Hierarchical Modeling and Analysis for Spatial Data

Hierarchical Modeling and Analysis for Spatial Data, Third Edition is the latest edition of this popular and authoritative text on Bayesian modeling and inference for spatial and spatial-temporal data. The text presents a comprehensive and up-to-date treatment of hierarchical and multilevel modeling for spatial and spatio-temporal data within a Bayesian framework. Over the past decade since the second edition, spatial statistics has evolved significantly driven by an explosion in data availability and advances in Bayesian computation. This edition reflects those changes, introducing new methods, expanded applications, and enhanced computational resources to support researchers and practitioners across disciplines, including environmental science, ecology, and public health.

Key features of the third edition:

- A dedicated chapter on state-of-the-art Bayesian modeling of large spatial and spatio-temporal datasets
- Two new chapters on spatial point pattern analysis, covering both foundational and Bayesian perspectives
- A new chapter on spatial data fusion, integrating diverse spatial data sources from different probabilistic mechanisms
- An accessible introduction to GPS mapping, geodesic distances, and mathematical cartography
- An expanded special topics chapter, including spatial challenges with finite population modeling and spatial directional data
- A thoroughly revised chapter on Bayesian inference, featuring an updated review of modern computational techniques
- A dedicated GitHub repository providing R programs and solutions to selected exercises, ensuring continued access to evolving software developments

With refreshed content throughout, this edition serves as an essential reference for statisticians, data scientists, and researchers working with spatial data. Graduate students and professionals seeking a deep understanding of Bayesian spatial modeling will find this volume an invaluable resource for both theory and practice.

MONOGRAPHS ON STATISTICS AND APPLIED PROBABILITY

Editors: F. Bunea, R. Henderson, L. Levina, N. Meinshausen, R. Smith

Recently Published Titles

For more information about this series please visit: https://www.crcpress.com/Chapman--HallCRC-Monographs-on-Statistics--Applied-Probability/book-series/CHMONSTAAPP

Hierarchical Modeling and Analysis for Spatial Data
Third Edition

Sudipto Banerjee, Alan E. Gelfand, and Bradley P. Carlin

CRC Press
Taylor & Francis Group
Boca Raton London New York

CRC Press is an imprint of the
Taylor & Francis Group, an **Informa** business

A CHAPMAN & HALL BOOK

Third edition published 2026
by CRC Press
2385 NW Executive Center Drive, Suite 320, Boca Raton FL 33431

and by CRC Press
4 Park Square, Milton Park, Abingdon, Oxon, OX14 4RN

CRC Press is an imprint of Taylor & Francis Group, LLC

© 2026 Sudipto Banerjee, Alan E. Gelfand and Bradley P. Carlin

First edition published by Taylor & Francis Group, LLC, 2003
Second edition published by Taylor & Francis Group, LLC 2015

Library of Congress Cataloging-in-Publication Data
Names: Banerjee, Sudipto author | Gelfand, Alan E., 1945- author | Carlin,
Bradley P. author
Title: Hierarchical modeling and analysis for spatial data / Sudipto
Banerjee, Alan E. Gelfand and Bradley P. Carlin.
Description: Third edition. | Boca Raton, FL : CRC Press, 2026. | Includes
bibliographical references and index. | Summary: "Hierarchical Modeling
and Analysis for Spatial Data, Third Edition is the 3rd edition of this
popular and authoritative text on Bayesian modeling and inference for
spatial and spatial-temporal data. The text presents a comprehensive
and up-to-date treatment of hierarchical and
multilevel modeling for spatial and spatio-temporal data within a
Bayesian framework. Over the past decade since the second edition,
spatial statistics has evolved significantly, driven by an explosion in
data availability and advances in Bayesian computation. This edition
reflects those changes, introducing new methods, expanded applications,
and enhanced computational resources to support researchers and
practitioners across disciplines, including environmental science,
ecology, and public health"-- Provided by publisher.
Identifiers: LCCN 2025015650 (print) | LCCN 2025015651 (ebook) | ISBN
9781032508559 hbk | ISBN 9781032513324 pbk | ISBN 9781003401728 ebk
Subjects: LCSH: Spatial analysis (Statistics) | Multilevel models
(Statistics)
Classification: LCC QA278.2 .B36 2026 (print) | LCC QA278.2 (ebook) | DDC
519.5--dc23/eng/20250425
LC record available at https://lccn.loc.gov/2025015650
LC ebook record available at https://lccn.loc.gov/2025015651

ISBN: 978-1-032-50855-9 (hbk)
ISBN: 978-1-032-51332-4 (pbk)
ISBN: 978-1-003-40172-8 (ebk)

DOI: 10.1201/9781003401728

Typeset in CMR10 font
by KnowledgeWorks Global Ltd.

Publisher's note: This book has been prepared from camera-ready copy provided by the authors.

TO SHARBANI, MARIASUN AND TANZY

Contents

Preface to the Third Edition

In the decade that has passed since the second edition of this book, we have witnessed substantial evolution in the spatial statistics landscape. We can now consider spatial statistics as a fairly mature field; there is little low-hanging fruit left to pursue. We have seen the remarkable growth in data collection, both spatial and spatial-temporal, helping us to learn about complex processes with datasets now of enormous size. This arises from the dramatic increase in data collection no longer designed. Rather, it is spatially referenced data exclusively observational and available from sources as diverse as global sampling efforts to satellite collection. According to the application, it may place us in the realm of so-called "big data". These are exciting times for modern statistical work, and in this regard, spatial analysis is a particularly important player due to increased appreciation of the information carried in spatial locations, perhaps across temporal scales, in learning about these complex processes.

We are witnessing an increased examination of complex space and space-time systems using such data, requiring synthesis of multiple sources of information (empirical, theoretical, physical, etc.), necessitating the development of multi-level models. We are seeing routine exemplification of the hierarchical framework

$$[data \mid process, parameters][process \mid parameters][parameters].$$

The role of the statistician continues to evolve in this landscape to that of an integral participant in team-based research: a participant in the framing of the questions to be investigated, the determination of data that needs to investigate these questions, the development of models to examine these questions, the development of strategies and algorithms to fit these models, and the analysis and summarization of the resultant inference under these specifications.

Applications abound, particularly in the environmental sciences and in ecology (two of our primary areas of exemplification), but also in public health, real estate, and many other fields. We believe this third edition moves forward in this spirit. The first edition was intended as a research monograph, presenting a state-of-the-art treatment of hierarchical modeling for spatial data at that time. It was a delightful success, far exceeding our expectations in terms of reception by the community and sales. With the passing of ten further years since the release of the previous edition, we find the need to offer a new edition, reflecting some of the new work that the decade has produced as well as updating the state of the art as we presented it in the second edition. We remain firmly in the realm of hierarchical or multi-level modeling specified within a Bayesian framework (though we do separate model specification from fitting and inference within the Bayesian setting). The result is an edition that has grown by nearly 100 pages (to the chagrin of the publishers but incorporating new material which we feel is important in the year 2025).

Rather than describing the contents chapter by chapter, we note the following new content. Due to the dramatic evolution in Bayesian computation for large spatial and spatial-temporal datasets, we now have a new chapter devoted to the "current" state of the art for model fitting in this setting. Due to the rapid pace of software development for spatial-temporal data analysis and its dissemination through web-based portals such as GitHub and freely accessible software repositories such as Comprehensive R Archive Network (CRAN) which hosts packages for use in the R statistical computing environment, we have now built a dedicated https://github.com/sudiptobanerjee/BGC_2023 page containing computer programs, primarily in R, to accompany this volume. The programs are arranged in folders corresponding to the chapters in this book and illustrate the implementation of the various models and methods in this text. Unlike in the previous two editions, where computer programs were supplied in the text itself, readers can now access these programs online. This website will also be updated as new upgrades to existing software become available.

We have also moved the solutions to selected exercises from the main text to the book's website. Earlier editions also included Appendices on linear algebra and numerical algorithms. With the considerable increase in the availability of such material in the form of online resources, new software, and an increasing number of textbooks on linear algebra and matrix computations for statisticians, machine learners, and data scientists, we have now removed the Appendices from the text and have added appropriate references in the Bibliography and in the book's companion website, which we intend to continue to update.

Due to the continuing growth of spatial point pattern application, we now present two chapters in this area, one foundational, the other purely Bayesian, jointly spanning nearly 100 pages. Due to the diversity of datasets being collected, we have now developed a chapter on spatial data fusion, acknowledging that the sources may be of any of the three basic spatial data types and the need for coherent/generative modeling to fuse sources arising from different probabilistic mechanisms. Due to increasing interest in how spatial data are mapped, we offer an accessible introduction to the mathematical concepts underlying how locations are determined using Global Positioning Systems (GPS), elements of the geometry of (and geodesic distances on) the surface of the earth, and the concepts behind mathematical cartography. Finally, due to the increasing diversity of applications, we now have an expanded special topics chapter (Chapter 17) where we present new material on spatial challenges with finite population modeling and on spatial directional data. Further, chapter by chapter, we have refreshed the material from the earlier editions. Most notably, this includes a thorough re-working of the chapter on the Basics of Bayesian inference, including an introductory review of current Bayesian computational methods (with references for further investigation). We believe this refreshing makes it more attractive to a reader in 2025!

The current edition would not have been possible without the assistance of the following colleagues, current and former post doctoral researchers, students and academic descendants (listed in alphabetical order by last name) who have collaborated with us in several new projects since the publication of the second edition: Jesus Abaurrea, Nada Abdallah, Luca Aiello, Matthew Aiello-Lammens, Pierfranceco Alaimo di Loro, Valentina Arputhasamy, Jesus Asin, Maria Asuncion Beamonte San Agustin, Laura Baracaldo, Thomas Belin, Anirban Bhattacharya, Ana Carmen Cebrian, Jorge Castillo-Mateo, Alec Chan-Golston, Xiang Chen, Michael Christensen, Abhirup Datta, Debangan Dey, Dipak Dey, Jeffrey Doser, Andrew Finley, Ian Frankenburg, Henry Frye, Leiwen Gao, Pilar Gargallo, Zeus Gracia Tabuenca, Caroline Groth, Rajarshi Guhaniyogi, Aritra Halder, Mark Handcock, Josh Hewitt, James Hodges, Michael Jerrett, Bokgyeong Kang, Miguel Lafuentes Blanco, Didong Li, Fan Li, Gianluca Mastroantonio, Marco Mingione, Rachel Mitchell, Jesper Moller, Lucia Paci, Jane Pan, Soumyakanti Pan, Fernando Perez-Cabello, Erica Porter, Luca Presicce, Beate Ritz, Eliana Rodrigues, Marcos Rodriguez, Manuel Salvador, Robert Schick, Erin Schliep, Elliot Shannon, Angie Shen, Shinichiro Shirota, Jasper Slingsby, Becky Tang, Wenpin Tang, Bradley Tomasek, Theresa Utlaut, Nina Virhs, Philip White, Kyle Wu, Lu Zhang, and Daniel Zhou.

In addition, we much appreciate the continuing support of Chapman & Hall/CRC in helping to bring this new edition to fruition, in particular the encouragement of the steadfast and indefatigable Rob Calver.

SUDIPTO BANERJEE   Los Angeles, California
ALAN E. GELFAND   Durham, North Carolina
BRADLEY P. CARLIN   Minneapolis, Minnesota
2025

About the Authors

Sudipto Banerjee is a professor of biostatistics and Senior Associate Dean for Academic Programs in the Fielding School of Public Health at the University of California, Los Angeles (UCLA). He holds joint appointments as a professor in the UCLA Department of Statistics and Data Science and as an affiliate faculty member in the UCLA Institute of Environment and Sustainability. Prof. Banerjee has authored over 200 research articles, 2 textbooks, 2 committee reports for the National Research Council of the National Academies, and an edited handbook on spatial epidemiology. He is well known for his research expertise and methodological advancements in Bayesian hierarchical modeling and inference for spatial-temporal data; theoretical and computational developments for Gaussian processes; environmental processes and their impacts on public health; spatial epidemiology; stochastic process models; statistical learning from physical and mechanistic systems; and survey sampling and survival analysis.

Alan E. Gelfand is The James B. Duke Professor Emeritus of Statistical Science at Duke University. He also has a secondary appointment as a professor of environmental science and policy in the Nicholas School. Author of more than 330 papers and 6 books, Prof. Gelfand is internationally known for his contributions to applied statistics, Bayesian computation, and Bayesian inference. For the past thirty years, his primary research focus has been on statistical modeling for spatial and space-time data. He has been involved in advanced methodology, using the Bayesian paradigm, to associate fully model-based inferences with spatial and space-time data. His chief areas of application include spatio-temporal environmental and ecological processes.

Bradley P. Carlin is a statistical researcher, methodologist, consultant, author, and instructor. He currently serves as Senior Director of Data Science and Statistics for PhaseV, a clinical trials technology and consulting firm. Prior to this, he spent 27 years on the faculty of the Division of Biostatistics at the University of Minnesota School of Public Health, serving as division head for 7 of those years. He was not involved in this third edition of this book, but contributed much to the development of the first two editions. Since a portion of that material remains in this new edition, he continues to act as a co-author.

Chapter 1

Overview of spatial data problems

1.1 Introduction to spatial data and models

Researchers in a wide array of scientific disciplines including, but not limited to, climatology, ecology, economics, environmental health, epidemiology, forestry, and several other areas within physical and biomedical sciences are increasingly faced with the task of analyzing data that are geographically referenced with respect to where variables have been measured or recorded. Such data are often presented as maps and may also be available over time in the form of longitudinal or other time series structures. Analyzing such data requires accounting for spatial, and possibly spatial-temporal, dependence. Furthermore, such data will often consist of multiple variables, and, depending upon the specific application and scientific questions being asked, one may wish to model only one variable as the dependent variable while treating other variables as predictors, or one may wish to model multiple variables jointly as dependent variables in which case it is necessary to model spatial associations for each of the dependent variables as well as possible associations among themselves.

Example 1.1 *In an epidemiological investigation, we might wish to analyze lung, breast, colorectal, and cervical cancer rates by county and year in a particular state, with smoking, mammography, and other important screening and staging information also available at some level. Public health professionals who collect such data are charged not only with surveillance, but also statistical* inference *tasks, such as* modeling *of trends and correlation* structures, estimation *of underlying model parameters*, hypothesis testing *(or comparison of competing models), and* prediction *of observations at unobserved times or locations.*

Example 1.2 *In ecology, one is often interested in the spatial distribution and abundance of species. Data collection varies according to whether interest is in plants or in animals. For the former, since plants don't move, collection can take the form of sampling at selected sites (usually with some design intentions) or perhaps opportunistically in the form of random encounters. For animals, collection can arise through, e.g., visual sighting, capture-recapture, camera traps, and tagging. A rich range of inference is sought. For plants, can we explain* presence/absence *or* abundance *at locations through environmental features? How can one model species jointly in order to capture* dependence/interactions *among species? How can one learn about change in distribution and abundance across time? For animals, as well there is interest in distribution and abundance over space and time. There is also interest in movement behavior in space and time as well as in behavioral response to disturbance.*

This monograph aims to present a practical, self-contained treatment of hierarchical modeling and data analysis for complex spatial (and spatiotemporal) data sets. Spatial statistics methods have been around for some time, with the landmark work by Cressie [1993] providing arguably the first comprehensive book in the area, and have been massively influenced by rapid developments in computational methods such as Markov chain Monte Carlo (MCMC) algorithms that allow fully Bayesian analyses of sophisticated multilevel models for complex geographically referenced data. This approach enables *generative/coherent* modeling, i.e., stochastic specifications that could actually produce the observed data. Further, it offers full inference for Gaussian and non-Gaussian spatial data, multivariate spatial data, spatiotemporal data, and stochastic approaches to address problems such as geographic and temporal misalignment of spatial data layers.

This book does not attempt to be fully comprehensive but does attempt to present a fairly thorough treatment of hierarchical Bayesian approaches for handling all of these problems. The mathematical level of the book is roughly comparable to that of Gelman et al. [2013] or Carlin and Louis [2008]. While we sometimes state results rather formally and attempt to not compromise on rigor, we do not pursue the presentation of such materials in the forms of theorems and proofs. For more mathematical treatments of spatial statistics, we refer to Cressie [1993], Chilés and Delfiner [2012], Wackernagel [2003], and Stein [1999a] for geostatistics, while van Lieshout [2019] and Kent and Mardia [2003] cover theoretical developments suitable for different types of spatial data analysis. Cressie and Wikle [2011] provide a comprehensive treatment of spatial-temporal data analysis, while texts such as Diggle and Ribeiro [2007], Bivand et al. [2013] Wikle et al. [2019], Moraga [2023], and Sahu [2021] focus on modeling spatial and spatial-temporal data with several examples in the R statistical programming environment. In fact, pleasantly enough, there has emerged a rich library of books on spatial statistics that are too extensive to be comprehensively reviewed here [see, e.g., Bailey and Gatrell, 1995, Haining, 2003, Schabenberger and Gotway, 2004, Fotheringham and Rogerson, 2009, Chun and Griffith, 2013, Fotheringham and Rogerson, 2013, Pebesma and Bivand, 2023, for a selection of texts written from diverse perspectives and catering to diverse data analytic needs]. A broad review of several areas in spatial and spatio-temporal analysis is provided in the handbook compiled by Gelfand et al. [2010].

Our primary focus is on the issues of *modeling* (where we offer rich, flexible classes of hierarchical structures to accommodate both static and dynamic spatial data), *computing* (both in terms of Markov chain Monte Carlo algorithms and methods for handling very large datasets in space and time), and *data analysis* (to illustrate the first two items in terms of inferential summaries and graphical displays). Reviews of both traditional spatial methods (Chapters 2, 3, and 4) and Bayesian methods (Chapter 5) attempt to ensure that previous exposure to either of these two areas is not required (though it will of course be helpful if available).

Following convention, we classify spatial data sets into one of three basic types:

- *point-referenced data* where $\mathbf{Y}(\mathbf{s})$ is a random vector at a location $\mathbf{s} \in \Re^r$, with $\mathbf{s}$ spanning D *continuously*, a fixed subset of $\Re^r$ that contains an r-dimensional rectangle of positive volume;

- *areal data* where D is again a fixed subset (of regular or irregular shape), but now partitioned into a *finite* number of areal units with well-defined boundaries; random observations arise at each unit;

- *point pattern data*, where now the number and locations of points in D are random; its index set, $\mathcal{S}$ gives the locations that form the spatial point pattern. In less common notation, $Y(\mathbf{s})$ itself can simply equal 1 for all $\mathbf{s} \in \mathcal{S}$ (indicating the occurrence of the event). Some additional covariate information associated with each point in $\mathcal{S}$ may be available, producing a *marked point pattern process*.

The first case is often referred to as *geocoded* or *geostatistical* data, names apparently arising from the long history of these types of problems in mining and other geological sciences. Figure 1.1 offers an example of this case, showing the locations of 114 air-pollution monitoring sites in three midwestern U.S. states (Illinois, Indiana, and Ohio). The plotting character indicates the 2001 annual average PM2.5 level (measured in ppb) at each site. PM2.5 stands for particulate matter less than 2.5 microns in diameter (a measure of the density of very small particles that can travel through the nose and windpipe and into the lungs, potentially damaging a person's health). Here we might be interested in a model of the geographic distribution of these levels that accounts for spatial correlation and perhaps

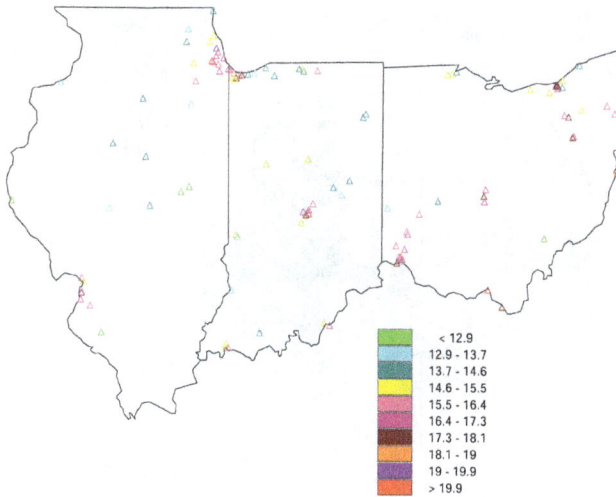

Figure 1.1 *Map of PM2.5 sampling sites over three midwestern U.S. states; plotting character indicates range of average monitored PM2.5 level over the year 2001.*

underlying covariates (regional industrialization, traffic density, and the like). The use of colors makes it somewhat easier to read, since the color allows the categories to be ordered more naturally, and helps sharpen the contrast between the urban and rural areas. Again, traditional analysis methods for point-level data like this are described in Chapter 2, while Chapter 6 introduces the corresponding hierarchical modeling approach.

The second case above (areal data) is frequently referred to as discrete spatial data since it works with only a finite set of spatial random variables. When areal units are identical in size and shape, it is often referred to as *lattice* data or *gridded* data to connote observations corresponding to corners or centroids of a checkerboard-like grid. Such data may arise from agricultural field trials (where the plots cultivated form a regular lattice) or in image restoration (where the data correspond to pixels on a screen, again in a regular lattice). However, more flexibly, areal data are summaries over an *irregular* lattice, like a collection of county or other regional boundaries, as in Figure 1.2. Here we have information on the percent of a surveyed population with household income falling below 200% of the federal poverty limit, for a collection of regions comprising Hennepin County, MN. Note that we have no information on any single household in the study area, only regional summaries for each region. Figure 1.2 is an example of a *choropleth map*, meaning that it uses shades of color (or greyscale) to classify values into a few broad classes (six in this case), like a histogram (bar chart) for nonspatial data. Choropleth maps are visually appealing (and therefore, also common), but provide a crude summary of the data, one that can be easily altered simply by manipulating the class cutoffs.

As with any map of the areal units, choropleth maps *do* show reasonably precise *boundaries* between the regions (i.e., a series of exact spatial coordinates that, when connected in the proper order, will trace out each region), and thus we also know which regions are adjacent to (touch) which other regions. Thus the "sites" $\mathbf{s} \in D$ in this case are actually the regions (or *blocks*) themselves, which in this text we will denote not by B_i, $i = 1, \ldots, n$, to avoid confusion between points $\mathbf{s}_i$ and blocks B_i. It may also be illuminating to think of the county centroids as forming the vertices of an irregular lattice, with two lattice points being connected if and only if the counties are "neighbors" in the spatial map, with physical adjacency being an obvious way to define a region's neighbors (but not the only way, e.g., between centroid distance offers another option).

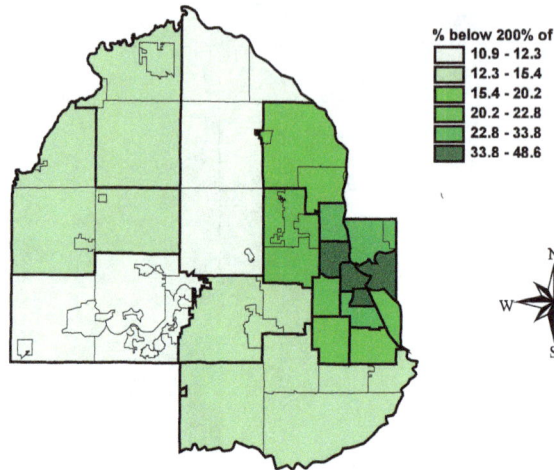

Figure 1.2 **ArcView** *map of percent of surveyed population with household income below 200% of the federal poverty limit, regional survey units in Hennepin County, MN.*

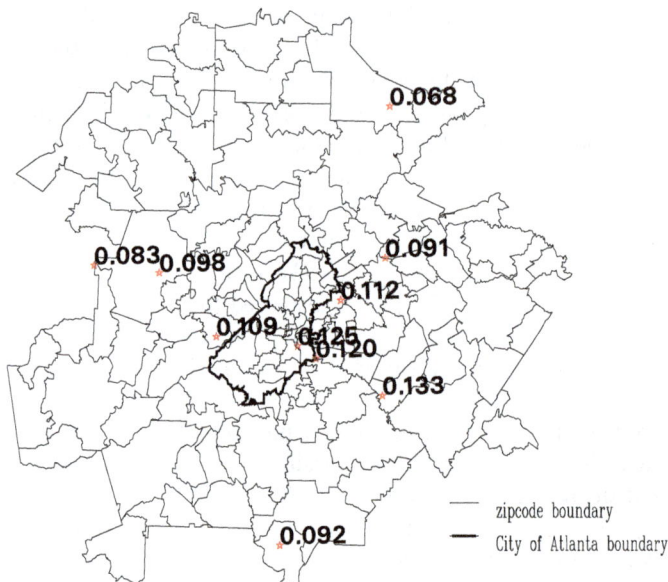

Figure 1.3 *Zip code boundaries in the Atlanta metropolitan area and 8-hour maximum ozone levels (ppm) at 10 monitoring sites for July 15, 1995.*

Some spatial data sets feature *both* point- and areal-level data, and require their simultaneous display and analysis. Figure 1.3 offers an example of this case. The first component of this data set is a collection of eight-hour maximum ozone levels at 10 monitoring sites in the greater Atlanta, GA, area for a particular day in July 1995. Like the observations in Figure 1.1, these were made at fixed monitoring stations for which exact spatial coordinates (say, latitude and longitude) are known. (That is, we assume the $Y(\mathbf{s}_i)$, $i = 1, \dots, 10$ are random, but the $\mathbf{s}_i$ are not.) The second component of this data set is the number of children in the area's zip codes (shown using the irregular subboundaries on the map) that

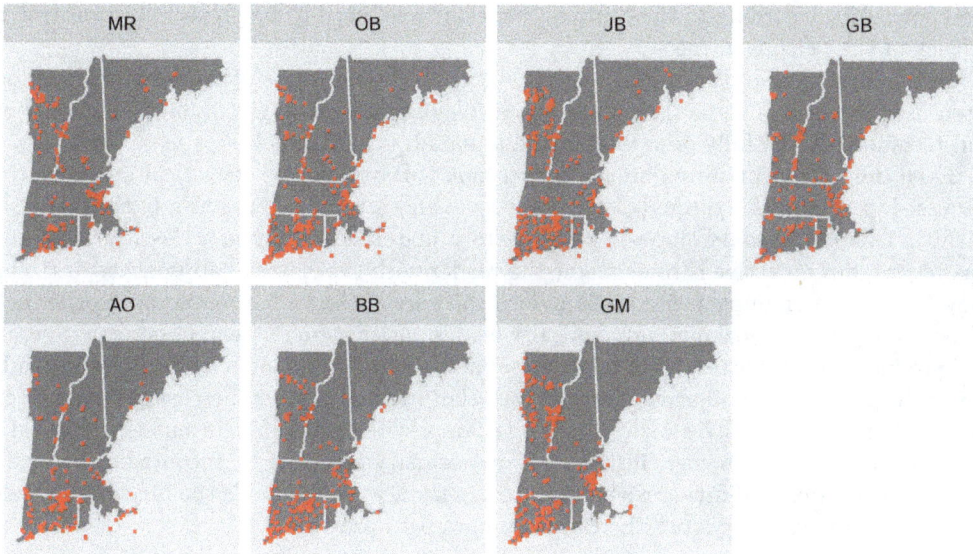

Figure 1.4 *The distribution of locations for seven invasive plant species across New England.*

reported at local emergency rooms (ERs) with acute asthma symptoms on the following day; confidentiality of health records precludes us from learning the precise address of any of the children. These are areal summaries that could be indicated by shading the zip codes, as in Figure 1.2. A natural question here is whether we can establish a connection between high ozone and subsequent high pediatric ER asthma visits. Since the data are misaligned (point-level ozone but areal-level ER counts), a formal statistical investigation of this question requires a preliminary *realignment* of the data; this is the subject of Chapter 4.

The third case above (spatial point pattern data) could be exemplified by residences of persons suffering from a particular disease, or by locations of a certain species of tree in a forest. In particular, Figure 1.4 shows the realization of the point pattern for each of seven of the most common invasive plant species across New England in the USA. Specifically, they are: multiflora rose (MR), oriental bittersweet (OB), Japanese barberry (JB), glossy buckthorn (GB), autumn olive (AO), burning bush (BB), and garlic mustard (GM). They are drawn from the Global Biodiversity Information Facility (GBIF) which is a data aggregator for biological collections worldwide. The number of points varies from species to species. Notably, fewer points occur in the north of New England, perhaps due to the colder temperatures. To enhance our understanding of the dependence/interaction among these point patterns, it would be attractive to model them jointly.

Here, only the locations $\mathbf{s}_i$ are thought of as random. One path of interest is in *explaining* the point pattern using spatially varying features (say as a covariate vector $\mathbf{X}(\mathbf{s})$) of the region D. An alternative path may address the occurrence of *clustering* or in *inhibition*. In the former, the process shows that points tend to aggregate, to be spatially closer to each other. In the latter, the process tends to discourage points from being too close to each other. Either is considered as departure from what is called complete spatial randomness where points are viewed as being distributed uniformly over D. In some cases the set of random points might be supplemented by, e.g., an associated species or disease type or an age or size, producing a *marked* point pattern). In contrast to areal data, where no individual points in the data set are supplied, here (and in point-referenced data as well) precise locations are known. So, according to the context, locations may need to be protected to protect the privacy of the persons in the set.

In the remainder of this initial section, we give a brief outline of the basic models most often used for each of these three data types. We only intend to give a flavor of the models and techniques to be fully described in the remainder of this book.

Even though our preferred inferential outlook is Bayesian, the statistical inference tools discussed in Chapters 2 through 4 for point-level and areal data are entirely classical. Similarly, in Chapter 8 we develop the point pattern material initially in a classical fashion. In this regard, our objective is to acquaint the reader with the classical approaches first, since they are more often implemented in standard software packages. Moreover, as in other fields of data analysis, classical methods typically enable easier model fitting, easier computation, and produce acceptable results in relatively simple settings. Further, classical methods often have interpretations as limiting cases of Bayesian methods under increasingly vague prior assumptions. Additionally, classical methods can provide insight for formulating and fitting hierarchical models. However, in the end, our primary intent is to encourage hierarchical model specification and fitting within a Bayesian framework to enable the full range of exact inference and uncertainty that the Bayesian paradigm provides.

1.1.1   Point-level models

In the case of point-level data, the location index $\mathbf{s}$ varies *continuously* over D, a fixed subset of $\Re^d$. Suppose we assume that the covariance between the random variables at two locations depends on the *distance* between the locations. One frequently used association specification is the exponential model. Here the covariance between measurements at two locations is an exponential function of the interlocation distance, i.e., $Cov(Y(\mathbf{s}_i), Y(\mathbf{s}_{i'})) \equiv C(d_{ii'}) = \sigma^2 e^{-\phi d_{ii'}}$ for $i \neq i'$, where $d_{ii'}$ is the distance between sites s_i and $s_{i'}$, and σ^2 and ϕ are positive parameters. The former is called the *spatial variance* or *partial sill*, the latter is called the *spatial decay*, and $1/\phi$ is called the *range* parameter. A plot of the covariance versus distance is called the *covariogram*. When $i = i'$, $d_{ii'}$ is of course 0, and $C(d_{ii'}) = Var(Y(\mathbf{s}_i))$ is often expanded to $\tau^2 + \sigma^2$, where $\tau^2 > 0$ is called a *nugget effect*, and $\tau^2 + \sigma^2$ is called the *sill*. Of course, while the exponential model is convenient and has some desirable properties, many other parametric models are commonly used; see Section 2.1 for further discussion of these and their relative merits.

Adding a joint distributional model to these variance and covariance assumptions then enables likelihood inference in the usual way. The most convenient approach would be to assume a multivariate *normal* (or *Gaussian*) distribution for the data. That is, suppose we are given observations $\mathbf{Y} \equiv \{Y(\mathbf{s}_i)\}$ at known locations $\mathbf{s}_i$, $i = 1, \ldots, n$. We then assume that

$$\mathbf{Y} \mid \mu, \boldsymbol{\theta} \sim N_n(\mu\mathbf{1}, \Sigma(\boldsymbol{\theta})), \tag{1.1}$$

where N_n denotes the n-dimensional normal distribution, μ, illustratively, is a constant mean level, $\mathbf{1}$ is a vector of ones (in practice, we would replace μ with regression in covariates $\mathbf{X}(\mathbf{s}_i)$), and $(\Sigma(\boldsymbol{\theta}))_{ii'}$ gives the covariance between $Y(\mathbf{s}_i)$ and $Y(\mathbf{s}_{i'})$. For the variance-covariance specification of the previous paragraph, we have $\boldsymbol{\theta} = (\tau^2, \sigma^2, \phi)^{\mathrm{T}}$, since the covariance matrix depends on the nugget, sill, and range.

In fact, the simplest choices for Σ are those corresponding to *isotropic* covariance functions, where we assume that the spatial correlation is a function solely of the distance $d_{ii'}$ between $\mathbf{s}_i$ and $\mathbf{s}_{i'}$. As mentioned above, exponential forms are particularly intuitive examples. In this case,

$$(\Sigma(\boldsymbol{\theta}))_{ii'} = \sigma^2 \exp(-\phi d_{ii'}) + \tau^2 I(i = i'), \ \sigma^2 > 0, \ \phi > 0, \ \tau^2 > 0, \tag{1.2}$$

where I denotes the indicator function (i.e., $I(i = i') = 1$ if $i = i'$, and 0 otherwise).

Many other choices are possible for $Cov(Y(\mathbf{s}_i), Y(\mathbf{s}_{i'}))$, e.g., the spherical, the Gaussian, and the Matérn (see Subsection 2.1.3 for a full discussion). In particular, while the Matérn requires the calculation of a modified Bessel function [Stein, 1999a, p. 51] illustrates its ability to capture a broader range of local correlation behavior. It includes the exponential as a special case. We shall say much more about point-level spatial methods and models in Chapters 2,3 and 6 and also provide illustrations using freely available statistical software.

1.1.2   Areal models

In models for areal data, the units of spatial indexing are the geographic regions or *blocks* (zip codes, counties, administrative units, etc.), and the data are typically a regional aggregate or summary (e.g., counts, sums, averages) of variables for each region. To introduce spatial association, one approach is to define a *neighborhood* structure based on the arrangement of the blocks in the map. Once the neighborhood structure is defined, two-dimensional models resembling one-dimensional autoregressive time series models are considered. Two common models that incorporate such neighborhood information are the *simultaneous* and *conditional* autoregression models (abbreviated as SAR and CAR), which were originally developed by Whittle [1954] and Besag [1974], respectively. The SAR model is computationally convenient for use with likelihood methods. By contrast, the CAR model is computationally convenient for Gibbs sampling used in conjunction with Bayesian model fitting. In this regard, it is often used to incorporate spatial correlation through a vector of spatially varying random effects $\boldsymbol{\phi} = (\phi_1, \ldots, \phi_n)^{\mathrm{T}}$. For example, writing Y_i as the value of a variable Y recorded in region i, we might assume $Y_i \overset{ind}{\sim} N(\mu + \phi_i, \sigma^2)$, and then model the spatial random effects as a CAR

$$\phi_i \mid \boldsymbol{\phi}_{(-i)} \sim N\left(\sum_{j=1}^{n} a_{ij}\phi_j \, , \, \tau_i^2\right), \qquad (1.3)$$

where $\boldsymbol{\phi}_{(-i)} = \{\phi_j : j \neq i\}$, τ_i^2 is the conditional variance, and the a_{ij} are constants (usually assumed known) such that $a_{ii} = 0$ for $i = 1, \ldots, n$. Letting $A = (a_{ij})$ and $M = Diag(\tau_1^2, \ldots, \tau_n^2)$, under certain conditions, we show in Chapter 4 that

$$p(\boldsymbol{\phi}) \propto \exp\{-\boldsymbol{\phi}^{\mathrm{T}} M^{-1}(I - A)\boldsymbol{\phi}/2\}, \qquad (1.4)$$

where I is a $n \times n$ identity matrix.

A common way to construct A and M is to let $A = \rho \, Diag(1/w_{i+})W$ and $M^{-1} = \tau^{-2} Diag(w_{i+})$. Here ρ is referred to as the *spatial correlation* parameter, and $W = (w_{ij})$ is a neighborhood matrix for the areal units, which can be defined as

$$w_{ij} = \begin{cases} 1 & \text{if subregions } i \text{ and } j \text{ share a common boundary, } i \neq j \\ 0 & \text{otherwise} \end{cases}. \qquad (1.5)$$

Thus $Diag(w_{i+})$ is a diagonal matrix with (i, i) entry equal to $w_{i+} = \sum_j w_{ij}$. We discuss areal models in greater detail in Chapters 4 and 3.

1.1.3   Point process models

Currently, the analysis of point process data follows two main paths. For the first, an intensity $\lambda(\mathbf{s})$ is specified over D to try to explain the point pattern, i.e., where the intensity is high, we expect to see more points, where it is low, we expect to see fewer points. So, this line of investigation focuses on modeling this intensity function. In its customary form, we employ so-called nonhomogeneous Poisson processes or logGaussian Cox processes as the models (see Chapters 8 and 9). For the former, we specify $\log \lambda(\mathbf{s}) = \mathbf{X}(\mathbf{s})^{\mathrm{T}} \boldsymbol{\beta}$. For the latter, we add a spatial random effects term, $w(\mathbf{s})$, modeled as a realization of a Gaussian process, to enable *local* adjustment to the intensity or to capture unmeasured spatial variables.

The second path focuses on clustering and inhibition, seeking to capture departure from a uniform distribution of point locations over D. Stochastically, such uniformity is described through a *homogeneous Poisson process*, which implies that the expected number of occurrences in region A is $\lambda|A|$, where λ is the *intensity* parameter of the process and $|A|$ is the area of A. To investigate this, in practice, plots of the data are useful, but the tendency of the human eye to see clustering or other structure in a point pattern realization renders a strictly graphical approach unreliable. Instead, statistics that measure clustering or inhibition, and perhaps even associated significance tests, are often used. A widely employed illustrative choice is *Ripley's K function*, given by

$$K(d) = \frac{1}{\lambda} \mathrm{E}[\text{number of points within } d \text{ of an arbitrary point}] \,, \qquad (1.6)$$

where, again, λ is the intensity of the process, i.e., the mean number of points per unit area. The usual estimator for K is given by

$$\widehat{K}(d) = n^{-2}|A| \sum_{i \neq j} \sum p_{ij}^{-1} I_d(d_{ij}) \,, \qquad (1.7)$$

where n is the number of points in A, d_{ij} is the distance between points i and j, p_{ij} is the proportion of the circle with center i and passing through j that lies within A, and $I_d(d_{ij})$ equals 1 if $d_{ij} < d$, and 0 otherwise.

We provide an extensive development for point processes in Chapter 8. Other useful texts focusing primarily on point processes and patterns include Diggle [2013], Lawson and Denison [2002], and Moller and Waagepetersen [2003] for treatments of spatial point processes and related methods in spatial cluster detection and modeling. A very broad and accessible volume, supplying both theory and application is Illian et al. [2007]. In the context of Bayesian analysis of spatial point patterns, we note the monograph by Gelfand and Schliep [2018].

1.2   Essential elements of spatial data science

Spatial data science is an increasingly conspicuous term broadly used to describe all scientific and technological issues pertaining to the collection, storage, processing, and analysis of spatial data. While our focus in this monograph pertains primarily to statistical modeling and analysis of spatial data, we provide a brief overview of some essential mathematical elements of spatial data science that are related to how such data are collected and mapped. Specifically, we offer some quick introduction to the mathematical concepts underlying how locations are determined using Global Positioning Systems (GPS), elements of the geometry of (and geodesic distances on) the surface of the earth, and the concepts behind mathematical cartography with formulas for some widely employed map projections.

1.2.1   *Mathematics of GPS*

The Global Positioning System (GPS) is a satellite-based navigation system that was developed by the United States Department of Defense (DoD) in the early 1970s with the primary aim of meeting U.S. military needs. It has since evolved into a system that can be accessed by both military and civilian users to obtain continuous positioning and timing information anywhere in the world under any weather conditions. The GPS receiver located on the Earth receives signals from multiple satellites whose positions and times are known. A method called *multilateration* uses the available information employing multiple equations to solve for the coordinates of the GPS receiver on the Earth. The location is approximated using the travel time of a signal from the satellite to the receiver and we need to account for the difference in time between the receiver's clock and the satellites' atomic clocks. This difference in time induces errors in precision because the satellites' are equipped with highly precise atomic clocks while the receiver's clock is much less accurate.

Multilateration finds the coordinates of the GPS receiver as a point of intersection of intersecting spheres using the locations and paths of the satellites as well as the time for the signal to travel from the satellite to the receiver [see Thompson, 1998, for a delightful exposition]. For example, with 4 satellites, we obtain the system

$$(x - A_1)^2 + (y - B_1)^2 + (z - C_1)^2 - (c(t_1 - d))^2 = 0$$
$$(x - A_2)^2 + (y - B_2)^2 + (z - C_2)^2 - (c(t_2 - d))^2 = 0$$
$$(x - A_3)^2 + (y - B_3)^2 + (z - C_3)^2 - (c(t_3 - d))^2 = 0$$
$$(x - A_4)^2 + (y - B_4)^2 + (z - C_4)^2 - (c(t_4 - d))^2 = 0 \, ,$$

where (x, y, z) denotes the Euclidean coordinates of the GPS receiver from which latitude and longitude can be derived, (A_i, B_i, C_i) are the coordinates of the satellites for $i = 1, \ldots, 4$, d is the difference in time between the receiver and the satellite's clocks, t_i is the travel time for the signal from the i-th satellite to the receiver, and $c = 299,792.458$ is the speed of light in kilometers per second. Given the high precision of the atomic clocks in the satellites, the variable d is assumed to be the same across the satellites, but needs to be estimated (not known) since the signal's speed changes as it enters the earth's atmosphere. A numerical solver, such as Newton-Raphson method, solves the above equations for (x, y, z, d).

To implement the above method, we will set up a 3-dimensional Euclidean coordinate system for the GPS receiver and the satellites that will allow us to calculate the geographic coordinates from (x, y, z). The most common one is the spherical coordinate system. This is a 3-dimensional Cartesian coordinate system (x, y, z), with the origin at the center of the earth, the z-axis along the North and South Poles, and the x-axis on the plane of the equator joining the center of the earth and the Greenwich meridian. Let $P_1 = (\theta_1, \lambda_1)$ and $P_2 = (\theta_2, \lambda_2)$ be any two points on the Earth, where θ_1 and λ_1 are the latitude and longitude, respectively, of the point P_1, while θ_2 and λ_2 are those for the point P_2. Using elementary trigonometry we derive the following relationships between (x, y, z) and (θ, λ):

$$x = R \cos\theta \cos\lambda, \quad y = R \cos\theta \sin\lambda \quad \text{and} \quad z = R \sin\theta \, , \tag{1.8}$$

where R is the radius of the earth (Figure 1.5). The positions for each of the GPS satellites, (A_i, B_i, C_i) are also expressible in the same coordinate frame, where R is replaced by their distance to the center of the earth and the angles (λ, θ) are obtained in similar manner to a point on the earth's surface by considering the satellite to be on the surface of a sphere concentric to the earth's surface. In other words, each satellite is located on a sphere with the same center as the earth and radius equal to its distance from the center. Given (x, y, z), the geographic coordinates of the GPS receiver on the Earth are

$$\lambda = \arctan\left(\frac{y}{x}\right) \quad \text{and} \quad \theta = \arcsin\left(\frac{z}{R}\right) \, .$$

Several other, more elaborate, models for GPS systems are available, including the use of more satellites for enhanced precision. Thompson [1998] offers a very accessible introduction drawing an analogy of this problem with a messenger from bringing a parcel from a car (satellite) to an individual (receiver) and passing through a pavement (space) at a given speed and then through gravel (earth's atmosphere) at a different speed. Strang and Borre [1997] is a comprehensive and accessible textbook on the subject.

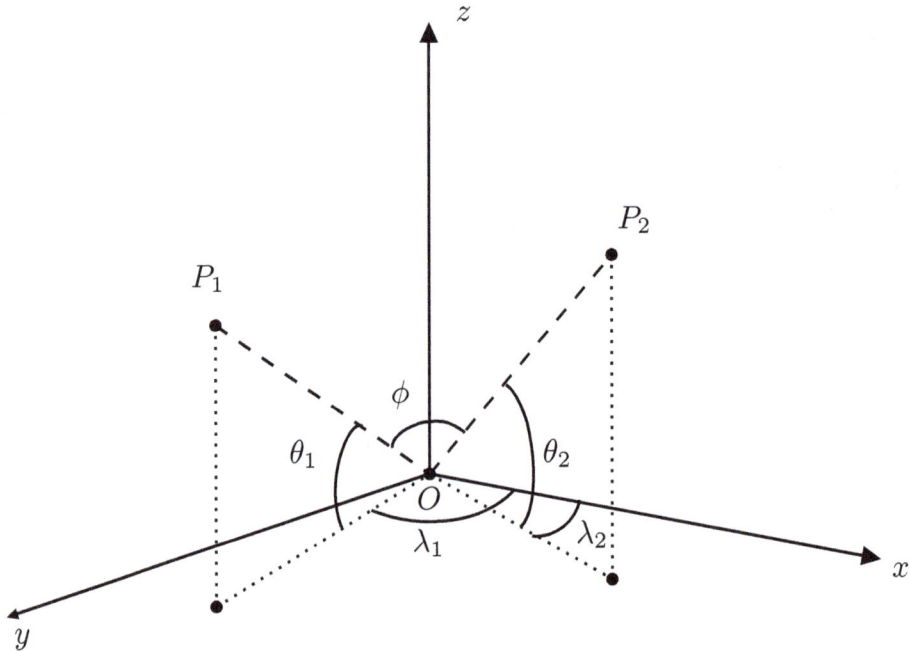

Figure 1.5 *Diagrams illustrating the spherical coordinate system with latitude and longitude.*

1.2.2   *Essentials of spherical trigonometry*

While the earth's shape is better approximated by an ellipsoid, a simpler spherical model often suffices even for data that span large areas of the earth's surface. The geometry of the sphere is simpler and most insights into distances and angles are obtained from considering spherical triangles. An elaborate study of spherical triangles, which comprises the fascinating subject of spherical trigonometry, is not our aim here and we refer to classic texts such as Todhunter [1863] and Van Brummelen [2017]. Statisticians familiar with the linear algebra of orthogonal projection matrices may find an accessible development of some of the basic identities in spherical trigonometry in Banerjee [2004].

Here, we offer some basic concepts of a spherical triangle and use simple vector algebra to provide quick derivations to some of the basic identities that help us understand geodesic distances and map projections. Geodesic distance is the shortest distance between two points. On a sphere, the geodesic distance is the distance along the great circle arc, which refers to the arc formed by the intersection of the sphere and the plane passing through the center of the earth and the two points on the sphere.

Figure 1.6 shows a spherical triangle ABC with vertices A, B, and C on the surface of the sphere and three "sides" a, b, and c. The sides are geodesic arcs joining the vertices. In fact, a spherical triangle has six angles: three side angles and three vertex angles (also called azimuthal angles). We place this sphere into an Euclidean coordinate system in $\mathbb{R}^3$ such that the origin **O** coincides with the center of the sphere. Note that we can take any of the vertices to act like our north pole and any two vertices to define the xz-plane along with the center. For illustration, let A be the north pole (on the z-axis) and let points A and B on the xz-plane. The y-axis is orthogonal to the xz-plane passes through the origin. The length of the vectors $\overrightarrow{OA}$, $\overrightarrow{OB}$, and $\overrightarrow{OC}$ are equal to the radius of the sphere.

Consider, for simplicity, the unit sphere (with radius one unit) and let $\overrightarrow{OA}$, $\overrightarrow{OB}$, and $\overrightarrow{OC}$ be three unit vectors drawn from the origin to the vertices of the spherical triangle. The arcs

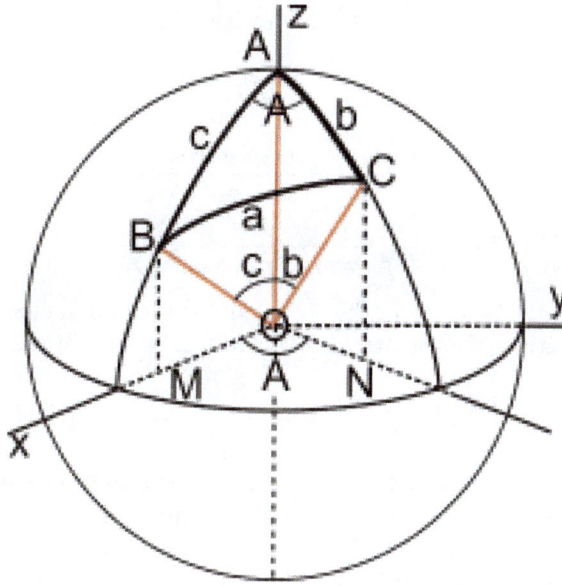

Figure 1.6 *Coordinate system for a spherical triangle.*

$\widehat{BC}$, $\widehat{CA}$, and $\widehat{AB}$ subtend angles of magnitude a, b, and c at the center, respectively. Our coordinate system with $\overrightarrow{OA}$ along the z-axis and $\overrightarrow{OB}$ in the xz-plane implies the following coordinates for the three vertices,

$$\overrightarrow{OA} = (0, 0, 1); \quad \overrightarrow{OB} = (\sin c, 0, \cos c), \quad \text{and} \quad \overrightarrow{OC} = (\sin b \cos A, \sin b \sin A, \cos b).$$
$$(1.9)$$

Each of the side angles can be expressed as simple dot products of the vectors as:

$$\cos a = \overrightarrow{OB} \cdot \overrightarrow{OC}; \quad \cos b = \overrightarrow{OC} \cdot \overrightarrow{OA}; \quad \text{and} \quad \cos c = \overrightarrow{OC} \cdot \overrightarrow{OA} . \qquad (1.10)$$

The *law of cosines* for spherical triangles are now derived by simply equating one of the dot products using the coordinates in (1.9). For example, for the side a, we get

$$\cos a = \cos b \cos c + \sin b \sin c \cos A . \qquad (1.11)$$

Analogous results for the other two sides can be obtained using their respective dot product expressions in (1.10). Better still, we could have worked with any other coordinate system since the relation only depends upon the sides and angles of the triangle. So, without any further calculation at all, we can imagine setting up (1.9) with the z-axis along $\overrightarrow{OB}$ with B as the north-pole or the z-axis along $\overrightarrow{OC}$ with C as the north pole to obtain (just cycling through the letters in (1.11)) $\cos b = \cos c \cos a + \sin c \sin a \cos B$ and $\cos c = \cos a \cos b + \sin a \sin b \cos C$ to complete the set of cosine rules for a spherical triangle.

The vertical angles of the spherical triangle are only slightly more complicated and are defined as the angles between the *planes* containing the two arcs meeting at that vertex. In fact, it is possible to entirely avoid further geometry and simply express the vertical angle in terms of the side angles from (1.11). This yields

$$\cos A = \frac{\cos a - \cos b \cos c}{\sin b \sin c} . \qquad (1.12)$$

One could go on, just for the sake of having fun with trigonometric identities, and write

$$\sin^2 A = 1 - \cos^2 A = 1 - \left(\frac{\cos a - \cos b \cos c}{\sin b \sin c} \right)^2$$

$$= \frac{(1 - \cos^2 b)(1 - \cos^2 c) - (\cos a - \cos b \cos c)^2}{\sin^2 b \sin^2 c}$$

$$\frac{\sin^2 A}{\sin^2 a} = \frac{1 - \cos^2 a - \cos^2 b - \cos^2 c + 2 \cos a \cos b \cos c}{\sin^2 a \sin^2 b \sin^2 c}.$$

Now comes a remarkable observation: the right-hand side is symmetric in a, b, and c and, hence, invariant to their cyclic permutations. Therefore, we obtain the *law of sines*

$$\frac{\sin^2 A}{\sin^2 a} = \frac{\sin^2 B}{\sin^2 b} = \frac{\sin^2 C}{\sin^2 c} \qquad (1.13)$$

Observe that these vertical angles are determined uniquely up to absolute values and their signs are determined by the direction and distance we move along the opposite side.

The above derivation is somewhat dull and offers no elegance beyond simple algebra. An alternative, perhaps more geometrically appealing, derivation relies upon computing the volume of the parallelepiped formed by the position vectors in (1.9). This volume, say V, is given by the scalar triple product (also called the box product) $\overrightarrow{OA} \cdot (\overrightarrow{OB} \times \overrightarrow{OC})$, where $\cdot$ is the dot product and $\times$ is the vector cross product. The scalar triple product is conveniently computed as the 3×3 determinant with $\overrightarrow{OA}$, $\overrightarrow{OB}$, and $\overrightarrow{OC}$ as its rows. Therefore,

$$V^2 = (\overrightarrow{OA} \cdot (\overrightarrow{OB} \times \overrightarrow{OC}))^2 = \begin{vmatrix} 0 & 0 & 1 \\ \sin c & 0 & \cos c \\ \sin b \cos A & \sin b \sin A & \cos b \end{vmatrix}^2 = (\sin b \sin c \sin A)^2.$$

This volume is invariant to the specific Euclidean coordinate system we used in (1.9) and we could have used any other system. For example, using a coordinate system with B as the north pole, i.e., the z-axis along $\overrightarrow{OB}$ would yield $V^2 = (\sin a \sin c \sin B)^2$ and setting up the z-axis along $\overrightarrow{OC}$ would yield $V^2 = (\sin a \sin b \sin C)^2$. Equating these expressions and dividing throughout by $\sin^2 a \sin^2 b \sin^2 c$ yields

$$\frac{\sin^2 A}{\sin^2 a} = \frac{\sin^2 B}{\sin^2 b} = \frac{\sin^2 C}{\sin^2 c} = \frac{V^2}{\sin^2(a) \sin^2(b) \sin^2(c)}. \qquad (1.14)$$

1.2.3   Calculating distances on the earth's surface

Distance computations are indispensable in spatial analysis. Inter-site distance computations are used in variogram analysis to assess the strength of spatial association. They help in setting starting values for non-linear least squares algorithms in classical analysis (more in Chapter 2) and in specifying priors on the range parameter in Bayesian modeling (more in Chapter 6), making them crucial for correct the interpretation of spatial range and the convergence of statistical algorithms. For data sets covering relatively small spatial domains, ordinary Euclidean distance offers an adequate approximation. However, for larger domains (say, the entire continental U.S.) the curvature of the earth causes distortions because of the difference in differentials in longitude and latitude (a unit increment in degree longitude is not the same length as a unit increment in degree latitude except at the equator).

Suppose we have two points on the surface of the earth, $B = (\lambda_1, \phi_1)$ and $C = (\lambda_2, \phi_2)$, where ϕ_1 and λ_1 are the latitude and longitude, respectively, of the point B, while ϕ_2 and

λ_2 are those for the point C. The problem is to find the shortest distance (*geodesic*) between the points. The geodesic is the arc of the great circle joining the two points. Thus geodesic distance is the length of the (shorter) arc of a *great circle* (i.e., a circle with radius equal to the radius of the earth). Recall that the length of the arc of a circle equals the angle subtended by the arc at the center multiplied by the radius of the circle. Therefore it suffices to find the angle subtended by the arc.

This angle can be found by a straightforward application of (1.11) for a sphere of radius R and by relating the three sides and the angle A of the spherical triangle ABC to latitudes and longitudes. Let vertex A be the north pole, B be the first point and C be the second point. The geodesic distance between B and C is the length of the side a in the triangle. The latitudes are $\phi_1 = \pi/2 - c$ and $\phi_2 = \pi/2 - b$, and the vertex angle $A = \lambda_1 - \lambda_2$ is the difference between the longitudes (measured on the equatorial plane from the prime-meridian). Note that the relation (1.11), while derived explicitly for a sphere of unit radius, remains exactly the same for a sphere of radius R. This is because each of the vectors in (1.9) are multiplied by R, which means that R multiplies both sides of the equation $\overrightarrow{OA} \cdot \overrightarrow{OB} = \|\overrightarrow{OA}\| \|\overrightarrow{OB}\| \cos a$ and cancels out in the expression for $\cos a$. Now, applying (1.11) and using the relation of the angles b, c, and A with latitude and longitude yields

$$\cos a = \sin \phi_1 \sin \phi_2 + \cos \phi_1 \cos \phi_2 \cos (\lambda_1 - \lambda_2) . \qquad (1.15)$$

The geodesic arc length is $D = Ra$, where a is the angle measure in *radians*.

1.2.4   Mathematics of map projections

A map projection is a systematic representation of all or part of the surface of the earth on a plane. This typically comprises lines delineating meridians (longitudes) and parallels (latitudes), as required by some definitions of the projection. A well-known fact from topology is that it is impossible to prepare a distortion-free flat map of a surface curving in all directions. Thus, the cartographer must choose the characteristic (or characteristics) that are to be shown accurately on the map. There is no single "best" projection for mapping. The purpose of the projection and the application at hand lead to projections that are appropriate. Even for a single application, there may be several appropriate projections, and choosing the "best" projection is subjective. Indeed there are an infinite number of projections that can be devised, and several hundred have been published.

Since the sphere cannot be flattened onto a plane without distortion, the general strategy for map projections is to use an intermediate surface that can be flattened. This intermediate surface is called a *developable surface* and the sphere is first projected onto this surface, which is then laid out as a plane. The three most commonly used surfaces are the cylinder, the cone and the plane itself. Using different orientations of these surfaces produce different classes of map projections. Figure 1.7 presents the basic geometry of wrapping a cylinder around the sphere along its equator. The points on the globe are projected onto the wrapping (or tangential) surface, which is then laid out to form the map. These projections may be performed in several ways, giving rise to different projections. In particular, Figure 1.7 demonstrates one way of projecting a point on the sphere on to the cylinder. The cylinder wraps around This maps the point (λ, ϕ) on the sphere, where λ is the longitude and ϕ is the latitude to the point $(\lambda, \sin \phi)$ on the plane formed by unfolding the cylinder and endowing the plane with an xy coordinate system with origin at the intersection of the equator and the Greenwich meridian.

Before the age of computers, the above orientations were used by cartographers in the physical construction of maps. With computational advances and digitizing of cartography, analytical formulae for projections were devised. Here, we briefly outline the underlying theory for equal-area and conformal (locally shape-preserving) maps. More detailed treatments are available in Pearson II [1990] and Bugayevskiy and Snyder [1995].

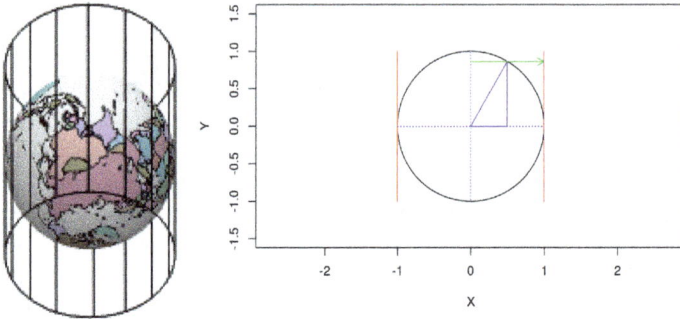

Figure 1.7 *A cylinder as the "developable" surface that wraps around the globe. The left panel shows the conceptual diagram, while the right panel presents one way of projecting a point on the sphere onto the cylinder (Lambert's projection). The green line is a perpendicular from the point on the sphere to the inner surface of the cylinder.*

The basic idea behind deriving equations for map projections is to consider a sphere with the geographical coordinate system (λ, ϕ) for longitude and latitude and to construct an appropriate (rectangular or polar) coordinate system (x, y) so that

$$x = f(\lambda, \phi),\ y = g(\lambda, \phi),$$

where f and g are appropriate functions to be determined, based upon the properties we want our map to possess. We will study map projections using differential geometry concepts, looking at infinitesimal patches on the sphere (so that curvature may be neglected and the patches are closely approximated by planes) and deriving a set of (partial) differential equations whose solution will yield f and g. Suitable initial conditions are set to create projections with desired geometric properties.

Thus, consider a small patch on the sphere formed by the infinitesimal quadrilateral, $ABCD$, given by the vertices,

$$A = (\lambda, \phi),\ B = (\lambda, \phi + d\phi),\ C = (\lambda + d\lambda, \phi),\ D = (\lambda + d\lambda, \phi + d\phi).$$

So, with R being the radius of the earth, the horizontal differential component along an arc of latitude is given by $|AC| = (R\cos\phi)d\lambda$ and the vertical component along a great circle of longitude is given by $|AB| = Rd\phi$. Since AC and AB are arcs along the latitude and longitude of the globe, they intersect each other at right angles. Therefore, the area of the patch $ABCD$ is given by $|AC||AB|$. Let $A'B'C'D'$ be the (infinitesimal) image of the patch $ABCD$ on the map. Then, $A' = (f(\lambda, \phi), g(\lambda, \phi))$, $B' = (f(\lambda, \phi + d\phi), g(\lambda, \phi + d\phi))$, $C' = (f(\lambda + d\lambda, \phi), g(\lambda + d\lambda, \phi))$ and $D' = (f(\lambda + d\lambda, \phi + d\phi), g(\lambda + d\lambda, \phi + d\phi))$. Therefore,

$$\overrightarrow{A'C'} = \left(\frac{\partial f}{\partial \lambda}, \frac{\partial g}{\partial \lambda}\right) d\lambda \text{ and } \overrightarrow{A'B'} = \left(\frac{\partial f}{\partial \phi}, \frac{\partial g}{\partial \phi}\right) d\phi.$$

If we seek an equal-area projection we need to equate the area of the patches $ABCD$ and $A'B'C'D'$. But note that the area of $A'B'C'D'$ is given by the area of the parallelogram formed by vectors $\overrightarrow{A'C'}$ and $\overrightarrow{A'B'}$. Treating them as vectors in the xy plane of an xyz system, we see that the area of $A'B'C'D'$ is the cross-product,

$$(\overrightarrow{A'C'}, 0) \times (\overrightarrow{A'B'}, 0) = \left(\frac{\partial f}{\partial \lambda}\frac{\partial g}{\partial \phi} - \frac{\partial f}{\partial \phi}\frac{\partial g}{\partial \lambda}\right) d\lambda d\phi.$$

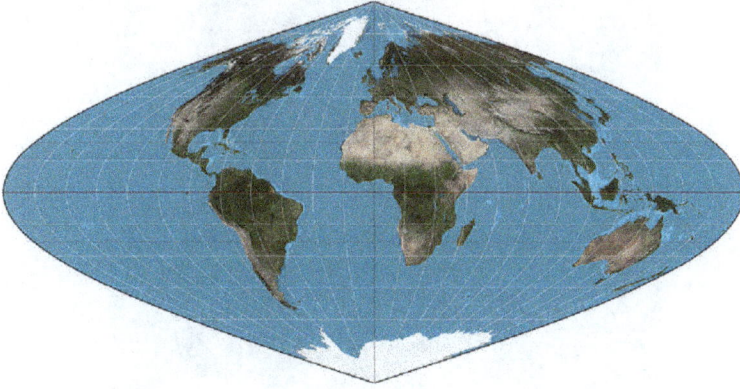

Figure 1.8 *The sinusoidal projection using the GeoCART software.*

Equating the above to $|AC||AB|$ renders a partial differential equation in f and g:

$$\left(\frac{\partial f}{\partial \lambda} \frac{\partial g}{\partial \phi} - \frac{\partial f}{\partial \phi} \frac{\partial g}{\partial \lambda} \right) = R^2 \cos \phi \,.$$

Note that this is the equation that must be satisfied by any equal-area projection. It is an underdetermined system and further conditions need to be imposed (that ensure other specific properties of the projection) to arrive at f and g.

Example 1.3 Equal-area maps are used for statistical displays of areal-referenced data. An easily derived equal-area projection is the sinusoidal projection. This is obtained by specifying $\partial g / \partial \phi = R$, which yields equally spaced straight lines for the parallels, and results in the following analytical expressions for f and g (with the 0 degree meridian as the central meridian):

$$f(\lambda, \phi) = R\lambda \cos \phi \,; \quad g(\lambda, \phi) = R\phi.$$

This is shown in Figure 1.8 ∎

Example 1.4 Another popular equal-area projection (with equally spaced straight lines for the meridians) is the Lambert cylindrical projection given by

$$f(\lambda, \phi) = R\lambda \,; \quad g(\lambda, \phi) = R \sin \phi \,.$$

This corresponds to the geometric construction shown in Figure 1.7. ∎

For conformal (angle-preserving) projections we set the angle $\angle(AC, AB)$ equal to $\angle(A'C', A'B')$. Since $\angle(AC, AB) = \pi/2$, $\cos(\angle(AC, AB)) = 0$, leading to

$$\frac{\partial f}{\partial \lambda} \frac{\partial f}{\partial \phi} + \frac{\partial g}{\partial \lambda} \frac{\partial g}{\partial \phi} = 0$$

or, equivalently, the Cauchy-Riemann equations of complex analysis,

$$\left(\frac{\partial f}{\partial \lambda} + i \frac{\partial g}{\partial \lambda} \right) \left(\frac{\partial f}{\partial \phi} - i \frac{\partial g}{\partial \phi} \right) = 0 \,.$$

A sufficient partial differential equation system for conformal mappings of the Cauchy-Riemman equations that simpler to use is

$$\frac{\partial f}{\partial \lambda} = \frac{\partial g}{\partial \phi} \cos \phi; \quad \frac{\partial g}{\partial \lambda} = \frac{\partial f}{\partial \phi} \cos \phi \,.$$

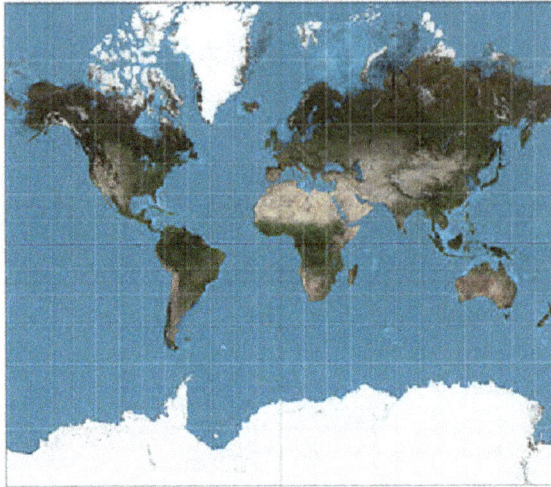

Figure 1.9 *The Mercator projection using the GeoCART software.*

Example 1.5 The Mercator projection shown in Figure 1.9 is a classical example of a conformal projection. It has the interesting property that rhumb lines (curves that intersect the meridians at a constant angle) are shown as straight lines on the map. This is particularly useful for navigation purposes. The Mercator projection is derived by letting $\partial g/\partial \phi = R\sec\phi$. After suitable integration, this leads to the analytical equations (with the 0 degree meridian as the central meridian),

$$f(\lambda, \phi) = R\lambda; \quad g(\lambda, \phi) = R\ln\tan\left(\frac{\pi}{4} + \frac{\phi}{2}\right) = \frac{R}{2}\ln\left(\frac{1 + \sin\phi}{1 - \sin\phi}\right),$$

where the last equality follows from an elementary trigonometric identity. ∎

As seen above, even the simplest map projections lead to complex transcendental equations relating latitude and longitude to positions of points on a given map. Therefore, rectangular grids have been developed for use by surveyors to designate each point by its distance from two perpendicular axes on a flat map. The y-axis usually coincides with a chosen central meridian, y increasing north, and the x-axis is perpendicular to the y-axis at a latitude of origin on the central meridian, with x increasing east. Frequently, the x and y coordinates are called "eastings" and "northings," respectively, and to avoid negative coordinates, may have "false eastings" and "false northings" added to them. The grid lines usually do not coincide with any meridians and parallels except for the central meridian and the equator. In Sections 1.2.5 and 1.2.6 we take a closer look at the Mercator projection of Example 1.5 and how it makes its way to a widely used grid in practice.

Spatial modeling of point-level data often requires computing distances between points on the earth's surface. One might wonder about a *planar* map projection, which would preserve distances between points. Unfortunately, the existence of such a map is precluded by Gauss' Theorema Eggregium in differential geometry [see, e.g., Guggenheimer, 1977, pp. 240–242]. Thus, while we have seen projections that preserve area and shapes, distances are always distorted. The *gnomonic* projection [Snyder, 1987, pp. 164–168] gives the correct distance from a single reference point, but is less useful for the spatial data analyst who needs to obtain complete inter-site distance matrices (since this would require not one but many such maps). Banerjee [2005] explores different strategies for computing distances on the earth and their impact on statistical inference. We present a brief summary in Section 1.3.

1.2.5 The Mercator and Transverse Mercator projections

Some further intuition for the Mercator projection is worthwhile, given its importance and its wide use in the Universal Transverse Mercator (UTM) projection. The equator is a circle with radius R and circumference $2\pi R$. The parallel at latitude ϕ is a circle of radius $R\cos\phi$ so its circumference is $2\pi R\cos\phi$. Therefore, on the sphere, the length of a parallel depends upon the latitude. However, the parallels all have the same length on the map. This means that the parallel with latitude ϕ on a map is stretched by a factor $1/\cos\phi$. Suppose we wish to travel along the diagonal of a very small square patch on the surface of the earth at latitude ϕ, say from its northeast corner to its southwest corner. Let hR be the length of the sides of this square patch (h representing a very small number yielding length as a fraction of R). The Mercator projection preserves angles, which means that we travel southwest both on the earth and on the map. Furthermore, since angles are preserved for all points, the image of a square patch is also a square on Mercator's map.

Now consider some basic properties of the small patch and how it translates to the map. The side length of the patch represents $\frac{hR}{2\pi R\cos(\phi)}$ of the length of the parallel at latitude ϕ, so its width on the map is $\frac{hR}{2\pi R\cos(\phi)}2\pi = h\sec(\phi)$. This means that its height on the map must also be $h\sec\theta$. When preparing a map we need to fix a scale. Let us assume that the equator will have length 2π on the map and let $F(\phi)$ be the distance from the equator to the parallel at latitude ϕ on the map. Consider the differential from ϕ to $\phi + h$, where $h = \Delta\phi$ so that the side length of the square patch is $hR = R\Delta\phi$. The corresponding height of the square patch on the map is $\Delta F(\phi) = F(\phi + h) - F(\phi) \approx h\sec\phi = \sec\phi\Delta\phi$. Thus $\frac{\Delta F}{\Delta\phi} \approx \sec(\phi)$. In the limit as $\Delta\phi \to 0$, we get $\frac{dF}{d\phi} = \sec(\phi)$. Taking the equator as height 0, we find $F(0) = 0$ and we obtain $F(\phi) = \int_0^\phi \sec(x)dx = \log\tan\left(\frac{\phi}{2} + \frac{\pi}{4}\right)$.

Different classes of map projections can be derived from different orientations (referred to as the "aspect" by cartographers) of the "developable" surface wrapping the sphere. One such widely used projection is the transverse Mercator projection. In the normal "aspect" of the Mercator projection given in Example 1.5, the cylinder stands vertically around the sphere touching it at the equator (Figure 1.7). The transverse Mercator orients the cylinder by rotating it horizontally 90 degrees and wrapping the earth to touch the Earth along a meridian instead of the equator. If a spherical model is assumed, which is what we do here as it usually suffices for most statistical applications, then this projection is called the Gauss-Lambert projection. If an ellipsoidal model is used, then we refer to it as the Gauss-Kruger projection.

We can derive the equations of the transverse Mercator projection by rotating the axes of the normal Mercator and using some spherical trigonometry. First, we need to choose a meridian that will be tangential to the cylinder. Recall that in the normal aspect we conveniently chose the equator to be tangential to the cylinder as it is a great circle. For the transverse aspect, we can choose any meridian, but for reference we choose Greenwich and 180 degrees East (extending the Greenwich meridian over the poles to the other side of the globe) as the point of contact. This ensures that the scale is unity on the meridian. We now seek functions $f(\lambda, \phi)$ and $g(\lambda, \phi)$ such that the projection is also conformal.

We supply a brief sketch for deriving the transverse Mercator formulas. For a rough visualization of the process, we refer to Figure 1.10. We set up a 2-dimensional euclidean coordinate system for the horizontally oriented cylinder with the origin at where the equator meets the Greenwich meridian. The y-axis goes north-south and the x-axis goes east-west with both axes being tangential to the Greenwich meridian and the equator at the point where they meet (we label this point O). Therefore, the y-axis runs in a direction parallel to the width of the cylinder and the x-axis runs parallel to the length (or height) of the cylinder. These axes are shown in the left panel of Figure 1.10 and they meet at the point O. The meridian of the point P on the sphere meets the equator (at a right angle) at the

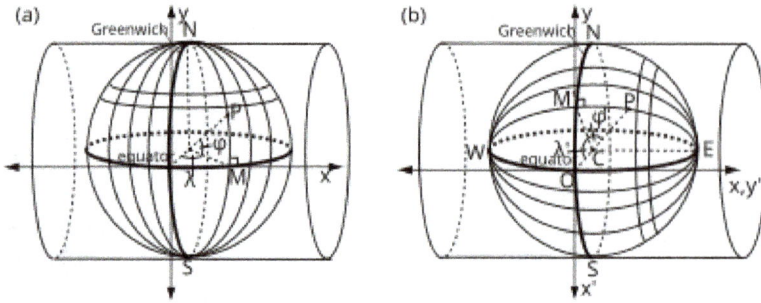

Figure 1.10 *The horizontal orientation of the cylinder and the coordinate system for the transverse Mercator projection.*

point M. If C is the center of the sphere, then ϕ is the angle between $\overrightarrow{CP}$ and $\overrightarrow{CM}$, while λ is the angle on the equatorial plane made by the triangles OCN and MCN.

We now introduce a new coordinate system (x', y') such that the x' and y' axes coincide with the original y axis and x axis, respectively. This is shown in the right panel of Figure 1.10. The tangential (Greenwich) meridian in the left panel is our new "equator" which defines a new (λ', ϕ') system of geographic coordinates. The point M' denotes the intersection of the meridian containing P with the new "equator" in the rotated coordinate system. With these new coordinate systems, we now see that x' and y' give us the normal Mercator projection in terms of λ' and ϕ' with a change in sign for the x' coordinate. Therefore,

$$x' = -R\lambda' ; \quad y' = R \ln \tan\left(\frac{\pi}{4} + \frac{\phi'}{2}\right) = \frac{R}{2} \ln\left(\frac{1 + \sin\phi'}{1 - \sin\phi'}\right) . \qquad (1.16)$$

What remains is to express (x', y') and (λ', ϕ') in terms of (x, y) and (λ, ϕ). The former is especially simple as it is given by a rotation of axes by 90 degrees. We see $x' = -y$ and $y' = x$. Using this along with (1.16) we obtain

$$y = R\lambda' ; \quad x = R \ln \tan\left(\frac{\pi}{4} + \frac{\phi'}{2}\right) = \frac{R}{2} \ln\left(\frac{1 + \sin\phi'}{1 - \sin\phi'}\right) . \qquad (1.17)$$

The relationship between the geographic coordinates need a bit more work. Figure 1.11 serves to illustrate the angles λ, ϕ, λ' and ϕ' in a single sphere. The points M and M' are the same as in Figure 1.10. We focus on the spherical triangle $NM'P$, which we relabel as ABC (to better connect with the notation in Section 1.2.2) such that two of the vertex angles are $A = \lambda$ and $B = \pi/2$, while the three sides are $a = \phi'$, $b = \pi/2 - \phi$ and $c = \pi/2 - \lambda'$.

Since $\sin B = 1$, the first equality in (1.13) (and assuming all angles are acute) yields

$$\sin\lambda / \sin\phi' = 1 / \sin(\pi/2 - \phi) \quad \text{or that} \quad \sin\phi' = \sin\lambda\cos\phi . \qquad (1.18)$$

This gives ϕ' in terms of λ and ϕ. Applying (1.11) to the sides $a = \phi'$, $b = \pi/2 - \phi$ render

$$\cos\phi' = \sin\phi\sin\lambda' + \cos\phi\cos\lambda'\cos\lambda \quad \text{and} \quad \sin\phi = \cos\phi'\sin\lambda' , \qquad (1.19)$$

where the last expression has simplified because $\cos B = 0$. Different, but equivalent, expressions for λ' in terms of λ and ϕ arise from different trigonometric identities. One popular expression is derived by eliminating $\cos\phi'$ from (1.19). Substituting $\cos\phi'$ in the second expression with its value in the first renders $\sin\phi = \sin\phi\sin^2\lambda' + \cos\phi\cos\lambda\cos\lambda'\sin\lambda'$. Dividing both sides by $\cos\phi$ and simplifying a bit further reveals $\tan\lambda' = \sec\lambda\tan\phi$. Substituting the values of ϕ' and λ' obtained as functions of ϕ and λ in (1.17) gives the formulas

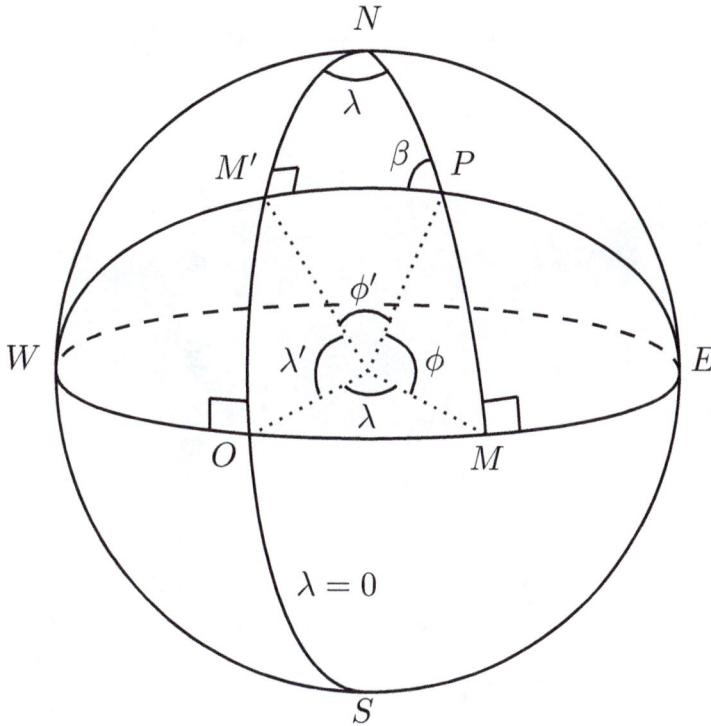

Figure 1.11 *Figure showing the relationships between (λ', ϕ') and (λ, ϕ) using spherical triangles.*

for the transverse Mercator projection as

$$x = \frac{R}{2} \ln \left(\frac{1 + \sin \lambda \cos \phi}{1 - \sin \lambda \cos \phi} \right) \; ; \quad y = R \arctan(\sec \lambda \tan \phi) \,. \tag{1.20}$$

1.2.6 Universal Transverse Mercator (UTM) projections

It remains to form a grid so that the transverse Mercator projection in (1.20) can be applied locally within each cell. One such popular grid, adopted by The National Imagery and Mapping Agency (NIMA) (formerly known as the Defense Mapping Agency) and used especially for military use throughout the world, is the Universal Transverse Mercator (UTM) grid; see Figure 1.12. The UTM divides the world into 60 north-south zones, each of width six degrees longitude. Starting with Zone 1 (between 180 degrees and 174 degrees west longitude), these are numbered consecutively as they progress eastward to Zone 60, between 174 degrees and 180 degrees east longitude. Within each zone, coordinates are measured north and east in meters, with northing values being measured continuously from zero at the Equator, in a northerly direction. Negative numbers for locations south of the Equator are avoided by assigning an arbitrary false northing value of 10,000,000 meters (as done by NIMA's cartographers). A central meridian cutting through the center of each 6 degree zone is assigned an easting value of 500,000 meters, so that values to the west of the central meridian are less than 500,000 while those to the east are greater than 500,000. In particular, the conterminous 48 states of the United States are covered by 10 zones, from Zone 10 on the west coast through Zone 19 in New England.

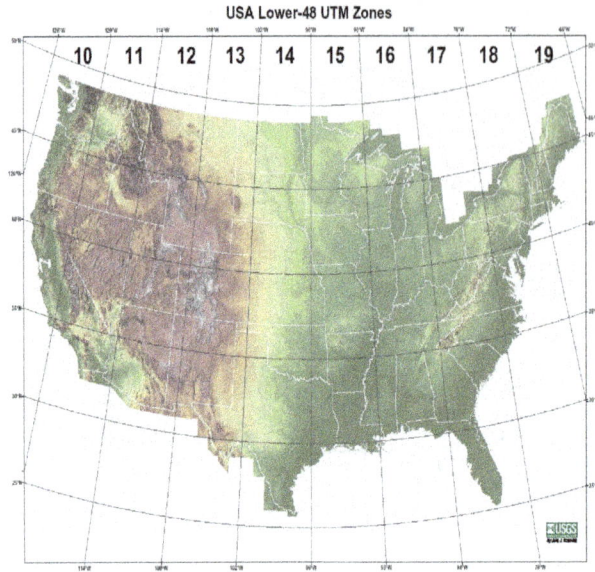

Figure 1.12 *Example of a UTM grid over the United States (figure courtesy of the U.S. Geological Survey).*

The UTM envisages overlaying a transparent grid on the map, allowing distances to be measured in meters at the map scale between any map point and the nearest grid lines to the south and west. The northing of the point is calculated as the sum of the value of the nearest grid line south of it and its distance north of that line. Similarly, its easting is the value of the nearest grid line west of it added to its distance east of that line. For instance, in Figure 1.13, the grid value of line A-A is 357,000 meters east, while that of line B-B is 4,276,000 meters north. Point P is 800 meters east and 750 meters north of the grid lines resulting in the grid coordinates of point P as north 4,276,750 and east 357,800.

1.3   Euclidean approximations to geodesic distances

While calculating (1.21) is straightforward, Euclidean metrics are popular due to their simplicity and easier interpretation. More crucially, statistical modeling of spatial correlations proceeds from *correlation functions* that are often valid only with Euclidean metrics. For example, using (1.21) to calculate the distances in general covariance functions may not result in a positive definite $\Sigma(\boldsymbol{\theta})$ in (1.1). We consider a few different approaches for computing distances on the earth using Euclidean metrics, classifying them as those arising from the classical spherical coordinates, and those arising from planar projections.

Equation (1.21) reveals that the relationship between the Euclidean distances and the geodetic distances is not just a matter of scaling. We cannot multiply one by a constant number to obtain the other. A simple scaling of the geographical coordinates results in a "naive Euclidean" metric obtained directly in degree units, and converted to kilometer units as $\|P_1 - P_2\|\pi R/180$. This may suffice for small domains but always overestimates the geodesic distance. It stretches distances by *flattening out* the meridians and parallels onto a plane. The accuracy deteriorates as the domain expands.

Banerjee [2005] also explores a more natural metric, which is along the "chord" joining the two points. This is simply the Euclidean metric $\|\mathbf{u}_2 - \mathbf{u}_1\|$, yielding a "burrowed through the earth" distance—the chordal length between P_1 and P_2. The slight underestimation of the geodesic distance is expected, since the chord "penetrates" the domain, producing a

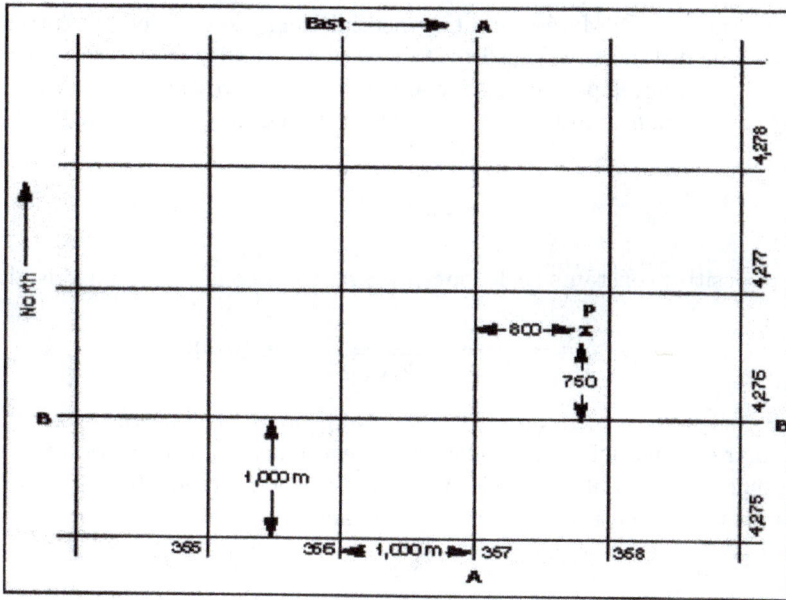

Figure 1.13 *Finding the easting and northing of a point in a UTM projection (figure courtesy of the U.S. Geological Survey).*

straight-line approximation to the geodetic arc. The first three rows of Table 1.1, compare the geodetic distance with the "naive Eucidean" and chordal metrics. The next three rows show distances computed by using three planar projections: the Mercator, the Sinusoidal and a centroid-based data projection, which is developed in Exercise 13. The first column corresponds to the distance between the farthest points in a spatially referenced data set comprising 50 locations in Colorado (we will revisit this dataset later in Chapter 6), while the next two present results for two differently spaced pairs of cities. The overestimation and underestimation of the "naive Euclidean" and "chordal" metrics respectively is clear, although the chordal metric excels even for distances over 2000 kms (New York and New Orleans). We find that the sinusoidal and centroid-based projections seem to be distorting distances much less than the Mercator, which performs even worse than the naive Euclidean.

Methods	Colorado	Chicago-Minneapolis	New York-New Orleans
geodetic	741.7	562.0	1897.2
naive Euclidean	933.8	706.0	2172.4
chord	741.3	561.8	1890.2
Mercator	951.8	773.7	2336.5
sinusoidal	742.7	562.1	1897.7
centroid-based	738.7	562.2	1901.5

Table 1.1 *Comparison of different methods of computing distances (in kms). For Colorado, the distance reported is the maximum inter-site distance for a set of 50 locations.*

This approximation of the chordal metric has an important theoretical implication for the spatial modeler. A troublesome aspect of geodetic distances is that they are *not* necessarily valid arguments for correlation functions defined on Euclidean spaces (see Chapter 2 for more general forms of correlation functions). However, the excellent approximation of the chordal metric (which is Euclidean) ensures that in most practical settings valid correlation functions in $\Re^3$ such as the Matérn and exponential yield positive definite correlation matrices with geodetic distances and ensures legitimate probability models for inference.

Schoenberg [1942] develops a necessary and sufficient representation for valid positive-definite functions on spheres in terms of normalized Legendre polynomials P_k of the form:

$$\psi(t) = \sum_{k=0}^{\infty} a_k P_k(\cos t),$$

where a_k's are positive constants such that $\sum_{k=0}^{\infty} a_k$ converges. An example is given by

$$\psi(t) = \frac{1}{\sqrt{1 + \alpha^2 - 2\alpha \cos t}}, \quad \alpha \in (0,1),$$

which can be easily shown to have the Legendre polynomial expansion $\sum_{k=0}^{\infty} \alpha^k P_k(\cos t)$. The chordal metric also provides a simpler way to construct valid correlation functions over the sphere using a sinusoidal composition of any valid correlation function in Euclidean space. To see this, consider a unit sphere ($R = 1$) and note that

$$\|\mathbf{u}_1 - \mathbf{u}_2\| = \sqrt{2 - 2\langle \mathbf{u}_1, \mathbf{u}_2 \rangle} = 2\sin(\phi/2).$$

Therefore, a correlation function $\rho(d)$ (suppressing the range and smoothness parameters) on the Euclidean space transforms to $\rho(2\sin(\phi/2))$ on the sphere, thereby *inducing* a valid correlation function on the sphere. This has some advantages over the Legendre polynomial approach of Schoenberg: (1) we retain the interpretation of the smoothness and decay parameters, (2) is simpler to construct and compute, and (3) builds upon a rich legacy of investigations (both theoretical and practical) of correlation functions on Euclidean spaces (again, see Chapter 2 for different correlation functions).

1.4   Software and datasets

This text extensively uses the R [R Core Team, 2021] software programming language (www.r-project.org) and environment for statistical computing and graphics. R is released under the GNU open-source license and can be downloaded for free from the Comprehensive R Archive Network (https://cran.r-project.org/). The capabilities of R are extended through "libraries" or "packages" that perform specialized tasks. These packages are also available from CRAN and can be downloaded and installed from within the R software environment.

The R statistical software environment today offers excellent interfaces with Geographical Information Systems (GIS) through a number of libraries (or packages). At the core of R's GIS capabilities is the maps library originally described by Becker et al. [1997] and Becker and Wilks [1998]. This map library contains the geographic boundary files for several maps, including county boundaries for every state in the U.S.

There are a variety of spatial packages in R that perform modeling and analysis for the different types of spatial data. For example, the gstat and geoR packages provide functions to perform traditional (classical) analysis for point-level data; the latter also offers simpler Bayesian models. The packages spBayes and spTimer have much more elaborate Bayesian functions, the latter focusing primarily on space-time data. The spdep package in R provides several functions for analyzing areal-level data, including basic descriptive statistics for areal data as well as fitting areal models using classical likelihood methods. Turning to point-process models, a popular spatial R package, spatstat, allows computation of K for any data set, as well as the approximate 95% intervals for it so the significance of departure from some theoretical model may be judged. However, full inference likely requires use of the R package Splancs, or perhaps a fully Bayesian approach with user-specific coding [also see Wakefield and Morris, 2001]. We will use a number of spatial and spatiotemporal datasets to illustrate the modeling and implementation of software. The number of packages performing spatial analysis is already too large to be discussed in this text. We refer the reader to the CRAN Task View (http://cran.r-project.org/web/views/Spatial.html) for an exhaustive list of such packages and brief descriptions regarding their capabilities.

This monograph emphasizes Bayesian inference using hierarchical models. Implementing these models rely upon Markov chain Monte Carlo (MCMC) algorithms that are implemented in several of the aforementioned R libraries. In addition, there are specialized Bayesian modeling environments built around the BUGS language [Gilks et al., 1994, Lunn et al., 2000, 2009, Goudie et al., 2020]. Notable Bayesian modeling environments offered through R packages include BRugs, R2OpenBUGS, RJAGS, rstan, and NIMBLE. Most of these packages implement MCMC algorithms including Gibbs sampling, Metropolis random walks, Hamiltonian Monte Carlo (notably by rstan) and slice sampling. Integrated Nested Laplace Approximation [Rue et al., 2009, Lindgren et al., 2011] is another very popular computational tool that relies upon posterior approximations using Gaussian Markov random fields and is implemented in R-INLA (hosted outside of CRAN). A host of computational resources, including Approximate Bayesian Computation [Csillery et al., 2012, Dutta et al., 2021] and variational inference [Fox and Roberts, 2012, Blei et al., 2017], are available from the CRAN Bayesian Task View.

1.5 Website accompanying this textbook

Computer programs and datasets that are used for illustrations in this text and can be shared publicly are available for free access and download at https://github.com/sudiptobanerjee/BGC_2023. Datasets are located in the "data" directory, while computer programs and resources are arranged according to chapters. Computer programs are often illustrated using HTML documents generated using the rmarkdown package.

1.6 Exercises

1. What sorts of areal unit variables can you envision that could be viewed as arising from point-referenced variables? What sorts of areal unit variables can you envision whose mean could be viewed as arising from a point-referenced surface? What sorts of areal unit variables fit neither of these scenarios?

2. What sorts of sensible properties should characterize association between point-referenced measurements? What sorts of sensible properties should characterize the association between areal unit measurements?

3. Suggest some regional-level covariates that might help explain the spatial pattern evident in Figure 1.2. (*Hint:* The roughly rectangular group of regions located on the map's eastern side is the city of Minneapolis, MN.)

4.(a) Suppose you recorded elevation and average daily temperature on a particular day for a sample of locations in a region. If you were given the elevation at a new location, how would you make a plausible estimate of the average daily temperature for that location?

  (b) Why might you expect a spatial association between selling prices of single-family homes in this region to be weaker than that between the observed temperature measurements?

5. For what sorts of point-referenced spatial data would you expect measurements across time to be essentially independent? For what sorts of point-referenced data would you expect measurements across time to be strongly dependent?

6. For point-referenced data, suppose the means of the variables are spatially associated. Would you expect the association between the variables themselves to be weaker than, stronger than, or the same as the association between the means?

7. For a spatial point pattern of observed locations for a plant species, what sorts of environmental predictors might be useful to explain this point pattern? Would you expect there to be spatial dependence in the set of locations? If so, what do you think modeling of this dependence might be able to capture?

8. We briefly review the *dot product* in 2-dimensional and 3-dimensional real Euclidean geometry. Let $\mathbf{u} = (u_1, \ldots, u_d)^{\mathrm{T}}$ and $\mathbf{v} = (u_1, \ldots, u_d)^{\mathrm{T}}$ be two $d \times 1$ vectors, where $d = 2$ or $d = 3$. The *dot-product* or *scalar product* is defined as $\mathbf{u} \cdot \mathbf{v} = \sum_{i=1}^{d} u_i v_i = \mathbf{v} \cdot \mathbf{u}$. If $\mathbf{u}$ and $\mathbf{v}$ are treated as $d \times 1$ column vectors, then $\mathbf{u} \cdot \mathbf{v} = \mathbf{u}^{\mathrm{T}}\mathbf{v}$ is a matrix product.

   (a) Explain why defining the length $\|\mathbf{u}\| = \sqrt{\mathbf{u} \cdot \mathbf{u}}$ is consistent with elementary geometry. *Hint:* Pythagoras.

   (b) Prove that $(\alpha\mathbf{u}) \cdot \mathbf{v} = \alpha(\mathbf{u} \cdot \mathbf{v}) = \mathbf{u} \cdot (\alpha\mathbf{v})$.

   (c) Prove that $\mathbf{u} \cdot (\mathbf{v} + \mathbf{w}) = \mathbf{u} \cdot \mathbf{v} + \mathbf{u} \cdot \mathbf{w}$.

   (d) If $\mathbf{u} \cdot \mathbf{v} = 0$, then prove the Pythagorean identity:

$$\|\mathbf{u} + \mathbf{v}\|^2 = \|\mathbf{u}\|^2 + \|\mathbf{v}\|^2 = \|\mathbf{u} + \mathbf{v}\|^2 \,.$$

   (e) Prove the Cauchy-Schwarz inequality for dot products: $|\mathbf{u} \cdot \mathbf{v}| \leq \|\mathbf{u}\|\|\mathbf{v}\|$.

   (f) Define the cosine function as

$$\cos\theta = \frac{\mathbf{u} \cdot \mathbf{v}}{\|\mathbf{u}\|\|\mathbf{v}\|} \,,$$

   where θ is the angle between $\mathbf{u}$ and $\mathbf{v}$. Explain why this definition of the cosine function is equivalent to that in elementary trigonometry. *Hint:* Expand $\|\mathbf{u} - \mathbf{v}\|^2 = (\mathbf{u} - \mathbf{v}) \cdot (\mathbf{u} - \mathbf{v})$ in terms of $\|\mathbf{u}\|$, $\|\mathbf{v}\|$ and $\mathbf{u} \cdot \mathbf{v}$; compare with the law of cosines in plane trigonometry.

   (g) Define $\mathbf{u}$ and $\mathbf{v}$ to be orthogonal if $\mathbf{u} \cdot \mathbf{v} = 0$. Explain why this makes geometric sense using the above definition of the cosine function.

9. This exercise computes the geodesic distance between two points using a different spherical coordinate system from the one used in (1.9). We form a three-dimensional Cartesian coordinate system (x, y, z), with the origin at the center of the earth, the z-axis along the North and South Poles, and the x-axis on the plane of the equator joining the center of the earth and the Greenwich meridian. Let $P_1 = (\theta_1, \lambda_1)$ and $P_2 = (\theta_2, \lambda_2)$, where θ_1 and λ_1 are the latitude and longitude, respectively, of the point P_1, while θ_2 and λ_2 are those for the point P_2. Using Figure 1.5 as a guide and recalling elementary trigonometry, prove the following relationships between (x, y, z) and latitude-longitude (θ, λ):

$$x \;=\; R\cos\theta\cos\lambda, \quad y = R\cos\theta\sin\lambda \quad \text{and} \quad z = R\sin\theta \,.$$

   Form the vectors $\mathbf{u}_1 = (x_1, y_1, z_1)$ and $\mathbf{u}_2 = (x_2, y_2, z_2)$ as the Cartesian coordinates corresponding to points P_1 and P_2. Hence, ϕ is the angle between $\mathbf{u}_1$ and $\mathbf{u}_2$. With Figure 1.14 as a guide, derive the geodesic distance formula on a sphere of radius R:

$$D = R\phi = R\arccos[\sin\theta_1\sin\theta_2 + \cos\theta_1\cos\theta_2\cos(\lambda_1 - \lambda_2)] \,. \qquad (1.21)$$

10.(a) Write your own R function that will compute the distance between 2 points P_1 and P_2 on the surface of the earth. The function should take the latitude and longitude of the P_i as input, and output the geodesic distance D given in (1.21). Use $R = 6371$ km.

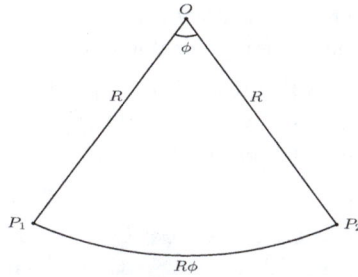

Figure 1.14 *Diagrams illustrating the geometry underlying the calculation of great circle (geodesic) distance.*

(b) Use your program to obtain the geodesic distance between Chicago (87.63W, 41.88N) and Minneapolis (93.22W, 44.89N), and between New York (73.97W, 40.78N) and New Orleans (90.25W, 29.98N).

11. A "naive Euclidean" distance may be computed between two points by simply applying the Euclidean distance formula to the longitude-latitude coordinates, and then multiplying by $(R\pi/180)$ to convert to kilometers. Find the naive Euclidean distance between Chicago and Minneapolis, and between New York and New Orleans, comparing your results to the geodesic ones in the previous problem.

12. The *chordal* ("burrowing through the earth") distance separating two points is given by the Euclidean distance applied to the cartesian spherical coordinate system given in Subsection 1.2.3. Find the chordal distance between Chicago and Minneapolis, and between New York and New Orleans, comparing your results to the geodesic and naive Euclidean ones above.

13. A two-dimensional projection, often used to approximate geodesic distances by applying Euclidean metrics sets up rectangular axes along the centroid of the observed locations, and scales the points according to these axes. Thus, with N locations having geographical coordinates $(\lambda_i, \theta_i)_{i=1}^{N}$, we first compute the centroid $(\bar{\lambda}, \bar{\theta})$ (the mean longitude and latitude). Next, two distances are computed. The first, d_X, is the geodesic distance (computed using (1.21)) between $(\bar{\lambda}, \theta_{\min})$ and $(\bar{\lambda}, \theta_{\max})$, where $\theta_{\min}$ and $\theta_{\max}$ are the minimum and maximum of the observed latitudes. Analogously, d_Y is the geodesic distance computed between $(\lambda_{\min}, \bar{\theta})$ and $(\lambda_{\max}, \bar{\theta})$. These actually scale the axes in terms of true geodesic distances. The projection is then given by

$$ x = \frac{\lambda - \bar{\lambda}}{\lambda_{\max} - \lambda_{\min}} d_X; \text{ and } y = \frac{\theta - \bar{\theta}}{\theta_{\max} - \theta_{\min}} d_Y . $$

Applying the Euclidean metric to the projected coordinates yields a good approximation to the inter-site geodesic distances. This projection is useful for entering coordinates in spatial statistics software packages that require two-dimensional coordinate input and uses Euclidean metrics to compute distances.

(a) Compute the above projection for Chicago and Minneapolis ($N = 2$) and find the Euclidean distance between the projected coordinates. Compare with the geodesic distance. Repeat this exercise for New York and New Orleans.

(b) When will the above projection fail to work?

14. Use the fields package to produce the inter-site distance matrix for the locations in the scallops data. Compute this matrix using the `rdist.earth` function, which yields the geodetic distances. Next project the data to UTM coordinates and use the `rdist`

function to compute the inter-site Euclidean distance matrix. Draw histograms of the inter-site distances and comment on any notable discrepancies.

15. We collect some basic identities regarding vector products in 3-dimensional Euclidean geometry. Let $\mathbf{u} = (u_1, u_2, u_3)^\mathsf{T}$ and $\mathbf{v} = (u_1, u_2, u_3)^\mathsf{T}$ be two 3×1 vectors. The *cross-product* or *vector product* is defined as the skew-symmetric linear transformation,

$$\mathbf{u} \times \mathbf{v} = \begin{bmatrix} u_2 v_3 - u_3 v_2 \\ u_3 v_1 - u_1 v_3 \\ u_1 v_2 - u_2 v_1 \end{bmatrix} = \mathbf{U}_\times \mathbf{v} , \quad \text{where} \quad \mathbf{U}_\times = \begin{bmatrix} 0 & -u_3 & u_2 \\ u_3 & 0 & -u_1 \\ -u_2 & u_1 & 0 \end{bmatrix} . \quad (1.22)$$

(a) Prove that $\mathbf{u} \times \mathbf{u} = \mathbf{0}$; $\mathbf{v} \times \mathbf{u} = -(\mathbf{u} \times \mathbf{v})$; and $(c\mathbf{u}) \times \mathbf{v} = \mathbf{u} \times (c\mathbf{v}) = c(\mathbf{u} \times \mathbf{v})$.

(b) Prove $\mathbf{u} \times (\mathbf{v} + \mathbf{w}) = \mathbf{u} \times \mathbf{v} + \mathbf{u} \times \mathbf{w}$. Derive the analogous expression for $(\mathbf{v} + \mathbf{w}) \times \mathbf{u}$ without explicitly using (1.22).

(c) Prove that the cross-product and dot-product are related by

$$\|\mathbf{u} \times \mathbf{v}\|^2 = \|\mathbf{u}\|^2 \|\mathbf{v}\|^2 - (\mathbf{u} \cdot \mathbf{v})^2 = \|\mathbf{u}\|^2 \|\mathbf{v}\|^2 \sin^2 \theta , \quad (1.23)$$

where θ is the angle between $\mathbf{u}$ and $\mathbf{v}$. *Hint:* Recall $\mathbf{u} \cdot \mathbf{v} = \|\mathbf{u}\| \|\mathbf{v}\| \cos \theta$.

(d) Prove that $\mathbf{u} \times \mathbf{v}$ is orthogonal to both $\mathbf{u}$ and $\mathbf{v}$. Prove that $\mathbf{u} \times \mathbf{v}$ is orthogonal to any vector in the plane spanned by $\mathbf{u}$ and $\mathbf{v}$.

(e) Explain why the value of $\|\mathbf{u} \times \mathbf{v}\|$ is the area of the parallelogram with sides $\mathbf{u}$ and $\mathbf{v}$.

(f) Prove the relationship of the cross product of 3 vectors with the dot product:

$$\mathbf{u} \times (\mathbf{v} \times \mathbf{w}) = (\mathbf{u} \cdot \mathbf{w})\mathbf{v} - (\mathbf{u} \cdot \mathbf{v})\mathbf{w} . \quad (1.24)$$

16. The scalar triple product of three vectors $\mathbf{u}$, $\mathbf{v}$ and $\mathbf{w}$ is defined as $\mathbf{u} \cdot (\mathbf{v} \times \mathbf{w})$.

(a) Prove that

$$\mathbf{u} \cdot (\mathbf{v} \times \mathbf{w}) = \det \left(\begin{bmatrix} u_1 & u_2 & u_3 \\ v_1 & v_2 & v_3 \\ w_1 & w_2 & w_3 \end{bmatrix} \right) = \det \left(\begin{bmatrix} u_1 & v_1 & w_1 \\ u_2 & v_2 & w_2 \\ u_3 & v_3 & w_3 \end{bmatrix} \right) = \det([\mathbf{u} \quad \mathbf{v} \quad \mathbf{w}]) . \quad (1.25)$$

(b) Prove that the scalar triple product is unchanged under a circular shift of its three operands $\mathbf{u}$, $\mathbf{v}$ and $\mathbf{w}$. That is,

$$\mathbf{u} \cdot (\mathbf{v} \times \mathbf{w}) = \mathbf{v} \cdot (\mathbf{w} \times \mathbf{u}) = \mathbf{w} \cdot (\mathbf{u} \times \mathbf{v}) \quad (1.26)$$

Hint: Recall properties of the determinant and use (1.25).

(c) Prove that $\mathbf{u} \cdot (\mathbf{v} \times \mathbf{w}) = 0$ if at least two of $\mathbf{u}$, $\mathbf{v}$ and $\mathbf{w}$ are equal.

(d) Prove that $|\mathbf{u} \cdot (\mathbf{v} \times \mathbf{w})|$ is the volume of the parallelepiped with sides $\mathbf{u}$, $\mathbf{v}$ and $\mathbf{w}$.

17. Exercises 15 and 16 sets the stage for a coordinate-free proof of the spherical law of sines. Consider, again, a spherical triangle with vertices A, B and C on the unit sphere, as in Section 1.2.2, but do not impose any coordinate system on them. Since a vertex angle is the angle between the normal to the planes meeting at that vertex, (1.23) yields

$$\sin^2 A = \frac{\|(\overrightarrow{OB} \times \overrightarrow{OA}) \times (\overrightarrow{OA} \times \overrightarrow{OC})\|^2}{\|\overrightarrow{OB} \times \overrightarrow{OA}\|^2 \|\overrightarrow{OA} \times \overrightarrow{OC}\|^2} ; \quad \sin^2 B = \frac{\|(\overrightarrow{OC} \times \overrightarrow{OB}) \times (\overrightarrow{OB} \times \overrightarrow{OA})\|^2}{\|\overrightarrow{OC} \times \overrightarrow{OB}\|^2 \|\overrightarrow{OB} \times \overrightarrow{OA}\|^2} ;$$

$$\sin^2 C = \frac{\|(\overrightarrow{OA} \times \overrightarrow{OC}) \times (\overrightarrow{OC} \times \overrightarrow{OB})\|^2}{\|\overrightarrow{OA} \times \overrightarrow{OC}\|^2 \|\overrightarrow{OC} \times \overrightarrow{OB}\|^2} . \quad (1.27)$$

The following steps lead us to rediscover the spherical law of sines in (1.13).

(a) Apply (1.24) to the cross product $(\overrightarrow{OB} \times \overrightarrow{OA}) \times (\overrightarrow{OA} \times \overrightarrow{OC})$ (denoting the first cross-product as $\vec{u}$ helps a bit) and the results in (1.25) and Exercise 16c to prove

$$(\overrightarrow{OB} \times \overrightarrow{OA}) \times (\overrightarrow{OA} \times \overrightarrow{OC}) = \det \left(\begin{bmatrix} \overrightarrow{OC} & \overrightarrow{OB} & \overrightarrow{OA} \end{bmatrix} \right) \overrightarrow{OA} . \qquad (1.28)$$

(b) Apply (1.23) to the denominator of $\sin^2 A$ in (1.27) and use (1.28) to derive

$$\left(\det \left(\begin{bmatrix} \overrightarrow{OC} & \overrightarrow{OB} & \overrightarrow{OA} \end{bmatrix} \right) \right)^2 = \| (\overrightarrow{OB} \times \overrightarrow{OA}) \times (\overrightarrow{OA} \times \overrightarrow{OC}) \|^2 = \sin^2 c \sin^2 b \sin^2 A .$$
$$(1.29)$$

(c) Repeat the above steps for $\sin^2 B$ in (1.27) to conclude

$$\left(\det \left(\begin{bmatrix} \overrightarrow{OA} & \overrightarrow{OC} & \overrightarrow{OB} \end{bmatrix} \right) \right)^2 = \| (\overrightarrow{OC} \times \overrightarrow{OB}) \times (\overrightarrow{OB} \times \overrightarrow{OA}) \|^2 = \sin^2 a \sin^2 c \sin^2 B .$$
$$(1.30)$$

(d) Finally, prove that $\left(\det \left(\begin{bmatrix} \overrightarrow{OC} & \overrightarrow{OB} & \overrightarrow{OA} \end{bmatrix} \right) \right)^2 = \left(\det \left(\begin{bmatrix} \overrightarrow{OA} & \overrightarrow{OC} & \overrightarrow{OB} \end{bmatrix} \right) \right)^2$ to conclude that (1.29) and (1.30) are equal and, hence,

$$\sin^2 b \sin^2 A = \sin^2 a \sin^2 B$$

This proves the first half of the spherical law of sines in (1.13). The second half follows by repeating the above arguments for $\sin^2 C$ using its definition in (1.27).

18. Develop a coordinate-free proof for the spherical law of cosines given in (1.11) by noting

$$\cos A = \frac{(\overrightarrow{OB} \times \overrightarrow{OA}) \cdot (\overrightarrow{OA} \times \overrightarrow{OC})}{\| \overrightarrow{OB} \times \overrightarrow{OA} \| \| \overrightarrow{OA} \times \overrightarrow{OC} \|} .$$

Hint: First verify the Cauchy-Binet identity:

$$(\vec{a} \times \vec{b}) \cdot (\vec{c} \times \vec{d}) = (\vec{a} \cdot \vec{c})(\vec{b} \cdot \vec{d}) - (\vec{a} \cdot \vec{d})(\vec{b} \cdot \vec{c}) .$$

To verify the above identity, define $\vec{x} = \vec{a} \times \vec{b}$. Simplify $\vec{x} \cdot (\vec{c} \times \vec{d})$ using (1.26) and (1.24) to obtain the right hand side. Apply this identity to the numerator in $\cos A$ and simplify.

19. Prove the following equivalent expressions in the Mercator projection of Example 1.5:

$$\ln \tan \left(\frac{\pi}{4} + \frac{\phi}{2} \right) = \frac{1}{2} \ln \left(\frac{1 + \sin \phi}{1 - \sin \phi} \right) .$$

Chapter 2

Basics of point-referenced data models

In this chapter we present the essential elements of spatial models and classical analysis for point-referenced data. As mentioned in Chapter 1, the fundamental concept underlying the theory is a stochastic process $\{Y(\mathbf{s}) : \mathbf{s} \in D\}$, where D is a fixed subset of r-dimensional Euclidean space. Such stochastic processes have a rich presence in the time series literature, where $r = 1$. In the spatial context, usually we encounter r to be 2 (say, northings and eastings) or 3 (e.g., northings, eastings, and altitude above sea level). For situations where $r > 1$, the process is often referred to as a *spatial process*. For example, $Y(\mathbf{s})$ may represent the level of a pollutant at site $\mathbf{s}$. While it is conceptually sensible to assume the existence of a pollutant level at all possible sites in the domain, in practice the data will be a partial realization of that spatial process. That is, it will consist of measurements at a finite set of locations, say $\{\mathbf{s}_1, \ldots, \mathbf{s}_n\}$, where there are monitoring stations. The problem facing the statistician is inference about the spatial process $Y(\mathbf{s})$ and prediction at new locations, based upon this partial realization. The remarkable feature of the models we employ here is that, despite only seeing the process, equivalently, the spatial surface at a finite set of locations, we can infer about the surface at an uncountable number of locations. The reason is that we specify association through *structured* dependence which enables this broad interpolation.

This chapter is organized as follows. We begin with a survey of the building blocks of point-level data modeling, including stationarity, isotropy, and variograms (and their fitting via traditional moment-matching methods). We defer theoretical discussion of the spatial (typically Gaussian) process modeling that enables likelihood (and Bayesian) inference in these settings to Chapters 3 and 6. Then, we illustrate helpful exploratory data analysis tools, as well as traditional classical methods, especially kriging (point-level spatial prediction). We view all of these activities in an exploratory fashion, i.e., as a prelude to fully model-based inference under a hierarchical model. We close with some short tutorials in R using easy-to-use and widely available point-level spatial statistical analysis packages.

The material we cover in this chapter is traditionally known as *geostatistics*, and could easily fill many more pages than we devote to it here. While we prefer the more descriptive term "point-level spatial modeling," we will at times still use "geostatistics" for brevity and perhaps consistency when referencing the literature. Chilés and Delfiner [2012] and Stein [1999a] offer more in-depth treatments of geostatistics.

2.1 Elements of point-referenced modeling

2.1.1 Stationarity

For our discussion we assume that our spatial process has a mean, say $\mu(\mathbf{s}) = \mathrm{E}(Y(\mathbf{s}))$, associated with it and that the variance of $Y(\mathbf{s})$ exists for all $\mathbf{s} \in D$. The process $Y(\mathbf{s})$ is said to be *Gaussian* if, for any $n \geq 1$ and any set of sites $\{\mathbf{s}_1, \ldots, \mathbf{s}_n\}$, $\mathbf{Y} = (Y(\mathbf{s}_1), \ldots, Y(\mathbf{s}_n))^{\mathrm{T}}$ has a multivariate normal distribution. The process is said to be *strictly stationary* (sometimes *strong* stationarity) if, for any given $n \geq 1$, any set of n sites $\{\mathbf{s}_1, \ldots, \mathbf{s}_n\}$ and any $\mathbf{h} \in \Re^r$,

DOI: 10.1201/9781003401728-2

the distribution of $(Y(\mathbf{s}_1), \ldots, Y(\mathbf{s}_n))$ is the same as that of $(Y(\mathbf{s}_1 + \mathbf{h}), \ldots, Y(\mathbf{s}_n + \mathbf{h}))$. Here D is envisioned as $\Re^r$ as well.

A less restrictive condition is given by *weak stationarity* (also called second-order stationarity). A spatial process is called weakly stationary if $\mu(\mathbf{s}) \equiv \mu$, i.e., it has a constant mean and $\text{Cov}(Y(\mathbf{s}), Y(\mathbf{s} + \mathbf{h})) = C(\mathbf{h})$ for all $\mathbf{h} \in \Re^r$ such that $\mathbf{s}$ and $\mathbf{s} + \mathbf{h}$ both lie within D. In fact, for stationarity as a second-order property we will need only the second property; $\text{E}(Y(\mathbf{s}))$ need not equal $\text{E}(Y(\mathbf{s}+\mathbf{h}))$. But since we will apply the definition only to a mean 0 spatial residual process, this distinction is not important for us. Weak stationarity implies that the covariance relationship between the values of the process at any two locations can be summarized by a covariance function $C(\mathbf{h})$, and this function depends only on the separation vector $\mathbf{h}$. Note that, with all variances assumed to exist, strong stationarity implies weak stationarity. The converse is not true in general, but it *does* hold for Gaussian processes; see Exercise 4.

We offer a simple illustration of a weakly stationary process that is not strictly stationary. It is easy to see in the one-dimensional case. Suppose the process $Y_t, t = 1, 2, \ldots$ consists of a sequence of independent variables such that for t odd, Y_t is a binary variable taking the values 1 and -1 each with probability .5 while for t even, Y_t is normal with mean 0 and variance 1. We have weak stationarity since $\text{Cov}(Y(t), Y(t')) = 0, t \neq t', = 1, t = t'$. That is, we only need to know the value of $t - t'$ to specify the covariance. However, clearly Y_1, Y_3 does not have the same distribution as Y_2, Y_4.

2.1.2 Variograms

There is a third type of stationarity called *intrinsic* stationarity. Here we assume $\text{E}[Y(\mathbf{s} + \mathbf{h}) - Y(\mathbf{s})] = 0$ and define

$$\text{E}[Y(\mathbf{s} + \mathbf{h}) - Y(\mathbf{s})]^2 = \text{Var}(Y(\mathbf{s} + \mathbf{h}) - Y(\mathbf{s})) = 2\gamma(\mathbf{h}) . \tag{2.1}$$

Equation (2.1) makes sense only if the left-hand side depends *solely* on $\mathbf{h}$ (so that the right-hand side can be written at all), and not the particular choice of $\mathbf{s}$. If this is the case, we say the process is *intrinsically stationary*. The function $2\gamma(\mathbf{h})$ is then called the *variogram*, and $\gamma(\mathbf{h})$ is called the *semivariogram*. We can offer some intuition behind the variogram, but it really arose simply as a result of its appearance in traditional kriging where one seeks the best linear unbiased predictor, as we clarify below. Behaviorally, at short distances (small $||\mathbf{h}||$), we would expect $Y(\mathbf{s} + \mathbf{h})$ and $Y(\mathbf{h})$ to be very similar, that is, $(Y(\mathbf{s} + \mathbf{h}) - Y(\mathbf{s}))^2$ to be small. As $||\mathbf{h}||$ grows larger, we expect less similarity between $Y(\mathbf{s} + \mathbf{h})$ and $Y(\mathbf{h})$, i.e., we expect $(Y(\mathbf{s} + \mathbf{h}) - Y(\mathbf{s}))^2$ to be larger. So, a plot of $\gamma(\mathbf{h})$ would be expected to increase with $||\mathbf{h}||$, providing some insight into spatial behavior. (The covariance function $C(\mathbf{h})$ is sometimes referred to as the *covariogram*, especially when plotted graphically.) Note that intrinsic stationarity defines only the first and second moments of the differences $Y(\mathbf{s} + \mathbf{h}) - Y(\mathbf{s})$. It says nothing about the joint distribution of a collection of variables $\{Y(\mathbf{s}_1), \ldots, Y(\mathbf{s}_n)\}$, and thus provides no likelihood.

In fact, it says nothing about the moments of the $Y(\mathbf{s})$'s, much less their distribution. It only describes the behavior of differences rather than the behavior of the data that we observe, clearly unsatisfying from the perspective of data analysis. The $Y(\mathbf{s})$'s need not have any moments. For example, we might have $Y(\mathbf{s}) = W(\mathbf{s}) + V$ where $W(\mathbf{s})$ is a collection of i.i.d. normal variables and, independently, V is a Cauchy random variable. Then $Y(\mathbf{s})$ is intrinsically stationary but the $Y(\mathbf{s})$'s have no moments. Even more disconcerting, the distribution of $Y(\mathbf{s}) - Y(\mathbf{s}')$ may be proper while the distribution of $Y(\mathbf{s})$ and of $Y(\mathbf{s}')$ may be improper. For instance, suppose the joint distribution, $f(Y(\mathbf{s}), Y(\mathbf{s}')) \propto \exp(-(Y(\mathbf{s}) - Y(\mathbf{s}'))^2/2)$. Then, $Y(\mathbf{s}) - Y(\mathbf{s}') \sim N(0, 1)$ and, in fact, $f(Y(\mathbf{s}) \mid Y(\mathbf{s}'))$ and $f(Y(\mathbf{s}') \mid Y(\mathbf{s}))$ are proper, $N(0, 1)$ but the joint distribution is improper. How could we employ a probability specification that could not possibly have generated the data we observe?

It is easy to see the relationship between the variogram and the covariance function:

$$
\begin{aligned}
2\gamma(\mathbf{h}) &= \operatorname{Var}\left(Y\left(\mathbf{s}+\mathbf{h}\right)-Y\left(\mathbf{s}\right)\right) = \operatorname{Var}(Y(\mathbf{s}+\mathbf{h})) + \operatorname{Var}(Y(\mathbf{s})) - 2\operatorname{Cov}(Y(\mathbf{s}+\mathbf{h}), Y(\mathbf{s})) \\
&= C(\mathbf{0}) + C(\mathbf{0}) - 2C(\mathbf{h}) = 2\left[C(\mathbf{0}) - C(\mathbf{h})\right] .
\end{aligned}
$$

We arrive at a rather important equation,

$$
\gamma(\mathbf{h}) = C(\mathbf{0}) - C(\mathbf{h}) . \tag{2.2}
$$

Equation (2.2) reveals that, given C, we are able to recover γ easily. But what about the converse? In general, can we recover C from γ? Here it turns out we need to assume a bit more: if the spatial process is *ergodic*, then $C(\mathbf{h}) \to 0$ as $\|\mathbf{h}\| \to \infty$, where $\|\mathbf{h}\|$ denotes the length of the $\mathbf{h}$ vector. This is an intuitively sensible condition, since it means that the covariance between the values at two points vanishes as the points become further separated in space. But taking the limit of both sides of (2.2) as $\|\mathbf{h}\| \to \infty$ yields $\lim_{\|\mathbf{h}\| \to \infty} \gamma(\mathbf{h}) = C(\mathbf{0})$. Using the dummy variable $\mathbf{u}$ to avoid confusion, we note

$$
C(\mathbf{h}) = C(\mathbf{0}) - \gamma(\mathbf{h}) = \lim_{\|\mathbf{u}\| \to \infty} \gamma(\mathbf{u}) - \gamma(\mathbf{h}) . \tag{2.3}
$$

In general, the limit on the right-hand side need not exist, but if it does, then the process is weakly (second-order) stationary with $C(\mathbf{h})$ as given in (2.3). We therefore have a way to determine the covariance function C from the semivariogram γ. Therefore, weak stationarity implies intrinsic stationarity, but the converse is not true; indeed, the next section offers examples of processes that are intrinsically stationary but not weakly stationary.

A valid variogram necessarily satisfies a negative definiteness condition. For any set of locations $\mathbf{s}_1, \ldots, \mathbf{s}_n$ and any set of constants $a_1, \ldots, a_n$ such that $\sum_i a_i = 0$, if $\gamma(\mathbf{h})$ is valid, then

$$
\sum_i \sum_j a_i a_j \gamma(\mathbf{s}_i - \mathbf{s}_j) \le 0 . \tag{2.4}
$$

To see this, note that

$$
\begin{aligned}
\sum_i \sum_j a_i a_j \gamma(\mathbf{s}_i - \mathbf{s}_j) &= \frac{1}{2} \mathrm{E}\left[\sum_i \sum_j a_i a_j (Y(\mathbf{s}_i) - Y(\mathbf{s}_j))^2\right] \\
&= -\mathrm{E}\left[\sum_i \sum_j a_i a_j Y(\mathbf{s}_i) Y(\mathbf{s}_j)\right] \\
&= -\mathrm{E}\left[\sum_i a_i Y(\mathbf{s}_i)\right]^2 \le 0 .
\end{aligned}
$$

We remark that, despite the suggestion of expression (2.2), there is no relationship between this result and the positive definiteness condition for covariance functions (see Subsection 3.1.2). Cressie [1993] discusses further necessary conditions for a valid variogram. Lastly, the condition (2.4) emerges naturally in ordinary kriging (see Section 2.5).

2.1.3   Isotropy

Another important related concept is that of isotropy. If the semivariogram function $\gamma(\mathbf{h})$ depends upon the separation vector only through its length $\|\mathbf{h}\|$, then we say that the

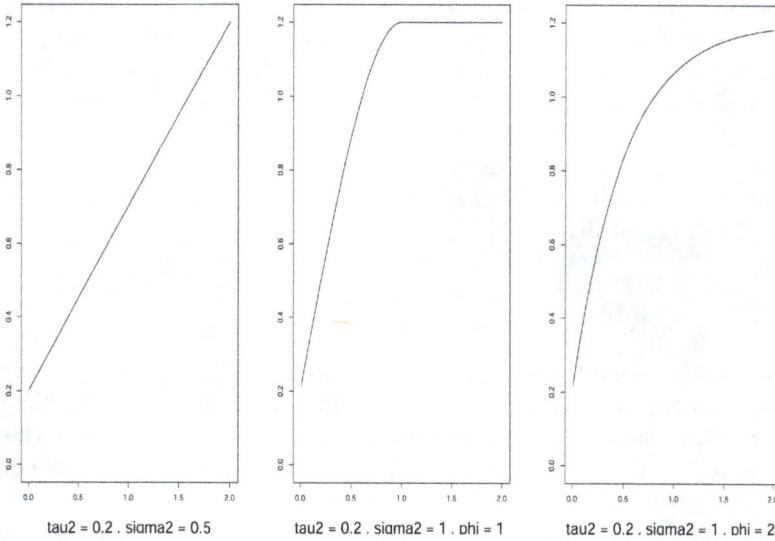

Figure 2.1 *Theoretical semivariograms for three models: (a) linear, (b) spherical, and (c) exponential.*

variogram is *isotropic*; that is, if $\gamma(\mathbf{h})$ is a real-valued function of a univariate argument, and can be written as $\gamma(\|\mathbf{h}\|)$. If not, we say it is *anisotropic*. Because of the foregoing issues with intrinsic stationarity, in the absence of a full probabilistic specification (as detailed in Chapter 3), we are reluctant to refer to associate an isotropic variogram with a stochastic process. Nonetheless, in the literature, we find terminology stating that, if a process is intrinsically stationary and isotropic, it is also called *homogeneous*.

Isotropic variograms are popular because of their simplicity, interpretability, and, in particular because a number of relatively simple parametric forms are available as candidates for the semivariogram. Denoting $\|\mathbf{h}\|$ by d for notational simplicity, we now consider a few of the more important such forms.

1. *Linear:*

$$\gamma(d) = \begin{cases} \tau^2 + \sigma^2 d & \text{if } d > 0, \ \tau^2 > 0, \ \sigma^2 > 0 \\ 0 & \text{otherwise} \end{cases}.$$

Note that $\gamma(d) \to \infty$ as $d \to \infty$, and so this semivariogram does not correspond to a weakly stationary process (although it is intrinsically stationary). This semivariogram is plotted in Figure 2.1(a) using the parameter values $\tau^2 = 0.2$ and $\sigma^2 = 0.5$.

2. *Spherical:*

$$\gamma(d) = \begin{cases} \tau^2 + \sigma^2 & \text{if } d \geq 1/\phi, \\ \tau^2 + \sigma^2 \left\{ \frac{3\phi d}{2} - \frac{1}{2}(\phi d)^3 \right\} & \text{if } 0 < d \leq 1/\phi, \\ 0 & \text{otherwise} \end{cases}.$$

The spherical semivariogram is valid in $r = 1, 2$, or 3 dimensions, but for $r \geq 4$ it fails to correspond to a spatial variance matrix that is positive definite (as required to specify a valid joint probability distribution). This variogram owes its popularity largely to the fact that it offers clear illustrations of the *nugget, sill,* and *range,* three characteristics traditionally associated with variograms. Specifically, consider Figure 2.1(b), which plots the spherical semivariogram using the parameter values $\tau^2 = 0.2$, $\sigma^2 = 1$, and $\phi = 1$. While $\gamma(0) = 0$ by definition, $\gamma(0^+) \equiv \lim_{d \to 0^+} \gamma(d) = \tau^2$; this quantity is the *nugget*. Next, $\lim_{d \to \infty} \gamma(d) = \tau^2 + \sigma^2$; this asymptotic value of the semivariogram is called the *sill*. (The sill minus the nugget, which is simply σ^2 in this case, is called the *partial sill*.)

Finally, the value $d = 1/\phi$ at which $\gamma(d)$ first reaches its ultimate level (the sill) is called the *range*. It is for this reason that many of the variogram models of this subsection are often parametrized through $R \equiv 1/\phi$. Confusingly, both R and ϕ are sometimes referred to as the *range* parameter, although ϕ is often more accurately referred to as the *decay* parameter.

Note that for the linear semivariogram, the nugget is τ^2 but the sill and range are both infinite. For other variograms (such as the next one we consider), the sill is finite but only reached asymptotically.

3. *Exponential:*

$$\gamma(d) = \begin{cases} \tau^2 + \sigma^2 \left(1 - \exp\left(-\phi d\right)\right) & \text{if } d > 0, \\ 0 & \text{otherwise} \end{cases}.$$

The exponential has an advantage over the spherical in that it is simpler in functional form while still being a valid variogram in all dimensions (and without the spherical's finite range requirement). However, note from Figure 2.1(c), which plots this semivariogram assuming $\tau^2 = 0.2$, $\sigma^2 = 1$, and $\phi = 2$, that the sill is only reached asymptotically; strictly speaking, the range $R = 1/\phi$ is infinite. In cases like this, the notion of an *effective range* is often used, i.e., the distance at which there is essentially no lingering spatial correlation. To make this notion precise, we must convert from γ scale to C scale (possible here since $\lim_{d\to\infty} \gamma(d)$ exists; the exponential is not only intrinsically but also weakly stationary). From (2.3) we have

$$\begin{aligned} C(d) &= \lim_{u\to\infty} \gamma(u) - \gamma(d) \\ &= \tau^2 + \sigma^2 - \left[\tau^2 + \sigma^2(1 - \exp(-\phi d))\right] \\ &= \sigma^2 \exp(-\phi d). \end{aligned}$$

Hence

$$C(t) = \begin{cases} \tau^2 + \sigma^2 & \text{if } d = 0 \\ \sigma^2 \exp(-\phi d) & \text{if } d > 0 \end{cases}. \tag{2.5}$$

If the nugget $\tau^2 = 0$, then this expression reveals that the correlation between two points d units apart is $\exp(-\phi d)$; note that $\exp(-\phi d) = 1^-$ for $d = 0^+$ and $\exp(-\phi d) = 0$ for $d = \infty$, both in concert with this interpretation.

A common definition of the *effective range*, d_0, is the distance at which this correlation is *negligible*, customarily taken as having dropped to only 0.05. Setting $\exp(-\phi d_0)$ equal to this value we obtain $t_0 \approx 3/\phi$, since $\log(0.05) \approx -3$. The range will be discussed in more detail in Subsection 3.1.2.

Finally, the form of (2.5) gives a clear example of why the nugget (τ^2 in this case) is often viewed as a "nonspatial effect variance," and the partial sill (σ^2) is viewed as a "spatial effect variance." That is, we have two variance components. Along with ϕ, a statistician would likely view fitting this model to a spatial data set as an exercise in estimating these three parameters. We shall return to the variogram model fitting in Subsection 2.1.4.

4. *Gaussian:*

$$\gamma(d) = \begin{cases} \tau^2 + \sigma^2 \left(1 - \exp\left(-\phi^2 d^2\right)\right) & \text{if } d > 0 \\ 0 & \text{otherwise} \end{cases}. \tag{2.6}$$

The Gaussian variogram is an analytic function and yields very smooth realizations of the spatial process. We shall say much more about process smoothness in Subsection 3.1.4.

5. *Powered exponential:*

$$\gamma(d) = \begin{cases} \tau^2 + \sigma^2 \left(1 - \exp\left(-|\phi d|^p\right)\right) & \text{if } d > 0 \\ 0 & \text{otherwise} \end{cases}. \tag{2.7}$$

Here $0 < p \leq 2$ yields a family of valid variograms. Note that both the Gaussian and the exponential forms are special cases.

6. *Rational quadratic:*

$$\gamma(d) = \begin{cases} \tau^2 + \frac{\sigma^2 d^2}{(\phi + d^2)} & \text{if } d > 0 \\ 0 & \text{otherwise} \end{cases}.$$

7. *Wave:*

$$\gamma(d) = \begin{cases} \tau^2 + \sigma^2 \left(1 - \frac{\sin(\phi d)}{\phi d}\right) & \text{if } d > 0 \\ 0 & \text{otherwise} \end{cases}.$$

Note this is an example of a variogram that is not monotonically increasing. The associated covariance function is $C(d) = \sigma^2 \sin(\phi d)/(\phi d)$. Bessel functions of the first kind include the wave covariance function and are discussed in detail in Subsections 3.1.2 and 6.4.

8. *Power law*

$$\gamma(d) = \begin{cases} \tau^2 + \sigma^2 d^\lambda & \text{of } d > 0 \\ 0 & d = 0 \end{cases}.$$

This generalizes the linear case and produces valid intrinsic (albeit not weakly) stationary semivariograms provided $0 \le \lambda < 2$.

9. *Matérn :* The variogram for the Matérn class is given by

$$\gamma(d) = \begin{cases} \tau^2 + \sigma^2 \left[1 - \frac{(2\sqrt{\nu}d\phi)^\nu}{2^{\nu-1}\Gamma(\nu)} K_\nu(2\sqrt{\nu}d\phi)\right] & \text{if } d > 0 \\ 0 & d = 0 \end{cases}. \qquad (2.8)$$

This class was originally suggested by Matérn [1986]. Interest in it was revived by Handcock and Stein [1993] and Handcock and Wallis [1994], who demonstrated attractive interpretations for ν as well as ϕ. In particular, $\nu > 0$ is a parameter controlling the smoothness of the realized random field (see Subsection 3.1.4) while ϕ is a spatial decay parameter. The function $\Gamma(\cdot)$ is the usual gamma function while K_ν is the modified Bessel function of order ν [see, e.g., Abramowitz and Stegun, 1965, Chapter 9].Note that special cases of the above are the exponential ($\nu = 1/2$) and the Gaussian ($\nu \to \infty$). At $\nu = 3/2$ we obtain a closed form as well, namely $\gamma(d) = \tau^2 + \sigma^2 \left[1 - (1 + \phi d)\exp(-\phi d)\right]$ for $t > 0$. In fact, we obtain a polynomial times exponential from for the Matérn for all ν of the form, $\nu = k + \frac{1}{2}$ with k a non-negative integer. The Matérn covariance function is often expressed as $\alpha = 2\sqrt{\nu}\phi$ along with $\eta = \sigma^2 \phi^{2\nu}$ and ν. This transformation is helpful in providing better behaved model fitting, particularly using Markov chain Monte Carlo (see Chapter 6 for further discussion).

   The covariance functions and variograms discussed in this subsection are conveniently summarized in Tables 2.1 and 2.2, respectively. An important point, that the reader may wonder about, is the fact that every presented covariance function and variogram has a discontinuity at zero. That is, the limit as $d \to 0$ for the covariance function is σ^2 not $\sigma^2 + \tau^2$ and for the variogram is τ^2, not 0. What is the reason? We elaborate in greater detail in the ensuing chapters. Here, we offer a simple explanation. Consider the form of the residual for a spatial model. We might write it as $r(\mathbf{s}) = w(\mathbf{s}) + \epsilon(\mathbf{s})$ where $w(\mathbf{s})$ is a spatial process, say a stationary Gaussian process with mean 0 and covariance function $\sigma^2 \rho(\mathbf{s} - \mathbf{s}')$ and $\epsilon(\mathbf{s})$ is a pure error process, i.e., the $\epsilon(\mathbf{s})$'s are i.i.d., say $N(0, \tau^2)$ with the ϵ's independent of the w's. Then, it is straightforward to compute $\text{Cov}(r(\mathbf{s}), r(\mathbf{s}')) = \sigma^2 \rho(\mathbf{s} - \mathbf{s}')$ while $\text{var}Y(\mathbf{s}) = \sigma^2 + \tau^2$, whence the discontinuity at 0 emerges. And, in fact, $\text{corr}(r(\mathbf{s}), r(\mathbf{s}'))$ is bounded by $\sigma^2/(\sigma^2 + \tau^2)$. The rationale for including the $\epsilon(\mathbf{s})$'s in the model is that we don't want to insist that all model error is spatial. Of course, we certainly want to include the w's in order to be able to capture a spatial story. Possible explanations for the pure error contribution include: (i) measurement error associated with the data collection at a given location, (ii) replication error to express the possibility that repeated

Model	Covariance function, $C(d)$
Linear	$C(d)$ does not exist
Spherical	$C(d) = \begin{cases} 0 & \text{if } d \geq 1/\phi \\ \sigma^2 \left[1 - \frac{3}{2}\phi d + \frac{1}{2}(\phi d)^3\right] & \text{if } 0 < d \leq 1/\phi \\ \tau^2 + \sigma^2 & d = 0 \end{cases}$
Exponential	$C(d) = \begin{cases} \sigma^2 \exp(-\phi d) & \text{if } d > 0 \\ \tau^2 + \sigma^2 & d = 0 \end{cases}$
Powered exponential	$C(d) = \begin{cases} \sigma^2 \exp(-\lvert\phi d\rvert^p) & \text{if } d > 0 \\ \tau^2 + \sigma^2 & d = 0 \end{cases}$
Gaussian	$C(d) = \begin{cases} \sigma^2 \exp(-\phi^2 d^2) & \text{if } d > 0 \\ \tau^2 + \sigma^2 & d = 0 \end{cases}$
Rational quadratic	$C(d) = \begin{cases} \sigma^2 \left(1 - \frac{d^2}{(\phi + d^2)}\right) & \text{if } d > 0 \\ \tau^2 + \sigma^2 & d = 0 \end{cases}$
Wave	$C(d) = \begin{cases} \sigma^2 \frac{\sin(\phi d)}{\phi d} & \text{if } d > 0 \\ \tau^2 + \sigma^2 & d = 0 \end{cases}$
Power law	$C(d)$ does not exist
Matérn	$C(d) = \begin{cases} \frac{\sigma^2}{2^{\nu-1}\Gamma(\nu)} (\phi d)^\nu K_\nu(\phi d) & \text{if } d > 0 \\ \tau^2 + \sigma^2 & d = 0 \end{cases}$
Matérn at $\nu = 3/2$	$C(d) = \begin{cases} \sigma^2 (1 + \phi d) \exp(-\phi d) & \text{if } d > 0 \\ \tau^2 + \sigma^2 & d = 0 \end{cases}$

Table 2.1 *Summary of common isotropic parametric covariance functions (covariograms).*

measurements at the same location might not provide identical observations, or (iii) micro-scale error to acknowledge that, though we never see observations closer to each other than the minimum pairwise distance in our sample, there might be a very fine scale structure which is represented as noise.

2.1.4   Variogram model fitting

Having seen a fairly large selection of models for the variogram, one might well wonder how we choose one of them for a given data set, or whether the data can really distinguish them (see Subsection 6.4 in this latter regard). Historically, a variogram model is chosen by plotting the *empirical semivariogram* [Matheron, 1963], a simple nonparametric estimate of the semivariogram, and then comparing it to the various theoretical shapes available from the choices in the previous subsection. The customary empirical semivariogram is

$$\hat{\gamma}(d) = \frac{1}{2N(d)} \sum_{(\mathbf{s}_i, \mathbf{s}_j) \in N(d)} [Y(\mathbf{s}_i) - Y(\mathbf{s}_j)]^2 \,, \tag{2.9}$$

where $N(d)$ is the set of pairs of points such that $\|\mathbf{s}_i - \mathbf{s}_j\| = d$, and $|N(d)|$ is the number of pairs in this set. Notice that, unless the observations fall on a regular grid, the distances between the pairs will all be different, so this will not be a useful estimate as it stands. Instead we would "grid up" the d-space into intervals $I_1 = (0, d_1), I_2 = (d_1, d_2)$, and so forth, up to $I_K = (d_{K-1}, d_K)$ for some (typically regular) grid $0 < d_1 < \cdots < d_K$. Representing the d values in each interval by the midpoint of the interval, we then alter our definition of $N(d)$ to

$$N(d_k) = \{(\mathbf{s}_i, \mathbf{s}_j) : \|\mathbf{s}_i - \mathbf{s}_j\| \in I_k\} \,, \; k = 1, \ldots, K \,.$$

model	Variogram, $\gamma(d)$
Linear	$\gamma(d) = \begin{cases} \tau^2 + \sigma^2 d & \text{if } d > 0 \\ 0 & d = 0 \end{cases}$
Spherical	$\gamma(d) = \begin{cases} \tau^2 + \sigma^2 & \text{if } d \geq 1/\phi \\ \tau^2 + \sigma^2 \left[\frac{3}{2}\phi d - \frac{1}{2}(\phi d)^3 \right] & \text{if } 0 < d \leq 1/\phi \\ 0 & d = 0 \end{cases}$
Exponential	$\gamma(d) = \begin{cases} \tau^2 + \sigma^2(1 - \exp(-\phi d)) & \text{if } d > 0 \\ 0 & d = 0 \end{cases}$
Powered exponential	$\gamma(d) = \begin{cases} \tau^2 + \sigma^2(1 - \exp(-|\phi d|^p)) & \text{if } d > 0 \\ 0 & d = 0 \end{cases}$
Gaussian	$\gamma(d) = \begin{cases} \tau^2 + \sigma^2(1 - \exp(-\phi^2 d^2)) & \text{if } d > 0 \\ 0 & d = 0 \end{cases}$
Rational quadratic	$\gamma(d) = \begin{cases} \tau^2 + \frac{\sigma^2 d^2}{(\phi + d^2)} & \text{if } d > 0 \\ 0 & d = 0 \end{cases}$
Wave	$\gamma(d) = \begin{cases} \tau^2 + \sigma^2(1 - \frac{\sin(\phi d)}{\phi d}) & \text{if } d > 0 \\ 0 & d = 0 \end{cases}$
Power law	$\gamma(d) = \begin{cases} \tau^2 + \sigma^2 d^\lambda & \text{if } d > 0 \\ 0 & d = 0 \end{cases}$
Matérn	$\gamma(d) = \begin{cases} \tau^2 + \sigma^2 \left[1 - \frac{(\phi d)^\nu}{2^{\nu-1}\Gamma(\nu)} K_\nu(\phi d) \right] & \text{if } d > 0 \\ 0 & d = 0 \end{cases}$
Matérn at $\nu = 3/2$	$\gamma(d) = \begin{cases} \tau^2 + \sigma^2 \left[1 - (1 + \phi d) \exp(-\phi d) \right] & \text{if } d > 0 \\ 0 & d = 0 \end{cases}$

Table 2.2 *Summary of common parametric isotropic variograms.*

Selection of an appropriate number of intervals K and the location of the upper endpoint t_K is reminiscent of similar issues in histogram construction. Journel and Huijbregts [2003] recommend bins wide enough to capture at least 30 pairs per bin.

Clearly (2.9) is nothing but a method of moments (MOM) estimate, the semivariogram analogue of the usual sample variance estimate s^2. While very natural, there is reason to doubt that this is the best estimate of the semivariogram. Certainly, it will be sensitive to outliers, and the sample average of the squared differences may be rather badly behaved since under a Gaussian distributional assumption for the $Y(\mathbf{s}_i)$, the squared differences will have a distribution that is a scale multiple of the heavily skewed χ_1^2 distribution. In this regard, Cressie and Hawkins [1980] proposed a robust estimate that uses sample averages of $|Y(\mathbf{s}_i) - Y(\mathbf{s}_j)|^{1/2}$; this estimate is available in several software packages (see Section 2.3 below). Perhaps more uncomfortable is the fact that (2.9) uses data differences, rather than the data itself. Also of concern is the fact that the components of the sum in (2.9) will be dependent within and across bins, and that $N(d_k)$ will vary across bins.

In any case, an empirical semivariogram estimate can be plotted, viewed, and an appropriately shaped theoretical variogram model can be fit to this "data." Since any empirical estimate naturally carries with it a significant amount of noise in addition to its signal, this fitting of a theoretical model has traditionally been as much art as science: in any given real data setting, any number of different models (exponential, Gaussian, spherical, etc.) may seem equally appropriate. Indeed, the fitting has historically been done "by eye," or at best by using trial and error to choose values of nugget, sill, and range parameters that provide a good match to the empirical semivariogram (where the "goodness" can be judged visually or by using some least squares or similar criterion); again see Section 2.3. More formally, we could treat this as a statistical estimation problem, and use nonlinear maximization routines to find a nugget, sill, and range parameters that minimize some goodness-of-fit criterion.

If we also have a distributional model for the data, we could use maximum likelihood (or restricted maximum likelihood, REML) to obtain sensible parameter estimates. In Chapter 5 and Chapter 6 we shall see that adopting the hierarchical Bayesian approach, it will often be easier and more intuitive to work directly with a covariance specification $C(d)$, rather than changing to a partial likelihood in order to introduce the semivariogram. In addition, we will gain full inference, e.g., posterior distributions for all unknowns of interest as well as more accurate assessment of uncertainty than appealing to arguably inappropriate asymptotics (see Chapter 3).

2.2 Anisotropy

Stationary correlation functions extend the class of correlation functions from isotropy where association only depends upon the distance to an association that depends upon the separation vector between locations. As a result, association depends upon direction. Here, we explore covariance functions that are stationary but not isotropic. (We defer the discussion of nonstationary covariance functions to Chapter 3.) A simple example is the class where we *separate* the components. That is, suppose we write the components of $\mathbf{s}$ as s_{lat}, s_{lon}, similarly for $\mathbf{h}$. Then, we can define $\mathrm{corr}(Y(\mathbf{s}+\mathbf{h}), Y(\mathbf{s})) = \rho_1(h_{lat})\rho_2(h_{lon})$ where ρ_1 and ρ_2 are valid correlation functions on $\Re^1$, a so-called *product* correlation function. Evidently, this correlation function is stationary but depends on direction. In particular, if we switch h_{lat} and h_{lon} we will get a different value for the correlation even though $||\mathbf{h}||$ is unchanged. A common choice is $\exp(\phi_1(|h_{lat}|) + \phi_2(|h_{lon}|))$.

The separable correlation function is usually extended to a covariance function by introducing the multiplier σ^2, as we have done above. This covariance function tends to be used, for instance, with Gaussian processes in computer model settings [Sacks et al., 1989, Kennedy and O'Hagan, 2002, Oakley and O'Hagan, 2004, Rasmussen and Williams, 2005, Bayarri et al., 2009, Santner et al., 2019]. In fact, Gaussian processes are extensively employed in machine learning applications and nonparametric regression modeling [see, e.g., Rasmussen and Williams, 2005, Ghosal and van der Vaart, 2017, for comprehensive accounts], where we seek a response surface over covariate space and, since the covariates live on their own spaces with their own scales, component-wise dependence seems appropriate. In the spatial setting, since latitude and longitude are on the same scale, it may be less suitable.

2.2.1 Geometric anisotropy

A commonly used class of stationary covariance functions is the geometric anisotropic covariance functions where we set

$$C(\mathbf{s}-\mathbf{s}') = \sigma^2\rho((\mathbf{s}-\mathbf{s}')^{\mathrm{T}}B(\mathbf{s}-\mathbf{s'})) . \tag{2.10}$$

In (2.10), B is positive definite with ρ a valid correlation function in $\Re^r$ (say, from Table 2.1). We would omit the range/decay parameter since it can be incorporated into B. When $r = 2$ we obtain a specification with three parameters rather than one. Contours of constant association arising from c in (2.10) are elliptical. In particular, the contour corresponding to $\rho = .05$ provides the range in each spatial direction. Ecker and Gelfand [1997] Ecker provide the details for Bayesian modeling and inference incorporating (2.10); see also Subsection 6.5.

Following the discussion in Subsection 3.1.2, we can extend geometric anisotropy to *product* geometric anisotropy. In the simplest case, we would set

$$C(\mathbf{s}-\mathbf{s}') = \sigma^2\, \rho_1((\mathbf{s}-\mathbf{s}')^{\mathrm{T}}B_1(\mathbf{s}-\mathbf{s}'))\, \rho_2((\mathbf{s}-\mathbf{s}')^{\mathrm{T}}B_2(\mathbf{s}-\mathbf{s}')) , \tag{2.11}$$

noting that c is valid since it arises as a product of valid covariance functions [see Ecker and Gelfand, 2003, for further details and examples]. Evidently, we can extend to a product of more than 2 geometric anisotropy forms and we can create rich directional range behavior. However, a challenge with (2.11) is that it introduces 7 parameters into the covariance function and it will be difficult to identify and learn about all of them unless we have many, many locations.

2.2.2 Other notions of anisotropy

In a more general discussion, Zimmerman [1993] suggests three different notions of anisotropy: *sill* anisotropy, *nugget* anisotropy, and *range* anisotropy. More precisely, working with a variogram $\gamma(\mathbf{h})$, let $\mathbf{h}$ be an arbitrary separation vector so that $\mathbf{h}/\|\mathbf{h}\|$ is a unit vector in $\mathbf{h}$'s direction. Consider $\gamma(a\mathbf{h}/\|\mathbf{h}\|)$. Let $a \to \infty$ and suppose $\lim_{a\to\infty} \gamma(a\mathbf{h}/\|\mathbf{h}\|)$ depends upon $\mathbf{h}$. This situation is naturally referred to as sill anisotropy. If we work with the usual relationship $\gamma(a\mathbf{h}/\|\mathbf{h}\|) = \tau^2 + \sigma^2\left(1 - \rho\left(a\frac{\mathbf{h}}{\|\mathbf{h}\|}\right)\right)$, then, in some directions, ρ must not go to 0 as $a \to \infty$. If this can be the case, then ergodicity assumptions (i.e., convergence assumptions associated with averaging) will be violated. If so, then perhaps the constant mean assumption, implicit for the variogram, does not hold. Alternatively, it is also possible that the constant nugget assumption fails.

Instead, let $a \to 0$ and suppose $\lim_{a\to 0} \gamma(a\mathbf{h}/\|\mathbf{h}\|)$ depends upon $\mathbf{h}$. This situation is referred to as nugget anisotropy. Since, by definition, ρ must go to 1 as $a \to 0$, this case says that the assumption of uncorrelated measurement errors with common variance may not be appropriate. In particular, a simple white noise process model with constant nugget for the nonspatial errors is not appropriate.

A third type of anisotropy is range anisotropy where the range depends upon direction. Zimmerman [1993] (1993) asserts that "this is the form most often seen in practice." Geometric anisotropy and the more general product geometric anisotropy from the previous subsections are illustrative cases. However, given the various constructive strategies offered in Subsection 3.1.2 to create more general stationary covariance functions, we can envision nongeometric range anisotropy, implying a general correlation function or variogram contours in $\Re^2$. However, due to the positive definiteness restriction on the correlation function, the extent of possible contour shapes is still rather limited.

Lastly, motivated by directional variograms (see Subsection 2.3.2), some authors propose the idea of nested models [see Zimmerman, 1993, and references therein]. That is, for each separation vector there is an associated angle with, say, the x-axis, which by symmetry considerations can be restricted to $[0, \pi)$. Partitioning this interval into a set of angle classes, a different variogram model is assumed to operate for each class. In terms of correlations, this would imply a different covariance function is operating for each angle class. But evidently this does not define a valid process model: the resulting covariance matrix for an arbitrary set of locations need not be positive definite.

This can be seen with as few as three points and two angle classes. Let $(\mathbf{s}_1, \mathbf{s}_2)$ belong to one angle class with $(\mathbf{s}_1, \mathbf{s}_3)$ and $(\mathbf{s}_2, \mathbf{s}_3)$ in the other. With exponential isotropic correlation functions in each class by choosing ϕ_1 and ϕ_2 appropriately we can make $\rho(\mathbf{s}_1 - \mathbf{s}_2) \approx 0$ while $\rho(\mathbf{s}_1 - \mathbf{s}_3) = \rho(\mathbf{s}_2 - \mathbf{s}_3) \approx 0.8$. A quick calculation shows that the resulting 3×3 covariance (correlation) matrix is not positive definite. So, in terms of being able to write proper joint distributions for the resulting data, nested models are inappropriate; they do not provide an extension of isotropy that allows for likelihood-based inference.

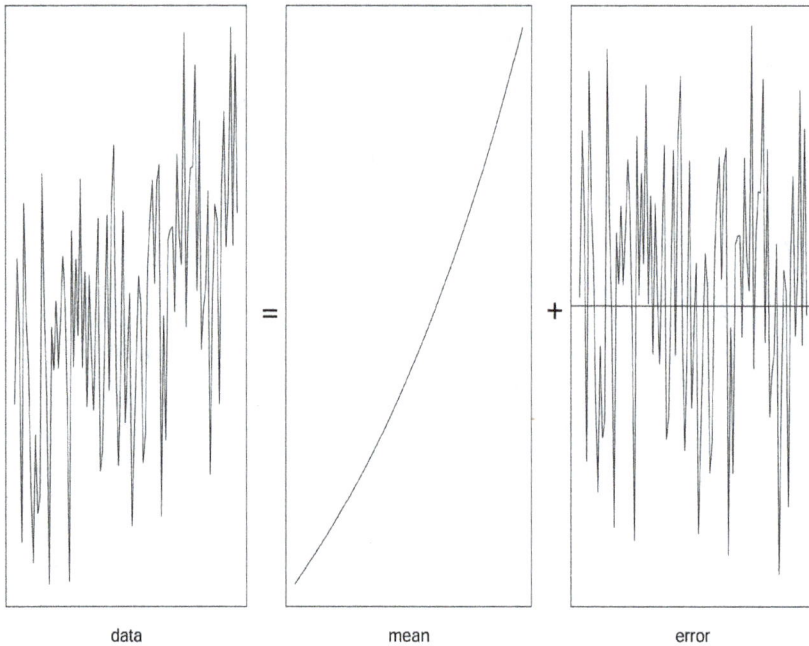

data   mean   error

Figure 2.2 *Illustration of the first law of geostatistics.*

2.3   Exploratory approaches for point-referenced data

2.3.1   Basic techniques

Exploratory data analysis (EDA) tools are routinely implemented in the process of analyzing one- and two-sample data sets, regression studies, generalized linear models and so on [see, e.g., Hoaglin and etc., 1985, Chambers, 1983, Hoaglin et al., 2000]. Similarly, such tools are appropriate for analyzing point-referenced spatial data.

For continuous spatial data, the starting point is the so-called "first law of geostatistics." Figure 2.2 illustrates this "law" in a one-dimensional setting. The data is partitioned into a mean term and an error term. The mean corresponds to global (or *first-order*) behavior, while the error captures local (or *second-order*) behavior through a covariance function. EDA tools are available to examine both first- and second-order behavior.

The law also clarifies that spatial association in the data, $Y(\mathbf{s})$, need not resemble spatial association in the residuals, $\epsilon(\mathbf{s})$. That is, spatial association in $Y(\mathbf{s})$ corresponds to looking at $\mathrm{E}(Y(\mathbf{s}) - \mu)(Y(\mathbf{s}') - \mu)$, while the spatial structure in $\epsilon(\mathbf{s})$ corresponds to looking at $\mathrm{E}(Y(\mathbf{s}) - \mu(\mathbf{s}))(Y(\mathbf{s}') - \mu(\mathbf{s}'))$. The difference between the former and the latter is $(\mu - \mu(\mathbf{s}))(\mu - \mu(\mathbf{s}'))$, which, if interest is in spatial regression, we would not expect to be negligible.

An initial exploratory display should be a simple map of the locations themselves. We need to assess how *regular* the arrangement of the points is and also whether there is a much larger maximum distance between points in some directions than in others. Next, some authors would recommend a stem-and-leaf display of the $Y(\mathbf{s})$. This plot is evidently nonspatial and is customarily for observations which are i.i.d. We expect both nonconstant mean and spatial dependence, but such a plot may at least suggest potential outliers. We might also consider a three-dimensional "drop line" scatterplot of $Y(\mathbf{s}_i)$ versus $\mathbf{s}_i$ or, alternatively, a three-dimensional surface plot or, perhaps, a contour plot as a *smoothed* summary. An example of such a plot is shown for a sample of 120 log-transformed home selling prices

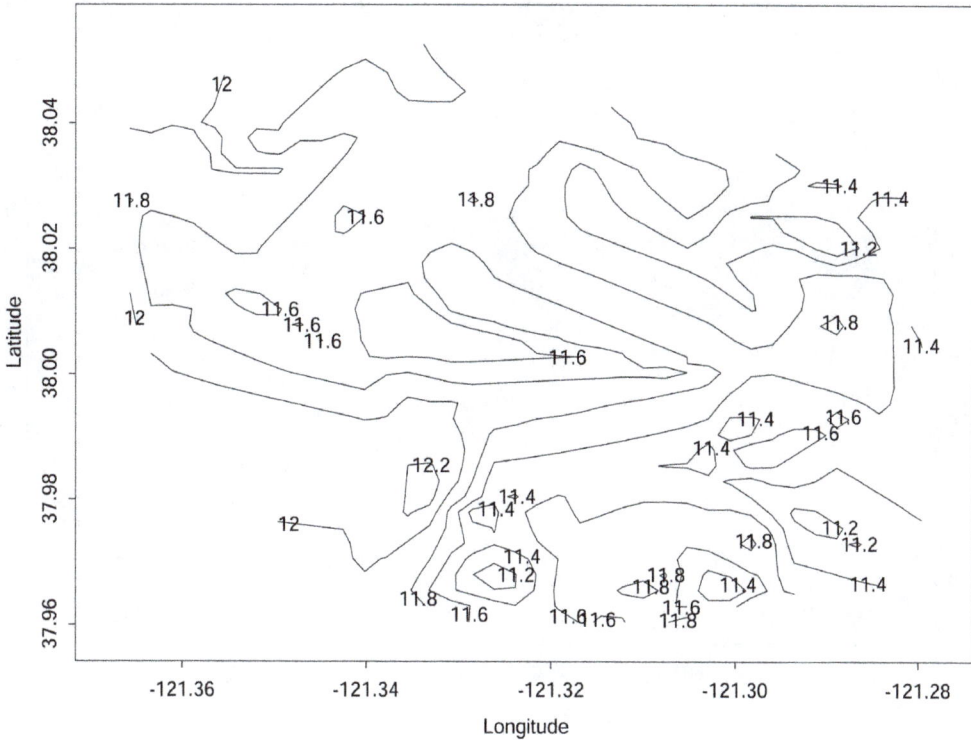

Figure 2.3 *Illustrative contour plot, Stockton real estate data.*

in Stockton, California, in Figure 2.3. We find the contour plot to be effective in revealing the entire spatial surface. However, as the preceding paragraph clarifies, such displays may be deceiving. They may show spatial patterns that will disappear after $\mu(\mathbf{s})$ is fitted, or perhaps vice versa. It seems more sensible to study spatial patterns in the residuals.

In exploring $\mu(\mathbf{s})$ we may have two types of information at location $\mathbf{s}$. One is purely geographic information, i.e., the geocoded location expressed in latitude and longitude or as projected coordinates such as eastings and northings (Subsection 1.2.4 above). The other will be features relevant for explaining the $Y(\mathbf{s})$ at $\mathbf{s}$. For instance, if $Y(\mathbf{s})$ is a pollution concentration, then elevation, temperature, and wind information at $\mathbf{s}$ could well be useful and important. If, instead, $Y(\mathbf{s})$ is the selling price of a single-family home at $\mathbf{s}$, then characteristics of the home (square feet, age, number of bathrooms, etc.) would be useful.

When the mean is described purely through geographic information, $\mu(\mathbf{s})$ is referred to as a *trend surface*. When $\mathbf{s} \in \Re^2$, the surface is usually developed as a low-dimensional bivariate polynomial in each coordinate. For data that is roughly on a grid (or can be assigned to row and column bins by overlaying a regular lattice on the points), we can make row and column box-plots looking for trend. Displaying these box-plots versus their center could clarify the existence and nature of such trend. In fact, median polishing [see, e.g., Hoaglin and etc., 1985] could be used to extract row and column effects, and also to see if a multiplicative trend surface term is useful [see Cressie, 1993, pp. 46–48 in this regard].

Figures 2.4 and 2.5 illustrate the row and column boxplot approach for a data set previously considered by Diggle and Ribiero [2002]. The response variable is the surface elevation ("height") at 52 locations on a regular grid within a 310-foot square (and where

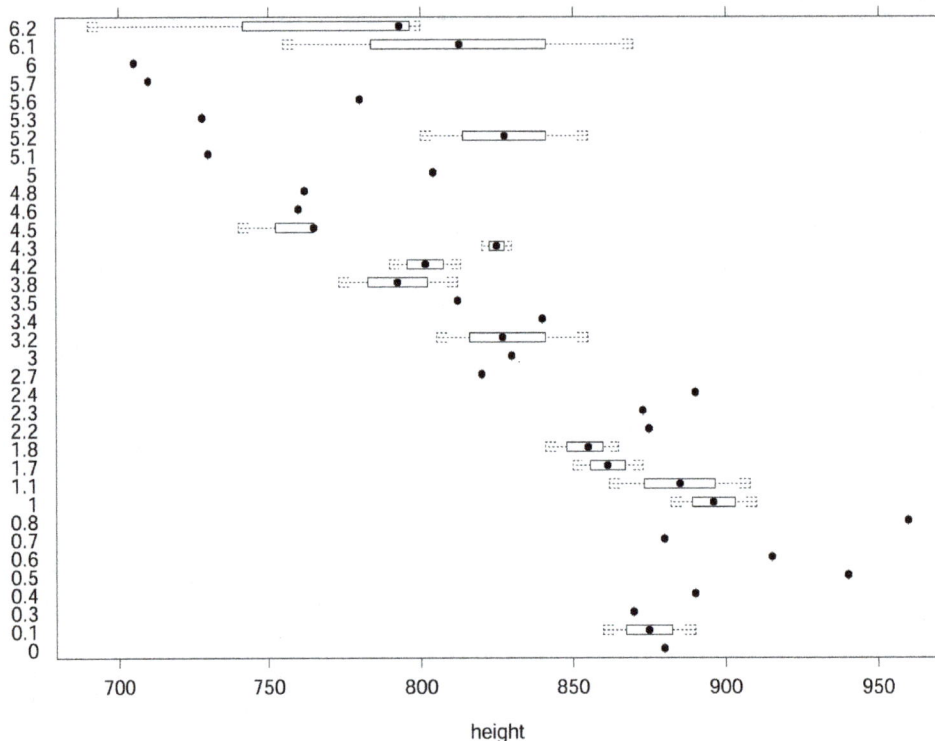

Figure 2.4 *Illustrative row box plots, Diggle and Ribeiro (2002) surface elevation data.*

the mesh of the grid is 50 feet). The plots reveal some evidence of spatial pattern as we move along the rows, but not down the columns of the regular grid.

To assess small-scale behavior, some authors recommend creating the *semivariogram cloud*, i.e., a plot of $(Y(\mathbf{s}_i) - Y(\mathbf{s}_j))^2$ versus $||\mathbf{s}_i - \mathbf{s}_j||$. Usually this cloud is too "noisy" to reveal very much; see, e.g., Figure 6.2. The empirical semivariogram (2.9) is preferable in terms of reducing some of the noise and can be a helpful tool in assessing the presence of spatial structure. Again, the caveat above suggests employing it for residuals (not the data itself) unless a constant mean is appropriate.

An empirical (nonparametric) covariance estimate, analogous to (2.9), is also available. Creating bins as in this earlier approach, we define

$$\widehat{C}(t_k) = \frac{1}{N_k} \sum_{(\mathbf{s}_i, \mathbf{s}_j) \in N(t_k)} (Y(\mathbf{s}_i) - \bar{Y})(Y(\mathbf{s}_j) - \bar{Y}) , \qquad (2.12)$$

where again $N(t_k) = \{(\mathbf{s}_i, \mathbf{s}_j) : ||\mathbf{s}_i - \mathbf{s}_j|| \in I_k\}$ for $k = 1, \ldots, K$, I_k indexes the kth bin, and there are N_k pairs of points falling in this bin. Equation (2.12) is a spatial generalization of a lagged autocorrelation in time series analysis. Since $\widehat{c}$ uses a common $\bar{Y}$ for all $Y(\mathbf{s}_i)$, it may be safer to employ (2.12) on the residuals. We note that the empirical covariance function is not used much in practice; the empirical variogram is much more common. Moreover, two further issues arise: first, how should we define $\widehat{C}(0)$, and second, regardless of this choice, we note that $\widehat{\gamma}(t_k)$ does *not* equal $\widehat{C}(0) - \widehat{C}(t_k)$, $k = 1, \ldots, K$. Details for both of these issues are left to Exercise 5. Again, since we are only viewing $\widehat{\gamma}$ and $\widehat{C}$ as exploratory tools, there seems to be no reason to pursue the latter further.

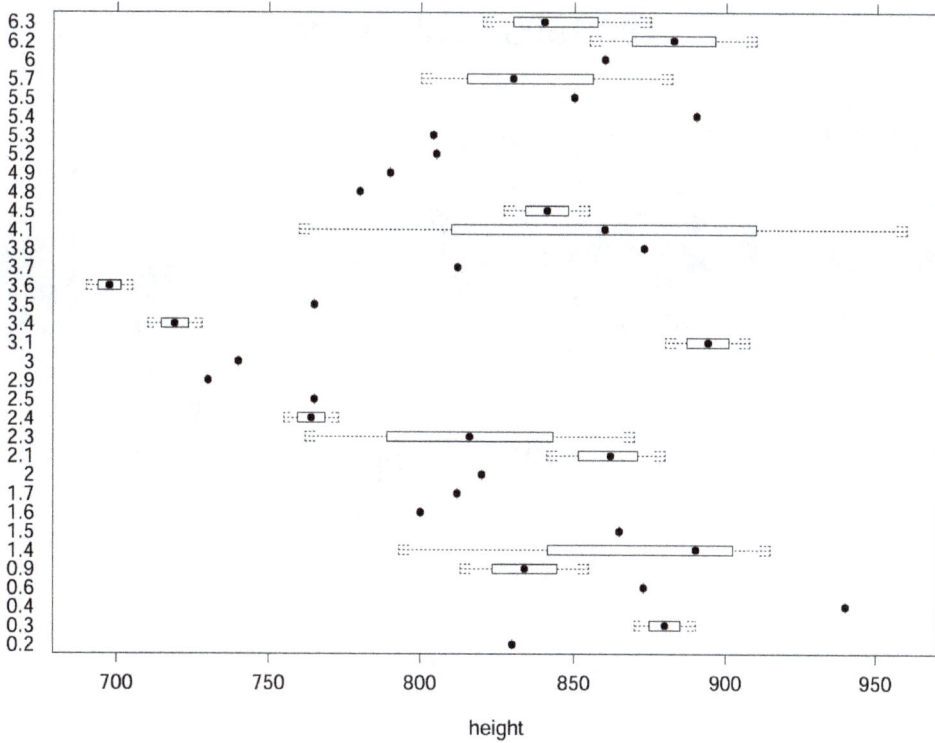

Figure 2.5 *Illustrative column box plots, Diggle and Ribeiro (2002) surface elevation data.*

Again, with a regular grid or binning, we can create "same-lag" scatterplots. These are plots of $Y(\mathbf{s}_i + h\mathbf{e})$ versus $Y(\mathbf{s}_i)$ for a fixed h and a fixed unit vector $\mathbf{e}$. Comparisons among such plots may reveal the presence of anisotropy and perhaps nonstationarity.

Lastly, suppose we attach a neighborhood to each point. We can then compute the sample mean and variance for the points in the neighborhood, and even a sample correlation coefficient using all pairs of data in the neighborhood. Plots of each of them versus location can be informative. The first may give some idea regarding how the mean structure changes across the study region. Plots of the second and third may provide evidence of nonstationarity. Implicit in extracting useful information from these plots is a roughly constant local mean. If $\mu(\mathbf{s})$ is to be a trend surface, this is plausible. But if $\mu(\mathbf{s})$ is a function of some geographic variables at $\mathbf{s}$ (say, home characteristics), then the use of residuals would be preferable.

2.3.2 Assessing anisotropy

We illustrate various EDA techniques to assess anisotropy using sampling of scallop abundance on the continental shelf off the coastline of the northeastern U.S. The data comes from a survey conducted by the Northeast Fisheries Science Center of the National Marine Fisheries Service. Figure 2.6 shows the sampling sites for 1990 and 1993. We see much more sampling in the southwest-to-northeast direction than in the northwest-to-southeast direction. Evidently, it is more appropriate to follow the coastline in searching for scallops.

1990 Scallop Sites

1993 Scallop Sites

Figure 2.6 *Sites sampled in the Atlantic Ocean for 1990 and 1993 scallop catch data.*

2.3.2.1 *Directional semivariograms and rose diagrams*

The most common EDA technique for assessing anisotropy involves use of directional semi-variograms. Typically, one chooses angle classes $\eta_i \pm \epsilon$, $i = 1, \ldots, L$ where ϵ is the half-width of the angle class and L is the number of angle classes. For example, a common choice of angle classes involves the four cardinal directions measured counterclockwise from the x-axis ($0°$, $45°$, $90°$, and $135°$), where ϵ is $22.5°$. Journel and Froidevaux [1982] display directional semivariograms at angles $35°$, $60°$, $125°$, and $150°$ in deducing anistropy for a tungsten deposit. While knowledge of the underlying spatial characteristics of region D is invaluable in choosing directions, often the choice of the number of angle classes and the directions seems to be arbitrary.

For a given angle class, the Matheron empirical semivariogram (2.9) can be used to provide a directional semivariogram for angle η_i. Theoretically, all types of anisotropy can be assessed from these directional semivariograms; however, in practice determining whether the sill, nugget, and/or range varies with direction can be difficult. In particular, it is unclear how much variability will arise in directional variograms generated under isotropy. Figure 2.7(a) illustrates directional semivariograms for the 1990 scallop data in the four cardinal directions. Note that the semivariogram points are connected only to aid comparison.

Possible conclusions from the figure are: the variability in the $45°$ direction (parallel to the coastline) is significantly less than in the other three directions and the variability perpendicular to the coastline ($135°$) is very erratic, possibly exhibiting sill anisotropy. We caution however that it is dangerous to read too much significance and interpretation

A   B

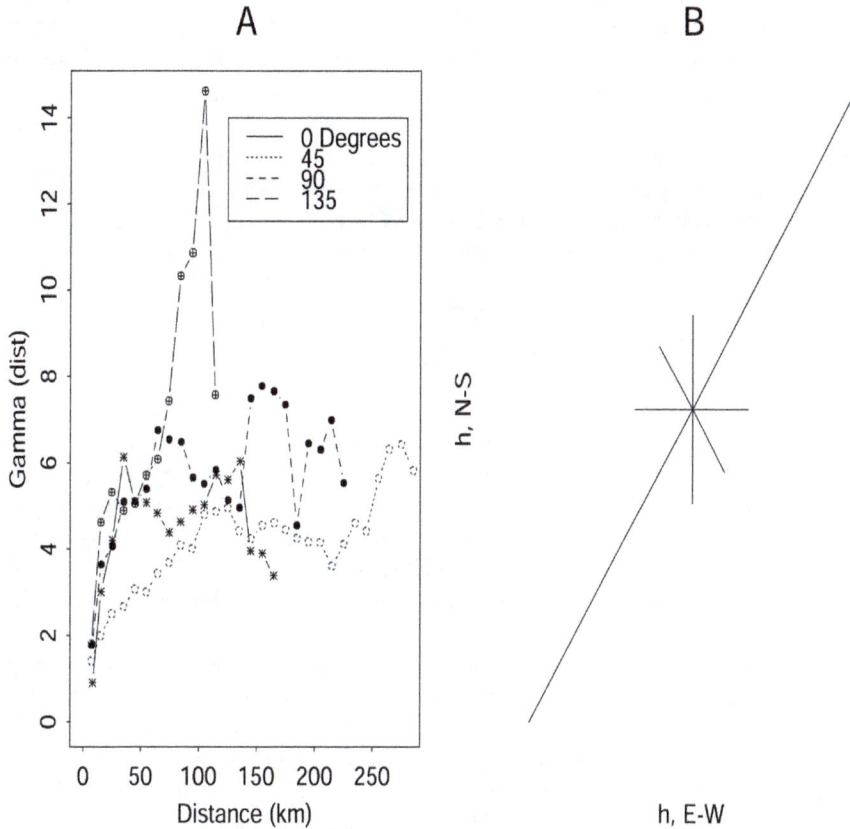

Figure 2.7 *Directional semivariograms (a) and a rose diagram (b) for the 1990 scallop data.*

into directional variograms. For example, from Figure 2.6, as noted above, we see far more sampling at greater distances in the southwest-northeast direction than in the northwest-southeast direction. Moreover, no sample sizes (and thus no assessments of variability) are attached to these pictures. Directional variograms from data generated under a simple isotropic model will routinely exhibit differences of the magnitudes seen in Figure 2.7(a). Furthermore, it seems difficult to draw any conclusions regarding the presence of geometric anisotropy from this figure.

A rose diagram[Isaaks and Srivastava, 1990, pp. 151–154] can be created from the directional semivariograms to evaluate geometric anisotropy. At an arbitrarily selected γ^*, for a directional semivariogram at angle η, the distance d^* at which the directional semivariogram attains γ^* can be interpolated. Then, the rose diagram is a plot of angle η and corresponding distance d^* in polar coordinates. If an elliptical contour describes the extremities of the rose diagram reasonably well, then the process exhibits geometric anisotropy. For instance, the rose diagram for the 1990 scallop data is presented in Figure 2.7(b) using the γ^* contour of 4.5. It is approximately elliptical, oriented parallel to the coastline ($\approx 45°$) with a ratio of major to minor ellipse axes of about 4.

2.3.2.2 *Empirical semivariogram contour (ESC) plots*

A more informative method for assessing anisotropy is a contour plot of the empirical semivariogram surface in $\Re^2$. Such plots are mentioned informally in Isaaks and Srivastava

[1990] (pp. 149–151) and in Haining [1990] (pp. 284–286); the former call them contour maps of the grouped variogram values, the latter an isarithmic plot of the semivariogram. Following Ecker and Gelfand [1999], we formalize such a plot here calling it an *empirical semivariogram contour* (ESC) plot. For each of the $\frac{N(N-1)}{2}$ pairs of sites in $\Re^2$, calculate h_x and h_y, the separation distances along each axis. Since the sign of h_y depends upon the arbitrary order in which the two sites are compared, we demand that $h_y \geq 0$. (We could alternatively demand that $h_x \geq 0$.) That is, we take $(-h_x, -h_y)$ when $h_y < 0$. These separation distances are then aggregated into rectangular bins B_{ij} where the empirical semivariogram values for the (i,j)th bin are calculated by

$$\gamma_{ij}^* = \frac{1}{2N_{B_{ij}}} \sum_{\{(k,l):(\mathbf{s}_k - \mathbf{s}_l) \in B_{ij}\}} (Y(\mathbf{s}_k) - Y(\mathbf{s}_l))^2, \qquad (2.13)$$

where $N_{B_{ij}}$ equals the number of sites in bin B_{ij}. Because we force $h_y \geq 0$ with h_x unrestricted, we make the bin width on the y-axis half of that for the x-axis. We also force the middle class on the x-axis to be centered around zero. Upon labeling the center of the (i,j)th bin by (x_i, y_j), a three-dimensional plot of γ_{ij}^* versus (x_i, y_j) yields an empirical semivariogram surface. Smoothing this surface using interpolation algorithms [e.g., Akima, 1978] produces a contour plot that we call the ESC plot. A symmetrized version of the ESC plot can be created by reflecting the upper left quadrant to the lower right and the upper right quadrant to the lower left.

The ESC plot can be used to assess departures from isotropy; isotropy is depicted by circular contours while elliptical contours capture geometric anisotropy. A rose diagram traces only one arbitrarily selected contour of this plot. A possible drawback to the ESC plot is the occurrence of sparse counts in extreme bins. However, these bins may be trimmed before smoothing if desired. Concerned that the use of geographic coordinates could introduce artificial anisotropy (since 1° latitude ≠ 1° longitude in the northeastern United States), we have employed a Universal Transverse Mercator (UTM) projection to kilometers in the E-W and N-S axes (see Subsection 1.2.4).

Figure 2.8 is the empirical semivariogram contour plot constructed using x-axis width of 30 kilometers for the 1993 scallop data. We have overlaid this contour plot on the bin centers with their respective counts. Note that using empirical semivariogram values in the row of the ESC plot for which $h_y \approx 0$ provides an alternative to the usual 0° directional semivariogram. The latter directional semivariograms are based on a polar representation of the angle and distance. For a chosen direction η and tolerance ϵ, the area for a class fans out as distance increases [see Figure 7.1 in Isaaks and Srivastava, 1990, p. 142]. Attractively, a directional semivariogram based on the rectangular bins associated with the empirical semivariogram in $\Re^2$ has bin area remaining constant as distance increases. In Figure 2.9, we present the four customary directional (polar representation) semivariograms for the 1993 scallop data. Clearly, the ESC plot is more informative, particularly in suggesting evidence of geometric anisotropy.

2.4   Spatial interpolation and kriging

An important aspect of spatial data analysis deals with fitting a spatial surface to a set of measurements over a finite set of locations. This exercise is loosely referred to as spatial interpolation and refers to the problem of estimating the value of the outcome at arbitrary locations (possibly new locations with no available measurements) using available measurements. In the subsequent chapters, we will deal, almost exclusively, with probabilistic spatial interpolation, or "kriging", so named by Matheron [1963] after the South African mining engineer D.G. Krige [Krige, 1951], whose seminal work on empirical geostatistical methods built the foundations of modern geostatistics. But before we embark upon such

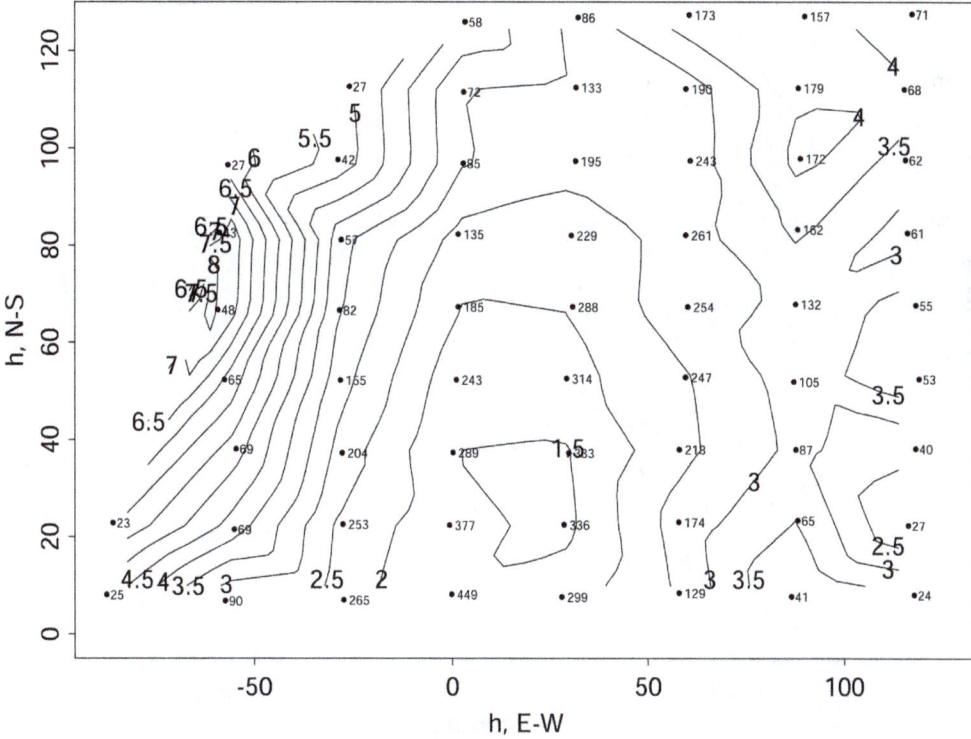

Figure 2.8 *ESC plot for the 1993 scallop data.*

developments, we offer a brief overview of spatial interpolation as an exercise in fitting surfaces to spatial data. We will, quite remarkably, see how the multivariate Gaussian distribution, which will in fact be a crucial component for building spatial processes later on, can be used as a *deterministic* spatial interpolator.

Consider measurements $\mathbf{Y} = (Y(\mathbf{s}_1), Y(\mathbf{s}_2), \ldots, Y(\mathbf{s}_n))^{\mathsf{T}}$ from a set of n fixed locations $\mathcal{S} = \{\mathbf{s}_1, \mathbf{s}_2, \ldots, \mathbf{s}_n\}$. We seek a function $f(\mathbf{s})$, defined for any arbitrary $\mathbf{s} \in \mathcal{D}$, such that $f(\mathbf{s}_i) = Y(\mathbf{s}_i)$ for each of the observed locations. We say that $f(\mathbf{s})$ *interpolates* the observed values, i.e., it represents a surface that passes exactly through the observed measurements. For now, we ignore the distribution of $\mathbf{Y}$. Suppose we restrict attention to generic linear interpolators of the form

$$f(\mathbf{s}) = \mu(\mathbf{s}) + \sum_{j=1}^{n} b_j(\mathbf{s})\beta_j \,, \qquad (2.14)$$

where $\mu(\mathbf{s})$ is any known function capturing trend, $b_i(\mathbf{s})$'s are a set of *basis functions* and β_i's are the associated coefficients. The condition for interpolation implies

$$\begin{pmatrix} Y(\mathbf{s}_1) \\ Y(\mathbf{s}_2) \\ \vdots \\ Y(\mathbf{s}_n) \end{pmatrix} = \begin{pmatrix} f(\mathbf{s}_1) \\ f(\mathbf{s}_2) \\ \vdots \\ f(\mathbf{s}_n) \end{pmatrix} = \begin{pmatrix} \mu(\mathbf{s}_1) \\ \mu(\mathbf{s}_2) \\ \vdots \\ \mu(\mathbf{s}_n) \end{pmatrix} + \begin{pmatrix} b_1(\mathbf{s}_1) & b_2(\mathbf{s}_1) & \ldots & b_n(\mathbf{s}_n) \\ b_1(\mathbf{s}_2) & b_2(\mathbf{s}_2) & \cdots & b_n(\mathbf{s}_2) \\ \vdots & \vdots & \ddots & \vdots \\ b_1(\mathbf{s}_n) & b_2(\mathbf{s}_n) & \ldots & b_n(\mathbf{s}_n) \end{pmatrix} \begin{pmatrix} \beta_1 \\ \beta_2 \\ \vdots \\ \beta_n \end{pmatrix} . \qquad (2.15)$$

We write (2.15) concisely as $\boldsymbol{\mu} + B\boldsymbol{\beta} = \mathbf{Y}$, where $\boldsymbol{\mu}$ is the $n \times 1$ vector with i-th element $\mu(\mathbf{s}_i)$, B is the $n \times n$ matrix with (i, j)th element $b_j(\mathbf{s}_i)$ and $\boldsymbol{\beta}$ is the $n \times 1$ vector with β_j as the j-th element.

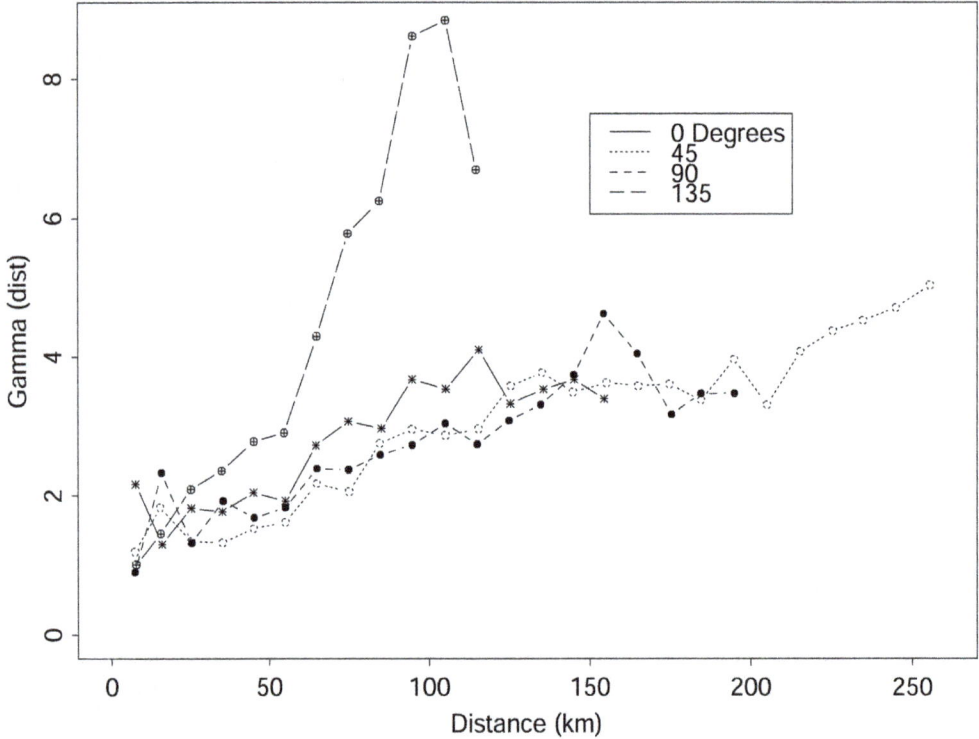

Figure 2.9 *Directional semivariograms for the 1993 scallop data.*

If $\hat{\boldsymbol{\beta}}$ is any solution for $B\boldsymbol{\beta} = \mathbf{Y} - \boldsymbol{\mu}$, then $f(\mathbf{s}) = \mu(\mathbf{s}) + \sum_{j=1}^{n} b_j(\mathbf{s})\hat{\beta}_j$ is a spatial interpolator. Matters are especially simple if the matrix B is nonsingular. Then, $\hat{\boldsymbol{\beta}} = B^{-1}(\mathbf{Y} - \boldsymbol{\mu})$ is the unique solution for (2.15) and we write (2.14) as

$$f(\mathbf{s}) = \mu(\mathbf{s}) + \sum_{j=1}^{n} b_j(\mathbf{s})\hat{\beta}_j = \mu(\mathbf{s}) + \mathbf{b}^{\mathrm{T}}(\mathbf{s})\hat{\boldsymbol{\beta}} = \mu(\mathbf{s}) + \mathbf{b}^{\mathrm{T}}(\mathbf{s})B^{-1}(\mathbf{Y} - \boldsymbol{\mu}) , \qquad (2.16)$$

where $\mathbf{b}(\mathbf{s})$ is the $n \times 1$ vector with j-th element $b_j(\mathbf{s})$. Clearly $f(\mathbf{s})$ in (2.16) interpolates the data because $\hat{\boldsymbol{\beta}}$ is obtained from the very conditions for interpolation in (2.15). Nevertheless, a direct verification is also instructive. Since B is nonsingular it satisfies $BB^{-1} = I$. Equating the i-th rows of both matrices yields $\mathbf{b}^{\mathrm{T}}(\mathbf{s}_i)B^{-1} = \mathbf{e}_i^{\mathrm{T}}$ because $\mathbf{b}^{\mathrm{T}}(\mathbf{s}_i)$ is the i-th row of B, where $\mathbf{e}_i^{\mathrm{T}}$ denotes the i-th row of the identity matrix. Plugging in $\mathbf{s} = \mathbf{s}_i$ renders

$$f(\mathbf{s}_i) = \mu(\mathbf{s}_i) + \mathbf{b}^{\mathrm{T}}(\mathbf{s}_i)B^{-1}(\mathbf{Y} - \boldsymbol{\mu}) = \mu(\mathbf{s}_i) + \mathbf{e}_i^{\mathrm{T}}(\mathbf{Y} - \boldsymbol{\mu}) = \mu(\mathbf{s}_i) + Y(\mathbf{s}_i) - \mu(\mathbf{s}_i) = Y(\mathbf{s}_i)$$

for each $i = 1, 2, \ldots, n$. The above framework yields a rich class of interpolating functions specified with suitable basis functions. Given a finite collection of spatial locations where the outcome has been observed, any choice of basis functions ensures a nonsingular B will yield a spatial interpolator of the form (2.16). The above calculations also reveal that the interpolation property holds for *any* choice of $\mu(\mathbf{s})$.

We can introduce spatial covariance functions into spatial interpolation. "Kriging" can be looked upon as a special case of the generic (2.14) by constructing the basis functions from a valid covariance function. Taking $b_j(\mathbf{s})$ to be a valid covariance function $C(\mathbf{s}, \mathbf{s}_j)$ implies that B is a positive definite matrix with elements $C(\mathbf{s}_i, \mathbf{s}_j)$. This is what we mean by a *valid* covariance function. Therefore, B is nonsingular so the resulting $f(\mathbf{s})$ interpolates the data for any $\mathbf{s}_i \in \mathcal{S}$. In this case, we write (2.16) as

$$f(\mathbf{s}) = \mu(\mathbf{s}) + \mathbf{c}^{\mathrm{T}}(\mathbf{s}; \boldsymbol{\theta}) C^{-1}(\boldsymbol{\theta})(\mathbf{Y} - \boldsymbol{\mu}) \,, \tag{2.17}$$

where $\mathbf{c}(\mathbf{s}; \boldsymbol{\theta})$ is the $n \times 1$ vector with $C(\mathbf{s}, \mathbf{s}_j)$ as its elements and $C(\boldsymbol{\theta})$ is the spatial covariance matrix with elements $C(\boldsymbol{\theta})$. The $f(\mathbf{s})$ constructed in (2.17) is a spatial interpolator over $\mathcal{S}$ for *all* values of $\boldsymbol{\theta}$ supporting the covariance function and for any choice of $\mu(\mathbf{s})$.

Equation 2.17, as derived above, emerges from some simple linear algebra. For our inferential purposes, it is more attractive to see how it arises from the joint probability law implied by a Gaussian spatial process model. Let us first recall a basic, but extremely useful, result on conditional densities derived from joint multivariate Gaussian distributions. Let $\mathbf{Y}_1 1$ and $\mathbf{Y}_2$ be $n_1 \times 1$ and $n_2 \times 1$ random vectors following the multivariate normal density

$$\begin{pmatrix} \mathbf{Y}_1 \\ \mathbf{Y}_2 \end{pmatrix} \sim N \left(\begin{pmatrix} \boldsymbol{\mu}_1 \\ \boldsymbol{\mu}_2 \end{pmatrix}, \begin{pmatrix} \Omega_{11} & \Omega_{12} \\ \Omega_{21} & \Omega_{22} \end{pmatrix} \right), \tag{2.18}$$

where $\mathrm{E}[\mathbf{Y}_1] = \boldsymbol{\mu}_1$ and $\mathrm{E}[\mathbf{Y}_2] = \boldsymbol{\mu}_2$ are $n_1 \times 1$ and $n_2 \times 1$ mean vectors respectively, and the $(n_1 + n_2) \times (n_1 + n_2)$ covariance matrix is presented as a 2×2 block matrix with $n_i \times n_j$ blocks denoted by Ω_{ij} and with $\Omega_{21} = \Omega_{12}^{\mathrm{T}}$ to ensure symmetry. The conditional distribution $p(\mathbf{Y}_1 \mid \mathbf{Y}_2)$ is normal with mean and variance,

$$\mathrm{E}[\mathbf{Y}_1 \mid \mathbf{Y}_2] = \boldsymbol{\mu}_1 + \Omega_{12}\Omega_{22}^{-1}(\mathbf{Y}_2 - \boldsymbol{\mu}_2) \quad \text{and} \quad \mathrm{Var}[\mathbf{Y}_1 \mid \mathbf{Y}_2] = \Omega_{11} - \Omega_{12}\Omega_{22}^{-1}\Omega_{21} \,, \tag{2.19}$$

respectively. If $Y(\mathbf{s})$ is a Gaussian spatial process with mean $\mu(\mathbf{s})$ and a covariance function $C(\cdot, \cdot; \boldsymbol{\theta})$, then $\mathbf{Y} = (Y(\mathbf{s}_1), Y(\mathbf{s}_2), \ldots, Y(\mathbf{s}_n))^{\mathrm{T}}$ is the observed outcomes with a $N(\boldsymbol{\mu}, C(\boldsymbol{\theta}))$ probability law implied by the Gaussian process. From (2.18), the joint distribution of $\mathbf{Y}$ and an unobserved $Y(\mathbf{s}_0)$ at a new location $\mathbf{s}_0$ is given by

$$\begin{pmatrix} \mathbf{Y} \\ Y(\mathbf{s}_0) \end{pmatrix} \sim N \left(\begin{bmatrix} \boldsymbol{\mu} \\ \mu(\mathbf{s}_0) \end{bmatrix}, \begin{bmatrix} C(\boldsymbol{\theta}) & \mathbf{c}(\mathbf{s}_0; \boldsymbol{\theta}) \\ \mathbf{c}(\mathbf{s}_0; \boldsymbol{\theta})^{\mathrm{T}} & C(\mathbf{s}_0, \mathbf{s}_0) \end{bmatrix} \right) \,,$$

where we have set $\mathbf{Y}_1 := \mathbf{Y}$, $\mathbf{Y}_2 := Y(\mathbf{s}_0)$, $\Omega_{11} := C(\boldsymbol{\theta})$, $\Omega_{12} = \mathbf{c}(\mathbf{s}_0; \boldsymbol{\theta})$, and $\Omega_{22} = C(\mathbf{s}_0, \mathbf{s}_0)$. Then, applying (2.19), we find

$$E[Y(\mathbf{s}_0) \mid \mathbf{Y}] = \mu(\mathbf{s}_0) + \mathbf{c}^{\mathrm{T}}(\mathbf{s}_0; \boldsymbol{\theta})C(\boldsymbol{\theta})^{-1}(\mathbf{Y} - \boldsymbol{\mu})$$

$$\text{and} \quad \mathrm{Var}[Y(\mathbf{s}_0) \mid \mathbf{Y}] = C(\mathbf{s}_0, \mathbf{s}_0) - \mathbf{c}^{\mathrm{T}}(\mathbf{s}_0; \boldsymbol{\theta})C(\boldsymbol{\theta})^{-1}\mathbf{c}(\mathbf{s}_0; \boldsymbol{\theta}) \,.$$

There are a couple of notable observations to be made here. First, the above conditional expectation is exactly $f(\mathbf{s}_0)$ in (2.17), but assuming a probability law enables us to quantify the uncertainty using the expression for the variance. This is called the "kriging" estimate and "kriging" variance for $Y(\mathbf{s}_0)$ at a new location $\mathbf{s}_0$. If $\mathbf{s}_0$ is the same as an observed location $\mathbf{s}_i$, then $\mathrm{E}[Y(\mathbf{s}_0) \mid \mathbf{Y}] = f(\mathbf{s}_i) = Y(\mathbf{s}_i)$ so the kriging estimator interpolates observed values. Second, it is easy to verify that $\mathrm{Var}[Y(\mathbf{s}_0) \mid \mathbf{Y}] = 0$ whenever $\mathbf{s}_0$ is one of the observed locations. This shows that "kriging" *deterministically* interpolates the data, i.e., it reproduces the value of the outcome at an observed location and there is no uncertainty of the estimator at that location. Finally, the "kriging" estimator retains its interpolating property irrespective of the value of $\boldsymbol{\theta}$ (as long as it is in a range that ensures the covariance function is positive definite). The process parameters $\boldsymbol{\theta}$ will not be known and have to be estimated. They can be fixed using eyeball estimates obtained from variograms, or can be formally estimated using likelihood-based or Bayesian methods; see Chapter 6.

It is also clear that the interpolation property of "kriging" holds irrespective of the presence or absence of a nugget. Consider the covariance function $C(\mathbf{s}, \mathbf{t}) = \sigma^2 \rho(\|\mathbf{s} - \mathbf{t}\|; \phi) + \tau^2 \delta_{\{\mathbf{s}=\mathbf{t}\}}$, where $\delta_{\{\mathbf{s}=\mathbf{t}\}} = 1$ if $\mathbf{s} = \mathbf{t}$ and 0 otherwise. Then $\boldsymbol{\theta} = \{\sigma^2, \phi, \tau^2\}$

and $C(\boldsymbol{\theta}) = \sigma^2 H(\phi) + \tau^2 I_n$, where $H(\phi)$ is a spatial correlation matrix with (i, j)-th entry $\rho(\|\mathbf{s}_i - \mathbf{s}_j\|; \phi)$. Irrespective of whether $\tau^2 = 0$ or not, the vector $\mathbf{c}(\mathbf{s}_i, \boldsymbol{\theta}) = (C(\mathbf{s}_i, \mathbf{s}_1; \boldsymbol{\theta}), C(\mathbf{s}_i, \mathbf{s}_2; \boldsymbol{\theta}), \ldots, C(\mathbf{s}_i, \mathbf{s}_n; \boldsymbol{\theta}))^{\mathrm{T}}$ and will equal the i-th row of $C(\boldsymbol{\theta})$ whenever $\mathbf{s}_i$ is an observed location and the kriging estimate will interpolate the value at $\mathbf{s}_i$.

We point out that some authors [e.g., Gneiting and Guttorp, 2010] distinguish spatial interpolation in the presence of a nugget from that without a nugget. The latter is sometimes referred to as *noiseless kriging*. The underlying idea here is to note that even if there is a measurement error in the process, for an unobserved location there are no measurements, hence no measurement error. Therefore, for a new location $\mathbf{s}_0$ the kriging estimator is

$$\mathrm{E}[Y(\mathbf{s}_0) \mid \mathbf{Y}] = \mu(\mathbf{s}_0) + \boldsymbol{\gamma}^{\mathrm{T}}(\mathbf{s}_0; \boldsymbol{\theta})C^{-1}(\boldsymbol{\theta})(\mathbf{Y} - \boldsymbol{\mu}) , \qquad (2.20)$$

where $\boldsymbol{\gamma}(\mathbf{s}_0; \boldsymbol{\theta})$ is the $n \times 1$ vector with i-th element $\sigma^2 \rho(\|\mathbf{s}_0 - \mathbf{s}_i\|; \phi)$. Simply plugging in $\mathbf{s}_0 = \mathbf{s}_i$, where $\mathbf{s}_i$ is an observed location, will not yield $Y(\mathbf{s}_i)$. This is because $\boldsymbol{\gamma}^{\mathrm{T}}(\mathbf{s}_i; \boldsymbol{\theta})$ does not equal the i-th row of $C(\boldsymbol{\theta})$—the i-th entry of the i-th row of $C(\boldsymbol{\theta})$ is $\sigma^2 + \tau^2$, while the i-th entry in $\boldsymbol{\gamma}^{\mathrm{T}}(\mathbf{s}_i; \boldsymbol{\theta})$ is simply σ^2. Thus, (2.20) is an interpolator only if $\tau^2 = 0$, which corresponds to noiseless "kriging;" see Section 2.6 for further details.

In the subsequent section, we briefly describe how kriging can be developed purely using the classical minimum mean-squared error approach to prediction. In fact, this is how it was originally formulated by Matheron [1963] using intrinsic stationarity. A linear predictor for $Y(\mathbf{s}_0)$ based on $\mathbf{Y}$ would take the form $\sum \ell_i Y(\mathbf{s}_i) + \delta_0$. The best linear prediction under squared error loss would minimize $\mathrm{E}[Y(\mathbf{s}_0) - (\sum \ell_i Y(\mathbf{s}_i) + \delta_0)]^2$ over δ_0 and the ℓ_i. Under the intrinsic stationarity, we would take $\sum \ell_i = 1$ in order that $\mathrm{E}[Y(\mathbf{s}_0) - \sum \ell_i Y(\mathbf{s}_i)] = 0$. This boils down to a fairly routine optimization for a quadratic form under a linear constraint and can be easily solved using Lagrange multipliers and, as we show below in Section 2.5, the optimal linear predictor can be seen to be a function of the variogram. Under weak stationarity, we can derive the covariance function from the variogram and the optimal linear predictor coincides with the right hand side of (2.20). Further details on deriving the kriging equations under intrinsic stationarity can be found, for example, in the texts by Cressie [1993] and Chilés and Delfiner [1999].

2.5   Classical spatial prediction

Here, we describe the classical (i.e., minimum mean-squared error) approach to spatial prediction or kriging in the point-referenced data setting. The problem is one of *optimal spatial prediction*: given observations of a random field $\mathbf{Y} = (Y(\mathbf{s}_1), \ldots, Y(\mathbf{s}_n))'$, how do we predict the variable Y at a site $\mathbf{s}_0$ where it has not been observed? In other words, what is the best predictor of the value of $Y(\mathbf{s}_0)$ based on the data $\mathbf{y}$?

As above, a linear predictor for $Y(\mathbf{s}_0)$ based on $\mathbf{y}$ would take the form $\sum \ell_i Y(\mathbf{s}_i) + \delta_0$. Using squared error loss, the best linear prediction would minimize $\mathrm{E}[Y(\mathbf{s}_0) - (\sum \ell_i Y(\mathbf{s}_i) + \delta_0)]^2$ over δ_0 and the ℓ_i. Under the intrinsic stationarity specification (2.1) we would take $\sum \ell_i = 1$ in order that $\mathrm{E}(Y(\mathbf{s}_0) - \sum \ell_i Y(\mathbf{s}_i)) = 0$. As a result, we would minimize $\mathrm{E}[Y(\mathbf{s}_0) - \sum \ell_i Y(\mathbf{s}_i)]^2 + \delta_0^2$, and clearly δ_0 would be set to 0. Now letting $a_0 = 1$ and $a_i = -\ell_i$ we see that the criterion becomes $\mathrm{E}[\sum_{i=0}^{n} a_i Y(\mathbf{s}_i)]^2$ with $\sum a_i = 0$. But from (2.4) this expectation becomes $-\sum_i \sum_j a_i a_j \gamma(\mathbf{s}_i - \mathbf{s}_j)$, revealing how, historically, the variogram arose in kriging within the geostatistical framework. Indeed, the optimal ℓ's can be obtained by solving this constrained optimization. In fact, it is a routine optimization of a quadratic from under a linear constraint and is customarily handled using Lagrange multipliers. The solution will be a function of $\gamma(\mathbf{h})$, in fact of the set of $\gamma_{ij} = \gamma(\mathbf{s}_i - \mathbf{s}_j)$ and $\gamma_{0j} = \gamma(\mathbf{s}_0 - \mathbf{s}_j)$ With an estimate of γ, one immediately obtains the so-called *ordinary kriging* estimate. Other than the intrinsic stationarity model (Subsection 2.1.2), no further distributional assumptions are required for the $Y(\mathbf{s})$'s.

To provide a bit more detail, restoring the ℓ_i's, we obtain

$$-\sum_i \sum_j a_i a_j \gamma(\mathbf{s}_i - \mathbf{s}_j) = -\sum_i \sum_j \ell_i \ell_j \gamma_{ij} + 2\sum_i \ell_i \gamma_{0i}$$

Adding the constraint, $\sum \ell_i = 1$, times the Lagrange multiplier, λ, we find that the partial derivative of this expression with regard to ℓ_i becomes $-\sum_j \ell_j \gamma_{ij} + \gamma_{0i} - \lambda = 0$. Letting Γ be the $n \times n$ matrix with entries $\Gamma_{ij} = \gamma_{ij}$ and $\boldsymbol{\gamma}_0$ be the $n \times 1$ vector with $(\boldsymbol{\gamma}_0)_i = \gamma_{0i}$, with $\boldsymbol{\ell}$ the $n \times 1$ vector of coefficients, we obtain the system of *kriging* equations, $\Gamma\boldsymbol{\ell} + \lambda\mathbf{1} = \boldsymbol{\gamma}_0$ and $\mathbf{1}^{\mathsf{T}}\boldsymbol{\ell} = 1$. The solution is $\boldsymbol{\ell} = \Gamma^{-1}\left(\boldsymbol{\gamma}_0 + \frac{(1 - \mathbf{1}^{\mathsf{T}}\Gamma^{-1}\boldsymbol{\gamma}_0)}{\mathbf{1}^{\mathsf{T}}\Gamma^{-1}\mathbf{1}}\mathbf{1}\right)$ and $\boldsymbol{\ell}^{\mathsf{T}}\mathbf{Y}$ becomes the best linear unbiased predictor (BLUP). Again, with $\gamma(\mathbf{h})$ unknown, we have to estimate it in order to calculate this estimator. Then, $\boldsymbol{\ell}$ is a function of the data and the estimator is no longer linear. Continuing in this regard, the usual estimator of the uncertainty in the prediction is the predictive mean square error (PMSE), $E(Y(\mathbf{s}_0) - \boldsymbol{\ell}^{\mathsf{T}}\mathbf{Y})^2$, rather than $\operatorname{var}(\boldsymbol{\ell}^{\mathsf{T}}\mathbf{Y})$. There is a closed-form expression for the former, i.e., $\boldsymbol{\gamma}_0^{\mathsf{T}}\Gamma^{-1}\boldsymbol{\gamma}_0 - (\boldsymbol{\ell}^{\mathsf{T}}\Gamma^{-1}\boldsymbol{\gamma}_0 - 1)^2/\boldsymbol{\ell}^{\mathsf{T}}\Gamma^{-1}\boldsymbol{\ell}$, which, also requires an estimate of $\gamma(\mathbf{h})$ in order to be calculated.

Let us now take a formal look at kriging in the context of Gaussian processes. Consider first the case where we have no covariates, but only the responses $Y(\mathbf{s}_i)$. This is developed by means of the following model for the observed data:

$$\mathbf{Y} = \mu\mathbf{1} + \boldsymbol{\epsilon}, \text{ where } \boldsymbol{\epsilon} \sim N(\mathbf{0}, \Sigma) .$$

For a spatial covariance structure having no nugget effect, we specify Σ as

$$\Sigma = \sigma^2 H(\phi) \text{ where } (H(\phi))_{ij} = \rho(\phi; d_{ij}) ,$$

where $d_{ij} = ||\mathbf{s}_i - \mathbf{s}_j||$, the distance between $\mathbf{s}_i$ and $\mathbf{s}_j$ and ρ is a valid correlation function on $\Re^r$ such as those in Table 2.1. For a model having a nugget effect, we instead set

$$\Sigma = \sigma^2 H(\phi) + \tau^2 I ,$$

where τ^2 is the nugget effect variance.

When covariate values $\mathbf{x} = (x(\mathbf{s}_1), \ldots, x(\mathbf{s}_n))^{\mathsf{T}}$ and $x(\mathbf{s}_0)$ are available for incorporation into the analysis, the procedure is often referred to as *universal kriging*, though we caution that some authors (e.g., Kaluzny et al., 1998) use the term "universal" in reference to the case where only latitude and longitude are available as covariates. The model now takes the more general form

$$\mathbf{Y} = X\boldsymbol{\beta} + \boldsymbol{\epsilon}, \text{ where } \boldsymbol{\epsilon} \sim N(\mathbf{0}, \Sigma) ,$$

with Σ being specified as above, either with or without the nugget effect. Note that ordinary kriging may be looked upon as a particular case of universal kriging with X being the $n \times 1$ matrix (i.e., column vector) $\mathbf{1}$, and $\boldsymbol{\beta}$ the scalar μ.

We now pose our prediction problem as follows: we seek the function $h(\mathbf{y})$ that minimizes the mean-squared prediction error,

$$E\left[(Y(\mathbf{s}_0) - h(\mathbf{y}))^2 \,\Big|\, \mathbf{y}\right] . \tag{2.21}$$

By adding and subtracting the conditional mean $E[Y(\mathbf{s}_0) \mid \mathbf{y}]$ inside the square, grouping terms, squaring and noting that the cross-product term equals zero, we obtain

$$E\left[(Y(\mathbf{s}_0) - h(\mathbf{y}))^2 \,\Big|\, \mathbf{y}\right] = E\left\{(Y(\mathbf{s}_0) - E[Y(\mathbf{s}_0) \mid \mathbf{y}])^2 \,\Big|\, \mathbf{y}\right\} + \{E[Y(\mathbf{s}_0) \mid \mathbf{y}] - h(\mathbf{y})\}^2 .$$

Since the second term on the right-hand side is nonnegative, we have

$$E\left[(Y(\mathbf{s}_0) - h(\mathbf{y}))^2 \,\Big|\, \mathbf{y}\right] \geq E\left\{(Y(\mathbf{s}_0) - E[Y(\mathbf{s}_0) \mid \mathbf{y}])^2 \,\Big|\, \mathbf{y}\right\}$$

for any function $h(\mathbf{y})$. Equality holds if and only if $h(\mathbf{y}) = \mathrm{E}[Y(\mathbf{s}_0) \mid \mathbf{y}]$, so it must be that the predictor $h(\mathbf{y})$ which minimizes the error is the conditional expectation of $Y(\mathbf{s}_0)$ given the data. This result is familiar since we know that the mean minimizes expected squared error loss. From a Bayesian point of view, this $h(\mathbf{y})$ is just the *posterior mean* of $Y(\mathbf{s}_0)$, and it is well known that the posterior mean is the Bayes rule (i.e., the minimizer of posterior risk) under squared error loss functions of the sort adopted in (2.21) above as our scoring rule. Note that the posterior mean is the best predictor under squared error loss regardless of whether we assume a Gaussian model and that, in general, it need not be linear; it need not be the BLUP which is the ordinary kriging predictor. However, under a Gaussian process model assumption, this posterior mean is linear, as we now clarify.

In particular, let us explicitly obtain the form of the posterior mean of $Y(\mathbf{s}_0)$. It will emerge as a function of all the population parameters ($\boldsymbol{\beta}, \sigma^2, \phi$, and τ^2). Again, we apply (2.18) with $\mathbf{Y}_1 = Y(\mathbf{s}_0)$ and $\mathbf{Y}_2 = \mathbf{y}$. It then follows that

$$\Omega_{11} = \sigma^2 + \tau^2, \quad \Omega_{12} = \boldsymbol{\gamma}^{\mathrm{T}}, \quad \text{and } \Omega_{22} = \Sigma = \sigma^2 H(\phi) + \tau^2 I \,,$$

where $\boldsymbol{\gamma}^{\mathrm{T}} = \left(\sigma^2 \rho(\phi; d_{01}), \ldots, \sigma^2 \rho(\phi; d_{0n})\right)$. Substituting these values into the mean and variance formulae above, we obtain

$$E[Y(\mathbf{s}_0) \mid \mathbf{y}] = \mathbf{x}_0^{\mathrm{T}} \boldsymbol{\beta} + \boldsymbol{\gamma}^{\mathrm{T}} \Sigma^{-1} (\mathbf{y} - X\boldsymbol{\beta}) \,, \tag{2.22}$$

$$\text{and } \mathrm{Var}[Y(\mathbf{s}_0) \mid \mathbf{y}] = \sigma^2 + \tau^2 - \boldsymbol{\gamma}^{\mathrm{T}} \Sigma^{-1} \boldsymbol{\gamma} \,. \tag{2.23}$$

We see that the posterior mean is a linear predictor. We remark that this solution assumes we have actually observed the covariate value $\mathbf{x}_0 = \mathbf{x}(\mathbf{s}_0)$ at the "new" site $\mathbf{s}_0$; we defer the issue of missing $\mathbf{x}_0$ for the moment.

Since, in practice, the model parameters are unknown, they must be estimated from the data. Here we would modify $h(\mathbf{y})$ to

$$\widehat{h(\mathbf{y})} = \mathbf{x}_0^{\mathrm{T}} \widehat{\boldsymbol{\beta}} + \widehat{\boldsymbol{\gamma}}^{\mathrm{T}} \widehat{\Sigma}^{-1} \left(\mathbf{y} - X\widehat{\boldsymbol{\beta}}\right) \,,$$

where $\widehat{\boldsymbol{\gamma}} = \left(\hat{\sigma}^2 \rho(\hat{\phi}; d_{01}), \ldots, \hat{\sigma}^2 \rho(\hat{\phi}; d_{0n})\right)^{\mathrm{T}}$, $\widehat{\boldsymbol{\beta}} = \left(X^{\mathrm{T}} \widehat{\Sigma}^{-1} X\right)^{-1} X^{\mathrm{T}} \widehat{\Sigma}^{-1} \mathbf{y}$, the usual weighted least squares estimator of $\boldsymbol{\beta}$, and $\widehat{\Sigma} = \hat{\sigma}^2 H(\hat{\phi})$. Thus $\widehat{h(\mathbf{y})}$ can be written as $\boldsymbol{\lambda}^{\mathrm{T}} \mathbf{y}$, where

$$\boldsymbol{\lambda} = \widehat{\Sigma}^{-1} \widehat{\boldsymbol{\gamma}} + \widehat{\Sigma}^{-1} X \left(X^{\mathrm{T}} \widehat{\Sigma}^{-1} X\right)^{-1} \left(\mathbf{x}_0 - X^{\mathrm{T}} \widehat{\Sigma}^{-1} \widehat{\boldsymbol{\gamma}}\right) \,. \tag{2.24}$$

The reader may be curious as to whether $\boldsymbol{\lambda}^{\mathrm{T}} \mathbf{Y}$ is the same as $\boldsymbol{\ell}^{\mathrm{T}} \mathbf{Y}$ assuming all parameters are known say, in the ordinary kriging case. That is, assuming we have $\gamma(\mathbf{h}) = c(\mathbf{0}) - c(\mathbf{h})$, we can write both linear predictors in terms of $c(\mathbf{h})$ or $\gamma(\mathbf{h})$ but will they agree? Immediately, we know that the answer is no because $\boldsymbol{\ell}^{\mathrm{T}} \mathbf{Y}$ depends only on $\gamma(\mathbf{h})$ or $c(\mathbf{h})$ while $\boldsymbol{\lambda}^{\mathrm{T}} \mathbf{Y}$ requires $\boldsymbol{\beta}$, hence μ in the constant mean case. In fact, it is a straightforward calculation to show that, if we replace μ by $\hat{\mu}$, the best linear unbiased estimator of μ, in $\boldsymbol{\lambda}$, we do obtain the ordinary kriging estimator. We leave this as an exercise.

If $\mathbf{x}_0$ is unobserved, we can estimate it and $Y(\mathbf{s}_0)$ jointly by iterating between this formula and a corresponding one for $\hat{\mathbf{x}}_0$, namely

$$\hat{\mathbf{x}}_0 = X^{\mathrm{T}} \boldsymbol{\lambda} \,,$$

which arises simply by multiplying both sides of (2.24) by X^{T} and simplifying. This is essentially an EM (expectation-maximization) algorithm [Dempster et al., 2018] with the calculation of $\hat{\mathbf{x}}_0$ being the E step and (2.24) being the M step.

In the classical framework, a lot of energy is devoted to the determination of the optimal estimates to plug into the above equations. Typically, restricted maximum likelihood

(REML) estimates are selected and shown to have certain optical properties. However, as we shall see in Chapter 6, how to perform the estimation is not an issue in the Bayesian setting. There, adopting prior distributions on the parameters produces the full posterior predictive distribution $p(Y(\mathbf{s}_0) \mid \mathbf{y})$. Any desired point or interval estimate (the latter to express our uncertainty in such prediction) as well as any desired probability statements may then be computed with respect to this distribution.

2.6 Noiseless kriging

It is natural to ask whether the kriging predictor (ordinary or posterior mean) will return the observed value at the $\mathbf{s}_i$'s where we observed the process. If so, we would refer to the predictor as an *exact* interpolator. The literature addresses this problem through a detailed inspection of the kriging equations. However, the answer is immediately clear once we look at the model specification. And, it is possibly counterintuitive; we obtain exact interpolation in the case of no nugget, rather than in the seemingly more flexible case where we add a nugget.

Analytically, one can determine whether or not the predictor in (2.22) will equal the observed value at a given $\mathbf{s}_i$. We leave as a formal exercise to verify that if $\tau^2 = 0$ (i.e., the no-nugget case, or so-called noiseless prediction,) then the answer is yes, while if $\tau^2 > 0$ then the answer is no.

However, we can illuminate the situation without formal calculation. There are two potential settings: (i) $Y(\mathbf{s}) = \mu(\mathbf{s}) + w(\mathbf{s})$ and (ii) $Y(\mathbf{s}) = \mu(\mathbf{s}) + w(\mathbf{s}) + \epsilon(\mathbf{s})$, the no nugget and nugget cases, respectively. Under (i), evidently, predicting $Y(\mathbf{s}_0)$ is the same as predicting $\mu(\mathbf{s}_0) + w(\mathbf{s}_0)$. There is only one surface for the process realization. Under (ii), predicting $Y(\mathbf{s}_0)$ is different from predicting $\mu(\mathbf{s}_0) + w(\mathbf{s}_0)$; we have two random surfaces where the former is everywhere discontinuous while the latter, for a continuous $\mu(\mathbf{s})$ and customary covariance functions, is continuous. Here, terminology can be confusing. Under (i), there is no noise, so we could refer to this case as noiseless kriging. However, under (ii), predicting $\mu(\mathbf{s}_0) + w(\mathbf{s}_0)$ could be referred to as noiseless interpolation. We ignore the terminology issue and address the exact interpolation question.

Now, suppose $\mathbf{s}_0$ is, in fact, one of the $\mathbf{s}_i$. Then, we need to distinguish the observed value at $\mathbf{s}_i$ say $Y_{obs}(\mathbf{s}_i)$ from a new or replicate value at $\mathbf{s}_i$, say $Y_{rep}(\mathbf{s}_i)$. Under (i) $f(Y_{rep}(\mathbf{s}_i) \mid \text{Data}) = f(\mu(\mathbf{s}_i) + w(\mathbf{s}_i) \mid \text{Data})$ and since, given the data, $\mu(\mathbf{s}_i) + w(\mathbf{s}_i) = Y_{obs}(\mathbf{s}_i)$, $f(Y_{rep}(\mathbf{s}_i) \mid \text{Data})$ is degenerate at $Y_{obs}(\mathbf{s}_i)$; we have exact interpolation. Under (ii), $Y_{rep}(\mathbf{s}_i) \sim Y_{obs}(\mathbf{s}_i)$ and $\mathrm{E}(Y_{rep}(\mathbf{s}_i) \mid \text{Data}) = \mathrm{E}(\mu(\mathbf{s}_i) + w(\mathbf{s}_i) \mid \text{Data}) \neq Y_{obs}(\mathbf{s}_i)$; we do not have exact interpolation.

2.7 Spatial data analysis using machine learning methods

While the subsequent developments in this text will focus on statistical models built from stochastic processes, it is worth remarking on developments in machine learning algorithms that focus upon minimizing an objective loss function (based on spatial interpolation or approximation) without, in general, attempting to account for spatial correlations. Ignoring spatial correlation may adversely impact inference. Saha et al. [2023] and Zhan and Datta [2024] present some of the first theoretical studies investigating the consequences of executing off-the-shelf machine learning algorithms ignoring spatial correlations in geospatial data analysis. These authors report that the performance of such machine learning algorithms is negatively affected in geospatial analysis when spatial correlations are not modeled. Using these algorithms for geospatial data can lead to inaccurate or inefficient inference.

In response to these challenges, novel machine learning algorithms have emerged that integrate traditional statistical principles to address geospatial correlation. Saha et al. [2023] introduce RF-GLS, a version of random forests that explicitly incorporates spatial correlation in the data using the generalized least squares (GLS) loss, a principle commonly employed in linear regression for dependent data. RF-GLS demonstrated superior performance compared to the naive application of random forests for geospatial data and was also applicable to time-series analysis. The study provided some of the first asymptotic consistency results for random forests for dependent data. A software implementation for RF-GLS is developed in Saha et al. [2022]. Finally, we comment on Zhan and Datta [2024] who develop NN-GLS, a neural network approach that uses a GLS loss to model spatial correlation. The NN-GLS is modeled as a graph neural networks (GNN), enabling the use of existing GNN implementations for NN-GLS, thereby offering greater scalability compared to random forest-based methods. The work also presents some of the first asymptotic results for neural networks for irregular geospatial data, including results on the consistency of NN-GLS estimators and a theoretical result quantifying the risks of using conventional neural networks for spatial data.

2.8 Computer tutorials

We provide computer programs to interpolate spatial surfaces, compute and plot variograms and conduct exploratory data analysis within the R statistical computing environment in the folder titled Chapter 2 from https://github.com/sudiptobanerjee/BGC_2023.

2.8.1 EDA and spatial data visualization in R

Figure 2.10 presents the spatial surface constructed by interpolating log metric tons of biomass over a region in Bartlett, New Hampshire, using data from the Bartlett Experimental Forest (BEF) of the U.S. Department of Agriculture and Forest Service. The full dataset is available in the spBayes package for the R statistical computing environment.

We will also illustrate our approaches using the so-called WEF forest inventory data from a long-term ecological research site in western Oregon. These data consist of a census of all trees in a 10-ha stand. Diameter at breast height (DBH) and tree height (HT) have been

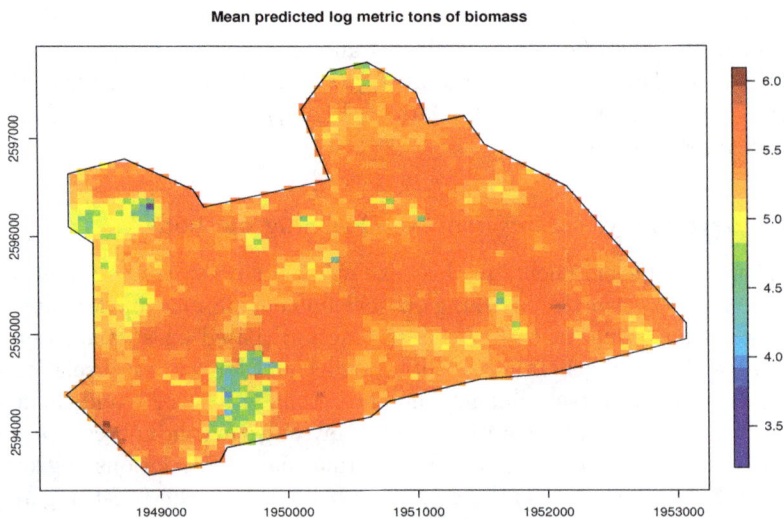

Figure 2.10 *Interpolated surface of observed log metric tons of biomass from the Bartlett Experimental Forest data.*

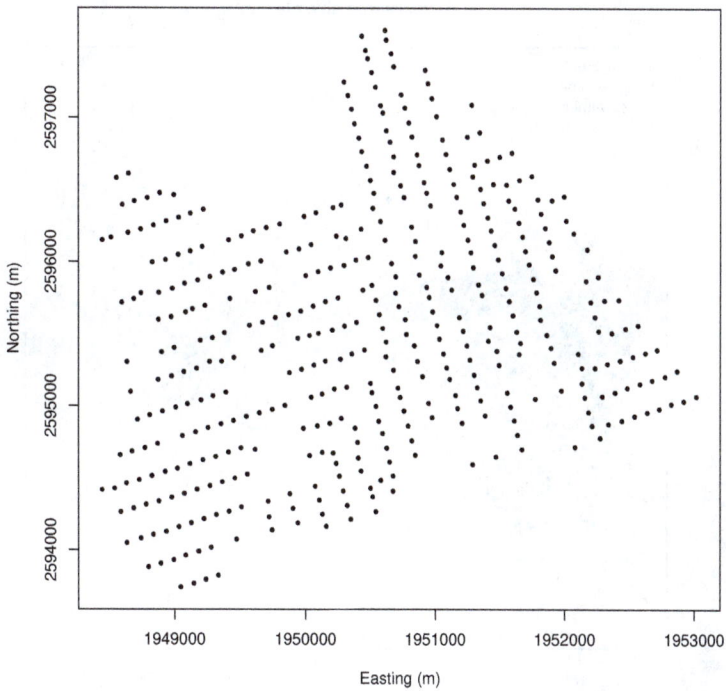

Figure 2.11 *Tree locations on the WEF.*

measured for all trees in the stand. For a subset of these trees, the distance from the center of the stem to the edge of the crown was measured at each of the cardinal directions. Rarely do we have the luxury of a complete census, but rather a subset of trees are sampled using a series of inventory plots. This sample is then used to draw inferences about parameters of interest (e.g., mean stand DBH or correlation between DBH and HT). We defer these analyses to subsequent example sessions. Here, we simply use these data to demonstrate some basics of spatial data manipulation, visualization, and exploratory analysis. Figure 2.11 presents the coordinates of the trees. Figure 2.12 displays the coordinates colored according to the four classes: sapling, poletimber, sawtimber, and large sawtimber. The figure uses four colors each with five shades.

If the data observations are well distributed over the domain, spatial patterns can often be detected by estimating a continuous surface using an interpolation function. Several packages provide suitable interpolators, including the akima for linear or cubic spline interpolation and MBA which provides efficient interpolation of large data sets with multilevel B-splines (see Figure 2.13 for interpolated DBH and Figure 2.14 for a 3-dimensional perspective plot).

2.8.2 Variogram analysis in R

Typical visual inspection of the forest inventory data suggests that there is some degree of spatial dependence in the distribution of DBH. This encourages further exploration using a variogram analysis to quantify the range of spatial dependence and obtain an idea about the sill and the nugget. The geoR package provides functions for computing and plotting variograms. It provides several of the variogram models in Table 2.2 and offers options for

Forestry tree size classes

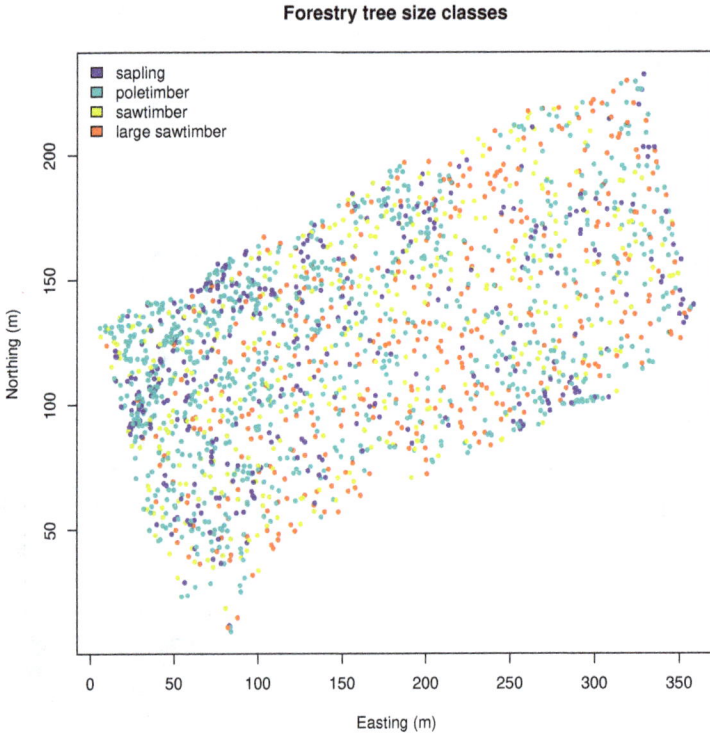

Figure 2.12 *Intervals based on previously defined tree size classes.*

nonlinear least squares and Bayesian methods to estimate the variogram model parameters from the data.

A typical exploratory analysis often proceeds by estimating and plotting variograms for the response variable as well as for the estimated residuals obtained from fitting an ordinary least squares regression. For illustrative purposes, we first fit an exponential variogram model to DBH, then fit a second variogram model to the residuals of a linear regression of DBH onto tree species. For the linear regression model, we regress the response DBH on the five tree species with one of the species taken as baseline and using four binary regressors for the four remaining species indicating if the tree belongs to that species. The resulting variograms are displayed in Figure 2.15. Here the upper and lower horizontal lines are the sill and nugget, respectively, and the vertical line is the effective range (i.e., that distance at which the correlation drops to 0.05).

We can also check for possible anisotropic patterns in spatial dependence, which can also be accomplished using functions in geoR and gstat packages. The resulting plot appears in Figure 2.16

2.9   Exercises

1. Consider the time series, $Y_t = X\sin(\omega t + \theta)$ (so X is the amplitude, ω is the frequency and θ is the phase) where X is distributed with mean 0 and variance 1 independent of $\theta \sim U(-\pi, \pi)$. Show that Y_t is weakly stationary.

2. Recalling Section 2.5, show that, under a constant mean assumption, i.e., $\mu(\mathbf{s}) = \mu$, the best linear unbiased estimator (BLUE), $\hat{\mu}$ of μ under a Gaussian model is $\frac{\mathbf{1}^{\mathrm{T}}\Sigma^{-1}\mathbf{Y}}{\mathbf{1}^{\mathrm{T}}\Sigma^{-1}\mathbf{1}}$. (In

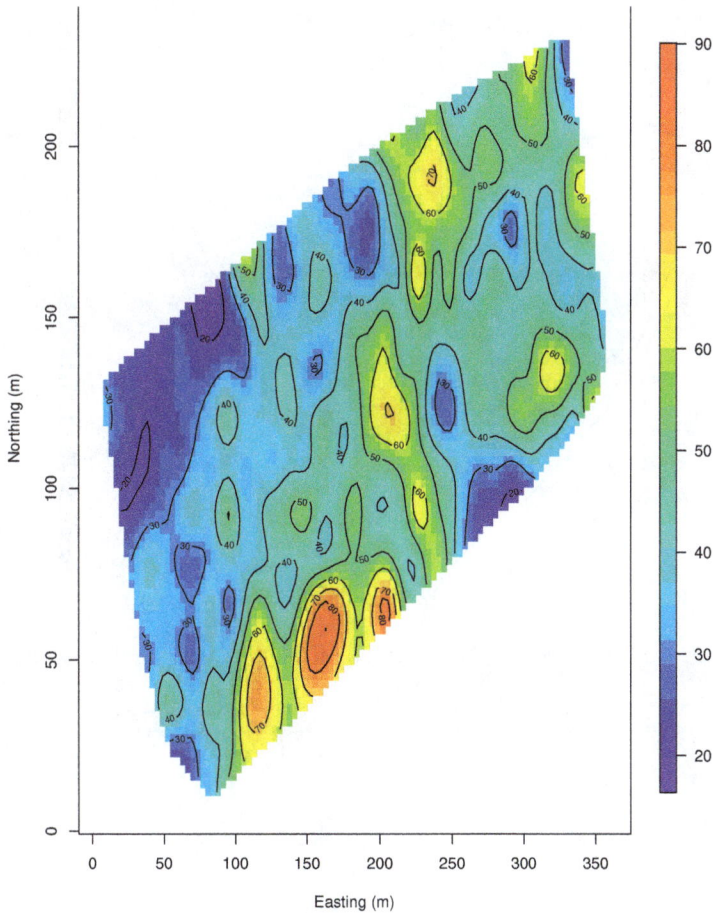

Figure 2.13 *Interpolation of DBH using Multilevel B-spline.*

fact, this only requires first and second moments and not normality.) Then, show that $\boldsymbol{\ell}^{\mathrm{T}}\mathbf{Y}$ is $\boldsymbol{\lambda}^{\mathrm{T}}\mathbf{Y}$ with $\hat{\mu}$ replacing μ.

3. For semivariogram models #2, 4, 5, 6, 7, and 8 in Subsection 2.1.3,

   (a) identify the nugget, sill, and range (or effective range) for each;

   (b) find the covariance function $C(t)$ corresponding to each $\gamma(t)$, provided it exists.

4. Prove that for Gaussian processes, strong stationarity is equivalent to weak stationarity.

5.(a) What is the issue with regard to specifying $\widehat{c}(0)$ in the covariance function estimate (2.12)?

   (b) Show either algebraically or numerically that regardless of how $\widehat{c}(0)$ is obtained, $\widehat{\gamma}(t_k) \neq \widehat{c}(0) - \widehat{c}(t_k)$ for all t_k.

6. The scallops data can be downloaded from https://github.com/sudiptobanerjee/BGC_2023/data.

   (a) Provide a descriptive summary of the scallops data.

   (b) Obtain rough estimates of the nugget, sill, and range.

   (c) Repeat the theoretical variogram fitting with an exponential variogram, and report your results.

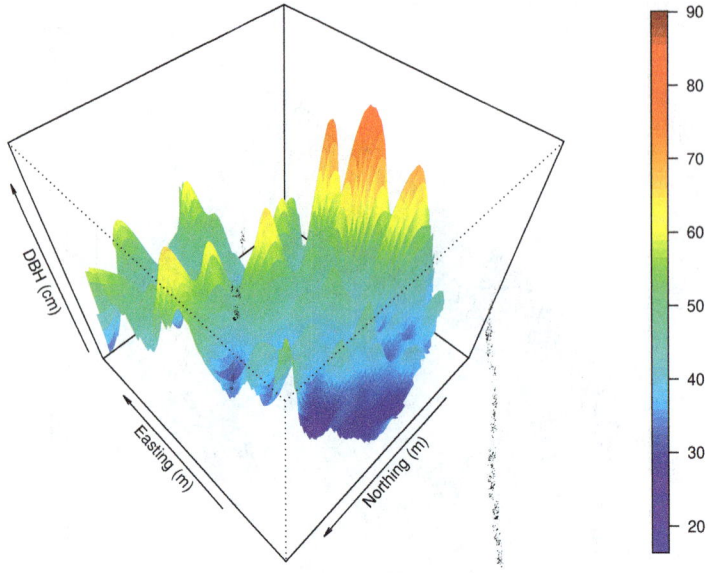

Figure 2.14 *Perspective plot of Multilevel B-spline interpolation of DBH.*

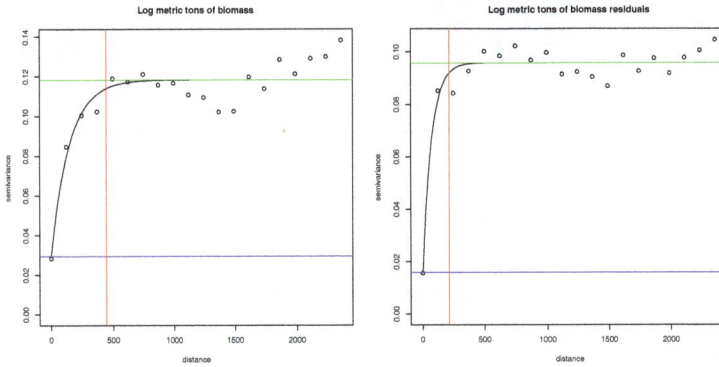

Figure 2.15 *Isotropic semivariograms for DBH and residuals of a linear regression of DBH onto tree species.*

7. Consider the `coalash` data frame in the `gstat` package in `R` and available from https://github.com/sudiptobanerjee/BGC_2023/data. This data comes from the Pittsburgh coal seam on the Robena Mine Property in Greene County, PA [Cressie, 1993, p. 32]. This data frame contains 208 coal ash core samples (the variable `coal` in the data frame) collected on a grid given by x and y planar coordinates (*not* latitude and longitude). Carry out the following tasks in `R`:

   (a) Plot the sampled sites embedded on a map of the region. Add contour lines to the plot.

   (b) Provide a descriptive summary (histograms, stems, quantiles, means, range, etc.) of the variable `coal` in the data frame.

   (c) Plot variograms and correlograms of the response and comment on the need for spatial analysis here.

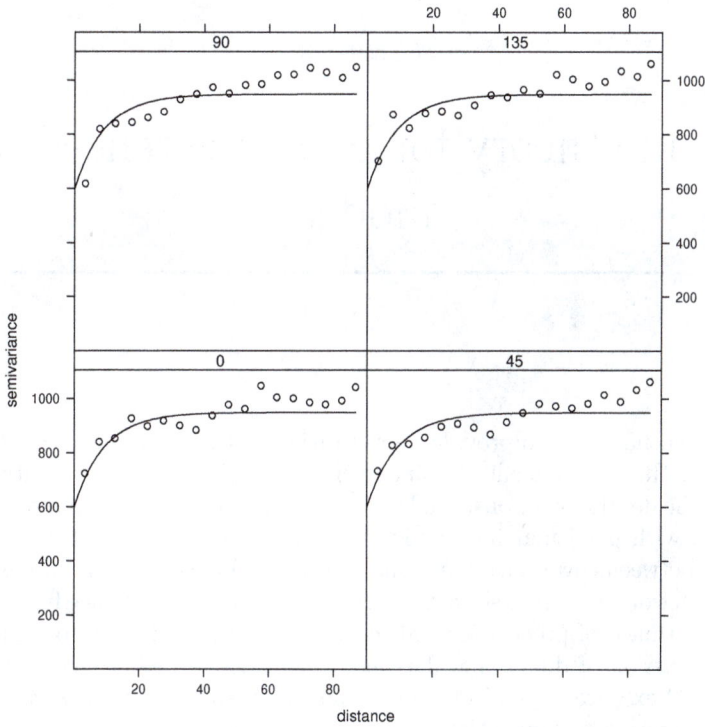

Figure 2.16 *Directional semivariograms for DBH.*

(d) If you think that there is a need for spatial analysis, arrive at your best estimates of the range, nugget, and sill.

8. Confirm expressions (2.22) and (2.23), and subsequently verify the form for $\boldsymbol{\lambda}$ given in (2.24).

9. Show that when using (2.22) to predict the value of the surface at one of the existing data locations $\mathbf{s}_i$, the predictor will equal the observed value at that location if and only if $\tau^2 = 0$. (That is, the usual Gaussian process is a spatial interpolator only in the "noiseless prediction" scenario.)

10. Recall that

$$
\begin{aligned}
\mathbf{Y} &= X\boldsymbol{\beta} + \boldsymbol{\epsilon}, \text{ where } \boldsymbol{\epsilon} \sim N\left(\mathbf{0}, \Sigma\right), \\
\text{and } \Sigma &= \sigma^2 H\left(\phi\right) + \tau^2 I, \text{ where } \left(H\left(\phi\right)\right)_{ij} = \rho\left(\phi; d_{ij}\right).
\end{aligned}
$$

Thus the dispersion matrix of $\widehat{\boldsymbol{\beta}}$ is given as $\mathrm{Var}(\widehat{\boldsymbol{\beta}}) = \left(X^{\mathrm{T}}\Sigma^{-1}X\right)^{-1}$. Thus $\widehat{\mathrm{Var}}(\widehat{\boldsymbol{\beta}}) = \left(X^{\mathrm{T}}\widehat{\Sigma}^{-1}X\right)^{-1}$ where $\widehat{\Sigma} = \widehat{\sigma}^2 H(\widehat{\phi}) + \widehat{\tau}^2 I$ and $X = [\mathbf{1}, \mathtt{long}, \mathtt{lat}]$. Given the estimates of the sill, range, and nugget (from the $\mathtt{nls}$ function), it is possible to estimate the covariance matrix $\widehat{\Sigma}$, and thereby get $\widehat{\mathrm{Var}}(\widehat{\boldsymbol{\beta}})$. Develop an $\mathtt{R}$ program to perform this exercise to obtain estimates of standard errors for $\widehat{\boldsymbol{\beta}}$ for the scallops data.

Hint: $\widehat{\tau}^2$ is the nugget; $\widehat{\sigma}^2$ is the partial sill (the sill minus the nugget). Finally, the correlation matrix $H(\widehat{\phi})$ can be obtained from the spherical covariance function, part of your solution to Exercise 3

Chapter 3

★ Some theory for point-referenced data models

The intent of this chapter is to provide a brief review of the basic theory of stochastic processes needed for the development of point-referenced spatial data or geostatistical models. We begin with the development of spatial stochastic processes built from independent increment processes, with particular interest in stationary spatial processes. We briefly discuss the connection between covariance functions and spectral measures. We discuss the validity of covariance functions as well as simple constructions of valid covariance functions. We then turn to the smoothness of process realizations as driven by stationary covariance functions with a brief discussion of directional derivative processes (anticipating a fuller discussion in Chapter 14). Next, we expand our development of isotropic covariance functions. We conclude with a section on nonstationary covariance specifications.

3.1 Formal modeling theory for spatial processes

When we write the collection of random variables $\{Y(\mathbf{s}) : \mathbf{s} \in D\}$ for some region of interest D or more generally $\{Y(\mathbf{s}) : \mathbf{s} \in \Re^r\}$, it is evident that we are envisioning a stochastic process indexed by $\mathbf{s}$. To capture spatial association it is also evident that these variables will be pairwise dependent with strength of dependence that is specified by their locations.

So, in fact, we have to determine the joint distribution for an uncountable number of random variables. In fact, we do this through the specification of arbitrary finite-dimensional distributions, i.e., for an arbitrary number of and choice of locations. This characterizes the stochastic process. More precisely, for the set of locations, $\{\mathbf{s}_1, \mathbf{s}_2, \ldots, \mathbf{s}_n\}$, let the finite-dimensional distribution be $P(Y(\mathbf{s}_1), Y(\mathbf{s}_2), \ldots, Y(\mathbf{s}_n) \in A)$ for A in a suitable σ-algebra of sets. In fact, without loss of generality, we can take $A = A_1 \times A_2 \times \ldots A_n$, a product set, and write $P(Y(\mathbf{s}_1) \in A_1, Y(\mathbf{s}_2) \in A_2, \ldots, Y(\mathbf{s}_n) \in A_n)$.

Consider the following two "consistency" conditions:

1. Under any permutation α of the indices $1, 2, \ldots, n$, say $\alpha_1, \alpha_2, \ldots, \alpha_n$,

$$P(Y(\mathbf{s}_{\alpha_1}) \in A_{\alpha_1}, Y(\mathbf{s}_{\alpha_2}) \in A_{\alpha_2}, \ldots, Y(\mathbf{s}_{\alpha_n}) \in A_n) = \\ P(Y(\mathbf{s}_1) \in A_1, Y(\mathbf{s}_2) \in A_2, \ldots, Y(\mathbf{s}_n) \in A_n).$$

That is, the permutation of the indices does not change the probability of events.

2. For any set of locations, $\{\mathbf{s}_1, \mathbf{s}_2, \ldots, \mathbf{s}_n\}$ consider an additional arbitrary location $\mathbf{s}_{n+1}$. If we marginalize the $n+1$ dimensional joint distribution specified for $Y(\mathbf{s}_1), Y(\mathbf{s}_2), \ldots, Y(\mathbf{s}_n), Y(\mathbf{s}_{n+1})$ over $Y(\mathbf{s}_{n+1})$, we obtain the n dimensional joint-distribution specified for $Y(\mathbf{s}_1), Y(\mathbf{s}_2), \ldots, Y(\mathbf{s}_n)$.

All stochastic processes on a continuous space satisfy these two conditions. Remarkably, a theorem due to Kolmogorov [see, e.g., Billingsley, 2012] informally states that the collection of finite dimensional probability measures satisfies (i) and (ii) if and only if a stochastic

DOI: 10.1201/9781003401728-3

process exists on the associated probability space having these finite dimensional distributions. Of course, characterizing the entire collection of finite dimensional distributions can be challenging. A convenient way to do this is by confining ourselves to Gaussian processes (possibly transformations) or to mixtures of such processes (a very rich class). That is, in this case, we can work with multivariate normal distributions and all that is required is a mean surface, $\mu(\mathbf{s})$ and a valid correlation function which provides the covariance matrix, as we discuss below.

Again, to clarify the inference setting, in practice we will only observe $Y(\mathbf{s})$ at a finite set of locations, $\mathbf{s}_1, \mathbf{s}_2, \ldots, \mathbf{s}_n$. Based upon $\{Y(\mathbf{s}_i), i = 1, \ldots, n\}$, we seek to infer about the mean, variability, and association structure of the process. We also seek to predict $Y(\mathbf{s})$ at arbitrary unobserved locations. Since our focus is on hierarchical modeling, often the spatial process is introduced through random effects at the second stage of the modeling specification. In this case, we still have the same inferential questions but now the process is never actually observed. It is latent and the data, modeled at the first stage, helps us to learn about the process. In this sense, we can make intuitive connections with familiar dynamic models [e.g., West and Harrison, 1997] where there is a latent state space model that is temporally updated. This reminds us of a critical difference between the one-dimensional time domain and the two-dimensional spatial domain: we have full order in the former, but only partial order in two or more dimensions.

The implications of this remark are substantial. Large sample analysis for time series usually lets time go to ∞. Asymptotics envision an increasing time domain. By contrast, large sample analysis for spatial process data usually envisions a fixed region with more and more points filling in this domain (so-called infill asymptotics). When applying increasing domain asymptotic results, we can assume that, as we collect more and more data, we can learn about temporal association at increasing distances in time. When applying infill asymptotic results for a fixed domain we can learn more and more about the association as the distance between points tends to 0. However, with a maximum distance fixed by the domain, we cannot learn about association (in terms of consistent inference) at increasing distances. The former remark indicates that we may be able to do an increasingly better job with regard to spatial prediction at a given location. However, we need not be doing better in terms of inferring about other features of the process. Learning about process parameters will be bounded; Fisher information does not go to ∞, Cramèr-Rao lower bounds and asymptotic variances do not go to 0.

It is worth remarking that the theoretical issues pertaining to asymptotic inference on spatial covariance parameters are different from inference on the realizations of the Gaussian process itself, which attracts significant attention in nonparametric modeling and functional data analysis with Gaussian processes [we refer the reader to Ghosal and van der Vaart, 2017, and references therein for a very comprehensive account of this topic]. There is, also, a substantial body of theoretical work on the identifiability and asymptotic properties of inference for Gaussian process parameters, especially in the Matérn covariance function. The text by Stein [1999a] is an excellent reference with regard to the rather deep issues pertaining to such inference. Stein [1999b], Zhang [2004], Loh [2005], Zhang and Zimmerman [2005], Anderes [2010], Tang et al. [2021], Li [2022], Li et al. [2023] and Li et al. [2024] for technical developments and discussion regarding such asymptotic results. Here, we view such concerns as providing encouragement for using a Bayesian framework for inference, since then we need not rely on any asymptotic theory for inference, but rather, under the implemented model, we obtain exact inference given whatever data we have observed. Of course, the fact that information is bounded implies that the data never overwhelms the prior, as is customarily assumed. There is no free lunch and prior sensitivity analysis may be needed.

In the ensuing subsections we turn to some technical discussion regarding the specification of spatial stochastic processes as well as covariance and correlation functions. However, we note that the above restriction to Gaussian processes enables several advantages. First, it allows a very convenient distribution theory. Joint marginal and conditional distributions are all immediately obtained from standard theory once the mean and covariance structure have been specified. In fact, as above, this is all we need to specify in order to determine all distributions. Also, as we shall see, in the context of hierarchical modeling, a Gaussian process assumption for spatial random effects introduced at the second stage of the model is very natural in the same way that independent random effects with variance components are customarily introduced in linear or generalized linear mixed models. From a technical point of view, as noted in Subsection 2.1.1, if we work with Gaussian processes and stationary models, strong stationarity is equivalent to weak stationarity. We will clarify these notions in the next subsection. Lastly, in most applications, it is difficult to criticize a Gaussian assumption. To argue this as simply as possible, in the absence of replication we have $\mathbf{Y} = (Y(\mathbf{s}_1), \ldots, Y(\mathbf{s}_n))^{\mathrm{T}}$, a single realization from an n-dimensional distribution. With a sample size of one, how can we criticize *any* multivariate distributional specification (Gaussian or otherwise)?

Strictly speaking, this last assertion is not quite true with a Gaussian process model. That is, the joint distribution is a multivariate normal with mean say $\mathbf{0}$, and a covariance matrix that is a parametric function of the parameters in the covariance function. As n grows large enough, the effective sample size will also grow. By linear transformation, we can obtain a set of approximately uncorrelated variables through which the adequacy of the normal assumption might be studied. We omit details.

3.1.1 Some basic stochastic process theory for spatial processes

A key remark here is that, when we develop a spatial stochastic process model, we can proceed along two paths. First, we can specify the process stochastically and obtain its induced covariance function, i.e., its induced dependence structure. A second path is to start with say, a Gaussian process and then specify a valid covariance function to overlay on the Gaussian process. This covariance function supplies the joint covariance matrix for any finite number of variables at any selected set of sites. In this subsection we follow the first path, in order to illuminate some of the rigor associated with formal stochastic process specification. In the subsequent subsections we define valid covariance functions and focus on their properties and examples.

To begin our development here, we start with independent increment processes. It will take a few paragraphs before we are able to explicitly connect them to spatial processes. A real-valued independent increment process, Z over say R^d (here, we are only interested in the case $d = 2$) is such that, for arbitrary disjoint sets A and B, where A and B belong to a suitable σ-algebra of sets, $Z(A)$ and $Z(B)$ are independent. In particular, we let $Z(d\mathbf{w})$ be the generator for these random variables in the sense that $Z(A) = \int_A Z(d\mathbf{w})$. Evidently, if A and B are disjoint, then $Z(A \bigcup B) = Z(A) + Z(B)$. We assume that the Z process has first and second moments and, for convenience that the first moment is 0. We set $\mathrm{E}(Z^2(A)) = \mathrm{Var}(A) = G(A)$, i.e., $G(A) = \int_A G(dw)$. Also, we see that $\mathrm{E}(Z(A)Z(B)) = \mathrm{E}(Z(A \bigcap B) + Z(A \bigcap B^C))(Z(A \bigcap B)Z(A^C \bigcap B)) = \mathrm{E}(Z^2(A \bigcap B)) = G(A \bigcap B)$ due to independence of increments.

Let $Z_f = \int f(\mathbf{w})Z(d\mathbf{w})$ for an appropriately measurable function f. Of course, $Z(A) = \int 1(\mathbf{w} \in A)Z(d\mathbf{w})$ is a special case. This suggests that we proceed to step functions as a way to study the behavior of Z_f for measurable f and the dependence between say Z_{f1} and Z_{f2}. If $f(\mathbf{w}) = \sum_l a_l 1(\mathbf{w} \in A_l)$, then $Z_f = \sum_l a_l Z(A_l)$. Because of the independent increments, $\mathrm{Var}(Z_f) = \sum a_l^2 G(A_l)$.

Next, consider $f_1(\mathbf{w}) = \sum_l a_l 1(\mathbf{w} \in A_l)$ and $f_2(\mathbf{w}) = \sum_k b_k 1(\mathbf{w} \in B_k)$. Then,

$$
\begin{aligned}
\mathrm{Cov}(Z_{f1}, Z_{f2}) &= \mathrm{E}(Z_{f1} Z_{f2}) = \sum_l \sum_k a_l b_k \mathrm{E}(Z(A_l) Z(B_k)) \\
&= \sum_l \sum_k a_l b_k \mathrm{E}(Z(A_l \textstyle\bigcap B_k)) = \sum_l \sum_k a_l b_k G(A_l \textstyle\bigcap B_k) .
\end{aligned}
$$

But, also, $\int f_1(\mathbf{w}) f_2(\mathbf{w}) G(d\mathbf{w}) = \sum_l \sum_k a_l b_k G(A_l \bigcap B_k)$. So, we have shown that $\mathrm{E}(Z_{f1} Z_{f2}) = \int f_1(\mathbf{w}) f_2(\mathbf{w}) G(d\mathbf{w})$.

Now, let us bring in the spatial process setting. Define

$$
Y(\mathbf{s}) = \int \psi(\mathbf{s}, \mathbf{w}) Z(d\mathbf{w}) ,
$$

i.e., $Y(\mathbf{s})$ is $Z_{\psi_\mathbf{s}}$ in our notation above. Here, we allow ψ to be a complex-valued function since we want to employ the particular form $\psi(\mathbf{s}, \mathbf{w}) = e^{i\mathbf{s}^\mathrm{T}\mathbf{w}}$. We also allow Z to be complex-valued, introducing the complex conjugate using an overline. So, now $G(A) = \mathrm{E}(Z(A)\overline{Z}(A)) = E|Z(A)|^2$. Then, we have defined a stochastic process and to calculate $\mathrm{Cov}(Y(\mathbf{s}), Y(\mathbf{s}'))$, we compute $\mathrm{E}(Y(\mathbf{s})\overline{Y}(\mathbf{s}')) = \int \psi(\mathbf{s}, \mathbf{w}) \overline{\psi}(\mathbf{s}', \mathbf{w}) G(d\mathbf{w})$, using the result of the previous paragraph.

With $\psi(\mathbf{s}, \mathbf{w}) = e^{i\mathbf{s}^\mathrm{T}\mathbf{w}}$, we obtain

$$
\mathrm{Cov}(Y(\mathbf{s}), Y(\mathbf{s}')) = \int e^{i\mathbf{s}^\mathrm{T}\mathbf{w}} e^{-i\mathbf{s}'^\mathrm{T}\mathbf{w}} G(d\mathbf{w}) = \int e^{i(\mathbf{s}-\mathbf{s}')^\mathrm{T}\mathbf{w}} G(d\mathbf{w}) .
$$

In other words, the association between $Y(\mathbf{s})$ and $Y(\mathbf{s}')$ depends only upon the separation vector $\mathbf{h} = \mathbf{s} - \mathbf{s}'$, not on the individual locations; $\mathrm{Cov}(Y(\mathbf{s}), Y(\mathbf{s}')) = C(\mathbf{s} - \mathbf{s}')$. Such a stochastic process is said to be *stationary*. In fact, we have another characterization, an elegant result which says that $Y(\mathbf{s})$ is a stationary stochastic process if and only if it can be represented in the form $Y(\mathbf{s}) = \int e^{i\mathbf{s}^\mathrm{T}\mathbf{w}} Z(d\mathbf{w})$ where Z is a possibly complex-valued, mean 0, independent increments process [Yaglom, 1987, Section 8]. We note the implicit parallel structure, $Y(\mathbf{s}) = \int e^{i\mathbf{s}^\mathrm{T}\mathbf{w}} Z(d\mathbf{w})$ and $C(\mathbf{s}) = \int e^{i\mathbf{s}^\mathrm{T}\mathbf{w}} G(d\mathbf{w})$.

Other choices for ψ appear in the literature. For instance, ψ might be a kernel function $K(\mathbf{s} - \mathbf{w})$ which is integrable over $\Re^2$. Then, we have $Y(\mathbf{s}) = \int K(\mathbf{s} - \mathbf{w}) Z(d\mathbf{w})$ and $\mathrm{Cov}(Y(\mathbf{s}), Y(\mathbf{s}')) = \int K(\mathbf{s} - \mathbf{w}) K(\mathbf{s}' - \mathbf{w}) G(d\mathbf{w})$. In the special case where $G(d\mathbf{w}) = \sigma^2 d\mathbf{w}$, we see that, after a simple change of variable, $\mathrm{Cov}(Y(\mathbf{s}), Y(\mathbf{s}')) = \sigma^2 \int K(\mathbf{s}-\mathbf{s}'+\mathbf{u}) K(\mathbf{u}) d\mathbf{u}$. Such a process construction is called *kernel convolution* and is discussed further in Section 3.2.2. A question we might ask is "how rich is the class of stationary process obtainable under kernel convolution?" We can illuminate this in the next subsection, after we characterize valid covariance functions through Bochner's Theorem.

We might also ask when the foregoing construction provides a Gaussian process? In particular, if $Z(\mathbf{s})$ is Brownian motion then $Y(\mathbf{s})$ is a Gaussian process. That is, Brownian motion is an independent increments process providing jointly normally distributed random variables. Informally, integration of such variables against a choice of ψ is like taking linear transformations of jointly normal variables; the resulting variable is normal, and a resulting set of variables is jointly normal. Let us add a few words about Brownian motion.

Brownian motion in one dimension is, again, an independent increments process, usually defined on $\Re^+$. In fact, for $t > 0$, we let $Z(t) \equiv Z((0, t])$, i.e., we convert a set function to a point function in order to define $\{Z(t) : t \in \Re^+\}$. We assume $Z(t) \sim N(0, \sigma^2 t)$ and $\mathrm{Cov}(Z(t), Z(t')) = \sigma^2 \min(t, t') = \frac{1}{2}\sigma^2(|t| + |t'| - |t - t'|)$. In other words, the spectral measure for Brownian motion in one dimension is $G(dt) = \sigma^2 |dt|$. We can also calculate that $\mathrm{E}(Z(t+h) - Z(t))^2 = \sigma^2|h|$. From this we can infer that process realizations are mean square continuous, i.e., $\lim_{h \to 0} \mathrm{E}(Z(t+h) - Z(t))^2 = 0$. However, process realizations are not mean square differentiable. (See Section 14.2 for more discussion on mean square smoothness.)

Moving to two dimensions, we would like to define $Z(\mathbf{s})$. Paralleling the one-dimensional case, we want $Z(\mathbf{s}) \sim N(0, \sigma^2 ||\mathbf{s}||)$. The easiest way to envision this is through circles. That is, $Z(\mathbf{s}) \sim Z(\mathbf{s}')$ if $||\mathbf{s}|| = ||\mathbf{s}'||$; marginal univariate distributions are common on circles. Again, imitating the one-dimensional case we want $\mathrm{Cov}(Z(\mathbf{s}), Z(\mathbf{s}')) = \frac{1}{2}\sigma^2 (||\mathbf{s}|| + ||\mathbf{s}'|| - ||\mathbf{s} - \mathbf{s}'||)$. We see that, even if $||\mathbf{s}'|| = ||\mathbf{s}''||$, $\mathrm{Cov}(Z(\mathbf{s}), Z(\mathbf{s}')) \neq (Z(\mathbf{s}), Z(\mathbf{s}''))$. Again, $\mathrm{E}(Z(\mathbf{s} + \mathbf{h}) - Z(\mathbf{s}))^2 = \sigma^2 ||\mathbf{h}||$ implying mean square continuity of process realizations. Moreover, we now see that this is a linear variogram (Section 2.1.3). That is, we have a stationary (in fact, isotropic) variogram associated with a nonstationary covariance function, illuminating our comment in that subsection regarding the fact that the linear variogram does not correspond to a weakly stationary process model.

Next, we clarify the connection between Brownian motion and white noise. Recall that white noise is usually defined as the process $V(\mathbf{s})$ where the $V(\mathbf{s})$ are i.i.d. $N(0, \sigma^2)$. Consider the finite differential $\dfrac{Z(\mathbf{s} + \mathbf{h}) - Z(\mathbf{s})}{||\mathbf{h}||}$. From the above, this variable is distributed as $N(0, \sigma^2 / ||\mathbf{h}||)$. As $||\mathbf{h}|| \to 0$, we do not obtain a limiting distribution. So, white noise is not the *derivative* of Brownian motion (though it is sometimes referred to as a *generalized* derivative). In fact, while $Y(\mathbf{s}) = \int \phi(\mathbf{s}, \mathbf{w}) Z(d\mathbf{w})$ defines a stochastic process, $Y(\mathbf{s}) = \int \phi(\mathbf{s}, \mathbf{w}) V(\mathbf{w}) d\mathbf{w}$ does not. Rather, we must work with finite-dimensional versions of this expression, $Y(\mathbf{s}) = \sum_{j=1}^{J} \phi(\mathbf{s}, \mathbf{s}_j) V(\mathbf{s}_j)$ to create a well-defined process. And, in no sense, should this finite version be viewed as an approximation to integration over $\Re^2$. The special case where $\phi(\mathbf{s}, \mathbf{w})$ is a kernel $K(\mathbf{s} - \mathbf{w})$ is discussed above, and a stationary covariance function results.

Finally, we mention the notion of fractional Brownian motion. Here, we define $\mathrm{Cov}(Z(\mathbf{s}), Z(\mathbf{s}')) = \frac{1}{2}\sigma^2 (||\mathbf{s}||^{2H} + ||\mathbf{s}'||^{2H} - ||\mathbf{s} - \mathbf{s}'||^{2H})$. $H \in (0, 1)$ is called the Hurst index. Fractional Brownian motion generalizes the foregoing specifications and allows for process with dependent increments. In particular, $H = 1/2$ is the usual Brownian motion. If H is greater than $1/2$ we have positively correlated increments while if it is less than $1/2$ we have negatively correlated increments. We leave this as an exercise.

3.1.2 *Covariance functions and spectra*

In order to specify a stationary process we must provide a valid covariance function. Here "valid" means that $C(\mathbf{h}) \equiv \mathrm{Cov}(Y(\mathbf{s}), Y(\mathbf{s} + \mathbf{h}))$ is such that for any finite set of sites $\mathbf{s}_1, \ldots, \mathbf{s}_n$ and for any $a_1, \ldots, a_n$,

$$Var\left[\sum_i a_i Y(s_i)\right] = \sum_{i,j} a_i a_j Cov(Y(\mathbf{s}_i), Y(\mathbf{s}_j)) = \sum_{i,j} a_i a_j C(\mathbf{s}_i - \mathbf{s}_j) \geq 0 \,,$$

with strict inequality if not all the a_i are 0. That is, we need $C(\mathbf{h})$ to be a positive definite function.

Verifying the positive definiteness condition is evidently not routine. Fortunately, we have *Bochner's Theorem* [see, e.g., Varadhan, 2001, Gikhman and Skorokhod, 2007], which provides a necessary and sufficient condition for $C(\mathbf{h})$ to be positive definite. This theorem is applicable for $\mathbf{h}$ in arbitrary r-dimensional Euclidean space, although our primary interest is in $r = 2$.

In general, for real-valued processes, Bochner's Theorem states that $C(\mathbf{h})$ is positive definite if and only if

$$C(\mathbf{h}) = \int \cos(\mathbf{w}^{\mathrm{T}} \mathbf{h}) \, G(d\mathbf{w}) \,, \tag{3.1}$$

where G is a bounded, positive, symmetric about 0 measure in $\Re^r$. Then $C(\mathbf{0}) = \int G d(\mathbf{w})$ becomes a normalizing constant, and $G(d\mathbf{w})/C(\mathbf{0})$ is referred to as the *spectral distribution* that induces $C(\mathbf{h})$. If $G(d\mathbf{w})$ has a density with respect to the Lebesgue measure, i.e., $G(d\mathbf{w}) = g(\mathbf{w})d\mathbf{w}$, then $g(\mathbf{w})/C(\mathbf{0})$ is referred to the as the *spectral density*. Evidently, (3.1) can be used to generate valid covariance functions; see Section 3.1.2.1 below. Of course, the behavioral implications associated with C arising from a given G will only be clear in special cases, and (3.1) will be integrable in closed form only in cases that are even more special.

Since $\exp{(i\mathbf{w}^{\mathrm{T}}\mathbf{h})} = \cos{(\mathbf{w}^{\mathrm{T}}\mathbf{h})} + i\sin{(\mathbf{w}^{\mathrm{T}}\mathbf{h})}$, where $i = \sqrt{-1}$, we have $C(\mathbf{h}) = \int \exp{(i\mathbf{w}^{\mathrm{T}}\mathbf{h})}\, G(d\mathbf{w})$. That is, the imaginary term disappears due to the symmetry of G around 0. In other words, $C(\mathbf{h})$ is a valid covariance function if and only if it is the characteristic function of an r-dimensional symmetric random variable (random variable with a symmetric distribution). We note that if G is not assumed to be symmetric about $\mathbf{0}$, $C(\mathbf{h}) = \int \exp{(i\mathbf{w}^{\mathrm{T}}\mathbf{h})}\, G(d\mathbf{w})$ still provides a valid covariance function (i.e., positive definite) but now for a complex-valued random process on $\Re^r$.

The Fourier transform of $C(\mathbf{h})$ is

$$\widehat{c}(\mathbf{w}) = \int \exp{(-i\mathbf{w}^{\mathrm{T}}\mathbf{h})}\ C(\mathbf{h})d\mathbf{h}\ . \tag{3.2}$$

Applying the inversion formula, $C(\mathbf{h}) = (2\pi)^{-r} \int \exp{(i\mathbf{w}^{\mathrm{T}}\mathbf{h})}\widehat{c}(\mathbf{w})d\mathbf{w}$, we see that $(2\pi)^{-r}\widehat{c}(\mathbf{w})/C(0) = g(\mathbf{w})$, the spectral density. Explicit computation of (3.2) is usually not possible except in special cases, but the fast Fourier transform (FFT) can be used for an approximate calculation. Expression (3.2) can be used to check whether a given $C(\mathbf{h})$ is valid: we simply compute $\widehat{c}(\mathbf{w})$ and check whether it is positive and integrable (so it is indeed a density up to normalization).

The one-to-one relationship between $C(\mathbf{h})$ and $g(\mathbf{w})$ enables the examination of spatial processes in the spectral domain rather than in the observational domain. Computation of $g(\mathbf{w})$ can often be expedited through fast Fourier transforms; g can be estimated using the so-called *periodogram*. Likelihoods can be obtained approximately in the spectral domain enabling inference to be carried out in this domain. See, e.g., Guyon [1995] or Stein [1999a] for a full development. Likelihood evaluation is much faster in the spectral domain. However, in this book we confine ourselves to the observational domain because of concerns regarding the accuracy associated with approximation in the spectral domain [e.g., the likelihood of Whittle, 1954], and with the ad hoc creation of the periodogram (e.g., how many low frequencies are ignored). We do however note that the spectral domain may afford the best potential for handling computation associated with large data sets.

Isotropic covariance functions, i.e., $C(\|\mathbf{h}\|)$, where $\|\mathbf{h}\|$ denotes the length of $\mathbf{h}$, are the most frequently adopted choice within the stationary class. There are various direct methods for checking the permissibility of isotropic covariance and variogram specifications [see, e.g., Armstrong and Diamond, 1984, Christakos, 1984, McBratney and Webster, 1986]. Again denoting $\|\mathbf{h}\|$ by t for notational simplicity, recall that Tables 2.1 and 2.2 provide the covariance function $C(t)$ and variogram $\gamma(t)$, respectively, for the widely encountered parametric isotropic choices that were initially presented in Section 2.1.3.

It is noteworthy that an isotropic covariance function that is valid in dimension r need not be valid in dimension $r + 1$. The intuition may be gleaned by considering $r = 1$ versus $r = 2$. For three points, in one-dimensional space, given the distances separating points 1 and 2 (d_{12}) and points 2 and 3 (d_{23}), then the distance separating points 1 and 3 d_{13} is either $d_{12} + d_{23}$ or $|d_{12} - d_{23}|$. But in two-dimensional space, given d_{12} and d_{23}, d_{13} can take any value in $\Re^+$ (subject to triangle inequality). With increasing dimension more sets of interlocation distances are possible for a given number of locations; it will be more difficult for a function to satisfy the positive definiteness condition. Armstrong and Jabin [1981] provide an explicit example that we defer to Exercise 7.

There are isotropic correlation functions that are valid in all dimensions. The Gaussian correlation function, $k\rho(\|h\|) = \exp(-\phi\|h\|^2)$ is an example. It is the characteristic function associated with r i.i.d. normal random variables, each with variance $1/(2\phi)$ for any r. More generally, the powered exponential, $\exp(-\phi\|h\|^\alpha)$, $0 < \alpha \leq 2$ (and hence the exponential correlation function) is valid for any r. Here, we note the general result that $C(\|\mathbf{h}\|)$ is a positive definite isotropic function on $\Re^r$ for all r if and only if it has the representation, $C(\|\mathbf{h}\|) = \int e^{-w\|\mathbf{h}\|^2} G(dw)$ where G is non-decreasing and bounded and $w \in R^+$. So, $C(\|\mathbf{h}\|)$ arises as a scale mixture of Gaussian correlation functions. G might be a c.d.f. on R^+ with a p.d.f., $g(w)$, i.e., $G(dw) = g(w)dw$.

Rather than seeking isotropic correlation functions that are valid in all dimensions, we might seek all valid isotropic correlation functions in a particular dimension r. Matérn [1986] provides the general result. The set of $C(\|h\|)$ of the form

$$C(\|h\|) = \int_0^\infty \left(\frac{2}{w\|\mathbf{h}\|}\right)^\alpha \Gamma(\nu+1) J_\nu(w\|\mathbf{h}\|) G(dw), \qquad (3.3)$$

where G is nondecreasing and integrable on $\Re^+$, J_ν is the Bessel function of the first kind of order ν, and $\nu = (r-2)/2$ provides all valid isotropic correlation functions on $\Re^r$.

When $r = 2$, $v = 0$ so that arbitrary correlation functions in two-dimensional space arise as scale mixtures of Bessel functions of order 0. In particular, $J_0(d) = \sum_{k=0}^\infty \frac{(-1)^k}{(k!)^2} \left(\frac{d}{2}\right)^{k/2}$. J_0 decreases from 1 at $d = 0$ and will oscillate above and below 0 with amplitudes and frequencies that are diminishing as d increases (see Figure 3.1). Typically, correlation functions that are monotonic and decreasing to 0 are chosen but, apparently, valid correlation functions can permit negative associations with w determining the scale in distance space. Such behavior might be appropriate in certain applications.

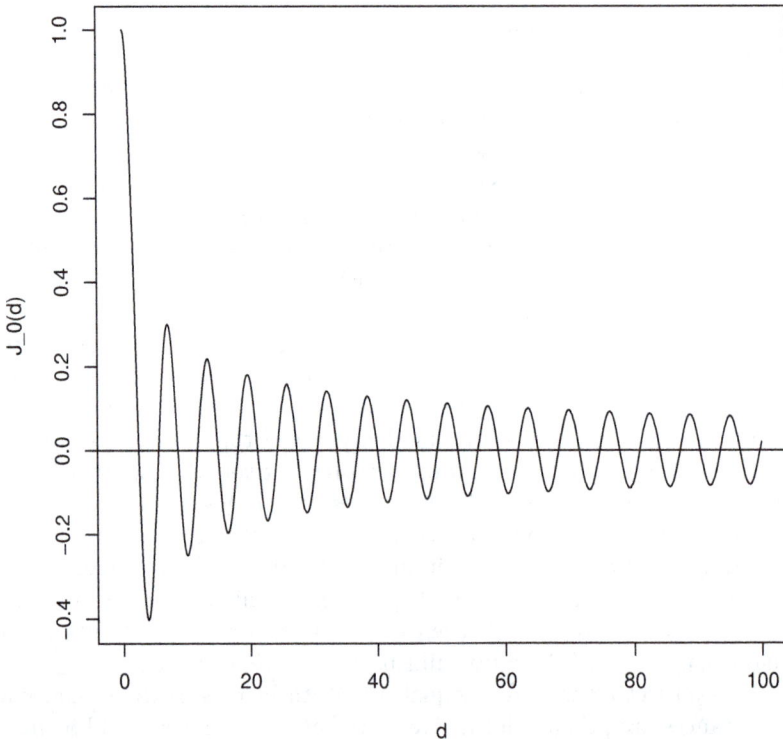

Figure 3.1 *A plot of $J_0(d)$ out to $d = 100$.*

The form in (3.3) at $\nu = 0$ was exploited in Shapiro and Botha [1991] and Ver Hoef and Barry [1998] to develop "nonparametric" variogram models and "black box" kriging. It was employed by Ecker and Gelfand [1997] to obtain flexible spatial process models within which to do inference from a Bayesian perspective (see Section 6.4).

If we confine ourselves to strictly monotonic isotropic covariance functions then we can introduce the notion of a range. As described above, the range is conceptualized as the distance beyond which association becomes negligible. If the covariance function reaches 0 in a finite distance, then we refer to this distance as the range. However, as Table 2.1 reveals, we customarily work with covariance functions that attain 0 asymptotically as $\|\mathbf{h}\| \to \infty$. In this case, it is common to define the range as the distance beyond which correlation is less than .05, and this is the definition we employ in the sequel. So if ρ is the correlation function, then writing the range as R we solve $\rho(R; \boldsymbol{\theta}) = 0.05$, where $\boldsymbol{\theta}$ denotes the parameters in the correlation function. Therefore, R is an implicit function of the parameter $\boldsymbol{\theta}$.

We do note that some authors define the range through the variogram, i.e., the distance at which the variogram reaches 95% of its sill. That is, we would solve $\gamma(R) = .95(\sigma^2 + \tau^2)$. Note, however, that if we rewrite this equation in terms of the correlation function we obtain $\tau^2 + \sigma^2(1 - \rho(R; \boldsymbol{\theta})) = 0.95(\tau^2 + \sigma^2)$, so that $\rho(R; \boldsymbol{\theta}) = .05\left(\frac{\sigma^2 + \tau^2}{\sigma^2}\right)$. Evidently, the solution to this equation is quite different from the solution to the above equation. In fact, this latter equation may not be solvable, e.g., if $\sigma^2/(\sigma^2 + \tau^2) \leq 0.05$, the case of a very weak "spatial story" in the model. As such, one might argue that a spatial model is inappropriate in this case. However, with σ^2 and τ^2 unknown, it seems safer to work with the former definition.

3.1.2.1 *More general isotropic correlation functions*

From Section 3.1.2, a correlation function $\rho(d, \phi)$ is valid only if it is positive definite in d, $\rho(0, \phi) = 1$, and $|\rho(d, \phi)| \leq 1$ for all d. From Bochner's Theorem (3.1), the characteristic function of a symmetric distribution in R^r satisfies these constraints. From Khinchin's Theorem above [e.g., Yaglom, 2004, p. 106] as well as (3.3), the class of all valid functions $\rho(d, \phi)$ in $\Re^r$ can be expressed as

$$\rho(d, \phi) = \int_0^\infty \Omega_r(zd)dG_\phi(z),\qquad(3.4)$$

where G_ϕ is nondecreasing integrable and $\Omega_r(x) = \left(\frac{2}{x}\right)^{\frac{r-2}{2}}\Gamma\left(\frac{r}{2}\right)J_{\left(\frac{r-2}{2}\right)}(x)$. Here again, $J_v(\cdot)$ is the Bessel function of the first kind of order v. For $r = 1, \Omega_1(x) = \cos(x)$; for $r = 2, \Omega_2(x) = J_0(x)$; for $r = 3, \Omega_3(x) = \sin(x)/x$; for $r = 4, \Omega_4(x) = \frac{2}{x}J_1(x)$; and for $r = \infty, \Omega_\infty(x) = \exp(-x^2)$. Specifically, $J_0(x) = \sum_{k=0}^\infty \frac{(-1)^k}{k!^2}\left(\frac{x}{2}\right)^{2k}$ and $\rho(d, \phi) = \int_0^\infty J_0(zd)dG_\phi(z)$ provides the class of all permissible correlation functions in $\Re^2$. Figure 3.1 provides a plot of $J_0(x)$ versus x, revealing that it is not monotonic. (This must be the case in order for $\rho(d, \phi)$ above to capture all correlation functions in $\Re^2$.) These more general isotropic covariance functions are revisited in much greater detail in Section 6.4.

3.1.3 *Constructing valid covariance functions*

We note that one can offer constructive strategies to build larger classes of correlation functions. Three approaches are mixing, products, and convolution. Mixing notes simply that if $C_1, \ldots, C_m$ are valid correlation functions in $\Re^r$ and if $\sum_{i=1}^m p_i = 1$, $p_i > 0$, then $C(\mathbf{h}) = \sum_{i=1}^m p_i C_i(\mathbf{h})$ is also a valid correlation function in $\Re^r$. This follows since $C(\mathbf{h})$ is the characteristic function associated with $\sum p_i f_i(\mathbf{x})$, where $f_i(\mathbf{x})$ is the (symmetric about 0) density in r-dimensional space associated with $C_i(\mathbf{h})$. In fact, the sum $\sum_{i=1}^\infty a_i C_i(\mathbf{h})$ yields a valid covariance function as well, provided the a_i are all greater than 0 and $\sum_{i=1}^\infty a_i <)\infty$.

Using product forms simply notes that again if $C_1, \ldots, C_m$ are valid in $\Re^r$, then $\prod_{i=1}^{m} C_i$ is a valid correlation function in $\Re^r$. This follows since $\prod_{i=1}^{m} C_i(\mathbf{h})$ is the characteristic function associated with $V = \sum_{i=1}^{m} V_i$ where the V_i are independent with V_i having characteristic function $C_i(\mathbf{h})$.

The use of products has attracted attention in the context of so-called *covariance tapering* [see, e.g., Furrer et al., 2006, Kaufman et al., 2008, and references therein]. The idea here is that, with covariance matrices which do not reach 0 until $||\mathbf{h}||$ reaches ∞, the resulting covariance matrices are never sparse. With a large number of locations, handling a large covariance matrix in terms of say inversion and determinant calculation can be very challenging (see Chapter 13). Introducing sparsity into this matrix can facilitates this computation. A naive thought might be to set to 0 all entries in the matrix that are smaller than some specified value, arguing that these are negligible. Unfortunately, this strategy can result in a covariance matrix which is no longer positive definite. As an alternative, suppose $C(\mathbf{h})$ is the covariance matrix you start with and suppose $\tilde{C}(\mathbf{h})$ is a covariance function with bounded support, i.e., it reaches 0 for all $\mathbf{h}$ with length $||\mathbf{h}|| < d_0$ for some $d_0 < \infty$. Then $C^*(\mathbf{h}) = \tilde{C}(\mathbf{h})C(\mathbf{h})$ is valid and produces sparsity at a distance greater than d_0. Of course, one might ask why not just use $\tilde{C}$ from the start?

We note that this connects to the issue of nested variograms or covariance functions which have been suggested in the literature [see. e.g., Hohn, 1988, Wackernagel, 2003]. The idea here would be to use different such functions in different portions of the region of interest. The problem is that, in doing this, the aggregated covariance matrix need not be positive definite; we have a model specification that can not possibly produce the data we are observing.

Convolution simply recognizes that if C_1 and C_2 are valid correlation functions in $\Re^r$, then $C_{12}(\mathbf{h}) = \int C_1(\mathbf{h} - \mathbf{t})C_2(\mathbf{t})d\mathbf{t}$ is a valid correlation function in $\Re^r$. The argument here is to look at the Fourier transform of $C_{12}(\mathbf{h})$. That is,

$$
\begin{aligned}
\widehat{c}_{12}(\mathbf{w}) &= \int \exp\left(-i\mathbf{w}^{\mathsf{T}}\mathbf{h}\right) C_{12}(\mathbf{h})d\mathbf{h} \\
&= \int \exp\left(-i\mathbf{w}^{\mathsf{T}}\mathbf{h}\right) \int C_1(\mathbf{h} - \mathbf{t})C_2(\mathbf{t})d\mathbf{t}d\mathbf{h} \\
&= \widehat{c}_1(\mathbf{w}) \cdot \widehat{c}_2(\mathbf{w}),
\end{aligned}
$$

where $\widehat{c}_i(\mathbf{w})$ is the Fourier transform of $C_i(\mathbf{h})$ for $i = 1, 2$. But then $C_{12}(\mathbf{h}) = (2\pi)^{-2} \int \exp\left(i\mathbf{w}^{\mathsf{T}}\mathbf{h}\right) \widehat{c}_1(\mathbf{w})\widehat{c}_2(\mathbf{w})d\mathbf{w}$. Now $\widehat{c}_1(\mathbf{w})$ and $\widehat{c}_2(\mathbf{w})$ are both symmetric about $\mathbf{0}$ since, up to a constant, they are the spectral densities associated with $C_1(\mathbf{h})$ and $C_2(\mathbf{h})$, respectively. Hence, $C_{12}(\mathbf{h}) = \int \cos \mathbf{w}^{\mathsf{T}}\mathbf{h}G(d\mathbf{w})$, where $G(d\mathbf{w}) = (2\pi)^{-2}\widehat{c}_1(\mathbf{w})\widehat{c}_2(\mathbf{w})d\mathbf{w}$.

Thus, from (3.1), $C_{12}(\mathbf{h})$ is a valid correlation function, i.e., G is a bounded, positive, symmetric about 0 measure on $\Re^2$. In fact, if C_1 and C_2 are isotropic then C_{12} is as well; we leave this verification as Exercise 9.

3.1.4 *Smoothness of process realizations*

How does one select among the various choices of correlation functions? Usual model selection criteria will typically find it difficult to distinguish, say, among one-parameter isotropic scale choices such as the exponential, Gaussian, or Cauchy. Ecker and Gelfand [1997] provide some graphical illustrations showing that, through suitable alignment of parameters, the correlation curves will be very close to each other. Of course, in comparing choices with parametrizations of differing dimensions (e.g., correlation functions developed using results from the previous section), we will need to employ a selection criterion that penalizes complexity and rewards parsimony.

An alternative perspective is to make the selection based on theoretical considerations. This possibility arises from the powerful fact that the choice of correlation function determines the smoothness of realizations from the spatial process. More precisely, a process realization is viewed as a random surface over the region. By choice of C we can ensure that these realizations will be almost surely continuous, or mean square continuous, or mean square differentiable, and so on. Of course, at best the process is only observed at finitely many locations. (At worst, it is never observed, e.g., when the spatial process is used as a second-stage model for random spatial effects.) So, it is not possible to "see" the smoothness of the process realization. Elegant theory, developed in Kent [1989], Stein [1999a] and Banerjee and Gelfand [2003] clarifies the relationship between the choice of the correlation function and such smoothness. We provide a bit of this theory below, with further discussion in Section 14.2. For now, the key point is that, according to the process being modeled, we may, for instance, anticipate surfaces to not be continuous (as with digital elevation models in the presence of gorges, escarpments, or other topographic features), or to be differentiable (as in studying land value gradients or temperature gradients). We can choose a correlation function to essentially ensure such behavior.

Of particular interest in this regard is the Matérn class of covariance functions. The parameter v (see Table 2.1) is, in fact, a smoothness parameter. In two-dimensional space, the greatest integer in v indicates the number of times process realizations will be mean square differentiable. Indeed, since $v = \infty$ corresponds to the Gaussian correlation function, the implication is that the use of the Gaussian correlation function results in process realizations that are mean square analytic, which may be too smooth to be appropriate in practice. That is, it is possible to predict $Y(\mathbf{s})$ perfectly for all $\mathbf{s} \in \Re^2$ based upon observing $Y(\mathbf{s})$ in an arbitrarily small neighborhood. Expressed in a different way, the use of the Matérn covariance function as a model enables the data to inform about v; we can learn about process smoothness despite observing the process at only a finite number of locations.

Hence, we follow Stein [1999a] in recommending the Matérn class as a general specification for building spatial models. The computation of this function requires the evaluation of a modified Bessel function. In fact, evaluation will be done repeatedly to obtain a covariance matrix associated with n locations, and then iteratively if a model is fit via MCMC methods. This may appear off-putting but, in fact, it is routinely available in most statistical computing packages and libraries. In fact, such computation can be done efficiently using expansions to approximate $K_v(\cdot)$ [Abramowitz and Stegun, 1965, p. 435], or working through the inversion formula below (3.2), which in this case becomes

$$2 \left(\frac{\phi \|\mathbf{h}\|}{2} \right)^{v} \frac{K_v(\phi(\|\mathbf{h}\|))}{\phi^{2v}\Gamma(v + \frac{r}{2})} = \int_{\Re^r} e^{i\mathbf{w}^{\mathrm{T}}\mathbf{h}}(\phi^2 + \|\mathbf{w}\|^2)^{-(v+r/2)}d\mathbf{w} , \tag{3.5}$$

where K_v is the modified Bessel function of order $\nu > 0$. We see that the Matérn covariance function arises as the characteristic function from a Cauchy spectral density. In fact, this is how Matérn came upon this covariance function when doing his thesis research (Matérn, 1960). In particular, the right side of (3.5) is readily approximated using fast Fourier transforms. Again, we revisit process smoothness in Section 14.2.

We conclude this subsection by returning to the question of how rich the class of stationary processes obtained using kernel mixing. From Section 3.2, we have the induced covariance function to be $C(\mathbf{s}) = \sigma^2 \int_{\Re^2} K(\mathbf{s} - \mathbf{w})K(\mathbf{w})d\mathbf{w}$. Using (3.2), we have the Fourier transform

$$\hat{c}(\mathbf{w}) = \int_{\Re^2} \int_{\Re^2} e^{i\mathbf{w}^{\mathrm{T}}\mathbf{s}} K(\mathbf{s} - \mathbf{w})K(\mathbf{w})d\mathbf{w}d\mathbf{s} = \overline{\hat{K}}(\mathbf{w})\hat{K}(\mathbf{w}) = |\hat{K}(\mathbf{w})|^2 .$$

In other words, K induces C, $K \Leftrightarrow \hat{K}$, and $C \Leftrightarrow \hat{c}$ by the one-to-one relationship between distributions and characteristic functions. So, if we start with C, is there a $\hat{K}$ that yields $\hat{c}$? Can $\hat{c}(\mathbf{w})$ be written as $|\hat{K}(\mathbf{w})|^2$, i.e. does $\hat{c}$ admit a "square root?" We have the elegant result [Yaglom, 1987] which says that a stationary random process can be defined by kernel mixing if and only if it has a spectral density.

Also, immediately, we can create an example of a stationary random process that does not arise from kernel mixing as follows (R. Wolpert, personal communication). In $\Re^1$, let V_1 and V_2 be independent $N(0,1)$ variables and set $Y(t) = Z_1\cos(t) + Z_2\sin(t)$. Then, it is easy to see that $\mathrm{E}(Y(t)) = 0$ and $\mathrm{Cov}(Y(t), Y(t')) = \cos(t - t')$. So, $Y(t)$ is, in fact, a stationary Gaussian process. But, directly, $C(h) = \int_{-\infty}^{\infty} e^{ihw} \frac{1}{2}[\delta_{-1}(w) + \delta_1(w)]dw$ where δ is the usual delta function, $\delta_c(x) = 1$ if $x = c$, $= 0$ otherwise. So, $\hat{c}(w) = \frac{1}{2}[\delta_{-1}(w) + \delta_1(w)]$ which does not admit a square root and, therefore, there is no kernel representation.

Finally, we can consider the foregoing in the context of the Matérn class of covariance functions above. The key point here is that, in (3.5), the Matérn covariance functions are one-to-one with Cauchy-type spectral densities. In particular, the smoothness parameter, $v > 0$, appears in $\hat{c}(\mathbf{w})$ in the exponent as $-(v + \frac{r}{2})$. Hence, taking a square root yields the power $-(\frac{v}{2} + \frac{r}{4}) = -(v' + \frac{r}{2})$ with $v' = \frac{v}{2} - \frac{r}{4}$. So, in order that $v' > 0$, we need $v > \frac{r}{2}$ to have a spectral density. In two dimensions, this implies that $v > 1$, i.e., only Matérn covariance functions that produce at least mean square differentiable realizations arise from kernel convolution. Since $k = \frac{1}{2}$ for the exponential, we can not create the exponential covariance from kernel mixing. Also noteworthy here is the implication that one should not employ Gaussian kernels in using kernel convolution since they produce Gaussian covariance functions, again, process realizations that are too smooth. See Paciorek and Schervish [2006] and Section 3.2.2 for further discussion in this regard.

3.1.5   Directional derivative processes

The previous section offered a discussion intended to clarify, for a spatial process, the connection between the correlation function and the smoothness of process realizations. When realizations are mean square differentiable, we can think about a directional derivative process. That is, for a given direction, at each location, we can define a random variable that is the directional derivative of the original process at that location in the given direction. The entire collection of random variables can again be shown to be a spatial process. We offer brief development below but note that, intuitively, such variables would be created through limits of finite differences. In other words, we can also formalize a finite difference process in a given direction. The value of formalizing such processes lies in the possibility of assessing where, in a region of interest, there are sharp gradients and in which directions. They also enable us to work at different scales of resolution. Application could involve land-value gradients away from a central business district, temperature gradients in a north-south direction as mentioned above, or perhaps the maximum gradient at a location and the direction of that gradient, in order to identify zones of rapid change (boundary analysis). Some detail on the development of directional derivative processes appears in Section 14.3.

3.2   Nonstationary spatial process models ⋆

Recognizing that isotropy is an assumption regarding spatial association that will rarely hold in practice, Section 2.2 proposed classes of covariance functions that were still stationary but anisotropic. However, we may wish to shed the stationarity assumption entirely and merely assume that $\mathrm{Cov}(Y(\mathbf{s}), Y(\mathbf{s}')) = C(\mathbf{s}, \mathbf{s}')$ where $C(\cdot, \cdot)$ is symmetric in its arguments. The choice of C must still be valid. Theoretical classes of valid nonstationary covariance functions can be developed [Rehman and Shapiro, 1996], but they are typically described through existence theorems, perhaps as functions in the complex plane.

We seek classes that are flexible but also offer attractive interpretation and are computationally tractable. To this end, we prefer constructive approaches. We first observe that nonstationarity can be immediately introduced through scaling and through marginalization of stationary processes.

For the former, suppose $w(\mathbf{s})$ is a mean 0, variance 1 stationary process with correlation function ρ. Then $v(\mathbf{s}) = \sigma(\mathbf{s})w(\mathbf{s})$ is a nonstationary process. In fact,

$$Var\ v(\mathbf{s}) = \sigma^2(\mathbf{s})$$
$$\text{and } Cov(v(\mathbf{s}), v(\mathbf{s}')) = \sigma(\mathbf{s})\sigma(\mathbf{s}')\rho(\mathbf{s} - \mathbf{s}')\ , \tag{3.6}$$

so $v(\mathbf{s})$ could be used as a spatial error process, replacing $w(\mathbf{s})$ in (6.1). Where would $\sigma(\mathbf{s})$ come from? Since the use of $v(\mathbf{s})$ implies heterogeneous variance for $Y(\mathbf{s})$ we could follow the familiar course in regression modeling of setting $\sigma(\mathbf{s}) = g(x(\mathbf{s}))\sigma$ where $x(\mathbf{s})$ is a suitable positive covariate and g is a strictly increasing positive function. Hence, $\text{Var}(Y(\mathbf{s}))$ increases in $x(\mathbf{s})$. Customary choices for $g(\cdot)$ are $(\cdot)$ or $(\cdot)^{\frac{1}{2}}$.

Instead, suppose we set $v(\mathbf{s}) = w(\mathbf{s}) + \delta z(\mathbf{s})$ with $z(\mathbf{s}) > 0$ and with δ being random with mean 0 and variance σ_δ^2. Then $v(\mathbf{s})$ is still a mean 0 process but now unconditionally, i.e., marginalizing over δ,

$$\text{Var}(v(\mathbf{s})) = \sigma_w^2 + z^2(\mathbf{s})\sigma_\delta^2$$
$$\text{and } \text{Cov}(v(\mathbf{s}), v(\mathbf{s}')) = \sigma_w^2\rho(\mathbf{s} - \mathbf{s}') + z(\mathbf{s})z(\mathbf{s}')\sigma_\delta^2\ . \tag{3.7}$$

(There is no reason to impose $\sigma_w^2 = 1$ here.) Again, this model for $v(\mathbf{s})$ can replace that for $w(\mathbf{s})$ as above. Now where would $z(\mathbf{s})$ come from? One possibility is that $z(\mathbf{s})$ might be a function of the distance from s to some externality in the study region. (For instance, in modeling land prices, we might consider distance from the central business district.) Another possibility is that $z(\mathbf{s})$ is an explicit function of the location, e.g., of latitude or longitude, of eastings or northings (after some projection). Of course, we could introduce a vector $\boldsymbol{\delta}$ and a vector $\mathbf{z}(\mathbf{s})$ such that $\boldsymbol{\delta}^{\mathrm{T}}\mathbf{z}(\mathbf{s})$ is a trend surface and then do a trend surface marginalization. In this fashion the spatial structure in the mean is converted to the association structure. And since $\mathbf{z}(\mathbf{s})$ varies with $\mathbf{s}$, the resultant association must be nonstationary.

In (3.6) the departure from stationarity is introduced in a multiplicative way, i.e., through scaling. The nonstationarity is really just in the form of a nonhomogeneous variance; the spatial correlations are still stationary. Similarly, through (3.7) it arises in an additive way. Now, the nonstationarity is really just arising by revising the mean structure with a regression term; the residual spatial dependence is still stationary. In fact, it is evident that we could create $v(\mathbf{s}) = \sigma(\mathbf{s})w(\mathbf{s}) + \delta z(\mathbf{s})$ yielding both types of departures from stationarity. But it is also evident that (3.6) and (3.7) are of limited value.

The foregoing suggests a simple strategy for developing nonstationary covariance structure using known functions. For instance, for a function $g(\mathbf{s})$ on $\Re^2$, $C(\mathbf{s}, \mathbf{s}') = \sigma^2 g(\mathbf{s})g(\mathbf{s}')$ is immediately seen to be a valid covariance function. Of course, it is not very interesting since, for locations $\mathbf{s}_1, \mathbf{s}_2, \ldots, \mathbf{s}_n$, the resulting joint covariance matrix is of rank 1 regardless of n! But then, the idea can evidently be extended by introducing more functions. This leads to the flexible class of nonstationary covariance functions introduced by Cressie and Johannesson [2008] to implement so-called fixed rank kriging for large spatial data sets. Specifically, let $C(\mathbf{s}, \mathbf{s}') = \mathbf{g}(\mathbf{s})^{\mathrm{T}} K \mathbf{g}(\mathbf{s}')$ where $\mathbf{g}(\mathbf{s})$ is an $r \times 1$ vector of known functions and K is an $r \times r$ positive definite matrix. Again, the validity of this C is immediate. There is no requirement that the functions in $\mathbf{g}(\mathbf{s})$ be orthogonal and standard classes such as smoothing splines, radial basis functions, or wavelets can be used. The challenges include the choice of r and the estimation of K (setting $K = I$ is not rich enough). For the latter, Cressie and Johannesson [2008] propose to obtain an empirical estimate of K, say $\hat{K}$ and then minimize a Frobenius norm[1] between $\hat{K}$ and $K(\boldsymbol{\theta})$ for some parametric class of positive definite matrices. Hence, this approach, when applied to kriging, will suffer the same problems regarding capturing uncertainty as noted in the context of ordinary kriging in Chapter 2. However, with the challenges regarding choice of r and specifying K as a

---

[1]The Frobenius norm between two matrices A and B is $||A - B||^2 \equiv \mathtt{tr}(A - B)^{\mathrm{T}}(A - B)$.

somewhat high dimensional unknown parametric matrix, at present there is no Bayesian version of this fixed rank kriging available.

Constructive approaches for building nonstationary processes have also been motivated by ideas from machine learning such as neural networks. Chen et al. [2022] devise building spatial processes using weights and activation functions analogous to neural networks. More specifically, these authors employ a $K \times 1$ vector of basis functions, $\boldsymbol{\phi}(\mathbf{s}) = (\phi_1(\mathbf{s}), \ldots, \phi_K(\mathbf{s}))^\mathsf{T}$ and construct a nested process with n layers, where the i-th layer is specified by $\mathbf{u}_i(\mathbf{s}) = A_i \psi_{i-1}(\mathbf{u}_{i-1}(\mathbf{s})) + \boldsymbol{\omega}_i(\mathbf{s})$ for $i = n, n-1, \ldots, 1$, $\psi_i(\cdot)$ is an activation function mapping a latent process $\mathbf{u}_i(\mathbf{s}) = A_i \psi_{i-1}(\mathbf{u}_{i-1}(\mathbf{s})) + \boldsymbol{\omega}_i(\mathbf{s})$, A_i is the matrix of weights and $\boldsymbol{\omega}_i(\mathbf{s})$ adjusts for bias. The first layer, i.e., $i = 1$, is specified as $\mathbf{u}_1(\mathbf{s}) = A_1 \mathbf{u}_0(\mathbf{s}) + \boldsymbol{\omega}_1(\mathbf{s})$, where $\mathbf{u}_0(\mathbf{s})$ is a vector consisting of predictors as well as basis functions. This is referred to as "deep kriging."

The above is an example of building a process as a hierarchical model. Such nested spatial processes can be constructed in different ways. For example, Bolin and Lindgren [2011] propose nested spatial processes using linear operators on spatial processes. Beginning with a spatially indexed white noise process, say $\epsilon(\cdot)$, a spatial process is constructed as the composition $\mathcal{L}_1 \circ \cdots \circ \mathcal{L}_n w_n(\cdot)$ with $\mathcal{L}_i w_i(\cdot) = w_{i-1}(\cdot)$ for $i = 1, \ldots, n$ with $w_0(\cdot) = \mathcal{L}_\epsilon \epsilon(\cdot)$ being a process arising from a linear operator on the spatially-indexed white noise process. Bolin and Lindgren [2011] build these linear operators using stochastic partial differential equations, while Sidén and Lindsten [2020] specify these operators using convolutional neural networks. Such deep processes can also be constructed using the composition of space-deformation functions (see Section 3.2.1) or by modeling the covariance kernel parameters themselves as spatial processes (see the discussion after (3.18) in Section 3.2.2). We do not attempt a comprehensive review of deep Gaussian processes in this text; Wikle and Zammit-Mangion [2023] offer a more detailed overview of different constructions.

3.2.1 Deformation

In what is regarded as a landmark paper in spatial data analysis, Sampson and Guttorp [1992] introduced an approach to nonstationarity through *deformation*. The basic idea is to transform the geographic region D to a new region G, a region such that stationarity and, in fact, isotropy holds on G. The mapping $\mathbf{g}$ from D to G is bivariate, i.e., if $\mathbf{s} = (\ell_1, \ell_2)$, $\mathbf{g}(\ell_1, \ell_2) = (g_1(\ell_1, \ell_2), g_2(\ell_1, \ell_2))$. If C denotes the isotropic covariance function on G we have

$$Cov(Y(\mathbf{s}), Y(\mathbf{s}')) = C(\|\mathbf{g}(\mathbf{s}) - \mathbf{g}(\mathbf{s}')\|) . \qquad (3.8)$$

Thus, from (3.8) there are two unknown functions to estimate, $\mathbf{g}$ and C. The latter is assumed to be a parametric choice from a standard class of covariance functions (as in Table 2.1). To determine the former is a challenging "fitting" problem. To what class of transformations shall we restrict ourselves? How shall we obtain the "best" member of this class? Sampson and Guttorp [1992] employ the class of thin plate splines and optimize a version of a two-dimensional nonmetric multidimensional scaling criterion [see, e.g., Mardia et al., 1979], providing an algorithmic solution. The solution is generally not well behaved, in the sense that $\mathbf{g}$ will be bijective, often folding over itself. Smith [1996] embedded this approach within a likelihood setting but worked instead with the class of radial basis functions. Damian et al. [2001] and Schmidt and O'Hagan [2003] have formulated fully Bayesian approaches to implement (3.8). The former still work with thin plate splines, but place priors over an identifiable parametrization (which depends upon the number of points, n being transformed). The latter elects not to model $\mathbf{g}$ directly but instead model the transformed locations. The set of n transformed locations is modeled as n realizations from a bivariate Gaussian spatial process (see Chapter 10) and a prior is placed on the process parameters.

That is, the $\mathbf{g}(\mathbf{s})$ surface arises as a random realization of a bivariate process over the $\mathbf{s}$ rather than through the values over $\mathbf{s}$ of an unknown bivariate transformation.

A fundamental limitation of the deformation approach is that implementation requires independent replications of the process in order to obtain an estimated sample covariance matrix for the set of $(Y(\mathbf{s}), \ldots, Y(\mathbf{s}_n))$. In practice, we rarely obtain i.i.d. replications of a spatial process. If we obtain repeated measurements at a particular location, they are typically collected across time. We would prefer to incorporate a temporal aspect in the modeling rather than attempting repairs (e.g., differencing and detrending) to achieve approximately i.i.d. observations. This is the focus of Chapter 12. Moreover, even if we are prepared to assume independent replications, we will require a rather large number of them to adequately estimate an $n \times n$ covariance matrix, even for a moderate size n, more than we would imagine in practice.

Spatial deformations have also motivated the construction of deep spatial models in machine-learning contexts for analyzing complex spatially dependent data exhibiting anisotropy and nonstationarity. For example, Zammit-Mangion et al. [2022] exploit a connection between the space deformation models of Sampson and Guttorp [1992] and feed-forward neural networks to build deep compositional spatial processes. Let $w(\cdot) \sim GP(0, C(\cdot, \cdot))$ be a Gaussian spatial process constructed over a domain $\mathcal{D} \subset \mathbb{R}^d$ and let $\Phi : \mathcal{D}_0 \to \mathcal{D}_n$ be a deterministic function that maps the domain $\mathcal{D}_0$ to a domain $\mathcal{D}_n$ such that $C(\mathbf{s}, \mathbf{s}') = K(\|\Phi(\mathbf{s}) - \Phi(\mathbf{s}')\|)$ is an isotropic covariance function on $\mathcal{D}_n$. Deep compositional models specify Φ to be a composition of several functions, $\Phi(\mathbf{s}) := \phi_n \circ \cdots \circ \phi_1(\mathbf{s})$. An immediate consideration for this composition to be well-defined is to appropriately specify the domain and range of the constituent functions. In general, we can define each $\phi_i(\cdot)$ to be a map from a subset $\mathcal{D}_{i-1}$ of $\mathbb{R}^{d_{i-1}}$ to another subset $\mathcal{D}_i$ of $\mathbb{R}^{d_i}$ with d_i being an integer specifying the dimension of the space over which $\phi_{i+1}(\cdot)$ is defined. For example, we can construct a feed-forward neural network such that $\phi_i(\mathbf{s}) = W_i \mathbf{f}_i(\mathbf{s})$, where $\mathbf{f}_i(\mathbf{s})$ is an $r_i \times 1$ vector of basis functions and W_i is a $d_i \times r_i$ matrix whose elements are the coefficients of the basis functions. This implies $\phi_{ij}(\mathbf{s}) = \sum_{k=1}^{r_i} w_{i,(j,k)} f_{ik}(\mathbf{s})$, where $w_{i,(j,k)}$ is the (j,k)-th element of W_i and $\phi_i(\mathbf{s}) = (\phi_{i1}(\mathbf{s}), \ldots, \phi_{id_i}(\mathbf{s}))^\mathsf{T}$ for $\mathbf{s} \in \mathcal{D}_{i-1}$. This framework is flexible and accommodates low-rank as well as full-rank specifications and also impose constraints on the elements of W_i to ensure injective maps. One could also specify the basis functions as thin-plate splines [as in Sampson and Guttorp, 1992] or, as in Schmidt and O'Hagan [2003], assume Gaussian process priors for each of the $\phi_n(\cdot)$. We refer the reader to Zammit-Mangion et al. [2022] and to Wikle and Zammit-Mangion [2023] for comprehensive accounts of these specifications.

3.2.2 *Nonstationarity through kernel mixing of process variables*

Kernel mixing as described above in the context of creating stationary processes. Here, we show that it provides a strategy for introducing non-stationarity while retaining clear interpretation and permitting analytic calculation. We look at two distinct approaches, one due to Higdon in a series of papers [e.g., Higdon, 1998, Higdon et al., 1999, Higdon, 2002a] and the other due to Fuentes [Fuentes and Smith, 2001, Fuentes, 2001, 2002b,a]. We note that kernel mixing has a long tradition in the statistical literature, especially in density estimation and regression modeling [Silverman, 2018]. Kernel mixing is often done with distributions and we will look at this idea in a later subsection. Here, we focus on kernel mixing of random variables.

Adding some detail to our earlier discussion, suppose we work with bivariate kernels starting with stationary choices of the form $K(\mathbf{s} - \mathbf{s}')$, e.g., $K(\mathbf{s} - \mathbf{s}') = \exp\{-\frac{1}{2}(\mathbf{s} - \mathbf{s}')^\mathsf{T} V(\mathbf{s} - \mathbf{s}')\}$. A natural choice for V would be diagonal with V_{11} and V_{22} providing componentwise scaling to the separation vector $\mathbf{s} - \mathbf{s}'$. Other choices of kernel function are available; specialization to versions based on Euclidean distance is immediate; again see, e.g., Silverman [2018]. Higdon [1998] and Higdon [2002a] let $z(\mathbf{s})$ be a white noise process, i.e., $\mathrm{E}[z(\mathbf{s})] = 0$,

$Var(z(\mathbf{s})) = \sigma^2$ and $Cov(z(\mathbf{s}), z(\mathbf{s}')) = 0$ and set

$$w(\mathbf{s}) = \int_{\Re^2} K(\mathbf{s} - \mathbf{t})z(\mathbf{t})dt . \tag{3.9}$$

Rigorously speaking, (3.9) is not defined. As we noted in Section 3.1.1, the convolution should be written as $w(\mathbf{s}) = \int K(\mathbf{s} - \mathbf{t})\mathcal{X}(dt)$ where $\mathcal{X}(\mathbf{t})$ is two-dimensional Brownian motion.

Reiterating earlier details, the process $w(\mathbf{s})$ is said to arise through *kernel convolution*. By change of variable, (3.9) can be written as

$$w(\mathbf{s}) = \int_{\Re^2} K(\mathbf{u})z(\mathbf{s} + \mathbf{u})du , \tag{3.10}$$

emphasizing that $w(\mathbf{s})$ arises as a kernel-weighted average of z's centered around $\mathbf{s}$. It is straightforward to show that $E[w(\mathbf{s})] = 0$, but also that

$$\begin{aligned} \text{Var}(w(\mathbf{s})) &= \sigma^2 \int_{\Re^2} k^2(\mathbf{s} - \mathbf{t})dt , \\ \text{and} \quad \text{Cov}(w(\mathbf{s}), w(\mathbf{s}')) &= \sigma^2 \int_{\Re^2} K(\mathbf{s} - \mathbf{t})K(\mathbf{s}' - \mathbf{t})dt . \end{aligned} \tag{3.11}$$

A simple change of variables $(\mathbf{t} \to \mathbf{u} = \mathbf{s}' - \mathbf{t})$, shows that

$$\text{Cov}(w(\mathbf{s}), w(\mathbf{s}')) = \sigma^2 \int_{\Re^2} K(\mathbf{s} - \mathbf{s}' + \mathbf{u})K(\mathbf{u})du , \tag{3.12}$$

i.e., $w(\mathbf{s})$ is stationary. In fact, (3.9) is a way of generating classes of stationary processes [see. e.g., Yaglom, 2004, Chapter 26] whose limitations we have considered above. We can extend (3.9) so that $z(\mathbf{s})$ is a mean 0 stationary spatial process with covariance function $\sigma^2 \rho(\cdot)$. Again $E[w(\mathbf{s})] = 0$ but now

$$\begin{aligned} \text{Var}(w(\mathbf{s})) &= \sigma^2 \int_{\Re^2} \int_{\Re^2} K(\mathbf{s} - \mathbf{t})K(\mathbf{s}' - \mathbf{t})\rho(\mathbf{t} - \mathbf{t}')dtdt' \\ \text{and} \quad \text{Cov}(w(\mathbf{s}), w(\mathbf{s}')) &= \sigma^2 \int_{\Re^2} \int_{\Re^2} K(\mathbf{s} - \mathbf{t})K(\mathbf{s}' - \mathbf{t}')\rho(\mathbf{t} - \mathbf{t}')dtdt' . \end{aligned} \tag{3.13}$$

Interestingly, $w(\mathbf{s})$ is still stationary. We now use the change of variables $(\mathbf{t} \to \mathbf{u} = \mathbf{s}' - \mathbf{t}, \mathbf{t}' \to \mathbf{u}' = \mathbf{s}' - \mathbf{t}')$ to obtain

$$\text{Cov}(w(\mathbf{s}), w(\mathbf{s}')) = \sigma^2 \int_{\Re^2} \int_{\Re^2} K(\mathbf{s} - \mathbf{s}' + \mathbf{u})K(\mathbf{u}')\rho(\mathbf{u} - \mathbf{u}')dudu' . \tag{3.14}$$

Note that (3.11) and (3.13) can be proposed as covariance functions. It is straightforward to argue that they are positive definite functions and so, can be attached for instance to a Gaussian process if we wish. However, the integrations in (3.12) and (3.14) will not be possible to do explicitly except in certain special cases [see, e.g., Ver Hoef and Barry, 1998]. Numerical integration across $\Re^2$ for (3.12) or across $\Re^2 \times \Re^2$ for (3.14) may be difficult, requiring nonlinear transformation to a bounded set. If we work with a subset $D \subset \Re^2$, we sacrifice stationarity. Monte Carlo integration is also not attractive here: we would have to sample from the standardized density associated with K. But since $\mathbf{s} - \mathbf{s}'$ enters into the argument, we would have to do a separate Monte Carlo integration for each pair of locations $(\mathbf{s}_i, \mathbf{s}_j)$.

An alternative is to replace (3.9) with a finite sum approximation, i.e., to define

$$w(\mathbf{s}) = \sum_{j=1}^{L} K(\mathbf{s} - \mathbf{t}_j)z(\mathbf{t}_j) \tag{3.15}$$

for locations $\mathbf{t}_j$, $j = 1, \ldots, L$. In the case of a white noise assumption for the z's,

$$\text{Var}(w(\mathbf{s})) = \sigma^2 \sum_{j=1}^{L} k^2(\mathbf{s} - \mathbf{t}_j)$$
$$\text{and } \text{Cov}(w(\mathbf{s}), w(\mathbf{s}')) = \sigma^2 \text{Var}(w(\mathbf{s})) = \sigma^2 \sum_{j=1}^{L} K(\mathbf{s} - \mathbf{t}_j) K(\mathbf{s}' - \mathbf{t}_j) . \tag{3.16}$$

In the case of spatially correlated z's,

$$\text{Var}(w(\mathbf{s})) = \sigma^2 \sum_{j=1}^{L} \sum_{j'=1}^{L} K(\mathbf{s} - \mathbf{t}_j) K(\mathbf{s} - \mathbf{t}_{j'}) \rho(\mathbf{t}_j - \mathbf{t}_{j'})$$
$$\text{and } \text{Cov}(w(\mathbf{s}), w(\mathbf{s}')) = \sigma^2 \sum_{j=1}^{L} \sum_{j'=1}^{L} K(\mathbf{s} - \mathbf{t}_j) K(\mathbf{s}' - \mathbf{t}_{j'}) \rho(\mathbf{t}_j - \mathbf{t}_{j'}) . \tag{3.17}$$

Expressions (3.16) and (3.17) can be calculated directly from (3.15) and, in fact, can be used to provide a limiting argument for expressions (3.11) and (3.13); see Exercise 12. Note that (3.15) provides a dimension reduction; we express the entire stochastic process of random variables through the finite collection of $z_{\mathbf{t}_j}$. Dimension reduction has been offered as a general way to handle the computational challenges associated with large datasets in space and time, see Chapter 13.

Note further that, while (3.16) and (3.17) are available explicitly, these forms reveal that the finite sum process in (3.15) is no longer stationary. While nonstationary specifications are the objective of this section, their creation through (3.15) is rather artificial as it arises from the arbitrary $\{\mathbf{t}_j\}$. It would be more attractive to modify (3.9) to achieve a class of nonstationary processes.

So, instead, suppose we allow the kernel in (3.9) to vary spatially. Notationally, we can write such an object as $K(\mathbf{s} - \mathbf{s}'; \mathbf{s})$. Illustratively, we might take $K(\mathbf{s} - \mathbf{s}'; \mathbf{s}) = \exp\{-\frac{1}{2}(\mathbf{s} - \mathbf{s}')^{\text{T}} V_{\mathbf{s}}(\mathbf{s} - \mathbf{s}')\}$. As above, we might take $V_{\mathbf{s}}$ to be diagonal with, if $\mathbf{s} = (\ell_1, \ell_2)$, $(V_{\mathbf{s}})_{11} = V(\ell_1)$ and $(V_{\mathbf{s}})_{22} = V(\ell_2)$. Higdon et al. [1999] adopt such a form with V taken to be a slowly varying function. One explicit choice would take $V_{\mathbf{s}} = V_{A_i}$ for $s \in A_i$ where the V_{A_i} are obtained through local range anisotropy. Another choice might take $V_{\mathbf{s}} = \eta(\mathbf{s})I$ where $\eta(\mathbf{s})$ might be a simple trend surface or a function of a local covariate $X(\mathbf{s})$ such as elevation.

We can insert $K(\mathbf{s} - \mathbf{s}'; \mathbf{s})$ into (3.9) in place of $K(\mathbf{s} - \mathbf{s}')$ with obvious changes to (3.11), (3.12), (3.13), and (3.14). Evidently, the process is now nonstationary. In fact, the variation in V provides insight into the departure from stationarity. For computational reasons Higdon et al. [1999] implement this modified version of (3.9) through a finite-sum analogous to (3.15). A particularly attractive feature of employing a finite sum approximation is dimension reduction. If $z(\mathbf{s})$ is white noise we have an approach for handling large data sets (see Section 13.1). That is, regardless of n, $\{w(\mathbf{s}_i)\}$ depends only on L latent variables z_j, $j = 1, \ldots, L$, and these variables are independent. Rather than fitting the model in the space of the $\{w(\mathbf{s}_i)\}$ we can work in the space of the z_ℓ.

Paciorek and Schervish [2006] substantially extend this work. First, they note that the general form $C(\mathbf{s}, \mathbf{s}') = \int_{\Re^2} K_{\mathbf{s}}(\mathbf{u}) K_{\mathbf{s}'}(\mathbf{u}) d\mathbf{u}$ is a valid covariance function, in fact for $\mathbf{s}, \mathbf{s}' \in \Re^r$ for an positive integer r. This is easily shown by direct calculation and is left as an exercise. More importantly, it means we do not need a process construction to justify such covariance functions. They also note that the use of Gaussian kernels in this construction (as well as the earlier stationary version) is unattractive for the same reasons noted above; the resulting covariance function will produce process realizations that are too smooth.

In fact, Paciorek and Schervish [2006] provide a much richer class of nonstationary covariance functions as follows. Let $Q(\mathbf{s}, \mathbf{s}') = (\mathbf{s} - \mathbf{s}')^{\text{T}} (\frac{V_{\mathbf{s}} + V_{\mathbf{s}'}}{2})^{-1} (\mathbf{s} - \mathbf{s}')$ and let ρ be any positive definite function on $\Re^p$. Then

$$C(\mathbf{s}, \mathbf{s}') = |V_{\mathbf{s}}|^{\frac{1}{2}} |V_{\mathbf{s}'}|^{\frac{1}{2}} \left| \frac{V_{\mathbf{s}} + V_{\mathbf{s}'}}{2} \right|^{-\frac{1}{2}} \rho(\sqrt{Q(\mathbf{s}, \mathbf{s}')}) \tag{3.18}$$

is a valid nonstationary correlation function on $\Re^r$. We can use the Matérn or other choices for ρ with choices for $V_\mathbf{s}$ as above. We can multiply by σ^2 to obtain a covariance function. Under conditions given in the Ph.D. thesis of Paciorek, process realizations under $C(\mathbf{s}, \mathbf{s}')$ inherit the smoothness properties of $\rho(\cdot)$. We leave as an exercise to show that if, in the general kernel convolution form above, we take $K_\mathbf{s}(\mathbf{u}) = |V_\mathbf{s}|^{-\frac{1}{2}}\exp(-(\mathbf{s} - \mathbf{u})^\top V_\mathbf{s}(\mathbf{s} - \mathbf{u}))$, then $C(\mathbf{s}, \mathbf{s}') = |V_\mathbf{s}|^{\frac{1}{2}}|V_{\mathbf{s}'}|^{\frac{1}{2}}|\frac{V_\mathbf{s}+V_{\mathbf{s}'}}{2}|^{-\frac{1}{2}}\exp(-Q(\mathbf{s}, \mathbf{s}'))$.

It is worth remarking here that the idea of modeling parameters in covariance kernels, as proposed in Paciorek and Schervish [2006], can be combined with multi-resolution spatial modeling [see, e.g., Nychka et al., 2015] to build deep spatial processes. We continue to add levels to our hierarchical model with the parameters in the spatial covariance kernel at each level itself being modeled as a Gaussian process. We construct a nested spatial process of depth n as follows. We begin with a Gaussian spatial process of interest $w(\mathbf{s}) := w_n(\mathbf{s})$ with $w_n(\cdot) \sim GP(0, C_n(\cdot, \cdot; \boldsymbol{\theta}_n(\cdot)))$, where $C_n(\cdot, \cdot; \boldsymbol{\theta}_n(\mathbf{s}))$ is a spatially varying covariance kernel and $\boldsymbol{\theta}_n(\cdot)$ is a collection of spatially varying covariance kernel parameters. We subsequently model each $\boldsymbol{\theta}_i(\cdot) \sim GP(\mathbf{0}, C_i(\cdot, \cdot; \boldsymbol{\theta}_{i-1}(\cdot)))$ for $i = n, n-1, \ldots, 1$ and $\boldsymbol{\theta}_1(\cdot) \sim GP(\mathbf{0}, C_1(\cdot, \cdot))$ with $C_1(\cdot, \cdot)$ being a covariance kernel that is completely specified. The validity of the space-varying covariance kernel at each depth ensures that the process of interest, $w_n(\cdot)$, is a valid stochastic process. These deep processes can serve as effective predictive models for nonstationary spatial data and have been explored in machine learning for computer models by Sauer et al. [2023a,b]. Evidently, the dimension of the parameter space explodes in these deep processes and for computational efficiency one resorts to sparsity-inducing processes such as nearest-neighbor Gaussian processes [see Section 13.6.2 and Sauer et al., 2023a, Coube-Sisqueille et al., in press].

Fuentes [2002a,b] offers a kernel mixing form that initially appears similar to (3.9) but is fundamentally different. Let

$$w(\mathbf{s}) = \int K(\mathbf{s} - \mathbf{t})z_{\boldsymbol{\theta}(\mathbf{t})}(\mathbf{s})d\mathbf{t} . \tag{3.19}$$

In (3.19), $K(\cdot)$ is as in (3.9) but $z_{\boldsymbol{\theta}}(\mathbf{s})$ denotes a mean 0 stationary spatial process with covariance function that is parametrized by $\boldsymbol{\theta}$. For instance $C(\cdot; \boldsymbol{\theta})$ might be $\sigma^2 \exp(-\phi \|\cdot\|^\alpha)$, a power exponential family with $\boldsymbol{\theta} = (\sigma^2, \phi, \alpha)$. In (3.19) $\boldsymbol{\theta}(\mathbf{t})$ indexes an uncountable number of processes. These processes are assumed independent across $\mathbf{t}$. Note that (3.19) is mixing an uncountable number of stationary spatial processes each at $\mathbf{s}$ while (3.9) is mixing a single process across all locations.

Formally, $w(\mathbf{s})$ has mean 0 and

$$\begin{aligned} \text{Var}(w(\mathbf{s})) &= \int_{\Re^2} k^2(\mathbf{s} - \mathbf{t})C(0; \boldsymbol{\theta}(\mathbf{t}))d\mathbf{t} \\ \text{and } \text{Cov}(w(\mathbf{s}), w(\mathbf{s}')) &= \int_{\Re^2} K(\mathbf{s} - \mathbf{t})K(\mathbf{s}' - \mathbf{t})C(\mathbf{s} - \mathbf{s}'; \boldsymbol{\theta}(\mathbf{t}))d\mathbf{t} . \end{aligned} \tag{3.20}$$

Expression (3.20) reveals that (3.19) defines a nonstationary process. Suppose k is very rapidly decreasing and $\boldsymbol{\theta}(\mathbf{t})$ varies slowly. Then $w(\mathbf{s}) \approx K(0)z_{\boldsymbol{\theta}(\mathbf{s})}(\mathbf{s})$. But also, if $\mathbf{s} - \mathbf{s}'$ is small, $w(\mathbf{s})$ and $w(\mathbf{s}')$ will behave like observations from a stationary process with parameter $\boldsymbol{\theta}(\mathbf{s})$. Hence, Fuentes [2002b] refers to the class of models in (3.19) as a nonstationary class that exhibits *local* stationarity.

In practice, one cannot work with (3.19) directly. Again, finite sum approximation is employed. Again, a finite set of locations $\mathbf{t}_1, \ldots, \mathbf{t}_L$ is selected and we set

$$w(\mathbf{s}) = \sum_j K(\mathbf{s} - \mathbf{t}_j)z_j(\mathbf{s}) , \tag{3.21}$$

writing $\boldsymbol{\theta}(\mathbf{t}_j)$ as j. Straightforwardly,

$$\begin{aligned} \text{Var}(w(\mathbf{s})) &= \sum_{j=1}^L k^2(\mathbf{s} - \mathbf{t}_j)C_j(0) \\ \text{and } \text{Cov}(w(\mathbf{s}), w(\mathbf{s}')) &= \sum_{j=1}^L K(\mathbf{s} - \mathbf{t}_j)K(\mathbf{s}' - \mathbf{t}_j)C_j(\mathbf{s} - \mathbf{s}') . \end{aligned} \tag{3.22}$$

It is worth noting that the discretization in (3.21) does not provide a dimension reduction. In fact, it is a dimension *explosion*; we need L $z_j(\mathbf{s})$'s for each $w(\mathbf{s})$.

In (3.21) it can happen that some $\mathbf{s}$'s may be far enough from each of the $\mathbf{t}_j$'s so that each $K(\mathbf{s}-\mathbf{t}_j) \approx 0$, whence $w(\mathbf{s}) \approx 0$. Of course, this cannot happen in (3.19) but we cannot work with this expression. A possible remedy was proposed by Banerjee et al. [2004]. Replace (3.21) with

$$w(\mathbf{s}) = \sum_{j=1}^{L} \alpha(\mathbf{s}, \mathbf{t}_j) z_j(\mathbf{s}) . \tag{3.23}$$

In (3.23), the $z_j(\mathbf{s})$ are as above, but $\alpha(\mathbf{s}, \mathbf{t}_j) = \gamma(\mathbf{s}, \mathbf{t}_j)/\sqrt{\sum_{j=1}^{L} \gamma^2(\mathbf{s}, \mathbf{t}_j)}$, where $\gamma(\mathbf{s}, \mathbf{t})$ is a decreasing function of the distance between $\mathbf{s}$ and $\mathbf{t}$, which may change with $\mathbf{s}$, i.e., $\gamma(\mathbf{s}, \mathbf{t}) = k_{\mathbf{s}}(\|\mathbf{s} - \mathbf{t}\|)$. (In the terminology of Higdon et al. [1999], $k_{\mathbf{s}}$ would be a spatially varying kernel function.) As a result, $\sum_{j=1}^{L} \alpha^2(\mathbf{s}, \mathbf{t}_j) = 1$, so regardless of where $\mathbf{s}$ is, not all of the weights in (3.23) can be approximately 0. Other standardizations for γ are possible; we have proposed this one because if all σ_j^2 are equal, then $\mathrm{Var}(w(\mathbf{s})) = \sigma^2$. That is, if each local process has the same variance, then this variance should be attached to $w(\mathbf{s})$. Furthermore, suppose $\mathbf{s}$ and $\mathbf{s}'$ are near to each other, whence $\gamma(\mathbf{s}, \mathbf{t}_j) \approx \gamma(\mathbf{s}', \mathbf{t}_j)$ and thus $\alpha(\mathbf{s}, \mathbf{t}_j) \approx \alpha(\mathbf{s}', \mathbf{t}_j)$. So, if in addition all $\phi_j = \phi$, then $\mathrm{Cov}(w(\mathbf{s}), w(\mathbf{s}')) \approx \sigma^2 \rho(\mathbf{s}-\mathbf{s}'; \phi)$. So, if the process is in fact stationary over the entire region, we obtain essentially the second-order behavior of this process.

The alternative scaling $\widetilde{\alpha}(\mathbf{s}, \mathbf{t}_j) = \gamma(\mathbf{s}, \mathbf{t}_j)/\sum_{j'} \gamma(\mathbf{s}, \mathbf{t}_{j'})$ gives a weighted average of the component processes. Such weights would preserve an arbitrary constant mean. However, since, in our context, we are modeling a mean 0 process, such preservation is not a relevant feature. Useful properties of the process in (3.23) are

$$
\begin{aligned}
\mathrm{E}\left(w\left(\mathbf{s}\right)\right) &= 0 , \\
\mathrm{Var}\left(w\left(\mathbf{s}\right)\right) &= \sum_{j=1}^{L} \alpha^2(\mathbf{s}, \mathbf{t}_j) \sigma_j^2 , \\
\text{and } \mathrm{Cov}(w(\mathbf{s}), w(\mathbf{s}')) &= \sum_{j=1}^{L} \alpha(\mathbf{s}, \mathbf{t}_j) \alpha(\mathbf{s}', \mathbf{t}_j) \sigma_j^2 \rho(\mathbf{s} - \mathbf{s}'; \phi_j) .
\end{aligned}
$$

We have clearly defined a proper spatial process through (3.23). In fact, for arbitrary locations $\mathbf{s}_1, \ldots, \mathbf{s}_n$, let $\mathbf{w}_\ell^{\mathrm{T}} = (w_\ell(\mathbf{s}_1)), \ldots, w_\ell(\mathbf{s}_n))$, $\mathbf{w}^{\mathrm{T}} = (w(\mathbf{s}_1), \ldots, w(\mathbf{s}_n))$, and let A_ℓ be diagonal with $(A_\ell)_{ii} = \alpha(\mathbf{s}_i, \mathbf{t}_\ell)$. Then $\mathbf{w} \sim N(\mathbf{0}, \sum_{\ell=1}^{L} \sigma_\ell^2 A_\ell R(\phi_\ell) A_\ell)$, where $(\Sigma(\phi_\ell))_{ii'} = \rho(\mathbf{s}_i - \mathbf{s}_{i'}, \phi_\ell)$. Note that $L = 1$ is permissible in (3.23); $w(\mathbf{s})$ is still a nonstationary process. Finally, Fuentes and Smith [2003] and Banerjee et al. [2004] offer some discussion regarding the precise number of and locations for the $\mathbf{t}_j$.

We conclude this subsection by noting that for a general nonstationary spatial process there is no sensible notion of a range. However, for the class of processes in (3.23) we can define a meaningful range. Under (3.23),

$$\mathrm{Corr}(w(\mathbf{s}), w(\mathbf{s}')) = \frac{\sum_{j=1}^{L} \alpha(\mathbf{s}, \mathbf{t}_j) \alpha(\mathbf{s}', \mathbf{t}_j) \sigma_j^2 \rho(\mathbf{s} - \mathbf{s}'; \phi_j)}{\sqrt{\left(\sum_{j=1}^{L} \alpha^2(\mathbf{s}, \mathbf{t}_j) \sigma_j^2\right) \left(\sum_{j=1}^{L} \alpha^2(\mathbf{s}', \mathbf{t}_j) \sigma_j^2\right)}} . \tag{3.24}$$

Suppose ρ is positive and strictly decreasing asymptotically to 0 as distance tends to ∞, as is usually assumed. If ρ is, in fact, isotropic, let d_ℓ be the range for the ℓth component process, i.e., $\rho(d_\ell, \phi_\ell) = .05$, and let $\widetilde{d} = \max_\ell d_\ell$. Then (3.24) immediately shows that at distance $\widetilde{d}$ between $\mathbf{s}$ and $\mathbf{s}'$, we have $\mathrm{Corr}(w(\mathbf{s}), w(\mathbf{s}')) \leq .05$. So $\widetilde{d}$ can be interpreted as a conservative range for $w(\mathbf{s})$. Normalized weights are not required in this definition. If ρ is only assumed stationary, we can similarly define the range in an arbitrary direction $\boldsymbol{\mu}$. Specifically, if $\boldsymbol{\mu}/\|\boldsymbol{\mu}\|$ denotes a unit vector in $\boldsymbol{\mu}$'s direction and if $d_{\boldsymbol{\mu},\ell}$ satisfies $\rho(d_{\boldsymbol{\mu},\ell}\boldsymbol{\mu}/\|\boldsymbol{\mu}\|\,;\phi_\ell) = .05$, we can take $\tilde{d}_{\boldsymbol{\mu}} = \max_\ell \tilde{d}_{\boldsymbol{\mu},\ell}$.

3.2.3 Mixing of process distributions

If a kernel $K(\cdot)$ is integrable and standardized to a density function and if f is also a density function, then

$$f_K(y) = \int K(y-x)f(x)dx \tag{3.25}$$

is a density function. (It is, of course, the distribution of $X + Y - X$ where $X \sim f$, $Y - X \sim k$, and X and $Y - X$ are independent). In (3.25), we can extend to allow $\mathbf{Y}$, a vector of dimension n. But recall that we have specified the distribution for a spatial process through arbitrary finite-dimensional distributions (see Section 3.1.1). This suggests that we can use (3.25) to build a process distribution.

Operating formally, let V_D be the set of all $V(\mathbf{s})$, $\mathbf{s} \in D$. Write $V_D = V_{0,D} + V_0 - V_{0,D}$ where $V_{0,D}$ is a realization of a mean 0 stationary Gaussian process over D, and $V_D - V_{0,D}$ is a realization of a white noise process with variance σ^2 over D. Write

$$f_K(V_D \mid \tau) = \int \frac{1}{\sigma}\,K\left(\frac{1}{\tau}(V_D - V_{0,D})\right)f(V_{0,D})dV_{0,D}\,. \tag{3.26}$$

Formally, f_K is the distribution of the spatial process $v(\mathbf{s})$. In fact, $v(\mathbf{s})$ is just the customary model for the residuals in a spatial regression, i.e., of the collection $v(\mathbf{s}) = w(\mathbf{s})+\epsilon(\mathbf{s})$ where $w(\mathbf{s})$ is a spatial process and $\epsilon(\mathbf{s})$ is a noise or nugget process.

Of course, in this familiar case, there is no reason to employ the form (3.26). However it does reveal how, more generally, a spatial process can be developed through "kernel mixing" of a process distribution. More importantly, it suggests that we might introduce an alternative specification for $V_{0,D}$. For example, suppose $f(V_{0,D})$ is a discrete distribution, say, of the form $\sum_\ell p_\ell \delta(v^*_{\ell,D})$ where $p_\ell \geq 0$, $\sum p_\ell = 1$, $\delta(\cdot)$ is the Dirac delta function, and $V^*_{\ell,D}$ is a surface over D. The sum may be finite or infinite. An illustration of the former arises when $f(V_{0,D})$ is a realization from a finite discrete mixture [Duan and Gelfand, 2003] or from a finite Dirichlet process; the latter arises under a general Dirichlet process and can be extended to more general Pitman-Yor processes [see, e.g., Gelfand et al., 2005a] for further details in this regard.

But then, if $K(\cdot)$ is Gaussian white noise, given $\{p_\ell\}$ and $\{v^*_{\ell,D}\}$, for any set of locations $\mathbf{s}_1,\ldots,\mathbf{s}_n$, if $\mathbf{V} = (v(\mathbf{s}_1),\ldots,v(\mathbf{s}_n))$,

$$f_K(\mathbf{V}) = \sum_\ell p_\ell N(\mathbf{v}^*_\ell,\sigma^2 I)\,, \tag{3.27}$$

where $\mathbf{v}^*_\ell = (v^*_\ell(\mathbf{s}_1),\ldots,v^*_\ell(\mathbf{s}_n))^\mathsf{T}$. So $v(\mathbf{s})$ is a continuous process that is non-Gaussian. But also, $\mathrm{E}[v(\mathbf{s}_i)] = \sum_\ell p_\ell v^*_\ell(\mathbf{s}_i)$ and

$$\begin{aligned} \mathrm{Var}\ v(\mathbf{s}_i) &= \textstyle\sum_\ell p_\ell v^{2*}_\ell(\mathbf{s}_i) - (\sum_\ell p_\ell v^*_\ell(\mathbf{s}_i))^2 \\ \text{and }\mathrm{Cov}(v(\mathbf{s}_i),v(\mathbf{s}_j)) &= \textstyle\sum_\ell p_\ell v^*_\ell(\mathbf{s}_i)v^*_\ell(\mathbf{s}_j) - (\sum_\ell p_\ell v^*_\ell(\mathbf{s}_i))(\sum_\ell p_\ell v^*_\ell(\mathbf{s}_j))\,. \end{aligned} \tag{3.28}$$

This last expression shows that $v(\mathbf{s})$ is *not* a stationary process. However, a routine calculation shows that if the $v^*_{\ell,D}$ are continuous surfaces, the $v(\mathbf{s})$ process is mean square continuous and almost surely continuous.

3.2.4   Covariates in the covariance function

The discussion at the beginning of Section 3.2 leads us to the more general question of introducing covariates into the covariance function in addition to bringing them into the mean. We note work here from Risser and Calder [2015], Schmidt et al. [2011] and Reich et al. [2011] along with the review paper of Schmidt and Guttorp [2020].

As above, an elementary way is to allow the variance of the process to change with the location as a function of covariates. As the variance must be positive, one can assume e.g. $\log \sigma(\mathbf{s}) = \mathbf{X}^{\mathrm{T}}(\mathbf{s})\boldsymbol{\gamma}$ for a suitable set of covariates. In fact, any one-to-one function $g(\cdot)$ from R^+ to R will work. Then, with a valid correlation function say $\rho(\mathbf{s} - \mathbf{s}')$, the induced covariance function is $C(\mathbf{s}, \mathbf{s}') = g^{-1}(\mathbf{s})\mathbf{X}^{\mathrm{T}}(\mathbf{s})\boldsymbol{\gamma}g^{-1}(\mathbf{X}^{\mathrm{T}}\boldsymbol{\gamma})\rho(\mathbf{s} - \mathbf{s}')$. This is just an extension of what was discussed around expression (3.6). To perform spatial interpolation one needs the values of the covariates at the unobserved locations of interest.

A natural extension of this follows by Reich et al. [2011]. We can extend the covariance function to one arising from a weighted sum of mean 0 Gaussian processes. That is, consider the spatial random effect in the form $\eta(\mathbf{s}) = \sum_{k=1}^{K} w_k(\mathbf{X}(\mathbf{s}))\omega_k(\mathbf{s})$. Here, the $\omega_k(\mathbf{s})$ are independent Gaussian processes with $\omega_k(\mathbf{s})$ having covariance function, say $C(\mathbf{s} - \mathbf{s}' : \theta_k)$. Full discussion of the properties of this process is supplied in Reich et al. [2011]. For identifiability, it is convenient to set the sum of the weights equal to 1. As an example, we could set $w_k(\mathbf{X}(\mathbf{s})) = \frac{e^{\mathbf{X}^{\mathrm{T}}(\mathbf{s})\boldsymbol{\gamma}_k}}{\sum_{k=1}^{K} e^{\mathbf{X}^{\mathrm{T}}(\mathbf{s})\boldsymbol{\gamma}_k}}$. Choice of K would use model comparison tools. Interpretation of the $\boldsymbol{\gamma}_k$ coefficient vectors is challenging.

We might opt for a simpler version of this idea. Suppose the goal is to bring covariates into the correlation function rather than in the variance function. Then, introduce the mean 0 Gaussian process, $w(\mathbf{s}) \equiv a v_1(\mathbf{s}) + \sqrt{1 - a^2} v_2(\mathbf{s})$. Here, $0 < a < 1$ and $v_1(\mathbf{s})$ and $v_2(\mathbf{s})$ are independent Gaussian processes with mean 0, variance 1, with say, exponential correlation functions having decay parameters ϕ_1 and ϕ_2, respectively. Then, we can specify a as $a(\mathbf{X}(\mathbf{s}))$ using say a probit or logit regression resulting in the valid correlation function, $\mathrm{Corr}(w(\mathbf{s}, w(\mathbf{s}') = a^2(\mathbf{X}(\mathbf{s}))e^{-\phi_1\|\mathbf{s}-\mathbf{s}'\|} + (1 - a^2(\mathbf{X}(\mathbf{s}))e^{-\phi_2\|\mathbf{s}-\mathbf{s}'\|}$. With p covariates, we have a p dimensional coefficient vector to estimate. If $\eta(\mathbf{s}) = \sigma^2 w(\mathbf{s})$, then we have a valid covariance function.

A different choice, elaborated in Schmidt et al. [2011] is to compute a distance function between the covariates at a pair of sites, say $d(\mathbf{X}(\mathbf{s}) - \mathbf{X}(\mathbf{s}'); \boldsymbol{\theta}))$ and then insert these distances into a valid correlation function, i.e. $\rho(d(\mathbf{X}(\mathbf{s}) - \mathbf{X}(\mathbf{s}'; \boldsymbol{\theta})); \psi)$. For example, we might consider a Mahalanobis-type distance, $\sqrt{(\mathbf{X}(\mathbf{s}) - \mathbf{X}(\mathbf{s}'))^{\mathrm{T}} V_\Psi (\mathbf{X}(\mathbf{s}) - \mathbf{X}(\mathbf{s}'))}$ with say V_Ψ say a diagonal matrix with entries supplying Ψ. Then, we could consider $e^{-\phi d(\mathbf{X}(\mathbf{s}) - \mathbf{X}(\mathbf{s}'); \boldsymbol{\theta}))}$. Here, we run into the deformation problem considered in Section 3.2.1. This covariance function is not valid, employing distances in covariate space rather than in Euclidean space.

A straightforward specification is through kernel convolution as in Section 3.2.2. Again, we envision the process as $w(\mathbf{s}) = \int_{R^2} K(\mathbf{s} - \mathbf{u})z(\mathbf{u})$. Again, we discretize the integration to $\sum_{j=1}^{L} K(\mathbf{s} - \mathbf{u}_j)z(\mathbf{u}_j$. Let independent $z(\mathbf{u}_j) \sim N(0, \sigma^2(\mathbf{X}^{\mathrm{T}}(\mathbf{u}_j)\boldsymbol{\gamma})$, employing a suitable link from $R^1 \text{to} R^+$. The induced non-stationary covariance function is $\mathrm{Cov}(w(\mathbf{s}), w(\mathbf{s}') = \sum_{j=1}^{L} K(\mathbf{s}-\mathbf{u}_j)K(\mathbf{s}'-\mathbf{u}_j)\sigma^2(\mathbf{X}^{\mathrm{T}}(\mathbf{u}_j)\boldsymbol{\gamma})$. Attractively, there is only one vector of γ coefficients to estimate.

In this spirit, Calder [2008] proposes a bivariate dynamic model for PM2:5 and PM10 using a discrete convolution approach. The kernel of the convolution is assumed to be Gaussian with a covariance matrix that depends on fixed values of wind direction and wind speed. Ideally, the values of the covariate should change with location and their association with the covariance structure should be estimated. So, in the univariate case, she would consider kernel convolution in the form of a kernel function with a Gaussian kernel having a variance which depends upon the covariates, as above.

Risser and Calder [2015] extend Calder's approach by employing the nonstationary covariance structure presented in Section 3.2.2, adopting Matérn correlation functions. We omit details except to note that the covariates are brought in to the $V_\mathbf{s}$ matrix in the form $\Phi + \Gamma \mathbf{X}(\mathbf{s})\mathbf{X}(\mathbf{s})^\mathsf{T}\Gamma^\mathsf{T}$ where Φ is a 2×2 error matrix and, with p covariates, Γ is a $2 \times p$ coefficient matrix. An explicit form is available when a Gaussian correlation function is used. In particular, following the text around expression (3.18), we can simplify to express $V_\mathbf{s}$ to be diagonal with diagonal entries introduced say, through probit regressions on $\mathbf{X}(\mathbf{s})$. This yields a model with $2 \times p$ regression coefficients. Interpretation of the covariance function can be made through anisotropy.

3.3   Exercises

1. Show that, if $\rho(\mathbf{s} - \mathbf{s}')$ is a valid correlation function then $e^{\sigma^2 \rho(\mathbf{s} - \mathbf{s}')}$ and $\sinh(\sigma^2 \rho(\mathbf{s} - \mathbf{s}'))$ is are valid covariance functions.

2. If $c(\mathbf{h}; \gamma)$ is valid for $\gamma \in \Gamma$ and ν is a positive measure on Γ, then $c_\nu(\mathbf{h} = \int_\Gamma c(\mathbf{h}; \gamma)\nu(d\gamma)$ is valid provided the integral exists for all $\mathbf{h}$

3. Suppose $Y(\mathbf{s}$ is a Gaussian process with mean surface $\mu(\mathbf{s})$ and covariance function $c(\mathbf{s}, \mathbf{s}')$. Let $Z(\mathbf{s})$ be the induced log Gaussian process, i.e., $Y(\mathbf{s}) = \log Z(\mathbf{s})$. Find the mean surface and covariance function for the $Z(\mathbf{s})$ process. If $Y(\mathbf{s})$ is stationary, is $Z(\mathbf{s})$ necessarily stationary?

4. Suppose $W(\mathbf{s})$ is a mean 0 stationary Gaussian process with correlation function $\rho(\mathbf{h})$ Let $Y(\mathbf{s}) = \frac{\sigma W(\mathbf{s})}{\sqrt{\lambda}}$. If $\lambda \sim Ga(\nu/2, \nu/2)$, when will the marginal process have a covariance function? If it does, what is the covariance function?

5. Some mixing results

   (a) If $Z \sim N(0, 1)$ and $V \sim Ga(r/2, r/2)$ independent of Z, show that $Y = \sigma z/\sqrt{V} \sim \sigma t_r$ where t_r is a t-distribution with r d.f. So, if $Z(\mathbf{s})$ is a mean 0 Gaussian process with covariance function/correlation function $\rho(\mathbf{s} - \mathbf{s}')$, then we will define $Y(\mathbf{s}) = \sigma Z(\mathbf{s})/sqrtV$ as a t-process with r d.f. Obtain the joint distribution of $Y(\mathbf{s})$ and $Y(\mathbf{s}')$. Obtain the covariance function of the process, provided $r > 2$.

   (b) If $Z \sim N(0, 1)$ and $V \sim \exp(1)$ independent of Z, show that $Y = \sigma Z/\sqrt{2V} \sim$ Laplace$(0, \sigma)$ where Laplace(μ, σ) is the Laplace (double exponential) distribution with location μ and scale σ. So, if $Z(\mathbf{s})$ is a mean 0 Gaussian process with covariance function/correlation function $\rho(\mathbf{s} - \mathbf{s}')$, then we will define $Y(\mathbf{s}) = \sigma Z(\mathbf{s})/\sqrt{2V}$ as a Laplace process. Obtain the joint distribution of $Y(\mathbf{s})$ and $Y(\mathbf{s}')$. Obtain the covariance function of the process.

6. Show, for fractional Brownian motion in two dimensions, if the Hurst index is greater than 1/2 we have positively correlated increments while if it is less than 1/2 we have negatively correlated increments.

7. Consider the *triangular* (or "tent") covariance function,

$$C(\|h\|) = \begin{cases} \sigma^2(1 - \|h\|/\delta) & \text{if } \|h\| \leq \delta, \ \sigma^2 > 0, \ \delta > 0, \\ 0 & \text{if } \|h\| > \delta \end{cases}.$$

It is valid in one dimension. (The reader can verify that it is the characteristic function of the density function $f(x)$ proportional to $[1 - \cos(\delta x)]/\delta x^2$.) Now in two dimensions, consider a 6×8 grid with locations $\mathbf{s}_{jk} = (j\delta/\sqrt{2}, k\delta/\sqrt{2})$, $j = 1, \ldots, 6$, $k = 1, \ldots, 8$. Assign a_{jk} to $\mathbf{s}_{jk}$ such that $a_{jk} = 1$ if $j + k$ is even, $a_{jk} = -1$ if $j + k$ is odd. Show that $Var[\Sigma a_{jk} Y(\mathbf{s}_{jk})] < 0$, and hence the triangular covariance function is *invalid* in two dimensions.

8. The *turning bands method* (Christakos, 1984; Stein, 1999a) is a technique for creating stationary covariance functions on $\Re^r$. Let $\mathbf{u}$ be a random unit vector on $\Re^r$ (by random we mean that the coordinate vector that defines $\mathbf{u}$ is randomly chosen on the surface of the unit sphere in $\Re^r$). Let $c(\cdot)$ be a valid stationary covariance function on $\Re^1$, and let $W(t)$ be a mean 0 process on $\Re^1$ having $c(\cdot)$ as its covariance function. Then for any location $\mathbf{s} \in \Re^r$, define

$$Y(\mathbf{s}) = W(\mathbf{s}^{\mathrm{T}}\mathbf{u}) \ .$$

Note that we can think of the process either conditionally given $\mathbf{u}$, or marginally by integrating with respect to the uniform distribution for $\mathbf{u}$. Note also that $Y(\mathbf{s})$ has the possibly undesirable property that it is constant on planes (i.e., on $\mathbf{s}^{\mathrm{T}}\mathbf{u} = k$).

   (a) If W is a Gaussian process, show that, given $\mathbf{u}$, $Y(\mathbf{s})$ is also a Gaussian process and is stationary.

   (b) Show that marginally $Y(\mathbf{s})$ is *not* a Gaussian process, but is stationary. [*Hint:* Show that $\mathrm{Cov}(Y(\mathbf{s}), Y(\mathbf{s}')) = \mathrm{E}_{\mathbf{u}}c((\mathbf{s} - \mathbf{s}')^{\mathrm{T}}\mathbf{u}).$]

   (c) If $c(\cdot)$ is isotropic, then so is $Cov(Y(\mathbf{s}), Y(\mathbf{s}'))$

9. (a) Based on (3.1), show that $c_{12}(\mathbf{h})$ is a valid correlation function; i.e., that G is a bounded, positive, symmetric about 0 measure on $\Re^2$.

   (b) Show further that if c_1 and c_2 are isotropic, then c_{12} is.

10. Show by direct calculation that $C(\mathbf{s}, \mathbf{s}') = \int_{Re^2} K_{\mathbf{s}}(\mathbf{u})K_{\mathbf{s}'}(\mathbf{u})d\mathbf{u}$ is a valid covariance function for $\mathbf{s}, \mathbf{s}' \in Re^p$ for any positive integer p.

11. Suppose we take $K_{\mathbf{s}}(\mathbf{u}) = |V_{\mathbf{s}}|^{-\frac{1}{2}}\exp(-(\mathbf{s} - \mathbf{u})^{\mathrm{T}}V_{\mathbf{s}}(\mathbf{s} - \mathbf{u}))$, then, if $C(\mathbf{s}, \mathbf{s}') = \int_{Re^2} K_{\mathbf{s}}(\mathbf{u})K_{\mathbf{s}'}(\mathbf{u})d\mathbf{u}$, show that

$$C(\mathbf{s}, \mathbf{s}') = |V_{\mathbf{s}}|^{\frac{1}{2}}|V_{\mathbf{s}'}|^{\frac{1}{2}}\left|\frac{V_{\mathbf{s}} + V_{\mathbf{s}'}}{2}\right|^{-\frac{1}{2}}\exp(-Q(\mathbf{s}, \mathbf{s}'))$$

12. (a) Derive the variance and covariance relationships given in (3.11).

    (b) Derive the variance and covariance relationships given in (3.13).

Chapter 4

Basics of areal data models

We now turn to some exploratory tools and modeling approaches that are customarily applied to data collected for areal units. Again, this literature is sometimes referred to as discrete spatial data modeling to reflect the fact that we are only specifying a joint model for a finite set of random variables. We have in mind general, possibly irregular geographic units, but of course include the special case of regular grids of cells (pixels). Indeed, many of the ensuing models have been proposed for regular lattices of points and parameters, and sometimes even for point-referenced data (see Chapter 13 on the problem of inverting very large matrices).

In the context of areal units the general inferential issues are the following:

(i) Is there spatial pattern? If so, how strong is it? Intuitively, "spatial pattern" suggests that measurements for areal units which are near to each other will tend to take more similar values than those for units far from each other. Though you might "know it when you see it," this notion is evidently vague and in need of quantification. Indeed, with independent measurements for the units, we expect to see *no pattern*, i.e., a completely random arrangement of larger and smaller values. But again, randomness will inevitably produce some patches of similar values.

(ii) Do we want to smooth the data? If so, how much? Suppose, for example, that the measurement for each areal unit is a count, say, a number of cancers. Even if the counts were independent, and perhaps even after population adjustment, there would still be extreme values, as in any sample. Are the observed high counts more elevated than would be expected by chance? If we sought to present a surface of expected counts we might naturally expect that the high values would tend to be pulled down, the low values to be pushed up. This is the notion of smoothing. No smoothing would present a display using simply the observed counts. Maximal smoothing would result in a shared common value for all units, clearly excessive. Suitable smoothing would fall somewhere in between, and take the spatial arrangement of the units into account.

How much smoothing is appropriate is not readily defined. In particular, for model-based smoothers such as we describe below, it is not evident what the extent of smoothing is, or how to control it. Specification of a utility function for smoothing [as attempted in Stern and Cressie, 1999] would help to address these questions but does not seem to be considered in practice.

(iii) For a new areal unit or set of units, how can we infer about what data values we expect to be associated with these units? That is, if we modify the areal units to new units, e.g., from zip codes to census block groups, what can we say about say, the cancer counts we expect for the latter given those for the former? This is the so-called *modifiable areal unit problem (MAUP)*, which historically (and in most GIS software packages) is handled by crude areal allocation. Sections 7.2 and 7.3 propose a model-based methodology for handling this problem.

DOI: 10.1201/9781003401728-4

As a matter of fact, in order to facilitate interpretation and better assess uncertainty, we will suggest model-based approaches to treat the above issues, as opposed to the more descriptive or algorithmic methods that have dominated the literature and are by now widely available in GIS software packages. We will also introduce further flexibility into these models by examining them in the context of regression. That is, we assume interest in explaining the areal unit responses and that we have available potential covariates to do this. These covariates may be available at the same or at different scales from the responses, but, regardless, we will now question whether there remains any spatial structure adjusted for these explanatory variables. This suggests that we may not try to model the data in a spatial way directly, but instead introduce spatial association through random effects. This will lead to versions of generalized linear mixed models [Breslow and Clayton, 1993]. We will generally view such models in the hierarchical fashion that is the primary theme of this text.

4.1   Exploratory approaches for areal data

We begin with the presentation of some tools that can be useful in the initial exploration of areal unit data. The primary concept here is a *proximity matrix*, W. Given measurements $Y_1, \ldots, Y_n$ associated with areal units $1, 2, \ldots, n$, the entries w_{ij} in W spatially connect units i and j in some fashion. (Customarily w_{ii} is set to 0.) Possibilities include binary choices, i.e., $w_{ij} = 1$ if i and j share some common boundary, perhaps a vertex (as in a regular grid). Alternatively, w_{ij} could reflect "distance" between units, e.g., a decreasing function of intercentroidal distance between the units (as in a county or other regional map). But distance can be returned to a binary determination. For example, we could set $w_{ij} = 1$ for all i and j within a specified distance. Or, for a given i, we could get $w_{ij} = 1$ if j is one of the K nearest (in distance) neighbors of i. The preceding choices suggest that W would be symmetric. However, for irregular areal units, this last example provides a setting where this need not be the case. Also, the w_{ij}'s may be standardized by $\sum_j w_{ij} = w_{i+}$. If $\widetilde{W}$ has entries $\widetilde{w}_{ij} = w_{ij}/w_{i+}$, then evidently $\widetilde{W}$ is row stochastic, i.e., $\widetilde{W}\mathbf{1} = \mathbf{1}$, but now $\widetilde{W}$ need not be symmetric.

As the notation suggests, the entries in W can be viewed as weights. More weight will be associated with j's closer (in some sense) to i than those farther away from i. In this exploratory context (but, as we shall see, more generally) W provides the mechanism for introducing spatial structure into our formal modeling.

Lastly, working with distance suggests that we can define distance bins, say, $(0, d_1], (d_1, d_2], (d_2, d_3]$, and so on. This enables the notion of *first-order neighbors* of unit i, i.e., all units within distance d_1 of i, *second-order neighbors*, i.e., all units more than d_1 but at most d_2 from i, *third-order neighbors*, and so on. Analogous to W we can define $W^{(1)}$ as the proximity matrix for first-order neighbors. That is, $w_{ij}^{(1)} = 1$ if i and j are first-order neighbors, and equal to 0 otherwise. Similarly we define $W^{(2)}$ as the proximity matrix for second-order neighbors; $w_{ij}^{(2)} = 1$ if i and j are second-order neighbors, and 0 otherwise, and so on to create $W^{(3)}, W^{(4)}$, etc.

Of course, the most obvious exploratory data analysis tool for lattice data is a map of the data values. Figure 4.1 gives the statewide average verbal SAT exam scores as reported by the College Board and initially analyzed by Wall [2004]. Clearly these data exhibit strong spatial pattern, with midwestern states and Utah performing best, and coastal states and Indiana performing less well. Of course, before jumping to conclusions, we must realize there are any number of spatial covariates that may help to explain this pattern; for instance, the percentage of eligible students taking the exam (Midwestern colleges have historically relied on the ACT exam, not the SAT, and only the strongest students in these states typically take the latter exam). Still, the map of these raw data show significant spatial pattern.

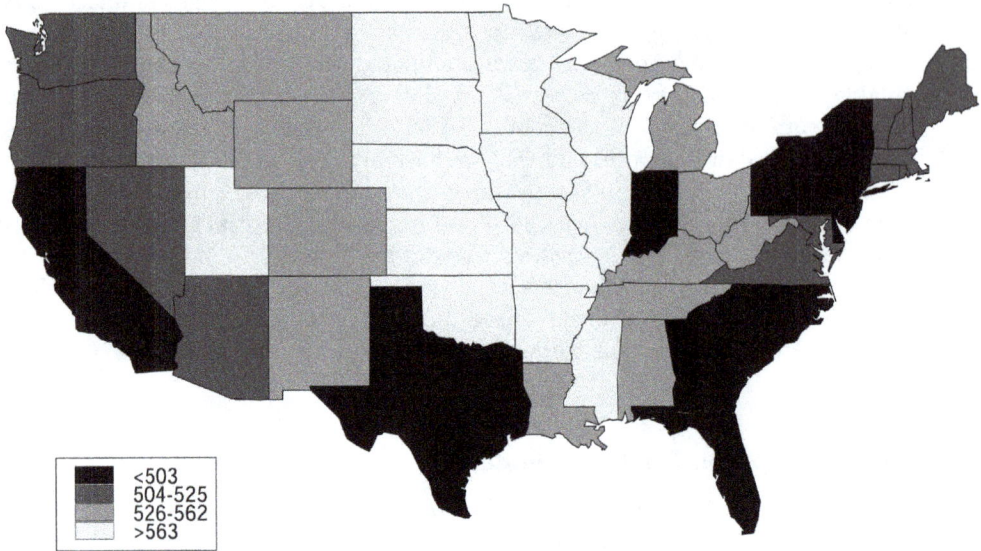

Figure 4.1 *Choropleth map of 1999 average verbal SAT scores, lower 48 U.S. states and the district of Columbia.*

4.1.1 Measures of spatial association

Two standard statistics that are used to measure the strength of spatial association among areal units are Moran's I and Geary's C [see, e.g., Ripley, 1981, Section 5.4]. These are spatial analogues of statistics for measuring association in time series, the lagged autocorrelation coefficient and the Durbin-Watson statistic, respectively. They can also be seen to be areal unit analogues of the empirical estimates for the correlation function and the variogram, respectively. Recall that, for point-referenced data, the empirical covariance function (2.12) and semivariogram (2.9), respectively, provide customary nonparametric estimates of these measures of association.

Moran's I takes the form

$$I = \frac{n \sum_i \sum_j w_{ij}(Y_i - \overline{Y})(Y_j - \overline{Y})}{\left(\sum_{i \neq j} w_{ij}\right) \sum_i (Y_i - \overline{Y})^2} . \tag{4.1}$$

I is not strictly supported on the interval $[-1, 1]$. It is evidently a ratio of quadratic forms in $\mathbf{Y}$, which provides the idea for obtaining approximate first and second moments through the delta method [see, e.g., Agresti, 2002, Chapter 14]. Moran shows under the null model where the Y_i are i.i.d., I is asymptotically normally distributed with mean $-1/(n-1)$ and a rather unattractive variance of the form

$$\text{Var}(I) = \frac{n^2(n-1)S_1 - n(n-1)S_2 - 2S_0^2}{(n+1)(n-1)^2 S_0^2} . \tag{4.2}$$

In (4.2), $S_0 = \sum_{i \neq j} w_{ij}$, $S_1 = \frac{1}{2}\sum_{i \neq j}(w_{ij} + w_{ji})^2$, and $S_2 = \sum_k (\sum_j w_{kj} + \sum_i w_{ik})^2$. We recommend the use of Moran's I as an exploratory measure of spatial association, rather than as a "test of spatial significance."

For the data mapped in Figure 4.1, we used the `moran.test` function in the spdep package in R to obtain a value for Moran's I of 0.6125, a reasonably large value. The associated standard error estimate of 0.0979 suggests very strong evidence against the null hypothesis of no spatial correlation in these data.

Geary's C takes the form

$$C = \frac{(n-1)\sum_i\sum_j w_{ij}(Y_i - Y_j)^2}{2\left(\sum_{i\neq j} w_{ij}\right)\sum_i(Y_i - \overline{Y})^2} \, . \tag{4.3}$$

C is never negative, and has mean 1 for the null model; *small* values (i.e., between 0 and 1) indicate *positive* spatial association. Also, C is a ratio of quadratic forms in $\mathbf{Y}$ and, like I, is asymptotically normal if the Y_i are i.i.d. We omit details of the distribution theory, recommending the interested reader to Cliff and Ord [1973] or [Ripley, 1981, p. 99].

Using the `geary.test` function on the SAT verbal data in Figure 4.1, we obtained a value of 0.3577 for Geary's C, with an associated standard error estimate of 0.0984. Again, the marked departure from the mean of 1 indicates strong positive spatial correlation in the data.

Convergence to asymptotic normality for a ratio of quadratic forms is extremely slow. We may believe the significant rejection of independence using the asymptotic theory for the example above because the results are so extreme. However, if one truly seeks to run a significance test using (4.1) or (4.3), our recommendation is a Monte Carlo approach. Under the null model the distribution of I (or C) is invariant to permutations of the Y_i's. The exact null distribution of I (or C) requires computing its value under all $n!$ permutation of the Y_i's, infeasible for n in practice. However, a Monte Carlo sample of say 1000 permutations, including the observed one, will position the observed I (or C) relative to the remaining 999, to determine whether it is extreme (perhaps via an empirical p-value). Again using `spatial.cor` function on our SAT verbal data, we obtained empirical p-values of 0 using both Moran's I and Geary's C; *no* random permutation achieved I or C scores as extreme as those obtained for the actual data itself.

A further display that can be created in this spirit is the *correlogram*. Working with say I, in (4.1) we can replace w_{ij} with the previously defined $w_{ij}^{(1)}$ and compute say $I^{(1)}$. Similarly, we can replace w_{ij} with $w_{ij}^{(2)}$ and obtain $I^{(2)}$. A plot of $I^{(r)}$ vs. r is called a correlogram and, if spatial pattern is present, is expected to decline in r initially and then perhaps vary about 0. Evidently, this display is a spatial analogue of a temporal lag autocorrelation plot [e.g., see Carlin and Louis, 2008, p. 181]. In practice, the correlogram tends to be very erratic and its information context is often not clear.

With large, regular grids of cells as we often obtain from remotely sensed imagery, it may be of interest to study spatial association in a particular direction (e.g., east-west, north-south, southwest-northeast, etc.). Now the spatial component reduces to one dimension and we can compute lagged autocorrelations (lagged appropriately to the size of the grid cells) in the specific direction. An analog of this was proposed for the case where the Y_i are binary responses (e.g., presence or absence of forest in the cell) by Agarwal et al. [2002]. In particular, Figure 4.2 shows rasterized maps of binary land use classifications for roughly 25,000 1 km $\times$ 1 km pixels in eastern Madagascar [see Agarwal et al., 2002, as well as Section 7.5 for further discussion].

While the binary map in Figure 4.2 shows spatial pattern in land use, we develop an additional display to provide quantification. For data on a regular grid or lattice, we calculate binary analogues of the sample autocovariances using the 1km $\times$ 1 km resolution with four illustrative directions: East (E), Northeast (NE), North (N), and Northwest (NW). In particular for any pair of pixels, we can identify say a Euclidean distance and direction between them by labeling one as X and the other as Y, creating a correlated binary pair. Then, we can go to the lattice and identify all pairs which share the same distance and direction. The collection all such (X,Y) pairs yields a 2 $\times$ 2 table of counts (with table

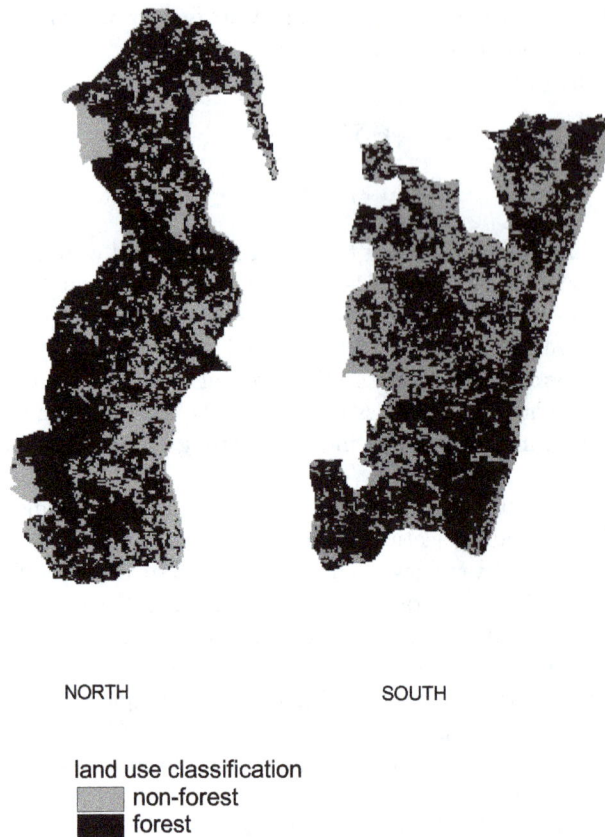

NORTH   SOUTH

land use classification
   non-forest
   forest

Figure 4.2 *Rasterized north and south regions (1 km × 1 km) with binary land use classification overlaid.*

cells labeled as X=0, Y=0; X=0, Y=1; X=1, Y=0, X=1, Y=1). The resultant log-odds ratio measures the association between pairs in that direction at that distance. (Note that if we followed the same procedure but reversed direction, e.g., changed from E to W, the corresponding log odds ratio would be unchanged.)

In Figure 4.3, we plot log odds ratio against direction for each of the four directions. Note that the spatial association is quite strong, requiring a distance of at least 40 km before it drops to essentially 0. This suggests that we would not lose much spatial information if we work with the lower (4 km × 4 km) resolution. In exchange we obtain a richer response variable (17 ordered levels, indicating the number of forested cells from 0 to 16) and a substantial reduction in the number of pixels (from 26,432 to 1,652 in the north region, from 24,544 to 1,534 in the south region) to facilitate model fitting.

4.1.2 *Spatial smoothers*

Recall from the beginning of this chapter that often a goal for, say, a choropleth map of the Y_i's is *smoothing*. Depending upon the number of classes used to make the map, there is already some implicit smoothing in such a display (although this is not *spatial* smoothing, of course).

The W matrix directly provides a spatial smoother; that is, we can replace Y_i by $\widehat{Y}_i = \sum_j w_{ij} Y_j / w_{i+}$. This ensures that the value for areal unit i "looks like" its neighbors, and

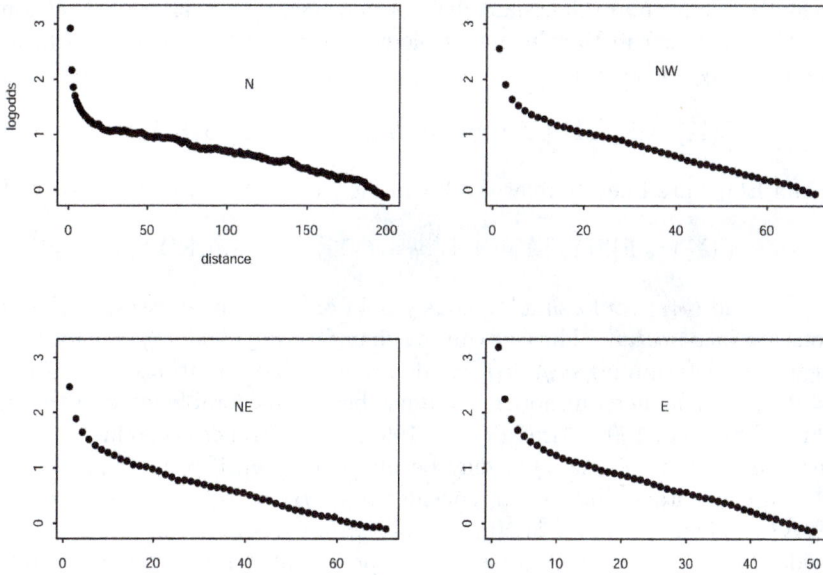

Figure 4.3 *Land use log-odds ratio versus distance in four directions.*

that the more neighbors we use in computing $\widehat{Y}_i$, the more smoothing we will achieve. In fact, $\widehat{Y}_i$ may be viewed as an unusual smoother in that it ignores the value actually observed for unit i. As such, we might revise the smoother to

$$\widehat{Y}_i^* = (1 - \alpha)Y_i + \alpha\widehat{Y}_i \,, \tag{4.4}$$

where $\alpha \in (0, 1)$. Working in an exploratory mode, various choices may be tried for α, but for any of these, (4.4) is a familiar *shrinkage* form. Thus, under a specific model with a suitable loss function, an optimal α could be sought. Finally, the form (4.4), viewed generally as a linear combination of the Y_j, is customarily referred to as a *filter* in the GIS literature. In fact, such software will typically provide choices of filters, and even a default filter to automatically smooth maps.

In Section 5.1 we will present a general discussion revealing how smoothing (shrinkage) emerges as a byproduct of the hierarchical models we propose to use to explain the Y_i. In particular, when W is used in conjunction with a stochastic model (as in Section 4.3), the $\widehat{Y}_i$ are updated across i and across Monte Carlo iterations as well. So the observed Y_i will affect the eventual $\widehat{Y}_i$; we achieve model-driven smoothing and a "manual" inclusion of Y_i as in (4.4) with some choice of α is unnecessary.

4.2 Brook's lemma and Markov random fields

A useful technical result for obtaining the joint distribution of the Y_i's in some of the models we discuss below is *Brook's Lemma* [Brook, 1964]. The usefulness of this lemma is exposed in Besag's seminal paper on conditionally autoregressive models [Besag, 1974].

It is clear that given $p(y_1, \ldots, y_n)$, the so-called *full conditional* distributions, $p(y_i \mid y_j, j \neq i)$, $i = 1, \ldots, n$, are uniquely determined. Brook's Lemma demonstrates the converse and, in fact, enables us to constructively retrieve the unique joint distribution determined by these full conditionals. But first, it is also clear that we cannot write down an arbitrary set of full conditional distributions and assert that they uniquely determine the joint distribution. To see this, let $Y_1 \mid Y_2 \sim N(\alpha_0 + \alpha_1 Y_2, \sigma_1^2)$ and let $Y_2 \mid Y_1 \sim N(\beta_0 + \beta_1 Y_1^3, \sigma_2^2)$, where N denotes the normal (Gaussian) distribution. intuitively, it seems that a mean for Y_1 given Y_2 which is linear in Y_2 is incompatible with a mean for Y_2 given Y_1 which is linear in Y_1^3. More formally, we see that

$$E(Y_1) = E[E(Y_1 \mid Y_2)] = E[\alpha_0 + \alpha_1 Y_2] = \alpha_0 + \alpha_1 E(Y_2) , \qquad (4.5)$$

i.e., $E(Y_1)$ and $E(Y_2)$ are linearly related. But in fact, it must also be the case that

$$E(Y_2) = E[E(Y_2 \mid Y_1)] = E[\beta_0 + \beta_1 Y_1] = \beta_0 + \beta_1 E(Y_1^3) . \qquad (4.6)$$

Equations (4.5) and (4.6) could simultaneously hold only in trivial cases, so the two mean specifications are *incompatible*. Thus we can say that $f(y_1 \mid y_2)$ and $f(y_2 \mid y_1)$ are incompatible with regard to determining $p(y_1, y_2)$. We do not examine conditions for compatibility of conditional distributions here, although there has been considerable work in this area [see, e.g., Arnold and Strauss, 1991, Arnold et al., 1999, and references therein].

Another point is that $p(y_1 \ldots, y_n)$ may be improper even if $p(y_i \mid y_j, j \neq i)$ is proper for all i. As an elementary illustration, consider $p(y_1, y_2) \propto \exp[-\frac{1}{2}(y_1 - y_2)^2]$. Evidently $p(y_1 \mid y_2)$ is $N(y_2, 1)$ and $p(y_2 \mid y_1)$ is $N(y_1, 1)$, but $p(y_1, y_2)$ is improper. Casella and George [1992] provide a similar example in a bivariate exponential (instead of normal) setting.

Brook's Lemma notes that

$$
\begin{aligned}
p(y_1, \ldots, y_n) \;=\; &\frac{p(y_1 \mid y_2, \ldots, y_n)}{p(y_{10} \mid y_2, \ldots, y_n)} \cdot \frac{p(y_2 \mid y_{10}, y_3, \ldots, y_n)}{p(y_{20} \mid y_{10}, y_3, \ldots, y_n)} \\
&\ldots \frac{p(y_n \mid y_{10}, \ldots, y_{n-1,0})}{p(y_{n0} \mid y_{10}, \ldots, y_{n-1,0})} \cdot p(y_{10}, \ldots, y_{n0}) ,
\end{aligned}
\qquad (4.7)
$$

an identity which is easily checked (Exercise 1). Here, $\mathbf{y}_0 = (y_{10}, \ldots, y_{n0})'$ is any fixed point in the support of $p(y_1, \ldots, y_n)$. Hence $p(y_1, \ldots, y_n)$ is determined by the full conditional distributions, since apart from the constant $p(y_{10}, \ldots, y_{n0})$ they are the only objects appearing on the right-hand side of (4.7). Hence the joint distribution is determined up to a proportionality constant. If $p(y_1, \ldots, y_n)$ is improper then this is, of course, the best we can do; if $p(y_1, \ldots, y_n)$ is proper then the fact that it integrates to 1 determines the constant. Perhaps most important is the constructive nature of (4.7): we can create $p(y_1, \ldots, y_n)$ simply by calculating the product of ratios. For more on this point see Exercise 2

When the number of areal units is very large (say, a regular grid of pixels associated with an image or a large number of small geographic regions), we do not seek to write down the joint distribution of the Y_i. Rather we prefer to work (and model) exclusively with the n corresponding full conditional distributions. In fact, from a spatial perspective, we would imagine that the full conditional distribution for Y_i would be more "local," that is, it should really depend only upon the neighbors of cell i. Adopting some definition of a neighbor structure (e.g., the one setting $W_{ij} = 1$ or 0 depending on whether i and j are adjacent or not), let ∂_i denote the set of neighbors of cell i.

Next suppose we specify a set of full conditional distributions for the Y_i such that

$$p(y_i \mid y_j, j \neq i) = p(y_i \mid y_j, j \in \partial_i) \qquad (4.8)$$

A critical question to ask is whether a specification such as (4.8) uniquely determines a joint distribution for $Y_1, \ldots Y_n$. That is, we do not need to see the explicit form of this distribution. We merely want to be assured that if, for example, we implement a Gibbs sampler (see Subsection 5.5.1.1) to simulate realizations from the joint distribution, then there is indeed a unique stationary distribution for this sampler.

The notion of using *local* specification to determine a joint (or global) distribution in the form (4.8) is referred to as a *Markov random field* (MRF). There is by now a substantial literature in this area, with Besag [1974] being a good place to start. Geman and Geman [1984] provide the next critical step in the evolution, while Kaiser and Cressie [2000] offer a more contemporary perspective and also provide further references [also see Rue and Held, 2005, and references therein].

A critical concept in this regard is that of a *clique*. A clique is a set of cells (equivalently, indices) such that each element is a neighbor of every other element. With n cells, depending upon the definition of the neighbor structure, cliques can possibly be of size 1, 2, and so on up to size n. A *potential function* (or simply *potential*) of order k is a function of k arguments that is exchangeable in these arguments. The arguments of the potential would be the values taken by variables associated with the cells for a clique of size k. For continuous Y_i, a customary potential on cliques of size $k = 2$ is $(Y_i - Y_j)^2$ when $i \sim j$. (We use the notation $i \sim j$ if i is a neighbor of j and j is a neighbor of i.) In fact, we may also view this potential as a sum of a potential on cliques of size $k = 1$, i.e., Y_i^2 with potential on cliques of size $k = 2$, i.e., $Y_i Y_j$. For, say, binary Y_i, a common potential on cliques of size $k = 2$ is

$$I(Y_i = Y_j) = Y_i Y_j + (1 - Y_i)(1 - Y_j) ,$$

where again $i \sim j$ and I denotes the indicator function. Next, we define a *Gibbs distribution* as follows: $p(y_1, \ldots, y_n)$ is a Gibbs distribution if it is a function of the Y_i only through potentials on cliques. That is,

$$p(y_1, \ldots, y_n) \propto \exp \left\{ \gamma \sum_k \sum_{\boldsymbol{\alpha} \in \mathcal{M}_k} \phi^{(k)}(y_{\alpha_1}, y_{\alpha_2}, \ldots, y_{\alpha_k}) \right\} . \tag{4.9}$$

Here, $\phi^{(k)}$ is a potential of order k, $\mathcal{M}_k$ is the collection of all subsets of size k from $\{1, 2, \ldots, n\}$, $\boldsymbol{\alpha} = (\alpha_1, \ldots, \alpha_k)'$ indexes this set, and $\gamma > 0$ is a scale (or "temperature") parameter. If we only use cliques of size 1, we see that we obtain an independence model, evidently not of interest. When $k = 2$, we achieve spatial structure. In practice, cliques with $k = 3$ or more are rarely used, introducing complexity with little benefit. So, throughout this book, only cliques of order less than or equal to 2 are considered.

Informally, the *Hammersley-Clifford Theorem* [Besag, 1974, Clifford, 1990, see] demonstrates that if we have an MRF, i.e., if (4.8) defines a unique joint distribution, then this joint distribution is a Gibbs distribution. That is, it is of the form (4.9), with all of its "action" coming in the form of potentials on cliques. [Cressie, 1993, pp. 417–18] offers a proof of this theorem, and mentions that its importance for spatial modeling lies in its limiting the complexity of the conditional distributions required, i.e., full conditional distributions can be specified locally. Geman and Geman [1984] provided essentially the converse of the Hammersley-Clifford theorem. If we begin with (4.9) we have determined an MRF. As a result, they argued that to sample a Markov random field, one could sample from its associated Gibbs distribution, hence coining the term "Gibbs sampler."

For continuous data on $\Re^1$, a common choice for the joint distribution is a pairwise difference form

$$p(y_1, \ldots, y_n) \propto \exp \left\{ -\frac{1}{2\tau^2} \sum_{i,j} (y_i - y_j)^2 I(i \sim j) \right\} . \tag{4.10}$$

Distributions such as (4.10) will be the focus of the next section. For the moment, we merely note that it is a Gibbs distribution on potentials of order 1 and 2 and that

$$p(y_i \mid y_j, j \neq i) = N \left(\sum_{j \in \partial_i} y_i / m_i , \tau^2 / m_i \right) , \tag{4.11}$$

where m_i is the number of neighbors of cell i. The distribution in (4.11) is clearly of the form (4.8) and shows that the mean of Y_i is the average of its neighbors, exactly the sort of local smoother we discussed in the section on spatial smoothers.

4.3   Conditionally autoregressive (CAR) models

Although they were introduced by Besag [1974] 50 years ago, conditionally autoregressive (CAR) models have enjoyed a dramatic increase in usage only in the 1990's and onward. This resurgence arises from their convenient employment in the context of Gibbs sampling and more general Markov chain Monte Carlo (MCMC) methods for fitting certain classes of hierarchical spatial models (seen, e.g., in Section 6.8.3).

4.3.1   The Gaussian case

We begin with the Gaussian (or *autonormal*) case. Suppose we set

$$Y_i \,|\, y_j, j \neq i \sim N\left(\sum_j b_{ij} y_j \,,\, \tau_i^2\right), \; i = 1, \ldots, n\,. \tag{4.12}$$

These full conditionals are compatible, so through Brook's Lemma we can obtain

$$p(y_1, \ldots, y_n) \propto \exp\left\{-\frac{1}{2}\mathbf{y}' D^{-1}(I - B)\mathbf{y}\right\}, \tag{4.13}$$

where $B = \{b_{ij}\}$ and D is diagonal with $D_{ii} = \tau_i^2$. Expression (4.13) suggests a joint multivariate normal distribution for $\mathbf{Y}$ with mean $\mathbf{0}$ and variance matrix $\Sigma_{\mathbf{y}} = (I - B)^{-1}D$.

But we are getting ahead of ourselves. First, we need to ensure that $D^{-1}(I - B)$ is symmetric. The resulting conditions are

$$\frac{b_{ij}}{\tau_i^2} = \frac{b_{ji}}{\tau_j^2} \quad \text{for all } i, j\,. \tag{4.14}$$

Evidently, from (4.14), B need not be symmetric. Returning to our proximity matrix W (which we assume to be symmetric), suppose we set $b_{ij} = w_{ij}/w_{i+}$ and $\tau_i^2 = \tau^2/w_{i+}$. Then (4.14) is satisfied and (4.12) yields $p(y_i | y_j, j \neq i) = N\left(\sum_j w_{ij} y_j / w_{i+}\,,\, \tau^2/w_{i+}\right)$. Also, (4.13) becomes

$$p(y_1, \ldots, y_n) \propto \exp\left\{-\frac{1}{2\tau^2}\mathbf{y}'(D_w - W)\mathbf{y}\right\}, \tag{4.15}$$

where D_w is diagonal with $(D_w)_{ii} = w_{i+}$.

Now a second aspect is noticed. $(D_w - W)\mathbf{1} = \mathbf{0}$, i.e., $\Sigma_{\mathbf{y}}^{-1}$ is singular, so that $\Sigma_{\mathbf{y}}$ does not exist and the distribution in (4.15) is improper. (The reader is encouraged to note the difference between the case of $\Sigma_{\mathbf{y}}^{-1}$ singular and the case of $\Sigma_{\mathbf{y}}$ singular. With the former we have a density function but one that is not integrable; effectively we have too many variables and we need a constraint on them to restore propriety. With the latter we have no density function but a proper distribution that resides in a lower dimensional space; effectively we have too *few* variables.) With a little algebra (4.15) can be rewritten as

$$p(y_1, \ldots, y_n) \propto \exp\left\{-\frac{1}{2\tau^2}\sum_{i \neq j} w_{ij}(y_i - y_j)^2\right\}. \tag{4.16}$$

This is a pairwise difference specification slightly more general than (4.10). But the impropriety of $p(\mathbf{y})$ is also evident from (4.16) since we can add any constant to all of the Y_i and

(4.16) is unaffected; the Y_i are not "centered." A constraint such as $\sum_i Y_i = 0$ would provide the needed centering. Thus we have a more general illustration of a joint distribution that is improper but has all full conditionals proper. The specification in (4.15) or (4.16) is often referred to as an *intrinsically autoregressive* (IAR) model.

As a result, $p(\mathbf{y})$ in (4.15) cannot be used as a model for data; data could not arise under an improper stochastic mechanism, and we cannot impose a constant center on randomly realized measurements. Hence, the use of an improper autonormal model must be relegated to a *prior* distributional specification. That is, it will be attached to random spatial effects introduced at the second stage of a hierarchical specification (again, see e.g., Section 6.8.3).

The impropriety in (4.15) can be remedied in an obvious way. Redefine $\Sigma_{\mathbf{y}}^{-1} = D_w - \rho W$ and choose ρ to make $\Sigma_{\mathbf{y}}^{-1}$ nonsingular. This is guaranteed if $\rho \in \left(1/\lambda_{(1)}, 1/\lambda_{(n)}\right)$, where $\lambda_{(1)} < \lambda_{(2)} < \cdots < \lambda_{(n)}$ are the ordered eigenvalues of $D_w^{-1/2} W D_w^{-1/2}$; see Exercise 5 Moreover, since $tr(D_w^{-1/2} W D_w^{-1/2}) = 0 = \sum_{i=1}^{n} \lambda_{(i)}$, $\lambda_{(1)} < 0$, $\lambda_{(n)} > 0$, and 0 belongs to $\left(1/\lambda_{(1)}, 1/\lambda_{(n)}\right)$.

Another elegant way to look at the propriety of CAR models is through the Gershgorin disk theorem [Golub and Van Loan, 2013, Horn and Johnson, 2012, Banerjee and Roy, 2014]. This famous theorem of linear algebra focuses on so-called *diagonal dominance* and, in its simplest form asserts that, for any symmetric matrix A, if all $a_{ii} > 0$ and $a_{ii} > \sum_{j \neq i} |a_{ij}|$, then A is positive definite. For instance, if $D_w^{-1}(I - B)$ is symmetric, then it is positive definite if, for each i, $\sum_{j \neq i} |B_{ij}| < 1$. With $D_w - \rho W$, a sufficient condition is that $|\rho| < 1$, weaker than the conditions created above. However, since we have positive definiteness for $\rho < 1$, impropriety at $\rho = 1$, this motivates us to examine the behavior of CAR models for ρ near 1, as we do below.

Returning to the unscaled situation, ρ can be viewed as an additional parameter in the CAR specification, enriching this class of spatial models. Furthermore, $\rho = 0$ has an immediate interpretation: the Y_i become independent $N(0, \tau^2/w_{i+})$. If ρ is not included, independence cannot emerge as a limit of (4.15). (Incidentally, this suggests a clarification of the role of τ^2, the variance parameter associated with the full conditional distributions: the magnitude of τ^2 should *not* be viewed as, in any way, *quantifying* the strength of spatial association. Indeed if all Y_i are multiplied by c, τ^2 becomes $c\tau^2$ but the strength of spatial association among the Y_i is clearly unaffected.) Lastly, $\rho \sum_j w_{ij} Y_j / w_{i+}$ can be viewed as a *reaction function*, i.e., ρ is the expected proportional "reaction" of Y_i to $\sum_j w_{ij} Y_j / w_{i+}$. (This interpretation is more common in the SAR literature (Section 4.4).)

With these advantages plus the fact that $p(\mathbf{y})$ (or the Bayesian posterior distribution, if the CAR specification is used to model constrained random effects) is now proper, is there any reason not to introduce the ρ parameter? In fact, the answer may be yes. Under $\Sigma_{\mathbf{y}}^{-1} = D_w - \rho W$, the full conditional $p(y_i | y_j, j \neq i)$ becomes $N\left(\rho \sum_j w_{ij} y_j / w_{i+}, \tau^2/w_{i+}\right)$. Hence we are modeling Y_i not to have a mean that is an average of its neighbors, but some *proportion* of this average. Does this enable any sensible spatial interpretation of the CAR model? Moreover, does ρ calibrate very well with any familiar interpretation of "strength of spatial association?" Fixing $\tau^2 = 1$ without loss of generality, we can simulate CAR realizations for a given n, W, and ρ. We can also compute for these realizations a descriptive association measure such as Moran's I or Geary's C. Here we do not present explicit details of the range of simulations we have conducted. However, for a 10×10 grid using a first-order neighbor system, when $\rho = 0.8$, I is typically 0.1 to 0.15; when $\rho = 0.9$, I is typically 0.2 to 0.25; and even when $\rho = 0.99$, I is typically at most 0.5. It thus appears that ρ can mislead with regard to the strength of association. Expressed in a different way, within a Bayesian framework, a prior on ρ that encourages a consequential amount of spatial association would place most of its mass near 1.

A related point is that if $p(\mathbf{y})$ is proper, the breadth of spatial pattern may be too limited. In the case where a CAR model is applied to random effects, an improper choice may actually enable a wider scope for posterior spatial pattern. As a result, we do not take a position with regard to propriety or impropriety in employing CAR specifications (though in the remainder of this text, we do sometimes attempt to illuminate relative advantages and disadvantages).

Referring to (4.12), we may write the entire system of random variables as

$$\mathbf{Y} = B\mathbf{Y} + \boldsymbol{\epsilon}, \quad \text{or equivalently,} \quad (I - B)\mathbf{Y} = \boldsymbol{\epsilon} \qquad (4.17)$$

In particular, the distribution for $\mathbf{Y}$ induces a distribution for $\boldsymbol{\epsilon}$. If $p(\mathbf{y})$ is proper then $\mathbf{Y} \sim N(\mathbf{0}, (I - B)^{-1}D)$ whence $\boldsymbol{\epsilon} \sim N(\mathbf{0}, D(I - B)^{\mathrm{T}})$, i.e., the components of $\boldsymbol{\epsilon}$ are not independent. Also, $\mathrm{Cov}(\boldsymbol{\epsilon}, \mathbf{Y}) = D$. The SAR specification in Section 4.4 reverses this specification, supplying a distribution for $\boldsymbol{\epsilon}$ which induces a distribution for $\mathbf{Y}$.

When $p(\mathbf{y})$ is proper we can appeal to standard multivariate normal distribution theory to interpret the entries in $\Sigma_{\mathbf{y}}^{-1}$. For example, $1/(\Sigma_{\mathbf{y}}^{-1})_{ii} = \mathrm{Var}(Y_i \mid Y_j, j \neq i)$. Of course with $\Sigma_{\mathbf{y}}^{-1} = D^{-1}(I - B)$, $(\Sigma_{\mathbf{y}}^{-1})_{ii} = 1/\tau_i^2$ providing immediate agreement with (4.12). But also, if $(\Sigma_{\mathbf{y}}^{-1})_{ij} = 0$, then Y_i and Y_j are conditionally independent given $Y_k, k \neq i, j$, a fact you are asked to show in Exercise 10 Hence if any $b_{ij} = 0$, we have conditional independence for that pair of variables. Connecting b_{ij} to w_{ij} shows that the choice of neighbor structure implies an associated collection of conditional independences. With first-order neighbor structure, all we are asserting is a spatial illustration of the local Markov property (Whittaker, 1990, p. 68).

We conclude this subsection with four remarks. First, one can directly introduce a regression component into (4.12), e.g., a term of the form $\mathbf{x}_i'\boldsymbol{\beta}$. Conditional on $\boldsymbol{\beta}$, this does not affect the association structure that ensues from (4.12); it only revises the mean structure. However, we omit details here (the interested reader can consult Besag, 1974), since we will only use the autonormal CAR as a distribution for spatial random effects. These effects are added onto the regression structure for the mean on some transformed scale (again, see Section 6.8.3).

We also note that in suitable contexts it may be appropriate to think of $\mathbf{Y}_i$ as a vector of dependent areal unit measurements or, in the context of random effects, as a vector of dependent random effects associated with an areal unit. This leads to the specification of multivariate conditionally autoregressive (MCAR) models, which is the subject of Section 11.1. From a somewhat different perspective, $\mathbf{Y}_i$ might arise as $(Y_{i1}, \ldots, Y_{iT})^{\mathrm{T}}$ where Y_{it} is the measurement associated with areal unit i at time t, $t = 1, \ldots, T$. Now we would of course think in terms of spatiotemporal modeling for Y_{it}. This is the subject of Section 12.7.

Thirdly, a (proper) CAR model can in principle be used for point-level data, taking w_{ij} to be, say, an inverse distance between points i and j. However, unlike the spatial prediction described in Section 2.5, now spatial prediction becomes *ad hoc*. That is, to predict at a new site Y_0, we might specify the distribution of Y_0 given $Y_1, \ldots, Y_n$ to be a normal distribution, such as a $N\left(\rho \sum_j w_{0j}y_j/w_{0+}, \ \tau^2/w_{0+}\right)$. Note that this determines the joint distribution of $Y_0, Y_1, \ldots, Y_n$. However, this joint distribution is *not* the CAR distribution that would arise by specifying the full conditionals for $Y_0, Y_1, \ldots, Y_n$ and using Brook's Lemma, as in constructing (4.15). In this regard, we can not "marginalize" a CAR model. That is, suppose we specify a CAR model for say n areal units and we want a CAR model for a subset of them, say the first m. If we consider the multivariate normal distribution with upper left $m \times m$ block $(D^{-1}(I - B))_m^{-1}$, the inverse of this matrix need not look anything like the CAR model for these m units.

Finally, Gaussian Markov random fields can introduce proximities more general than those that we have discussed here. In particular, working with a regular lattice, there is much scope for further theoretical development. For instance, [Rue and Held, 2005, p. 114] describe the derivation of the following model weights based on the forward difference analogue of penalizing the derivatives of a surface used to specify the thin plate spline. They consider twelve neighbors of a given point. The north, east, south, and west neighbors each receive a weight of $+8$, the northeast, southeast, southwest, and northwest neighbors, each receive a weight of -2 and the "two away" north, east, south, and west neighbors, each receive a weight of -1. Thus, the $w_{i+} = 20$. These weights would possibly viewed as unusual with regard to spatial smoothing, in particular the negative values, but, again, they do have a probabilistic justification through the two-dimensional random walk on the lattice. Moreover, they do play a role in Markov random field approximation to Gaussian processes. Some care needs to be taken with regard to edge specifications. See further discussion in Rue and Held [2005].

4.3.2 The non-Gaussian case

If one seeks to model the data directly using a CAR specification, then in many cases a normal distribution would not be appropriate. Binary response data and sparse count data are two examples. In fact, one might imagine any exponential family model as a first-stage distribution for the data. Here, we focus on the case where the Y_i are binary variables and present the so-called autologistic CAR model [(historically, the Ising model; see Brush, 1967]. This model has received attention in the literature [Heikkinen and Hogmander, 1994, Hogmander and Møller, 1995, Hoeting et al., 2000].. Ignoring covariates for the moment, as we did with the CAR models above, consider the joint distribution

$$p(y_1, y_2, ..., y_n; \psi) \propto \exp(\psi \sum_{i,j} w_{ij} 1(y_i = y_j))$$

$$= \exp(\psi \sum_{i,j} w_{ij}(y_i y_j + (1 - y_i)(1 - y_j))). \qquad (4.18)$$

We immediately recognize this specification as a Gibbs distribution with a potential on cliques of order $k = 2$. Moreover, this distribution is always proper since it can take on only 2^n values. However, we will assume that ψ is an unknown parameter (how would we know it in practice?) and hence we will need to calculate the normalizing constant $c(\psi)$ in order to infer about ψ. But, computation of this constant requires summation over all of the 2^n possible values that $(Y_1, Y_2, ..., Y_n)$ can take on. Even for moderate sample sizes this will present computational challenges. Hoeting et al. [2000] propose approximations to the likelihood using a pseudo-likelihood and a normal approximation.

From (4.18) we can obtain the full conditional distributions for the Y_i's. In fact, $P(Y_i = 1 \mid y_j, j \neq i) = e^{\psi S_{i,1}}/(e^{\psi S_{i,0}} + e^{\psi(S_{i,1})})$ where $S_{i,1} = \sum_{j \sim i} 1(y_j = 1)$ and $S_{i,0} = \sum_{j \sim i} 1(y_j = 0)$ and $P(Y_i = 0 | y_j, j \neq i) = 1 - P(Y_i = 1 | y_j, j \neq i)$. That is, $S_{i,1}$ is the number of neighbors of i that are equal to 1 and $S_{i,0}$ is the number of neighbors of i that are equal to 0. We can see the role that ψ plays; larger values of ψ place more weight on matching. This is most easily seen through $\log \frac{P(Y_i=1 \mid y_j, j \neq i)}{P(Y_i=0 \mid y_j, j \neq i)} = \psi(S_{i,1} - S_{i,0})$. Since the full conditional distributions take on only two values, there are no normalizing issues with them.

Bringing in covariates is natural on the log scale, i.e.,

$$\log \frac{P(Y_i = 1 \mid y_j, j \neq i)}{P(Y_i = 0 \mid y_j, j \neq i)} = \psi(S_{i,1} - S_{i,0}) + \mathbf{X}_i^{\mathsf{T}} \boldsymbol{\beta}. \qquad (4.19)$$

Solving for $P(Y_i = 1 \mid y_j, j \neq i)$, we obtain

$$P(Y_i = 1 \mid y_j, j \neq i) = \frac{\exp\{\psi(S_{i,1} - S_{i,0}) + \mathbf{X}_i^{\mathsf{T}} \boldsymbol{\beta}\}}{1 + \exp\{(S_{i,1} - S_{i,0}) + \mathbf{X}_i^{\mathsf{T}} \boldsymbol{\beta}\}}.$$

Now, to update both ψ and $\boldsymbol{\beta}$, we will again need the normalizing constant, now $c(\psi, \boldsymbol{\beta})$. In fact, we leave as an exercise, the joint distribution of $(Y_1, Y_2, ..., Y_n)$ up to a constant.

The case where Y_i can take on one of several categorical values presents a natural extension to the autologistic model. If we label the (say) L possible outcomes as simply $1, 2, ..., L$, then we can define the joint distribution for $(Y_1, Y_2, \ldots, Y_n)$ exactly as in (4.18), i.e.

$$p(y_1, y_2, ..., y_n; \psi) \propto \exp\left(\psi \sum_{i,j} w_{ij} 1(y_i = y_j)\right) \qquad (4.20)$$

with w_{ij} as above. The distribution in (4.20) is referred to as a *Potts model* [Potts, 1952]. Now the distribution takes on L^n values; now, the calculation of the normalizing constant is even more difficult. Because of this challenge, fitting Potts models to data is rare in the literature; rather, it is customary to run a forward simulation using the Potts model since this only requires implementing a routine Gibbs sampler, updating the Y_i's (see Chapter 5) for a fixed ψ. However, one nice data analysis example is the allocation model in Green and Richardson [2002]. There, the Potts model is employed as a random effects specification, in a disease mapping context as an alternative to a CAR model.

4.4   Simultaneous autoregressive (SAR) models

Returning to (4.17), suppose that instead of letting $\mathbf{Y}$ induce a distribution for $\boldsymbol{\epsilon}$, we let $\boldsymbol{\epsilon}$ induce a distribution for $\mathbf{Y}$. Imitating usual autoregressive time series modeling, suppose we take the ϵ_i to be independent innovations. For a little added generality, assume that $\boldsymbol{\epsilon} \sim N\left(0, \tilde{D}\right)$ where $\tilde{D}$ is diagonal with $\left(\tilde{D}\right)_{ii} = \sigma_i^2$. (Note $\tilde{D}$ has no connection with D in Section 4.3; the B we use below may or may not be the same as the one we used in that section.) Analogous to (4.12), now $Y_i = \sum_j b_{ij} Y_j + \epsilon_i$, $i = 1, 2, ..., n$, with $\epsilon_i \sim N\left(0, \sigma_i^2\right)$ or, equivalently, $(I - B)\mathbf{Y} = \boldsymbol{\epsilon}$ with $\boldsymbol{\epsilon}$ distributed as above. Therefore, if $(I - B)$ is full rank,

$$\mathbf{Y} \sim N\left(\mathbf{0}, (I - B)^{-1} \tilde{D} \left((I - B)^{-1}\right)^{\mathrm{T}}\right). \qquad (4.21)$$

Also, $\text{Cov}(\boldsymbol{\epsilon}, \mathbf{Y}) = \tilde{D}(I - B)^{-1}$. If $\tilde{D} = \sigma^2 I$ then (4.21) simplifies to $\mathbf{Y} \sim N\left(\mathbf{0}, \sigma^2 \left[(I - B)(I - B)^{\mathrm{T}}\right]^{-1}\right)$. In order that (4.21) be proper, $I - B$ must be full rank but not necessarily symmetric. Two choices are most frequently discussed in the literature [e.g., Griffith, 1988]. The first assumes $B = \rho W$, where W is a so-called contiguity matrix, i.e., W has entries that are 1 or 0 according to whether or not unit i and unit j are direct neighbors (with $w_{ii} = 0$). So W is our familiar first-order neighbor proximity matrix. Here ρ is called a *spatial autoregression parameter* and, evidently, $Y_i = \rho \sum_j Y_j I(j \in \partial_i) + \epsilon_i$, where ∂_i denotes the set of neighbors of i. In fact, any symmetric proximity matrix can be used and, paralleling the discussion below (4.15), $I - \rho W$ will be nonsingular if $\rho \in \left(\frac{1}{\lambda_{(1)}}, \frac{1}{\lambda_{(n)}}\right)$ where now $\lambda_{(1)} < \cdots < \lambda_{(n)}$ are the ordered eigenvalues of W. As a weaker conclusion, if W is symmetric, we can apply the diagonal dominance result from Section 4.3.1. Now, if $\rho \sum_{j \neq i} w_{ij} < 1$ for each i, i.e., $\rho < \min \frac{1}{w_{i+}}$, we have positive definiteness, hence nonsingularity.

Alternatively, W can be replaced by $\widetilde{W}$ where now, for each i, the ith row has been normalized to sum to 1. That is, $\left(\tilde{W}\right)_{ij} = w_{ij}/w_{i+}$. Again, $\widetilde{W}$ is not symmetric, but it is row stochastic, i.e., $\widetilde{W}\mathbf{1} = \mathbf{1}$. If we set $B = \alpha\widetilde{W}$, α is called a *spatial autocorrelation parameter* and, were W a contiguity matrix, now $Y_i = \alpha \sum_j Y_j I(j \in \partial_i)/w_{i+} + \epsilon_i$. With a very regular grid the w_{i+} will all be essentially the same and thus α will be a multiple of ρ. But, perhaps more importantly, with $\widetilde{W}$ row stochastic the eigenvalues of $\widetilde{W}$ are all less than or equal to 1 (i.e., $\max |\lambda_i| = 1$). Thus $I - \alpha \widetilde{W}$ will be nonsingular if $\alpha \in (-1, 1)$, justifying referring to α as an autocorrelation parameter; see Exercise 9.

A SAR model is customarily introduced in a regression context, i.e., the *residuals* $\mathbf{U} = \mathbf{Y} - X\boldsymbol{\beta}$ are assumed to follow a SAR model, rather than $\mathbf{Y}$ itself. But then, following (4.17), if $\mathbf{U} = B\mathbf{U} + \boldsymbol{\epsilon}$, we obtain the attractive form

$$\mathbf{Y} = B\mathbf{Y} + (I - B) X\boldsymbol{\beta} + \boldsymbol{\epsilon} \,. \tag{4.22}$$

Expression (4.22) shows that $\mathbf{Y}$ is modeled through a component that provides a spatial weighting of neighbors and a component that is a usual linear regression. If B is the zero matrix we obtain an ordinary least squares (OLS) regression; if $B = I$ we obtain a purely spatial model.

We note that from (4.22) the SAR model does not introduce any spatial effects; the errors in (4.22) are independent. Expressed in a different way, if we modeled $\mathbf{Y} - X\boldsymbol{\beta}$ as $\mathbf{U} + \mathbf{e}$ with $\mathbf{e}$ independent errors, we would have $\mathbf{U} + \mathbf{e} = B\mathbf{U} + \boldsymbol{\epsilon} + \mathbf{e}$ and $\boldsymbol{\epsilon} + \mathbf{e}$ would result in a redundancy. Equivalently, if we write $\mathbf{U} = B\mathbf{U} + \boldsymbol{\epsilon}$, we see, from the distribution of $B\mathbf{U}$, that both terms on the right side are driven by the same variance component, σ_ϵ^2. As a result, in practice a SAR specification is not used in conjunction with a GLM. To introduce $\mathbf{U}$ as a vector of spatial adjustments to the mean vector, a transformed scale creates redundancy between the independent Gaussian error in the definition of the U_i and the stochastic mechanism associated with the conditionally independent Y_i.

We briefly note the somewhat related spatial modeling approach of Langford et al. [2002]. Rather than modeling the residual vector $\mathbf{U} = B\mathbf{U} + \boldsymbol{\epsilon}$, they propose that $\mathbf{U} = \tilde{B}\boldsymbol{\epsilon}$ where $\boldsymbol{\epsilon} \sim N\left(\mathbf{0}, \sigma^2 I\right)$, i.e., that $\mathbf{U}$ be modeled as a spatially motivated linear combination of independent variables. This induces $\Sigma_U = \sigma^2 \tilde{B}\tilde{B}^{\mathrm{T}}$. Thus, the U_i and hence the Y_i will be dependent and given $\tilde{B}$, $\mathrm{Cov}\,(Y_i, Y_{i'}) = \sigma^2 \sum_j \tilde{b}_{ij}\tilde{b}_{i'j}$. If $\tilde{B}$ arises through some proximity matrix W, the more similar rows i and i' of W are, the stronger the association between Y_i and $Y_{i'}$. However, the difference in nature between this specification and that in (4.22) is evident. To align the two, we would set $(I - B)^{-1} = \tilde{B}$, i.e. $B = I - \tilde{B}^{-1}$ (assuming $\tilde{B}$ is of full rank). $I - \tilde{B}^{-1}$ would not appear to have any interpretation through a proximity matrix.

Perhaps the most important point to note with respect to SAR models is that they are well suited to maximum likelihood estimation but not at all for MCMC fitting of Bayesian models. That is, the log likelihood associated with (4.22) (assuming $\tilde{D} = \sigma^2 I$) is

$$\frac{1}{2} \log \left| \sigma^{-1} \left(I - B\right) \right| - \frac{1}{2\sigma^2} \left(\mathbf{Y} - X\boldsymbol{\beta}\right)^{\mathrm{T}} \left(I - B\right) \left(I - B\right)^{\mathrm{T}} \left(\mathbf{Y} - X\boldsymbol{\beta}\right) \,. \tag{4.23}$$

Though B will introduce a regression or autocorrelation parameter, the quadratic form in (4.23) is quick to calculate (requiring no inverse) and the determinant can usually be calculated rapidly using diagonally dominant, sparse matrix approximations [see, e.g., Pace and Barry, 1997a,b]. Thus maximization of (4.23) can be done iteratively but, in general, efficiently.

Also, note that while the form in (4.23) can certainly be extended to a full Bayesian model through appropriate prior specifications, the absence of a hierarchical form with random effects implies straightforward Bayesian model fitting as well. Indeed, the general spatial slice Gibbs sampler [Agarwal and Gelfand, 2005] can easily handle this model. However, suppose we attempt to introduce SAR random effects in some fashion. Unlike CAR random effects that are defined through full conditional distributions, the full conditional distributions for the SAR effects have no convenient form. For large n, computation of such distributions using a form such as (4.21) will be expensive.

SAR models as in (4.22) are frequently employed in the spatial econometrics literature. With point-referenced data, B is taken to be ρW where W is the matrix of inter-point distances. Likelihood-based inference can be implemented in the spdep package in R. An illustrative example is provided in Exercise 12.

4.4.1   SAR versus CAR models

[Cressie, 1993, pp. 408–10] credits Brook [1964] with being the first to make a distinction between the CAR and SAR models, and offers a comparison of the two. To begin with, we may note from (4.13) and (4.21) that, under propriety, the two forms are equivalent if and only if

$$(I - B)^{-1}D = (I - \tilde{B})^{-1}\tilde{D}((I - \tilde{B})^{-1})^{\mathrm{T}} ,$$

where we use the tilde to indicate matrices in the SAR model. Cressie then shows that any SAR model can be represented as a CAR model (since D is diagonal, we can straightforwardly solve for B), but gives a counterexample to prove that the converse is not true. Since all SAR models are proper while we routinely employ improper CAR models, it is not surprising that the latter is a larger class.

We can add a bit more clarification here, following ideas in Ver Hoef et al. [2018]. In particular, when $I - B$ and $I - \tilde{B}$ are both full rank, Ver Hoef et al. [2018] prove that any positive definite covariance matrix can be written in the form $(I - B)^{-1}D$ for a unique choice of B and D. They also prove that any positive definite covariance matrix can be written in the form $(I - \tilde{B})^{-1}\tilde{D}(I - \tilde{B}^{-1})^{\mathrm{T}}$ for (non-unique) choices of $\tilde{B}$ and $\tilde{D}$. So, any SAR model can be written as a unique CAR model, and any CAR model can be written as a non-unique SAR model. Again, in the CAR setting, the usual conditions for $I - B$ to be full rank set $B = \rho W$ as above, where W is the neighbor weight matrix and ρ is suitably constrained. Similarly, in the SAR setting above, $\tilde{B} = \tilde{\rho}W$ for a suitably constrained $\tilde{\rho}$ yields $I - \tilde{B}$ to be full rank.

In addition, for the "proper" CAR and SAR models that include spatial correlation parameters ρ, Wall [2004] shows that the correlations between neighboring regions implied by these two models can be rather different; in particular, the first-order neighbor correlations increase at a slower rate as a function of ρ in the CAR model than they do for the SAR model. (As an aside, she notes that these correlations are not even monotone for $\rho < 0$, another reason to avoid negative spatial correlation parameters.) Also, correlations among pairs can switch in nonintuitive ways. For example, when working with the adjacency relationships generated by the lower 48 contiguous U.S. states, she finds that when $\rho = .49$ in the CAR model, $Corr(Alabama, Florida) = .20$ and $Corr(Alabama, Georgia) = .16$. But when ρ increases to .975, we instead get $Corr(Alabama, Florida) = .65$ and $Corr(Alabama, Georgia) = .67$, a slight reversal in ordering.

4.4.2   STAR models

In the literature, SAR models have frequently been extended to handle spatiotemporal data. The idea is that in working with proximity matrices, we can define neighbors in time as well as in space. Figure 4.4 shows a simple illustration with 9 areal units, 3 temporal units for each areal unit yielding $i = 1, \ldots, 9$, $t = 1, 2, 3$, labeled as indicated.

The measurements Y_{it} are spatially associated at each fixed t. But also, we might seek to associate, say, Y_{i2} with Y_{i1} and Y_{i3}. Suppose we write Y as the 27×1 vector with the first nine entries at $t = 1$, the second nine at $t = 2$, and the last nine at $t = 3$. Also let

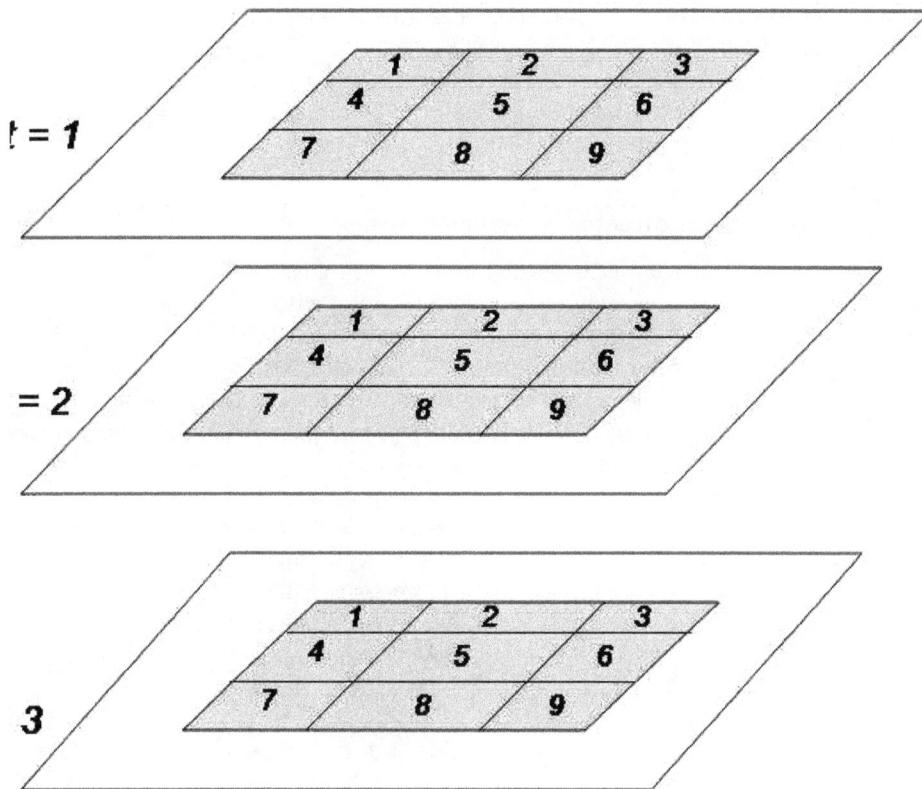

Figure 4.4 *Illustration of spatiotemporal areal unit setting for STAR model.*

$W_S = BlockDiag(W_1, W_1, W_1)$, where

$$W_1 = \begin{pmatrix} 0 & 1 & 0 & 1 & 0 & 0 & 0 & 0 & 0 \\ 1 & 0 & 1 & 0 & 1 & 0 & 0 & 0 & 0 \\ 0 & 1 & 0 & 0 & 0 & 1 & 0 & 0 & 0 \\ 1 & 0 & 0 & 0 & 1 & 0 & 1 & 0 & 0 \\ 0 & 1 & 0 & 1 & 0 & 1 & 0 & 1 & 0 \\ 0 & 0 & 1 & 0 & 1 & 0 & 0 & 0 & 1 \\ 0 & 0 & 0 & 1 & 0 & 0 & 0 & 1 & 0 \\ 0 & 0 & 0 & 0 & 1 & 0 & 1 & 0 & 1 \\ 0 & 0 & 0 & 0 & 0 & 1 & 0 & 1 & 0 \end{pmatrix}.$$

Then W_S provides a spatial contiguity matrix for the Y's. Similarly, let $W_T = \begin{pmatrix} 0 & W_2 & 0 \\ W_2 & 0 & W_2 \\ 0 & W_2 & 0 \end{pmatrix}$, where $W_2 = I_{9\times 9}$. Then W_T provides a *temporal* contiguity matrix for the Y's. But then, in our SAR model, we can define $B = \rho_s W_S + \rho_t W_T$. In fact, we can also introduce $\rho_{ST} W_S W_T$ into B and note that

$$W_S W_T = \begin{pmatrix} 0 & W_1 & 0 \\ W_1 & 0 & W_1 \\ 0 & W_1 & 0 \end{pmatrix}.$$

In this way, we introduce association across both space and time. For instance Y_{21} and Y_{41} affect the mean of Y_{12} (as well as affecting Y_{11}) from W_S by itself. Many more possibilities exist. Models formulated through such more general definitions of B are referred to as *spatiotemporal autoregressive* (STAR) models. See Pace et al. [2000] for a full discussion and development. The interpretation of the ρ's in the above example measures the relative importance of first-order spatial neighbors, first-order temporal neighbors, and first-order spatiotemporal neighbors.

4.5   Areal models using directed acyclic graphs

Areal models can be looked upon as special cases of graphical models [Lauritzen, 1996], where the graph $\mathcal{G} = \{\mathcal{V}, \mathcal{E}\}$ represents the underlying map with each region corresponding to a vertex in the set $\mathcal{V}$ and each edge $(i,j) \in \mathcal{E}$ representing the relation $i \sim j$, i.e., i is a neighbor of j. Auto-regression models, such as CAR and SAR, build a joint probability distribution for random variables over the vertices by regressing one variable, say Y_i, on a subset of the remaining variables while conforming to the conditional dependencies or independence implied by the presence or absence of edges.

While CAR models are built upon undirected graphs, where the edges have no direction and represent conditional independence of the nodes given all other nodes, it is possible to build areal models using directed acyclic graphs. Directed acyclic graphs or *DAGs* have been used in the spatial literature for modeling large spatial datasets [Datta et al., 2016a, we will revisit DAGs in the context of nearest neighbor Gaussian processes in Chapter 13] and for generating image textures [Cressie and Davidson, 1998]. Datta et al. [2019] propose areal models using a directed acyclic graph. We specify distributions of the Y_i's as

$$Y_1 = \epsilon_1, \ Y_2 = b_{21}Y_1 + \epsilon_2, \ \ldots, \ Y_n = b_{n1}Y_1 + \cdots + b_{n,n-1}Y_{n-1} + \epsilon_n, \qquad (4.24)$$

where the ϵ_i's are independent $N(0, \tau_i^2)$ errors. The specification in (4.24) is similar to a SAR model except that $B = (b_{ij})$ in (4.24) is a strictly lower triangular matrix. Let F be the diagonal matrix with $1/\tau_1^2, \ldots, 1/\tau_n^2$ on the diagonal. Then $\mathbf{Y} \sim N(\mathbf{0}, Q^{-1})$ where the precision matrix $Q = L^{\mathrm{T}}FL$ and $L = I - B$.

In fact, switching from the specification in the SAR model to a strictly lower triangular matrix B is not restrictive because any precision matrix Q can be expressed as $PQP^{\mathrm{T}} = (I - B)^{\mathrm{T}}F(I - B)$ for some permutation matrix P, strictly lower-triangular matrix B, and diagonal matrix F with positive diagonal entries. If Q is positive-definite, then $P = I$ and the above factorization is essentially the usual Cholesky factorization [Banerjee and Roy, 2014]. Datta et al. [2019] extend this result for rank deficient Q. Hence, any multivariate normal distribution can be expressed as in (4.24) under certain orderings of the areal units.

For low rank distributions this will be equivalent to setting some of the τ_i's to zero. In fact, the lower triangular B has several advantages over the SAR model. First, L is lower triangular with ones on the diagonal, ensuring that $L^{\mathrm{T}}FL$ is positive definite as long as all τ_i's are positive. Next, $\det(L^{\mathrm{T}}FL)$ is simply $\prod_{i=1}^{n} \tau_i$ and the quadratic form $\mathbf{Y}^{\mathrm{T}}L^{\mathrm{T}}FL\mathbf{Y}$ is

$$\mathbf{Y}^{\mathrm{T}}L^{T}FL\mathbf{Y} = (1/\tau_1^2)Y_1^2 + \sum_{i=2}^{n}(1/\tau_i^2)\left(Y_i - \sum_{\{j<i\}} Y_j b_{ij}\right)^2.$$

This requires $O(k+s)$ floating point operations (FLOPs) where s is the number of non-zero entries of B. Hence, if B is sparse, the joint density of $\mathbf{Y}$ is evaluated cheaply.

To complete the specification in (4.24), we need to fully specify the matrices B and F. Datta et al. [2019] propose a class of directed acyclic graphical autoregression (DAGAR) model by considering an ordered sequence of regions (nodes in $\mathcal{V}$) and defining the neighbor sets for each region i as $N(i) = \{j < i, j \sim i\}$ for $i > 1$; $N(1)$ is empty. Therefore, $N(i)$ consists of only geographic neighbors of region i that *precede* i in the ordered sequence of regions. We set $b_{ij} = 0$ for all $j \notin N(i)$ and the constraint $j < i$ ensures that B is lower triangular. This reduces (4.24) to

$$Y_1 = \epsilon_1, \quad Y_i = \sum_{j \in N(i)} Y_j b_{ij} + \epsilon_i, \quad \text{for} \quad i = 2, \ldots, n. \tag{4.25}$$

This specification is analogous to auto-regressive models for time series. In fact, if Y_i denotes the response at time i, $N(i)$ includes all time points less than i up to a lag of r, and $b_{ij} = b_{i-j}$, then (4.25) simply denotes the autoregressive model of order r. In a time-series context, where i and j denote time points, assigning the weights b_{ij} based on the temporal lag seems natural, but for irregular areal datasets, enumeration of the areal units does not have any physical interpretation. A reasonable simplification is to assign equal weights to all the neighbors, i.e., letting $b_{ij} = b_i$ for all $j \in N(i)$.

A natural specification is $b_i = 1/n_{<i}$ and $(1/\tau_i^2) \propto n_{<i}$ where $n_{<i} = |N(i)|$ denotes the cardinality of the neighbor set for $i > 1$ and $n_{<1} = 0$. This specification is similar to the CAR model except that DAGAR uses only the directed neighbors $N(i)$ instead of all neighbors. However, since $n_{<1} = 0$, this choice of b_{ij} does not provide a well-defined specification for τ_1^2. Either we define τ_1^2 in a manner inconsistent with the definition of τ_i^2 for $i > 1$ or we define $\tau_1^2 = 0$ which yields an improper distribution for $\mathbf{Y}$. Instead, Datta et al. [2019] use a spanning tree construction to specify the nonzero elements of B and F as

$$b_{ij} = \frac{\rho}{1 + (n_{<i} - 1)\rho^2} \quad \text{for} \quad j \in N(i), \quad \frac{1}{\tau_i^2} = \frac{1 + (n_{<i} - 1)\rho^2}{1 - \rho^2}, \quad \text{for} \quad i = 1, \ldots, n. \tag{4.26}$$

The specifications in (4.26) retain $b_{ij} = b_i$ for all $j \in N(i)$, thereby assigning equal weights to all the directed neighbors. The conditional precision, $1/\tau_i^2$, increases with the number of directed neighbors.

Datta et al. [2019] show that ρ acts as an interpretable spatial autocorrelation parameter in the above graphical autoregression structure and is not afflicted by the strange behavior of the spatial autocorrelation parameter in SAR and CAR models exposed by Wall [2004]. Unlike the CAR model, DAGAR's covariance matrix is always positive definite and, hence, it can be regarded as a model either for the observations or for the random effects with dependence structure derived from a graph. Also, the Cholesky factor has the same level of sparsity as the undirected graph ensuring scalability for analyzing very large areal datasets. Datta et al. [2019] extend (4.24) and (4.26) to any permutation of the original ordering.

4.6 Computer tutorials

We provide the computer programs to outline the use of some GIS functions in R for obtaining neighborhood (adjacency) matrices, computing Moran's and Geary's statistic and fitting CAR and SAR models using traditional maximum likelihood techniques and mapping the results for certain classes of problems to execute other exploratory approaches in the folder titled Chapter 4 from https://github.com/sudiptobanerjee/BGC_2023.

A convenient illustration is offered by the SIDS (sudden infant death syndrome) data, analyzed by Cressie and Read [1985], Cressie and Chan [1989], [Cressie, 1993, Section 6.2] and [Kaluzny et al., 1999, Section 5.3] and available from the spdep package. This dataset contains counts of SIDS deaths from 1974 to 1978 and counts from 1979 to 1983 along with related covariate information for the 100 counties in the U.S. state of North Carolina. In particular, we show how to produce maps such as the one displayed in Figure 4.5 that are useful to compare the raw data with the spatially smoothed rates. Although the spatial autocorrelation in the data was found to be modest, the fitted values from the SAR model clearly show the smoothing. Both maps have the same color scheme.

a) Raw Freeman–Tukey transformed SIDS rates

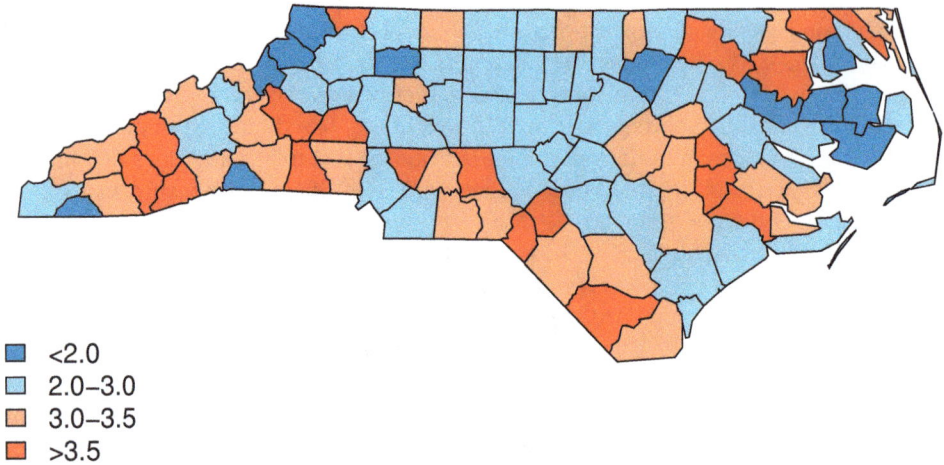

- ■ <2.0
- ■ 2.0–3.0
- ■ 3.0–3.5
- ■ >3.5

b) Fitted SIDS rates from SAR model

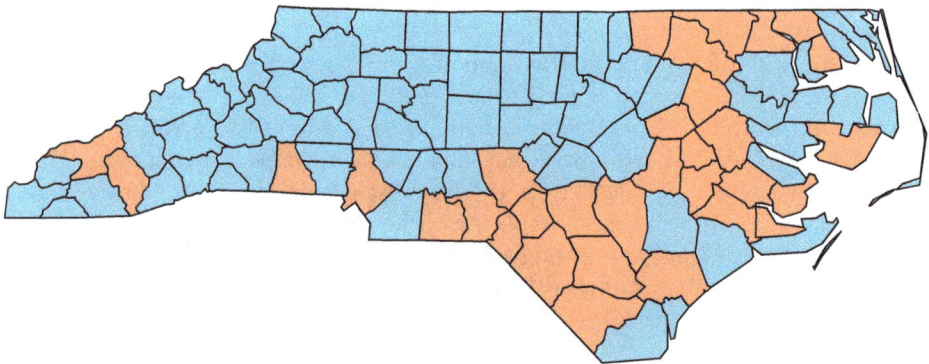

Figure 4.5 *Unsmoothed raw (a) and spatially smoothed fitted (b) rates, North Carolina SIDS data.*

4.7   Exercises

1. Verify Brook's Lemma, equation (4.7).

2.(a) To appreciate how Brook's Lemma works, suppose Y_1 and Y_2 are both binary variables, and that their joint distribution is defined through conditional logit models. That is,

$$\log \frac{P(Y_1 = 1 \mid Y_2)}{P(Y_1 = 0 \mid Y_2)} = \alpha_0 + \alpha_1 Y_2 \quad \text{and} \quad \log \frac{P(Y_2 = 1 \mid Y_1)}{P(Y_2 = 0 \mid Y_1)} = \beta_0 + \beta_1 Y_1 \ .$$

Obtain the joint distribution of Y_1 and Y_2.

(b) This result can be straightforwardly extended to the case of more than two variables, but the details become increasingly clumsy. Illustrate this issue in the case of *three* binary variables, Y_1, Y_2, and Y_3.

3. Returning to (4.13) and (4.14), let $B = ((b_{ij}))$ be an $n \times n$ matrix with positive elements; that is, $b_{ij} > 0$, $\sum_j b_{ij} \leq 1$ for all i, and $\sum_j b_{ij} < 1$ for at least one i. Let $D = Diag\left(\tau_i^2\right)$ be a diagonal matrix with positive elements τ_i^2 such that $D^{-1}\left(I - B\right)$ is symmetric; that is, $b_{ij}/\tau_i^2 = b_{ji}/\tau_j^2$, for all i, j. Show that $D^{-1}\left(I - B\right)$ is positive definite.

4. Looking again at (4.13), obtain a simple sufficient condition on B such that the CAR specification with precision matrix $D^{-1}\left(I - B\right)$ is a pairwise difference specification, as in (4.16).

5. Show that, for W symmetric, $\Sigma_{\mathbf{y}}^{-1} = D_w - \rho W$ is positive definite (thus resolving the impropriety in (4.15)) if $\rho \in \left(1/\lambda_{(1)}, 1/\lambda_{(n)}\right)$, where $\lambda_{(1)} < \lambda_{(2)} < \cdots < \lambda_{(n)}$ are the ordered eigenvalues of $D_w^{-1/2} W D_w^{-1/2}$.

6. Show that if all entries in W are nonnegative and $D_w - \rho W$ is positive definite with $0 < \rho < 1$, then all entries in $(D_w - \rho W)^{-1}$ are nonnegative.

7. Under a proper CAR model for $\mathbf{Y}$, i.e., with $\Sigma_{\mathbf{y}} = D_w - \rho W$, obtain the correlation and covariance between Y_i and Y_j.

8. Obtain the joint distribution, up to the normalizing constant, for $(Y_1, Y_2, ..., Y_n)$ under (4.19). Hint: You might try to guess it but Brook's lemma can be used as well.

9. Recalling the SAR formulation using the scaled adjacency matrix $\widetilde{W}$ just below (4.21), prove that $I - \alpha \widetilde{W}$ will be nonsingular if $\alpha \in (-1, 1)$, so that α may be sensibly referred to as an "autocorrelation parameter."

10. In the setting of Subsection 4.3.1, if $(\Sigma_{\mathbf{y}}^{-1})_{ij} = 0$, then show that Y_i and Y_j are conditionally independent given $Y_k, k \neq i, j$.

11. The file https://github.com/sudiptobanerjee/BGC_2023/data/state-sat.dat gives the 1999 state average SAT data (part of which is mapped in Figure 4.1).

(a) Use the `spautolm` function to fit the SAR model of Section 4.4, taking the verbal SAT score as the response Y and the percent of eligible students taking the exam in each state as the covariate X. Do this analysis twice: first using binary weights and then using row-normalized weights. Is the analysis sensitive to these choices of weights? Is knowing X helpful in explaining Y?

(b) Using the `maps` library in R, draw choropleth maps similar to Figure 4.1 of both the fitted verbal SAT scores and the spatial residuals from the SAR model. Is there evidence of spatial correlation in the response Y once the covariate X is accounted for?

(c) Repeat your SAR model analysis above, again using `spautolm` but now assuming the CAR model of Section 4.3. Compare your estimates with those from the SAR model and interpret any changes.

(d) One might imagine that the percentage of eligible students taking the exam should perhaps affect the variance of our model, not just the mean structure. To check this, refit the SAR model replacing your row-normalized weights with weights equal to the reciprocal of the percentage of students taking the SAT. Is this model sensible?

12. Consider the data https://github.com/sudiptobanerjee/BGC_2023/data/Columbus. dat available within the spdep package (but with possibly different variable names). These data record crime information for 49 neighborhoods in Columbus, OH, during 1980. Variables measured include NEIG, the neighborhood id value (1–49); HOVAL, its mean housing value (in $1,000); INC, its mean household income (in $1,000); CRIME, its number of residential burglaries and vehicle thefts per thousand households; OPEN,

a measure of the neighborhood's open space; PLUMB, the percentage of housing units without plumbing; DISCBD, the neighborhood centroid's distance from the central business district; X, an x-coordinate for the neighborhood centroid (in arbitrary digitizing units, not polygon coordinates); Y, the same as X for the y-coordinate; AREA, the neighborhood's area; and PERIM, the perimeter of the polygon describing the neighborhood.

(a) Use `spdep` in `R` to construct adjacency matrices for the neighborhoods of Columbus based upon centroid distances less than

    i. 25% of the maximum intercentroidal distances;
   ii. 50% of the maximum intercentroidal distances;
  iii. 75% of the maximum intercentroidal distances.

(b) For each of the three spatial neighborhoods constructed above, use the `spautolm` function to fit SAR models with CRIME as the dependent variable, and HOVAL, INC, OPEN, PLUMB, and DISCBD as the covariates. Compare your results and interpret your parameter estimates in each case.

(c) Repeat your analysis by setting $B = \rho W$ in equation (4.22) with W_{ij} the Euclidean distance between location i and location j.

(d) Repeat part (b) for CAR models. Compare your estimates with those from the SAR model and interpret them.

Chapter 5

Basics of Bayesian inference

We provide a brief review of hierarchical Bayesian modeling and computing for readers less familiar with these topics. In one chapter we can only scratch the surface of this field and readers may well wish to consult one of the many textbooks (classics or more recent) on the subject, either as preliminary work or on an as-needed basis. First, we mention the texts stressing Bayesian theory, including DeGroot [2004], Box and Tiao [1992], Berger [1993], Bernardo and Smith [2000], Robert [2007]. These books tend to focus on foundations and decision theory, rather than computation or data analysis. Gelman et al. [2013] provides a very comprehensive treatment of the field spanning methods, computation and data analysis. O'Hagan et al. [1994], Carlin and Louis [2008] and Hoff [2009] present treatments that would perhaps be deemed at the same level as Gelman et al. [2013], while more introductory texts include Lee [2012] and Albert [2009] with the latter emphasizing Bayesian data analysis using R packages.

5.1 Introduction to hierarchical modeling and Bayes' Theorem

The Bayesian approach provides a coherent probabilistic framework for analyzing data by building a joint probability model on the set of "unknowns" and "knowns." We offer a heuristic development that will suffice for subsequent developments. The Bayesian approach to statistical inference seeks to provide full uncertainty quantification of unknown quantities of interest conditional upon known (or seen) quantities. Denoting $\mathcal{U}$ to be a collection of unknowns and $\mathcal{K}$ a collection of "knowns," we build a joint probability model for $\mathcal{U}$ and $\mathcal{K}$. *Bayes' Theorem* or *Bayes' Rule* is a simple consequence of expressing the joint distribution in two different ways. Using somewhat informal notation,

$$
\begin{aligned}
P(\mathcal{U}, \mathcal{K}) = P(\mathcal{K}) \times P(\mathcal{U} \mid \mathcal{K}) = P(\mathcal{U}) \times P(\mathcal{K} \mid \mathcal{U}) \\
\implies P(\mathcal{U} \mid \mathcal{K}) \propto P(\mathcal{U}) \times P(\mathcal{K} \mid \mathcal{U}) ,
\end{aligned}
\tag{5.1}
$$

where the last proportionality follows from the fact that $P(\mathcal{K})$, which is the marginal distribution of $\mathcal{K}$, does not depend on $\mathcal{U}$ and, therefore, can be treated as a constant when evaluating $P(\mathcal{U} \mid \mathcal{K})$. Probabilistic inference arrives at this distribution by specifying $P(\mathcal{K} \mid \mathcal{U})$ (the likelihood) and $P(\mathcal{U})$ (the prior). We treat probability measure as a tool to quantify uncertainty. The distribution over $\mathcal{K}$ models uncertainties associated with the data (e.g., how it is sampled from a population) while the distribution on $\mathcal{U}$ models the uncertainty associated with everything that we have not observed or measured in our experiment. This is the essence of Bayes' Theorem attributed to the Reverend Thomas Bayes, an 18th-century nonconformist minister and part-time mathematician [Bayes, 1763, Barnard, 1958]. We apply Bayes' Theorem derived in (5.1) to construct hierarchical models.

  A quick word about notation in the subsequent sections. We typically denote vectors with boldface (e.g., $\mathbf{y}$, $\boldsymbol{\beta}$) when we need to carry out linear algebra operations on them, but we do not find it necessary to use boldface for *sets* of parameters when there is no apparent

DOI: 10.1201/9781003401728-5

linear algebra operation required. Hence, we denote $p(\theta)$ to denote the joint probability distribution of a collection of parameters $\theta = \{\theta_1, \ldots, \theta_k\}$ without ambiguity.

Consider the customary setting where we specify a distribution $f(\mathbf{y} \mid \theta)$ for the observed data $\mathbf{y} = (y_1, \ldots, y_n)^{\mathrm{T}}$ given a set of unknown parameters $\theta = \{\theta_1, \ldots, \theta_k\}$. Bayesian inference treats observed data as $\mathcal{K}$ and parameters as $\mathcal{U}$ when applying (5.1). Therefore,

$$p(\theta \mid \mathbf{y}) \propto \underbrace{p(\theta)}_{\text{Prior}} \times \underbrace{p(\mathbf{y} \mid \theta)}_{\text{Likelihood}} \quad . \tag{5.2}$$

We specify a *prior* distribution to model θ, which itself can depend on other parameters. Therefore, we can write $p(\theta \mid \lambda)$, where λ is a set of hyperparameters as the prior distribution for θ. For instance, y_i might be the empirical mammography rate in a sample of women aged 40 and over from county i, θ_i the underlying true mammography rate for all such women in this county, and λ a parameter controlling how these true rates vary across counties. If λ is known, inference concerning θ is based on its *posterior* distribution,

$$p(\theta \mid \mathbf{y}, \lambda) = \frac{p(\mathbf{y}, \theta \mid \lambda)}{p(\mathbf{y} \mid \lambda)} = \frac{p(\mathbf{y}, \theta \mid \lambda)}{\int p(\mathbf{y}, \theta \mid \lambda) \, d\theta} = \frac{p(\mathbf{y} \mid \theta) p(\theta \mid \lambda)}{\int p(\mathbf{y} \mid \theta) p(\theta \mid \lambda) \, d\theta} \quad . \tag{5.3}$$

Bayesian inference takes into account contributions from the data (in the form of the likelihood) and any available external knowledge or opinion (in the form of the prior π) to the posterior. Since, in practice, λ will not be known, a second stage (or *hyperprior*) distribution $h(\lambda)$ will often be required, and (5.3) will be replaced with

$$p(\theta \mid \mathbf{y}) = \frac{p(\mathbf{y}, \theta)}{p(\mathbf{y})} = \frac{\int p(\mathbf{y} \mid \theta) p(\theta \mid \lambda) h(\lambda) \, d\lambda}{\int p(\mathbf{y} \mid \theta) p(\theta \mid \lambda) h(\lambda) \, d\theta d\lambda} \quad . \tag{5.4}$$

It is worth noting the *hierarchical* structure implicit in (5.4), i.e., three levels of distributional specification, typically with primary interest in the θ level. As the title of this book suggests, such hierarchical specification is the focus of the inference methodology in this volume. Rather than integrating over λ, we might replace λ by an estimate $\hat{\lambda}$ obtained by maximizing the marginal distribution $p(\mathbf{y} \mid \lambda) = \int p(\mathbf{y} \mid \theta) \times p(\theta \mid \lambda) d\theta$, viewed as a function of λ. Inference could then proceed based on the *estimated* posterior distribution $p(\theta \mid \mathbf{y}, \hat{\lambda})$, obtained by plugging $\hat{\lambda}$ into equation (5.3). This approach is referred to as *empirical Bayes* analysis [see, e.g., Berger, 1993, Maritz and Lwin, 2019, Carlin and Louis, 2008, Efron, 2012, for details regarding empirical Bayes methodology and applications] see Berger [1993], Maritz and Lwin [2019].

Bayesian predictive inference is carried out the same principles. Here, we introduce a predictive random variable that represents the unknown values of the outcome. Now $\mathcal{U}$ represents the predictive random variable and the posterior $P(\mathcal{U} \mid \mathcal{K})$ is the *posterior predictive distribution*,

$$p(\tilde{\mathbf{y}} \mid \mathbf{y}) = \int p(\tilde{\mathbf{y}} \mid \theta, \mathbf{y}) \times p(\theta \mid \mathbf{y}) d\theta \,, \tag{5.5}$$

where $\tilde{\mathbf{y}}$ denotes the predictive random variable. The density $p(\tilde{\mathbf{y}} \mid \theta, \mathbf{y})$ is often referred to as the *conditional predictive density*. Note that (5.5) is the expectation of the conditional probability density of the predictive random variable with respect to the posterior distribution of the parameters, i.e., $\mathrm{E}_{\theta \mid \mathbf{y}}[p(\tilde{\mathbf{y}} \mid \theta, \mathbf{y})]$, where $\mathrm{E}_{\theta \mid \mathbf{y}}[\cdot]$ is the expectation with respect to $p(\theta \mid \mathbf{y})$. Rather than imputing point estimates of parameters, Bayesian predictive inference averages over the entire posterior distribution of the parameters. The posterior distribution captures uncertainty about the model parameters while the conditional predictive density captures the uncertainty associated with measuring the predictive random variable.

The appeal of (5.5) as a predictive tool emanates from its flexibility. Note that (5.5) is a valid probability distribution for any valid conditional predictive density $p(\tilde{\mathbf{y}} \mid \theta, \mathbf{y})$. This is ensured as long as $\tilde{\mathbf{y}}$ and $\mathbf{y}$ have a valid joint distribution conditional on the shared parameters θ. In fact, in several data analysis settings, the predictive random variable $\tilde{\mathbf{y}}$ is assumed conditionally independent of the observed data $\mathbf{y}$ given parameters θ, i.e., $p(\tilde{\mathbf{y}} \mid \theta, \mathbf{y}) = p(\tilde{\mathbf{y}} \mid \theta)$, we could reasonably model the predictive random variable using the same probability model as that of the likelihood. Looking ahead at what will follow in this book, in spatial models such an assumption will not hold for the data and the predictive random variable because of a shared underlying process. In such cases, conditional independence between $\tilde{\mathbf{y}}$ and $\mathbf{y}$ will hold if one conditions on the underlying process in addition to parameters θ. We will discuss such models later in the text.

This text will pursue "full" Bayesian inference with hierarchical models and avoid empirical Bayes. The fully Bayesian inferential paradigm offers potentially attractive advantages over the classical approach through its more philosophically sound foundation, its unified approach to data analysis, and its ability to formally incorporate prior opinion or external empirical evidence into the results via the prior distribution π. Data analysts, formerly reluctant to adopt the Bayesian approach due to skepticism concerning its philosophy and a lack of necessary computational tools, are now turning to it with increasing regularity since classical methods emerge as both theoretically and practically inadequate to handle the challenges of today's complex modeling landscape. In our context, modeling the θ_i as random (instead of fixed) effects allows us to induce specific (e.g., spatial) correlation structures among them, hence among the observed data y_i as well. Hierarchical Bayesian methods now enjoy broad application in the analysis of spatial data, as the remainder of this book reveals.

5.1.1 Illustrations of Bayes' Theorem

Here we consider a few basic examples of Bayesian models.

Example 5.1 Suppose we are observing a data value Y from a binomial distribution, $Bin(n, \theta)$, with density proportional to

$$p(y \mid \theta) \propto \theta^y (1 - \theta)^{n-y} . \tag{5.6}$$

The $Beta(\alpha, \beta)$ distribution offers a conjugate prior for this likelihood since its density is proportional to (5.6) as a function of θ, namely

$$p(\theta) \propto \theta^{\alpha-1} (1 - \theta)^{\beta-1} . \tag{5.7}$$

Using Bayes' Rule (5.3), it is clear that

$$\begin{aligned} p(\theta \mid y) &\propto \theta^{y+\alpha-1} (1 - \theta)^{n-y+\beta-1} \\ &\propto Beta(y + \alpha, \, n - y + \beta) , \end{aligned} \tag{5.8}$$

another Beta distribution. For example, consider a setting where $n = 10$ and we observe $Y = y_{obs} = 7$. Choosing $\alpha = \beta = 1$ (i.e., a uniform prior for θ), the posterior is a $Beta(y_{obs} + 1, n - y_{obs} + 1) = Beta(8, 4)$ distribution. ∎

Example 5.2 Consider the following hierarchical model for n independent and identically distributed observations with mean μ and variance σ^2,

$$y_i \mid \mu, \sigma^2 \overset{iid}{\sim} N(\mu, \sigma^2), \; i = 1, 2, \ldots, n; \quad \mu \mid \theta, \sigma^2 \sim N(\theta, \sigma^2/n_0); \tag{5.9}$$

where θ, σ^2, n_0, a_0 and b_0 are all known scalars. Therefore, the only unknown is μ so we compute its posterior distribution given the data $\mathbf{y} = (y_1, \ldots, y_n)^{\mathrm{T}}$ as,

$$
p(\mu \mid \mathbf{y}, \sigma^2, \theta) \propto N\left(\mu \,\middle|\, \theta, \frac{\sigma^2}{n_0}\right) \times \prod_{i=1}^{n} N(y_i \mid \mu, \sigma^2) \propto N\left(\mu \,\middle|\, \theta, \frac{\sigma^2}{n_0}\right) \times N\left(\bar{y} \,\middle|\, \theta, \frac{\sigma^2}{n}\right)
$$

$$
\propto \exp\left\{-\frac{n_0}{2\sigma^2}(\mu - \theta)^2\right\} \times \exp\left\{-\frac{n}{2\sigma^2}(\mu - \bar{y})^2\right\}
$$

$$
\propto \exp\left\{-\frac{n_0 + n}{2\sigma^2}\left(\mu - \frac{n_0\theta + n\bar{y}}{n_0 + n}\right)^2\right\} \propto N\left(\mu \,\middle|\, \frac{n_0\theta + n\bar{y}}{n_0 + n}, \frac{\sigma^2}{n_0 + n}\right),
$$

$$
(5.10)
$$

where $\bar{y} = (1/n)\sum_{i=1}^{n} y_i$ is the sample mean. ∎

Example 5.2 offers very appealing intuition. Without any data, our prior belief on the population mean μ is quantified by mean θ and its precision σ^2/n_0. In fact, the scalar n_0 can be imagined as a prior "sample size" that tells us how precise our belief on μ is (there is no need for n_0 to be an integer). After we have observed a sample of size n, the belief on μ is updated and expressed by the posterior distribution. The mean is a weighted average of the prior mean and the sample mean, while the precision (reciprocal of the variance) in our belief is proportional to the "new sample size" $n_0 + n$. Also note that the prior chosen leads to a posterior distribution for μ that is available in closed form, and is a member of the same distributional family as the prior. Such a prior is referred to as a *conjugate* prior. We will often use such priors in our work, since, when they are available, conjugate families are convenient and still allow a variety of shapes wide enough to capture our prior beliefs.

We turn to an example where both μ and σ^2 are unknown. Here, we will again use conjugate priors to arrive at analytically tractable distributions. The algebra to derive the joint posterior distribution again follows from writing down the joint distribution of the data and all unknown parameters and, subsequently, keeping track of only expressions involving the unknown parameters. This is only slightly more complicated than the previous example.

Example 5.3 We model observations y_i randomly sampled from a Normal distribution with unknown mean μ and unknown variance σ^2 using the following hierarchical model,

$$
y_i \stackrel{iid}{\sim} N(\mu, \sigma^2),\ i = 1, 2, \ldots, n; \quad \mu \sim N(\theta, \sigma^2/n_0); \quad \sigma^2 \sim IG(a_0, b_0), \quad (5.11)
$$

where θ, n_0, a_0 and b_0 are known scalars and σ^2 is distributed as the Inverse-Gamma random variable with shape a_0 and scale b_0. The joint posterior distribution for $\{\mu, \sigma^2\}$ conditional on data $\mathbf{y} = (y_1, \ldots, y_n)^{\mathrm{T}}$ is calculated as

$$
p(\mu, \sigma^2 \mid \mathbf{y}) \propto IG(\sigma^2 \mid a, b) \times N\left(\mu \,\middle|\, \theta, \frac{\sigma^2}{n_0}\right) \times \prod_{i=1}^{n} N(y_i \mid \mu, \sigma^2)
$$

$$
\propto \left(\frac{1}{\sigma^2}\right)^{a+1+\frac{n+1}{2}} \times \exp\left\{-\frac{1}{\sigma^2}\left(b + \frac{1}{2}Q(\mu, \theta, \mathbf{y})\right)\right\},
$$

where $Q(\mu, \theta, \mathbf{y}) = n_0(\mu - \theta)^2 + \sum_{i=1}^{n}(y_i - \mu)^2$. We further simplify $Q(\mu, \theta, \mathbf{y})$ by writing $\sum_{i=1}^{n}(y_i - \mu)^2 = n(\mu - \bar{y})^2 + (n-1)s^2$, where $\bar{y} = \sum_{i=1}^{n} y_i/n$ is the sample mean and $s^2 = \sum_{i=1}^{n}(y_i - \bar{y})^2/(n-1)$ is the sample variance. We "complete the square" to collect the parameter μ into a single quadratic term (we follow our hunch to see if a Normal distribution emerges for $p(\mu \mid \mathbf{y}, \sigma^2)$ as in Example 5.2). This yields

$$Q(\mu, \theta, \mathbf{y}) = n_0(\mu - \theta)^2 + n(\mu - \bar{y})^2 + (n-1)s^2$$

$$= (n_0 + n)\left(\mu - \frac{n_0\theta + n\bar{y}}{n_0 + n}\right)^2 + \frac{n_0 n}{n_0 + n}(\bar{y} - \theta)^2 + (n-1)s^2$$

$$= (n_0 + n)\left(\mu - \frac{n_0\theta + n\bar{y}}{n_0 + n}\right)^2 + f(\theta, \bar{y}),$$

where $f(\theta, \bar{y}) = (n-1)s^2 + \frac{n_0 n}{n_0 + n}(\bar{y} - \theta)^2$. The joint posterior distribution factorizes as

$$p(\mu, \sigma^2 \mid \mathbf{y}) \propto \underbrace{IG(\sigma^2 \mid a_1, b_1)}_{p(\sigma^2 \mid \mathbf{y})} \times \underbrace{N\left(\mu \;\middle|\; \frac{n_0\theta + n\bar{y}}{n_0 + n}, \frac{\sigma^2}{n_0 + n}\right)}_{p(\mu \mid \sigma^2, \mathbf{y})}, \qquad (5.12)$$

where $a_1 = a_0 + \frac{n}{2}$ and $b_1 = b_0 + \frac{f(\theta, \bar{y})}{2}$. ∎

We conclude this section with a famous example from a landmark paper by Lindley and Smith [1972] on the Bayesian hierarchical linear model.

Example 5.4 *(the general linear model).* Let $\mathbf{y}$ be an $n \times 1$ vector of correlated observations, X an $n \times p$ matrix of covariates. Since the elements of $\mathbf{y}$ are correlated, we now construct a hierarchical model using a multivariate normal likelihood,

$$\mathbf{y} \mid \boldsymbol{\beta}, \Sigma \;\sim\; N_n(X\boldsymbol{\beta}, \Sigma), \qquad \boldsymbol{\beta} \sim N_p(A\boldsymbol{\alpha}, V).$$

Here $\boldsymbol{\beta}$ is a $p \times 1$ vector of regression coefficients and Σ is a $p \times p$ covariance matrix that is assumed known. It can be shown [now a classic result, first published by Lindley and Smith, 1972], that the marginal distribution $p(\mathbf{y})$ is $N_n(\mathbf{y} \mid XA\boldsymbol{\alpha}, \Sigma + XVX^{\mathrm{T}})$. The easiest way to derive this (without calculus) is to simply write down the hierarchical model in terms of the linear models corresponding to the likelihood and the prior:

$$\mathbf{y} = X\boldsymbol{\beta} + \mathbf{u}; \quad \text{and} \quad \boldsymbol{\beta} = A\boldsymbol{\alpha} + \mathbf{v}, \qquad (5.13)$$

where $\mathbf{u} \sim N_n(\mathbf{0}, \Sigma)$ and $\mathbf{v} \sim N_p(\mathbf{0}, V)$ are independently distributed. The marginal distribution of $\mathbf{y}$ is now easily obtained by "eliminating" $\boldsymbol{\beta}$ from these equations by simply substituting the second expression in (5.13) into the first. This yields $\mathbf{y} = XA\boldsymbol{\alpha} + X\mathbf{v} + \mathbf{u}$ from which the marginal distribution follows easily.

The posterior distribution of $\boldsymbol{\beta} \mid \mathbf{y}$ is derived by noting the joint distribution from (5.13)

$$\begin{bmatrix} \mathbf{y} \\ \boldsymbol{\beta} \end{bmatrix} \sim N_{n+p}\left(\begin{bmatrix} XA\boldsymbol{\alpha} \\ A\boldsymbol{\alpha} \end{bmatrix}, \begin{bmatrix} \Sigma + XVX^{\mathrm{T}} & XV \\ VX^{\mathrm{T}} & V \end{bmatrix}\right). \qquad (5.14)$$

The posterior distribution of $\boldsymbol{\beta}$ given $\mathbf{y}$ is derived from the familiar formulas of the conditional distribution that yield $p(\boldsymbol{\beta} \mid \mathbf{y})$ to be a multivariate Gaussian distribution with mean $A\boldsymbol{\alpha} + VX^{\mathrm{T}}(\Sigma + XVX^{\mathrm{T}})^{-1}(\mathbf{y} - XA\boldsymbol{\alpha})$ and variance $V - VX^{\mathrm{T}}(\Sigma + XVX^{\mathrm{T}})^{-1}XV$. This expression shows how the prior mean and variance are updated by information from $\mathbf{y}$.

Alternate expressions for the mean and variance of $p(\boldsymbol{\beta} \mid \mathbf{y})$ arise from working with Gaussian densities. Retaining only terms involving $\boldsymbol{\beta}$ (everything else is treated as constant),

$$p(\boldsymbol{\beta} \mid \mathbf{y}) \propto N(\boldsymbol{\beta} \mid A\boldsymbol{\alpha}, V) \times N(\mathbf{y} \mid X\boldsymbol{\beta}, \Sigma) \propto \exp\left\{-\frac{1}{2}(Q_1(\boldsymbol{\beta}) + Q_2(\boldsymbol{\beta}, \mathbf{y}))\right\},$$

where $Q_1(\boldsymbol{\beta}) = (\boldsymbol{\beta} - A\boldsymbol{\alpha})^\mathsf{T} V^{-1}(\boldsymbol{\beta} - A\boldsymbol{\alpha})$ and $Q_2(\boldsymbol{\beta}, \mathbf{y}) = (\mathbf{y} - X\boldsymbol{\beta})^\mathsf{T} \Sigma^{-1}(\mathbf{y} - X\boldsymbol{\beta})$. We now simplify the sum of these two quadratic forms using a matrix completion of squares

$$
\begin{aligned}
Q_1(\boldsymbol{\beta}) + Q_2(\boldsymbol{\beta}, \mathbf{y}) &= (\boldsymbol{\beta} - A\boldsymbol{\alpha})^\mathsf{T} V^{-1}(\boldsymbol{\beta} - A\boldsymbol{\alpha}) + (\mathbf{y} - X\boldsymbol{\beta})^\mathsf{T} \Sigma^{-1}(\mathbf{y} - X\boldsymbol{\beta}) \\
&= \boldsymbol{\beta}^\mathsf{T} V^{-1} \boldsymbol{\beta} - 2\boldsymbol{\beta}^\mathsf{T} V^{-1} A\boldsymbol{\alpha} + \boldsymbol{\alpha}^\mathsf{T} A^\mathsf{T} V^{-1} A\boldsymbol{\alpha} \\
&\quad + \boldsymbol{\beta}^\mathsf{T} X^\mathsf{T} \Sigma^{-1} X\boldsymbol{\beta} - 2\boldsymbol{\beta}^\mathsf{T} X^\mathsf{T} \Sigma^{-1} \mathbf{y} + \mathbf{y}^\mathsf{T} \Sigma^{-1} \mathbf{y} \\
&= \boldsymbol{\beta}^\mathsf{T} \left(V^{-1} + X^\mathsf{T} \Sigma^{-1} X \right) \boldsymbol{\beta} - 2\boldsymbol{\beta}^\mathsf{T} \left(V^{-1} A\boldsymbol{\alpha} + X^\mathsf{T} \Sigma^{-1} \mathbf{y} \right) \\
&\quad + \boldsymbol{\alpha}^\mathsf{T} A^\mathsf{T} V^{-1} A\boldsymbol{\alpha} + \mathbf{y}^\mathsf{T} \Sigma^{-1} \mathbf{y} \\
&= \boldsymbol{\beta}^\mathsf{T} M^{-1} \boldsymbol{\beta} - 2\boldsymbol{\beta}^\mathsf{T} m + c
\end{aligned}
\tag{5.15}
$$

where $M^{-1} = V^{-1} + X^\mathsf{T} \Sigma^{-1} X$, $m = A\boldsymbol{\alpha} + X^\mathsf{T} \Sigma^{-1} \mathbf{y}$ and $c = \boldsymbol{\alpha}^\mathsf{T} A^\mathsf{T} V^{-1} A\boldsymbol{\alpha} + \mathbf{y}^\mathsf{T} \Sigma^{-1} \mathbf{y}$. Therefore, $p(\boldsymbol{\beta} \mid \mathbf{y}) = N_p(\boldsymbol{\beta} \mid Mm, M)$, where $m = X^\mathsf{T} \Sigma^{-1} \mathbf{y} + V^{-1} A\boldsymbol{\alpha}$ and $M^{-1} = V^{-1} + X^\mathsf{T} \Sigma^{-1} X$. Thus $\mathrm{E}(\boldsymbol{\beta} \mid \mathbf{y}) = Mm$ provides a point estimate for $\boldsymbol{\beta}$, with variability captured by the associated variance matrix M. ∎

Equating the expression for the posterior variance $\mathrm{Var}(\boldsymbol{\beta} \mid \mathbf{y})$ given by M with that derived from (5.14) yields the venerable Sherman-Woodbury-Morrison identity

$$
\left(V^{-1} + X^\mathsf{T} \Sigma^{-1} X \right)^{-1} = V - V X^\mathsf{T} (\Sigma + X V X^\mathsf{T})^{-1} X V
\tag{5.16}
$$

Both these expressions for the posterior variance offer insights. In particular, for a vague prior we may set $V^{-1} = O$ (the matrix of zeros) so that $M^{-1} = X \Sigma^{-1} X$ and $m = X^\mathsf{T} \Sigma^{-1} \mathbf{y}$. In the simple case where $\Sigma = \sigma^2 I_p$, the posterior becomes $N_p \left(\boldsymbol{\beta} \mid \hat{\boldsymbol{\beta}}, \sigma^2 (X^\mathsf{T} X)^{-1} \right)$ where $\hat{\boldsymbol{\beta}} = (X^\mathsf{T} X)^{-1} X^\mathsf{T} \mathbf{y}$. This reveals that Bayesian inference with "flat priors" for the regression coefficients are formally equivalent to the usual likelihood approach. On the other hand, the posterior variance on the right-hand side of (5.16) is not of use when $V^{-1} = O$, but explicitly yields the shrinkage in variance from V (for the prior) by the amount $V X^\mathsf{T} (\Sigma + X V X^\mathsf{T})^{-1} X V$. This is indeed "shrinking" the prior variance because the matrix subtracted from V is positive definite.

5.2   Bayesian inference

Bayesian inference is conceptually straightforward because once we have computed (or obtained an estimate of) the posterior, inference comes down merely to extracting features of this distribution. The posterior summarizes everything we know about the model parameters in light of the data. In the remainder of this section, we shall assume for simplicity that the posterior $p(\theta \mid \mathbf{y})$ itself (and not merely an estimate of it) is available either analytically or we can draw samples from it.

Bayesian methods for estimation are also reminiscent of corresponding maximum likelihood methods. This should not be surprising, since likelihoods form part of the Bayesian specification; we have even seen that a normalized (i.e., standardized) likelihood can be thought of a posterior when this is possible. However, when we turn to hypothesis testing (Bayesians prefer the term *model comparison*), the approaches have little in common. Bayesians (and many likelihoodists) have a deep and abiding antipathy toward p-values, for a long list of reasons we shall not go into here; the interested reader may consult [Berger, 1993, Sec. 4.3.3], [Kass and Raftery, 1995, Sec. 8.2] or [Carlin and Louis, 2008, Sec. 2.3.3].

5.2.1   *Point estimation*

Suppose for the moment that θ is univariate. Given the posterior $p(\theta \mid \mathbf{y})$, a natural Bayesian point estimate of θ would be some measure of centrality. Three familiar choices are the posterior mean,

$$\hat{\theta} = \mathrm{E}(\theta \mid \mathbf{y}) \, ,$$

the posterior median,

$$\hat{\theta} \, : \, \int_{-\infty}^{\hat{\theta}} p(\theta \mid \mathbf{y}) d\theta = 0.5 \, ,$$

and the posterior mode,

$$\hat{\theta} \, : \, p(\hat{\theta} \mid \mathbf{y}) = \sup_{\theta} p(\theta \mid \mathbf{y}) \, .$$

Notice that the lattermost estimate is typically easiest to compute, since it does not require any integration: we can replace $p(\theta \mid \mathbf{y})$ by its unstandardized form, $f(\mathbf{y} \mid \theta)p(\theta)$, and get the same answer (since these two differ only by a multiplicative factor of $m(\mathbf{y})$, which does not depend on θ). Indeed, if the posterior exists under a flat prior $p(\theta) = 1$, then the posterior mode is nothing but the maximum likelihood estimate (MLE).

Note that for symmetric unimodal posteriors (e.g., a normal distribution), the posterior mean, median, and mode will all be equal. However, for multimodal or otherwise nonnormal posteriors, the mode will often be an unsatisfying choice of centrality measure (consider for example the case of a steadily decreasing, one-tailed posterior; the mode will be the very first value in the support of the distribution—hardly central!). By contrast, the posterior mean, though arguably the most commonly used, will sometimes be overly influenced by heavy tails (just as the sample mean $\bar{y}$ is often nonrobust against outlying observations). As a result, the posterior median will often be the best and safest point estimate. It is also the most difficult to compute (since it requires both an integration and a rootfinder), but this difficulty is somewhat mitigated for posterior estimates obtained via Markov chain Monte Carlo (MCMC), as developed in Section 5.5.

5.2.2 *Interval estimation*

The posterior allows us to make any desired probability statements about θ. By inversion, we can infer about any quantile. For example, suppose we find the $\alpha/2$- and $(1-\alpha/2)$-quantiles of $p(\theta \mid \mathbf{y})$, that is, the points q_L and q_U such that

$$\int_{-\infty}^{q_L} p(\theta \mid \mathbf{y}) d\theta = \alpha/2 \text{ and } \int_{q_U}^{\infty} p(\theta \mid \mathbf{y}) d\theta = 1 - \alpha/2 \, .$$

Then clearly $P(q_L < \theta < q_U \mid \mathbf{y}) = 1 - \alpha$; our confidence that θ lies in (q_L, q_U) is $100 \times (1-\alpha)\%$. Thus this interval is a $100 \times (1-\alpha)\%$ *credible set* (or simply *Bayesian confidence interval*) for θ. This interval is relatively easy to compute, and enjoys a direct interpretation ("the probability that θ lies in (q_L, q_U) is $(1-\alpha)$") which the usual frequentist interval does not.

The interval just described is often called the *equal tail* credible set, for the obvious reason that is obtained by chopping an equal amount of support $(\alpha/2)$ off the top and bottom of $p(\theta \mid \mathbf{y})$. Note that for symmetric unimodal posteriors, this equal tail interval will be symmetric about this mode (which we recall equals the mean and median in this case). It will also be optimal in the sense that it will have shortest length among sets C satisfying

$$1 - \alpha \leq P(C \mid \mathbf{y}) = \int_C p(\theta \mid \mathbf{y}) d\theta \, . \tag{5.17}$$

Note that any such set C could be thought of as a $100 \times (1-\alpha)\%$ credible set for θ. For posteriors that are not symmetric and unimodal, a better (shorter) credible set can be obtained by taking only those values of θ having posterior density greater than some cutoff $k(\alpha)$, where this cutoff is chosen to be as large as possible while C continues to satisfy equation (5.17). This *highest posterior density* (HPD) confidence set will always be of optimal length, but will typically be significantly more difficult to compute. The equal tail interval emerges as HPD in the symmetric unimodal case since then, it captures the "most likely" values of θ. Fortunately, many of the posteriors we will be interested in will be (at least approximately) symmetric unimodal, so the much simpler equal tail interval will often suffice. In fact, it is the routine choice in practice.

5.2.3   Bayes factors

Bayesian inference (point or interval) is quite straightforward given the posterior distribution, or an estimate thereof. By contrast, hypothesis testing is less straightforward for two reasons. First, there is less agreement among Bayesians as to the proper approach to the problem. For years, posterior probabilities and Bayes factors were considered the only appropriate method. But these methods are only suitable with fully proper priors, and for relatively low-dimensional models. With the recent proliferation of very complex models, employing at least partly improper priors, other methods have come to the fore. Second, solutions to hypothesis testing questions often involve not just the posterior $p(\theta \mid \mathbf{y})$, but also the *marginal* distribution, $m(\mathbf{y})$. Unlike the case of posterior and the predictive distributions, marginal distributions are not easily calculated.

Bayesian hypothesis testing can be regarded as a model choice problem, replacing the customary two hypotheses H_0 and H_A by two candidate parametric models M_1 and M_2 having respective parameter vectors $\boldsymbol{\theta}_1$ and $\boldsymbol{\theta}_2$.[1] Under prior densities $\pi_i(\boldsymbol{\theta}_i)$, $i = 1, 2$, the marginal distributions of $\mathbf{y}$ are found by integrating out the parameters,

$$p(\mathbf{y} \mid M_i) = \int f(\mathbf{y} \mid \boldsymbol{\theta}_i, M_i)\pi_i(\boldsymbol{\theta}_i)d\boldsymbol{\theta}_i \;, \; i = 1, 2 \;. \tag{5.18}$$

Bayes' Theorem (5.3) may then be applied to obtain the posterior probabilities $P(M_1 \mid \mathbf{y})$ and $P(M_2 \mid \mathbf{y}) = 1 - P(M_1 \mid \mathbf{y})$ for the two models. The quantity commonly used to summarize these results is the *Bayes factor*, BF, which is the ratio of the posterior odds of M_1 to the prior odds of M_1, given by Bayes' Theorem as

$$BF \;\; = \;\; \frac{P(M_1 \mid \mathbf{y})/P(M_2 \mid \mathbf{y})}{P(M_1)/P(M_2)} \tag{5.19}$$

$$= \;\; \frac{\left[\frac{p(\mathbf{y} \mid M_1)P(M_1)}{p(\mathbf{y})}\right] \Big/ \left[\frac{p(\mathbf{y} \mid M_2)P(M_2)}{p(\mathbf{y})}\right]}{P(M_1)/P(M_2)}$$

$$= \;\; \frac{p(\mathbf{y} \mid M_1)}{p(\mathbf{y} \mid M_2)} \;, \tag{5.20}$$

the ratio of the observed marginal densities for the two models. If the two models are *a priori* equally probable (i.e., $P(M_1) = P(M_2) = 0.5$), then $BF = P(M_1 \mid \mathbf{y})/P(M_2 \mid \mathbf{y})$, the posterior odds of M_1.

Consider the case where both models share the same parametrization (i.e., $\boldsymbol{\theta}_1 = \boldsymbol{\theta}_2 = \theta$), and both hypotheses are simple (i.e., $M_1 : \theta = \theta^{(1)}$ and $M_2 : \theta = \theta^{(2)}$). Then $\pi_i(\theta)$ consists of a point mass at $\theta^{(i)}$ for $i = 1, 2$, and so from (5.18) and (5.20) we have

$$BF = \frac{f(\mathbf{y} \mid \theta^{(1)})}{f(\mathbf{y} \mid \theta^{(2)})} \;,$$

which is nothing but the likelihood ratio between the two models. Hence, in the simple-versus-simple setting, the Bayes factor is precisely the odds in favor of M_1 over M_2 *given solely by the data.*

---

[1] Attractively, this perspective removes the nesting of hypotheses in customary frequentist analysis.

Bayes factors have received substantial attention in statistical inference from diverse perspectives and we do not attempt a comprehensive review. We direct the reader to a landmark paper by Kass and Raftery [1995] who offer a comprehensive account of Bayes' factors for comparing models, while Morey et al. [2016] offer perspectives from the philosophies of scientific hypothesis testing. Foundational treatments are available in Jeffreys [1998] and Good [1979]. [Gelman et al., 2013, section 7.4] offers a nice summary of Bayes factors, where they may be useful and where they are not, in the context of practical Bayesian data analysis.

5.2.4 *The importance of computing the posterior distribution*

We will not pursue formal decision-theoretic Bayesian hypothesis testing in the subsequent chapters. In fact, for complex dependent hierarchical models involving spatial and spatial-temporal processes, formal Bayesian hypothesis testing and Bayes' factors are more of a distraction. Instead, in Section 5.6 we will turn to model choice using criteria that assign a score or value to each model and help us select among different models. However, in order to develop such scores, we will need to make sure that we can compute the posterior distribution. Unfortunately, barring a few restricted classes of models, the posterior distribution is usually analytically intractable. We will provide accounts of methods that conduct inference by drawing samples from the posterior distribution or approximate the posterior distribution.

But before getting into computations, it will be instructive to investigate a very important class of models for which we are able to derive the posterior distributions in closed-form—the Bayesian linear regression model with conjugate priors. This example will not only help elucidate some common algebraic manipulations in Gaussian linear models that will be used in subsequent chapters, but will also help us illustrate the principles of sampling based Bayesian inference that will be the mainstay in subsequent chapters.

5.3 The conjugate Bayesian linear regression model: gory details

5.3.1 *Posterior distribution of model parameters*

We extend Example 5.3 to a regression model over $i = 1, \ldots, n$ independent observations,

$$y_i \mid \boldsymbol{\beta}, \sigma^2 \overset{ind}{\sim} N(\mathbf{x}_i^{\mathsf{T}}\boldsymbol{\beta}, \sigma^2) \; ; \quad \boldsymbol{\beta} \mid \sigma^2 \sim N(M_0\mathbf{m}_0, \sigma^2 M_0) \; ; \quad \sigma^2 \sim IG(a_0, b_0) \, , \qquad (5.21)$$

where y_i is an outcome of interest, $\mathbf{x}_i$ is a $p \times 1$ vector of design variables corresponding to y_i, $\boldsymbol{\beta}$ is the corresponding $p \times 1$ vector of slopes and σ^2 is the variance parameter of the model for the data. The above model is often written in terms of the posterior density

$$p(\boldsymbol{\beta}, \sigma^2 \mid \mathbf{y}) \propto p(\boldsymbol{\beta}, \sigma^2, \mathbf{y}) = IG(\sigma^2 \mid a_0, b_0) \times N(\boldsymbol{\beta} \mid M_0\mathbf{m}_0, \sigma^2 M_0) \times \prod_{i=1}^{n} N(y_i \mid \mathbf{x}_i^{\mathsf{T}}\boldsymbol{\beta}, \sigma^2) \, .$$

This is a Normal-Inverse-Gamma (NIG) family with density,

$$\underbrace{\frac{NIG(\boldsymbol{\beta}, \sigma^2 \mid M_0\mathbf{m}_0, M_0, a_0, b_0)}{p(\boldsymbol{\beta}, \sigma^2)}} = \underbrace{\frac{IG(\sigma^2 \mid a_0, b_0)}{p(\sigma^2)}} \times \underbrace{\frac{N\left(\boldsymbol{\beta} \mid M_0\mathbf{m}_0, \sigma^2 M_0\right)}{p(\boldsymbol{\beta} \mid \sigma^2)}}$$

$$= \frac{b_0^{a_0}}{(2\pi)^{p/2}(\det(M_0))^{1/2}\Gamma(a_0)} \left(\frac{1}{\sigma^2}\right)^{a_0+1+\frac{p}{2}} \exp\left(-\frac{1}{\sigma^2}\left\{b_0 + \frac{1}{2}Q(\boldsymbol{\beta}; \mathbf{m}_0, M_0)\right\}\right) \, ,$$

$$(5.22)$$

where $Q(\beta; \mathbf{m}_0, M_0) = (\beta - M_0\mathbf{m}_0)^{\mathrm{T}}M_0^{-1}(\beta - M_0\mathbf{m}_0)$. The marginal distribution of σ^2 is already specified above, while that of β is obtained as

$$\int NIG(\beta, \sigma^2 \mid M_0\mathbf{m}_0, M_0, a_0, b_0)d\sigma^2 = t_{2a_0}\left(\beta \,\middle|\, M_0\mathbf{m}_0, \frac{b_0}{a_0}M_0\right), \qquad (5.23)$$

where $t_\nu(\mathbf{x} \mid \boldsymbol{\mu}, \Sigma) = \dfrac{\Gamma\left(\frac{\nu+p}{2}\right)}{(\nu\pi)^{p/2}\Gamma(\frac{\nu}{2})\,|\Sigma|^{1/2}}\left(1 + \dfrac{(\mathbf{x}-\boldsymbol{\mu})^{\mathrm{T}}\Sigma^{-1}(\mathbf{x}-\boldsymbol{\mu})}{\nu}\right)^{-\frac{\nu+p}{2}}$ is a multivariate t-density in $\mathbf{x}$ with mean $\boldsymbol{\mu}$ and degrees of freedom ν.

The NIG family is a conjugate prior for the Gaussian linear model in (5.21) and yields

$$p(\beta, \sigma^2 \mid \mathbf{y}) = \underbrace{\frac{IG(\sigma^2 \mid a^*, b^*)}{p(\sigma^2 \mid \mathbf{y})}}_{} \times \underbrace{\frac{N(\beta \mid M\mathbf{m}, \sigma^2 M)}{p(\beta \mid \sigma^2, \mathbf{y})}}_{} = \underbrace{\frac{NIG(\beta, \sigma^2 \mid M\mathbf{m}, M, a^*, b^*)}{p(\beta, \sigma^2 \mid \mathbf{y})}}_{},$$

$$(5.24)$$

where $\mathbf{y} = (y_1, y_2, \ldots, y_n)^{\mathrm{T}}$, X is $n \times p$ with i-th row $\mathbf{x}_i^{\mathrm{T}}$, $\mathbf{m} = \mathbf{m}_0 + X^{\mathrm{T}}\mathbf{y}$, $M^{-1} = M_0^{-1} + X^{\mathrm{T}}X$, $a^* = a_0 + \frac{n}{2}$, and $b^* = b_0 + \frac{1}{2}(\mathbf{y}^{\mathrm{T}}\mathbf{y} + \mathbf{m}_0^{\mathrm{T}}M_0\mathbf{m}_0 - \mathbf{m}^{\mathrm{T}}M\mathbf{m})$. The marginal posteriors are $IG(\sigma^2 \mid a^*, b^*)$ and $t_\nu(\beta \mid \boldsymbol{\mu}, \Sigma)$ with $\nu = 2a^*$, $\boldsymbol{\mu} = M\mathbf{m}$ and $\Sigma = (b^*/a^*)M$.

We can easily adapt (5.21) to allow for dependence among the elements of $\mathbf{y}$, which is more appropriate for spatial data analysis. We modify (5.21) as

$$\mathbf{y} \mid \beta, \sigma^2, X \sim N(X\beta, \sigma^2 V_y); \quad \beta \mid \sigma^2 \sim N(M_0\mathbf{m}_0, \sigma^2 M_0); \quad \sigma^2 \sim IG(a_0, b_0), \qquad (5.25)$$

where V_y is an $n \times n$ positive definite covariance matrix that, for now, we assume as fixed. The posterior density is now

$$p(\beta, \sigma^2 \mid \mathbf{y}) \propto \underbrace{NIG(\beta, \sigma^2 \mid M_0\mathbf{m}_0, a_0, b_0)}_{p(\beta, \sigma^2)} \times \underbrace{N(\mathbf{y} \mid X\beta, \sigma^2 V_y)}_{p(\mathbf{y} \mid \beta, \sigma^2)}$$

$$\propto \left(\frac{1}{\sigma^2}\right)^{a_0+1}\exp\left\{-\frac{b_0}{\sigma^2}\right\} \times \left(\frac{1}{\sigma^2}\right)^{\frac{n+p}{2}}\exp\left\{-\frac{1}{2\sigma^2}(Q_1(\beta) + Q_2(\beta, \mathbf{y}))\right\}$$

$$(5.26)$$

where $Q_1(\beta) = (\beta - M_0\mathbf{m}_0)^{\mathrm{T}}M_0^{-1}(\beta - M_0\mathbf{m}_0)$ and $Q_2(\beta, \mathbf{y}) = (\mathbf{y} - X\beta)^{\mathrm{T}}V_y^{-1}(\mathbf{y} - X\beta)$. We "complete the square," familiar from (5.15) in Example 5.4, to obtain

$$Q_1(\beta) + Q_2(\beta, \mathbf{y}) = (\beta - M_0\mathbf{m}_0)^{\mathrm{T}}M_0^{-1}(\beta - M_0\mathbf{m}_0) + (\mathbf{y} - X\beta)^{\mathrm{T}}V_y^{-1}(\mathbf{y} - X\beta)$$

$$= \beta^{\mathrm{T}}M^{-1}\beta - 2\beta^{\mathrm{T}}\mathbf{m} + c \qquad (5.27)$$

where $M^{-1} = M_0^{-1} + X^{\mathrm{T}}V_y^{-1}X$, $\mathbf{m} = \mathbf{m}_0 + X^{\mathrm{T}}V_y^{-1}\mathbf{y}$ and $c = \mathbf{m}_0^{\mathrm{T}}M_0\mathbf{m}_0 + \mathbf{y}^{\mathrm{T}}V_y^{-1}\mathbf{y}$. The $p \times p$ matrix M^{-1} is positive definite (hence so is M), $\mathbf{m}$ is $p \times 1$ and c is a scalar. The identity we use is

$$\beta^{\mathrm{T}}M^{-1}\beta - 2\beta^{\mathrm{T}}\mathbf{m} = (\beta - M\mathbf{m})^{\mathrm{T}}M^{-1}(\beta - M\mathbf{m}) - \mathbf{m}^{\mathrm{T}}M\mathbf{m}, \qquad (5.28)$$

which is easy to verify by expanding the right-hand side. Applying (5.28) to (5.27) yields

$$Q_1(\beta) + Q_2(\beta, \mathbf{y}) = (\beta - M\mathbf{m})^{\mathrm{T}}M^{-1}(\beta - M\mathbf{m}) - \mathbf{m}^{\mathrm{T}}M\mathbf{m} + c$$

$$= (\beta - M\mathbf{m})^{\mathrm{T}}M^{-1}(\beta - M\mathbf{m}) + c^*, \qquad (5.29)$$

where $c = \mathbf{m}_0^{\mathsf{T}} M_0 \mathbf{m}_0 + \mathbf{y}^{\mathsf{T}} V_y^{-1} \mathbf{y}$ and $c^* = c - \mathbf{m}^{\mathsf{T}} M \mathbf{m} = \mathbf{m}_0^{\mathsf{T}} M_0 \mathbf{m}_0 + \mathbf{y}^{\mathsf{T}} V_y^{-1} \mathbf{y} - \mathbf{m}^{\mathsf{T}} M \mathbf{m}$.
Arranging the terms in (5.26) we get

$$p(\boldsymbol{\beta}, \sigma^2 \mid \mathbf{y}) \propto \left(\frac{1}{\sigma^2}\right)^{a_0 + \frac{n}{2} + 1} \exp\left\{-\frac{1}{\sigma^2}\left(b_0 + \frac{c^*}{2}\right)\right\}$$

$$\times \left(\frac{1}{\sigma^2}\right)^{\frac{p}{2}} \exp\left\{-\frac{1}{2\sigma^2}(\boldsymbol{\beta} - M\mathbf{m})^{\mathsf{T}} M^{-1}(\boldsymbol{\beta} - M\mathbf{m}) + c^*\right\} \quad (5.30)$$

$$\propto \underbrace{IG\left(\sigma^2 \mid a^*, b^*\right)}_{p(\sigma^2 \mid \mathbf{y})} \times \underbrace{N(\boldsymbol{\beta} \mid M\mathbf{m}, \sigma^2 M)}_{p(\boldsymbol{\beta} \mid \sigma^2, \mathbf{y})} = NIG(\boldsymbol{\beta}, \sigma^2 \mid M\mathbf{m}, M, a^*, b^*),$$

where $a^* = a_0 + \frac{n}{2}$ and $b^* = b_0 + \frac{c^*}{2} = b_0 + \frac{1}{2}\left\{\mathbf{m}_0^{\mathsf{T}} M_0 \mathbf{m}_0 + \mathbf{y}^{\mathsf{T}} V_y^{-1} \mathbf{y} - \mathbf{m}^{\mathsf{T}} M \mathbf{m}\right\}$. The joint
posterior for (5.25) is of the form (5.24) with the values of M, $\mathbf{m}$, a^* and b^* computed as
above. Note that (5.21) is a special case of (5.25) with $V_y = I_n$ (the $n \times n$ identity matrix).

5.3.2 Posterior predictive distributions

Turning to predictive inference, let us apply (5.5) to the Bayesian linear regression setting.
First consider (5.21) (i.e., (5.25) with $V_y = I_n$). Let $\tilde{X}$ be an $\tilde{n} \times p$ matrix of predictors
available to us and we wish to predict the outcome corresponding to $\tilde{X}$. We denote the
predictive random variable by the $\tilde{n} \times 1$ vector $\tilde{\mathbf{y}} = (\tilde{y}_1, \ldots, \tilde{y}_{\tilde{n}})^{\mathsf{T}}$. Since the measurements are
independently distributed, it is reasonable to derive the conditional predictive density from
the model $\tilde{y}_i \overset{ind}{\sim} N(\tilde{\mathbf{x}}_i^{\mathsf{T}} \boldsymbol{\beta}, \sigma^2)$ for $i = 1, \ldots, \tilde{n}$, i.e, $\tilde{\mathbf{y}} \sim N(\tilde{X}\boldsymbol{\beta}, \sigma^2 I_{\tilde{n}})$. Therefore, conditionally
on $\{\boldsymbol{\beta}, \sigma^2\}$ the predictive random variable $\tilde{\mathbf{y}}$ is conditionally independent of the data $\mathbf{y}$, i.e.,
$p(\tilde{\mathbf{y}} \mid \boldsymbol{\beta}, \sigma^2, \mathbf{y}) = p(\tilde{\mathbf{y}} \mid \boldsymbol{\beta}, \sigma^2)$. The posterior predictive distribution is

$$p(\tilde{\mathbf{y}} \mid \mathbf{y}) = \int \underbrace{N(\tilde{\mathbf{y}} \mid \tilde{X}\boldsymbol{\beta}, \sigma^2 I_{\tilde{n}})}_{p(\tilde{\mathbf{y}} \mid \boldsymbol{\beta}, \sigma^2)} \times \underbrace{NIG(\boldsymbol{\beta}, \sigma^2 \mid M\mathbf{m}, M, a^*, b^*)}_{p(\boldsymbol{\beta}, \sigma^2 \mid \mathbf{y})} d\beta d\sigma^2 = t_{\tilde{\nu}}(\tilde{\mathbf{y}} \mid \tilde{\boldsymbol{\mu}}, \tilde{\Sigma}),$$

$$\text{(5.31)}$$

where $t_{\tilde{\nu}}(\tilde{\mathbf{y}} \mid \tilde{\boldsymbol{\mu}}, \tilde{\Sigma})$ is the multivariate Student's t distribution given below (5.24) with
$\nu = 2a^*$, $\tilde{\boldsymbol{\mu}} = \tilde{X} M \mathbf{m}$ and $\tilde{\Sigma} = (b^*/a^*)(I_{\tilde{n}} + \tilde{X} M \tilde{X}^{\mathsf{T}})$. This derivation should come as no
surprise: it follows from mixing a normal distribution with an NIG prior analogous to how
the posterior was derived in (5.24). The posterior predictive distribution simply uses the
NIG posterior (updated from observed data) as a "prior" to update our Normal model for
the predictive random variable. In fact, we derive (5.31) using only linear models.

Conditional on the data $\mathbf{y}$ and the variance σ^2, we construct two normal linear models,

$$\tilde{\mathbf{y}} = \tilde{X}\boldsymbol{\beta} + \tilde{\mathbf{u}}, \ \tilde{\mathbf{u}} \sim N(\mathbf{0}, \sigma^2 I_n); \quad \text{and} \quad \boldsymbol{\beta} = M\mathbf{m} + \mathbf{v}, \ \mathbf{v} \sim N(\mathbf{0}, \sigma^2 M), \quad (5.32)$$

where M and $\mathbf{m}$ are as in the posterior distribution of $\boldsymbol{\beta}$, and $\tilde{\mathbf{u}}$ and $\mathbf{v}$ are independent
random variables with the indicated distributions. The first equation is the normal linear
model corresponding to the conditional predictive density $p(\tilde{\mathbf{y}} \mid \boldsymbol{\beta}, \sigma^2)$ while the second
equation represents $p(\boldsymbol{\beta} \mid \mathbf{y}, \sigma^2)$. Integrating out $\boldsymbol{\beta}$ yields $p(\tilde{\mathbf{y}} \mid \mathbf{y}, \sigma^2)$ and amounts to
eliminating $\boldsymbol{\beta}$ from (5.32). Substituting the expression for $\boldsymbol{\beta}$ from the second equation into
the first yields

$$\tilde{\mathbf{y}} = \tilde{X}(M\mathbf{m} + \mathbf{v}) + \tilde{\mathbf{u}} = XM\mathbf{m} + M\mathbf{v} + \tilde{\mathbf{u}} \implies \tilde{\mathbf{y}} \mid \sigma^2 \sim N\left(\tilde{X}M\mathbf{m}, \sigma^2(I_{\tilde{n}} + \tilde{X}M\tilde{X}^{\mathsf{T}})\right).$$

Combining the above with the posterior distribution $p(\sigma^2 \mid \mathbf{y}) = IG(\sigma^2 \mid a^*, b^*)$ yields

$$p(\tilde{\mathbf{y}}, \sigma^2 \mid \mathbf{y}) = NIG(\tilde{\mathbf{y}}, \sigma^2 \mid \tilde{X}M\mathbf{m}, I_{\tilde{n}} + \tilde{X}M\tilde{X}^{\mathsf{T}}, a^*, b^*). \quad (5.33)$$

Applying (5.23) to the above NIG density produces the multivariate t density in (5.31).

Next, we turn to predictive inference for the dependent model (5.25). Here, we will need to ensure that the joint distribution $p(\mathbf{y}, \tilde{\mathbf{y}} \mid \boldsymbol{\beta}, \sigma^2)$ is well-defined. This, in fact, is crucial because without a valid joint model for the observed data and the predictive random variable there is no valid conditional predictive distribution, hence no valid posterior predictive distribution in (5.5). We specify $p(\mathbf{y}, \tilde{\mathbf{y}} \mid \boldsymbol{\beta}, \sigma^2)$ through the joint linear model for $(\mathbf{y}^{\mathsf{T}}, \tilde{\mathbf{y}}^{\mathsf{T}})^{\mathsf{T}}$,

$$\begin{bmatrix} \mathbf{y} \\ \tilde{\mathbf{y}} \end{bmatrix} = \begin{bmatrix} X \\ \tilde{X} \end{bmatrix} \boldsymbol{\beta} + \begin{bmatrix} \mathbf{u} \\ \tilde{\mathbf{u}} \end{bmatrix}, \quad \begin{bmatrix} \mathbf{u} \\ \tilde{\mathbf{u}} \end{bmatrix} \sim N \left(\begin{bmatrix} \mathbf{0}_{n \times 1} \\ \mathbf{0}_{\tilde{n} \times 1} \end{bmatrix}, \sigma^2 \begin{bmatrix} V_y & V_{y,\tilde{y}} \\ V_{y,\tilde{y}}^{\mathsf{T}} & V_{\tilde{y}} \end{bmatrix} \right), \tag{5.34}$$

where the variance-covariance matrix in the normal distribution must be positive definite to ensure a proper joint distribution. This yields the conditional predictive density,

$$\tilde{\mathbf{y}} \mid \boldsymbol{\beta}, \sigma^2, \mathbf{y} \sim N \left(\tilde{X}\boldsymbol{\beta} + V_{y,\tilde{y}}^{\mathsf{T}} V_y^{-1} (\mathbf{y} - X\boldsymbol{\beta}), \sigma^2 \left[V_{\tilde{y}} - V_{y,\tilde{y}}^{\mathsf{T}} V_y^{-1} V_{y,\tilde{y}} \right] \right). \tag{5.35}$$

The positive-definiteness of the joint variance-covariance matrix $V = \begin{bmatrix} V_y & V_{y,\tilde{y}} \\ V_{y,\tilde{y}}^{\mathsf{T}} & V_{\tilde{y}} \end{bmatrix}$ ensures the positive-definiteness of the Schur's complement [see, e.g., Harville, 2000, Banerjee and Roy, 2014] $V_{\tilde{y}|y} = V_{\tilde{y}} - V_{y,\tilde{y}}^{\mathsf{T}} V_y^{-1} V_{y,\tilde{y}}$, which, in turn, ensures that (5.35) is a proper density.

Note that the joint model (5.34) does not affect the posterior distribution $p(\boldsymbol{\beta}, \sigma^2 \mid \mathbf{y})$ in (5.30), while (5.35) is written as the model $\tilde{\mathbf{y}} = \boldsymbol{\mu}_{\tilde{y}|y} + \tilde{\mathbf{u}}$, where $\boldsymbol{\mu}_{\tilde{y}|y} = \tilde{X}\boldsymbol{\beta} + V_{y,\tilde{y}}^{\mathsf{T}} V_y^{-1} (\mathbf{y} - X\boldsymbol{\beta})$ and $\tilde{\mathbf{u}} \sim N \left(\mathbf{0}, \sigma^2 V_{\tilde{y}|y} \right)$. We can now proceed analogously to the derivation of (5.33) from (5.32), now using $p(\boldsymbol{\beta}, \sigma^2 \mid \mathbf{y})$ defined in (5.30) to obtain

$$p(\tilde{\mathbf{y}}, \sigma^2 \mid \mathbf{y}) = NIG(\tilde{\mathbf{y}}, \sigma^2 \mid \tilde{\mathbf{m}}, \tilde{M}, a^*, b^*), \tag{5.36}$$

where $\tilde{\mathbf{m}} = AM\mathbf{m} + V_{y,\tilde{y}}^{\mathsf{T}} V_y^{-1} \mathbf{y}$, $\tilde{M} = V_{\tilde{y}} + AMA^{\mathsf{T}}$ with $A = \tilde{X} - V_{y,\tilde{y}}^{\mathsf{T}} V_y^{-1} X$, and M, $\mathbf{m}$, a^* and b^* are as in (5.30). The posterior predictive density is $p(\tilde{\mathbf{y}} \mid \mathbf{y}) = t_{2a^*}(\tilde{\mathbf{y}} \mid \tilde{\mathbf{m}}, (b^*)/a^* \tilde{M})$.

5.4   Exact sampling from joint distributions

In Section 5.3, we derived closed forms for $p(\sigma^2 \mid \mathbf{y})$, $p(\boldsymbol{\beta} \mid \sigma^2, \mathbf{y})$ and $p(\tilde{\mathbf{y}} \mid \mathbf{y}, \sigma^2)$. While the marginal posterior distribution $p(\boldsymbol{\beta} \mid \mathbf{y})$ from (5.25) is also available in closed form as a $t_\nu(\boldsymbol{\mu}, \Sigma)$ density with $\nu = 2a^*$, $\boldsymbol{\mu} = M\mathbf{m}$ and $\Sigma = (b^*/a^*)M$, it will be more convenient to conduct inference using samples drawn from the joint posterior distribution $p(\boldsymbol{\beta}, \sigma^2 \mid \mathbf{y})$.

Before turning to the specific Bayesian linear regression model in (5.21) or (5.25), let us briefly consider drawing exact samples from a joint distribution using marginal and conditional sampling. Consider 2 random variables X and Y with joint density function $(X, Y) \sim f(X, Y)$. We write the density at (x, y) as $f_{X,Y}(x, y) = f_X(x) f_{Y|X}(y \mid x)$, where $X \sim f_X(\cdot)$ and $Y \mid (X = x) \sim f_{Y|X}(\cdot \mid x)$.

Suppose we can draw samples from f_X and $f_{Y|X}$ for any given $X = x$. How can we draw N samples from the above joint distribution? We first draw N samples $X_i^* = x_i \sim f_X$ for $i = 1, \ldots, N$ from the marginal distribution of X. Then, for each drawn value x_i, we draw $Y_i^* \sim f_{Y|X}(\cdot \mid x_i)$ for $i = 1, \ldots, N$. This is presented in Algorithm 1.

Algorithm 1 Composition sampling

1: **for** $(i$ in 1:N$)$, **do**
2:     Draw $X_i^* = x_i \sim f_X$
3:     Draw $Y_i^* \sim f_{Y|X}(\cdot \mid x_i)$
4: **end for**
5: **return** $\{X_i^*, Y_i^*\}_{i=1,\ldots,N}$

We offer some heuristics on why the distribution of the simulated random variables is $(X_i^*, Y_i^*) \overset{ind}{\sim} f_{X,Y}(\cdot, \cdot)$. It is clear that $P(X_i^* \leq u) = \int_{-\infty}^u f_X(x)dx$ because $X_i^* \sim f_X(\cdot)$, i.e., X_i^* has the same distribution as X. And $P(Y_i^* \leq u \mid X_i^* = x) = \int_{-\infty}^u f_{Y|X}(y \mid x)dy$ because $Y_i^* \mid (X_i^* = x) \sim f_{Y|X}(\cdot \mid x)$. Therefore,

$$P(X_i^* \leq u, Y_i^* \leq v) = \int_{-\infty}^u f_X(x) \int_{-\infty}^v f_{Y|X}(y \mid x)dydx$$
$$= \int_{-\infty}^u \int_{-\infty}^v f_X(x)f_{Y|X}(y \mid x)dydx$$
$$= P(X \leq u, Y \leq v).$$

Hence, $(X_i, Y_i) \overset{d}{=} (X, Y)$. Since each draw is independent across i, the samples are independent. We can apply this method to draw samples from a joint posterior distribution,

$$p(\theta_1, \theta_2 \mid \mathbf{y}) = p(\theta_1 \mid \mathbf{y}) \times p(\theta_2 \mid \theta_1, \mathbf{y}).$$

In conjugate models, it is often easy to draw samples from $p(\theta_1 \mid \mathbf{y})$ and from $p(\theta_2 \mid \theta_1, \mathbf{y})$. We first draw N samples $\theta_{1(i)} \sim p(\theta_1 \mid \mathbf{y})$ and, for each drawn value $\theta_{1(i)}$, we draw $\theta_{2(i)} \sim p(\theta_2 \mid \theta_{1(i)}, \mathbf{y})$. Algorithm 2 presents the steps.

Algorithm 2 Composition sampling from posteriors

1: **for** $(i$ in $1 : N)$, **do**
2:     Draw $\theta_{1(i)} \sim p(\theta_1 \mid \mathbf{y})$
3:     Draw $\theta_{2(i)} \sim p\left(\theta_2 \mid \theta_{1(i)}, \mathbf{y}\right)$
4: **end for**
5: **return** $\left\{\theta_{1(i)}, \theta_{2(i)}\right\}_{i=1,\ldots,N}$

The remarkable impact of the above sampling scheme, which we call *composition sampling*, is that it replaces integration with Monte Carlo simulation. The above discussion shows that generating samples from the joint distribution $p(\theta_1, \theta_2 \mid \mathbf{y})$ effectively and *automatically* produces samples from the marginal distributions $p(\theta_1 \mid \mathbf{y})$ and $p(\theta_2 \mid \mathbf{y})$ [this remarkable fact and its significant applicability in Bayesian inference is the essence of the paper by Gelfand and Smith, 1990]. Therefore, we have completely avoided the integration,

$$p(\theta_2 \mid \mathbf{y}) = \int p(\theta_2 \mid \theta_1, \mathbf{y})p(\theta_1 \mid \mathbf{y})d\theta_1 = \int p(\theta_2, \theta_1 \mid \mathbf{y})d\theta_1,$$

by simulating draws from the left hand side. Since N is arbitrary, we can make it as large as we wish. So, importantly, through this sampling we can learn *arbitrarily well* about features of the joint and marginal distributions.

It should, then, come as no surprise that the same principle can be applied to evaluate the posterior predictive distribution in (5.5). We avoid the integration there by drawing samples from $p(\tilde{\mathbf{y}} \mid \mathbf{y})$, where $\tilde{\mathbf{y}}$ represents the predictive random variable and $\mathbf{y}$ the observed data. Given that we have drawn N samples $\theta_{(i)} \sim p(\theta \mid \mathbf{y})$ for $i = 1, \ldots, N$, we simply draw one value of the predictive random variable $\tilde{\mathbf{y}}$ from its conditional predictive distribution $p(\tilde{\mathbf{y}} \mid \theta, \mathbf{y})$, i.e., we draw $\tilde{\mathbf{y}}_{(i)} \sim p(\tilde{\mathbf{y}} \mid \mathbf{y}, \theta_{(i)})$ for $i = 1, \ldots, N$. The resulting samples are from $p(\tilde{\mathbf{y}} \mid \mathbf{y})$. This is summarized in the following Algorithm.

Algorithm 3 Sampling from posterior predictive distribution

1: Draw $\theta_{(1)}, \ldots, \theta_{(N)} \sim p(\theta \mid \mathbf{y})$
2: **for** $(i$ in $1 : N)$, **do**
3:      Draw $\tilde{\mathbf{y}}_{(i)} \sim p(\tilde{\mathbf{y}}_{(i)} \mid \theta_{(i)}, \mathbf{y})$
4: **end for**
5: **return** $\left\{ \tilde{\mathbf{y}}_{(i)} \right\}_{i=1,\ldots,N}$

Sampling from the joint posterior distribution in (5.24) is achieved by first sampling $\sigma^2 \sim IG(a^*, b^*)$ and then drawing one instance of $\boldsymbol{\beta} \sim N(M\mathbf{m}, \sigma^2 M)$ for each sampled value of σ^2. This yields marginal posterior samples from $p(\boldsymbol{\beta} \mid \mathbf{y})$, which is a multivariate t distribution that is available in closed form, but we do not need to work with its density function. We summarize this in the following Algorithm.

Algorithm 4 Bayesian inference for linear regression

1: Compute M, $\mathbf{m}$, a^* and b^* in (5.24)
2: **for** $(i$ in $1 : N)$, **do**
3:      Draw $\sigma^2_{(i)} \sim IG(a^*, b^*)$
4:      Draw $\boldsymbol{\beta}_{(i)} \sim N\left(M\mathbf{m}, \sigma^2_{(i)} M\right)$
5: **end for**
6: **return** $\left\{ \sigma^2_{(i)}, \boldsymbol{\beta}_{(i)} \right\}_{i=1,\ldots,N}$

Bayesian predictive inference follows by applying Algorithm 3 with the posterior samples of $\boldsymbol{\beta}$ and σ^2 drawn using Algorithm 4. For each drawn $\boldsymbol{\beta}_{(i)}$ and $\sigma^2_{(i)}$, $i = 1, \ldots, N$ we draw $\tilde{\mathbf{y}}_{(i)} \sim N(\tilde{X}\boldsymbol{\beta}_{(i)}, \sigma^2_{(i)})$. The resulting $\{\tilde{\mathbf{y}}_{(i)}\}_{i=1,\ldots,N}$ are samples from the posterior predictive distribution given in (5.31). We have sampled from the multivariate t distribution given there without worrying about its mathematical form.

The aforementioned discussion illustrates some basic concepts in simulation-based inference using conditional and marginal distributions. The application to Bayesian linear regression is especially useful because we had access to a marginal posterior distribution $(\sigma^2 \mid \mathbf{y})$ and a conditional posterior distribution $(\boldsymbol{\beta} \mid \sigma^2, \mathbf{y})$. This enables us to draw Monte Carlo samples *exactly* from the joint posterior distribution. By "exactly" we mean that every drawn sample (from $i = 1, \ldots, N$) is from the joint distribution. We do not need to undergo several iterations in order to "converge" to the joint posterior. However, more general hierarchical models, including several we will be encountering for spatial data analysis, may not afford us such luxuries and we will need to turn to iterative (as opposed to exact) Markov chain Monte Carlo algorithms to sample from the joint posterior distribution. These algorithms require tuning and time to converge to the desired joint posterior distribution (after they "burn in" with initial sampling), but once they have converged we can use the posterior samples and avoid possibly high-dimensional integration. We discuss these next.

5.5   Bayesian computation

The exact sampling methods outlined in Section 5.4 are restricted by the analytical tractability of the marginal posterior distributions for certain parameters (e.g., $p(\theta_1 \mid \mathbf{y})$ in Algorithm 2). In general, we do not have closed forms for the marginal posterior distributions for *any* of the posterior distributions and we need more versatile computational methods.

In fact, a closer look at (5.3) reveals that the computational challenge in applying Bayesian methods emanate from the integration required to do inference which is generally not tractable in closed form and must be approximated numerically. Here the emergence of inexpensive, high-speed computing equipment and software played a key role, enabling the application of Markov chain Monte Carlo (MCMC) methods, such as the Metropolis-Hastings algorithm [Metropolis et al., 1953, Hastings, 1970] and the Gibbs sampler[Geman and Geman, 1984, Gelfand and Smith, 1990] 1990).

We provide a brief introduction to Bayesian computing while referring the reader to more expansive treatments in a number of textbooks. Almost all textbooks on Bayesian statistics include discussions on computation [e.g., Carlin and Louis, 2008, Christensen et al., 2011, Gelman et al., 2013] . The explosion in Bayesian activity and computing power in the past twenty or so years has caused a similar explosion in the number of books in this area. One of the earliest comprehensive treatment was by Tanner [1997], with books by Gilks et al. [1995], Chen et al. [2000], Gamerman and Lopes [2006] and Amaral Turkman et al. [2019] offering updated and expanded discussions that are primarily Bayesian in focus. Also significant are the computing books by Liu [2002] and Robert and Casella [2004], which, while not specifically Bayesian, still emphasize Markov chain Monte Carlo methods typically used in modern Bayesian analysis. Perhaps the comprehensive summary of this activity appears in Brooks et al. [2011].

5.5.1   Markov chain Monte Carlo methods

Without doubt the most widely used computing tools in Bayesian practice today are Markov chain Monte Carlo (MCMC) methods. This is due to their ability (in principle) to enable inference from posterior distributions of arbitrarily large dimension, essentially by reducing the problem to one of recursively addressing a series of lower-dimensional (often unidimensional) problems. Like traditional Monte Carlo methods, MCMC methods work by producing not a closed form for the posterior (of a feature of interest) in (5.3), but a *sample* of values $\{\theta^{(g)}, g = 1, \ldots, G\}$ from this distribution. In this sense, we revert to the most basic of statistical ideas; in order to learn about a distribution/population, and sample from it. While sampling obviously does not carry as much information as the closed form itself, a histogram or kernel density estimate based on such a sample is typically sufficient for reliable inference; moreover, as noted above, such inference can be made arbitrarily accurate merely by increasing the Monte Carlo sample size G. However, unlike traditional Monte Carlo methods, MCMC algorithms produce *correlated* samples from this posterior, since they arise as iterative draws from a particular Markov chain, the stationary distribution of which is the same as the posterior.

The convergence of the Markov chain to the correct stationary distribution can be guaranteed for an enormously broad class of posteriors, explaining the popularity of MCMC. But this convergence is also the source of most of the difficulty in actually implementing MCMC procedures, for two reasons. First, it forces us to make a decision about when it is safe to stop the sampling algorithm and summarize its output, an area known in the business as *convergence diagnosis*. Second, it clouds the determination of the quality of the estimates produced (since they are based not on i.i.d. draws from the posterior, but on correlated samples. This is sometimes called the *variance estimation* problem since a common goal here is to estimate the Monte Carlo variances (equivalently standard errors) associated with our MCMC-based posterior estimates.

In the remainder of this section, we introduce the three most popular MCMC algorithms, the Gibbs sampler, the Metropolis-Hastings algorithm and the slice sampler. We then return to the convergence diagnosis and variance estimation problems.

5.5.1.1   The Gibbs sampler

Suppose our model features k parameters, $\theta = \{\theta_1, \ldots, \theta_k\}$. To implement the Gibbs sampler, we must assume that samples can be generated from each of the *full* or *complete* conditional distributions $\{p(\theta_i \mid \theta_{j \neq i}, \mathbf{y}), \ i = 1, \ldots, k\}$ in the model. Such samples might be available directly (say, if the full conditionals were familiar forms, like normals and gammas) or indirectly (say, via a rejection sampling approach). In this latter case two popular alternatives are the adaptive rejection sampling (ARS) algorithm of Gilks and Wild [1992], and the Metropolis algorithm described in the next subsection. In either case, under mild conditions, the collection of full conditional distributions uniquely determine the joint posterior distribution, $p(\theta \mid \mathbf{y})$, and hence all marginal posterior distributions $p(\theta_i \mid \mathbf{y})$, $i = 1, \ldots, k$.

Given arbitrary starting values $\{\theta_2^{(0)}, \ldots, \theta_k^{(0)}\}$, Gibbs sampling proceeds as below.

Algorithm 5 Gibbs sampling

1: Initialize $\{\theta_2^{(0)}, \ldots, \theta_k^{(0)}\}$
2: **for** $(t \text{ in } 1 : T)$, **do**
3:      **for** $(j \text{ in } 1 : k)$ **do**
4:          Draw $\theta_j^{(t)}$ from $p\left(\theta_j \,\middle|\, \theta_1^{(t)}, \ldots, \theta_{j-1}^{(t)}, \theta_{j+1}^{(t-1)}, \ldots, \theta_k^{(t-1)}, \mathbf{y}\right)$;
5:      **end for**
6: **end for**
7: **return** $\left(\theta_1^{(1:T)}, \ldots, \theta_k^{(1:T)}\right)$

Under mild regularity conditions [see, e.g., Geman and Geman, 1984, Robert and Casella, 2004, Gamerman and Lopes, 2006, Amaral Turkman et al., 2019, and, in particular, the comprehensive treatment with references in the latter text], one can show that the k-tuple obtained at iteration t, $(\theta_1^{(t)}, \ldots, \theta_k^{(t)})$, converges in distribution to a draw from the true joint posterior distribution $p(\theta_1, \ldots, \theta_k \mid \mathbf{y})$.

As an important aside, the above still holds if we *block* the components of θ, replacing univariate updating with block updating. If such sampling is convenient it will typically produce faster convergence of the Gibbs sampling (although good choices of blocking will be dependent upon the joint distribution so may not be clear).

A formal proof is not our intent here, but a bit of heuristics cannot hurt [see a very accessible exposition by Casella and George, 1992, that also includes an explanation, more rigorous than what we offer here, of why the Gibbs sampler works]. The key point to note is that if the resulting Markov chain has reached its stationary distribution, then it stays there in the next update. For clarity consider the case with $k = 2$ in Algorithm 5 with $\theta = (\theta_1, \theta_2)$ and we want to draw samples from the joint posterior distribution $p(\theta \mid \mathbf{y})$. Using somewhat informal notation, suppose that $(\theta_1^{(t)}, \theta_2^{(t)})$ are in fact samples from the true joint posterior distribution $p\left(\theta_1^{(t)}, \theta_2^{(t)} \mid \mathbf{y}\right)$, which means that $\theta_2^{(t)} \sim p(\theta_2 \mid \mathbf{y})$ (the true marginal posterior). In the transition to time $t + 1$ in the Gibbs sampler, we generate $\theta_1^{(t+1)} \sim p(\theta_1 \mid \theta_2^{(t)}, \mathbf{y})$ so $(\theta_1^{(t+1)}, \theta_2^{(t)}) \sim p(\theta_1, \theta_2 \mid \mathbf{y})$ and, hence, $\theta_1^{(t+1)} \sim p(\theta_1 \mid \mathbf{y})$. Next, we draw $\theta_2^{(t+1)} \sim p(\theta_2 \mid \theta_1^{(t+1)}, \mathbf{y})$ and this implies that $(\theta_1^{(t+1)}, \theta_2^{(t+1)}) \sim p(\theta_1, \theta_2 \mid \mathbf{y})$.

As a historical footnote, we add that Geman and Geman [1984] chose the name "Gibbs sampler" because the distributions used in their context (image restoration, where the parameters were actually the colors of pixels on a screen) were Gibbs distributions (as previously seen in equation (4.9)). These were in turn named after J.W. Gibbs, a 19th-century American physicist and mathematician generally regarded as one of the founders of modern thermodynamics and statistical mechanics. While Gibbs distributions form an exponential family on potentials (as in Section 4.2), it becomes clear that they are better suited for sampling in Bayesian applications. This follows because the form of the joint posterior distribution is known, up to the normalizing constant, immediately from the model specification. Since this joint distribution uniquely determines the marginal distributions, the forms of the conditional distributions required for Gibbs sampling are immediately determined, again up to normalizing constants. In some cases these distributions are standard statistical ones (normal, gamma, etc.). In most modern hierarchical modeling contexts they are not so we turn to other methods to administer the sampling.

5.5.1.2 The Metropolis-Hastings algorithm

The Gibbs sampler is easy to understand and implement but requires the ability to readily sample from each of the full conditional distributions, $p(\theta_i \mid \boldsymbol{\theta}_{j\neq i}, \mathbf{y})$. Unfortunately, when the prior distribution $p(\theta)$ and the likelihood $f(\mathbf{y} \mid \theta)$ are not a conjugate pair, one or more of these full conditionals may not be available in closed form. Even in this setting, however, $p(\theta_i \mid \theta_{j\neq i}, \mathbf{y})$ *will* be available up to a proportionality constant, since it is proportional to the portion of $f(\mathbf{y} \mid \theta) \times p(\theta)$ that involves θ_i.

The *Metropolis algorithm* (or *Metropolis-Hastings algorithm*) is a rejection algorithm that attacks precisely this problem, since it requires only a function proportional to the distribution to be sampled, at the cost of requiring a rejection step from a particular *candidate* density. Like the Gibbs sampler, this algorithm was not developed by statistical data analysts for this purpose, but by statistical physicists working on the Manhattan Project in the 1940s seeking to understand the particle movement theory underlying the first atomic bomb (one of the coauthors on the original Metropolis et al. (1953) paper was Edward Teller, who is often referred to as "the father of the hydrogen bomb"). In this regard, it was used to implement forward simulation of realizations (scenarios) under a complex, high dimensional model.

Suppose we wish to generate from a joint posterior distribution distribution $p(\theta \mid \mathbf{y})$ for a possibly vector-valued parameter θ. The Metropolis algorithm requires us to specify a *symmetric* probability density function $q(x \mid z) = q(z \mid x)$ so that q is a valid density function in x and in z on the same domain as θ. This density is called a *proposal density* because we "propose" a value of θ at iteration t by drawing a random value $\theta^* \sim q(\cdot \mid \theta^{(t-1)})$. We then accept the proposed draw as the value of $\theta^{(t)}$ with probability $\min\left\{1, p(\theta^* \mid \mathbf{y})/p(\theta^{(t-1)} \mid \mathbf{y})\right\}$ by simulating a coin toss. If the proposal is rejected, then we set $\theta^{(t)} = \theta^{(t-1)}$ and propose a new draw followed by the accept-reject step to move to $\theta^{(t+1)}$. Algorithm 6 describes the Metropolis sampler given a starting value $\theta^{(0)}$ at iteration $t = 0$.

Algorithm 6 Metropolis random walk algorithm

1: Initialize $\theta^{(0)}$ and choose symmetric proposal distribution $q(\cdot \mid \cdot)$.
2: **for** $(t$ in $1 : T)$ **do**
3:     $\theta^{(t)} = \theta^{(t-1)}$
4:     Draw $\theta^* \sim q(\cdot \mid \theta^{(t-1)})$;
5:     Set $\theta^{(t)} = \theta^*$ with probability $\alpha = \min\left\{1, \exp[\log p(\theta^* \mid \mathbf{y}) - \log p(\theta^{(t-1)} \mid \mathbf{y})]\right\}$.
6: **end for**
7: **return** $(\theta^1, \ldots, \theta^{\mathrm{T}})$

Let us offer some intuition [deferring to texts such as Robert and Casella, 2004, Gamerman and Lopes, 2006, Amaral Turkman et al., 2019, for further details]. An important concept underlying Algorithm 6 is "time reversibility," which, again using informal notation, essentially means that $P(\theta^{(t)} = a, \theta^{(t+1)} = b) = P(\theta^{(t)} = b, \theta^{(t+1)} = a)$. The Metropolis random walk ensures "time reversibility" when $\theta^{(t)} \sim p(\theta \mid \mathbf{y})$, i.e., once the sample is drawn from the true posterior distribution, because (assuming below that all divisions are legal)

$$P(\theta^{(t)} = a, \theta^{(t+1)} = b) = p(a \mid \mathbf{y})q(b \mid a) \min \left(1, \frac{p(b \mid \mathbf{y})}{p(a \mid \mathbf{y})}\right)$$

$$= \min \left(p(a \mid \mathbf{y})q(b \mid a), q(b \mid a)p(b \mid \mathbf{y})\right) = \min \left(p(a \mid \mathbf{y})q(a \mid b), q(a \mid b)p(b \mid \mathbf{y})\right)$$

$$= p(b \mid \mathbf{y})q(a \mid b) \min \left(1, \frac{p(a \mid \mathbf{y})}{p(b \mid \mathbf{y})}\right) = P(\theta^{(t)} = b, \theta^{(t+1)} = a) , \quad (5.37)$$

where $p(a \mid \mathbf{y})$ and $p(b \mid \mathbf{y})$ denote the posterior density $p(\theta \mid \mathbf{y})$ evaluated at $\theta = a$ and $\theta = b$. Note the use of symmetry, $q(a \mid b) = q(b \mid a)$, in the 3rd equality in (5.37).

Time reversibility implies that if $\theta^{(t)} \sim p(\theta \mid \mathbf{y})$, then $\theta^{(t+1)} \sim p(\theta \mid \mathbf{y})$. This is straightforward: Let $f_{t,t+1}(a, b)$ be the joint density of $(\theta^{(t)}, \theta^{(t+1)})$ evaluated at (a, b) and let $f_{t+1}(b)$ be the marginal density of $\theta^{(t+1)}$ evaluated at $\theta^{(t+1)} = b$. Then,

$$f_{t+1}(b) = \int f_{t,t+1}(a, b)da = \int f_{t,t+1}(b, a)da = f_t(b) = p(b \mid \mathbf{y}),$$

where the second equality comes from time-reversibility and the last equality comes from our assumption that $\theta^{(t)} \sim p(\theta \mid \mathbf{y})$. Then under generally the same mild conditions as those supporting the Gibbs sampler, a draw $\theta^{(t)}$ converges in distribution to a draw from the true posterior density $p(\theta \mid \mathbf{y})$.

Recall that the steps of the Gibbs sampler were fully determined by the statistical model under consideration (since full conditional distributions for well-defined models are unique). By contrast, the Metropolis algorithm affords substantial flexibility through the selection of the candidate density q. This flexibility can be a blessing and a curse: while theoretically we are free to pick almost any candidate density, in practice only a "good" choice will result in sufficiently many candidate acceptances. A simple and widely employed practical choice for the proposal density for sampling vector-valued $\theta \in \mathbb{R}^p$ is the multivariate normal density, i.e., $q(x \mid z) = N(x \mid z, \Sigma)$, where Σ is the fixed covariance matrix. Therefore, in Algorithm 6,

$$q(\theta^* \mid \theta^{(t-1)}) = N(\theta^* \mid \theta^{(t-1)}, \Sigma). \qquad (5.38)$$

The symmetry of the density ensures that $q(\theta^* \mid \theta^{(t-1)}) = q(\theta^{(t-1)} \mid \theta^*)$ and is "self correcting" in the sense that candidates are always centered around the current value of the chain. This proposal is often referred to as a "random walk" proposal density.

An effective Metropolis algorithm should sample from the entire support of the posterior distribution. The variance of the proposal distribution affects the acceptance rate. High variances result in some proposed draws of θ^* that are distant from $\theta^{(t-1)}$ and in low probability regions of the posterior distribution. This leads to $p(\theta^* \mid \mathbf{y})/p(\theta^{(t-1)} \mid \mathbf{y})$ being small implying lower values of the acceptance probability α in Algorithm 6. With too few proposed values being accepted, the chain will move slowly and struggle to explore the posterior. On the other hand, small variances will yield θ^* to be sampled close to $\theta^{(t-1)}$, which means $p(\theta^* \mid \mathbf{y})/p(\theta^{(t-1)} \mid \mathbf{y}) \approx 1$ and, hence, $\alpha \approx 1$. This too yields undesirable chains because the chain moves only very short distances and will, again, be impracticably slow in exploring the posterior. A good proposal density should offer the flexibility to control, or "tune," the acceptance rates. It should ensure that we do not accept all, or nearly all, of the proposed values drawn because the chain will "baby-step" around the parameter space, leading to high acceptance but also high autocorrelation in the sampled chain and slow exploration of the support for the distribution. This is often the result of an overly narrow proposal density. On the other hand, an overly dispersed candidate density will also struggle, proposing leaps to places far from the bulk of the support of the posterior, leading to high rejection and, again, high autocorrelation.

The Gaussian proposal density is attractive because the variance in (5.38) is explicitly modeled by Σ. Increasing or decreasing the variance components (diagonal elements) of Σ increases or decreases the variability so Σ acts as a tuning parameter that is easily controlled. Here we might try to mimic the posterior variance by setting Σ equal to an empirical estimate of the true posterior variance, derived from a preliminary sampling run. While intuition may suggest values of Σ so that roughly 50% of the candidates are accepted, theoretical results [e.g., Gelman et al., 1996b] indicate even lower acceptance rates (25% to 40%) are optimal, but this result varies with the dimension and the true posterior correlation structure of θ. As a result, the choice of Σ is often done *adaptively*. For instance, in one dimension (setting $\Sigma = \sigma$, and thus avoiding the issue of correlations among the elements of θ), a common trick is to simply pick some initial value of σ, and then keep track of the empirical proportion of candidates that are accepted. If this fraction is too high (75% to 100%), we simply increase σ; if it is too low (0 to 20%), we decrease it. Since adaptation infinitely often can actually disturb the chain's convergence to the desired stationary distribution, the simplest approach is to allow this adaptation only during the burn-in period, a practice sometimes referred to as *pilot adaptation*. A more involved alternative is to allow adaptation at *regeneration points* which, once defined and identified, break the Markov chain into independent sections [see, e.g., Mykland et al., 1995, Hobert et al., 2002, Mira and Sargent, 2003, for discussions of the use of regeneration in practical MCMC settings].

The Gaussian proposal distribution has support over $\mathbb{R}^p$ and is an especially convenient choice for parameters having the same support. Often, however, the support of θ will be restricted to perhaps positive values or bounded domains. In such cases we can still use a Gaussian proposal density is applied to a transformation of θ that will take it to the support of the Gaussian density. For example, considering a real-valued scalar parameter $\theta > 0$ (such as a variance), we can use a logarithm transformation $\tilde{\theta} = \log\theta$ and use the posterior $p(\tilde{\theta} \mid \mathbf{y})$ which is obtained from the standard change of variable adjusted for the Jacobian. Another common example is a scalar parameter θ that has bounded support, say an interval (a, b) inherited from a Uniform prior, and we wish to transform to $\tilde{\theta} \in \mathbb{R}$. One could then use the transformation $\tilde{\theta} = \log\left((b - \theta)/(\theta - a)\right)$. For vector-valued θ, we apply element-wise transformations as needed to make sure that all elements are mapped to the real line congruous with the support of the Gaussian density.

A different option is to adapt the Metropolis algorithm as devised by Hastings [1970], where we no longer require that q be symmetric in its arguments. Here, we modify the calculation of the acceptance ratio to $\alpha = \min\left\{1, \dfrac{p(\theta^* \mid \mathbf{y})q(\theta^{(t-1)} \mid \theta^*)}{p(\theta^{(t-1)} \mid \mathbf{y})q(\theta^* \mid \theta^{(t-1)})}\right\}$. Calculations similar to (5.37) reveal that time-reversibility is satisfied with this modified α, which implies, as for the Metropolis, that if $\theta^{(t)} \sim p(\theta \mid \mathbf{y})$ then $\theta^{(t+1)} \sim p(\theta \mid \mathbf{y})$. Therefore, the posterior distribution is the *stationary distribution* of the Markov chain and once we start sampling from $p(\theta \mid \mathbf{y})$, all subsequent samples are drawn from it.

Algorithm 7 Metropolis-Hastings algorithm

1: Initialize $\theta^{(0)}$ and choose symmetric proposal distribution $q(\cdot \mid \cdot)$.

2: **for** $(t$ in $1 : T)$ **do**

3:     $\theta^{(t)} = \theta^{(t-1)}$

4:     Draw $\theta^* \sim q(\cdot \mid \theta^{(t-1)})$;

5:     Set $\theta^{(t)} = \theta^*$ with probability $\alpha = \min\left\{1, \dfrac{p(\theta^* \mid \mathbf{y})q(\theta^{(t-1)} \mid \theta^*)}{p(\theta^{(t-1)} \mid \mathbf{y})q(\theta^* \mid \theta^{(t-1)})}\right\}$.

6: **end for**

7: **return** $\left(\theta^1, \ldots, \theta^T\right)$

5.5.1.3   Metropolis-Adjusted Langevin algorithm

The random walk Metropolis (RWM) algorithm is very popular among practitioners because it is general and easy to implement. In addition, the RWM is quite robust to the choice of the tuning (scaling) parameters. Unfortunately, this simplicity often leads to performance that scales poorly with increasing dimensions and increasing complexity of the target density. Even when the proposal is optimally chosen, RWM relies on local moves that lead to slow mixing, especially in high dimensions. The proposal distribution of a RWM is randomly exploring the interesting parts of the posterior density without considering its structure.

So, we are seeking better proposal distributions that incorporate the structure of the target density, leading to faster mixing. Using gradient information of the target function can help improve MCMC convergence. Methods using this information are especially useful in high dimensions where a simple random walk has many more directions to wander off and tracing a narrow high probability density shell will be very difficult. The simplest method for incorporating gradient information is the Metropolis-adjusted Langevin algorithm (MALA) [Roberts and Rosenthal, 1998a] which simulates a reversible Langevin diffusion whose stationary distribution is a target distribution, say $p(\boldsymbol{\theta})$. In this section (and the next) we use boldface ($\boldsymbol{\theta}$) to indicate that the parameters reside in an Euclidean vector space.

In particular, let $p(\boldsymbol{\theta} \mid \mathbf{y})$ be a continuous and differentiable posterior density in $\mathbb{R}^p$. Gradient-based methods exploit the gradient of the logarithm of the target density, written

$$\nabla_{\boldsymbol{\theta}} \log p(\boldsymbol{\theta} \mid \mathbf{y}) = \nabla_{\boldsymbol{\theta}} \log p(\mathbf{y} \mid \boldsymbol{\theta}) + \nabla_{\boldsymbol{\theta}} \log p(\boldsymbol{\theta}). \tag{5.39}$$

The gradient is often available in closed form, and it does not require the knowledge of the normalizing constant. The gradient informs about the direction and the rate of increase of a given function. For instance, for a given value $\boldsymbol{\theta}^{(r)}$ and $\epsilon > 0$, the update $\boldsymbol{\theta}^{(r+1)} \leftarrow \boldsymbol{\theta}^{(r)} + \epsilon \nabla_{\boldsymbol{\theta}} \log p(\boldsymbol{\theta}^{(r)} \mid \mathbf{y})$, leads to an increase of $p(\boldsymbol{\theta} \mid \mathbf{y})$, for ϵ small enough. This corresponds to the well-known gradient ascent method.

Incorporating the gradient in an MCMC procedure is an intuitive and appealing idea. It will push the Markov chain towards values with higher density. Further, a strong theoretical justification exists for gradient-adjusted Metropolis Hastings proposals, based on Langevin diffusions [Roberts and Tweedie, 1996].

Let $\mathbf{B}^{(t)}$ be p-dimensional standard Brownian motion. Consider a continuous-time stochastic process $\boldsymbol{\theta}^{(t)}$ satisfying the following stochastic differential equation:

$$d\boldsymbol{\theta}^{(t)} = \frac{1}{2} \nabla_{\boldsymbol{\theta}} \log p(\boldsymbol{\theta}^{(t)} \mid \mathbf{y}) dt + d\mathbf{B}^{(t)}. \tag{5.40}$$

A key result is that the stationary distribution of this Langevin diffusion is the posterior density $p(\boldsymbol{\theta} \mid \mathbf{y})$.

In practice, we need to consider discrete approximations of the Langevin diffusion, for example using the so-called Euler method. This leads to the following discrete-time stochastic process

$$\boldsymbol{\theta}^{(r+1)} = \boldsymbol{\theta}^{(r)} + \frac{\epsilon^2}{2} \nabla_{\boldsymbol{\theta}} \log p(\boldsymbol{\theta}^{(r)} \mid \mathbf{y}) + \epsilon \mathbf{z}^{(r)} \tag{5.41}$$

for any chosen discretization step $\epsilon > 0$, and with i.i.d. $\mathbf{z}^{(r)} \sim N_p(0, \mathbb{I}_p)$. This discrete approximation is no longer guaranteed to converge to $p(\boldsymbol{\theta} \mid \mathbf{y})$. There is a delicate trade-off between the accuracy of this approximation (as $\epsilon \to 0$) and the sampling efficiency, increasing as ϵ grows. So, this issue is solved by treating the above distribution as a proposal density of a Metropolis-Hastings algorithm.

The Metropolis adjusted Langevin algorithm (MALA) therefore can be seen as a specific Metropolis Hastings algorithm with proposal distribution

$$\boldsymbol{\theta}^* \mid \boldsymbol{\theta} \sim N_p \left(\boldsymbol{\theta} + \frac{\sigma_p^2}{2} \nabla_{\boldsymbol{\theta}} \log p(\boldsymbol{\theta} \mid \mathbf{y}), \sigma_p^2 \mathbb{I}_p \right) \tag{5.42}$$

where σ_p^2 is a tuning parameter, which depends on dimension and is chosen with care. In fact, convergence can be further improved by modifying the tuning parameter within the chain using an adaptive MCMC algorithm [Roberts and Rosenthal, 2007].

We note that this proposal distribution is not symmetric as in the random walk Metropolis case. Therefore the acceptance probability ratio takes into account also the proposal densities, as in Metropolis-Hastings, namely $\alpha = \min\left(1, \frac{p(\boldsymbol{\theta}^* \mid \mathbf{y})q(\boldsymbol{\theta} \mid \boldsymbol{\theta}^*)}{p(\boldsymbol{\theta} \mid \mathbf{y})q(\boldsymbol{\theta}^* \mid \boldsymbol{\theta})}\right)$. There is strong theoretical and empirical evidence showing that a much faster mixing compensates for the price paid for computing the gradient.

5.5.1.4 Hamiltonian Monte Carlo (HMC)

The local transitions that define Markov chains often adversely affect the convergence of MCMC algorithms such as the Gibbs sampler, Metropolis, and Metropolis-Hastings [Neal, 1994]. For example, Gibbs samplers are often afflicted by high auto-correlations in the posterior samples, which results in very slow mixing of the Markov chains. With a Metropolis or Metropolis-Hastings algorithm, it can be challenging to find effective proposal distributions that suppress slow-moving random walks. A substantial literature exists on developing variants of such algorithms that help the Markov chain explore the target posterior distribution more effectively [Liu, 2002, Robert and Casella, 2004, Amaral Turkman et al., 2019]. Here, we discuss one such variant based on Hamiltonian dynamics.

Hamiltonian Monte Carlo (HMC), also known as hybrid Monte Carlo, modifies the Metropolis algorithm to suppress random walk behavior by introducing an auxiliary variable that replaces the step of generating a random draw from a proposal distribution with a deterministic simulation of Hamiltonian dynamics [Duane et al., 1987, Neal, 2011, Hoffman and Gelman, 2014]. The idea is conceptually simple and can be looked upon as a means to modify Algorithm 6 by replacing the symmetric proposal distribution q with a deterministic transition. We assume our parameters are in $\mathbb{R}^d$ (hence indicated in boldface).

Our objective remains, as in Algorithm 6, to draw samples from the posterior distribution $p(\boldsymbol{\theta} \mid \mathbf{y})$. If $\boldsymbol{\theta} \in \mathbb{R}^d$, we augment $\boldsymbol{\theta}$ with an auxiliary variable $\mathbf{r} \in \mathbb{R}^d$ and sample from

$$p(\boldsymbol{\zeta} \mid \mathbf{y}) := p(\boldsymbol{\theta}, \mathbf{r} \mid \mathbf{y}) \propto p(\boldsymbol{\theta} \mid \mathbf{y}) \times N(\mathbf{r} \mid \mathbf{0}, \mathbb{I}) \propto \exp\left\{\mathcal{L}(\boldsymbol{\theta}) - \frac{\mathbf{r}^{\mathrm{T}}\mathbf{r}}{2}\right\}, \qquad (5.43)$$

where $\boldsymbol{\zeta} := (\boldsymbol{\theta}^{\mathrm{T}}, \mathbf{r}^{\mathrm{T}})^{\mathrm{T}} \in \mathbb{R}^{2d}$ and $\mathcal{L}(\boldsymbol{\theta}) = \log p(\boldsymbol{\theta} \mid \mathbf{y})$. This augmentation may appear somewhat unnecessary at first given that we are interested in sampling from $p(\boldsymbol{\theta} \mid \mathbf{y})$. In fact, perhaps somewhat curiously, the auxiliary variable $\mathbf{r}$ is independent of the data $\mathbf{y}$ and $\boldsymbol{\theta}$ is conditionally independent of $\mathbf{r}$ given the data. This would tempt us to simply simulate $\mathbf{r} \sim N(\mathbf{0}, \mathbb{I})$ and then $\boldsymbol{\theta} \sim p(\boldsymbol{\theta} \mid \mathbf{y})$, which would yield $(\boldsymbol{\theta}, \mathbf{r})$ from (5.43), but that would defeat our purpose of improving the efficiency of sampling from $p(\boldsymbol{\theta} \mid \mathbf{y})$. Instead, the purpose of the auxiliary variable is to simplify and improve sampling from the joint posterior density $p(\boldsymbol{\zeta} \mid \mathbf{y})$ in (5.43). And, as we discussed in Section 5.4, sampling from this joint posterior density delivers the samples of $\boldsymbol{\theta}$ from $p(\boldsymbol{\theta} \mid \mathbf{y})$; the samples of $\mathbf{r}$ are ignored and play no role in subsequent inference.

For circumventing the choice of the proposal density q in the Metropolis algorithms (Algorithms 6 or (7)) to draw $\boldsymbol{\zeta}$ from (5.43), we could consider the following move at iteration t with state $\boldsymbol{\theta}^{(t)}$. We draw an $\mathbf{r}^* \sim N(\mathbf{0}, \mathbb{I})$ and execute a *deterministic* move at iteration $t+1$,

$$F : (\boldsymbol{\theta}^{(t)}, \mathbf{r}^*) \longrightarrow (\tilde{\boldsymbol{\theta}}, \tilde{\mathbf{r}}),$$

where F is a deterministic map that "proposes" the new candidate $(\tilde{\boldsymbol{\theta}}, \tilde{\mathbf{r}})$. It remains for us to construct a rule for evaluating the map $F : (\boldsymbol{\theta}^{(t)}, \mathbf{r}^*) \longrightarrow (\tilde{\boldsymbol{\theta}}, \tilde{\mathbf{r}})$ so that the time-reversibility property of the algorithm is retained and $p(\boldsymbol{\theta} \mid \mathbf{y})$ remains the stationary distribution of the Markov chain.

So what properties would be required of F for us to ensure time reversibility? Since F is deterministic, its proposed value $\tilde{\boldsymbol{\zeta}} = (\tilde{\boldsymbol{\theta}}^{\mathrm{T}}, \tilde{\mathbf{r}}^{\mathrm{T}})^{\mathrm{T}}$ could be looked upon as a move with probability one, i.e., $q(\tilde{\boldsymbol{\zeta}} \mid \boldsymbol{\zeta}^{(t)}) = 1$. But this, by itself, would not suffice and we would also need a deterministic inverse map that ensures $q(\boldsymbol{\zeta}^{(t)} \mid \tilde{\boldsymbol{\zeta}}) = 1$. This would ensure symmetry of the proposal and a Metropolis algorithm would ensure time reversibility as

$$P(\boldsymbol{\zeta}^{(t)} = \mathbf{a}, \boldsymbol{\zeta}^{(t+1)} = \mathbf{b}) = p(\mathbf{a} \mid \mathbf{y}) \min\left(1, \frac{p(\mathbf{b} \mid \mathbf{y})}{p(\mathbf{a} \mid \mathbf{y})}\right)$$

$$= \min\left(p(\mathbf{a} \mid \mathbf{y}), p(\mathbf{b} \mid \mathbf{y})\right) = p(\mathbf{b} \mid \mathbf{y}) \min\left(1, \frac{p(\mathbf{a} \mid \mathbf{y})}{p(\mathbf{b} \mid \mathbf{y})}\right) = P(\boldsymbol{\zeta}^{(t)} = \mathbf{b}, \boldsymbol{\zeta}^{(t+1)} = \mathbf{a}) .$$

where $p(\mathbf{a} \mid \mathbf{y})$ and $p(\mathbf{b} \mid \mathbf{y})$ denote the values of $p(\boldsymbol{\zeta} \mid \mathbf{y})$ at $\boldsymbol{\zeta} = \mathbf{a}$ and $\boldsymbol{\zeta} = \mathbf{b}$, respectively. However, treating this deterministic map as proposing candidates with probability one is not very meaningful. After all, if both $q(\tilde{\boldsymbol{\zeta}} \mid \boldsymbol{\zeta}^{(t)}) = 1$ and $q(\boldsymbol{\zeta}^{(t)} \mid \tilde{\boldsymbol{\zeta}}) = 1$, then wouldn't the chain simply be oscillating between $\boldsymbol{\zeta}^{(t)}$ and $\tilde{\boldsymbol{\zeta}}$? The matter of using deterministic proposals in Metropolis algorithms is more subtle and our best bet is to look at physical systems with the joint posterior as a target left invariant by movement in its arguments.

HMC achieves this by simulating transitions with F as a solution to a differential equations system from Hamiltonian mechanics [Duane et al., 1987]. Here, we treat $\boldsymbol{\theta}$ as the spatial location of a moving object in $\mathbb{R}^d$ and $\mathbf{r}$ is its momentum vector (also in $\mathbb{R}^d$). Analyzing dynamical systems proceeds by treating the location of the particle as a function of time, $\boldsymbol{\theta}(t)$ with elements $\theta_i(t) \in \mathbb{R}$. The momentum is derived from $\boldsymbol{\theta}(t)$ as the mass of the object multiplied by the velocity vector, $(d\boldsymbol{\theta}(t)/dt)$, so $\mathbf{r}(t)$ is also changing over time. These two quantities are now coupled into the system

$$\frac{d\theta_i(t)}{dt} = \frac{\partial H}{\partial r_i} \quad \text{and} \quad \frac{dr_i(t)}{dt} = -\frac{\partial H}{\partial \theta_i} \text{ for } i = 1, \ldots, d , \qquad (5.44)$$

where $H(\boldsymbol{\theta}, \mathbf{r}) = -\log p(\boldsymbol{\theta} \mid \mathbf{y}) + \mathbf{r}^{\mathrm{T}}\mathbf{r}/2$ is the sum of the potential energy, $-\log p(\boldsymbol{\theta} \mid \mathbf{y})$, and kinetic energy, $\mathbf{r}^{\mathrm{T}}\mathbf{r}/2$, of the object at location $\boldsymbol{\theta}$. Equation 5.44 leaves H unchanged even as $(\boldsymbol{\theta}(t), \mathbf{r}(t))$ change over time. Since $p(\boldsymbol{\zeta} \mid \mathbf{y}) \propto \exp(-H(\boldsymbol{\zeta}))$ in (5.44), the joint posterior distribution also remains invariant and simulating *exact* Hamiltonian dynamics would ensure that we are traveling along equal-probability contours.

In practice we simulate (5.44) using deterministic "leapfrog" integrator steps

$$\mathbf{r}\left(t + \frac{\epsilon}{2}\right) = \mathbf{r}(t) + \frac{\epsilon}{2}\nabla\mathcal{L}(\boldsymbol{\theta}); \quad \boldsymbol{\theta}(t + \epsilon) = \boldsymbol{\theta}(t) + \epsilon\mathbf{r}\left(t + \frac{\epsilon}{2}\right);$$

$$\mathbf{r}(t + \epsilon) = \mathbf{r}\left(t + \frac{\epsilon}{2}\right) + \frac{\epsilon}{2}\nabla\mathcal{L}(\theta(t + \epsilon)) . \qquad (5.45)$$

Algorithm 8 presents the function for "leapfrogging" from (θ, r) to $(\tilde{\theta}, \tilde{r})$.

Algorithm 8 Leapfrog integrator

1: **function** LEAPFROG$(\mathcal{L}, \boldsymbol{\theta}, \mathbf{r}, \epsilon)$
2:    Set $\tilde{\mathbf{r}} \leftarrow \mathbf{r} + (\epsilon/2)\nabla\mathcal{L}(\boldsymbol{\theta})$
3:    Set $\tilde{\boldsymbol{\theta}} \leftarrow \boldsymbol{\theta} + \epsilon\tilde{\mathbf{r}}$
4:    Set $\tilde{\mathbf{r}} \leftarrow \tilde{\mathbf{r}} + (\epsilon/2)\nabla\mathcal{L}(\boldsymbol{\theta})$
5: **return** $\left(\tilde{\theta}, \tilde{r}\right)$
6: **end function**

The leapfrog integrator generates an approximation of the system in (5.44), but, unlike the exact system, it does not leave $p(\boldsymbol{\zeta} \mid \mathbf{y})$ unchanged. This requires an additional Metropolis step after L leapfrog steps, where we accept $(\boldsymbol{\theta}^{(t+1)}, \mathbf{r}^{(t+1)}) = (\tilde{\boldsymbol{\theta}}, \tilde{\mathbf{r}})$ with probability

$$
\min\left(1, \frac{p(\tilde{\boldsymbol{\theta}}, \tilde{\mathbf{r}} \mid \mathbf{y})}{p(\boldsymbol{\theta}^{(t)}, \mathbf{r}^* \mid \mathbf{y})}\right) = \min\left(1, \frac{\exp\left\{\mathcal{L}(\tilde{\boldsymbol{\theta}}) - \frac{\tilde{\mathbf{r}}^{\mathrm{T}}\tilde{\mathbf{r}}}{2}\right\}}{\exp\left\{\mathcal{L}(\boldsymbol{\theta}^{(t)}) - \frac{\mathbf{r}^{*\mathrm{T}}\mathbf{r}^*}{2}\right\}}\right)
$$

to ensure time-reversibility. Algorithm 9 summarizes HMC with L "leapfrog" steps.

Algorithm 9 Hamiltonian Monte Carlo (HMC) algorithm

1: Initialize $\boldsymbol{\theta}^{(0)}$; specify $\mathcal{L}$, L, ϵ and T.
2: **for** (t in $1:T$) **do**
3:     Draw $\mathbf{r}^* \sim N(\mathbf{0}, \mathbb{I}_d)$
4:     Set $\boldsymbol{\theta}^{(t)} \leftarrow \boldsymbol{\theta}^{(t-1)}$, $\tilde{\boldsymbol{\theta}} \leftarrow \boldsymbol{\theta}^{(t-1)}$, and $\tilde{\mathbf{r}} \leftarrow \mathbf{r}^*$
5:     **for** (l in $1:L$) **do**
6:         Set $(\tilde{\boldsymbol{\theta}}, \tilde{\mathbf{r}}) \leftarrow \text{Leapfrog}(\mathcal{L}, \tilde{\boldsymbol{\theta}}, \tilde{\mathbf{r}}, \epsilon)$ (call Algorithm 8)
7:     **end for**
8:     Set $\boldsymbol{\theta}^{(t)} \leftarrow \tilde{\boldsymbol{\theta}}$ with probability $\min\left(1, \dfrac{\exp\left\{\mathcal{L}(\tilde{\boldsymbol{\theta}}) - \frac{\tilde{\mathbf{r}}^{\mathrm{T}}\tilde{\mathbf{r}}}{2}\right\}}{\exp\left\{\mathcal{L}(\boldsymbol{\theta}^{(t)}) - \frac{\mathbf{r}^{*\mathrm{T}}\mathbf{r}^*}{2}\right\}}\right)$
9: **end for**
10: **return** $\left(\boldsymbol{\theta}^1, \ldots, \boldsymbol{\theta}^T\right)$

We provide some further intuition on the time-reversibility of the simple HMC algorithm. The key to this result is that the leapfrog integrator in (5.45) preserves volumes. The update for each coordinate depends only on the other coordinates, which ensures that the leapfrog updates keeps the volume of a region unchanged after mapping that region to a new region. To be slightly more precise, let $\mathcal{D}$ be a small region in the $\boldsymbol{\zeta} = (\boldsymbol{\theta}^{\mathrm{T}}, \mathbf{r}^{\mathrm{T}})^{\mathrm{T}}$ space and suppose the L leapfrog steps map $\mathcal{D}$ to a region $\tilde{\mathcal{D}}$. Then $\mathcal{D}$ and $\tilde{\mathcal{D}}$ both have the same volume. We write the transition probability from $\mathcal{D}$ to $\tilde{\mathcal{D}}$ as

$$
T(\mathcal{D} \to \tilde{\mathcal{D}}) = (\delta V)\min\left(1, \frac{\exp(-H(\tilde{\mathcal{D}}))}{\exp(-H(\mathcal{D}))}\right) = (\delta V)\min\left(1, \exp\left(-H(\tilde{\mathcal{D}}) + H(\mathcal{D})\right)\right),
$$

(5.46)

where δV is the volume of $\mathcal{D}$ and $\tilde{\mathcal{D}}$, $H(\mathcal{D}) = -\log\left(\int_{\mathcal{D}} p(\boldsymbol{\zeta} \mid \mathbf{y})d\boldsymbol{\zeta}\right)$ and $p(\boldsymbol{\zeta} \mid \mathbf{y})$ is the joint posterior density in (5.43). If $\boldsymbol{\zeta}^{(t)} = (\boldsymbol{\theta}^{(t)}, \mathbf{r}^{(t)})$ is drawn from the joint density $p(\boldsymbol{\zeta} \mid \mathbf{y})$ in (5.43), then $P(\boldsymbol{\zeta}^{(t)} \in \mathcal{D}) = \int_{\mathcal{D}} p(\boldsymbol{\zeta} \mid \mathbf{y})d\boldsymbol{\zeta} = \exp(-H(\mathcal{D}))$. Therefore,

$$
\begin{aligned}
P(\boldsymbol{\zeta}^{(t)} \in \mathcal{D}, \boldsymbol{\zeta}^{(t+1)} \in \tilde{\mathcal{D}}) &= P(\boldsymbol{\zeta}^{(t)} \in \mathcal{D})T(\mathcal{D} \to \tilde{\mathcal{D}}) \\
&= \exp(-H(\mathcal{D}))(\delta V)\min\left(1, \exp\left(-H(\tilde{\mathcal{D}}) + H(\mathcal{D})\right)\right) \\
&= (\delta V)\min\left(\exp(-H(\mathcal{D})), \exp\left(-H(\tilde{\mathcal{D}})\right)\right) \\
&= P(\boldsymbol{\zeta}^{(t)} \in \tilde{\mathcal{D}}, \boldsymbol{\zeta}^{(t+1)} \in \mathcal{D}),
\end{aligned}
$$

(5.47)

where the last equality follows from the symmetry in the expression above it.

An important advantage of HMC over the usual Metropolis algorithms is that it avoids the issue of tuning a proposal density to obtain desirable acceptance rates. Learning from

Hamiltonian dynamics and taking steps informed by first-order gradient information makes HMC especially viable for exploring high-dimensional posteriors due to accelerated convergence, whereas Metropolis random walks are difficult to tune and Gibbs samplers suffer from high auto-correlations. However, as with almost any MCMC algorithm, there are caveats. Most notably, HMC can be sensitive to the step size ϵ and the number of leapfrog steps L. Hoffman and Gelman [2014] introduced the "No-U-Turn Sampler" (NUTS), that extends HMC using a recursive algorithm (with an additional Gibbs sampling step) to obviate some of the drawbacks of the HMC sampler in Algorithm 9 by eliminating the need to set L and automating the choice of ϵ using a method called primal-dual averaging. NUTS can thus be used with no hand-tuning at all and has taken a giant step in making MCMC suited for automatic inference engines.

5.5.1.5 Slice sampling

Another alternative to the Metropolis-Hastings algorithm that is still quite general is *slice sampling* [Neal, 2003]. In its most basic form, suppose we seek to sample a univariate $\theta \sim f(\theta) \equiv h(\theta)/\int h(\theta)d\theta$, where $h(\theta)$ is known. Suppose we add a so-called *auxiliary variable* U such that $U \mid \theta \sim Unif(0, h(\theta))$. Then the joint distribution of θ and U is $p(\theta, u) \propto 1 \cdot I(U < h(\theta))$, where I denotes the indicator function. If we run a Gibbs sampler drawing from $U \mid \theta$ followed by $\theta \mid U$ at each iteration, we can obtain samples from $p(\theta, u)$, and hence from the marginal distribution of θ, $f(\theta)$. Sampling from $\theta \mid u$ requires a draw from a uniform distribution for θ over the set $S_U = \{\theta : U < h(\theta)\}$.

Figure 5.1 reveals why this approach is referred to as slice sampling. U "slices" the nonnormalized density, and the resulting "footprint" on the axis provides S_U. If we can enclose S_U in an interval, we can draw θ uniformly on this interval and simply retain it only if $U < h(\theta)$ (i.e., if $\theta \in S_U$). If θ is instead multivariate, S_U is more complicated and now we would need a bounding rectangle.

Note that if $h(\theta) = h_1(\theta)h_2(\theta)$ where, say, h_1 is a standard density that is easy to sample, while h_2 is nonstandard and difficult to sample, then we can introduce an auxiliary variable U such that $U \mid \theta \sim U(0, h_2(\theta))$. Now $p(\theta, u) = h_1(\theta)I(U < h_2(\theta))$. Again $U \mid \theta$ is routine to sample, while to sample $\theta \mid U$ we would now draw θ from $h_1(\theta)$ and retain it only if θ is such that $U < h_2(\theta)$.

Slice sampling incurs problems similar to rejection sampling in that we may have to draw many θ's from h_1 before we are able to retain one. On the other hand, it has an advantage over the Metropolis-Hastings algorithm in that it always samples from the exact full conditional $p(\theta \mid u)$. As noted above, Metropolis-Hastings does not, and thus slice sampling would be expected to converge more rapidly. Nonetheless, overall comparison of computation time may make one method a winner in some cases, and the other a winner in other cases. We do remark that slice sampling is attractive for fitting a large range of point-referenced spatial data models following Agarwal and Gelfand [2005]. In fact, it has become useful in the context of spatial point pattern data for fitting so-called log Gaussian Cox processes. See Section 8.4.3 and the papers by Murray et al. [2010a] and Murray and Adams [2010].

5.5.1.6 Melding algorithms for posterior sampling

All of the aforementioned algorithms can be executed with the joint posterior distribution as the target to sample from. The choice of the specific algorithm depends upon a range of factors including ease of implementation, speed of convergence and overall computational efficiency. While such choices are not always easy to determine and may often require computational explorations with different algorithms, MCMC affords substantial flexibility in that they can be combined with each other without disturbing the target posterior. This is possible because each of them leaves the target posterior density (the stationary

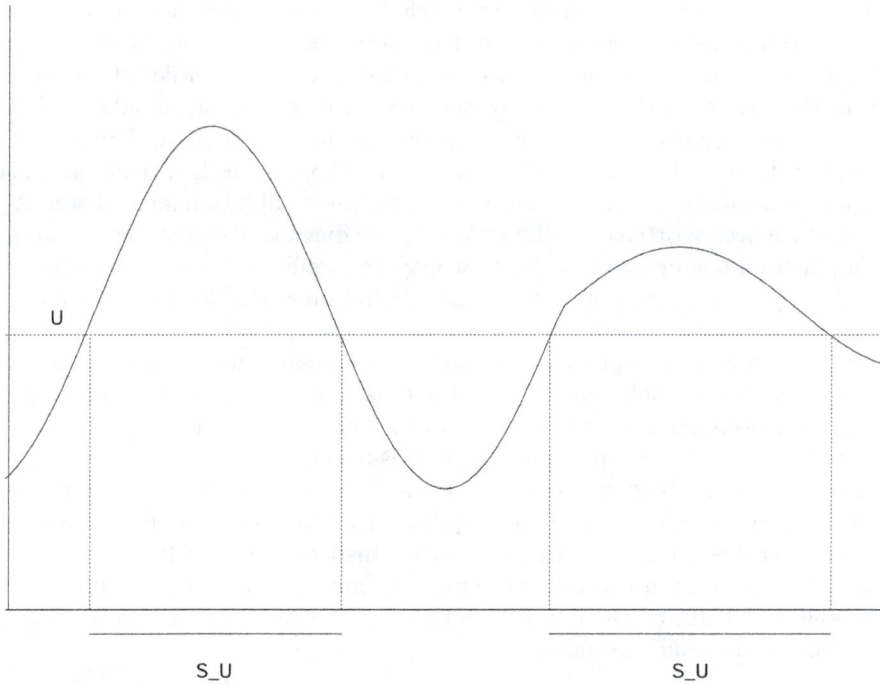

Figure 5.1 *Illustration of slice sampling. For this bimodal distribution, S_U is the union of two disjoint intervals.*

distribution of the Markov chain) invariant. For example, Gibbs samplers are especially useful for models where the full conditional distributions are available for exact sampling. However, in designing a Gibbs sampler for a complex hierarchical model we could encounter some full conditional distributions that are not available for exact sampling. In such cases, we could sample from that full conditional distribution using any other algorithm that maintains that full conditional as its stationary distribution. To be more specific, if for some j in Algorithm 5, we cannot exactly draw $\theta_j^{(t)} \sim p\left(\theta_j^{(t)} \mid \theta_1^{(t)}, \ldots, \theta_{j-1}^{(t)}, \theta_{j+1}^{(t-1)}, \ldots, \theta_k^{(t-1)}\right)$, then we can invoke either the Metropolis algorithms (Algorithms 6 or 7) or the HMC algorithm (Algorithm 9) or even slice sampling to draw from the full conditional in the Gibbs. What is even more appealing is that we need to execute only one draw using the Metropolis or HMC steps (i.e., $T = 1$) within the Gibbs sampler. Since all of these algorithms keep the stationary distribution invariant, melding these algorithms results in a new MCMC algorithm with the same target posterior distribution.

5.5.1.7   Convergence diagnosis

Diagnosing convergence in MCMC computation amounts to deciding when it is safe to stop the algorithm and summarize the output. This means that we must make a guess as to the iteration t_0 after which all output may be thought of as coming from the true stationary distribution of the Markov chain (i.e., the true posterior distribution). The most common approach here is to run a few (say, $m = 3$ or 5) *parallel* sampling chains, initialized at disparate starting locations that are overdispersed with respect to the true posterior. These chains are then plotted on a common set of axes, and their *trace plots* are then viewed to see if there is an identifiable point t_0 after which all m chains seem to be "overlapping" (traversing essentially the same part of θ-space).

There are obvious problems with this approach. First, since the posterior is unknown at the outset, there is no reliable way to ensure that the m chains are "initially overdispersed," as required for a convincing diagnostic. We might use extreme quantiles of the prior $p(\theta)$ and rely on the fact that the support of the posterior is typically a subset of that of the prior, but this requires a proper prior and in any event is perhaps doubtful in high-dimensional or otherwise difficult problems. Second, it is hard to see how to automate such a diagnosis procedure, since it requires a subjective judgment call by a human viewer. A great many papers have been written on various convergence diagnostic statistics that summarize MCMC output from one or many chains that may be useful when associated with various stopping rules; see Cowles and Carlin [1996] and Mengersen et al. [1999] for reviews of many such diagnostics.

Gelman and Rubin [1992] propose a statistic that remains widely used to assess convergence. Here, we run a small number (m) of parallel chains with different starting points that are "initially overdispersed" with respect to the true posterior. (Of course, since we do not know the true posterior before beginning there is technically no way to ensure this; still, the rough location of the bulk of the posterior may be discernible from known ranges, the support of the (proper) prior, or perhaps a preliminary posterior mode-finding algorithm.) Running the m chains for $2N$ iterations each, we then try to see whether the variation within the chains for a given parameter of interest λ approximately equals the total variation across the chains during the latter N iterations. Specifically, we monitor convergence by the estimated *scale reduction factor*,

$$\sqrt{\hat{R}} = \sqrt{\left(\frac{N-1}{N} + \frac{m+1}{mN} \frac{B}{W} \right) \frac{df}{df - 2}} \, , \qquad (5.48)$$

where B/N is the variance between the means from the m parallel chains, W is the average of the m within-chain variances, and df is the degrees of freedom of an approximating t density to the posterior distribution. Equation (5.48) is the factor by which the scale parameter of the t density might shrink if sampling were continued indefinitely; the authors show it must approach 1 as $N \to \infty$.

The approach is fairly intuitive and is applicable to output from any MCMC algorithm. However, it focuses only on detecting bias in the MCMC estimator and does not necessarily reflect the *accuracy* of the resulting posterior estimate. It is also an inherently univariate quantity, meaning it must be applied to each parameter (or parametric function) of interest in turn, although Brooks and Gelman [1998] extend the Gelman and Rubin approach in three important ways, one of which is a multivariate generalization for simultaneous convergence diagnosis of every parameter in a model.

While the Gelman-Rubin-Brooks and other formal diagnostic approaches remain popular, in practice very simple checks often work just as well and may even be more robust against "pathologies" (e.g., multiple modes) in the posterior surface that may easily fool some diagnostics. For instance, sample auto-correlations in any of the observed chains can inform about whether slow traversing of the posterior surface is likely to impede convergence. Sample cross-correlations (i.e., correlations between two different parameters in the model) may identify ridges in the surface (say, due to collinearity between two predictors) that will again slow convergence; such parameters may need to be updated in multivariate blocks, or one of the parameters dropped from the model altogether. Combined with a visual inspection of a few sample trace plots, the user can at least get a good feeling for whether posterior estimates produced by the sampler are likely to be reliable. However, as a caveat, all convergence diagnostics explore solely MCMC output. They never compare proximity of the output to the truth since the truth is not known (if it was, we wouldn't be implementing an MCMC algorithm).

As a final comment here, the foregoing diagnostics apply to a scalar parameter when in fact, in practice $\boldsymbol{\theta}$ will be perhaps of very high dimension. Further, if we are interested in functions of the parameters, in principle, the number of diagnostics is uncountable. So, as above, in practice, the best we can do is to consider convergence diagnosis for a few unknowns of interest.

5.5.1.8 Variance estimation

An obvious criticism of Monte Carlo methods generally is that no two analysts will obtain the same answer, since the components of the estimator are random. This makes assessment of the variance of these estimators is an important point. Combined with a central limit theorem, the result would be an ability to test whether two Monte Carlo estimates were significantly different. For example, suppose we have a single chain of N post-burn-in samples of a parameter of interest λ, so that our basic posterior mean estimator becomes $\hat{E}(\lambda \mid \mathbf{y}) = \hat{\lambda}_N = \frac{1}{N}\sum_{t=1}^{N}\lambda^{(t)}$. Assuming the samples comprising this estimator are independent, a variance estimate for it would be given by

$$\widehat{\mathrm{Var}}_{iid}(\hat{\lambda}_N) = s_\lambda^2/N = \frac{1}{N(N-1)}\sum_{t=1}^{N}(\lambda^{(t)} - \hat{\lambda}_N)^2 \, , \tag{5.49}$$

i.e., the sample variance, $s_\lambda^2 = \frac{1}{N-1}\sum_{t=1}^{N}(\lambda^{(t)} - \hat{\lambda}_N)^2$, divided by N. But while this estimate is easy to compute, it would very likely be an *underestimate* due to positive autocorrelation in the MCMC samples. One can resort to *thinning*, which is simply retaining only every kth sampled value, where k is the approximate lag at which the autocorrelations in the chain become insignificant. However, Maceachern and Berliner [1994] show that such thinning from a stationary Markov chain always increases the variance of sample mean estimators, and is thus suboptimal. This is intuitively reminiscent of Fisher's view of sufficiency: it is never a good idea to throw away information (in this case, $(k-1)/k$ of our MCMC samples) just to achieve approximate independence among those that remain.

A better alternative is to use all the samples but in a more sophisticated way. One such alternative uses the notion of *effective sample size*, or *ESS* [Kass et al., 1998, p. 99]. *ESS* is defined as

$$ESS = N/\kappa(\lambda) \, ,$$

where $\kappa(\lambda)$ is the *autocorrelation time* for λ, given by

$$\kappa(\lambda) = 1 + 2\sum_{k=1}^{\infty}\rho_k(\lambda) \, , \tag{5.50}$$

where $\rho_k(\lambda)$ is the autocorrelation at lag k for the parameter of interest λ. We may estimate $\kappa(\lambda)$ using sample autocorrelations estimated from the MCMC chain. The variance estimate for $\hat{\lambda}_N$ is then

$$\widehat{\mathrm{Var}}_{ESS}(\hat{\lambda}_N) = s_\lambda^2/ESS(\lambda) = \frac{\kappa(\lambda)}{N(N-1)}\sum_{t=1}^{N}(\lambda^{(t)} - \hat{\lambda}_N)^2 \, .$$

Note that unless the $\lambda^{(t)}$ are uncorrelated, $\kappa(\lambda) > 1$ and $ESS(\lambda) < N$, so that $\widehat{\mathrm{Var}}_{ESS}(\hat{\lambda}_N) > \widehat{\mathrm{Var}}_{iid}(\hat{\lambda}_N)$, in concert with intuition. That is, since we have fewer than N effective samples, we expect some inflation in the variance of our estimate.

In practice, the autocorrelation time $\kappa(\lambda)$ in (5.50) is often estimated simply by cutting off the summation when the magnitude of the terms first drops below some "small" value (say, 0.1). This procedure is simple but may lead to a biased estimate of $\kappa(\lambda)$.

[Gilks et al., 1995, pp. 50–51] recommend an *initial convex sequence estimator* mentioned by Geyer [1992] which, while still output-dependent and slightly more complicated, actually yields a consistent (asymptotically unbiased) estimate here.

A somewhat simpler (though also more naive) method of estimating $\text{Var}(\hat{\lambda}_N)$ is through *batching*. Here we divide our single long run of length N into m successive batches of length k (i.e., $N = mk$), with batch means $B_1, \ldots, B_m$. Clearly $\hat{\lambda}_N = \bar{B} = \frac{1}{m}\sum_{i=1}^{m} B_i$. We then have the variance estimate

$$\widehat{\text{Var}}_{batch}(\hat{\lambda}_N) = \frac{1}{m(m-1)} \sum_{i=1}^{m} (B_i - \hat{\lambda}_N)^2 \, , \tag{5.51}$$

provided that k is large enough so that the correlation between batches is negligible, and m is large enough to reliably estimate $\text{Var}(B_i)$. It is important to verify that the batch means are indeed roughly independent, say, by checking whether the lag 1 autocorrelation of the B_i is less than 0.1. If this is not the case, we must increase k (hence N, unless the current m is already quite large), and repeat the procedure.

Regardless of which of the above estimates, $\hat{V}$, is used to approximate $\text{Var}(\hat{\lambda}_N)$, a 95% confidence interval for $\text{E}(\lambda \mid \mathbf{y})$ is then given by

$$\hat{\lambda}_N \pm z_{.025}\sqrt{\hat{V}} \, ,$$

where $z_{.025} = 1.96$, the upper .025 point of a standard normal distribution. If the batching method is used with fewer than 30 batches, it is a good idea to replace $z_{.025}$ with $t_{m-1,.025}$, the upper .025 point of a t distribution with $m-1$ degrees of freedom.

5.5.2   *Approximate Bayesian Computation (ABC)*

A further model fitting tool in the Bayesian arsenal is approximate Bayesian computation (henceforth ABC). ABC is intended for the Bayesian model setting $[\mathbf{y} \mid \boldsymbol{\theta}][\boldsymbol{\theta}]$ where $[\mathbf{y} \mid \boldsymbol{\theta}]$ is intractable but given $\boldsymbol{\theta}$ you can sample $\mathbf{y} \sim [\mathbf{y} \mid \boldsymbol{\theta}]$. ABC comes into play in the setting where the likelihood is implicit, from which we can simulate samples but we do not have access to an explicit expression for the likelihood. These models are often described as generative models. Such models are at the core of so-called *forward simulation* [Lynch and Vaandrager, 1995], a model-checking approach, i.e., comparing simulated realizations under the model with the observed, with a long history in the literature. Nowadays, simulations from implementations of generative models have increasingly been used to give training data sets for supervised machine learning purposes. Typically, in this setting, the likelihood arises as an intractable multidimensional integral. Alternative examples include cases where the normalizing constant for the distribution of the data is an intractable function of the parameters or the likelihood is a product of terms where at least one is unknown. The basic outline of what subsequently became known as ABC was introduced by J et al. [1999] for addressing an application in population genetics.

The ABC formulation insists that $[\boldsymbol{\theta}]$ is proper. If so, then if $\boldsymbol{\theta}^* \sim [\boldsymbol{\theta}]$ and $\mathbf{y}^* \sim [\mathbf{y} \mid \boldsymbol{\theta}]$, evidently, $\mathbf{y}^*$ is a realization under this model. How does this connect us to obtaining draws from the posterior, $[\boldsymbol{\theta} \mid \mathbf{y}_{obs}]$ where $\mathbf{y}_{obs}$ is the observed data? Suppose we take a situation too simple to be useful in practice namely the support for $\mathbf{y}$ is finite. Then, suppose we draw a $\boldsymbol{\theta}^*$ followed by a $\mathbf{y}^*$ such that $\mathbf{y}^* = \mathbf{y}_{obs}$. It is clear that $\boldsymbol{\theta}^* \sim [\boldsymbol{\theta} \mid \mathbf{y}^*] \propto [\mathbf{y}^* \mid \boldsymbol{\theta}][\boldsymbol{\theta}] = [\mathbf{y}_{obs} \mid \boldsymbol{\theta}][\boldsymbol{\theta}] \propto [\boldsymbol{\theta} \mid \mathbf{y}_{obs}]$.

When the support for $\mathbf{y}$ is continuous, $P(\mathbf{y}^* = \mathbf{y}_{obs}) = 0$. So, suppose we choose a distance function $d(\cdot, \cdot)$ and a lower dimensional and simpler vector of *summary* statistics $\mathbf{s}(\mathbf{y})$. Then, if $d(\mathbf{s}(\mathbf{y}^*), \mathbf{s}(\mathbf{y}_{obs})) \leq \epsilon$, we retain $\boldsymbol{\theta}^*$. The resulting target distribution for inference under ABC becomes

$$[\boldsymbol{\theta}, \mathbf{s} \mid \mathbf{s}(\mathbf{y}_{obs})]_\epsilon \propto [\mathbf{s} \mid \boldsymbol{\theta}][\boldsymbol{\theta}]1(d(\mathbf{s}, \mathbf{s}(\mathbf{y}_{obs})) \leq \epsilon) \tag{5.52}$$

and the ABC posterior for $\boldsymbol{\theta}$ is

$$[\boldsymbol{\theta} \mid \mathbf{s}(\mathbf{y}_{obs})]_\epsilon = \int [\boldsymbol{\theta}, \mathbf{s} \mid \mathbf{s}(\mathbf{y}_{obs})]_\epsilon d\mathbf{s} \tag{5.53}$$

So, the attractiveness of ABC is that, if we can draw $(\mathbf{y}^*, \boldsymbol{\theta}^*)$ under the model of interest, we can immediately either retain $\boldsymbol{\theta}^*$ as an observation from $[\boldsymbol{\theta} \mid \mathbf{s}(\mathbf{y}_{obs})]_\epsilon$ or discard it and then make a new draw from the model of interest. By repeating this sufficiently often, we can obtain draw an arbitrarily large sample from the ABC posterior, $[\boldsymbol{\theta} \mid \mathbf{s}(\mathbf{y}_{obs})]_\epsilon$.

As a result, the game here consists of choosing a suitable set of summary statistics and a suitable ϵ. Evidently, the smaller ϵ the more difficult it is to retain a $\boldsymbol{\theta}$ but the closer $[\boldsymbol{\theta} \mid \mathbf{s}(\mathbf{y}_{obs})]_\epsilon$ is to $[\boldsymbol{\theta} \mid \mathbf{s}(\mathbf{y}_{obs})]$. Further, the choice of $\mathbf{s}$ determines how close $[\boldsymbol{\theta} \mid \mathbf{s}(\mathbf{y}_{obs})]$ is to $[\boldsymbol{\theta} \mid \mathbf{y}_{obs}]$.

Given an $\mathbf{s}(\mathbf{y})$, empirically, we can choose ϵ to achieve a desired acceptance rate for the drawn $\boldsymbol{\theta}^*$'s. Therefore, the literature is filled with dimension reduction strategies to select $\mathbf{s}(\mathbf{y})$. If there were a set of sufficient statistics, there would be no loss of information with such summary statistics; only the choice of ϵ would matter. However, for the models to which ABC is applied in practice, this will not be the case. The choice of these statistics is a dimension reduction problem and is model specific. We do not go into detail here but two frequent approaches are regression adjustment methods and best subset selection methods. Two useful review papers are Blum et al. [2013] and Beaumont [2019].

As a final remark, it has been suggested to employ ABC within MCMC Marjoram et al. [2003]. ABC-MCMC differs from traditional MCMC in that the calculation of the ratio of likelihoods for new and old parameter values in the Hastings ratio is replaced by a step in which we simulate a single dataset, $\mathbf{y}^*$, using $\boldsymbol{\theta}^*$, and then proceed to calculate the rest of the ratio only if $d(\mathbf{s}(\mathbf{y}^*), \mathbf{s}(\mathbf{y}_{obs})) \leq \epsilon$. Thus, the intractable likelihood has been replaced by a simulation step, recovering computational tractability. However, ABC-MCMC has been shown to mix relatively poorly, compared to traditional MCMC, in the tails of the posterior.

5.5.3 *Amortized Bayesian computation using neural networks*

Approximate Bayesian Computation (ABC), discussed in Section 5.5.2, approximates the posterior by repeatedly sampling from the joint distribution of the parameters $\boldsymbol{\theta}$ and data $\mathbf{y}$. If the resulting data set is sufficiently similar to the actually observed data set, then the generated value of $\boldsymbol{\theta}$ is retained as a sample from the desired posterior; otherwise, it is rejected. While ABC's performance may benefit from stricter similarity criteria for the generated and observed data, it may also be encumbered by prohibitive rejection rates.

Related ideas in machine learning offer neural estimation and amortized inference [Radev et al., 2020, 2023b,a, Sainsbury-Dale et al., 2024, Zammit-Mangion et al., 2024] that are generative in nature, like ABC, but circumvent the rejection steps by training a neural network to arrive at optimal Bayesian estimates. We simulate realizations from the joint distribution of the parameters and the data to build and train specialized neural networks. The trained network is then used to deliver rapid statistical inference from the observed data. The term "amortization," which is popularly used in asset management to refer to the exercise of gradually writing off the initial cost of (an asset) over a period of time, is applied in the sense that a potentially time consuming simulation-based training phase of the neural network is rewarded (or "written off") with extremely rapid inference using the trained network.

Unlike in likelihood-based inference where the parameters of a model are estimated from every new dataset, amortized inference attempts to train a neural network using a large number of simulated datasets. We seek to train a (possibly deep) neural machine $f_\phi(\mathbf{y}; \mathcal{M})$ for the task of estimate $\hat{\theta}(\mathbf{y}; \phi) = f_\phi(\mathbf{y}; \mathcal{M})$, where $\mathcal{M}$ is a Bayesian model (consisting of a prior and likelihood), $\mathbf{y}$ is any realized data from $\mathcal{M}$ and $\hat{\boldsymbol{\theta}}(\mathbf{y}; \phi)$ is a neural estimator of our parameter of interest $\boldsymbol{\theta}$. Training the network amounts to optimizing neural parameter,

$$\phi^* = \arg\min_\phi \mathrm{E}_{\boldsymbol{\theta},\mathbf{y}}[L(\boldsymbol{\theta}, \hat{\boldsymbol{\theta}}(\mathbf{y}; \phi))] \,,$$

for some specified loss function $L(\cdot, \cdot)$. In a Bayesian setting, we generate a large number of simulated random variables $\{\boldsymbol{\theta}^{(i)}, \mathbf{y}^{(i)}\} \sim \mathcal{M}$ for $i = 1, \ldots, M$ and find $\phi^* = \arg\min \frac{1}{M} \sum_{i=1}^M L(\boldsymbol{\theta}^{(i)}, \hat{\boldsymbol{\theta}}(\mathbf{y}^{(i)}; \phi))$. Once the neural network is trained, we achieve generative inference for any new dataset $\tilde{\mathbf{y}}$ by computing $\hat{\boldsymbol{\theta}} = \hat{\boldsymbol{\theta}}(\tilde{Y}, \phi^*)$ using the optimal ϕ^* obtained from the simulated data.

Rather than obtain neural estimates of some statistic of interest, amortized inference can also be used to generate samples from the posterior distribution. This is carried out in practice by training an invertible neural network that approximates the true posterior using the concept of a normalizing flow [see Kobyzev et al., 2021, and references therein for a comprehensive account]. Normalizing flows [and related ideas in transport maps Marzouk et al., 2016, Katzfuss and Schäfer, 2024] are invertible and differentiable mappings that transform a simple probability distribution (e.g., a standard normal) into a possibly more complex distribution. We offer a brief overview of how training a net works[see Radev et al., 2020, 2023a,b, for details].

Let $p(\boldsymbol{\theta} \mid \mathbf{y}) \propto p(\boldsymbol{\theta}) \times p(\mathbf{y} \mid \boldsymbol{\theta})$ be the posterior distribution for our Bayesian model. We seek to approximate this posterior distribution using an invertible neural network $p_\phi(\boldsymbol{\theta} \mid \mathbf{y}) \approx p(\boldsymbol{\theta} \mid \mathbf{y})$, where ϕ is a set of neural parameters that need to be trained using simulated output from $p(\boldsymbol{\theta}, \mathbf{y})$. We implement a normalizing flow that maps $\boldsymbol{\theta} \sim p_\phi(\boldsymbol{\theta} \mid \mathbf{y})$ to a standard Gaussian variable $\mathbf{z} \sim N(\mathbf{0}, \mathbb{I})$, $\boldsymbol{\theta}$ and $\mathbf{z}$ are both $p \times 1$ vectors, such that $\mathbf{z} = f_\phi(\boldsymbol{\theta}; \mathbf{y})$, where $f_\phi(\boldsymbol{\theta}; \mathbf{y}) : \mathbb{R}^p \to \mathbb{R}^p$ is an invertible and differentiable map with $\boldsymbol{\theta} = f_\phi^{-1}(\mathbf{z}; \mathbf{y})$. We seek to train the neural parameters $\hat{\phi}$ so that $f_\phi^{-1}(\mathbf{z}; \mathbf{y})$ is distributed as our target posterior $p(\boldsymbol{\theta} \mid \mathbf{y})$. Radev et al. [2020] show that this is achieved by

$$\hat{\phi} = \arg\min_\phi \mathrm{E}_{[\mathbf{y}]}[KL(p(\boldsymbol{\theta} \mid \mathbf{y}), p_\phi(\boldsymbol{\theta} \mid \mathbf{y}))] = \arg\max_\phi \mathrm{E}_{[\boldsymbol{\theta},\mathbf{y}]}[\log p_\phi(\boldsymbol{\theta} \mid \mathbf{y})] \,, \tag{5.54}$$

where $KL(p(\boldsymbol{\theta} \mid \mathbf{y}), p_\phi(\boldsymbol{\theta} \mid \mathbf{y}))$ is the Kullback-Leibler divergence between densities, the first expectation is with respect to the marginal distribution $p(\mathbf{y})$ while the second is with respect to the joint distribution $p(\boldsymbol{\theta}, \mathbf{y})$. The last equality follows by writing $KL(p(\boldsymbol{\theta} \mid \mathbf{y}), p_\phi(\boldsymbol{\theta} \mid \mathbf{y})) = \mathrm{E}_{[\boldsymbol{\theta} \mid \mathbf{y}]}[\log p(\boldsymbol{\theta} \mid \mathbf{y}) - \log p_\phi(\boldsymbol{\theta} \mid \mathbf{y})]$ and noting that the first term inside the expectation, $\log p(\boldsymbol{\theta} \mid \mathbf{y})$ does not depend upon ϕ and, therefore, does not affect the optimal solution in (5.54).

The normalizing flow allows us to reformulate the optimization problem using the standard change of variable from $\boldsymbol{\theta}$ to $\mathbf{z} = f_\phi(\boldsymbol{\theta}; \mathbf{y})$ into

$$\hat{\phi} = \arg\max_\phi \mathrm{E}_{[\boldsymbol{\theta},\mathbf{y}]} \left[\log p(f_\phi(\boldsymbol{\theta}; \mathbf{y})) + \log \left(\left| \det \left(\frac{\partial f_\phi(\boldsymbol{\theta}; \mathbf{y})}{\partial \boldsymbol{\theta}} \right) \right| \right) \right] \,, \tag{5.55}$$

where $p(f_\phi(\boldsymbol{\theta}; \mathbf{y}))$ is the p-dimensional standard normal density with argument $\mathbf{z} = f_\phi(\boldsymbol{\theta}; \mathbf{y})$. The optimization can proceed using M simulated realizations $\{\boldsymbol{\theta}^{(m)}, \mathbf{y}^{(m)}\} \overset{ind}{\sim} p(\boldsymbol{\theta}, \mathbf{y})$ and then solving a Monte Carlo version of (5.55),

$$\hat{\phi} = \arg\min_\phi \frac{1}{M} \sum_{m=1}^M \left(\frac{\|f_\phi(\boldsymbol{\theta}^{(m)}; \mathbf{y}^{(m)})\|^2}{2} - \log \left(\left| \det \left(\frac{\partial f_\phi(\boldsymbol{\theta}; \mathbf{y}^{(m)})}{\partial \boldsymbol{\theta}} \bigg|_{\boldsymbol{\theta}=\boldsymbol{\theta}^{(m)}} \right) \right| \right) \right) \,. \tag{5.56}$$

The objective function in (5.56) is computed using the form of the specific neural network we use (including the partial derivatives Jacobian $\det\left(\partial f_\phi(\boldsymbol{\theta}; \mathbf{y}^{(m)})/\partial\boldsymbol{\theta}\right)$). Radev et al. [2020]

offers some specific examples of composing invertible neural networks so that the Jacobian matrix is triangular, hence the determinant is easy to compute. The optimization in (5.56) can be performed using any stochastic gradient descent algorithm. Algorithm 10 presents a simplified version of amortized Bayesian inference. An expanded version of this algorithm introduces a summary neural network that maps the data $\mathbf{y}$ to a low-dimensional sufficient statistics $\tilde{\mathbf{y}} := h_\psi(\mathbf{y})$ and we consider jointly optimizing the neural parameters ϕ and ψ. These details as well as the explicit specifications of the neural networks are developed in Radev et al. [2020]. Other examples of amortized inference by training neural networks with applications to spatial models are discussed in Zammit-Mangion et al. [2024] who consider forward and reverse KL-divergence for training, while Sainsbury-Dale et al. [2024] offer an excellent exposition to the amortized inference that includes examples of spatial modeling.

Algorithm 10 Amortized Bayesian inference using normalizing flow neural networks

---

1: **Network Training Module** *(online learning with batch size M):*
2: **while** $\hat{\phi}$ has not converged **do**
3:     Sample $N \sim \mathcal{U}(N_{\min}, N_{\max})$.
4:     **for** $m = 1, \ldots, M$ **do**
5:         Sample $\boldsymbol{\theta}^{(m)} \sim p(\boldsymbol{\theta})$.
6:         **for** $i = 1, \ldots, N$ **do**
7:             Sample $\boldsymbol{y}_i^{(m)} \sim p(\mathbf{y} \mid \boldsymbol{\theta}^{(m)})$
8:         **end for**
9:         Compute $\boldsymbol{z}^{(m)} = f_\phi(\boldsymbol{\theta}^{(m)}; \boldsymbol{y}_{1:N}^{(m)})$.
10:     **end for**
11:     Update neural network parameters $\hat{\phi}$ from (5.56).
12: **end while**

13: **Inference Module** *(given observed or test data $\tilde{\mathbf{y}}_{1:N}$):*
14: **for** $l = 1, \ldots, L$ **do**
15:     Sample (normalizing flow): $\mathbf{z}^{(l)} \sim \mathcal{N}_D(\mathbf{0}, \mathbf{I})$.
16:     Compute $\boldsymbol{\theta}^{(l)} = f_{\hat{\phi}}^{-1}(\mathbf{z}^{(l)}; \tilde{\mathbf{Y}}_{1:N})$.
17: **end for**
18: Return $\{\boldsymbol{\theta}^{(l)}\}_{l=1}^L$ as a sample from $p(\boldsymbol{\theta} \mid \tilde{\mathbf{Y}}_{1:N})$.

5.5.4 Variational Bayes

Variational methods have their origins in the 18th century with the work of Euler, Lagrange, and others on the calculus of variations [Gelfand and Fomin, 1963]. The subject explores mappings that take a functional, instead of a variable, as input and returns the value of the functional as the output. Many problems can be expressed in terms of an optimization problem in which the quantity being optimized is a functional. The solution is obtained by exploring all possible function to find the one that maximizes (or minimizes) the functional. Usually no closed-form solution can be found. Therefore, variational methods naturally focus on approximations to the optimal solutions.

Variational inference is widely employed in machine learning to approximate posterior densities for Bayesian models as an alternative to MCMC when speed takes precedence over accuracy. Variational inference for Bayesian models is referred to as *Variational Bayes* (VB) [Jordan et al., 1999, Wainwright and Jordan, 2008, Blei et al., 2017, and references therein]. Blei et al. [2017], in particular, offer an excellent review of variational inference from a statistician's perspective. Variational Bayes, like MCMC, also attempts to circumvent high-dimensional integrals in the posterior distribution. While MCMC replaces such integration with simulations, VB reformulates the problem in terms of optimization.

Consider how variational optimization can be applied to the Bayes inference problem. Let $\mathbf{y}$ denote the observed variables and $\boldsymbol{\theta}$ denote the unobserved parameters. We assume a prior distribution $p(\boldsymbol{\theta})$ for parameter $\boldsymbol{\theta}$. Then the marginal likelihood $p(\mathbf{y}) = \int p(\mathbf{y} \mid \boldsymbol{\theta}) p(\boldsymbol{\theta}) \, d\boldsymbol{\theta}$ can be bounded below using any distribution over the parameter $\boldsymbol{\theta}$. To see how, let $q(\boldsymbol{\theta})$ be any probability density function on $\boldsymbol{\theta}$. Then,

$$\log p(\mathbf{y}) = \log\left(\frac{p(\mathbf{y}, \boldsymbol{\theta})}{p(\boldsymbol{\theta} \mid \mathbf{y})}\right) = \log p(\mathbf{y}, \boldsymbol{\theta}) - \log p(\boldsymbol{\theta} \mid \mathbf{y}) = \log\left(\frac{p(\mathbf{y}, \boldsymbol{\theta})}{q(\boldsymbol{\theta})}\right) + \log\left(\frac{q(\boldsymbol{\theta})}{p(\boldsymbol{\theta} \mid \mathbf{y})}\right).$$

Multiplying both sides by $q(\boldsymbol{\theta})$ and integrating with respect to $\boldsymbol{\theta}$, we obtain

$$
\begin{aligned}
\log p(\mathbf{y}) &= \int q(\boldsymbol{\theta}) \log\left(\frac{p(\mathbf{y}, \boldsymbol{\theta})}{q(\boldsymbol{\theta})}\right) d\boldsymbol{\theta} + \int q(\boldsymbol{\theta}) \log\left(\frac{q(\boldsymbol{\theta})}{p(\boldsymbol{\theta} \mid \mathbf{y})}\right) d\boldsymbol{\theta} \\
&= \text{ELBO}(q) + KL(q, p(\cdot \mid \mathbf{y})) \geq \text{ELBO}(q),
\end{aligned}
$$

where $\text{ELBO}(q) = \text{E}[\log p(\mathbf{y}, \boldsymbol{\theta})] - \text{E}[\log q(\boldsymbol{\theta})]$ is the *evidence lower bound* with $\text{E}[\cdot]$ denoting expectation with respect to q, and $KL(q, p(\cdot \mid \mathbf{y}))$ is the Kullback-Leibler (KL) divergence between $q(\boldsymbol{\theta})$ to $p(\boldsymbol{\theta} \mid \mathbf{y})$. Since $KL(q, p)$ is always nonnegative, $\text{ELBO}(q)$ is a lower bound for the log marginal likelihood. Given data $\mathbf{y}$ the sum of $\text{ELBO}(q)$ and $KL(q, p)$ is a constant. Therefore, to find a $q(\boldsymbol{\theta})$ that approximates $p(\boldsymbol{\theta} \mid \mathbf{y})$ well, we can either maximize $\text{ELBO}(q)$ or minimize $KL(q, p)$.

Variational Bayes focuses on maximizing $\text{ELBO}(q)$. We begin by specifying a class of densities $\mathcal{Q}$ from which we select the optimal q. Perhaps the most conspicuous class in machine learning applications is $\mathcal{Q} = \left\{ q(\boldsymbol{\theta}) : q(\boldsymbol{\theta}) = \prod_{j=1}^{m} q_i(\boldsymbol{\theta}_j) \right\}$, where we look at densities that assume independence across the components of $\boldsymbol{\theta} = (\boldsymbol{\theta}_1, \ldots, \boldsymbol{\theta}_m)$ and each $\boldsymbol{\theta}_i$ can be scalar or vector. This class offers substantial tractability in optimizing ELBO and leads to so called *mean field approximations* of the posterior. A crucial implication of the mean field class is that we can carry out the optimization in a coordinate-wise manner, which gives a tractable form of the optimal $q_j(\boldsymbol{\theta}) \in \mathcal{Q}$ for $j = 1, \ldots, m$.

More specifically, the mean-field class allows us to easily rewrite $\text{ELBO}(q)$ in terms of the j-th variational factor $q_j(\boldsymbol{\theta}_j)$ for a specific j as follows

$$
\begin{aligned}
\text{ELBO}(q) &= \int \prod_{i=1}^{m} q_i(\boldsymbol{\theta}_i) \log\left(\frac{p(\mathbf{y}, \boldsymbol{\theta})}{\prod_{i=1}^{m} q(\boldsymbol{\theta}_i)}\right) = \int \prod_{i=1}^{m} q_i(\boldsymbol{\theta}_i) \left(\log p(\mathbf{y}, \boldsymbol{\theta}) - \sum_{i=1}^{m} \log q_i(\boldsymbol{\theta}_i)\right) d\boldsymbol{\theta} \\
&= \int q_j(\boldsymbol{\theta}_j) \left\{ \int \prod_{i \neq j} q_i(\boldsymbol{\theta}_i) \log p(\mathbf{y}, \boldsymbol{\theta}) \, d\boldsymbol{\theta}_i \right\} d\boldsymbol{\theta}_j - \int q_j(\boldsymbol{\theta}_j) \log q_j(\boldsymbol{\theta}_j) \, d\boldsymbol{\theta}_j + \text{const} \\
&= \text{E}_j\left[\text{E}_{-j}\left[\log p(\mathbf{y}, \boldsymbol{\theta}_j, \boldsymbol{\theta}_{-j})\right]\right] - \text{E}_j[\log q_j(\boldsymbol{\theta}_j)] + \text{const} := \text{ELBO}(q_j),
\end{aligned}
$$

$$(5.57)$$

where $\boldsymbol{\theta}_{-j}$ is the vector $\boldsymbol{\theta}$ will all components except $\boldsymbol{\theta}_j$, $\text{E}_j[\cdot]$ is the expectation with respect to $q_j(\boldsymbol{\theta}_j)$, $\text{E}_{-j}[\cdot]$ is the expectation with respect to $\prod_{i=1 \neq j}^{m} q_i(\boldsymbol{\theta}_i)$ and const absorbs all quantities not involving $q_j(\boldsymbol{\theta}_j)$. Equation (5.57) reveals that $\text{ELBO}(q_j)$ is the negative Kullback-Leibler divergence between $q_j(\boldsymbol{\theta}_j)$ and a density given by $\log q_i^*(\boldsymbol{\theta}_i) = \text{E}_{-j}[\log p(\mathbf{y}, \boldsymbol{\theta}_j, \boldsymbol{\theta}_{-j})] + \text{constant}$. Therefore, we maximize $\text{ELBO}(q_j)$ when we set

$$q_j(\boldsymbol{\theta}_j) = q_j^*(\boldsymbol{\theta}_j) = \frac{\exp\left\{\text{E}_{-j}[\log p(\mathbf{y}, \boldsymbol{\theta}_j, \boldsymbol{\theta}_{-j})]\right\}}{\int \exp\left\{\text{E}_{-j}[\log p(\mathbf{y}, \boldsymbol{\theta}_j, \boldsymbol{\theta}_{-j})]\right\} d\boldsymbol{\theta}_j} \propto \exp\left\{\text{E}_{-j}[\log p(\boldsymbol{\theta}_j \mid \boldsymbol{\theta}_{-j}, \mathbf{y})]\right\},$$

$$(5.58)$$

where the last expression is a simple consequence of the fact that the full conditional distribution $p(\boldsymbol{\theta}_j \mid \boldsymbol{\theta}_{-j}, \mathbf{y})$ is proportional to the joint distribution $p(\mathbf{y}, \boldsymbol{\theta}_j, \boldsymbol{\theta}_{-j})$. This draws resemblance with the Gibbs sampler. While the Gibbs sampler draws samples from the full conditionals, the mean-field variational algorithm computes an optimal approximation based on the full conditional distribution.

Equation (5.58) represents a set of consistent conditions for the maximum of the lower bound subject to the factorization constraint. However, it does not represent an explicit solution because the right-hand side of (5.58) depends on the expectation computed with respect to the other parameters $\boldsymbol{\theta}_j$. So we must initialize the distribution of all the $\boldsymbol{\theta}_j$ and then cycle through them iteratively. Each parameter's distribution is updated in turn with a revised function given by (5.58) and evaluated using the current estimate of the distribution function for all other parameters. Convergence is guaranteed because the bound is convex with respect to each of the factors $q_j(\boldsymbol{\theta}_j)$ [Attias, 2000]. Algorithm 11 presents the widely employed coordinate ascent variational inference (CAVI) algorithm.

Algorithm 11 Coordinate ascent variational inference (CAVI) algorithm

---

1: **Input:** A joint probability model $p(\mathbf{y}, \boldsymbol{\theta})$ up to a proportionality constant; data $\mathbf{y}$.
2: **Output:** A variational mean field approximation $q(\boldsymbol{\theta}) = \prod_{i=1}^{m} q(\boldsymbol{\theta}_i)$
3: Initialize: $q_j(\boldsymbol{\theta}_j)$ for $j = 1, \ldots, m$
4: **while** ELBO has not converged **do**
5:     **for** j in $1:m$ **do**
6:         Set $q_j(\boldsymbol{\theta}_j) \propto \exp\{\mathrm{E}_{-j}[\log p(\boldsymbol{\theta}_j \mid \boldsymbol{\theta}_{-j}, \mathbf{y})]\}$
7:     **end for**
8:     Compute ELBO$(q) = \mathrm{E}[\log p(\mathbf{y}, \boldsymbol{\theta})] - \mathrm{E}[\log q(\boldsymbol{\theta})]$ and check for convergence.
9: **end while**
10: **return** $q(\boldsymbol{\theta})$

Algorithm 11 is best illustrated with a relatively simple example. We apply the VB algorithm to a Bayesian linear regression model with the conjugate Normal-Inverse-Gamma prior specified in (5.21). The posterior distribution is accessible in closed form, which helps us assess the VB method against an analytical benchmark. Letting $\mathbf{y}$ be an $n \times 1$ vector of outcomes, we write $\mathbf{y} = X\boldsymbol{\beta} + \boldsymbol{\epsilon}$, where X is the $n \times p$ matrix of predictors, $\boldsymbol{\beta}$ is the slope vector of regression coefficients and $\boldsymbol{\epsilon}$ is an $n \times 1$ vector of Gaussian errors, $\boldsymbol{\epsilon} \sim N(\mathbf{0}, \sigma^2 \mathbb{I}_n)$. In (5.21) we assume that $M_0 \mathbf{m}_0 = \boldsymbol{\mu}_\beta$, $M_0 = V_\beta$, $a_0 = a$, and $b_0 = b$ are known constants.

We apply (5.58) to the above model using $\boldsymbol{\theta} = (\boldsymbol{\beta}, \sigma^2)$ with the mean field approximation $q(\boldsymbol{\beta}, \sigma^2) = q_\beta(\boldsymbol{\beta}) q_{\sigma^2}(\sigma^2)$. Assuming that the approximate variational distributions for all the parameters are known at iteration t, Algorithm 11 involves the following iterative solutions: (i) the variational factor for $\boldsymbol{\beta}$ is $q^{(t+1)}(\boldsymbol{\beta}) \sim N\left(\boldsymbol{\mu}^*, (\zeta^2)^{(t+1)} V^*\right)$, where $(\zeta^2)^{(t+1)} = \left[\int d\sigma^2 q^{(t)}(\sigma^2)/\sigma^2\right]^{-1}$, $V^* = (V_\beta^{-1} + X^\mathrm{T} X)^{-1}$ and $\boldsymbol{\mu}^* = V^*(V_\beta^{-1} \boldsymbol{\mu}_\beta + X^\mathrm{T} \mathbf{y})$; and (ii) the variational factor for σ^2 is $q^{(t+1)}(\sigma^2) \sim IG\left(a^* + \frac{p}{2}, \frac{2b^* + p \times (\zeta^2)^{(t+1)}}{2}\right)$, where $a^* = a + \frac{n}{2}$ and $b^* = b + \frac{1}{2}\left(\boldsymbol{\mu}_\beta^\mathrm{T} V_\beta^{-1} \boldsymbol{\mu}_\beta + \mathbf{y}^\mathrm{T}\mathbf{y} - \boldsymbol{\mu}^{*\mathrm{T}} V^{*-1} \boldsymbol{\mu}^*\right)$. Notice that this algorithm only needs the starting value for $(\zeta^2)^{(0)} = \mathrm{E}^{(0)}(1/\sigma^2)$. We do not have to calculate the expectation by specifying $q^{(0)}(\sigma^2)$, but give an initial value to $(\zeta^2)^{(0)}$ directly.

Thus using the distribution of σ^2 at iteration $t+1$, we find

$$(\zeta^2)^{(t+2)} = \left\{\int d\sigma^2 \frac{q^{(t+1)}(\sigma^2)}{\sigma^2}\right\}^{-1} = \frac{2b^* + p \times (\zeta^2)^{(t+1)}}{2a^* + p}. \tag{5.59}$$

Defining $\lim_{t \to +\infty} (\zeta^2)^{(t)} = \zeta^2$ and taking limit on both sides of (5.59), we obtain $\zeta^2 = \frac{b^*}{a^*}$. So when $t \to \infty$,

$$\frac{2b^* + p \times (\zeta^2)^{(t+1)}}{2} \to \frac{2b^* + p\zeta^2}{2} = \frac{b^*}{a^*}\left(a^* + \frac{p}{2}\right).$$

The approximate posterior distributions are, for $\boldsymbol{\beta}$, a multivariate normal centered at $\boldsymbol{\mu}^*$ with variance $\frac{b^*}{a^*}V^*$, and for σ^2, an Inverse Gamma with parameters $a^* + \frac{p}{2}$ and $\frac{b^*}{a^*}\left(a^* + \frac{p}{2}\right)$. The joint posterior distribution for $\boldsymbol{\beta}$ and σ^2 with the conjugate NIG prior is $NIG(\boldsymbol{\beta}, \sigma^2 \mid \boldsymbol{\mu}^*, V^*, a^*, b^*)$. The exact marginal posterior distributions are $p(\boldsymbol{\beta} \mid \mathbf{y}) \sim t_{2a^*}\left(\boldsymbol{\mu}^*, \frac{b^*}{a^*}V^*\right)$ and $p(\sigma^2 \mid \mathbf{y}) \sim IG(a^*, b^*)$. Both the true and the VB estimated marginal posterior distribution for $\boldsymbol{\beta}$ have the same mean $\boldsymbol{\mu}^*$ and scale parameter $\frac{b^*}{a^*}V^*$. However, the posterior variance of $\boldsymbol{\beta}$ estimated from VB is smaller because the t distribution has a heavier tail than the normal distribution. It is easier to see this when $p = 1$. The variance of the *Student t* distribution is $\frac{a^*}{a^*-1} > 1$, while it is 1 for a standard normal distribution.

A similar situation arises for the marginal posterior distribution of σ^2. An inverse gamma random variable's mean and mode can be estimated as the ratio of its scale and shape parameter when the latter is large. Here, when the sample size n, and thus $a^* = a + \frac{n}{2}$, are large enough, the approximate posterior mean and mode of σ^2 from VB are the same as the exact true posterior because $\frac{b^*}{a^*}\left(a^* + \frac{p}{2}\right) / \left(a^* + \frac{p}{2}\right) = \frac{b^*}{a^*}$. But the approximate posterior variance of σ^2 from VB is smaller due to a larger shape parameter: $a^* + p/2 > a^*$. For both parameters, when sample size $n \to \infty$, the posterior variance estimates from VB have limits which equal the true values, i.e., $\frac{a^*}{a^*-1} \to 1$ and $\frac{a^*+p/2}{a^*} \to 1$. For data sets with reasonable sample sizes, the difference between the VB estimate and the true posterior is very small. Thus the VB approach offers a very good approximation for the true posterior distribution in this simple model.

We conclude our brief account of VB with a further remark. While in the above example we focused on variational approximations for model parameters, the same principle applies to predictive variational inference, where we seek to approximate $p(\boldsymbol{\beta}, \sigma^2, \tilde{\mathbf{y}} \mid \mathbf{y})$. Now $\boldsymbol{\theta} = (\boldsymbol{\beta}, \sigma^2, \tilde{\mathbf{y}})$, where we have appended the $\tilde{n} \times 1$ predictive random variable $\tilde{\mathbf{y}}$ to the set of unknowns. We could approximate the posterior predictive density $p(\tilde{\mathbf{y}} \mid \mathbf{y})$ in (5.5) by simply replacing the posterior $p(\boldsymbol{\beta}, \sigma^2 \mid \mathbf{y})$ with its mean field variational approximation, $p(\tilde{\mathbf{y}} \mid \mathbf{y}) \approx \mathrm{E}_{q^*}[p(\tilde{\mathbf{y}} \mid \mathbf{y}, \boldsymbol{\beta}, \sigma^2)]$, where the expectation is taken with respect to the optimal variational approximation $q_{\boldsymbol{\beta}}^*(\boldsymbol{\beta})q_{\sigma^2}^*(\sigma^2)$. Alternatively, we could specify the mean field class as $q(\boldsymbol{\theta}) = q_{\boldsymbol{\beta}}(\boldsymbol{\beta})q_{\tilde{y}}(\tilde{\mathbf{y}})q_{\sigma^2}(\sigma^2)$ or, yet another alternative, use a variational factor jointly for $\boldsymbol{\beta}$ and $\tilde{\mathbf{y}}$, i.e., $q(\boldsymbol{\theta}) = q_{\gamma}(\boldsymbol{\gamma})q_{\sigma^2}(\sigma^2)$, where $\boldsymbol{\gamma} = (\boldsymbol{\beta}^\mathsf{T}, \tilde{\mathbf{y}}^\mathsf{T})^\mathsf{T}$. In either case, we can again apply (5.58) and Algorithm 11 to arrive at an optimal $q^*(\boldsymbol{\theta})$ for the ELBO function. We refer to Blei et al. [2017] for a more comprehensive account of VB methods.

5.5.4.1   *Integrated nested Laplace approximation*

The integrated nested Laplace approximation (INLA) approach to Bayesian inference circumvents the challenges of programming, fine-tuning, and running computationally challenging MCMC algorithms [Illian et al., 2012, 2013]. As a result, INLA has become a popular method for parameter inference and model fitting. for the class of latent Gaussian models [Rue et al., 2009].

The main aim of INLA is to approximate the posterior distributions; it is not a sampling scheme like the algorithms discussed above. INLA will approximate, marginally, the posterior distribution of the latent field as well as the model parameters. The approximate (marginal) posterior distributions can then be used to obtain estimates of means, standard deviations, etc.

We provide some details of the Laplace approximation, first generally, and then as used in obtaining posterior inference. Laplace approximations are used to approximate $\int_{-\infty}^{\infty} g(\mathbf{v})d\mathbf{v}$

where $g(\cdot) > 0$. To begin, write $h(\mathbf{v}) = \log g(\mathbf{v})$ and expand $h(\mathbf{v})$ in a Taylor series as $h(\mathbf{v}) \approx h(\mathbf{v}^*) + (\mathbf{v} - \mathbf{v}^*)^\mathsf{T} \nabla h(\mathbf{v})|_{\mathbf{v}=\mathbf{v}^*} - (\mathbf{v} - \mathbf{v}^*)^\mathsf{T} H(\mathbf{v}^*)(\mathbf{v} - \mathbf{v}^*)$. Here, $\mathbf{v}^*$ is the mode and H is the Hessian for h. Then, plugging these values in and integrating out the d-dim multivariate normal for $\mathbf{v}$ yields the approximation to the integral, $(2\pi)^{d/2}|H(\mathbf{v}^*)|^{.5}e^{h(\mathbf{v}^*)}$.

In practice, using the Laplace approximation, and thus, INLA, to obtain posterior inference, begins with writing the joint posterior distribution $f(\boldsymbol{\theta} \mid \mathbf{y}) \propto e^{\log(f(\mathbf{y} \mid \boldsymbol{\theta})f(\boldsymbol{\theta}))}$. The Laplace approximation entails using a second-order expansion of $\log(f(\mathbf{y} \mid \boldsymbol{\theta})f(\boldsymbol{\theta}))$ to create what is essentially a multivariate normal density approximation for the posterior. In a hierarchical setting this can be extended to

$$\Pi_i f(y_i \mid z_i, \boldsymbol{\theta}_y)f(\mathbf{z} \mid \boldsymbol{\theta}_z)f(\boldsymbol{\theta})$$

where, typically, the $\mathbf{z}$ are latent variables and $f(\mathbf{z} \mid \boldsymbol{\theta}_z)$ is $MVN(\mathbf{0}, \Sigma_z)$. Now, the posterior of interest is $f(\boldsymbol{\theta}, \mathbf{z} \mid \mathbf{y}) \propto \Pi_i f(y_i \mid z_i, \boldsymbol{\theta}_y)f(\mathbf{z} \mid \boldsymbol{\theta}_z)f(\boldsymbol{\theta})$. This approach requires that we can assume $\boldsymbol{\theta}$ is low dimensional in order to obtain posteriors for each component of $\boldsymbol{\theta}$. More importantly, $\boldsymbol{\theta}$ must be low dimensional to obtain $f(\mathbf{z} \mid \mathbf{y}) = \int f(\mathbf{z} \mid \boldsymbol{\theta}, \mathbf{y})f(\boldsymbol{\theta} \mid \mathbf{y})d\boldsymbol{\theta}$.

The first step is to obtain the Laplace approximation for the full conditional, $f(\mathbf{z} \mid \boldsymbol{\theta}, \mathbf{y})$ for a given $\boldsymbol{\theta}$, which requires first computing the mode of the full conditional, $\mathbf{z}^*$, and Hessian. Let $\tilde{f}(\mathbf{z} \mid \boldsymbol{\theta}, \mathbf{y})$ denote this approximation. Next, $f(\boldsymbol{\theta} \mid \mathbf{y}) \propto \frac{f(\mathbf{y}, \mathbf{z}, \boldsymbol{\theta})}{f(\mathbf{z} \mid \boldsymbol{\theta}, \mathbf{y})}$ regardless of $\mathbf{z}$. Then, our approximation to the posterior of $\boldsymbol{\theta}$ can be written $\tilde{f}(\boldsymbol{\theta} \mid \mathbf{y}) \propto \frac{f(\mathbf{y}, \mathbf{z}, \boldsymbol{\theta})}{f(\mathbf{z} \mid \boldsymbol{\theta}, \mathbf{y})}|_{\mathbf{z}=\mathbf{z}^*}$. Lastly, $\tilde{f}(\mathbf{z} \mid \mathbf{y}) = \int \tilde{f}(\mathbf{z} \mid \boldsymbol{\theta}, \mathbf{y})\tilde{f}(\boldsymbol{\theta} \mid \mathbf{y})d\boldsymbol{\theta}$, which can be approximate using a grid over $\boldsymbol{\theta}$-space and replacing the integral by a sum approximation. From this, it is easy to see the computational challenge when the dimension of $\boldsymbol{\theta}$ gets large, say, more than 8 to 10.

In Section 9.4.2.4 we present further detail linking the use of INLA in the context of spatial model fitting through a Markov Random Field approximation Markov Random Field approximation to a Matérn Gaussian random field. In that section the application is to fitting logGaussian Cox process models but the application is broader to general geostatistical models.

5.6 Bayesian model choice

Model choice essentially requires specification of the *utility* for a model. This is a challenging exercise and, in practice, is not often considered. Off-the-shelf criteria are typically adopted. For instance, do we see our primary goal for the model to be an explanation or, alternatively, prediction? Utility for the former places emphasis on the parameters and in fact, calculates a criterion over the parameter space. An approximate yet very easy-to-use model choice tool of this type, known as the Deviance Information Criterion (DIC), has gained popularity. Utility for the latter places us in the space of the observations, considering the performance of *predictive* distributions. One such choice is the posterior predictive criterion due to Gelfand and Ghosh [1998], while Gneiting and Raftery [2007] have considered the problem of model choice from the standpoint of proper scoring rules recommending (continuous) rank probability scores (CRPS). In any event, Bayesian model selection (and model selection in general) is always a contentious issue; there is rarely a unanimously agreed upon criterion. In this subsection, we limit our attention to Bayes factors, the DIC, the Gelfand-Ghosh criterion and the (continuous) rank probability score (CRPS). A further important point in this regard is the notion of cross-validation or hold-out data, i.e., data that is not used to fit the model but rather only for model validation or comparison. See the discussion on the use of hold-out samples at the end of this section.

It is worth noting that model comparison can be considered in the parameter space (as in the ensuing Sections 5.6.1–5.6.3) or in the data space (section 5.6.4). It seems that the former approaches are much more common in the literature but nowadays there seems to be a movement toward the latter approach. We prefer model choice in the data space. That is, in general, parameters are artificial constructs associated with a model; they are not real. So, model assessment with regard to parameter recovery seems unsatisfying. However, model performance in the data space is real. The data is what we observe; posterior predictive performance relative to what we have observed seems a more meaningful way to assess performance.

As an aside, it is important to distinguish model adequacy from model selection. The latter is an easier problem. In what follows, a criterion is proposed and the model which performs best under that criterion is selected. So, model selection becomes a *relative* decision. On the other hand, model adequacy again requires computing a chosen criterion but then, assessing whether, typically, it is too large for the model to be viewed as adequate. So, model adequacy/checking becomes an *absolute* decision. With a complex model and a selected criterion it will be very difficult to calibrate the magnitude of the criterion for the observed data with regard to adequacy. As a result, Bayesian model checking takes us to approaches which we discuss in Section 5.6.5.

5.6.1   Information criterion from Bayes factors

Given that Bayesian hypothesis testing is, in fact, formulated as a model selection problem between two competing models (see Section 5.2.3), it is reasonable to consider developing model selection criteria based on Bayes factors. A popular "shortcut" method is the *Bayesian Information Criterion* (BIC) (also known as the *Schwarz Criterion*), the change in which across the two models is given by

$$\Delta BIC = W - (p_2 - p_1) \log n \,, \tag{5.60}$$

where p_i is the number of parameters in model $M_i, i = 1, 2$, and

$$W = -2 \log \left[\frac{\sup_{M_1} f(\mathbf{y} \mid \theta)}{\sup_{M_2} f(\mathbf{y} \mid \theta)} \right] \,,$$

the usual likelihood ratio test statistic. Schwarz [1978] showed that for nonhierarchical (two-stage) models and large sample sizes n, BIC approximates $-2 \log BF$. An alternative to BIC is the *Akaike Information Criterion* [Akaike, 1998, AIC], which alters (5.60) slightly to

$$\Delta AIC = W - 2(p_2 - p_1) \,. \tag{5.61}$$

Both AIC and BIC are *penalized likelihood ratio* model choice criteria since both have second terms that act as a penalty, correcting for differences in size between the models (to see this, think of M_2 as the "full" model and M_1 as the "reduced" model). Evidently, using a penalty which depends upon the sample size, BIC criticizes differences in model dimension more strongly than AIC does.

The more serious (and aforementioned) limitation in using Bayes factors or their approximations is that they are not appropriate under noninformative priors. To see this, note that if $\pi_i(\boldsymbol{\theta}_i)$ is improper, then $p(\mathbf{y} \mid M_i) = \int f(\mathbf{y} \mid \boldsymbol{\theta}_i, M_i)\pi_i(\boldsymbol{\theta}_i)d\boldsymbol{\theta}_i$ necessarily is as well, and so BF as given in (5.20) is not well defined. While several authors [see, e.g., O'Hagan, 1995, Berger and Pericchi, 1996] have attempted to modify the definition of BF to repair this deficiency, we suggest more informal yet general approaches described below.

5.6.2   Model selection using information criterion

Bayesian inference for spatial data primarily proceeds from developing models that introduce spatial dependence and accommodate other features as deemed plausible by the investigator. Once a model is fit to the data, we should evaluate the quality of the fit of the model. There are various approaches for this and we defer to the text by Gelman et al. [2013] for a comprehensive account that is relevant to practical data analysis [also see Gelman et al., 2014, and references therein for an article focusing on predictive information criteria, which we will follow here].

Here we will consider the predictive information criterion that assigns a score to each model. Such a score rewards a model for fitting the data well and penalizes the model for being overly complex without delivering substantive inferential gains. Let $y = (y_1, y_2, \ldots, y_n)^{\mathrm{T}}$ be an $n \times 1$ vector of outcome variables that are modeled independently conditional on parameters θ as $p(y \mid \theta) = \prod_{i=1}^{n} p(y_i \mid \theta)$. This conditional independence given a parameter value is essential for building valid predictive information criteria and may appear, at first, to be somewhat at odds with spatial data where the outcomes are dependent. However, in the hierarchical models we explore in the subsequent chapters, we will introduce a latent process in θ so that the observations are independent and conditional on the process.

An ideal measure of a model's predictive efficiency would be based upon how well it predicts new data realized from a true data-generating process but that were not used in training the model. Let y be an $n \times 1$ vector of observations realized from a true model, say f_{true}, and let $\tilde{y}$ denote an $n \times 1$ predictive random variable representing a hypothetically realized $n \times 1$ dataset $\tilde{y} \sim f$ from the true data generating process. Inference for this predictive random variable is achieved by computing the posterior predictive distribution, $p(\tilde{\mathbf{y}} \mid \mathbf{y})$ in (5.5). A measure evaluating the predictive fit corresponding to the i-th element, $\tilde{y}_i$, of $\tilde{\mathbf{y}}$ is given by $\log p(\tilde{y}_i \mid \mathbf{y}) = \log \mathrm{E}_{\theta \mid \mathbf{y}}[p(\tilde{y}_i \mid \theta)]$, where $\mathrm{E}_{\theta \mid \mathbf{y}}[\cdot]$ is the expectation with respect to the posterior distribution $p(\theta \mid \mathbf{y})$. The true underlying distribution now enters the picture to account for the uncertainty in $\tilde{y}$. This leads to the expected log predictive density, denoted $\mathrm{ELPD}_i = \mathrm{E}_f[\log p(\tilde{y}_i \mid \mathbf{y})]$, where $\mathrm{E}_f[\cdot]$ is the expectation with respect to the true data generating model. For the full $\tilde{\mathbf{y}}$ we simply add $\sum_{i=1}^{n} \mathrm{ELPD}_i$.

Different modifications of the ELPD adjusted for (estimated) model complexity leads to different predictive information criteria such as the Akaike Information Criteria (AIC), Bayesian Information Criteria (BIC), Deviance Information Criteria (DIC) and Widely Applicable Information Criteria (WAIC). We provide some additional details on the latter two, which are, arguably, among the two most widely used scores used in Bayesian inference.

In practice, we do not know θ and we do not know f. The uncertainty about the former is accounted by the posterior distribution over which we average. For the unknown data generating model, we use the empirical approximation using the values of the sampled data y itself. This means calculating $\sum_{i=1}^{n} \log p(\tilde{y}_i \mid \mathbf{y})$ at the value $\tilde{y}_i = y_i$, which yields the *log point-wise predictive density* defined as,

$$\mathrm{LPPD} = \sum_{i=1}^{n} \log \mathrm{E}_{\theta \mid \mathbf{y}}[p(y_i \mid \theta)] \approx \sum_{i=1}^{n} \log \left(\frac{1}{L} \sum_{l=1}^{L} p(y_i \mid \theta^{(l)}) \right) , \qquad (5.62)$$

where the last expression is the Monte Carlo approximation of the LPPD using posterior samples $\theta^{(l)} \sim p(\theta \mid \mathbf{y})$. WAIC [Watanabe and Opper, 2010, Gelman et al., 2014] compensates (5.62) for model complexity by approximating the effective number of parameters in the model to adjust for complexity using the posterior variance function $\mathbb{V}_{\theta \mid \mathbf{y}}[\cdot]$,

$$p_{\mathrm{WAIC}} = \sum_{i=1}^{n} \mathbb{V}_{\theta \mid \mathbf{y}}[\log p(y_i \mid \theta)] \approx \sum_{i=1}^{n} \left(\frac{1}{L-1} \sum_{l=1}^{L} (v_{il} - \bar{v}_i)^2 \right) , \qquad (5.63)$$

where $v_{il} = \log p(y_i \mid \theta^{(l)})$, $\theta^{(l)} \sim p(\theta \mid \mathbf{y})$ and $\bar{v}_i = (1/L) \sum_{l=1}^{L} v_{il}$. The WAIC is defined by

$$\mathrm{WAIC} = -2 \times (\mathrm{LPPD} - p_{\mathrm{WAIC}}) . \qquad (5.64)$$

This is computed using the Monte Carlo approximations in (5.62) and (5.63). Alternate definitions of p_{WAIC} are possible; we refer the reader to an excellent exposition by Gelman et al. [2014] for details.

As mentioned earlier, the LPPD defined above is not the only way to compute a predictive information criteria. In fact, the AIC [Akaike, 1998], defines ELPD as $\mathrm{E}_f[\log p(\tilde{\mathbf{y}} \mid \hat{\theta}(\mathbf{y}))]$, where $\hat{\theta}(\mathbf{y})$ is the maximum likelihood estimate and approximates this as $\log p(\mathbf{y} \mid \hat{\theta}(\mathbf{y}))$, while the penalty is simply the number of parameters so that $\text{AIC} = \log p(\mathbf{y} \mid \hat{\theta}(\mathbf{y})) - k$.

Spiegelhalter et al. [2002] propose a generalization of the AIC, where ELPD is now defined as $\log p(\mathbf{y} \mid \hat{\theta}_{\text{BAYES}})$ with $\hat{\theta}_{\text{BAYES}}$ being a posterior estimate (mean or median) of θ. Defining $D(\theta) = -2 \times \log p(\mathbf{y} \mid \theta)$ and penalty $p_D = \bar{D} - D(\bar{\theta})$, where $\bar{\theta} = \mathrm{E}_{\theta \mid \mathbf{y}}[\theta]$ is the posterior mean of θ and $\bar{D} = \mathrm{E}_{\theta \mid \mathbf{y}}[D(\theta)]$ is the posterior mean of $D(\theta)$, we define the *deviance information criteria* as

$$\text{DIC} = \mathrm{E}_{\theta \mid \mathbf{y}}[D(\theta)] + p_D \,. \tag{5.65}$$

The *fit* of a model is measured by the posterior expectation of the deviance, which is the first term, and the *complexity* of a model by the effective number of parameters p_D (which may well be less than the total number of model parameters, due to the borrowing of strength across random effects). For Gaussian models, one can show that a reasonable definition of p_D is the expected deviance minus the deviance evaluated at the posterior expectations. Smaller values of DIC indicate a better-fitting model.

5.6.3 Posterior predictive loss criterion

An alternative to the above metrics that is also easily implemented using output from the posterior simulation is the *posterior predictive loss* (performance) approach of Gelfand and Ghosh [1998]. To elucidate further, we revisit the point-wise predictive distribution [see, e.g., Gelman et al., 2013] defined as

$$p(Y_{\ell,rep} \mid \mathbf{y}) = \mathrm{E}_{\theta \mid \mathbf{y}}\left[p(Y_{\ell,rep} \mid \theta)\right] = \int p(Y_{\ell,rep} \mid \theta) \times p(\theta \mid \mathbf{y}) \, d\theta \tag{5.66}$$

for each unit i. We sample one value of $Y_{\ell,rep} \sim p(Y_{\ell,rep} \mid \theta)$ for each posterior sample of θ. An ideal measure of a model's performance is its out-of-sample predictive performance for new data produced from the true data-generating process. This, unfortunately, is impracticable (except in synthetic experiments that simulate the population) because we do not know the true data generating process. Therefore, the predictive distribution (5.66) encodes the distribution of alternative data that would be replicated from the fitted model.

Focusing on prediction, in particular, with regard to replicates of the observed data, $Y_{\ell,rep}$, $\ell = 1, \ldots, n$, the selected models are those that perform well under a *balanced* loss function. Roughly speaking, this loss function penalizes actions both for departure from the corresponding observed value ("fit") as well as for departure from what we expect the replicate to be ("smoothness"). The loss puts weights k and 1 on these two components, respectively, to allow for adjustment of relative regret for the two types of departure. It can be shown that for squared error loss, the resulting criterion becomes

$$D_k = \frac{k}{k+1} \underbrace{\sum_{\ell=1}^{n}(y_\ell - \mu_\ell)^2}_{G} + \underbrace{\sum_{\ell=1}^{n} \sigma_\ell^2}_{P} = \frac{k}{k+1}G + P \,, \tag{5.67}$$

where $\mu_\ell = \mathrm{E}[Y_{\ell,rep} \mid y]$ and $\sigma_\ell^2 = \mathbb{V}(Y_{\ell,rep} \mid y)$ are the posterior predictive mean and variance computed for each replicate, respectively. The components of D_k have natural interpretations. G is a goodness-of-fit term, while P is a penalty term. To clarify, we are seeking to penalize complexity and reward parsimony, just as other penalized likelihood criteria do. For a poor model we expect large predictive variance and poor fit. As the model improves, we expect to do better on both terms. But as we start to overfit, we will continue to do better with regard to goodness of fit, but also begin to inflate the variance (as we introduce multicollinearity). Eventually the resulting increased predictive variance penalty will exceed the gains in goodness of fit. As we sort through a collection of models, the one with the smallest D_k is preferred. When $k = \infty$ (so that $D_k = D_\infty = G + P$), we will write D_∞ simply as D for brevity. Lower values of D indicate preferred models.

Two remarks are appropriate. First, we may report the first and second terms (excluding $k/(k+1)$) on the right side of (5.67), rather than reducing to the single number D_k. Second, in practice, ordering of models is typically insensitive to the particular choice of k.

The quantities μ_ℓ and σ_ℓ^2 that are required to compute (5.67) can be readily computed from posterior samples. If under model m we have parameters $\theta^{(m)}$, then

$$p(y_{\ell,rep} \mid \mathbf{y}) = \int p(y_{\ell,rep} \mid \theta^{(m)})\, p(\theta^{(m)} \mid \mathbf{y})\, d\theta^{(m)} . \tag{5.68}$$

Hence each posterior realization (say, θ^*) can be used to draw a corresponding $y_{\ell,rep}$ from $p(y_{\ell,rep} \mid \theta^{(m)} = \theta^*)$. The resulting $y_{\ell,rep}^*$ has marginal distribution $p(y_{\ell,rep} \mid \mathbf{y})$. With samples from this distribution we can obtain μ_ℓ and σ_ℓ^2. Hence development of D_k requires an extra level of simulation, one-for-one with the posterior samples.

More general loss functions can be used, including the so-called deviance loss (based upon $p(y_\ell \mid \theta^{(m)})$), again yielding two terms for D_k with corresponding interpretation and predictive calculation. This enables the application to, say, binomial or Poisson likelihoods.

Finally, we note that Gneiting and Raftery [2007] suggest modifying (5.67) to propose

$$GRS = -\sum_{\ell=1}^{n} \frac{(y_\ell - \mu_\ell)^2}{\sigma_\ell^2} - \sum_{\ell=1}^{n} \log \sigma_\ell^2 , \tag{5.69}$$

as a proper scoring rule. Higher values of GRS indicate better models.

5.6.4 Model assessment using hold-out data

An important point is that all of the foregoing criteria evaluate model performance based upon the data used to fit the model. That is, they use the data twice with regard to model comparison. Arguably, a more attractive way to compare models is to partition the dataset into a fitting (or learning) set and a validation or "hold out" set and apply the criterion to the hold-out data after fitting the model to the fitting dataset. This enables us to see how a model will perform with *new* data; using the fitted data to compare models will provide too optimistic an assessment of model performance. Of course, how much data to retain for fitting and how much to use for hold out is not well defined; it depends upon the modeling and the amount of available data. Rough suggestions in the literature propose holding out as much as 20−30%. We don't think the amount is a critical issue.

In fact, what we prefer and recommend is so-called K-fold cross-validation. That is, we repeat the hold-out exercise K times, to avoid chance bias in any particular hold-out set, and then look at model performance across the K replicated hold-out sets.

Following our words at the start of this section, when working with point referenced spatial data, prediction is the primary goal. This is the whole intent of the kriging development in Chapter 3. If we are concerned with the predictive performance of a model, then holding out data becomes natural. In this regard, we have two potential perspectives. The first makes a comparison between an observed value, e.g., an observation at a spatial location and an estimate of this observation, obtained through say kriging. Applied to hold-out data, it leads to criteria such as predicted mean square or absolute error, i.e., the sum of squared differences or absolute differences across the hold-out data.

The second perspective compares an observed value with its associated posterior predictive distribution (See Chapter 6). Conceptually, it can be argued that it is preferable to compare a held-out observation to an *entire* distribution rather than to just the held out value. For instance, using this distribution, we can create a posterior $1 - \alpha$ predictive interval for a held-out observation. If we do this across many observations, we can obtain an *empirical* coverage probability (proportion of times the predictive interval contained the observed value) which can be compared with the *nominal* $1 - \alpha$ coverage probability. Rough agreement supports the model. Empirical under-coverage suggests that the model is too optimistic with regard to uncertainty, intervals are too short. However, over-coverage is also not desirable. It indicates that predictive intervals are wider than need be, uncertainty is overestimated.

For model choice, comparing a predictive distribution with an observation takes us into the realm of probabilistic calibration or forecasting, making forecasts for the future and providing suitable measures of the uncertainty associated with them. Probabilistic forecasting has become routine in such applications as weather and climate prediction, computational finance, and macroeconomic forecasting. In our context, the goodness of a predictive distribution relative to an observation is measured by how concentrated the distribution is around the observation. Bypassing all of the elegant theory regarding proper scoring rules [see Gneiting and Raftery, 2007], the proposed measure is the continuous rank probability score (CRPS), the squared integrated distance between the predictive distribution and the degenerate distribution at the observed value,

$$CRPS(F, y) = \int_{-\infty}^{\infty} \left(F(u) - 1(u \geq y) \right)^2 du \, , \qquad (5.70)$$

where F is the predictive distribution and y is the observed value. For us, $Y(\mathbf{s}_0)$ is the observation and F is the posterior predictive distribution for $Y(\mathbf{s}_0)$. With a collection of such hold-out observations and associated predictive distributions, we would sum the CRPS over these observations to create the model comparison criterion. Recall that, under model fitting using posterior samples, we will not have F explicitly but, rather, a sample from F. Fortunately, Gneiting and Raftery [2007] present a convenient alternative form for (5.70), providing F has a first moment:

$$CRPS(F, y) = \frac{1}{2} \mathrm{E}_F |Y - Y'| + \mathrm{E}_F |Y - y| \qquad (5.71)$$

where Y and Y' are independent replicates from F. With samples from F, we have immediate Monte Carlo integrations to compute (5.71).

Finally, we do not recommend a choice among the model comparison approaches discussed. Again, the information criteria approaches work in the parameter space with the likelihood, while the predictive approaches work in the space of the data with posterior predictive distributions. The former addresses comparative explanatory performance, while the latter addresses comparative predictive performance. So, if the objective is to use the model for explanation, we may prefer the former; if instead the objective is prediction, we may prefer a predictive criterion. In different terms, we can argue that, since, with areal unit data, we are most interested in explanation, information criteria approaches will be attractive. In fact, the notion of hold-out data is not clear for this setting. By contrast, with point-referenced data, we can easily hold-out some locations; predictive validation/comparison seems the more attractive path.

5.6.5   In-sample model adequacy

From a formal Bayesian perspective, if we look at the identity, $f(\mathbf{y} \mid \boldsymbol{\theta})f(\boldsymbol{\theta}) = f(\boldsymbol{\theta} \mid \mathbf{y})f(\mathbf{y})$, we can interpret the left side as the generative perspective for the model with the right side as the inferential perspective for the model. In particular, the first term on the right side provides the posterior, as we have already discussed. The second term on the right side can be viewed with regard to model adequacy. In fact, it is a density ordinate so large values suggest that a model is adequate, i.e., the data falls within a likely portion of the marginal distribution for the data under the model. Then, small values of $f(\mathbf{y})$ would criticize the model. While this is an elegant way to interpret the Bayesian model specification, it is completely infeasible in practice. The calculation of $f(\mathbf{y})$ when $\mathbf{y}$ is high dimensional is challenging, perhaps infeasible, but even if one can, more problematic is calibrating $f(\mathbf{y})$. For a density over high dimensional space, how can one assess whether the ordinate at the observed data is large or small? In this regard, assessment of the magnitude of this value could be facilitated by standardizing, using the maximum value or an average value of this density but again, such standardizing will be infeasible with high dimensional $\mathbf{y}$'s. As noted at the beginning of Section 5.6, any model adequacy measure will suffer a calibration problem in the sense that the measure yields an absolute criterion, unlike model comparison which enables relative comparison.

In the absence of a hold-out data set (Section 5.6.4) how can we investigate model adequacy? What has emerged are in-sample predictive model checks, in particular, posterior model checks [Gelman et al., 1996a, henceforth GMS] and prior model checks [Dey et al., 1998, henceforth DGSV].

GMS propose a posterior predictive strategy. They define a discrepancy measure as a function of data and parameters, treating both as unknown in one case, inserting the observed data in the other. They then compare the resulting posterior distributions given the observed data. DGSV also study the posterior distributions of various discrepancy measures given the observed data. However, they compare a particular choice with what is expected under the model, not with what would be expected under the model **and** the observed data. GMS dismiss prior predictive checking, arguing that the prior predictive distribution treats the prior as a true "population distribution" whereas the posterior predictive distribution treats the prior as an outmoded first guess. However, it can be argued that model checking should examine the acceptability of the model actually fitted to the data. Therefore, model parameter values must be generated from the prior prescribed under the model. In this regard, GMS can be criticized for using the data twice. The observed data, through the posterior, suggest values of the parameter which are likely under the model. Then, to assess adequacy, the observed data is checked against data generated using such parameter values, apparently making it difficult to criticize the model.

We briefly further describe both the GMS and DGSV approaches. Both employ Monte Carlo tests in looking at discrepancy measures, $D(\mathbf{y}; \boldsymbol{\theta})$ which, for instance, in the context of a general linear model, might take the form of a raw residual, $Y_i - g(\mathbf{X}_i^{\mathsf{T}}\boldsymbol{\beta} + \mathbf{w}_i^{\mathsf{T}}\boldsymbol{\alpha})$. Here, g is the link function, $\mathbf{X}_i^{\mathsf{T}}\boldsymbol{\beta}$ is the fixed effects component, and $\mathbf{w}_i^{\mathsf{T}}\boldsymbol{\alpha}$ is the random effects component. We let $\boldsymbol{\theta}$ denote the entire collection of parameters.

GMS looks at draws from $[D(\mathbf{y}; \boldsymbol{\theta}) \mid \mathbf{y}_{obs}]$ and compares with draws from $[D(\mathbf{y}_{obs}; \boldsymbol{\theta}) \mid \mathbf{y}_{obs}]$. The problem is evident; the data is used twice. Draws of $\mathbf{y}$ from $[\mathbf{y}_{obs}; \boldsymbol{\theta} \mid \mathbf{y}_{obs}]$ will look too much like $\mathbf{y}_{obs}$; discrepancies, $D(\mathbf{y}, \boldsymbol{\theta})$, will look too much like $D(\mathbf{y}_{obs}; \boldsymbol{\theta})$. The model checking will not be critical enough.

DGSV create draws from $[D(\mathbf{y}, \boldsymbol{\theta}) \mid \mathbf{y}]$ by sampling $\boldsymbol{\theta}$ from the prior, then sampling $\mathbf{y}$ under the model given $\boldsymbol{\theta}$. Fitting the model enables draws from $[\boldsymbol{\theta} \mid \mathbf{y}]$ and, hence, with a collection $\mathbf{y}_l^*$, draws from $[D(\mathbf{y}_l^*, \boldsymbol{\theta}) \mid \mathbf{y}_l^*]$. Then, comparison is made between $[D(\mathbf{y}_l^*, \boldsymbol{\theta}) \mid \mathbf{y}_l^*]$ and $[D(\mathbf{y}_{obs}; \boldsymbol{\theta}) \mid \mathbf{y}_{obs}]$. This is an "apples vs. apples" comparison which uses the data only once. That is, DGSV compare the observed discrepancy with the discrepancies you expect under the model; GMS compares the observed discrepancy with what you expect under the model and the observed data. However, the computational demand required for DGSV is evident; one must fit and sample for every $\mathbf{y}_l^*$.

We do not present further details of these model-checking strategies here but note that GMS has emerged as far more commonly used since it is easier to implement (immediately after model fitting). However, again, it doesn't criticize the model well enough and uses the data twice (once to fit, once to check). DGSV is more computationally demanding but is formally coherent and uses the data only once. We elaborate on both approaches in the context of model checking for spatial point patterns in Section 9.2.5.1. However, our recommendation throughout this text is to adopt an out-of-sample, equivalently cross-validation approach, when available, for both model checking and model comparison.

5.7   Computer tutorials

Computer programs illustrating Bayesian data analysis using different libraries in R and its interfaces with Bayesian modeling environments such as NIMBLE, Stan and JAGS are supplied in the folder titled Chapter 5 from https://github.com/sudiptobanerjee/BGC_2023.

5.8   Exercises

1. During her senior year in high school, Minnesota basketball sensation Carolyn Kieger scored at least 30 points in 9 consecutive games, helping her team win 7 of those games. The data for this remarkable streak are shown in Table 5.1. Notice that the rest of the team *combined* managed to outscore Kieger on only 2 of the 9 occasions.

A local press report on the streak concluded (apparently quite sensibly) that Kieger was primarily responsible for the team's relatively good win-loss record during this period. A natural statistical model for testing this statement would be the *logistic regression* model,

$$Y_i \overset{ind}{\sim} Bernoulli(p_i),$$
$$\text{where} \quad \text{logit}(p_i) = \beta_0 + \beta_1 x_{1i} + \beta_2 x_{2i}.$$

Here, Y_i is 1 if the team won game i and 0 if not, x_{1i} and x_{2i} are the corresponding points scored by Kieger and the rest of the team, respectively, and the logit transformation is

Game	Points scored by Kieger	Rest of team	Game outcome
1	31	31	W, 62–49
2	31	16	W, 47–39
3	36	35	W, 71–64
4	30	42	W, 72–48
5	32	19	L, 64–51
6	33	37	W, 70–49
7	31	29	W, 60–37
8	33	23	W, 56–45
9	32	15	L, 57–47

Table 5.1 *Carolyn Kieger prep basketball data.*

defined as $\text{logit}(p_i) \equiv \log(p_i/(1-p_i))$, so that

$$p_i = \frac{\exp(\beta_0 + \beta_1 x_{1i} + \beta_2 x_{2i})}{1 + \exp(\beta_0 + \beta_1 x_{1i} + \beta_2 x_{2i})} .$$

(a) Using vague (or even flat) priors for the β_j, $j = 0, 1, 2$, fit this model to the data. Obtain posterior summaries for the β_j parameters, as well as a DIC score and effective number of parameters p_D. Also investigate MCMC convergence using trace plots, autocorrelations, and crosscorrelations (the latter from the "Correlations" tool under the "Inference" menu). Is this model acceptable, numerically or statistically?

(b) Fit an appropriate two-parameter reduction of the model in part (a). Center the remaining covariate(s) around their own mean to reduce crosscorrelations in the parameter space, and thus speed MCMC convergence. Is this model an improvement?

(c) Fit one additional two-parameter model, namely,

$$\text{logit}(p_i) = \beta_0 + \beta_1 z_i ,$$

where $z_i = x_{1i}/(x_{1i}+x_{2i})$, the *proportion* of points scored by Kieger in game i. Again investigate convergence behavior, the β_j posteriors, and model fit relative to those in parts (a) and (b).

(d) For this final model, look at the estimated posteriors for the p_i themselves, and interpret the striking differences among them. What does this suggest might still be missing from our model?

2. Show that (5.38) is indeed a symmetric proposal density, as required by the conditions of the Metropolis algorithm.

3. Suppose now that θ is univariate but confined to the range $(0, \infty)$, with density proportional to $h(\theta)$.

(a) Find the Metropolis acceptance ratio r assuming a Gaussian proposal density (5.38). Is this an efficient generation method?

(b) Find the Metropolis acceptance ratio r assuming a Gaussian proposal density for $\eta \equiv \log \theta$. (Hint: Don't forget the Jacobian of this transformation!)

(c) Finally, find the Metropolis-Hastings acceptance ratio r assuming a $Gamma(a, b)$ proposal density for θ.

4. Consider the conjugate Bayesian linear model in (5.21) with $M_0 m_0 = \mu$, i.e., the prior density for the regression slopes is $N(\beta \mid \mu, \sigma^2 M_0)$.

(a) Derive the joint posterior density and show that it is of the form

$$p(\beta, \sigma^2 \mid \mathbf{y}) = IG(\sigma^2 \mid a^*, b^*) \times N(\beta \mid M\mathbf{m}, \sigma^2 M) , \qquad (5.72)$$

where $\mathbf{m} = M_0^{-1}\mu + X^{\mathsf{T}}\mathbf{y}$, $M^{-1} = M_0^{-1} + X^{\mathsf{T}}X$, $a^* = a + n/2$ and $b^* = b_0 + (1/2)(\mu^{\mathsf{T}}M_0^{-1}\mu + \mathbf{y}^{\mathsf{T}}\mathbf{y} - \mathbf{m}^{\mathsf{T}}M\mathbf{m})$.

(b) Assume that $M_0^{-1} \to O$ (the matrix of zeros) and the elements of μ are finite. Then show that the density in (5.72) reduces to

$$p(\beta, \sigma^2 \mid \mathbf{y}) = IG(\sigma^2 \mid a^*, b^*) \times N(\beta \mid \mathbf{m}, \sigma^2 M) , \qquad (5.73)$$

where $\mathbf{m} = X^{\mathsf{T}}\mathbf{y}$, $M^{-1} = X^{\mathsf{T}}X$, $a^* = a + n/2$ and $b^* = b_0 + (1/2)(\mathbf{y}^{\mathsf{T}}\mathbf{y} - \mathbf{m}^{\mathsf{T}}M\mathbf{m})$.

(c) Show that $\hat{\beta} = M\mathbf{m}$ in (5.73) is the ordinary least-squares estimate for β and b^* simplifies to

$$b^* = b_0 + \frac{1}{2}(\mathbf{y}^{\mathsf{T}}\mathbf{y} - \mathbf{m}^{\mathsf{T}}M\mathbf{m}) = b_0 + \frac{(n-p)s^2}{2} ,$$

where $(n-p)s^2 = \sum_{i=1}^{n}(\mathbf{y} - X\hat{\beta})^{\mathsf{T}}(\mathbf{y} - X\hat{\beta})$.

(d) If $a_0 \to -p/2$ and $b_0 \to 0$ in (5.73), then prove that

$$p(\boldsymbol{\beta}, \sigma^2 \mid \mathbf{y}) = IG\left(\sigma^2 \left| \frac{n-p}{2}, \frac{(n-p)s^2}{2} \right.\right) \times N\left(\boldsymbol{\beta} \left| \hat{\boldsymbol{\beta}}, \sigma^2 (X^{\mathrm{T}} X)^{-1} \right.\right) . \qquad (5.74)$$

Explain how Bayesian inference from (5.74) numerically reproduces classical inference from least squares regression.

5. Guo and Carlin [2004] (2004) consider a joint analysis of the AIDS longitudinal and survival data originally analyzed separately by Goldman et al. [1996] and [Carlin and Louis, 2008, Sec. 8.1] . These data compare the effectiveness of two drugs, didanosine (ddI) and zalcitabine (ddC), in both preventing death and improving the longitudinal CD4 count trajectories in patients with late-stage HIV infection. The joint model used is one due to Henderson et al. [2000], which links the two submodels using bivariate Gaussian random effects. Specifically,

Longitudinal model: For data $y_{i1}, y_{i2}, \ldots, y_{in_i}$ from the ith subject at times $s_{i1}, s_{i2}, \ldots, s_{i,n_i}$, let

$$y_{ij} = \mu_i(s_{ij}) + W_{1i}(s_{ij}) + \epsilon_{ij} , \qquad (5.75)$$

where $\mu_i(s) = \mathbf{x}_{1i}^{\mathrm{T}}(s)\boldsymbol{\beta}_1$ is the mean response, $W_{1i}(s) = \mathbf{d}_{1i}^{\mathrm{T}}(s)\mathbf{U}_i$ incorporates subject-specific random effects (adjusting the main trajectory for any subject), and $\epsilon_{ij} \sim N(0, \sigma_\epsilon^2)$ is a sequence of mutually independent measurement errors. This is the classic longitudinal random effects setting of Laird and Ware [1982].

Survival model: Letting t_i is time to death for subject i, we assume the parametric model,

$$t_i \sim \text{Weibull}\left(p, \mu_i(t)\right),$$

where $p > 0$ and

$$\log(\mu_i(t)) = \mathbf{x}_{2i}^{\mathrm{T}}(t)\boldsymbol{\beta}_2 + W_{2i}(t) .$$

Here, $\boldsymbol{\beta}_2$ is the vector of fixed effects corresponding to the (possibly time-dependent) explanatory variables $\mathbf{x}_{2i}(t)$ (which may have elements in common with $\mathbf{x}_{1i}$), and $W_{2i}(t)$ is similar to $W_{1i}(s)$, including subject-specific covariate effects and an intercept (often called a *frailty*).

The specific joint model studied by Guo and Carlin [2004] assumes

$$\begin{aligned} W_{1i}(s) &= U_{1i} + U_{2i}\, s , \quad \text{and} & (5.76) \\ W_{2i}(t) &= \gamma_1 U_{1i} + \gamma_2 U_{2i} + \gamma_3 (U_{1i} + U_{2i}\, t) + U_{3i} , & (5.77) \end{aligned}$$

where $(U_{1i}, U_{2i})^{\mathrm{T}} \overset{iid}{\sim} N(\mathbf{0}, \Sigma)$ and $U_{3i} \overset{iid}{\sim} N(0, \sigma_3^2)$, independent of the $(U_{1i}, U_{2i})^{\mathrm{T}}$. The γ_1, γ_2, and γ_3 parameters in the model (5.77) measure the association between the two submodels induced by the random intercepts, slopes, and fitted longitudinal value at the event time $W_{1i}(t)$, respectively.

(a) Fit the version of this model with $U_{3i} = 0$ for all i, as well as the further simplified version that sets $\gamma_3 = 0$ ("Model XI"). Which models fits better according to the DIC criterion?

(b) For your chosen model, investigate and comment on the posterior distributions of γ_1, γ_2, $\beta_{1,3}$ (the relative effect of ddI on the overall CD4 slope), and $\beta_{2,2}$ (the relative effect of ddI on survival).

(c) For each drug group separately, estimate the posterior distribution of the median survival time of a hypothetical patient with covariate values corresponding to a male who is AIDS-negative and intolerant of AZT at study entry. Do your answers change if you fit only the survival portion of the model (i.e., ignoring the longitudinal information)?

Chapter 6

Hierarchical modeling for univariate spatial data

Having reviewed the basics of inference and computing under the hierarchical Bayesian modeling paradigm, we now turn our attention to its application in the setting of univariate point referenced and areal unit data. The spatial models discussed in the previous chapters will be of interest, but now they may be introduced in either the first-stage specification, to directly model the data in a spatial fashion, *or* in the second-stage specification, to model spatial structure in the random effects. We begin with models for point-level data, then proceed to areal data models.

There is a substantial body of literature focusing on spatial prediction from a Bayesian perspective. Early work includes Le and Zidek [1992], Handcock and Stein [1993], Brown et al. [1994], Handcock and Wallis [1994], Victor De Oliveira and Short [1997], Ecker and Gelfand [1997], Diggle et al. [1998], and Gaudard et al. [1999]. The work of Woodbury [1989], Abrahamsen [1993], and several papers by Omre and colleagues [Omre, 1987, 1988, Omre and Halvorsen, 1989, Omre et al., 1989, Hjort et al., 1994] is partially Bayesian in the sense that prior specification of the mean parameters and covariance function are elicited; however, no distributional assumption is made for the $Y(\mathbf{s})$. Over the course of the 20 years since the publication of the first edition of this book, there has been an explosion of Bayesian spatial work, too many papers to cite here. Rather, the reader will find references to them sprinkled throughout the ensuing chapters of this new edition.

6.1 Stationary spatial process models

The basic model we will work with is

$$Y(\mathbf{s}) = \mu(\mathbf{s}) + w(\mathbf{s}) + \epsilon(\mathbf{s}) , \tag{6.1}$$

where $\mu(\mathbf{s}) = \mathbf{x}^{\mathrm{T}}(\mathbf{s})\boldsymbol{\beta}$. The residual is partitioned into two pieces, one endowed with spatial structure and the other not so. That is, the $w(\mathbf{s})$ are assumed to be realizations from a zero-centered stationary Gaussian spatial process (see Section 3.1), capturing residual spatial association, while the $\epsilon(\mathbf{s})$ are uncorrelated pure error terms. A point which is often ignored in the literature is that (6.1) is well-defined for an individual point $\mathbf{s}$. but not for a process over say $D \subset R^2$. The issue is that, while $\{w(\mathbf{s}) : \mathbf{s} \in D\}$ as a Gaussian process is well-defined over D, $\{\epsilon(\mathbf{s}) : \mathbf{s} \in D\}$ is not, formally, a process. Brownian motion is well-defined over D but is not differentiable so $\epsilon(\mathbf{s})$, as a process, does not exist. Sometimes $\{\epsilon(\mathbf{s}) : \mathbf{s} \in D\}$ is referred to as a pseudo-derivative of Brownian motion but, in fact, is properly defined only for a countably infinite set of $\mathbf{s}$'s. However, practically, this presents no problem since we always work with only a finite set of $\mathbf{s}$'s.

Thus, for the covariance functions presented in Chapter 2, the $w(\mathbf{s})$ introduce the partial sill (σ^2) and range (ϕ) parameters, while the $\epsilon(\mathbf{s})$ add the nugget effect (τ^2). Valid correlation functions were discussed in Section 2.1 as well as Chapter 3. More specifically, supplying the

DOI: 10.1201/9781003401728-6

correlation function as a function of the separation between sites yields a *stationary* model. If this dependence is captured only through the distance $||\mathbf{s}_i - \mathbf{s}_j||$, we obtain *isotropy*. Again, the most common such forms (exponential, Matérn , etc.) were presented in Subsection 2.1.3 and Tables 2.1 and 2.2.

Several interpretations can be attached to $\epsilon(\mathbf{s})$ and its associated variance τ^2. For instance, $\epsilon(\mathbf{s})$ can be viewed as a pure error term, as opposed to the spatial error term $w(\mathbf{s})$. That is, we would not necessarily insist that the residual error be entirely spatially structured. Correspondingly, the nugget τ^2 is a variance component of $Y(\mathbf{s})$, as is σ^2. In other words, while $w(\mathbf{s} + \mathbf{h}) - w(\mathbf{s}) \to 0$ as $\mathbf{h} \to 0$ (if process realizations are continuous; see Subsection 3.1.4 and Section 14.2), $[w(\mathbf{s} + \mathbf{h}) + \epsilon(\mathbf{s} + \mathbf{h})] - [w(\mathbf{s}) - \epsilon(\mathbf{s})]$ will not. We are proposing residuals that are not spatially continuous, but not because the spatial process is not smooth. Instead, it is because we envision additional variability associated with the observed process, $Y(\mathbf{s})$. This could be viewed as measurement error (as might be the case with data from a monitoring device) or more generally as "noise" associated with replication of measurement at location $\mathbf{s}$ (as might be the case with the sale of a single-family home at $\mathbf{s}$, in which case $\epsilon(\mathbf{s})$ would capture the effect of the particular seller, buyer, realtors, and so on).

Another view of τ^2 is that it represents *microscale* variability, i.e., variability at distances smaller than the smallest inter-location distance in the data. In this sense, arguably, $\epsilon(\mathbf{s})$ could also be viewed as a spatial process, but with very rapid decay in association, i.e., with very small range. The dependence between the $\epsilon(\mathbf{s})$ would only matter at very high resolution. In this regard, [Cressie, 1993, pp. 112–113] suggests that $\epsilon(\mathbf{s})$ and τ^2 may themselves be partitioned into two pieces, one reflecting pure error and the other reflecting microscale spatial error. In practice, we will rarely know much about the latter (and the data can not inform about it since we never observe the process at finer spatial resolution), so in this book we employ $\epsilon(\mathbf{s})$ to represent only the former.

One of the simplest examples of a Bayesian model for kriging arises by specifying $w(\mathbf{s})$ in (6.1) as a spatial process with an isotropic covariance function. Suppose we have data $Y(\mathbf{s}_i)$, $i = 1, \ldots, n$, and let $\mathbf{Y} = (Y(\mathbf{s}_1), \ldots, Y(\mathbf{s}_n))^{\mathsf{T}}$. The basic Gaussian isotropic kriging models of Section 2.5 are a special case of the general linear model. The problem boils down to the appropriate definition of the Σ matrix. For example, in the case with a nugget effect,

$$\Sigma = \sigma^2 H(\phi) + \tau^2 I,$$

where H is a correlation matrix with $H_{ij} = \rho(\mathbf{s}_i - \mathbf{s}_j; \phi)$ and ρ is a valid isotropic correlation function on $\Re^2$ indexed by a parameter (or parameters) ϕ. The set of unknown parameters is $\{\boldsymbol{\beta}, \sigma^2, \tau^2, \phi\}$, a Bayesian solution requires an appropriate prior distribution. Parameter estimates may then be obtained from the posterior distribution, which by (5.3) is

$$p\left(\boldsymbol{\beta}, \sigma^2, \tau^2, \phi \mid \mathbf{y}\right) \propto p\left(\mathbf{y} \mid \boldsymbol{\beta}, \sigma^2, \tau^2, \phi\right) \times p(\boldsymbol{\beta}, \sigma^2, \tau^2, \phi), \tag{6.2}$$

where

$$\mathbf{Y} \mid \boldsymbol{\beta}, \sigma^2, \tau^2, \phi \sim N\left(X\boldsymbol{\beta}, \ \sigma^2 H(\phi) + \tau^2 I\right). \tag{6.3}$$

Modeling point-referenced data using (6.2) and (6.3) yields legitimate probabilistic inference long as $H(\phi)$ is positive definite. We will return to this hierarchical model in Section 6.3. While isotropic correlation models are easy choices that ensure this, they are by no means the only choice. We now take a brief peek into how any valid covariance function will still yield valid posterior distributions.

6.2 Modeling point-referenced data

As per the title, we emphasize the role of hierarchical modeling throughout this book. In particular, we emphasize the generic model

$$[\text{data} \mid \text{process}] \times [\text{process} \mid \text{parameters}] \times [\text{parameters}], \tag{6.4}$$

which is more flexible than might first appear since the nature of the data and the nature of the process are not specified. This rich framework emphasizes our goal of learning about a process (typically, fairly complex) that is driving the data we are observing, a process that we seek to better understand. This framework occurs throughout the duration of this book.

Readers will recall from Chapter 3 the rich possibilities available to us for constructing valid spatial covariance models that relax assumptions of isotropy or even stationarity. In fact, any valid specification for the process $w(\mathbf{s})$ in (6.1) delivers valid probabilistic inference. Furthermore, keeping an eye toward modeling point-referenced spatial-temporal data (see Chapter 12) and also more general applications of Gaussian processes [Rasmussen and Williams, 2005], it may be more appealing to treat the stochastic process as a tool to introduce dependence among any finite collection of random variables.

Formally, we can define our process as an uncountable set of random variables, say $\{w(\ell) : \ell \in \mathcal{L}\}$, over a domain of interest $\mathcal{L}$. This uncountable set is endowed with a probability law specifying the joint distribution for any finite subset of random variables. Spatial processes are usually constructed assuming $\mathcal{L} \subseteq \Re^d$ (usually $d = 2$ or 3) or, perhaps, as a subset of points on a sphere or ellipsoid. Spatiotemporal settings, which are considered in detail in Chapter 12, may treat $\mathcal{L} = \mathcal{S} \times \mathcal{T}$, where $\mathcal{S} \subset \Re^d$ and $\mathcal{T} \subset [0, \infty)$ are the space and time domains, respectively, and $\ell = (\mathbf{s}, t)$ is a space-time coordinate with spatial location $\mathbf{s} \in \mathcal{S}$ and time point $t \in \mathcal{T}$ [see, e.g., Stein, 2005, Gneiting and Guttorp, 2010, Cressie and Wikle, 2011, and Section 12.3 for more elaborate developments on spatiotemporal covariance models].

We now denote our covariance function $\text{Cov}\{w(\ell), w(\ell')\} = K_\theta(\ell, \ell')$ for any two points ℓ and ℓ' in $\mathcal{L}$, where K_θ denotes any valid real-valued covariance function on $\mathcal{L} \times \mathcal{L}$. If $\mathcal{U}$ and $\mathcal{V}$ are finite sets comprising n and m points in $\mathcal{L}$, respectively, then $K_\theta(\mathcal{U}, \mathcal{V})$ denotes the $n \times m$ matrix whose (i, j)-th element is evaluated using the covariance function $K_\theta(\cdot, \cdot)$ between the i-th point $\mathcal{U}$ and the j-th point in $\mathcal{V}$. If $\mathcal{U}$ or $\mathcal{V}$ comprises a single point, $K_\theta(\mathcal{U}, \mathcal{V})$ is a row or column vector, respectively. A valid spatiotemporal covariance function ensures that $K_\theta(\mathcal{U}, \mathcal{U})$ is positive definite for any finite set $\mathcal{U}$, which we will denote simply as K_θ if the context is clear. A customary specification models $\{w(\ell) : \ell \in \mathcal{L}\}$ as a zero-centered Gaussian process, denoted as $w(\ell) \sim GP(0, K_\theta(\cdot, \cdot))$. For any finite collection $\mathcal{U} = \{\ell_1, \ell_2, \ldots, \ell_n\}$ in $\mathcal{L}$, the $n \times 1$ random vector $\mathbf{w} = (w(\ell_1)), w(\ell_2), \ldots, w(\ell_n))^\mathrm{T}$ is distributed as $N(\mathbf{0}, K_\theta)$, where $K_\theta = K_\theta(\mathcal{U}, \mathcal{U})$. If X is $n \times p$ with (i, j)-th element recording the fixed value of predictor j at ℓ_i, $x_j(\ell_i)$, then the hierarchical model in (6.2) and (6.3) is

$$p(\theta, \tau, \boldsymbol{\beta}, w \mid \mathbf{y}) \propto p(\theta, \boldsymbol{\beta}, \tau) \times N(\mathbf{w} \mid \mathbf{0}, K_\theta) \times N(\mathbf{y} \mid X\boldsymbol{\beta} + \mathbf{w}, D_\tau), \quad (6.5)$$

where $\mathbf{y} = (y(\ell_1), y(\ell_2), \ldots, y(\ell_n))^\mathrm{T}$ is the $n \times 1$ vector of observed outcomes and $D_\tau = \tau^2 I_n$.

For Gaussian likelihoods, one can integrate out the random effects $\mathbf{w}$ from (6.5) and work with the posterior

$$p(\theta, \boldsymbol{\beta}, \tau \mid \mathbf{y}) \propto p(\theta, \boldsymbol{\beta}, \tau) \times N(\mathbf{y} \mid X\boldsymbol{\beta}, K_\theta + D_\tau). \quad (6.6)$$

This reduces the parameter space to $\{\tau^2, \theta, \boldsymbol{\beta}\}$ by excluding the possibly high-dimensional vector $\mathbf{w}$. These are known as *collapsed* models and are often preferred for model fitting since iterative algorithms such as MCMC are much more efficient and converge much more quickly as they explore the much smaller parameter space. Furthermore, inference on $\mathbf{w}$ is not lost in (6.6). Once posterior samples of $\{\theta, \boldsymbol{\beta}, \tau\}$ are drawn from (6.6), we draw posterior samples of $\mathbf{w} \sim p(\mathbf{w} \mid \boldsymbol{\beta}, \theta, \tau)$ from its full conditional distribution derived from (6.5), which is of the form $N(M\mathbf{m}, M)$, where $\mathbf{m} = X^\mathrm{T} D_\tau^{-1} \mathbf{y}$ and $M^{-1} = K_\theta^{-1} + D_\tau^{-1}$. This completes the posterior sampling from $p(\theta, \tau, \boldsymbol{\beta}, \mathbf{w} \mid \mathbf{y})$ without explicitly sampling from (6.5).

Fitting the collapsed model still requires matrix computations involving $K_\theta + D_\tau$, which is $n \times n$. Thus, it does not obviate the computational problems associated with large or massive values of n. In fact, fitting either (6.5) or (6.6) to large spatial datasets incurs a substantial computational expense from the size of K_θ. Since θ is unknown, each iteration of the model fitting algorithm will involve decomposing or factorizing K_θ, which typically requires $\sim n^3$ floating point operations (flops) and memory requirements in the order of $\sim n^2$. Geostatistical data are almost never observed on regular grids and the configuration of points are typically highly irregular. The covariance models effective for inference do not, in general, result in any computationally exploitable structure for K_θ, which makes the matrix computations prohibitive for large values of n. These are referred to as "big-n" or "high-dimensional" problems in geostatistics and are the main topic of Chapter 13.

6.2.1   Model selection

In adapting Bayesian model selection metrics as described, for example, in Sections 5.6.2 and 5.6.3, we compute either point-wise predictive densities or posterior predictive distributions for replicated data. Such metrics are developed using the independence of the data points in the likelihood function, which implies that the likelihood $\prod_{i=1}^{n} N(y(\ell_i) \mid \mathbf{x}(\ell_i)^{\mathrm{T}}\boldsymbol{\beta} + \mathbf{w}, \tau^2)$ in (6.5) should be used for computing such metrics.

For example, in calculating the LPPD in (5.62), we compute

$$\text{LPPD} = \sum_{i=1}^{n} \log \mathbb{E}_{\Omega \mid \mathbf{y}}[p(y(\ell_i) \mid \Omega)] \approx \sum_{i=1}^{n} \log \left(\frac{1}{L} \sum_{l=1}^{L} p(y(\ell_i) \mid \Omega^{(l)}) \right), \qquad (6.7)$$

where $\Omega = \{\boldsymbol{\beta}, \mathbf{w}, \tau\}$, $p(y(\ell_i) \mid \Omega) = N(y(\ell_i) \mid \mathbf{x}(\ell_i)^{\mathrm{T}}\boldsymbol{\beta} + \mathbf{w}, \tau^2)$ is the normal density evaluated at the value of the response $y(\ell_i)$ given Ω, and the last expression is the Monte Carlo approximation of the LPPD using posterior samples $\Omega^{(l)} \sim p(\Omega \mid \mathbf{y})$. We compensate (5.62) for model complexity using the posterior variance function $\mathbb{V}_{\Omega \mid y}[\cdot]$,

$$p_{\text{WAIC}} = \sum_{i=1}^{n} \mathbb{V}_{\Omega \mid y}[\log p(y(\ell_i) \mid \Omega)] \approx \sum_{i=1}^{n} \left(\frac{1}{L-1} \sum_{l=1}^{L} (v_{il} - \bar{v}_i)^2 \right), \qquad (6.8)$$

where $v_{il} = \log p(y(\ell_i) \mid \Omega^{(l)})$, $\Omega^{(l)} \sim p(\Omega \mid \mathbf{y})$ and $\bar{v}_i = (1/L)\sum_{l=1}^{L} v_{il}$. The WAIC is then calculated as $\text{WAIC} = -2 \times (\text{LPPD} - p_{\text{WAIC}})$ using the Monte Carlo approximations in (5.62) and (5.63). Similarly, for DIC, we compute the deviance $D(\Omega) = -2 \times \sum_{i=1}^{n} \log p(y(\ell_i) \mid \Omega)$ and penalty $p_D = \bar{D} - D(\bar{\Omega})$, where $\bar{\Omega} = \mathbb{E}_{\Omega \mid y}[\Omega]$.

For the posterior predictive loss approach in Section 5.6.3, we compute

$$p(Y_{rep}(\ell_i) \mid \mathbf{y}) = \mathbb{E}_{\Omega \mid y}[p(Y_{rep}(\ell_i) \mid \Omega)] = \int p(Y_{rep}(\ell_i) \mid \Omega) \times p(\Omega \mid \mathbf{y})\, d\Omega \qquad (6.9)$$

for each spatial location $\ell_i \in \mathcal{L}$. We sample one value of $Y_{rep}(\ell_i) \sim p(Y_{rep}(\ell_i) \mid \Omega)$ for each posterior sample of Ω. The key point, again, is to use (6.5) with $\Omega = \{\boldsymbol{\beta}, \mathbf{w}, \tau\}$. The D_k and P, or the GRS, are then calculated as described in (5.67) or (5.68), respectively.

6.2.2   Predictive inference for point-referenced data

Posterior predictive inference for point-referenced data includes inference on (i) the latent process and (ii) the values of the response or outcome at arbitrary locations. Let $\tilde{\mathcal{L}} = \{\tilde{\ell}_1, \tilde{\ell}_2, \ldots, \tilde{\ell}_{\tilde{n}}\}$ be a set of $\tilde{n}$ locations where we wish to predict $Y(\ell)$. Let $\tilde{\mathbf{Y}} = (\tilde{Y}(\tilde{\ell}_1), \ldots, \tilde{Y}(\tilde{\ell}n))^{\mathrm{T}}$ be the $\tilde{n} \times 1$ random vector with i-th element $\tilde{Y}(\tilde{\ell}_i)$ representing the unknown value of the outcome at location $\tilde{\ell}_i$ and let $\tilde{\mathbf{w}}$ be the $\tilde{n} \times 1$ vector with elements $w(\tilde{\ell}_i)$. Spatial prediction extends the joint distribution $p(\theta, \mathbf{w}, \boldsymbol{\beta}, \tau, \mathbf{y})$ to include $\tilde{\mathbf{Y}}$ as

$$p(\theta, \tau, \boldsymbol{\beta}, \mathbf{w}, \mathbf{y}, \tilde{\mathbf{w}}, \tilde{\mathbf{Y}}) = p(\theta, \tau, \boldsymbol{\beta}) \times p(\mathbf{w} \mid \theta) \times p(\tilde{\mathbf{w}} \mid \mathbf{w}, \theta) \times p(\mathbf{y} \mid \boldsymbol{\beta}, \mathbf{w}, \tau) \times p(\tilde{\mathbf{Y}} \mid \boldsymbol{\beta}, \tilde{\mathbf{w}}, \tau).$$
$$(6.10)$$

The factorization in (6.10) relies upon conditional independence between $\tilde{\mathbf{Y}}$ and w given $\tilde{\mathbf{w}}$ and $\boldsymbol{\beta}$. The density $p(\tilde{\mathbf{w}} \mid \mathbf{w}, \theta)$ is derived from the joint distribution $p(\mathbf{w}, \tilde{\mathbf{w}} \mid \theta)$ specified by the process, while $p(\tilde{\mathbf{Y}} \mid \boldsymbol{\beta}, \tilde{\mathbf{w}}, \tau)$ is the likelihood function applied over $\tilde{\mathcal{L}}$.

Predictive inference evaluates the posterior predictive distribution $p(\tilde{\mathbf{Y}}, \tilde{\mathbf{w}} \mid \mathbf{y})$. This is the joint posterior distribution for the outcomes and the spatial effects at locations in $\tilde{\mathcal{L}}$. This distribution is easily derived from (6.10) as

$$p(\tilde{\mathbf{Y}}, \tilde{\mathbf{w}}, \boldsymbol{\beta}, \mathbf{w}, \theta, \tau \mid \mathbf{y}) \propto p(\boldsymbol{\beta}, \mathbf{w}, \theta, \tau \mid \mathbf{y}) \times p(\tilde{\mathbf{w}} \mid \mathbf{w}, \theta) \times p(\tilde{\mathbf{Y}} \mid \boldsymbol{\beta}, \tilde{\mathbf{w}}, \tau) . \tag{6.11}$$

We draw samples from (6.11) by first collecting the posterior samples from $p(\boldsymbol{\beta}, \mathbf{w}, \theta, \tau \mid \mathbf{y})$. For each drawn value of $\{\boldsymbol{\beta}, \mathbf{w}, \theta, \tau\}$, we make one draw of the $\tilde{n} \times 1$ vector $\tilde{\mathbf{w}}$ from $p(\tilde{\mathbf{w}} \mid \mathbf{w}, \theta)$ and then, using this sampled $\tilde{\mathbf{w}}$, we make one draw of $\hat{\mathbf{Y}}$ from $p(\tilde{\mathbf{Y}} \mid \boldsymbol{\beta}, \tilde{\mathbf{w}}, \tau)$. The resulting samples of $\tilde{\mathbf{w}}$ and $\tilde{\mathbf{Y}}$ are draws from the desired posterior predictive distribution $p(\tilde{\mathbf{w}}, \tilde{\mathbf{Y}} \mid \mathbf{y})$. This yields posterior inference for the latent spatial process $\tilde{\mathbf{w}}$ and the outcome $\tilde{\mathbf{Y}}$ at arbitrary locations since $\mathcal{L}$ can be any finite collection of samples. Summarizing these distributions by computing their sample means, standard errors, and the 2.5-th and 97.5-th quantiles yields point estimates with associated uncertainty quantification.

It is worth pausing here for a moment to appreciate how the spatial covariance function induces the conditional distribution $p(\tilde{\mathbf{w}} \mid \mathbf{w}, \theta)$. Since there is one underlying random field over the entire domain, the covariance function specifies the joint distribution $p(\mathbf{w}, \tilde{\mathbf{w}} \mid \theta)$ as $N\left(\mathbf{0}, K_\theta(L \cup \tilde{L})\right)$, where $K_\theta(L \cup \tilde{L}) = \begin{bmatrix} K_\theta(\mathcal{L}, \mathcal{L}) & K_\theta(\mathcal{L}, \tilde{\mathcal{L}}) \\ K_\theta(\tilde{\mathcal{L}}, \mathcal{L}) & K_\theta(\tilde{\mathcal{L}}, \tilde{\mathcal{L}}) \end{bmatrix}$. Hence, the conditional distribution $p(\tilde{\mathbf{w}} \mid \mathbf{w}, \theta)$ is $N(\tilde{\mathbf{w}} \mid C\mathbf{w}, F_\theta)$, where $C = K_\theta(\tilde{\mathcal{L}}, \mathcal{L})K_\theta^{-1}$ and $F_\theta = K_\theta(\tilde{\mathcal{L}}, \tilde{\mathcal{L}}) - K_\theta(\tilde{\mathcal{L}}, \mathcal{L})K_\theta^{-1}K_\theta(\mathcal{L}, \tilde{\mathcal{L}})$. In the next section, we provide some further details on exact distribution theory achieved in conjugate Bayesian linear models by fixing the range parameter and the ratio of the nugget to the partial sill.

6.2.3   Conjugate Bayesian linear geostatistical models

Let us consider incorporating spatial dependence in a Bayesian linear regression model. Recall the conjugate model discussed in Section 5.3,

$$\mathbf{y} \mid \boldsymbol{\beta}, \sigma^2 \sim N(X\boldsymbol{\beta}, \sigma^2 V_y) ; \quad \boldsymbol{\beta} \mid \sigma^2 \sim N(\boldsymbol{\beta} \mid \boldsymbol{\mu}_\beta, \sigma^2 V_\beta) ; \quad \sigma^2 \sim IG(a_\sigma, b_\sigma) , \tag{6.12}$$

where $\mathbf{y}$ is an $n \times 1$ vector of observations of the dependent variable, X is $n \times p$ (assumed to be of rank p) consisting of independent variables (covariates or predictors) and its first column is usually taken to be the intercept, V_y is a fixed (i.e., known) $n \times n$ positive definite matrix, $\boldsymbol{\mu}_\beta$, V_β, a_σ and b_σ are assumed to be fixed hyper-parameters specifying the prior distributions on the regression slopes $\boldsymbol{\beta}$ and the scale σ^2. This model is easily tractable and the posterior distribution is

$$p(\boldsymbol{\beta}, \sigma^2 \mid \mathbf{y}) = \underbrace{IG(\sigma^2 \mid a_\sigma^*, b_\sigma^*)}_{p(\sigma^2 \mid \mathbf{y})} \times \underbrace{N(\boldsymbol{\beta} \mid M\mathbf{m}, \sigma^2 M)}_{p(\boldsymbol{\beta} \mid \sigma^2, \mathbf{y})} , \tag{6.13}$$

where $a_\sigma^* = a_\sigma + n/2$, $b_\sigma^* = b_\sigma + (1/2)\left\{ \boldsymbol{\mu}_\beta^{\mathrm{T}} V_\beta^{-1} \boldsymbol{\mu}_\beta + \mathbf{y}^{\mathrm{T}} V_y^{-1} \mathbf{y} - \mathbf{m}^{\mathrm{T}} M\mathbf{m} \right\}$, $M^{-1} = V_\beta^{-1} + X^{\mathrm{T}} V_y^{-1} X$ and $\mathbf{m} = V_\beta^{-1} \boldsymbol{\mu}_\beta + X^{\mathrm{T}} V_y^{-1} \mathbf{y}$.

It is easy to accommodate spatial dependence in (6.12). Consider (6.6) with the customary specification $D_\tau = \tau^2 I$ and let $K_\theta = \sigma^2 H(\phi)$, where $H(\phi)$ is a correlation matrix whose entries are given by a correlation function $\rho(\phi; \ell_i, \ell_j)$. Thus, $\theta = \{\sigma^2, \phi\}$, where σ^2 is the spatial variance component and ϕ is a spatial decay parameter controlling the rate at which the spatial correlation decays with separation between points. This is equivalent to setting $V_y = H(\phi) + \delta^2 I$ in (6.12), where $\delta^2 = \tau^2/\sigma^2$ is the ratio between the "noise" variance and "spatial" variance. If we assume that ϕ and δ^2 are fixed and that the prior on $\{\boldsymbol{\beta}, \sigma^2\}$ are as in (6.12), then we can draw samples of $\{\boldsymbol{\beta}, \sigma^2\}$ by first sampling $\sigma^2 \sim IG(a_\sigma^*, b_\sigma^*)$ and then sampling $\boldsymbol{\beta} \sim N(M\mathbf{m}, \sigma^2 M)$ for each sampled σ^2.

We will return to the issue of fixing $\{\phi, \delta^2\}$ shortly, but before that it is instructive to note that the spatial process $w(\cdot)$ itself can be introduced in (6.12), which would enable us to directly sample the spatial random effects w from their marginal posterior $p(\mathbf{w} \mid \mathbf{y})$. Here, we write the joint distribution of y and w as a linear model,

$$
\underbrace{\begin{bmatrix} \mathbf{y} \\ \boldsymbol{\mu}_\beta \\ \mathbf{0} \end{bmatrix}}_{\mathbf{y}_*} = \underbrace{\begin{bmatrix} X & I_n \\ I_p & O \\ O & I_n \end{bmatrix}}_{X_*} \underbrace{\begin{bmatrix} \boldsymbol{\beta} \\ \mathbf{w} \end{bmatrix}}_{\boldsymbol{\gamma}} + \underbrace{\begin{bmatrix} \boldsymbol{\eta}_1 \\ \boldsymbol{\eta}_2 \\ \boldsymbol{\eta}_3 \end{bmatrix}}_{\boldsymbol{\eta}} , \tag{6.14}
$$

where $\boldsymbol{\eta} \sim N(\mathbf{0}, \sigma^2 V_{y_*})$ and $V_{y_*} = \begin{bmatrix} \delta^2 I_n & O & O \\ O & V_\beta & O \\ O & O & H(\phi) \end{bmatrix}$. If we assume that δ^2 and ϕ are fixed at known values, then V_{y_*} is fixed. We have a conjugate Bayesian linear regression model $\mathbf{y}_* = X_* \boldsymbol{\gamma} + \boldsymbol{\eta}$, where $\boldsymbol{\gamma}$ has a flat prior and $\sigma^2 \sim IG(a_\sigma, b_\sigma)$. Thus,

$$
p(\boldsymbol{\gamma}, \sigma^2 \mid \mathbf{y}) = \underbrace{IG(\sigma^2 \mid a_\sigma^*, b_\sigma^*)}_{p(\sigma^2 \mid \mathbf{y})} \times \underbrace{N(\boldsymbol{\gamma} \mid M_* \mathbf{m}_*, \sigma^2 M_*)}_{p(\boldsymbol{\gamma} \mid \sigma^2, \mathbf{y})} , \tag{6.15}
$$

where $a_\sigma^* = a_\sigma + n/2$, $b_\sigma^* = b_\sigma + (1/2)\{\mathbf{y}_*^{\mathsf{T}} V_{y_*}^{-1} \mathbf{y}_* - \mathbf{m}_*^{\mathsf{T}} M_* \mathbf{m}_*\}$, $M_*^{-1} = X_*^{\mathsf{T}} V_{y_*}^{-1} X_*$ and $\mathbf{m}_* = X_*^{\mathsf{T}} V_{y_*}^{-1} \mathbf{y}_*$. The posterior mean of $\boldsymbol{\gamma}$ is $\hat{\boldsymbol{\gamma}} = M_* \mathbf{m}_* = \left(X_*^{\mathsf{T}} V_{y_*}^{-1} X_*\right)^{-1} X_*^{\mathsf{T}} V_{y_*}^{-1} \mathbf{y}_*$, which is the generalized least squares estimate obtained from the augmented linear system in (6.14). Sampling from the posterior proceeds analogous to that described below (6.13).

Predictive inference can also be cast into an augmented linear regression model. The predictive model for $\tilde{\mathbf{Y}}$ can be written as a spatial regression

$$
\tilde{\mathbf{Y}} = \tilde{X}\boldsymbol{\beta} + \tilde{\mathbf{w}} + \tilde{\boldsymbol{\epsilon}} ; \quad \tilde{\mathbf{w}} = C\mathbf{w} + \boldsymbol{\omega} , \tag{6.16}
$$

where $\tilde{X}$ is $\tilde{n} \times p$ comprising predictors observed at locations in $\tilde{\mathcal{L}}$ and $\tilde{\boldsymbol{\epsilon}} \sim N(\mathbf{0}, \tilde{D}_\tau)$, where $\tilde{\boldsymbol{\epsilon}}$ is $\tilde{n} \times 1$ with elements $\epsilon(\tilde{\ell}_i)$ and $\tilde{D}_\tau = \tau^2 I_{\tilde{n}}$. The second equation in (6.16) models the spatial effects $\tilde{\mathbf{w}}$ across the unobserved locations in $\tilde{\mathcal{L}}$ in terms of those from the observed locations in $\mathcal{L}$. Since there is one underlying random field over the entire domain, the covariance function specifies the $\tilde{n} \times n$ coefficient matrix C. In particular, if $\mathbf{w} \sim N(\mathbf{0}, \sigma^2 H(\phi))$, then $C = H(\phi; \tilde{\mathcal{L}}, \mathcal{L}) H(\phi)^{-1}$ and $\boldsymbol{\omega} \sim N(\mathbf{0}, \sigma^2 F_\theta)$, where $F_\theta = H(\phi; \tilde{\mathcal{L}}, \tilde{\mathcal{L}}) - H(\phi; \tilde{\mathcal{L}}, \mathcal{L}) H(\phi)^{-1} H(\phi, \mathcal{L}, \tilde{\mathcal{L}})$, $H(\phi; \mathcal{U}, \mathcal{V})$ is the correlation matrix with $\rho(\phi; \ell_i, \ell_j)$ being the value of its (i, j)-th element, $\ell_i \in \mathcal{U}$ and $\ell_j \in \mathcal{V}$. The model for the data and the predictions is combined into

$$
\underbrace{\begin{bmatrix} \mathbf{y} \\ \boldsymbol{\mu}_\beta \\ \mathbf{0} \\ \mathbf{0} \\ \mathbf{0} \end{bmatrix}}_{\mathbf{y}_*} = \underbrace{\begin{bmatrix} X & I_n & O & O \\ I_p & O & O & O \\ O & C & -I_{\tilde{n}} & O \\ \tilde{X} & O & I_{\tilde{n}} & -I_{\tilde{n}} \end{bmatrix}}_{X_*} \underbrace{\begin{bmatrix} \boldsymbol{\beta} \\ \mathbf{w} \\ \tilde{\mathbf{w}} \\ \tilde{\mathbf{Y}} \end{bmatrix}}_{\boldsymbol{\gamma}} + \underbrace{\begin{bmatrix} \boldsymbol{\eta}_1 \\ \boldsymbol{\eta}_2 \\ \boldsymbol{\eta}_3 \\ \boldsymbol{\eta}_4 \\ \boldsymbol{\eta}_5 \end{bmatrix}}_{\boldsymbol{\eta}} , \tag{6.17}
$$

where $\boldsymbol{\eta} \sim N\left(\mathbf{0}, \sigma^2 V_\eta\right)$ and $V_\eta = \begin{bmatrix} \delta^2 I_n & O & O & O & O \\ O & V_\beta & O & O & O \\ O & O & H(\phi) & O & O \\ O & O & O & F_\theta & O \\ O & O & O & O & \delta^2 I_{\tilde{n}} \end{bmatrix}$.

If locations where predictions are sought are fixed by study design, then fitting (6.17) using the Bayesian conjugate framework can be beneficial. Equipped with an efficient computer program to calculate generalized least squares estimates from linear regression models, we can immediately obtain the posterior mean of $\hat{\gamma}$ from (6.17). The advantage of this formulation is that an efficient least squares algorithm to solve (6.17) that can exploit the sparsity of the design matrix X_* will immediately deliver inference on the regression slopes (β), the spatial process ($\mathbf{w}$) at observed points, the interpolated process ($\tilde{w}$) at unobserved points, and the predicted response ($\tilde{\mathbf{Y}}$) all at once. On the other hand, this is not necessary for sampling from the joint posterior distribution and would not be efficient for predictions if we wish to predict at new locations after training the model. In such cases, it will be more helpful to first store the posterior samples of $\{\beta, \mathbf{w}, \sigma^2\}$. Then, for each stored sample of the parameters we draw one sample of $\tilde{\mathbf{w}} \sim N(C\mathbf{w}, F_\theta)$ followed by one draw of $\tilde{\mathbf{Y}} \sim N(\tilde{X}\beta + \tilde{\mathbf{w}}, \tilde{D}_\tau)$. The resulting $\{\tilde{\mathbf{w}}, \tilde{\mathbf{Y}}\}$ will be the desired posterior predictive samples for the latent spatial process and the unobserved outcomes.

From the preceding account, we see that fixing the spatial range decay parameter ϕ and the noise-to-spatial variance ratio δ^2 casts the Bayesian geostatistical model into a conjugate framework that will allow inference on $\{\beta, w, \sigma^2\}$. Note that multiplying the posterior samples of σ^2 by the fixed quantity δ^2 fetches us the posterior samples of τ^2. Therefore, we neglect uncertainty in ϕ and, partially, for one of the variance components due to fixing their ratio. This, however, provides the computational advantage that inference can be carried out without resorting to expensive iterative algorithms such as MCMC that require several iterations before sampling from the posterior distribution. This computational benefit becomes especially relevant when handling massive spatial data. Furthermore, fixing the values of δ^2 and ϕ is not entirely unreasonable given that these parameters are weakly identified by the data [Zhang, 2004, Tang et al., 2021] and difficult to learn from the posterior. Nevertheless, the inference will depend upon these fixed parameters so we discuss a practical approach to fix ϕ and δ^2 at reasonable values.

Choosing ϕ and δ^2

We can set values for ϕ and δ^2 by conducting some simple spatial exploratory data analysis using the "variogram." Several practical algorithms exist for empirically calculating the variogram (or semivariogram) from observations using finite sample moments. As one example, Finley et al. [2019] investigate the impact of tree cover and the occurrence of forest fires on forest height. They first fit an ordinary linear regression of the form $y_{FH} = \beta_0 + \beta_1 x_{\text{tree}} + \beta_2 x_{\text{fire}} + \epsilon$ and then compute a variogram for the residuals from the ordinary linear regression.

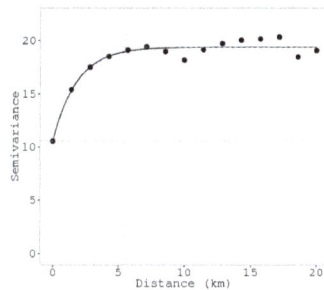

Figure 6.1 *Variogram of the residuals from non-spatial regression indicates strong spatial pattern.*

Figure 6.1 depicts the variogram, which informs about the process parameters. The lower horizontal line represents the "nugget" or the micro-scale variation captured by the measurement error variance component τ^2. The top horizontal line represents the "sill" (or ceiling) which is the total variation captured by $\sigma^2 + \tau^2$. Therefore, the difference between the two horizontal lines is called the "partial sill" and is captured by σ^2. Finally, the vertical line represents the distance beyond which the variogram flattens or the covariance tends to zero. One can provide "eye-ball" estimates for these quantities and, in particular, fix the values of ϕ and $\delta^2 = \tau^2/\sigma^2$. Fixing these values from the variogram yields the desired highly accessible conjugate framework and the models can be estimated without resorting to Markov chain Monte Carlo (MCMC) as described earlier. Note that instead of $\{\phi, \delta^2\}$, we could also have fixed ϕ and any one of the variance components, σ^2 or τ^2, which would also yield a conjugate model with exact distribution theory. The one slight advantage of fixing δ^2 is that we will get the posterior samples of both σ^2 and τ^2, the latter obtained simply as $\sigma^2\delta^2$. The above crude estimates can be improved using a K-fold cross-validation. We split the data randomly into K different folds. Let $S[k]$ be the k-th folder of observed points and let $S[-k]$ denote the observed points outside of $S[k]$. For each k, we compute the predictive mean $\mathrm{E}[y(S[k]) \mid y(S[-k])]$. We then compute the "Root Mean Square Predictive Error" (RMSPE) and choose the value of $\{\phi, \delta^2\}$ corresponding to the smallest RMSPE from a grid of candidate values. The range of the grid is based on the interpretation of the hyper-parameters. We suggest a reasonably wide range for δ^2 (e.g., $[0.001, 1000]$), which accommodates one variance component substantially dominating the other in either direction. For the spatial decay ϕ we suggest a lower bound of $\dfrac{3}{\text{maximum inter-site distance}}$, which, based on the exponential covariance function indicates that the spatial correlation drops below 0.05 at the maximum inter-site distance, and an upper bound that can be initially set as 100 times of the lower bound. After initial fitting, we can shrink the range and refine the grid of the candidate values for more precise estimators.

It is worth pointing out that the above ideas have led to developments in Bayesian predictive stacking [Wolpert, 1992, Clyde and Iversen, 2013, Le and Clarke, 2017, Yao et al., 2018, 2021, 2022, Zhang et al., 2025, Pan et al., 2025] for conjugate geostatistical models. Unlike fixing $\{\phi, \delta^2\}$ at a prescribed value, Bayesian predictive stacking averages, or stacks, the closed-form conditional conjugate posterior distributions described in Section 6.2.3 over a set of fixed values of $\{\phi, \delta^2\}$. Zhang et al. [2025] devise stacking of means and predictive densities in a manner that is computationally efficient without resorting to iterative algorithms such as MCMC and can exploit the benefits of parallel computations. Zhang et al. [2025] provide some theoretical and empirical insights into inference for geostatistical inference in an infill asymptotic setting showing that stacked inference is comparable to full sampling-based Bayesian inference at a significantly lower computational cost. Pan et al. [2025] extend stacked Bayesian inference for exponential families.

6.2.4 More on prior specifications

We now depart from the conjugate priors discussed above. Typically, independent priors are chosen for the different parameters, i.e.,

$$p(\boldsymbol{\theta}) = p(\boldsymbol{\beta})p(\sigma^2)p(\tau^2)p(\phi) \,,$$

and natural candidates are multivariate normal for $\boldsymbol{\beta}$ and inverse gamma for σ^2 and τ^2. Specification for ϕ of course depends upon the choice of ρ function; in the simple exponential case where $\rho(\mathbf{s}_i - \mathbf{s}_j; \phi) = \exp(-\phi\|\mathbf{s}_i - \mathbf{s}_j\|)$ (and ϕ is thus univariate), a gamma might seem sensible.

However, at this point, we need to devote a few paragraphs to a more careful development of prior specifications. First, let us consider the setting with no nugget so $\Sigma = \sigma^2 H(\phi)$. Let's focus on the Matérn class of covariance functions. An elegant result due to Zhang [2004], which requires a somewhat sophisticated analysis of equivalent measures for stochastic processes, tells us that for the Matérn covariance function with smoothness parameter ν, the product $\sigma^2 \phi^{2\nu}$ can be identified but not the individual parameters. For instance, with the exponential, we can identify $\sigma^2 \phi$ but not the range or the variances themselves. Only if we fix one can we identify the other. Implications for inference become apparent. Kriging will only be sensitive to the product. Which parameter are we more interested in learning about? Likely, it is the spatial variance, especially in the interest of comparison with the pure error variance. The two variance components inform about the relative strength of the spatial story vs. the pure error story, suggesting greater interest in σ^2. Furthermore, generally, learning regarding the spatial range is weak. So, practically, we recommend a very informative prior for ϕ and a relatively vague prior for σ^2. For the former, rather than a Gamma, we often employ a uniform over a specified interval or a discretized uniform over a finite set of points.

Still care must be taken with regard to the latter. Here, we note the work of Berger et al. [2001]. Again, in the case of a spatial model with no nugget, with the exponential covariance function, they consider the class of *objective* priors of the form $p(\boldsymbol{\beta}, \sigma^2, \phi) \propto \frac{p(\phi)}{(\sigma^2)^\alpha}$ which implies a flat prior for the regression coefficients, $\boldsymbol{\beta}$. They demonstrate that, with say a uniform prior for ϕ, an improper posterior arises for $\alpha < 2$. The implication for us is that if we adopt an inverse Gamma prior $IG(\epsilon, \epsilon)$ prior for σ^2, this corresponds to the case of $\alpha = 1 + \epsilon$. For small ϵ, we have a specification that yields a posterior which is close to improper. While we may not explicitly see problems with our MCMC chains, we know that sampling from a nearly improper posterior can yield poorly behaved MCMC inference. Hence, our recommendation is to always use $IG(a, b)$ priors with $a \geq 1$ (implying $\alpha \geq 2$). Again, these priors are still quite vague since, with $a = 1$, we have no mean, hence no variance, and with $a = 2$, we have no variance. The foregoing theory has not been extended to the case when we bring in the nugget, τ^2. However, practical experience suggests that the foregoing problems only worsen and thus the same cautions should be taken.

Since we will often want to make inferential statements about the parameters separately, we will need to obtain *marginal* posterior distributions. For example, a point estimate or credible interval for $\boldsymbol{\beta}$ arises from

$$
\begin{aligned}
p(\boldsymbol{\beta} \mid \mathbf{y}) &= \int \int \int p(\boldsymbol{\beta}, \sigma^2, \tau^2, \phi \mid \mathbf{y}) \, d\sigma^2 d\tau^2 d\phi \\
&\propto p(\boldsymbol{\beta}) \int \int \int f(\mathbf{y} \mid \boldsymbol{\theta}) \, p(\sigma^2) p(\tau^2) p(\phi) d\sigma^2 d\tau^2 d\phi \,.
\end{aligned}
$$

In principle this is simple, but in practice there will be no closed form for the above integrations. As such, we will often resort to MCMC or other numerical integration techniques, as described in Section 5.5.

6.3   Details on Bayesian inference for isotropic models

Here, we return to the isotropic covariance models and note that expression (6.3) can be recast as a hierarchical model of this form by writing the first-stage specification as $\mathbf{Y}$ conditional not only on $\boldsymbol{\theta}$, but also on the vector of spatial random effects $\mathbf{W} = (w(\mathbf{s}_1), \ldots, w(\mathbf{s}_n))^{\mathsf{T}}$. That is,

$$
\mathbf{Y} \mid \boldsymbol{\theta}, \mathbf{W} \sim N(X\boldsymbol{\beta} + \mathbf{W}, \tau^2 I) \,. \tag{6.18}
$$

The $Y(\mathbf{s}_i)$ are conditionally independent given the $w(\mathbf{s}_i)$. The second-stage specification is for $\mathbf{W}$, namely, $\mathbf{W} \mid \sigma^2, \phi \sim N(\mathbf{0}, \sigma^2 H(\phi))$ where $H(\phi)$ is as above. This is the *process* model. Here, it is quite simple, a process merely introduced to capture spatial dependence. In later examples it will become richer as we incorporate more process features into its specification. Lastly, the model specification is completed by adding priors for $\boldsymbol{\beta}$ and τ^2 as well as for σ^2

and ϕ, the latter two of which may be viewed as hyperparameters. The parameter space is now augmented from $\boldsymbol{\theta}$ to $(\boldsymbol{\theta}, \mathbf{W})$, and its dimension is increased by n.

Regardless, the resulting $p(\boldsymbol{\theta} \mid \mathbf{y})$ is the same, but we have the choice of using Gibbs sampling (or some other MCMC method) to fit the model either as $f(\mathbf{y} \mid \boldsymbol{\theta})p(\boldsymbol{\theta})$, or as $f(\mathbf{y} \mid \boldsymbol{\theta}, \mathbf{W})p(\mathbf{W} \mid \boldsymbol{\theta})p(\boldsymbol{\theta})$. The former is the result of marginalizing the latter over $\mathbf{W}$. Generally, we would prefer to work with the former. Apart from the conventional wisdom that we should do as much marginalization in closed form as possible before implementing an MCMC algorithm (i.e., in as low a dimension as possible), the matrix $\sigma^2 H(\phi) + \tau^2 I$ is typically better behaved than $\sigma^2 H(\phi)$. To see this, note that if, say, $\mathbf{s}_i$ and $\mathbf{s}_j$ are very close to each other, $\sigma^2 H(\phi)$ will be close to singular while $\sigma^2 H(\phi) + \tau^2 I$ will not. Determinant and inversion calculations will also tend to be better behaved for the marginal model form than the conditional model form.

Interest is often in the spatial surface that involves $\mathbf{W} \mid \mathbf{y}$, as well as prediction for $W(\mathbf{s}_0) \mid \mathbf{y}$ for various choices of $\mathbf{s}_0$. At first glance it would appear that fitting the conditional model here would have an advantage, since realizations essentially from $p(\mathbf{W} \mid \mathbf{y})$ are directly produced in the process of fitting the model. However, since $p(\mathbf{W} \mid \mathbf{y}) = \int p(\mathbf{W} \mid \boldsymbol{\theta}, \mathbf{y})p(\boldsymbol{\theta} \mid \mathbf{y})d\boldsymbol{\theta}$, posterior realizations of $\mathbf{W}$ can be obtained one for one via *composition* sampling using posterior realizations of $\boldsymbol{\theta}$. Specifically, if the values $\boldsymbol{\theta}^{(g)}$ are draws from an MCMC algorithm with stationary distribution $p(\boldsymbol{\theta} \mid \mathbf{y})$, then corresponding draws $\mathbf{W}^{(g)}$ from $p(\mathbf{W} \mid \boldsymbol{\theta}^{(g)}, \mathbf{y})$ will have marginal distribution $p(\mathbf{W} \mid \mathbf{y})$, as desired. Thus we need not generate the $\mathbf{W}^{(g)}$ within the Gibbs sampler itself, but instead obtain them immediately given the output of the smaller, marginal sampler. Note that marginalization over $\mathbf{W}$ is only possible if the hierarchical form has a first-stage Gaussian specification, as in (6.18). We return to this matter in Section 6.2.

Next we turn to prediction of the response Y at a new value s_0 with associated covariate vector $\mathbf{x}(\mathbf{s}_0)$; this predictive step is the Bayesian "kriging" operation. Denoting the unknown value at that point by $Y(\mathbf{s}_0)$ and using the notations $Y_0 \equiv Y(\mathbf{s}_0)$ and $\mathbf{x}_0 \equiv \mathbf{x}(\mathbf{s}_0)$ for convenience, the solution in the Bayesian framework simply amounts to finding the predictive distribution,

$$p(y_0 \mid \mathbf{y}, X, \mathbf{x}_0) = \int p(y_0, \boldsymbol{\theta} \mid \mathbf{y}, X, \mathbf{x}_0)\, d\boldsymbol{\theta} = \int p(y_0 \mid \mathbf{y}, \boldsymbol{\theta}, \mathbf{x}_0)\, p(\boldsymbol{\theta} \mid \mathbf{y}, X)\, d\boldsymbol{\theta}\ ,$$

where $p(y_0 \mid \mathbf{y}, \boldsymbol{\theta}, \mathbf{x}_0)$ has a conditional normal distribution arising from the joint multivariate normal distribution of Y_0 and the original data $\mathbf{Y}$; see (2.22) and (2.23).

In practice, MCMC methods may again be readily used to obtain estimates of (6.19). Suppose we draw (after burn-in, etc.) our posterior sample $\boldsymbol{\theta}^{(1)}, \boldsymbol{\theta}^{(2)}, \ldots, \boldsymbol{\theta}^{(G)}$ from the posterior distribution $p(\boldsymbol{\theta} \mid \mathbf{y}, X)$. Then the above predictive integral may be computed as a Monte Carlo mixture of the form

$$\widehat{p}(y_0 \mid \mathbf{y}, X, \mathbf{x}_0) = \frac{1}{G} \sum_{g=1}^{G} p\left(y_0 \mid \mathbf{y}, \boldsymbol{\theta}^{(g)}, \mathbf{x}_0\right)\ . \tag{6.19}$$

In practice we typically use composition sampling to draw, one for one for each $\boldsymbol{\theta}^{(g)}$, a $y_0^{(g)} \sim p\left(y_0 \mid \mathbf{y}, \boldsymbol{\theta}^{(g)}, \mathbf{x}_0\right)$. The collection $\left\{y_0^{(1)}, y_0^{(2)}, \ldots, y_0^{(G)}\right\}$ is a sample from the posterior predictive density, and so can be fed into a histogram or kernel density smoother to obtain an approximate plot of the density, bypassing the mixture calculation (6.19). A point estimate and credible interval for the predicted Y_0 may be computed in the same manner as in the estimation case above.

Next suppose that we want to predict at a *set* of m sites, denoted, say, by $S_0 = \{\mathbf{s}_{01}, \mathbf{s}_{02}, \ldots, \mathbf{s}_{0m}\}$. We could individually predict at each of these points "independently" using the above method. But *joint* prediction may also be of interest since it enables realizations from the same random spatial surface. As a result, it allows the estimation of posterior associations among the m predictions. We may form an unobserved vector $\mathbf{Y}_0 = (Y(\mathbf{s}_{01}), \ldots, Y(\mathbf{s}_{0m}))^{\mathrm{T}}$ with associated design matrix X_0 having rows $\mathbf{x}(\mathbf{s}_{0j})^{\mathrm{T}}$, and compute its joint predictive density as

$$
\begin{aligned}
p\left(\mathbf{y}_0 \mid \mathbf{y}, X, X_0\right) &= \int p\left(\mathbf{y}_0 \mid \mathbf{y}, \boldsymbol{\theta}, X_0\right) p\left(\boldsymbol{\theta} \mid \mathbf{y}, X\right) d\boldsymbol{\theta} \\
&\approx \frac{1}{G} \sum_{g=1}^{G} p\left(\mathbf{y}_0 \mid \mathbf{y}, \boldsymbol{\theta}^{(g)}, X_0\right),
\end{aligned}
$$

where again $p\left(\mathbf{y}_0 \mid \mathbf{y}, \boldsymbol{\theta}^{(j)}, X_0\right)$ is available from standard conditional normal formulae. We could also use composition to obtain, one for one for each $\boldsymbol{\theta}^{(g)}$, a collection of $\mathbf{y}_0^{(g)}$ and make any inferences we like based on this sample, either jointly or componentwise.

Often we are interested in not only the variables $Y(\mathbf{s})$, but also in functions of them, e.g., $\log Y(\mathbf{s})$ (if $Y(\mathbf{s}) > 0$), $I(Y(\mathbf{s}) > c)$, and so on. These functions are random variables as well. More generally we might be interested in functions $g(\mathbf{Y}_D)$ where $\mathbf{Y}_D = \{Y(\mathbf{s}) : \mathbf{s} \in D\}$. These include, for example, $(Y(\mathbf{s}_i) - Y(\mathbf{s}_j))^2$, which enter into the variogram, linear transformations $\sum_i \ell_i Y(\mathbf{s}_i)$, which include filters for spatial prediction at some locations, and finite differences in specified directions, $[Y(\mathbf{s} + h\mathbf{u}) - Y(\mathbf{s})]/\mathbf{h}$, where $\mathbf{u}$ is a particular unit vector (see Subsection 14.3).

Functions of the form $g(\mathbf{Y}_D)$ also include block averages, i.e. $Y(A) = \frac{1}{|A|} \int_A g(Y(\mathbf{s})) d\mathbf{s}$. Block averages are developed in much more detail in Chapter 4. The case where $g(\mathbf{Y}_D) = I(Y(\mathbf{s}) \leq c)$ leads to the definition of the spatial CDF (SCDF) as in Section 17.2. Integration of a process or of a function of a process yields a new random variable, i.e., the integral is random and is usually referred to as a stochastic integral. An obvious but important point is that $E_A g(Y(\mathbf{s})) \neq g(E_A Y(\mathbf{s}))$ if g is not linear. Hence modeling $g(Y(\mathbf{s}))$ is not the same as modeling $g(Y(A))$. See Wakefield and Salway [2001] for further discussion.

6.4   More general isotropic correlation functions, revisited

In Section 3.1.2.1 we noted the general characterization result for all valid isotropic correlation functions in R^r. That is, from Khinchin's Theorem [Yaglom, 2004, p. 106] as well as (3.3), this class of functions $\rho(d, \phi)$ in $\Re^r$ can be expressed as

$$
\rho(d, \phi) = \int_0^\infty \Omega_r(zd) dG_\phi(z), \tag{6.20}
$$

where G_ϕ is nondecreasing integrable and $\Omega_r(x) = \left(\frac{2}{x}\right)^{\frac{r-2}{2}} \Gamma\left(\frac{r}{2}\right) J_{\left(\frac{r-2}{2}\right)}(x)$. Repeating, $J_v(\cdot)$ is the Bessel function of the first kind of order v. For $r = 1, \Omega_1(x) = \cos(x)$; for $r = 2, \Omega_2(x) = J_0(x)$; for $r = 3, \Omega_3(x) = \sin(x)/x$; for $r = 4, \Omega_4(x) = \frac{2}{x} J_1(x)$; and for $r = \infty, \Omega_\infty(x) = \exp(-x^2)$. Specifically, $J_0(x) = \sum_{k=0}^\infty \frac{(-1)^k}{k!^2} \left(\frac{x}{2}\right)^{2k}$ and $\rho(d, \phi) = \int_0^\infty J_0(zd) dG_\phi(z)$ provides the class of all permissible correlation functions in $\Re^2$. Figure 3.1 provides a plot of $J_0(x)$ versus x, revealing that it is not monotonic. (This must be the case in order for $\rho(d, \phi)$ above to capture all correlation functions in $\Re^2$.)

In practice, a convenient simple choice for $G_\phi(z)$ is a step function that assigns positive mass (jumps or weights) w_ℓ at points (nodes) ϕ_ℓ, $\ell = 1, ..., p$ yielding, with $\mathbf{w} = (w_1, w_2, ..., w_p)$,

$$
\rho(d, \phi, \mathbf{w}) = \sum_{\ell=1}^{p} w_\ell \Omega_n(\phi_\ell d). \tag{6.21}
$$

The forms in (6.21) are referred to as *nonparametric* variogram models in the literature to distinguish them from standard or parametric forms for $\rho(d, \phi)$, such as those given in Table 2.2. This is a separate issue from selecting a parametric or nonparametric methodology for parameter estimation. 1992, Shapiro and Botha [1991], and Cherry [1996] use a step function for G_ϕ. Barry and Ver Hoef [1996] employ a mixture of piecewise linear variograms in R^1 and piecewise-planar models for sites in $\Re^2$. Hall et al. [1994] transform the problem from choosing ϕ_ℓ's and w_ℓ's in (6.21) to determining a kernel function and its associated bandwidth. Lele [1995] proposes iterative spline smoothing of the variogram yielding a ρ which is not obviously of the form (6.20). Most of these *nonparametric* models are fit to some version of the empirical semivariogram (2.9).

Sampson and Guttorp [1992] fit their model, using $\Omega_\infty(x)$ in (6.21), to the semivariogram cloud rather than to the smoothed Matheron semivariogram estimate. Their example involves a data set with 12 sites yielding only 66 points in the semivariogram cloud, making this feasible. Application of their method to a much larger (hence "noisier") data set would be expected to produce a variogram mixing hundreds and perhaps thousands of Gaussian forms. The resulting variogram will follow the semivariogram cloud too closely to be plausible.

Working in $\Re^2$, where again $\Omega_2(x) = J_0(x)$, under the Bayesian paradigm we can introduce (6.21) directly into the likelihood but keep p small (at most 5), allowing random w_ℓ or random ϕ_ℓ. This offers a compromise between the rather limiting standard parametric forms (Table 2.1) that specify two or three parameters for the covariance structure, and the above nonparametric methods that are based upon a practically implausible (and potentially overfitting) mixture of hundreds of components. Moreover, by working with the likelihood, inference is conditioned upon the observed $\mathbf{y}$, rather than on a summary such as a smoothed version of the semivariogram cloud.

Returning to (6.20), when $n = 2$ we obtain

$$\rho(d, \phi) = \int_0^\infty \sum_{k=0}^\infty \frac{(-1)^k}{k!^2} \left(\frac{zd}{2}\right)^{2k} dG_\phi(z) . \tag{6.22}$$

Only if z is bounded, i.e., if G_ϕ places no mass on say $z > \phi_{max}$, can we interchange summation and integration to obtain

$$\rho(d, \phi) = \sum_{k=0}^\infty \frac{(-1)^k}{k!^2} \left(\frac{d}{2}\right)^{2k} \delta_{2k} , \tag{6.23}$$

where $\delta_{2k} = \int_0^{\phi_{\max}} z^{2k} dG_\phi(z)$. The simplest such choice for G_ϕ puts discrete mass w_ℓ at a finite set of values $\phi_\ell \in (0, \phi_{\max})$, $\ell = 1, ..., p$ resulting in a finite mixture of Bessels model for $\rho(d, \phi)$, which in turn yields

$$\gamma(d_{ij}) = \tau^2 + \sigma^2 \left(1 - \sum_{\ell=1}^p w_\ell J_0(\phi_\ell d_{ij})\right) . \tag{6.24}$$

Under a Bayesian framework for a given p, if the w_ℓ's are each fixed to be $\frac{1}{p}$ with ϕ_ℓ's unknown (hence random), they are constrained by $0 < \phi_1 < \phi_2 < \cdots < \phi_p < \phi_{max}$ for identifiability. The result is an equally weighted mixture of random curves. If a random mixture of fixed curves is desired, then the w_ℓ's are random and the ϕ_ℓ's are systematically chosen to be $\phi_\ell = \left(\frac{\ell}{p+1}\right) \phi_{\max}$. We examine $p = 2, 3, 4, 5$ for fixed nodes and $p = 1, 2, 3, 4, 5$ for fixed weights. Mixture models using random w_ℓ's and random ϕ_ℓ's might be considered but, in our limited experience, the posteriors have exhibited weak identifiability in the parameters and thus are not recommended.

(a)

(b)

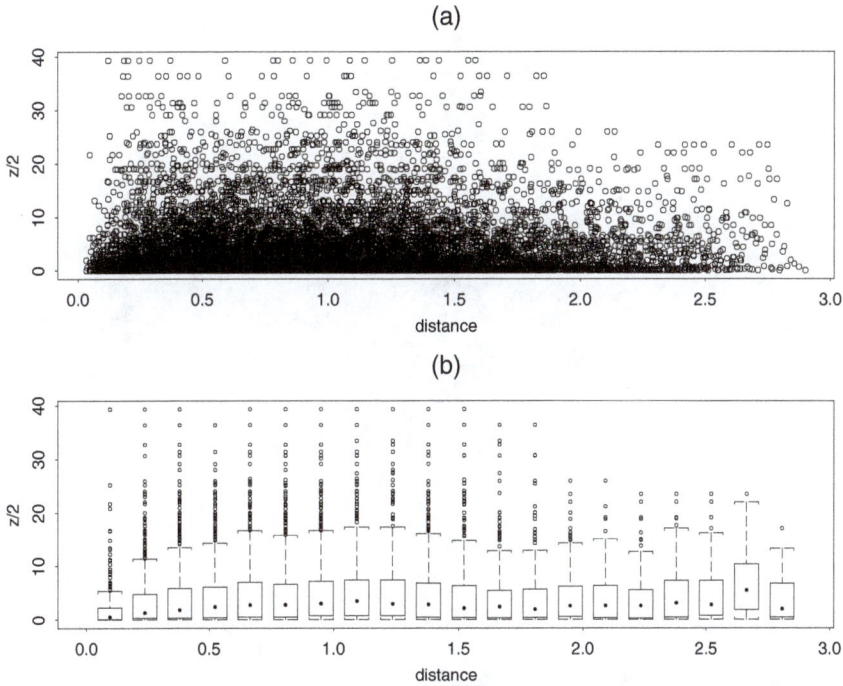

Figure 6.2 *Semivariogram cloud (a) and boxplot produced from 0.05 lag (b), 1993 scallop data.*

In choosing ϕ_{max}, we essentially determine the maximum number of sign changes we allow for the dampened sinusoidal Bessel correlation function over the range of d's of interest. For, say, $0 \leq d \leq d^{\max}$ where $d^{\max}$ is the maximum of the $d_{ij} = ||\mathbf{s}_i - \mathbf{s}_j||$, the larger ϕ is, the more sign changes $J_0(\phi d)$ will have over this range. This suggests making ϕ_{max} very large. However, as noted earlier in this section, we seek to avoid practically implausible ρ and γ, which would arise from an implausible $J_0(\phi d)$. For illustration, the plot in Figure 3.1 above allows several sign changes, to show the longer-term stability of its oscillation. Letting κ be the value of x where $J_0(x) = 0$ attains its kth sign change (completes its $\frac{k-1}{2}$ period) we set $\kappa = \phi_{\max} d^{\max}$, thus determining $\phi_{\max}$. We reduce the choice of $\phi_{\max}$ to choosing the maximum number of Bessel periods allowable. For a given p, when the ϕ's are random, the posterior distribution for ϕ_p will reveal how close to $\phi_{\max}$ the data encourages ϕ_p to be.

Example 6.1 We return to the 1990 log-transformed scallop data, originally presented in Subsection 2.3.2. In 1990, 148 sites were sampled in the New York Bight region of the Atlantic Ocean, which encompasses the area from the tip of Long Island to the mouth of the Delaware River. These data have been analyzed by Ecker and Heltshe [1994], Ecker and Gelfand [1997], Ecker and Gelfand [1999] , Kaluzny et al. [1999] and others. Figure 6.2 shows the semivariogram cloud (panel a) together with boxplots (panel b) formed from the cloud using the arbitrary lag $\delta = 0.05$. The 10,731 pairs of points that produce the semivariogram cloud do not reveal any distinct pattern. In a sense, this shows the folly of fitting a curve to this data: we have a weak signal and a great deal of noise.

However, the boxplots and the Matheron empirical semivariograms each based on lag $\delta = 0.05$ (Figure 6.3) clearly, exhibit spatial dependence, in the sense that when separation distances are small, the spatial variability tends to be less. Here the attempt is to remove the noise to see whatever signal there may be. Of course, the severe skewness revealed by the boxplots (and expected from squared differences) raises the question of whether the bin averages are an appropriate summary [expression (2.9); see Ecker and Gelfand, 1997, in this regard].. Clearly such displays and attempts to fit an empirical variogram must be viewed as part of the exploratory phase of our data analysis.

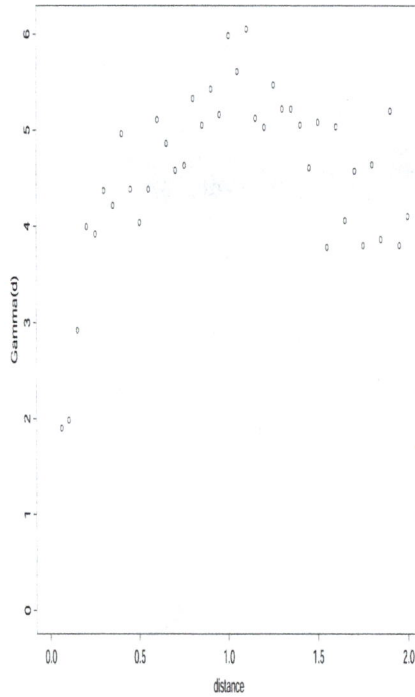

Figure 6.3 *Matheron empirical semivariograms for lag $\delta = 0.05$.*

For the choice of ϕ_{max} in the nonparametric setup, we selected seven sign changes, or three Bessel periods. With $d_{ij}^{max} = 2.83$ degrees, ϕ_{max} becomes 7.5. A sensitivity analysis with two Bessel mixtures ($p = 2$) having a fixed weight w_1 and random nodes was undertaken. Two, four, and five Bessel periods revealed little difference in results as compared with three. However, when one Bessel period was examined ($\phi_{max} = 3$), the model fit poorly and in fact ϕ_p was just smaller than 3. This is an indication that more flexibility (i.e., a larger value of ϕ_{max}) is required.

Several of the parametric models from Tables 2.1 and 2.2 and several nonparametric Bessel mixtures with different combinations of fixed and random parameters were fit to the 1990 scallop data. (Our analysis here parallels that of Ecker and Gelfand [1997], although our results are not identical to theirs since they worked with the 1993 version of the data set.) Figure 6.4 shows the posterior mean of each respective semivariogram, while Table 6.1 provides the value of model choice criteria for each model along with the independence model, $\Sigma_{\mathbf{Y}} = (\tau^2 + \sigma^2)I$. Here we use the model selection criterion (5.67), as described in Subsection 5.6 [Gelfand and Ghosh, 1998]. However, since we are fitting variograms, we work somewhat less formally using $Z_{ij,obs} = (Y(\mathbf{s}_i) - Y(\mathbf{s}_j))^2/2$. Since Z_{ij} is distributed as a multiple of a χ_1^2 random variable, we use a loss associated with a gamma family of distributions, obtaining a $D_{k,m}$ value of

$$(k+1) \sum_{i,j} \left\{ \log\left(\frac{\lambda_{ij}^{(m)} + kz_{ij,obs}}{k+1}\right) - \frac{\log(\lambda_{ij}^{(m)}) + k\log(z_{ij,obs})}{k+1} \right\}$$

$$+ \sum_{i,j} \left(\log(\lambda_{ij}^{(m)}) - \mathrm{E}(\log(z_{ij,rep}) \mid \mathbf{y}, m) \right) \qquad (6.25)$$

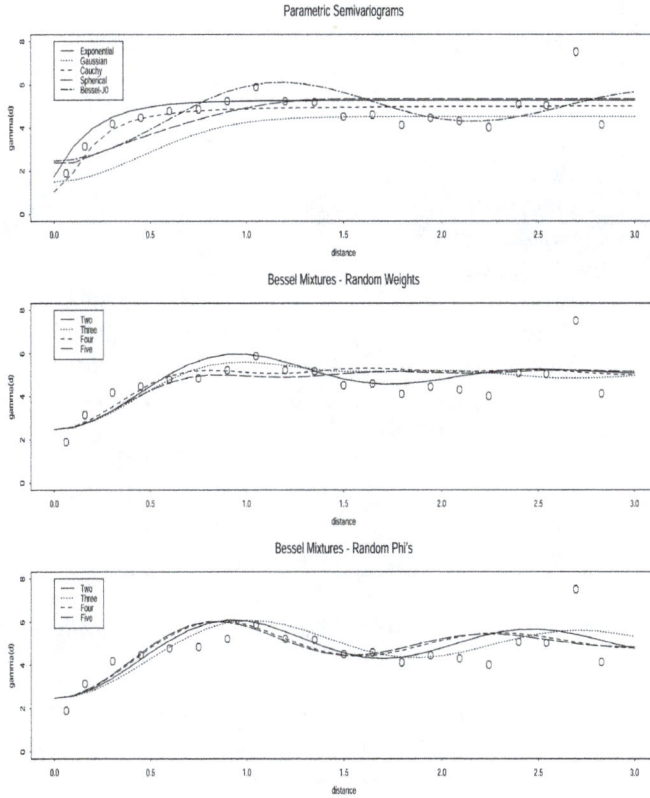

Figure 6.4 *Posterior means for various semivariogram models.*

for model m, where $\lambda_{ij}^{(m)} = \mathrm{E}(z_{ij,rep} \mid \mathbf{y}, m)$. The concavity of the log function ensures that both summations on the right-hand side of (6.25) are positive. (As an aside, in theory $z_{ij,obs} > 0$ almost surely, but in practice we may observe some $z_{ij} = 0$ as, for example, with the log counts in the scallop data example. A correction is needed and can be achieved by adding ϵ to $z_{ij,obs}$ where ϵ is, say, one-half of the smallest possible positive $z_{ij,obs}$.)

Setting $k = 1$ in (6.25), we note that of the Bessel mixtures, the five-component model with fixed ϕ's and random weights is best according to the $D_{1,m}$ statistic. Here, given $\phi_{max} = 7.5$, the nodes are fixed to be $\phi_1 = 1.25, \phi_2 = 2.5, \phi_3 = 3.75, \phi_4 = 5.0$, and $\phi_5 = 6.25$. One would expect that the fit measured by the $G_{1,m}$ criterion should improve with increasing p. However, the models do not form a nested sequence in p, except in some instances (e.g., the $p = 2$ model is a special case of the $p = 5$ model). Thus, the apparent poorer fit of the four-component fixed ϕ model relative to the three-component model is indeed possible. The random ϕ Bessel mixture models were all very close and, as a class, these models fit as well or better than the best parametric model. Hence, modeling mixtures of Bessel functions appears more sensitive to the choice of fixed ϕ's than to fixed weights. ∎

6.5 Modeling geometric anisotropy

Anistropy was introduced in Section 2.2, in the form of the geometric, sill, and nugget anisotropy, to refer to particular cases of stationarity. In any event, we have $\mathrm{Cov}\left(Y\left(\mathbf{s} + \mathbf{h}\right), Y\left(\mathbf{s}\right)\right) = C\left(\mathbf{h}; \phi\right).$ The most prominent, tractable, and interesting case in

Model	$G_{1,m}$	P_m	$D_{1,m}$
Parametric			
exponential	10959	13898	24857
Gaussian	10861	13843	24704
Cauchy	10683	13811	24494
spherical	11447	13959	25406
Bessel	11044	14037	25081
independent	11578	16159	27737
Semiparametric			
fixed ϕ_ℓ, random w_ℓ:			
two	11071	13968	25039
three	10588	13818	24406
four	10934	13872	24806
five	10567	13818	24385
random ϕ_ℓ, fixed w_ℓ:			
two	10673	13907	24580
three	10677	13959	24636
four	10636	13913	24549
five	10601	13891	24492

Table 6.1 *Model choice for fitted variogram models, 1993 scallop data.*

applications is *geometric anisotropy*. This refers to the situation where the coordinate space can be linearly transformed to an isotropic space. A linear transformation may correspond to the rotation or stretching of the coordinate axes. Thus in general,

$$\rho\left(\mathbf{h};\phi\right) = \rho_0\left(\|L\mathbf{h}\|;\phi\right) ,$$

where l is a $d \times d$ matrix describing the linear transformation. Of course, if L is the identity matrix, this reduces to the isotropic case.

We assume a second-order stationary normal model for $\mathbf{Y}$, arising from the customary model, $Y(\mathbf{s}) = \mu + w(\mathbf{s}) + \epsilon(\mathbf{s})$ as in (6.1). This yields $\mathbf{Y} \sim N(\mu\mathbf{1}, \Sigma(\boldsymbol{\alpha}))$, where $\boldsymbol{\alpha} = (\tau^2, \sigma^2, B)^{\mathrm{T}}$, $B = L^{\mathrm{T}}L$, and

$$\Sigma(\boldsymbol{\alpha}) = \tau^2 I + \sigma^2 H((\mathbf{h}^{\mathrm{T}}B\mathbf{h})^{\frac{1}{2}}) . \tag{6.26}$$

In (6.26), the matrix H has (i,j)th entry $\rho((\mathbf{h}_{ij}^{\mathrm{T}}B\mathbf{h}_{ij})^{\frac{1}{2}})$ where ρ is a valid correlation function and $\mathbf{h}_{ij} = \mathbf{s}_i - \mathbf{s}_j$. Common forms for ρ would be those in Table 2.2. In (6.26), τ^2 is the semiovariogram nugget and $\tau^2 + \sigma^2$ is the sill. The variogram is $2\gamma(\tau^2, \sigma^2, (\mathbf{h}^{\mathrm{T}}B\mathbf{h})^{\frac{1}{2}}) = 2(\tau^2 + \sigma^2(1 - \rho((\mathbf{h}^{\mathrm{T}}B\mathbf{h})^{\frac{1}{2}})))$.

Turning to $\Re^2$, B is 2×2 and the orientation of the associated ellipse, ω, is related to B by

$$\cot(2\omega) = \frac{b_{11} - b_{22}}{2b_{12}} . \tag{6.27}$$

The range in the direction η, where η is the angle $\mathbf{h}$ makes with the x-axis and which we denote as r_η, is determined by the relationship

$$\rho(r_\eta(\widetilde{\mathbf{h}}_\eta^{\mathrm{T}}B\widetilde{\mathbf{h}}_\eta)^{\frac{1}{2}}) = 0.05 , \tag{6.28}$$

where $\widetilde{\mathbf{h}}_\eta = (\cos\eta, \sin\eta)$ is a unit vector in direction η.

The *ratio of anisotropy* [Journel and Huijbregts, 2003, pp. 178–181], also called the *ratio of affinity* [Journel and Froidevaux, 1982, p. 228], which here we denote as λ, is the ratio of the major axis of the ellipse to the minor axis, and is related to B by

$$\lambda = \frac{r_\omega}{r_{(\pi-\omega)}} = \left(\frac{\widetilde{\mathbf{h}}^{\mathrm{T}}_{(\pi-\omega)} B \widetilde{\mathbf{h}}_{(\pi-\omega)}}{\widetilde{\mathbf{h}}^{\mathrm{T}}_\omega B \widetilde{\mathbf{h}}_\omega} \right)^{\frac{1}{2}}, \tag{6.29}$$

where again $\widetilde{\mathbf{h}}_\eta$ is the unit vector in direction η. Since (6.27), (6.28), and (6.29) are functions of B, posterior samples (hence inference) for them is straightforward given posterior samples of $\boldsymbol{\alpha}$.

A customary prior distribution for a positive definite matrix such as B is Wishart(R,p), where

$$\pi(b) \propto |B|^{\frac{p-n-1}{2}} \exp\left(-\frac{1}{2} tr(pBR^{-1}) \right), \tag{6.30}$$

so that $\mathrm{E}(B) = R$ and $p \geq n$ is a precision parameter in the sense that $\mathrm{Var}(B)$ increases as p decreases. In $\Re^2$, the matrix $R = \begin{bmatrix} R_{11} & R_{12} \\ R_{12} & R_{22} \end{bmatrix}$. Prior knowledge is used to choose R, but we choose the prior precision parameter, p, to be as small as possible, i.e., $p = 2$.

A priori, it is perhaps easiest to assume that the process is isotropic, so we set $R = \delta I$ and then treat δ as fixed or random. For δ random, we model $p(B,\delta) = p(B \mid \delta)p(\delta)$, where $p(B \mid \delta)$ is the Wishart density given by (6.30) and $p(\delta)$ is an inverse gamma distribution with mean obtained from a rough estimate of the range and infinite variance (i.e., shape paramater equal to 2).

However, if we have prior evidence suggesting geometric anisotropy, we could attempt to capture it using (6.27), (6.28), or (6.29) with $\widetilde{\mathbf{h}}^{\mathrm{T}}_\eta R \widetilde{\mathbf{h}}_\eta$ replacing $\widetilde{\mathbf{h}}^{\mathrm{T}}_\eta B \widetilde{\mathbf{h}}_\eta$. For example, with a prior guess for ω, the angle of orientation of the major axis of the ellipse, a prior guess for λ, the ratio of major to minor axis (say, from a rose diagram), and a guess for the range in a specific direction (say, from a directional semivariogram), then (6.27), (6.28), and (6.29) provides a system of three linear equations in three unknowns to solve for R_{11}, R_{12}, and R_{22}. Alternatively, from three previous directional semivariograms, we might guess the range in three given directions, say, r_{η_1}, r_{η_2}, and r_{η_3}. Now, using (6.28), we again arrive at three linear equations with three unknowns in R_{11}, R_{12}, and R_{22}. One can also use an empirical semivariogram in $\Re^2$ constructed from prior data to provide guesses for R_{11}, R_{12}, and R_{22}. By computing a 0° and 90° directional semivariogram based on the ESC plot with rows where $h_y \approx 0$ for the former and columns where $h_x \approx 0$ in the latter, we obtain guesses for R_{11} and R_{22}, respectively. Finally, R_{12} can be estimated by examining a bin where neither $h_x \approx 0$ nor $h_y \approx 0$. Equating the empirical semivariogram to the theoretical semivariogram at the associated (x_i, y_j), with R_{11} and R_{22} already determined, yields a single equation to solve for R_{12}.

Example 6.2 Here we return again to the log-transformed sea scallop data of Subsection 2.3.2, and reexamine it for geometric anisotropy. Previous analyses [e.g. Ecker and Heltshe, 1994] have detected geometric anisotropy with the major axes of the ellipse oriented parallel to the coastline ($\approx$ 50° referenced counterclockwise from the x-axis). [Kaluzny et al., 1999, p. 90] suggest that λ, the ratio of the major axis to the minor axis, is approximately 3. The 1993 scallop catches with 147 sites were analyzed in Ecker and Gelfand [1997] under isotropy. Referring back to the ESC plot in Figure 2.8, a geometrically anisotropic model seems reasonable. Here we follow Ecker and Gelfand [1999] and illustrate with a Gaussian correlation form, $\rho((\mathbf{h}^{\mathrm{T}} B \mathbf{h})^{\frac{1}{2}}) = \exp(-\mathbf{h}^{\mathrm{T}} B \mathbf{h})$.

We can use the 1990 scallop data to formulate isotropic and geometrically anisotropic prior specifications for R, the prior mean for B. The first has $R = \delta I$ with fixed $\widehat{\delta} = 0.0003$,

	Isotropic prior	Geometrically anisotropic prior		
	fixed $\widehat{\psi} = 0.0003$	ω, λ and $r_{50°}$	three ranges	ESC plot
τ^2	1.29	1.43	1.20	1.33
	(1.00, 1.64)	(1.03, 1.70)	(1.01, 1.61)	(0.97, 1.73)
σ^2	2.43	2.35	2.67	2.58
	(1.05, 5.94)	(1.27, 5.47)	(1.41, 5.37)	(1.26, 5.67)
sill	3.72	3.80	3.87	3.91
	(2.32, 7.17)	(2.62, 6.69)	(2.66, 6.76)	(2.39, 7.09)
μ	2.87	2.55	3.14	2.90
	(2.16, 3.94)	(1.73, 3.91)	(2.24, 3.99)	(2.14, 4.02)
ω	55.3	64.4	57.2	60.7
	(26.7, 80.7)	(31.9, 77.6)	(24.5, 70.7)	(46.7, 75.2)
λ	2.92	3.09	3.47	3.85
	(1.59, 4.31)	(1.77, 4.69)	(1.92, 4.73)	(2.37, 4.93)

Table 6.2 *Posterior means and 95% interval estimates for a stationary Gaussian model with Gaussian correlation structure under various prior specifications.*

i.e., a prior isotropic range of 100 km. Another has $\widehat{\delta} = 0.000192$, corresponding to a 125-km isotropic prior range to assess the sensitivity of choice of $\widehat{\delta}$, and a third has δ random. Under prior geometric anisotropy, we can use $\omega = 50°$, $\lambda = 3$, and $r_{50°} = 125$ km to obtain a guess for R. Solving (6.27), (6.28), and (6.29) gives $R_{11} = 0.00047$, $R_{12} = -0.00023$, and $R_{22} = 0.00039$. Using the customary directional semivariograms with the 1990 data, another prior guess for R can be built from the three prior ranges $r_{0°} = 50$ km, $r_{45°} = 125$ km, and $r_{135°} = 30$ km. Via (6.28), we obtain $R_{11} = 0.012$, $R_{12} = -0.00157$, and $R_{22} = 0.00233$. Using the ESC plot for the 1990 data, we use all bins where $h_x = h_{long} \approx 0$ (90° semivariogram) to provide $R_{22} = 0.0012$, and bins where $h_y = h_{lat} \approx 0$ (0° semivariogram) to provide $R_{11} = 0.00053$. Finally, we pick three bins with large bin counts (328, 285, 262) and along with the results of the 0° and 90° ESC plot directional semivariograms, we average the estimated R_{12} for each of these three bins to arrive at $R_{12} = -0.00076$.

The mean and 95% interval estimates for the isotropic prior specification with $\widehat{\delta} = 0.0003$, and the three geometrically anisotropic specifications are presented in Table 6.2. Little sensitivity to the prior specifications is observed as expected, given that we use the smallest allowable prior precision. The posterior mean for the angle of orientation, ω, is about 60° and the ratio of the major ellipse axis to the minor axis, λ, has a posterior mean of about 3 to 3.5. Furthermore, the value 1 is not in any of the three 95% interval estimates for λ, indicating that isotropy is inappropriate.

We next present posterior inference associated with the ESC plot-based prior specification. Figure 6.5 shows the posteriors for the nugget in panel (a), the sill in panel (b), the angle of orientation in panel (c), and the ratio of the major axis to the minor axis in panel (d). Figure 6.6 shows the mean posterior range plotted as a function of angle with associated individual 95% intervals. This plot is much more informative in revealing departure from isotropy than merely examining whether the 95% interval for λ contains 1. Finally, Figure 6.7 is a plot of the contours of the posterior mean surface of the semivariogram. Note that it agrees with the contours of the ESC plot given in Figure 2.8 reasonably well. ∎

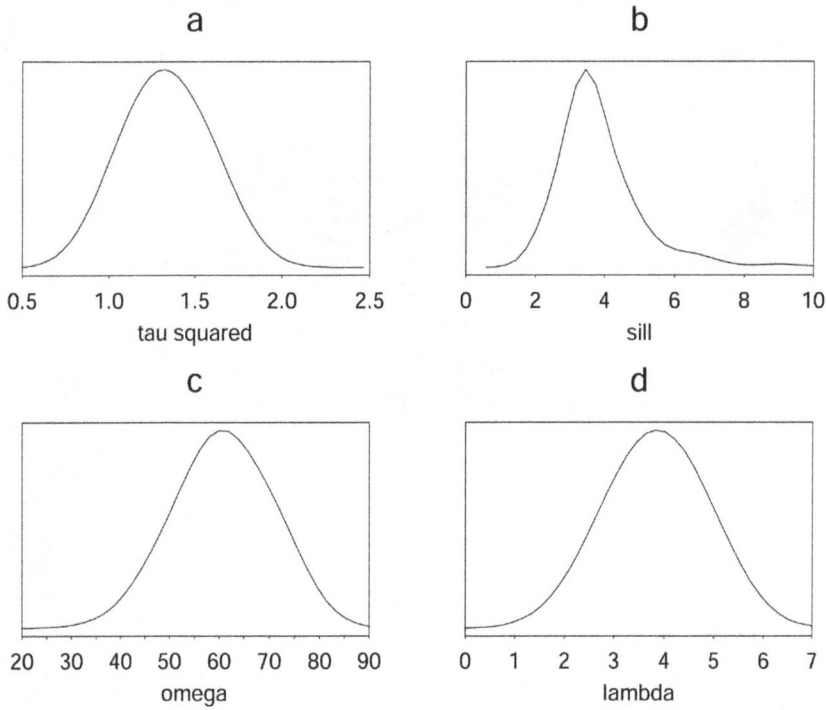

Figure 6.5 *Posterior distributions under the geometrically anisotropic prior formed from the ESC plot.*

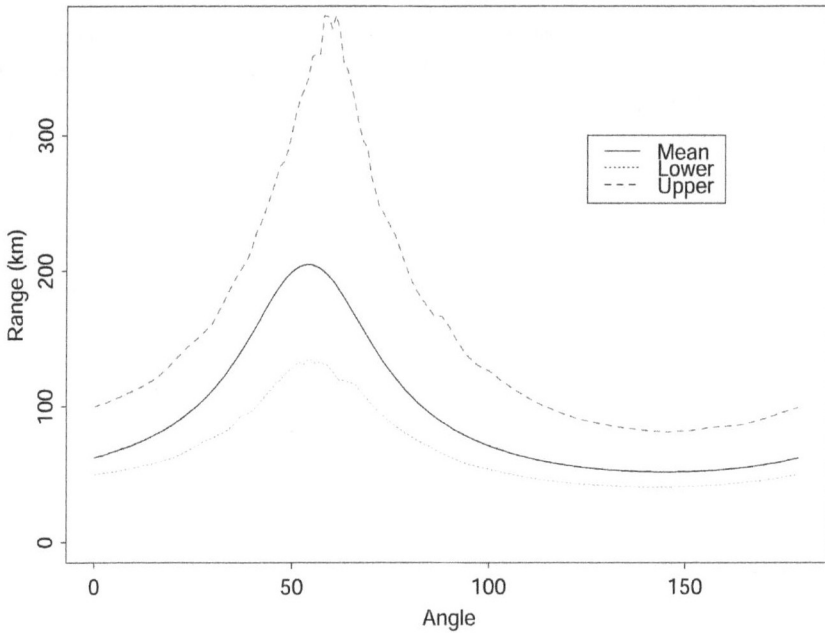

Figure 6.6 *Posterior range as a function of angle for the geometrically anisotropic prior formed from the ESC plot.*

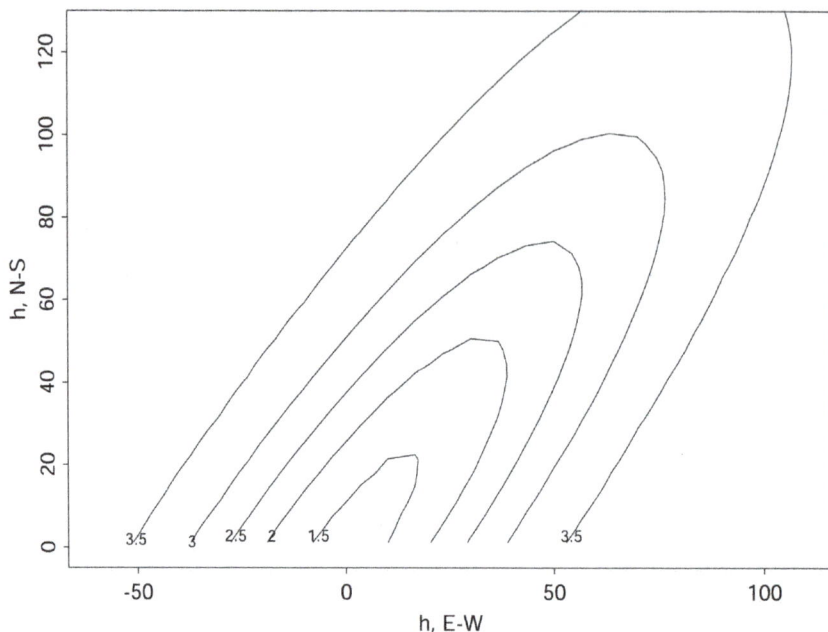

Figure 6.7 *Contours of the posterior mean semivariogram surface for the geometrically anisotropic prior formed from the ESC plot.*

6.6   Generalized linear spatial process modeling

In some point-referenced data sets we obtain measurements $Y(\mathbf{s})$ that would not naturally be modeled using a normal distribution; indeed, they need not be continuous. Most commonly, $Y(\mathbf{s})$ might be a binary variable, e.g., indicating whether or not measurable rain fell at location $\mathbf{s}$ in the past 24 hours or whether the tumor was present at a location in a tissue scan. In an aggregated data context examining species range and richness, $Y(\mathbf{s})$ might indicate presence or absence of a particular species at $\mathbf{s}$ (although here, strictly speaking $\mathbf{s}$ is not a point, but really an area that is sufficiently small to be thought of as a point within the overall study area). In this regard, a count variable at $\mathbf{s}$ might provide the number of distinct species observed at that location.

Following Diggle et al. [1998], we formulate a hierarchical model analogous to those in Section 6.1, but with the Gaussian model for $Y(\mathbf{s})$ replaced by another suitable member of the class of exponential family models. Assume the observations $Y(\mathbf{s}_i)$ are conditionally independent given $\boldsymbol{\beta}$ and $w(\mathbf{s}_i)$ with distribution,

$$f(y(\mathbf{s}_i) \mid \boldsymbol{\beta}, w(\mathbf{s}_i), \gamma) = h(y(\mathbf{s}_i), \gamma) \exp\{\gamma[y(\mathbf{s}_i)\eta(\mathbf{s}_i) - \psi(\eta(\mathbf{s}_i))]\}\,, \qquad (6.31)$$

where $g(\eta(\mathbf{s}_i)) = \mathbf{x}^{\mathrm{T}}(\mathbf{s}_i)\boldsymbol{\beta} + w(\mathbf{s}_i)$ for some link function g, and γ is a dispersion parameter. We presume the $w(\mathbf{s}_i)$ to be spatial random effects coming from a Gaussian process, as in Section 6.1. The second-stage specification is $\mathbf{W} \sim N(\mathbf{0}, \sigma^2 H(\phi))$ as before. Were the $w(\mathbf{s}_i)$ i.i.d., we would have a customary generalized linear mixed effects model [Breslow and Clayton, 1993]. Hence (6.31) is still a generalized linear mixed model, but now with spatial structure in the random effects.

Note that, although we have defined a process for $w(\mathbf{s})$ we have not created a process for $Y(\mathbf{s})$. That is, using conditional independence, what we have done is to create a joint distribution $f(y(\mathbf{s}_1), \ldots, y(\mathbf{s}_n) \mid \boldsymbol{\beta}, \sigma^2, \boldsymbol{\phi}, \gamma)$, namely,

$$\int \left(\prod_{i=1}^{n} f(y(\mathbf{s}_i) \mid \boldsymbol{\beta}, w(\mathbf{s}_i), \gamma) \right) p(\mathbf{W} \mid \sigma^2, \boldsymbol{\phi}) d\mathbf{W} . \tag{6.32}$$

The class of distributions that can support a stochastic process is limited, characterized through mixtures of elliptical distributions which, of course, includes Gaussian processes.

We have an opportunity to make another important point here. As above, a frequent first stage spatial specification is a binary response model. That is, at every location $\mathbf{s}$, there is a binary variable $Y(\mathbf{s})$. The resulting surface is frequently referred to as a binary map (DeOliveira, 2000). In this case, we can usefully consider two hierarchical specifications to model the binary map. The first sets $Y(\mathbf{s}) = 1$ or 0 according to whether $Z(\mathbf{s}) \geq 0$ or < 0. Then, we model $Z(\mathbf{s}) = \mathbf{x}(\mathbf{s})^{\mathrm{T}}\boldsymbol{\beta} + w(\mathbf{s}) + \epsilon(\mathbf{s})$. That is $Z(\mathbf{s})$ is our usual geostatistical model, (6.1). So, $Z(\mathbf{s})$ is a Gaussian process and determines $Y(\mathbf{s})$. In particular, $P(Y(\mathbf{s}) = 1) = P(Z(\mathbf{s}) \geq 0) = \Phi(\mathbf{x}(\mathbf{s})^{\mathrm{T}}\boldsymbol{\beta} + w(\mathbf{s}))$. As an alternative, resembling (6.31), let $P(Y(\mathbf{s}) = 1) \equiv p(\mathbf{s})$. Now, adopt a link function to take $p(\mathbf{s})$ to $\Re^1$, say $\Phi^{-1}(\cdot)$ and set $\Phi^{-1}(p(\mathbf{s})) = \mathbf{x}(\mathbf{s})^{\mathrm{T}}\boldsymbol{\beta} + w(\mathbf{s})$. Though they appear different, these two models are equivalent; rather, now $Y(\mathbf{s}) \mid p(\mathbf{s})$ is a random mechanism while $Y(\mathbf{s}) \mid Z(\mathbf{s})$ is a deterministic mechanism. We have exchanged the binary first-stage stochastic specification for a pure error Gaussian specification. But, this also clarifies why it is not sensible to add a pure error term to the specification for $p(\mathbf{s})$. Such an error term would be redundant, as is immediately evident were we to add a corresponding additional pure error term to the specification for $Z(\mathbf{s})$. In fact, with such an additional error term, we would obtain an unidentified MCMC model fitting which we would see with poorly behaved convergence. This point is evidently true for whatever first stage generalized linear model specification is used in (6.31).

We also note an important consequence of modeling with spatial random effects (which incidentally is relevant for Sections 6.8 and 6.9 as well). Introducing these effects in the (transformed) mean, as below (6.31), encourages the means of the spatial variables at proximate locations to be close to each other (adjusted for covariates). Though marginal spatial dependence is induced between, say, $Y(\mathbf{s})$ and $Y(\mathbf{s}')$, the observed $Y(\mathbf{s})$ and $Y(\mathbf{s}')$ need *not* be close to each other. This would be the case even if $Y(\mathbf{s})$ and $Y(\mathbf{s}')$ had the same mean. As a result, second-stage spatial modeling is attractive when spatial explanation in the *mean* is of interest. Direct (first-stage) spatial modeling is appropriate to encourage proximate *observations* to be close.

Turning to computational issues, note that (6.32) cannot be integrated in closed form; we cannot marginalize over $\mathbf{W}$. Unlike the Gaussian case, an MCMC algorithm will have to update $\mathbf{W}$ as well as $\boldsymbol{\beta}, \sigma^2, \boldsymbol{\phi}$, and γ. This same difficulty occurs with simulation-based model fitting of standard generalized linear mixed models [again see, e.g., Breslow and Clayton, 1993]. In fact, the $w(\mathbf{s}_i)$ would likely be updated using a Metropolis step with a Gaussian proposal, or through adaptive rejection sampling (since their full conditional distributions will typically be log-concave); see Exercise 4.

Example 6.3 Non-Gaussian point-referenced spatial model. Here we consider a real estate data set, with observations at 50 locations in Baton Rouge, LA. The response $Y(\mathbf{s})$ is a binary variable, with $Y(\mathbf{s}) = 1$ indicating that the price of the property at location $\mathbf{s}$ is "high" (above the median price for the region), and $Y(\mathbf{s}) = 0$ indicating that the price is "low." Observed covariates include the house's age, total living area, and other area in the property. We fit the model given in (6.31) where $Y(\mathbf{s}) \sim Bernoulli(p(\mathbf{s}))$ and g is the logit link.

Table 6.3 provides the parameter estimates and Figure 6.8 shows the image plot with overlaid contour lines for the posterior mean surface of the latent $w(\mathbf{s})$ process. These are obtained by assuming vague priors for $\boldsymbol{\beta}$, a Uniform$(0, 10)$ prior for ϕ, and an Inverse

Parameter	50%	(2.5%, 97.5%)
intercept	−1.096	(−4.198, 0.4305)
living area	0.659	(−0.091, 2.254)
age	0.009615	(−0.8653, 0.7235)
ϕ	5.79	(1.236, 9.765)
σ^2	1.38	(0.1821, 6.889)

Table 6.3 *Parameter estimates (posterior medians and upper and lower .025 points) for the binary spatial model.*

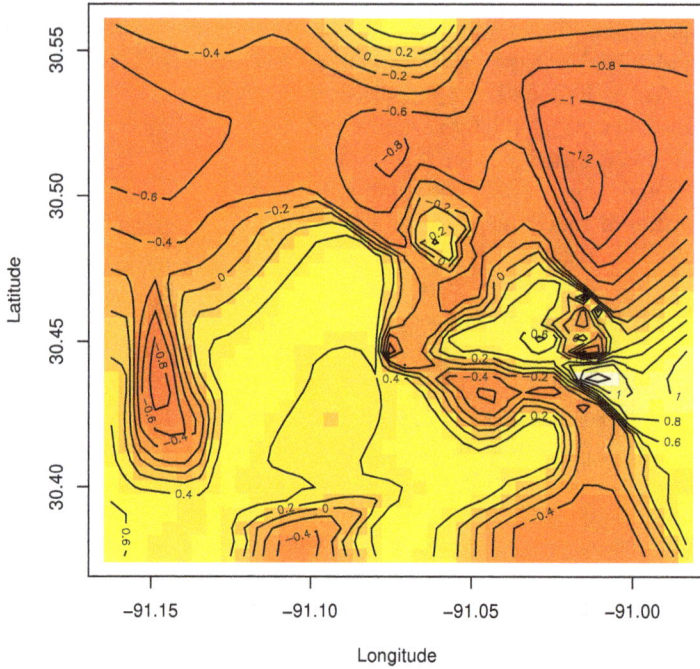

Figure 6.8 *Image plot of the posterior median surface of the latent spatial process $w(\mathbf{s})$, binary spatial model.*

Gamma$(0.1, 0.1)$ prior for σ^2. The image plot reveals negative residuals (i.e., lower prices) in the northern region, and generally positive residuals (higher prices) in the south-central region, although the southeast shows some lower price zones. The distribution of the contour lines indicates smooth flat stretches across the central parts, with downward slopes toward the north and southeast. The covariate effects are generally uninteresting, though living area seems to have a marginally significant effect on price class. ∎

6.7   Fitting Bayesian models for point-referenced data

Computer programs illustrating Bayesian data analysis for point-referenced data using different libraries in R and its interfaces with Bayesian modeling environments such as NIMBLE, Stan and JAGS are supplied in the folder titled Chapter 6 from https://github.com/sudiptobanerjee/BGC_2023. Here we present some brief illustrations using code from the website.

6.7.1   Gaussian spatial regression models

We fit a Gaussian spatial regression model to forest inventory data from the U.S. Department of Agriculture Forest Service, Bartlett Experimental Forest (BEF), Bartlett, NH. This dataset holds 1991 and 2002 forest inventory data for 437 plots. In our illustration, we use log-transformed total tree biomass as the outcome and regress it on five predictors: slope, elevation, and tasseled cap brightness (TC1), greenness (TC2), and wetness (TC3) components from spring, summer, and fall 2002 Landsat images. We subsequently make prediction of biomass for every image pixel across the BEF.

We obtained estimates of the partial sill, σ^2, nugget, τ^2, and decay parameter ϕ based upon some empirical semivariogram plots. We fixed ϕ and the nugget to partial sill ratio $\delta^2 = \tau^2/\sigma^2$ and applied the conjugate Bayesian linear regression model in Section 6.2.3 to a BEF subset consisting of 415 observations. We carry out exact inference by sampling directly from the posterior; no MCMC algorithm is needed. The following estimates are based upon 1000 posterior samples. Again, no burn-in is needed as this is exact sampling from the true posterior. A summary of the results is presented below (rounded up to 3 significant digits).

	2.5%	25%	50%	75%	97.5%
Intercept	-0.372	0.636	1.161	1.644	2.701
Elevation	0.000	0.000	0.000	0.001	0.001
Slope	-0.016	-0.011	-0.009	-0.006	-0.002
TC1	-0.002	0.007	0.010	0.015	0.023
TC2	-0.002	0.003	0.006	0.008	0.013
TC3	0.010	0.017	0.021	0.025	0.032
σ^2	0.072	0.078	0.082	0.086	0.095
τ^2	0.014	0.016	0.016	0.017	0.019

Table 6.4 *Posterior quantiles of regression parameters, the partial sill σ^2 and the nugget τ^2 from the conjugate Bayesian regression analysis of the BEF data with τ^2/σ^2 and ϕ fixed at values estimated from a semivariogram analysis.*

Next, we relax our model and assume that all the covariance function parameters are unknown. Hence, we do not fix ϕ, nor do we assume that the ratio $\delta^2 = \tau^2/\sigma^2$ is fixed. Instead, we assign individual priors on σ^2, τ^2 and ϕ and fit a collapsed model using MCMC. The regression coefficients are updated from their normal full conditional distributions, while ϕ, σ^2 and τ^2 are updated using Metropolis steps and require tuning parameters. A flat prior for β is used by default. Their posterior estimates are not very different from those obtained above so we do not present them again. Below is a summary of the estimates of σ^2, τ^2 and ϕ:

	2.5%	25%	50%	75%	97.5%
σ^2	0.072	0.078	0.082	0.086	0.095
τ^2	0.014	0.016	0.016	0.017	0.019
ϕ	0.005	0.008	0.010	0.012	0.017

Table 6.5 *Posterior quantiles of the partial sill σ^2, the nugget τ^2 and the correlation decay ϕ from a Bayesian regression analysis of the BEF data with all covariance parameters unknown.*

We obtain the posterior samples of the marginalized regression coefficients and the spatial effects from their full conditional distributions after collecting the posterior samples of $\{\sigma^2, \tau^2, \phi\}$. Figure 6.9 shows the resulting trace plots.

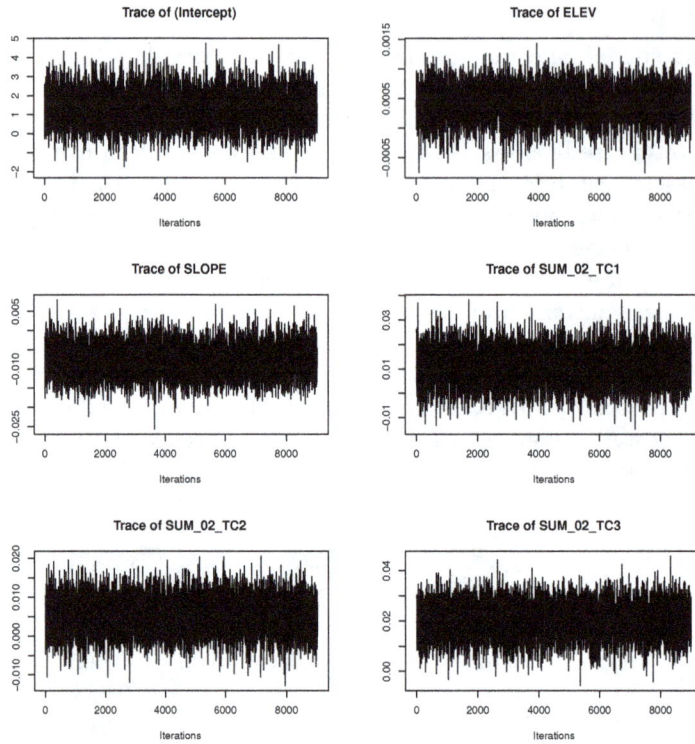

Figure 6.9 *MCMC trace plots of β in the Bayesian regression analysis of the BEF data with all covariance parameters unknown.*

We also obtain the posterior mean and standard deviation for the spatial effects. We plot the posterior means and the the residual means from an Ordinary Least Squares (OLS) model representing a linear model without spatial effects. These are presented in Figure 6.10

6.7.1.1   Prediction

Turning to prediction or Bayesian kriging, we sample from the posterior predictive distribution of every pixel across the BEF. We construct the prediction design matrix for the entire grid extent and extract the coordinates of the BEF bounding polygon vertices. With access to each pixel's posterior predictive distribution, we can map any summary statistics of interest. In Figure 6.11 we compare the log metric tons of biomass interpolated over the observed plots to that of the pixel-level prediction.

6.7.1.2   Model selection

To compare several alternative models with varying degrees of richness, we use DIC (Section 5.6.2) and GPD scores ((5.67) in Section 5.6.3). The results are displayed in Tables 6.6 and 6.7.

6.7.2   Non-Gaussian spatial GLM

We also fit Poisson and binomial models using the log and logit link function, respectively. Here we illustrate a Poisson generalized linear mixed model with spatially dependent random effects. We consider a simulated dataset with 50 locations inside the unit square. We generate

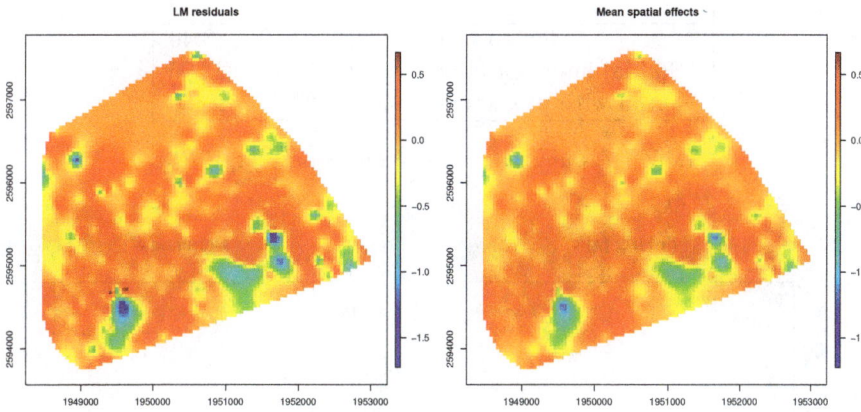

Figure 6.10 *Interpolated surface of the OLS model residuals and the mean of the random spatial effects posterior distribution.*

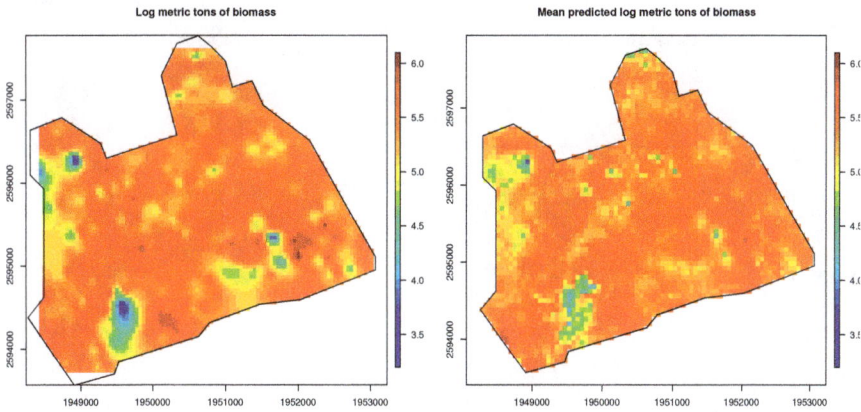

Figure 6.11 *Interpolated surface of observed log metric tons of biomass and the posterior predictive mean of each pixels.*

a latent Gaussian spatial random field $w(\mathbf{s})$ using an exponential covariance function with $\sigma^2 = 2$ and $\phi = 3/0.5$ (so the spatial range is 0.5). Finally, the outcome in each location is generated from a Poisson distribution with intensity $\exp(\beta_0 + w(\mathbf{s}_i))$. These coefficients and the Cholesky square root of the parameters' estimated covariances are used as starting values and Metropolis sampler tuning values. In addition to the regression coefficients we specify starting values for the spatial correlation decay ϕ and variance σ^2 as well as the

	D	$D(\Omega)$	p_D	DIC
Non spatial	-538.50	-545.40	6.90	-531.60
Spatial intercept only	-1202.50	-1416.80	214.30	-988.20
Spatial with predictors	-1084.90	-1266.80	182.00	-902.90

Table 6.6 *Candidate model comparison using DIC.*

	G	P	D
Non spatial	41.00	42.40	83.40
Spatial intercept only	2.80	18.20	21.00
Spatial with predictors	5.00	22.30	27.30

Table 6.7 *Candidate model comparison using GPD.*

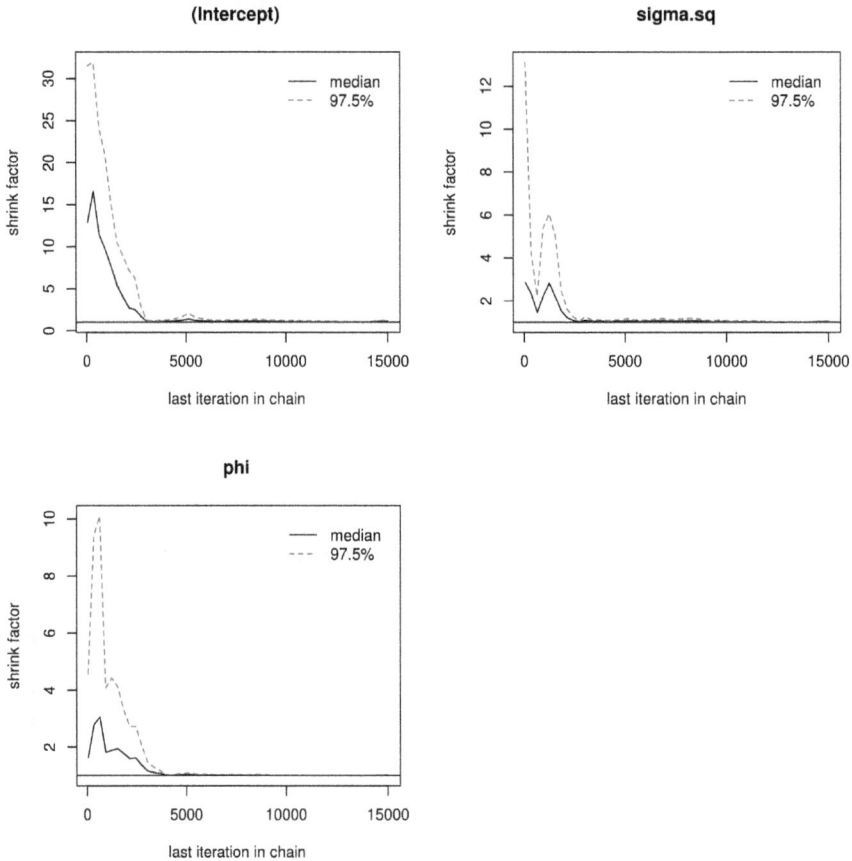

Figure 6.12 *MCMC chain convergence diagnostics.*

random spatial effects **w**. Posterior inference is based on three MCMC chains each of length 15,000 iterations.

The convergence of multiple chains can be assessed using diagnostics detailed in Gelman and Rubin [1992]. We calculate the "potential scale reduction factor" for each parameter, along with the associated upper and lower confidence limits. Approximate convergence is diagnosed when the upper confidence limit is close to 1. We can also plot Gelman and Rubin's shrink factor versus the number of MCMC samples, here again convergence is diagnosed when the upper confidence limit remains close to 1. The estimated potential scale reduction factors were 1.09 (upper confidence limit 1.20) for the intercept, 1.02 (upper confidence limit 1.05) for σ^2 and 1.01 (upper confidence limit 1.02) for ϕ. Figure 6.12 suggests we should discard the first $\sim$10,000 samples as burn-in prior to summarizing the parameters' posterior distributions.

	50%	2.5%	97.5%
Intercept	0.18	-1.44	0.82
σ^2	1.58	0.87	3.41
ϕ	9.89	3.69	23.98

Table 6.8 *Parameter estimates for the intercept, σ^2 and ϕ from a Bayesian spatial GLM analysis of a simulated data set.*

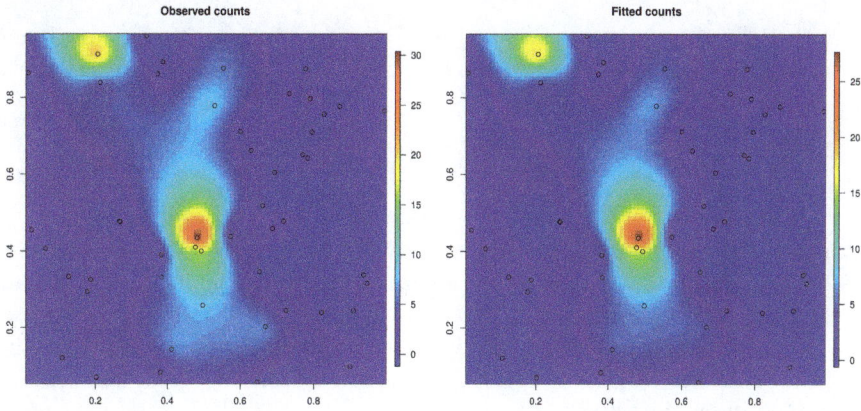

Figure 6.13 *Observed and estimated counts.*

Given the post-burn in samples, we are also able to generate surfaces of the estimated counts. Given the posterior samples of the model parameters, we use composition sampling to draw from the posterior predictive distribution of any new location. These realization can then be summarized to assess predictive performance. For example Figure 6.14 illustrates the median prediction over the domain.

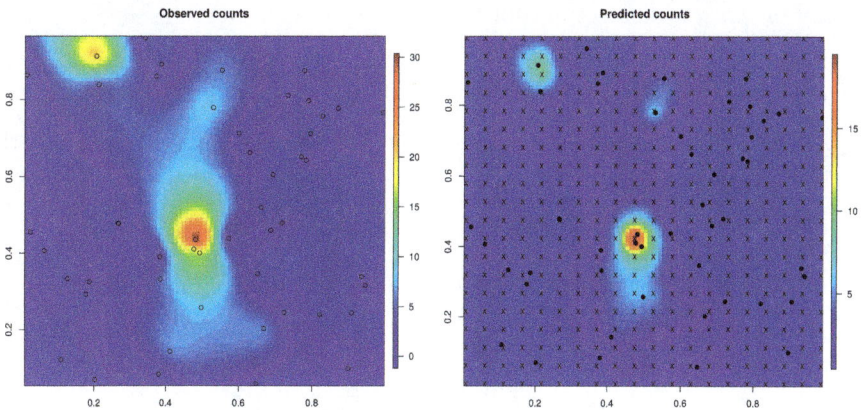

Figure 6.14 *Observed and predicted counts.*

6.8   Areal data models

6.8.1   Disease mapping

An area of strong biostatistical and epidemiological interest is that of *disease mapping*. Here we typically have count data of the following sort:

$$Y_i \;=\; \text{observed number of cases of disease in county } i, \; i = 1, \ldots, I$$
$$E_i \;=\; \text{expected number of cases of disease in county } i, \; i = 1, \ldots, I$$

The Y_i are thought of as random variables, while the E_i are thought of as fixed and known functions of n_i, the number of persons at risk for the disease in county i. As a simple starting point, we might assume that

$$E_i = n_i \bar{r} \equiv n_i \left(\frac{\sum_i y_i}{\sum_i n_i} \right) \equiv \sum_i y_i \frac{\sum_i n_i}{\sum_i n_i} ;,$$

i.e., $\bar{r}$ is the overall disease rate in the entire study region. The second equivalence interprets E_i as scaling the total number of cases by the proportion of the population at risk in county i. These E_i thus correspond to a kind of "null hypothesis," where we expect a constant disease rate in every county. This process is called *internal standardization* since it centers the data (some counties will have observed rates higher than expected, and some less) but uses only the observed data to do so.

     Internal standardization is "cheating" (or at least "empirical Bayes") in some sense, since, evidently, the E_i are not fixed but are functions of the data. A better approach might be to make reference to an existing standard table of age-adjusted rates for the disease (as might be available for many types of cancer). Then after stratifying the population by age group, the E_i emerge as

$$E_i = \sum_j n_{ij} r_j \,,$$

where n_{ij} is the person-years at risk in area i for age group j (i.e., the number of persons in age group j who live in area i times the number of years in the study), and r_j is the disease rate in age group j (taken from the standard table). This process is called *external standardization*. In either case, in its simplest form, a disease map is just a display (in color or greyscale) of the raw disease rates overlaid on the areal units.

6.8.2   Traditional models and frequentist methods

If E_i is not too large (i.e, the disease is rare or the regions i are sufficiently small), the usual model for the Y_i is the Poisson model,

$$Y_i \mid \eta_i \sim Po(E_i \eta_i) \,,$$

where η_i is the true *relative risk* of disease in region i. The maximum likelihood estimate (MLE) of η_i is readily shown to be

$$\hat{\eta}_i \equiv SMR_i = \frac{Y_i}{E_i} \,,$$

the *standardized morbidity (or mortality) ratio* (SMR), i.e., the ratio of observed to expected disease cases (or deaths). Note that $\mathrm{Var}(SMR_i) = \mathrm{Var}(Y_i)/E_i^2 = \eta_i/E_i$, and so we might take $\widehat{\mathrm{Var}}(SMR_i) = \hat{\eta}_i/E_i = Y_i/E_i^2$. This in turn permits the calculation of traditional confidence intervals for η_i (although this is a bit awkward since the data are discrete), as well as hypothesis tests.

Example 6.4 To find a confidence interval for η_i, arguably, it would be preferable to work on the log scale, i.e., to assume that $\log SMR_i$ is roughly *normally* distributed. Using the delta method (Taylor series expansion), one can find that

$$\mathrm{Var}[\log(SMR_i)] \approx \frac{1}{SMR_i^2}\mathrm{Var}(SMR_i) = \frac{E_i^2}{Y_i^2} \times \frac{Y_i}{E_i^2} = \frac{1}{Y_i} \, .$$

An approximate 95% CI for $\log \eta_i$ is thus $\log SMR_i \pm 1.96/\sqrt{Y_i}$, and so (transforming back) an approximate 95% CI for η_i is

$$\left(SMR_i \exp(-1.96/\sqrt{Y_i}) \, , \; SMR_i \exp(1.96/\sqrt{Y_i}) \right) \, .$$

∎

Example 6.5 Suppose we wish to test whether the true relative risk in county i is elevated or not, i.e.,

$$H_0 : \eta_i = 1 \; \text{ versus } \; H_A : \eta_i > 1 \, .$$

Under the null hypothesis, $Y_i \sim Po(E_i)$, the p-value for this test is

$$p = Pr(X \geq Y_i \mid E_i) = 1 - Pr(X < Y_i \mid E_i) = 1 - \sum_{x=0}^{Y_i-1} \frac{\exp(-E_i)E_i^x}{x!} \, .$$

This is the (one-sided) p-value; if it is less than 0.05 we would typically reject H_0 and conclude that there is a statistically significant excess risk in the county i. ∎

6.8.3 *Hierarchical Bayesian methods*

The methods of the previous section are fine for detecting extra-Poisson variability (overdispersion) in the observed rates, but what if we seek to *estimate* and *map* the underlying relative risk surface $\{\eta_i, i = 1, \ldots, I\}$? In this case we might naturally think of a *random effects* model for the η_i, since we would likely want to assume that all the true risks come from a common underlying distribution. Random effects models also allow the procedure to "borrow strength" across the various counties in order to come up with an improved estimate for the relative risk in each.

  The random effects here, however, can be high-dimensional and are couched in a nonnormal (Poisson) likelihood. Thus, as in the previous sections of this chapter, the most natural way of handling this more complex model is through hierarchical Bayesian modeling, as we now describe.

6.8.3.1 *Poisson-gamma model*

As a simple initial model, consider

$$Y_i \mid \eta_i \; \overset{ind}{\sim} \; Po(E_i\eta_i) \, , i = 1, \ldots, I,$$
$$\text{and } \eta_i \; \overset{iid}{\sim} \; G(a, b) \, ,$$

where $G(a, b)$ denotes the *gamma* distribution with mean $\mu = a/b$ and variance $\sigma^2 = a/b^2$; note that this is the gamma parametrization used by the `WinBUGS` package. Solving these two equations for a and b we get $a = \mu^2/\sigma^2$ and $b = \mu/\sigma^2$. Suppose we set $\mu = 1$ (the "null" value) and $\sigma^2 = (0.5)^2$ (a fairly large variance for this scale). Figure 6.15(a) shows a sample of size 1000 from the resulting $G(4, 4)$ prior; note the vertical reference line at $\eta_i = \mu = 1$.

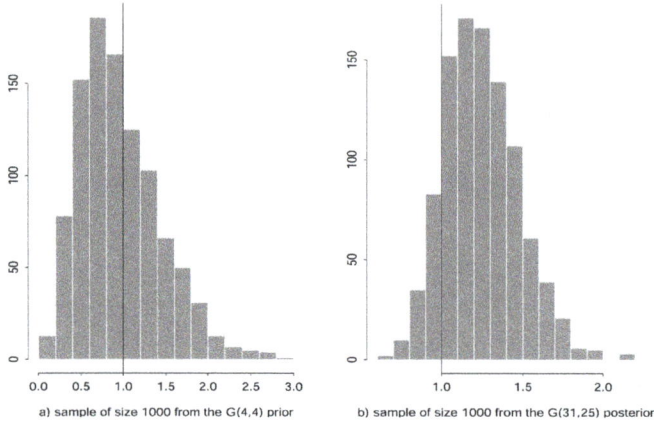

Figure 6.15 *Samples of size 1000 from a Gamma(4,4) prior (a) and a Gamma(27+4, 21+4) posterior (b) for η_i.*

Inference about $\boldsymbol{\eta} = (\eta_1, \ldots, \eta_I)'$ is now based on the resulting posterior distribution, which in the Poisson-gamma emerges in closed form (thanks to the conjugacy of the gamma prior with the Poisson likelihood) as $\prod_i p(\eta_i \mid y_i)$, where $p(\eta_i \mid y_i)$ is $G(y_i + a, E_i + b)$. Thus a suitable point estimate of η_i might be the posterior mean,

$$\mathrm{E}(\eta_i \mid \mathbf{y}) \; = \; \mathrm{E}(\eta_i \mid y_i) \; = \; \frac{y_i + a}{E_i + b} \; = \; \frac{y_i + \frac{\mu^2}{\sigma^2}}{E_i + \frac{\mu}{\sigma^2}} \tag{6.33}$$

$$= \; \frac{E_i \left(\frac{y_i}{E_i} \right)}{E_i + \frac{\mu}{\sigma^2}} + \frac{\left(\frac{\mu}{\sigma^2} \right) \mu}{E_i + \frac{\mu}{\sigma^2}}$$

$$= \; w_i \, SMR_i + (1 - w_i)\mu \, , \tag{6.34}$$

where $w_i = E_i / [E_i + (\mu/\sigma^2)]$, so that $0 \le w_i \le 1$. Thus the Bayesian point estimate (6.34) is a *weighted average* of the the data-based SMR for region i, and the prior mean μ. This estimate is approximately equal to SMR_i when w_i is close to 1 (i.e., when E_i is big, so the data are strongly informative, or when σ^2 is big, so the prior is weakly informative). On the other hand, (6.34) will be approximately equal to μ when w_i is close to 0 (i.e., when E_i is small, so the data are sparse, or when σ^2 is small, so that the prior is highly informative).

As an example, suppose in county i we observe $y_i = 27$ disease cases, when only $E_i = 21$ were expected. Under our $G(4, 4)$ prior we obtain a $G(27 + 4, 21 + 4) = G(31, 25)$ posterior distribution; Figure 6.15(b) shows a sample of size 1000 drawn from this distribution. From (6.33) this distribution has mean $31/25 = 1.24$ (consistent with the figure), indicating slightly elevated risk (24%). However, the posterior probability that the true risk is bigger than 1 is $P(\eta_i > 1 \mid y_i) = .863$, which we can derive exactly (say, using `1 - pgamma(25,31)` in R), or estimate empirically as the proportion of samples in Figure 6.15(b) that are greater than 1. In either case, we see substantial but not overwhelming evidence of risk elevation in this county.

If we desired a $100 \times (1 - \alpha)\%$ confidence interval for η_i, the easiest approach would be to simply take the upper and lower $\alpha/2$-points of the $G(31, 25)$ posterior, since the resulting interval $\left(\eta_i^{(L)}, \eta_i^{(U)} \right)$, would be such that $P \left[\eta_i \in \left(\eta_i^{(L)}, \eta_i^{(U)} \right) \mid y_i \right] = 1 - \alpha$, by definition of the posterior distribution. This is the so-called *equal-tail credible interval* mentioned in Subsection 5.2.2. In our case, taking $\alpha = .05$ we obtain $(\eta_i^{(L)}, \eta_i^{(U)}) = (.842, 1.713)$, again indicating no "significant" elevation in risk for this county. (In R the appropriate commands here are `qgamma(.025, 31)/25` and `qgamma(.975, 31)/25`.)

Finally, in a "real" data setting we would obtain not 1 but I point estimates, interval estimates, and posterior distributions, one for each county. Such estimates would often be summarized in a choropleth map, say, in R or `ArcView`. Full posteriors are obviously difficult to summarize spatially, but posterior means, variances, or confidence limits are easily mapped in this way. We shall explore this issue in the next subsection.

6.8.3.2 Poisson-lognormal models

The gamma prior of the preceding section is very convenient computationally, but suffers from a serious defect: it fails to allow for spatial correlation among the η_i. To do this we would need a *multivariate* version of the gamma distribution; such structures exist but are awkward both conceptually and computationally. Instead, the usual approach is to place some sort of multivariate *normal* distribution on the $\psi_i \equiv \log \eta_i$, the *log*-relative risks.

Thus, consider the following augmentation of our basic Poisson model:

$$Y_i \mid \psi_i \overset{ind}{\sim} Po\left(E_i\, e^{\psi_i}\right),$$
$$\text{where } \psi_i = \mathbf{x}_i'\boldsymbol{\beta} + \theta_i + \phi_i. \tag{6.35}$$

The $\mathbf{x}_i$ are explanatory spatial covariates, having parameter coefficients $\boldsymbol{\beta}$. The covariates are *ecological*, or county (not individual) level, which may lead to problems of ecological bias (to be discussed later). However, the hope is that they will explain some (perhaps all) of the spatial patterns in the Y_i.

Before, we discuss modeling for the θ_i and ϕ_i, we digress to return to the internal standardization problem mentioned at the beginning of this section. It is evident in (6.35) that, with internal standardization, we do not have a valid model. That is, the data appears on both sides of the Poisson specification. This issue is generally ignored in the literature but the reader might appreciate the potential benefit of an alternative parametrization which writes $\text{E}(Y_i) = n_i p_i$. Here, again, n_i is the number of people at risk in say county i and p_i is the incidence rate in county i. Now, we would model p_i and avoid the internal standardization problem. In fact, this is the customary Poisson parametrization where we expect large n_i and small p_i. And, if interest is in the η_i's, the SMR's, after posterior inference on p_i, we get posterior inference for η_i for free since, given the data, $\eta_i = n_i p_i/E_i$. Why isn't this model more widely used? We suspect it is the fact that, in principle, we would have to introduce say a logit or probit link to model the p_i which, perhaps, makes the model more difficult to fit. (The difficulty with retaining the log link is that, upon inversion, we might find probabilities greater than 1. However, when the p_i are small this need not be a problem.) In the sequel, we stay with the customary parametrization but encourage the reader to consider the alternative.

Returning to (6.35), the θ_i capture region-wide *heterogeneity* via an ordinary, exchangeable normal prior,

$$\theta_i \overset{iid}{\sim} N(0, 1/\tau_h), \tag{6.36}$$

where τ_h is a precision (reciprocal of the variance) term that controls the magnitude of the θ_i. These random effects capture extra-Poisson variability in the log-relative risks that varies "globally," i.e., over the entire study region.

Finally, the ϕ_i are the parameters that make this a spatial model by capturing regional *clustering*. That is, they model extra-Poisson variability in the log-relative risks that varies "locally," so that nearby regions will have more similar rates. A plausible way to attempt this might be to try a point-referenced model on the parameters ϕ_i. For instance, writing $\phi = (\phi_1, \ldots, \phi_I)'$, we might assume that

$$\phi \mid \mu, \boldsymbol{\lambda} \sim N_I(\mu, H(\boldsymbol{\lambda})),$$

where N_I denotes the I-dimensional normal distribution, μ is the (stationary) mean level, and $(H(\boldsymbol{\lambda}))_{ii'}$ gives the covariance between ϕ_i and $\phi_{i'}$ as a function of some hyperparameters $\boldsymbol{\lambda}$. The standard forms given in Table 2.1 remain natural candidates for this purpose. The issue here is that it is not clear what are appropriate inter-areal unit distances. Should we use centroid to centroid? Does this make sense with units of quite differing sizes and irregular shapes? A further challenge arises when we have many areal units. Then, the so-called big N problem emerges (see Chapter 13) with regard to high dimensional matrix computations. Moreover, with areal unit data, we are not interested in interpolation, that is, kriging is the usual benefit for introducing a covariance function. As a result, it is customary in hierarchical analyses of areal data to return to neighbor-based notions of proximity, and ultimately, to return to CAR specifications for $\boldsymbol{\phi}$ (Section 4.3). In the present context (with the CAR model placed on the elements of $\boldsymbol{\phi}$ rather than the elements of $\mathbf{Y}$), we will write

$$\boldsymbol{\phi} \sim CAR(\tau_c) , \qquad (6.37)$$

where by this notation we mean the improper CAR (IAR) model in (4.16) with y_i replaced by ϕ_i, τ^2 replaced by $1/\tau_c$, and using the 0-1 (adjacency) weights w_{ij}. Thus τ_c is a *precision* (not variance) parameter in the CAR prior (6.37), just as τ_h is a precision parameter in the heterogeneity prior (6.36).

6.8.3.3 CAR models and their difficulties

Compared to point-level (geostatistical) models, CAR models are very convenient computationally, since our method of finding the posterior of $\boldsymbol{\theta}$ and $\boldsymbol{\phi}$ is itself a conditional algorithm, the Gibbs sampler. Recall from Subsection 5.5.1.1 that this algorithm operates by successively sampling from the *full conditional* distribution of each parameter (i.e., the distribution of each parameter given the data and every other parameter in the model). So for example, the full conditional of ϕ_i is

$$p(\phi_i \mid \phi_{j \neq i}, \boldsymbol{\theta}, \boldsymbol{\beta}, \mathbf{y}) \propto Po(y_i \mid E_i e^{\mathbf{x}_i'\boldsymbol{\beta} + \theta_i + \phi_i}) \times N(\phi_i \mid \bar{\phi}_i , 1/(\tau_c m_i)) , \qquad (6.38)$$

meaning that we do not need to work with the joint distribution of $\boldsymbol{\phi}$ at all. The conditional approach also eliminates the need for any matrix inversion.

While computationally convenient, CAR models have various theoretical and computational challenges, some of which have already been noted in Section 4.3, and others of which emerge now that we are using the CAR as a distribution for the random effects $\boldsymbol{\phi}$, rather than the data $\mathbf{Y}$ itself. We consider two of these.

1. Impropriety: Recall from the discussion surrounding (4.15) that the IAR prior we selected in (6.37) above is *improper*, meaning that it does not determine a legitimate probability distribution (one that integrates to 1). That is, the matrix $\Sigma^{-1} = (D_w - W)$ is singular, and thus its inverse does not exist.

As mentioned in that earlier discussion, one possible fix for this situation is to include a "propriety parameter" ρ in the precision matrix, i.e., $\Sigma^{-1} = (D_w - \rho W)$. Taking $\rho \in \left(1/\lambda_{(1)}, 1/\lambda_{(n)}\right)$, where $\lambda_{(1)}$ and $\lambda_{(n)}$ are the smallest and largest eigenvalues of $D_w^{-1/2} W D_w^{-1/2}$, respectively, ensures the existence of $\Sigma_{\mathbf{y}}$. Alternatively, using the scaled adjacency matrix $\widetilde{W} \equiv Diag(1/w_{i+})W$, Σ^{-1} can be written as $M^{-1}(I - \alpha \widetilde{W})$ where M is diagonal. One can then show [Carlin and Banerjee, 2003, Gelfand and Vounatsou, 2003] that if $|\alpha| < 1$, then $(I - \alpha \widetilde{W})$ will be positive definite, resolving the impropriety problem without eigenvalue calculation.

This immediately raises the question of whether or not to include a propriety parameter in the CAR model. On the one hand, adding say ρ, in addition to supplying propriety, enriches the spatial model. Moreover, it provides a natural interpretation and, when $\rho = 0$.

we obtain an independence model. So, why would we not include ρ? One reason is that the interpretation is not quite what we want. Recognizing that the inclusion of ρ makes $E(\phi_i = \rho \sum_{j \sim i} w_{ij}\phi_j$, we see that it isn't necessarily sensible that we expect ϕ_i to be only a portion of the average of its neighbors. Why is that an appropriate smoother? A further difficulty with this fix (discussed in more detail near the end of Subsection 4.3.1) is that this new prior typically does not deliver enough spatial similarity unless ρ is quite close to 1, thus getting us very close to the same problem again! Indeed in model fitting, ρ always is very close to 1; it is as though, as a spatial random effects model, the data wants ρ to be 1. Some authors [Carlin and Banerjee, 2003] recommend an informative prior that insists on larger ρ's or α's (say, a $Beta(18, 2)$), but this is controversial since there will typically be little true prior information available regarding the magnitude of α.

Our recommendation is to work with the intrinsic CAR specification, i.e., ignore the impropriety of the standard CAR model (6.37) and continue! After all, we are only using the CAR model as a *prior*; the *posterior* will generally still emerge as proper, so Bayesian inference may still proceed. This is the usual approach, but it also requires some care, as follows: this improper CAR prior is a *pairwise difference prior* Besag et al. [1995] that is identified only up to an additive constant. Thus to identify an intercept term β_0 in the log-relative risk, we must add the constraint $\sum_{i=1}^{I} \phi_i = 0$. Note that, in implementing a Gibbs sampler to fit (6.35), this constraint can be imposed *numerically* by recentering each sampled ϕ vector around its own mean following each Gibbs iteration, so-called centering on the fly. Note that we will prefer to implement this centering during the course of model fitting rather than transforming to the lower dimensional proper distribution, in order to retain the convenient conditional distributions associated with the pairwise difference CAR.

As a brief digression here, we note that the simple model, $Y_i = \phi_i + \theta_i$ where, again, the θ_i are i.i.d. error terms and the ϕ's are an intrinsic CAR is a legitimate data model. That is, $Y_i \mid \phi_i \sim N(\phi_i, \sigma_h^2)$. (This setting is analogous to placing an improper prior on μ in the model, $Y_i = \mu + \theta_i$.) In fact, we do not need an intercept in this model. If we were to include one, we would have to center the ϕ's at each model fitting iteration. Contrast this specification with $Y_i = \mathbf{X}_i^\mathsf{T}\boldsymbol{\beta} + \phi_i$ which can not be a data model as we see from the conditional distribution of $Y_i \mid \boldsymbol{\beta}$.

2. Selection of τ_c and τ_h: Clearly the values of these two prior precision parameters will control the amount of extra-Poisson variability allocated to "heterogeneity" (the θ_i) and "clustering" (the ϕ_i). But they cannot simply be chosen to be arbitrarily large, since then the ϕ_i and θ_i would be *unidentifiable*: note that we see only a single Y_i in each county, yet we are attempting to estimate *two* random effects for each i! Eberly and Carlin [2000] investigate convergence and Bayesian learning for this data set and model, using fixed values for τ_h and τ_c.

Suppose we decide to introduce third-stage priors (*hyperpriors*) on τ_c and τ_h. Under the intrinsic CAR specification, the distribution for the ϕ_i's is improper. What power of τ_c should be introduced into the multivariate "Gaussian distribution" for the ϕ's to yield, with the hyperprior, the full conditional distribution for τ_c? Evidently, with an improper distribution, there is no *correct* answer; the power is not determined. This problem has been considered in the literature; see, e.g. Hodges [2013]. A natural way to think about this issue is to recognize the effective dimension of the intrinsic CAR model. For instance, with a connected set of areal units, we have one more ϕ than we need; the intrinsic CAR is proper in the $n-1$ dimensional space where we set one of the ϕ's equal to the sum of the remainder. Thus, the centering on the fly constraint makes the distribution proper so the suggestion is to use the power $(n-1/2$ for τ_c. However, in some areal unit settings, we have singleton units, units with no neighbors (perhaps say, two units not connected to the rest). Here, the recommendation is to deduce from n the number of "islands" in the data. With no singletons, we have one island, hence $n - 1$, with one singleton, we have two islands, hence, $n - 2$, etc. Of course, with a proper CAR, this problem disappears.

These hyperpriors can not be arbitrarily vague due to the identifiability challenge above. Still, the gamma offers a conjugate family here, so we might simply take

$$\tau_h \sim G(a_h, b_h) \text{ and } \tau_c \sim G(a_c, b_c) .$$

To make this prior "fair" (i.e., equal prior emphasis on heterogeneity and clustering), it is tempting to simply set $a_h = a_c$ and $b_h = b_c$, but this would be incorrect for two reasons. First, the τ_h prior (6.36) uses the usual *marginal* specification, while the τ_c prior (6.37) is specified *conditionally*. Second, τ_c is multiplied by the number of neighbors m_i before playing the role of the (conditional) prior precision. Bernardinelli et al. [1995b] note that the prior marginal standard deviation of ϕ_i is roughly equal to the prior conditional standard deviation divided by 0.7. Thus a scale that delivers

$$sd(\theta_i) = \frac{1}{\sqrt{\tau_h}} \approx \frac{1}{0.7\sqrt{\bar{m}\tau_c}} \approx sd(\phi_i) \tag{6.39}$$

where $\bar{m}$ is the average number of neighbors that may offer a reasonably "fair" specification. Of course, it is fundamentally unclear how to relate the marginal variance of a proper joint distribution with the conditional variance of an improper joint distribution.

Example 6.6 As an illustration of the Poisson-lognormal model (6.35), consider the data displayed in Figure 6.16. These data from Clayton and Kaldor [1987] are the observed (Y_i) and expected (E_i) cases of lip cancer for the $I = 56$ districts of Scotland during the period 1975–1980. One county-level covariate x_i, the percentage of the population engaged in agriculture, fishing or forestry (AFF), is also available (and also mapped in Figure 6.16). Modeling the log-relative risk as

$$\psi_i = \beta_0 + \beta_1 x_i + \theta_i + \phi_i , \tag{6.40}$$

we wish to investigate a variety of vague, proper, and arguably "fair" priors for τ_c and τ_h, find the estimated posterior of β_1 (the AFF effect), and find and map the fitted relative risks $E(\psi_i \mid \mathbf{y})$.

Recall that Y_i cannot inform about θ_i or ϕ_i, but only about their sum $\xi_i = \theta_i + \phi_i$. Making the reparameterization from $(\boldsymbol{\theta}, \boldsymbol{\phi})$ to $(\boldsymbol{\theta}, \boldsymbol{\xi})$, we have the joint posterior,

$$p(\boldsymbol{\theta}, \boldsymbol{\xi} \mid \mathbf{y}) \propto L(\boldsymbol{\xi}; \mathbf{y}) p(\boldsymbol{\theta}) p(\boldsymbol{\xi} - \boldsymbol{\theta}).$$

This means that

$$p(\theta_i \mid \theta_{j \neq i}, \boldsymbol{\xi}, \mathbf{y}) \propto p(\theta_i) \, p(\xi_i - \theta_i \mid \{\xi_j - \theta_j\}_{j \neq i}) .$$

Since this distribution is free of the data $\mathbf{y}$, the θ_i are *Bayesianly unidentified* (and so are the ϕ_i). But this does not preclude *Bayesian learning* (i.e., prior to posterior movement) about θ_i. No Bayesian learning would instead require

$$p(\theta_i \mid \mathbf{y}) = p(\theta_i) , \tag{6.41}$$

in the case where both sides are proper (a condition not satisfied by the CAR prior). Note that (6.41) is a much stronger condition than Bayesian unidentifiability, since the data have no impact on the *marginal* (not merely the *conditional*) posterior distribution.

Recall that, though they are unidentified, the θ_i and ϕ_i are interesting in their own right, as is

$$\alpha = \frac{sd(\boldsymbol{\phi})}{sd(\boldsymbol{\theta}) + sd(\boldsymbol{\phi})} ,$$

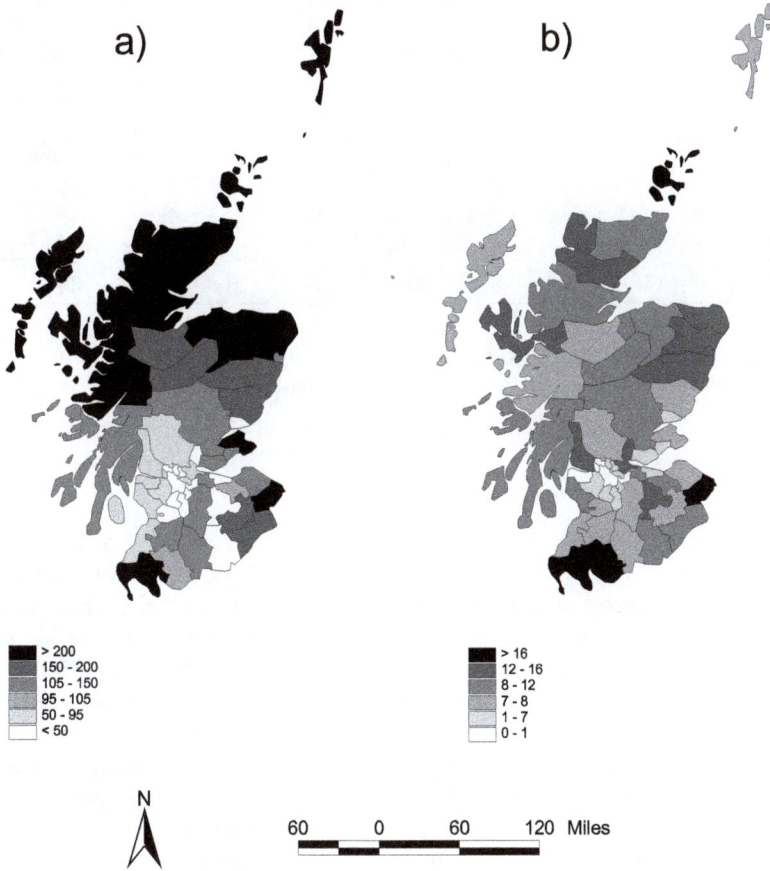

Figure 6.16 *Scotland lip cancer data: (a) crude standardized mortality ratios (observed / expected × 100); (b) AFF covariate values.*

where $sd(\cdot)$ is the empirical marginal standard deviation function. That is, α is the proportion of the variability in the random effects that is due to clustering (hence $1 - \alpha$ is the proportion due to unstructured heterogeneity). Recall we need to specify vague but proper prior values τ_h and τ_c that lead to acceptable convergence behavior, yet still allow Bayesian learning. This prior should also be "fair," i.e., lead to $\alpha \approx 1/2$ *a priori*.

We also experiment with the use of vague priors for τ_c and τ_h, as suggested by Best et al. [1999]. Posterior moments (mean, sd) and convergence (lag 1 autocorrelation) summaries for α, β_1, ξ_1, and ξ_{56} are given in Table 6.9. Besides the Best et al. [1999] prior, two priors inspired by equation (6.39) are also reported [see Carlin and Pérez, 2000]. The AFF covariate appears significantly different from 0 under all 3 priors, although convergence is *very* slow (very high values for l1acf). The excess variability in the data seems mostly due to clustering ($E(\alpha \mid \mathbf{y}) > .50$), but the posterior distribution for α does *not* seem robust to changes in the prior. Finally, convergence for the ξ_i (reasonably well identified) is rapid; convergence for the ψ_i (not shown) is virtually immediate.

Of course, a full analysis of these data would also involve a map of the posterior means of the raw and estimated SMR's. ∎

Priors for τ_c, τ_h	Posterior for α			Posterior for β		
	mean	sd	l1acf	mean	sd	l1acf
G(1.0, 1.0), G(3.2761, 1.81)	.57	.058	.80	.43	.17	.94
G(.1, .1), G(.32761, .181)	.65	.073	.89	.41	.14	.92
G(.1, .1), G(.001, .001)	.82	.10	.98	.38	.13	.91

Priors for τ_c, τ_h	Posterior for ξ_1			Posterior for ξ_{56}		
	mean	sd	l1acf	mean	sd	l1acf
G(1.0, 1.0), G(3.2761, 1.81)	.92	.40	.33	−.96	.52	.12
G(.1, .1), G(.32761, .181)	.89	.36	.28	−.79	.41	.17
G(.1, .1), G(.001, .001)	.90	.34	.31	−.70	.35	.21

Table 6.9 *Posterior summaries for the spatial model with Gamma hyperpriors for τ_c and τ_h, Scotland lip cancer data; "sd" denotes standard deviation while "l1acf" denotes lag 1 sample autocorrelation.*

6.8.4 Extending the CAR model

From the above, the reader may glean the idea that it might be easier to just work with the CAR model in (6.35), discarding the heterogeneity component. In a sense, this merely asserts that we are interested in a spatial model, we are interested in implementing spatial smoothing, and since we can not separate the spatial structure from the heterogeneity, we adopt an appropriate spatial specification. In fact, we see the connection here with the discussion of the redundant pure error term in Section 6.6. With conditionally independent first-stage Poisson counts, adding a heterogeneity model is redundant.

Instead, we might extend the CAR model to include two variance components to attempt to capture both heterogeneity and spatial dependence. A version was presented in the work of Leroux et al. [2000] and developed further in MacNab and Dean [2000]. The key point is that the precision parameter in the intrinsic CAR model represents both overdispersion and spatial association. Letting $Q = D_W − W$, Letting $Q = D_W − W$. We would like a covariance form, we would like a covariance form, $\Sigma = \sigma^2 Q^{-1} + \tau^2 I$. Of course, Q^{-1} doesn't exist. So, instead, define $\Sigma = \omega^2 A^{-1}$ where $A = \lambda Q + (1 − \lambda)I$ where $\lambda \in (0, 1)$ so that A^{-1} exists and, in the likelihood, $\Sigma^{-1} = \frac{\lambda}{\omega^2}Q + \frac{1-\lambda}{\omega^2}I$, resembling the covariance matrix components associated with (6.35). The key difference is that, now, we introduce just a single set of random effect with covariance matrix Σ rather than two sets of random effects, separating the dependence structure.

For this new multivariate normal prior for, say, the set say the set ξ, we can immediately calculate the conditional normal distributions. In fact, $E(\xi_i \mid \{\xi_j, j \neq i\}) = \frac{\lambda}{1-\lambda+\lambda w_{i+}} \sum_{j \sim i} \xi_j$ and $Var(\xi_i \mid \{\xi_j, j \neq i\}) = \frac{\omega^2}{1-\lambda+\lambda w_{i+}}$. When $\lambda = 1$, we have just an intrinsic CAR prior; when $\lambda = 0$, we have an independence model. And, as with the proper CAR, we see that the expected value of ξ_i is only a portion of the average of its neighbors if $\lambda \neq 1$.

A more recent variation on the basic CAR model is offered in Riebler et al. [2016]. It sets $Q = \lambda R_*^- + (1−\lambda)I_A$, where R_* is the scaled spatial neighborhood matrix [see Riebler et al., 2016] and $\lambda \in [0, 1)$. In the literature, this prior has been referred to as the BYM2 spatial prior. It has a scaled spatially structured component and its variance matrix represents a weighted average of the variances of the structured and unstructured components, which facilitates the interpretation of the model.

6.9 General linear areal data modeling

By analogy with Section 6.6, the areal unit measurements Y_i that we model need not be restricted to counts, as in our disease mapping setting. They may also be binary events (say, presence or absence of a particular facility in region i), or continuous measurements (say, population density, i.e., a region's total population divided by its area).

Again formulating a hierarchical model, Y_i may be described using a suitable first-stage member of the exponential family. Now given $\boldsymbol{\beta}$ and ϕ_i, analogous to (6.31) the Y_i are conditionally independent with density,

$$f(y_i \mid \boldsymbol{\beta}, \phi_i, \gamma) = h(y_i, \gamma) \exp\{\gamma[y_i \eta_i - \psi(\eta_i)]\}, \tag{6.42}$$

where $g(\eta_i) = \mathbf{x}_i^{\mathrm{T}} \boldsymbol{\beta} + \phi_i$ for some link function g with γ a dispersion parameter. The ϕ_i will be spatial random effects coming from a CAR model; the pairwise difference, intrinsic (IAR) form is most commonly used. As a result, we have a generalized linear mixed model with spatial structure in the random effects. Paralleling previous sections, it makes no sense to introduce independent heterogeneity effects, θ_i. Again, we are in a situation where the stochastic mechanism in f replaces these independent errors. Again, we recommend fitting models of the form (6.42) having only a spatial random effect. Computation is more stable and a "balanced" (or "fair") prior specification (as mentioned in connection with (6.39) above) is not an issue.

6.10 Comparison of point-referenced and areal data models

We conclude this chapter with a brief summary and comparison between point-referenced data models and areal unit data models. First, the former are defined with regard to an uncountable number of random variables. The process specification determines the n-dimensional joint distribution for the $Y(\mathbf{s}_i)$, $i = 1, \ldots, n$. For areal units, we envision only a single, finite, n-dimensional distribution for the Y_i, $i = 1, \ldots, n$, which we write down to begin with.

Next, with point-referenced data, we model association directly. For example, if $\mathbf{Y} = (Y(\mathbf{s}_1), \ldots, Y(\mathbf{s}_n))^{\mathrm{T}}$ we specify the variance-covariance matrix of $\mathbf{Y}$, say $\Sigma_{\mathbf{Y}}$ using isotropy (or anisotropy), stationarity (or nonstationarity), and so on. With areal data $\mathbf{Y} = (Y_1, \ldots, Y_n)^{\mathrm{T}}$ and CAR (or SAR) specifications, we instead model $\Sigma_{\mathbf{Y}}^{-1}$ directly. For instance, with CAR models, Brook's Lemma enables us to reconstruct $\Sigma_{\mathbf{Y}}^{-1}$ from the conditional specifications; $\Sigma_{\mathbf{Y}}^{-1}$ provides conditional association structure (as in Section 4.3) but says nothing about *unconditional* association structure. When $\Sigma_{\mathbf{Y}}^{-1}$ is full rank, the transformation to $\Sigma_{\mathbf{Y}}$ is very complicated and very nonlinear. Positive conditional association can become a negative unconditional association. If the CAR is defined through distance-based w_{ij}'s there need not be any corresponding distance-based order to the unconditional associations [see Besag and Kooperberg, 1995, Hrafnkelsson and Cressie, 2003, Wall, 2004, for further discussion].

With regard to formal specification, in the most commonly employed point-level Gaussian case, the process is specified through a valid covariance function. With CAR modeling, the specification is instead done through Markov random fields (Section 4.2) employing the Hammersley-Clifford Theorem to ensure a unique joint distribution.

Explanation is a common goal of point-referenced data modeling, but often an even more important goal is spatial prediction or interpolation (i.e., kriging). This may be done at at new points, or for block averages (see Section 7.1). With areal units, again a goal is explanation, but now often is supplemented by smoothing. Here the interpolation problem is infrequent and, in any event, is to new areal units, the so-called modifiable areal unit problem (MAUP) as discussed in Sections 7.2 and 7.3.

Finally, with spatial processes, likelihood evaluation requires the computation of a quadratic form involving $\Sigma_{\mathbf{Y}}^{-1}$ and the determinant of $\Sigma_{\mathbf{Y}}$. (With spatial random effects, this evaluation is deferred to the second stage of the model, but is still present.) With an increasing number of locations, such computation becomes very expensive (computing time is greater than order n^2), and may also become unstable, due to the enormous number of floating point operations required. We refer to this situation as a "big N" problem (and devote Chapter 13 to it). On the other hand, with CAR modeling the likelihood (or the second-stage model for the random effects, as the case may be) is written down immediately, since this model parametrizes $\Sigma_{\mathbf{Y}}^{-1}$ (rather than $\Sigma_{\mathbf{Y}}$). Full conditional distributions needed for MCMC sampling are immediate, and there is no big n problem. Also, for SAR models (though not of much interest in the context of hierarchical modeling because of the fact that the local conditional distributions are not available in simple, closed forms) the quadratic form is directly evaluated, while the determinant is usually evaluated efficiently (even for very large n) using sparse matrix methods [see, e.g., Pace and Barry, 1997a,b].

6.11   Computer tutorials

Computer programs illustrating Bayesian analysis of univariate spatial data using different libraries in R and its interfaces with Bayesian modeling environments are supplied in the folder titled Chapter 6 from https://github.com/sudiptobanerjee/BGC_2023.

6.12   Exercises

1. Derive the forms of the full conditionals for $\boldsymbol{\beta}, \sigma^2, \tau^2, \phi$, and $\mathbf{W}$ in the exponential kriging model (6.1) and (6.3).

2. Assuming the likelihood in (6.3), suppose that ϕ is fixed and we adopt the prior $p(\boldsymbol{\beta}, \sigma^2, \tau^2) \propto 1/(\sigma^2 \tau^2)$, a rather standard noninformative prior often chosen in non-spatial analysis settings. Show that the resulting posterior $p(\boldsymbol{\beta}, \sigma^2, \tau^2 \mid \mathbf{y})$ is *improper*.

3. Derive the form of $p(y_0 \mid \mathbf{y}, \boldsymbol{\theta}, \mathbf{x}_0)$ in (6.19) via the usual conditional normal formulae; i.e., following [Guttman, 1982, pp. 69-72], if $\mathbf{Y} = (\mathbf{Y}_1^{\mathrm{T}}, \mathbf{Y}_2^{\mathrm{T}})^{\mathrm{T}} \sim N(\boldsymbol{\mu}, \boldsymbol{\Sigma})$ where

$$\boldsymbol{\mu} = \begin{pmatrix} \boldsymbol{\mu}_1 \\ \boldsymbol{\mu}_2 \end{pmatrix} \text{ and } \boldsymbol{\Sigma} = \begin{pmatrix} \Sigma_{11} & \Sigma_{12} \\ \Sigma_{21} & \Sigma_{22} \end{pmatrix} ,$$

   then $\mathbf{Y}_2 \mid \mathbf{Y}_1 \sim N(\boldsymbol{\mu}_{2.1}, \boldsymbol{\Sigma}_{2.1})$, where

$$\boldsymbol{\mu}_{2.1} = \boldsymbol{\mu}_2 + \Sigma_{21}\Sigma_{11}^{-1}(\mathbf{Y}_1 - \boldsymbol{\mu}_1) \text{ and } \boldsymbol{\Sigma}_{2.1} = \Sigma_{22} - \Sigma_{21}\Sigma_{11}^{-1}\Sigma_{12} .$$

4. In expression (6.31), if $g(\theta) = \theta$ and the prior on $\boldsymbol{\beta}$ is a proper normal distribution,

   (a) Show that the full conditional distributions for the components of $\boldsymbol{\beta}$ are log-concave.

   (b) Show that the full conditional distributions for the $w(\mathbf{s}_i)$ are log-concave.

5. The lithology data set is available from https://github.com/sudiptobanerjee/BGC_2023 and consists of measurements taken at 118 sample sites in the Radioactive Waste Management Complex region of the Idaho National Engineering and Environmental Laboratory. At each site, bore holes were drilled and measurements were taken to determine the elevation and thickness of the various underground layers of soil and basalt. Understanding the spatial distribution of variables like these is critical to predicting the fate and transport of groundwater and the (possibly harmful) constituents carried thereinsee Leecaster [2002] (2002) for full details.

   For this problem, consider only the variables Northing, Easting, Surf Elevation, Thickness, and A-B Elevation, and only those records for which full information is available (i.e., extract only those data rows without an "NA" for any variable).

(a) Produce image plots of the variables `Thickness`, `Surf Elevation`, and `A-B Elevation`. Add contour lines to each plot and comment on the descriptive topography of the region.

(b) Taking `log(Thickness)` as the response and `Surf Elevation` and `A-B Elevation` as covariates, fit a univariate Gaussian spatial model with a nugget effect, using the exponential and Matérn covariance functions. You may start with flat priors for the covariate slopes, Inverse Gamma$(0.1, 0.1)$ priors for the spatial and nugget variances, and a Gamma$(0.1, 0.1)$ prior for the spatial range parameter. Modify the priors and check for their sensitivity to the analysis.

(c) Perform Bayesian kriging on a suitable grid of values and create image plots of the posterior mean residual surfaces for the spatial effects. Overlay the plots with contour lines and comment on the consistency with the plots from the raw data in part (a).

(d) Repeat the above for a purely spatial model (without a nugget) and compare this model with the spatial+nugget model using a model choice criterion (say, DIC).

6. The real estate data set available from https://github.com/sudiptobanerjee/BGC_2023/data consists of information regarding 70 sales of single-family homes in Baton Rouge, LA, during the month of June 1989. It is customary to model the log-selling price.

(a) Obtain the empirical variogram of the raw log-selling prices.

(b) Fit an ordinary least squares regression to log-selling price using living area, age, other area, and number of bathrooms as explanatory variables. Such a model is usually referred to as a *hedonic* model.

(c) Obtain the empirical variogram of the residuals to the least squares fit.

(d) Using an exponential spatial correlation function, attempt to fit the model $Y(\mathbf{s}) = \mathbf{x}^{\mathrm{T}}(\mathbf{s})\boldsymbol{\beta} + W(\mathbf{s}) + \epsilon(\mathbf{s})$ as in equation (6.1) to the log-selling prices.

(e) Predict the actual selling price for a home at location (longitude, latitude) = $(-91.1174, 30.506)$ that has characteristics LivingArea $= 938$ sqft, OtherArea $= 332$sqft, Age $= 25$yrs, Bedrooms $= 3$, Baths $= 1$, and HalfBaths $= 0$. (*Reasonability check:* The actual log selling price for this location turned out to be 10.448.)

(f) Fit the above model in a Bayesian framework. Begin with the following prior specification: a flat prior for $\boldsymbol{\beta}$, Gamma$(0.1,0.1)$ priors for $1/\sigma^2$ and $1/\tau^2$, and a Uniform$(0,10)$ prior for ϕ. Also investigate prior robustness by experimenting with other choices.

(g) Obtain samples from the predictive distribution for log-selling price and selling price for the particular location mentioned above. Summarize this predictive distribution.

(h) Compare the classical and Bayesian inferences.

(i) *(advanced):* Hold out the first 20 observations in the data file, and fit the nonspatial (i.e., without the $W(\mathbf{s})$ term) and spatial models to the observations that remain. For both models, compute

- $\sum_{j=1}^{20}(Y(\mathbf{s}_{0j}) - \hat{Y}(\mathbf{s}_{0j}))^2$
- $\sum_{j=1}^{20} \mathrm{Var}(Y(\mathbf{s}_{0j}) \mid \mathbf{y})$
- the proportion of predictive intervals for $Y(\mathbf{s}_0)$ that are correct
- the proportion of predictions that are within 10% of the true value
- the proportion of predictions that are within 20% of the true value

Discuss the differences in predictive performance.

7. Suppose we have the Poisson-Gamma model, i.e., independent $Y_i \sim Po(E_i \eta_i), i = 1, 2, ..., I$ with independent $\eta_i \sim Gamma(a, b)$ and $\mu = \frac{a}{b}$, $\sigma^2 = \frac{a}{b^2}$. We are interested in the posterior mean estimate for the rate r_i at areal unit i, $\mathrm{E}(r_i \mid \mathbf{y})$. Show that it is a weighted average of $\hat{r}_i = \frac{Y_i}{n_i}$, with n_i the population at risk, and the prior mean on r_i. What are the revised weights?

8. Suppose $Z_i = Y_i/n_i$ is the observed disease *rate* in each county, and we adopt the model $Z_i \overset{ind}{\sim} N(\eta_i, \sigma^2)$ and $\eta_i \overset{iid}{\sim} N(\mu, \tau^2)$, $i = 1, \ldots, I$. Find $\mathrm{E}(\eta_i \mid y_i)$, and express it as a weighted average of Z_i and μ. Interpret your result as the weights vary.

9. In fitting model (6.35) with priors for the θ_i and ϕ_i given in (6.36) and (6.37), suppose we adopt the hyperpriors $\tau_h \sim G(a_h, b_h)$ and $\tau_c \sim G(a_c, b_c)$. Find closed form expressions for the full conditional distributions for these two parameters.

10. The full conditional (6.38) does *not* emerge in closed-form, since the CAR (normal) prior is not conjugate with the Poisson likelihood. However, prove that this full conditional *is* log-concave, meaning that the necessary samples can be generated using the adaptive rejection sampling (ARS) algorithm of Gilks and Wild [1992] (1992).

11. Confirm algebraically that, taken together, the expressions

$$\phi_i \mid \boldsymbol{\phi}_{j \neq i} \sim N(\phi_i \mid \bar{\phi}_i,\ 1/(\tau_c m_i))\ ,\ i = 1, \ldots, I$$

are equivalent to the (improper) joint specification

$$p(\phi_1, \ldots, \phi_I) \propto \exp\left\{ -\frac{\tau_c}{2} \sum_{i\,adj\,j} (\phi_i - \phi_j)^2 \right\}\ ,$$

i.e., the version of (4.16) corresponding to the usual, adjacency-based CAR model (6.37).

12. The Minnesota Department of Health is charged with investigating the possibility of geographical clustering of the rates for the late detection of colorectal cancer in the state's 87 counties. For each county, the late detection rate is simply the number of regional or distant case detections divided by the total cases observed in that county. Information on several potentially helpful covariates is also available. The most helpful is the county-level estimated proportions of persons who have been screened for colorectal cancer, as estimated from telephone interviews available biannually between 1993 and 1999 as part of the nationwide Behavioral Risk Factor Surveillance System (BRFSS).

(a) Model the log-relative risk using (6.40), fitting the heterogeneity plus clustering (CAR) model to these data. Find and summarize the posterior distributions of α (the proportion of excess variability due to clustering) and β_1 (the screening effect). Does MCMC convergence appear adequate in this problem?

(b) Since the screening covariate was estimated from the BRFSS survey, we should really account for survey measurement error, since this may be substantial for rural counties having few respondents. Replace the observed covariate x_i in the log-relative risk model (6.40) by T_i, the true (unobserved) rate of colorectal screening in county i. Following Xia and Carlin [1998], we further augment our hierarchical model with

$$T_i \overset{iid}{\sim} N(\mu_0, 1/\lambda) \quad \text{and} \quad x_i \mid T_i \overset{ind}{\sim} N(T_i, 1/\delta)\ ,$$

That is, x_i is acknowledged as an imperfect (albeit unbiased) measure of the true screening rate T_i. Treat the measurement error precision λ and prior precision δ either as known "tuning parameters," or else assign them gamma hyperprior distributions, and recompute the posterior for β_1. Observe and interpret any changes.

(c) A more realistic errors-in-covariates model might assume that the precision of x_i given T_i should be proportional to the survey sample size r_i in each county. Write down (but do not fit) a hierarchical model that would address this problem.

Chapter 7

Spatial misalignment

In this chapter we tackle the problem of spatial misalignment. It is more and more the case that different spatial data layers are collected at different scales. For example, we may have one layer at point level, another at point level but at different locations, yet another for one set of areal units and a last over a different set of areal units. Standard GIS software can routinely create overlays and themes with such layers but this is primarily descriptive. Here we seek a formal inferential framework to deal with such misalignment. As a canonical example, consider an environmental justice setting where we seek to assess whether one group is adversely affected by say, an environmental contaminant, compared with another. So, we might record exposure levels at monitoring stations, we might collect adverse health outcomes at the scale of zip or postcodes, and we might learn about population groups at risk through census data at the census tract scale. How might we assemble these layers to assess inequities?

As a result, two types of problems arise with such data settings. For the first, we seek an analysis of the spatial data at a different scale of spatial resolution than it was originally collected. For the second, we would be interested in a regression setting where we seek to use some set of spatially referenced variables to explain another where the variables have been collected at different spatial scales. For example, we might wish to obtain the spatial distribution of some variable at the county level, even though it was originally collected at the census tract level. We might have a very low-resolution global climate model for weather prediction, and seek a regional model or even more locally (i.e., at higher resolution). For areal unit data, our purpose might be simply to understand the distribution of a variable at a new level of spatial aggregation (the so-called *modifiable areal unit problem*, or MAUP). For data modeled at point level through a spatial process we would envision block averaging (Section 7.1.1.2 below) at different spatial scales, (the so-called *change of support problem*, or COSP), again possibly for connection with another variable observed at a particular scale. For either type of data, our goal in the first case is typically one of spatial *interpolation*, while in the second it is one of spatial *regression*.

In addition to our presentation here, we encourage the reader to look at the review paper by Gotway and Young [2002]. These authors give nice discussions of (as well as both traditional and Bayesian approaches for) the MAUP and COSP, spatial regression, and the *ecological fallacy*. This last term refers to the fact that relationships observed between variables measured at the ecological (aggregate) level may not accurately reflect (and will often overstate) the relationship between these same variables measured at the individual level. Discussion of this problem dates at least to Robinson [1950] [see Wakefield, 2003, 2004, for more modern treatments of this difficult subject]. As a simple graphical illustration, Figure shows the incidence of low birth weight at three spatial scales, county, zipcode, and census tract, for the state of North Carolina. Visually, the nature of the spatial structure changes substantially according to scale.

As in previous sections, we group our discussion according to whether the data is suitably modeled using a spatial process as opposed to a CAR or SAR model. Here the former

DOI: 10.1201/9781003401728-7

• Spatial scale

    • Now: Counties,

       • Next: ZCTA and
         beyond.

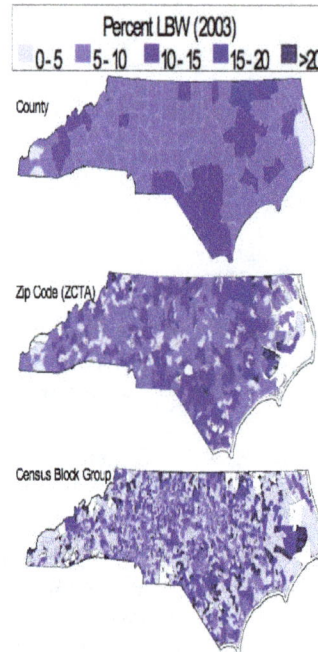

Figure 1. Spatial pattern in percent of low birthweight births in North Carolina.

Figure 7.1 *Incidence of low birth weight at three spatial scales, county, zipcode, and census tract, for the state of North Carolina.*

assumption leads to more general modeling, since point-level data may be naturally aggregated to block level, but the reverse procedure may or may not be possible; e.g., if the areal data are counts or proportions, what would the point-level variables be? However, since block-level summary data are quite frequent in practice (often due to confidentiality restrictions), methods associated with such data are also of importance. We thus consider point-level and block-level modeling.

7.1   Point-level modeling

7.1.1   Gaussian process models

Consider a univariate variable that is modeled through a spatial process. In particular, assume that it is observed either at points in space or over areal units (e.g., counties or zip codes), which we will refer to as *block* data. The *change of support problem* is concerned with inference about the values of the variable at points or blocks different from those at which it has been observed.

7.1.1.1   Motivating data set

A solution to the change of support problem is required in many health science applications, particularly spatial and environmental epidemiology. To illustrate, consider again the data set of ozone levels in the Atlanta, GA metropolitan area, originally reported by Tolbert et al. [2000a]. Ozone measures are available at between 8 and 10 fixed monitoring sites during the 92 summer days (June 1 through August 31) of 1995. Similar to Figure 1.3 (which shows 8-hour maximum ozone levels), Figure 7.2 shows the 1-hour daily maximum ozone

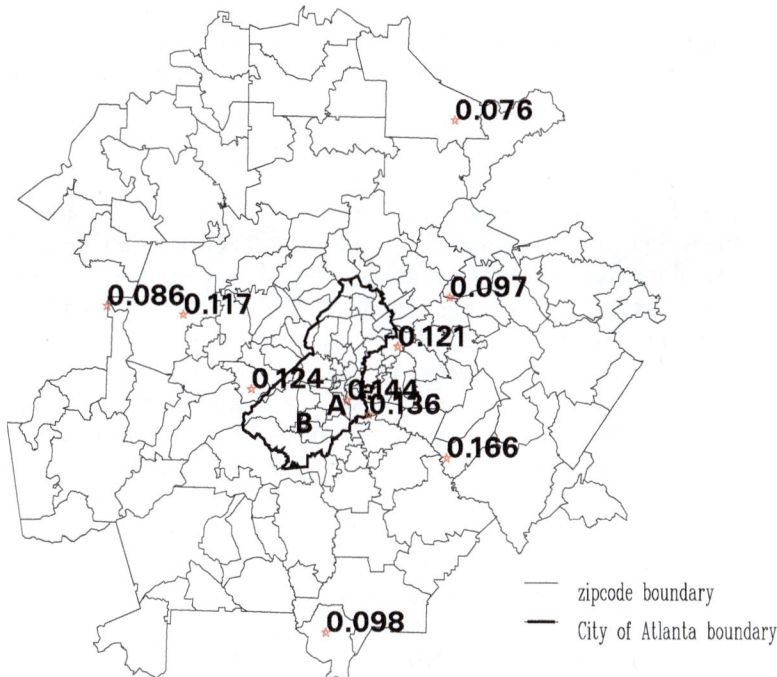

Figure 7.2 *Zip code boundaries in the Atlanta metropolitan area and 1-hour daily maximum ozone levels at the 10 monitoring sites for July 15, 1995.*

measures at the 10 monitoring sites on July 15, 1995, along with the boundaries of the 162 zip codes in the Atlanta metropolitan area. Here we might be interested in predicting the ozone level at different points on the map (say, the two points marked **A** and **B**, which lie on opposite sides of a single city zip), or the average ozone level over a particular zip (say, one of the 36 zips falling within the city of Atlanta, the collection of which are encircled by the dark boundary on the map). The latter problem is of special interest, since, in this case, relevant health outcome data are available only at the zip level. In particular, for each day and zip, we have the number of pediatric ER visits for asthma, as well as the total number of pediatric ER visits. Thus an investigation of the relationship between ozone exposure and pediatric asthma cannot be undertaken until the mismatch in the support of the two variables is resolved. Situations like this are relatively common in health outcome settings, since personal privacy concerns often limit the access of statisticians to data other than at the areal or block level.

In many earth science and population ecology contexts, presence/absence is typically recorded at essentially point-referenced sites while relevant climate layers are often down-scaled to grid cells at some resolution. A previous study of the Atlanta ozone data by Carlin et al. [1999] realigned the point-level ozone measures to the zip level by using an `ARC/INFO` universal kriging procedure to fit a smooth ozone exposure surface, and subsequently took the kriged value at each zip centroid as the ozone value for that zip. But this approach uses a single centroid value to represent the ozone level in the entire zip, and fails to properly capture variability and spatial association by treating these kriged estimates as observed values.

7.1.1.2   Model assumptions and analytic goals

Consider a realization of a spatial process over a surface D. We would like to formalize averaging the realization over D or over a subset $B \subset D$. Conceptually, we can write this as an integral,

$$Y(B) = |B|^{-1} \int_B Y(\mathbf{s}) d\mathbf{s} \,, \qquad (7.1)$$

where $|B|$ denotes the area of B [see, e.g., Cressie, 1993] and refer to (7.1) as a block average. The integration in (7.1) is not the usual integration of a function. Rather, it is an average over random variables which are a realization of a stochastic process, hence a random or stochastic integral. This raises the question of whether such an integral exists. This is an analogous question to defining a Reimann integral over B where we need $Y(\mathbf{s})$ to be continuous (a.e.) over D. Here, we need the process realization to be continuous (a.e.) over D. Following to the discussion below (6.1), if we remove the pure error term with a spatial process that yields continuous realizations (a.e.) and the mean structure is also continuous (a.e.) then the integral will exist. However, if, as a model, $Y(\mathbf{s})$ includes a pure error term, then the $Y(\mathbf{s})$ surface is everywhere discontinuous and the integral does not exist. Instead, suppose we consider an increasingly finer finite gridding of B (or D) and, for a given grid, we compute the average of the process values over the centroids of the grid cells. The limit of this average as grid cell size tends to 0 can be well defined. We will take this as the definition of the block average over B but, for convenience of notation, we will continue to write it as an integral in the form (7.1). Further, $Y(B)$ is a conceptual object; we can never observe it. We can only observe approximations to it in the form of a grid cell approximation or, as we do below, with a Monte Carlo approximation. As a random variable, we can obtain the distributional properties of $Y(B)$, as we do below.

We note that the assumption of an underlying spatial process is only appropriate for block data which can be sensibly viewed as an averaging over continuous-valued point data; examples of this would include rainfall, pollutant level, temperature, and elevation.

It would be inappropriate for most proportions. For instance, if $Y(B)$ is envisioned as the proportion of college-educated persons in B, then we would envision the support for $Y(B)$ to be continuous on $[0, 1]$. However, even were we to conceptualize an individual at every point in B, the $Y(\mathbf{s})$ surface would be a binary map and our definition of $Y(B)$ above through a limit on grid size would not yield continuous support for $Y(B)$. Nonetheless, such approximation for the distribution of $Y(B)$ is frequently adopted.

In general, we envision four possibilities. First, starting with point data $Y(\mathbf{s}_1), \ldots, Y(\mathbf{s}_I)$, we seek to predict at new locations, i.e., to infer about $Y(\mathbf{s}_1'), \ldots, Y(\mathbf{s}_K')$ (points to points). Second, starting with point data, we seek to predict at blocks, i.e., to infer about $Y(B_1), \ldots, Y(B_K)$ (points to blocks). Third, starting with block data $Y(B_1), \ldots, Y(B_I)$, we seek to predict at a set of locations, i.e., to infer about $Y(\mathbf{s}_1'), \ldots, Y(\mathbf{s}_K')$ (blocks to points). Finally, starting with block data, we seek to predict new blocks, i.e., to infer about $Y(B_1'), \ldots, Y(B_K')$ (blocks to blocks).

All of this prediction may be collected under the umbrella of kriging, as in Sections 2.5 and 6.1. Our kriging here will be implemented within the Bayesian framework enabling full inference (a posterior predictive distribution for every prediction of interest, joint distributions for all pairs of predictions, etc.) and avoiding asymptotics. We will however use rather noninformative priors so that our results will roughly resemble those of a likelihood analysis.

Inference about blocks through averages as in (7.1) is not only formally attractive but demonstrably preferable to *ad hoc* approaches. One such approach would be to average over the observed $Y(\mathbf{s}_i)$ in B. But this presumes there is at least one observation in any B, and ignores the information about the spatial process in the observations outside of B. Another *ad hoc* approach would be to simply predict the value at some central point of B. But this value has larger variability than (and may be biased for) the block average.

In the next section, we develop the methodology for spatial data at a single time point; the general spatiotemporal case is similar and described in Section 12.2. Example 7.1 applies our approaches to the Atlanta ozone data are pictured in Figure 7.2.

7.1.2 *Methodology for the point-level realignment*

We start with a stationary Gaussian process specification for $Y(\mathbf{s})$ having a continuous mean function $\mu(\mathbf{s}; \boldsymbol{\beta})$ and covariance function $C(\mathbf{s} - \mathbf{s}'; \boldsymbol{\theta}) = \sigma^2 \rho(\mathbf{s} - \mathbf{s}'; \boldsymbol{\phi})$ yielding continuous realizations, so that $\boldsymbol{\theta} = (\sigma^2, \boldsymbol{\phi})^{\mathrm{T}}$. Here μ is a regression model with coefficient vector $\boldsymbol{\beta}$, which we write as $\mu(\mathbf{s}; \boldsymbol{\beta})$, while σ^2 is the process variance and $\boldsymbol{\phi}$ denotes the parameters associated with the stationary correlation function ρ. With point data observed at sites $\mathbf{s}_1, ..., \mathbf{s}_I$, let $\mathbf{Y}_s^{\mathrm{T}} = (Y(\mathbf{s}_1), ..., Y(\mathbf{s}_I))$. Then

$$\mathbf{Y}_s \mid \boldsymbol{\beta}, \boldsymbol{\theta} \sim N(\boldsymbol{\mu}_s(\boldsymbol{\beta}), \sigma^2 H_s(\boldsymbol{\phi})) , \tag{7.2}$$

where $\boldsymbol{\mu}_s(\boldsymbol{\beta})_i = \mu(\mathbf{s}_i; \boldsymbol{\beta})$ and $(H_s(\boldsymbol{\phi}))_{ii'} = \rho(\mathbf{s}_i - \mathbf{s}_{i'}; \boldsymbol{\phi})$.

Given a prior on $\boldsymbol{\beta}, \sigma^2$, and $\boldsymbol{\phi}$, models such as (7.2) are straightforwardly fit using simulation methods as described in Section 6.1, yielding posterior samples $(\boldsymbol{\beta}_g^*, \boldsymbol{\theta}_g^*)$, $g = 1, ..., G$ from $f(\boldsymbol{\beta}, \boldsymbol{\theta} \mid \mathbf{Y}_s)$.

Then prediction at a set of new locations $\mathbf{Y}_{s'}^{\mathrm{T}} = (Y(\mathbf{s}_1'), ..., Y(\mathbf{s}_K'))$ is really just multiple kriging; we require only the predictive distribution,

$$f(\mathbf{Y}_{s'} \mid \mathbf{Y}_s) = \int f(\mathbf{Y}_{s'} \mid \mathbf{Y}_s, \boldsymbol{\beta}, \boldsymbol{\theta}) f(\boldsymbol{\beta}, \boldsymbol{\theta} \mid \mathbf{Y}_s) d\boldsymbol{\beta} d\boldsymbol{\theta} . \tag{7.3}$$

By drawing $\mathbf{Y}_{s',g}^* \sim f(\mathbf{Y}_{s'} \mid \mathbf{Y}_s, \boldsymbol{\beta}_g^*, \boldsymbol{\theta}_g^*)$ we obtain a sample from (7.3) via composition which provides any desired inference about $\mathbf{Y}_{s'}$ and its components.

Under a Gaussian process,

$$f\left(\begin{pmatrix} \mathbf{Y}_s \\ \mathbf{Y}_{s'} \end{pmatrix} \middle| \boldsymbol{\beta}, \boldsymbol{\theta}\right) = N\left(\begin{pmatrix} \boldsymbol{\mu}_s(\boldsymbol{\beta}) \\ \boldsymbol{\mu}_{s'}(\boldsymbol{\beta}) \end{pmatrix}, \sigma^2 \begin{pmatrix} H_s(\boldsymbol{\phi}) & H_{s,s'}(\boldsymbol{\phi}) \\ H_{s,s'}^{\mathrm{T}}(\boldsymbol{\phi}) & H_{s'}(\boldsymbol{\phi}) \end{pmatrix}\right) , \tag{7.4}$$

with entries defined as in (7.2). Hence, $\mathbf{Y}_{s'} \mid \mathbf{Y}_s, \boldsymbol{\beta}, \boldsymbol{\theta}$ is distributed as

$$\begin{aligned} N\big(&\boldsymbol{\mu}_{s'}(\boldsymbol{\beta}) + H_{s,s'}^{\mathrm{T}}(\boldsymbol{\phi}) H_s^{-1}(\boldsymbol{\phi})(\mathbf{Y}_s - \boldsymbol{\mu}_s(\boldsymbol{\beta})) , \\ &\sigma^2 [H_{s'}(\boldsymbol{\phi}) - H_{s,s'}^{\mathrm{T}}(\boldsymbol{\phi}) H_s^{-1}(\boldsymbol{\phi}) H_{s,s'}(\boldsymbol{\phi})]\big) . \end{aligned} \tag{7.5}$$

Sampling from (7.5) requires the inversion of $H_s(\boldsymbol{\phi}_g^*)$, which will already have been done in sampling $\boldsymbol{\phi}_g^*$, and then the square root of the $K \times K$ covariance matrix in (7.5).

For a single set B, it is straightforward to show (see Exercise 6.2) that $Y(B) \sim N(\mu_B(\boldsymbol{\beta}), \sigma_B^2(\boldsymbol{\theta}))$ where $\mu_B(\boldsymbol{\beta}) = \frac{1}{|B|} \int_B \mu(\mathbf{s}, \boldsymbol{\beta}) d\mathbf{s}$ and $\sigma_B^2(\boldsymbol{\theta}) = \sigma^2 \frac{1}{|B|^2} \int_B \int_B \rho(\mathbf{s} - \mathbf{s}'; \boldsymbol{\phi}) d\mathbf{s} d\mathbf{s}'$. Since $\rho < 1$, we immediately see that $\mathrm{Var}(Y(B)) < \mathrm{Var}(Y(\mathbf{s}))$ for any $\mathbf{s} \in B$ (and, in fact, this holds for the average of any finite set of points in B).

Now, turning to prediction for $\mathbf{Y}_B^{\mathrm{T}} = (Y(B_1), ..., Y(B_K))$, the vector of averages over blocks $B_1, ..., B_K$, we again require the predictive distribution, which is now

$$f(\mathbf{Y}_B \mid \mathbf{Y}_s) = \int f(\mathbf{Y}_B \mid \mathbf{Y}_s; \boldsymbol{\beta}, \boldsymbol{\theta}) f(\boldsymbol{\beta}, \boldsymbol{\theta} \mid \mathbf{Y}_s) d\boldsymbol{\beta} d\boldsymbol{\theta} . \tag{7.6}$$

Under a Gaussian process, we now have

$$f\left(\begin{pmatrix} \mathbf{Y}_s \\ \mathbf{Y}_B \end{pmatrix} \middle| \boldsymbol{\beta}, \boldsymbol{\theta}\right) = N\left(\begin{pmatrix} \boldsymbol{\mu}_s(\boldsymbol{\beta}) \\ \boldsymbol{\mu}_B(\boldsymbol{\beta}) \end{pmatrix}, \sigma^2 \begin{pmatrix} H_s(\boldsymbol{\phi}) & H_{s,B}(\boldsymbol{\phi}) \\ H_{s,B}^{\mathrm{T}}(\boldsymbol{\phi}) & H_B(\boldsymbol{\phi}) \end{pmatrix}\right) , \tag{7.7}$$

where

$$(\boldsymbol{\mu}_B(\boldsymbol{\beta}))_k \;=\; \mathrm{E}(Y(B_k) \mid \boldsymbol{\beta}) \;=\; |B_k|^{-1} \int_{B_k} \mu(\mathbf{s}; \boldsymbol{\beta}) d\mathbf{s}\,,$$

$$(H_B(\boldsymbol{\phi}))_{kk'} \;=\; |B_k|^{-1} |B_{k'}|^{-1} \int_{B_k} \int_{B_{k'}} \rho(\mathbf{s} - \mathbf{s}'; \boldsymbol{\phi}) d\mathbf{s}' d\mathbf{s}\,,$$

$$\text{and} \quad (H_{s,B}(\boldsymbol{\phi}))_{ik} \;=\; |B_k|^{-1} \int_{B_k} \rho(\mathbf{s}_i - \mathbf{s}'; \boldsymbol{\phi}) d\mathbf{s}'\,.$$

Analogously to (7.5), $\mathbf{Y}_B \mid \mathbf{Y}_s, \boldsymbol{\beta}, \boldsymbol{\theta}$ is distributed as

$$N\left(\boldsymbol{\mu}_B(\boldsymbol{\beta}) + H_{s,B}^{\mathrm{T}}(\boldsymbol{\phi}) H_s^{-1}(\boldsymbol{\phi})(\mathbf{Y}_s - \boldsymbol{\mu}_s(\boldsymbol{\beta}))\,,\right.$$
$$\left. \sigma^2 \left[H_B(\boldsymbol{\phi}) - H_{s,B}^{\mathrm{T}}(\boldsymbol{\phi}) H_s^{-1}(\boldsymbol{\phi}) H_{s,B}(\boldsymbol{\phi})\right]\right)\,. \tag{7.8}$$

The major difference between (7.5) and (7.8) is that in (7.5), given $(\boldsymbol{\beta}_g^*, \boldsymbol{\theta}_g^*)$, numerical values for all of the entries in $\boldsymbol{\mu}_{s'}(\boldsymbol{\beta})$, $H_{s'}(\boldsymbol{\phi})$, and $H_{s,s'}(\boldsymbol{\phi})$ are immediately obtained. In (7.8) every analogous entry requires an integration as above. Anticipating irregularly shaped B_k's, Riemann approximation to integrate over these regions may be awkward. Instead, noting that each such integration is an expectation with respect to a uniform distribution, we propose Monte Carlo integration. In particular, for each B_k we propose to draw a set of locations $\mathbf{s}_{k,\ell}$, $\ell = 1, 2, ..., L_k$, distributed independently and uniformly over B_k. Here L_k can vary with k to allow for very unequal $|B_k|$.

Hence, we replace $(\boldsymbol{\mu}_B(\boldsymbol{\beta}))_k$, $(H_B(\boldsymbol{\phi}))_{kk'}$, and $(H_{s,B}(\boldsymbol{\phi}))_{ik}$ with

$$(\widehat{\boldsymbol{\mu}}_B(\boldsymbol{\beta}))_k \;=\; L_k^{-1} \sum_{\ell} \mu(\mathbf{s}_{k,\ell}; \boldsymbol{\beta})\,,$$

$$(\widehat{H}_B(\boldsymbol{\phi}))_{kk'} \;=\; L_k^{-1} L_{k'}^{-1} \sum_{\ell} \sum_{\ell'} \rho(\mathbf{s}_{k\ell} - \mathbf{s}_{k'\ell'}; \boldsymbol{\phi})\,, \tag{7.9}$$

$$\text{and} \quad (\widehat{H}_{s,B}(\boldsymbol{\phi}))_{ik} \;=\; L_k^{-1} \sum_{\ell} \rho(\mathbf{s}_i - \mathbf{s}_{k\ell}; \boldsymbol{\phi})\,.$$

In our notation, the "hat" denotes a Monte Carlo integration that can be made arbitrarily accurate and has nothing to do with the data $\mathbf{Y}_s$. Note also that the same set of $\mathbf{s}_{k\ell}$'s can be used for each integration and with each $(\boldsymbol{\beta}_g^*, \boldsymbol{\theta}_g^*)$; we need only obtain this set once. In obvious notation we replace (7.7) with the $(I + K)$-dimensional multivariate normal distribution $\widehat{f}\left((\mathbf{Y}_s, \mathbf{Y}_B)^{\mathrm{T}} \,\middle|\, \boldsymbol{\beta}, \boldsymbol{\theta}\right)$ with entries using (7.9).

As in Section 7.1.1.2, we can not *observe* $Y(B)$ and we can not *sample* $Y(B)$; we can only approximate it. If we define $\widehat{Y}(B_k) = L_k^{-1} \sum_{\ell} Y(\mathbf{s}_{k\ell})$, then $\widehat{Y}(B_k)$ is a Monte Carlo integration for $Y(B_k)$ as given in (7.1). With an obvious definition for $\widehat{\mathbf{Y}}_B$, it is apparent that

$$\widehat{f}((\mathbf{Y}_s, \mathbf{Y}_B)^{\mathrm{T}} \mid \boldsymbol{\beta}, \boldsymbol{\theta}) = f\left((\mathbf{Y}_s, \widehat{\mathbf{Y}}_B)^{\mathrm{T}} \mid \boldsymbol{\beta}, \boldsymbol{\theta}\right) \tag{7.10}$$

where (7.10) is interpreted to mean that the approximate joint distribution of $(\mathbf{Y}_s, \mathbf{Y}_B)$ is the exact joint distribution of $\mathbf{Y}_s, \widehat{\mathbf{Y}}_B$. In practice, we will work with $\widehat{f}$, converting to $\widehat{f}(\mathbf{Y}_B \mid \mathbf{Y}_s, \boldsymbol{\beta}, \boldsymbol{\theta})$ to sample $\mathbf{Y}_B$ rather than sampling the $\widehat{Y}(B_k)$'s through the $Y(\mathbf{s}_{k\ell})$'s. But, evidently, we are sampling $\widehat{\mathbf{Y}}_B$ rather than $\mathbf{Y}_B$.

As a technical point, we might ask when $\widehat{\mathbf{Y}}_B \xrightarrow{P} \mathbf{Y}_B$. An obvious sufficient condition is that realizations of the $Y(\mathbf{s})$ process are almost surely continuous. In the stationary case, Kent [1989] provides sufficient conditions on $C(\mathbf{s} - \mathbf{t}; \boldsymbol{\theta})$ to ensure this. Alternatively, Stein [1999a] defines $Y(\mathbf{s})$ to be a *mean square continuous* if $\lim_{\mathbf{h} \to 0} \mathrm{E}(Y(\mathbf{s} + \mathbf{h}) - Y(\mathbf{s}))^2 = 0$ for all $\mathbf{s}$. But then $Y(\mathbf{s} + \mathbf{h}) \xrightarrow{P} Y(\mathbf{s})$ as $\mathbf{h} \to 0$, which is sufficient to guarantee that $\widehat{\mathbf{Y}}_B \xrightarrow{P} \mathbf{Y}_B$.

Stein notes that if $Y(\mathbf{s})$ is stationary, we only require $C(\cdot; \boldsymbol{\theta})$ continuous at $\mathbf{0}$ for mean square continuity. (See Subsection 3.1.4 and Section 14.2 for further discussion of smoothness of process realizations.)

This leads us to an opportunity to make an important distinction between estimation and prediction in the spatial context. Suppose we assume a constant mean surface over the bounded set B say at height μ. Then we could consider $\hat{Y}(B)$ as an *estimator* of μ as well as a *predictor* of $Y(B)$. We note that $\hat{Y}(B) \to Y(B)$ in probability and, in fact, if the process covariance function is continuous at 0, we have mean square convergence, as we argued above. However, $\hat{Y}(B)$ need not converge in probability to μ. It suffices to look at $\mathrm{Var}(\hat{Y}(B))$ which, for a sample of size L, from (7.6), is $\sigma^2 \frac{1}{L^2} \sum_{\ell} \sum_{\ell'} \rho(\mathbf{s}_\ell - \mathbf{s}_{\ell'})$. In this double sum, the "diagonal" terms contribute σ^2/L and so will tend to 0 as $\ell \to \infty$. However, the remaining $L(L-1)$ terms in the double sum, when divided by L^2 need not tend to 0. For instance, with an isotropic covariance function that reaches 0 only when distance goes to ∞, we will have every entry in the remaining double sum bigger than $\rho(d_{max})$, where d_{max} is the largest pairwise distance between the locations. Therefore, the remaining sum is greater than $\frac{L(L-1)}{L^2} \rho(d_{max})$. And, as $L \to \infty$, this term will not tend to 0 since d_{max} is bounded because B is bounded. The result holds more generally, even with covariance functions having bounded support since, as $L \to \infty$, the proportion of points that will stay within a subset such that the maximum distance over the subset will be less than the upper support bound for the covariance function will also tend to ∞. In summary, we have consistent prediction, but need not have consistency for estimation.

A related block averaging is associated with a binary process. Binary processes are routinely created from continuous processes using indicator functions. In any event, suppose $Z(\mathbf{s})$ is either 1 or 0. For instance, with presence/absence data, we observe an indicator of whether a particular species is present at location $\mathbf{s}$. Now, let $Z(B) = \frac{1}{|B|} \int_B Z(\mathbf{s}) d\mathbf{s}$. How shall we interpret $Z(B)$? If we think of $Z(\mathbf{s})$ as a light bulb that is either on (1) or off (0) at each location, then $Z(B)$ is the proportion of light bulbs that are on in areal unit B. Evidently, while every $Z(\mathbf{s})$ is either 1 or 0, $Z(B)$ will be in $(0,1)$. This reminds us of the foregoing issue of scaling from blocks to points. However, suppose we define the binary variable $U(B)$ as the result of choosing a location at random in B and taking the value of the binary variable at that location. Then, we see that $P(U(B) = 1) = \mathrm{E}(Z(B))$. These issues, in the context of modeling presence/absence data, are considered in detail in Gelfand et al.(2005) an further in Gelfand and Shirota [2019].

Returning to the prediction problem, starting with block data $\mathbf{Y}_B^{\mathrm{T}} = (Y(B_1), \ldots, Y(B_I))$, analogous to (7.2) the likelihood is well-defined as

$$f(\mathbf{Y}_B \mid \boldsymbol{\beta}, \boldsymbol{\theta}) = N(\boldsymbol{\mu}_B(\boldsymbol{\beta}), \sigma^2 H_B(\boldsymbol{\phi})). \qquad (7.11)$$

Hence, given a prior on $\boldsymbol{\beta}$ and $\boldsymbol{\theta}$, the Bayesian model is completely specified. As above, evaluation of the likelihood requires integrations. So, we replace (7.11) with

$$\widehat{f}(\mathbf{Y}_B \mid \boldsymbol{\beta}, \boldsymbol{\theta}) = N(\widehat{\boldsymbol{\mu}}_B(\boldsymbol{\beta}), \sigma^2 \widehat{H}_B(\boldsymbol{\phi})). \qquad (7.12)$$

Simulation-based fitting is now straightforward, as below (7.2), albeit somewhat more time consuming due to the need to calculate $\widehat{\boldsymbol{\mu}}_B(\boldsymbol{\beta})$ and $\widehat{H}_B(\boldsymbol{\phi})$.

To predict for $\mathbf{Y}_{s'}$ we require $f(\mathbf{Y}_{s'} \mid \mathbf{Y}_B)$. As above, we only require $f(\mathbf{Y}_B, \mathbf{Y}_{s'} \mid \boldsymbol{\beta}, \boldsymbol{\theta})$, which has been given in (7.7). Using (7.10) we now obtain $\widehat{f}(\mathbf{Y}_{s'} \mid \mathbf{Y}_B, \boldsymbol{\beta}, \boldsymbol{\theta})$ to sample $\mathbf{Y}_{s'}$. Note that $\widehat{f}$ is used in (7.12) to obtain the posterior samples and again to obtain the predictive samples. Equivalently, the foregoing discussion shows that we can replace $\mathbf{Y}_B$ with $\widehat{\mathbf{Y}}_B$ throughout. To predict for new blocks $B_1', \ldots, B_K'$, let $\mathbf{Y}_{B'}^{\mathrm{T}} = (Y(B_1'), \ldots, Y(B_K'))$. Now we require $f(\mathbf{Y}_{B'} \mid \mathbf{Y}_B)$, which in turn requires $f(\mathbf{Y}_B, \mathbf{Y}_{B'} \mid \boldsymbol{\beta}, \boldsymbol{\theta})$. The approximate distribution $\widehat{f}(\mathbf{Y}_B, \mathbf{Y}_{B'} \mid \boldsymbol{\beta}, \boldsymbol{\theta})$ employs Monte Carlo integrations over the B_k''s as well as the B_i's, and yields $\widehat{f}(\mathbf{Y}_{B'} \mid \mathbf{Y}_B, \boldsymbol{\beta}, \boldsymbol{\theta})$ to sample $\mathbf{Y}_{B'}$. Again $\widehat{f}$ is used to obtain both the posterior and predictive samples.

Note that in all four prediction cases, we can confine ourselves to an $(I+K)$-dimensional multivariate normal. Moreover, we have only an $I \times I$ matrix to invert repeatedly in the model fitting and a $K \times K$ matrix whose square root is required for the predictive sampling.

For the modifiable areal unit problem (i.e., prediction at new blocks using data for a given set of blocks), suppose we take as our point estimate for a generic new set B_0 the posterior mean,

$$\mathrm{E}(Y(B_0) \mid \mathbf{Y}_B) = E\{\mu(B_0; \boldsymbol{\beta}) + \mathbf{H}_{B,B_0}^{\mathsf{T}}(\boldsymbol{\phi}) H_B^{-1}(\boldsymbol{\phi})(\mathbf{Y}_B - \boldsymbol{\mu}_B(\boldsymbol{\beta})) \mid \mathbf{Y}_B\} \,,$$

where $\mathbf{H}_{B,B_0}(\boldsymbol{\phi})$ is $I \times 1$ with i^{th} entry equal to $\mathrm{Cov}(Y(B_i), Y(B_0) \mid \boldsymbol{\theta})/\sigma^2$. If $\mu(\mathbf{s}; \boldsymbol{\beta}) \equiv \mu_i$ for $\mathbf{s} \in B_i$, then $\mu(B_0; \boldsymbol{\beta}) = |B_0|^{-1} \sum_i |B_i \cap B_0| \mu_i$. But $\mathrm{E}(\mu_i \mid \mathbf{Y}_B) \approx Y(B_i)$ to a first-order approximation, so in this case $\mathrm{E}(Y(B_0) \mid \mathbf{Y}_B) \approx |B_0|^{-1} \sum_i |B_i \cap B_0| Y(B_i)$, the areally weighted estimate.

Example 7.1 We now use the foregoing approach to perform point-point and point-block inference for the Atlanta ozone data pictured in Figure 7.2. Recall that the target points are those marked **A** and **B** on the map, while the target blocks are the 36 Atlanta city zips. The differing block sizes suggest the use of a different L_k for each k in equation (7.9). Conveniently, our GIS (`ARC/INFO`) can generate random points over the whole study area, and then allocate them to each zip. Thus L_k is proportional to the area of the zip, $|B_k|$. Illustratively, our procedure produced 3743 randomly chosen locations distributed over the 36 city zips, and average L_k of nearly 104.

Suppose that log-ozone exposure $Y(\mathbf{s})$ follows a second-order stationary spatial Gaussian process, using the exponential covariance function $C(\mathbf{s}_i - \mathbf{s}_{i'}; \boldsymbol{\theta}) = \sigma^2 e^{-\phi \|\mathbf{s}_i - \mathbf{s}_{i'}\|}$. A preliminary exploratory analysis of our data set suggested that a constant mean function $\mu(\mathbf{s}_i; \boldsymbol{\beta}) = \mu$ is adequate for our data set. We place the customary flat prior on μ, and assume that $\sigma^2 \sim IG(a, b)$ and $\phi \sim G(c, d)$. We chose $a = 3$, $b = 0.5$, $c = 0.03$, and $d = 100$, corresponding to fairly vague priors. We then fit this three-parameter model using an MCMC implementation, sampling μ and σ^2 via Gibbs steps and ϕ through Metropolis-Hastings steps with a $G(3, 1)$ candidate density. Convergence of the sampling chains was virtually immediate. We obtained the following posterior medians and 95% equal-tail credible intervals for the three parameters: for μ, 0.111 and (0.072, 0.167); for σ^2, 1.37 and (1.18, 2.11); and for ϕ, 1.62 and (0.28, 4.13).

Figure 7.3 maps summaries of the posterior samples for the 36 target blocks (city zips) and the 2 target points (A and B); specifically, the posterior medians, $q_{.50}$, upper and lower .025 points, $q_{.975}$ and $q_{.025}$, and the lengths of the 95% equal-tail credible intervals, $q_{.975} - q_{.025}$. The zip-level medians show a clear spatial pattern, with the highest predicted block averages occurring in the southeastern part of the city near the two high observed readings (0.144 and 0.136), and the lower predictions in the north apparently the result of smoothing toward the low observed value in this direction (0.076). The interval lengths reflect spatial variability, with lower values occurring in larger areas (which require more averaging) or in areas nearer to observed monitoring stations (e.g., those near the southeastern, northeastern, and western city boundaries). Finally, note that our approach allows sensibly differing predicted medians for points A and B, with A being higher due to the slope of the fitted surface. Previous centroid-based analyses [like that of Carlin et al., 1999] would instead implausibly impute the same fitted value to both points, since both lie within the same zip. ∎

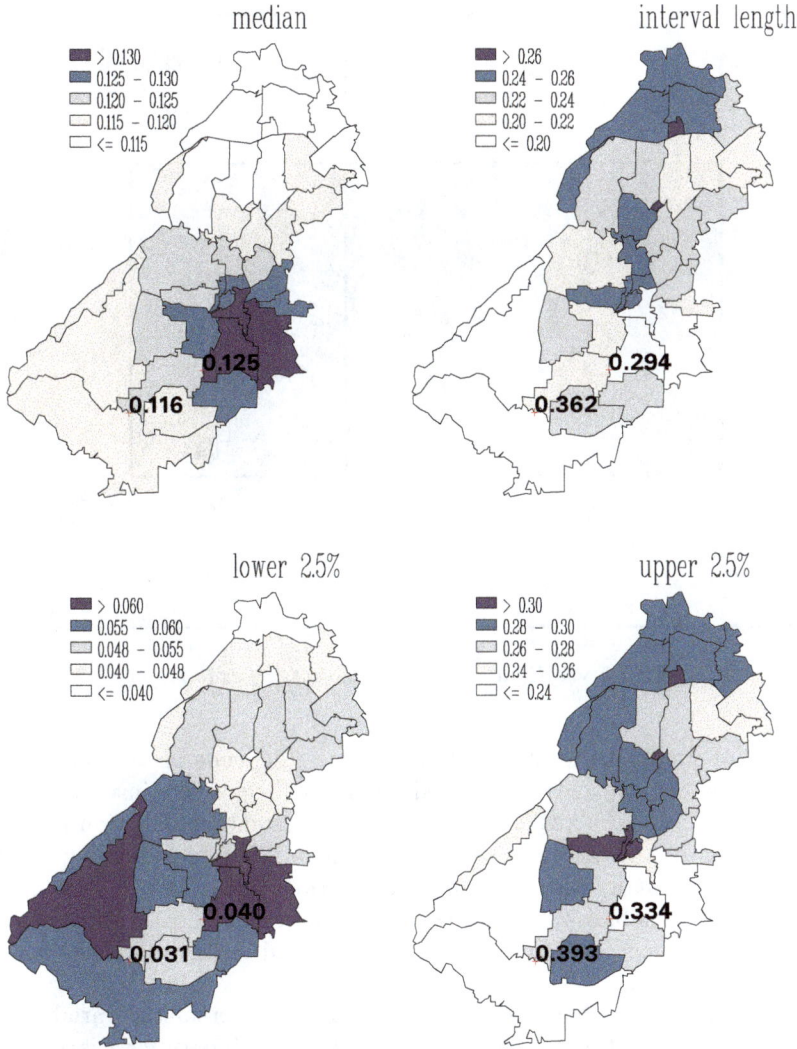

Figure 7.3 *Posterior point-point and point-block summaries, static spatial model, Atlanta ozone data for July 15, 1995.*

7.2 Nested block-level modeling

We now turn to the case of variables available (and easily definable) only as block-level summaries. For example, it might be that disease data are known at the county level, but hypotheses of interest pertain to sociodemographically depressed census tracts. We refer to regions on which data are available as "source" zones and regions for which data are needed as "target" zones.

As mentioned earlier, the block-block interpolation problem has a rich literature and is often referred to as the *modifiable areal unit problem* [see, e..g., Cressie, 1996]. In the case of an *extensive* variable (i.e., one whose value for a block can be viewed as a sum of sub-block values, as in the case of population, disease counts, productivity, or wealth), areal weighting offers a simple imputation strategy. While rather naive, such allocation proportional to area has a long history and is routinely available in GIS software.

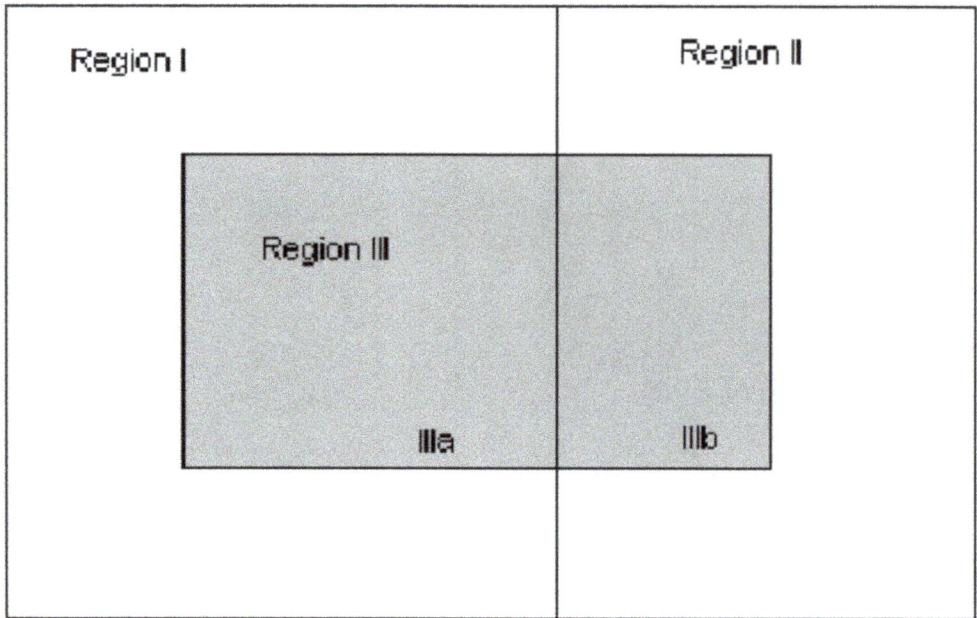

Figure 7.4 *Regional map for motivating example.*

The validity of simple areal interpolation obviously depends on the spatial variable in question being more or less evenly distributed across each region. For instance, Tobler [1979] introduced the so-called *pycnophylactic* approach. He assumed population density to be a continuous function of location, and proposed a simple "volume preserving" (with regard to the observed areal data) estimator of that function. This method is appropriate for continuous outcome variables but is harder to justify for count data, especially counts of human populations, since people do not generally spread out continuously over an areal unit; they tend to cluster.

Flowerdew and Green [1989] presented an approach wherein the variable of interest is count data uses information about the distribution of a binary covariate in the target zone to help estimate the counts. Their approach applies Poisson regression iteratively, using the EM algorithm, to estimate target zone characteristics. While subsequent work [Flowerdew and Green, 1992] extended this EM approach to continuous (typically normally distributed) outcome variables, neither of these papers reflects a fully inferential approach to the population interpolation problem.

In this section we follow Mugglin and Carlin [1998] and focus on the setting where the target zonation of the spatial domain D is a refinement of the source zonation, a situation we term *nested* misalignment. In the data setting we describe below, the source zones are U.S. census tracts, while the target zones (and the zones on which covariate data are available) are U.S. census block groups.

7.2.1 *Methodology for nested block-level realignment*

Consider the diagram in Figure 7.4. Assume that a particular rectangular tract of land is divided into two regions (I and II), and spatial variables (say, disease counts) y_1 and y_2 are known for these regions (the source zones). But suppose that the quantity of interest is Y_3, the unknown corresponding count in Region III (the target zone), which is comprised of subsections (IIIa and IIIb) of Regions I and II.

As already mentioned, a crude way to approach the problem is to assume that disease counts are distributed evenly throughout Regions I and II, and so the number of affected individuals in Region III is just

$$y_1 \left[\frac{area(IIIa)}{area(I)} \right] + y_2 \left[\frac{area(IIIb)}{area(II)} \right] . \tag{7.13}$$

This simple areal interpolation approach is available within many GIS packages. However, (7.13) is based on an assumption that is likely to be unviable, and also offers no associated estimate of uncertainty.

Let us now assume that the entire tract can be partitioned into smaller subsections, where on each subsection we can measure some other variable that is correlated with the disease count for that region. For instance, if we are looking at a particular tract of land, in each subsection we might record whether the land is predominantly rural or urban in character. We do this in the belief that this variable affects the likelihood of disease. Continuous covariates could also be used (say, the median household income in the subsection). Note that the subsections could arise simply as a refinement of the original scale of aggregation (e.g., if disease counts were available only by census tract, but covariate information arose at the census block group level), or as the result of overlaying a completely new set of boundaries (say, a zip code map) onto our original map. The statistical model is easier to formulate in the former case, but the latter case is, of course, more general and is the one motivated by modern GIS technology (to which we return in Section 7.3).

To facilitate our discussion in the former case, we consider a data set on the incidence of leukemia in Tompkins County, New York, that was originally presented and analyzed by Waller et al. [1994] and available on the web at https://www.stats.ox.ac.uk/pub/datasets/csb/. As seen in Figure 7.5, Tompkins County, located in west-central New York state, is roughly centered around the city of Ithaca, NY. The county is divided into 23 census tracts, with each tract further subdivided into between 1 and 5 block groups, for a total of 51 such subregions. We have leukemia counts available at the tract level, and we wish to predict them at the block group level with the help of population counts and covariate information available on this more refined scale. In this illustration, the two covariates we consider are whether the block group is coded as "rural" or "urban," and whether or not the block group centroid is located within 2 kilometers of a hazardous chemical waste site. There are two waste sites in the county, one in the northeast corner and the other in downtown Ithaca, near the county's center. (For this data set, we in fact have leukemia counts at the block group level, but we use only the tract totals in the model-fitting process, reserving the refined information to assess the accuracy of our results.) In this example, the unequal population totals in the block groups will play the weighting role that unequal areas would have played in (7.13).

Figure 7.6 shows a census tract-level disease map produced by the GIS `MapInfo`. The data record the block group-level population counts n_{ij} and covariate values u_{ij} and w_{ij}, where u_{ij} is 1 if block group j of census tract i is classified as urban, 0 if rural, and w_{ij} is 1 if the block group centroid is within 2 km of a waste site, 0 if not. Typical of GIS software, `MapInfo` permits the allocation of the census tract totals to the various block groups proportional to block group area or population. We use our hierarchical Bayesian method to incorporate the covariate information, as well as obtain variance estimates to accompany the block group-level point estimates.

As in our earlier disease mapping discussion (Subsection 6.8.1), we introduce a first-stage Poisson model for the disease counts,

$$Y_{ij} \mid m_{k(i,j)} \stackrel{ind}{\sim} Po(E_{ij} m_{k(i,j)}), \ i = 1, \dots, I, \ j = 1, \dots, J_i \,,$$

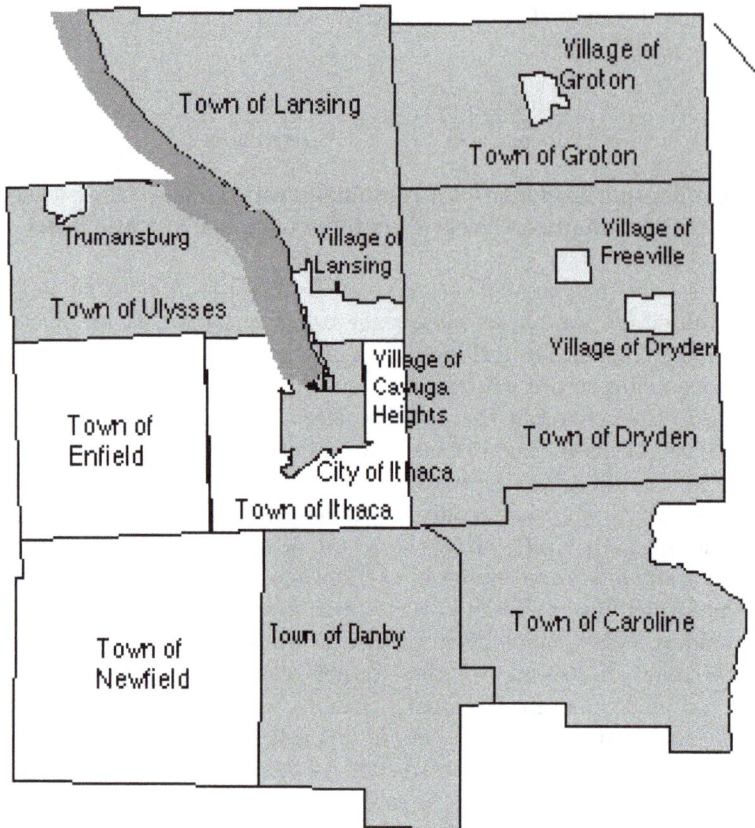

Figure 7.5 *Map of Tompkins County, NY.*

where $I = 23$, J_i varies from 1 to 5, Y_{ij} is the disease count in block group j of census tract i, and E_{ij} is the corresponding "expected" disease count, computed as $E_{ij} = n_{ij}\lambda$ where n_{ij} is the population count in the cell and λ is the overall probability of contracting the disease. This "background" probability could be estimated from our data; here we take $\lambda = 5.597 \times 10^{-4}$, the crude leukemia rate for the 8-county region studied by Waller et al. [1994] an area that includes Tompkins County. Hence, $m_{k(i,j)}$ is the relative risk of contracting leukemia in the block group (i, j), and $k = k(i, j) = 1, 2, 3$, or 4 depending on the covariate status of the block group. Specifically, we let

$$k(i,j) = \begin{cases} 1, & \text{if } (i,j) \text{ is rural, not near a waste site} \\ 2, & \text{if } (i,j) \text{ is urban, not near a waste site} \\ 3, & \text{if } (i,j) \text{ is rural, near a waste site} \\ 4, & \text{if } (i,j) \text{ is urban, near a waste site} \end{cases} .$$

Defining $\mathbf{m} = (m_1, m_2, m_3, m_4)$ and again adopting independent and minimally informative gamma priors for these four parameters, we seek estimates of $p(m_k \mid \mathbf{y})$, where $\mathbf{y} = (y_{1.}, \ldots, y_{I.})$, and $y_{i.} = \sum_{j=1}^{J_i} y_{ij}$, the census tract disease count totals. We also wish to obtain block group-specific mean and variance estimates $\mathrm{E}[Y_{ij} \mid \mathbf{y}]$ and $\mathrm{Var}[Y_{ij} \mid \mathbf{y}]$, to be plotted in a disease map at the block group (rather than census tract) level. Finally, we may

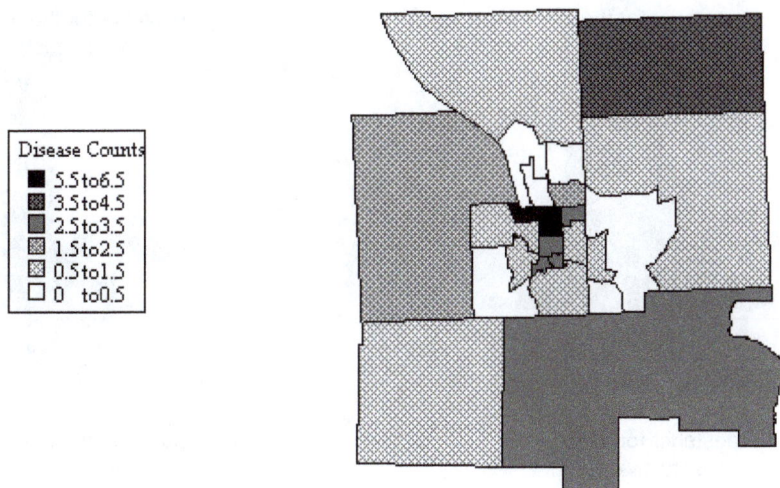

Figure 7.6 *GIS map of disease counts by census tract, Tompkins County, NY.*

also wish to estimate the distribution of the total disease count in some conglomeration of block groups (say, corresponding to some village or city).

By the conditional independence of the block group counts we have $Y_{i.} \mid \mathbf{m} \overset{ind}{\sim} Po(\sum_{k=1}^{4} s_k m_k)$, $i = 1, \ldots, I$, where $s_k = \sum_{j:k(i,j)=k} E_{ij}$, the sum of the expected cases in block groups j of region i corresponding to covariate pattern k, $k = 1, \ldots, 4$. The likelihood $L(\mathbf{m}; \mathbf{y})$ is then the product of the resulting $I = 23$ Poisson kernels. After multiplying this by the prior distribution term $\prod_{k=1}^{4} p(m_k)$, we can obtain forms proportional to the four full conditional distributions $p(m_k \mid m_{l \neq k}, \mathbf{y})$, and sample these sequentially via univariate Metropolis steps.

Once again it is helpful to reparameterize to $\delta_k = \log(m_k)$, $k = 1, \ldots, 4$, and perform the Metropolis sampling on the log scale. We specify reasonably vague $Gamma(a, b)$ priors for the m_k by taking $a = 2$ and $b = 10$ (similar results were obtained with even less informative Gamma priors unless a was quite close to 0, in which case convergence was unacceptably poor). For this "base prior," convergence obtains after 200 iterations, and the remaining 1800 iterations in 5 parallel MCMC chains are retained as posterior samples from $p(\mathbf{m} \mid \mathbf{y})$.

A second reparametrization aids in interpreting our results. Suppose we write

$$\delta_{k(i,j)} = \theta_0 + \theta_1 u_{ij} + \theta_2 w_{ij} + \theta_3 u_{ij} w_{ij} , \qquad (7.14)$$

so that θ_0 is an intercept, θ_1 is the effect of living in an urban area, θ_2 is the effect of living near a waste site, and θ_3 is the urban/waste site interaction. This reparametrization expresses the log-relative risk of disease as a linear model, a common approach in spatial disease mapping [Besag et al., 1991, Waller et al., 1997] . A simple 1-1 transformation converts our $(m_1^{(g)}, m_2^{(g)}, m_3^{(g)}, m_4^{(g)})$ samples to $(\theta_0^{(g)}, \theta_1^{(g)}, \theta_2^{(g)}, \theta_3^{(g)})$ samples on the new scale, which in turn allows direct investigation of the main effects of urban area and waste site proximity, as well as the effect of interaction between these two. Figure 7.7 shows the histograms of the posterior samples for θ_i, $i = 0, 1, 2, 3$. We note that θ_0, θ_1, and θ_3 are not significantly different from 0 as judged by the 95% BCI, while θ_2 is "marginally significant" (in a Bayesian sense) at this level. This suggests a moderately harmful effect of residing within 2 km of a waste site, but no effect of merely residing in an urban area (in this case, the city of Ithaca). The preponderance of negative $\theta_3^{(g)}$ samples is somewhat surprising; we

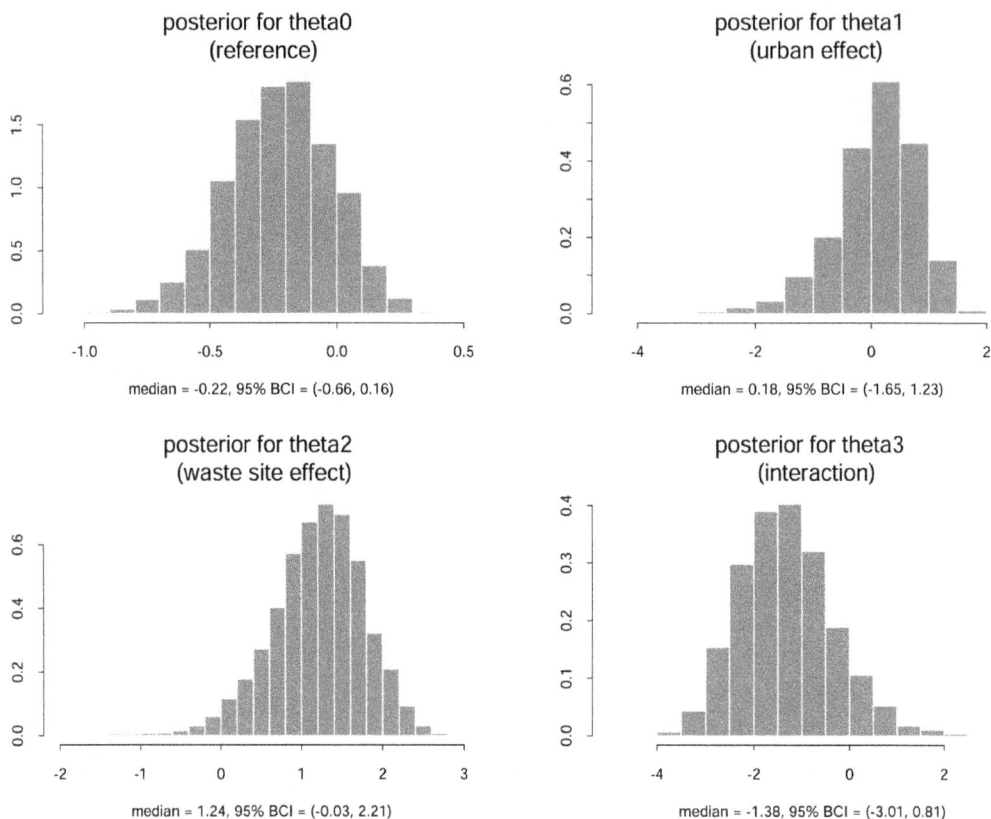

Figure 7.7 *Posterior histograms of sampled log-relative risk parameters, Tompkins County, NY, data set.*

might have expected living near an urban waste site to be associated with an increased (rather than decreased) risk of leukemia. This is apparently the result of the high leukemia rate in a few rural block groups *not* near a waste site (block groups 1 and 2 of tract 7, and block group 2 of tract 20), forcing θ_3 to adjust for the relatively lower overall rate near the Ithaca waste site.

7.2.2 Individual block group estimation

To create the block group-level estimated disease map, for those census tracts having $J_i > 1$, we obtain a conditional binomial distribution for Y_{ij} given the parameters $\mathbf{m}$ and the census tract totals $\mathbf{y}$, so that

$$\mathrm{E}(Y_{ij} \mid \mathbf{y}) = \mathrm{E}[\mathrm{E}(Y_{ij} \mid \mathbf{m}, \mathbf{y})] \approx \frac{y_{i\cdot}}{G} \sum_{g=1}^{G} p_{ij}^{(g)} , \qquad (7.15)$$

where p_{ij} is the appropriate binomial probability arising from conditioning a Poisson random variable on the sum of itself and a second, independent Poisson variable. For example, for $p_{11}^{(g)}$ we have

$$p_{11}^{(g)} = \frac{1617 m_1^{(g)}}{(1617 + 702) m_1^{(g)} + (1526 + 1368) m_3^{(g)}} ,$$

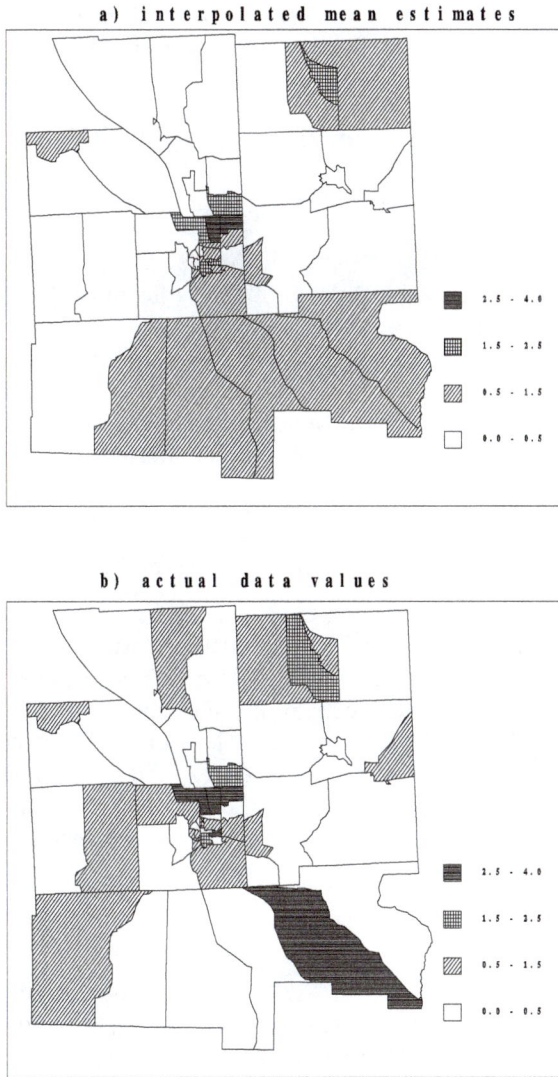

Figure 7.8 *GIS maps of interpolated (a) and actual (b) block group disease counts, Tompkins County, NY.*

as determined by the covariate patterns in the first four rows of the data set. Note that when $J_i = 1$ the block group total equals the known census tract total, hence no estimation is necessary.

The resulting collection of estimated block group means $E(Y_{ij} \mid \mathbf{y})$ are included in the data set on our webpage, along with the actual case counts y_{ij}. (The occasional noninteger values of y_{ij} in the data are not errors, but arise from a few cases in which the precise block group of occurrence is unknown, resulting in fractional counts being allocated to several block groups.) Note that, like other interpolation methods, the sum of the estimated cases in each census tract is the same as the corresponding sum for the actual case counts. The GIS maps of the $E(Y_{ij} \mid \mathbf{y})$ and the actual y_{ij} shown in Figure 7.8 reveal two pockets of elevated disease counts (in the villages of Cayuga Heights and Groton).

To get an idea of the variability inherent in the posterior surface, we might consider mapping the estimated posterior variances of our interpolated counts. Since the block group-level variances do not involve aggregation across census tracts, these variances may be easily estimated as $\mathrm{Var}(Y_{ij} \mid \mathbf{y}) = \mathrm{E}(Y_{ij}^2 \mid \mathbf{y}) - [\mathrm{E}(Y_{ij} \mid \mathbf{y})]^2$, where the $\mathrm{E}(Y_{ij} \mid \mathbf{y})$ are the estimated means (already calculated), and

$$\begin{aligned}
\mathrm{E}(Y_{ij}^2 \mid \mathbf{y}) &= \mathrm{E}[\mathrm{E}(Y_{ij}^2 \mid \mathbf{m}, \mathbf{y})] = \mathrm{E}[y_{i.}p_{ij}(1-p_{ij}) + y_{i.}^2 (p_{ij})^2] \\
&\approx \frac{1}{G} \sum_{g=1}^{G} \left[y_{i.}p_{ij}^{(g)}(1 - p_{ij}^{(g)}) + y_{i.}^2 (p_{ij}^{(g)})^2 \right] ,
\end{aligned} \tag{7.16}$$

where p_{ij} is again the appropriate binomial probability for block group (i,j) [see Mugglin and Carlin, 1998, for more details].

We remark that most of the census tracts are composed of homogeneous block groups (e.g., all rural with no waste site nearby); in these instances the resulting binomial probability for each block group is free of $\mathbf{m}$. In such cases, posterior means and variances are readily available without any need for mixing over the Metropolis samples, as in equations (7.15) and (7.16).

7.2.3   Aggregate estimation: Block groups near the Ithaca, NY waste site

In order to assess the number of leukemia cases we expect in those block groups within 2 km of the Ithaca waste site, we can sample the predictive distributions for these blocks, sum the results, and draw a histogram of these sums. Twelve block groups in five census tracts fall within these 2-km radii: all of the block groups in census tracts 11, 12, and 13, plus two of the three (block groups 2 and 3) in tract 6 and three of the four (block groups 2, 3, and 4) in tract 10. Since the totals in census tracts 11, 12, and 13 are known to our analysis, we need only samples from two binomial distributions, one each for the conglomerations of near-waste site block groups within tracts 6 and 10. Defining the sum over the twelve block groups as Z, we have

$$Z^{(g)} = Y_{6,(2,3)}^{(g)} + Y_{10,(2,3,4)}^{(g)} + y_{11,.} + y_{12,.} + y_{13,.} \ .$$

A histogram of these values is shown in Figure 7.9. The estimated median value of 10 happens to be exactly equal to the true value of 10 cases in this area. The sample mean, 9.43, is also an excellent estimate. Note that the minimum and maximum values in Figure 7.9, $Z = 7$ and $Z = 11$, are imposed by the data structure: there must be at least as many cases as the total is known to have occurred in census tracts 11, 12, and 13 (which is 7), and there can be no more than the total number known to have occurred in tracts 6, 10, 11, 12, and 13 (which is 11).

Finally, we may again compare our results to those produced by a GIS under either area-based or population-based interpolation. The former produces a mean estimate of 9.28, while the latter gives 9.59. These are close to the Bayesian mean 9.43, but neither approach produces an associated confidence interval, much less a full graphical display of the sort given in Figure 7.9.

7.3   Nonnested block-level modeling

The approach of the previous section [see also Mugglin and Carlin, 1998, Mugglin et al., 1999] offered a hierarchical Bayesian method for interpolation and smoothing of Poisson responses with covariates in the nested case. In the remainder of this section, we develop a framework for hierarchical Bayesian interpolation, estimation, and spatial smoothing over *nonnested* misaligned data grids. In Subsection 7.3.1 we summarize a data set collected in response to possible contamination resulting from the former Feed Materials Production Center (FMPC) in southwestern Ohio with the foregoing analytic goals. In Subsection 7.3.2

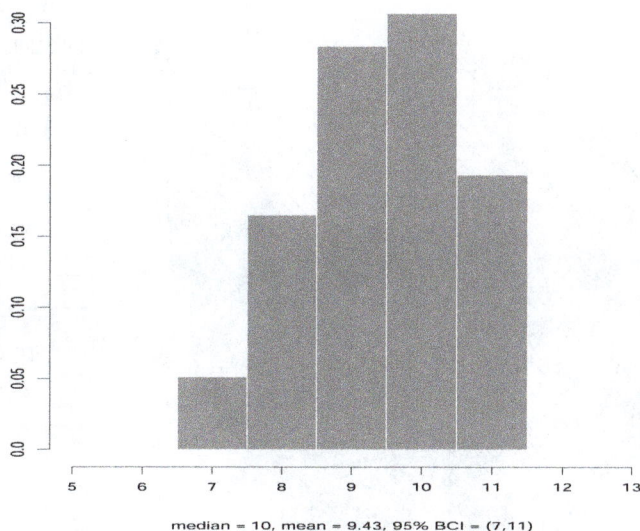

median = 10, mean = 9.43, 95% BCI = (7,11)

Figure 7.9 *Histogram of sampled disease counts, total of all block groups having centroid within 2 km of the Ithaca, NY, waste site.*

we develop the theory of our modeling approach in a general framework, as well as our MCMC approach and a particular challenge that arises in its implementation for the FMPC data. Finally in Example 7.2 we set forth the conclusions resulting from our analysis of the FMPC data.

7.3.1   Motivating data set

Risk-based decision making is often used for prioritizing cleanup efforts at U.S. Superfund sites. Often these decisions will be based on estimates of the past, present, and future potential health impacts. These impact assessments usually rely on the estimation of the number of outcomes, and the accuracy of these estimates will depend heavily on the ability to estimate the number of individuals at risk. Our motivating data set is connected with just this sort of risk assessment.

In the years 1951–1988 near the town of Ross in southwestern Ohio, the former Feed Materials Production Center (FMPC) processed uranium for weapons production. Draft results of the Fernald Dosimetry Reconstruction Project, sponsored by the Centers for Disease Control and Prevention (CDC), indicated that during production years the FMPC released radioactive materials (primarily radon and its decay products and, to a lesser extent, uranium and thorium) from the site. Although radioactive liquid wastes were released, the primary exposure to residents of the surrounding community resulted from breathing radon decay products. The potential for increased risk of lung cancer was thus the focus of intense local public interest and subsequent public health studies [see Devine et al., 1998].

Estimating the number of adverse health outcomes in the population (or in subsets thereof) requires estimation of the number of individuals at risk. Population counts, broken down by age and sex, are available from the U.S. Census Bureau according to federal census block groups, while the areas of exposure interest are dictated by both direction and distance from the plant. Rogers and Killough [1997] construct an exposure "windrose," which consists of 10 concentric circular bands at 1-kilometer radial increments divided into 16 compass sectors (N, NNW, NW, WNW, W, etc.). Through the overlay of such a windrose onto U.S. Geological Survey (USGS) maps, provide counts of the number of "structures"

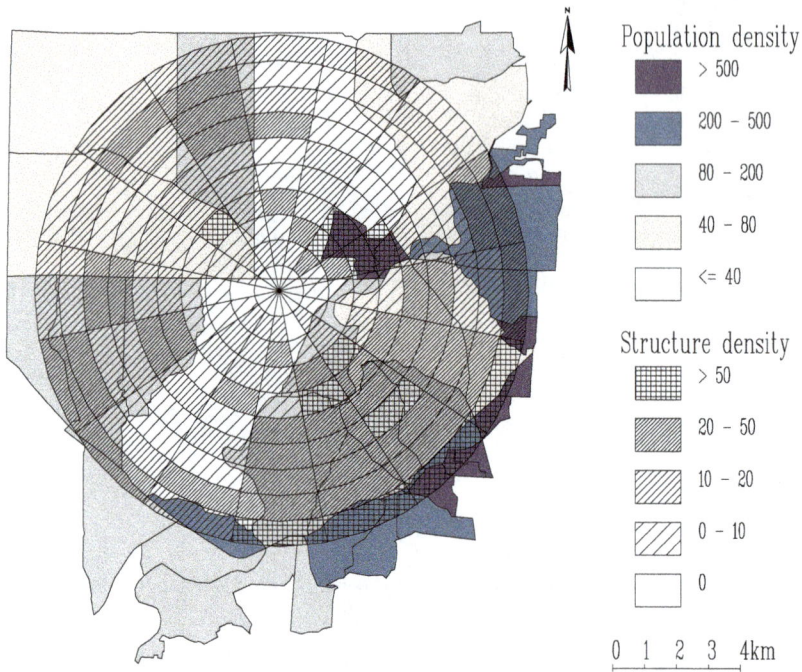

Figure 7.10 *Census block groups and 10-km windrose near the FMPC site, with 1990 population density by block group and 1980 structure density by cell (both in counts per km^2).*

(residential buildings, office buildings, industrial building complexes, warehouses, barns, and garages) within each subdivision (*cell*) of the windrose.

Figure 7.10 shows the windrose centered at the FMPC. We assign numbers to the windrose cells, with 1 to 10 indexing the cells starting at the plant and running due north, then 11 to 20 running from the plant to the north-northwest, and so on. Structure counts are known for each cell; the hatching pattern in the figure indicates the areal density (structures per square kilometer) in each cell.

Also shown in Figure 7.10 are the boundaries of 39 Census Bureau block groups, for which 1990 population counts are known. These are the source zones for our interpolation problem. Shading intensity indicates the population density (persons per square kilometer) for each block group. The intersection of the two (nonnested) zonation systems results in 389 regions we call *atoms,* which can be aggregated appropriately to form either cells or block groups.

The plant was in operation for 38 years, raising concern about the potential health risks it caused. Efforts to assess the impact of the FMPC on cancer morbidity and mortality require the analysis of this misaligned data set; in particular, it is necessary to interpolate gender- and age group-specific population counts to the windrose exposure cells. These numbers of persons at risk could then be combined with cell-specific dose estimates obtained by Killough et al. [1996] et al. and estimates of the cancer risk per unit dose to obtain expected numbers of excess cancer cases by cell.

In fact, such an expected death calculation was made by Devine et al. [1998] using traditional life table methods operating on the Rogers and Killough [1997] cell-level population estimates (which were in turn derived simply as proportional to the structure counts). However, these estimates were only for the total population in each cell; sex- and age group-specific counts were obtained by "breaking out" the totals into subcategories using a standard table (i.e., the *same* table in each cell, regardless of its true demographic

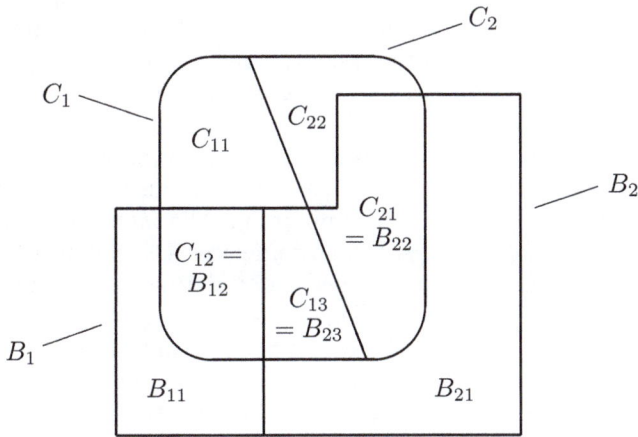

Figure 7.11 *Illustrative representation of areal data misalignment.*

makeup). In addition, the uncertainty associated with the cell-specific population estimates was quantified in an ad hoc way.

7.3.2 Methodology for nonnested block-level realignment

We confine our model development to the case of two misaligned spatial grids. Given this development, the extension to more than two grids will be conceptually apparent. The additional computational complexity and bookkeeping detail will also be evident.

Let the first grid have regions indexed by $i = 1, ..., I$, denoted by B_i, and let $S_B = \bigcup_i B_i$. Similarly, for the second grid we have regions C_j, $j = 1, ..., J$ with $S_C = \bigcup_j C_j$. In some applications $S_B = S_C$, i.e., the B-cells and the C-cells offer different partitions of a common region. Nested misalignment (e.g., where each C_j is contained entirely in one and only one B_i) is evidently a special case. Another possibility is that one data grid contains the other; say, $S_B \subset S_C$. In this case, there will exist some C cells for which a portion lies outside of S_B. In the most general case, there is no containment and there will exist B-cells for which a portion lies outside of S_C and C-cells for which a portion lies outside of S_B. Figure 7.11 illustrates this most general situation.

Atoms are created by intersecting the two grids. For a given B_i, each C-cell which intersects B_i creates an atom (which possibly could be a union of disjoint regions). There may also be a portion of B_i which does not intersect with any C_j. We refer to this portion as the *edge* atom associated with B_i, i.e., a B-edge atom. In Figure 7.11, atoms B_{11} and B_{21} are B-edge atoms. Similarly, for a given C_j, each B-cell which intersects with C_j creates an atom, and we analogously determine C-edge atoms (atoms C_{11} and C_{22} in Figure 7.11). It is crucial to note that each non-edge atom can be referenced relative to an appropriate B-cell, say B_i, and denoted as B_{ik}. It also can be referenced relative to an appropriate C cell, say C_j, and denoted by $C_{j\ell}$. Hence, there is a one-to-one mapping within $S_B \bigcap S_C$ between the set of ik's and the set of $j\ell$'s, as shown in Figure 7.11 (which also illustrates our convention of indexing atoms by area, in descending order). Formally we can define the function c on nonedge B-atoms such that $c(B_{ik}) = C_{j\ell}$, and the *inverse* function b on C-atoms such that $b(C_{j\ell}) = B_{ik}$. For computational purposes we suggest the creation of "look-up" tables to specify these functions. (Note that the possible presence of both types of edge cell precludes a single "ij" atom numbering system since such a system could index cells on either S_B or S_C, but not their union.)

Without loss of generality we refer to the first grid as the *response* grid, that is, at each B_i we observe a response Y_i. We seek to explain Y_i using a variety of covariates. Some of these covariates may, in fact, be observed on the response grid; we denote the value of this vector for B_i by $\mathbf{W}_i$. But also, some covariates are observed on the second or *explanatory* grid. We denote the value of this vector for C_j by $\mathbf{X}_j$.

We seek to explain the observed Y's through both $\mathbf{X}$ and $\mathbf{W}$. The misalignment between the $\mathbf{X}$'s and Y's is the obstacle to standard regression methods. What levels of $\mathbf{X}$ should be assigned to Y_i? We propose a fully model-based approach in the case where the Y's and X's are aggregated measurements. As always, the advantage of a model-based approach implemented within a Bayesian framework is full inference both with regard to estimation of model parameters and prediction using the model.

The assumption that the Y's are aggregated measurements means Y_i can be envisioned as $\sum_k Y_{ik}$, where the Y_{ik} are unobserved or latent and the summation is over all atoms (including perhaps an edge atom) associated with B_i. To simplify, we assume that the X's are also scalar aggregated measurements, i.e., $X_j = \sum_\ell X_{j\ell}$ where the summation is over all atoms associated with C_j. As for the $\mathbf{W}$'s, we assume that each component is either an aggregated measurement or an *inheritable* measurement. For component r, in the former case $W_i^{(r)} = \sum_k W_{ik}^{(r)}$ as with Y_i; in the latter case $W_{ik}^{(r)} = W_i^{(r)}$.

In addition to (or perhaps in place of) the $\mathbf{W}_i$ we will introduce B-cell random effects μ_i, $i = 1, ..., I$. These effects are employed to capture spatial association among the Y_i's. The μ_i can be given a spatial prior specification. A Markov random field form Besag [1974], Bernardinelli and Montomoli [1992], as described below, is convenient. Similarly, we will introduce C-cell random effects ω_j, $j = 1, ..., J$ to capture spatial association among the X_j's. It is assumed that the latent Y_{ik} inherits the effect μ_i and that the latent $X_{j\ell}$ inherit the effect ω_j.

For aggregated measurements that are counts, we assume the latent variables are conditionally independent Poissons. As a result, the observed measurements are Poissons as well and the conditional distribution of the latent variables given the observed is a product multinomial. We note that it is not required that the Y's be counted data. For instance, with aggregated measurements that are continuous, a convenient distributional assumption is conditionally independent gammas, in which case the latent variables would be rescaled to product Dirichlet. An alternative choice is normal, whereupon the latent variables would have a distribution that is a product of conditional multivariate normals. In this section we detail the Poisson case.

As mentioned above, the area naturally plays an important role in the allocation of spatial measurements. Letting $|A|$ denote the area of region A, if we apply the standard assumption of allocation proportional to the area to the $X_{j\ell}$ in a stochastic fashion, we would obtain

$$X_{j\ell} \mid \omega_j \sim Po(e^{\omega_j}|C_{j\ell}|) , \qquad (7.17)$$

assumed independent for $\ell = 1, 2, ..., L_j$. Then $X_j \mid \omega_j \sim Po(e^{\omega_j}|C_j|)$ and $(X_{j1}, X_{j2}, ..., X_{j,L_j} \mid X_j, \omega_j) \sim Mult(X_j; q_{j1}, ..., q_{j,L_j})$ where $q_{j\ell} = |C_{j\ell}|/|C_j|$.

Such strictly area-based modeling cannot be applied to the Y_{ik}'s since it fails to connect the Y's with the X's (as well as the $\mathbf{W}$'s). To do so we again begin at the atom level. For nonedge atoms, we use the previously mentioned look-up table to find the $X_{j\ell}$ to associate with a given Y_{ik}. It is convenient to denote this $X_{j\ell}$ as X_{ik}'. Ignoring the $\mathbf{W}_i$ for the moment, we assume

$$Y_{ik} \mid \mu_i, \theta_{ik} \sim Po\left(e^{\mu_i}|B_{ik}| h(X_{ik}'/|B_{ik}| ; \theta_{ik})\right) , \qquad (7.18)$$

independent for $k = 1, \ldots, K_i$. Here h is a preselected parametric function, the part of the model specification that adjusts an expected proportional-to-area allocation according to X_{ik}'. Since (7.17) models expectation for $X_{j\ell}$ proportional to $|C_{j\ell}|$, it is natural to use the *standardized* form $X_{ik}'/|B_{ik}|$ in (7.18). Particular choices of h include $h(z ; \theta_{ik}) = z$ yielding

$Y_{ik} \mid \mu_i \sim Po(e^{\mu_i} X'_{ik})$, which would be appropriate if we choose not to use $|B_{ik}|$ explicitly in modeling $E(Y_{ik})$. In our FMPC implementation, we actually select $h(z \; ; \theta_{ik}) = z + \theta_{ik}$ where $\theta_{ik} = \theta/(K_i|B_{ik}|)$ and $\theta > 0$; see equation (7.23) below and the associated discussion.

If B_i has no associated edge atom, then

$$Y_i \mid \mu_i, \boldsymbol{\theta}, \{X_{j\ell}\} \sim Po\left(e^{\mu_i} \sum_k |B_{ik}| \, h(X'_{ik}/|B_{ik}| \; ; \theta_{ik})\right) . \qquad (7.19)$$

If B_i has an edge atom, say B_{iE}, since there is no corresponding $C_{j\ell}$, there is no corresponding X'_{iE}. Hence, we introduce a latent X'_{iE} whose distribution is determined by the nonedge atoms that are neighbors of B_{iE}. Paralleling equation (7.17), we model X'_{iE} as

$$X'_{iE} \mid \omega_i^* \sim Po(e^{\omega_i^*}|B_{iE}|) , \qquad (7.20)$$

thus adding a new set of random effects $\{\omega_i^*\}$ to the existing set $\{\omega_j\}$. These two sets together are assumed to have a single CAR specification. An alternative is to model $X'_{iE} \sim Po\left(|B_{iE}| \left(\sum_{N(B_{iE})} X'_t / \sum_{N(B_{iE})} |B_t|\right)\right)$, where $N(B_{iE})$ is the set of neighbors of B_{iE} and t indexes this set. Effectively, we multiply $|B_{iE}|$ by the overall count per unit area in the neighboring nonedge atoms. While this model is somewhat more data-dependent than the (more model-dependent) one given in (7.20), we remark that it can actually lead to better MCMC convergence due to the improved identifiability in its parameter space: the spatial similarity of the structures in the edge zones is being modeled directly, rather than indirectly via the similarity of the ω_i^* and the ω_j.

Now, with an X'_{ik} for all ik, (7.18) is extended to all B-atoms and the conditional distribution of Y_i is determined for all i as in (7.19). But also $Y_{i1}, ..., Y_{ik_i} \mid Y_i, \mu_i, \theta_{ik}$ is distributed Multinomial$(Y_i; q_{i1}, ..., q_{ik_i})$, where $q_{ik} = |B_{ik}| \, h(X'_{ik}/|B_{ik}| \; ; \theta_{ik}) / \sum_k |B_{ik}| \, h(X'_{ik}/|B_{ik}| \; ; \theta_{ik})$.

To capture the spatial nature of the B_i we may adopt an IAR model for the μ_i, i.e.,

$$p(\mu_i \mid \mu_{i', i' \neq i}) = N\left(\sum_{i'} w_{ii'} \mu_{i'}/w_{i.} \, , \, 1/(\lambda_\mu w_{i.})\right) \qquad (7.21)$$

where $w_{ii} = 0$, $w_{ii'} = w_{i'i}$ and $w_{i.} = \sum_{i'} w_{ii'}$. Below, we set $w_{ii'} = 1$ for $B_{i'}$ a neighbor of B_i and $w_{ii'} = 0$ otherwise, the standard "0-1 adjacency" form.

Similarly we assume that

$$f(\omega_j \mid \omega_{j', j' \neq j}) = N\left(\sum_{j'} v_{jj'} \omega_{j'}/v_{j.} \, , \, 1/(\lambda_\omega v_{j.})\right) .$$

We adopt a proper Gamma prior for λ_μ and also for λ_ω. When $\boldsymbol{\theta}$ is present we require a prior that we denote by $f(\boldsymbol{\theta})$. The choice of $f(\boldsymbol{\theta})$ will likely be vague but its form depends upon the adopted parametric form of h.

The entire specification can be given a representation as a graphical model, as in Figure 7.12. In this model the arrow from $\{X_{j\ell}\} \to \{X'_{ik}\}$ indicates the inversion of the $\{X_{jl}\}$ to $\{X'_{ik}\}$, augmented by any required edge atom values X'_{iE}. The $\{\omega_i^*\}$ would be generated if the X'_{iE} are modeled using (7.20). Since the $\{Y_{ik}\}$ are not observed but are distributed as multinomial given the fixed block group totals $\{Y_i\}$, this is a predictive step in our model, as indicated by the arrow from $\{Y_i\}$ to $\{Y_{ik}\}$ in the figure. In fact, as mentioned above the further predictive step to impute Y'_j, the Y total associated with X_j in the jth target zone, is of key interest. If there are edge atoms C_{jE}, this will require a model for the associated Y'_{jE}. Since there is no corresponding B-atom for C_{jE} a specification such as (7.18) is not

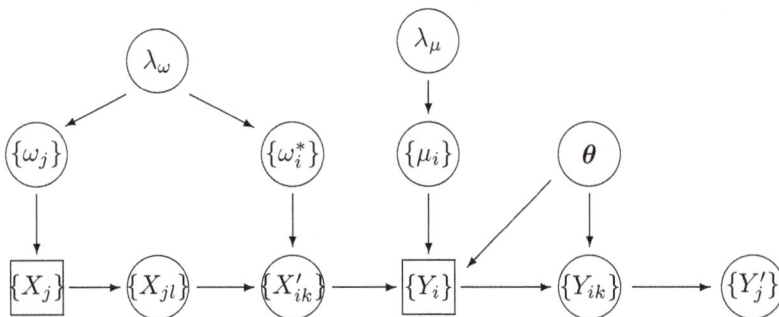

Figure 7.12 *Graphical version of the model, with variables as described in the text. Boxes indicate data nodes, while circles indicate unknowns.*

appropriate. Rather, we can imitate the above modeling for X'_{iE} using (7.20) by introducing $\{\mu^*_j\}$, which along with the μ_i follow the prior in (7.21). The $\{\mu^*_j\}$ and $\{Y'_{jE}\}$ would add two consecutive nodes to the right side of Figure 7.12, connecting from λ_μ to $\{Y'_j\}$.

The entire distributional specification overlaid on this graphical model has been supplied in the foregoing discussion and (in the absence of C_{jE} edge atoms, as in Figure 7.10) takes the form

$$
\begin{aligned}
\prod_i & f(Y_{i1}, ..., Y_{ik_i} \mid Y_i, \boldsymbol{\theta}) \prod_i f(Y_i \mid \mu_i, \boldsymbol{\theta}, \{X'_{ik}\}) \, f(\{X'_{ik}\} \mid \omega^*_i, \{X_{j\ell}\}) \\
& \times \prod_j f(X_{j1}, ..., X_{jL_j} \mid X_j) \prod_j f(X_j \mid \omega_j) \\
& \times f(\{\mu_i\} \mid \lambda_\mu) \, f(\lambda_\mu) \, f(\{\omega_j\}, \{\omega^*_i\} \mid \lambda_\omega) \, f(\lambda_\omega) \, f(\boldsymbol{\theta}) \,.
\end{aligned}
\tag{7.22}
$$

Bringing in the $\mathbf{W}_i$ merely revises the exponential term in (7.18) from $\exp(\mu_i)$ to $\exp(\mu_i + \mathbf{W}^{\mathsf{T}}_{ik}\boldsymbol{\beta})$. Again, for an inherited component of $\mathbf{W}_i$, say, $W^{(r)}_i$, the resulting $W^{(r)}_{ik} = W^{(r)}_i$. For an aggregated component of $\mathbf{W}_i$, again, say, $W^{(r)}_i$, we imitate (7.17) assuming $W^{(r)}_{ik} \mid \mu^{(r)}_i \sim Po(e^{\mu^{(r)}_i}|B_{ik}|)$, independent for $k = 1, ..., K_i$. A spatial prior on the $\mu^{(r)}_i$ and a Gaussian (or perhaps flat) prior on $\boldsymbol{\beta}$ completes the model specification.

Finally, on the response grid, for each B_i rather than observing a single Y_i we may observe Y_{im}, where $m = 1, 2, ..., M$ indexes levels of factors such as sex, race, or age group. Here we seek to use these factors, in an ANOVA fashion, along with the X_j (and $\mathbf{W}_i$) to explain the Y_{im}. Ignoring $\mathbf{W}_i$, the resultant change in (7.18) is that Y_{ikm} will be Poisson with μ_i replaced by μ_{im}, where μ_{im} has an appropriate ANOVA form. For example, in the case of sex and age classes, we might have a sex main effect, an age main effect, and a sex-age interaction effect. In our application these effects are not nested within i; we include only a spatial overall mean effect indexed by i.

Regarding the MCMC implementation of our model, besides the usual concerns about appropriate choice of Metropolis-Hastings candidate densities and acceptability of the resulting convergence rate, one issue deserves special attention. Adopting the identity function for h in (7.18) produces the model $Y_{ik} \sim Po(e^{\mu_i}(X'_{ik}))$, which in turn implies $Y_{i.} \sim Po(e^{\mu_i}(X'_{i.}))$. Suppose however that $Y_{i.} > 0$ for a particular block group i, but in some MCMC iteration no structures are allocated to any of the atoms of the block group. The result is a flawed probabilistic specification. To ensure $h > 0$ even when $z = 0$, we revised our model to $h(z \,;\, \theta_{ik}) = z + \theta_{ik}$ where $\theta_{ik} = \theta/(K_i|B_{ik}|)$ with $\theta > 0$, resulting in

$$
Y_{ik} \sim Po\left(e^{\mu_i}\left(X'_{ik} + \frac{\theta}{K_i}\right)\right) \,.
\tag{7.23}
$$

This adjustment eliminates the possibility of a zero-valued Poisson parameter but does allow for the possibility of a nonzero population count in a region where there are no structures observed. When conditioned on $Y_{i\cdot}$, we find $(Y_{i1}, \ldots, Y_{iK_i} \mid Y_{i\cdot}) \sim \text{Mult}(Y_{i\cdot}\,;\, p_{i1}, \ldots, p_{iK_i})$, where

$$p_{ik} = \frac{X'_{ik} + \theta/K_i}{X'_{i\cdot} + \theta} \ \text{ and } \ Y_{i\cdot} \sim Po\left(e^{\mu_i}(X'_{i\cdot} + \theta)\right) . \qquad (7.24)$$

Our basic model then consists of (7.23) to (7.24) together with

$$\mu_i \stackrel{iid}{\sim} N\left(\eta_\mu, 1/\tau_\mu\right), \ X_{jl} \sim Po\left(e^{\omega_j}|C_{jl}|\right) \Rightarrow X_{j\cdot} \sim Po\left(e^{\omega_j}|C_j|\right) ,$$
$$(X_{j1}, \ldots, X_{jL_j} \mid X_{j\cdot}) \sim \text{Mult}(X_{j\cdot}\,;\, q_{j1}, \ldots, q_{jL_j}), \text{ where } q_{jl} = |C_{jl}|/|C_j|, \qquad (7.25)$$
$$X'_{iE} \sim Po\left(e^{\omega_i^*}|B_{iE}|\right) , \ \text{ and } \ (\omega_j, \omega_i^*) \sim \text{CAR}(\lambda_\omega) ,$$

where X'_{iE} and ω_i^* refer to edge atom structure counts and log relative risk parameters, respectively. While θ could be estimated from the data, in our implementation we simply set $\theta = 1$; [Mugglin et al., 2000, Sec. 6] discuss the impact of alternate selections.

Example 7.2 *(FMPC data analysis).* We turn now to the particulars of the FMPC data analysis, examining two different models in the context of the misaligned data as described in Section 7.3.1. In the first case we take up the problem of total population interpolation, while in the second we consider age- and sex-specific population interpolation.

7.3.2.1 *Total population interpolation model*

We begin by taking $\eta_\mu = 1.1$ and $\tau_\mu = 0.5$ in (7.25). The choice of mean value reflects the work of Rogers and Killough [1997], who found population per household (PPH) estimates for four of the seven townships in which the windrose lies. Their estimates ranged in value from 2.9 to 3.2, hence our choice of $\eta_\mu = 1.1 \approx \log(3)$. The value $\tau_\mu = 0.5$ is sufficiently small to make the prior for μ_i large enough to support all feasible values of μ_i (two prior standard deviations in either direction would enable PPH values of 0.18 to 50.8).

For $\boldsymbol{\omega} = \{\omega_j, \omega_i^*\}$ we adopted a CAR prior and fixed $\lambda_\omega = 10$. We did not impose any centering of the elements of $\boldsymbol{\omega}$ around 0, allowing them to determine their own mean level in the MCMC algorithm. Since most cells have four neighbors, the value $\lambda_\omega = 10$ translates into a conditional prior standard deviation for the ω's of $\sqrt{1/(10 \cdot 4)} = .158$, hence a marginal prior standard deviation of roughly $.158/.7 \approx .23$ [Bernardinelli et al., 1995a]. In any case, we found $\lambda_\omega < 10$ too vague to achieve MCMC convergence. Typical posterior medians for the ω's ranged from 2.2 to 3.3 for the windrose ω_j's and from 3.3 to 4.5 for the edge ω_i^*s.

Running 5 parallel sampling chains, acceptable convergence obtains for all parameters within 1,500 iterations. We discarded this initial sample and then continued the chains for an additional 5,000 iterations each, obtaining a final posterior sample of size 25,000. From the resulting samples, we can examine the posterior distributions of any parameters we wish. It is instructive first to examine the distributions of the imputed structure counts X_{jl}. For example, consider Figure 7.13, which shows the posterior distributions of the structure counts in cell 106 (the sixth one from the windrose center in the SE direction), for which $L_j = 4$. The known cell total $X_{106,\cdot}$ is 55. Note that the structure values indicated in the histograms are integers. The vertical bars in each histogram indicate how the 55 structures would be allocated if imputed proportionally to the area. In this cell, we observe good general agreement between these naively imputed values and our histograms, but the advantage of assessing variability from the full distributional estimates is immediately apparent.

Population estimates per cell for cells 105 through 110 (again in the SE direction, from the middle to the outer edge of the windrose) are indicated in Figure 7.14. Vertical bars here represent estimates calculated by multiplying the number of structures in the cell by a fixed (map-wide) constant representing population per household (PPH), a method roughly

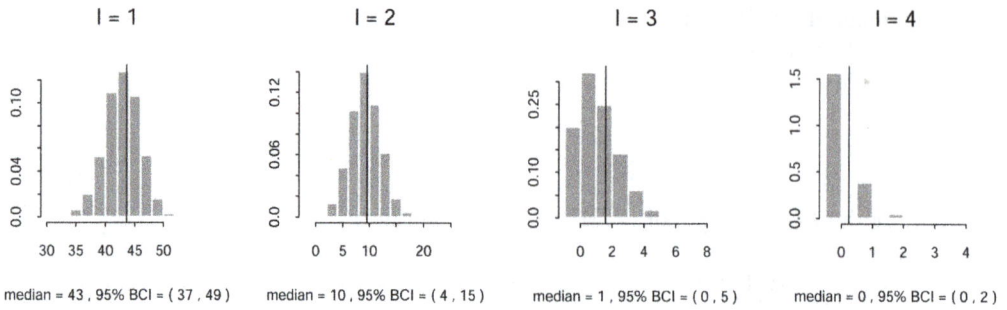

Figure 7.13 *Posterior distributions of structure estimates for the four atoms of cell 106 (SE6). Vertical bars represent structure values if imputed proportionally to the area. Here and in the next figure, "median" denotes posterior median, and "95% BCI" denotes the equal-tail Bayesian confidence interval.*

equivalent to that employed by Rogers and Killough [1997], who as mentioned above actually used four different PPH values. Our reference lines use a constant value of 3 (the analog of our prior mean). While cells 105 and 106 indicate good general agreement in these estimates, cells 107 through 110 display markedly different population estimates, where our estimates are substantially higher than the constant-PPH estimates. This is typical of cells toward the outer edge of the southeast portion of the windrose, since the suburbs of Cincinnati encroach on this region. We have population data only (no structures) in the southeastern edge atoms, so our model must estimate both the structures and the population in these regions. The resulting PPH is higher than a map-wide value of 3 (one would expect suburban PPH to be greater than rural PPH) and so the CAR model placed on the $\{\omega_j, \omega_i^*\}$ parameters induces a spatial similarity that can be observed in Figure 7.14.

We next implement the $\{Y_{i\cdot}\} \rightarrow \{Y_{ik}\}$ step. From the resulting $\{Y_{ik}\}$ come the $\{Y_{j\cdot}'\}$ cell totals by appropriate reaggregation. Figure 7.15 shows the population densities by atom $(Y_{ik}/|B_{ik}|)$, calculated by taking the posterior medians of the population distributions for each atom and dividing by atom area in square kilometers. This figure clearly shows the encroachment by suburban Cincinnati on the southeast side of our map, with some spatial smoothing between the edge cells and the outer windrose cells. Finally, Figure 7.16 shows population densities by cell $(Y_{j\cdot}'/|C_j|)$, where the atom-level populations have been aggregated to cells before calculating densities. Posterior standard deviations, though not shown, are also available for each cell. While this figure, by definition, provides less detail than Figure 7.15, it provides information at the scale appropriate for combination with the exposure values of Killough et al. [1996]. Moreover, the scale of aggregation is still fine enough to permit the identification of the locations of Cincinnati suburban sprawl, as well as the communities of Ross (contained in cells ENE 4-5 and NE 4), Shandon (NW 4-5), New Haven (WSW 5-6), and New Baltimore (SSE 4-5).

7.3.2.2 Age and sex effects

Recall from Section 7.3.1 that we seek population counts not only by cell but also by sex and age group. This is because the dose resulting from a given exposure will likely differ depending on gender and age and because the risk resulting from that dose can also be affected by these factors. Again we provide results only for the year 1990; the extension to other timepoints would of course be similar. Population counts at the block group level by sex and age group are provided by the U.S. Census Bureau. Specifically, age is recorded as counts in 18 quinquennial (5-year) intervals: 0–4, 5–9, . . . , 80–84, and 85+. We consider an additive extension of our basic model (7.23)–(7.25) to the sex- and age group-specific case;

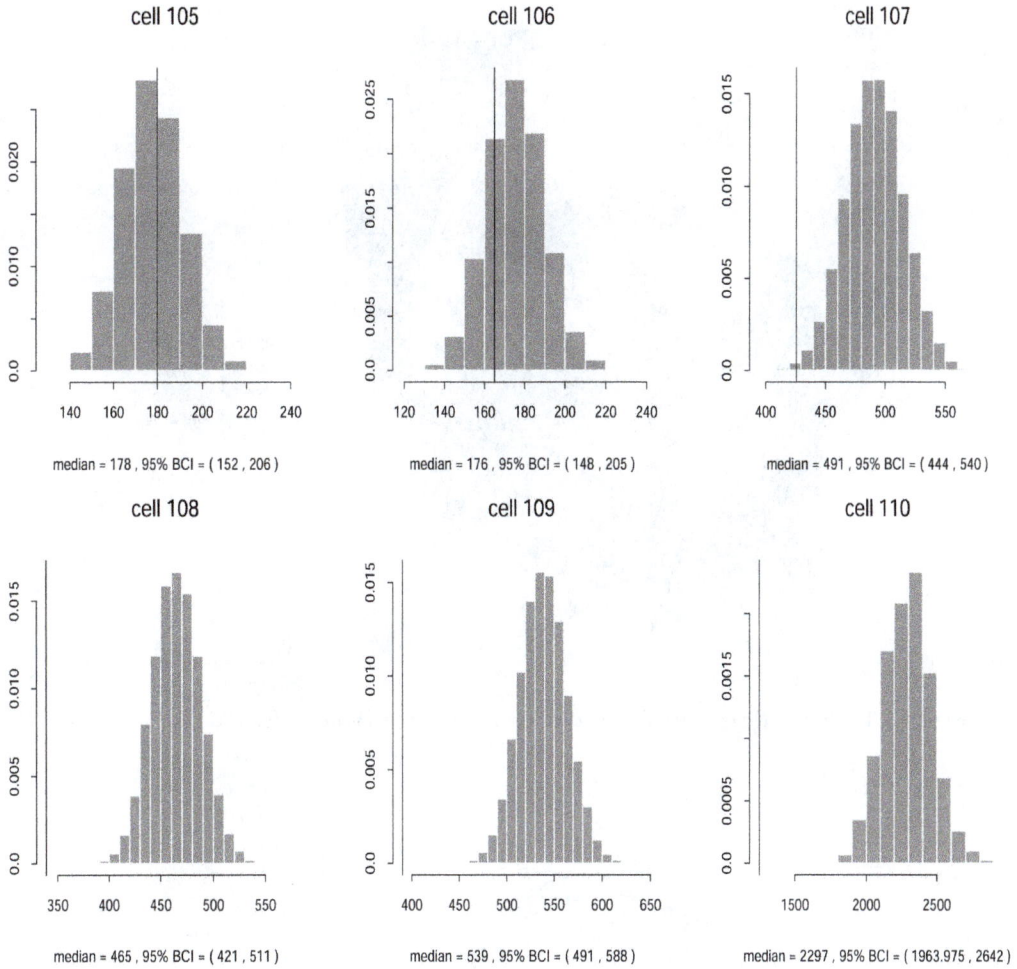

Figure 7.14 *Posterior distributions of populations in cells 105 to 110. Vertical bars represent estimates formed by multiplying structures per cell by a constant population per household (PPH) of 3.0.*

see Mugglin et al. [2000] for results from a slightly more complex additive-plus-interaction model.

We start with the assumption that the population counts in atom k of block group i for gender g at age group a is Poisson-distributed as

$$Y_{ikga} \sim Po\left(e^{\delta_{iga}}\left(X'_{ik} + \frac{\theta}{K_i}\right)\right), \quad \text{where} \quad \delta_{iga} = \mu_i + g\alpha + \sum_{a=1}^{17} \beta_a I_a \,,$$

$g=0$ for males and 1 for females, and I_a is a 0–1 indicator for age group a ($a = 1$ for ages 5-9, $a = 2$ for 10-14, etc.). The μ_i are block group-specific baselines (in our parametrization, they are the logs of the fitted numbers of males in the 0–4 age bracket), and α and the $\{\beta_a\}$ function as main effects for sex and age group, respectively. Note the α and $\{\beta_a\}$ parameters are not specific to any one block group, but rather apply to all 39 block groups in the map.

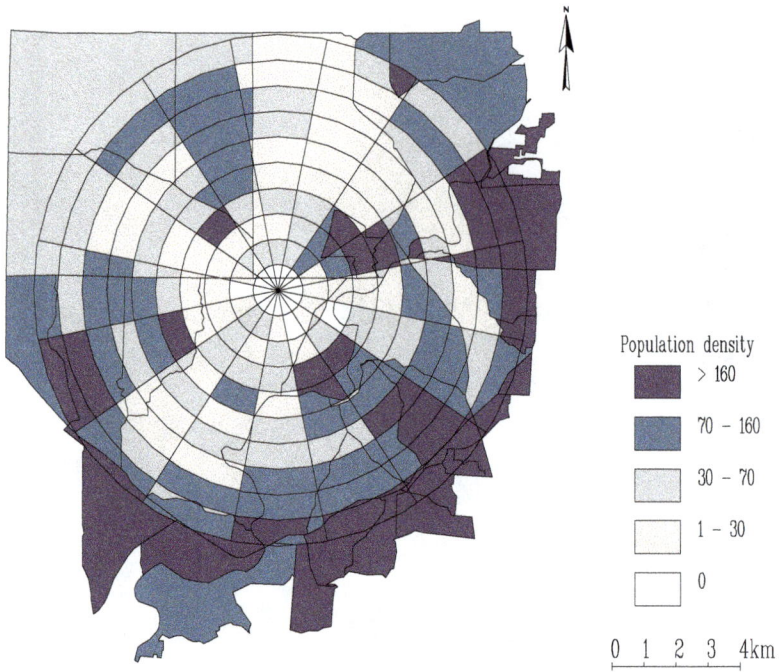

Figure 7.15 *Imputed population densities* (persons/km^2) *by atom for the FMPC region.*

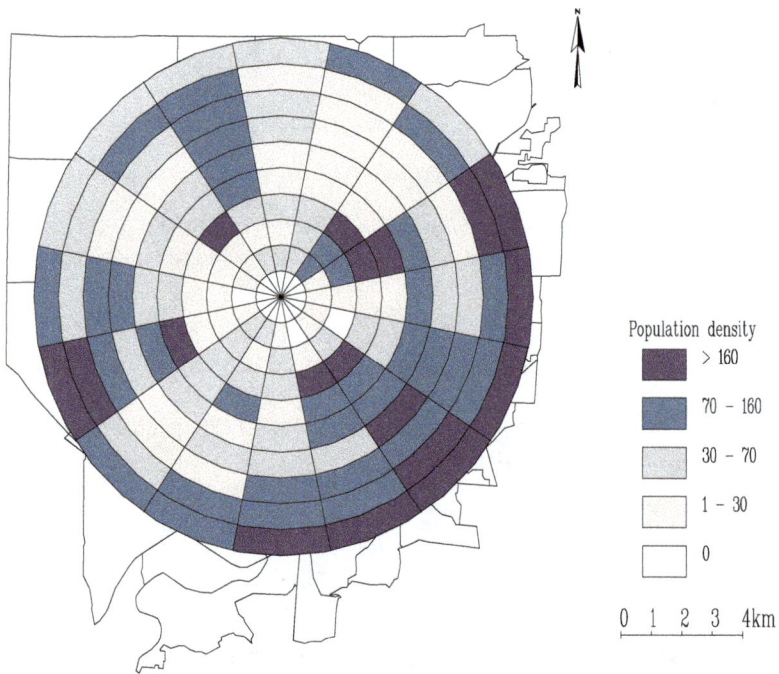

Figure 7.16 *Imputed population densities* (persons/km^2) *by cell for the FMPC windrose.*

Effect	Parameter	Median	2.5%	97.5%
Gender	α	0.005	−0.012	0.021
Ages 5–9	β_1	0.073	0.033	0.116
Ages 10–14	β_2	0.062	0.021	0.106
Ages 15–19	β_3	−0.003	−0.043	0.041
Ages 20–24	β_4	−0.223	−0.268	−0.177
Ages 25–29	β_5	−0.021	−0.063	0.024
Ages 30–34	β_6	0.137	0.095	0.178
Ages 35–39	β_7	0.118	0.077	0.160
Ages 40–44	β_8	0.044	0.001	0.088
Ages 45–49	β_9	−0.224	−0.270	−0.180
Ages 50–54	β_{10}	−0.404	−0.448	−0.357
Ages 55–59	β_{11}	−0.558	−0.609	−0.507
Ages 60–64	β_{12}	−0.627	−0.677	−0.576
Ages 65–69	β_{13}	−0.896	−0.951	−0.839
Ages 70–74	β_{14}	−1.320	−1.386	−1.255
Ages 75–79	β_{15}	−1.720	−1.797	−1.643
Ages 80–84	β_{16}	−2.320	−2.424	−2.224
Ages 85+	β_{17}	−2.836	−2.969	−2.714

Table 7.1 *Quantiles and significance of gender and age effects for the age-sex additive model.*

With each μ_i now corresponding only to the number of baby boys (not the whole population) in block group i, we expect its value to be decreased accordingly. Because there are 36 age-sex divisions, we modified the prior mean η_μ to $-2.5 \approx \log(3/36)$. We placed vague independent $N(0, 10^2)$ priors on α and the βs, and kept all other prior values the same as in Section 7.3.2.1. Convergence of the MCMC algorithm obtains in about 1,500 iterations. (The slowest parameters to converge are those pertaining to the edge atoms, where we have no structure data. Some parameters converge much faster: the α and β_a parameters, for example, converge by about 500 iterations.) We then ran 5,000 iterations for each of the 5 chains, resulting in a final sample of 25,000.

Population interpolation results are quite similar to those outlined in Section 7.3.2.1, except that population distributions are available for each cell at any combination of age and sex. While we do not show these results here, we do include a summary of the main effects for age and sex. Table 7.1 shows the posterior medians and 2.5% and 97.5% quantiles for the α and β_a parameters. Among the β_a parameters, we see a significant negative value of β_4 (ages 20–24), reflecting a relatively small group of college-aged residents in this area. After a slight increase in the age distribution for ages 30–44, we observe increasingly negative values as a increases, indicating the expected decrease in population with advancing age. ∎

7.4   A data assimilation example

Here, we offer a brief discussion of an example from Wikle and Berliner (2005) regarding wind vector data. The data consists of monitoring observations at one areal scale along with computer model output at a different areal scale. The objective is to fuse the data to provide inference at a third spatial scale. So, this is a misalignment or change of support problem. However, this problem also falls under the classification of data assimilation which is the topic of Chapter 16. In particular, the two sources of data are daily wind satellite data and computer model output from a weather center over the period 15 September 1996-29 June 1997. There are satellite-based wind estimates from a NASA Scatterometer (NSCAT) at 0.5 degree spatial resolution, not on a regular grid, along with the National Center for Environmental Prediction (NCEP) analysis of wind direction at 2.5 degree resolution on a regular grid. The goal is to predict surface streamfunction at a resolution of 1.0 degrees. Adopting our usual paradigm, [Data | Process, Parameters][Process | Parameters][Parameters] we have measurement data, Z, from the two sources and we let Y denote the true underlying process which operates at point level.

Let $A_i, i = 1, 2, .., n_a$ denote the .5 degree resolution grid, $B_j, j = 1, 2, ..., n_b$ the grid at 1.0 degree resolution and $C_k, k = 1, 2, ..., n_c$, the sets at 2.5 degree resolution. Though the scales are nested the sets need not be. We let $\mathbf{Z}_A = (Z(A_1), ..., Z(A_{n_a}))^\mathrm{T}$ denote the, observations on the subgrid and $\mathbf{Z}_C = (Z(C_1), ..., Z(C_{n_c}))^\mathrm{T}$ denote the observations on the supergrid.

Next, denote by $Y_D = \{Y(\mathbf{s}) : \mathbf{s} \in D\}$ the true spatial process and employ block averaging to obtain the true process values at the three spatial scales. That is, in general $Y(S) = \int_S Y(\mathbf{s})d\mathbf{s}$. Hence, $\mathbf{Y}_A = (Y(A_1), ..., Y(A_{n_a}))^\mathrm{T}$ is the true subgrid wind vector process, $\mathbf{Y}_C = (Y(C_1), ..., Y(C_{n_c}))^\mathrm{T}$ is the true supergrid process and $\mathbf{Y}_B = (Y(B_1), ..., Y(B_{n_b}))^\mathrm{T}$ is the true process on the desired prediction grid. A simple measurement error model is introduced for the data, i.e., $\mathbf{Z}_A = \mathbf{Y}_A + \mathbf{e}_A$, where $\mathbf{e}_A \sim N(0, \sigma_a^2 I_{n_a})$ and $\mathbf{Z}_C = \mathbf{Y}_C + \mathbf{e}_C$, where $\mathbf{e}_C \sim N(0, \sigma_c^2 I_{n_c})$. Wikle and Berliner treat the two measurement error models as independent.

The remaining challenge is to align the $Y(A_i)$'s and, $Y(C_k)$ with the $Y(B_j)$'s. This requires creating the highest resolution partition (in the spirit of the previous section) by intersecting all of the sets in A, B, and C. We omit the details and encourage the reader to consult the Wikle and Berliner paper for full details.

7.5 Misaligned regression modeling

The methods of the preceding sections allow us to realign spatially misaligned data. The results of such methods may be interesting in and of themselves, but in many cases our real interest in data realignment will be as a precursor to fitting *regression* models relating the (newly realigned) variables.

For instance, Agarwal et al. [2002] apply the ideas of Section 7.2 in a rasterized data setting. Such data are common in remote sensing, where satellites can collect data (say, land use) over a pixelized surface, which is often fine enough so that town or other geopolitical boundaries can be (approximately) taken as the union of a collection of pixels.

The focal area for the Agarwal et al. [2002] study is the tropical rainforest biome within Toamasina (or Tamatave) Province of Madagascar. This province is located along the east coast of Madagascar and includes the greatest extent of tropical rainforest in the island nation. The aerial extent of Toamasina Province is roughly 75,000 square km. Four georeferenced GIS coverages were constructed for the province: town boundaries with associated 1993 population census data, elevation, slope, and land cover. Ultimately, the total number of towns was 159, and the total number of pixels was 74,607. For analysis at a lower resolution, the above 1-km raster layers are aggregated into 4-km pixels.

Figure 7.17 shows the town-level map for the 159 towns in the Madagascar study region. In fact, there is an escarpment in the western portion where the climate differs from the rest of the region. It is a seasonally dry grassland/savanna mosaic. Also, the northern part is expected to differ from the southern part, since the north has fewer population areas with large forest patches, while the south has more villages with many smaller forest patches and more extensive road development, including commercial routes to the national capital west of the study region. The north and south regions with a transition zone were created as shown in Figure 7.17.

The joint distribution of land use and population count is modeled at the pixel level. Let L_{ij} denote the land use value for the jth pixel in the ith town and let P_{ij} denote the

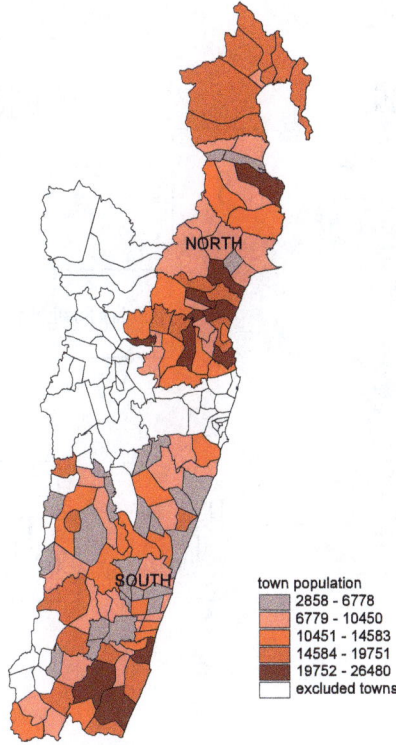

Figure 7.17 *Northern and southern regions within the Madagascar study region, with population overlaid.*

population count for the jth pixel in the ith town. Again, the L_{ij} are observed but only $P_{i\cdot} = \sum_j P_{ij}$ are observed at the town level. Collect the L_{ij} and P_{ij} into town-level vectors $\mathbf{L}_i$ and $\mathbf{P}_i$, and overall vectors $\mathbf{L}$ and $\mathbf{P}$.

Covariates observed at each pixel include an elevation, E_{ij}, and a slope, S_{ij}. To capture spatial association between the L_{ij}, pixel-level spatial effects φ_{ij} are introduced; to capture spatial association between the $P_{i\cdot}$, town-level spatial effects δ_i are introduced. That is, the spatial process governing land use may differ from that for population.

The joint distribution, $p(\mathbf{L}, \mathbf{P} \mid \{E_{ij}\}, \{S_{ij}\}, \{\varphi_{ij}\}, \{\delta_i\})$ is specified by factoring it as

$$p(\mathbf{P} \mid \{E_{ij}\}, \{S_{ij}\}, \{\delta_i\}) \, p(\mathbf{L} \mid \mathbf{P}, \{E_{ij}\}, \{S_{ij}\}, \{\varphi_{ij}\}) \,. \qquad (7.26)$$

Conditioning is done in this fashion in order to explain the effect of population on land use. Causality is *not* asserted; the conditioning could be reversed. (Also, implicit in (7.26) is a marginal specification for $\mathbf{L}$ and a conditional specification for $\mathbf{P} \mid \mathbf{L}$.)

Turning to the first term in (7.26), the P_{ij} are assumed conditionally independent given the E's, S's, and δ's. In fact, we assume $P_{ij} \sim \text{Poisson}(\lambda_{ij})$, where

$$\log \lambda_{ij} = \beta_0 + \beta_1 E_{ij} + \beta_2 S_{ij} + \delta_i \,. \qquad (7.27)$$

Thus $\mathrm{P}_{i\cdot} \sim \text{Poisson}(\lambda_{i\cdot})$, where $\log \lambda_{i\cdot} = \log \sum_j \lambda_{ij} = \log \sum_j \exp(\beta_0 + \beta_1 E_{ij} + \beta_2 S_{ij} + \delta_i)$. In other words, the P_{ij} inherits the spatial effect associated with $P_{i\cdot}$. Also, $\{P_{ij}\} \mid P_{i\cdot} \sim$ Multinomial$(P_{i\cdot}; \{\gamma_{ij}\})$, where $\gamma_{ij} = \lambda_{ij}/\lambda_{i\cdot}$.

In the second term in (7.26), conditional independence of the L_{ij} given the P's, E's, S's, and φ's is assumed. To facilitate computation, we aggregate to 4 km $\times$ 4 km resolution. (The

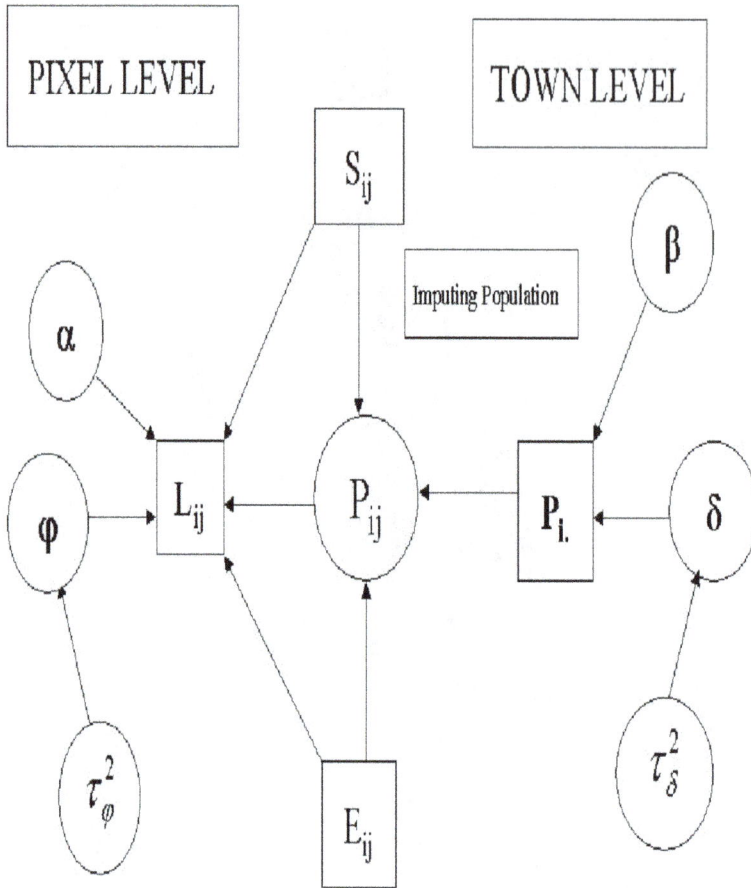

Figure 7.18 *Graphical representation of the land use-population model.*

discussion regarding Figure 4.2 in Subsection 4.1.1 supports this.) Since L_{ij} lies between 0 and 16, it is assumed that $L_{ij} \sim$Binomial$(16, q_{ij})$, i.e., that the sixteen 1 km $\times$ 1 km pixels that comprise a given 4 km $\times$ 4 km pixel are i.i.d. Bernoulli random variables with q_{ij} such that

$$\log\left(\frac{q_{ij}}{1 - q_{ij}}\right) = \alpha_0 + \alpha_1 E_{ij} + \alpha_2 S_{ij} + \alpha_3 P_{ij} + \varphi_{ij} \; . \qquad (7.28)$$

For the town-level spatial effects, a conditionally autoregressive (CAR) prior is assumed using only the adjacent towns for the mean structure, with variance τ_δ^2, and similarly for the pixel effects using only adjacent pixels, with variance τ_φ^2.

To complete the hierarchical model specification, priors for $\boldsymbol{\alpha}, \boldsymbol{\beta}, \tau_\delta^2$, and τ_φ^2 (when the φ_{ij} are included) are required. Under a binomial, with proper priors for τ_δ^2 and τ_φ^2, a flat prior for $\boldsymbol{\alpha}$ and $\boldsymbol{\beta}$ will yield a proper posterior. For τ_δ^2 and τ_φ^2, inverse Gamma priors may be adopted. Figure 7.18 offers a graphical representation of the full model.

We now present a brief summary of the data analysis. At the 4 km x 4 km pixel scale, two versions of the model in (7.28) were fit, one with the φ_{ij} (Model 2) and one without them (Model 1). Models 1 and 2 were fitted separately for the northern and southern regions. The results are summarized in Table 7.2, point (posterior median) and interval (95% equal tail) estimate. The population-count model results are little affected by the inclusion of the φ_{ij}. For the land-use model this is not the case. Interval estimates for the fixed effects

Model:	M_1		M_2	
Region:	North	South	North	South
Population model parameters:				
β_1	−.577	−.245	−.592	−.176
(elev)	(−.663,−.498)	(−.419,−.061)	(−.679,−.500)	(−.341,.019)
β_2	.125	−.061	.127	−.096
(slope)	(.027,.209)	(−.212,.095)	(.014,.220)	(−.270,.050)
τ_{δ^2}	1.32	1.67	1.33	1.71
	(.910,2.04)	(1.23,2.36)	(.906,1.94)	(1.22,2.41)
Land use model parameters:				
α_1	.406	−.081	.490	.130
(elev)	(.373,.440)	(−.109,−.053)	(.160,.857)	(−.327,.610)
α_2	.015	.157	.040	−.011
(slope)	(−.013,.047)	(.129,.187)	(−.085,.178)	(−.152,.117)
α_3	−5.10	−3.60	−4.12	−8.11
$(\times 10^{-4})$	(−5.76,−4.43)	(−4.27,−2.80)	(−7.90,−.329)	(−14.2,−3.69)
τ_{φ^2}	—	—	6.84	5.85
			(6.15,7.65)	(5.23,6.54)

Table 7.2 *Parameter estimation (point and interval estimates) for Models 1 and 2 for the northern and southern regions.*

coefficients are much wider when the φ_{ij} are included. This is not surprising from the form in (7.28). Though the P_{ij} are modeled and are constrained by summation over j and though the ϕ_{ij} are modeled dependently through the CAR specification, since neither is observed, strong collinearity between the P_{ij} and ϕ_{ij} is expected, inflating the variability of the α's.

Specifically, for the population count model in (7.27), in all cases the elevation coefficient is significantly negative; higher elevation yields a smaller expected population. Interestingly, the elevation coefficient is more negative in the north. The slope variable is intended to provide a measure of the differential in elevation between a pixel and its neighbors. However, a crude algorithm is used within the ARC/INFO software for its calculation, diminishing its value as a covariate. Indeed, a higher slope would typically encourage a lower expected population. While this is roughly true for the south under either model, the opposite emerges for the north. The inference for the town-level spatial variance component τ_δ^2 is consistent across all models. Homogeneity of spatial variance for the population model is acceptable.

Turning to (7.28), in all cases the coefficient for the population is significantly negative. There is a strong relationship between land use and population size; increased population increases the chance of deforestation, in support of the primary hypothesis for this analysis. The elevation coefficients are mixed with regard to significance. However, for both Models 1 and 2, the coefficient is always at least .46 larger in the north. Elevation more strongly encourages forest cover in the north than in the south. This is consistent with the discussion of the preceding paragraph but, apparently, the effect is weaker in the presence of the population effect. Again, the slope covariate provides inconsistent results; but is insignificant in the presence of spatial effects. Inference for the pixel-level spatial variance component does not criticize homogeneity across regions. Note that τ_φ^2 is significantly larger than τ_δ^2. Again, this is expected. With a model having four population parameters to explain 3186

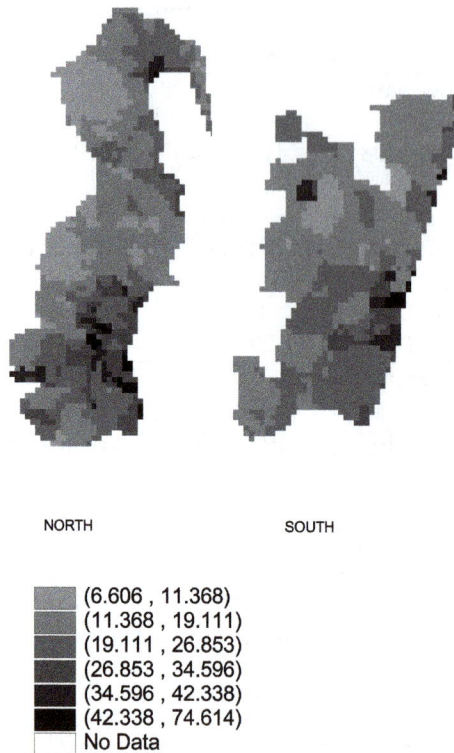

NORTH   SOUTH

(6.606 , 11.368)
(11.368 , 19.111)
(19.111 , 26.853)
(26.853 , 34.596)
(34.596 , 42.338)
(42.338 , 74.614)
No Data

Figure 7.19 *Imputed population (on the square root scale) at the pixel level for north and south regions.*

$q'_{ij}s$ as opposed to a model having three population parameters to explain 115 $\lambda'_i s$, we would expect much more variability in the $\varphi'_{ij}s$ than in the $\delta'_i s$. Finally, Figure 7.19 shows the imputed population at the 4 km × 4 km pixel level.

The approach of Section 7.3 will be difficult to implement with more than two mutually misaligned areal data layers, due mostly to the multiple labeling of atoms and the needed higher-way look-up table. However, the approach of this section suggests a simpler strategy for handling this situation. First, rasterize all data layers to a common scale of resolution. Then, build a suitable latent regression model at that scale, with conditional distributions for the response and explanatory variables constrained by the observed aggregated measurements for the respective layers.

Zhu et al. [2003] consider regression in the point-block misalignment setting, illustrating with the Atlanta ozone data pictured in Figure 7.2. Recall that in this setting the problem is to relate several air quality indicators (ozone, particulate matter, nitrogen oxides, etc.) and a range of sociodemographic variables (age, gender, race, and a socioeconomic status surrogate) to the response, pediatric emergency room (ER) visit counts for asthma in Atlanta, GA. Here the air quality data is collected at fixed monitoring stations (point locations) while the sociodemographic covariates and response variable is collected by zip code (areal summaries). In fact, the air quality data is available as daily averages at each monitoring station, and the response is available as daily counts of visits in each zip code. Zhu et al. [2003] use the methods of Section 7.1 to realign the data, and then fit a Poisson regression model on this scale. Since the data also involves a temporal component, we defer further details until Subsection 12.7.4.

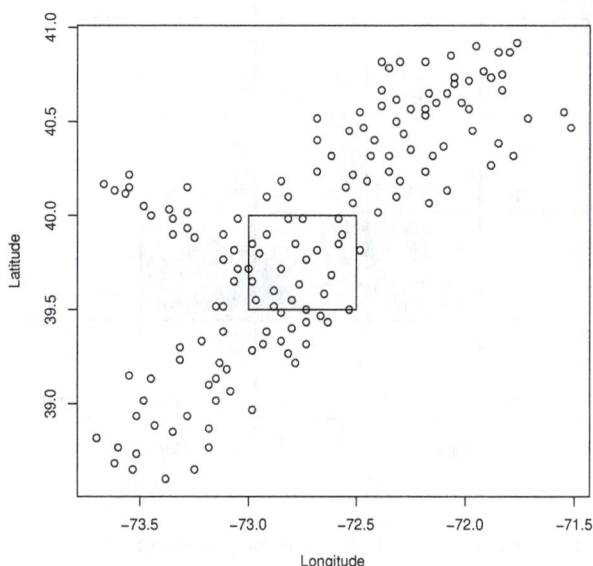

Figure 7.20 *Scallop data from 1993, with rectangle over which a block average is desired.*

7.6 Exercises

1. Suppose we estimate the average value of some areal variable $Y(B)$ over a block B by the predicted value $Y(\mathbf{s}^*)$, where $\mathbf{s}^*$ is some central point of B (say, the population-weighted centroid). Prove that $\mathrm{Var}(Y(\mathbf{s}^*)) \geq \mathrm{Var}(Y(B))$ for any $\mathbf{s}^*$ in B. Is this result still true if $Y(\mathbf{s})$ is nonstationary?

2. Derive the form for $H_B(\phi)$ given below (7.7). (*Hint:* This may be easiest to do by gridding the B_k's, or through a limiting Monte Carlo integration argument.)

3. Suppose g is a differentiable function on $\Re^+$, and suppose $Y(\mathbf{s})$ is a mean-zero stationary process. Let $Z(\mathbf{s}) = g(Y(\mathbf{s}))$ and $Z(B) = \frac{1}{|B|} \int_B Z(\mathbf{s}) d\mathbf{s}$. Approximate $\mathrm{Var}(Z(B))$ and $\mathrm{Cov}(Z(B), Z(B'))$. (*Hint:* Try the delta method here.)

4. Define a process (for convenience, on $\Re^1$) such that $\widehat{Y}(B)$ defined as above (7.10) does *not* converge almost surely to $Y(B)$.

5. Consider the scallop data sites formed by the rectangle having opposite vertices (73.0W, 39.5N) and (72.5W, 40.0N) (refer to Figure 7.20). This rectangle includes 20 locations; the full scallop data are provided in https://github.com/sudiptobanerjee/BGC_2023/data/myscallops.dat, which includes our transformed variable, log(tcatch+1).

   (a) Krige the block average of log(tcatch+1) for this region by simulating from the posterior predictive distribution given all of the 1993 data. Adopt the model and prior structure in Example 7.1, and use equation (7.6) implemented through (7.12) to carry out the generation.

   (b) Noting the caveats regarding vague priors mentioned just below equation (6.3), change to a more informative prior specification on the spatial variance components. Are your findings robust to this change?

6. Suppose that Figure 7.21 gives a (nested) subdivision of the region in Figure 7.4, where we assume the disease count in each subsection is Poisson-distributed with parameter m_1 or m_2, depending on which value (1 or 2) a subregional binary measurement assumes. Suppose further that these Poisson variables are independent given the covariate. Let the

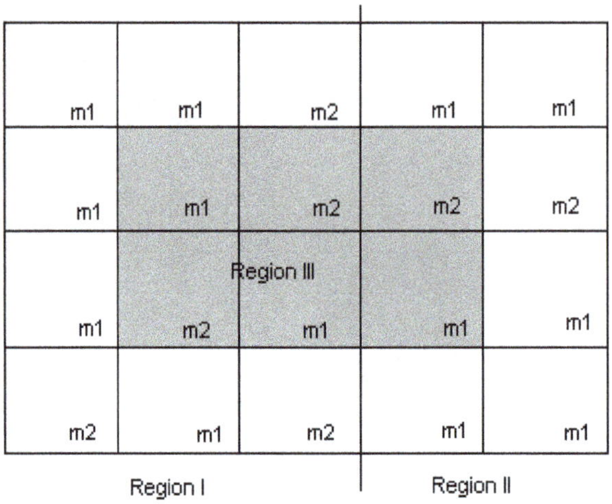

Figure 7.21 *Subregional map for motivating example.*

observed disease counts in Region I and Region II be $y_1 = 632$ and $y_2 = 311$, respectively, and adopt independent $Gamma(a, b)$ priors for m_1 and m_2 with $a = 0.5$ and $b = 100$, so that the priors have mean 50 (roughly the average observed count per subregion) and variance 5000.

(a) Derive the full conditional distributions for m_1 and m_2, and obtain estimates of their marginal posterior densities using MCMC or some other approach. (*Hint:* To improve the numerical stability of your algorithm, you may wish to transform to the log scale. That is, reparametrize to $\delta_1 = \log(m_1)$ and $\delta_2 = \log(m_2)$, remembering to multiply by the Jacobian $(\exp(\delta_i), i = 1, 2)$ for each transformation.)

(b) Find an estimate of $E(Y_3 \mid \mathbf{y})$, the predictive mean of Y_3, the total disease count in the shaded region. (*Hint:* First estimate $E(Y_{3a} \mid \mathbf{y})$ and $E(Y_{3b} \mid \mathbf{y})$, where Y_{3a} and Y_{3b} are the subtotals in the left (Region I) and right (Region II) portions of Region III.)

(c) Obtain a sample from the posterior predictive distribution of Y_3, $p(y_3 \mid \mathbf{y})$. Is your answer consistent with the naive one obtained from equation (7.13)?

7. For the Tompkins County data, available from https://github.com/sudiptobanerjee/BGC_2023/data/tompkins.dat with supporting information on **StatLib** at lib.stat.cmu.edu/datasets/csb/, obtain smoothed estimates of the underlying block group-level relative risks of disease by modifying the log-relative risk model (7.14) to

$$\delta_{k(i,j)} = \theta_0 + \theta_1 u_{ij} + \theta_2 w_{ij} + \theta_3 u_{ij} w_{ij} + \phi_k \,,$$

where we assume

(a) $\phi_k \overset{iid}{\sim} N(0, 1/\tau)$ (*global* smoothing), and

(b) $\boldsymbol{\phi} \sim CAR(\lambda)$, i.e., $\phi_k \mid \phi_{k' \neq k} \sim N\left(\bar{\phi}_k, \frac{1}{\lambda n_k}\right)$ (*local* smoothing).

Do your estimates significantly differ? How do they change as you change λ?

8. For the FMPC data and model in Section 7.3,

(a) Write an explicit expression for the full Bayesian model, given in shorthand notation in equation (7.22).

(b) For the full conditionals for μ_i, X_{ji}, and X'_{iE}, show that the Gaussian, multinomial, and Poisson (respectively) are sensible choices as Metropolis-Hastings proposal densities, and give the rejection ratio in each case.

Chapter 8

Basics of point pattern data modeling

8.1 Introduction

In Chapter 1 we offered point patterns as a third type of spatial data that is collected. Of the three data types, in our view spatial and space-time point patterns continue to be the least developed in terms of Bayesian modeling, fitting, and application. For the sake of keeping chapter size more manageable, we split our development of point pattern work into two chapters. In this chapter we develop the basic ideas, theory, and models which are commonly used in analyzing spatial point pattern data. The next chapter (Chapter 9) is devoted to presenting all of the tools needed for fully Bayesian analysis of such data using these models. This includes the generic approach, the required simulation methodology, the model fitting and model assessment, space-time point patterns, and examples. We can not be comprehensive; rather, we seek to open the door to the range of challenges that arise with such data.

There is a consequential formal theoretical literature and there is by now a substantial body of exploratory tools. We shall explore both of these aspects in Sections 8.2 and 8.3 . In Section 8.4 we look at basic modeling specifications. Here, it is evident that the modeling side, in particular, the hierarchical approach through full Bayesian modeling has received much less attention. In Section 8.5 we extend the class of models to Neyman-Scott and Gibbs processes, again, with an eye toward Bayesian model fitting. In Section 8.6 we consider marked-point processes.

Examples of spatial point patterns arise in various contexts. For instance, in looking at ecological processes, we may be interested in the pattern of occurrences of species, e.g., the pattern of trees in a forest, say junipers and pinions, so-called species distribution. What sort of pattern does each of the species reveal? Do the two species present *different* patterns? Are there environmental/habitat features which explain the observed patterns? Do the species respond differently to the available environment? In epidemiology, in particular so-called spatial epidemiology, we seek to find patterns in disease cases, perhaps different patterns for cases vs. controls. With breast cancer cases where a woman may elect one of say, two treatment options—mastectomy or radiation—do point patterns differ according to option? In *syndromic surveillance*, we seek to identify disease outbreaks. Here, we would be looking for clustering of cases. Additionally, we might be looking at the evolution of a pattern over time. Examples include outbreaks involving e-coli, H1N1, swine flu, bird flu, and bovine tuberculosis. Another variant is to investigate the point pattern of invasive species. Here, we see a picture of locations that is not yet in equilibrium. We might be able to gather subsequent time-slices of the pattern but our primary inferential objective might be to anticipate where the species will eventually appear given where it is now. Again, environmental/habitat features that explain where it currently resides and how it has gotten there can help to address this question. Finally, we might look at the evolution/growth of a city, i.e., urban development. We could look at the pattern of development of single-family homes or of commercial property over time. Also for a city, we could look at the point pattern of property sales, perhaps if there is "change" in this point pattern over time.

DOI: 10.1201/9781003401728-8

Again, point patterns consider the setting where the randomness is associated with the locations of the points themselves. We are immediately led to attempt to clarify what we mean by "no spatial pattern." What do we mean by a *uniform* distribution of points? This is referred to as complete spatial randomness and will be developed formally in Section 8.2. Moreover, it is at the heart of several of the tools in Section 8.3. We can further imagine a collection of point patterns, each indexed by a level of a variable, a so-called "mark," resulting in a *marked* point pattern. Conceptually, marks may be discrete or continuous. With discrete marks, natural interest would be in comparing point patterns across marks, for example, the spatial distributions for different species. Continuous marks naturally arise for example for the selling price of a residential property or for the diameter at the breast height of a tree. More generally and, in particular, with continuous marks, we may be more interested in the joint distribution of marks and locations. Joint modeling is formalized through a conditional × marginal specification. The form $[marks|points][points]$ leads us back to geostatistical analysis where we have a *mark* at each location and, typically, we seek to infer about spatially continuous phenomena given observations at a fixed finite set of locations. In marked point pattern analysis, the randomness in locations, along with associated marks, is analyzed. This leads us into the world of preferential sampling (See Chapter 16). Conditioning in the opposite direction, i.e., $[points|marks][marks]$ seeks to tell an entirely different story, how point pattern behavior varies across marks. We illuminate this point further in Section 8.6.

For a specified, bounded region D, we will denote the realization as $\mathcal{S} = \{\mathbf{s}_i, i = 1, 2..., n\}$ where, again, both n and the $\mathbf{s}_i$ are random. Below, we will consider the role of D more carefully. Are we seeing a finite realization of an infinite point pattern as a result of imposing the restriction to D (in which case, we might need to worry about edge effects and the shape of D might matter). Or, are we seeing a finite point pattern associated with a specified D (conceptually, perhaps an island or a forest or within the limits of a city)? The modeling for these two settings will not be the same and, it may be argued that, in practice, the choice between these two settings is arbitrary. In fact, we will focus more on the second case since, arguably, it is better suited to application. Further, it allows more flexible modeling with more straightforward model fitting and analysis.

Again, we need not have variables at locations, just the pattern of points provided by the locations. We seek to extract relatively rudimentary features of the patterns. Complete spatial randomness or spatial homogeneity is a place to start, an assumption which we expect to criticize on several accounts. The first is because, realistically, it can't be imagined to be operating in practice. A second is because we seek to shed light on where there is departure from randomness and what its nature might be. Third, such departure can result from environmental features in which case we would like to develop regression models to explain why we observed the pattern that we did. Fourth, alternatively, the pattern may reveal possibly a form of clustering or attraction, possibly of inhibition or repulsion, perhaps regular or systematic behavior which, again, we would seek to explain.

Figure 8.1 shows realizations of *spatial homogeneity* for six samples each of 30 points. The plots reveal that the eye can not easily assess complete randomness; it tends to "see" structure. Here, the analogy is with seeing functional relationships in a scatterplot. The eye will tend to respond to artifacts in the randomness. By contrast, Figure 8.2 shows clustering and systematic pattern. Now, still with a small number of points, we can see real structure, real departure from homogeneity.

We can take this a bit further in Figure 8.3 where we show *intensity* surface which can be used to generate point patterns. The surface is supplied in both a perspective plot and a contour plot. We formalize the notion of an intensity in the next section but, for now, we merely conceptualize it as a surface that is such that in regions where it is high we expect to see more points and in regions where it is low we expect to see fewer points. Realizations of point patterns from this intensity surface are shown in Figure 8.4, with some contours

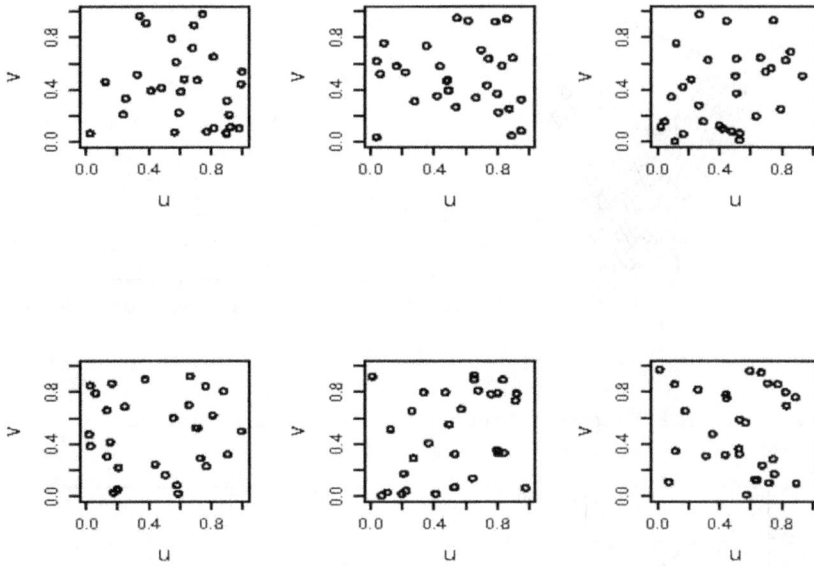

Figure 8.1 *The panels depict* spatial homogeneity *for six samples each of 30 points. The plots reveal that the eye cannot easily assess complete randomness and tends to look for structure.*

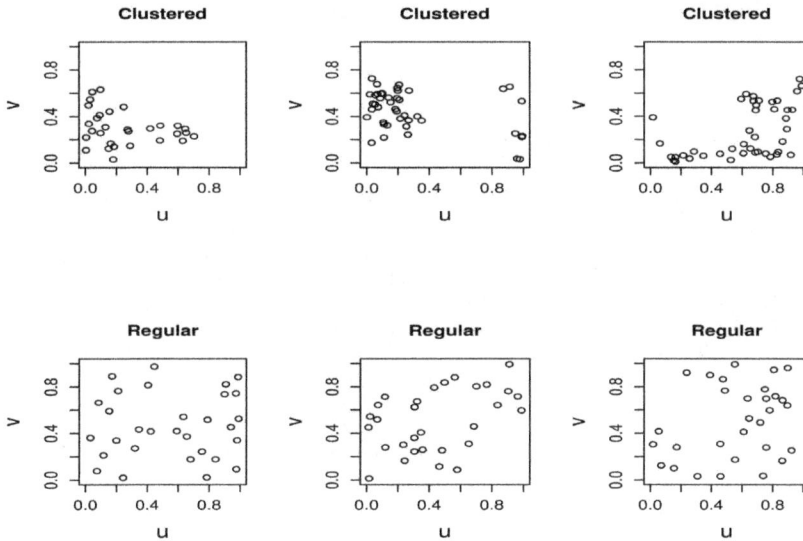

Figure 8.2 *Clustering and systematic (regular) pattern.*

overlaid (dashed lines) to reveal the nature of resultant point patterns and their inherent variability.

A noteworthy remark with regard to point patterns is that often we are prevented from seeing them. Particularly with regard to public health data, points may be aggregated to geographic units, e.g., census units, zip/postcodes, and counties, for confidentiality/privacy reasons. That is, we may be denied the opportunity to analyze the data at point level resolution. Such aggregation returns us to counts associated with areal units which may be analyzed using versions of the discrete spatial data methodology presented in Chapter 4.

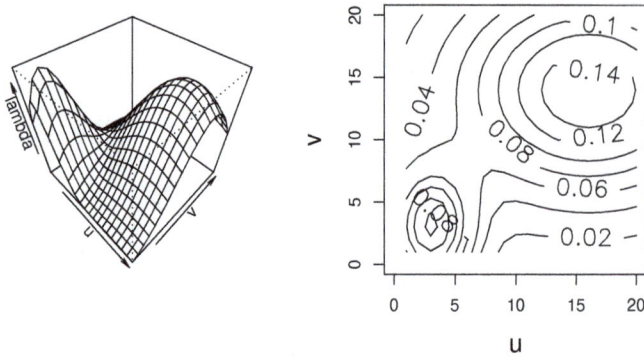

Figure 8.3 *Intensity surface used to generate point patterns.*

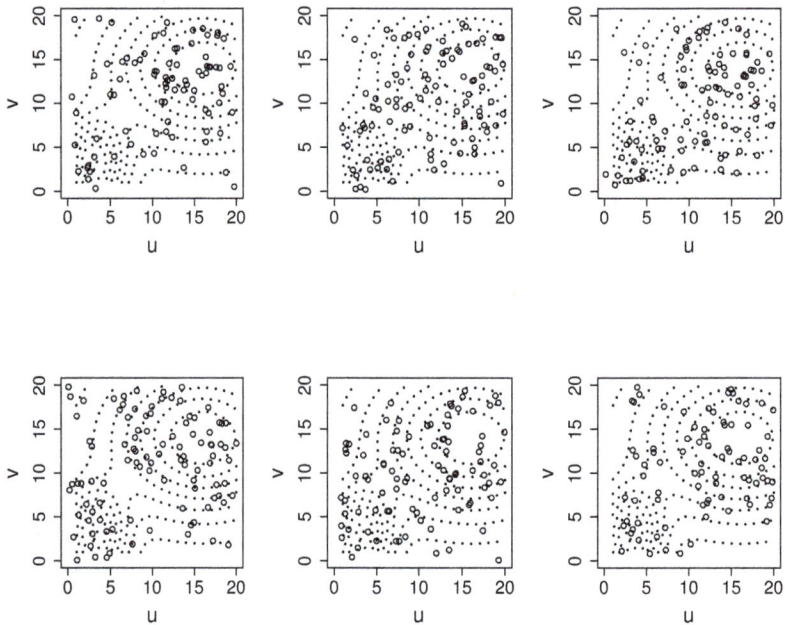

Figure 8.4 *Realizations from the intensity surface in Figure 8.3 with overlaid contours are shown as dashed lines.*

In this regard, we will again confront the ecological fallacy, offered in Section 8.4.3.1. Aggregating points with covariate marks to counts over units, with an associated explanatory variable for the units typically does not arise through simple scaling as we elaborate in Section 8.4.

Another limitation with regard to seeing point patterns over a specified region is sampling effort. Time and resources may preclude sampling all of D. As a result, we will only observe a partial realization of the actual point pattern. Can we learn about the actual point pattern realization or the intensity that generated it? This objective, learning about the actual intensity in the presence of thinning/degradation due to effort, is discussed in Section 9.3.6.2 with an example in Section 9.8.2.1.

There are various books that devote all or a portion of their space to point patterns. In particular, Cressie [1993] presents a fairly detailed, formal discussion at a fairly high technical level. Peter Diggle has made substantial contributions through his website and his book [Diggle, 2013] which is accessible and now a classic but is not broad. Moller and Waagepetersen [2003] offers a model-driven perspective that is likelihood based and fairly technical while Møller, in a set of online notes [Møller, 2012], presents some Bayesian work for point patterns. Baddeley et al. [2015] provide a full text to go with their very rich and useful spatstat package [Baddeley et al., 2005]. Waller and Gotway [2004] provide an easy read, focusing primarily on what would be viewed as exploratory tools, with little on modeling.

An attractive book with broad coverage is by Illian et al. [2007]. It offers a fair bit of technical development along with a data analytic perspective, paying attention to modeling. It has an engaging feel but, by its own admission, emphasizes descriptive (summary) statistics for features of the process model in order to reveal the nature of departures from complete spatial randomness. In fairness, they note that many processes for point patterns are described constructively (in terms of such features), avoiding likelihood specification or making it infeasible. They also indicate that they would prefer to go beyond the exploratory stage when possible. Evidently, consistent with the fully model-based character of this book, leading to posterior inference, we will focus on specifications where this is possible. We also note Gelfand et al. [2010] which presents a review of the field, devoting 161 pages spanning 7 chapters to theory and inference for point patterns. The recent monograph from Gelfand and Schliep [2018] considers modeling, model fitting, and inference exclusively from a Bayesian perspective.

8.2   Theoretical development

We focus on point patterns over $D \subset R^2$. Much theoretical work is over a subregion of R^d since many of the features of interest are geometric in nature, e.g., the distance between points or number of points in a sphere, and these can be formulated for general d dimensional locations. Of course, there is substantial literature for point patterns over an interval on R^1 (usually specified as $(0, T]$). These are sometimes referred to as *counting processes*, emphasizing the number of events as well as their locations, typically times, [see, e.g., Andersen et al., 1996, Fleming and Harrington, 1991]. Working in R^1 offers the advantage of a well-defined "history." We can build models to capture what we expect in the future, given what we have seen thus far. That is, we can view the points as arriving sequentially and model them accordingly. Alternatively, we might view the entire pattern retrospectively, once we have reached time T. If so, we might consider the event times as conditionally independent given some distribution over $(0, T]$. This latter approach is widely adopted for point patterns over R^2 since we have no ordering of the points. It is the approach below through nonhomogeneous Poisson processes and, more generally, Cox processes. However, in Section 8.5 we clarify that, in order to achieve inhibition or clustering behavior, we might work with pairwise interaction specifications which remove this assumption.

There are settings in R^1 where the scale is not time but, rather, some other continuous classification such as size. For instance, in population demography with regard to trees, we can consider a point pattern of tree diameters (usually referred to as diameter at breast height—DBH) or of basal areas, observed in some regions of interest. In fact, over time, we may see several censuses of such diameters. This could enable assessment of forest dynamics, for example, whether the number of individuals is changing as well as whether the distribution of size is changing. See, e.g., Ghosh et al. [2012] in this regard.

Returning to R^2, we consider a bounded, connected subset D. Again and henceforth, we denote a random realization of a point pattern by $\mathcal{S}$ with elements $\mathbf{s}_1, ..., \mathbf{s}_n$. Here, $\mathcal{S}$ is random and so are any features we calculate from it. A probabilistic model for $\mathcal{S} \in D$ must place a distribution over all possible realizations in D. Evidently, this is where the modeling challenge emerges. In practice, it will be easier to specify features/functionals of this distribution than it will be to specify the distribution. Perhaps, this is not surprising since the required distribution must be over a countable set in order to provide the number of points and then, jointly over the continuous domain, D, in order to locate the set of points.

However, continuing in a formal sense, one path for probabilistic modeling for $\mathcal{S}$ is to specify two ingredients. One is the distribution for $N(D)$, the number of points in D. Evidently, this is a distribution over the set $n \in \{0, 1, .., \infty\}$. The second is, for any n, a multivariate density over D^n, for any n, say $f(\mathbf{s}_1, \mathbf{s}_2, ..., \mathbf{s}_n)$. We will call f a *location density* and, since points are unordered/unlabeled, f must be symmetric in its arguments. The implication is, with $\partial \mathbf{s}$ denoting an arbitrarily small circular neighborhood around $\mathbf{s}$,

$$P(N(\partial \mathbf{s}_1) = 1, N(\partial \mathbf{s}_2) = 1, \ldots, N(\partial \mathbf{s}_n) = 1 \mid N(D) = n)$$
$$\approx f(\mathbf{s}_1, \mathbf{s}_2, ..., \mathbf{s}_n) \prod_i |\partial \mathbf{s}_i| \,, \tag{8.1}$$

where $|\partial \mathbf{s}|$ denotes the area of $\partial \mathbf{s}$. An additional implication is that the likelihood will take the form:

$$L(\mathcal{S}) = P(N(D) = n)n! f(\mathbf{s}_1, \mathbf{s}_2, ..., \mathbf{s}_n). \tag{8.2}$$

The first term on the right side of (8.2) appears because the number of points is random. The $n!$ appears in (8.2) because the unordered points can be assigned to the n locations in $n!$ ways. A further thought at this point is that, according to the above, we need to specify f consistently over all $\mathcal{S}$. This sort of consistency condition reminds us of the requirements for providing a model for point-referenced data back in Chapter 2. If we think of $N(D)$ fixed at n, then we only need to provide a finite n-dimensional distribution for f, as with the discrete spatial data models in Chapter 4. In fact, both cases are discussed below in attempting to supply stochastic models for $\mathcal{S}$.

Further, we can define a stationary point pattern model. If $f(\mathbf{s}_1, \mathbf{s}_2, ..., \mathbf{s}_n) = f(\mathbf{s}_1 + \mathbf{h}, \mathbf{s}_2 + \mathbf{h}, ..., \mathbf{s}_n + \mathbf{h})$ for all n, $\mathbf{s}_i$, and $\mathbf{h}$, we say that the point process is stationary. This condition would naturally be proposed over R^2, in which case it would hold, suitably, over D. Note that stationarity is a model property, not a model specification.

8.2.1 Counting measure

Given a point pattern, arguably the simplest feature to think about is the number of points in a specified set. To this end, analogous to the definition of $N(D)$, we introduce count variables, $N(B)$ where, for set B, $N(B)$ is the number of points in set B. That is, $N(B) = \sum_{\mathbf{s}_i \in \mathcal{S}} 1(\mathbf{s}_i \in B)$. Note that $N(B)$ is computed by looking at the points in $\mathcal{S}$ individually, which we will call a *first* order property. This is opposed to looking at objects based upon say, pairs of points, which we will call a second order property below. Since the number of sets B is uncountable, formally, we will need to specify a counting measure over a σ-algebra of measurable sets. Moreover, since the point pattern is random, so is the $N(B)$. We need a random counting measure in order to specify a joint model for an uncountable number of random counts. We can think of $N(B) = \int_B N(d\mathbf{s}) = \int_B N(s)ds$. In this notation, $N(d\mathbf{s})$ is the generator for counts in sets B (analogous to the generators in Section 3.1.1) and $N(\mathbf{s})$ is the Radon-Nikodym derivative of $N(d\mathbf{s})$ with respect to the Lebesgue measure. We develop this measure through finite-dimensional distributions, i.e., the joint distribution for a finite set of count variables. Initially, we do this through Poisson processes, as is customarily done due to the convenient distribution theory. However, in Section 8.5 we consider more general cluster/mixture and Gibbs process models, where $N(B)$ is induced rather than modeled directly. We also briefly digress to consider first and second-order moment measures.

We recall the definition of a Poisson process over a set D, driven by the intensity function $\lambda(\mathbf{s})$. For $B \subseteq D$, $N(B) \sim \text{Po}(\lambda(B))$ where $\lambda(B) = \int_B \lambda(\mathbf{s}) d\mathbf{s}$.[1] So, we have specified the first-order moment measure, $\text{E}(N(B))$; different from above, this is not a *counting* measure. In addition, if B_1 and B_2 are disjoint, then $N(B_1)$ and $N(B_2)$ are independent; see [Cressie, 1993, p. 620] or [Illian et al., 2007, p. 118]. This definition clarifies the need for a bounded set; otherwise, for some B's the integral could be ∞. We also note that we can define the random Poisson measure induced by $\lambda(\mathbf{s})$. We may view this as $\lim_{\partial\mathbf{s}\to 0} \frac{N(\partial\mathbf{s})}{|\partial\mathbf{s}|} = N(\mathbf{s})$ or equivalently, $N(B) = \int_B N(\mathbf{s}) d\mathbf{s}$.

Evidently, $\text{E}(N(B)) = \text{Var}(N(B)) = \lambda(B)$. The independence of disjoint sets immediately implies that $f(\mathbf{s}_1, \mathbf{s}_2, \ldots, \mathbf{s}_n) = \Pi_i f(\mathbf{s}_i) = \Pi_i \lambda(\mathbf{s}_i)/\lambda(D)$ where $\lambda(D) = \int_D \lambda(\mathbf{s}) d\mathbf{s}$. In fact, $P(N(\partial\mathbf{s}) = 1) \approx \text{E}(N(\partial\mathbf{s})) = \lambda(\partial\mathbf{s}) \approx \lambda(\mathbf{s})|\partial\mathbf{s}|$ which is usually written as $\lambda(\mathbf{s})d\mathbf{s}$.

This formalization allows us to specify the notion of complete spatial randomness (CSR), equivalently, spatial homogeneity. This arises when $\lambda(\mathbf{s}) = \lambda$, i.e., we have a constant surface over D; this is referred to as a homogeneous Poisson process (HPP). Evidently, $\lambda(B) = \lambda|B|$ where $|B|$ is the area of B, that is, the expected number of points in set B is proportional to the area of B. The total number of points expected over D is $\lambda|D|$.

From a different perspective, we note that stationarity of the process implies that $\lambda(\mathbf{s}) = \lambda$ for all $\mathbf{s}$ and thus, $\lambda(B) = \lambda|B|$ for all $B \subseteq D$. This is evident since stationarity implies that $f(\mathbf{s}) = f(\mathbf{s} + \mathbf{h})$ for all $\mathbf{s}$ and $\mathbf{h}$. In different terms, if $\lambda(B) = \lambda(B + \mathbf{h})$ for all $B \subseteq D$ and $\mathbf{h}$, then standard real analysis shows that $\lambda(\mathbf{s})$ is constant, i.e., the unique measure satisfying this condition is proportional to Lebesgue measure. It is also clear that $f(\mathbf{s}_1, \mathbf{s}_2, ..., \mathbf{s}_n) = 1/|D|^n$.

It is important to emphasize that the HPP is only one (arguably, the simplest nontrivial) stationary process specification. It is the one that specifies a constant intensity with conditionally independent locations. More general models include the stationary Gibbs processes which we consider in Section 8.5.3. Also, as we shall see below, with stationary processes, one can ask questions regarding so-called *typical* points. A typical point is one of the locations in a point pattern realization, $\mathcal{S}$. Under stationarity, the questions address issues such as the probability of no other points in $\mathcal{S}$ within a specified distance of a typical point or the expected number of points in $\mathcal{S}$ within a specified distance from a typical point. We discuss these ideas further in Section 8.3. Of course, stationarity, hence, a constant intensity would be appropriate only for certain types of data collection. An illustrative application would be to physical processes in a homogeneous environment, for example, interacting particle models.

In most environmental data collection settings, it is anticipated that the environment will not be homogeneous and that the heterogeneity of the environment affects the first-order intensity of the process. Within the Poisson process setting, this suggests the more general case when $\lambda(\mathbf{s})$ is not constant which we refer to the model as a nonhomogeneous Poisson process (NHPP), in some literature, an *inhomogeneous* Poisson process. So, the NHPP is not stationary. Again, there are nonstationary process beyond the NHPP; the NHPP assumes conditionally independent locations with local density, $f(\mathbf{s}) = \lambda(\mathbf{s})/\lambda(D)$.

When $\lambda(\mathbf{s})$ is random we refer to the NHPP as a Cox process. For us, we should take care with regard to this distinction since, from a Bayesian inference perspective, whenever $\lambda(\mathbf{s})$ is unknown, it will be assumed to be random. The Cox processes notion thinks of $\lambda(\mathbf{s})$ as a realization of a stochastic process. For example, $\log\lambda(\mathbf{s})$ might be a realization from a Gaussian process, say restricted to D. Then, the Cox process is, in fact, the marginal process obtained by integrating over the randomness in $\lambda(\mathbf{s})$. But, clearly, it is a hierarchical (two-stage) specification, i.e., draw $\lambda(\mathbf{s})$ and then draw $\mathcal{S}$ given $\lambda(\mathbf{s})$; so is naturally amenable to Bayesian inference.

---

[1]It is worth noting that this is a double integral over two dimensional locations and, for circular regions it may be more easily calculated using polar coordinates.

We now formalize the notion of counting measure. For any set $B \subset D$, let $N(B)$ count the number of points in B. Then, $N(B) \in \{0, 1, 2, ...\}$. We say that $N(B)$ is a counting measure if $N(B) < \infty, \forall B \in \mathcal{B}$ where $\mathcal{B}$ is a σ–algebra of sets intersected with D. Counting measure satisfies the usual countable additivity property thus enabling $N(B_1 \cup B_2)$ and $N(B_1 \cap B_2)$. In particular, we can argue that a realization of a point pattern is equivalent to a realization of a counting measure. That is, given the point pattern, we can immediately assign $N(B)$ to all of the measurable sets. But, conversely (and informally), given $N(B)$ for a σ–algebra of sets, by selecting a suitable increasing or decreasing sequence of sets we can identify the location of a point through the change in counting measure. Again, $\{N(\cdot)\}$ over $\mathcal{B}$ is random with a distribution induced by that for $\mathcal{S}$ in the set of all possible point pattern realizations over D. Conversely, a distribution specified for the $\{N(\cdot)\}$ over all possible point pattern realizations over D would induce a distribution for $\mathcal{S}$ over all possible point pattern realizations in D. In any event, we refer to either collection of variables as a *spatial point process*. Evidently, a NHPP is an example of a spatial point process. In fact, it is an example where the distribution over the space of point patterns and the distribution over the σ-algebra of sets can both be written down explicitly. In more general cases, we specify the process model through a location density, $f(\mathbf{s}_1, ..., \mathbf{s}_n)$ with n random but not generated first. Further, in some cases, such as in Section 8.5.1, we specify the process constructively, without an explicit location density.

8.2.2 *Moment measures*

Next, we turn to the notion of moment measures, also product densities as in Illian et al. [2007]. We start with first order properties, i.e., the first moment measure, $\{E(N(B)) : B \in \mathcal{B}\}$. With regard to this collection of sets, given $\lambda(\mathbf{s})$, we can compute $E(N(B))$ in the usual way, i.e., $E(N(B)) = \int_B \lambda(\mathbf{s})d\mathbf{s}$. However, given that the collection, $\{E(N(B)) : B \in \mathcal{B}\}$, is a measure, can we extract the *first-order* intensity? The approach is to take limits in the form, $\lambda(\mathbf{s}) = \lim_{|\partial\mathbf{s}| \to 0} \frac{E(N(\partial\mathbf{s}))}{|\partial\mathbf{s}|}$, where, as above, $\partial\mathbf{s}$ is a neighborhood of $\mathbf{s}$. The intuition is that $E(N(\partial\mathbf{s})) = \int_{\partial\mathbf{s}} \lambda(\mathbf{s}')d\mathbf{s}' \approx \lambda(\mathbf{s})|\partial\mathbf{s}|$. An analogy may be made with obtaining the probability density function given the cumulative distribution function. That is, we "build up" from $\lambda(\mathbf{s})$, "scale down" from $\{E(N(B)) : B \in \mathcal{B}\}$. Also, we note that, if $f(\mathbf{s}_1, ..., \mathbf{s}_n) = \Pi_i f(\mathbf{s}_i)$, then $\lambda(\mathbf{s}) = f(\mathbf{s})\lambda(D)$ and this conclusion does not depend upon an NHPP assumption. In fact, by the conditional independence of the locations, we see that, given $N(D) = n$, $N(B) \sim Bi(n, P(B))$ where $P(B) = \int_B f(\mathbf{s})d\mathbf{s}$. So, $E(N(B)) = E(E(N(B)|N(D) = n)) = E(nP(B)) = E(n \int_B f(\mathbf{s})d\mathbf{s}) = \int_B E(n)f(\mathbf{s})d\mathbf{s} = \int_B \lambda(D)f(\mathbf{s})d\mathbf{s}$ and thus, $\lambda(\mathbf{s}) = f(\mathbf{s})\lambda(D)$.

A remarkably elegant result for computing certain expectations with regard to point patterns is Campbell's Theorem. We start with a first-order version whose proof is left as an exercise. It will often be of interest to compute features associated with individual points, e.g., as above, whether or not they are in a given set or whether or not they are within a specified distance from a given point. If the feature is denoted by $g(\mathbf{s})$, then we might be interested in the value of this feature summed over the points in $\mathcal{S}$, i.e., $\sum_{\mathbf{s}_i \in \mathcal{S}} g(\mathbf{s}_i)$. Campbell's Theorem provides the expectation of this variable, i.e., with regard to region D,

$$E_{\mathcal{S} \cap D}\left(\sum_{\mathbf{s}_i \in \mathcal{S} \cap D} g(\mathbf{s}_i) \right) = \int_D g(\mathbf{s})\lambda(\mathbf{s})d\mathbf{s}. \tag{8.3}$$

Evidently, applied to the indicator the indicator of whether $\mathbf{s}_i \in B$, we obtain $E(N(B)) = \int_B \lambda(\mathbf{s})d\mathbf{s}$ which, as a result, enables the proof of the theorem. We will return to Campbell's Theorem in the next chapter where it provides a useful tool for implementing fully Bayesian inference for spatial point patterns.

Second order properties refer to properties of the model that consider the points in pairs. For second-order properties, consider $\gamma(B_1 \times B_2) \equiv E_{\mathcal{S}} \sum_{\mathbf{s},\mathbf{s}' \in \mathcal{S}} 1(\mathbf{s} \in B_1, \mathbf{s}' \in B_2)$.

Define $\gamma(\mathbf{s}, \mathbf{s}')$, the *second order intensity* through $\gamma(B_1 \times B_2) = \int_{B_1} \int_{B_2} \gamma(\mathbf{s}, \mathbf{s}')d\mathbf{s}'d\mathbf{s}$. As a result, if B_1 and B_2 are disjoint, $\mathrm{E}_{\mathcal{S}}(N(B_1)N(B_2)) = \int_{B_1} \int_{B_2} \gamma(\mathbf{s}, \mathbf{s}')d\mathbf{s}'d\mathbf{s}$. Hence, with sufficiently small sets, $\gamma(\mathbf{s}, \mathbf{s}') = \lim_{|\partial\mathbf{s}|\to 0, |\partial\mathbf{s}'|\to 0} \frac{\mathrm{E}(N(\partial\mathbf{s})N(\partial\mathbf{s}'))}{|\partial\mathbf{s}||\partial\mathbf{s}'|}$. Again, we build up from $\gamma(\mathbf{s}, \mathbf{s}')$, scale down from $\{\mathrm{E}(N(B_1)N(B_2)) : B_1, B_2 \in \mathcal{B}\}$. Analogous to the first-order case, if $f(\mathbf{s}_1, ..., \mathbf{s}_n) = \Pi_i f(\mathbf{s}_i)$, then $\gamma(\mathbf{s}, \mathbf{s}') = \lambda(\mathbf{s})\lambda(\mathbf{s}') = \lambda^2(D)f(\mathbf{s})f(\mathbf{s}')$. We leave this proof as an exercise.

The *pair correlation function*, $g(\mathbf{s}, \mathbf{s}')$, sometimes referred to as the *reweighted* second-order intensity, is defined as $\gamma(\mathbf{s}, \mathbf{s}')/\lambda(\mathbf{s})\lambda(\mathbf{s}')$. When $\lambda(\mathbf{s}) = \lambda$, $g(\mathbf{s}, \mathbf{s}')$ simplifies to $\gamma(\mathbf{s}, \mathbf{s}')/\lambda^2$. In fact, $g(\mathbf{s}, \mathbf{s}') = 1$ under CSR. Furthermore, when $g(\mathbf{s}, \mathbf{s}') > 1$, attraction is implied, while $g(\mathbf{s}, \mathbf{s}') < 1$ implies repulsion. Under stationarity, $\gamma(\mathbf{s}, \mathbf{s}') = \gamma(\mathbf{s} - \mathbf{s}')$. In this context, isotropy means $\gamma(\mathbf{s}, \mathbf{s}') = \gamma(||\mathbf{s} - \mathbf{s}'||)$.

More generally, we have that

$$\mathrm{E}[N(B_1)N(B_2)] = \mathrm{E}[N(B_1)]\mathrm{E}[N(B_2)] + \mathrm{Var}[N(B_1 \cap B_2)]$$

since $\mathrm{Cov}(N(B_1), N(B_2)) = \mathrm{Var}(N(B_1 \cap B_2))$. This result simplifies to $\mathrm{E}(N(B_1)N(B_2)) = \mathrm{E}(N(B_1))\mathrm{E}(N(B_2))$ when B_1 and B_2 are disjoint.

There is a bivariate version of Campbell's Theorem as follows. Suppose g is a feature which is a function of two arguments, $g(\mathbf{s}, \mathbf{s}')$, e.g., the distance between $\mathbf{s}$ and $\mathbf{s}'$. Suppose we are interested in the value of this feature summed over pairs of point in $\mathcal{S}$, $\sum_{\mathbf{s}_i, \mathbf{s}_j \in \mathcal{S}, i \neq j} g(\mathbf{s}_i, \mathbf{s}_j)$. Then, Campbell's theorem provides the expectation of this variable over $\mathcal{S}$, i.e., again, with restriction to D

$$\mathrm{E}_{\mathcal{S} \cap D}\left(\sum_{\mathbf{s}_i, \mathbf{s}_j \in \mathcal{S} \cap D, i \neq j} g(\mathbf{s}_i, \mathbf{s}_j)\right) = \int_D \int_D g(\mathbf{s}, \mathbf{s}')\gamma(\mathbf{s}, \mathbf{s}')d\mathbf{s}d\mathbf{s}'. \tag{8.4}$$

This result can again be built up from indicator functions and, in fact, when $g(\mathbf{s}, \mathbf{s}') = 1(\mathbf{s} \in B_1, \mathbf{s}' \in B_2)$, we obtain

$$\mathrm{E}(N(B_1)N(B_2)) = \int_{B_1} \int_{B_2} \gamma(\mathbf{s}, \mathbf{s}')d\mathbf{s}'d\mathbf{s} + \int_{B_1 \cap B_2} \lambda(\mathbf{s})d\mathbf{s} .$$

If the spatial point process is stationary, then $\gamma(\mathbf{s}, \mathbf{s}') = \gamma(\mathbf{s} - \mathbf{s}')$. If $\gamma(\mathbf{s}, \mathbf{s}') = \gamma(||\mathbf{s} - \mathbf{s}'||)$, we say that the spatial point process is isotropic. We link an isotropic γ to the K-function in Section 8.3.3 below.

Further insight can be obtained by noticing that, if $\partial\mathbf{s}$ is sufficiently small, $P(N(\partial\mathbf{s}) > 1)$ will be negligible so $\mathrm{E}(N(\partial\mathbf{s})) \approx P(N(\partial\mathbf{s}) = 1) \approx P(N(\partial\mathbf{s}) > 0)$. Similarly, $\mathrm{E}(N(\partial\mathbf{s})N(\partial\mathbf{s}')) \approx P(N(\partial\mathbf{s}) > 0, N(\partial\mathbf{s}') > 0)$. But, since $\mathrm{E}(N(\partial\mathbf{s})N(\partial\mathbf{s}')) \approx \gamma(\mathbf{s}, \mathbf{s}')|\partial\mathbf{s}||\partial\mathbf{s}'|$, we find that $\gamma(\mathbf{s}, \mathbf{s}') \approx P(N(\partial\mathbf{s}) > 0, N(\partial\mathbf{s}') > 0)/|\partial\mathbf{s}||\partial\mathbf{s}'|$. That is, $\gamma(\mathbf{s}, \mathbf{s}')|\partial\mathbf{s}||\partial\mathbf{s}'|$ is the probability of a point of $\mathcal{S}$ in $\partial\mathbf{s}$ and a point of $\mathcal{S}$ in $\partial\mathbf{s}'$, with an evident limiting interpretation as an intensity. For instance, if γ is isotropic, $\gamma(||\mathbf{s} - \mathbf{s}'||)$ can be interpreted loosely as the density for inter-point distances.

We conclude this subsection with a brief discussion of the conditional or Papangelou intensity. Consider the notation, $\lambda(\mathbf{s}|\mathcal{S})$ where $\mathbf{s}$ is a fixed location and $\mathcal{S}$ is a realization of the point process. How might we interpret this intensity function for a given location $\mathbf{s}$ and a given realization $\mathcal{S}$? Here, we have $\lambda(B|\mathcal{S}) = \int_B \lambda(d\mathbf{s}|\mathcal{S}) = \int_B \lambda(\mathbf{s}|\mathcal{S})d\mathbf{s}$ where $\lambda(d\mathbf{s}|\mathcal{S})$ is a generator and $\lambda(\mathbf{s}|\mathcal{S})$ is a Radon-Nikodym derivative. Further, for an arbitrarily small neighborhood around $\mathbf{s}$, $\lambda(\partial\mathbf{s}|\mathcal{S}) \approx \lambda(\mathbf{s}|\mathcal{S})|\partial\mathbf{s}|$. But also, $\lambda(\partial\mathbf{s}|\mathcal{S}) \approx P(N(\partial\mathbf{s}) = 1 \mid \mathcal{S})$. That is, $\lambda(\partial\mathbf{s}|\mathcal{S})$ is roughly the probability that there is a point of $\mathcal{S}$ in $\partial\mathbf{s}$ and the rest of the $\mathbf{s}_i \in \mathcal{S}$ lie outside of $\partial\mathbf{s}$. Formally, this suggests that, with n random, we view

$$\lambda(\partial\mathbf{s}|\mathcal{S}) = \int_{\partial\mathbf{s}} \frac{f(\mathbf{u}, \mathbf{s}_1, ..., \mathbf{s}_n)}{f(\mathbf{s}_1, ..., \mathbf{s}_n)}d\mathbf{u} \approx \frac{f(\mathbf{s}, \mathbf{s}_1, ..., \mathbf{s}_n)}{f(\mathbf{s}_1, ..., \mathbf{s}_n)}|\partial\mathbf{s}|. \tag{8.5}$$

So, in the limit $\lambda(\mathbf{s}|\mathcal{S}) = \dfrac{f(\mathbf{s}, \mathbf{s}_1, ..., \mathbf{s}_n)}{f(\mathbf{s}_1, ..., \mathbf{s}_n)}$, presuming the denominator is not zero. It may be shown that $\lambda(\mathbf{s}) = \mathrm{E}_{\mathcal{S}}(\lambda(\mathbf{s}|\mathcal{S}))$ [see Exercise 5 at the end of this chapter or refer to Møller and Waagepetersen, 2007]. Evidently, for conditionally independent locations $\lambda(\mathbf{s}|\mathcal{S}) = f(\mathbf{s}) = \lambda(\mathbf{s})/\lambda(D)$. The conditional intensity will also take a convenient explicit form for Gibbs processes (pairwise interaction processes) as we show in Section 8.5.3 which also facilitates the simulation of these processes (and will connect with the Markov random field theory presented in Chapter 4). In general, $\lambda(\mathbf{s}|\mathcal{S})$ is random since $\mathcal{S}$ is random. We leave for an exercise the proof that $\mathrm{E}_{\mathcal{S}}(\lambda(\mathbf{s}|\mathcal{S})) = \lambda(\mathbf{s})$.

Returning to the second-moment measure or product intensity, using the foregoing infinitesimal argument, we can view $\gamma(\mathbf{s}, \mathbf{s}')/\lambda(\mathbf{s}')$ as the intensity at $\mathbf{s}$ given a point at $\mathbf{s}'$. In particular, under stationarity, we can write $\gamma(\mathbf{s} - \mathbf{s}')/\lambda = \pi(\mathbf{s} - \mathbf{s}')$ where $\pi(\mathbf{s} - \mathbf{s}')|\partial\mathbf{s}| \approx P(N(\partial\mathbf{s}) > 0)|N(\partial\mathbf{s}') > 0)$, i.e., the probability that there is a point in $\partial\mathbf{s}'$ given there is a point in $\partial\mathbf{s}$.

8.3   Diagnostic tools

Here, we take up what has, historically, been the "bread and butter" of point pattern data analysis and is incorporated into most of the software for analyzing point pattern data. Arguably, at present, the best in this regard is spatstat [Baddeley et al., 2005], which we use in conjunction with several of the examples below and in Chapter 9. We look at approaches for studying departure from spatial homogeneity. In particular, we look the traditional distance-based approaches, G-functions, F-functions, and K-functions. We also look at empirical intensity estimates. Again, we view this work as exploratory. Hence, our review is brief; there are many other texts that develop this material in full detail [see, e.g., Illian et al., 2007]. A useful perspective here is to recognize that all of these exploratory tools (and many further ones which we do not present here) are descriptive or summary measures from a sample to investigate a process feature. They are nonparametric in nature; in the above cases, they are based upon first or second order properties, taking forms that are analogues of empirical cdf's, typically offering no associated uncertainty.

In other words, if you assume less, say only the existence of second-order moment measures then you return less with regard to inference. As noted in Section 8.1, from the Bayesian point of view, we will require the full probabilistic specification for the model. Then, if we can fit the model, we will enjoy the full Bayesian benefit—posterior inference for any model features of interest. These features will merely be functionals of the specification. Hence, we experience the usual pluses and minuses of working within the Bayesian paradigm. On the plus side, fitting models with sampling-based methods yields posterior samples for model unknowns which can be converted to Monte Carlo approximations for these features—the focus of the next chapter. On the minus side, MCMC model fitting can be challenging for many of these models. Some of the process specifications include intractable normalizing constants which are functions of the model parameters. For others, specifications are provided constructively, not yielding an explicit likelihood. We discuss these issues in detail in the next chapter.

8.3.1   Exploratory data analysis; investigating complete spatial randomness

The most elementary way to investigate complete spatial randomness (CSR) proceeds from the definition of an HPP. In particular, if we look at a collection of cells in D, all of equal area, mutually disjoint, not necessarily exhaustive, then, given λ, the observed counts for these cells are i.i.d. Hence, we can compute $\bar{N}$, the mean of the cell counts. As well, we can compute S_N^2, the sample variance of the counts. Then, if we look at $S_N^2/\bar{N}$, under CSR we would expect this to be near one; substantial departure would criticize CSR. Other functions of $\bar{N}$ and S_N^2 have been considered in the literature [see, e.g., Cressie, 1993, p. 590] with hypothesis tests proposed. However, confining our diagnostics to these two moment estimates seem to drastically discard information in the point pattern.

In the same spirit, given a collection of counts, we can examine the i.i.d. Poisson assumption through a chi-square test, i.e., comparing, respectively, across the collection of cells, the observed number of 0's, 1's, etc., with the expected number of 0's, 1's, etc. We would do some sort of aggregation for the large counts, develop an empirical estimate of λ (e.g., total count divided by total area), and employ a χ^2 distribution with degrees of freedom one less than the total number of cells for the χ^2 statistic [again, see Cressie, 1993, pp. 688-689].

8.3.2 G and F functions

If we confine ourselves to stationary processes, we can think in terms of typical points and notions like the probability of being within distance d of a typical point or the expected number of points within distance d of a typical point. In what follows, we formally define these process features and then provide the customary sample estimates for them. The recurring theme here is that, under CSR, these notions have simple explicit forms. The goal of the exploratory data analysis is to see if the sample estimate supports CSR or not. It is important to appreciate that these conceptual quantities are only sensible for stationary processes. For nonstationary processes, we will have to focus on estimating the associated first and, possibly, second-order characteristics (Section 8.3.3).

Let us view the process over all of $\Re^2$, i.e., a countably infinite point pattern. Consider the random variable $N(\mathbf{s}, d; \mathcal{S})$, where $\mathbf{s} \in \mathcal{S}$, $\partial_d \mathbf{s}$ is a circle of radius d centered at $\mathbf{s}$, and N counts the number of points in the circle from $\mathcal{S}$, excluding $\mathbf{s}$. By stationarity, $N(\mathbf{s}, d; \mathcal{S}) \sim N(\mathbf{0}, d; \mathcal{S} - \mathbf{s})$, where $\mathcal{S} - \mathbf{s}$ is the translation of $\mathcal{S}$ by $\mathbf{s}$. This distributional result is equivalent to saying that every point in $\mathcal{S}$ is a *typical* point, in the sense that each one can be viewed as equivalent to $\mathbf{0}$ under translation.

Under restriction to a bounded set D, consider

$$\mathrm{E}_{\mathcal{S}} \left(\sum_{\mathbf{s}_i \in \mathcal{S}, \mathcal{S} \in D} 1(N(\mathbf{s}_i, d; \mathcal{S}) > 0) \right) = \lambda |D| P(N_D(\mathbf{s}, d; \mathcal{S}) > 0). \tag{8.6}$$

The right side of (8.6) follows by noting that the expectation over $\mathcal{S}$ can be calculated in two stages, over $\mathcal{S}$ given $N(D)$ and then over $N(D)$. Here, $N_D(\mathbf{s}, d; \mathcal{S})$ is the count under restriction of the random $\mathcal{S}$ to D and, using the left side of (8.6), we see an obvious Monte Carlo integration for it. Moreover, it is clear that this integration arises in two stages. First, $\mathcal{S}$ is sampled, then N is calculated, given $\mathcal{S}$. Again, note that to obtain (8.6) requires restriction to a bounded set D and enables a Monte Carlo integration for $P(N_D(\mathbf{s}, d; \mathcal{S}) > 0)$, not for $P(N(\mathbf{s}, d; \mathcal{S}) > 0) \geq P(N_D(\mathbf{s}, d; \mathcal{S}) > 0)$. Empirical estimation of the latter requires an *edge correction* (see below).

In the literature, $P(N(\mathbf{s}, d; \mathcal{S}) > 0)$ is denoted by $G(d)$. That is, this probability does not depend upon $\mathbf{s}$, consistent with the notion of a typical point. We see that $G(d)$ increases in d and, in fact, can be viewed as a cdf in distance d. An alternative to $G(d)$ in the literature is $F(d)$ where now $N(\mathbf{s}, d; \mathcal{S})$ would assume $\mathbf{s}$ is not in $\mathcal{S}$. We might distinguish these two definitions of event N by subscript, say N_G and N_F, (sometimes, N^- and N). More importantly, $G(d)$ need not equal $F(d)$. For instance, inhibition might preclude two points in $\mathcal{S}$ from being within distance d of each other. Comparison of estimates of G and F may be informative.

More informally, think of $G(d)$ as the "nearest neighbor" distribution, i.e., the c.d.f. of the nearest neighbor distance, event to event, i.e., at an observed event, $G(d) =$

$Pr(\texttt{nearest event} \leq d)$. Similarly, think of $F(d)$ as the "empty space" distribution, i.e., for an arbitrary location, the c.d.f. of the nearest neighbor distance, point to event, $F(d) = Pr(\texttt{nearest event} \leq d)$. Under CSR, $G(d) = F(d) = 1 - \exp(-\lambda\pi d^2)$. That is, given d, the outcome $\texttt{nearest event} \leq d$ occurs if there is at least one event within a circle of radius d. With constant intensity λ, we know that the number of events in this circle follows a $\texttt{Po}(\lambda\pi d^2)$ distribution from which the expression for $G(d) = F(d)$ follows. The above reveals that, if $X \sim G$, then $Z = \pi X^2 \sim \texttt{Exp}(\lambda)$ from which it is easy to obtain the mean and variance of the distribution G. We leave this as an exercise. Moreover, since $2\pi\lambda X^2 \sim \chi_2^2$, we note that G places a lot of mass on small distances. We expect to see some clustering under CSR.

The empirical c.d.f. for G, $\hat{G}(d)$, arises from the n nearest neighbor distances (nearest neighbor distance for $\mathbf{s}_1$, for $\mathbf{s}_2$, etc.). Denote this set by $\{d_1, d_2, ..., d_n\}$. The empirical c.d.f. for F is different since the number of "points" is arbitrary. That is, $\hat{F}(d)$ is the empirical c.d.f. arising from the m nearest neighbor distances associated with a randomly selected set of m points in D. Evidently, m is arbitrary and $\hat{G} \neq \hat{F}$ though, as above, we will be interested in looking at the difference.

With restriction to D, it is clear that we will need an edge correction if, for $\mathbf{s}_i$, $d > b_i$, where b_i is the *distance* from $\mathbf{s}_i$ to edge of D. Introducing this into the empirical c.d.f., we take

$$\hat{G}(d) = \frac{\sum_i I(d_i \leq d < b_i)}{\sum_i I(d < b_i)} \tag{8.7}$$

with a similar form for $\hat{F}$. The rationale for (8.7) is that, if $d > b_i$, then the event $\{d_i < d\}$ is not observed. Evidently, (8.7) is only sensible when d is not large.

Comparison of $\hat{G}$ and $\hat{F}$ with $G = F$ under CSR, is usually through a customary theoretical Q-Q plot.[2] Shorter tails suggest clustering/attraction, i.e., nearest neighbor distances are shorter than expected. Longer tails suggest inhibition/repulsion, i.e., nearest neighbor distances are longer than expected. Another potentially useful function is the J function, $J(d) = \frac{1-G(d)}{1-F(d)}$. This function avoids comparison with CSR (though it equals 1 in that case), rather bringing, upon reflection, the interpretation of clustering for $J(d) < 1$ and inhibition for $J(d) > 1$. $\hat{J}(d) = \frac{1-\hat{G}(d)}{1-\hat{F}(d)}$ is the customary estimate of $J(d)$. Clearly, diagnostic tools, the F and G functions provide more information than the foregoing χ^2 test. We do note that, technically, $\hat{G}(d)$ is not exactly an empirical c.d.f. since the d_i's are not independent. The same is true for $\hat{F}$. For formal testing of CSR, using say a Cramer—Von Mises test statistic, this would be problematic and perhaps a version of a Monte Carlo test would be preferred. However, with EDA intentions, perhaps this issue can be ignored.

8.3.3 The K function

Again, for a stationary process, continuing with the notation from Section 8.3.2, now consider $\mathrm{E}(N(\mathbf{s}, d; \mathcal{S}))$ which is the expected number of points in $\partial_d\mathbf{s}$, a circle of radius d, centered at $\mathbf{s}$, when $\mathbf{s} \in \mathcal{S}$ but not including $\mathbf{s}$. Using the foregoing notation and a similar calculation, we have

$$\mathrm{E}_{\mathcal{S}}\left(\sum_{\mathbf{s}_i \in \mathcal{S}, \mathcal{S} \in D} N(\mathbf{s}_i, d; \mathcal{S})\right) = \lambda|D|\mathrm{E}(N_D(\mathbf{s}, d; \mathcal{S})). \tag{8.8}$$

We have formalized an expectation of interest. That is, with respect to the set D, the right side is the expected number of points from a random $\mathcal{S}$ within distance d of a point in $\mathcal{S}$. Again, the left side of the equality motivates a natural Monte Carlo integration. Again,

---

[2]Note that $\hat{G}$ and $\hat{F}$ are free of λ while the expression for $G = F$ requires it. However, it is simple to obtain quantiles of G with say $\lambda = 1$ to compare with the ordered d_i.

empirical estimation of $\mathrm{E}(N(\mathbf{s}, d; \mathcal{S})) \geq \mathrm{E}(N_D(\mathbf{s}, d; \mathcal{S}))$ requires edge correction (see below). $\mathrm{E}(N(\mathbf{s}, d; \mathcal{S})$ does not depend on $\mathbf{s}$. E and P with respect to $N(\mathbf{s}, d; \mathcal{S})$ are referred to as Palm characteristics, features of so-called Palm distributions [see Daley and Vere-Jones, 2003]. In the literature, we find $\mathrm{E}(N(\mathbf{s}, d; \mathcal{S})) \equiv \lambda K(d)$.

The K function was introduced by Ripley [1977] and its estimator [there are several in the literature by now; see Illian et al., 2007, p. 231] is a widely used descriptive statistic. Informally, rather than nearest neighbor distance, the K function considers the *expected number* of points within distance d of an arbitrary point. Under stationarity, this expectation is the same for any point. Under CSR we can calculate it explicitly.

$$K(d) = (\lambda)^{-1}\mathrm{E}(\text{number of points within } d \text{ of an arbitrary point})$$

The scaling by $1/\lambda$, along with stationarity, enables us to have $K(d)$ free of λ. Under CSR, $K(d) = \lambda\pi d^2/\lambda = \pi d^2$, i.e., the area of a circle of radius d. A customary estimate of $K(d)$ is

$$\hat{K}(d) = (\hat{\lambda})^{-2}\sum_i\sum_{j\neq i}1(d_{ij} \equiv ||\mathbf{s}_i - \mathbf{s}_j|| \leq d)/n \tag{8.9}$$

which may be written as $(n\hat{\lambda})^{-1}\sum_i r_i$ where r_i is the number of $\mathbf{s}_j$ within d of $\mathbf{s}_i$. Here, $\hat{\lambda} = n/|D|$ where $|D|$ is the area of D.

Once again, an edge correction is needed when $\mathbf{s}_i$ is too near the boundary of D. More precisely, in (8.9), we place $\frac{1}{w_{ij}}$ inside the double sum where w_{ij} is the conditional probability that an event is in D given that it is exactly distance d_{ij} from $\mathbf{s}_i$. This probability is approximated as the proportion of the circumference of the circle centered at $\mathbf{s}_i$ with radius $||\mathbf{s}_i - \mathbf{s}_j||$ that lies within D.

As with G and F, we compare $\hat{K}(d)$ with $K(d) = \pi d^2$. Regularity/inhibition implies $\hat{K}(d) < \pi d^2$; clustering implies $\hat{K}(d) > \pi d^2$. A plot which has been proposed in the literature [see, e.g., Cressie, 1993] is $L(d)$ vs d where $L(d) = \sqrt{\frac{\hat{K}(d)}{\pi}} - d$. Evidently, $L(d) = 0$ under CSR, suggesting that we look for peaks and valleys in the plot. For instance, a peak at distance d would suggest clustering at that distance.

Under isotropy, we can connect $K(d)$ to $\gamma(d)$, where $\gamma(d)$ is the second moment measure, presented in the previous section. The relationship is $\gamma(d) = \frac{\lambda^2 K^{\mathrm{T}}(d)}{2\pi d}$. We offer a simple proof. Recall that $\gamma(\mathbf{s}, \mathbf{s}')|\partial\mathbf{s}||\partial\mathbf{s}'|$ is approximately the probability that there is a point of $\mathcal{S}$ in each of $\partial\mathbf{s}$ and $\partial\mathbf{s}'$. Hence, given there is a point in $\mathcal{S}$ at $\mathbf{s}$, in the isotropic case, $\gamma(r)/\lambda$ is the conditional probability of a point in $\mathcal{S}$ at distance r from $\mathbf{s}$ given there is a point at $\mathbf{s}$ (from the end of Section 8.2.2). If we sweep out the area of the circle of radius d around $\mathbf{s}$ using concentric circles of increasing radius r centered at $\mathbf{s}$, we have that $\int_0^d 2\pi r\gamma(r)dr/\lambda$ is the expected number of points within distance d of $\mathbf{s}$. That is, $\frac{2\pi}{\lambda}\int_0^d r\gamma(r)dr = \mathrm{E}(N(\mathbf{s}, d; \mathcal{S})) = \lambda K(d)$. So, $\frac{\lambda^2 K(d)}{2\pi} = \int_0^d r\gamma(r)dr$. Differentiating both sides with respect to d yields $\frac{\lambda^2 K'(d)}{2\pi} = d\gamma(d)$ from which the result follows.

As an aside, in the literature, we find the inhomogeneous K function associated with a general $\lambda(\mathbf{s})$ rather than a constant λ (Baddeley et al., 2001). It is defined through the pair correlation function, $g(\mathbf{s}, \mathbf{s}') = \frac{\gamma(\mathbf{s}, \mathbf{s}')}{\lambda(\mathbf{s})\lambda(\mathbf{s}')}$. In particular, if $g(\mathbf{s}, \mathbf{s}') = g_o(||\mathbf{s} - \mathbf{s}'||)$, the associated inhomogeneous K function, $K_o(d)$, satisfies $K_o(d) = 2\pi\int_0^d rg_o(r)dr$, analogous to the calculation of the preceding paragraph. In the case of an NHPP, it may be shown that $K_o(d)$ is still equal to πd^2. The modified version of (8.9) would replace $\hat{\lambda}^2$ with $\hat{\lambda}(\mathbf{s})\hat{\lambda}(\mathbf{s}')$, moved to the denominator in the respective terms of the summation.

Returning to the homogeneous case, the awkwardness in estimating $\gamma(d)$ compared with the ease of estimation and interpretation of $K(d)$ has led to the dominant usage of the latter. However, Illian et al. [2007] prefer the use of an estimator of $\gamma(d)$, in fact, of the pair

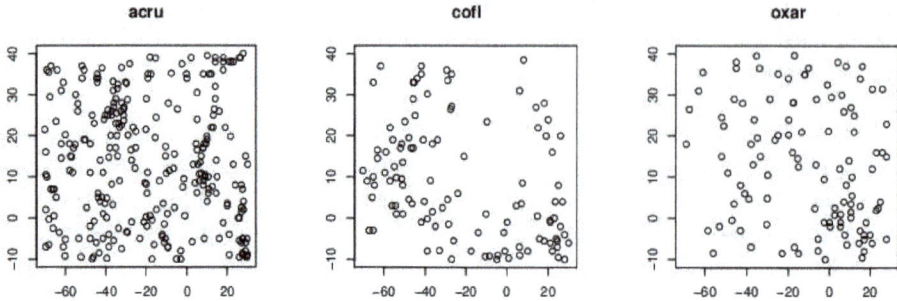

Figure 8.5 *The observed point patterns for the three species ACRU, COFL and OXAR in the Duke Forest dataset.*

correlation coefficient, $\gamma(d)/\lambda^2$. The analogue of (8.9) for γ is given by

$$\frac{1}{2\pi d} \sum_{\mathbf{s}_i, \mathbf{s}_j \in \mathcal{S}_{obs}} h(d - d_{ij})w_{ij} \tag{8.10}$$

where $d_{ij} = \|\mathbf{s}_i - \mathbf{s}_j\|$, h is a kernel function, and again, w_{ij} is as above. We see that (8.10) is a customary kernel estimator, based upon the d_{ij}'s, adjusted for edge effects. Of course, the d_{ij} are not independent.

We illustrate the EDA using the foregoing diagnostic tools. In particular, we work with a set of data from Duke Forest which considers three tree species prevalent in Duke Forest in North Carolina. The three species are ACRU: *Acer rubrum* (red maple), COFL: *Cornus florida* (flowering dogwood), and OXAR: *Oxydendrum arboretum* (sourwood) and the point patterns of their locations are from the Blackwood region in Duke Forest. Indeed, in Figure 8.5 we show the observed point pattern for each species. The region is rectangular and the coordinates shown are local in meters.

Figure 8.6 provides the theoretical Q-Q plots for the G function and the F function (adding $m = 100$ points) for the three species in the Duke Forest data. There does not seem to be strong evidence for departure from CSR in these plots (perhaps a bit for COFL). Figure 8.7 presents the theoretical Q-Q plots for the J function (top) and the L function (bottom) for the three species in the Duke Forest dataset. Compared to the F and G plots, the J plots here are a bit more revealing. For COFL, there seems to be considerable evidence of clustering and, to a lesser extent for ACRU. Finally, the lower panel in Figure 8.7 provides the L plots, taken from the K functions. Here, for all species, we see evidence of clustering with peaks in the vicinity of $d = 6$ to 8 meters.

8.3.4 Empirical estimates of the intensity

For nonstationary processes, moving away from CSR leads to interest in estimating the first-order intensity, $\lambda(\mathbf{s})$. In the absence of a model for $\lambda(\mathbf{s})$, we briefly discuss empirical estimates. The first is the analogue of a histogram, and the second of a kernel density estimate.

Imagine a refined grid over D. Then, as above, $\lambda(\partial\mathbf{s}) = \int_{\partial\mathbf{s}} \lambda(\mathbf{s})d\mathbf{s} \approx \lambda(\mathbf{s})|\partial\mathbf{s}|$. So, for grid cell A_l, assume $\lambda(\mathbf{s})$ is constant over A_l. Then, the natural estimate is $N(A_l)/|A_l|$. Evidently, a picture of this estimate will reveal a two-dimensional step surface which we might call a tile surface. Its appearance will resemble a two-dimensional histogram but the area under the surface will be the number of points in the pattern.

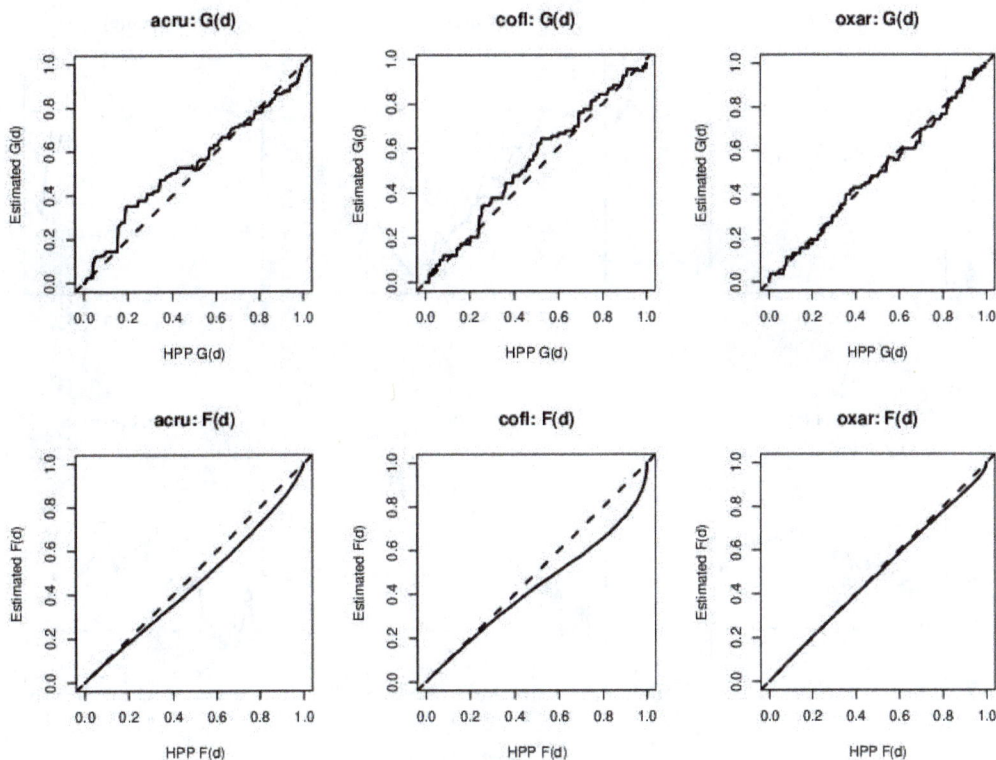

Figure 8.6 *The theoretical Q-Q plots for the G function (top) and the F function (bottom) for the three species in the Duke Forest dataset.*

Kernel density estimates are widely used, providing a smoothing of a histogram. In the same spirit, a kernel intensity estimate takes the form

$$\hat{\lambda}_\tau(\mathbf{s}) = \sum_i h(\|\mathbf{s} - \mathbf{s}_i\|/\tau)/\tau^2, \mathbf{s} \in D. \tag{8.11}$$

In (8.11), h is a radially symmetric bivariate pdf (usually a bivariate normal) while τ is "bandwidth" which controls the smoothness of $\hat{\lambda}_\tau(\mathbf{s})$. The power τ^2 in the denominator reflects the fact that the scaling of the bandwidth operates over R^2. Finally, note that we don't divide by n, as with kernel density estimates. The reason is that we *cumulate* intensity rather than *normalizing* density.

The second order intensity, γ, is defined in general but would be difficult to estimate in the absence of replications. Furthermore, the K function is not defined for a nonstationary situation except in a special case. That special case was noted in Section 8.2.2, a so-called *intensity reweighted stationary process*. This process has the feature that the pair correlation function, $\gamma(\mathbf{s}_1, \mathbf{s}_2)/\lambda(\mathbf{s}_1)\lambda(\mathbf{s}_2)$ (where $\lambda(\mathbf{s})$ is strictly positive) is stationary, in fact, isotropic, say $g(d)$ where $d = \|\mathbf{s}_1 - \mathbf{s}_2\|$. The associated intensity reweighted K function would become $K_{rew}(d) = 2\pi \int_0^d rg(r)dr$ (following the argument of the previous subsection).

As a result the natural extension of the estimate $K(d)$ becomes

$$\hat{K}_{rew}(d) = |D|^{-1} \sum_i \sum_{j \neq i} I(d_{ij} \equiv \|\mathbf{s}_i - \mathbf{s}_j\| \leq d)/\hat{\lambda}(\mathbf{s}_i)\hat{\lambda}(\mathbf{s}_j)$$

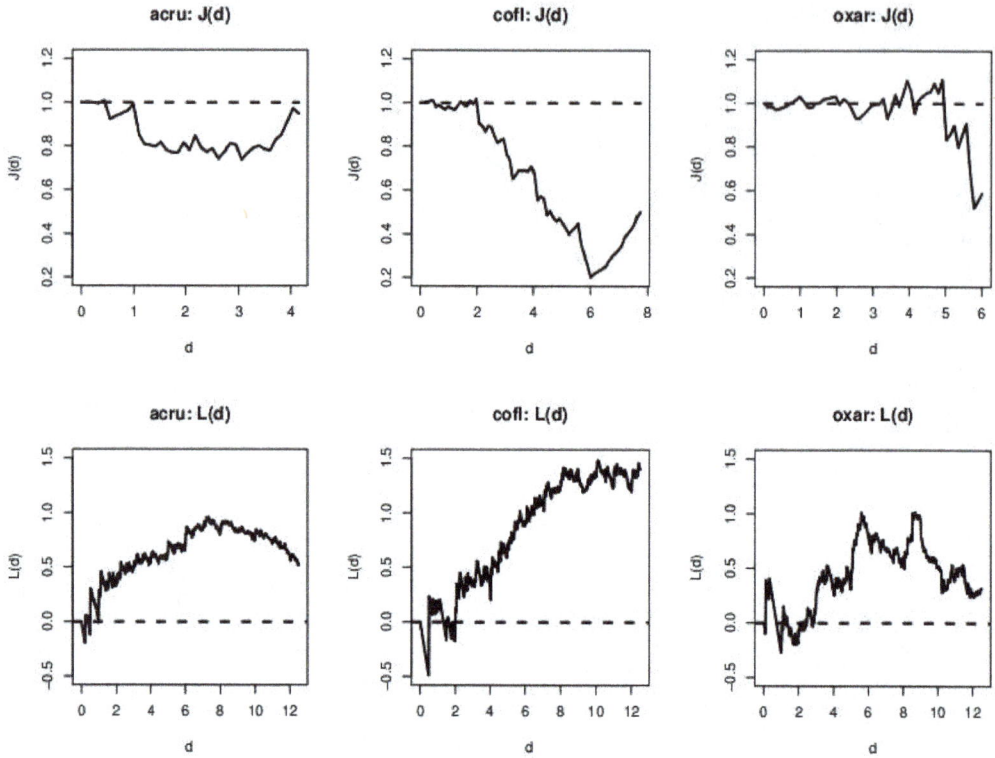

Figure 8.7 *The theoretical Q-Q plots for the J function (top) and the L function (bottom) for the three species in the Duke Forest dataset.*

with say a kernel estimate of the intensity. Similarly, the natural estimate of $\gamma(\mathbf{s}_1, \mathbf{s}_2)$ would be $\hat{g}(d)\hat{\lambda}(\mathbf{s}_1)\hat{\lambda}(\mathbf{s}_2)$ where

$$\hat{g}(d) = \frac{1}{2\pi d|D|} \sum_{\mathbf{s}_i, \mathbf{s}_j \in \mathcal{S}_{obs}} h(d - d_{ij}) w_{ij} / \hat{\lambda}(\mathbf{s}_i)\hat{\lambda}(\mathbf{s}_j).$$

It is evident [and noted in Baddeley et al., 2000], that these estimates will typically be unstable and badly biased due to the intensity estimates in the denominators. A further concern is that the stationarity of the second-order reweighted intensity is restrictive.

To illustrate, we consider a tropical rainforest tree dataset consisting of the locations of 3605 trees (left panel, Figure 8.8(a)). Provided covariates include elevation and slope (center and right panels, Figure 8.8(a)) and suggest that a constant intensity is not operating. This data has been analyzed in Møller and Waagepetersen [2007] as well as in the `spatstat` R package [Baddeley and Turner, 2005]. Here, in Figure 8.8(b), we present the associated kernel (without and with edge correction) and tiled intensity estimates. We see the very high peak associated with the very high concentration of points in the northeastern part of the region.

8.4   Modeling point patterns; NHPPs and Cox processes

We first confine ourselves to modeling of point patterns using NHPP's, deferring alternative specifications to Section 8.5. Arguably, NHPP's are the most widely used models for point

(a)

(b)

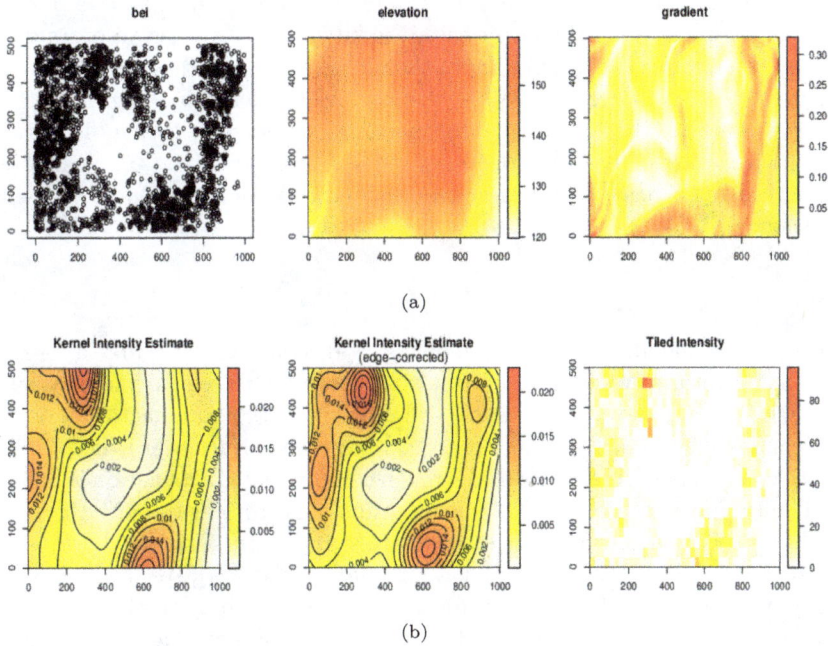

Figure 8.8 *Tropical rainfall tree data. Panel (a) shows the number of locations (left), elevation (center), and slope (right). Panel (b) presents the associated kernel and tiled intensity estimates.*

patterns. We begin by providing the NHPP likelihood, developed from the joint density in two distinct ways. In fact, the likelihood has already been alluded to in Section 8.2 through expression (8.2) and the discussion of the conditionally independent form of the location density in Section 8.2.1.

First, from Section 8.2.1, given $N(D) = n$, the location density,

$$f(\mathbf{s}_1, \mathbf{s}_2, ..., \mathbf{s}_n | N(D) = n) = \prod_i \frac{\lambda(\mathbf{s}_i)}{(\lambda(D))^n} ,$$

where, again $\lambda(D) = \int_D \lambda(\mathbf{s}) d\mathbf{s}$. So, the "joint density,"

$$f(\mathbf{s}_1, \mathbf{s}_2, ..., \mathbf{s}_n, N(D) = n) = \prod_i \frac{\lambda(s_i)}{(\lambda(D))^n} \times (\lambda(D))^n \frac{\exp(-\lambda(D))}{n!} .$$

Thus, the likelihood becomes

$$L(\lambda(\mathbf{s}), \mathbf{s} \in D; \mathbf{s}_1, ..., \mathbf{s}_n) = \prod_i \lambda(\mathbf{s}_i) \exp(-\lambda(D)) \tag{8.12}$$

Alternatively, partition D into a fine grid. The Poisson assumption implies that the likelihood will be a product over the grid cells, that is, $\prod_l \exp(-\lambda(A_l))(\lambda(A_l))^{N(A_l)}$. Regardless of the grid, the product of the exponential terms is $\exp(-\lambda(D))$. Moreover, as the grid becomes finer, $N(A_l) = 1$ or 0 according to whether there is a $\mathbf{s}_i$ in A_l or not. In the limit, we obtain (8.12).

A key point to note in (8.12) is that the likelihood is a function of the entire intensity surface. We will have an uncountable dimensional model unless we provide a parametric form for $\lambda(\mathbf{s})$. In the latter case, we can replace $\lambda(\mathbf{s})$ with its associated parameters and L becomes a function of those parameters. In the former case we have *nonparametric* specification; for example, $\lambda(\mathbf{s})$ could be a realization of a log Gaussian process, i.e., $\lambda(\mathbf{s}) = \exp(Z(\mathbf{s}))$ where $Z(\mathbf{s})$ is a Gaussian process. In the latter case, the integral in (8.12) is an ordinary integral, i.e., an integral with respect to a function. In the nonparametric case, with a GP, the integral is with regard to a random process realization; we have a stochastic integral. Such integrals have been discussed in Chapter 7 but may be more challenging to handle here because, for example, with a log Gaussian process, $\lambda(D)$ is not linear in the process. Strategy for *approximation* of $\lambda(D)$ has to be considered with care. We provide further discussion on this issue below.

8.4.1 Parametric specifications

Armed with the likelihood, we turn to modeling $\lambda(\mathbf{s})$. One form that has appeared in the literature is $\lambda(\mathbf{s}) = \sigma\lambda_0(\mathbf{s})$ with $\lambda_0(\mathbf{s})$ known, σ unknown. This seems, patently, silly; where would λ_0 come from? How would we know the intensity up to scale factor?

A more common choice is to assume that $\lambda(s)$ is a tiled surface over a grid. This requires specifying λ_l for A_l. Such tiling naturally occurs when covariate information is only available at say grid cell scale because then, it will be impossible to learn about the intensity at higher resolution. With such tiling, to incorporate spatial structure into λ, rather than the Gaussian process specification that we mentioned above for $\lambda(\mathbf{s})$, we might now employ a Markov random field model for the λ_l's. In fact, this will take us to the realm of discrete spatial data and disease mapping models, the subjects of Chapters 4 and 6, with elaboration below.

A convenient choice for λ would be a parametric function, $\lambda(\mathbf{s}; \theta)$. The challenge here is to specify a rich enough class. Simple polynomial forms will not be flexible enough; they will suffer the phenomenon of the "tail wagging the dog." That is, if they are high in some places they will be forced to be low in others. Moreover, they must be nonnegative over D. More flexible choices will be through the use of basis functions, e.g., a collection of two-dimensional basis functions (usually orthonormal), yielding $\lambda(\mathbf{s}; \boldsymbol{\theta})$ of the form, $\sum_{k=1}^{K} a_k g_k(\mathbf{s})$, where the g_k are a set of K basis functions [see, e.g., Fahrmeir et al., 2013]. The nonnegativity requirement may be handled through suitable basis functions or perhaps through exponentiation.

In this spirit, we might write $\lambda(\mathbf{s}; \theta) = \lambda f(\mathbf{s}; \theta)$ where f is a bivariate density function truncated to D. Such a form has the implicit and appealing interpretation that $\lambda = \lambda(D)$ along with the assurance that $f \geq 0$. However, to create sufficiently rich choices for f, we would turn to mixture models, e.g., $f(\mathbf{s}) = \sum_{k=1}^{K} p_k f_k(\mathbf{s})$. Fitting such models introduces challenges. First, identifying the components is a well-known difficulty [see Celeux et al., 2000]. Second, there is the potentially awkward restriction of the component densities to D. Normalizing these densities to D when D is a general geographic region will require numerical integration. Additionally, the component-wise normalizing constants will be functions of the respective component parameters.

The foregoing treats the locations as explanatory variables in the specification for $\lambda(\mathbf{s})$. So, as an alternative, $\lambda(\mathbf{s})$ may be viewed as a trend surface form. We could write $\log\lambda(\mathbf{s})$ as a trend surface in latitude and longitude or in eastings and northings. Of course, this will still result in the limitations of a polynomial form but working on the log scale mitigates the nonnegativity concern.

In practice, it is most typical to specify $\log\lambda(\mathbf{s}) = \mathbf{X}^{\mathrm{T}}(\mathbf{s})\boldsymbol{\gamma}$. That is, we envision spatial covariates to drive the point pattern and we express this through a regression model on the log scale. For example, with regard to the distribution of a species, we might expect more points where elevation is lower or where temperature is higher. If so, we will need to calculate $\int_D e^{\mathbf{X}^{\mathrm{T}}(\mathbf{s})\boldsymbol{\gamma}} d\mathbf{s}$ to obtain (8.12). This integral can be problematic since rarely will $\mathbf{X}(\mathbf{s})$ be given in functional form—consider, for example elevation as a covariate. A numerical integration will be needed.

In this regard, as noted above, often, tiling is imposed with such covariates in the sense that the covariate surface may only be resolved to subregions, e.g., population density at census scales or, with regard to climate variables, mean annual temperature or a drought index at some grid box scale. Now, the issue of the ecological fallacy may emerge (See Chapter 7). More precisely, for any subregion B, we would need $\int_B e^{\mathbf{X}^{\mathrm{T}}(\mathbf{s})\boldsymbol{\gamma}} d\mathbf{s}$. What we have is $e^{\mathbf{X}^{\mathrm{T}}(B)\boldsymbol{\gamma}}$ which can be quite different unless we are comfortable with $\mathbf{X}^{\mathrm{T}}(\mathbf{s})$ essentially constant for $\mathbf{s} \in B$, say $\mathbf{X}^{\mathrm{T}}(B)$. If so, then we only require a simple rescaling, $|B|e^{\mathbf{X}^{\mathrm{T}}(\mathbf{s})\boldsymbol{\gamma}} = \int_B e^{\mathbf{X}^{\mathrm{T}}(\mathbf{s})\boldsymbol{\gamma}} d\mathbf{s}$. In the absence of finer covariate resolution, we can not do better with regard to the ecological fallacy.

8.4.2 Nonparametric specifications

Returning to the above, suppose we take $\lambda(\mathbf{s})$ to be a process realization. Writing $\lambda(\mathbf{s}) = g(\mathbf{X}(\mathbf{s})^{\mathrm{T}}\boldsymbol{\gamma})\lambda_0(\mathbf{s})$, we insist that $g(\cdot) \geq 0$ and can think of $\lambda_0(\mathbf{s})$ as the *error* process, a realization of a positive stochastic process which, in the interest of centering, might naturally have mean 1. We recall, from Section 8.2.1, that this specification is a Cox process. In fact, conditional on $\{\lambda_0(\mathbf{s}), \mathbf{s} \in D\}$ (and $\boldsymbol{\gamma}$), we have a NHPP. Suppose $\lambda(\mathbf{s}) = \exp(Z(\mathbf{s}))$ where $Z(\mathbf{s})$ is a realization from a spatial Gaussian process with mean say $\mathbf{X}^{\mathrm{T}}(\mathbf{s})\boldsymbol{\gamma}$ and covariance function $\sigma^2\rho(\cdot)$. Then, we refer to this specification as a log Gaussian Cox process (LGCP), noting that it provides a prior for $\lambda(\mathbf{s})$, analogous to the parametric case, $\lambda(\mathbf{s}; \boldsymbol{\theta})$, where we have a prior on $\boldsymbol{\theta}$. As earlier, we might write this as $\mathbf{X}^{\mathrm{T}}(\mathbf{s})\boldsymbol{\gamma} + w(\mathbf{s})$ so that $\log\lambda_0(\mathbf{s}) = w(\mathbf{s})$.[3] Conditional on $w(\mathbf{s})$, we immediately have the first and second moments for this process. Calculation of the marginal product densities is left as an exercise. In fact, if $g(\mathbf{X}^{\mathrm{T}}(\mathbf{s})\boldsymbol{\gamma}) = \lambda$, marginally, the process is stationary.

With a Cox process, the further challenge is the evaluation of the likelihood which requires the stochastic integration, $\int_D e^{\mathbf{X}^{\mathrm{T}}(\mathbf{s})\boldsymbol{\gamma}+w(\mathbf{s})} d\mathbf{s}$. Evidently, such an integral does not have a closed-form expression. Furthermore, approximation is tricky. Suppose we can dispense with the ecological fallacy concerns as in the previous subsection. In particular, suppose $\mathbf{X}(\mathbf{s})$ is tiled to M subregions, $B_i, i = 1, 2, ..., M$. Then,

$$\int_D e^{\mathbf{X}^{\mathrm{T}}(\mathbf{s})\boldsymbol{\gamma}+w(\mathbf{s})} d\mathbf{s} = \sum_{m=1}^{M} \int_{B_m} e^{\mathbf{X}^{\mathrm{T}}(\mathbf{s})\boldsymbol{\gamma}+w(\mathbf{s})} d\mathbf{s} = \sum_{m=1}^{M} e^{\mathbf{X}^{\mathrm{T}}(B_m)\boldsymbol{\gamma}} \int_{B_m} e^{w(\mathbf{s})} d\mathbf{s}.$$

The required stochastic integrations are clear, one for each B_m. For a given B_m, if we divide the integral by $|B_m|$, we can view the integral as the expectation of $\exp(w(\mathbf{s}))$ with respect to a uniform distribution over B_m. Here, we find ourselves in similar territory to the block averaging discussed in Section 7.1. Hence, it is natural to attempt a Monte Carlo approximation to the integral, drawing random $\mathbf{s}_j, j = 1, 2, ..., J$ over B_m and taking the approximation, $|B_m| \sum_j \exp(w(\mathbf{s}_j))/J$. A critical difference between the Monte Carlo approximation proposed here and that of Section 7.1 is that there we integrate the process directly while here we integrate a nonlinear function of the process. For the former, we can argue that the behavior of the so-called *quadratic variation* is such that the difference between the stochastic integral and the Monte Carlo approximation tends to 0 in probability as $J \to \infty$. With a nonlinear function, this need not be the case.

The customary fix is to work with a finite-dimensional process. One version would simply replace $\int_{B_m} e^{w(\mathbf{s})} d\mathbf{s}$ with e^{ϕ_m} where $\{\phi_i\}$ follow a CAR model (Chapter 3 but see below). Then, $\lambda(D)$ is approximated by $\sum_{m=1}^{M} e^{\mathbf{X}^{\mathrm{T}}(B_m)\boldsymbol{\gamma}+\phi_m}$. An alternative is to specify the ϕ_m as the value of a realization of a mean 0 GP at a set of *representative points*, i.e., at a

---

[3]It is customary in the literature to make $w(\mathbf{s})$ have mean 0 but, to provide $\lambda_0(\mathbf{s})$ with mean 1 implies making $\mathrm{E}(w(\mathbf{s})) = -\sigma^2/2$

suitable point within each of the B_m's. Now $\{\phi_m\}$ will follow a familiar multivariate normal distribution. There is substantial literature on likelihood approximation in the Cox process case. See, for example, Wolpert and Ickstadt [1998], Benes et al. [2003] and the book of Moller and Waagepetersen [2003].

8.4.3   Bayesian modeling details for NHPPs and Cox processes

While the next chapter develops Bayesian modeling fitting and inference in general for spatial point patterns, it is straightforward for the setting we have described thus far. For parametric cases, we write the likelihood as $L(\boldsymbol{\theta}; \mathbf{s}_1, ..., \mathbf{s}_n)$ with a prior on a finite-dimensional $\boldsymbol{\theta}$ as usual. The parametric model may be a trend surface so that the $\boldsymbol{\theta}$ are associated coefficients. However, as noted above, it would be customary to include covariates in $\lambda(\mathbf{s}; \boldsymbol{\theta})$, so that some of the $\boldsymbol{\theta}$ become regression coefficients. In any event, with a prior on $\boldsymbol{\theta}$, the specification is complete. The result, with regard to model fitting, is to replace conditioning on λ_D with conditioning on $\boldsymbol{\theta}$. Again, we will need to calculate $\int_D \lambda(s; \theta) ds$. According to the specification for $\lambda(\mathbf{s})$, perhaps this can be done explicitly. If not, we would likely resort to numerical integration.

For the nonparametric case, again, $\lambda(\mathbf{s}) = \exp(Z(\mathbf{s}))$ where $Z(\mathbf{s}) = \mathbf{X}^{\mathrm{T}}(\mathbf{s})\boldsymbol{\gamma} + w(\mathbf{s})$ is a realization of a GP, i.e., a GP prior on $\log \lambda_D$. Following the remark at the end of the previous subsection, we can only work with a finite-dimensional distribution. We replace λ_D with $\lambda(\mathbf{s}_l^*)$ where the $\mathbf{s}_l^*$ are centroids associated with a suitably fine grid over D and yield a tiled or step surface. This introduces a multivariate normal prior for λ on the Z scale. Still, we need $\lambda(\mathbf{s})$ at every $\mathbf{s} \in D$ and we achieve this by using the $Z(\mathbf{s}_l^*)$ to create a tiled surface over D. This converts the likelihood from $L(\lambda_D; s_1, ... s_n)$ to $L(\boldsymbol{\theta}, \{w(\mathbf{s}_l^*)\}, \{w(\mathbf{s}_i), i = 1, 2, .., n\}; \mathbf{s}_1, ..., \mathbf{s}_n)$ where $\boldsymbol{\theta}$ denotes the remaining model parameters. Evidently, this is an approximation. However, the finite-dimensional set of $Z(\mathbf{s}_l^*)$'s allows a convenient numerical integration.

Prior specification for the covariance function of $Z(\mathbf{s})$, say σ^2 and ϕ with an exponential covariance choice, is not apparent nor is the potential to learn much about them. Recall the challenges articulated in Chapter 3 in the geostatistical case. Here, we are in a much more difficult setting. In the geostatistical case we observe values on the process surface, perhaps up to nugget error, to learn about the dependence structure associated with the surface. Here, we never see any realizations associated with the intensity surface; we only see a point pattern resulting from it. How much information is there in the point pattern regarding σ^2 and ϕ? Evidently, this is not a Bayesian issue; any fitting package will be challenged in this regard. Practically, we will have to depend upon the covariates, $\mathbf{X}(\mathbf{s})$, to inform about the intensity and hope that the errors are small. In the absence of useful covariates, e.g., say a constant λ as above, the $\lambda_0(\mathbf{s})$ will reflect any random clustering and gaps in the observed point pattern and can not possibly produce an estimated intensity which is roughly constant. (The same would be true for a kernel intensity estimator.) So, in our experience, we need very informative priors for σ^2 and ϕ. In fact, we typically fix ϕ to be relatively small compared with the size of D. Then, at least we have σ^2 identified [recall the discussion in Chapter 6 following the work of Zhang, 2004]. Even so, Bayesian learning may be limited and it will still be helpful to adopt an informative appropriately centered inverse gamma or log normal prior for σ^2. A convenient way to do this is through EDA based upon minimum contrast estimation using the K function or the paired correlation function [see, e.g., Diggle, 2013, Waagepetersen and Guan, 2009].

Model fitting requires updating $\boldsymbol{\gamma}, \sigma^2$ (and perhaps ϕ) along with the the w's associated with $\mathbf{s}_l^*$ as well as the w's associated with the $\mathbf{s}_i$. We use standard Gaussian proposals for the γ's and, depending upon the prior, an inverse gamma or log normal proposal for σ^2. With m representative points, the latter requires updating an $m + n$ dimensional latent multivariate normal random variable. There are many MCMC algorithms for implementing this sort of updating. A particularly efficient choice is elliptical slice sampling as developed in the work of Murray et al. [2010a] and Murray and Adams [2010]. See Section 9.4.2 for full elaboration of model fitting details for HPP and LGCP models.

8.4.3.1   The "poor man's" version; revisiting the ecological fallacy

Due to the computational demands of fitting a fully Bayesian LGCP model, it is sometimes the case that these models are fitted by aggregating the points to cells in a grid. We refer to this as the "poor man's" version of model fitting. In particular, we overlay a regular grid on D after which we work with the likelihood arising from the Poisson counts associated with the grid. Effectively, we reduce the problem to the disease mapping setting of Chapters 4 and 6. In fact, we now model the λ_i's associated with the grid cells. As in the disease mapping case, we can introduce covariates, now at the scale of the grid. Moreover, we could now introduce a CAR model for the spatial random effects, replacing the Gaussian process. The resultant specification for the intensity for grid cell i would become

$$\log\lambda_i = \mathbf{X}_i^{\mathsf{T}}\boldsymbol{\beta} + \phi_i. \tag{8.13}$$

Such a model is routine to fit, in fact, easily using WinBUGS as discussed in Section 6.8. An obvious difference between this setting and that for disease mapping is the fact that, for the latter, the areal units are specified. For the poor man's version we have no such provision. We can select any number of grid cells, any sizes, orientations, etc. Evidently, there will be a concern with regard to sensitivity to the choice of the grid. We know that the appearance of the pattern with regard to the grid cells will be scale dependent. So, we recommend taking the effort to fit at the highest resolution, the scale of the point pattern itself.

In fact, such grid cell approximation reminds us of the foregoing concern, the ecological fallacy. With no spatial random effects and intensity over D, $\lambda(\mathbf{s}) = \exp(\mathbf{X}(\mathbf{s})^{\mathsf{T}}\boldsymbol{\beta})$, we would have $\lambda_i = \int_{A_i} \lambda(\mathbf{s})d\mathbf{s} = \int_{A_i} \exp(\mathbf{X}(\mathbf{s})^{\mathsf{T}}\boldsymbol{\beta})d\mathbf{s})$. Clearly, this need not at all be close to $\exp(\int_{A_i} \mathbf{X}(\mathbf{s})^{\mathsf{T}}\boldsymbol{\beta})d\mathbf{s} = \exp(\mathbf{X}_i^{\mathsf{T}}\boldsymbol{\beta})$ which would arise using (8.13). Bringing in spatial random effects only compounds the problem. Now, ignoring the covariate term, we seek $\int_{A_i} \lambda(\mathbf{s})d\mathbf{s} = \int_{A_i} \exp(Z(\mathbf{s})d\mathbf{s}$ which evidently, is not $\exp(\int_{A_i} Z(\mathbf{s})d\mathbf{s}) = \exp(Z(A_i))$. Moreover, starting with a Gaussian process, the collection $\{Z(A_i)\}$ is a set of block averages, with a joint normal random distribution that has no connection with a CAR Model for the set $\{\phi_i\}$.

8.5   More general point pattern models

8.5.1   Cluster processes

Many observed point process realizations exhibit patterns of clustering. The notion of *clustering* or even a *cluster* is not well defined. We would recognize it as a variation in point density across region D. Informally, we would think of it as groups of points with interpoint distances that are shorter than the average distance across the pattern. In a sense it is easier to formulate models for spatial clustering than it may be to assert it or explain it for observed point patterns. We offer the family of Neyman Scott models in Section 8.5.1 and, alternatively, shot noise processes in Section 8.5.2. However, recall that, even HPP's will exhibit clustering (as we noted from the behavior of the G function in Section 8.3.2); $G(d)$ places a lot of mass on small distances. Moreover, as remarked in Diggle et al. [2007], there is a "fundamental ambiguity between heterogeneity and clustering," the first due to the spatial variation in the first-order intensity, $\lambda(\mathbf{s})$, the second due to potential stochastic dependence between the point locations, arising from the (stationary) second-order intensity $\gamma(\|\mathbf{s} - \mathbf{s}'\|)$. These two phenomena are "difficult to disentangle." Indeed, detection of

clusters in an observed point pattern is a very challenging problem. There are various algorithmic cluster detection procedures [see, e.g., Illian et al., 2007, p. 273 for some discussion]
but it will not be possible to assess how well they work since there is no notion of "truth."

The other side of the coin is inhibition or repulsion. Many physical processes are assumed theoretically to discourage points from being too close to each other. In ecological
processes, competition for resources may discourage plants from being too proximate with
each other. In a sense, such behavior is more well-defined than clustering and, as a result, it
is simpler to specify models to capture this. We consider the family of Gibbs process models
in Section 8.5.3.

In this regard, an important take home message emerges below. If you want to model
clustering, it may be best to use the Neyman-Scott or shot noise processes of Sections 8.5.1.1
and 8.5.2. If you want to model inhibition or repulsion, it may be best to use the Gibbs
processes of Section 8.5.3.

8.5.1.1   Neyman-Scott processes

Neyman-Scott processes offer one clustering strategy for enriching NHPP specifications. At
the first stage, 'parents' are generated using an HPP (more generally, an NHPP). At the
second stage, 'children' are generated, and associated with respective parents. In fact this
typically requires two stochastic stages: first generate the number of children associated
with each parent, then generate locations for each child relative to its parent. Typically,
the parents are removed (sigh!) and the point pattern realization is that of the entire set of
children.

In particular, suppose we generate *parent* events from a NHPP with $\lambda(\mathbf{s})$ say K, and
their locations say $\boldsymbol{\mu}_k, k = 1, 2, ..., K$. Next, suppose each parent produces a random (but
i.i.d.) number of offspring, N_k, where the N_k are i.i.d. according to a distribution on the
integers, say g. Typically, $g = \text{Po}(\delta)$ but, in some cases, it might make sense to have g be a
mixture of say two Poisson distributions to allow for *small* and *large* numbers of offspring.
Next, we need to locate the offspring relative to the parent. For the kth parent, suppose this
is done by assigning positions according to i.i.d. draws from a bivariate density, $f(\mathbf{s}; \boldsymbol{\mu}_k)$,
i.e., a density centered at $\boldsymbol{\mu}_k$ (usually with radial symmetry, usually a Gaussian). The case
where the bivariate density in $N(\boldsymbol{\mu}_k, \sigma^2 I)$ is referred to as the (modified) Thomas process.
Again, only the offspring are retained to yield the point pattern.

Note that, in the above specification for the Neyman-Scott process, conditional on the
number of parents, K, and their locations $\boldsymbol{\mu}_k, k = 1, 2, ..., K$, we can combine the steps
of generating the number of children and their locations. That is, generate N i.i.d $\sim g_K$
and generate $\mathbf{s}_1, \mathbf{s}_2, ..., \mathbf{s}_N$ i.i.d $\sim \sum_{k=1}^{K} \frac{1}{K} f(\mathbf{s}; \boldsymbol{\mu}_k, \Sigma)$ (yielding conditionally independent
locations). Here, we envision g_K as the distribution of the sum of K i.i.d. variates from g
(this distribution is easy to obtain when g is a member of the exponential family). This
allows us to connect Neyman-Scott processes to mixture models for the intensity for the
locations, as in Section 8.4.1, i.e., the earlier λf formulation for the intensity surface, with
λ being the expected value parameter for g_K and f being the mixture of f_k's, here with
uniform weights across components rather than random weights. With Gaussian choices for
the f_k, we might work with $\sigma^2 I_{2 \times 2}$ as the covariance matrix. In fact, we could introduce a
more general Σ but would likely work with a common choice across the k's. So, with regard
to generalization and application, it may be useful to interpret a Neyman-Scott process
through mixtures.

As a simple special case, we note the compound Poisson process. This process arises by
setting the variance of the bivariate offspring density to a point mass at 0. Then, all of the
children cluster at the $\boldsymbol{\mu}_k$. We obtain a count at each $\boldsymbol{\mu}_k$ which can be interpreted as a
'mark' at that location.

Again, the case above where $\Sigma = \sigma^2 I_{2\times 2}$ is referred to as the (modified) Thomas process in the literature [Illian et al., 2007, p. 377]. Here the density of distance, r, of offspring from the parent is evidently a Rayleigh, i.e. $\frac{r}{\sigma^2}e^{-\frac{r^2}{2\sigma^2}}, r \geq 0$. We can obtain the distribution of the distance, d between two random points in the same cluster. It is easy to see that $d \sim 2r$.

As another example, consider the Matérn process. Here, we restrict the N_k offspring of the parent at $\boldsymbol{\mu}_k$ to be uniformly distributed in a circle of radius R around $\boldsymbol{\mu}_k$. R is a model parameter and the density of distance from the parent is $\frac{2r}{R^2}, 0 \leq r \leq R$. Given R, we can obtain the distribution of the distance between two random points in the same cluster. It is

$$f_{interpoint}(d) = \frac{4d}{\pi R^2}\left(\cosh\frac{d}{2R} - \frac{d}{2R}\sqrt{1 - \frac{d^2}{4R^2}}\right), 0 \leq d \leq 2R.$$ We leave this as an exercise.

It is known that the set of Neyman-Scott processes is equivalent to the set of Cox processes [see, e.g., Cressie, 1993, p. 663]. This should not be surprising since both begin with a first-stage NHPP and then introduce process realizations to provide the second stage.

In extending this modeling idea a bit, we can imagine that there are parents outside of D who have children inside D and parents inside D who have some children outside of D. Careful investigation of this situation is taken up in Chakraborty et al. [2010] which we review in Section 9.8 of the next chapter. The goal there is to consider measurement error in point patterns so that there is an observed point pattern along with a true unobserved point pattern (i.e., with the true locations). With a simple measurement error model for observed varying around true, say a bivariate normal distribution, the challenge falls to modeling the latent true point pattern. Now the above setting emerges. Due to measurement error, the true value can be in D and the observed value outside or vice versa.

8.5.2 Shot noise processes

We offer a few words regarding the so-called shot noise process. The basic idea is, again, to define a process in two stages. For a region D, draw a point pattern $\mathcal{S}$ over D, say from a homogeneous or nonhomogeneous Poisson process. Then, assign a random mass to each sampled location. The process realization at $\mathbf{s}$ becomes $Y(\mathbf{s}) = \sum_{\mathbf{s}_i \in \mathcal{S}} h(\mathbf{s} - \mathbf{s}_i; m(\mathbf{s}_i))$. Forms for h include $h(\mathbf{s} - \mathbf{s}_i; m(\mathbf{s}_i)) = f(\mathbf{s} - \mathbf{s}_i)m(\mathbf{s}_i)$ where f is a density over D and $m(\mathbf{s}_i)$ is a positive random variable. The m's might be i.i.d., free of $\mathbf{s}$ or a regression on some covariate say $X(\mathbf{s})$ over D or a process realization over D, like a log Gaussian process or a Gamma process. So, $m(\mathbf{s}_i)$ denotes the contribution to $Y(\mathbf{s})$ from the point at $\mathbf{s}_i$ and $Y(\mathbf{s})$ accumulates the "shots" arising from the realization, $\mathcal{S}$. In essence, if we have a realization of a marked point process (Section 8.6), $\mathcal{S}_M = \{\mathbf{s}_i, m(\mathbf{s}_i)\}$, the *impulse* function h provides what is referred to as a shot noise random field over all of D.

We also can make connections to implementation of constructions such as kernel convolution (see Section 3.8.2) and predictive processes see Section 13.4. In both cases the marks are Gaussian, independent in the former, from a GP in the latter. For kernel convolution, f is a suitable kernel function; for predictive processes, it arises from the correlation function for the process of interest (see Section 13.4). The wrinkle added by the shot noise process is that the $\mathbf{s}_i$'s are a realization from a point process.

Generically, we can compute $E_{\mathcal{S}}Y(\mathbf{s})$ as a single integral over $\mathcal{S}$, given $m(\cdot)$, i.e. $\int h(\mathbf{s} - \mathbf{s}'; m(\mathbf{s}'))\lambda(\mathbf{s}')d\mathbf{s}'$ or as a double integral over the randomness in m using Campbell's theorem.

We introduce shot noise processes here as models for intensities. That is, we view $Y(\mathbf{s})$ as creating a realization of a random intensity, $\lambda(\mathbf{s})$. We supply a Cox process that is an alternative to a log Gaussian Cox process. More specifically, suppose, as in Section 8.4.2, we write $\lambda(\mathbf{s}) = e^{X^{\mathrm{T}}(\mathbf{s})\beta}\lambda_0(\mathbf{s})$ where the exponential term is the usual deterministic specification we have used before and now, $\lambda_0(\mathbf{s})$ is a mean 1 shot noise process so that $\lambda(\mathbf{s})$ is *centered* around the deterministic component. Suppose we adopt the form $\lambda_0(\mathbf{s}) = $

$\sum_{\mathbf{s}_i \in \mathcal{S}} f(\mathbf{s} - \mathbf{s}_i) m(\mathbf{s}_i)$, with $\mathcal{S}$ drawn from an HPP(λ) and $m(\mathbf{s}_i)$ a constant, m. From Campbell's theorem, we have $E(\lambda_0(\mathbf{s})) = m\lambda = 1$ so $m = 1/\lambda$ (and, then, of course, $E(\lambda_0(D)) = |D|$).

The likelihood arises in two stages, following the process definition. That is, for $\mathcal{S}_{obs}$, we have

$$L(\beta, \lambda_0(\mathbf{s}), \mathbf{s} \in D; \mathcal{S}_{obs}) = e^{-\int_D e^{X^T(\mathbf{s})\beta} \lambda_0(\mathbf{s}) d\mathbf{s}} \Pi_i e^{X^T(\mathbf{s}_i)\beta} \lambda_0(\mathbf{s}_i). \qquad (8.14)$$

Approximation to the stochastic integral would be done using representative points as described in Section 8.4. However, note that the likelihood in (8.14) is conditional on $\mathcal{S}$, a realization from an HPP(λ), i.e., $\lambda_0(\mathbf{s})$ is a function of $\mathcal{S}$. That is, we first draw $\mathcal{S}$ given λ and then we draw $\mathcal{S}_{obs}$ given β and $\mathcal{S}$. In the context of this book, such hierarchical specification is routine and causes no problems.

A version considered in, e.g., Møller and Waggepetersen (2007) extends the point pattern realization to a Poisson process over $D \times R^+$, yielding a point pattern which we denote as $\{(\mathbf{s}_i, \lambda_i)\}$. Then, $\lambda(\mathbf{s}) = \sum_{\{(\mathbf{s}_i, \lambda_i)\}} \lambda_i f(\mathbf{s} - \mathbf{s}_i)$. We see that $\lambda(\mathbf{s}) = \sum_i \lambda_i(\mathbf{s})$ so that realizations from $\lambda(\mathbf{s})$ arise as a superposition or union of independent Poisson processes with intensities $\tilde{\lambda}_i(\mathbf{s}) = \lambda_i f(\mathbf{s} - \mathbf{s}_i)$ and so provides a natural clustering process.

In fact, we can see that the Neyman-Scott process can be viewed as a special case of the shot noise process. Simply, let the $\mathbf{s}_i$ play the role of the $\boldsymbol{\mu}_i$, the cluster centers, drawn from an HPP or an NHPP over D and set the λ_i all to a constant, say δ. Then locally, at $\boldsymbol{\mu}_i$, we expect δ offspring.

8.5.3   Gibbs processes

Having invested a large part of a chapter (Chapter 4) in the development of Markov random fields and, in particular, on the formalization of Gibbs distributions, it is attractive that here, in the context of developing point process models that are not Poisson, we can revisit these specifications. Specifically, we can define the probability density of a point process over a bounded set D as the Radon-Nikodym derivative for the process measure with regard to an HPP with unit intensity over D. In particular, paralleling Chapter 4, we say that the point process model over D is a finite Gibbs process if, for n locations, its location density is

$$f(\mathcal{S}) = \exp(-Q(\mathcal{S})). \qquad (8.15)$$

Here,

$$Q(\mathbf{s}_1, \mathbf{s}_2, \ldots, \mathbf{s}_n) = c_0 + \sum_{i=1}^n h_1(\mathbf{s}_i) + \sum_{i \neq j} h_2(\mathbf{s}_i, \mathbf{s}_j) + \cdots + h_n(\mathbf{s}_1, \mathbf{s}_2, \ldots, \mathbf{s}_n). \qquad (8.16)$$

In (8.16), the h's are potentials of order $1, 2, \ldots, n$, respectively, each symmetric in its arguments. Here, c_0 plays the role of a *normalizing* constant to make $f(\mathcal{S})$ integrate to 1 over $\times D^n$. With potentials only of order 1, we obtain an NHPP with $\lambda(\mathbf{s}) = e^{-c_0 - h_1(\mathbf{s})}$. Higher order potentials capture/control interaction. As with Markov random fields, we only look at pairwise interactions, i.e., we only include h_1 and h_2. To guarantee integrability, we must take $h_2 \geq 0$. This implies that we can only capture inhibition. In other words, if we require $Q(\mathbf{s}_1, \mathbf{s}_2) \geq c_0 + h_1(\mathbf{s}_1) + h_1(\mathbf{s}_2)$, this means for pairs of points at a given distance, $f(\mathbf{s}_1, \mathbf{s}_2)$ puts less mass under the Gibbs specification than with the corresponding NHPP; we encourage inhibition. If $h_1(\mathbf{s})$ is constant, we have a homogeneous Gibbs process.

In this regard, the most common form for h_2 is $\phi(\|\mathbf{s} - \mathbf{s}'\|)$, e.g., $\phi(\|\mathbf{s} - \mathbf{s}'\|) = -\|\mathbf{s} - \mathbf{s}'\|^2/\tau^2$. Conveniently, the Papangelou conditional intensity has a simple form in this case,

$$\lambda(\mathbf{s}|\mathcal{S}) = \exp(-(h_1(\mathbf{s}) + \sum_{i=1}^n \phi(\|\mathbf{s} - \mathbf{s}_i\|))). \qquad (8.17)$$

We have the intensity for the new point adjusted by its interaction with all of the points in the given $\mathcal{S}$. Attractively, the unknown normalizing constant cancels from the conditional intensity, which we leave as an exercise.

Forward simulation of realizations from a Gibbs process is straightforward using MCMC, with birth and death in order to have n random, as presented in Moller and Waagepetersen [2003] [also see Illian et al., 2007]. Intuitively, this should be the case since we have the full conditional intensities in (8.17) to use in order to develop a Gibbs sampler. However, Bayesian model fitting for Gibbs processes, using MCMC, requires this constant since it will be a function of the parameters in h_1 and h_2. Unfortunately, this constant will be intractable (as an n-dimensional integral over $\times D^n$). A clever auxiliary variables approach to dealing with this challenge has been developed by Møller and collaborators and is presented in Section 9.3.5.

We conclude this subsection by presenting some specific examples of Gibbs processes which have received attention in the literature. They are specified through $\phi(d)$ where, again, d is an interpoint distance. The Strauss process [Strauss, 1975] sets $h_2(d) = -\log\gamma$ if $d \leq R$ and 0 otherwise. Since $h_2 \geq 0$ is required for integrability we have $0 \leq \gamma \leq 1$. Specifying $h_1(\mathbf{s}) = -log\beta$ provides a constant first-order intensity, resulting in a homogeneous Strauss process. The finite point process density for the homogeneous Strauss process is then

$$f(\mathcal{S}; \boldsymbol{\theta}) = e^{-c_0(\beta,\gamma)} \, \beta^{N(D)} \gamma^{\mathbf{s}_R(\mathcal{S})}, \tag{8.18}$$

where $N(D)$ is the number of points in $\mathcal{S} \cap D$ and $\mathbf{s}_R(\mathcal{S})$ counts the number of pairs of points $(\mathbf{s}_i, \mathbf{s}_j) \subset \mathcal{S}$ with $||\mathbf{s}_i - \mathbf{s}_j|| \leq R$, the so-called close pairs function. We note that, given R, $\mathbf{s}_R(\mathcal{S})$ is a *sufficient* statistic. Viewed as a function of R, it is useful (Section 9.2.5.1) for model checking. The Papangelou conditional intensity takes the form

$$\lambda(\mathbf{s}|\mathcal{S}; \boldsymbol{\theta}) = \beta \, \gamma^{s_R(\mathcal{S}\cup\{s\})-s_R(\mathcal{S}\setminus\{s\})}. \tag{8.19}$$

Again, the unknown normalizing constant cancels out.

The Strauss process sets $\phi(d) = \beta, d \leq d_0, = 0, d > d_0$. We see that, when $\beta > 0$, $e^{-\phi(d)} \leq 1$ for all d implying inhibition or repulsion. That is, with $\beta > 0$, the interaction term downweights patterns with more points close to each other. Choosing $\beta < 0$ implies clustering but raises the integrability challenge; the resulting density for $\mathcal{S}$ will not be integrable.

The hardcore Strauss process is an extreme case, setting $\phi(d) = \infty, d \leq R, = 0, d > R$. Now, the density is 0 for all $\mathcal{S}$ having a pair of points less than R apart. Given these two examples, we can imagine other choices for $\phi(d)$ [see, e.g., Illian et al., 2007, Section 3.6] with the only constraint being that $\phi(d) \geq 0$.

8.6 Marked point processes

Marked point processes add considerable vitality to the investigation of point patterns. That is, each point carries the extra information of a mark which captures a feature of whatever object was observed at that point. Jointly, we may be better able to understand the process yielding the points. In other words, one could ignore the marks (essentially marginalizing over the marks) but it is evident that this will sacrifice potentially important information. Marks are often discrete such as providing labels for different types of cancers over a point pattern of cancer cases (See Section 8.6.4.2) or labels for different species of trees in a forest. In this case, the data is often referred to as *multi-type*. In fact they can even index time points, leading to space-time point patterns (See Section 9.7 in the next chapter.) In each of these settings, we would be interested in seeing differences between the point patterns; aggregating them would lose this opportunity. Figure 8.9 shows kernel intensity estimates for the three tree species in the Duke Forest data viewed as three different marks.

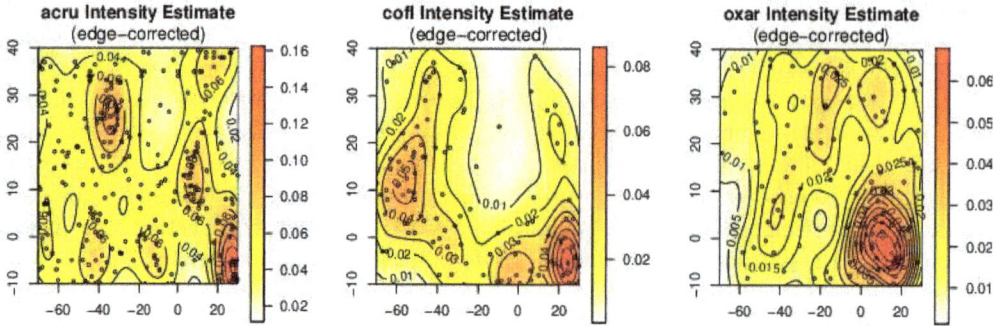

Figure 8.9 *Kernel intensity estimates for the three tree species in the Duke Forest dataset viewed as three different marks.*

We clearly see differences in the intensities. With a fuller analysis, we might try to explain these differences using local environmental features such as soil type, soil moisture, or light availability (exposed canopy area).

Continuous marks may also be of interest. For trees, we may record a height or a basal area. For an earthquake, the mark may be its strength, say on the Richter scale. As noted in the introduction, the emphasis here is on looking at both the location and the mark as random, with appropriate modeling. This contrasts with the usual geostatistical analysis (Chapter 2), where only the feature at a location is viewed as random. (See, Section 8.6.3 for further discussion in this regard.)

8.6.1   Model specifications

From a mathematical perspective, a mark is merely viewed as adding an extra coordinate to the observation, i.e., we observe $(\mathbf{s}, m)$ as a point over $D \times M$ where M is the support set for the marks. If the marks are continuous and univariate, M will be some subset of R^1 and the marked point pattern is equivalent to a point pattern over $D \times M$. If the marks are discrete, M will be a set of labels, say $l, l = 1, 2, ..., L$ and the overall point pattern can be viewed as a set of L point patterns, each over D. We remark that the notation $(\mathbf{s}, m)$ is often modified according to interpretation of the marked point process. Sometimes we might write $\mathcal{S}$ with $m(\mathbf{s}))$, $\mathbf{s} \in \mathcal{S}$, other times, $\mathcal{S}_m$. The former suggests drawing locations and then assigning labels, the latter suggests selecting labels and then drawing locations. (see Section 8.6.3).

If we follow a product space representation for a marked point process, then a marked point process is really just a point process over this product space. So, we can adopt much of the earlier theory in this chapter. For instance, $N(B \times A)$ is the number of points with locations in $B \subseteq D$ and marks in $A \subseteq M$. Defining a random counting measure leads us defining count random variables on a σ-algebra of sets over $D \times M$. In turn, this suggests a Poisson marked point process where $N(B \times A) \sim Po(\lambda(B \times A))$ for a suitable intensity measure $\lambda(B \times A)$, with independence of the count variables over disjoint product sets.

Stationarity assumes that, for any n, $\{\mathbf{s}_i, m_i, i = 1, 2, .., n\} \sim \{\mathbf{s}_i + \mathbf{h}, m_i, i = 1, 2, .., n\}$. This definition says that points are translated but marks remain the same. It may be a sensible assumption for marks that are size features. Applied to a Poisson marked point process, it simplifies the intensity measure to $\lambda(B \times A) = \lambda|B|\nu(A)$ where $\nu(\cdot)$ is a probability distribution over M. The marginal point process is an HPP with intensity λ. Conditional on locations, the marks are i.i.d. according to ν.

The process may have an intensity function, i.e., $\lambda(\mathbf{s}, m)$ such that $E(N(B \times A)) = \lambda(B \times A) = \int_B \int_A \lambda(\mathbf{s}, m) d\mu(m) d\mathbf{s}$. If the marks are continuous, we usually take $\mu(m)$ to be the Lebesgue measure. If the marks are discrete/categorical, we take $\mu(m)$ to be counting measure and write $E(N(B \times A)) = \lambda(B \times A) = \int_B \sum_{l \in A} \lambda(\mathbf{s}, l) d\mathbf{s}$. In fact, more naturally, we would write $\lambda(\mathbf{s}, l)$ as $\lambda_l(\mathbf{s})$.

For continuous marks, integrating over m yields $\lambda(\mathbf{s}) = \int_M \lambda(\mathbf{s}, m) dm$, the intensity for the point process of locations. In fact, $f(m|\mathbf{s}) = \frac{\lambda(\mathbf{s}, m)}{\lambda(\mathbf{s})}$ is the conditional density for the mark at location $\mathbf{s}$. For categorical marks, the marginal intensity is $\lambda(\mathbf{s}) = \sum_{l=1}^L \lambda_l(\mathbf{s})$. Now, the conditional probability for mark l at location $\mathbf{s}$ is $\frac{\lambda_l(\mathbf{s})}{\lambda(\mathbf{s})}$. In epidemiological settings, we would likely investigate the relative risk or relative intensity for mark l' to mark l, $\frac{\lambda_{l'}(\mathbf{s})}{\lambda_l(\mathbf{s})}$ with an interest in how it varies over D.

More general marked point process models can be developed following the earlier ideas for unmarked point processes. One possibility would be thinning, using mark-dependent thinning (Section 9.3.6.2), e.g., with discrete marks, introducing a thinning surface $p_{l,D}$ for each mark l.

We can also envision random field mark processes. Such processes assume that the realization of the point pattern S over D is independent of the realization of the stochastic process $m_D = \{m(\mathbf{s}) : \mathbf{s} \in D\}$. This model would be appropriate if we wanted to assign marks that exhibited spatial structure, e.g., marks are more similar at locations closer to each other than at locations more distant from each other. The special case where m_D is white noise, i.e., random i.i.d. assignment of marks to locations, is a null model that we would hope to reject. Such processes can be readily sampled if both S and m_D can be sampled. Dependence structure for the joint field over $D \times M$ is very complicated even if m_D is deterministic.

Another option is to extend Gibbs process models. This can be specified most conveniently with discrete marks. In particular, following Section 8.5.3, we can introduce a Gibbs process for each l. Following Section 8.2.2, we can obtain a Papangelou conditional intensity for each l. Now, we might investigate relative conditional intensities. A special case could consider the pairwise interaction terms to be common over l with only the first-order intensities dependent on l. It can be shown that such a process is equivalent to generating a realization from an unmarked Gibbs process with this common intensity and then labeling with probabilities according to the conditional mark probabilities above, $q_l(\mathbf{s}) = \frac{\lambda_l(\mathbf{s})}{\lambda(\mathbf{s})}$.

8.6.2 Bayesian model fitting for marked point processes

We offer some brief words here regarding the fitting of marked point processes. Fitting for continuous marks will follow model fitting for unmarked point process models (as elaborated in Chapter 9), working with the product space representation, employing one of the foregoing models over $D \times M$. In the literature, the predominance of such examples employs an NHPP with intensity $\lambda(\mathbf{s}, m)$ over $D \times M$. Extended to a LGCP, the foregoing approximate fitting using representative points can be directly carried over. Fitting say a Gibbs process with marks is extremely challenging and beyond our scope here.

With discrete marks, we need to model an intensity for each mark. This raises the question of whether the point patterns are dependent. For instance, in ecological processes, we can imagine symbiotic relationships which will encourage approximate co-occurrence of points. We could also imagine competitive processes such that the presence of an individual from one species would discourage the presence of an individual from another species.

Formally, these ideas are different from modeling attraction/clustering or inhibition/repulsion associated with a single species, i.e., with a single intensity. Again, these latter objectives motivate pairwise interaction processes which were discussed in Section 8.5.3.

Returning to dependence between patterns, suppose our modeling assumes the points are conditionally independent given their intensity, as with an NHPP. So, to introduce dependence we must do it for the intensities which, in turn, will impart marginal dependence to the point patterns. If we work with parametric intensities, say $\lambda(\mathbf{s}; \boldsymbol{\theta}_1)$ and $\lambda(\mathbf{s}; \boldsymbol{\theta}_2)$, then we would need to introduce a prior for $(\boldsymbol{\theta}_1, \boldsymbol{\theta}_2)$ which makes them dependent. How to do this depends upon the nature of the specification for the λ's. For instance, with regression coefficients, we might assume them to be exchangeable and add another hierarchical level to the model. If we work with nonparametric forms, employing say log Gaussian Cox processes, specified through say $w_1(\mathbf{s})$ and $w_2(\mathbf{s})$ (Section 8.4.2), we can make these dependent using elementary coregionalization as developed in Chapter 10.

8.6.3   Modeling clarification

A brief clarification of some modeling issues alluded to in the introduction to this section may prove useful here. Suppose we obtain data in the form $(\mathbf{s}_i, m_i), i = 1, 2, ..., n$, where the $\mathbf{s}_i$'s are observed locations and we think of $m_i = L(\mathbf{s}_i)$ as a discrete label, say from $l = 1, 2, ..., M$. So, we think of the $L(\cdot)$'s as marks, indicating which mark was assigned to each of the observed points. If we ignored the marks, under an NHPP model, we know the joint distribution of $(n, \{s_1, s_2, ..., s_n\}) | \lambda_D$ where $\lambda_D = \{\lambda(\mathbf{s}) : \mathbf{s} \in D\}$. From a Bayesian perspective, we only need to model λ_D to complete the specification. Adopting this perspective, with marks as above, we imagine a point pattern for each label/mark value and would extend to $\lambda_{l,D}$, the intensity associated with each label value. Then, assuming marks are also random, we would assign a prior on labels say $p_l, l = 1, 2, ...L$. In this fashion, we model the joint distribution of location and label as [location|label][label] and we would assume the pairs $(\mathbf{s}_i, L(\mathbf{s}_i))$ are conditionally independent given the $\lambda_l(\mathbf{s})$'s. Under this modeling, we have specified $[\mathcal{S}|\, L = l; \{\lambda_l(\mathbf{s})\}]$.

As in Section 8.6.1, the cumulative intensity is $\lambda(\mathbf{s}) = \sum_l \lambda_l(\mathbf{s})$, which, we note, has nothing to do with the p_l's, and

$$f(\mathbf{s}) = \frac{\lambda(\mathbf{s})}{\lambda(D)} = \sum_l \frac{\lambda_l(\mathbf{s})}{\sum_l \lambda_l(D)}$$

is the marginal location density. Turning to the joint distribution, we have $f_l(\mathbf{s}) p_l$ where $f_l(\mathbf{s}) = \lambda_l(\mathbf{s})/\lambda_l(D)$, the location density associated with mark l. We interpret $f_l(\mathbf{s}) p_l$ as drawing a label $L = l$ and then locating the label at $\mathbf{s}$ given $L = l$. That is, the draw $(\mathbf{s}, L = l)$ creates the event $(\mathbf{s}, L(\mathbf{s}) = l)$. Note that this has nothing to do with the joint intensity $\lambda(\mathbf{s}, m)$ discussed in Section 8.7.1 which adopts counting measure for m when m is discrete. It therefore yields $\lambda(\partial \mathbf{s}, \{l\}) \approx \lambda(\mathbf{s}, l)|\partial \mathbf{s}|$ and thus $\lambda(\mathbf{s}, l) = \lambda_l(\mathbf{s})$, again, free of the p_l's.

Now, consider conditioning in the opposite direction. That is, we draw a location and then assign a label to the location. Again, the draw $(\mathbf{s}, L = l)$ creates the event $(\mathbf{s}, L(\mathbf{s}) = l)$. Using Bayes' Theorem, $P(L = l|\mathbf{s}) = f_l(\mathbf{s}) p_l / \tilde{f}(\mathbf{s})$. We have the familiar rescaling of the prior weights with $\tilde{f}(\mathbf{s}) = \sum_l p_l f_l(\mathbf{s})$, a mixture density having nothing to do with the foregoing marginal location density $f(\mathbf{s})$.

Alternatively, we can imagine that the modeling situation is reversed. The label is viewed as the response at a location; now we would be modeling the joint distribution as [label|location][location]. The model for location would now have a single λ_D and the distribution for label given location would be a multinomial trial with location-specific probabilities. In the case of two labels, we might adopt a logit model, i.e., a model for $\log \frac{P(L(\mathbf{s})=1)}{P(L(\mathbf{s})=2)}$. In general, the joint distribution becomes $P(L = l|\mathbf{s}) f(\mathbf{s})$ where, as usual, $f(\mathbf{s}) = \lambda(\mathbf{s})/\lambda(D)$. Again, this means that we draw a location $\mathbf{s}$ and then assign a label $L = l$ to the location, creating the event $(\mathbf{s}, L(\mathbf{s}) = l)$. Turning to Bayes' Theorem,

$f(\mathbf{s}|L = l) = P(L = l|\mathbf{s})f(\mathbf{s})/\int_D P(L = l|\mathbf{s})f(\mathbf{s})d\mathbf{s}$ and, in fact $f(\mathbf{s}|L = l) = f_l(\mathbf{s})$, the location density associated with mark l. Thus, $\lambda_l(\mathbf{s}) = c_l P(L = l|\mathbf{s})f(\mathbf{s})$ where the constant c_l can not be identified; we can only learn about the location density for mark l but not the intensity for this mark. A last calculation shows that $\lambda(\mathbf{s}) = \sum_l \lambda_l(\mathbf{s}) = \frac{\sum_l c_l P(L=l|\mathbf{s})}{\lambda(D)}\lambda(\mathbf{s})$. Hence, $\sum_l c_l P(L = l|\mathbf{s}) = \lambda(D)$ but the c_l is not determined.

In summary, note the fundamental difference between the two joint modeling scenarios. In the first case, it is most natural that the L's have nothing to do with locations. There is a single distribution for them and given a realization (label), we have an associated intensity which provides the joint distribution for the points having that label. In the second case, we formalize an uncountable collection of $L(\mathbf{s})$'s with a single intensity for the observed points. In other words, conceptually, the joint distributions for the first case live in a different space from the joint distributions for the second case. See, also, the chapter by A. Baddeley in Gelfand et al. [2010, Chapter 21] in this regard.

This leads to the more general question of whether covariates are spatially referenced or not. Typically, variables such as sex or species type are not spatially referenced. They would more naturally be marks and we would find ourselves in the first modeling scenario, seeking to compare point patterns. If we turn them into response variables, this substantially changes the problem. It would presume that an individual can exist at every location whence, we can imagine a label for every location. Expressed in different terms, we note that a point pattern can be expressed using an indicator variable $V(\mathbf{s})$ which takes the value 1 if there is a point in the pattern at $\mathbf{s}$ and takes the value 0 otherwise. That is, the pattern is $\{V(\mathbf{s}) = 1, \mathbf{s} \in D\}$. V is equal to 1 for only a finite set of locations and is 0 for an uncountable number of locations. This clarifies the impossibility of modeling $P(V(\mathbf{s}) = 1)$ from a point pattern and is another way of distinguishing the foregoing order of conditioning in the modeling. It is intimately connected to the distinction between modeling for presence-only data vs. modeling presence-absence data; this issue forms the basis for Section 9.8.2 as well as discussion in Section 16.5. For presence-only data, we see a point pattern of locations; for presence-absence data, we see a collection of Bernoulli trials at a given set of locations.

Other covariates are naturally spatially referenced such as elevation or aspect. In this case, using an NHPP, we would naturally insert them into the model for the intensity of the point pattern. That is, they may illuminate where points are more or less likely to be. Denoting such a covariate by $X(\mathbf{s})$, the intensity would become $\lambda(X(\mathbf{s}) : \boldsymbol{\theta})$, as discussed in Section 8.4.1. Alternatively, $X(\mathbf{s})$ can be a response variable. In this case, we would build a customary point-referenced spatial model (as in Chapter 2). Presumably, we would be interested in interpolating $X(\mathbf{s})$ over D. For instance, with elevation, this would be the goal of stochastic modeling for a digital /elevation/terrain map (for more information than you need about such maps see http://en.wikipedia.org/wiki/Digital_elevation_model).

8.6.4 Enriching intensities

So, we can imagine modeling the intensity using covariates of both types. For instance, with mark l and covariate $X(\mathbf{s})$, we might specify the $\lambda(\mathbf{s})$'s using the form:

$$\lambda_l(X(\mathbf{s}); \boldsymbol{\theta}) = \mu + \alpha_l + \beta_l X(\mathbf{s}). \tag{8.20}$$

Here, we briefly present the development and extension of (8.20) as discussed Shengde et al. (2009). To provide concrete motivation, we consider marks for colon and rectum cancer. Colon and rectum cancer share many risk factors and are often tabulated together as "colorectal cancer" in published summaries. However, recent work indicating that exercise, diet, and family history may have differential impacts on the two cancers encourages analyzing them separately. The data is from the Minnesota Cancer Surveillance System from 1998–2002 over the 16-county Twin Cities (Minneapolis-St. Paul) metro and exurban area.

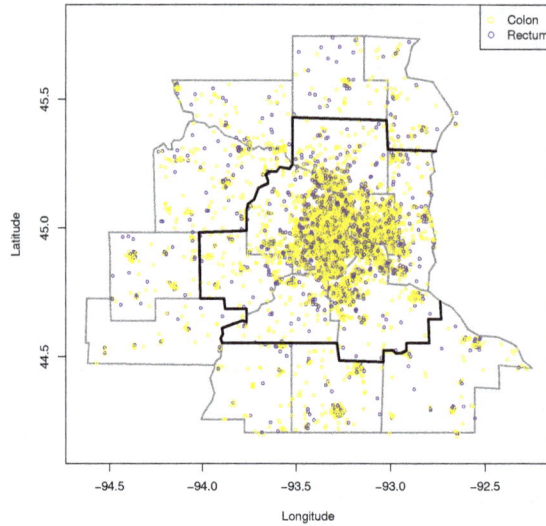

Figure 8.10 *Jittered residential locations of colon (light circle) and rectum (dark circle) cancer cases, Twin Cities metro and exurban counties, 1998–2002.*

The data consist of two marked point patterns, one for each cancer type and we expect association between the cancer types.

Figure 8.10 shows the 7 counties comprising the Twin Cities metro area as those encircled by the dark boundary; also shown are 9 adjacent, exurban counties. Within these 16 counties, we have 6544 individuals for analysis. Figure 8.10 plots the approximate locations of the cancers after the addition of a random "jitter" to protect patient confidentiality (explaining why some of the cases appear to lie outside of the spatial domain). The physiological adjacency of the colon and the rectum suggests positive dependence in these point patterns; persons with rectum cancer beyond stage 1 (i.e., regional or distant) are at risk for colon cancer due to metastasis. Moreover, the two cancers likely share unmodeled spatially-varying risk factors (such as local health care quality or availability), also suggesting positive dependence.

We assume a nonhomogeneous Poisson process with intensity function $\lambda(\mathbf{s})$ for all $\mathbf{s} \in D$. Let $\mathbf{X}(\mathbf{s})$ be a vector of location-specific covariates corresponding to a disease case observed at $\mathbf{s}$. For us, a key component of $\mathbf{X}(\mathbf{s})$ is the indicator of whether the case is in the metro area or not. However, in other contexts, we could envision information such as elevation, climate, exposure to pollutants, and so on to be relevant. We model $\lambda(\mathbf{s}) = r(\mathbf{s})\pi(\mathbf{s})$ where $r(\mathbf{s})$ is the population density surface at location $\mathbf{s}$. In practice, we may create such a surface using GIS tools and census data, or we may just work with areal unit population counts, letting $r(\mathbf{s}) = n(A)/|A|$ if $\mathbf{s} \in A$, where, as usual, $n(A)$ is the number of persons residing in A and $|A|$ is the area of A.

Returning to our framework, $r(\mathbf{s})$ serves as an offset and $\pi(\mathbf{s})$ is interpreted as a population adjusted (or *relative*) intensity, which we model on the log scale yielding

$$\pi(\mathbf{s}) = \exp(\mathbf{X}(\mathbf{s})'\boldsymbol{\beta} + w(\mathbf{s})), \qquad (8.21)$$

where $w(\mathbf{s})$ is a zero-centered stochastic process, and $\boldsymbol{\beta}$ is an unknown vector of regression coefficients. If $w(\mathbf{s})$ is taken to be a Gaussian process, we have the previously-discussed log Gaussian Cox process (LGCP). We follow the "representative points" approach resulting in the replacement of w_D with a finite set, say $w^* = \{w(\mathbf{s}_j^*), j = 1, 2, \ldots, m\}$. Then we revise the NHPP likelihood to

$$L(\beta, w^*, w(\mathbf{s}_1), ...w(\mathbf{s}_n); S)p(w^*, w(\mathbf{s}_i), ...w(\mathbf{s}_n))p(\boldsymbol{\beta}) \,. \qquad (8.22)$$

Now, we only need to work with an $(n + m)$-dimensional random variable to handle the w's, hence their prior is just an $(n+m)$-dimensional multivariate normal distribution. Note that, in (8.22), we will require that $\mathbf{X}(\mathbf{s})$ be available at each of the above $\mathbf{s}$'s. To this point, we are essentially following Section 8.4.2.

8.6.4.1   Introducing non-spatial covariate information

Next, we introduce nonspatial covariates which we think of as being of two types (though the distinction will depend upon the application). One type of covariate provides marks. For us, this covariate is cancer type (colon vs. rectum), and we are interested in whether the two cancer intensity patterns differ. The second type of covariate we view as an "auxiliary" variable that provides additional information associated with intensity. For us, age and cancer stage are examples of such covariates. Clearly patient age is associated with cancer intensity, but the strength of this association may differ across cancers. We wish to adjust the intensity to reflect patient age, analogous to the age standardization used in aggregated areal data settings.

In general, we view these latter covariates as continuous and introduce a second argument into the definition of the intensity, yielding a surface in $\mathbf{s}$ and $\mathbf{v}$ over the product space $D \times \mathcal{V}$ (i.e., the geographic space by the auxiliary covariate space).[4] Here, we are following the path of Section 8.7.1. Therefore, we generalize to

$$\pi(\mathbf{s}, \mathbf{v}) = \exp(\beta_0 + \mathbf{X}(\mathbf{s})'\boldsymbol{\beta} + \mathbf{v}'\boldsymbol{\alpha} + (\mathbf{v} \otimes \mathbf{X}(\mathbf{s}))'\boldsymbol{\gamma} + w(\mathbf{s})) \,, \qquad (8.23)$$

where the Kronecker product $\mathbf{v} \otimes \mathbf{X}(\mathbf{s})$ denotes the set of all the first-order multiplicative interaction terms between $\mathbf{X}(\mathbf{s})$ and $\mathbf{v}$. When a particular interaction term is not of interest, the corresponding coefficient in $\boldsymbol{\gamma}$ is set to zero. This development is essentially that of Section 8.7.2 where here we have made the intensity $\lambda(\mathbf{s}, m)$ explicit. That is, this expression envisions a conceptual intensity value at each $(\mathbf{s}, \mathbf{v})$ combination. The interaction terms between spatial and non-spatial covariates provide the ability to adjust the spatial intensity by individual risk factors. If we fix $\mathbf{v}$ in (8.23), we can view $\lambda(\mathbf{s}, \mathbf{v}) = r(\mathbf{s})\pi(\mathbf{s}, \mathbf{v})$ as the intensity associated with level $\mathbf{v}$. If we *integrate* over $\mathbf{v}$ (see below), we obtain the (cumulative) marginal intensity $\lambda(\mathbf{s})$ associated with $\pi(\mathbf{s}, \mathbf{v})$. Note that $\pi(\mathbf{s}, \mathbf{v})$ is a Cox process.

Now, introducing marks $k = 1, 2, \ldots, K$, a general additive form for the log relative intensity is

$$\log \pi_k(\mathbf{s}, \mathbf{v}) = \beta_{0k} + \mathbf{X}'(s)\boldsymbol{\beta}_k + \mathbf{v}'\boldsymbol{\alpha}_k + (\mathbf{v} \otimes \mathbf{X}(\mathbf{s}))'\boldsymbol{\gamma}_k + w_k(\mathbf{s}) \,. \qquad (8.24)$$

We can immediately interpret the terms on the right side of (8.24). The global mark effect is captured with the β_{0k}. Therefore, there is no intercept in $\mathbf{X}(\mathbf{s})$ and we have mark-varying coefficients for the spatially-referenced covariates, reflecting the possibility that these covariates can differentially affect the intensity surfaces of the marks. Similarly, we have mark-varying coefficients for the nuisance variables. We also have mark-varying coefficients for the interaction terms, reflecting possibly different effects of the non-spatial covariates over spatial domains. Finally, we allow the spatial random effects to vary with mark, i.e., a different Gaussian process realization for each k. Dependence in the $w_k(\mathbf{s})$ surfaces may be expected (say, increased intensity at $\mathbf{s}$ for one marked outcome encourages increased intensity for another at that $\mathbf{s}$), suggesting the need for a *multivariate* Gaussian process over

---

[4]In the case of a discrete valued covariate, any integrals over $\mathbf{v}$ in our development are replaced by sums.

the w_k. Both separable and nonseparable forms for the associated cross-covariance function can be conveniently specified through *coregionalization* as in Chapter 10.

Reduced models of (8.24) are immediately available, including e.g. $w_k(\mathbf{s}) = w(\mathbf{s})$, $\boldsymbol{\beta}_k = \boldsymbol{\beta}$, and $\boldsymbol{\alpha}_k = \boldsymbol{\alpha}$. Another interesting reduced model obtains by setting $\boldsymbol{\gamma}_k = 0$, leading to

$$\log \pi_k(\mathbf{s}, \mathbf{v}) = \beta_{0k} + \mathbf{X}'(\mathbf{s})\boldsymbol{\beta}_k + \mathbf{v}'\boldsymbol{\alpha}_k + w_k(\mathbf{s}) . \tag{8.25}$$

This separable form enables us to directly study the effect of the marks on spatial intensity. Specifically, the intensity associated with (8.24) is

$$\lambda_k(s, v) = \exp(\beta_{0k} + \mathbf{v}'\boldsymbol{\alpha}_k) \times r(\mathbf{s}) \exp(\mathbf{X}'(\mathbf{s})\boldsymbol{\beta}_k + w_k(\mathbf{s})) . \tag{8.26}$$

We see a factorization into nonspatial nuisance and spatial covariate terms. Presuming the former is integrable over $\mathbf{v}$, the latter, up to a constant, is the "marginal spatial intensity."

Integration of $\lambda_k(\mathbf{s}, \mathbf{v})$, based upon (8.24), can be computed analytically in most cases. When $\mathbf{v}$ is categorical, the likelihood integral involves only integration over the spatial domain D. When $\mathbf{v}$ is continuous, simple algebra shows

$$\int_{\mathcal{V}} \lambda_k(\mathbf{s}, \mathbf{v}) dv ds = r(\mathbf{s}) \exp(\beta_{0k} + \mathbf{X}(\mathbf{s})'\boldsymbol{\beta}_k + w_k(\mathbf{s}))$$
$$\times \int_{\mathcal{V}} \exp(\mathbf{v}'\boldsymbol{\alpha}_k + (\mathbf{v} \otimes \mathbf{X}(\mathbf{s}))'\boldsymbol{\gamma}_k) d\mathbf{v} .$$

Suppose, for instance, that there is only one component in $\mathbf{X}(\mathbf{s})$ and one component in $\mathbf{v}$ having range (v_l, v_u). Provided $\alpha_k + X(\mathbf{s})\gamma_k \neq 0$, the marginal intensity $\lambda_k(\mathbf{s})$ is then $int_{\mathcal{V}} \lambda_k(\mathbf{s}, v) dv ds$, which is equal to

$$r(\mathbf{s}) \exp(\beta_{0k} + \beta_k X(\mathbf{s}) + w_k(\mathbf{s})) \times \int_{\mathcal{V}} \exp(v(\alpha_k + X(\mathbf{s})\gamma_k)) dv$$
$$= r(\mathbf{s}) \exp(\beta_{0k} + \beta_k X(\mathbf{s}) + w_k(\mathbf{s}))$$
$$\times \frac{1}{\alpha_k + X(\mathbf{s})\gamma_k} \left[\exp(v_u(\alpha_k + z(\mathbf{s})\gamma_k)) - \exp(v_l(\alpha_k + X(\mathbf{s})\gamma_k)) \right] .$$

Turning to the revised likelihood associated with (8.24), let $\{(\mathbf{s}_{ki}, \mathbf{v}_{ki}), i = 1, 2, ... n_k\}$ be the locations and nuisance covariates associated with the n_k points having mark k. The likelihood becomes

$$\prod_k \exp\left(-\int_D \int_{\mathcal{V}} \lambda_k(\mathbf{s}, \mathbf{v}) d\mathbf{v} ds \right) \times \prod_k \prod_{\mathbf{s}_{ki}, \mathbf{v}_{ki}} \lambda_k(\mathbf{s}_{ki}, \mathbf{v}_{ki}) . \tag{8.27}$$

Using the calculations above, the double integral becomes

$$\int_D \int_{\mathcal{V}} \lambda_k(\mathbf{s}, v) dv d\mathbf{s} = \int_D \left(r(\mathbf{s}) \exp(\beta_{0k} + \beta_k X(\mathbf{s}) + w_k(\mathbf{s})) \right.$$
$$\left. \times \frac{1}{\alpha_k + X(\mathbf{s})\gamma_k} \left[\exp(v_u(\alpha_k + X(\mathbf{s})\gamma_k)) - \exp(v_l(\alpha_k + X(\mathbf{s})\gamma_k)) \right] \right) d\mathbf{s} ,$$

provided that the set $\{\mathbf{s} : \alpha_k + X(\mathbf{s})\gamma_k = 0\}$ has Lebesgue measure zero. Hence the difficulty in the likelihood evaluation is the same as in the basic likelihood and can be treated with approximation, as above. In this regard, note that we bound the components of $\mathbf{v}$ in order to integrate explicitly over $\mathbf{v}$. We do not have a stochastic integration with regard to $\mathcal{V}$ as we have over D.

	total	late=0	late=1	metro	non-metro
all	6544	2606(40%)	3938(60.2%)	5481(83.8%)	1063(16.2%)
colon	4857	1855(38%)	3002(61.8%)	4079(84%)	778(16%)
rectum	1687	751(44.5%)	751(55.5%)	1402(83.1%)	285(16.9%)
ratio	2.88	2.47	4.0	2.91	2.73

Table 8.1 *Table of colorectum cancer patients' characteristics in metro and adjacent area of Minnesota. Percentages across appropriate columns are given in parentheses, and "ratio" gives the ratio of colon to rectum cases.*

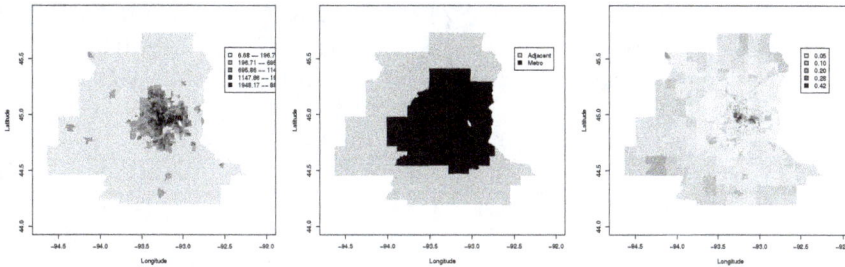

Figure 8.11 *Left, population density by tract; middle, metro/non-metro area; right, poverty rate by tract.*

8.6.4.2 Results of the analysis

Previous studies suggest that covariates related to a patient's socioeconomic status (SES) may be related to the patient's risk factors through its impact on diet, health care quality, or propensity to seek care. While our dataset lacks any individual-level SES measures, from census data we have several related tract-level variables: percentage of farm population, percentage of rural population, percentage of people with less than high school education, percentage of minority population, and poverty rate. A preliminary population-adjusted nonspatial Poisson regression analysis of our data on these covariates revealed only poverty rate and the metro indicator as significant predictors.

We consider two location-specific covariates: $z_1(\mathbf{s})$, the metro area indicator, and $z_2(\mathbf{s})$, the poverty rate in the census tract containing $\mathbf{s}$. We also employ two non-location-specific covariates: v_1, cancer stage (set to 1 if the cancer is diagnosed "late" (regional or distant stage) and 0 otherwise), and v_2, the patient's age at diagnosis. The population density $r(\mathbf{s})$ we use for standardization is available at the 2000 census tract level, meaning that we assume population density is constant across any tract.

Table 8.1 breaks down the data by stage and metro/non-metro area. We see that 38% of colon cancer cases were diagnosed at an early stage, while 44.5% of rectum cancer cases were. In total, colon cancer is nearly three times as prevalent as rectum cancer in both the metro and non-metro areas. A fact not revealed by the table is that there are 72 individuals who contribute *both* a colon and a rectum tumor. Since this is only around 1% of the total of 6544 individuals, we do not explicitly model this particular kind of dependence, but rather "lump it in" with the bivariate dependence modeled by ρ.

Figure 8.11 shows tract-level maps of population density, $r(\mathbf{s})$, and our two location-specific covariates, $z_1(\mathbf{s})$ and $z_2(\mathbf{s})$. Not surprisingly, the central metro areas are the most populated. The poverty rate is fairly uniform except for high rates in a concentrated portion of the central metro.

We fit our model, using independent Inverse Gamma(2, 0.5) priors for σ_1^2 and σ_2^2, and a $Unif(-0.999, 0.999)$ prior for ρ. The scale of the spatial decay parameter ϕ is determined by the distance function employed. In this application, we started with a $Unif(130, 390)$

model	p_D	DIC
GLM (no residuals)	11.8	1194.4
Univariate spatial residuals	72.0	692.4
Bivariate spatial residuals	80.2	688.8

Table 8.2 *Model comparison using effective model size p_D and DIC score. GLM refers to generalized linear model having no random effects.*

colon	BSR	USR	GLM
intercept	-8.76 (-9.12,-8.44)	-8.75 (-9.25,-8.40)	-8.91 (-8.99,-8.83)
metro	-0.23 (-0.49,0.04)	-0.19 (-0.42,0.06)	-0.21 (-0.29,-0.14)
poverty	-2.01 (-2.47,-1.55)	-1.90 (-2.36,-1.47)	-0.26 (-0.61,0.09)
age	0.36 (0.31,0.40)	0.36 (0.31,0.40)	0.32 (0.28,0.36)
late	0.48 (0.42,0.54)	0.48 (0.42,0.54)	0.48 (0.43,0.54)
metro*age	-0.06 (-0.11,-0.02)	-0.06 (-0.11,-0.02)	-0.06 (-0.11,-0.02)
rectum–colon	BSR	USR	GLM
intercept	-0.86 (-1.08,-0.65)	-0.84 (-1.00,-0.68)	-0.84 (-1.01,-0.69)
metro	0.02 (-0.21,0.26)	-0.07 (-0.22,0.08)	-0.07 (-0.22,0.09)
poverty	0.14 (-0.70,0.98)	-0.24 (-1.06,0.52)	-0.22 (-1.00,0.49)
age	-0.18 (-0.26,-0.10)	-0.18 (-0.26,-0.10)	-0.18 (-0.25,-0.11)
late	-0.26 (-0.37,-0.15)	-0.26 (-0.38,-0.15)	-0.26 (-0.37,-0.15)
metro*age	0.06 (-0.03,0.15)	0.05 (-0.03,0.15)	-0.01 (-0.08,0.07)
ρ	0.98(0.95,0.99)	–	–
ϕ	195	195	–
σ_1^2	0.95(0.57,1.48)	0.76(0.43,1.33)	–
σ_2^2	0.75(0.41,1.33)		–

Table 8.3 *Parameter estimates for the model with metro indicator and poverty rate as the spatial covariates, and stage and age as individual covariates. The estimates for rectum are relative effects to colon cancer. BSR=bivariate spatial residual model, USR=univariate spatial residual model, GLM=no random effects model.*

prior for ϕ, so that the effective range lies between one-fourth and three-fourths of the maximal distance between any two knots. As expected, ϕ is only weakly identified, so a fairly informative prior is needed for satisfactory MCMC behavior. For simplicity, we simply fix the range parameter at $\phi = 195$, so that the effective range is roughly half of the maximal distance. A random-walk Metropolis-Hastings algorithm is used to draw posterior samples.

Table 8.2 compares the effective model size and DIC score of three models. It can be seen that the no-random effect model (GLM) is unacceptably bad, and the model with a single set of spatial residuals is not much worse than the bivariate residual model. This suggests that the two sets of residuals are fairly similar, and that ρ is close to 1.

Table 8.3 shows parameter estimates from some of our models. We parameterize so that the top rows concern the fixed effects for colon cancers, $\boldsymbol{\beta}_1$, but the second set of rows give the *differential* effect in the rectum cancer group, $\boldsymbol{\Delta} \equiv \boldsymbol{\beta}_2 - \boldsymbol{\beta}_1$. Thus, any 95% Bayesian confidence intervals that exclude 0 in this part of the table suggest a variable that has a significantly different impact on the two cancers.

In general, the effects of the non-spatial covariates are fairly similar across the models considered. We find that in the metro area there are relatively fewer cases of both colon and rectum cancer. This is consistent with statewide patterns of colorectal cancer occurrence in Minnesota, where higher age-adjusted rates are often found in non-metro areas. However, there is no significant change in this relationship in the rectum group relative to the colon group. Turning to the non-location-specific covariates, age is significantly associated with increasing colon cancer, but a somewhat surprising relative *decrease* in rectum cancer. This difference (-0.18) is statistically significant, but not large enough in magnitude to make the overall age effect in the rectum group negative. A look at the data bears this out, with rectum cancers arising in a somewhat younger population; our preliminary Poisson regression also concurs, though here the relative decrease in the rectum group is not significant. Late detection provides another interesting difference between the colon and rectum groups: while there are significantly more cases diagnosed late than early, the effect of late diagnosis is significantly smaller in the rectum group (point estimate -0.26). Thus public health interventions to encourage screening and early detection of colorectal cancer will have a significantly greater impact on prevention for the colon than for the rectum. The metro-age interaction shows that the effect of age on colon cancer is significantly less pronounced in the metro area; a smaller "age adjustment" to the colon cancer intensity process is needed in the metro area. This effect is largely absent for rectum cancer, but this difference is not quite statistically significant. Finally, the estimate of ρ is very close to 1, indicating very similar spatial residual patterns. This is perhaps a surprisingly strong association, but believable given that these are *residual* surfaces, which account (at least conceptually) for important missing covariates, which could be spatial (e.g., local screening percentage, other sociodemographic factors) or nonspatial (e.g., the physiological adjacency of the colon and the rectum).

8.7 Exercises

1. If $\mathcal{S}_1$ is a realization from a NHPP with intensity $\lambda_1(\mathbf{s})$ and $\mathcal{S}_2$ is a realization from a NHPP with intensity $\lambda_2(\mathbf{s})$, independent of $\mathcal{S}_1$, show that the joint realization comes from a NHPP with intensity $\lambda_1(\mathbf{s}) + \lambda_2(\mathbf{s})$.

2. If $f(\mathbf{s}_1, ..., \mathbf{s}_n) = \Pi_i f(\mathbf{s}_i)$, show that the second order intensity satisfies $\gamma(\mathbf{s}, \mathbf{s}') = \lambda(\mathbf{s})\lambda(\mathbf{s}')$.

3. (a) **Stationary spatial point process**: For a spatial point process, suppose $N(B + \mathbf{h}) \sim N(B), \forall B \in \mathcal{B}$ and $\forall \mathbf{h}$. Show that the second-order intensity, $\gamma(\mathbf{s}, \mathbf{s}')$ satisfies $\gamma(\mathbf{s}, \mathbf{s} + \mathbf{h}) = \gamma(\mathbf{h})$.
   (b) **Isotropic spatial point process**: For a spatial point process, suppose $N(B) \sim N(PB), \forall B \in \mathcal{B}$ where P is an orthogonal matrix and $PB = \{\mathbf{s}^* = P\mathbf{s} : \mathbf{s} \in B\}$. Then $\gamma(\mathbf{s}, \mathbf{s}') = \gamma(||\mathbf{s} - \mathbf{s}'||)$.

4. Prove Campbell's Theorem. That is, if $g(\mathbf{s})$ is a point feature and we are interested in $\sum_{\mathbf{s}_i \in \mathcal{S}} g(\mathbf{s}_i)$, then $E_{\mathcal{S}}(\sum_{\mathbf{s}_i \in \mathcal{S}} g(\mathbf{s}_i) = \int g(\mathbf{s})\lambda(\mathbf{s})d\mathbf{s}$. Hint: Show it is true for an indicator function.

5. Show that, for $\mathbf{s} \in D$, a bounded region, for the conditional intensity, $\lambda(\mathbf{s}|\mathcal{S})$, $E_{\mathcal{S}}(\lambda(\mathbf{s}|\mathcal{S}) = \lambda(\mathbf{s})$. (Hint: If you can not prove this in general, show it for the NHPP.)

6. Show that all nonstationary point processes that arise from $p(\mathbf{s})$ thinning of a stationary point process are such that the second-order reweighted intensity functions (or pair correlation functions) for the original stationary and the reweighted nonstationary process are identical.

7. For the homogeneous Gibbs process in Section 8.5.3, show that the Papangelou conditional intensity is given by (8.17). What does it become for the Strauss process? For the hardcore process?

8. Recalling the discussion of the F and G functions from Section 8.3.2, show that, if $X \sim G$, then $Z = \pi X^2 \sim \text{Exp}(\lambda)$. Hence, obtain the mean and variance of X.

9. For a log Gaussian Cox process with log intensity, $\log\lambda(\mathbf{s}) = X^{\mathrm{T}}(\mathbf{s})\beta + z(\mathbf{s})$ where $z(\mathbf{s})$ is a mean 0 Gaussian process with covariance function $C(\mathbf{s}, \mathbf{s}')$, obtain the marginal first and second order product densities.

10. For a Matérn process with restriction of offspring to a circle of radius R, consider the distribution of the distance between two random points in the same cluster. Show that it is $f_{interpoint}(d) = \frac{4d}{\pi R^2}\left(\cosh\frac{d}{2R} - \frac{d}{2R}\sqrt{1 - \frac{d^2}{4R^2}}\right), 0 \le d \le 2R.$

Chapter 9

Bayesian analysis of point pattern models

9.1 Introduction

Here, we consider model fitting and inference exclusively within the Bayesian framework. Our general approach follows a simulation-based strategy predicated upon the ability to draw realizations (point patterns) under the specified model after model fitting. Customarily, model fitting is done using a Markov chain Monte Carlo algorithm. However, in Section 9.5 we discuss approximate Bayesian computation (ABC) model fitting which only requires simulation under the model and thus, fits in well with our general approach. Also, INLA (see references below) is now becoming a widely employed fitting software. However, in either case, full inference (beyond that for explicit functions of model parameters) entails simulation as we develop in the ensuing sections.

This chapter also contains discussion of spatio-temporal point pattern analysis for both continuous and discrete time and concludes with some additional topics.

With posterior samples of model parameters, we can obviously do arbitrary inference on the model parameters. With posterior predictive samples of point patterns, we can perform arbitrary inference on any process features. With prior and/or predictive samples of point patterns, we can study model adequacy and model comparison, as well as make prior-posterior comparison. Important Bayesian contributions for analyzing spatial point patterns have been made by Møller and colleagues [see, e.g., Moller and Waagepetersen, 2003, Møller and Waagepetersen, 2007] and references therein. In addition, there has been a recent strand which considers Poisson process models, focusing on a rich range of specifications for the intensity [see, e.g., Kottas and Sansó, 2007, Taddy and Kottas, 2012]. Some recent Bayesian work has employed integrated nested Laplace approximation (INLA) [Rue et al., 2009] for inference. Bayesian point pattern analysis using INLA can be found in Illian et al. [2012], King et al. [2012a], Illian et al. [2013]. Work that follows our inference paradigm is suggested in a conference address by Møller [Møller, 2012]. In particular, for model adequacy, we propose an examination of Bayesian residuals (drawing on the work of Baddeley et al. [2005, 2008]), both realized and predictive, as well as empirical coverage, and prior predictive checks through Monte Carlo tests. For model selection, we use predictive mean square error, empirical coverage, and ranked probability scores. We consider both in-sample and, when possible, out-of-sample approaches. We can demonstrate that *prior* predictive model checking is needed for an effective assessment of model adequacy. Bayesian model checking work, primarily validating intensities, is found in Taddy [2010], Zhou et al. [2015], and Xiao et al. [2015].

9.2 The generic approach

At a high level, we consider the generic model form $[\mathcal{S} \mid \boldsymbol{\theta}][\boldsymbol{\theta}]$. We observe $\mathcal{S}_{obs}$. Upon model fitting, we obtain posterior samples $\boldsymbol{\theta}_l^*$ from $[\boldsymbol{\theta} \mid \mathcal{S}_{obs}]$. These enable inference about a function of $\boldsymbol{\theta}$, $b(\boldsymbol{\theta})$, assuming it is available explicitly. Using composition sampling, we can create posterior predictive samples $\mathcal{S}_l^*$ from $[\mathcal{S} \mid \mathcal{S}_{obs}]$ by drawing $\mathcal{S}_l^*$ from $[\mathcal{S} \mid \boldsymbol{\theta}_l^*]$.

DOI: 10.1201/9781003401728-9

Using these point pattern samples, we create posterior samples of any function h of $\mathcal{S}$ as $\{h(\mathcal{S}_l^*), l = 1, 2, \ldots, L\}$ from $[h(\mathcal{S}) \mid \mathcal{S}_{obs}]$. **So, our story is: for a given model, if we can fit it and if we can sample realizations from it, we can carry out arbitrary inference.**

A primary feature we are trying to infer about is a (random) surface, i.e., an intensity. It is important to appreciate that the intensity surface, $\lambda(\mathbf{s})$, reflects observed data points but also unobserved points. That is, a point pattern provides more information than just the locations of the observed points. *Absence* at other locations is informative [Baddeley et al., 2005]. This leads to a consequential challenge: we never observe a point on this surface. We can see the analogue with density estimation. In fact, we have empirical kernel intensity estimates, as shown in Section 8.3.4. However, unlike density estimation, here, the number of points is random. This implies that comparison of point patterns must be done with care.

To see the challenge in its simplest form, consider a homogeneous Poisson process (HPP) setting. From Section 8.3.4 of the previous chapter, any observed point pattern will provide an empirical intensity estimate which is a two-dimensional step surface, not at all close to a flat surface. Similarly, a kernel intensity estimate will not be close to flat unless a very large bandwidth is employed. How far from a constant surface can we be and still believe that a constant intensity is operating? In this regard, the null hypotheses for the HPP, $H_o : \lambda(\mathbf{s}) = \lambda$ seems *untenable*; it would never be operating in practice except perhaps under a simulation setting. Rather, it is more sensible to compare inference under an HPP model with that from other models.

In general, with spatial point patterns, it is hard to criticize models and to choose between models. For instance, a regression based intensity using spatially referenced covariates, say an NHPP, and a clustering-based intensity using say a Neyman Scott process, may equally well explain peaks in an intensity. Often, the choice will be made according to the nature of the process generating the observed data. There is not a large literature here (but we do review some below and the `spatstat` package [Baddeley and Turner, 2005] offers some tools).

Returning to the general inference framework, let us elaborate posterior study of features. Posterior distributions of interest might include: for arbitrary sets A and B, $[N(A) \mid \mathcal{S}_{obs}]$, $[N(A), N(B) \mid \mathcal{S}_{obs}]$, $[N(A) \mid N(B), \mathcal{S}_{obs}]$, and $[\frac{N(A)}{N(D)} \mid \mathcal{S}_{obs}]$. We might also seek the posterior for the G and K functions under a given model. This is novel territory since the literature only considers empirical estimates of these function (Section 8.3.3). Comparison of the posterior distribution for G or K with the associated empirical estimate could be informative. Comparison, if appropriate, with the G or K functions under an HPP could also be informative. Here, comparison is through formal inference rather than exploratory comparison.

Turning to model assessment, residuals are a commonly used tool. In particular, Baddeley et al. [2005, 2008] develop various notions of residuals for point patterns. For example, they define a *raw* residual, analogous to the standard residual from a regression model, as

$$R_{\hat{\boldsymbol{\theta}}}(B) \equiv N(B) - \int_B \hat{\lambda}(\mathbf{s} \mid \mathcal{S})d\mathbf{s} \tag{9.1}$$

for $B \subseteq D$, where $\hat{\lambda}(\mathbf{s} \mid \mathcal{S}) \equiv \lambda(\mathbf{s} \mid \mathcal{S}; \hat{\boldsymbol{\theta}})$ is the estimated Papangelou conditional intensity function. In the Bayesian setting, we would work with the *realized* residual, which considers the posterior of (9.1), employing $\lambda(\mathbf{s} \mid \mathcal{S}, \boldsymbol{\theta})$.

More generally, Baddeley et al. [2005] define the h-weighted innovation measure as

$$I(B, h, \lambda) \equiv \sum_{\mathbf{s}_i \in \mathcal{S} \cap B} h(\mathbf{s}_i, \mathcal{S} \backslash \mathbf{s}_i) - \int_B h(\mathbf{s}, \mathcal{S} \backslash \mathbf{s}) \lambda(\mathbf{s} \mid \mathcal{S})d\mathbf{s}. \tag{9.2}$$

These innovations have mean 0 under the true model, as can be seen using (9.7), developed below. Choices of h include $h(\mathbf{s}, \mathcal{S} \backslash \mathbf{s}) = 1/\lambda(\mathbf{s} \mid \mathcal{S})$ which defines the inverse λ residuals, in the spirit of Stoyan and Grabarnik [1991]. With $h(\mathbf{s}, \mathcal{S} \backslash \mathbf{s}) = 1/\sqrt{\lambda(\mathbf{s} \mid \mathcal{S})}$, we obtain an analogue of the Pearson residual from Poisson regression. Estimators are obtained by inserting an estimator of $\lambda(\mathbf{s} \mid \mathcal{S})$. A final residual, which we shall not consider here, is the pseudoscore residual, which sets $h(\mathbf{s}, \mathcal{S} \backslash \mathbf{s}) = \frac{\partial}{\partial \boldsymbol{\theta}} \log\{\lambda(\mathbf{s} \mid \mathcal{S})\}$, where $\boldsymbol{\theta}$ denotes the parameters of the intensity function λ.

Again, from a Bayesian perspective, the posterior distribution of $\int_B h(\mathbf{s}, \mathcal{S} \backslash \mathbf{s}) \lambda(\mathbf{s} \mid \mathcal{S}) ds$ and $I(B, h, \lambda)$ would be studied. In particular, these innovations are of the form $t(\mathcal{S}, \boldsymbol{\theta})$ and so their posteriors can be obtained through simulation as detailed below. We can use the posterior mean, $\mathrm{E}(\int_B h(\mathbf{s}, \mathcal{S} \backslash \mathbf{s}) \lambda(\mathbf{s} \mid \mathcal{S}) ds \mid \mathcal{S}_{obs})$, to obtain a point estimate and can also examine whether 0 falls in a given credible interval. Baddeley et al. [2005, 2008] provide formulas for the variance calculations of residuals and innovations.

With regard to validation, under a given model, consider credible intervals created from these innovation distributions developed over many sets. If the model is true, should we expect to achieve empirical coverage of 0 at roughly the nominal level? For the raw/realized innovations, the answer is clearly no. The raw innovations compare an observed count with the posterior distribution for the *expectation* of that count. Though we hope the expectations are close to the raw innovations, the credible intervals provide coverage for the expected counts rather than for the counts themselves. Thinking of the regression analogue, the raw innovations are akin to employing the distribution $[y - \mu_y \mid \text{Data}]$ when we should be employing the distribution for the *predictive* innovations, $[y - y_{\text{pred}} \mid \text{Data}]$.

So instead, we adopt *predictive residuals*,

$$R_{\text{pred}}(B) = N_{\text{obs}}(B) - N_{\text{pred}}(B), \tag{9.3}$$

where, as above, posterior samples $\mathcal{S}_l^*$ supply the draws $N_l^*(B)$, hence the posterior predictive distribution of $N_{\text{pred}}(B)$ and, in turn, of $R_{\text{pred}}(B)$.

Finally, for an h-scaled innovation as in (9.2), Baddeley et al. [2005] define the smoothed innovation field $r(\mathbf{u}; \boldsymbol{\theta})$ at location $\mathbf{u} \in D$ as

$$
\begin{aligned}
r(\mathbf{u}; \boldsymbol{\theta}) &= e(\mathbf{u}) \int_D k(\mathbf{u} - \mathbf{v}) dI(\mathbf{v}, \mathbf{h}, \boldsymbol{\theta}) \\
&= e(\mathbf{u}) \left[\sum_{\mathbf{s}_i \in \mathcal{S}} k(\mathbf{u} - \mathbf{s}_i) \mathbf{h}(\mathbf{s}_i, \mathcal{S} \backslash \{\mathbf{s}_i\}) - \int_D k(\mathbf{u} - \mathbf{v}) \mathbf{h}(\mathbf{v}, \mathcal{S}) \lambda(\mathbf{v} \mid \mathcal{S}; \boldsymbol{\theta}) d\mathbf{v} \right],
\end{aligned} \tag{9.4}
$$

where $k(\mathbf{s})$ is a probability density on $\mathbb{R}^2$ used as a smoothing kernel and $e(\mathbf{u}) \equiv 1/\int_D k(\mathbf{u} - \mathbf{v}) d\mathbf{v}$ is an edge correction. This field puts positive atoms at each $\mathbf{s}_i \in \mathcal{S}$ and a negative value elsewhere and then smooths using the kernel. So, positive values indicate locations where the empirical intensity was higher than the intensity of the fitted model while negative values indicate areas where the intensity of the fitted model was higher.

Baddeley et al. [2005] estimate $\boldsymbol{\theta}$ to obtain a residual field, $r(\mathbf{u}; \hat{\boldsymbol{\theta}})$. For us, for the NHPP and LGCP models, with a posterior distribution for $\lambda(\mathbf{s}; \boldsymbol{\theta})$, we can obtain a posterior distribution for $r(\mathbf{u}; \boldsymbol{\theta})$. Additionally, we can create a plot showing those regions that have a credible interval (for the smoothed innovation) which contains 0, as well as those regions that have a credible interval above or below 0.

9.2.1   Some details

Often the parametric function $b(\boldsymbol{\theta})$ is not available explicitly (see below). Then, we need to find an $h(\mathcal{S})$ such that $\mathrm{E}(h(\mathcal{S}) \mid \boldsymbol{\theta}) = b(\boldsymbol{\theta})$. As a result, in order to obtain a $b(\boldsymbol{\theta}_l^*)$, for each $\boldsymbol{\theta}_l^*$, we need to generate replicates, i.e., for each l, the set $\{\mathcal{S}_{lb}^*, b = 1, 2, \ldots, B\}$. These replicates provide a Monte Carlo integration for $b(\boldsymbol{\theta}_l^*)$, i.e., $\frac{1}{B}\sum_b h(\mathcal{S}_{lb}^*)$.

The most general objects of interest would take the form $t(\mathcal{S}, \boldsymbol{\theta})$, leading to the posterior, $[t(\mathcal{S}, \boldsymbol{\theta}) \mid \mathcal{S}_{obs}]$. With t available explicitly (as it will be in practice), we can use the $\{\boldsymbol{\theta}_l^*, \mathcal{S}_l^*\}$ to create posterior draws.

Examples of $b(\boldsymbol{\theta})$ include $\mathrm{E}(N(A) \mid \boldsymbol{\theta})$ and $\mathrm{E}(N(A)N(B) \mid \boldsymbol{\theta})$. They also include $\lambda(\mathbf{s}; \boldsymbol{\theta})$, $\gamma(d; \boldsymbol{\theta})$, $\lambda(A; \boldsymbol{\theta})$, $g(d; \boldsymbol{\theta})$ (the pair correlation function), $G(d; \boldsymbol{\theta})$, and $K(d; \boldsymbol{\theta})$. To be more explicit, $\mathrm{E}(N(A) \mid \mathcal{S}_{\mathrm{obs}}) \approx \frac{1}{L}\sum_{l=1}^{L}\sum_{\mathbf{s}_{li}^* \in \mathcal{S}_l^*} 1(\mathbf{s}_{li}^* \in A)$. This suggests how we could create model-based Bayesian intensity estimates in the setting where we do not have an explicit form for $\lambda(\mathbf{s})$. (With an explicit form, we would directly obtain the posterior mean $\mathrm{E}(\lambda(\mathbf{s}) \mid \mathcal{S}_{obs})$ from posterior samples and plot this as a surface using a fine grid of $\mathbf{s}$.) Taking $A = \partial\mathbf{s}$ yields the Bayes estimator for $\lambda(\partial\mathbf{s}) \approx \lambda(\mathbf{s})|\partial\mathbf{s}|$, hence for $\lambda(\mathbf{s})$. Therefore, again with a fine grid of $\mathbf{s}$, a Bayesian estimator of the intensity surface results. The size of $\partial\mathbf{s}$ can be viewed as analogous to a bandwidth selection for a kernel intensity estimate. In fact, recall the usual kernel smoothing yields kernel intensity estimate, $\lambda_\tau(\mathbf{s}) = \frac{1}{\tau^2}\sum_{\mathbf{s}_i \in \mathcal{S}} h(\|\mathbf{s} - \mathbf{s}_i\|/\tau)$ discussed in Section 8.3.3. We can contrast the model-based posterior estimate of the intensity with the empirical kernel intensity estimate.

Turning things around, if we can write λ as a parametric function, $\lambda(\mathbf{s}; \boldsymbol{\theta})$ (say for an NHPP but not for a LGCP), posterior samples of $\boldsymbol{\theta}$ yield an estimate of $\lambda(\mathbf{s}; \boldsymbol{\theta})$. Then, numerical integration would enable a posterior estimate of $\lambda(A; \boldsymbol{\theta})$. Details for the G and K functions, as parametric functions, are presented below.

Examples of $h(\mathcal{S})$'s include: $N(A)$, $(N(A), N(B))$, $\frac{N(A)}{N(D)}$ along with the foregoing predictive residuals , e.g., the distribution $[N_{obs}(A) - N(A) \mid \mathcal{S}_{obs}]$. We could also estimate conditional events with distribution $[N(A) \mid N(B) = m; \mathcal{S}_{obs}]$. Examples of $t(\mathcal{S}, \boldsymbol{\theta})$ include: realized residuals, e.g., the distribution of $[N(A) - \lambda(A; \boldsymbol{\theta}) \mid \mathcal{S}_{obs}]$ and the inhomogeneous K function $K_{inhom}(d; \boldsymbol{\theta})$ [Baddeley et al., 2000] discussed in Section 8.3.3.

9.2.2   Campbell's Theorem and the GNZ result

The main theoretical tool we employ here is Campbell's Theorem (Section 8.2.2), which we recall here. It provides the expectation of the summation over $\mathcal{S} \cap D$ of a function $h(\mathbf{s}_i)$ (restriction to D ensures that realizations of $\mathcal{S}$ are finite so that expectations exist). It states that

$$\mathrm{E}_{\mathcal{S} \cap D}\left(\sum_{\mathbf{s}_i \in \mathcal{S} \cap D} h(\mathbf{s}_i)\right) = \int_D h(\mathbf{s})\lambda(\mathbf{s})\, d\mathbf{s}. \tag{9.5}$$

For example, letting $h(\mathbf{s}) = 1(s \in A)$ for some set $A \subset D$, Campbell's Theorem says that $\sum_{\mathbf{s}_i \in \mathcal{S}} 1(\mathbf{s}_i \in A)$ is an unbiased estimator for $\int_D 1(s \in A)\lambda(\mathbf{s})\, d\mathbf{s} = \int_A \lambda(\mathbf{s})\, d\mathbf{s} = \lambda(A)$ based on $\mathcal{S}$, i.e., $N(A)$ is an unbiased estimator of $\lambda(A)$, as we already know. However, more generally, $\sum_{\mathbf{s}_i \in \mathcal{S} \cap D} h(\mathbf{s}_i)$ is an unbiased estimator of $\int_D h(\mathbf{s})\lambda(\mathbf{s})d\mathbf{s}$. More usefully, (9.5) suggests Monte Carlo integration to obtain the right side. With samples $\mathcal{S}_l^*$ from $[\mathcal{S} \cap D \mid \lambda(\mathbf{s})]$, or, more generally, $[\mathcal{S} \cap D \mid \boldsymbol{\theta}]$, averaging the sum on the left side of (9.5) over these replicates provides a Monte Carlo integration for the left side, hence for the right side. That is, we can compute integrations relative to $\lambda(\mathbf{s})$ even if we can not obtain $\lambda(\mathbf{s})$ explicitly!

The utility of this approach for Bayesian inference is immediately evident. Now, $\lambda(\mathbf{s})$ is random and so, for a given h as above, the right side of (9.5) is random. However, with posterior (prior) predictive samples of point patterns, we can directly create a Monte Carlo integration for the expectation on the left side, hence a Monte Carlo integration for the posterior (prior) mean of the integral on the right side. Again, the right side of (9.5) is viewed as $b(\boldsymbol{\theta})$ which typically will not be available explicitly. So, for example, to obtain $\mathrm{E}(\int_D h(\mathbf{s})\lambda(\mathbf{s})ds \mid \mathcal{S}_{obs})$, posterior draws of $\lambda_l(\mathbf{s})$ would provide posterior predictive draws,

$\mathcal{S}_l^*$, to which we apply the sum on the left side of (9.5). Averaging over these draws produces the posterior mean for the right side.

We also recall the bivariate version of Campbell's Theorem (Section 8.2.2). For h, a function of two points in $\mathcal{S}$:

$$\mathrm{E}_{\mathcal{S} \cap D}\Big(\sum_{\substack{\mathbf{s}_i, \mathbf{s}_j \in \mathcal{S} \cap D \\ i \neq j}} h(\mathbf{s}_i, \mathbf{s}_j) \Big) = \int_D \int_D h(\mathbf{s}, \mathbf{s}') \gamma(\mathbf{s}, \mathbf{s}') \, d\mathbf{s} \, d\mathbf{s}'. \tag{9.6}$$

(9.6) is useful for exploring second-order properties of a point process, e.g., the second-order intensity which is defined through $h(\mathbf{s}, \mathbf{s}') = 1(\mathbf{s} \in A, \mathbf{s}' \in B)$. Another application arises for the Strauss process (Section 8.6.3) where we consider the "close pairs" function, $s_R(\mathcal{S}) = \sum_{\mathbf{s}_i, \mathbf{s}_j \in \mathcal{S} \subset D} 1(\|\mathbf{s}_i - \mathbf{s}_j\| \leq R)$. Applying Campbell's Theorem, we find $\mathrm{E}_{\mathcal{S} \cap D}(s_R(\mathcal{S})) = \int_D \int_D 1(\|\mathbf{u} - \mathbf{v}\| \leq R) \gamma(\mathbf{u}, \mathbf{v}) d\mathbf{u} d\mathbf{v} = \int \int_{\|\mathbf{u}-\mathbf{v}\| \leq d} \gamma(\mathbf{u}, \mathbf{v}) d\mathbf{u} d\mathbf{v}$. Therefore, it enables similar Monte Carlo integration for the posterior mean of the right side.

A more general result which can also be useful is the Georgii-Nguyen-Zessin (GNZ) formula [Georgii, 1976, Xanh and Zessin, 1979], which applies to h of the form $h(\mathbf{s}; \mathcal{S} \backslash \{\mathbf{s}\})$ and gives the equality

$$\mathrm{E}_{\mathcal{S} \cap D}\Big(\sum_{\mathbf{s}_i \in \mathcal{S}} h(\mathbf{s}_i, \mathcal{S} \backslash \{\mathbf{s}_i\}) \Big) = \mathrm{E}_{\mathcal{S} \cap D}\Big(\int_D h(\mathbf{s}, \mathcal{S} \backslash \mathbf{s}) \lambda(\mathbf{s} \mid \mathcal{S}) d\mathbf{s} \Big), \tag{9.7}$$

where $\lambda(\mathbf{s} \mid \mathcal{S})$ is the Papangelou conditional intensity. Again, Monte Carlo integration enables a posterior mean for the right side. Here, a choice for $h(\mathbf{s}_i \mid \mathcal{S} \backslash \mathbf{s}_i)$ might be $1(\min_{\mathbf{s}_j \in \mathcal{S} \backslash \mathbf{s}_i} \|\mathbf{s}_i - \mathbf{s}_j\| \leq d)$ for a given d which connects to the G function (Section 8.3.2).

Returning to Campbell's Theorem, it was noted above that by summing over the indicator function $1(\mathbf{s}_i \in A)$, we provide an unbiased estimator for $\mathrm{E}(N(A)) = \lambda(A; \boldsymbol{\theta})$ whose usual Bayes estimate is $\mathrm{E}(\lambda(A; \boldsymbol{\theta}) \mid \mathcal{S}_{obs})$. If $\lambda(A; \boldsymbol{\theta})$ is available explicitly, a Monte Carlo integration for $\mathrm{E}(\lambda(A; \boldsymbol{\theta}) \mid \mathcal{S}_{obs})$ is $\frac{1}{L} \sum_l \lambda(A; \boldsymbol{\theta}_l^*)$. When we cannot calculate $\lambda(A; \boldsymbol{\theta})$, we note that $\mathrm{E}(\lambda(A; \boldsymbol{\theta}) \mid \mathcal{S}_{obs}) = \mathrm{E}(N(A) \mid \mathcal{S}_{obs}) \approx \frac{1}{L} \sum_{l=1}^{L} \sum_{\mathbf{s}_{li}^* \in \mathcal{S}_l^*} 1(\mathbf{s}_{li}^* \in A)$, providing a Monte Carlo integration. Of course, the elements of the set $\{ \sum_{\mathbf{s}_{li}^* \in \mathcal{S}_l^*} 1(\mathbf{s}_{li}^* \in A) \}_{l=1}^{L}$ provide posterior samples of $N(A)$.

Summarizing, we may be interested in inference on $b(\boldsymbol{\theta})$ based upon $[b(\boldsymbol{\theta}) \mid \mathcal{S}_{obs}]$. With posterior samples, $\{\boldsymbol{\theta}_l^*\}$, we obtain $\{b(\boldsymbol{\theta}_l^*)\}$ for such inference, as usual. If interest is in $[h(\mathcal{S}) \mid \mathcal{S}_{obs}]$, then the set $\{\mathcal{S}_l^*\}$ provides the set $\{h(\mathcal{S}_l^*)\}$ for inference. For $[t(\mathcal{S}, \boldsymbol{\theta}) \mid \mathcal{S}_{obs}]$ with t available explicitly, we can use $\{\boldsymbol{\theta}_l^*, \mathcal{S}_l^*\}$. Again, if $b(\boldsymbol{\theta})$ is not available explicitly, the strategy is then to find $h(\mathcal{S})$ such that $\mathrm{E}(h(\mathcal{S}) \mid \boldsymbol{\theta}) = b(\boldsymbol{\theta})$.

With regard to the G and K functions, they are parametric functions of the form $G(d; \boldsymbol{\theta})$ and $K(d; \boldsymbol{\theta})$, respectively. Except in special cases, they are not available in closed form, leading to the use of empirical, rather than model-based, estimation. However, we have expressions (8.6) and (8.8) from Section 8.3.2. For each of these, the summation on the left side is over terms of the form $h(\mathbf{s}, \mathcal{S} \backslash \mathbf{s})$. Hence the GNZ result in (9.7) applies, enabling Monte Carlo integrations with respect to the Papangelou conditional intensity. In other words, these Monte Carlo integrations can provide the posterior mean of the G and K functions under a specified model given $\mathcal{S}_{obs}$. We obtain model based estimates for G and K rather than the empirical estimates supplied in Sections 8.3.2 and 8.3.3. Comparison between the model based estimate and the empirical estimate can be enlightening with regard to the adequacy of the model. The same argument can be applied to the inhomogeneous K function [Baddeley et al., 2000] (Section 8.3.3). However, there we have a form $t(\mathcal{S}, \boldsymbol{\theta})$ to which we would apply the results.

The discussion of residuals above provides another set of quantities of interest to which we can apply the foregoing. For example, suppose $h(\mathbf{u}; \mathcal{S} \backslash \mathbf{u}) = 1(\mathbf{u} \in B)$. This yields

$E_{\mathcal{S}} N(\mathcal{S} \cap B) = \int_B E_{\mathcal{S}} \lambda(\mathbf{u} \mid \mathcal{S}) d\mathbf{u}$. In turn, this suggests $N(\mathcal{S} \cap B) - \int_B \lambda(\mathbf{s} \mid \mathcal{S}) d\mathbf{s}$, the realized *innovation* residuals, which have mean 0 [Baddeley et al., 2005]. Suppose $h(\mathbf{u}; \mathcal{S} \backslash \mathbf{u}) = 1(\mathbf{u} \in B)/\lambda(\mathbf{u} \mid \mathcal{S})$. This yields $E_{\mathcal{S}}(\sum_{\mathbf{s}_i \in \mathcal{S}} 1(\mathbf{s}_i \in B)/\lambda(\mathbf{s}_i \mid \mathcal{S} \backslash \mathbf{s}_i)) = |B|$ [Stoyan and Grabarnik, 1991], the so-called "inverse" residuals. Application to other *scaled* residuals is clear.

Consider the GNZ applied to a LGCP with random intensity $\log \lambda(\mathbf{s}) = z(\mathbf{s})$ where $z(\mathbf{s})$ is a GP. Then, given D, $E_{\mathcal{S} \cap D}(\sum_{\mathbf{s}_i \in \mathcal{S} \cap D} h(\mathbf{s}_i; (\mathcal{S} \cap D) \backslash \mathbf{s}_i) = \int_D h(\mathbf{s}) E(\lambda(\mathbf{s})) d\mathbf{s}$ if h depends only on $\mathbf{s}$. Again, with restriction to D, a finite point pattern and posterior samples immediately provide a Monte Carlo integration yielding a Bayes estimate of the right side.

Suppose $h(\mathbf{u}; \mathcal{S} \backslash \mathbf{u}) = 1(\mathbf{u} \in \partial \mathbf{s})$ yielding $E_{\mathcal{S}} N(\mathcal{S} \cap \partial \mathbf{s}) = \int_{\partial \mathbf{s}} E_{\mathcal{S}} \lambda(\mathbf{u} | \mathcal{S}) d\mathbf{u}$. The left side is $\int_{\partial \mathbf{s}} \lambda(\mathbf{u}) d\mathbf{u} \approx \lambda(\mathbf{s}) |\partial \mathbf{s}|$. The right side is $E_{\mathcal{S}}(\int_{\partial \mathbf{s}} \lambda(\mathbf{u} \mid \mathcal{S}) d\mathbf{u}) = E_{\mathcal{S}}(\lambda(\partial \mathbf{s} \mid \mathcal{S})) \approx E_{\mathcal{S}}(\lambda(\mathbf{s} \mid \mathcal{S})) |\partial \mathbf{s}|$. So, marginally, $\lambda(\mathbf{s}) \approx E_{\mathcal{S}}(\lambda(\partial \mathbf{s}) \mid \mathcal{S})$.

Continuing with the GNZ result, now suppose $h_D(\mathcal{S}) \equiv \sum_{\mathbf{s}_i \in \mathcal{S} \cap D} h(\mathbf{s}_i; (\mathcal{S} \cap D) \backslash \mathbf{s}_i)$. Then, $E_{\mathcal{S} \cap D \mid \boldsymbol{\theta}}(h(\mathcal{S})) = E_{\mathcal{S} \cap D \mid \boldsymbol{\theta}}(\int_D h(\mathbf{s}; (\mathcal{S} \cap D) \backslash \mathbf{s}) \lambda(\mathbf{s} \mid \mathcal{S}) d\mathbf{s}) \equiv b_{h_D}(\boldsymbol{\theta})$. In order to achieve a normalization, we propose to work with $\bar{h}_D(\mathcal{S}) \equiv h_D(\mathcal{S})/N(\mathcal{S} \cap D)$. If $N(\mathcal{S} \cap D) = 0$, then $h_D(\mathcal{S}) = 0$ and we define $\frac{0}{0} = 1$. So, we consider $E_{\mathcal{S} \cap D \mid \boldsymbol{\theta}}(\bar{h}(\mathcal{S})) \equiv b_{\bar{h}_D}(\boldsymbol{\theta})$. Evidently, we need a different version of the GNZ result which we offer in the next subsection.

9.2.3 *An iterated expectation version*

We consider a different way of calculating expectations which suggests a different way of developing Monte Carlo integrations. We imagine that our model can provide a realization $\mathcal{S}$ over $\mathbb{R}^2$ which induces $\mathcal{S} \cap D$ over D with an associated finite $N(\mathcal{S} \cap D)$. Alternatively, suppose, given D, we first generate $N(\mathcal{S} \cap D) = n$ and then we locate $\mathcal{S}$ over D given $N(\mathcal{S} \cap D) = n$, assuming the $\mathbf{s}_i$ are exchangeable. This is the *generative* view that we have noted before for, e.g., a NHPP or a cluster process as opposed to a *modeling* or mechanistic view, e.g., a Gibbs process.

Regardless, there is a joint distribution $[\mathcal{S} \cap D, N(\mathcal{S} \cap D)]$, hence a conditional times marginal version $[\mathcal{S} \cap D \mid N(\mathcal{S} \cap D)][N(\mathcal{S} \cap D)]$. We may not be able to write down these densities explicitly but, formally, we can calculate the expectation iteratively (and we can obtain expectations explicitly in certain cases as we show below).

That is,

$$E_{\mathcal{S} \cap D}(\sum_{\mathbf{s}_i \in \mathcal{S} \cap D} h(\mathbf{s}_i; (\mathcal{S} \cap D) \backslash \mathbf{s}_i)) = E_{N(\mathcal{S} \cap D)} E_{\mathcal{S} \cap D \mid N(\mathcal{S} \cap D)} \sum_{\mathbf{s}_i \in \mathcal{S} \cap D} h(\mathbf{s}_i; (\mathcal{S} \cap D) \backslash \mathbf{s}_i)$$

$$= E_{N(\mathcal{S} \cap D)}(N(\mathcal{S} \cap D) E_{\mathcal{S} \cap D \mid N(\mathcal{S} \cap D)}(h(\mathbf{s}, (\mathcal{S} \cap D) \backslash \mathbf{s})).$$

That is, using the normalized form (and defining $0/0 = 1$), $E_{\mathcal{S} \cap D}(\sum_{\mathbf{s}_i \in \mathcal{S} \cap D} h(\mathbf{s}_i; (\mathcal{S} \cap D) \backslash \mathbf{s}_i))/N(\mathcal{S} \cap D) = E_{\mathcal{S} \cap D} h(\mathbf{s}; (\mathcal{S} \cap D) \backslash \mathbf{s})$. Attractively, we remove $N(\mathcal{S} \cap D)$ from the second expectation on the right side. With this notation, $E_{\mathcal{S} \cap D \mid \boldsymbol{\theta}} \bar{h}_D(\mathcal{S}) = b_{\bar{h}_D}(\boldsymbol{\theta})$ where $b_{\bar{h}_D}(\boldsymbol{\theta}) = E_{\mathcal{S} \cap D \mid \boldsymbol{\theta}} h(\mathbf{s}; (\mathcal{S} \cap D) \backslash \mathbf{s})$. Below, we propose choices for $h(\mathbf{s}; (\mathcal{S} \cap D) \backslash \mathbf{s})$ which are of interest. As usual, a Bayes estimate for $b_{\bar{h}_D}(\boldsymbol{\theta})$ is $E(b_{\bar{h}_D}(\boldsymbol{\theta}) \mid \mathcal{S}_{obs})$. With posterior samples, $\{\boldsymbol{\theta}_l^*\}$, a Monte Carlo integration for the posterior mean is $\frac{1}{L} \sum_l b_{h_D}(\boldsymbol{\theta}_l^*)$. Of course, typically, we can not calculate $b_{\bar{h}_D}(\boldsymbol{\theta})$ explicitly. However, from above, $E_{\mathcal{S} \cap D \mid \mathcal{S}_{obs}} \bar{h}(\mathcal{S}) = E_{\boldsymbol{\theta} \mid \mathcal{S}_{obs}} E_{\mathcal{S} \cap D \mid \boldsymbol{\theta}} \bar{h}(\mathcal{S}) = E_{\boldsymbol{\theta} \mid \mathcal{S}_{obs}} b_{\bar{h}}(\boldsymbol{\theta})$. So, a direct Monte Carlo integration becomes $\frac{1}{L} \bar{h}(\mathcal{S}_l^*)$.

Suppose we want posterior samples of $b_{\bar{h}_D}(\boldsymbol{\theta})$. Obviously, they are $\{b_{\bar{h}_D}(\boldsymbol{\theta}_l^*)\}$. But again, we cannot calculate $b_{\bar{h}_D}(\boldsymbol{\theta}_l^*)$. With a sample of point patterns $\{\mathcal{S}_{lb}^*, b = 1, 2, \ldots, B\}$ from $[\mathcal{S} | \boldsymbol{\theta}_l^*]$, we have Monte Carlo integrations for $\{b_{\bar{h}_D}(\boldsymbol{\theta}_l^*)\}$. Altogether, we need a *nested* sampling of point patterns to obtain the desired posterior samples.

9.2.4   Model adequacy and model comparison

Out of sample model assessment using a fitting/training sample and an independent vali-dation/test sample is now becoming standard practice. With point pattern data, such an approach is little discussed and may not be available. With a conditionally independent location distribution, as with NHPPs and LGCPs, the answer is yes. However, with, e.g., an inhibition model, holding out points will alter the geometry of the point pattern. It will change the nature of the inter-point distances, hence the interaction structure. This will be true in general for a point pattern model where there is dependence between the locations of the points as with a Gibbs process.

For models with conditionally independent locations, we develop training and test datasets. Suppose we decide to administer 20% holdout. We can not simply remove 20% of the data at random. This will *fix* the size of the point pattern rather than allowing it to be random. Rather, the p-thinning approach, as in Section 9.3.6.2, can be applied to create appropriate training and test data. Letting p denote the retention probability, p-thinning proceeds by independently, point-by-point, removing $\mathbf{s}_i \in \mathcal{S}$ with probability $1 - p$. This produces a training point pattern $\mathcal{S}_{train}$ and test point pattern $\mathcal{S}_{test}$, which are independent, conditional on $\lambda(\mathbf{s})$. In fact, $\mathcal{S}_{train}$ has intensity $p\lambda(\mathbf{s})$, $\mathcal{S}_{test}$ has intensity $(1-p)\lambda(\mathbf{s})$, and the revised validation intensity compared with the fitting intensity is $\lambda_{test}(\mathbf{s}) = \left(\frac{1-p}{p}\right)\lambda_{train}(\mathbf{s})$.

9.2.5   Model adequacy through empirical coverage

When cross-validation is possible, consider the following strategy. For a given set B, pos-terior predictive point patterns will supply the posterior predictive distribution of $N(B)$. The predictive residuals discussed in Section 9.2 should be centered around zero for an adequate model. If we look at many subregions B_k, we expect the empirical coverage to be roughly the nominal level of coverage if the model is adequate. How shall we create a set $\{B_k\}$? [Baddeley et al., 2005, section 11.1] propose to analyze a set of residuals over disjoint partitions B_k of the domain, similar to quadrat counting [Diggle, 2013]. With an irregular domain D, division into disjoint subregions of similar size can be time-consuming and is, in fact, unnecessary. We suggest to draw random subregions uniformly over D and then evaluate the residuals or innovations in each subregion. Moreover, there is no reason to require the B_k be disjoint. If not, this allows us to draw as many B_k as desired, subject to the requirement that each B_k has the same area. Denote the area of each B_k by $q|D|$ where $q \in (0,1)$ such that q represents the size of each B_k relative to D. For various q's we can evaluate the innovation or residual measures on each of the B_k's and obtain the observed empirical coverage of 0. Intuition (and practical experience) suggests that larger regions will validate better than smaller ones.

In the sequel, we take the shape of each B_k to be a square but, depending upon D, there may be some reason to choose the shape more carefully. The use of squares sometimes limits the placement of the B_k when q is large and also access to the edges of D. Furthermore, with randomly placed, overlapping B_k, it can be hard to identify regions where the model fits poorly. Disjoint B_k alleviate this problem but, with regard to empirical coverage, the success rate of Bernoulli trials based upon random B_k's will suffice.

9.2.5.1   In-sample model adequacy

When we can not develop a test data set, how can we investigate model adequacy? Evidently, the foregoing empirical coverage approach can be implemented in-sample. However, we would not expect it to criticize the model very well. This leads to the work on posterior model checks [Gelman et al., 1996a, henceforth GMS] and prior model checks [Dey et al., 1998, henceforth DGSV] introduced in Section 5.2. As noted there, GMS is more common and easier to do. However, it doesn't criticize the model well enough and uses the data twice (once to fit, once to check). DGSV is more computationally demanding but is formally coherent and uses the data only once.

We now elaborate on both the GMS and DGSV approaches in the context of spatial point patterns. Both employ Monte Carlo tests in looking at discrepancy measures, $D(\mathcal{S}; \boldsymbol{\theta})$ which, for instance, might take the form of a realized residual, $N(A) - \lambda(A; \boldsymbol{\theta})$. GMS looks at draws from $[D(\mathcal{S}; \boldsymbol{\theta}) \mid \mathcal{S}_{obs}]$ and compares with draws from $[D(\mathcal{S}_{obs}; \boldsymbol{\theta}) \mid \mathcal{S}_{obs}]$. The double use of the data is evident. Draws of $\mathcal{S}_l^*$ from $[\mathcal{S}; \boldsymbol{\theta} \mid \mathcal{S}_{obs}]$ will look too much like $\mathcal{S}_{obs}$; discrepancies, $D(\mathcal{S}_l^*, \boldsymbol{\theta})$, will look too much like $D(\mathcal{S}_{obs}; \boldsymbol{\theta})$. The model checking will not be critical enough.

DGSV creates draws from $[D(\mathcal{S}, \boldsymbol{\theta}) \mid \mathcal{S}]$ by sampling $\boldsymbol{\theta}$ from the prior, then sampling $\mathcal{S}$ under the model given $\boldsymbol{\theta}$. Fitting the model enables draws from $[\boldsymbol{\theta} \mid \mathcal{S}]$ and, hence, with a collection $\mathcal{S}_l^*$, draws from $[D(\mathcal{S}_l^*, \boldsymbol{\theta}) \mid \mathcal{S}_l^*]$. Then, comparison is made between $[D(\mathcal{S}_l^*, \boldsymbol{\theta}) \mid \mathcal{S}_l^*]$ and $[D(\mathcal{S}_{obs}; \boldsymbol{\theta}) \mid \mathcal{S}_{obs}]$. This is an "apples vs. apples" comparison which uses the data only once. That is, DGSV compares the observed discrepancy with the discrepancies you expect under the model; GMS compares the observed discrepancy with what you expect under the model **and** the observed data. The computational demand required for DGSV is evident; one must fit and sample for every $\mathcal{S}_l^*$.

In-sample, our empirical coverage model adequacy check also suffers the GMS problem; it will not be critical enough. Again, for a collection of B_k's, we look at the set $\{[N_{obs}(B_k) - N(B_k) \mid \mathcal{S}_{obs}]\}$ and check empirical coverage relative to nominal coverage. We see that the $\mathcal{S}_l^*$'s will be too similar to $\mathcal{S}_{obs}$ (using noninformative priors) so the $N(B_k)$ that we generate given $\mathcal{S}_{obs}$ will tend to look too much like $N_{obs}(B_k)$, since the latter is a function of $\mathcal{S}_{obs}$. So, we assert that there is no role for empirical coverage here unless we can do it out-of-sample. In-sample empirical coverage will be inadequate to criticize the model.

As an alternative, it is better to generate $N_l^*(B)$ through $\mathcal{S}_l^*$'s from the marginal distribution rather than from the posterior distribution. Now, we can run a Monte Carlo test comparison between $[N_{obs}(B_k) - N(B_k) \mid \mathcal{S}_{obs}]$ and $\{[N_l^*(B_k) - N(B_k) \mid \mathcal{S}_l^*]\}$. Unfortunately, this demands a lot of comparison. For each B_k, we need to compare an "observed" posterior distribution vs. say 99 generated posterior distributions, say using quantiles. This reveals the issue/challenge of lots of simultaneous inference.

Let's consider a simpler checking function approach which can be expected to supply model criticism through the prior predictive framework without requiring the computation associated with DGSV. Consider $h(\mathcal{S})$, a function only of the point pattern. For instance, in assessing the adequacy of an HPP or a Strauss process, given a radius R, suppose we consider the statistic, $\mathbf{s}_R(\mathcal{S})$ discussed in Section 8.5.3. We can implement a Monte Carlo test for $\mathbf{s}_R(\mathcal{S}_{obs})$ and the set $\{\mathbf{s}_R(\mathcal{S}_l^*), l = 1, 2, \ldots, L\}$ where the $\mathcal{S}_l^*$'s are generated under the model. If there is interaction between the points in $\mathcal{S}$, then, as we run through a set of R's (motivated by the size of the region), these Monte Carlo tests should criticize the HPP model but potentially support Strauss process models in the vicinity of a suitable R. However, we may need a large number of points in the point pattern to extract criticism.

9.2.6 Model comparison

Model selection tools are notably lacking for point pattern models. A typical analysis uses ad hoc tests of the homogeneity and independence assumptions of CSR but, having decided which assumption to relax, there is no clear procedure for comparing models. Often model comparison is not even considered; as we noted earlier, a model is adopted on mechanistic or behavioral grounds. Lack of fit using the methods described above can eliminate some models but will not help when choosing among adequately fitting models.

Again, following Section 5.2.3.4, we would argue that model comparison should be done in predictive space since parameters are constructs within models and need not mean the same thing across models. So, then the question is, "What would we be predicting?" A natural choice would focus on the distribution, $[N(A) \mid S_{obs}]$ for $A \subset D$. In particular, we would compare $N_{obs}(A)$ with $[N(A) \mid S_{obs}; M_j]$ for each model, $j = 1, 2, \ldots, J$. Here, for model j with parameters $\boldsymbol{\theta}_j$, we obtain posterior samples, $\boldsymbol{\theta}_{j,l}^*$ and then $S_{j,l}^*$. Again, we would want to do this out of sample through p-thinning, as with NHPPs, LGCPs, and cluster processes, which are superpositions of NHPPs. (See Section 9.3.6 with regard to the above.)

As for criteria, we can look at predictive mean square error (PMSE), perhaps normalized by the expected number (the usual loss function for Poisson counts). We can look at ranked probability scores (RPS) [Gneiting and Raftery, 2007] following Section 5.2.3.4. The RPS arises from proper scoring rules and offers a useful metric for assessing the performance of a predictive distribution. Again, it compares an entire distribution (in this case a posterior predictive distribution) to the observed value. The more concentrated the distribution is around the observation, the smaller the RPS. That is, RPS prefers models which provide predictions that are concentrated around the observed value. For count data, the RPS is appropriate [Epstein, 1969]. Specifically, the RPS compares the posterior predictive distribution for a cell count with the degenerate distribution associated with the observed cell count using a sum of squares over the set of support values $\{0, 1, 2, \ldots\}$.

We propose choosing subregions B_k uniformly over D, with each B_k having the same size and potentially overlapping other $B_{k'}$. In fact, we can use the same B_k as in the Monte Carlo assessment above. We obtain $N(B_k)$ from the hold-out dataset and compare with $[N(B_k) \mid S_{fitted}]$ using posterior predictive point patterns. For any B_k, we can write the RPS as $\mathrm{RPS}(B_k) = \sum_{n=0}^{\infty} [F_{N(B_k) \mid S_{fitted}}(n) - 1[n \geq N_{obs}(B_k)]]^2$. We would average over k to compare models. That is, model selection would choose the model with the smallest average RPS.

If cross-validation is available, we would employ the RPS with our hold-out data, comparing observed counts in subsets to posterior predictive distributions for these counts. If holding out data is not possible, we would examine these same metrics in-sample.

Finally, we remark that we *cannot* use diagnostics like G, F, K, and K_{inhom} to compare models. For example, we can't say that G for one model is "better" than G for another model? That is, posterior distributions, e.g., $[G(d : \boldsymbol{\theta}_j) \mid S_{obs}; M_j]$, can criticize say CSR which has known distance functions when CSR is nested within the fitted model. With a set of models say $\{M_j\}$, using the G function, we can compare, e.g., $[G(d : \boldsymbol{\theta}_j) \mid S_{obs}; M_j]$ with the empirical estimate $\hat{G}(d)$. Since the latter is a *nonparametric* estimate, such comparison could be used to criticize M_j.

9.3 Simulating point patterns

As the previous section has shown, simulating realizations under a given point pattern model is a critical feature of our model-fitting approach. We begin by discussing methods for simulating point patterns under various model specifications. Algorithms along with R software available for simulating point patterns will be given for some models.

9.3.1 *Homogeneous Poisson Process (HPP)*

To start, recall the HPP, the simplest stationary process with complete spatial randomness discussed in Section 8.4. The HPP assumes a constant intensity function, $\lambda(\mathbf{s}) = \lambda$ for all $\mathbf{s} \in D$. We assume $D \in \mathbb{R}^2$, but the approaches are easily generalizable to domains with higher dimension. Additionally, for convenience, we assume D to be the unit square, such that $D = [0, 1] \times [0, 1]$, more generally a product set.

Let $|D|$ be the area of the spatial domain. A simple two stage process to simulate from an HPP with $\lambda(\mathbf{s}) = \lambda$ for all $\mathbf{s} \in D$ begins by first drawing $N \sim Po(\lambda|D|)$ to obtain the random number of points in the region. Then, each point, $\mathbf{s}_i$ for $i = 1, 2, \ldots, N$, is randomly *located* independently and uniformly over D. That is, letting $\mathbf{s}_i = (s_{i1}, s_{i2})$, we draw $s_{i1} \sim Unif(0, 1)$ and $s_{i1} \sim Unif(0, 1)$. Collectively, $\mathcal{S} = \{\mathbf{s}_i; i = 1, \ldots, N\}$ specifies the realized point pattern of the HPP. An extension, using rejection sampling, readily accommodates irregular domains. That is, if D is not a product set, we embed it into a rectangle. Then we draw points uniformly over the rectangle, retaining those that fall in D.

9.3.2   Nonhomogeneous Poisson Process (NHPP)

We immediately extend the simulation to the NHPP where $\lambda(\mathbf{s})$ is location-specific and varies with $\mathbf{s}$. This method dates to [Lewis and Shedler, 1979]. Letting $\lambda_{max} = \max\limits_{\mathbf{s} \in D} \lambda(\mathbf{s})$, we use the HPP simulation procedure by first drawing $N_{max} \sim Po(\lambda_{max}|D|)$ and then randomly locating the points in D. Let $\mathcal{S}_{max}$ denote the observed spatial point pattern from the HPP. The realization of the NHPP is obtained using a thinning procedure where each point, $\mathbf{s}_i \in \mathcal{S}_{max}$, is retained with probability $\frac{\lambda(\mathbf{s}_i)}{\lambda_{max}}$. That is, for each $\mathbf{s}_i$, $i = 1, 2, \ldots, N_{max}$, we sample an independent Bernoulli random variable with $p(\mathbf{s}_i) = \frac{\lambda(\mathbf{s}_i)}{\lambda_{max}}$ where a 1 means $\mathbf{s}_i$ is retained, 0 otherwise. The number, N, and collection of points retained, $\mathcal{S}$, yields the realization from the NHPP with intensity $\lambda(\mathbf{s})$. The proof is straightforward; we only need to show that, for any set $A \in D$, $N(A) \sim Po(\lambda(A))$.

Two issues to consider when simulating from an NHPP are the following. First, it is common for NHPPs to be specified using spatial covariates. For example, a spatial point process of the locations of tree species might be driven by climate variables such as temperature and precipitation, or elevation. Here, the intensity surface might be specified in log-linear form where $\log\lambda(\mathbf{s}) = \mathbf{x}^{\mathrm{T}}(\mathbf{s})\boldsymbol{\gamma}$ where $\mathbf{x}(\mathbf{s})$ is a vector of covariates for location $\mathbf{s}$, which is, in theory, observable for all $\mathbf{s} \in D$. In practice, spatial covariates are often only available across a collection of grid cells that form a disjoint partition of the spatial domain. Therefore, we approximate the continuous intensity surface, $\lambda(\mathbf{s})$, with a piecewise constant intensity. For locations simulated randomly within the domain using the NHPP simulation procedure, two locations within the same grid cell will have the same probability of being retained but independent Bernoulli random variables are sampled for each.

Second, the procedure outlined above assumes λ_{max} can be computed directly. That is, $\lambda(\mathbf{s})$, and, therefore, $\mathbf{x}(\mathbf{s})$ need to be available everywhere in order to obtain the maximum over the intensity surface. In practice, again, we will have to rely on the disjoint partition of the spatial domain in order to approximate the maximum intensity to be used in the simulation procedure.

9.3.3   Log Gaussian Cox Process (LGCP)

Recall that the LGCP, introduced in Section 8.4.2, is an extension of the NHPP where the intensity function, $\lambda(\mathbf{s})$, is a stochastic process. Simulating a realization of a LGCP requires first simulating a realization of the stochastic process, namely, a realization of the Gaussian process. Then, conditioning on this realization, we have an NHPP with intensity $\lambda(\mathbf{s})$ and can follow the procedure above using thinning to obtain our realization from the LGCP.[1] Due to the multiple stages of random sampling, i.e., the Gaussian process realization, the number of points, and the location of points, we provide additional details for this approach.

To begin, we must first obtain a realization of the Gaussian process, which we refer to as $z(\mathbf{s})$, that is defined over $\mathbf{s} \in D$. We will assume $z(\mathbf{s})$ is a mean zero GP with valid spatial covariance function $C(\mathbf{s}, \mathbf{s}')$. To obtain a realization of the GP, we first need to define a set of representative points, $\mathcal{U}$, at which we will *observe* $z(\mathbf{s})$. In theory, this set of representative

---

[1]To be clear, the resulting point pattern realization is *given* the GP realization. We are not attempting to create point patterns *marginalized* over the GP.

points can be chosen in any way; for example, randomly sampling a set of locations in D. Since, however, we will be using this realization of the GP to aid in simulating from the point process, it is often more efficient to discretize the domain into a collection of grid cells that form a disjoint partition of D and use, for example, the centroid of each grid as a representative point. The realization of the LCGP is sensitive to this discretization and a finer resolution of the surface will result in a better approximation of the true Gaussian process.[2]

Let $\mathcal{U}$ denote the set of representative points such that $\mathbf{u}_j \in \mathcal{U}$ and $|\mathbf{u}_j|$ is the area of the grid cell j with centroid $\mathbf{u}_j$. Then, we obtain a finite realization of the GP at the set of representative points by drawing $\mathbf{z} \sim MVN(\mathbf{0}, \mathbf{C})$ where $\mathbf{C}$ is the covariance matrix between the set of representative points. The realization of $\mathbf{z}$ at the grid cell centroids results in a tiled surface over D, and, in turn, a tiled surface for the intensity. That is, $\lambda_0(\mathbf{u}_j) = e^{z(\mathbf{u}_j)}$ denotes the *baseline* intensity for the entire grid cell containing $\mathbf{u}_j$ for each $\mathbf{u}_j \in \mathcal{U}$.

Then, we add the regressors, $\mathbf{x}^{\mathrm{T}}(\mathbf{s})\boldsymbol{\gamma}$, resulting in

$$\lambda(\mathbf{u}_j) = \exp\left\{\mathbf{x}^{\mathrm{T}}(\mathbf{u}_j)\boldsymbol{\gamma}\right\} \lambda_0(\mathbf{u}_j) = \exp\left\{\mathbf{x}^{\mathrm{T}}(\mathbf{u}_j)\boldsymbol{\gamma} + z(\mathbf{u}_j)\right\} .$$

Now, conditional on the realization $\mathbf{z}$, $\lambda(\mathbf{u}_j)$ fully specifies the intensity of an NHPP. Therefore, we can obtain a realization of the LGCP using the NHPP simulation procedure where $\lambda_{max} = \max_{\mathcal{U}} \sim \lambda(\mathbf{u})$. That is, we draw $N_{max} \sim Po(\lambda_{max}|D|)$ and randomly locate the N_{max} points in D, where $\mathcal{S}_{max}$ denotes the collection of points. The thinning procedure introduced above is again employed to obtain the simulated spatial point pattern under the specified LGCP. Letting $\mathbf{u}_i$ denote the centroid of the grid cell containing simulated point $\mathbf{s}_i \in \mathcal{S}_{max}$, $\mathbf{s}_i$ is retained with probability $p(\mathbf{u}_i) = \frac{\lambda(\mathbf{u}_i)}{\lambda_{max}}$. It is important to note that each sample point $\mathbf{s}_i \in \mathcal{S}_{max}$ is retained independently of all other points, even if there are multiple $\mathbf{s}_i$'s located in the same grid cell. Again, the number and locations of the points retained makes up the realization of the point pattern $\mathcal{S}$ from the LGCP.

9.3.4 Cluster Processes

We now turn from Poisson processes and the conditional independence assumption to the simulation of point patterns that exhibit dependence in the form of spatial clustering (Section 8.5.1). In general, clustering in spatial point patterns can be thought of as groups or *clusters* of points that exhibit inter-point distances smaller than the average distance between points for a point pattern of the same size under CSR across the domain. Two common families of cluster processes discussed in the previous chapter (Section 8.5.1) are the Neyman Scott process and the shot noise processes. The simulation procedures for each of these cluster processes are directly generative and outlined in detail below.

The Neyman Scott process can be described as a *parent-offspring* process. The two-stage procedure for simulating this family of cluster processes begins by randomly simulating a set of *parent* nodes/locations within the spatial domain. Then, for each parent, *offspring* are simulated randomly and independently around the parent node, which requires first sampling randomly the number of offspring for each parent and then randomly locating them within D. A simulated realization of the Neyman-Scott process is the result of the collection of offspring from this two-stage process; that is, the parents are removed.

More formally, the Neyman-Scott process requires a specification of the parent intensity and an offspring intensity, which could be specified conditionally given the parent. A common choice for the parent intensity might be an NHPP, where $\lambda(\mathbf{s})$ is defined over D

---

[2]The downside with increasing the grid resolution is that as the dimension of $\mathcal{U}$ increases, we have to sample from an increasingly higher dimensional multivariate normal distribution which becomes increasingly computationally expensive.

and parent nodes can be simulated according to the procedure outlined above. Note that a simpler HPP or a more general LGCP would also be suitable models for the parent process.

Next, for each parent node, we need to generate the number and locations of the offspring. Assuming K parent nodes were generated with locations $\boldsymbol{\mu}_1, \boldsymbol{\mu}_2, \ldots, \boldsymbol{\mu}_K$, one approach is to simulate N_k, the number of offspring for parent k where $k = 1, 2, \ldots, K$, according to, e.g., $N_k \overset{iid}{\sim} Po(\delta)$, and then locate them by simulating i.i.d. samples from the bivariate density $f(\mathbf{s}; \boldsymbol{\mu}_k)$. Customarily, the offspring of parent $\boldsymbol{\mu}_k$ are located using the bivariate Gaussian density $N(\boldsymbol{\mu}_k, \tau^2 \mathbf{I})$ resulting in a Thomas process [Illian et al., 2008] and Section 8.5.1. Alternatively, a Matérn process can be used in which the offspring are simulated uniformly within a circle of radius R centered at $\boldsymbol{\mu}_k$. The resulting collection of offspring from all K parent nodes yields the simulated point pattern from Neyman-Scott process.

For shot noise processes (Section 8.5.2), like the LGCP, the intensity is stochastic. Here, consider the form for a shot noise process is $\lambda(\mathbf{s}) = e^{\mathbf{x}^{\mathrm{T}}(\mathbf{s})\boldsymbol{\gamma}} \lambda_0(\mathbf{s})$ where $e^{\mathbf{x}^{\mathrm{T}}(\mathbf{s})\boldsymbol{\gamma}}$ is borrowed from the NHPP specified with covariates $\mathbf{x}(\mathbf{s})$ and $\lambda_0(\mathbf{s})$ is a mean 1 shot process. As discussed in Section 8.5.2, a simple yet flexible form for the shot process is $\lambda_0(\mathbf{s}) = \sum_{\mathbf{s}_i \in \mathcal{S}_{shot}} f(\mathbf{s} - \mathbf{s}_i) m(\mathbf{s}_i)$. Here, f is a unimodal density over D centered at 0, $m(\mathbf{s}_i) \geq 0$, and the summation is over the collection of points in the shot process. $m(\mathbf{s}_i)$ is referred to as the "shot" and the density f spreads the influence of the shot at $\mathbf{s}_i$ on $\lambda(\mathbf{s})$ according to the distance between $\mathbf{s}$ and $\mathbf{s}_i$.

To simulate from the shot noise process, first simulate $\mathcal{S}_{shot} \sim HPP(\lambda)$ to obtain a realization of points for the shot process. If we assume a fixed constant for the shot such that $m(\mathbf{s}_i) = m$, then, by letting $m = 1/\lambda$, $\mathrm{E}(\lambda_0(\mathbf{s})) = 1$ and thus, $\mathrm{E}(\lambda_0(D)) = |D|$. Now, using the simulated points of the shot process, compute $\lambda_0(\mathbf{s}) = \sum_{\mathbf{s}_i \in \mathcal{S}_{shot}} f(\mathbf{s} - \mathbf{s}_i) m(\mathbf{s}_i)$ and finally, $\lambda(\mathbf{s}) = \exp\{\mathbf{x}^{\mathrm{T}}(\mathbf{s})\boldsymbol{\gamma}\} \lambda_0(\mathbf{s})$ is the resulting shot noise intensity for $\mathbf{s} \in D$.

Conditional on $\lambda(\mathbf{s})$, we find ourselves with an NHPP intensity from which we need to simulate a point pattern. So, here we just follow the path above for doing so. The resulting collection of points retained yields a simulated point pattern from the shot noise process.

9.3.5   Gibbs Process

Gibbs processes offer models for point patterns when there is dependence between points. As noted in the previous chapter (Section 8.5.3), Gibbs processes are usually used to capture negative dependence, or *inhibition*, but can be used to model the clustering of points as well. Depending on the process specifications, such as the strength of inhibition or attraction of the point pattern or whether the number of points in the process is fixed or random, different simulation algorithms have been proposed in the literature.

For example, consider the Strauss process from the previous chapter (again, Section 8.5.3), which exhibits inhibition through a penalty based on the distance between pairwise points. Here, it is possible for two points to lie closer than distance R apart like in a hard-core process but can be unlikely. The Strauss process is intuitive from a simulation perspective. First, generate a realization, $\mathcal{S}$, from an HPP within the domain D. If no pair of points is less than some threshold distance R apart, we accept the realization. Otherwise, for each close pair of points, we have an independent Bernoulli trial to decide whether to reject or accept the pair. If *any* pairs of points are rejected, the entire point pattern is rejected. Letting $0 \leq p \leq 1$ denote the probability of acceptance of a pair of points and $s_R(\mathcal{S})$, the number of close pairs of points in the point pattern $\mathcal{S}$, the realization $\mathcal{S}$ is accepted based on a Bernoulli trial with probability $p^{s_R(\mathcal{S})}$. When $p = 1$, $\mathcal{S}$ will be accepted with probability 1 and thus, denotes an HPP. When $p = 0$, this results in a hard-core process such that $\mathcal{S}$ is rejected when $s_R(\mathcal{S}) > 0$. Therefore, for $0 < p < 1$, the Strauss process spans the range between complete spatial randomness and the hard-core process. Note that both increases in R and decreases in p will decrease the acceptance probability of candidate realizations of $\mathcal{S}$, making the simulation procedure outlined above increasingly inefficient.

An alternative approach for simulating from a more general Gibbs process uses a Markov chain Monte Carlo (MCMC) algorithm. In particular, MCMC iterative procedures can be used as a simulation algorithm where the MCMC algorithm is based on a spatial birth-death process [Illian et al., 2008, Baddeley et al., 2015]. The algorithm begins with an initial point pattern and after a series of iterations where points are added (births) and removed (deaths), the algorithm will tend to an equilibrium state. Once approximate equilibrium is reached, any instantaneous snapshot of the spatial patterns of points can be assumed to follow the probability law of a Gibbs process. That is, given $Q(\mathcal{S})$ (as defined in (8.16)) the joint location density for a particular point pattern realization, $\mathcal{S}$, is proportional to $e^{-Q(\mathcal{S})}$. To obtain a collection of realizations of a Gibbs process, one continues iterating through the birth-death process where the number of iterations between retained snapshots of the spatial patterns of points is large enough such that dependence between observations is negligible.

We offer a few more details on the MCMC approach for simulating Gibbs processes. The MCMC algorithm consists of random births and deaths of points, ultimately reaching an equilibrium state. Assume $\mathcal{S}$ is the initialized point pattern for the birth-death process. Additionally, assume a repulsive Gibbs process where d_0 denotes the threshold distance as discussed above. Then, for any small window of time, Δt, each point, independently, has probability $m\Delta t$ of mortality (i.e. being removed from the point pattern). Additionally, letting b denote the birth rate per unit area and per unit time, for some small spatial region of area Δa, a new point is "born", independently, with probability $bp^k \Delta t \Delta a$, where p is the probability of acceptance and k is the number of points in the current point pattern $\mathcal{S}$ that lie closer than the threshold R of the proposed birth point. Again, $p = 0$ implies a hard-core process. Iterating between random births and deaths in this fashion creates the sequence of spatial point patterns.

As opposed to the long-run approximations from the birth-death MCMC procedure, *perfect simulation* provides an alternative method for simulating from Gibbs processes [Propp and Wilson, 1996, 1998]. The benefit of perfect simulation is that it solves the problem of having to assess when approximate equilibrium of the target distribution has been reached. Perfect simulation is a Markov chain simulation algorithm such that the exact equilibrium is attained when the algorithm completes; it is therefore able to return perfect simulations from the target distribution. Perfect simulation based on coupling from the past (CFTP) repeatedly generates upper and lower Markov chains starting increasingly further back in time until the pair of chains merge at time 0. Then, the chains return a *perfect* simulation from the specified target distribution. While CFTP is the most common perfect sampler, other perfect samplers exist in the literature [Møller, 2001, Berthelsen and Møller, 2002].

Pairwise interaction processes are also addressed in Berthelsen and Møller [2003], where the dominated CFTP perfect simulation and path sampling [Gelman and Meng, 1998] are combined for likelihood based inference and non-parametric Bayesian inference. Häggström et al. [1999] extend the perfect simulation algorithm using a two component Gibbs sampler for a bivariate point process model in order to obtain exact samples from a mixture model. They investigate the algorithm in terms of computational feasibility through simulation and find that exact samples from the target distribution can be obtained as long as the rate of the underlying Poisson processes is small or moderate. Perfect simulation algorithms are further utilized for simulating from multivariate discrete and continuous target distributions by Møller [1999].

The `spatstat` package in R includes functions for simulating Gibbs processes using various simulation algorithms. Versions of these simulation algorithms include a Metropolis-Hastings algorithm, which reaches approximate convergence to the underlying Gibbs process, as well as perfect simulation, which guarantees convergence but often at the expense of large computation time. In general, as the strength of inhibition in the Gibbs process increases, the computation time needed to reach convergence also increases, thus, requiring more advanced algorithms. *Simulated tempering* [Marinari and Parisi, 1992, Geyer and Thompson, 1995] is one such approach for processes with strong inhibition where auxiliary variables are introduced into the algorithm to decrease autocorrelation. The `spatstat` package, using a Metropolis-Hastings algorithm, provides a method for simulating point patterns using tempering. Details of the simulated tempering algorithm, as they pertain to simulation and inference for hard-core Gibbs processes, can be found in Mase et al. [2001].

9.3.6   Other considerations for constructing and simulating point patterns

The procedures outlined above yield realizations of spatial point patterns under various explicit model specifications. Other considerations for the construction and simulation of point patterns include irregularly shaped spatial domains, and domains of higher dimension, as well as possible process enhancement through thinning procedures, displacement, censoring, and superposition. In this subsection we give a brief overview of these ideas, highlighting how they might be used in practice to better specify the underlying process and mechanism for data collection.

9.3.6.1   Irregular spatial domains

In the procedures outlined above, we assumed the spatial domain was the unit square, which enabled us to sample locations by independently drawing from two standard uniform distributions. In practice, the spatial domain can take any shape (and be of any dimension). Here, we offer a few comments regarding simulating point patterns for general domains. It is trivial to extend the sampling to rectangular spatial domains with non-unit-length widths and heights by applying appropriate shifts and/or scalings of the uniform distributions above.

For irregularly shaped domains, e.g., state or county boundaries, lakes, national forests, even after shifts, scalings, or rotations, a realization of the spatial point process can not be obtained by drawing from two independent uniform distributions. In this case, an efficient approach for simulating the point pattern entails embedding the irregular domain of interest, D, into a rectangular domain, D^* such that $D \subset D^*$. Then, under any of the process models outlined above, we can simulate points under D^* and use rejection sampling to retain only those in $D \cap D^*$. Note, here that we reject each point with $p = 1$ if it is outside D, retain it with $p = 1$ if it is in D. While D^* can be defined to be any rectangle such that $D \subset D^*$, in practice it should be specified as small as possible to obtain maximum efficiency in sampling points (i.e., retaining a high percentage of points.) Under an NHPP model, for example, $N^*_{max} \sim Po(\lambda_{max}|D^*|)$, and points are located uniformly over D^*. Points that are not contained in D are rejected, and those in D are retained according to their specified probability using an appropriate Bernoulli trial.

9.3.6.2   Thinned processes

Thinned processes are obtained through two surfaces, namely, the intensity associated with the point process, $\lambda(\mathbf{s})$ along with a thinning surface, $p(\mathbf{s})$, with realized values $0 \leq p(\mathbf{s}) \leq 1$. Here, $\lambda(\mathbf{s})$ might be specified using any of the point process models discussed above. The thinning process $p(\mathbf{s})$ can take various forms. A simple thinning process, known as p-thinning assumes $p(\mathbf{s}) = p$ for all $\mathbf{s} \in D$ such that each point in the spatial point pattern is retained independently with probability p. Other versions might view $p(\mathbf{s})$ as say a population density surface so that the thinning corresponds to a population density thinning. That is, if $\lambda(\mathbf{s})$ is an intensity associated with the incidence of locations of a particular disease, then $p(\mathbf{s})\lambda(\mathbf{s})$ provides a population (at risk) adjustment to the intensity surface. A more stochastic thinning process could view $p(\mathbf{s})$ arising as a cdf transformation of a Gaussian process.

With regard to simulating thinned processes, a spatial point pattern is first simulated according to $\lambda(\mathbf{s})$. The thinning process is applied at the second stage of the simulation, analogous to the second stage in the NHPP simulation procedure, where each point simulated in the first stage is retained independently with associated probability $p(\mathbf{s})$. The collection of retained points yields the simulated thinned spatial point pattern [Diggle, 2013].

Simple p-thinning has also been proposed as a tool for model validation with spatial point patterns (Section 9.2.4). The thinning is used to create training and test datasets [Leininger and Gelfand, 2017]. Recall that in traditional model validation, a training set, say, consisting of 80% of the data is randomly partitioned from the test set consisting of the remaining 20%. In spatial point pattern analyses, p-thinning entails using a Bernoulli random variable with probability p of being retained for each data point in $\mathcal{S}$, yielding the training point pattern $\mathcal{S}_{train}$. The random collection of points not retained creates the test point pattern, $\mathcal{S}_{test}$. Under proper rescaling by the ratio $\frac{1-p}{p}$ to the fitted intensity surface obtained from training the modeling using $\mathcal{S}_{train}$, model cross-validation can be conducted using $\mathcal{S}_{test}$.

We note that this p-thinning procedure can be applied to any spatial point process model with a conditionally independent location density, e.g., an NHPP, a LGCP, even more general Cox processes, with care. The resulting induced intensity will agree with that of the underlying process of the spatial point pattern up to scaling. With, e.g., pairwise dependence between locations to prescribe the joint location density, as in a Gibbs process, p-thinning can not be applied. Such thinning will reduce the size of the point pattern and, as a result, change the interpoint distance structure.

Again, richer specifications of $p(\mathbf{s})$ are also common and often take a more process-based approach for point process modeling and simulation. For example, assume that a given spatial domain provides a suitable habitat for a particular species and the distribution of this species is defined according to a spatial process with intensity $\lambda(\mathbf{s})$. In addition, based on some biotic or abiotic environmental conditions, e.g., soil moisture or a high abundance of predator species, there may be subregions within the spatial domain that are less likely to contain the particular species. The thinning process, $p(\mathbf{s})$, offers an additional stochastic modification that is independent of the primary process and can be employed to capture such deviations to the primary spatial process.

Therefore, given realizations from the primary process, $\lambda(\mathbf{s})$, using the simulation approaches outlined above, each point is retained according to a realization from the thinning process with respective probabilities $p((s)$. The *operating* intensity for the observed point pattern becomes $p(\mathbf{s})\lambda(\mathbf{s})$.

A different version arises when a point pattern is observed in the presence of a detection function as with distance sampling [Farr et al., 2021, Martino et al., 2021]. In this setting, usually a linear sampling transect in space is followed and the locations of observations seen on either side of the transect are recorded. Evidently, some points will be missed as the transect is followed. The detection function provides the probability that an observation at perpendicular distance d to the trajectory will be detected.

A different type of thinning reflects availability or sampling effort with regard to finding a species at location $\mathbf{s}$. Here, $p(\mathbf{s})$ will be 1 or 0 according to whether or not, say the location was available for the species or whether or not the location was visited . A degraded point pattern results (see Chakraborty et al. [2011]). Section 9.8.2 describes thinning in the context of presence-only data for species distributions. For a given bounded region D, there is a finite set of species locations which can be considered as a point pattern, i.e., the full census of individuals in D. However, in practice, money and time considerations imply that sampling effort over D will be sparse. The observed point pattern of locations will be a degraded/censored point pattern of the full census of locations due to the fact that some regions have never been sampled. The *operating* intensity for those regions is censored to 0.

Lastly, we note that sampling from and applying the thinning process, $p(\mathbf{s})$ may also require a disjoint partition of the spatial domain and a collection of representative points. For example, in the thinning by population density example above, the population density surface will be at an areal unit scale, e.g., block or census tracts.

9.3.6.3 Censoring

The term *censoring* of a point pattern is often used in practice; it emerges as a special case of thinning. Consider spatial point patterns as observational data in ecological studies, such as a collection of locations where a particular bird call was heard and identified, or the locations of observed grass species across a field site. In each of these examples, based on the sampling effort of the individuals who are collecting the data or the rate of missed detections, both of which could vary across space, the observed spatial point pattern would be a censored, degraded, or partially-realized version of the true point process. When censoring is homogenous across the spatial region, the observed point pattern would resemble that of a p-thinned point pattern as discussed above. Imperfect detection might be an example of a constant censoring process such that the ability to detect a species is the same across the entire spatial region. On the other hand, censoring that might vary across space could be the result of varying sampling efforts across a region. For regions within the domain with zero sampling effort, $p = 0$. Nonhomogeneous censoring would mimic that of a $p(\mathbf{s})$-thinned process.

As a further example, the Breeding Bird Survey (BBS) is a massive citizen science survey database of birds seen and identified across North America.[3] It is common for birders to revisit the same locations, resulting in high sampling effort in regions such as nature areas or parks. Therefore, the observed spatial point pattern of a particular species of interest might exhibit clustering of points in these high traffic areas and few to no points in other areas. From an inferential standpoint for spatial point pattern analysis, the effects of both homogenous and nonhomogeneous censoring can be dramatic, as they could lead to vastly underestimating the abundance of a species or misinformation with regard to its spatial distribution. To simulate a censored point pattern, the procedures outlined above for simulating p-thinning and $p(\mathbf{s})$-thinning can be applied.

9.3.6.4 Superposition

The superposition, or aggregation, of two or more component spatial point processes offers a natural way of enriching the specification of a point process. For instance, let $\mathcal{S}_1$ and $\mathcal{S}_2$ each denote a spatial point pattern with intensity functions, $\lambda_1(\mathbf{s})$ and $\lambda_2(\mathbf{s})$, respectively. Then, their superposition $\mathcal{S} = \mathcal{S}_1 \cup \mathcal{S}_2$ also forms a spatial point pattern. Simulating from a superposition spatial point process entails simulating from each of the component processes using the procedures outlined above and combining them onto one underlying space. Realizations of a marked spatial point pattern where the marks are discrete labels can be easily simulated by a superposition point process. Here, letting $\lambda_l(\mathbf{s})$ define the intensity of the process for label l where $l = 1, 2, \ldots, L$, a superposition point process could be obtained by $\bigcup_{l=1}^{L} \mathcal{S}_l$.

Various specifications of the component spatial point processes offer useful flexibility in building a marginal point pattern from the superposition point process. As a special case, when each component process is a Poisson point process, the resulting marginal superposition point process is also a Poisson process with intensity $\lambda(\mathbf{s}) = \sum_{l=1}^{L} \lambda_l(\mathbf{s})$. An early

---

[3]https://www.pwrc.usgs.gov/bbs/

version of this considers a point pattern of disease incidence that resulted from a *background* intensity surface, essentially the baseline process, with an overlay of an intensity for points during a high risk season or a disease outbreak setting [Kelsall and Diggle, 1995]. If $\lambda_B(\mathbf{s})$ is the baseline intensity and $\lambda_E(\mathbf{s})$ is the elevated risk intensity, then an interesting ratio here would be $\frac{\lambda_E(\mathbf{s})}{\lambda_B(\mathbf{s})+\lambda_E(\mathbf{s})}$.

Going further, Diggle [2013] shows that the superpositioned point process of a homogeneous Poisson process and a Poisson cluster process has second-order properties that are equivalent to pure Poisson cluster processes. However, interestingly, the nearest neighbor properties between the Poisson cluster process and the superpositioned process remain quite different. Similarly, the superposition of inhibition processes, such as two Gibbs processes, can also yield point processes with complex second-order and nearest-neighbor properties. While each Gibbs process will itself exhibit inhibition between points, within a specified radius, the superposition point process need not retain minimum distance behavior between points or number of points.

Lastly, shot noise process realizations can also naturally be developed by additive superposition. That is, the shot noise process is the superposition of the realizations of each of the random kernels. Recall the shot noise process $\lambda(\mathbf{s}) = e^{\mathbf{X}^T(\mathbf{s})\gamma}\lambda_0(\mathbf{s})$ where $\lambda_0(\mathbf{s})$ is the shot process such that $\lambda_0(\mathbf{s}) = \sum_{\mathbf{s}_i \in \mathcal{S}_{shot}} f(\mathbf{s} - \mathbf{s}_i)m(\mathbf{s}_i)$ and $\mathcal{S}_{shot}$ is the realization of the shot locations. Let $\lambda_{\mathbf{s}_i}(\mathbf{s})$ denote the random kernel for shot $\mathbf{s}_i \in \mathcal{S}$ such that $\lambda_{\mathbf{s}_i}(\mathbf{s}) = e^{\mathbf{X}^T(\mathbf{s})\gamma}f(\mathbf{s} - \mathbf{s}_i)m(\mathbf{s}_i)$. Realizations from each random kernel, $\lambda_{\mathbf{s}_i}(\mathbf{s})$, can be easily simulated, from which their superposition yields the realization of the shot noise process. Lastly, the shot noise process can be defined by the summation $\lambda(\mathbf{s}) = \sum_{\mathbf{s}_i \in \mathcal{S}_{shot}} \lambda_{\mathbf{s}_i}(\mathbf{s})$.

9.3.6.5   *Displacement*

Displacement is a point process operation that entails random relocations of the point pattern, obtained by randomly moving points of the point process to other locations within the defined domain. For each iteration of the operation, each point $\mathbf{s}_i \in \mathcal{S}$ is randomly and independently displaced, such that $\mathbf{s}_i \to \mathbf{s}_i + \mathbf{h}_i$ and the collection of displaced points, $\mathbf{s}_i + \mathbf{h}_i$ yields the new realization of the point process. There are many ways of specifying methods of displacement, such as rectangular or radial displacement. In simulation, care needs to be taken to control for the edges of the domain to ensure that displaced points remain in D. Toroidial boundaries are one choice, such that when points disappear off the right edge, they re-appear at the left. Adding a buffer zone is another choice, such that the point pattern is embedded in a larger region and the maximum distance of the rectangular or radial shift is less than the size of the buffer [Thapa and Bossler, 1992, Caspary and Scheuring, 1993].

Most commonly, displacement is used in testing hypotheses pertaining to the underlying processes of observed point patterns. Monte Carlo randomization tests, for example, can be used to test claims regarding distributional assumptions of a point pattern by comparing values of a test statistic, such as the K function, $K(d)$, for some distance, d, to assess clustering. The test statistic is first computed using the observed point pattern and compared to values of the statistic computed using a random and independent sequence of displaced point patterns. We note here that the function $rshift()$ in the R package `spatstat` conducts random shifts of point patterns that could be used in this capacity.

Another interesting application of displacement would be for a superposition of component point patterns. For example, consider the superposition of locations of male adults and yearlings of a species of birds, or two different species of trees across a forest with continuous marks denoting the diameter at breast height. There may be scientific insight to understanding not only the distribution of both adults and yearlings or the different tree species and sizes, but also in the their interaction. The continuous marks could be treated as a third dimension of the point pattern where the domain is defined as $D \times M$, $M \in \mathbb{R}^1$.

In this particular case, random and independent displacements of the points could be used to test whether or not there is clustering or inhibition between species of tree (or adults and yearlings) and whether or not these patterns are similar for different sizes of trees.

9.3.6.6 *Measurement error in locations*

In many applications, the observed point pattern $\mathcal{S}$ may not represent the collection of *true* locations of the points. For example, a forest inventory analysis may rely on remotely sensed data from, say, light detection and ranging technology [LiDAR, Lim et al., 2003], which yields spatial location of a species of tree in a forest. Due to measurement error of LIDAR, whether known or unknown, the observed point pattern may differ from the true locations of the trees. These discrepancies are often referred to as map positional error in geographic information systems (GIS) applications. In practice, ignoring measurement error could have important impacts on model inference, especially since these data products often transcend map projections and scalings.

From a modeling perspective, we can account for this uncertainty by defining the model hierarchically and adding a measurement error model to the data level. This would suggest looking at the observed location as $\mathbf{s}_{obs} = \mathbf{s}_{true} + \boldsymbol{\epsilon}(\mathbf{s})$ where $\boldsymbol{\epsilon}(\mathbf{s})$ is a bivariate noise process capturing the error in measurement. Here, the collection of points $\mathbf{s}_i \in \mathcal{S}$ is treated as one realization. Barber et al. [2006] develop models that address map positional error in order to infer the true location of feature coordinates. (See also Cressie and Kornak [2003]). In addition, they consider multiple realizations of the point pattern from different sources with possibly varying accuracy and propose model-averaging strategies to improve estimation of the true locations of the points. Model-based inference also yields uncertainty quantification where model averaging yields predictions with smaller uncertainty than any individual model. The hierarchical form of the model can be naturally handled within the Bayesian framework; both model fitting and simulation are easily attainable under the general specification of the latent process generating the *true* point pattern along with a measurement error specification.

9.4   Some details for Bayesian model fitting

9.4.1 *General comments on fitting point pattern models*

To provide some context, we begin with a few words regarding the general fitting of spatial point process models. The literature on such model fitting is fairly rich by now and is presented in the references at the beginning of the preceding chapter. A broad tool is the minimum contrast method as advocated by Diggle [Diggle, 2013]. This is essentially a method of moments idea. First, a summary function, typically the K function, is computed from the observed point pattern. Second, the theoretical expected value of this summary statistic under the point process model is derived (if possible, as an algebraic expression involving the parameters of the model) or estimated from simulations of the model. Then the model is fitted by finding the optimal parameter values for the model to give the closest match between the theoretical and empirical curves. With a single parameter, the criterion becomes $\int (\hat{K}(d) - K(d))^2 dd$. With additional parameters, often powers are introduced.

More generally, for a model with parameter vector $\boldsymbol{\theta}$, the minimum contrast estimator arises as the value of $\boldsymbol{\theta}$ which minimizes $\sum_d |\hat{K}_d(\mathcal{S})^a - K_{\boldsymbol{\theta}}(d)^a|^b$ where $K_{\boldsymbol{\theta}}$ is the theoretical K function, $\hat{K}_d(\mathcal{S})$ is the empirical estimator for the K function, and a and b are user-specified parameters [Diggle, 2013]. The summation replaces the integral by using a set of radii d for the K function.

From our perspective, likelihood-based methods are more attractive and have recently become more common practice [Baddeley and Turner, 2005]. For instance, with the homogeneous Poisson process (HPP), we have a closed form likelihood and the MLE is straightforward. For the nonhomogeneous process (NHPP), the Berman-Turner device [Berman and Turner, 1992] is attractive. It connects the NHPP log-likelihood to a weighted Poisson regression log-likelihood using quadrature to do numerical integration. It is also well suited for Bayesian model fitting of the NHPP and is detailed below. As mentioned in Section 8.4, for the log Gaussian Cox process (LGCP), we have a stochastic integral in the likelihood which can never be obtained explicitly. Customarily, we resort to numerical integration using so-called "representative points". Since we need likelihood evaluation in Bayesian model fitting, it is employed there as well. Recently, [Gonçalves and Gamerman, 2018] proposed a fitting strategy that avoids numerical integration of the stochastic integral.

For Neyman-Scott clustering processes, likelihood-based inference is computationally very demanding and is not straightforward to implement. Hence, the method of minimum contrast is the usual choice; implementation with a Cauchy distribution kernel is offered in spatstat. For Markov and Gibbs processes, a common approach is to employ the pseudo-maximum likelihood using the Papangelou conditional intensity. That is, the pseudo-likelihood is expressed through $\Pi_i\lambda(\mathbf{s}_i \mid \mathcal{S}/\mathbf{s}_i)$. Again, the spatstat package [Baddeley and Turner, 2005] is a very useful piece of software which offers likelihood fitting for many spatial point pattern models.

9.4.2 Bayesian computational strategies for log Gaussian Cox processes

Bayesian inference for Cox processes is natural since the observed point pattern depends on the latent random process. In the case of the LGCP, the point pattern depends on the latent Gaussian process. Using Bayes' Theorem, we can derive the inverse relationship and obtain inference for the latent random process given the observed point pattern. Guttorp and Thorarinsdottir [2012] offer a nice overview of Bayesian inference for point processes, focusing attention on Poisson processes, doubly stochastic processes (e.g., the LGCP), as well as cluster processes.

We begin with a brief review of Bayesian model fitting for the HPP. We then continue with the extension of Bayesian inference to the NHPP using the Berman-Turner device [Berman and Turner, 1992]. Building on these process specifications, we address Bayesian modeling fitting of the LGCP. In particular, due to its wide usage, we introduce a collection of Markov chain Monte Carlo methods for obtaining Bayesian inference for the LGCP. Then, we discuss integrated nested Laplacian approximation (INLA) as a computationally efficient approach for obtaining approximate Bayesian inference.

As above, let D denote the spatial domain and $\mathcal{S}$ the collection of observed spatial points, $\mathbf{s}_1, \mathbf{s}_2, \ldots, \mathbf{s}_n$ such that $\mathbf{s}_i \in D$ for all $i = 1, 2, \ldots, n$. The intensity function of the HPP is $\lambda(\mathbf{s}) \equiv \lambda$ over the entire domain, D. Bayesian posterior inference for the parameter λ requires specification of the likelihood function and prior distribution. Here, it is convenient and computationally efficient to assign a conjugate Gamma prior distribution to λ such that $\lambda \sim Gamma(\alpha, \beta)$ where α is the shape parameter and β is the rate parameter. The posterior distribution of λ given the data, $\mathcal{S}$ can be written

$$[\lambda \mid \mathcal{S}] \propto L(\lambda; \mathcal{S})[\lambda]$$
$$\propto e^{-\lambda(|D|+\beta)}\lambda^{n+\alpha-1}$$

where $|D|$ denotes the area of D. This posterior follows a Gamma distribution with shape $n + \alpha$ and rate $|D| + \beta$. As such, direct samples from the posterior distribution can be obtained efficiently.

In the context of the NHPP and the LGCP, our likelihood function takes the form

$$L((\lambda(\mathbf{s}), \mathbf{s} \in D); \mathcal{S}) = e^{-\lambda(D)} \prod_{i=1}^{n} \lambda(\mathbf{s}_i)$$

where $\lambda(D) = \int_D \lambda(\mathbf{s})d\mathbf{s}$. Again, for NHPPs, this is a regular integral, whereas for the LGCP, this integral is stochastic, requiring further computational considerations. Except for the simplest forms where the true intensity $\lambda(\mathbf{s})$ varies as a tiled surface across D, numerical integration will yield an approximation to $\lambda(D)$. The Berman-Turner device [Berman and Turner, 1992], using numerical quadrature, offers a simple and efficient approach for the numerical integration. Here, we discuss it in the context of spatial point patterns and as a method of approximating $\int_D \lambda(\mathbf{s})d\mathbf{s}$. We assume the intensity function of the NHPP is specified with parameter vector $\boldsymbol{\gamma}$. For example, the intensity could take a log-linear form where $\log\lambda_{\boldsymbol{\gamma}}(\mathbf{s}) = \mathbf{x}(\mathbf{s})'\boldsymbol{\gamma}$ with $\mathbf{x}(\mathbf{s})$ denoting a vector of covariates for location $\mathbf{s}$ and $\boldsymbol{\gamma}$ a vector of coefficients needing to be estimated.

To begin, we will approximate the integral using the weighted sum $\sum_{j=1}^m w_j \lambda_{\boldsymbol{\gamma}}(\mathbf{u}_j)$ where the $w_j > 0$ are the quadrature weights that sum to $|D|$ and $\mathbf{u}_j \in D$ for $j = 1, 2, \dots, m$ are the quadrature points. We can approximate the log-likelihood of the NHPP using these quadrature weights and points as

$$\log L(\boldsymbol{\gamma}; \mathcal{S}) \approx \sum_{i=1}^n \log\lambda_{\boldsymbol{\gamma}}(\mathbf{s}_i) - \sum_{j=1}^m w_j \lambda_{\boldsymbol{\gamma}}(\mathbf{u}_j).$$

A natural and convenient choice for quadrature points includes both the observation points $\mathbf{s}_i \in \mathcal{S}$ along with other randomly sampled points within D. Additional comments with regard to the quadrature points and weights are given below. Letting 1_j denote an indicator variable such that $1_j = 1$ if point $\mathbf{u}_j$ is a point in $\mathcal{S}$ and 0 otherwise, we can write the log likelihood as

$$\log L(\boldsymbol{\gamma}; \mathcal{S}) \approx \sum_{j=1}^m 1_j \log\lambda_{\boldsymbol{\gamma}}(\mathbf{u}_j) - w_j \lambda_{\boldsymbol{\gamma}}(\mathbf{u}_j).$$

By defining $y_j = 1_j/w_j$, we can conveniently rewrite this as

$$\log L(\boldsymbol{\gamma}; \mathcal{S}) \approx \sum_{y=1}^m (y_j \log\lambda_{\boldsymbol{\gamma}}(\mathbf{u}_j) - \lambda_{\boldsymbol{\gamma}}(\mathbf{u}_j))w_j,$$

which now takes the form of a weighted log likelihood of independent Poisson variables $Y_j \sim Po(\lambda_{\boldsymbol{\gamma}}(\mathbf{u}_j))$. Standard generalized linear model software can be employed to obtain approximate MLE estimates under this specification.

In obtaining inference within the Bayesian framework, we can utilize this approximation of the log likelihood in a Metropolis-Hastings algorithm for sampling from the posterior distribution of $\boldsymbol{\gamma}$. In practice, note that inference in either paradigm requires evaluation of $\lambda_{\boldsymbol{\gamma}}(\mathbf{u}_j)$ for all $\mathbf{u}_j$. That is, for all observed points in $\mathcal{S}$ as well as the quadrature points. Under a loglinear speciation of the NHPP where $\log\lambda_{\boldsymbol{\gamma}}(\mathbf{s}) = \mathbf{x}(\mathbf{s})^{\mathsf{T}}\boldsymbol{\gamma}$, this requires obtaining the covariates $\mathbf{x}(\mathbf{u}_j)$ for all $j = 1, 2, \dots, m$. Now, sampling from the posterior distribution of $\boldsymbol{\gamma}$ given $\mathcal{S}$ will require a Metropolis-Hastings algorithm. Assuming a vector form for $\boldsymbol{\gamma}$, these could be updated individually or as a block. In either case, the standard Metropolis-Hastings ratio comprised of likelihood, prior, and proposal can be employed where the approximate likelihood is used instead of the true likelihood.

One final consideration when using numerical quadrature in practice is the choice of quadrature points and weights. In general, numerical integration provides more accurate approximations an integral when the resolution of the discretized surface is high. Discretizing the spatial region using a fine-resolution grid is one possible choice to ensure adequate spatial coverage of the domain. In such a case, weights can then be computed as the ratio of the area of each grid cell relative to $|D|$. For an NHPP specified above with location-specific covariates, the resolution of the grid need only be as fine as the scale at which the covariates are observed. In ecological applications, for example, variables such as soil moisture or canopy cover may only be observable at a specified scale, e.g., hectare scale, in which case the resolution of the tiled surface of $\lambda_\gamma(\mathbf{s})$ will only be at this scale.

Under the LGCP specification, $\lambda(\mathbf{s})$ is now random so direct application of numerical integration techniques are not possible. Here, in order to complete the numerical integration, we will need to condition on a realization of the Gaussian process at a set of representative points. For example, if we obtain a realization of the Gaussian process at all quadrature locations, $\mathbf{u}_j, j = 1, 2, \ldots, m$, we can evaluate the approximate log-likelihood using the approach above. Note, however, that in addition to the possible hyperparameters in the mean and covariance function of the Gaussian process, the realization of the Gaussian process at the representative points is now also part of the joint posterior distribution. Common specification of the Gaussian process assumes a mean $-\sigma^2/2$ (this choice is made in order that, for $\lambda(\mathbf{s}) = \exp\{\mathbf{x}(\mathbf{s})^\mathsf{T}\gamma\}\exp\{z(\mathbf{s})\}$, we have $\mathrm{E}[\exp\{z(\mathbf{s})\}] = 1$), variance σ^2, and correlation specified using an appropriate covariance kernel (e.g., Matérn covariance function), with parameter(s) ϕ. Collectively, our MCMC sampling algorithm requires iterative sampling of the parameters γ, σ^2, ϕ, as well as obtaining the needed realizations of the Gaussian process, $z(\mathbf{s})$.

While this may seem straightforward, sampling from the Gaussian process can pose computational challenges. First, using standard geostatistical models with spatial random effects, such as Gaussian processes, MCMC algorithms are much more computational efficient when the models are specified marginally (Section 6.1.1.1), that is, when the Gaussian process is integrated out. Even when inference is obtained conditionally, a linear model form with normal errors enables conjugate, direct sampling of the spatial random effects. Here, under an LGCP, we do not have closed-form conditional distributions for posterior sampling and cannot marginalize over the Gaussian process since we require realizations for numerical integration.

In general, Markov chain Monte Carlo methods are common for fitting LGCPs. MCMC can be computationally challenging and often requires precise tuning in order to obtain proper mixing and convergence. Some advanced sampling algorithms that mitigate these challenges are discussed in Chapter 5. As an alternative to MCMC, approximate model fitting and inference for LGCPs can be obtained using integrated nested Laplace approximation (INLA) [Rue et al., 2009], also discussed in Chapter 5. In the remainder of this section, we will visit some of the common and more recent approaches proposed in the literature for Bayesian model inference of LGCPs. In this regard, we are briefly reviewing contemporary Bayesian hierarchical model fitting as well as supplementing some of the Bayesian model fitting methods discussed in Chapter 5 but without the emphasis on accommodating large datasets (Chapter 13).

9.4.2.1 *Elliptical slice sampling*

Slice sampling was discussed in Section 5.5.1.5. Here, we consider elliptical slice sampling, proposed by Murray et al. [2010b], as an efficient way to obtain samples of a multivariate latent Gaussian random variable in a generalized linear model when the likelihood function prohibits the posterior distribution of the latent random variable to be sampled directly. The elliptical slice sampler is suitable for obtaining samples from a posterior distribution that is proportional to the product of a multivariate Gaussian distribution and a likelihood function that connects the Gaussian prior to the data. Sampling from the posterior distribution of the random variable $z(\mathbf{s})$ of the LGCP is an example of this form, where the log intensity is specified as $\log\lambda(\mathbf{s}) = \mathbf{x}^\mathsf{T}(\mathbf{s})\gamma + z(\mathbf{s})$ and $z(\mathbf{s})$ is modeled as a Gaussian process.

Let $\mathbf{z}$ denote a realization of the Gaussian process at a collection of representative spatial locations defining a lattice over the spatial region. These points, representing grid cells, should be specified at a sufficiently fine resolution such that with low probability, two points in the observed point pattern are in the same grid cell and the error in using a piecewise constant approximation for the continuous intensity surface is negligible. For a finite set of locations, $\mathbf{z}$ follows a multivariate normal distribution with covariance Σ capturing the spatial correlation across locations. In specifying the algorithm, it is assumed that $\mathbf{z}$ is mean $\mathbf{0}$. However, this can easily be generalized to non-zero mean distributions by a shift through a change of variables.

Let $\mathbf{z}$ denote the current value of the process realization at the specified set of locations. Step one of the algorithm entails sampling from the prior distribution of $\mathbf{z}$ to define the *ellipse* for the algorithm. More precisely, $\boldsymbol{\nu} \sim MVN(\mathbf{0}, \Sigma)$. Then, the threshold for the log-likelihood used in the accept/reject step of the algorithm is obtained by sampling $u \sim Unif[0, 1]$. Now, to draw the initial sample, $\mathbf{z}_c$, we first sample $\theta \sim Unif[0, 2\pi]$ and define the bracket $[\theta_{min}, \theta_{max}] = [\theta - 2\pi, \theta]$. Then, the candidate value, $\mathbf{z}_c$, is obtained by setting $\mathbf{z}_c = \mathbf{z}\cos\theta + \boldsymbol{\nu}\sin\theta$. This draw of $\mathbf{z}_c$ is accepted as a sample from the posterior distribution if $\log L(\mathbf{z}_c; \mathcal{S}, \boldsymbol{\gamma}) - \log L(\mathbf{z}; \mathcal{S}, \boldsymbol{\gamma}) > \log u$ where $\log L(\mathbf{z}_c; \mathcal{S}, \boldsymbol{\gamma})$ and $\log L(\mathbf{z}; \mathcal{S}, \boldsymbol{\gamma})$ are the log likelihoods evaluated at $\mathbf{z}_c$ and $\mathbf{z}$, respectively, given the data, $\mathcal{S}$, and other parameters, $\boldsymbol{\gamma}$. If $\mathbf{z}_c$ is rejected, we shrink the bracket above such that if $\theta < 0$, $\theta_{min} = \theta$, or if $\theta \geq 0$, $\theta_{max} = \theta$. A new value of θ is then sampled such that $\theta \sim Unif[\theta_{min}, \theta_{max}]$ and we obtain our new candidate value of $\mathbf{z}$ as above. We continue through this procedure of shrinking the bracket and obtaining new draws of θ and candidate values of $\mathbf{z}$ until we accept the draw based on the log-likelihoods and threshold draw. Therefore, each iteration of the MCMC algorithm results in a unique sample of $\mathbf{z}$.

Computationally, the most expensive part of the elliptical slice sampler outlined above for obtaining realizations of the Gaussian process in the LGCP model is in sampling $\boldsymbol{\nu}$. For m representative points, this cost is $\mathcal{O}(m^3)$. Conditional on the realization of $\mathbf{z}$, and, thus, the intensity surface $\lambda(\mathbf{s})$, the observations of the point pattern are independent and computing the log likelihood potentially costs $\mathcal{O}(m + n)$ if the representative points and the observed points are distinct. If the observed points are *projected* to the nearest representative points, then the cost becomes $\mathcal{O}(m)$.

This sampler can easily be embedded into an MCMC algorithm where other model parameters are updated using traditional Gibbs sampling or Metropolis-Hastings algorithms. One benefit of the elliptical slice sampler is that it allows $\mathbf{z}$ to be sampled as a block. This is an advantage in terms of mixing over algorithms that sample each component separately, especially when the random effects $\mathbf{z}$ exhibit strong dependence. Additionally, the step-size within the algorithm is controlled by the bracket $[\theta_{min}, \theta_{max}]$. Whereas other algorithms require tuning to obtain appropriate step sizes, the elliptical slice sampler properly controls the step-size within each iteration of the algorithm to more efficiently draw likely candidate values of the latent Gaussian random variable.

9.4.2.2 *Metropolis-Adjusted Langevin algorithm*

A second MCMC method for inference for the LGCP uses the Metropolis-Adjusted Langevin algorithm (MALA) [Moller and Waagepetersen, 2003] as discussed in Section 5.5.1.3. Traditional Metropolis-Hastings algorithms tend to perform poorly in terms of mixing and convergence when proposal distributions and tuning parameters are poorly chosen, and the problem is compounded when parameters and latent processes are highly dependent [Girolami and Calderhead, 2011]. The Metropolis-Hastings algorithm with a Langevin-type proposal offers an efficient approach when the gradient of the transition density can be written down explicitly. MALA simulates the posterior distribution based on a linear transformation of the variable(s) of interest. Proposals of the new states for the (transformed) parameters or random variables are obtained using the gradient of the target posterior density and they are accepted or rejected using the Metropolis-Hastings algorithm [Roberts and Rosenthal, 1998a, Roberts and Stramer, 2002]. For traditional Metropolis-Hastings algorithms, the target acceptance rate for a random walk proposal and Gaussian target distribution has been shown to be 0.234 [Roberts and Rosenthal, 1998b]. Roberts and Rosenthal [1998a], Roberts et al. [2001] found 0.57 to be the optimal acceptance rate for MALA and Andrieu and Thoms [2008] developed an algorithm to find the tuning parameter to achieve this optimal rate.

Let $\mathcal{U}$ denote the set of grid cells in the discretization such that $\mathbf{u}_j, j = 1, 2, \ldots, m$ are the grid cell centroids and $|\mathbf{u}_j|$ is the area of grid cell j. Here, $\lambda(\mathbf{u}_j)$ is the piecewise constant intensity for grid cell j. Let $\mathbf{z} = (z(\mathbf{u}_1), z(\mathbf{u}_2), \ldots, z(\mathbf{u}_m))$ denote the realization of the Gaussian process at the set of m locations. Now, we are interested in obtaining samples from $\pi(\mathbf{z} \mid \mathcal{S}, \boldsymbol{\theta})$ where $\mathcal{S}$ is the observed spatial point pattern and $\boldsymbol{\theta}$ are all other parameters. Let $\mathbf{z} \sim MVN(-\frac{\sigma^2}{2}\mathbf{1}, \Sigma)$ where Σ is a spatial covariance matrix with variance σ^2 and spatial dependence specified according to a spatial covariance function. The transformation of $\mathbf{z}$ is such that $\tilde{\mathbf{z}} = \Sigma^{-1/2}(\mathbf{z} + \frac{\sigma^2}{2}\mathbf{1})$. Now, MALA is used to obtain samples from $\pi(\tilde{\mathbf{z}} \mid \mathcal{S}, \boldsymbol{\theta}) \propto \pi(\mathcal{S} \mid \mathbf{z}, \boldsymbol{\theta})\pi(\tilde{\mathbf{z}})$.

The Langevin-type proposal distribution for $\tilde{\mathbf{z}}$ is

$$p(\tilde{\mathbf{z}}_{cand} \mid \tilde{\mathbf{z}}) \sim MVN(\tilde{\mathbf{z}} + \frac{1}{2}\nabla\log(\pi(\tilde{\mathbf{z}} \mid \mathcal{S}, \boldsymbol{\theta})), h^2\mathbf{I})$$

where $\nabla\log(\pi(\tilde{\mathbf{z}} \mid \mathcal{S}, \boldsymbol{\theta}))$ is the gradient of the posterior density of interest, h is a tuning parameter, and $\mathbf{I}$ is the identity matrix. Modifications to the proposed covariance matrix have been proposed and are discussed below.

Now,

$$\log(\pi(\tilde{\mathbf{z}} \mid \mathcal{S}, \boldsymbol{\theta})) = constant - \frac{1}{2}||\tilde{\mathbf{z}}|| + \sum_{\mathbf{u}_j \in \mathcal{U}} (n_{\mathbf{u}_j}\mathbf{z}(\mathbf{u}_j) - |\mathbf{u}_j|\exp(\mathbf{x}^{\mathsf{T}}(\mathbf{u}_j)\boldsymbol{\gamma} + z(\mathbf{u}_j)))$$

where $n_{\mathbf{u}_j}$ is the number of observed points in grid cell j. The Discrete Fourier Transform (DFT) can then be used to compute the gradient, $\nabla\log(\pi(\tilde{\mathbf{z}} \mid \mathcal{S}, \boldsymbol{\theta}))$.

See Moller and Waagepetersen [2003], Møller and Waagepetersen [2007] for further discussion of MCMC with Langevin-Hastings updates for LGCPs as well as applications to environmental spatial point patterns. Modifications to the specification of the proposal covariance matrix that further improves convergence of the MCMC algorithm have been proposed by Girolami and Calderhead [2011]. They demonstrate the improvements of these methods through exhaustive simulations and applications for LGCPs as well as other models with intractable likelihoods. In particular, these modified specifications can increase efficiency with increases in both the dimension of the parameter space and correlation in the latent spatial processes.

The modifications to the proposal distribution of the MALA algorithm arise from using an approximation to the Fisher information matrix evaluated at the maximum likelihood estimate of the transformed process $\tilde{\mathbf{z}}$. Convergence can be further improved by modifying the tuning parameter h within the chain using an adaptive MCMC algorithm [Roberts and Rosenthal, 2007]. Details and an application using the modified proposals for the MALA algorithm are given in Diggle [2013]. Lastly, code for fitting LGCPs using the algorithm outlined above is provided in Moller and Waagepetersen [2003]. The lgcp package in R [Taylor and Diggle, 2014] also includes functions for fitting LGCPs using MALA.

9.4.2.3   Hamiltonian Monte Carlo

Hamiltonian Monte Carlo (HMC) [Neal, 1993a,b, 1996, Liu, 2002] algorithms, as discussed in Section 5.5.1.3, pose an alternative to Metropolis-Hastings algorithms for Bayesian inference of LGCPs. Here, we offer an elaboration of HMC in a different form from Section 5.5.1.3.

In general, in order to learn about a high dimensional density, we must efficiently explore the parameter space for $\boldsymbol{\theta}$, and can not stay near the mode(s) since there is very little mass concentrated at these locations. Of course, there are also regions, away from the modes, where there is little mass. So, we need the informal concept of a "typical set." Such a set is a high mass set within the parameter space. That is, random walk Metropolis is a "guess and check" style approach where proposals are drawn from a specified distribution. High acceptance is not favorable for exploring the parameter space as it is too local, and low acceptance means the candidates are being drawn too far from the mass of the density.

This suggests that information about $p(\boldsymbol{\theta})$ and the gradient, or rather, a moving direction, should be leveraged to make efficient proposals with regard to exploring the density. Whereas MALA suggests that the direction should be back towards the mode, in high dimension, the mode contains very little mass. Instead, the goal here is to choose a direction to move to (or stay in) the foregoing typical set. This approach for a probabilistic specification with (*mode, gradient, typical set*), is analogous to a physical system with (*planet, gravitational field, orbit*). For a physical system, staying in orbit requires momentum to be added or subtracted in order to appropriately offset the gravitational pull. The HMC approach moves from probability densities to energies associated with the physical system, where latent momentum is added for each component of $\boldsymbol{\theta}$. The energy is specified through the *Hamiltonian* and Hamilton's equations are used to control the behavior of the associated physical system.

Here, we describe the simplest version of the HMC approach. First, for $\boldsymbol{\theta}$ of p-dimensions, the latent joint distribution is specified in $2p$-dimensions. We introduce p auxiliary parameters, $\boldsymbol{\phi}$, an associated vector of momentums. We set $\pi(\boldsymbol{\theta}, \boldsymbol{\phi}) = e^{-H(\boldsymbol{\theta}, \boldsymbol{\phi})}$ where H is the Hamiltonian, or *energy*, at $(\boldsymbol{\theta}, \boldsymbol{\phi})$ The Hamiltonian is defined $H(\boldsymbol{\theta}, \boldsymbol{\phi}) = P(\boldsymbol{\theta}) + K(\boldsymbol{\phi}, \boldsymbol{\theta})$, where $P(\boldsymbol{\theta})$ is the potential energy and $K(\boldsymbol{\phi}, \boldsymbol{\theta})$ is the kinetic energy. In the density space, this is written $\pi(\boldsymbol{\theta}, \boldsymbol{\phi}) \propto e^{-P(\boldsymbol{\theta})} \times \exp(-K(\boldsymbol{\phi}, \boldsymbol{\theta}))$, which can be viewed as a conditional times a marginal form. The potential energy function is already determined through the density function for $\boldsymbol{\theta}$ which we wish to sample. So, we are left to choose the kinetic energy function, equivalently, the (conditional) density for the momentums. Every choice of kinetic energy, $K(\boldsymbol{\phi}, \boldsymbol{\theta})$, yields a new Hamiltonian transition that will interact differently with a given target distribution for $\boldsymbol{\theta}$. So, careful tuning is required to "stay in orbit'.' The Hamiltonian equations provide an approach to achieve this equilibrium, since Hamilton's equations maintain energy, equivalently, orbit. That is, they maintain a level set for $H(\boldsymbol{\theta}, \boldsymbol{\phi})$.

In implementation, the algorithm proceeds by drawing from $[\boldsymbol{\phi} \mid \boldsymbol{\theta}]$. The simplest choice is so-called *Euclidean-Gaussian* kinetic energy, $K(\boldsymbol{\phi} \mid \boldsymbol{\theta}) \propto \boldsymbol{\phi}^{\mathrm{T}} M^{-1} \boldsymbol{\phi}$ implying a multivariate normal for the momentums, i.e., $\boldsymbol{\phi} \sim MVN(\mathbf{0}, M)$ Theory shows that a good choice for M is $\Sigma_{\boldsymbol{\theta}}$, which can be learned from the MCMC samples, bringing in some conditional dependence. A draw from $[\boldsymbol{\phi} \mid \boldsymbol{\theta}]$ implies a draw, h, from $[H(\boldsymbol{\theta}, \boldsymbol{\phi}) \mid \boldsymbol{\theta}]$, which yields a fixed energy.

Then, to maintain the energy/orbit, we explore the level set determined by h through random trajectories over the level set. The level set is the set of $(\boldsymbol{\theta}^*, \boldsymbol{\phi}^*)$ such that $H(\boldsymbol{\theta}^*, \boldsymbol{\phi}^*) = h$. This set is determined by solutions to Hamilton's equations which don't exist in closed form. Therefore, the main obstruction to implementing the Hamiltonian Monte Carlo method generates the Hamiltonian trajectories themselves. Aside from a few trivial examples, we cannot solve Hamilton's equations exactly and any implementation must instead solve them numerically.

In practice, the proposal step in HMC is more costly than the traditional proposal draws in usual MCMC, making the computational cost of a single iteration higher. The benefit, however, is a more efficient exploration of the space for the high dimensional distribution, making it more efficient overall. The HMC algorithm is built into the STAN software packages [Carpenter et al., 2017], which is easily accessible and adaptable to a wide range of models. As with all MCMC algorithms, HMC can still struggle to capture isolated local maxima that do not fall in the typical set. See Girolami and Calderhead [2011] for additional details and modifications to the proposal covariance matrix that further improves convergence of the MCMC algorithm.

9.4.2.4   Integrated nested Laplace approximation

As noted in Section 5.5.1.5, the integrated nested Laplace approximation (INLA) approach to Bayesian inference for LGCP models circumvents the challenges of programming, fine-tuning, and running computationally challenging MCMC algorithms [Illian et al., 2012, 2013]. Some technical details were presented in Section 5.5.1.5. For the LGCP, INLA will approximate, marginally, the posterior distribution of the latent field as well as the model parameters. The approximate (marginal) posterior distributions can then be used to obtain estimates of means, standard deviations, etc.

The speed of model inference using INLA relies on approximating the continuous Gaussian process with a discrete Gaussian Markov random field (GMRF). Recall the LGCP specification where the intensity function of the spatial point pattern is written as $\log\lambda(\mathbf{s}) = \mathbf{x}^{\mathrm{T}}(\mathbf{s})\boldsymbol{\gamma} + z(\mathbf{s})$ and $z(\mathbf{s})$ is a Gaussian process. INLA requires the Gaussian process, which is capturing the spatial dependence in the process not accounted for the covariates, to be approximated by a GMRF.

The Matérn covariance function was discussed at length in Chapters 2 and 3. With regard to INLA, it has attracted consequential attention these days in the spatial community due to an important paper of Lindgren et al. [2011]. This paper demonstrated that a Gaussian random field with a Matérn covariance function arises as the solution to a particular stochastic partial differential equation (SPDE). The authors further show that, using an approximate stochastic weak solution to this stochastic partial differential equation, for some Gaussian fields in the Matérn class, an explicit link, through triangulation, can be made between these Gaussian fields and Gaussian Markov random fields (see Chapter 4). The consequence is that modeling can be done by using the former but computations can be done using the latter. This Markov random field approximation has enabled convenient implementation of spatial model fitting using INLA.

We explore the SPDE a little further here. It is a fractional SPDE taking the form

$$(\kappa^2 - \Delta)^{\alpha/2} Z(\mathbf{s}) = W(\mathbf{s}) \tag{9.8}$$

where, in general, $\mathbf{s} \in R^r$, Δ is the Laplacian, $\alpha = \nu + r/2$, $W(\mathbf{s})$ is spatial white noise with variance 1, and $Z(\mathbf{s})$ is the solution we seek. Here, $(\kappa^2 - \Delta)^{\alpha/2}$ is a pseudo-differential operator. Whittle [1954] obtained the solution using the Fourier transform definition of the fractional Laplacian in R^r. The reader unfamiliar with the world of SPDE's would surely ask what any of the above means!

Let's try to help a bit. Recall that, in R^2, the Laplacian of a function $f(\mathbf{s})$, $\Delta f(\mathbf{s}) = \sum_{j=1}^{2} \partial^2 f(\mathbf{s})/\partial^2 x_j$ where $\mathbf{s} = (x_1, x_2)$. Let us calculate $\Delta f(\mathbf{s})$ through its Fourier transform. Assuming differentiation can be brought under the integration, straightforward calculation yields $\Delta f(\mathbf{s}) = \int_{R^r} -||\mathbf{w}||^2 \mathcal{F}(f(\mathbf{s})) d\mathbf{w}$ where $\mathcal{F}(f(\mathbf{s}))$ is a function of $\mathbf{w}$ denoting the Fourier transform of $f(\mathbf{s})$. This suggests that we replace Δ with $-\Delta$ in order to make the integral positive and write $\mathcal{F}(-\Delta f(\mathbf{s})) = ||\mathbf{w}||^2 \mathcal{F}(f(\mathbf{s}))$. It further suggests considering fractional powers, e.g., $\mathcal{F}(-\Delta f(\mathbf{s})^\alpha) = ||\mathbf{w}||^{2\alpha} \mathcal{F}(f(\mathbf{s}))$ and finally $\mathcal{F}(\kappa^2 - \Delta f(\mathbf{s})^\alpha) = ||\kappa^2 + \mathbf{w}||^{2\alpha} \mathcal{F}(f(\mathbf{s}))$. In fact, from expression (3.5) in Chapter 3, using the inversion formula below expression (3.2) in Chapter 3, we see an illustration of the above applied to $f(\mathbf{s})$ as the Matérn correlation function. Finally, replacing $f(\mathbf{s})$ with $Z(\mathbf{s})$ and equating to $W(\mathbf{s})$ yields the fractional SPDE above. Altogether, we see that the solution will be a Gaussian field and that the solution will have a Matérn correlation function.

The `inla` package in R contains functions for fitting spatial point process models using INLA under various model specifications. For example, the (log)intensity function can be specified with a variety of location-specific covariates and/or spatially structured random effects observed on different spatial scales. Some of these model specifications are discussed in detail in Illian et al. [2013]. Multiple alternatives also exist within the `inla` packages for modeling the spatial dependence in the latent GMRF, such as first- and second-order random walks or conditional autoregressive models. Extensions that allow for an irregular grid for the GMRF and neighborhood dependence structure [Simpson et al., 2016, Lindgren et al., 2011] are also available.

Comparisons between MCMC and INLA approaches to inference for LGCP models can be found in Taylor and Diggle [2014], Renner et al. [2015]. In general, there are advantages and disadvantages to each of these approaches. MCMC, and, thus, fully Bayesian approaches benefit from producing full Bayesian inference; namely, full joint posterior distributions which can be used to obtain parameter inference (means, credible intervals) and to obtain full posterior predictive distributions. In addition, the inference is obtained for the Gaussian process with full a covariance matrix, rather than from a process approximated by a GMRF. Overwhelmingly, computational efficiency in obtaining inference is the main benefit of INLA. In particular, model comparison and validation greatly benefit from the rapid computation time of INLA, making it a convenient method for assessing a collection of candidate models. As above, INLA will struggle, however, when the dimension of θ grows large in a point process model because of the challenge of the above numerical integration over θ.

9.5   Computational strategies for inhibition and clustering processes

Arguably, the most widely adopted class of spatial point pattern models is the nonhomogeneous Poisson processes (NHPP) or, more generally, the log Gaussian Cox processes (LGCP) (see Moller and Waagepetersen [2003] and references therein). As developed in the previous chapter, such models assume conditionally independent event locations given the process intensity. However, there we noted that in many applications, we find evidence of clustering or of inhibition. So, here we focus on inhibition and refer to associated models as repulsive spatial point processes. Most common in this setting are Gibbs point processes (here, denoted as GPP) [see, e.g., Illian et al., 2008, and Section 8.5.3]. These processes specify the joint location density, up to the normalizing constant, in the form of a Gibbs distribution, introducing potentials on cliques of order 1 but also potentials on cliques of higher order, which capture the interaction. The most familiar example in the literature is the Strauss process and its extreme version, the hardcore process [Strauss, 1975, Kelly and Ripley, 1976]. An attractive alternative is the determinantal point process (here, denoted as DPP). Though these processes have some history in the mathematics and physics communities, they have only recently come to the attention of the statistical community thanks, most notably, to recent efforts of Jesper Møller and colleagues. See, for instance, Lavancier et al. [2015].

Approximate Bayes computation (ABC) to fit both classes of models is discussed in detail in Shirota and Gelfand [2017]. Here, we confine ourselves to ABC model fitting for GPPs as an alternative to Markov chain Monte Carlo (MCMC) model fitting as discussed above and in greater detail in Møller et al. [2006], Affandi et al. [2013, 2014], Goldstein et al. [2015]. The MCMC algorithms are complex and implementation can often result in poorly behaved chains with concerns regarding posterior convergence. Here, we demonstrate much simpler model fitting using ABC. ABC is particularly promising for GPPs since these processes allow straightforward simulation of point pattern realizations given parameter values. Additionally, such simulation facilitates posterior inference as well as consideration of model adequacy and model comparison, as we have argued in previously in this chapter.

ABC methods are now attracting considerable attention [Pritchard et al., 1999, Beaumont et al., 2002, Marjoram et al., 2003, Sisson and Fan, 2011, Marin et al., 2012]. The scope of ABC applications is also increasing, e.g., population genetics [Beaumont et al., 2002], multidimensional stochastic differential equations [Picchini, 2013], macroparasite populations

[Drovandi and Pettitt, 2011] and the evolution of HIV-AIDS [Blum and Tran, 2010]. As for spatial statistical applications, Erhardt and Smith [2012] implemented ABC for max-stable processes in order to model spatial extremes and Soubeyrand et al. [2013] applied ABC with functional summary statistics to fit a cluster and marked spatial point process.

The Gibbs point process offers a mechanistic model with interpretable parameters and has been used for modeling repulsive point patterns in environmental science and biology [Stoyan and Penttinen, 2000, Mattfeldt et al., 2007, King et al., 2012b, Goldstein et al., 2015]). The main challenge for fitting models using these processes is that likelihoods have intractable normalizing constants which depend on parameters. Hence, likelihood inference is difficult [Moller and Waagepetersen, 2003] and standard Bayesian inference using Markov chain Monte Carlo (MCMC)) is not directly available. Maximum pseudo-likelihood estimation was proposed in Besag [1975, 1986] and Jensen and Møller [1991]. These estimators show poor performance in the presence of strong inhibition [e.g., Huang and Ogata, 1999]. In the Bayesian framework, a clever auxiliary variable MCMC strategy was developed by Møller et al. [2006] (Section 9.3.5) and extended by Murray et al. [2006]. However, perfect simulation within the MCMC algorithm is needed along with approximations.

Here, following Shirota and Gelfand [2017], we show how to implement the ABC algorithm for fitting GPP's based on ABC-MCMC proposed by Marjoram et al. [2003]. We include a discussion about how to choose summary statistics, kernel functions, and tune user-specific parameters. Again, the attractiveness of ABC for repulsive point processes rests in the fact that it is straightforward to generate realizations under these point processes given parameter values. This enables the ABC presumption: randomly draw parameters and then randomly draw point patterns given parameters. Further, as has been our theme for the entire monograph, with posterior inference achieved for the model parameters, we can use composition sampling to draw posterior predictive point patterns, enabling posterior inference about features of point patterns realized under the models. In addition, through these posterior samples of point patterns, we can propose model assessment for repulsive point processes, following the discussion in Leininger and Gelfand [2017] and Section 9.2.

Briefly reviewing, for GPPs, since $c_0(\boldsymbol{\theta})$ cancels out of the Papangelou conditional intensity, the pseudo-likelihood, in log form, $\log PL(\mathcal{S} \mid \boldsymbol{\theta}) = -\int_D \lambda(u \mid \mathcal{S}, \boldsymbol{\theta})du + \sum_i \log \lambda(\mathbf{s}_i \mid \mathcal{S}, \boldsymbol{\theta})$, has been proposed [Besag, 1975] yielding the maximum pseudo-likelihood estimator. Although the maximum pseudo-likelihood estimator is consistent [see Jensen and Møller, 1991], the performance of the maximum pseudo-likelihood estimator is poor in the case of a small number of points and strong repulsion [Huang and Ogata, 1999].

The pseudo-likelihood can be used for MCMC in the Bayesian framework [e.g., King et al., 2012b]. Møller et al. [2006] and Berthelsen and Møller [2008] proposed an auxiliary variable MCMC method (AVM) where, conveniently, $c_0(\boldsymbol{\theta})$ cancels out of the Hastings ratio. The challenge is to obtain the conditional density of the auxiliary variable. A partially ordered Markov model is used to approximate this density. A similar approach is the exchange algorithm proposed by Murray et al. [2006]. Both algorithms require perfect simulation from the likelihood given $\boldsymbol{\theta}$ for each MCMC iteration. Although, perfect simulation is available for GPPs, this step can be computationally burdensome and obtaining a good acceptance rate is difficult.

More recently, Liang et al. [2009b] proposed the double MCMC algorithm which does not require perfect simulation from the likelihood. It only requires simulation from the Markov transition kernel and is faster than the AVM and exchange algorithms but convergence to the stationary distribution is not guaranteed. Goldstein et al. [2015] implement this algorithm, with an application, for a class of GPP models.

For ABC, we need to simulate realizations of a GPP given parameter values. This is usually based on a birth-and-death algorithm [e.g., Geyer and Møller, 1994, Moller and Waagepetersen, 2003, Illian et al., 2008]. An alternative simulation algorithm to generate the point pattern is "dominated coupling from the past" [Kendall and Møller, 2000] as implemented by Berthelsen and Møller [2002] and Berthelsen and Møller [2003]. This algorithm can be called as a default setting in `spatstat` [Baddeley and Turner, 2005].

9.5.1 Approximate Bayesian Computation for Repulsive Point Processes

Let $\mathcal{S}_{obs}$ be the observed point pattern and $\mathcal{S}^*$ be a simulated point pattern. For a Bayesian model of the form $\pi(\mathcal{S} \mid \boldsymbol{\theta})\pi(\boldsymbol{\theta})$, ABC consists of three steps: (1) generate $\boldsymbol{\theta} \sim \pi(\boldsymbol{\theta})$, (2) generate $\mathcal{S}^* \sim \pi(\mathcal{S} \mid \boldsymbol{\theta})$, (3) compare summary statistics for the generated $\mathcal{S}^*$, $\mathbf{T}(\mathcal{S}^*)$, with those of the observed data, $\mathbf{T}(\mathcal{S}_{obs})$, and accept $\boldsymbol{\theta}$ if $\Psi(\mathbf{T}(\mathcal{S}), \mathbf{T}(\mathcal{S}_{obs})) < \epsilon$ for a selected *kernel(distance)* measure Ψ. Accepted $\boldsymbol{\theta}$ are samples from the approximate posterior distribution, $\pi_\epsilon(\boldsymbol{\theta} \mid \mathbf{T}(\mathcal{S}_{obs}))$. Approximation error relative to the exact posterior distribution $\pi(\boldsymbol{\theta} \mid \mathcal{S}_{obs})$ comes from the choice of $\mathbf{T}(\cdot)$, Ψ, and ϵ. If $\mathbf{T}(\cdot)$ is a sufficient statistic for $\boldsymbol{\theta}$, then $\pi(\boldsymbol{\theta} \mid \mathbf{T}(\mathcal{S}_{obs})) = \pi(\boldsymbol{\theta} \mid \mathcal{S}_{obs})$ and, given Ψ, the only approximation error is from ϵ. Since sufficient statistics are not usually available, the selection of informative summary statistics $\mathbf{T}(\cdot)$ is critically important. Small values of ϵ are desired but require more simulation of $\boldsymbol{\theta} \sim \pi(\boldsymbol{\theta})$ and $\mathcal{S} \sim \pi(\mathcal{S} \mid \boldsymbol{\theta})$ in order to retain $\mathcal{S}$. Again, with regard to simulation of $\mathcal{S}$ for the GPP, we can utilize perfect simulation.

9.5.1.1 Summary Statistics

The Strauss process was discussed in Section 8.5.3. It is a Gibbs point process (GPP) with density often written as

$$\pi(\mathcal{S}) = \beta^{N(\mathcal{S})}\gamma^{s_R(\mathcal{S})}/c(\beta,\gamma), \quad \mathcal{S} \subset D \tag{9.9}$$

where $\beta > 0$, $0 \leq \gamma \leq 1$, $N(\mathcal{S})$ is the number of points, and $c(\beta,\gamma)$ is the normalizing constant. Here,

$$s_R(\mathcal{S}) = \sum_{\{\mathbf{s}_i,\mathbf{s}_j\}\subseteq\mathcal{S}\subset D} 1(\|\mathbf{s}_i - \mathbf{s}_j\| \leq R) \tag{9.10}$$

is the number of *R-close* pairs of points in $\mathcal{S}$. Given R, $N(\mathcal{S})$ and $s_R(\mathcal{S})$ are sufficient statistics for (β,γ). γ is an interaction parameter indicating the degree of repulsion. Large values of γ suggest weak repulsion while small values of γ indicate strong repulsion. $\gamma = 0$ provides the hardcore Strauss process which does not allow the occurrence of any points within the interaction radius R. $\gamma = 1$ provides a homogeneous Poisson process.

Hence, the appropriate summary statistics would be $\mathbf{T} = (\log N(\mathcal{S}), K_R(\mathcal{S}))$. In practice, R is not known but we can choose a radius R though profile pseudo likelihoods. Alternatively, creating a set of R values yields a set of summary statistics, indexed by R.

In other repulsive point process settings, second-order summary statistics would emerge as potential summary statistics because they illuminate clustering or inhibition behavior. For instance, with a stationary point process, the K function with radius d, $K(d)$, the *expected* number of the remaining points in the pattern within distance d from a typical point, is a useful measure. The empirical estimator of K_d provides such a choice. The actual summary statistics would employ a set of d's.

The L function [Besag and Diggle, 1977]; $\hat{L}_d(\mathcal{S}) = \sqrt{\hat{K}(d;\mathcal{S})/\pi}$ is often preferred because the fluctuations of the estimated K function increase with increasing d while the root transformation stabilizes these fluctuations [e.g., Illian et al., 2008]. Again, a set of D's would be employed.

9.5.2 Explicit specification of an ABC algorithm

The ABC algorithm presented here is based on a semi-automatic approach proposed by [Fearnhead and Prangle, 2012]. They argue that the optimal choice of $\mathbf{T}(\mathcal{S}_{obs})$ is $\mathrm{E}(\boldsymbol{\theta} \mid \mathcal{S}_{obs})$

and then discuss how to construct $E(\boldsymbol{\theta} \mid \mathcal{S}_{obs})$. They consider a linear regression approach to construct the summary statistics through a pilot run. In our setting, we generate L sets of $\{\boldsymbol{\theta}_\ell, \mathcal{S}_\ell\}_{\ell=1}^L$. Then, we implement a linear regression for $E(\boldsymbol{\theta}_\ell \mid \mathcal{S}_{obs}) = \mathbf{a} + \mathbf{b}^{\mathrm{T}}\boldsymbol{\eta}(\mathcal{S}_\ell, \mathcal{S}_{obs})$ where $\boldsymbol{\eta}(\mathcal{S}_\ell, \mathcal{S}_{obs})$ is a vector of functions of the summary statistics constructed from the simulated and observed point patterns.[4] Following above, we take $\boldsymbol{\eta}(\mathcal{S}, \mathcal{S}_{obs}) = (\eta_1(\mathcal{S}, \mathcal{S}_{obs}), \boldsymbol{\eta}_2(\mathcal{S}, \mathcal{S}_{obs}))$ where $\eta_1(\mathcal{S}, \mathcal{S}_{obs})$ and the $M \times 1$ vector $\boldsymbol{\eta}_2$ are

$$\eta_1(\mathcal{S}, \mathcal{S}_{obs}) = \log n(\mathcal{S}) - \log n(\mathcal{S}_{obs}), \quad \text{and}$$

$$\eta_{2,d}(\mathcal{S}, \mathcal{S}_{obs}) = \left| \sqrt{\hat{K}_d(\mathcal{S})} - \sqrt{\hat{K}_d(\mathcal{S}_{obs})} \right|^2. \tag{9.11}$$

for $d = 1, 2, \ldots M$.

After obtaining $\hat{\mathbf{a}}$ and $\hat{\mathbf{b}}$ by least squares, we can calculate $\hat{\boldsymbol{\theta}}^* = \hat{\mathbf{a}} + \hat{\mathbf{b}}^{\mathrm{T}}\boldsymbol{\eta}(\mathcal{S}^*, \mathcal{S}_{obs})$ for any simulated $\mathcal{S}^*$. We set $\hat{\boldsymbol{\theta}}_{obs} = \hat{\mathbf{a}}$ and specify our distance function for the ABC through $\Psi(\hat{\boldsymbol{\theta}}^*, \hat{\boldsymbol{\theta}}_{obs})$ with Ψ specified below. To facilitate the regression, one can take a log transformation of the parameter vector, e.g., $\boldsymbol{\theta} = (\log \beta, \log \gamma)$ for the Strauss process.

Given the results of the pilot run, the approach proposed by Fearnhead and Prangle [2012] implements the ABC-MCMC algorithm by Marjoram et al. [2003]. ABC-MCMC is a straightforward extension of the standard MCMC framework to ABC; convergence to the approximate posterior distribution, $\pi_\epsilon(\boldsymbol{\theta} \mid \mathbf{T}(\mathcal{S}_{obs}))$, is guaranteed. Specifically, with t denoting iterations and $q(\cdot \mid \cdot)$ denoting a proposal density,

1. Let $t = 1$.

2. Generate $\boldsymbol{\theta}^* \sim q(\boldsymbol{\theta} \mid \boldsymbol{\theta}^{(t-1)})$ and $\mathcal{S}^* \sim \pi(\mathcal{S} \mid \boldsymbol{\theta}^*)$ and calculate $\hat{\boldsymbol{\theta}}^* = \hat{\mathbf{a}} + \hat{\mathbf{b}}\boldsymbol{\eta}(\mathcal{S}^*, \mathcal{S}_{obs})$. Repeat this step until $\Psi(\hat{\boldsymbol{\theta}}^*, \hat{\boldsymbol{\theta}}_{obs}) < \epsilon$ where $\hat{\boldsymbol{\theta}}_{obs} = \hat{\mathbf{a}}$ and $\Psi(\hat{\boldsymbol{\theta}}^*, \hat{\boldsymbol{\theta}}_{obs})$ is defined below.

3. Calculate the acceptance ratio $\alpha = \min\left\{1, \frac{\pi(\boldsymbol{\theta}^*)q(\boldsymbol{\theta}^{(t-1)} \mid \boldsymbol{\theta}^*)}{\pi(\boldsymbol{\theta}^{(t-1)})q(\boldsymbol{\theta}^* \mid \boldsymbol{\theta}^{(t-1)})}\right\}$. If $u < \alpha$ where $u \sim Unif(0,1)$, retain $\boldsymbol{\theta}^{(t)} = \boldsymbol{\theta}^*$, otherwise $\boldsymbol{\theta}^{(t)} = \boldsymbol{\theta}^{(t-1)}$. Return to step 2 and $t \to t+1$.

As a distance measure, we use the componentwise sum of quadratic loss for the log of the parameter vector, i.e., $\Psi(\hat{\boldsymbol{\theta}}_\ell, \hat{\boldsymbol{\theta}}_{obs}) = \sum_j (\hat{\theta}_{\ell,j} - \hat{\theta}_{obs,j})^2/\hat{\mathrm{Var}}(\hat{\theta}_j)$ where $\hat{\mathrm{Var}}(\hat{\theta}_j)$ is the sample variance of j-th component of $\hat{\boldsymbol{\theta}}$. To choose an acceptance rate ϵ, through the pilot run, we obtain the empirical percentiles of $\{\Psi(\hat{\boldsymbol{\theta}}_\ell, \hat{\boldsymbol{\theta}}_{obs})\}_{\ell=1}^L$ and then select ϵ according to these percentiles. Step 2 is the most computationally demanding. We need to simulate the proposed point pattern $\mathcal{S}^* \sim \pi(\mathcal{S} \mid \boldsymbol{\theta}^*)$ until $\Psi(\hat{\boldsymbol{\theta}}^*, \hat{\boldsymbol{\theta}}_{obs}) < \epsilon$.

With regard to choice of a number of summary statistics, when M is large more information is obtained but, if too large, overfitting, relative to the number of points in the point pattern, results. As a strategy for this selection, we specify M with equally spaced d's, and implement a lasso [Tibshirani, 1996]. We determine the penalty parameter for the lasso by cross-validation and preserve the regression coefficients corresponding to the optimal penalty by using `glmnet` [Friedman et al., 2010]. A simpler alternative is to fit using several choices of M and assess the sensitivity of posterior inference.

In a frequentist analysis, the minimum contrast estimator is often used to fit models using the K function. The minimum contrast estimator requires the analytical form for the functional statistics which are not necessarily available for repulsive point processes. ABC does not require analytical expressions for the functional statistics because the approach compares the "estimated" K function for observed and simulated point patterns. However, if analytical forms for the functional summary statistics are available, the minimum contrast estimator or composite likelihood estimators [Baddeley and Turner, 2000, Guan, 2006] would

---

[4] Fearnhead and Prangle [2012] implement linear regression for each component of $\boldsymbol{\theta}$. However, with a small number of parameters, we keep the notation as linear regression for multivariate responses.

be available and easy to implement. Furthermore, software for these estimators has already been developed [Baddeley and Turner, 2005].

As a final comment here, Shirota and Gelfand [2017] compared this proposed ABC-MCMC algorithm with the straightforward exchange algorithm of Murray et al. (2006), mentioned above, for a Strauss point process. They considered inefficiency factors (IF) for parameters, i.e., the ratio of the numerical variance of the estimate from the MCMC samples relative to that from hypothetical uncorrelated samples, using both model fitting approaches. They found that IFs for the exchange algorithm tend to be an order of magnitude greater than those from the ABC-MCMC algorithm. Also, the ABC-MCMC algorithm allows simple parallelization, which is not possible for the exchange algorithm. So, computationally, the ABC-MCMC algorithm can be much faster.

9.5.3 *Neyman Scott process and shot noise processes*

In a sense, ABC is easier for the Neyman Scott and the shot noise processes than for Gibbs processes because simulation of samples is so straightforward. See [Soubeyrand et al., 2013] in this regard. That is, the Neyman Scott processes are defined in a generative fashion (Section 8.5.1) while the shot noise processes (Section 8.5.2) are generated in two straightforward stages. That is, first a realization of an HPP(λ) yields the shot noise intensity, followed by a realization of an NHPP given the shot noise intensity. Shot noise process realizations can also be developed by additive superposition, i.e., summing up realizations over each of the random kernels. We note that Soubeyrand et al. [2013] illustrate the use of ABC for a Thomas process.

The question to ask here is, in general, what should be the choice of summary statistics for spatial point patterns? Again, usual measures are the K function or the pair correlation function. For ABC we only require the empirical K function or the empirical partial correlation function. $\hat{K}(d)$ was given in Section 8.3.3. The empirical correlation function $\hat{g}(d)$ is derived from the relationship, $g(d) = K'(d)/2\pi d$ using either a smoothing spline for $K'(d)$ or else a direct kernel estimate for $y(d)$. Whichever is selected, the infinity of statistics is reduced to a single measure for ABC comparison by a weighted integration, $\int_{d \leq d_o} w(d)(\hat{K}(d) - \hat{K}_{obs}(d))^2 dd$ for a suitable maximum distance, d_o . See Soubeyrand et al. [2013] for further details.

9.6   Examples

9.6.1   *An NHPP and LGCP example*

Returning to the tropical rainforest tree data in Section 8.3.4, we consider four models here. In all cases we employ the two covariates, elevation and slope, resulting in three regression coefficients. We fit an NHPP model, we fit an LGCP model, and we fit two versions of the poor man's model, one using a 20×10 grid, the other using a 100×50 grid. In Figure 9.1(a) we show the results of the NHPP fit. All three coefficients are significant but the estimated intensity surface Figure 9.1(b) is poor; the regression form can not capture the extreme peaks (compare with the left panel in Figure 8.8(a) or Figure 8.11 below). For the LGCP, in Figure 9.2 we see that, again all three regression coefficients are significant. We also provide a comparison of the prior to posterior for σ^2 to show the Bayesian learning. Figure 9.3 shows the point pattern and the intensity surface estimates. Three posterior mean intensity surface estimates are provided, $\lambda(\mathbf{s})$, $\lambda_0(\mathbf{s})$, and $z_0(\mathbf{s}) = \log\lambda_0(\mathbf{s})$. Now, we see the substantial local GP adjustment and we do a much better job with the estimated $\lambda(\mathbf{s})$. Finally, we show the results of the poor man's fitting for the 20×10 and 100×50 grids in Figure 9.4. Panels (a) and (b) correspond to the 20×10 grid and panels (c) and (d) correspond to the

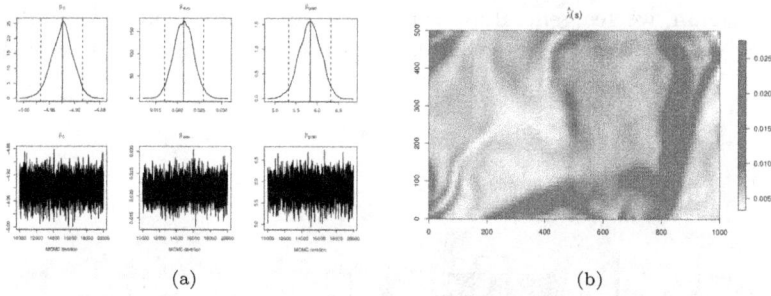

(a)   (b)

Figure 9.1 *Estimates from the NHPP model. Left panel shows the estimated posterior distributions for the three coefficients along with their MCMC chains. Right panel depicts the estimated intensity surface.*

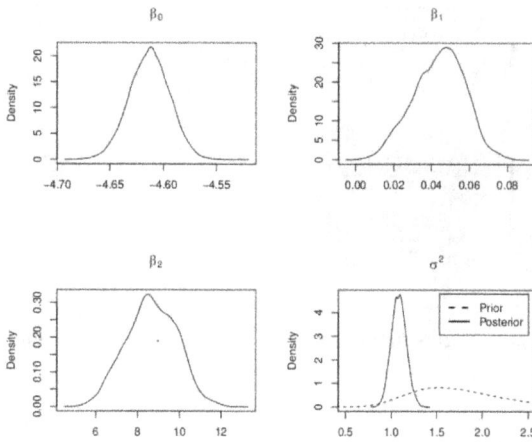

Figure 9.2 *Posterior estimates for the three regression coefficients and σ^2 from the LGCP model.*

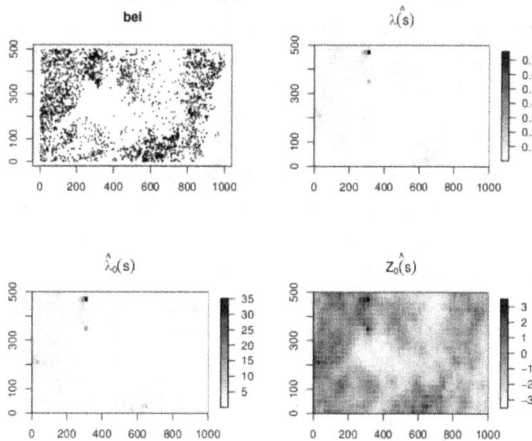

Figure 9.3 *Estimates from the LGCP model: Point pattern and the intensity estimates.*

100×50 grid. Again, we see general agreement regarding the coefficients while the higher resolution poor man's estimated intensity surfaces are very similar to those of the LGCP.

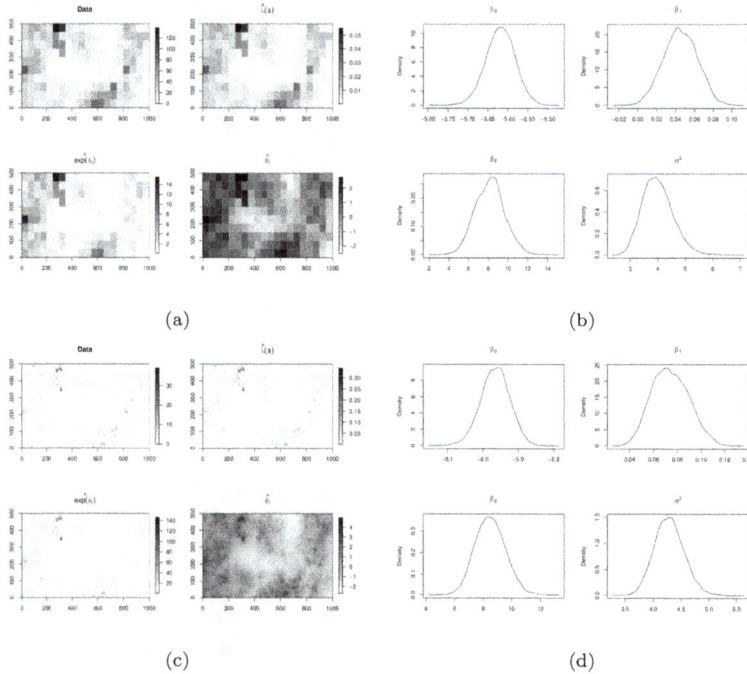

Figure 9.4 *Results for the poor man's fitting for the 20×10 ((a) and (b)) and 100×50 ((c) and (d)) grids.*

9.6.2 Modeling earthquake epicenters with a LGCP

We next consider a dataset of earthquake epicenter locations. The data comes from USGS Earthquake Hazards Program, which is publicly available.[5] The data in this analysis consist of 1035 earthquakes of magnitude ≥ 2.5 occurring in 2017 with epicenters in northern Oklahoma and southern Kansas, as depicted in Figure 9.5.

The intensity function for the LGCP is specified as

$$\log\lambda(\mathbf{s}) = \mu + z(\mathbf{s}).$$

That is, we have no regressors but $z(\mathbf{s})$ is a Gaussian process with mean $-\sigma^2/2$ and covariance specified by the exponential covariance function where $\mathrm{Cov}(z(\mathbf{s}), z(\mathbf{s}')) = \sigma^2 \exp^{-\|\mathbf{s}-\mathbf{s}'\|/\phi}$, where σ^2 is the variance and ϕ is the range parameter.

A collection of representative points was used for model fitting and inference as discussed in Section 9.4.2. The locations of the earthquake epicenters were projected from latitude and longitude to UTM coordinates. Distances between points were computed using Euclidean distance of the UTM coordinates, given in kilometers (km). The maximum distance between epicenters was 390 km. The locations of the representative points were defined on a 12×12 km grid as shown in Figure 9.6. A total of 529 representative points were used for model fitting and inference. Specifically, realizations of the Gaussian process were obtained at these representative points and used for numerical integration when evaluating the likelihood.

---

[5]https://earthquake.usgs.gov/earthquakes/

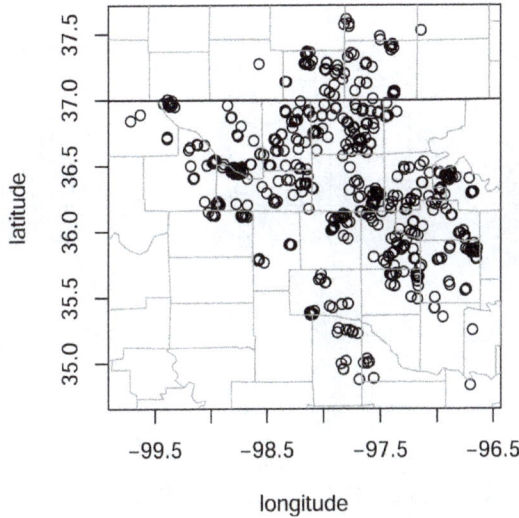

Figure 9.5 *Epicenters of earthquakes with magnitude greater than 2.5 during 2017 across northern Oklahoma and southern Kansas.*

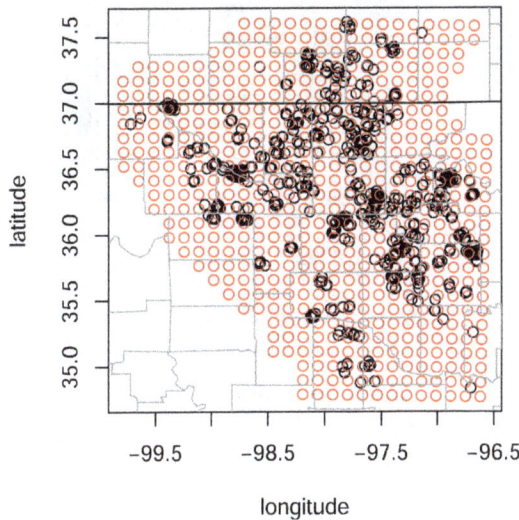

Figure 9.6 *The grid of representative points used for model fitting and inference. The points were located on a 12×12 km grid using UTM coordinates, however they are projected here to latitude and longitude.*

Inference was implemented within a Bayesian framework. Independent prior distributions were assigned to each of the three parameters, μ, σ^2, and ϕ. Specifically, a noninformative prior was assigned to μ such that $\mu \sim N(0, 100^2)$. For the covariance parameters, $\sigma^2 \sim IG(2, 2)$ and $\phi \sim Unif(8, 60)$. The parametrization of the inverse gamma distribution is highly noninformative; σ^2 has non-finite variance. The specification of the uniform distribution is based on the spatial domain of the data, and controls the effective range to be between approximately 24 and 180 km. These values were chosen such that the effective range is at least two times the resolution of the grid of representative points and less than approximately half the maximum distance over the domain.

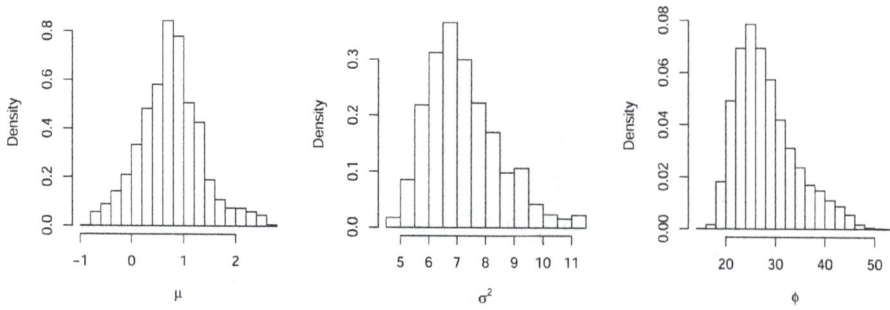

Figure 9.7 *Posterior distributions of the parameters μ, σ^2, and ϕ.*

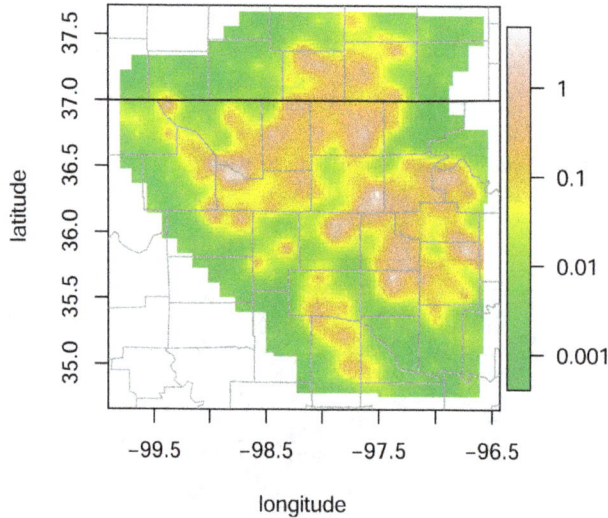

Figure 9.8 *Posterior mean intensity surface.*

Model fitting was carried out using an MCMC algorithm . Posterior samples were obtained for each of the three parameters using Metropolis-Hastings algorithms. Samples of the Gaussian process, $\mathbf{z}(\mathbf{s})$, at the set of representative points were obtained using the elliptical slice sampler [Murray et al., 2010b, and Section 9.4.2.1]. The chain was run for 100,000 iterations. The first half of the chain was disregarded as burn-in and every 10th iteration post-burn-in was retained for posterior inference in order to reduce dependence in the retained samples. Marginal posterior distributions for the parameters μ, σ^2, and ϕ are shown in Figure 9.7.

The posterior mean intensity surface is shown in Figure 9.8, highlighting areas of increased earthquake activity. Figure 9.9 shows two simulated realizations of the spatial point pattern using samples from the posterior distribution of the model parameter given the data following the procedure outlined in Section 9.3.3. By generating a collection of realizations of the point pattern using the posterior distribution of the model parameters, we also obtained samples from the posterior predictive distribution of $N(D)$, the number of points in the domain. The posterior mean of $N(D)$ was 1041.2 and the 90% credible interval was (987.5, 1109.1). Recall that the true number of earthquakes in the dataset was 1035, and, thus, our credible interval captures the true number of events.

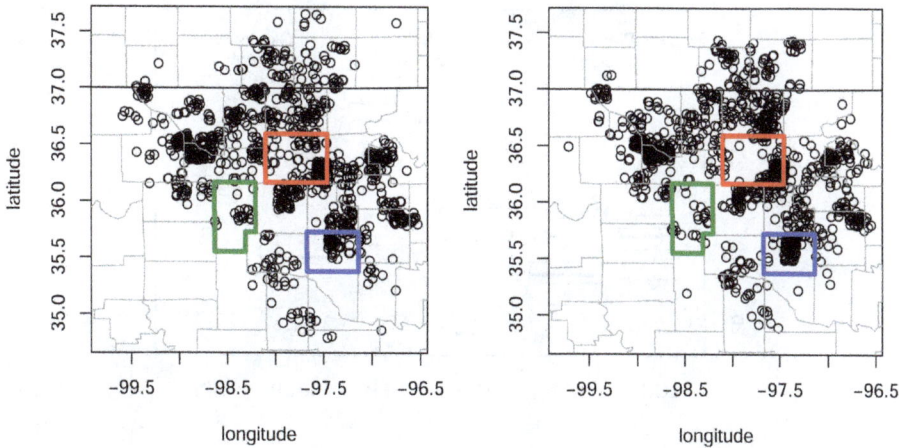

Figure 9.9 *Two realizations of the spatial point pattern from the posterior distribution. Blaine County (green), Garfield County (red), and Oklahoma County (blue) are also highlighted.*

The numbers of observed events for the three counties, Blaine County (Figure 9.9, green), Garfield County (Figure 9.9, red), and Oklahoma County (Figure 9.9, blue), were 15, 92, and 54, respectively. We obtained posterior predictive distributions of $N(B)$, $N(G)$, and $N(O)$, the marginal distribution of the number of events in Blaine, Garfield, and Oklahoma County. The posterior means and 90% credible intervals, from the LGCP model, for these random variables are given in Table 9.1. Each credible interval captured the true number of events for the county. In addition, we obtained posterior predictive distributions of $N(B)$, $N(G)$, and $N(O)$ conditional on the number of events in Oklahoma. For each conditional distribution, we set the number of events in Oklahoma equal to 930, which was the number of observed events in the state. The posterior mean and 90% credible intervals for $N(B)$, $N(G)$, and $N(O)$ are also given in Table 9.1. The marginal and conditional means for each county are similar. However, these intervals indicate that the conditional distributions have smaller variance, as expected given a fixed total.

To assess the predictive performance of the model, we created training and test spatial point patterns using p-thinning, discussed in Section 9.2.4. We thinned the observed point pattern using $p(\mathbf{s}) = 0.5$ for all $\mathbf{s} \in \mathcal{S}$, which resulted in a training dataset of 500 points and a test dataset of 535 points. We refitted the LGCP model to the training dataset, along with an HPP model for comparison. The intensity of the HPP model was specified as $\log\lambda(\mathbf{s}) = \mu$ for all $\mathbf{s} \in D$. The same prior distributions were assigned to each of the parameters; $\mu \sim N(0, 100^2)$ for the HPP. Both models were fitted using MCMC.

Figure 9.10 illustrates the posterior predictive distributions of residuals, R_{pred} as outlined in Section 9.2. For the three counties, Blaine, Garfield, and Oklahoma, we obtain samples from the posterior predictive distribution of the residuals under the two models. For Blaine county, this is computed as $R_{pred}(B) = N_{obs}(B) - N_{pred}(B)$. The distributions

	Observed	Marginal	Conditional
$N(B)$	15	16.6 (8.0 26.1)	15.8 (8.0, 25.0)
$N(G)$	92	92.9 (71.8 115.1)	91.5 (69.9, 111.0)
$N(O)$	54	53.2 (37.0, 69.0)	54.5 (39.0, 68.1)

Table 9.1 *Posterior mean (90% CI) of the posterior predictive distributions of $N(B)$, $N(G)$, and $N(O)$ from the LGCP model.*

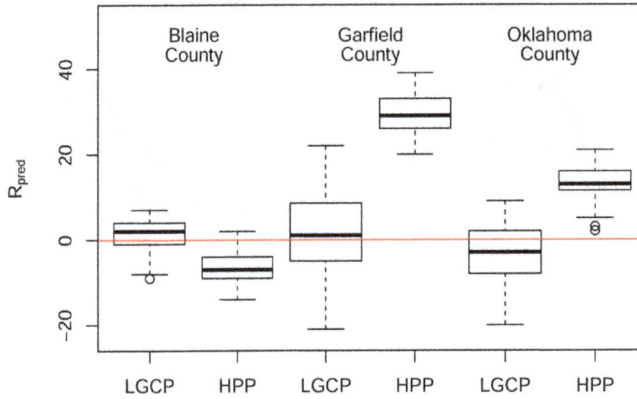

Figure 9.10 *Posterior predictive distributions of residuals R_{pred} for the LGCP and HPP models for the three counties, Blaine, Garfield, and Oklahoma, shown in Figure 9.9.*

for the other two counties were obtained analogously. Figure 9.10 shows boxplots of the predictive distribution under each model for the three counties. The predictive distributions of the residuals under the LGCP are all centered at approximately 0, whereas the predictive distributions under the HPP are far from 0. The HPP model overestimates the number of events in Blaine County while underestimating the number of events in both Garfield and Oklahoma County.

We also compared the LGCP and HPP models based on empirical coverage. For radial distances 6kms, 12kms, and 20kms, we randomly generated 400 points within the domain and defined circular regions. Then, using the test point pattern, we computed the true number of events within each circular region. We obtained the predictive distribution of the number of events within each circular region for each model and computed the 90% credible interval of the residual, R_{pred}. Empirical coverage was computed as the proportion of regions having 90% credible intervals of R_{pred} containing 0. The empirical coverage for the HPP model was 0.915, 0.585, and 0.350 for the three radial distances, indicating that the HPP is unable to capture the clustering behavior in the data for moderate or large regions. Under the LGCP model, empirical coverage was 0.945, 0.880, and 0.895, much closer to the nominal level for all distances.

Figure 9.11 consider the G function, showing four estimates. One is the empirical $G(d)$ function with edge correction. The second arises as the estimated $G(d)$ function under the HPP. That is, we insert the posterior mean for λ into the explicit expression for $G(d)$. The third and fourth show two estimators arising under the LGCP model. Formally, the G function is only applicable to stationary processes but since our model is "marginally" stationary, we compute it for the LGCP. The one in blue was obtained by applying the empirical estimator of $G(d)$ in Section 8.3.2 to posterior point pattern samples under the model and averaging the resulting functions. The one with a dashed line is the fully model-based estimator arising from using formal Bayesian edge correction [Gelfand and Schliep, 2018, Section 3.2.3]. The two versions of the estimator are indistinguishable, suggesting that the more computationally demanding edge-corrected version may not be worth the effort to obtain. More importantly, due to the clustering in earthquake locations (departure from complete spatial randomness), the LGCP function is above the HPP function for all distances. Finally, we see that the LGCP G function lies below the customary empirical G function. This does not criticize the LGCP model since the latter is conditionally nonstationary; comparison between the curves is not meaningful. The empirical G function only criticizes the HPP as a choice of stationary model.

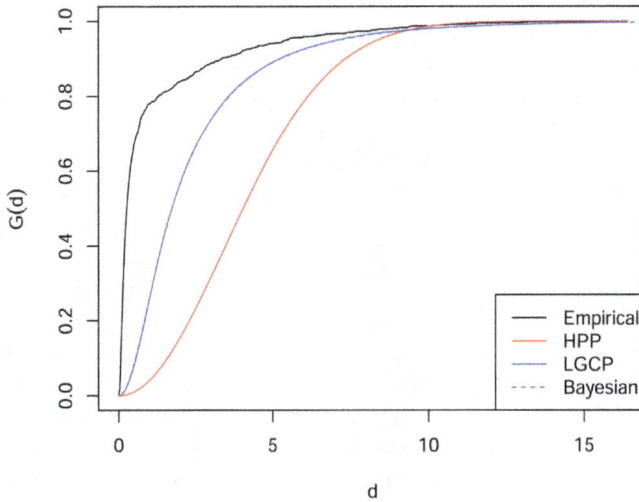

Figure 9.11 *The empirical G function as well as the estimated G function under the HPP and LGCP models. Also shown is the estimated G function with Bayesian edge correction.*

9.7   Space-time point patterns

We often encounter space-time point patterns. Customarily, we do not imagine that a point pattern occurs instantaneously. Rather, what occurs instantaneously are single events with an associated time of occurrence. That is, the time of occurrence of the event would be a continuous variable, e.g., when a cancer case was diagnosed, when a house was built. The data is in the form of $s_i, t_i, i = 1, 2, \ldots n$. In this sense, the times are continuous marks leading us back to the discussion in Section 8.6. In this framework, an interval of time is required to reveal an entire point pattern, e.g., the point pattern of disease cases over a week in a county or the point pattern of single-family homes built in a city during a given year. This raises the issue of whether points are viewed as 'births' and remain in the pattern until 'death.' This perspective would be sensible for say trees or houses but not for cancer cases or locations of traffic accidents. This raises the question of whether we view the pattern cumulatively or differentially, with associated modeling implications. That is, in principle, we can move from one view to the other but model specification needs to reflect the choice.

This is the modeling perspective that we consider here. However, we note an alternative view of space-time point pattern data. Rather than thinking in terms of each point having both a location and an arrival time, we can imagine that the point pattern is indexed by time t. That is, the point pattern at time t is viewed as a *snapshot* and, in fact, conceptually, we have such a snapshot evolving over time within a window of interest. An example of this setting is presented in Chapter 16 where, at time t, there is a point pattern (latent) of whale locations over Cape Cod Bay in Massachusetts. That is, the point pattern of locations changes with time as individuals move around the Bay (and, in fact, there will be births and deaths in the sense that individuals move in and out of the Bay). The point pattern is not viewed as arising cumulatively; we don't have to *integrate* over a time window to see a point pattern. With regard to modeling, with NHPPs or LGCPs we instead imagine a dynamic intensity subscripted by t, e.g., $\lambda_t(\mathbf{s})$.

Adding time to the investigation of point patterns can be critical with respect to adequate understanding of them. Point patterns can appear quite different over disjoint time windows and explaining these differences may be a vital aspect of the analysis. In different words, aggregating the patterns over time may remove this story. Furthermore, point pattern data collection is often very time consuming. Consider, for example, collecting complete inventories over reasonably large regions. Adding time can lead to infeasible demand in terms of a number of hours and cost, especially for a slowly evolving process whence a long time period may be needed. In Section 9.7.3.1, we consider one source of such data which is well-collected—new home construction over a specified geographic region. Methodology for space-time point patterns is underdeveloped and Bayesian analysis for such patterns even more so. However, some relatively recent work includes Gonçalves and Gamerman [2018] and Shirota and Banerjee [2019].

9.7.1 Space-time Poisson process models

So, when we consider adding time to the modeling, we have to determine whether we will view it as continuous or discrete. In the case of continuous time, we introduce a second argument to the intensity function, writing $\lambda(\mathbf{s}, t)$. We focus on space-time Poisson process models. Richer discussion is available in the book of Daley and Vere-Jones (2008). Under an NHPP specification over say $D \times (0, T]$, we observe $(\mathbf{s}, t)$'s and time is viewed as a continuous mark. Then, the expectation for a realization of a point pattern in the time interval $[t_1, t_2]$ is $\lambda_{[t_1, t_2]}(\mathbf{s}) = \int_{t_1}^{t_2} \lambda(\mathbf{s}, t) dt$. The intensity associated with say $B \times [t_1, t_2]$, with $B \in D$ and $0 \leq t_1 < t_2 \leq T$ is $\lambda(B \times [t_1, t_2]) = \int_B \int_{t_1}^{t_2} \lambda(\mathbf{s}, t) dt d\mathbf{s}$. Note that, though $\lambda(\cdot, \cdot)$ is a function over a subset of R^3, it makes no sense to consider Euclidean distance in three dimensions. The scale for time has nothing to do with the scale for space.

As with the spatial NHPP, we assume $N(B \times [t_1, t_2]) \sim Po(\lambda(B \times [t_1, t_2]))$ where $B \subseteq D$. In addition, given $N(D \times (0, T])$, we assume the points are scattered independently over $D \times (0, T]$ with density $\lambda(\mathbf{s}, t)/\lambda(D \times (0, T])$ where $\lambda(D \times (0, T]) = \int_D \int_0^T \lambda(\mathbf{s}, t) dt d\mathbf{s}$. Hence, $\lambda_{[t_1, t_2]}(\mathbf{s})$ is the intensity for an NHPP over D for events in the time window, $[t_1, t_2]$. Similarly, if we integrate over the set B, we obtain the intensity for a one dimensional point pattern in time over $(0, T]$ for events restricted to B. Special cases of $\lambda(\mathbf{s}, t)$ lead to time stationarity with $\lambda(\mathbf{s}, t) = \lambda(\mathbf{s})$, spatial stationarity with $\lambda(\mathbf{s}, t) = \lambda(t)$, or space-time stationarity with $\lambda(\mathbf{s}, t) = \lambda$. With space-time stationarity, we can obtain the naive estimate of λ, i.e., $\hat{\lambda} = N(D \times (0, T]/T|D|$. If we retain stationarity but not an HPP, then we can develop analogues of the second order intensity, K-functions, etc. as in Daley and Vere-Jones [2003] or Diggle and Gabriel [2010]. A relatively convenient class of Cox process models arises when $\lambda(\mathbf{s}, t)$ is a log space-time Gaussian process. Also, the notion of a Gibbs process can be extended, allowing pairwise spatial interaction, scaled by a function in time. Extending Section 8.2.2, the Papangelou conditional intensity takes the form, $\lambda(\mathbf{s}, t \mid \mathcal{S}_t) = \exp(\alpha(t) + \sum_{i=1}^{N_t} h(\mathbf{s}, \mathbf{s}_i))$ where $\mathcal{S}_t$ is the point pattern up to time t and N_t is the number of events in this pattern, i.e., in $(0, T]$.

More explicitly, if time is continuous, then we can imagine a parametric $\lambda(\mathbf{s}, t; \boldsymbol{\theta})$, say

$$\log \lambda(\mathbf{s}, t; \boldsymbol{\theta}) = \mathbf{X}(\mathbf{s}, t)^{\mathrm{T}} \boldsymbol{\theta}. \tag{9.12}$$

In (9.12), some components of the $\mathbf{X}$'s may be indexed only by space, e.g., elevation or by time, e.g., a global geographic covariate. Again, we can move to the nonparametric case, by introducing a spatio-temporal Gaussian process, $w(\mathbf{s}, t)$. That is,

$$\log \lambda(\mathbf{s}, t) = \mathbf{X}(\mathbf{s}, t)^{\mathrm{T}} \boldsymbol{\theta} + w(\mathbf{s}, t). \tag{9.13}$$

We have flexibility in the specification of $w(\mathbf{s}, t)$. For instance, $w(\mathbf{s}, t) = w(\mathbf{s}) + g(t)$, where $w(\mathbf{s})$ is a Gaussian process over D and $g(t)$ might be a specified function of time. Alternatively, $g(t)$ might be a stochastic process over $(0, T]$, e.g., Brownian motion, as in Section 9.7.3 below.

9.7.2 Dynamic models for discrete time data

In case of discrete t's, say equally spaced, we introduce intensities $\lambda_t(\mathbf{s})$, $t = 1, 2, ..., T$. In this regard, time is viewed as a discrete mark, inviting comparison of intensities across time. If time is discrete, then $\lambda_t(\mathbf{s})$ can be modeled dynamically, using parametric and/or nonparametric specifications. More precisely, we envision a customary dynamic model with conditionally independent first stage and time-evolving second stage (e.g., West and Harrison, 1997). That is, the point pattern at time t

$$\mathcal{S}_t \sim \text{NHPP}(\lambda_t(\mathbf{s})), \tag{9.14}$$

and the patterns are independent over t with intensities, $\{\lambda_t(\mathbf{s}) : s \in D\}$. Then, we add a transition specification for $\{\lambda_t(\mathbf{s}) : s \in D\} \mid \{\lambda_{t-1}(\mathbf{s}) : s \in D\}$. Here, we have many modeling options. At the simplest level, we would have a parametric family for the λ_t's indexed by say, $\boldsymbol{\theta}_t$, e.g., $\boldsymbol{\gamma}_t$ with $\log \lambda_t(\mathbf{s}) = \mathbf{X}_t(\mathbf{s})^{\mathsf{T}} \boldsymbol{\gamma}_t$, and the transition would take the form $[\boldsymbol{\theta}_t \mid \boldsymbol{\theta}_{t-1}]$ with an evolutionary equation for $\boldsymbol{\theta}_t$. A more complex transition could take the form of an integro-difference equation,

$$\lambda_t(\mathbf{s}) = \int_D K(\mathbf{s}, \mathbf{s}') \lambda_{t-1}(\mathbf{s}') d\mathbf{s}' \tag{9.15}$$

where $K(\cdot, \cdot)$ is a kernel or *propagator* function. Practically, the integral would have to be discretized over space in order to be evaluated.

With nonparametric specification, we find ourselves introducing log Gaussian processes, e.g.,

$$\log \lambda_t(\mathbf{s}) = \mathbf{X}(\mathbf{s})^{\mathsf{T}} \boldsymbol{\gamma}_t + w_t(\mathbf{s}) \tag{9.16}$$

where $w_t(\mathbf{s})$ is a mean 0 Gaussian process. In addition to dynamics in $\boldsymbol{\gamma}_t$, we can also introduce dynamics in the $w_t(\mathbf{s})$, analogous to Section 12.4.

9.7.3 Space-time Cox process models using stochastic PDEs

Here, we turn to the use of a stochastic differential equation (SDE) to provide a Cox process model for space-time point patterns with continuous time, drawing upon ideas in Duan et al. [2009]. Let D again be a fixed region and let $\mathcal{S}_T$ denote the observed space-time point pattern within D over the time interval $[0, T]$. Denote the stochastic intensity by $\lambda(\mathbf{s}, t), \mathbf{s} \in D, t \in [0, T]$. In practice, we may only know the spatial coordinates of all the points whereas the time coordinates are censored to time intervals in $[0, T]$.[6] Provided that $\lambda(\mathbf{s}, t)$ is integrable over $[0, T]$, the integrated process intensity is $\lambda(\mathbf{s}, T) = \int_0^T \lambda(t, \mathbf{s}) dt$. As above, we may observe the point pattern in subintervals of $[0, T]$: $[t_1 = 0, t_2), \ldots, [t_{J-1}, t_J = T]$. These data constitute a series of discrete-time spatio-temporal point patterns, which we denote by $\mathcal{S}_{[t_1=0,t_2)}, \ldots, \mathcal{S}_{[t_{j-1}, t_N=T]}$. The integrated process also provides stochastic intensities for these point patterns

$$\Delta \lambda_j(\mathbf{s}) = \lambda(\mathbf{s}, t_j) - \lambda(\mathbf{s}, t_{j-1}) = \int_{t_{j-1}}^{t_j} \lambda(\mathbf{s}, t) dt.$$

When the time intervals are sufficiently small, we may use the approximation

$$\Delta \lambda_j(\mathbf{s}) = \lambda(\mathbf{s}, t_j) - \lambda(\mathbf{s}, t_{j-1}) = \int_{t_{j-1}}^{t_j} \lambda(\mathbf{s}, \tau) d\tau \approx \lambda(\mathbf{s}, t_{j-1})(t_j - t_{j-1}).$$

---

[6]For example, in our house construction data below, we only have the geo-coded locations of the newly constructed houses within a year. The exact time when the construction of a new house starts is not available.

As a concrete example, below we consider a house construction dataset from Irving, TX, as discussed in Duan et al. [2009]. Let $\mathcal{S}_j = \mathcal{S}_{[t_{j-1}, t_j)} = x_j$ be the observed set of locations of new houses built in region D and period $j = [t_{j-1}, t_j)$. We supply a Cox process model for the $\mathcal{S}_j$ by specifying $\lambda(\mathbf{s}, t)$ through a stochastic differential equation. We do this in order to introduce a mechanistic modeling scheme with parameters that convey physical meanings in the mechanism described by a stochastic differential equation. Here, we use the logistic equation

$$\frac{\partial \lambda(\mathbf{s}, t)}{\partial t} = r(\mathbf{s}, t) \lambda(\mathbf{s}, t) \left[1 - \frac{\lambda(\mathbf{s}, t)}{K(\mathbf{s})} \right],$$

where $K(\mathbf{s})$ is the "carrying capacity" (assuming it to be time-invariant) and $r(\mathbf{s}, t)$ is the "growth rate." Spatially and/or temporally varying *parameters*, such as growth rate and carrying capacity, can be modeled by spatio-temporal processes. In practice, the logistic equation finds applications in population growth in ecology, product and technology diffusion in economics, and urban development. The last is our context, with growth rate and carrying capacity being readily interpretable surfaces over space and time and over space, respectively.

Let the initial point pattern be $\mathcal{S}_0$ and the intensity be $\lambda_0(\mathbf{s}) = \lambda(\mathbf{s}, 0) = \int_{-\infty}^{0} \lambda(\tau, \mathbf{s}) \, d\tau$. The hierarchical model for the space-time point patterns becomes

$$\begin{aligned} \mathcal{S}_j \mid \Delta \lambda_j &\sim \text{Poisson Process}(D, \Delta \lambda_j), \quad j = 1, \ldots, J \\ \mathcal{S}_0 \mid \lambda_0 &\sim \text{Poisson Process}(D, \lambda_0), \end{aligned} \tag{9.17}$$

where we suppress the indices $\mathbf{s}$ and t again for the periods $t_1, \ldots, t_J$. Note that the intensity $\Delta \lambda_j$ for $\mathcal{S}_j$ must be positive. Therefore, we model the growth rate r as a log-process, that is

$$\log r(\mathbf{s}, t) = \mu_r(\mathbf{s}; \beta_r) + \zeta(\mathbf{s}, t), \quad \zeta \sim GP(0, \varrho(\mathbf{s} - \mathbf{s}', t - t'; \varphi_r)). \tag{9.18}$$

The J spatial point patterns are conditionally independent given the space-time intensity so the likelihood is

$$\prod_{j=1}^{J} \left\{ \exp \left(-\int_D \Delta \lambda_j(s) \, ds \right) \prod_{i=1}^{n_j} \Delta \lambda_j(x_{ji}) \right\}$$

$$\times \exp \left(-\int_D \lambda_0(s) \, ds \right) \prod_{i=1}^{n_0} \lambda_0(x_{0i}). \tag{9.19}$$

This likelihood introduces the familiar stochastic integral, $\int_D \Delta \lambda_j(s) \, ds$, which we approximate in model fitting using a Riemann sum with representative points. As before, we divide the geographical region D into M cells and assume the intensity is homogeneous within each cell. Let $\Delta \lambda_j(m)$ and $\lambda_0(m)$ denote this average intensity in cell m. Let the area of cell m be $A(m)$. Then, the likelihood becomes

$$\prod_{j=1}^{J} \left[\exp \left(-\sum_{m=1}^{M} \Delta \lambda_j(m) A(m) \right) \prod_{m=1}^{M} \Delta \lambda_j(m)^{n_{jm}} \right]$$

$$\times \exp \left(-\sum_{m=1}^{M} \lambda_0(m) A(m) \right) \prod_{m=1}^{M} \lambda_0(m)^{n_{0m}} \tag{9.20}$$

where n_{jm} is the number of point in cell m in period j. We approximate the parameter processes $r(\mathbf{s}, t_j)$ and $K(\mathbf{s})$ accordingly as $r(m, t_j)$ and $K(m)$, which are homogeneous in each cell m.

9.7.3.1   An example: Modeling the House Construction Data for Irving, TX

Our house construction dataset consists of the geo-coded locations and years of the newly constructed residential houses in Irving, TX from 1901 to 2002. Irving started to develop in the early 1950's and the outline of the city was already in its current shape by the late 1960's. The city became almost fully developed by the early 1970's with little new construction afterward. We use the data from years 1951–1966 to fit our model and hold out the last three years (1967, 1968 and 1969) for prediction and model validation.

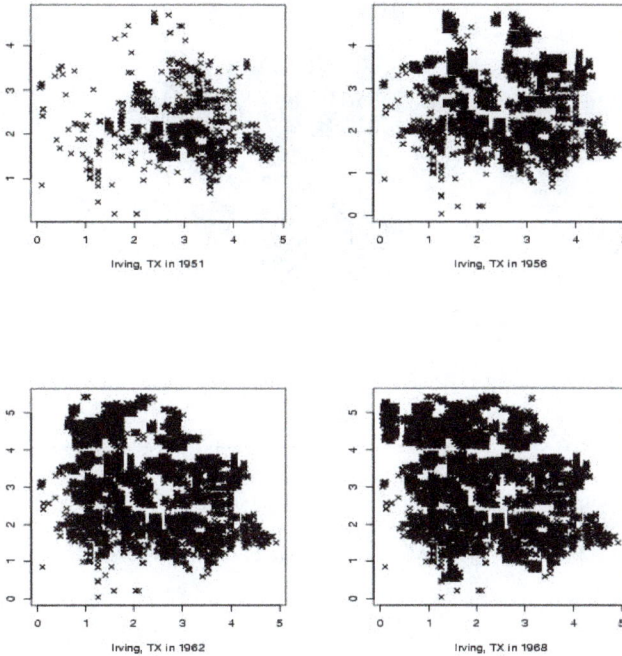

Figure 9.12 *Growth of residential houses in Irving, TX.*

Figure 9.12 shows our study region D as a square of 5.6×5.6 square miles with Irving, TX in the middle. This region is geographically disconnected from other major urban areas in Dallas County, which enables us to isolate Irving for analysis. In fact, the figure shows the growth of residential housing from 1951 to 1966. We divide the region into 100 (10×10) equally spaced grid cells as shown in Figure 9.13. Within each cell, we model the point pattern with a homogeneous Poisson process given $\Delta\lambda_j (m)$. The corresponding $\lambda_0 (m)$, $K (m)$ and $r (m, j)$ are collected into vectors $\lambda_0, K,$ and r which are modeled as follows.

$$\begin{aligned}
\log \lambda_0 &= \mu_\lambda + \theta_\lambda, \ \theta_\lambda \sim N(0, C_\lambda) \\
\log K &= \mu_K + \theta_K, \ \theta_K \sim N(0, C_K) \\
\log r &= \mu_r + \zeta, \ \zeta \sim N(0, C_r)
\end{aligned}$$

where the spatial covariance matrix C_λ and C_K are constructed using the Matérn class covariance function with distances between the centroids of the cells. The smoothness parameter ν is set to be $3/2$. The variances $\sigma_\lambda^2, \sigma_K^2$ and range parameters ϕ_λ and ϕ_K are to be estimated. The spatio-temporal log growth rate r is assumed to have a separable covariance matrix $C_r = \sigma_r^2 \Sigma_{\mathbf{s}} \otimes \Sigma_t$, where the spatial correlation $\Sigma_{\mathbf{s}}$ is also constructed as a Matérn class function of the distances between cell centroids with the smoothness parameter again being set to $3/2$. The temporal correlation Σ_t is of exponential form. The variance σ_r^2, spatial and temporal correlation parameters ϕ_r and α_r are to be estimated.

We use vague priors for the parameters in the mean function: $\pi(\mu_\lambda), \ \pi(\mu_K), \ \pi(\mu_r) \overset{ind}{\sim} N(0, 10^8)$. We use natural conjugate priors for the precision parameters (inverse of variances) of r and λ_0: $\pi(1/\sigma_\lambda^2), \ \pi(1/\sigma_K^2), \ \pi(1/\sigma_r^2) \overset{ind}{\sim} Gamma(1, 1)$. The temporal correlation

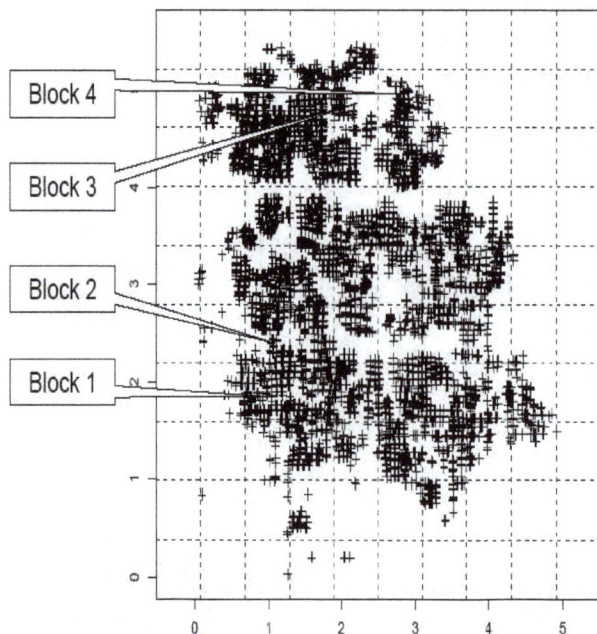

Figure 9.13 *The gridded study region encompassing Irving, TX.*

parameter of r also has a vague log-normal prior: $\pi\left(\alpha_r\right) \sim log\text{-}N\left(0, 10^8\right)$. Again, the spatial range parameters ϕ_λ ϕ_K and ϕ_r are only weakly identified (Zhang, 2004) so we use informative, discrete priors for them. In particular, we have chosen 40 values (from 1.1 to 5.0) and assume uniform priors on them for ϕ_λ ϕ_K and ϕ_r.

9.7.3.2 Results of the data analysis

Posterior inference (mean and 95% equal tail credible intervals) are presented in Table 9.2. Figure 9.14 shows the posterior mean growth curves and 95% Bayesian predictive bound for the intensity in the four blocks (marked as blocks 1, 2, 3 and 4). Comparing with the observed number of houses in the four blocks from 1951 to 1966, we can see the estimated curves fit the data very well.

In Figure 9.15 we display the posterior mean intensity surface for the year 1966 and the predictive mean intensity surfaces for the years 1967, 1968 and 1969. We also overlay the actual point patterns of the new homes constructed in those four years on the intensity surfaces. Figure 9.15 shows that our model can forecast the major areas of high intensity, hence high growth very well.

9.8 Additional topics

9.8.1 Measurement error in point patterns

Here, we consider the setting where the observed locations are measured with error and we seek to assess the resultant effect on the object of our interest, the intensity function, under an NHPP. Intuitively, adding noise will "blur" the intensity surface, making detection of its features more difficult. In fact, it is quite likely that, in recording locations, measurement error is introduced due to the accuracy of the measuring instrument as well as factors influencing detection of event occurrences within the region.

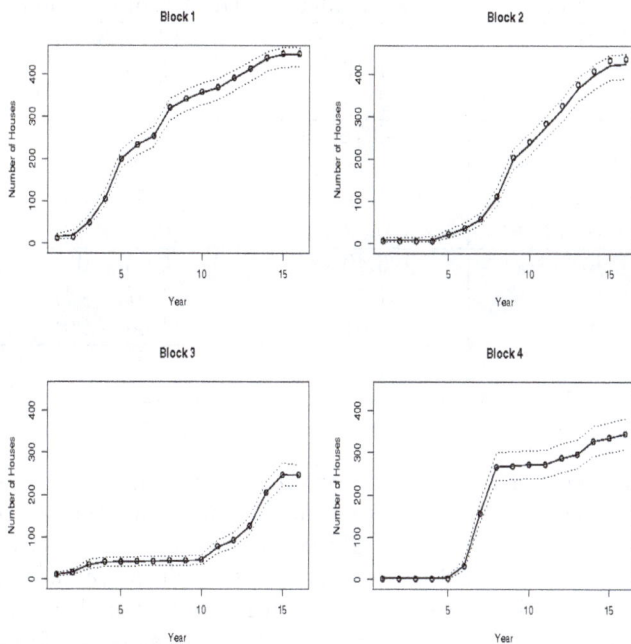

Figure 9.14 *Mean growth curves and their corresponding 95% predictive intervals (dotted lines) for the intensity for the four blocks marked in Figure 9.13.*

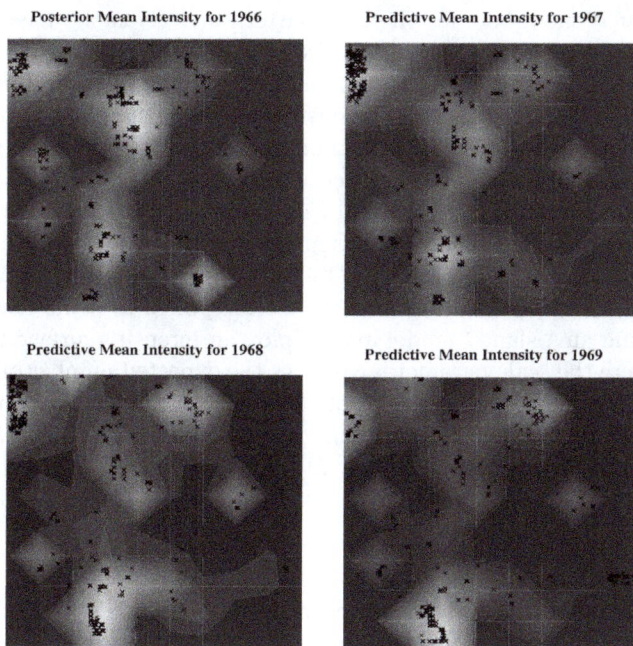

Figure 9.15 *Posterior and predictive mean intensity surfaces for the years 1966, 1967, 1968, and 1969.*

Model Parameters	Posterior Mean	95% Equal-tail Interval
μ_λ	2.78	(2.15, 3.40)
σ_λ	1.77	(1.49, 2.11)
ϕ_λ	3.03	(2.70, 3.20)
μ_r	-2.76	(-3.24, -2.29)
σ_r	2.48	(2.32, 2.68)
ϕ_r	4.09	(3.70, 4.30)
α_r	0.52	(0.43, 0.62)
μ_K	6.49	(5.93, 7.01)
σ_K	1.17	(1.02, 1.44)
ϕ_K	1.91	(1.60, 2.20)

Table 9.2 *Posterior inference for the house construction data.*

Modeling point patterns using intensities requires restriction to a bounded subset of the plane. As a result, such noise can push locations in and out of the study domain. We are not only observing a noisy version of the original realization, but it is also possible that we are missing some of the true events and also observing some which are not truly in the study region. For patterns having high event aggregation near the boundary of the region, this problem can be quite significant. Thus, measurement error results in a form of censoring to yield the actual dataset.

There is a small previous literature on degraded point patterns. A general description is that the observed pattern is a random transformation of the true pattern, as in Diggle [1993] who viewed the transformation as a conditionally independent random deformation of the true pattern. Chakraborty and Gelfand [2010] look at the problem as a two-stage specification—what is the model for the true pattern and given the true pattern, what is the model for the random transformation. Work in this spirit appears in Lund and Rudemo [2000] and Lund et al. [1999]. There, the distinction is made between inferring about the properties of the true point pattern and reconstructing the true point pattern.

Within a fully hierarchical framework, there is no need to separate the point process parameter estimation and pattern reconstruction problems. Both can be addressed through suitable posterior inference for fairly general Cox processes. Here, we consider the scenario where events can only occur inside D so when we talk about a shift of a location due to noise, it can only throw a point from D to D^c. But, since no event is allowed to take place outside D, each of the noisy locations observed corresponds to some true location in D. We term this setting an "island" model and employ an intensity surface which is a scaled mixture model where the scale parameter captures the expected number of points in D. We can remove this restriction by assuming that our region of interest is actually a subset of a bigger region of possible event findings (e.g., mapping tree locations in a specific part of a forest). Now events outside can also enter D because of noise; we refer to this case as a "subregion" model with details presented in Chakraborty and Gelfand [2010].

9.8.1.1 Modeling details

In order to simplify notation, in this subsection we label true locations by $\mathbf{x}$ and observed locations by $\mathbf{y}$. We consider measurement error in additive form. For a bounded study domain D and a true location $\mathbf{x}$, we assume the recorded location $\mathbf{y} = \mathbf{x} + \boldsymbol{\epsilon}$, where $\boldsymbol{\epsilon}$ is the measurement error. Further, we assume conditionally independent homogeneous displacements in the MEM scenario as employed in Diggle (1993). We note that, in some contexts, we might imagine that the error variability has spatial structure. Unfortunately, since the

x's are latent, this specification produces a complicated posterior full conditional for x and overall model fitting that is unstable.

Under the island model we assume our study region D contains the support of the true point process i.e. $P(\mathbf{x} \in D^c) = 0$ for any event location $\mathbf{x}$. So now we can only have (i) $\mathbf{x} \in D, \mathbf{y} \in D$, (ii) $\mathbf{x} \in D, \mathbf{y} \in D^c$. So, again, in a bounded region D, we assume n observed event locations $(\mathbf{y}_1, \mathbf{y}_2, ..., \mathbf{y}_n)$, which are a noisy version of a set of m actual locations $(\mathbf{x}_1, \mathbf{x}_2, ..., \mathbf{x}_m)$ representing the complete realization of the point pattern in D. Now, m is unknown but $m \geq n$, that is, $(m - n)$ of these locations fell outside of D. For $\mathbf{x} \in D$ we adopt the Gaussian noise distribution, $\mathbf{y} \mid \mathbf{x}, \beta \sim N(\mathbf{x}, \Omega)$.

We model $\lambda(\mathbf{s}) = \lambda f(\mathbf{s})$, $\mathbf{s} \in D$, as discussed in Section 8.4.1. Here, for illustration and convenience, we choose f_D as Gaussian mixture distribution restricted to D. With regard to the number of mixture components, K, evidently more components can allow us to to tease out more features of the $\lambda(\cdot)$ surface but, within our measurement error framework will lead to poorly behaved computation. Practically, we can make an empirical choice based upon the observed point pattern or we can do a model comparison across various choices for the number of components.

Relabeling the $\mathbf{x}$'s so that, for $i = 1, 2, ..., n$, $\mathbf{x}_i$ is the true location corresponding to $\mathbf{y}_i$ with the last $(m - n)$ $\mathbf{x}$'s corresponding to $\mathbf{y}$ locations outside D, we obtain the following model:

$$
\begin{aligned}
\mathbf{y}_i &\sim N_2(\mathbf{x}_i, \Omega), \ i = 1, 2, ..., n \\
f(\mathbf{x}_1, \mathbf{x}_2, ..., \mathbf{x}_m) &= NHPP(\lambda(.)) \\
\lambda(\mathbf{x}) &= \lambda f_D(\mathbf{x})
\end{aligned}
\tag{9.21}
$$

where $\lambda f_D(\mathbf{x}) = \lambda \sum_{k=1}^{K} q_k N_D(\mathbf{x} \mid \boldsymbol{\mu}_k, \Sigma_k)$, $N_{2,D}$ denotes the restriction of the bivariate normal density to D and q_k are the mixing weights.

Details on the MCMC model fitting and computation is supplied in Chakraborty and Gelfand [2010]. We do note that the likelihood has two parts, one from the observed locations $(\mathbf{y}_1, \mathbf{y}_2, ..., \mathbf{y}_n)$ (say L_1), another from the unobserved $\mathbf{y}$'s known to be in D^c (say (L_2)). Upon associating the $\mathbf{x}_i$'s with $\mathbf{y}_i$'s, the likelihood takes the form $L = L_1 L_2$ (as in Lund and Rudemo, 2000, and in Lund et al., 1999) where $L_1 = \prod_{i=1}^{n} \phi_2(\mathbf{y}_i \mid \mathbf{x}_i, \Omega)$ and $L_2 = \prod_{i=n+1}^{m} \bar{\Phi}_2(D; \mathbf{x}_i, \Omega)$, with ϕ_2 being the bivariate Gaussian pdf and $\bar{\Phi}_2$ its integral, in this case, over D, respectively, with appropriate parameters.

In writing the NHPP prior we assume that the first n of the $\mathbf{x}$'s are identified with the observed $\mathbf{y}$'s. In fact, there are $\frac{m!}{(m-n)!}$ possible matchings which have been collapsed into a single case. So, the prior density is, in fact,

$$
\pi(\mathbf{x}_{1:n}, \mathbf{x}_{n+1:m}) = \frac{m!}{(m-n)!} \lambda^m \prod_{i=1}^{m} f_D(\mathbf{x}_i) \frac{\exp(-\lambda)}{m!}
\tag{9.22}
$$

In the sequel we assume Ω for the measurement error process is known, obtained in some fashion, following our earlier discussion.

Hence, the full posterior for the model parameters becomes

$$
\pi(m, \mathbf{x}_{1:m}, \lambda, \boldsymbol{\mu}_{1:K}, \Sigma_{1:K}, q_{1:K} \mid \mathbf{y}_{1:n})
\tag{9.23}
$$

$$
\propto \binom{m}{n} \exp(-\lambda) \frac{\lambda^m}{m!} \prod_{i=1}^{m} f_D(\mathbf{x}_i \mid \boldsymbol{\mu}_{1:K}, \Sigma_{1:K}, q_{1:K})
$$

$$
\prod_{i=1}^{n} \phi_2(\mathbf{y}_i \mid \mathbf{x}_i) \prod_{i=n+1}^{m} \bar{\Phi}_2(D \mid \mathbf{x}_i) \pi(\lambda, \boldsymbol{\mu}_{1:K}, \Sigma_{1:k}, q_{1:K})
$$

$$
\tag{9.24}
$$

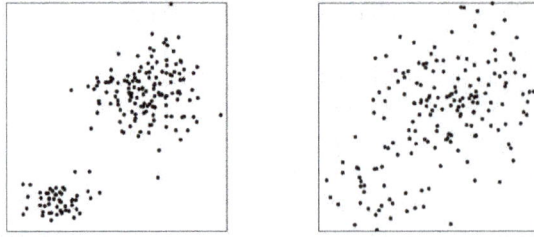

Figure 9.16 *Original (left) and perturbed (right) point patterns.*

We offer a simulation example to illustrate our methodology. We take f to be a 2-component normal mixture distribution (see Table 9.3) within a unit square and contaminate it with Gaussian noise having a dispersion matrix as $\begin{pmatrix} 0.023 & 0.002 \\ 0.002 & 0.019 \end{pmatrix}$. In Figure 9.16 we show the original and perturbed datasets. There were 199 points initially in the window, but after noise addition only 177 are left, roughly 11% loss of the points. Apart from the increased spread in the noisy pattern, the bimodality of the intensity essentially disappears. We fit the Island model as well as a noiseless NHPP. Included in Table 9.3 is the comparison between the models while Figure 9.17 provides a comparison of the estimated intensities. We see the benefit of the measurement error model. As expected, the estimation of the Σ's along with q and λ was severely affected by the noise. The effect on the μ's is noteworthy. Fitting a mixture model directly to that data has likely caused the μ's to shift a bit in order to adjust for the overlap. In Table 9.3 we can see that the 95% credible interval for $\mu_1^{(1)}$ produced by the noiseless NHPP excludes the true value (parameters that noticeably differ are in **bold**).

9.8.2   Presence-only data application

Learning about species distributions is a long-standing issue in ecology with, by now, an enormous literature. The focus of the work here is on the so-called *presence-only* setting, drawing on the work of Chakraborty et al. [2010]. Analysis of presence-only data has seen growing popularity in recent years due to increased availability of such records from museum databases and other non-systematic surveys. One model-based strategy for presence-only data has attempted to implement a presence/absence approach. All of this work depends upon drawing so-called *background samples*, a random sample of locations in the region

Parameters	Simulated Model	Island Model	Noiseless NHPP
$\mu_1^{(1)}$	0.64	0.6291, (0.5855, 0.6689)	**0.6053**, (0.5697, 0.6389)
$\mu_2^{(1)}$	0.61	0.5965, (0.5656, 0.6291)	0.5821, (0.5524, 0.6123)
$\mu_1^{(2)}$	0.25	0.2454, (0.1713, 0.3243)	0.2546, (0.2002, 0.3069)
$\mu_2^{(2)}$	0.14	0.1575, (0.0889, 0.2292)	0.1694, (0.1282, 0.2125)
q	0.71	0.7238, (0.6138, 0.8140)	**0.8150**, (0.7461, 0.8771)
λ	200	200.7096, (168.4630, 238.9672)	**177.7389**, (152.5286, 204.9620)
$\Sigma_{11}^{(1)}$	0.016	0.0153, (0.0068, 0.0275)	**0.0339**, (0.0263, 0.0432)
$\Sigma_{12}^{(1)}$	0.0007	0.0003, (−0.0060, 0.0081)	0.0037, (−0.0019, 0.0098)
$\Sigma_{22}^{(1)}$	0.018	0.0116, (0.0040, 0.0206)	**0.0271**, (0.0207, 0.0352)
$\Sigma_{11}^{(2)}$	0.007	0.0105, (0.0038, 0.0176)	**0.0175**, (0.0091, 0.0306)
$\Sigma_{12}^{(2)}$	0.0005	0.0004, (−0.0050, 0.0072)	−0.0033, (−0.0093, 0.0017)
$\Sigma_{22}^{(2)}$	0.002	0.0037, (0.0015, 0.0060)	**0.0096**, (0.0051, 0.0172)

Table 9.3 *Comparison of models with and without measurement error in case of bivariate Gaussian mixture intensity. Point estimates with 95% interval estimates in parentheses.*

(a)   (b)

(c)   (d)

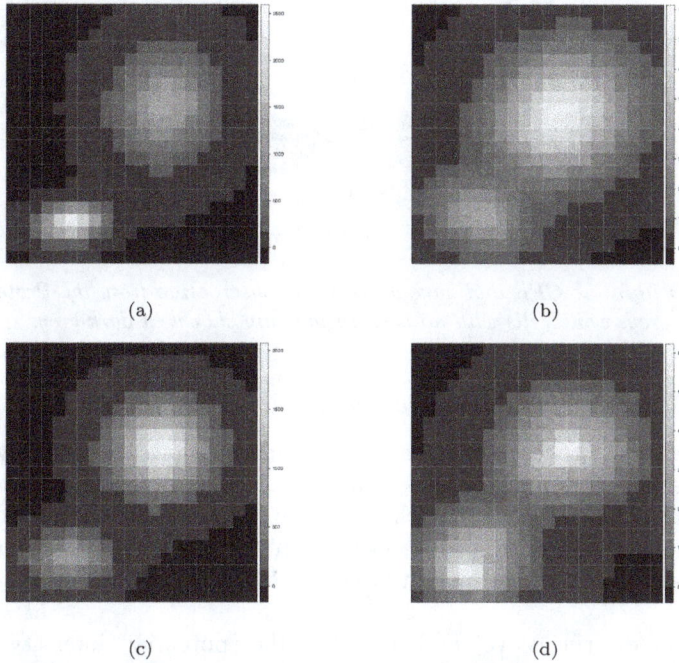

Figure 9.17 *Model Analysis :: (a) actual intensity surface, (b) its estimate based on noiseless NHPP, (c) posterior intensity estimate from Island model, (d) uncertainty of estimated intensity.*

with known environmental features. Early work here characterized these samples as pseudo-absences and fitted a logistic regression to the observed presences and these pseudo-absences. Since presence/absence is unknown for these samples, work (Ward et al., 2009) shows how to adjust the resulting logistic regression to account for this. All of this work is non-spatial and requires the choice of an *arbitrary* number of background samples. Perhaps, most importantly, as we argue below, this approach conditions in the wrong direction. We assert that the observed presence should be viewed as a *marked point pattern*, with the mark indicating presence (see the work of Warton and Shepherd, 2010, in this regard). We do not have a point pattern of absences; pseudo-absences create an unobserved and artificial pattern of absences.

We model presence-only data as a point pattern under a Cox process specification with associated intensity is given as a regression in terms of the available environments across the region. We employ a hierarchical model to introduce spatial structure for the intensity surface through spatial random effects. We do not assume any background or pseudo-absence samples; rather, we assume that the covariates we employ are available as surfaces over the region in order to interpolate an intensity over the entire region. We acknowledge that the observed point pattern is degraded/biased through anthropogenic processes, e.g., human intervention to transform the landscape and non-uniform (in fact, often very irregular) sampling effort.

We work with presence-only data collected from the Cape Floristic Region (CFR) in South Africa (Figure 9.18). The region is divided into approximately $37,000$ grid cells, each one minute by one minute (roughly $1.55km \times 1.85km$). Covariate information is only available at grid cell level so we model the intensity as a tiled surface over these cells. We illustrate with potential and degraded intensities for six species.

Figure 9.18 *Cells within the CFR that have at least one observation from the Protea Atlas dataset are shown in light grey, while cells with no observations are shown in dark grey.*

9.8.2.1 Probability model for presence locations

Again, we assume a Cox process model for the set of presence locations. We have to introduce degradation caused by sampling bias as well as by land transformation. As a result, we conceptualize a *potential* intensity, i.e., the intensity in the absence of degradation, as well as a *realized* (or effective) intensity that operates in the presence of degradation. Further, we tile the intensity to reflect our inability to explain it at spatial resolution finer than our grid cells.

We imagine three surfaces over D. Let $\lambda(\mathbf{s})$ be the "potential" intensity surface, i.e., a positive function which is integrable over D. $\lambda(\mathbf{s})$ is the intensity in the absence of degradation. Let $\int_D \lambda(\mathbf{s})d\mathbf{s} = \lambda(D)$. Then, $f(s) = \lambda(\mathbf{s})/\lambda(D)$ gives the potential density over D. Modeling for $\lambda(\mathbf{s})$ is given below. Next, we envision an availability surface, $U(\mathbf{s})$, a binary surface over D such that $U(\mathbf{s}) = 1$ or 0 according to whether location $\mathbf{s}$ is untransformed by land use or not. That is, assuming no sampling bias, $\lambda(\mathbf{s})U(\mathbf{s})$ can only be $\lambda(\mathbf{s})$ or 0 according whether $\mathbf{s}$ is available or not. Let A_i denote the geographical region corresponding to cell i. Then, if we average $U(s)$ over A_i, we obtain $u_i = \int_{A_i} U(\mathbf{s})d\mathbf{s}/|A_i|$, where u_i is the proportion of cell i that is transformed and $|A_i|$ is the area of cell i. In our setting u_i is known, through remote sensing, for all grid cells. Similarly, we envision a sampling effort surface over D which we denote as $T(\mathbf{s})$. $T(\mathbf{s})$ is also a binary surface and $T(\mathbf{s})U(\mathbf{s}) = 1$ indicates that location $\mathbf{s}$ is both available and sampled. Now, we can set $q_i = \int_{A_i} T(\mathbf{s})U(\mathbf{s})d\mathbf{s}/|A_i|$ and interpret q_i as the probability that a randomly selected location in A_i was available and sampled. Thus, we can capture availability and sampling effort at areal unit scale.

Hence, $\lambda(\mathbf{s})U(\mathbf{s})T(\mathbf{s})$ becomes the degradation at location $\mathbf{s}$. This implies that in regions where no locations were sampled, the operating intensity for the species is 0. If $T(\mathbf{s})$ is viewed as random, with $p(\mathbf{s}) = P(T(\mathbf{s}) = 1) \in [0, 1]$, then $p(s)$ gives the local probability of sampling. This is analogous to $p(\mathbf{s})$ thinning discussed in Section 9.3.6.2.

To go forward, we assume that $\lambda(\mathbf{s})$ is independent of $T(\mathbf{s})U(\mathbf{s})$. That is, the potential intensity for a species is independent of the degradation process. Then, omitting the details, we can write $\int_{A_i} \lambda(\mathbf{s})T(\mathbf{s})U(\mathbf{s})d\mathbf{s} = \lambda_i q_i$ where $\lambda_i = \int_{A_i} \lambda(\mathbf{s})d\mathbf{s}$ is the cumulative intensity associated with cell A_i and, again, $q_i = \frac{1}{|A_i|}\int_{A_i} T(\mathbf{s})U(\mathbf{s})d\mathbf{s}$. It is not sensible to imagine that sampling effort is independent of land transformation. For instance, if $U(\mathbf{s}) = 0$ then $T(\mathbf{s}) = 0$. Hence, if we define $q_i = u_i p_i$, then $p_i = \frac{\int_{A_i} T(\mathbf{s})U(\mathbf{s})d\mathbf{s}}{\int_{A_i} U(\mathbf{s})d\mathbf{s}}$, i.e., p_i is the conditional probability that a randomly selected location in cell i is sampled given it is available. In our application below we set p_i equal to 1 or 0 which we interpret as $T(\mathbf{s}) = U(\mathbf{s}) \forall \mathbf{s} \in A_i$ or $T(\mathbf{s}) = 0 \forall \mathbf{s} \in A_i$, respectively. In particular, we set $p_i = 1$ if cell i was sampled for any species in our dataset; otherwise, we set $p_i = 0$. For the CFR, this sets $p_i = 1$ for the $10, 158$ grid cells (28%) that have been visited.

To model the potential intensity surface $\lambda(\cdot)$, we employ a log Gaussian Cox process (Section 8.4.2). We expect the environmental covariates, say $\mathbf{x}(\mathbf{s})$ to influence the intensity and model the mean of the GP as a linear combination of them. Then for any location $\mathbf{s} \in D$, we have

$$\log \lambda(\mathbf{s}) \;=\; \mathbf{x}^{\mathrm{T}}(\mathbf{s})\beta + w(\mathbf{s}) \tag{9.25}$$

with $w(\cdot)$, a zero-mean stationary, isotropic GP over D; in the sequel we use the exponential covariance function.

Suppose we have n_i presence locations $(\mathbf{s}_{i,1}, \mathbf{s}_{i,2}, ..., \mathbf{s}_{i,n_i})$ within cell i for $i = 1, 2, ..., I$. Following the discussion above, $U(\mathbf{s}_{i,j})T(\mathbf{s}_{i,j}) \equiv 1, 0 \le j \le n_i, 1 \le i \le I$. Then the likelihood function becomes

$$L(\lambda(\cdot); \{\mathbf{s}_{i,j}\}) \;\propto\; \exp\left(-\int_D \lambda(\mathbf{s})U(\mathbf{s})T(\mathbf{s})\,d\mathbf{s}\right)\prod_{i=1}^{I}\prod_{j=1}^{n_i}\lambda(\mathbf{s}_{i,j}) \tag{9.26}$$

Fortunately, we have a natural approximation to handle the stochastic integral in (9.26) by recalling that the dataset is gathered at the scale of grid cells in the CFR. That is, though we have geo-coded locations for the observed sites, with covariate information at grid cell level, we only attempt to explain the point pattern at grid cell level. In particular, let D denote our CFR study domain where D is divided into $I = 36,907$ grid cells of equal area. For each cell $i = 1, 2, 3, ..., I$, we are given information on l covariates as $x_i = (x_{i1}, x_{i2}, ..., x_{il})$. We also have cell-level information about land availability across D, as a proportion of of the area of the cell (Figure 9.18). Following the previous subsection, we denote this by u_i. For many cells $n_i = 0$ primarily because 72% were actually unsampled. Additionally, a computational advantage accrues to working at the grid cell level; we can work with a product Poisson likelihood approximation rather than the point pattern likelihood in (9.26). That is, we assume $\lambda(\cdot)$ is a tiled surface such that for cell i, the height is $\Delta\lambda(\mathbf{s}_i)$ where Δ is the area of the cell and $\mathbf{s}_i$ is the centroid. Then, given the set $\{\lambda(\mathbf{s}_i), i = 1, 2, ..., I$, the n_i are independent and $n_i \sim \mathrm{Po}(\Delta\lambda(s_i)q_i)$. For any cell with $q_i = 0$ (which, by definition, can happen if either $p_i = 0$ or $u_i = 0$) there is no contribution from A_i in the product Poisson likelihood. Altogether, the posterior distribution takes the form

$$\pi(\lambda(\mathbf{s}_{1:m}), \beta, \theta \mid \mathbf{n}, \mathbf{x}, \mathbf{u}, \mathbf{q}) \propto \exp\left\{-\sum_{i=1}^{I}\lambda(\mathbf{s}_i)\Delta_i q_i\right\}\prod_{i=1}^{m}\lambda^{n_i}(\mathbf{s}_i)$$
$$\times \phi_m(\log \lambda(\mathbf{s}_{1:m}) \mid \beta, \mathbf{x}, \theta)\pi(\beta)\pi(\theta) \tag{9.27}$$

where ϕ_m denotes the m dimensional Gaussian density and θ the parameters in the covariance function of $w(\cdot)$ in (9.25).

The primary computational challenge in working with the CFR is handling the model fitting for $37,000$ grid cells, the familiar 'large n' problem for Gaussian process, see Chapter 13. We employ the predictive process approximation for Gaussian random fields as in Section 13.4, with bias correction. With grid cells, an alternative is a parallelization scheme in conjunction with a CAR model as described in Chakraborty et al. (2010). The joint set of locations, $(\mathbf{s}_{1:I}, \mathbf{s}_{1:r}^0)$, partition the spatial covariance matrix as $\sigma^2 R_{n+r}(\phi) = \sigma^2\begin{pmatrix} R_I(\phi) & R_{r,I}(\phi) \\ R_{I,r}(\phi) & R_r(\phi) \end{pmatrix}$, where the entries of R_{r+I} are exponential correlation terms with decay parameter ϕ. We rewrite $\lambda_{0,i} = \lambda(\mathbf{s}_i)\Delta$, which denotes the expected species count in cell i under the potential intensity. Now the hierarchical model becomes

$$
\begin{aligned}
n_i \mid \lambda_{0,i} &\stackrel{ind}{\sim} \text{Poi}(\lambda_{0,i} q_i), i = 1, 2, \ldots I \\
\log \lambda(\mathbf{s}_i) &= x_i^{\mathrm{T}} \beta + \tilde{w}(\mathbf{s}_i) + \epsilon_i^* \\
\tilde{w}(\mathbf{s}_{1:I}) &= R_{I,r}(\phi) R_r^{-1}(\phi)\, w(\mathbf{s}_{1:r}^0) \\
w(\mathbf{s}_{1:r}^0) &\sim N_r(\mathbf{0}_r, R_r(\phi)) \\
\epsilon_i^* &\stackrel{ind}{\sim} N(0, \sigma^2(1 - (R_{I,r}(\phi) R_r^{-1}(\phi) R_{r,I}(\phi))_{ii})) \\
\pi(\beta, \phi, \sigma^2) &= \pi(\beta)\pi(\phi)\pi(\sigma^2)
\end{aligned}
\tag{9.28}
$$

Posterior inference addresses two principal objectives: to understand the effect of environmental variables on species distribution and to construct maps of the potential and realized intensities over the entire study region. Posterior samples of β help us to infer whether a particular factor has a significant impact (positive or negative) on species intensity. With regard to displays of intensity surfaces, since, in our CFR application, $p_i = 1$ (i.e., $T(\mathbf{s}) = U(\mathbf{s})$ for all $\mathbf{s}$ in cell i) or $p_i = 0$ (i.e., $T(\mathbf{s}) = 0$ for all $\mathbf{s}$ in cell i) and since only 28% of cells were sampled, the $\lambda_i p_i$ surface will be 0 for 72% of cells, primarily capturing the (lack of) sampling effort. The $\lambda_i u_i$ surface reveals the effect of transformation. Since few cells are completely transformed, most $\lambda_i u_i > 0$. Of course, the $\lambda(\mathbf{s})$ surface is most interesting since it offers insight into the expected pattern of presences over all of D. Posterior draws of $\lambda_{1:I}$ can be used to infer about the potential intensity, displaying say the posterior mean surface. We can also learn about the potential density f in this discretized setting as $f_i = \lambda_i / \sum_{k=1}^I \lambda_k$, and the corresponding density under transformation as $f_{u,i} = \lambda_i u_i / \sum_{k=1}^I \lambda_k u_k$.

We consider six species within the Proteaceae family, ranging from prevalent to somewhat rare. Our point pattern for each species is drawn from the Protea Atlas data set (Rebelo et al, 2002). They are: *Protea aurea* (PRAURE) at 603 locations, *Protea cynaroides* (PRCYNA) at 8172 locations, *Leucadendron salignum* (LDSG) at 22949 locations, *Protea mundii* (PRMUND) at 764 locations, *Protea punctata* (PRPUNC) at 2148 locations, *Protea repens* (PRREPE) at 14574 locations.

In earlier work [Gelfand et al., 2005c] , 18 environmental explanatory variables were considered, available at the 1 minute by 1 minute resolution. Based upon these analyses, the six most important were selected as covariates for the intensity function. They are: mean annual precipitation (MAP), July (winter) minimum temperature (MIN07), January (summer) maximum temperature (MAX01), potential evapotranspiration (EVAP), summer soil moisture days (SMDSUM), and percent of the grid cell with low fertility soil (FERT). There is considerable spatial variation in these covariates across the region

Table 9.4 provides the posterior mean covariate effects for all species along with the associated 95% equal tail credible intervals in parentheses. Most of the coefficients are significantly different from 0 and also, the direction of significance varies with species.

Together, Figures 9.19 and 9.20 show the posterior mean intensity surfaces (potential and transformed) for the six species. Evidently there is strong spatial pattern and the pattern varies with species, i.e., the nature of local adjustment to the regression is species dependent. Comparison between the transformed and potential for each species is illuminating. Differentials of multiple orders of magnitude in expected cell counts are seen across many grid cells.

9.8.3 Spatial self-exciting processes

Self-exciting processes (also referred to as Hawkes processes) have attracted increasing interest [Reinhart, 2018]. Broadly, they capture the setting where the rate of events at time

Species	EVAP	MAX01	MIN07	MAP	SMDSUM	FERT
PRAURE	−4.909	2.702	−0.301	−1.222	−0.049	0.501
	(−6.506,−3.057)	(1.574,3.678)	(−0.967,0.425)	(−1.975,−0.423)	(−0.816,0.711)	(0.034,0.967)
PRCYNA	−2.447	1.268	−1.032	−0.833	0.552	0.802
	(−2.981,−1.859)	(0.853,1.619)	(−1.314,−0.469)	(−1.021,−0.626)	(0.255,0.830)	(0.642,0.957)
LDSG	0.721	−0.420	0.137	−0.376	0.488	0.099
	(0.373,1.085)	(−0.658,−0.181)	(−0.011,0.295)	(−0.513,−0.237)	(0.304,0.673)	(0.045, 0.152)
PRMUND	−0.219	0.028	−0.609	−0.199	1.082	0.507
	(−2.724,1.429)	(−1.163,1.702)	(−1.039,−0.055)	(−1.024,0.510)	(0.277,1.809)	(0.101,0.929)
PRPUNC	2.076	−1.590	−1.722	0.363	0.535	0.186
	(1.031,3.096)	(−2.290,−0.921)	(−2.048,−1.409)	(0.082,0.662)	(0.052,1.079)	(−0.014,0.381)
PRREPE	1.690	−1.205	−0.275	0.124	0.094	0.224
	(1.243,2.124)	(−1.498,−0.907)	(−0.431,−0.110)	(−0.011,0.278)	(−0.112,0.320)	(0.152,0.295)

Table 9.4 *Posterior mean of covariate effects with central 95% credible interval in parenthesis.*

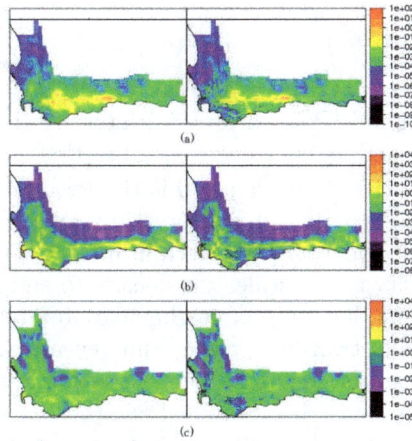

Figure 9.19 *Intensity maps for (a)* Protea aurea, *(b)* Protea cynaroides, *and (c)* Leucadendron salignum, *potential (left) and transformed (right). Values are cellwise expected frequency for the corresponding species.*

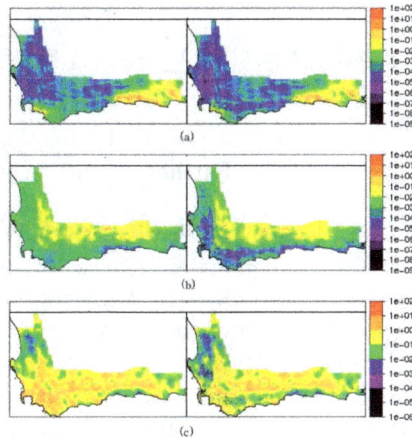

Figure 9.20 *Intensity maps for (a)* Protea mundi, *(b)* Protea punctate, *and (c)* Protea repens, *potential (left) and transformed (right). Values are cellwise expected frequency for the corresponding species.*

t may depend on the history of events at times preceding t, allowing events to trigger new events. Probabilistically, they are a temporal process which extends the usual Poisson process point pattern modeling on R^1. Specifically, given that a Poisson process is creating conditionally independent events in a time window $(0, T]$, the notion of self-exciting suggests that the occurrence of an event in the past encourages more events after that event than would be captured by a customary Poisson process. We offer a brief development in the simplest case. The idea is to propose a conditional intensity for time t, formally defined as

$$\lambda^*(t) \equiv \lambda(t \mid \mathcal{H}_t) = \lim_{\Delta t \to 0} \mathrm{E}(N(t, t + \Delta t)/\Delta t. \qquad (9.29)$$

Here, $\mathcal{H}_t$ is the history of events up to time t and, as usual, N is a random counting measure denoting the number of events in an interval. The practical form for the conditional intensity is

$$\lambda(t \mid \mathcal{H}_t) = \mu(t) + \int_0^T h(t - u) N(du) = \mu(t) + \sum_{t_j < t} h(t - t_j). \qquad (9.30)$$

Here, $\mu(t) \geq 0$ is the background (Poisson process) intensity (typically $\mu(t) = \mu$ providing an HPP as the background), and $h(\cdot)$ is a decreasing function and the sum supplements the intensity according to where the t_j are. $h() \geq 0$ is the *triggering* function which captures the nature of self-excitation. Typically, it is decreasing, reflecting decreasing contribution to the intensity as we look at events further back in history. Typically, h is written as αf where f is a decreasing density and α scales the density to an intensity.

Initial applications were to the study of earthquakes and aftershocks. More recent applications include crime events triggering more crime events and online posting where a particular post may encourage a flurry of subsequent posts. Spatial applications also consider the above but also the spread of infectious diseases.

The spatio-temporal form extends (9.29) over a domain d with again $t \in (0, T]$ to

$$\lambda(\mathbf{s}, t \mid \mathcal{H}_t) = \lim_{|\partial \mathbf{s}| \to 0, \Delta t \to 0} \mathrm{E}\left(\frac{N((\mathbf{s}, \partial \mathbf{s}) \times [t, t + \Delta t))}{|\partial \mathbf{s}| \Delta t}\right). \qquad (9.31)$$

Here, $(\mathbf{s}, \partial \mathbf{s})$ is a ball centered at $\mathbf{s}$ with radius $\partial \mathbf{s}$. Similarly, the practical form is

$$\lambda(\mathbf{s}, t \mid \mathcal{H}_t) = \mu(\mathbf{s}, t) + \sum_{t_j < t} h(t - t_j, \mathbf{s} - \mathbf{s}_j). \qquad (9.32)$$

Here, $\mathbf{s}_j$ is the spatial location of the event at time t_j. In practice, the triggering function is often taken as separable in space and time. Further, often $\mu(\mathbf{s}, t) = \mu(\mathbf{s})$ to focus on spatial variation in the background intensity.

Finally, marks can be added, analogous to Section 8.6.6. The common example in the literature is the magnitude of the earthquake aftershock, m_j at time t_j and location $\mathbf{s}_j$. Stochastically, they would be introduced through $\lambda(\mathbf{s}, t, m) = \lambda(\mathbf{s}, t) \mid \mathcal{H}_t \times f(m \mid \mathbf{s}, t)$ where $f(m \mid \mathbf{s}, t)$ is the conditional density of the mark at time t and location $\mathbf{s}$ given the history of the process up to t [see, e.g., Mohler et al., 2011, for discussion and a crime illustration]. For a full development of the foregoing spatio-temporal self-exciting processes the reader is referred to the review paper by [Reinhart, 2018, pp. 299-318]. Bayesian fitting of such models can be implemented following the development in Rasmussen [2013].

A very different spatial perspective views the spatial aspect as a special case of a self-exciting network [Linderman and Adams, 2014]. That is, communication is considered among a finite set of spatial locations analogous to nodes in a graph. The data sequence now becomes $\{(t_j, \mathbf{s}_{t_j}), j = 1, 2, ..., n\}$ where $\mathbf{s}_{t_j}$ is the spatial location of the event at time t_j. Now, the conditional intensity takes the form, $\lambda^*(t) = \sum_{k=1}^K \lambda_k^*(t)$ where there are K

spatial nodes and $\lambda_k^*(t)$ is the conditional intensity associated with the kth node. In particular, $\lambda_k^*(t) = \sum_{t_j < t} h(t - t_j; d(\mathbf{s}_k, \mathbf{s}_{t_j}))$. As above, h is usually written as a constant time a density, i.e., $\alpha_{\mathbf{s}_k, \mathbf{s}_{t_j}} f(t - t_j; d(\mathbf{s}_k, \mathbf{s}_{t_j}))$. A matrix of α's emerges. An application would view the network as a collection of monitoring stations over a region which can communicate with each other, which can mutually excite each other.

9.8.4 Scan statistics

Scan statistics are widely used to detect hotspots or local clustering in point patterns. They find application to epidemiology where there is interest in disease clustering but also in surveillance problems, e.g., syndromic, environmental or military settings. A primary reason for their wide usage is the availability of a free user-friendly software package, SaTScan, developed by Martin Kulldorff (http://www.satscan.org/). See also Kulldorff [1997]. However, this material does not usually find its way into point pattern textbooks because it is ad hoc in its nature, because it is not clearly an analysis of the pattern of the points, and because it is primarily algorithmic in its implementation. We offer just a brief treatment here.

The basic idea is as follows: define a window (subregion of the region of interest); scan the region using this window; compute the value of some statistic for each stop of the window; determine which windows (if any) give significantly large values of the statistic. Hence, the ad hoc aspect of the problem emerges. First, the notion of a cluster is not well-defined. How many points, how concentrated, what shape? Also, the choice of window is arbitrary. What shape, what size, what orientation? And, there is also a substantial computational challenge. With a large region and with a relatively small window, the statistic must be computed over many windows. Significance testing is usually done through Monte Carlo tests (recall the Monte Carlo tests in conjunction wit the significance of Moran's I and Geary's C in Section 3.1) so random datasets must be generated over the region of interest and the scanning process must be conducted for each Monte Carlo replicate. Such computation does not scale for large datasets while data mining settings are natural candidates for such procedures.

To provide a bit more detail, the setting is, in fact, one of a marked point pattern where each point has an observed mark and a baseline mark. That is, we have $\mathbf{s}$ with mark $c(\mathbf{s})$, an observed count for that location and $b(\mathbf{s})$, a baseline or expected count (based say on the population at risk at that location). Hence, it emerges that the points are, in fact, small areal units relative to the region D and the randomness may be viewed primarily with regard to the observed mark at the location rather than with regard to the observed location. In fact, the customary model assumes, at $\mathbf{s}_i$, that $c(\mathbf{s}_i) \sim Po(q_i b(\mathbf{s}_i))$, independently over i. So, of interest are large q's and we see similarity with the modeling for the disease mapping problem in Chapters 3 and 5. There r_i played the role of a relative risk or rate for an areal unit, with interest in elevated r's. Here, q_i is a relative risk, with interest as well in large q's. In Chapter 3, we modeled spatially dependent r_i's using Markov random fields (CAR models). Here, there is no spatial structure attached to the q_i's.

Now, consider the spatial scan window, say W. Assume q_i is constant over all points $\mathbf{s}_i$ in W, say equal to q_{in} and that it is constant for all points $\mathbf{s}_i$ in D but outside of W, say q_{out}. Under these clearly dubious assumptions along with the independent Poisson assumption, a likelihood ratio test can be created to test the null hypothesis that $q_{in} = q_{out}$ against the alternative that $q_{in} > q_{out}$ (elevated risk). It takes a simple form. Let $c(W) = \sum_{\mathbf{s}_i \in W} c(\mathbf{s}_i)$, similarly $b(W) = \sum_{\mathbf{s}_i \in W} b(\mathbf{s}_i)$, with analogous statistics $c(W^C)$ and $b(W^C)$ where W^C is the complement of W relative to D. Then the log of the likelihood ratio is $c(W) \log(\frac{c(W)}{b(W)}) + c(W^C) \log(\frac{c(W^C)}{b(W^C)})$. Again, we seek large values of this statistic. Evidently, there is the issue of multiple testing over many W's. Clearly there will be no tractable test statistic when implementing repeated calculations over overlapping regions, hence the foregoing Monte Carlo testing. There is a Bayesian version of this They introduce conjugate Gamma priors and for the q's and obtain posterior probabilities for the two hypotheses. Scan statistics can be created for variables other than counts, using alternative specifications to the Poisson, creating different likelihood ratio statistics.

9.9   Exercises

1. Justify the *rejection* method approach for obtaining samples of point patterns under a NHPP. That is, let $\lambda_{max} = \max_s \lambda(s), s \in D$. Generate a point pattern from a HPP using the constant intensity λ_{max}. For each point generated, do a rejection step, i.e., for s_i, draw $U_i \sim U(0,1)$ and retain s_i if $U_i < \lambda(s_i)/\lambda_{max}$. Argue that the remaining points come from an NHPP with intensity $\lambda(s)$. Hint: Obtain the distribution of the retained number of points in A, $N(A)$ for any $A \subset D$.

2. Suppose we have a NHPP with intensity $\lambda(\mathbf{s})$ and we implement p-thinning to a point pattern from this process (Section 8.5). Show that the thinned pattern is a realization from an NHPP with intensity $p\lambda(\mathbf{s})$.

3. Generate a random point pattern over the unit square from the intensity $\lambda(s) = \lambda(x,y) = \{200\exp(-3(x-\frac{1}{3})^2 - 4(y-\frac{2}{3})^2) + 300\exp(-2(x-\frac{2}{3})^2 - 2(y-\frac{1}{3})^2)\}I(x \in (0,1), y \in (0,1))$

   (a) How many points do you expect in the unit square?

   (b) Test for complete spatial randomness using a quadrat based χ-square test.

   (c) Obtain $\hat{G}(w)$. Obtain $\hat{F}(w)$ for $m = 100$. Obtain the theoretical $Q - Q$ plot of $\hat{G}(w)$ vs. G for complete spatial randomness. Obtain the theoretical $Q - Q$ plot of $\hat{F}(w)$ vs. G for complete spatial randomness.

   (d) Obtain $\hat{K}(d)$. Plot $\hat{L}(d) = \sqrt{\frac{\hat{K}(d)}{\pi}} - d$ vs. d.

4. For the Lansing Woods data, available at http://www.spatstat.org/, you will find six species. Choose a couple and, for each, assume a stationary, in fact, isotropic model, and provide the exploratory data analysis from Section 8.3. That is, estimate $G(d)$, $F(d)$, $K(d)$, and $\gamma(d)$. Compare with corresponding plots assuming CSR. Plot $J(d)$ and $L(d)$ vs. d. Fit NHPP and LGCP models to these choices within a Bayesian framework.

5. For the Strauss-Ripley redwood saplings dataset (redwood) available at http://www.spatstat.org/, fit a Gibbs process model, in fact, a Strauss process within a Bayesian framework.

Chapter 10

Multivariate spatial modeling for point-referenced data

In this chapter we take up the problem of multivariate spatial modeling. Spatial data is often *multivariate* in the sense that multiple outcomes are measured at each spatial unit. As in the univariate case, the spatial units can be referenced by points or by areal units. Examples of multivariate point-referenced data abound in the sciences. For instance, at a particular environmental monitoring station, levels of several pollutants would typically be measured (e.g., ozone, nitric oxide, carbon monoxide, $PM_{2.5}$, etc.). We anticipate both dependence between measurements at a particular location, and association between measurements across locations. Multivariate areal data are conspicuous in public health, where each county or administrative unit supplies counts or rates for a number of diseases. Again, we expect dependence between diseases within each county as well as across counties.

Statistical modeling and analysis will depend upon the type of spatial referencing— point-referenced or areal. Models for point-referenced spatial data usually proceed by constructing a multivariate spatial process that will enable interpolation and prediction (or kriging) over the entire domain. For areal data, modeling focuses upon multivariate versions of CAR models.

We first treat multivariate point-referenced data and turn to multivariate areal data in the next chapter. Analysis of multivariate point-referenced data can proceed using either a *conditioning approach*, along the lines of the way misalignment was treated in Chapter 7 (e.g., X followed by $Y \mid X$) or a *joint approach* that directly models the joint distribution of the outcomes variables. Both these approaches are based upon extensions of univariate kriging to multivariate contexts, where the conditional approach is also referred to as *kriging with external drift* [see, e.g., Royle and Berliner, 1999], while the joint approach is referred to as *co-kriging* [see Wackernagel, 2003].

10.1 Joint modeling in classical multivariate Geostatistics

Classical multivariate geostatistics begins, as with much of geostatistics, with early work of Matheron [1973] and Matheron [1979]. The basic ideas here include cross-variograms and cross-covariance functions, intrinsic coregionalization, and co-kriging. The emphasis is on prediction. A thorough discussion of the work in this area is provided in Wackernagel [2003]. To add generality to the conditional modeling approach, one could envision a latent multivariate spatial process defined over locations in a region. For example, in ambient air quality assessment, we seek to jointly model multiple contaminants at a fixed set of monitoring sites. Inference focuses upon three major aspects: (i) estimation of associations among the contaminants, (ii) estimation of the strength of spatial association for each contaminant, and (iii) prediction of the contaminants at arbitrary locations.

Let $\mathbf{Y}(\mathbf{s}) = (Y_1(\mathbf{s}), Y_2(\mathbf{s}), \dots, Y_p(\mathbf{s}))^{\mathsf{T}}$ be a $p \times 1$ vector, where each $Y_i(\mathbf{s})$ represents an outcome of interest referenced by $\mathbf{s} \in \mathcal{D}$. We seek to capture the association both within

DOI: 10.1201/9781003401728-10

components of $\mathbf{Y}(\mathbf{s})$ and across $\mathbf{s}$. The joint second-order (weak) stationarity hypothesis defines the cross-variogram as

$$\gamma_{ij}(\mathbf{h}) = \frac{1}{2}E(Y_i(\mathbf{s}+\mathbf{h}) - Y_i(\mathbf{s}))(Y_j(\mathbf{s}+\mathbf{h}) - Y_j(\mathbf{s})). \qquad (10.1)$$

Implicitly, we assume $E(Y(\mathbf{s}+\mathbf{h}) - Y(\mathbf{s})) = 0$ for all $\mathbf{s}$ and $\mathbf{s}+\mathbf{h} \in \mathcal{D}$. Clearly, $\gamma_{ij}(\mathbf{h})$ is an even function, i.e. $\gamma_{ij}(\mathbf{h}) = \gamma_{ij}(-\mathbf{h})$ and, as a consequence of the Cauchy-Schwarz inequality, satisfies $|\gamma_{ij}(\mathbf{h})|^2 \leq \gamma_{ii}(\mathbf{h})\gamma_{jj}(\mathbf{h})$.

The cross-covariance function is defined as

$$C_{ij}(\mathbf{h}) = E[(Y_i(\mathbf{s}+\mathbf{h}) - \mu_i)(Y_j(\mathbf{s}) - \mu_j)] \qquad (10.2)$$

where, for each i, a constant mean μ_i is assumed for $Y_i(\mathbf{s})$. The associated $p \times p$ matrix $C(\mathbf{h})$ is called the cross-covariance matrix. Note that it need not be symmetric (think of a setting where you might expect $C_{ij}(\mathbf{h}) \neq C_{ji}(\mathbf{h})$). Of course, $C(\mathbf{h})$ need not be positive definite but, in the limit, as $\|h\| \to 0$, it becomes positive definite.

Using standard properties of covariances, we see that the cross-covariance function must satisfy $|C_{ij}(\mathbf{h})|^2 \leq C_{ii}(\mathbf{0})C_{jj}(\mathbf{0})$. However, $|C_{ij}(\mathbf{h})|$ need not be bounded above by $C_{ij}(\mathbf{0})$ because the maximum value of $C_{ij}(\mathbf{h})$ need not occur at $\mathbf{0}$. A particular example is the so-called *spatial delay models* [see Wackernagel, 2003]. Consider the bivariate setting with $p = 2$ and assume that

$$Y_2(\mathbf{s}) = aY_1(\mathbf{s}+\mathbf{h}_0) + \epsilon(\mathbf{s}) ,$$

where $Y_1(\mathbf{s})$ is a spatial process with stationary covariance function $C(\mathbf{h})$, and $\epsilon(\mathbf{s})$ is a pure error process with variance τ^2. Then, the associated cross covariance function has $C_{11}(\mathbf{h}) = C(\mathbf{h}), C_{22}(\mathbf{h}) = a^2C(\mathbf{h})$ and $C_{12} = C(\mathbf{h}+\mathbf{h}_0)$. Similarly, $|C_{ij}(\mathbf{h})|^2$ need not be bounded above by $C_{ii}(\mathbf{h})C_{jj}(\mathbf{h})$. The matrix $C(\mathbf{h})$ of direct and cross-covariances with $C_{ij}(\mathbf{h})$ as its (i, j)-th entry is called the cross-covariance matrix. It need not be positive definite at any $\mathbf{h}$ though as $\mathbf{h} \to \mathbf{0}$, it converges to a positive definite matrix, the (local) covariance matrix associated with $\mathbf{Y}(\mathbf{s})$. We will discuss cross-covariance matrices in greater detail in Section 10.2.

Analogous to the relationship between the variogram and the covariance function, we find a familiar connection between the cross-variogram and the cross-covariance function. The latter determines the former and it is easy to show that

$$\gamma_{ij}(\mathbf{h}) = C_{ij}(\mathbf{0}) - \frac{1}{2}(C_{ij}(\mathbf{h}) + C_{ij}(-\mathbf{h})). \qquad (10.3)$$

If we decompose $C_{ij}(\mathbf{h})$ as $\frac{1}{2}(C_{ij}(\mathbf{h}) + C_{ij}(-\mathbf{h})) + \frac{1}{2}(C_{ij}(\mathbf{h}) - C_{ij}(-\mathbf{h}))$, then the cross-variogram only captures the even term of the cross-covariance function, suggesting that it may be inadequate in certain modeling situations. Such concerns led to the proposal of the pseudo cross-variogram Clark et al. [1989], Myers [1992], Cressie [1993]. In particular, Clark et al. [1989] proposed $\pi_{ij}^c(\mathbf{h}) = E(Y_i(\mathbf{s}+\mathbf{h}) - Y_j(\mathbf{h}))^2$ and Myers [1991] subsequently offered a mean-corrected version, $\pi_{ij}^m(\mathbf{h}) = var(Y_i(\mathbf{s}+\mathbf{h}) - Y_j(\mathbf{h}))$. It is easy to show that $\pi_{ij}^c(\mathbf{h}) = \pi_{ij}^m(\mathbf{h}) + (\mu_i - \mu_j)^2$. The psuedo cross-variogram is not constrained to be an even function. However, the assumption of stationary cross increments is unrealistic, certainly with variables measured on different scales and even with rescaling of the variables. A further limitation is the restriction of the pseudo cross-variogram to be positive. Despite the unattractiveness of "apples and oranges" comparison across components, Cressie and Wikle [1998] demonstrate successful multivariate kriging or co-kriging, using $\pi_{ij}^m(\mathbf{h})$.

10.1.1   Co-kriging

Co-kriging in classical multivariate geostatistics refers to spatial prediction at a new location that uses not only information from observations of the outcome variable being considered, say $Y_i(\mathbf{s})$, but also information from the measurements of the other outcome variables, i.e. the other components in $\mathbf{Y}(\mathbf{s})$. Early presentations of co-kriging are available in Journel and Huijbregts [2003] and Matheron [1979], while Myers [1982] presents a general and more comprehensive treatment using matrices. Corsten [1989] and Stein and Corsten [1991] develop cokriging within a linear regression framework. Again, a detailed review can be found in Wackernagel [2003].

It is instructive to distinguish between prediction of a single variable as above and joint prediction of several variables at a new location [Myers, 1982]. Consider the joint second order stationarity model in (10.1) above. Suppose we seek to predict $Y_1(\mathbf{s}_0)$, i.e., the first element in $\mathbf{Y}(\mathbf{s})$, at a new location $\mathbf{s}_0$. An unbiased estimator based upon $\mathbf{Y} = (\mathbf{Y}(\mathbf{s}_1), \mathbf{Y}(\mathbf{s}_2), ..., \mathbf{Y}(\mathbf{s}_n))^\mathsf{T}$ would assume the form $\hat{Y}_1(\mathbf{s}_0) = \sum_{i=1}^n \sum_{l=1}^p \lambda_{il} Y_l(\mathbf{s}_i)$ where we have the constraints that $\sum_{i=1}^n \lambda_{il} = 0, l \neq 1, \sum_{i=1}^n \lambda_{i1} = 1$. On the other hand, if we were interested in predicting $\mathbf{Y}(\mathbf{s}_0)$, we would now write $\hat{\mathbf{Y}}(\mathbf{s}_0) = \sum_{i=1}^n \Lambda_i \mathbf{Y}(\mathbf{s}_i)$. The unbiasedness condition is $\sum_{i=1}^n \Lambda_i = I$. Moreover, now, what should we take as the "optimality" condition? One choice is to choose the set $\{\Lambda_{0i}, 1 = 1, 2, ..., n\}$ with associated estimator $\hat{\mathbf{Y}}_0(\mathbf{s}_0)$ such that for any other unbiased estimator, $\tilde{\mathbf{Y}}(\mathbf{s}_0)$, $E(\tilde{\mathbf{Y}}(\mathbf{s}_0) - \mathbf{Y}(\mathbf{s}_0))(\tilde{\mathbf{Y}}(\mathbf{s}_0) - \mathbf{Y}(\mathbf{s}_0))^\mathsf{T} - E(\hat{\mathbf{Y}}_0(\mathbf{s}_0) - \mathbf{Y}(\mathbf{s}_0))(\hat{\mathbf{Y}}_0(\mathbf{s}_0) - \mathbf{Y}(\mathbf{s}_0))^\mathsf{T}$ is non-negative definite (Ver-Hoef and Cressie, 1993). Alternatively, one could minimize $\operatorname{tr} E(\hat{\mathbf{Y}}(\mathbf{s}_0) - \mathbf{Y}(\mathbf{s}_0))(\hat{\mathbf{Y}}(\mathbf{s}_0) - \mathbf{Y}(\mathbf{s}_0))^\mathsf{T} = E(\hat{\mathbf{Y}}(\mathbf{s}_0) - \mathbf{Y}(\mathbf{s}_0))^\mathsf{T}(\hat{\mathbf{Y}}(\mathbf{s}_0) - \mathbf{Y}(\mathbf{s}_0))$ (Myers, 1982).

Returning to the individual prediction case, minimization of predictive mean square error, $E(Y_1(\mathbf{s}_0) - \hat{Y}_1(\mathbf{s}_0))^2$ amounts to a quadratic optimization subject to linear constraints and the solution can be obtained using Lagrange multipliers. As in the case of univariate kriging, the solution can be written as a function of a cross-variogram specification. In fact, Ver-Hoef and Cressie (1993) show that $\pi_{ij}(\mathbf{h})$ above emerges in computing predictive mean square error, which suggests that it is a natural cross variogram for co-kriging. But, altogether, given the concerns noted regarding $\gamma_{ij}(\mathbf{h})$ and $\pi_{ij}(\mathbf{h})$, it seems preferable to assume the existence of second moments for the multivariate process, captured through a *valid* cross-covariance function (as defined in (10.2)). In fact, to introduce a likelihood, to implement full inference, and to do prediction with accurate uncertainty assessment, in the sequel we will work exclusively with Gaussian process models defined through cross-covariance functions.

We next turn to a brief discussion of *valid* cross-variograms. The matter is not as clear as in the univariate setting. Wackernagel [2003] induces valid cross-variograms from valid cross-covariance functions (see Section 10.2). Myers [1982] and Ver Hoef and Cressie [1993] assume second order stationarity and also a finite cross-covariance in order to bring $\gamma_{ij}(\mathbf{h})$ into the optimal co-kriging equations. Rehman and Shapiro [1996] define a *permissible* cross variogram $\gamma_{ij}(\mathbf{h})$ as one that meets the following conditions: (i) the $\gamma(\mathbf{h})$ are continuous except possibly at the origin, (ii) $\gamma_{ij}(\mathbf{h}) \geq 0, \forall \mathbf{h} \in \mathcal{D}$, (iii) $\gamma_{ij}(\mathbf{h}) = \gamma(-\mathbf{h}), \forall \mathbf{h} \in \mathcal{D}$, and (iv) the functions, $-\gamma_{ij}(\mathbf{h})$, are conditionally non-negative definite, the usual condition for individual variograms.

In fact, this directly renders an explicit solution to the individual co-kriging problem if we assume a multivariate Gaussian spatial process. The preceding developments show that such a process specification only requires supplying mean surfaces for each component of the outcome $\mathbf{Y}(\mathbf{s})$ and a valid cross-covariance function. For simplicity, assume $\mathbf{Y}(\mathbf{s})$ is centered to have mean $\mathbf{0}$. The cross-covariance function provides $\Sigma_\mathbf{Y}$, the $np \times np$ covariance matrix for the data $\mathbf{Y} = (\mathbf{Y}(\mathbf{s}_1)^\mathsf{T}, \mathbf{Y}(\mathbf{s}_2)^\mathsf{T}, ..., \mathbf{Y}(\mathbf{s}_n)^\mathsf{T})^\mathsf{T}$. In addition, it provides the $np \times 1$ vector, $\mathbf{c}_0$ which is blocked as vectors $\mathbf{c}_{0j}, j = 1, 2, .., n$ with l-th element $c_{0j,l} = \operatorname{Cov}(Y_1(\mathbf{s}_0), Y_l(\mathbf{s}_j))$. Then, from the multivariate normal distribution of $\mathbf{Y}, Y_1(\mathbf{s}_0)$, we obtain the co-kriging estimate,

$$\mathrm{E}(Y_1(\mathbf{s}_0) \mid \mathbf{Y}) = \mathbf{c}_0^\mathsf{T} \Sigma_\mathbf{Y}^{-1} \mathbf{Y}. \tag{10.4}$$

The associated variance, $\text{var}(Y_1(\mathbf{s}_0) \mid \mathbf{Y})$ is also immediately available, i.e., $\text{var}(Y_1(\mathbf{s}_0) \mid \mathbf{Y}) = C_{11}(\mathbf{0}) - \mathbf{c}_0^{\mathsf{T}} \Sigma_{\mathbf{Y}}^{-1} \mathbf{c}_0$.

In particular, consider the special case of the $p \times p$ cross-covariance matrix, $C(\mathbf{h}) = \rho(\mathbf{h})T$, where $\rho(\cdot)$ is a valid correlation function and T is the local positive definite covariance matrix. Then, $\Sigma_{\mathbf{Y}} = R \otimes T$, where R is the $n \times n$ matrix with (i,j)-th entry $\rho(\mathbf{s}_i - \mathbf{s}_j)$ and $\otimes$ denotes the Kronecker product. This specification also yields $\mathbf{c}_0 = \mathbf{r}_0 \otimes \mathbf{t}_{*1}$, where $\mathbf{r}_0$ is $n \times 1$ with entries $\rho(\mathbf{s}_0 - \mathbf{s}_j)$ and $\mathbf{t}_{*1}$ is the first column of T. Then, (10.4) becomes $t_{11} \mathbf{r}_0^{\mathsf{T}} R^{-1} \tilde{\mathbf{Y}}_1$ where t_{11} is the $(1,1)$-th element of T and $\tilde{\mathbf{Y}}_1$ is the vector of observations associated with the first component of the $\mathbf{Y}(\mathbf{s}_j)$'s. This specification is known as the *intrinsic* multivariate correlation and is discussed in greater generality in Section 10.1.2. In other words, under an intrinsic specification, only observations on the first component are used to predict the first component at a new location. We leave this as an exercise but also, see Helterbrand and Cressie (1994) and Wackernagel (2003) in this regard.

In all of the foregoing work, inference assumes the cross-covariance or the cross-variogram to be known. In practice, a parametric model is adopted and data-based estimates of the parameters are plugged in. A related issue here is whether the data is available for each variable at all sampling points (so-called *isotopy*—not to be confused with isotropy), some variables share some sample locations (partial *heterotopy*), or the variables have no sample locations in common (entirely *heterotopic*). Similarly, in the context of prediction, if any of the $Y_l(\mathbf{s}_0)$ are available to help predict $Y_1(\mathbf{s}_0)$, we refer to this as "collocated cokriging." The challenge with heterotopy in classical work is that the empirical cross-variograms can not be computed and empirical cross-covariances, though they can be computed, do not align with the sampling points used to compute the empirical direct covariances. Furthermore, the value of the cross-covariances at $\mathbf{0}$ can not be computed.[1]

10.1.2   *Intrinsic Multivariate Correlation and Nested Models*

A somewhat different approach to multivariate spatial models is based upon the so called *structural analysis* approach, where typically more than one variogram model is used to accommodate different spatial scales. For example, following Grzebyk and Wackernagel [1994], we might deploy a *nested* variogram model written as

$$\gamma(\mathbf{h}) = t_1 \gamma_1(\mathbf{h}) + t_2 \gamma_2(\mathbf{h}) + \cdots + t_r \gamma_r(\mathbf{h}) .$$

With three spatial scales, corresponding to a nugget, fine scale dependence, and long range dependence, respectively, we might write $\gamma(\mathbf{h}) = t_1 \gamma_1(\mathbf{h}) + t_2 \gamma_2(\mathbf{h}) + t_3 \gamma_3(\mathbf{h})$ where $\gamma_1(\mathbf{h}) = 0$ if $\|\mathbf{h}\| = 0$, $= 1$ if $\|\mathbf{h}\| > 0$, while $\gamma_2(\cdot)$ reaches a sill equal to 1 very rapidly and $\gamma_3(\cdot)$ reaches a sill equal to 1 much more slowly.

The nested variogram model corresponds to the spatial process $\sqrt{t_1} w_1(\mathbf{s}) + \sqrt{t_2} w_2(\mathbf{s}) + \sqrt{t_3} w_3(\mathbf{s})$—a linear combination of independent processes. Why not, then, use this idea to build a multivariate version of a nested variogram model? Journel and Huijbregts (see 1978) propose to do this using the specification

$$w_l(\mathbf{s}) = \sum_{r=1}^{m} \sum_{j=1}^{p} a_{rj}^{(l)} w_{rj}(\mathbf{s}) \text{ for } l = 1, \ldots, p ,$$

where the $w_{rj}(\mathbf{s})$ are independent process replicates across j and, for each r, the process has correlation function $\rho_r(\mathbf{h})$ and variogram $\gamma_r(\mathbf{h})$ (with sill 1). In the case of isotropic ρ's, this

---

[1] The empirical cross-variogram imitates the usual variogram (Chapter 3), creating bins and computing averages of cross-products of differentials within the bins. Similar words apply to the empirical cross-covariance.

implies that we have a different range for each r but a common range for all components given r.

The representation in terms of independent processes can now be given in terms of the $p \times 1$ vector process $\mathbf{w}(\mathbf{s}) = [w_l(\mathbf{s})]_{l=1}^p$, formed by collecting the $w_l(\mathbf{s})$'s into a column for $l = 1, \ldots, p$. We write the above linear specification as $\mathbf{w}(\mathbf{s}) = \sum_{r=1}^m A_r \mathbf{w}_r(\mathbf{s})$, where each A_r is a $p \times p$ matrix with (l, j)-th element $a_{rj}^{(l)}$ and $\mathbf{w}_r(\mathbf{s}) = (w_{r1}(\mathbf{s}), \ldots, w_{rp}(\mathbf{s}))^{\mathsf{T}}$ are $p \times 1$ vectors that are independent replicates from a spatial process with a correlation function $\rho_r(\mathbf{h})$ and variogram $\gamma_r(\mathbf{h})$ for $r = 1, 2, .., p$.

Letting $C_r(\mathbf{h})$ be the $p \times p$ cross covariance matrix and $\Gamma_r(\mathbf{h})$ denote the $p \times p$ matrix of direct and cross variograms associated with $\mathbf{w}(\mathbf{s})$, we have $C_r(\mathbf{h}) = \rho_r(\mathbf{h})T_r$ and $\Gamma_r(\mathbf{h}) = \gamma_r(\mathbf{h})T_r$. Here, T_r is positive definite with $T_r = A_r A_r^{\mathsf{T}} = \sum_{j=1}^p \mathbf{a}_{rj} \mathbf{a}_{rj}^{\mathsf{T}}$, where $\mathbf{a}_{rj}$ is the j-th column vector of A_r. Finally, the cross-covariance and cross variogram nested model representations take the form $C(\mathbf{h}) = \sum_{r=1}^m \rho_r(\mathbf{h})T_r$ and $\Gamma(\mathbf{h}) = \sum_{r=1}^m \gamma_r(\mathbf{h})T_r$.

The case $m = 1$ is called the intrinsic correlation model, the case $m > 1$ is called the intrinsic multivariate correlation model. Vargas-Guzmán et al. [2002] allow the $w_{rj}(\mathbf{s})$ to be dependent. Such modeling is natural when scaling is the issue, i.e., we want to introduce spatial effects to capture dependence at different scales (and, thus, m has nothing to do with p). When we have prior knowledge about these scales, such modeling will be successful. However, to find datasets that inform about such scaling may be challenging.

10.2 Some Theory for Cross Covariance Functions

Using the generic notation $\mathbf{Y}(\mathbf{s})$ to denote a $p \times 1$ vector of random variables at location $\mathbf{s}$, we seek flexible, interpretable, and computationally tractable models to describe the process $\{\mathbf{Y}(\mathbf{s}) : \mathbf{s} \in D\}$. As in the univariate setting, a well-defined multivariate process must ensure that for every finite set of locations $\mathcal{S} = \{\mathbf{s}_1, \mathbf{s}_2, \ldots, \mathbf{s}_n\}$, the vector $\mathbf{Y} = (\mathbf{Y}(\mathbf{s}_1)^{\mathsf{T}}, \mathbf{Y}(\mathbf{s}_2)^{\mathsf{T}}, \ldots, \mathbf{Y}(\mathbf{s}_n)^{\mathsf{T}})^{\mathsf{T}}$ has a valid joint distribution.

The key ingredient to supply a well-defined multivariate spatial process is the *cross-covariance* function associated with $\mathbf{Y}(\mathbf{s})$. Given its importance in multivariate spatial modeling, we provide some formal theories regarding the validity and properties of these functions. Let $\mathcal{D} \subset \Re^d$ be a connected subset of the d-dimensional Euclidean space and let $\mathbf{s} \in \mathcal{D}$ represent a generic point in $\mathcal{D}$. Consider a vector-valued spatial process $\{\mathbf{w}(\mathbf{s}) \in \Re^p : \mathbf{s} \in \mathcal{D}\}$, where $\mathbf{w}(\mathbf{s})$ is $p \times 1$ with components $w_j(\mathbf{s})$. For convenience, assume that $\mathrm{E}[\mathbf{w}(\mathbf{s})] = \mathbf{0}$.

The *cross-covariance function* is a matrix-valued function, say $\mathbf{C}(\mathbf{s}, \mathbf{s}')$, defined for any pair of locations $(\mathbf{s}, \mathbf{s}') \in \mathcal{D} \times \mathcal{D}$ and yielding the $p \times p$ matrix whose (i, j)-th element is the cross-covariance function

$$C_{ij}(\mathbf{s}, \mathbf{s}') = \mathrm{Cov}(w_i(\mathbf{s}), w_j(\mathbf{s}')) = \mathrm{E}[w_i(\mathbf{s})w_j(\mathbf{s}')] . \tag{10.5}$$

Note that the cross-covariance function in (10.2) arises as a special case of (10.5) when each $w_i(\mathbf{s}) = Y_i(\mathbf{s}) - \mu_i$ and $C_{ij}(\mathbf{s}, \mathbf{s}')$ is a function purely of $\mathbf{h} = \mathbf{s} - \mathbf{s}'$. Using more compact notation, as is customary in multivariate statistics, we write the cross-covariance matrix as

$$C(\mathbf{s}, \mathbf{s}') = \mathrm{Cov}(\mathbf{w}(\mathbf{s}), \mathbf{w}(\mathbf{s}')) = \mathrm{E}[\mathbf{w}(\mathbf{s})\mathbf{w}^{\mathsf{T}}(\mathbf{s}')] . \tag{10.6}$$

For example, if $\mathbf{w}(\mathbf{s})$ is a bivariate process with components $w_1(\mathbf{s})$ and $w_2(\mathbf{s})$, then the cross-covariance matrix for $\mathbf{w}(\mathbf{s})$ is 2×2:

$$C(\mathbf{s}, \mathbf{s}') = \begin{pmatrix} \mathrm{Cov}(w_1(\mathbf{s}), w_1(\mathbf{s}')) & \mathrm{Cov}(w_1(\mathbf{s}), w_2(\mathbf{s}')) \\ \mathrm{Cov}(w_2(\mathbf{s}), w_1(\mathbf{s}')) & \mathrm{Cov}(w_2(\mathbf{s}), w_2(\mathbf{s}')) \end{pmatrix} .$$

In general, a cross-covariance matrix $C(\mathbf{s}, \mathbf{s}')$ is not required to be symmetric (hence, positive definite) since $\mathrm{Cov}(w_i(\mathbf{s}), w_j(\mathbf{s}'))$ is not necessarily equal to $\mathrm{Cov}(w_j(\mathbf{s}), w_i(\mathbf{s}'))$.

Neither is it required that $C(\mathbf{s}, \mathbf{s}') = C(\mathbf{s}', \mathbf{s})$. However, since $\text{Cov}(w_i(\mathbf{s}), w_j(\mathbf{s}')) = \text{Cov}(w_j(\mathbf{s}'), w_i(\mathbf{s}))$, the cross-covariance matrix must satisfy

$$C(\mathbf{s}, \mathbf{s}') = C(\mathbf{s}', \mathbf{s})^{\mathrm{T}} \tag{10.7}$$

In other words, the cross-covariance matrix evaluated at $(\mathbf{s}, \mathbf{s}')$ is the transpose of the cross-covariance matrix evaluated at $(\mathbf{s}', \mathbf{s})$. A second condition for $C(\mathbf{s}, \mathbf{s}')$ is necessitated by the fact that for any finite collection of locations $\mathcal{S}$, we must have

$$\text{Var}\left\{\sum_{i=1}^{n} \mathbf{a}_i^{\mathrm{T}} \mathbf{w}(\mathbf{s}_i)\right\} = \sum_{i=1}^{n} \sum_{j=1}^{n} \mathbf{a}_i^{\mathrm{T}} C(\mathbf{s}_i, \mathbf{s}_j) \mathbf{a}_j > 0 \tag{10.8}$$

for every nonzero $\mathbf{a}_i \in \Re^d$.

The cross-covariance matrix completely determines the joint dispersion structure implied by the spatial process. To be precise, for any n and any arbitrary collection of sites $\mathcal{S} = \{\mathbf{s}_1, \dots, \mathbf{s}_n\}$ the $np \times 1$ vector of realizations $\mathbf{w} = (\mathbf{w}(\mathbf{s}_1)^{\mathrm{T}}, \mathbf{w}(\mathbf{s}_2)^{\mathrm{T}}, \dots, \mathbf{w}(\mathbf{s}_n)^{\mathrm{T}})^{\mathrm{T}}$ will have the variance-covariance matrix $\Sigma_{\mathbf{w}}$, which is an $nm \times nm$ block matrix whose (i, j)-th block is precisely the $p \times p$ cross-covariance matrix $\mathbf{C}(\mathbf{s}_i, \mathbf{s}_j)$. The conditions (10.7) and (10.8) ensure that $\Sigma_{\mathbf{w}}$ is symmetric and positive-definite. The conditions (10.7) and (10.8) also imply that $C(\mathbf{s}, \mathbf{s})$ is always symmetric and positive definite. In fact, it is precisely the variance-covariance matrix for the elements of $\mathbf{w}(\mathbf{s})$ within site $\mathbf{s}$. Again, since our primary focus in this text is Gaussian process models (or mixtures of such processes), specification of $C(\mathbf{s}, \mathbf{s}')$ is all we need to provide all finite dimensional distributions.

We say that $\mathbf{w}(\mathbf{s})$ is *stationary* if $C(\mathbf{s}, \mathbf{s}') = C(\mathbf{s}' - \mathbf{s})$, i.e. the cross-covariance function depends only upon the separation of the sites, while we say that the cross-covariance matrix is *isotropic* if $\mathbf{C}(\mathbf{s}, \mathbf{s}') = \mathbf{C}(\|\mathbf{s}' - \mathbf{s}\|)$, i.e. it depends only upon the distance between the sites. Note that for stationary processes we write the cross-covariance matrix as $C(\mathbf{h}) = C(\mathbf{s}, \mathbf{s} + \mathbf{h})$. From (10.7) it is immediate that

$$C(-\mathbf{h}) = C(\mathbf{s} + \mathbf{h}, \mathbf{s}) = C^{\mathrm{T}}(\mathbf{s}, \mathbf{s} + \mathbf{h}) = C^{\mathrm{T}}(\mathbf{h}).$$

Thus, for a stationary process, a symmetric cross-covariance functions is equivalent to having $C(-\mathbf{h}) = C(\mathbf{h})$ (i.e., an even function). For isotropic functions,

$$C(\mathbf{h}) = C(\|\mathbf{h}\|) = C(\| - \mathbf{h}\|) = C(-\mathbf{h}) = C^{\mathrm{T}}(\mathbf{h}),$$

hence the cross-covariance function is even and the matrix is necessarily symmetric.

What more can we say about functions that will satisfy (10.7) and (10.8)? The primary characterization theorem for cross-covariance functions [Cramér, 1940, Yaglom, 1987] says that real-valued functions, say $C_{ij}(\mathbf{h})$, will form the elements of a valid cross-covariance matrix $C(\mathbf{h})$ if and only if each $C_{ij}(\mathbf{h})$ has the cross-spectral representation

$$C_{ij}(\mathbf{h}) = \int \exp(2\pi i \mathbf{t}^{\mathrm{T}} \mathbf{h}) d(F_{ij}(\mathbf{t})), \quad \text{where} \quad i = \sqrt{-1}, \tag{10.9}$$

with respect to a positive definite measure $F(\cdot)$, i.e., where the cross-spectral matrix $M(B) = [F_{ij}(B)]_{i,j=1}^{p}$ is positive definite for any Borel subset $B \subseteq \Re^d$. The representation in (10.9) can be considered the most general representation theorem for cross-covariance functions. It is the analogue of Bochner's Theorem for covariance functions and has been employed by several authors to construct classes of valid cross-covariance functions.

Essentially, one requires a choice of the $F_{ij}(\mathbf{t})$'s. Matters simplify when $F_{ij}(\mathbf{t})$ is assumed to be square-integrable ensuring that a spectral density function $f_{ij}(\mathbf{t})$ exists such that $d(F_{ij}(\mathbf{t})) = f_{ij}(\mathbf{t}) d\mathbf{t}$. Now, one simply needs to ensure that the $p \times p$ matrix constructed with $f_{ij}(\mathbf{t})$ as its (i,j)-th entry is positive definite for all $\mathbf{t} \in \Re^d$. Corollaries of the above representation lead to the approaches proposed in Gaspari and Cohn [1999] and in Majumdar and Gelfand [2007] for constructing valid cross-covariance functions as convolutions of covariance functions of stationary random fields (see Section 10.8 later). For isotropic settings we use the notation $||\mathbf{s}' - \mathbf{s}||$ for the distance between sites $\mathbf{s}$ and $\mathbf{s}'$. The representation in (10.9) can be viewed more broadly in the sense that, working in the complex plane, if the matrix valued measure $M(\cdot)$ is Hermitian non negative definite, then we obtain a valid cross-covariance matrix in the complex plane. Rehman and Shapiro [1996] use this broader definition to obtain permissible cross variograms. Grzebyk and Wackernagel [1994] employ the induced complex covariance function to create a bilinear model of coregionalization.

As in the univariate case, it is evident that not every matrix $C(\mathbf{s}, \mathbf{s}')$ which we might propose will be *valid*. Consistent with our objective of using multivariate spatial process models in an applied context, we prefer constructive approaches for such cross-covariance functions. The next three sections describe approaches based upon separability, coregionalization, moving averages, and convolution. Nonstationarity can be introduced following the univariate approaches in Section 3.2; however, no details are presented here [see, e.g., Gelfand et al., 2004b, Sec. 4].

10.3   Separable models

Perhaps the most obvious specification of a valid cross-covariance function for a p-dimensional $\mathbf{Y}(\mathbf{s})$ is to let ρ be a valid correlation function for a univariate spatial process, let T be a $p \times p$ positive definite matrix, and let

$$C(\mathbf{s}, \mathbf{s}') = \rho(\mathbf{s}, \mathbf{s}') \cdot T \ . \tag{10.10}$$

In (10.10), $T \equiv (T_{ij})$ is interpreted as the covariance matrix associated with $\mathbf{Y}(\mathbf{s})$, and ρ attenuates association as $\mathbf{s}$ and $\mathbf{s}'$ become farther apart. The covariance matrix for $\mathbf{Y}$ resulting from (10.10) is easily shown to be

$$\Sigma_{\mathbf{Y}} = H \otimes T \ , \tag{10.11}$$

where $(H)_{ij} = \rho(\mathbf{s}_i, \mathbf{s}_j)$ and $\otimes$ denotes the Kronecker product. $\Sigma_{\mathbf{Y}}$ is evidently positive definite since H and T are. In fact, $\Sigma_{\mathbf{Y}}$ is convenient to work with since $|\Sigma_{\mathbf{Y}}| = |H|^p |T|^n$ and $\Sigma_{\mathbf{Y}}^{-1} = H^{-1} \otimes T^{-1}$. This means that updating $\Sigma_{\mathbf{Y}}$ requires working with a $p \times p$ and an $n \times n$ matrix, rather than one that is $np \times np$. Moreover, if we permute the rows of $\mathbf{Y}$ to $\widetilde{\mathbf{Y}}$ where $\widetilde{\mathbf{Y}}^{\mathrm{T}} = (Y_1(\mathbf{s}_1), \ldots, Y_1(\mathbf{s}_n), Y_2(\mathbf{s}_1), \ldots, Y_2(\mathbf{s}_n), \ldots, Y_p(\mathbf{s}_1), \ldots, Y_p(\mathbf{s}_n))$, then $\Sigma_{\widetilde{\mathbf{Y}}} = T \otimes H$.

In fact, working in the fully Bayesian setting, additional advantages accrue to (10.10). With $\boldsymbol{\phi}$ and T *a priori* independent and an inverse Wishart prior for T, the full conditional distribution for T, that is, $p(T|\mathbf{W}, \boldsymbol{\phi})$, is again an inverse Wishart [e.g., Banerjee et al., 2000]. If the Bayesian model is to be fitted using a Gibbs sampler, updating T requires a draw of a $p \times p$ matrix from a Wishart distribution, substantially faster than updating the $np \times np$ matrix $\Sigma_{\mathbf{Y}}$.

What limitations are associated with (10.10)? Clearly $C(\mathbf{s}, \mathbf{s}')$ is symmetric, i.e., $\mathrm{Cov}(Y_\ell(\mathbf{s}_i), Y_{\ell'}(\mathbf{s}_{i'})) = \mathrm{Cov}(Y_{\ell'}(\mathbf{s}_i), Y_\ell(\mathbf{s}_{i'}))$ for all i, i', ℓ, and ℓ'. Moreover, it is easy to check that if ρ is stationary, the *generalized* correlation, also referred to as the *coherence* in the time series literature [see, e.g., Wei, 2005] is such that

$$\frac{\mathrm{Cov}(Y_\ell(\mathbf{s}), Y_{\ell'}(\mathbf{s} + \mathbf{h}))}{\sqrt{\mathrm{Cov}(Y_\ell(\mathbf{s}), Y_\ell(\mathbf{s} + \mathbf{h}))\mathrm{Cov}(Y_{\ell'}(\mathbf{s}), Y_{\ell'}(\mathbf{s} + \mathbf{h}))}} = \frac{T_{\ell\ell'}}{\sqrt{T_{\ell\ell} T_{\ell'\ell'}}} \ , \tag{10.12}$$

regardless of $\mathbf{s}$ and $\mathbf{h}$. Also, if ρ is isotropic and strictly decreasing, then the spatial range (see Section 2.1.3) is identical for each component of $\mathbf{Y}(\mathbf{s})$. This must be the case since only one correlation function is introduced in (10.10). This seems the most unsatisfying restriction, since if, e.g., $\mathbf{Y}(\mathbf{s})$ is a vector of levels of different pollutants at $\mathbf{s}$, then why should the range for all pollutants be the same? In any event, some preliminary marginal examination of the $Y_\ell(\mathbf{s}_i)$ for each ℓ, $\ell = 1, \ldots, p$, might help to clarify the feasibility of a common range.

Additionally, (10.10) implies that, for each component of $\mathbf{Y}(\mathbf{s})$, correlation between measurements tends to 1 as the distance between measurements tends to 0. For some variables, including those in our illustration, such an assumption is appropriate. For others it may not be, in which case microscale variability (captured through a nugget) is a possible solution. Formally, suppose independent $\boldsymbol{\epsilon}(\mathbf{s}) \sim N\left(\mathbf{0}, Diag(\boldsymbol{\tau}^2)\right)$, where $Diag(\boldsymbol{\tau}^2)$ is a $p \times p$ diagonal matrix with (i,i) entry τ_i^2, are included in the modeling. That is, we write $\mathbf{Y}(\mathbf{s}) = \mathbf{V}(\mathbf{s}) + \boldsymbol{\epsilon}(\mathbf{s})$ where $\mathbf{V}(\mathbf{s})$ has the covariance structure in (10.10). An increased computational burden results, since the full conditional distribution for T is no longer an inverse Wishart, and likelihood evaluation requires working with an $np \times np$ matrix.

In a sequence of papers by Le and Zidek and colleagues, it was proposed that $\Sigma_\mathbf{Y}$ be taken as a random covariance matrix drawn from an inverse Wishart distribution centered around (10.11). In other words, an extra hierarchical level is added to the modeling for $\mathbf{Y}$. In this fashion, we are not specifying a spatial process for $\mathbf{Y}(\mathbf{s})$; rather, we are creating a joint distribution for $\mathbf{Y}$ with a flexible covariance matrix. Indeed, the resulting $\Sigma_\mathbf{Y}$ will be nonstationary. In fact, the entries will have no connection to the respective $\mathbf{s}_i$ and $\mathbf{s}_j$. This may be unsatisfactory since we expect to obtain many inconsistencies with regard to distance between points and corresponding association across components. We may be able to obtain the posterior distribution of, say, $Corr(Y_\ell(\mathbf{s}_i), Y_\ell(\mathbf{s}_j))$, but there will be no notion of a range.

The form in (10.10) was presented in Mardia and Goodall [1993] who used it in conjunction with maximum likelihood estimation. Banerjee and Gelfand [2002] discuss its implementation in a fully Bayesian context, as we outline in the next subsection.

10.4   Conjugate Bayesian multivariate spatial regression

We briefly show how the separable covariance structure mentioned above can be exploited to build a conjugate multivariate spatial regression model with analytically tractable posterior distributions. This development is similar in spirit to Section 6.2.3, where we discussed exact Bayesian inference with closed-form posterior distributions for univariate spatial data.,

Let $\mathcal{S} = \{\mathbf{s}_1, \mathbf{s}_2, \ldots, \mathbf{s}_n\}$ be a finite set of n locations in our spatial domain $\mathcal{D}$ and let $y_j(\mathbf{s}_i)$ be the value of our measurement on variable j at location $\mathbf{s}_i$, where there are q different variables indexed by $j = 1, 2, \ldots, q$ each of which yields n measurements indexed by $i = 1, 2, \ldots, n$. Let $\mathbf{x}(\mathbf{s}_i) = (x_1(\mathbf{s}_i), \ldots, x_p(\mathbf{s}_i))^\mathrm{T}$ be a $p \times 1$ vector of predictors whose values are assumed to be known and fixed for each $i = 1, 2, \ldots, n$. A general multivariate regression model posits

$$y_j(\mathbf{s}_i) = \mathbf{x}(\mathbf{s}_i)^\mathrm{T}\boldsymbol{\beta}_j + \epsilon_j(\mathbf{s}_i), \quad j = 1, \ldots, q; \ i = 1, 2, \ldots, n, \tag{10.13}$$

where $\boldsymbol{\beta}_j$ is the $p \times 1$ slope vector capturing the impact of $\mathbf{x}(\mathbf{s}_i)$ on $y_j(\mathbf{s}_i)$, and $\epsilon_j(\mathbf{s}_i)$ is the measurement or residual error corresponding to $y_j(\mathbf{s}_i)$. Thus, (10.13) appears as a collection of univariate regression models for $j = 1, \ldots, q$ variables.

Multivariate regression enriches (10.13) by introducing dependencies on the $\epsilon_j(\mathbf{s}_i)$'s so that associations among the q variables and among the measurements can be captured. In order to achieve such dependence, it is convenient to write (10.13) as a matrix-variate regression model

$$Y = XB + E \quad E \sim MN(O, H, \Sigma), \tag{10.14}$$

where $Y = (y_j(\mathbf{s}_i))$ is the $n \times q$ matrix of measurements $y_j(\mathbf{s}_i)$, $X = (\mathbf{x}_1, \ldots, \mathbf{x}_n)^{\mathsf{T}}$ is $n \times p$ with i-th row $\mathbf{x}(\mathbf{s}_i)^{\mathsf{T}}$, and $E = (\epsilon_j(\mathbf{s}_i))$ is the $n \times q$ matrix of errors assumed to follow the matrix-variate normal probability law

$$p(E \mid H, \Sigma) = MN(E \mid O, H, \Sigma) = \frac{\exp\left[-\frac{1}{2}\mathrm{Tr}(E^{\mathsf{T}}H^{-1}E\Sigma^{-1})\right]}{(2\pi)^{nq/2}(\det(H))^{q/2}(\det(\Sigma))^{n/2}}, \tag{10.15}$$

where $Tr(\cdot)$ is the trace function of a matrix, and H and Σ are $n \times n$ and $q \times q$ positive definite matrices modeling the associations among the rows (measurements) and columns (variables) of E, respectively. If the errors are independent across the n measurements then $H = I_n$, the $n \times n$ identity matrix. If the q variables are independent of each other, hence (10.13) represents q independent univariate regression models, then $\Sigma = I_q$. For the purposes of this section, we will assume that H is a fixed positive definite matrix and parameters to be estimated are $\{B, \Sigma\}$.

A Bayesian multivariate regression model is built using the conjugate Matrix-Normal Inverse-Wishart (MNIW) distribution on $\{B, \Sigma\}$,

$$p(B, \Sigma \mid C, V_\beta, \nu, S) = IW(\Sigma \mid \nu, S) \times MN(B \mid C, V_\beta, \Sigma)$$

$$\propto \frac{\exp\left(-\frac{1}{2}\mathrm{Tr}(S\Sigma^{-1})\right)}{(\det(\Sigma))^{\frac{\nu+q+1}{2}}} \times \frac{\exp\left[-\frac{1}{2}\mathrm{Tr}((B-C)^{\mathsf{T}}V_\beta^{-1}(B-C)\Sigma^{-1})\right]}{(2\pi)^{pq/2}(\det(V_\beta))^{q/2}(\det(\Sigma))^{p/2}}, \tag{10.16}$$

which is the hierarchical specification $\Sigma \sim IW(\nu, S)$ and $B \mid \Sigma \sim MN(C, V_\beta, \Sigma)$, where C is the prior mean of B and V_β is the prior variance-covariance matrix for the rows of B. The prior incorporates the same Σ as in the likelihood of (10.14) to ensure conjugacy.

The posterior distribution of $\{B, \Sigma\} \mid Y$ is an MNIW distribution,

$$p(B, \Sigma \mid Y) = \underbrace{IW(\Sigma \mid \nu + n, S_*)}_{p(\Sigma \mid Y)} \times \underbrace{MN(B \mid \Omega M, \Omega, \Sigma)}_{p(B \mid \Sigma, Y)}, \tag{10.17}$$

where $\Omega^{-1} = V_\beta^{-1} + X^{\mathsf{T}}H^{-1}X$, $M = V_\beta^{-1}C + X^{\mathsf{T}}H^{-1}Y$ and $S_* = S + C^{\mathsf{T}}V_\beta^{-1}C + Y^{\mathsf{T}}H^{-1}Y - M^{\mathsf{T}}\Omega M$. Bayesian inference proceeds by sampling directly from the posterior distribution $p(B, \Sigma \mid Y)$. Each draw of $\Sigma \sim p(\Sigma \mid Y) = IW(\nu, S_*)$ is followed by one draw of $B \sim p(B \mid \Sigma, Y) = MN(B \mid \Omega M, \Omega, \Sigma)$.

Posterior predictive inference is also available in closed form. Let $\tilde{\mathcal{S}} = \{\tilde{\mathbf{s}}_1, \tilde{\mathbf{s}}_2, \ldots, \tilde{\mathbf{s}}_{n'}\}$ be a set of n' spatial locations such that $\tilde{Y} = (Y_j(\tilde{\mathbf{s}}_i))$ is an $n' \times q$ matrix of unobserved values of the j outcomes that we wish to predict using a given $n' \times p$ matrix of predictors $\tilde{X} = (x_j(\tilde{\mathbf{s}}_i))$, where $x_j(\tilde{\mathbf{s}}_i)$ is the value of the j-th predictor at location $\tilde{\mathbf{s}}_i$. We extend (10.14) to include the predictive model so that

$$\begin{bmatrix} Y \\ \tilde{Y} \end{bmatrix} = \begin{bmatrix} X \\ \tilde{X} \end{bmatrix} B + \begin{bmatrix} E \\ \tilde{E} \end{bmatrix}, \quad \begin{bmatrix} E \\ \tilde{E} \end{bmatrix} \sim MN\left(\begin{bmatrix} O_{n \times q} \\ \tilde{O}_{n' \times q} \end{bmatrix}, \begin{bmatrix} H & J \\ J^{\mathsf{T}} & \tilde{H} \end{bmatrix}, \Sigma\right). \tag{10.18}$$

The $(n + n') \times (n + n')$ spatial correlation matrix $\begin{bmatrix} H & J \\ J^{\mathsf{T}} & \tilde{H} \end{bmatrix}$, which models joint associations among the rows of Y and $\tilde{Y}$, is constructed over the $n + n'$ locations in $\mathcal{S} \cup \tilde{\mathcal{S}}$. An easy specification is $H = R_\theta + \frac{1-\alpha}{\alpha}I_n$, where $R_\theta = (\rho_\theta(\mathbf{s}_i, \mathbf{s}_j))$, $\rho_\theta(\mathbf{s}, \mathbf{s}')$ is any valid spatial correlation function, and $\alpha = \frac{\sigma^2}{\sigma^2+\tau^2}$, where σ^2 and τ^2 are two variance components corresponding to the spatial process and measurement error or random noise, respectively. If the spatial variance is much larger than the noise variance, i.e., $\tau^2/\sigma^2 \approx 0$, then $\alpha \approx 1$ and $H \approx R_\theta$. At the other extreme, if $\sigma^2/\tau^2 \approx 0$ then $\alpha \approx 0$ in which case the weight on I_n becomes much larger than on R_θ. This situation, however, is less relevant since spatial modeling is pursued if exploratory data analysis indicates that $\sigma^2/\tau^2 > 0$ so that the spatial correlation function as well as the variation due to random noise contribute to H. Thus, $\tilde{H} = \tilde{R}_\theta + \frac{1-\alpha}{\alpha}I_{n'}$, where $\tilde{R}_\theta = (\rho_\theta(\tilde{\mathbf{s}}_i, \tilde{\mathbf{s}}_j))$ for $i, j = 1, 2, \ldots, n'$, and $J = (\rho_\theta(\mathbf{s}_i, \tilde{\mathbf{s}}_j))$ for $i = 1, 2, \ldots, n$ and $j = 1, 2, \ldots, n'$. Any valid spatial correlation function $\rho_\theta(\mathbf{s}, \mathbf{s}')$ will ensure the positive definiteness of $\begin{bmatrix} H & J \\ J^{\mathrm{T}} & \tilde{H} \end{bmatrix}$.

Posterior predictive inference uses the posterior samples of $\{B, \Sigma\}$ drawn from (10.17). For each drawn value of B and Σ, we draw one instance of $\tilde{Y}$ from the posterior conditional predictive density

$$p(\tilde{Y} \mid Y, B, \Sigma) = MN(\tilde{Y} \mid \tilde{X}B + J^{\mathrm{T}}H^{-1}(Y - XB), \tilde{H} - J^{\mathrm{T}}H^{-1}J, \Sigma). \qquad (10.19)$$

Repeating this for all the posterior samples of $\{B, \Sigma\}$ produces the desired samples of $\tilde{Y}$ from the posterior predictive distribution $p(\tilde{Y} \mid Y)$. No iterative algorithms are required and one only needs to sample from matrix-normal and inverse-Wishart distributions, which are standard distributions offered in statistical computing packages.

10.5   Spatial prediction, interpolation, and regression

Multivariate spatial process modeling is required when we are analyzing several point-referenced data layers, when we seek to explain or predict for one layer given the others, or when the layers are not all collected at the same locations. The last of these is a type of spatial misalignment that can also be viewed as a missing data problem, in the sense that we are missing observations to completely align all of the data layers. For instance, in monitoring pollution levels, we may observe some pollutants at one set of monitoring sites, and other pollutants at a different set of sites. Alternatively, we might have data on temperature, elevation, and wind speed, but all at different locations.

More formally, suppose we have a conceptual response $Z(\mathbf{s})$ along with a conceptual vector of covariates $\mathbf{x}(\mathbf{s})$ at each location $\mathbf{s}$. However, in the sampling, the response and the covariates are observed at possibly different locations. To set some notation, let us partition our set of sites into three mutually disjoint groups: let S_Z be the sites where only the response $Z(\mathbf{s})$ has been observed, S_X the set of sites where only the covariates have been observed, S_{ZX} the set where both $Z(\mathbf{s})$ and the covariates have been observed, and finally S_U the set of sites where no observations have been taken.

In this context we can formalize three types of inference questions. One concerns $Y(\mathbf{s})$ when $\mathbf{s} \in S_X$, which we call *interpolation*. The second concerns $Y(\mathbf{s})$ for $\mathbf{s}$ belonging to S_U, which we call *prediction*. Evidently, prediction and interpolation are similar but interval estimates will be at least as tight for the latter compared with the former. The last concerns the functional relationship between $X(\mathbf{s})$ and $Y(\mathbf{s})$ at an arbitrary site $\mathbf{s}$, along with other covariate information at $\mathbf{s}$, say $\mathbf{U}(\mathbf{s})$. We capture this through $E[Y(\mathbf{s})|X(\mathbf{s}), \mathbf{U}(\mathbf{s})]$, and refer to it as *spatial regression*. Figure 10.1 offers a graphical clarification of the foregoing definitions.

In the usual stochastic regressors setting one is interested in the relationship between $Y(\mathbf{s})$ and $X(\mathbf{s})$ where the pairs $(X(\mathbf{s}_i), Y(\mathbf{s}_i))$, $i = 1, \ldots, n$ (suppressing $\mathbf{U}(\mathbf{s}_i)$) are independent. For us, they are dependent with the dependence captured through a spatial characterization. Still, one may be interested in the regression of $Y(\mathbf{s})$ on $X(\mathbf{s})$ at an arbitrary $\mathbf{s}$. Note that there is no notion of a conditional spatial process, $Y(\mathbf{s}) \mid X(\mathbf{s})$, associated with the bivariate spatial process $(X(\mathbf{s}), Y(\mathbf{s}))$; how would one define the joint distribution of $Y(\mathbf{s}_i) \mid X(\mathbf{s}_i)$ and $Y(\mathbf{s}_{i'}) \mid X(\mathbf{s}_{i'})$?

We also note that our modeling structure here differs considerably from that of Diggle et al. [1998]. These authors specify a univariate spatial process in order to introduce unobserved spatial effects (say, V(**s**)) into the modeling, after which the $Y(\mathbf{s})$'s are conditionally

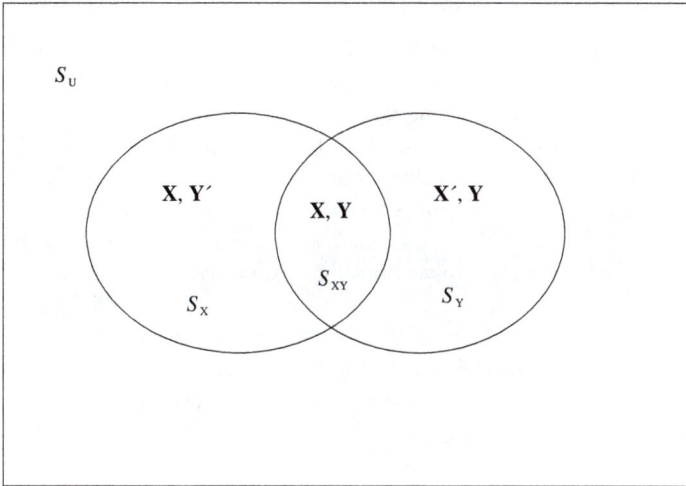

Figure 10.1 *A graphical representation of the S sets. Interpolation applies to locations in S_X, prediction applies to locations in S_U, and regression applies to all locations.* $\mathbf{X}_{aug} = (\mathbf{X}, \mathbf{X}')$, $\mathbf{Y}_{aug} = (\mathbf{Y}, \mathbf{Y}')$.

independent given the $V(\mathbf{s})$'s. In other words, the $V(\mathbf{s})$'s are intended to capture spatial association in the means of the $Y(\mathbf{s})$'s. For us, the $X(\mathbf{s})$'s are also modeled through a spatial process, but they are observed and introduced as an explanatory variable with a regression coefficient. Hence, along with the $Y(\mathbf{s})$'s, we require a bivariate spatial process.

Here, we provide a full Bayesian examination of the foregoing questions. In Subsection 10.5.1 we study the case where $Y(\mathbf{s})$ is Gaussian, but in Subsection 10.5.3 we allow the response to be binary.

The Gaussian interpolation problem is addressed from an empirical Bayes perspective in a series of papers by Zidek and coworkers. For instance, Le and Zidek [1992] and Brown et al. [1994] develop a Bayesian interpolation theory (both spatial and temporal) for multivariate random spatial data. Le et al. [1997] extend this methodology to account for misalignment, i.e., where possibly not all monitored sites measured the same set of pollutants (data missing by design). Their method produces the joint predictive distribution for several locations and different time points using all available data, thus allowing for simultaneous temporal and spatial interpolation without assuming the random field to be stationary. Their approach provides a first-stage multivariate normal distribution for the observed data. However, this distribution does not arise from a spatial Gaussian process.

Framing multivariate spatial prediction (often referred to as *cokriging*) in the context of linear regression dates at least to Corsten [1989] and Stein and Corsten [1991]. In this work, the objective is to carry out predictions for a possible future observation. Stein and Corsten [1991] advocate looking at the prediction problem under a regression setup. They propose trend surface modeling of the point source response using polynomials in the coordinates. Typically in trend surface analysis [Cressie, 1993], spatial structure is modeled through the mean but observations are assumed to be independent. Instead, Stein and Corsten [1991] retain familiar spatial dependence structure but assume the resultant covariances and cross-covariances (and hence the dispersion matrix) are known. In this context, Stein et al. [1991] use restricted maximum likelihood to estimate unknown spatial dependence structure parameters.

10.5.1   Regression in the Gaussian case

The regression problem posed here is solely to learn about the conditional distribution for $Y(\mathbf{s}_0) \mid X(\mathbf{s}_0)$. We are not interested in kriging. In this regard, the reader might ask why we do not proceed as in Chapter 6, specifying a univariate Gaussian process model for $Y(\mathbf{s})$, our usual geostatistical model; why do we consider a bivariate process model here? Two reasons are as follows. First, we might be interested in an inverse regression problem, to learn about $X(\mathbf{s}_0)$ for a given $Y(\mathbf{s}_0)$. To do so, requires modeling $X(\mathbf{s})$ to be random, hence introducing a process model for the X's as well as for the Y's, a bivariate process model. Second, and perhaps more importantly, we may have missingness, equivalently, misalignment. That is, we have some X's without associated Y's and vice versa. We would like to use all of the data rather than just the *complete* cases. We need a model for the missing data we need a bivariate model in order to do model based imputation (see, e.g., Little and Rubin [2002]).

So, to start, assume a single covariate with no misalignment, and let $\mathbf{X} = (X(\mathbf{s}_1), \dots, X(\mathbf{s}_n))^{\mathrm{T}}$ and $\mathbf{Y} = (Y(\mathbf{s}_1), \dots, Y(\mathbf{s}_n))^{\mathrm{T}}$. be the measurements on the covariates and the response, respectively. Supposing that $X(\mathbf{s})$ is continuous and that is it meaningful to model it in a spatial fashion, our approach is to envision (perhaps after a suitable transformation) a bivariate Gaussian spatial process, $\mathbf{W}(\mathbf{s}) = (X(\mathbf{s}), Y(\mathbf{s}))^{\mathrm{T}}$ with mean $\boldsymbol{\mu}(\mathbf{s}) = (\mu_X(\mathbf{s}), \mu_Y(\mathbf{s}))^{\mathrm{T}}$ and a separable cross-covariance function as in (10.10). Since $\rho(\mathbf{s}, \mathbf{s}) \equiv 1$, this specification implies that the joint distribution of $Y(\mathbf{s})$ and $X(\mathbf{s})$ at any location $\mathbf{s}$ is

$$\mathbf{W}(\mathbf{s}) = \left(\begin{array}{c} X(\mathbf{s}) \\ Y(\mathbf{s}) \end{array} \right) \sim \mathrm{N}\left(\boldsymbol{\mu}(\mathbf{s}), T\right). \tag{10.20}$$

With misalignment, let $\mathbf{X}$ be the vector of observed $X(\mathbf{s})$'s at the sites in $S_{XY} \cup R_X$, while $\mathbf{Y}$ will be the vector of $Y(\mathbf{s})$'s the sites in $S_{XY} \cup S_Y$. If we let $\mathbf{X}'$ denote the vector of missing X observations in S_Y and $\mathbf{Y}'$ the vector of missing Y observations in S_X, then in the preceding discussion we can replace $\mathbf{X}$ and $\mathbf{Y}$ by the augmented vectors $\mathbf{X}_{aug} = (\mathbf{X}, \mathbf{X}')$ and $\mathbf{Y}_{aug} = (\mathbf{Y}, \mathbf{Y}')$; see Figure 10.1 for clarification. After permutation to line up the X's and Y's, they can be collected into a vector $\mathbf{W}_{aug}$. In the Bayesian model specification, $\mathbf{X}'$ and $\mathbf{Y}'$ are viewed as latent (unobserved) vectors. In implementing a Gibbs sampler for model-fitting, we update the model parameters given $\mathbf{X}'$ and $\mathbf{Y}'$ (i.e., given $\mathbf{W}_{aug}$), and then update $(\mathbf{X}', \mathbf{Y}')$ given $\mathbf{X}$, $\mathbf{Y}$, and the model parameters.

The latter updating is routine since the associated full conditional distributions are normal. Such augmentation proves computationally easier with regard to bookkeeping since we retain the convenient Kronecker form for $\Sigma_{\mathbf{W}}$. That is, it is easier to marginalize over $\mathbf{X}'$ and $\mathbf{Y}'$ after simulation than before. For convenience of notation, we suppress the augmentation in the sequel.

In what we have called the prediction problem, it is desired to predict the outcome of the response variable at some unobserved site. Thus we are interested in the posterior predictive distribution $p(y(\mathbf{s}_0)|\mathbf{y}, \mathbf{x})$. We note that $x(\mathbf{s}_0)$ is also not observed here. On the other hand, the interpolation problem may be regarded as a method of imputing missing data. Here the covariate $x(\mathbf{s}_0)$ is observed but the response is "missing." Thus our attention shifts to the posterior predictive distribution For the regression problem, the distribution of interest is $p(E[Y(\mathbf{s}_0) \mid x(\mathbf{s}_0)] \mid x(\mathbf{s}_0), \mathbf{y}, \mathbf{x})$.

For simplicity, suppose $\boldsymbol{\mu}(\mathbf{s}) = (\mu_1, \mu_2)^{\mathrm{T}}$, independent of the site coordinates. (With additional fixed site-level covariates for $Y(\mathbf{s})$, say $\mathbf{U}(\mathbf{s})$, we would replace μ_2 with $\mu_2(\mathbf{s}) = \boldsymbol{\alpha}^{\mathrm{T}} \mathbf{U}(\mathbf{s})$.) Then, from (10.20), for the pair $(X(\mathbf{s}), Y(\mathbf{s}))$, $p(y(\mathbf{s})|x(\mathbf{s}), \beta_0, \beta_1, \sigma^2)$ is $\mathrm{N}\left(\beta_0 + \beta_1 x(\mathbf{s}), \sigma^2\right)$. That is, $E[Y(\mathbf{s}) \mid x(\mathbf{s})] = \beta_0 + \beta_1 x(\mathbf{s})$, where

$$\beta_0 = \mu_2 - \frac{T_{12}}{T_{11}} \mu_1, \ \ \beta_1 = \frac{T_{12}}{T_{11}}, \ \text{and} \ \sigma^2 = T_{22} - \frac{T_{12}^2}{T_{11}}. \tag{10.21}$$

So, given samples from the joint posterior distribution of (μ_1, μ_2, T, ϕ), we directly have samples from the posterior distributions for the parameters in (10.21), and thus from the posterior distribution of $E[Y(\mathbf{s}) \mid x(\mathbf{s})]$.

Rearrangement of the components of $\mathbf{W}$ as below (10.11) yields

$$
\begin{pmatrix} \mathbf{X} \\ \mathbf{Y} \end{pmatrix} \sim N \left(\begin{pmatrix} \mu_1 \mathbf{1} \\ \mu_2 \mathbf{1} \end{pmatrix} , \; T \otimes H(\phi) \right) , \tag{10.22}
$$

which simplifies the calculation of the conditional distribution of $\mathbf{Y}$ given $\mathbf{X}$.

Assuming an inverse Wishart prior for T, completing the Bayesian specification requires a prior for $\mu_1, \mu_2,$ and ϕ. For (μ_1, μ_2), for convenience, we would take a vague but proper bivariate normal prior. A suitable prior for ϕ depends upon the choice of $\rho(h; \phi)$. Then we use a Gibbs sampler to simulate the necessary posterior distributions. The full conditionals for μ_1 and μ_2 are in fact Gaussian distributions, while that of the T matrix is inverted Wishart as already mentioned. The full conditional for the ϕ parameter finds ϕ arising in the entries in H, and so is not available in closed form. Metropolis or slice sampling can be employed for its updating.

Under the above framework, interpolation presents no new problems. Let $\mathbf{s}_0$ be a new site at which we would like to predict the variable of interest. We first modify the $H(\phi)$ matrix forming the new matrix H^* as follows:

$$
H^*(\phi) = \begin{pmatrix} H(\phi) & \mathbf{h}(\phi) \\ \mathbf{h}(\phi)^{\mathrm{T}} & \rho(0; \phi) \end{pmatrix} , \tag{10.23}
$$

where $\mathbf{h}(\phi)$ is the vector with components $\rho(\mathbf{s}_0 - \mathbf{s}_j; \phi)$, $j = 1, 2, \ldots, n$. It then follows that

$$
\mathbf{W}^* \equiv (\mathbf{W}(\mathbf{s}_0), \ldots, \mathbf{W}(n))^{\mathrm{T}} \sim N \left(\mathbf{1}_{n+1} \otimes \begin{pmatrix} \mu_1 \\ \mu_2 \end{pmatrix} , \; H^*(\phi) \otimes T \right) . \tag{10.24}
$$

Once again a simple rearrangement of the above vector enables us to arrive at the conditional distribution $p(\mathbf{y}(\mathbf{s}_0)|\mathbf{x}(\mathbf{s}_0), \mathbf{y}, \mathbf{x}, \boldsymbol{\mu}, T, \phi)$ as a Gaussian distribution. The predictive distribution for the interpolation problem, $p(\mathbf{y}(\mathbf{s}_0)|\mathbf{y}, \mathbf{x})$, can now be obtained by marginalizing over the parameters, i.e.,

$$
p(y(\mathbf{s}_0)|\mathbf{y}, \mathbf{x}) = \int p(y(\mathbf{s}_0)|x(\mathbf{s}_0), \mathbf{y}, \mathbf{x}, \boldsymbol{\mu}, T, \phi) \, p(\boldsymbol{\mu}, T, \phi|x(\mathbf{s}_0), \mathbf{y}, \mathbf{x}) . \tag{10.25}
$$

For prediction, we do not have $x(\mathbf{s}_0)$. But this does not create any new problems, as it may be treated as a latent variable and incorporated into $\mathbf{x}'$. This only results in an additional draw within each Gibbs iteration, and is a trivial addition to the computational task.

10.5.2   Avoiding the symmetry of the cross-covariance matrix

In the spirit of Le and Zidek [1992], we can avoid the symmetry in separable cross-covariances noted above (10.12). Instead of directly modeling $\Sigma_{\mathbf{W}} = H(\phi) \otimes T$, we can add a further hierarchical level, by assuming that $\Sigma_{\mathbf{W}} \mid \phi, T$ follows an inverted Wishart distribution with mean $H(\phi) \otimes T$. All other specifications remain as before. Note that the marginal model (i.e., marginalizing over $\Sigma_{\mathbf{W}}$) is no longer Gaussian. However, using standard calculations, the resulting cross-covariance matrix is a function of $\rho(\mathbf{s} - \mathbf{s}'; \phi)$, retaining desirable spatial interpretation. Once again we resort to the Gibbs sampler to arrive at the posteriors, although in this extended model, the number of parameters has increased substantially, since the elements of $\Sigma_{\mathbf{W}}$ are being introduced as new parameters.

The full conditionals for the means μ_1 and μ_2 are still Gaussian and it is easily seen that the full conditional for $\Sigma_\mathbf{W}$ is inverted Wishart. The full conditional distribution for ϕ is now proportional to $p(\Sigma_\mathbf{W} \mid \phi, T)p(\phi)$; a Metropolis step may be employed for its updating. Also, the full conditional for T is no longer an inverse Wishart and a Metropolis step with an inverse Wishart proposal is used to sample the T matrix. All told, this is indeed a much more computationally demanding proposition since we now have to deal with the $2n \times 2n$ matrix $\Sigma_\mathbf{W}$ with regard to sampling, inversion, determinants, and so on.

10.5.3   Regression in a probit model

Now suppose we have binary response from a point-source spatial dataset. At each site, $Z(\mathbf{s})$ equals 0 or 1 according to whether we observed "failure" or "success" at that particular site. Thus, a realization of the process can be partitioned into two disjoint subregions, one for which $Z(\mathbf{s}) = 0$, the other $Z(\mathbf{s}) = 1$, and is called a *binary map* [Oliveira, 2000, 2020]. Again, the process is only observed at a finite number of locations. Along with this binary response we have a set of covariates observed at each site. We follow the latent variable approach for probit modeling as in, e.g., Oliveira [2000]. Let $Y(\mathbf{s})$ be a latent spatial process associated with the sites and let $X(\mathbf{s})$ be a process that generates the values of a particular covariate, in particular, one that is misaligned with $Z(\mathbf{s})$ and is sensible to model in a spatial fashion. For the present we assume $X(\mathbf{s})$ is univariate but extension to the multivariate case is apparent. Let $Z(\mathbf{s}) = 1$ if and only if $Y(\mathbf{s}) > 0$. We envision our bivariate process $\mathbf{W}(\mathbf{s}) = (X(\mathbf{s}), Y(\mathbf{s}))^\mathsf{T}$ distributed as in (10.20), but where now $\boldsymbol{\mu}(\mathbf{s}) = (\mu_1, \mu_2 + \boldsymbol{\alpha}^\mathsf{T}\mathbf{U}(\mathbf{s}))^\mathsf{T}$, with $\mathbf{U}(\mathbf{s})$ regarded as a $p \times 1$ vector of fixed covariates. Note that the conditional variance of $Y(\mathbf{s})$ given $X(\mathbf{s})$ is not identifiable. Thus, without loss of generality, we set $T_{22} = 1$, so that the T matrix has only two parameters.

Now, we formulate a probit regression model as follows:

$$P(Z(\mathbf{s}) = 1 \mid x(\mathbf{s}), \mathbf{U}(\mathbf{s}), \boldsymbol{\alpha}, \mu_1, \mu_2, T_{11}, T_{12})$$
$$= \Phi\left([\beta_0 + \beta_1 X(\mathbf{s}) + \boldsymbol{\alpha}^\mathsf{T}\mathbf{U}(\mathbf{s})] / \sqrt{1 - \frac{T_{12}^2}{T_{11}}} \right). \tag{10.26}$$

Here, as in (10.21), $\beta_0 = \mu_2 - (T_{12}/T_{11})\mu_1$, and $\beta_1 = T_{12}/T_{11}$.

The posterior of interest is $p(\mu_1, \mu_2, \boldsymbol{\alpha}, T_{11}, T_{12}, \phi, \mathbf{y} \mid \mathbf{x}, \mathbf{z})$, where $\mathbf{z} = (z(\mathbf{s}_1), \ldots, z(\mathbf{s}_n))^\mathsf{T}$ is a vector of 0's and 1's. The fitting again uses MCMC. Here, $\mathbf{X} = (X(\mathbf{s}_1), \ldots, X(\mathbf{s}_n))^\mathsf{T}$ and $\mathbf{Y} = (Y(\mathbf{s}_1), \ldots, Y(\mathbf{s}_n))^\mathsf{T}$ as in Subsection 10.5.1, except that $\mathbf{Y}$ is now unobserved, and introduced only for computational convenience. Analogous to (10.22),

$$\begin{pmatrix} \mathbf{X} \\ \mathbf{Y} \end{pmatrix} \sim \mathrm{N}\left(\begin{pmatrix} \mu_1 \mathbf{1} \\ \mu_2 \mathbf{1} + \mathbf{U}\boldsymbol{\beta} \end{pmatrix}, T \otimes H(\phi) \right), \tag{10.27}$$

where $\mathbf{U} = (U(\mathbf{s}_1), \ldots, U(\mathbf{s}_n))^\mathsf{T}$.

From (10.27), the full conditional distribution for each latent $Y(\mathbf{s}_i)$ is a univariate normal truncated to a set of the form $\{Y(\mathbf{s}_i) > 0\}$ or $\{Y(\mathbf{s}_i) < 0\}$. The full conditionals for μ_1 and μ_2 are both univariate normal, while that of $\boldsymbol{\beta}$ is multivariate normal with the appropriate dimension. For the elements of the T matrix, we may simulate first from a Wishart distribution (as mentioned in Subsection 10.5.1) and then proceed to scale it by T_{22}, or we may proceed individually for T_{12} and T_{11} using Metropolis-Hastings over a restricted convex subset of a hypercube [Chib and Greenberg, 1998]. Finally, ϕ can be simulated using a Metropolis step, as in Subsection 10.5.1. Misalignment is also treated as in Subsection 10.5.1, introducing appropriate latent $\mathbf{X}'$ and $\mathbf{Y}'$.

With posterior samples from $p(\mu_1, \mu_2, \boldsymbol{\alpha}, T_{11}, T_{12}, \phi \mid \mathbf{x}, \mathbf{z})$, we immediately obtain samples from the posterior distributions for β_0 and β_1. Also, given $x(\mathbf{s}_0)$, (10.26) shows how to obtain samples from the posterior for a particular probability, such as $p(P(Z(\mathbf{s}_0) =$

$1 \mid x(\mathbf{s}_0), \mathbf{U}(\mathbf{s}_0), \boldsymbol{\alpha}, \mu_1, \mu_2, T_{11}, T_{12} \mid x(\mathbf{s}_0), \mathbf{x}, \mathbf{z}$ at an unobserved site $\mathbf{s}_0$, clarifying the regression structure. Were $x(\mathbf{s}_0)$ not observed, we could still consider the chance that $Z(\mathbf{s}_0)$ equals 1. This probability, $P(Z(\mathbf{s}_0) = 1 \mid \mathbf{U}(\mathbf{s}_0), \boldsymbol{\alpha}, \mu_1, \mu_2, T_{11}, T_{12})$, arises by averaging over $X(\mathbf{s}_0)$, i.e.,

$$\int P(Z(\mathbf{s}_0) = 1 \mid x(\mathbf{s}_0), \mathbf{U}(\mathbf{s}_0), \boldsymbol{\alpha}, \mu_1, \mu_2, T_{11}, T_{12}) \, p(x(\mathbf{s}_0) \mid \mu_1, T_{11}) \, dx(\mathbf{s}_0) . \qquad (10.28)$$

In practice, we would replace the integration in (10.28) by a Monte Carlo integration. Then, plugging into this Monte Carlo integration, the foregoing posterior samples would yield essentially posterior realizations of (10.28).

Both the prediction problem and the interpolation problem may be viewed as examples of *indicator kriging* (e.g., Solow, 1986, Oliveira, 2000). For the prediction case we seek $p(z(\mathbf{s}_0) \mid \mathbf{x}, \mathbf{z})$; realizations from this distribution arise if we can obtain realizations from $p(y(\mathbf{s}_0) \mid \mathbf{x}, \mathbf{z})$. But

$$p(y(\mathbf{s}_0) \mid \mathbf{x}, \mathbf{z}) = \int p(y(\mathbf{s}_0) \mid \mathbf{x}, \mathbf{y}) \, p(\mathbf{y} \mid \mathbf{x}, \mathbf{z}) d \mathbf{y} . \qquad (10.29)$$

Since the first distribution under the integral in (10.29) is a univariate normal, as in Subsection 10.5.1, the posterior samples of $\mathbf{Y}$ immediately provide samples of $Y(\mathbf{s}_0)$. For the interpolation case we seek $p(z(\mathbf{s}_0) \mid x(\mathbf{s}_0), \mathbf{x}, \mathbf{z})$. Again we only need realizations from $p(y(\mathbf{s}_0) \mid x(\mathbf{s}_0), \mathbf{x}, \mathbf{z})$, but

$$p(y(\mathbf{s}_0) \mid x(\mathbf{s}_0), \mathbf{x}, \mathbf{z}) = \int p(y(\mathbf{s}_0) \mid x(\mathbf{s}_0), \mathbf{x}, \mathbf{y}) \, p(\mathbf{y} \mid x(\mathbf{s}_0), \mathbf{x}, \mathbf{z}) d \mathbf{y} . \qquad (10.30)$$

As with (10.29), the first distribution under the integral in (10.30) is a univariate normal.

10.5.4 *Examples*

Example 10.1 *(Gaussian model).* Our examples are based upon an ecological dataset collected over a west-facing watershed in the Negev Desert in Israel. The species under study is called an isopod, and builds its residence by making burrows. Some of these burrows thrive through the span of a generation while others do not. We study the following variables at each of 1129 sites. The variable "dew" measures time in minutes (from 8 a.m.) to evaporation of the morning dew. The variables "shrub" and "rock" density are percentages (the remainder is sand) characterizing the environment around the burrows. In our first example we try to explain shrub density (Y) through dew duration (X). In our second example we try to explain burrow survival (Z) through shrub density, rock density, and dew duration, treating only the last one as random and spatial. We illustrate the Gaussian case for the first example with 694 of the sites offering both measurements, 204 sites providing only the shrub density, and 211 containing only the dew measurements.

The spatial locations are displayed in Figure 10.2 using rescaled planar coordinates after UTM projection. The rectangle in Figure 10.2 is roughly 300 km by 250 km. Hence the vector $\mathbf{X}$ consists of $694 + 211 = 905$ measurements, while the vector $\mathbf{Y}$ consists of $694 + 204 = 898$ measurements. For these examples we take the exponential correlation function, $\rho(h; \phi) = e^{-\phi h}$. We assign a vague inverse gamma specification for the parameter ϕ, namely an IG(2, 1/0.024). This prior has infinite variance and suggests a range $(3/\phi)$ of 125 km, which is roughly half the maximum pairwise distance in our region. We found little inference sensitivity to the mean of this prior. The remaining prior specifications are all rather noninformative, i.e., a $N\left(\mathbf{0}, Diag(10^5, 10^5)\right)$ prior for (μ_1, μ_2) and an $IW\left(2, Diag(0.001, 0.001)\right)$ for T. That is, $E(T_{11}) = E(T_{22}) = 0.001$, $E(T_{12}) = 0$, and the variances of the T_{ij}'s do not exist.

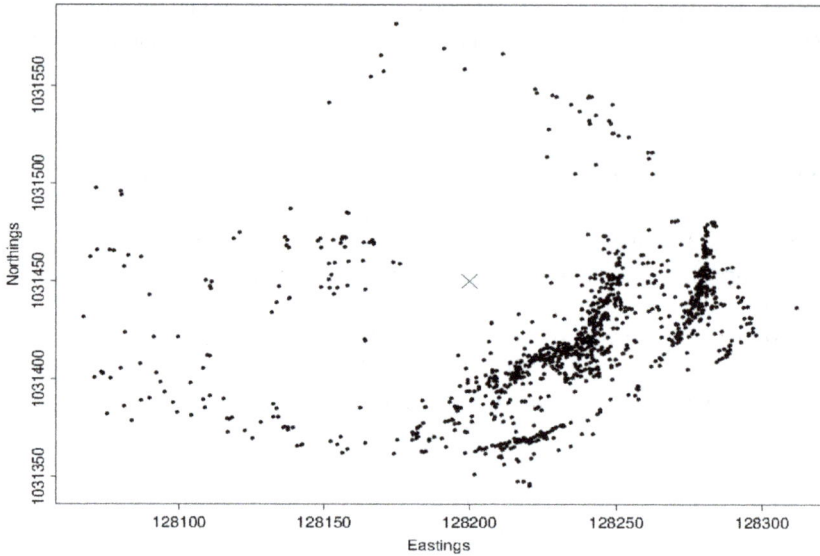

Figure 10.2 *Spatial locations of the isopod burrows data. The axes represent the eastings and the northings on a UTM projection.*

Parameter	Quantiles		
	2.5%	50%	97.5%
μ_1	73.118	73.885	74.665
μ_2	5.203	5.383	5.572
T_{11}	95.095	105.220	117.689
T_{12}	−4.459	−2.418	−0.528
T_{22}	5.564	6.193	6.914
$T_{12}/\sqrt{T_{11}T_{22}}$ (nonspatial corr. coef.)	−0.171	−0.095	−0.021
β_0 (intercept)	5.718	7.078	8.463
β_1 (slope)	−0.041	−0.023	−0.005
σ^2	5.582	6.215	6.931
ϕ	0.0091	0.0301	0.2072

Table 10.1 *Posterior quantiles for the shrub density/dew duration example.*

Table 10.1 provides the 95% credible intervals for the regression parameters and the decay parameter ϕ. The significant negative association between dew duration and shrub density is unexpected but is evident on a scatterplot of the 714 sites having both measurements. The intercept β_0 is significantly high, while the slope β_1 is negative. The maximum distance in the sample is approximately 248.1 km, so the spatial range, computed from the point estimate of $3/\phi$ from Table 10.1, is approximately 99.7 km, or about 40% of the maximum distance.

In Figure 10.3(a) we show the relative performances (using posterior density estimates) of prediction, interpolation, and regression at a somewhat central location $\mathbf{s}_0$, indicated by an "×" in Figure 10.2. The associated $X(\mathbf{s}_0)$ has a value 73.10 minutes. Regression (dotted line), since it models the means rather than predicting a variable, has substantially smaller variability than prediction (solid line) or interpolation (dashed line). In Figure 10.3(b), we "zoom in" on the latter pair. As expected, interpolation has less variability due to the specification of $x(\mathbf{s}_0)$. It turns out that in all cases, the observed value falls within the associated

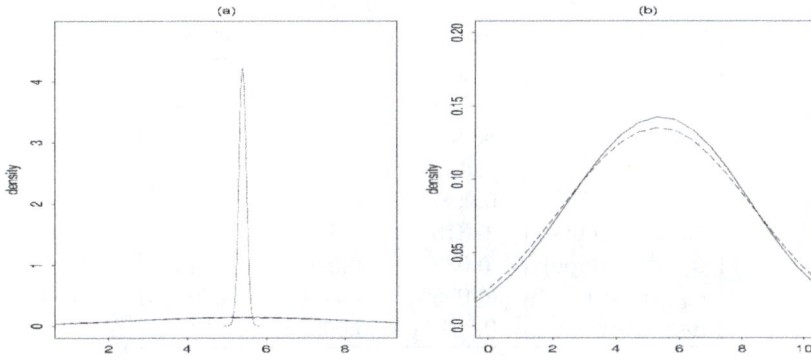

Figure 10.3 *Posterior distributions for inference at the location s_0, denoted by "$\times$" in the previous figure. Line legend: dotted line denotes $p\left(E[Y(\mathbf{s}_0)|x(\mathbf{s}_0)] \mid \mathbf{x}, \mathbf{y}\right)$ (regression); solid line denotes $p\left(y(\mathbf{s}_0) \mid x(\mathbf{s}_0), \mathbf{x}, \mathbf{y}\right)$ (prediction); and dashed line denotes $p\left(y(\mathbf{s}_0) \mid \mathbf{x}, \mathbf{y}\right)$ (interpolation).*

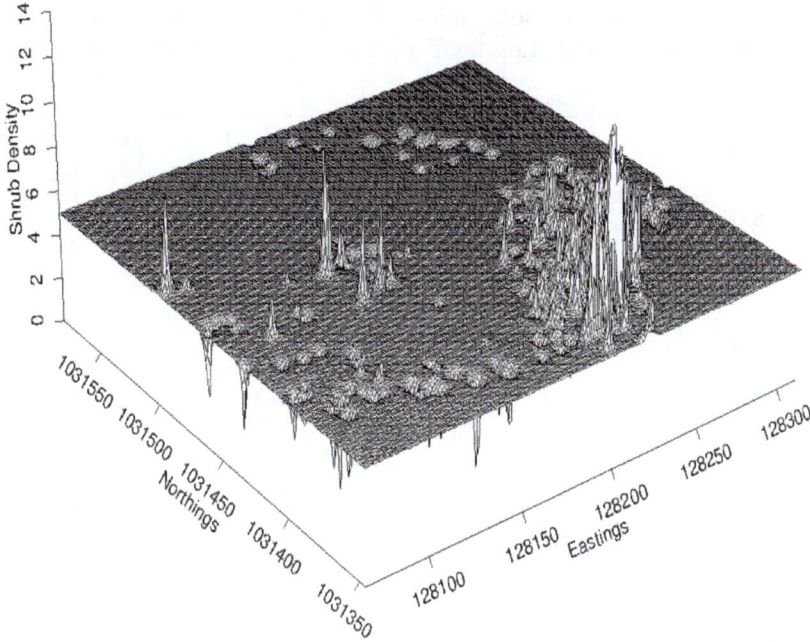

Figure 10.4 *For the Gaussian analysis case, a three-dimensional surface plot of $E\left(Y(\mathbf{s})|\mathbf{x}, \mathbf{y}\right)$ over the isopod burrows regional domain.*

intervals. Finally, in Figure 10.4 we present a three-dimensional surface plot of $E(Y(\mathbf{s})|\mathbf{x}, \mathbf{y})$ over the region. This plot reveals the spatial pattern in shrub density over the watershed. Higher measurements are expected in the eastern and particularly the southeastern part of the region, while relatively fewer shrubs are found in the northern and western parts. ∎

Example 10.2 *(Probit model).* Our second example uses a smaller data set, from the same region as Figure 10.2, which has 246 burrows of which 43 do not provide the dew measurements. Here the response is binary, governed by the success ($Y = 1$) or failure ($Y = 0$) of a burrow at a particular site. The explanatory variables (dew duration, shrub density, and rock density) relate, in some fashion, to water retention. Dew measurements are taken as the X's in our modeling with shrub and rock density being U_1 and U_2, respectively. The

| | Quantiles | | |
Parameter	2.5%	50%	97.5%
μ_1	75.415	76.095	76.772
μ_2	0.514	1.486	2.433
T_{11}	88.915	99.988	108.931
T_{12}	0.149	0.389	0.659
ϕ	0.0086	0.0302	0.2171
β_0 (intercept)	0.310	1.256	2.200
β_1 (dew slope)	0.032	0.089	0.145
α_1 (shrub)	−0.0059	−0.0036	−0.0012
α_2 (rock)	−0.00104	−0.00054	−0.00003

Table 10.2 *Posterior quantiles for the burrow survival example.*

prior specifications leading to the probit modeling again have vague bivariate normal priors for (μ_1, μ_2) and also for $\boldsymbol{\beta}$, which is two-dimensional in this example. For ϕ we again assign a noninformative inverse gamma specification, the IG(2, 0.024). We generate T_{11} and T_{12} through scaling a Wishart distribution for T with prior $IW\left(2, Diag(0.001, 0.001)\right)$.

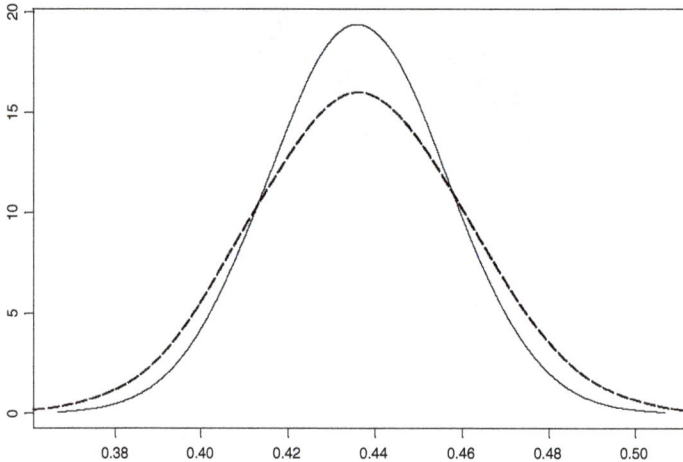

Figure 10.5 *Estimated posterior densities for the probit data analysis: solid line indicates $P\left(Z(s_0) = 1 \mid X(s_0), \mathbf{U}(s_0), \boldsymbol{\alpha}, \beta_0, \beta_1\right)$, while dashed line indicates $P\left(Z(s_0) = 1 \mid \mathbf{U}(s_0), \boldsymbol{\alpha}, \mu_1, \mu_2, T_{11}, T_{12}\right)$.*

In Table 10.2, we present the 95% credible intervals for the parameters in the model. The positive coefficient for dew is expected. It is interesting to note that shrub and rock density seem to have a negative impact on the success of the burrows. This leads us to believe that although high shrub and rock density may encourage the hydrology, it is perhaps not conducive to the growth of food materials for the isopods, or encourages predation of the isopods. The spatial range parameter again explains about 40% of the maximum distance. Figure 10.5 presents the density estimates for the posteriors $p\left(P(Z(\mathbf{s}_0) = 1 \mid x(0), \mathbf{U}(s_0), \boldsymbol{\alpha}, \beta_0, \beta_1) \mid x(0), \mathbf{x}, \mathbf{z}\right)$ and $p\left(P(Z(\mathbf{s}_0) = 1 \mid \mathbf{U}(\mathbf{s}_0), \boldsymbol{\alpha}, \mu_1, \mu_2, T_{11}, T_{12}) \mid \mathbf{x}, \mathbf{z}\right)$, with $\mathbf{s}_0$ being a central location and $x(\mathbf{s}_0) = 74.7$ minutes (after 8 a.m.), to compare the performance of interpolation and prediction. As expected, interpolation provides a slightly tighter posterior distribution. ∎

10.5.5 Conditional modeling

The spatial regression model in Section 10.5.1 uses a bivariate Gaussian process to model the variables $Y(\mathbf{s})$ and $X(\mathbf{s})$ *jointly*. Alternatively, one could proceed using a *conditional approach*, where we first model one variable and then, conditional upon the first, model the second. Again, consider two spatial variables $Y(\mathbf{s})$ and $X(\mathbf{s})$ observed over a finite set of locations $\mathcal{S} = \{\mathbf{s}_1, \mathbf{s}_2, \ldots, \mathbf{s}_n\}$. Let $\mathbf{Y}$ and $\mathbf{X}$ represent $n \times 1$ vectors of the observed $Y(\mathbf{s}_i)$'s and $X(\mathbf{s}_i)$'s respectively. The conditional approach specifies the distribution of $\mathbf{X}$ and, subsequently, the conditional distribution of $\mathbf{Y}$ given $\mathbf{X}$. This is attractive in that valid specification of these two distributions yields a legitimate joint distributions.

How do we model $\mathbf{Y} \mid \mathbf{X}$? It would be attractive to imagine a conditional *process* $Y(\mathbf{s}) \mid X(\mathbf{s})$ but this, in general, is not well defined for an arbitrary collection of variables. In fact, recall that, practically, we attempt such definition through finite dimensional distributions. Yet, it is meaningless to talk about the joint distribution of $Y(\mathbf{s}_i) \mid X(\mathbf{s}_i)$ and $Y(\mathbf{s}_j) \mid X(\mathbf{s}_j)$ for two distinct locations $\mathbf{s}_j$ and $\mathbf{s}_i$. This reveals the impossibility of specifying conditioning that would yield a consistent (in the sense of Section 3.1) definition of a process. In fact, Cressie and Wikle [2011] consider multivariate normal distributions associated with say m locations for $Y(\mathbf{s})$ and say n locations for $X(\mathbf{s})$. However, such specifications do not necessarily yield a bivariate spatial process and, thus, it may not be possible to develop kriging. To fix such issues, Cressie and Zammit-Mangion [2016] develop a conditional approach to construct multivariate spatial processes by extending bivariate models to multivariate settings using networks of spatial variables. More general conditional approaches based on graphical dependence structures among spatial variables lead to *graphical Gaussian processes* discussed in Section 10.10.

A special case of the conditional approach does produce a valid bivariate process model. Suppose that $X(\mathbf{s})$ is a univariate Gaussian spatial process with mean $\mu_X(\mathbf{s})$ and covariance function $C_X(\cdot; \boldsymbol{\theta}_X)$ indexed by process parameters $\boldsymbol{\theta}_X$. Therefore, $\mathbf{X} \sim N(\boldsymbol{\mu}_X \Sigma_X(\boldsymbol{\theta}_X))$, where $\boldsymbol{\mu}_X$ is $n \times 1$ with $\mu_X(\mathbf{s}_i)$ as its i-th entry and $\Sigma_X(\boldsymbol{\theta}_X)$ is an $n \times n$ spatial covariance matrix with entries $C_X(\mathbf{s}_i, \mathbf{s}_j; \boldsymbol{\theta}_X)$. Let $e(\mathbf{s})$ be another Gaussian process, independent of $X(\mathbf{s})$, with zero mean and covariance function $C_e(\cdot; \boldsymbol{\theta}_e)$ indexed by process parameters $\boldsymbol{\theta}_e$. Then, for any finite collection of n locations, suppose that

$$Y(\mathbf{s}_i) = \beta_0 + \beta_1 X(\mathbf{s}_i) + e(\mathbf{s}_i), \quad \text{for} \quad i = 1, 2, \ldots, n. \tag{10.31}$$

The joint distribution between $\mathbf{X}$ and $\mathbf{Y}$ is

$$\begin{pmatrix} \mathbf{X} \\ \mathbf{Y} \end{pmatrix} \sim N \left(\begin{pmatrix} \boldsymbol{\mu}_X \\ \boldsymbol{\mu}_Y \end{pmatrix}, \begin{pmatrix} \Sigma_X(\boldsymbol{\theta}_X) & \beta_1 \Sigma_X(\boldsymbol{\theta}_X) \\ \beta_1 \Sigma_X(\boldsymbol{\theta}_X) & \Sigma_e(\boldsymbol{\theta}_e) + \beta_1^2 \Sigma_X(\boldsymbol{\theta}_X) \end{pmatrix} \right), \tag{10.32}$$

where $\boldsymbol{\mu}_Y = \beta_0 \mathbf{1} + \beta_1 \boldsymbol{\mu}_X$ and $\Sigma_e(\boldsymbol{\theta}_e)$ is the $n \times n$ variance-covariance matrix for the $e(\mathbf{s}_i)$'s. Note that the above joint distribution arises from a legitimate bivariate spatial process $\mathbf{W}(\mathbf{s}) = (X(\mathbf{s}), Y(\mathbf{s}))^{\mathsf{T}}$, with mean $\boldsymbol{\mu}_{\mathbf{W}}(\mathbf{s}) = \begin{pmatrix} \mu_X(\mathbf{s}) \\ \beta_0 + \beta_1 \mu_X(\mathbf{s}) \end{pmatrix}$ and cross-covariance

$$C_{\mathbf{W}}(\mathbf{s}, \mathbf{s}') = \begin{pmatrix} C_X(\mathbf{s}, \mathbf{s}') & \beta_1 C_X(\mathbf{s}, \mathbf{s}') \\ \beta_1 C_X(\mathbf{s}, \mathbf{s}') & \beta_1^2 C_X(\mathbf{s}, \mathbf{s}') + C_e(\mathbf{s}, \mathbf{s}') \end{pmatrix}, \tag{10.33}$$

where we have suppressed the dependence of $C_X(\mathbf{s}, \mathbf{s}')$ and $C_e(\mathbf{s}, \mathbf{s}')$ on $\boldsymbol{\theta}_X$ and $\boldsymbol{\theta}_e$ respectively. Equation (10.31) implies that $\mathrm{E}[Y(\mathbf{s}) \mid X(\mathbf{s})] = \beta_0 + \beta_1 X(\mathbf{s})$ for any arbitrary location $\mathbf{s}$, thereby specifying a well-defined spatial regression model for an arbitrary $\mathbf{s}$.

An adaptation of (10.31) produces asymmetric cross-covariance matrices. For clarity, suppose that $X(\mathbf{s})$ is a Gaussian process with mean $\mu_X(\mathbf{s})$ and a stationary covariance function $C_X(\mathbf{s}, \mathbf{s}') = C_X(\mathbf{h})$, where $\mathbf{h} = \mathbf{s} - \mathbf{s}'$, and consider the following regression model

$$Y(\mathbf{s}) = \beta_0 + \beta_1 X(\mathbf{s} + \mathbf{r}) + \epsilon(\mathbf{s}), \tag{10.34}$$

where $\mathbf{r}$ is some *fixed* location. This model, often called a *spatial delay*, again defines a legitimate bivariate process with a stationary but *asymmetric* cross-covariance function

$$C_{\mathbf{W}}(\mathbf{h}) = \begin{pmatrix} C_X(\mathbf{h}) & \beta_1 C_X(\mathbf{r}+\mathbf{h}) \\ \beta_1 C_X(\mathbf{r}-\mathbf{h}) & \beta_1^2 C_X(\mathbf{h}) + C_e(\mathbf{h}) \end{pmatrix}.$$

While the introduction of the asymmetry may seem attractive, it is unclear when a spatial delay arises in practice. Put differently, why would we want to regress $Y(\mathbf{s})$ on $X(\mathbf{s}+\mathbf{r})$? Wackernagel [2003] draws an analogy with time-series, where it has been observed that in some cases that the effect of one variable on another is not instantaneous. The time for the second variable to react to the first causes a lag or "delay" in the correlations between the time series. Here is another explanation: the Cauchy-Schwarz inequality ensures that $\text{Cov}(X(\mathbf{s}), Y(\mathbf{s}+\mathbf{h})) = C_{XY}(\mathbf{h})$ attains an extremum that is proportional to that of $C_X(\mathbf{h})$. It may happen that the extremum (maximum in the case of a positively correlated variable pair and minimum otherwise) is shifted away in a direction $\mathbf{r}$. In that case, $C_{XY}(\mathbf{h})$ attains an extremum which is proportional to that of $C_{XX}(\mathbf{r}+\mathbf{h})$. Wackernagel [2003] suggests plotting the empirical cross-covariance function over appropriately constructed bins to diagnose spatial lag and, hence, ascertain $\mathbf{r}$. This, however, is awkward in practice. In any case, the modeling is rather restrictive and is perhaps feasible only in the bivariate setting. Later in the chapter, we discuss richer and more flexible nonstationary spatially-varying cross-covariance modeling for jointly modeling several spatial outcomes, which accommodates asymmetric cross-covariances without requiring one to estimate spatial lags or delays.

The conditional approach is not bereft of problems. As already mentioned, it will not generally produce multivariate process models, which precludes predictions at arbitrary locations for all the outcomes. The example above is one of the few instances where a legitimate bivariate process is obtained [see Royle and Berliner, 1999, for other examples]. Another issue with the conditional approach is the order of the hierarchy. Specifying $p(\mathbf{X})p(\mathbf{Y} \mid \mathbf{X})$ $p(\mathbf{Y})$ yields a joint distribution different from specifying $p(\mathbf{Y})p(\mathbf{X} \mid \mathbf{Y})$. How do we decide upon the ordering? In certain applications, there may be a natural ordering based upon a causal relationship. For example, Cressie and Zammit-Mangion [2016] build multivariate spatial covariance models using graphical networks among multiple variables and exploit the graph structure to reduce the number of possible models and select models based on possible causative links in the network. When such structures are not relevant or available, this ambiguity with regard to the ordering is undesirable and practically inefficient as we move to more than two outcomes. For example, with three outcomes, we will have six models emerging from the order of the outcomes in the hierarchy and with four outcomes we have twenty-four models. Clearly, when the information regarding the order of the hierarchy is lacking, we have an explosion in the number of models.

Finally, we note that the specification associated with $X(\mathbf{s})$ and $Y(\mathbf{s})$ in (10.31) can obviously be extended so that both $\mathbf{X}(\mathbf{s})$ and $\mathbf{Y}(\mathbf{s})$ are vectors. We can create a joint process model through such conditioning. However, in this case, we would need multivariate process models for both $\mathbf{X}(\mathbf{s})$ and $\mathbf{Y}(\mathbf{s})$; we are back to our original joint modeling problem.

10.5.6 *Spatial regression with kernel averaged predictors*

In formulating a spatial regression model, we have argued that it can be useful to assume that $X(\mathbf{s})$ is random and in the previous sections, we have shown how to implement spatial regression under a bivariate process model for $\mathbf{Z}(\mathbf{s}) = (Y(\mathbf{s}), X(\mathbf{s}))^{\mathsf{T}}$. However, in spatial applications, *neighboring* $X(\mathbf{s}')$ can be expected to inform about $Y(\mathbf{s})$ particularly when the distance between $\mathbf{s}$ and $\mathbf{s}'$ is small. For instance, precipitation can affect the water table (the depth at which soil and pour spaces become completely saturated with water) not only where the precipitation fell but also at surrounding locations due to run off and changes in slope from uneven ground surfaces. Similarly, the concentration of ozone is affected by pollutants, ultraviolet rays, and temperature within a neighborhood of location $\mathbf{s}$. Thus, a mean specification which only includes $X(\mathbf{s})$ and not neighboring $X(\mathbf{s}')$ may not adequately capture the process. Here, drawing upon Heaton and Gelfand [2011], we develop a kernel averaged predictor regression model to address this issue.

A mean specification which incorporates neighboring predictors might take the form $E(Y(\mathbf{s})) = \beta_0(\mathbf{s}) + \int_D \beta(\mathbf{s}, \mathbf{u}; \boldsymbol{\theta}) X(\mathbf{u}) d\mathbf{u}$ where $\beta(\mathbf{s}, \mathbf{s}'; \boldsymbol{\theta})$ is a coefficient model capturing the effect of $X(\mathbf{s}')$ on $Y(\mathbf{s})$ with parameters $\boldsymbol{\theta}$. With, say n observations $\mathbf{Z} = (\mathbf{Z}(\mathbf{s}_1)^{\mathsf{T}}, \ldots, \mathbf{Z}(\mathbf{s}_n)^{\mathsf{T}})^{\mathsf{T}}$, we might consider the conditional distribution, $Y(\mathbf{s}_i) \mid X(\mathbf{s}_1), \ldots, X(\mathbf{s}_n)$. In the Gaussian setting, $[Y(\mathbf{s}_i) \mid X(\mathbf{s}_1), \ldots, X(\mathbf{s}_n)]$ contains n coefficients $\beta(\mathbf{s}_i, \mathbf{s}_1), \ldots, \beta(\mathbf{s}_i, \mathbf{s}_n)$ which describe the loading of $X(\mathbf{s}_j)$ on $Y(\mathbf{s}_i)$ for $j = 1, \ldots, n$. Such a choice is unattractive because the explained relationship between the response and predictor depends on the number of sampling locations and the arrangement of those locations in D. It also fails to explicitly capture the idea of local $X(\mathbf{s}')$ informing about $Y(\mathbf{s})$.

Rather, we seek a *kernel averaged* predictor, $\tilde{X}(\mathbf{s})$ based upon $\{X(\mathbf{s}') : \mathbf{s}' \in D\}$ under an appropriate choice for $\beta(\mathbf{s}, \mathbf{s}'); \boldsymbol{\theta})$. What we are doing differs from what is usually referred to as employing functional covariates (see Baíllo and A. Grané, 2009). Functional covariates envision a function at location $\mathbf{s}$, say, $X(\mathbf{s}, t)$ for $t \in (0, T]$ and seek to reduce this to a single covariate at $\mathbf{s}$ to explain $Y(\mathbf{s})$. Usually, this is achieved through some integration of the function wherein the integration is over t rather than over $\mathbf{s}$ as we seek. Also, we do not seek to build process models for $\beta(\mathbf{s}, \mathbf{s}'; \boldsymbol{\theta})$. We provide process modeling for $(Y(\mathbf{s}), X(\mathbf{s}))$ but specify β as a parametric function. Specifying the latter using processes will provide poorly-identified, over-fitted models.

Let $Y(\mathbf{s})$ and $X(\mathbf{s})$ denote a univariate response variable and a single covariate at location $\mathbf{s} \in D$. Furthermore, assume $X(\mathbf{s})$ follows a Gaussian process of the form,

$$X(\mathbf{s}) = \mu_X(\mathbf{s}) + \sigma_X w_X(\mathbf{s}), \qquad (10.35)$$

where $\mu_X(\mathbf{s})$ is the mean surface at location $\mathbf{s}$ and $w_X(\mathbf{s})$ is a mean 0, variance 1 GP with correlation function $\rho_X(\mathbf{s}, \mathbf{s}'; \boldsymbol{\phi}_X)$. Here, we imitate the specification above. Notice that (10.35) defines a purely spatial covariate process. In some applications, however, covariates are measured with error. In these cases, $X(\mathbf{s})$ can be thought of as the "true" underlying covariate process and the observed covariate is $H(\mathbf{s}) = X(\mathbf{s}) + \epsilon_X(\mathbf{s})$ where $\epsilon_X(\mathbf{s})$ is an *iid* white noise process. Distributional results for $H(\mathbf{s})$ will differ from those of $X(\mathbf{s})$ by an additive nugget variance term only so, for simplicity, below, we assume $X(\mathbf{s})$ is observed. Moreover, we still want to use $X(\mathbf{s})$ to explain $Y(\mathbf{s})$, i.e., we want the regression of $Y(\mathbf{s})$ to be on the true $X(\mathbf{s})$.

We define the unobserved kernel-averaged local covariate at $\mathbf{s}$ as

$$\tilde{X}(\mathbf{s}) \equiv \frac{1}{K(\mathbf{s}; \boldsymbol{\xi})} \int_D K(\mathbf{s}, \mathbf{u}; \boldsymbol{\xi}) X(\mathbf{u}) d\mathbf{u}, \qquad (10.36)$$

where $K(\mathbf{s}, \mathbf{s}'; \boldsymbol{\xi})$ is a kernel defining a weight on the distance between $\mathbf{s}$ and $\mathbf{s}'$ with parameters $\boldsymbol{\xi}$ and $0 < K(\mathbf{s}; \boldsymbol{\xi}) = \int_D K(\mathbf{s}, \mathbf{u}; \boldsymbol{\xi}) d\mathbf{u} < \infty$. The parameter $\boldsymbol{\xi}$, most commonly, consists of *scale* parameters such as the entries of a covariance matrix but can also include *location* parameters as in Higdon [1998]. Choices for K are discussed below.

Because a valid GP was defined for $X(\mathbf{s})$, $\tilde{X}(\mathbf{s})$ also is a valid GP with mean $\tilde{\mu}_X(\mathbf{s}) = K(\mathbf{s}; \boldsymbol{\xi})^{-1} \int_D K(\mathbf{s}, \mathbf{u}; \boldsymbol{\xi}) \mu_X(\mathbf{u}) d\mathbf{u}$ and

$$\mathrm{Cov}(\tilde{X}(\mathbf{s}), \tilde{X}(\mathbf{s}')) = \sigma_X^2 \rho_{\tilde{X}}(\mathbf{s}, \mathbf{s}') \quad \text{where} \qquad (10.37)$$

$$\rho_{\tilde{X}}(\mathbf{s}, \mathbf{s}') = \frac{1}{K(\mathbf{s}; \boldsymbol{\xi}) K(\mathbf{s}'; \boldsymbol{\xi})} \int_D \int_D K(\mathbf{s}, \mathbf{u}; \boldsymbol{\xi}) K(\mathbf{s}', \mathbf{v}; \boldsymbol{\xi}) \rho_X(\mathbf{u}, \mathbf{v}) d\mathbf{v} d\mathbf{u} \,.$$

Not only are $X(\mathbf{s})$ and $\tilde{X}(\mathbf{s})$ marginally Gaussian processes, but a valid bivariate GP is induced for the pair $(X(\mathbf{s}), \tilde{X}(\mathbf{s}))^{\mathsf{T}}$. Specifically, $(X(\mathbf{s}), \tilde{X}(\mathbf{s}))^{\mathsf{T}}$ follows a bivariate GP with mean $(\mu_X(\mathbf{s}), \tilde{\mu_X}(\mathbf{s}))^{\mathsf{T}}$ and

$$\text{Cov}\left(\left(\begin{array}{c} X(\mathbf{s}) \\ \tilde{X}(\mathbf{s}) \end{array}\right), \left(\begin{array}{c} X(\mathbf{s}') \\ \tilde{X}(\mathbf{s}') \end{array}\right)\right) = \sigma_X^2 \left(\begin{array}{cc} \rho_X(\mathbf{s}, \mathbf{s}') & \rho_{X,\tilde{X}}(\mathbf{s}, \mathbf{s}') \\ \rho_{\tilde{X},X}(\mathbf{s}, \mathbf{s}') & \rho_{\tilde{X}}(\mathbf{s}, \mathbf{s}') \end{array}\right), \tag{10.38}$$

where

$$\rho_{X,\tilde{X}}(\mathbf{s}, \mathbf{s}') = \frac{1}{K(\mathbf{s}'; \boldsymbol{\xi})} \int_D K(\mathbf{s}', \mathbf{u}; \boldsymbol{\xi}) \rho_X(\mathbf{s}, \mathbf{u}) d\mathbf{u} \text{ and}$$

$$\rho_{\tilde{X},X}(\mathbf{s}, \mathbf{s}') = \frac{1}{K(\mathbf{s}; \boldsymbol{\xi})} \int_D K(\mathbf{s}, \mathbf{u}; \boldsymbol{\xi}) \rho_X(\mathbf{u}, \mathbf{s}') d\mathbf{u}$$

for any other location $\mathbf{s}' \in D$.

Next, consider the linear model defined by,

$$Y(\mathbf{s}) \mid X(\mathbf{s}), \tilde{X}(\mathbf{s}) = \beta_0 + \beta_1 \tilde{X}(\mathbf{s}) + \sigma_Y w_Y(\mathbf{s}) + \epsilon_Y(\mathbf{s}), \tag{10.39}$$

where $w_Y(\mathbf{s})$ is defined analogously to $w_X(\mathbf{s})$ in (10.35) but with correlation function ρ_Y and $\epsilon_Y(\mathbf{s})$ is a Gaussian white noise process with variance τ_Y^2. Intuitively, the kernel $K(\mathbf{s}, \mathbf{u}; \boldsymbol{\xi})$ describes how the effect of the covariate $X(\mathbf{s}), \mathbf{s} \in D$ propagates to the response.

Notice that the bivariate GP for $(X(\mathbf{s}), \tilde{X}(\mathbf{s}))^{\mathsf{T}}$ along with (10.39) provide a joint specification of a trivariate GP for $\mathbf{Z}(\mathbf{s}) = (X(\mathbf{s}), \tilde{X}(\mathbf{s}), Y(\mathbf{s}))^{\mathsf{T}}$. Specifically, $\mathbf{Z}(\mathbf{s})$ follows a valid trivariate GP with mean

$$\boldsymbol{\mu}(\mathbf{s}) = (\mu_X(\mathbf{s}), \tilde{\mu}_X(\mathbf{s}), \beta_0 + \beta_1 \tilde{\mu}_X(\mathbf{s}))^{\mathsf{T}}$$

and cross-covariance,

$$\sigma_X^2 \left(\begin{array}{ccc} \rho_X(\mathbf{s}, \mathbf{s}') & \rho_{X,\tilde{X}}(\mathbf{s}, \mathbf{s}') & \beta_1 \rho_{X,\tilde{X}}(\mathbf{s}, \mathbf{s}') \\ \rho_{\tilde{X},X}(\mathbf{s}, \mathbf{s}') & \rho_{\tilde{X}}(\mathbf{s}, \mathbf{s}') & \beta_1 \rho_{\tilde{X}}(\mathbf{s}, \mathbf{s}') \\ \beta_1 \rho_{\tilde{X},X}(\mathbf{s}, \mathbf{s}') & \beta_1 \rho_{\tilde{X}}(\mathbf{s}, \mathbf{s}') & \frac{\sigma_Y^2}{\sigma_X^2} \rho_Y(\mathbf{s}, \mathbf{s}') + \beta_1^2 \rho_{\tilde{X}}(\mathbf{s}, \mathbf{s}') \end{array}\right) \tag{10.40}$$

The joint distribution of $\mathbf{Z}(\mathbf{s})$ is useful for evaluating the properties of (10.39) such as induced correlations between $Y(\mathbf{s})$ and $(X(\mathbf{s}), \tilde{X}(\mathbf{s}))$. Note also that the induced bivariate process model for $(X(\mathbf{s}), Y(\mathbf{s}))$ is not a usual bivariate process specification; the kernel appears in the covariance structure.

We focus on the *local* spatial regression in terms of the conditional distribution of $Y(\mathbf{s})$ given $\tilde{X}(\mathbf{s})$ in the form of the conditional mean

$$E[Y(\mathbf{s}) \mid \tilde{X}(\mathbf{s})] = \beta_0 + \beta_1 \tilde{X}(\mathbf{s}) .$$

Thus, given $\tilde{X}(\mathbf{s})$, concepts such as R^2, mean square error, variable selection, shrinkage, etc. are applicable. Of course, all are random and would be averaged over the distribution of $\tilde{X}(\mathbf{s})$ in order to interpret them. Evidently, the potentially complex relationship between $\{X(\mathbf{s}) : \mathbf{s} \in D\}$ and $Y(\mathbf{s})$ is captured through a single parameter (β_1). However, from (10.39), we see that, effectively, we are introducing a coefficient weighting of the entire surface $X(\mathbf{s})$ to explain $Y(\mathbf{s})$. That is, the coefficient of $X(\mathbf{s}')$ is $\beta_1 K(\mathbf{s}, \mathbf{s}'; \boldsymbol{\xi})/K(\mathbf{s}; \boldsymbol{\xi})$. The normalization by $K(\mathbf{s}; \boldsymbol{\xi})$ identifies β_1.

The choice of K with the data informing about $\boldsymbol{\xi}$ enables a fairly rich regression specification while attractively reducing to a simple linear regression model in $\tilde{X}(\mathbf{s})$. Examples of kernels are given in Table 10.3. We note that the computation of $K(\mathbf{s}; \boldsymbol{\xi})$ for general regions D varies with $\mathbf{s}$ and can be computationally expensive when done repeatedly over MCMC

Kernel	$K(\mathbf{s}, \mathbf{s}'; \boldsymbol{\xi})$	Parameters ($\boldsymbol{\xi}$)
Uniform	$\mathbf{I}_{\{\|\mathbf{s}-\mathbf{s}'\| \leq \xi\}}$	ξ
Epanechnikov	$(\xi^2 - \|\mathbf{s} - \mathbf{s}'\|^2)\mathbf{I}_{\{\|\mathbf{s}-\mathbf{s}'\| \leq \xi\}}$	ξ
Component Wise Gaussian	$\prod_{i=1}^{d} \xi_i^{-1} \exp\{-(s_i - s_i')^2/(2\xi_i^2)\}$	$\xi_i, i = 1, \ldots, d$
Oriented Gaussian	$\|\Xi\|^{-1/2} \exp\left\{-\frac{(\mathbf{s}'-\mathbf{s})^\mathrm{T}\Xi^{-1}(\mathbf{s}'-\mathbf{s})}{2}\right\}$	$\Xi = \{\xi_{ij}\}$

Table 10.3 *Examples of kernel functions $K(\mathbf{s}, \mathbf{s}'; \boldsymbol{\xi})$ and their parameters. $\mathbf{I}_A$ is an indicator for the set A.*

iterations. Below, the kernel is taken to be $K(\mathbf{s}, \mathbf{s}'; \xi) = I(\{\|\mathbf{s} - \mathbf{s}'\| \leq \xi g(\phi_X)\})$ where $I(A)$ is an indicator for the set A, $\|\cdot\|$ denotes Euclidean distance, $\xi \in (0, 1)$ and $g(\phi_X)$ is the effective spatial range associated with ρ_X; for example, from Chapter 2, the exponential correlation function has $g(\phi_X) \approx 3/\phi_X$ where ϕ_X is the spatial decay parameter. Intuitively, the kernel $K(\mathbf{s}, \mathbf{s}'; \xi)$ is a disk centered at location $\mathbf{s}$ with radius $r = \xi g(\phi_X)$. The scale parameter r can be loosely interpreted as a hard threshold "decay" parameter in that the effect of $X(\mathbf{s}')$ on $Y(\mathbf{s})$ is negligible if $\|\mathbf{s}-\mathbf{s}'\| \geq r$. For very small r, $\tilde{X}(\mathbf{s}) \approx X(\mathbf{s})$. For large r, $\tilde{X}(\mathbf{s}) \approx \tilde{X}(\mathbf{s}')$ for all $\mathbf{s}, \mathbf{s}'$ yielding a highly collinear regression.

With the choice of $K(\mathbf{s}, \mathbf{s}'; \xi)$ above, r and $\boldsymbol{\phi}_X$ are strongly associated parameters; this is evident from the forms in (10.40). Plausible values for scale parameters of kernels depend on ϕ_X as well as D. For the K above, parameterizing the scale parameter as $r = \xi g(\boldsymbol{\phi}_X)$ where $\xi \in (0, 1)$ removes this dependency such that, a priori, ξ and ϕ_X can be taken as independent and also restricts K to be within the effective spatial range of ρ_X.

Next, consider what (10.39) implies about using $X(\mathbf{s})$ instead of $\tilde{X}(\mathbf{s})$ in the conditional mean for $Y(\mathbf{s}) \mid X(\mathbf{s}), \tilde{X}(\mathbf{s})$ assuming $\mathrm{E}(Y(\mathbf{s}) \mid X(\mathbf{s}), \tilde{X}(\mathbf{s})) = \beta_0 + \beta_1\tilde{X}(\mathbf{s})$. Using the fact that $\mathbf{Z}(\mathbf{s}) = (X(\mathbf{s}), \tilde{X}(\mathbf{s}), Y(\mathbf{s}))^\mathrm{T}$ follows a trivariate Gaussian process with known mean and covariance given by (10.40), multivariate normal theory gives that $Y(\mathbf{s}) \mid X(\mathbf{s})$ is also normally distributed with mean,

$$\mathrm{E}[Y(\mathbf{s}) \mid X(\mathbf{s})] = \beta_0 + \beta_1(\tilde{\mu}_X(\mathbf{s}) + \rho_{X,\tilde{X}}(\mathbf{s}, \mathbf{s})(X(\mathbf{s}) - \mu_X(\mathbf{s}))), \qquad (10.41)$$

variance,

$$\mathrm{Var}(Y(\mathbf{s}) \mid X(\mathbf{s})) = \tau_Y^2 + \sigma_Y^2 + \beta_1^2\sigma_X^2(\rho_{\tilde{X}}(\mathbf{s}, \mathbf{s}) - \rho_{X,\tilde{X}}^2(\mathbf{s}, \mathbf{s})), \qquad (10.42)$$

and covariance,

$$\mathrm{Cov}(Y(\mathbf{s}), Y(\mathbf{s}') \mid X(\mathbf{s}), X(\mathbf{s}')) = \sigma_Y^2\rho_Y(\mathbf{s}, \mathbf{s}') + \beta_1^2\sigma_X^2\rho_{\tilde{X}}(\mathbf{s}, \mathbf{s}') . \qquad (10.43)$$

So, when the model given by (10.39) holds, then the change in $Y(\mathbf{s})$ as a result of a unit change in $X(\mathbf{s})$ is $\beta_1\rho_{X,\tilde{X}}(\mathbf{s}, \mathbf{s})$ as opposed to β_1 when using $\tilde{X}(\mathbf{s})$. Hence, under the trivariate Gaussian process model, i.e., when the assumptions of (10.39) hold, using $X(\mathbf{s})$ implies that the effect of the covariate on $Y(\mathbf{s})$ is shrunk towards zero; $|\beta_1\rho_{X,\tilde{X}}(\mathbf{s}, \mathbf{s})| \leq |\beta_1|$ because $0 \leq \rho_{X,\tilde{X}}(\mathbf{s}, \mathbf{s}) \leq 1$. This result is not surprising in that if other $X(\mathbf{s}')$ in D besides $X(\mathbf{s})$ affect $Y(\mathbf{s})$ then the effect due to $X(\mathbf{s})$ is expected to diminish. The amount of shrinkage is determined by the kernel parameters, $\boldsymbol{\xi}$, as well as ϕ_X. For example, if $Y(\mathbf{s})$ is ozone and $X(\mathbf{s})$ is temperature, the implication would be that using $X(\mathbf{s})$ in the model could, potentially, underestimate the change in $Y(\mathbf{s})$ as a result from a unit change in temperature, leading to underestimation of the production of ozone.

A second consequence of using $X(\mathbf{s})$ instead of $\tilde{X}(\mathbf{s})$ when (10.39) holds is that the percent of variation in $Y(\mathbf{s})$ explained by $X(\mathbf{s})$ is less than the percent of variation in $Y(\mathbf{s})$ explained by $\tilde{X}(\mathbf{s})$ for many common covariance functions as detailed by the following result from Heaton and Gelfand [2011]:

Let $\rho^2_{Y|X}$ and $\rho^2_{Y|\tilde{X}}$ be the population coefficient of determination for the linear model defined by (10.41) and (10.39), respectively. If ρ_X is an isotropic, log-concave correlation function then $\rho^2_{Y\,|\,X} \leq \rho^2_{Y\,|\,\tilde{X}}$.

The class of log-concave covariance functions includes the powered exponential, $\rho(\mathbf{s}, \mathbf{s}') = \exp\{-\phi\|\mathbf{s} - \mathbf{s}'\|^\alpha \; 0 \leq \alpha \leq 2$, by direct calculation. Also included are closed form Matérn models, i.e., those with smoothness parameter ν of the form $\nu = k + 1/2$ for $k \in \{0, 1, 2, \dots\}$ again by direct calculation with an indication that this is the case for arbitrary ν [see Majumdar and Gelfand, 2007]. Also, it is easy to argue that convolution of covariance functions produces valid covariance functions. In fact, convolution of log-concave functions produces log-concave functions. Thus, for such covariance functions, using $X(\mathbf{s})$ instead of $\tilde{X}(\mathbf{s})$ results in less variation in $Y(\mathbf{s})$ explained. Additionally, notice that marginalizing over $\tilde{X}(\mathbf{s})$ adds extra spatial variation to $Y(\mathbf{s})$. To see this, notice that the covariance given by (10.43) has the added variance term $\beta_1^2 \sigma_X^2 \rho_{\tilde{X}}(\mathbf{s})$.

Model fitting, efficient computation, kernel selection, and illustrative examples are presented in Heaton and Gelfand [2011] and we encourage the interested reader to look at this paper for further discussion and details.

10.6   Coregionalization models ⋆

10.6.1   *Coregionalization models and their properties*

We now consider a constructive modeling strategy to add flexibility to (10.10) while retaining interpretability and computational tractability. Our approach is through the *linear model of coregionalization* (LMC), as for example in Grzebyk and Wackernagel [1994] and Wackernagel [2003]. The term "coregionalization" is intended to denote a model for measurements that covary jointly over a region.

The most basic coregionalization model, the so-called *intrinsic specification*, dates at least to Matheron [1982]. It arises as $\mathbf{Y}(\mathbf{s}) = A\mathbf{w}(\mathbf{s})$ where the components of $\mathbf{w}(\mathbf{s})$ are i.i.d. spatial processes. If the $w_j(\mathbf{s})$ have mean 0 and are stationary with variance 1 and correlation function $\rho(h)$, then $E(\mathbf{Y}(\mathbf{s}))$ is $\mathbf{0}$ and the cross-covariance matrix, $\Sigma_{\mathbf{Y}(\mathbf{s}),\mathbf{Y}(\mathbf{s}')} \equiv C(\mathbf{s} - \mathbf{s}') = \rho(\mathbf{s} - \mathbf{s}')AA^{\mathrm{T}}$. Letting $AA^{\mathrm{T}} = T$ immediately reveals the equivalence between this simple intrinsic specification and the separable covariance specification as in Section 10.3 above. As in Subsection 2.1.2, the term "intrinsic" is taken to mean that the specification only requires the first and second moments of differences in measurement vectors and that the first-moment difference is $\mathbf{0}$ and the second moments depend on the locations only through the separation vector $\mathbf{s} - \mathbf{s}'$. In fact here $E(\mathbf{Y}(\mathbf{s}) - \mathbf{Y}(\mathbf{s}')) = \mathbf{0}$ and $\frac{1}{2}\Sigma_{\mathbf{Y}(\mathbf{s}) - \mathbf{Y}(\mathbf{s}')} = G(\mathbf{s} - \mathbf{s}')$ where $G(\mathbf{h}) = C(\mathbf{0}) - C(\mathbf{h}) = T - \rho(\mathbf{s} - \mathbf{s}')T = \gamma(\mathbf{s} - \mathbf{s}')T$ where γ is a valid variogram. Of course, as in the $p = 1$ case, we need not begin with a covariance function but rather just specify the process through γ and T. A more insightful interpretation of "intrinsic" is that given in equation (10.12).

We assume A is full rank and, for future reference, we note that A can be assumed to be the lower triangular. This, lower-triangular specification for A does impose certain conditional independence constraints. To see how, consider the bivariate case, two locations $\mathbf{s}_1$ and $\mathbf{s}_2$ and suppose that the elements of A are fixed. Then, for $i = 1, 2$, $Y_1(\mathbf{s}_i) = a_{11}w_1(\mathbf{s}_i)$ and $Y_2(\mathbf{s}_i) = a_{21}w_1(\mathbf{s}_i) + a_{22}w_2(\mathbf{s}_i)$. Since the process $w_1(\mathbf{s})$ completely determines $Y_1(\mathbf{s})$, we can write

$$
\begin{aligned}
\mathrm{Cov}&\{Y_1(\mathbf{s}_1), Y_2(\mathbf{s}_2) \mid Y_1(\mathbf{s}_2)\} \\
&= \mathrm{Cov}\{a_{11}w_1(\mathbf{s}_1), a_{21}w_1(\mathbf{s}_2) + a_{22}w_2(\mathbf{s}_2) \mid w_1(\mathbf{s}_2)\} \\
&= a_{11}a_{22}\mathrm{Cov}\{w_1(\mathbf{s}_1), w_2(\mathbf{s}_2) \mid w_1(\mathbf{s}_2)\} = 0 ,
\end{aligned} \qquad (10.44)
$$

where the last equality follows because the process $w_1(\cdot)$ is independent of $w_2(\cdot)$. This reveals that $Y_1(\mathbf{s}_1)$ and $Y_2(\mathbf{s}_2)$ will be independent conditional upon $Y_1(\mathbf{s}_2)$. In terms of the dispersion matrix, this conditional independence implies that the Σ_Y^{-1} will have redundant zeroes. In theory, we can easily obviate restrictions such as in (10.44) by specifying A to be a non-triangular square root of T obtained by spectral decomposition. One example is a spectrally decomposed matrix, which is thereafter modeled using a set of eigenvalues and *Givens* angles. One could set $A = P\Lambda^{1/2}$, or the symmetric square-root $A = P\Lambda^{1/2}P'$, where $T = P\Lambda P'$ is the spectral decomposition for T. This requires further parametrization for the orthogonal matrix P, such as in terms of the $p(p-1)/2$ *Givens* angles $\theta_{ij}(\mathbf{s})$ for $i = 1,\ldots,p-1$ and $j = i+1,\ldots,p$ (e.g., Daniels and Kass, 1999). Specifically, $P = \prod_{i=1}^{p-1}\prod_{j=i+1}^{p}G_{ij}(\theta_{ij})$ where i and j are distinct and $G_{ij}(\theta_{ij})$ is almost the $p \times p$ identity matrix except that its i-th and j-th diagonal elements are replaced by $\cos(\theta_{ij})$ and $\pm\sin(\theta_{ij})$ respectively. Given P for any $\mathbf{s}$, the θ_{ij}'s are unique within range $(-\pi/2, \pi/2)$. These may be further modeled by means of Gaussian processes on a suitably transformed function, say $\tilde{\theta}_{ij} = \log(\frac{\pi/2+\theta_{ij}}{\pi/2-\theta_{ij}})$. For further reference, see Daniels and Kass [1999] and Kang and Cressie [2011].

While the conditional independence in (10.44) from the triangular specification for A may seem theoretically unnecessary, it is only an *a priori* assumption that does not carry over to posterior inference. Our ultimate interest resides with the cross-covariance function, which is robustly estimated using any bijective mapping with some square-root matrix. Furthermore, the number of parameters to be estimated in the Given's angle specification is the same as that for the triangular Cholesky. In practical settings, these specifications matter little but Cholesky decompositions are numerically more stable than the spectral decomposition. The former is also less expensive, requiring $O(m^3/3)$ flops as compared to more than $O(4m^3/3)$ flops required by the latter. Hence, we opt for the Cholesky decomposition in our subsequent data analysis. No additional richness accrues, at least from a practical standpoint, to a more general A.

A more general LMC arises if again $\mathbf{Y}(\mathbf{s}) = A\mathbf{w}(\mathbf{s})$ but now the $w_j(\mathbf{s})$ are independent but no longer identically distributed. In fact, let the $w_j(\mathbf{s})$ process have mean μ_j, variance 1, and correlation function $\rho_j(h)$. Then $E(\mathbf{Y}(\mathbf{s})) = A\boldsymbol{\mu}$ where $\boldsymbol{\mu}^{\mathsf{T}} = (\mu_1, \cdots, \mu_p)$ and the cross-covariance matrix associated with $\mathbf{Y}(\mathbf{s})$ is now

$$\Sigma_{\mathbf{Y}(\mathbf{s}),\mathbf{Y}(\mathbf{s}')} \equiv C(\mathbf{s} - \mathbf{s}') = \sum_{j=1}^{p} \rho_j(\mathbf{s} - \mathbf{s}')T_j , \qquad (10.45)$$

where $T_j = \mathbf{a}_j\mathbf{a}_j^{\mathsf{T}}$ with $\mathbf{a}_j$ the jth column of A. Note that $\sum_j T_j = T$. More importantly, we note that such a linear combination produces stationary spatial processes. We return to this point in Section 10.7.4.

The one-to-one relationship between T and lower triangular A is standard. For future use, when $p = 2$ we have

$$a_{11} = \sqrt{T_{11}} , \quad a_{21} = \frac{T_{12}}{\sqrt{T_{11}}} \quad \text{and} \quad a_{22} = \sqrt{T_{22} - \frac{T_{12}^2}{T_{11}}} .$$

When $p = 3$ we add

$$a_{31} = \frac{T_{13}}{\sqrt{T_{11}}} , \quad a_{32} = \frac{T_{11}T_{23} - T_{12}T_{13}}{\sqrt{T_{11}T_{22} - T_{12}^2}\sqrt{T_{11}}} \quad \text{and}$$

$$a_{33} = \sqrt{T_{33} - \frac{T_{13}^2}{T_{11}} - \frac{(T_{11}T_{23} - T_{12}T_{13})^2}{T_{11}(T_{11}T_{22} - T_{12}^2)}} .$$

Lastly, if we introduce monotonic isotropic correlation functions, we will be interested in the range associated with $Y_j(\mathbf{s})$. An advantage to (10.45) is that each $Y_j(\mathbf{s})$ has its own range. In particular, for $p = 2$ the range for $Y_1(\mathbf{s})$ solves $\rho_1(d) = 0.05$, while the range for $Y_2(\mathbf{s})$ solves the weighted average correlation,

$$\frac{a_{21}^2 \rho_1(d) + a_{22}^2 \rho_2(d)}{a_{21}^2 + a_{22}^2} = 0.05 . \tag{10.46}$$

Since ρ_1 and ρ_2 are monotonic the left side of (10.46) is decreasing in d. Hence, solving (10.46) is routine. If we have $p = 3$, we need in addition the range for $Y_3(\mathbf{s})$. We require the solution of

$$\frac{a_{31}^2 \rho_1(d) + a_{32}^2 \rho_2(d) + a_{33}^2 \rho_3(d)}{a_{31}^2 + a_{32}^2 + a_{33}^2} = 0.05 . \tag{10.47}$$

The left side of (10.47) is again decreasing in d. The form for general p is clear.

In practice, the ρ_j are parametric classes of functions. Hence the range d is a parametric function that is not available explicitly. However, within a Bayesian context, when models are fitted using simulation-based methods, we obtain posterior samples of the parameters in the ρ_j's, as well as A. Each sample, when inserted into the left side of (10.46) or (10.47), enables a solution for a corresponding d. In this way, we obtain posterior samples of each of the ranges, one-for-one with the posterior parameter samples.

Extending in a different fashion, we can define a process having a general *nested* covariance model (see, e.g., Wackernagel, 2003) as

$$\mathbf{Y}(\mathbf{s}) = \sum \mathbf{Y}^{(u)}(\mathbf{s}) = \sum_{u=1}^{r} A^{(u)} \mathbf{w}^{(u)}(\mathbf{s}) , \tag{10.48}$$

where the $\mathbf{Y}^{(u)}$ are independent intrinsic LMC specifications with the components of $\mathbf{w}^{(u)}$ having correlation function ρ_u. The cross-covariance matrix associated with (10.48) takes the form

$$C(\mathbf{s} - \mathbf{s}') = \sum_{u=1}^{r} \rho_u(\mathbf{s} - \mathbf{s}')T^{(u)} , \tag{10.49}$$

with $T^{(u)} = A^{(u)}(A^{(u)})^\mathrm{T}$. The $T^{(u)}$ are full rank and are referred to as *coregionalization matrices*. Expression (10.49) can be compared to (10.45). Note that r need not be equal to p, but $\boldsymbol{\Sigma}_{\mathbf{Y}(\mathbf{s})} = \sum_u T^{(u)}$. Also, work of Vargas-Guzmán et al. [2002] allows the $\mathbf{w}^{(u)}(\mathbf{s})$, hence the $\mathbf{Y}^{(u)}(\mathbf{s})$ in (10.48), to be dependent.

Returning to the more general LMC, in applications we introduce (10.45) as a spatial random effects component of a general multivariate spatial model for the data. That is, we assume

$$\mathbf{Y}(\mathbf{s}) = \boldsymbol{\mu}(\mathbf{s}) + \mathbf{v}(\mathbf{s}) + \boldsymbol{\epsilon}(\mathbf{s}) , \tag{10.50}$$

where $\boldsymbol{\epsilon}(\mathbf{s})$ is a white noise vector, i.e., $\boldsymbol{\epsilon}(\mathbf{s}) \sim N(\mathbf{0}, D)$ where D is a $p \times p$ diagonal matrix with $(D)_{jj} = \tau_j^2$. In (10.50), $\mathbf{v}(\mathbf{s}) = A\mathbf{w}(\mathbf{s})$ following (10.45) as above, but further assuming that the $w_j(\mathbf{s})$ are mean-zero Gaussian processes. Lastly $\boldsymbol{\mu}(\mathbf{s})$ arises from $\mu_j(\mathbf{s}) = \mathbf{X}_j^\mathrm{T}(\mathbf{s})\boldsymbol{\beta}_j$. Each component can have its own set of covariates with its own coefficient vector.

As in Section 6.1, (10.50) can be viewed as a hierarchical model. At the first stage, given $\{\boldsymbol{\beta}_j, j = 1, \cdots, p\}$ and $\{\mathbf{v}(\mathbf{s}_i)\}$, the $\mathbf{Y}(\mathbf{s}_i), i = 1, \cdots, n$ are conditionally independent with $\mathbf{Y}(\mathbf{s}_i) \sim N(\boldsymbol{\mu}(\mathbf{s}_i) + \mathbf{v}(\mathbf{s}_i), D)$. At the second stage, the joint distribution of $\mathbf{v} \equiv (\mathbf{v}(\mathbf{s}_1), \cdots, \mathbf{v}(\mathbf{s}_n))^\mathrm{T}$ is $N(\mathbf{0}, \sum_{j=1}^{p} H_j \otimes T_j)$, where H_j is $n \times n$ with $(\mathbf{H}_j)_{ii'} = \rho_j(\mathbf{s}_i - \mathbf{s}_{i'})$. Concatenating the $\mathbf{Y}(\mathbf{s}_i)$ into an $np \times 1$ vector $\mathbf{Y}$ (and similarly $\boldsymbol{\mu}(\mathbf{s}_i)$ into $\boldsymbol{\mu}$), we can marginalize over $\mathbf{v}$ to obtain

$$p(\mathbf{Y} \mid \{\boldsymbol{\beta}_j\}, D, \{\rho_j\}, T) = N\left(\boldsymbol{\mu}, \sum_{j=1}^{p}(H_j \otimes T_j) + I_{n \times n} \otimes D\right). \tag{10.51}$$

Prior distributions on $\{\boldsymbol{\beta}_j\}$, $\{\tau_j^2\}$, T, and the parameters of the ρ_j complete the Bayesian hierarchical model specification.

10.6.2 Unconditional and conditional Bayesian specifications

10.6.2.1 Equivalence of likelihoods

In Section 10.5.5 we briefly discussed bivariate spatial modeling using a conditional specification. Having discussed the LMC in the previous section in a fairly general context, it is worth exploring the connections between the conditional approach and the LMC in more general settings. The LMC of the previous section can be developed through a conditional approach rather than a joint modeling approach. This idea has been elaborated in, e.g., Royle and Berliner [1999] and Berliner [2000], who refer to it as a hierarchical modeling approach to multivariate spatial modeling and prediction.

In the context of say $\mathbf{v}(\mathbf{s}) = A\mathbf{w}(\mathbf{s})$ where the $w_j(\mathbf{s})$ are mean-zero Gaussian processes, by taking A to be lower triangular the equivalence and associated reparametrization are easy to see. Upon permutation of the components of $\mathbf{v}(\mathbf{s})$ we can, without loss of generality, write

$$p(\mathbf{v}(\mathbf{s})) = p(v_1(\mathbf{s})) \times p(v_2(\mathbf{s}) \mid v_1(\mathbf{s})) \times \cdots \times p(v_p(\mathbf{s}) \mid v_1(\mathbf{s}), \cdots, v_{p-1}(\mathbf{s})).$$

In the case of $p = 2$, $p(v_1(\mathbf{s}))$ is clearly $N(0, T_{11})$, i.e. $v_1(\mathbf{s}) = \sqrt{T_{11}}w_1(\mathbf{s}) = a_{11}w_1(\mathbf{s})$, $a_{11} > 0$. But

$$v_2(\mathbf{s}) \mid v_1(\mathbf{s}) \sim N\left(\frac{T_{12}v_1(\mathbf{s})}{T_{11}}, T_{22} - \frac{T_{12}^2}{T_{11}}\right), \quad \text{i.e.} \quad N\left(\frac{a_{21}}{a_{11}}v_1(\mathbf{s}), a_{22}^2\right).$$

In fact, from the previous section we have $\Sigma_{\mathbf{v}} = \sum_{j=1}^{p} H_j \otimes T_j$. If we permute the rows of $\mathbf{v}$ to $\tilde{\mathbf{v}} = \left(\mathbf{v}^{(1)}, \mathbf{v}^{(2)}\right)^{\mathrm{T}}$, where $\mathbf{v}^{(l)} = (v_l(\mathbf{s}_1), \dots, v_l(\mathbf{s}_n))^{\mathrm{T}}$ for $l = 1, 2$, then $\Sigma_{\mathbf{v}} = \sum_{j=1}^{p} T_j \otimes H_j$. Again with $p = 2$ we can calculate $E(\mathbf{v}^{(2)} \mid \mathbf{v}^{(1)}) = \frac{a_{21}}{a_{11}}\mathbf{v}^{(1)}$ and $\Sigma_{\mathbf{v}^{(2)} \mid \mathbf{v}^{(1)}} = a_{22}^2 H_2$. But this is exactly the mean and covariance structure associated with variables $\{v_2(\mathbf{s}_i)\}$ given $\{v_1(\mathbf{s}_i)\}$, i.e., with $v_2(\mathbf{s}_i) = \frac{a_{21}}{a_{11}}v_1(\mathbf{s}_i) + a_{22}w_2(\mathbf{s}_i)$. Note that as in Subsection 10.5, there is no notion of a *conditional* process here. Again there is only a joint distribution for $\mathbf{v}^{(1)}, \mathbf{v}^{(2)}$ given any n and any $\mathbf{s}_1, \cdots, \mathbf{s}_n$, hence a conditional distribution for $\mathbf{v}^{(2)}$ given $\mathbf{v}^{(1)}$.

Suppose we write $v_1(\mathbf{s}) = \sigma_1 w_1(\mathbf{s})$ where $\sigma_1 > 0$ and $w_1(\mathbf{s})$ is a mean 0 spatial process with variance 1 and correlation function ρ_1 and we write $v_2(\mathbf{s}) \mid v_1(\mathbf{s}) = \alpha v_1(\mathbf{s}) + \sigma_2 w_2(\mathbf{s})$ where $\sigma_2 > 0$ and $w_2(\mathbf{s})$ is a mean 0 spatial process with variance 1 and correlation function ρ_2. The parametrization $(\alpha, \sigma_1, \sigma_2)$ is obviously equivalent to (a_{11}, a_{12}, a_{22}), i.e., $a_{11} = \sigma_1$, $a_{21} = \alpha\sigma_1$, $a_{22} = \sigma_2$ and hence to T, i.e., to (T_{11}, T_{12}, T_{22}), that is, $T_{11} = \sigma_1^2$, $T_{12} = \alpha\sigma_1^2$, $T_{22} = \alpha^2\sigma_1^2 + \sigma_2^2$.

Extension to general p is straightforward but notationally messy. We record the transformations for $p = 3$ for future use. First, $v_1(\mathbf{s}) = \sigma_1 w_1(\mathbf{s})$, $v_2(\mathbf{s}) \mid v_1(\mathbf{s}) = \alpha^{(2|1)}v_1(\mathbf{s}) + \sigma_2 w_2(\mathbf{s})$ and $v_3(\mathbf{s}) \mid v_1(\mathbf{s}), v_2(\mathbf{s}) = \alpha^{(3|1)}v_1(\mathbf{s}) + \alpha^{(3|2)}v_2(\mathbf{s}) + \sigma_3 w_3(\mathbf{s})$. Then $a_{11} = \sigma_1$, $a_{21} = \alpha^{(2|1)}\sigma_1$, $a_{22} = \sigma_2$, $a_{31} = \alpha^{(3|1)}\sigma_1$, $a_{32} = \alpha^{(3|2)}\sigma_2$ and $a_{33} = \sigma_3$. But also $a_{11} = \sqrt{T_{11}}$,

$$a_{21} = \frac{T_{12}}{\sqrt{T_{11}}}, \quad a_{22} = \sqrt{T_{22} - \frac{T_{12}^2}{T_{11}}}, \quad a_{31} = \frac{T_{13}}{\sqrt{T_{11}}}, \quad a_{32} = \sqrt{\frac{T_{11}T_{23} - T_{12}T_{13}}{T_{11}(T_{11}T_{22} - T_{12}^2)}}, \quad \text{and } a_{33} =$$

$$\sqrt{T_{33} - \frac{T_{13}^2}{T_{11}} - \frac{(T_{11}T_{23} - T_{12}T_{13})^2}{T_{11}(T_{11}T_{12} - T_{12}^2)}}.$$

Advantages to working with the conditional form of the model are certainly computational and possibly mechanistic or interpretive. For the former, with the "σ, α" parametrization, the likelihood factors and thus, with a matching prior factorization, models can be fitted componentwise. Rather than the $pn \times pn$ covariance matrix involved in working with $\mathbf{v}$ we obtain p covariance matrices each of dimension $n \times n$, one for $\mathbf{v}^{(1)}$, one for $\mathbf{v}^{(2)} \mid \mathbf{v}^{(1)}$, etc. Since likelihood evaluation with spatial processes is more than an order n^2 calculation, there can be substantial computational savings in using the conditional model. If there is some natural chronology or perhaps causality in events, then this would determine a natural order for conditioning and hence suggest natural conditional specifications. For example, in the illustrative commercial real estate setting of Example 10.3, we have the income (I) generated by an apartment block and the selling price (P) for the block. A natural modeling order here is I, then P given I.

10.6.2.2   Equivalence of prior specifications

Working in a Bayesian context, it is appropriate to ask about the choice of parametrization with regard to prior specification. Suppose we let $\boldsymbol{\phi}_j$ be the parameters associated with the correlation function ρ_j. Let $\boldsymbol{\phi}^{\mathrm{T}} = (\boldsymbol{\phi}_1, \cdots, \boldsymbol{\phi}_p)$. Then the distribution of $\mathbf{v}$ depends upon T and $\boldsymbol{\phi}$. Suppose we assume *a priori* that $p(T, \boldsymbol{\phi}) = p(T)p(\boldsymbol{\phi}) = p(T)\prod_j p(\boldsymbol{\phi}_j)$. Then reparametrization, using obvious notation, to the $(\boldsymbol{\sigma}, \boldsymbol{\alpha})$ space results on a prior $p(\boldsymbol{\sigma}, \boldsymbol{\alpha}, \boldsymbol{\phi}) = p(\boldsymbol{\sigma}, \boldsymbol{\alpha})\prod_j p(\boldsymbol{\phi}_j)$.

Standard prior specification for T would of course be an inverse Wishart, while standard modeling for $(\sigma^2, \boldsymbol{\alpha})$ would be a product inverse gamma by normal form. In the present situation, when will they agree? We present the details for the $p = 2$ case. The Jacobian from $T \to (\sigma_1, \sigma_2, \alpha)$ is $|\mathbf{J}| = \sigma_1^2$, hence in the reverse direction it is $1/T_{11}$. Also $|T| = T_{11}T_{22} - T_{12}^2 = \sigma_1^2\sigma_2^2$ and

$$T^{-1} = \frac{1}{T_{11}T_{22} - T_{12}^2}\begin{pmatrix} T_{22} & -T_{12} \\ -T_{12} & T_{11} \end{pmatrix} = \frac{1}{\sigma_1^2\sigma_2^2}\begin{pmatrix} \alpha^2\sigma_1^2 + \sigma_2^2 & -\alpha\sigma_1^2 \\ -\alpha\sigma_1^2 & \sigma_1^2 \end{pmatrix}.$$

After some manipulation we have the following result:

Result 1: $T \sim IW_2(\nu, (\nu'D)^{-1})$; that is,

$$p(T) \propto |\mathbf{T}|^{-\frac{\nu+3}{2}}\exp\left\{-\frac{1}{2}tr(\nu'D\mathbf{T}^{-1})\right\},$$

where $D = \mathrm{Diag}(d_1, d_2)$ and $\nu' = \nu - 3$ if and only if

$$\sigma_1^2 \sim IG\left(\frac{\nu-1}{2}, \frac{d_1}{2}\right), \; \sigma_2^2 \sim IG\left(\frac{\nu+1}{2}, \frac{d_2}{2}\right), \; \text{and} \; \alpha \mid \sigma_2^2 \sim N\left(0, \frac{\sigma_2^2}{d_1}\right).$$

Note also that the prior in $(\boldsymbol{\sigma}, \boldsymbol{\alpha})$ space factors into $p(\sigma_1^2)p(\sigma_2^2, \alpha)$ to match the likelihood factorization.

This result is obviously order dependent. If we condition in the reverse order, σ_1^2, σ_2^2, and α no longer have the same meanings. In fact, writing this parametrization as $(\tilde{\sigma}_1^2, \tilde{\sigma}_2^2, \tilde{\alpha})$, we obtain equivalence to the above inverse Wishart prior for T if and only if $\tilde{\sigma}_1^2 \sim IG\left(\frac{\nu+1}{2}, \frac{d_1}{2}\right)$, $\tilde{\sigma}_2^2 \sim IG\left(\frac{\nu-1}{2}, \frac{d_2}{2}\right)$, and $\tilde{\alpha} \mid \tilde{\sigma}_1^2 \sim N\left(0, \frac{\tilde{\sigma}_1^2}{d_1}\right)$.

The result can be extended to $p > 2$ but the expressions become messy. However, if $p = 3$ we have:

Result 2: $T \sim IW_3(\nu, (\nu'D)^{-1})$, that is,

$$p(T) \propto |\mathbf{T}|^{-\frac{\nu+4}{2}}\exp\left\{-\frac{1}{2}tr(\nu'D\mathbf{T}^{-1})\right\},$$

where now where $D = Diag(d_1, d_2, d_3)$ and $\nu' = \nu - 3 + 1$ if and only if $\sigma_1^2 \sim IG\left(\frac{\nu-2}{2}, \frac{d_1}{2}\right)$, $\sigma_2^2 \sim IG\left(\frac{\nu}{2}, \frac{d_2}{2}\right)$, $\sigma_3^2 \sim IG\left(\frac{\nu+2}{2}, \frac{d_3}{2}\right)$, $\alpha^{(2|1)} \mid \sigma_2^2 \sim N\left(0, \frac{\sigma_2^2}{d_1}\right)$, $\alpha^{(3|1)} \mid \sigma_3^2 \sim N\left(0, \frac{\sigma_3^2}{d_1}\right)$, and $\alpha^{(3|2)} \mid \sigma_3^2 \sim N\left(0, \frac{\sigma_3^2}{d_2}\right)$. Though there is a one-to-one transformation from T-space to $(\boldsymbol{\sigma}, \boldsymbol{\alpha})$-space, a Wishart prior with nondiagonal D implies a nonstandard prior on $(\boldsymbol{\sigma}, \boldsymbol{\alpha})$-space. Moreover, it implies that the prior in $(\boldsymbol{\sigma}, \boldsymbol{\alpha})$-space will not factor to match the likelihood factorization.

Returning to the model in (10.50), the presence of white noise in (10.50) causes difficulties with the attractive factorization of the likelihood under conditioning. Consider again the $p = 2$ case. If

$$\begin{aligned} Y_1(\mathbf{s}) &= \mathbf{X}_1^{\mathrm{T}}(\mathbf{s})\boldsymbol{\beta}_1 + v_1(\mathbf{s}) + \epsilon_1(\mathbf{s}) \\ \text{and } Y_2(\mathbf{s}) &= \mathbf{X}_2^{\mathrm{T}}(\mathbf{s})\boldsymbol{\beta}_1 + v_2(\mathbf{s}) + \epsilon_2(\mathbf{s})\,, \end{aligned} \tag{10.52}$$

then the conditional form of the model writes

$$\begin{aligned} Y_1(\mathbf{s}) &= \mathbf{X}_1^{\mathrm{T}}(\mathbf{s})\boldsymbol{\beta}_1 + \sigma_1 w_1(\mathbf{s}) + \tau_1 u_1(\mathbf{s}) \\ \text{and } Y_2(\mathbf{s}) \mid Y_1(\mathbf{s}) &= \mathbf{X}_2^{\mathrm{T}}(\mathbf{s})\boldsymbol{\beta}_2 + \alpha Y_1(\mathbf{s}) + \sigma_2 w_2(\mathbf{s}) + \tau_2 u_2(\mathbf{s})\,. \end{aligned} \tag{10.53}$$

In (10.53), $w_1(\mathbf{s})$ and $w_2(\mathbf{s})$ are as above with $u_1(\mathbf{s})$, $u_2(\mathbf{s}) \sim N(0,1)$, independent of each other and the $w_l(\mathbf{s})$. But then, unconditionally, $Y_2(\mathbf{s})$ equals

$$\begin{aligned} & \mathbf{X}_2^{\mathrm{T}}(\mathbf{s})\widetilde{\boldsymbol{\beta}}_2 + \alpha\left(\mathbf{X}_1^{\mathrm{T}}(\mathbf{s})\boldsymbol{\beta}_1 + \sigma_1 w_1(\mathbf{s}) + \tau_1 u_1(\mathbf{s})\right) + \sigma_2 w_2(\mathbf{s}) + \tau_2 u_2(\mathbf{s}) \\ & = \mathbf{X}_2^{\mathrm{T}}(\mathbf{s})\widetilde{\boldsymbol{\beta}}_2 + \mathbf{X}_1^{\mathrm{T}}(\mathbf{s})\alpha\boldsymbol{\beta}_1 + \alpha\sigma_1 w_1(\mathbf{s}) + \sigma_2 w_2(\mathbf{s}) + \alpha\tau_1 u_1(\mathbf{s})) + \tau_2 u_2(\mathbf{s})\,. \end{aligned} \tag{10.54}$$

In attempting to align (10.54) with (10.52) we require $\mathbf{X}_2(\mathbf{s}) = \mathbf{X}_1(\mathbf{s})$, whence $\boldsymbol{\beta}_2 = \widetilde{\boldsymbol{\beta}}_2 + \alpha\boldsymbol{\beta}_1$. We also see that $v_2(\mathbf{s}) = \alpha\sigma_1 w_1(\mathbf{s}) + \sigma_2 w_2(\mathbf{s})$. But, perhaps most importantly, $\epsilon_2(\mathbf{s}) = \alpha\tau_1 u_1(\mathbf{s}) + \tau_2 u_2(\mathbf{s})$. Hence $\epsilon_1(\mathbf{s})$ and $\epsilon_2(\mathbf{s})$ are not independent, violating the white noise modeling assumption associated with (11.3). If we have a white noise component in the model for $Y_1(\mathbf{s})$ and also in the conditional model for $Y_2(\mathbf{s}) \mid Y_1(\mathbf{s})$ we do not have a white noise component in the unconditional model specification. Obviously, the converse is true as well.

If $\mathbf{u}_1(\mathbf{s}) = 0$, i.e., the $Y_1(\mathbf{s})$ process is purely spatial, then, again with $\mathbf{X}_2(\mathbf{s}) = \mathbf{X}_1(\mathbf{s})$, the conditional and marginal specifications agree up to reparametrization. More precisely, the parameters for the unconditional model are $\boldsymbol{\beta}_1$, $\boldsymbol{\beta}_2$, τ_2^2 with T_{11}, T_{12}, T_{22}, $\boldsymbol{\phi}_1$, and $\boldsymbol{\phi}_2$. For the conditional model we have $\boldsymbol{\beta}_1$, $\boldsymbol{\beta}_2$, τ_2^2 with σ_1, σ_2, α, $\boldsymbol{\phi}_1$, and $\boldsymbol{\phi}_2$. We can appeal to the equivalence of (T_{11}, T_{12}, T_{22}) and $(\sigma_1, \sigma_2, \alpha)$ as above. Also note that if we extend (10.52) to $p > 2$, in order to enable conditional and marginal specifications to agree, we will require a common covariate vector and that $u_1(\mathbf{s}) = u_2(\mathbf{s}) = \cdots = u_{p-1}(\mathbf{s}) = 0$, i.e., that all but one of the processes is purely spatial.

10.7 Spatially varying coefficient models

In Section 12.1 we introduce a spatially varying coefficient process in the evolution equation of the spatiotemporal dynamic model. Similarly, in Section 15.4 we consider multiple spatial frailty models with regression coefficients that were allowed to vary spatially. Here, we develop this topic more fully to amplify the scope of possibilities for such modeling. In particular, in the spatial-only case, we denote the value of the coefficient at location $\mathbf{s}$ by $\beta(\mathbf{s})$. This coefficient can be resolved at either areal unit or point level. With the former, the $\beta(\mathbf{s})$ surface consists of "tiles" at various heights, one tile per areal unit. The former are, perhaps, more natural with areal data (see Section 11.1). For the latter, we achieve a more flexible spatial surface.

Using tiles, concern arises regarding the arbitrariness of the scale of resolution, the lack of smoothness of the surface, and the fact that interpolation of the surface is not resolved to individual locations. When working with point-referenced data it will be more attractive to allow the coefficients to vary by location, to envision a particular coefficient, a spatial surface. For instance, in our example below we also model the (log) selling price of single-family houses. Customary explanatory variables include the age of the house, the square feet of living area, the square feet of other area, and the number of bathrooms. If the region of interest is a city or greater metropolitan area, it is evident that the capitalization rate (e.g., for age) will vary across the region. In some parts of the region older houses will be more valued than in other parts. By allowing the coefficient of age to vary with location, we can remedy the foregoing concerns. With practical interest in mind (say, real estate appraisal), we can predict the coefficient for arbitrary properties, not just for those that sold during the period of investigation. Similar issues arise in modeling environmental exposure to a particular pollutant where covariates might include temperature and precipitation.

One possible approach would be to model the spatial surface for the coefficient parametrically. In the simplest case this would require the rather arbitrary specification of a polynomial surface function; surfaces too limited or inflexible might result. More flexibility could be introduced using a spline surface over two- dimensional space; see, e.g., Luo and Wahba (1998) and references therein. However, this requires the selection of a spline function and determination of the number of and locations of the knots in the space. Also, with multiple coefficients, a multivariate specification of a spline surface is required.

The approach we adopt here is arguably more natural and at least as flexible. We model the spatially varying coefficient surface as a realization from a spatial process. For multiple coefficients we employ a multivariate spatial process model.

To clarify interpretation and implementations, we first develop our general approach in the case of a single covariate, hence two spatially varying coefficient processes, one for the "intercept" and one for the "slope." We then turn to the case of multiple covariates. Recall that, even when fitting a simple linear regression, the slope and intercept are almost always strongly (and, usually, inversely) correlated. (This is intuitive if one envisions overlaying random lines that are likely relative to a fixed scatterplot of the data points.) So, if we extend to a process model for each, it seems clear that we would want the processes to be dependent (and the same reasoning would extend to the case of multiple covariates. Hence, we employ a multivariate process model. Indeed we present a further generalization to build a spatial analogue of a multilevel regression model [see, e.g., Goldstein, 2003]. We also consider flexible spatiotemporal possibilities (anticipating Chapter 12). The previously mentioned real estate setting provides site level covariates whose coefficients are of considerable practical interest and a data set of single-family home sales from Baton Rouge, LA, enables illustration. Except for regions exhibiting special topography, we anticipate that a spatially varying coefficient model will prove more useful than, for instance, a trend surface model. That is, incorporating a polynomial in latitude and longitude into the mean structure would not be expected to serve as a surrogate for allowing a coefficient for, say, age or living area of a house to vary across the region.

10.7.1 Approach for a single covariate

Recall the usual Gaussian stationary spatial process model as in (6.1),

$$Y(\mathbf{s}) = \mu(\mathbf{s}) + w(\mathbf{s}) + \epsilon(\mathbf{s}) , \qquad (10.55)$$

where $\mu(\mathbf{s}) = \mathbf{x}(\mathbf{s})^{\mathrm{T}}\beta$ and $\epsilon(\mathbf{s})$ is a white noise process, i.e., $\mathrm{E}(\epsilon(\mathbf{s})) = 0$, $\mathrm{Var}(\epsilon(\mathbf{s})) = \tau^2$, $\mathrm{Cov}(\epsilon(\mathbf{s}), \epsilon(\mathbf{s}')) = 0$, and $w(\mathbf{s})$ is a second-order stationary mean-zero process independent of the white noise process, i.e., $\mathrm{E}(w(\mathbf{s})) = 0$, $\mathrm{Var}(w(\mathbf{s})) = \sigma^2$, $\mathrm{Cov}(w(\mathbf{s}), w(\mathbf{s}')) = \sigma^2\rho(\mathbf{s}, \mathbf{s}'; \phi)$, where ρ is a valid two-dimensional correlation function.

Letting $\mu(\mathbf{s}) = \beta_0 + \beta_1 x(\mathbf{s})$, write $w(\mathbf{s}) = \beta_0(\mathbf{s})$ and define $\tilde{\beta}_0(\mathbf{s}) = \beta_0 + \beta_0(\mathbf{s})$. Then $\beta_0(\mathbf{s})$ can be interpreted as a random spatial adjustment at location $\mathbf{s}$ to the overall intercept β_0. Equivalently, $\tilde{\beta}_0(\mathbf{s})$ can be viewed as a random intercept process. For an observed set of locations $\mathbf{s}_1, \mathbf{s}_2, \ldots, \mathbf{s}_n$ given $\beta_0, \beta_1, \{\beta_0(\mathbf{s}_i)\}$ and τ^2, the $Y(\mathbf{s}_i) = \beta_0 + \beta_1 x(\mathbf{s}_i) + \beta_0(\mathbf{s}_i) + \epsilon(\mathbf{s}_i), i = 1, \ldots, n$, are conditionally independent. Then $L(\beta_0, \beta_1, \{\beta_0(\mathbf{s}_i)\}, \tau^2; \mathbf{y})$, the first-stage likelihood, is

$$(\tau^2)^{-\frac{n}{2}} \exp\left\{ -\frac{1}{2\tau^2} \sum (Y(\mathbf{s}_i) - (\beta_0 + \beta_1 x(\mathbf{s}_i) + \beta_0(\mathbf{s}_i))^2 \right\} . \tag{10.56}$$

In obvious notation, the distribution of $\mathbf{B}_0 = (\beta_0(\mathbf{s}_1), \ldots, \beta_0(\mathbf{s}_n))^{\mathrm{T}}$ is

$$f(\mathbf{B}_0 \mid \sigma_0{}^2, \phi_0) = N(\mathbf{0}, \sigma_0{}^2 H_0(\phi_0)) , \tag{10.57}$$

where $(H_0(\phi_0))_{ij} = \rho_0(\mathbf{s}_i - \mathbf{s}_j; \phi_0)$. For all of the discussion and examples below, we adopt the Matérn correlation function, (2.8). With a prior on $\beta_0, \beta_1, \tau^2, \sigma_0^2$, and ϕ_0, specification of the Bayesian hierarchical model is completed. Under (10.56) and (10.57), we can integrate over $\mathbf{B}_0$, obtaining $L(\beta_0, \beta_1, \tau^2, \sigma_0{}^2, \phi_0; \mathbf{y})$, the marginal likelihood, as

$$|\sigma_0^2 H_0(\phi_0) + \tau^2 I|^{-\frac{1}{2}} \times \exp\left\{ -\frac{1}{2} (\mathbf{y} - \beta_0 \mathbf{1} - \beta_1 \mathbf{x})^{\mathrm{T}} (\sigma_0^2 H_0(\phi_0) + \tau^2 I)^{-1} (\mathbf{y} - \beta_0 \mathbf{1} - \beta_1 \mathbf{x}) \right\} , \tag{10.58}$$

where $\mathbf{x} = (x(\mathbf{s}_1), \ldots, x(\mathbf{s}_n))^{\mathrm{T}}$.

Following Gelfand et al. [2003], the foregoing development immediately suggests how to formulate a spatially varying coefficient model. Suppose we write

$$Y(\mathbf{s}) = \beta_0 + \beta_1 x(\mathbf{s}) + \beta_1(\mathbf{s}) x(\mathbf{s}) + \epsilon(\mathbf{s}) . \tag{10.59}$$

In (10.59), $\beta_1(\mathbf{s})$ is a second-order stationary mean-zero Gaussian process with variance σ_1^2 and correlation function $\rho_1(\cdot; \phi_1)$. Also, let $\tilde{\beta}_1(\mathbf{s}) = \beta_1 + \beta_1(\mathbf{s})$. Now $\beta_1(\mathbf{s})$ can be interpreted as a random spatial adjustment at location $\mathbf{s}$ to the overall slope β_1. Equivalently, $\tilde{\beta}_1(\mathbf{s})$ can be viewed as a random slope process. In effect, we are employing an uncountable dimensional function to explain the relationship between $x(\mathbf{s})$ and $Y(\mathbf{s})$. This model might be characterized as *locally linear*; however, it is difficult to imagine a more flexible specification for the relationship between $x(\mathbf{s})$ and $Y(\mathbf{s})$.

Expression (10.59) yields obvious modification of (10.56) and (10.57). In particular, the resulting in marginalized likelihood becomes

$$L(\beta_0, \beta_1, \tau^2, \sigma_1{}^2, \phi_1; \mathbf{y}) = |\sigma_1{}^2 D_x H_1(\phi_1) D_x + \tau^2 I|^{-\frac{1}{2}} \times \exp\left\{ -\frac{1}{2} Q \right\} , \tag{10.60}$$

where $Q = (\mathbf{y} - \beta_0 \mathbf{1} - \beta_1 \mathbf{x})^{\mathrm{T}} (\sigma_1^2 D_x H_1(\phi_1) D_x + \tau^2 I)^{-1} (\mathbf{y} - \beta_0 \mathbf{1} - \beta_1 \mathbf{x})$ and D_x is diagonal with $(D_x)_{ii} = x(\mathbf{s}_i)$. With $\mathbf{B}_1 = (\beta_1(\mathbf{s}_1), \ldots, \beta_1(\mathbf{s}_n))^{\mathrm{T}}$ we can sample $f(\mathbf{B}_1 \mid \mathbf{y})$ and $f(\beta_1(\mathbf{s}_{new}) \mid \mathbf{y})$ via composition.

Note that (10.59) provides a heterogeneous, nonstationary process for the data regardless of the choice of covariance function for the $\beta_1(\mathbf{s})$ process, since $\mathrm{Var}(Y(\mathbf{s}) \mid \beta_0, \beta_1, \tau^2, \sigma_1{}^2, \phi_1) = x^2(\mathbf{s}) \sigma_1{}^2 + \tau^2$ and $\mathrm{Cov}(Y(\mathbf{s}), Y(\mathbf{s}') \mid \beta_0, \beta_1, \tau^2, \sigma_1{}^2, \phi_1) = \sigma_1{}^2 x(\mathbf{s}) x(\mathbf{s}') \rho_1(\mathbf{s} - \mathbf{s}'; \phi_1)$. As a result, we observe that in practice, (10.59) is sensible only if we have $x(\mathbf{s}) > 0$. In fact, centering and scaling, which is usually advocated for better behaved model fitting, is inappropriate here. With centered $x(\mathbf{s})$'s we would find the likely untenable behavior that $\mathrm{Var}(Y(\mathbf{s}))$ decreases and then increases in $x(\mathbf{s})$. Worse, for an essentially central $x(\mathbf{s})$ we would find $Y(\mathbf{s})$ essentially independent of $Y(\mathbf{s}')$ for any $\mathbf{s}'$. Also, scaling the $x(\mathbf{s})$'s accomplishes nothing. $\beta_1(\mathbf{s})$ would be inversely rescaled since the model only identifies $\beta_1(\mathbf{s}) x(\mathbf{s})$.

This leads to concerns regarding possible approximate collinearity of $\mathbf{x}$, the vector of $x(\mathbf{s}_i)$'s, with the vector $\mathbf{1}$. Expression (10.60) shows that a badly behaved likelihood will arise if $\mathbf{x} \approx c\mathbf{1}$. But, we can reparametrize (10.59) to $Y(\mathbf{s}) = \beta_0' + \beta_1'\tilde{x}(\mathbf{s}) + \beta_1(\mathbf{s})x(\mathbf{s}) + \epsilon(\mathbf{s})$ where $\tilde{x}(\mathbf{s})$ is centered and scaled with obvious definitions for β_0' and β_1'. Now $\tilde{\beta}_1(\mathbf{s}) = \beta_1'/s_x + \beta_1(\mathbf{s})$ where s_x is the sample standard deviation of the $x(\mathbf{s})$'s.

As below (10.58), we can draw an analogy with standard longitudinal linear growth curve modeling, where $Y_{ij} = \beta_0 + \beta_1 x_{ij} + \beta_{1i} x_{ij} + \epsilon_{ij}$, i.e., a random slope for each individual. For growth curve models, we consider both a population level growth curve (marginalizing over the random effects) as well as these individual level growth curves. In this regard, here, $Y(\mathbf{s}) = \beta_0 + \beta_1 x(\mathbf{s}) + \epsilon(\mathbf{s})$ provides the global growth curve while (10.61) below also introduces the local growth curves [see Banerjee and Johnson, 2006, for spatially-varying growth curves]. We will discuss the richness afforded by spatially-varying growth curves in Section 10.7.2 below.

The general specification incorporating both $\beta_0(\mathbf{s})$ and $\beta_1(\mathbf{s})$ would be

$$Y(\mathbf{s}) = \beta_0 + \beta_1 x(\mathbf{s}) + \beta_0(\mathbf{s}) + \beta_1(\mathbf{s})x(\mathbf{s}) + \epsilon(\mathbf{s}). \tag{10.61}$$

Expression (10.61) parallels the usual linear growth curve modeling by introducing both an intercept process and a slope process. The model in (10.61) requires a bivariate process specification in order to determine the joint distribution of $\mathbf{B}_0$ and $\mathbf{B}_1$. It also partitions the total error into intercept process error, slope error, and pure error. A noteworthy remark is that we can fit the bivariate spatial process model in (10.61) without ever observing it. That is, we only observe the $Y(\mathbf{s})$ process. This demonstrates the power of hierarchical modeling with structured dependence.

10.7.2   Spatially-varying growth curves

Following Banerjee and Johnson [2006] we take a closer look at the richness afforded by spatially-varying coefficient processes in the context of growth curves. Growth curves are widely employed in longitudinal data analysis [Diggle et al., 2013]. If Y_{it} is the response recorded at a location indexed by i in time t, then a growth curve model that recognizes the spatial indexing but does not introduce spatial dependence using a process is

$$Y_{it} = \mathbf{x}_{it}^{\mathsf{T}}\boldsymbol{\beta} + f_i(t) + \epsilon_{it}, \; i = 1, \ldots, N \tag{10.62}$$

where $\mathbf{x}_{it}$ is a vector of covariates specific to the measurements at the location indexed by i and time t, $f_i(t)$ is a function capturing the growth through time and $\epsilon_{it} \overset{ind}{\sim} N(0, \tau_i^2)$ is measurement error with variance specific to the index i.

Equation (10.62) includes several growth curve models according to how $f_i(t)$ is specified. If no variation in growth patterns is expected across spatial locations, a *uniform* growth curve model, where $f_i(t)$ does not depend upon i will suffice. Linear functions yield models such as

$$f_i(t) = \alpha_0 + \alpha_1 t, \; \boldsymbol{\alpha} \sim N(\mathbf{0}, \text{Diag}(\sigma_0^2, \sigma_1^2)); \quad f_i(t) = \alpha_0 + \alpha_1 t, \; \boldsymbol{\alpha} \sim N(\mathbf{0}, \Lambda); \tag{10.63}$$

$$f_i(t) = \alpha_{0i} + \alpha_{1i} t, \; \boldsymbol{\alpha}_i \overset{i.i.d}{\sim} N(\mathbf{0}, \Lambda), \tag{10.64}$$

where $\boldsymbol{\alpha} = (\alpha_0, \alpha_1)^{\mathsf{T}}$ in the first two models in (10.63), while in the third model given by (10.64) we introduce coefficients $\boldsymbol{\alpha}_i = (\alpha_{0i}, \alpha_{1i})^{\mathsf{T}}$ specific to the location index. In the first model, α_0 and α_1 are independent normal coefficients, in the second they are correlated with covariance matrix Λ, and in the third model $\boldsymbol{\alpha}_i = (\alpha_{0i}, \alpha_{1i})^{\mathsf{T}}$ are i.i.d. bivariate normal distributions with covariance matrix Λ. The third model allows independent variation across sites and correlation between the intercept and slope *within* each site, but does not account for correlation *among* the sites.

If we expect neighboring locations to exhibit similar growth patterns, perhaps because they share similar topographic and environmental conditions, then we may wish to introduce spatial dependence among the growth curves. Let N be the number of locations and let us now denote the response recorded in location $\mathbf{s}$ at time t as $Y_t(\mathbf{s})$. Also, $\mathbf{x}_t(\mathbf{s})$ is the associated vector of covariates, and we now write our model as

$$Y_t(s) = \mathbf{x}_t(s)^{\mathrm{T}}\boldsymbol{\beta} + f(\mathbf{s},t) + \epsilon_t(\mathbf{s}), \tag{10.65}$$

where $f(\mathbf{s},t) = \alpha_0(\mathbf{s}) + \alpha_1(\mathbf{s})t$ and $\epsilon_t(\mathbf{s})$ are i.i.d. $N(0,\tau^2)$ for any finite collection of spatooal locations. Unlike in (10.64), where $\boldsymbol{\alpha}_i$ indexed the site i, here $\alpha_0(\mathbf{s})$ and $\alpha_1(\mathbf{s})$ are spatial processes introducing dependence among locations.

Modeling the intercept and slope as spatial processes enable us to estimate the random fields associated with the intercept and slope and to carry out predictive inference for growth patterns at arbitrary locations. We can model spatial association for the intercept and slope processes and the correlation between them using a bivariate Gaussian process $\boldsymbol{\alpha}(\mathbf{s}) = (\alpha_0(s), \alpha_1(s))^{\mathrm{T}}$, denoted by $GP(\boldsymbol{\mu}(\cdot), C_\alpha(\cdot,\cdot))$. Here, $\boldsymbol{\mu}(\mathbf{s}) = (\mu_0(\mathbf{s}), \mu_1(\mathbf{s}))^{\mathrm{T}}$ is the vector of process mean functions, and $C_\alpha(\cdot,\cdot)$ is a 2×2 cross-covariance function for $\boldsymbol{\alpha}(\mathbf{s})$. Equivalently, we may define $\tilde{\alpha}_0(\mathbf{s}) = \alpha_0(\mathbf{s}) - \mu_0(\mathbf{s})$, and $\tilde{\alpha}_1(\mathbf{s}) = \alpha_1(\mathbf{s}) - \mu_1(\mathbf{s})$. These yield a corresponding zero-centered bivariate Gaussian process, $(\tilde{\alpha}_0(\mathbf{s}), \tilde{\alpha}_1(\mathbf{s}))^{\mathrm{T}} \sim GP(\mathbf{0}, C_\alpha(\cdot,\cdot))$. In particular, when the process means are constant across sites, say $\mu_0(\mathbf{s}) = \mu_0$ and $\mu_1(\mathbf{s}) = \mu_1$, the growth function is the sum of a non-spatial component and a spatially-varying component, $f(\mathbf{s},t) = g(t) + \tilde{f}(\mathbf{s},t)$, where $\tilde{f}(\mathbf{s},t) = \tilde{\alpha}_0(\mathbf{s}) + \tilde{\alpha}_1(\mathbf{s})t$ and $g(t) = \mu_0 + \mu_1 t$.

While the cross-covariance function $C_\alpha(\mathbf{s},\mathbf{s}')$ can be constructed using a variety of methods [see, e.g., Genton and Kleiber, 2015, for a comprehensive review], the linear model of regionalization introduced earlier in Section 10.6.1 allows us to build models of increasing richness in covariance structures. For example, Banerjee and Johnson [2006] construct and evaluate the following three regionalization models,

$$C_\alpha(\mathbf{s} - \mathbf{s}') = \begin{pmatrix} \sigma_0^2 \rho_0(\mathbf{s} - \mathbf{s}') & 0 \\ 0 & \sigma_1^2 \rho_1(\mathbf{s} - \mathbf{s}') \end{pmatrix}; \quad C_\alpha(\mathbf{s} - \mathbf{s}') = \rho(\mathbf{s} - \mathbf{s})\Lambda; \tag{10.66}$$

$$C_\alpha(\mathbf{s} - \mathbf{s}') = A C_v(\mathbf{s} - \mathbf{s}') A^{\mathrm{T}}; \text{ where } A \text{ is lower-triangular.} \tag{10.67}$$

The first model in (10.66) assumes that $\alpha_0(s)$ and $\alpha_1(s)$ are independent processes with possibly different correlation structures. The second model associates $\alpha_0(s)$ and $\alpha_1(s)$ through an "intrinsic" specification so that $\Lambda = C_\alpha(\mathbf{0})$ is a 2×2 variance-covariance matrix between the processes (within each site), and $\rho(\mathbf{s} - \mathbf{s}')$ is a common spatial correlation function capturing spatial association. This is precisely the separable modeling, which factors out the cross-covariance function into a spatial and non-spatial component. Separability has nicer interpretations and offers significant computational benefits in likelihood evaluations, but imposes a common spatial correlation function (hence a common spatial range) for both the intercept and slope processes. Since there is no reason, a priori, to make such an assumption, this may be undesirable. Finally, the third model given by (10.67) drops separability and lets A remain an unspecified lower-triangular matrix while $C_v(\mathbf{s} - \mathbf{s}')$ is diagonal with elements $\rho_0(\mathbf{s} - \mathbf{s}')$ and $\rho_1(\mathbf{s} - \mathbf{s}')$. The nicer interpretation of separability is somewhat lost, but we achieve richer modeling with process-specific spatial ranges. Recall that AA^T is interpreted as the covariance matrix between the intercept and slope processes. Its diagonal elements play the roles of spatial variance parameters (similar to σ_0^2 and σ_1^2 in the first model) and its off-diagonals capture the association between the intercept and slope processes.

Banerjee and Johnson [2006] extend the above framework to multi-resolution spatial growth curves that would be appropriate when measurements are recorded at subplots within plots, as is often the case in agronomy and forestry. This is different from the models in (10.63) and (10.64), where we do not need to introduce dependence among measurements in the subplots. Now we distinguish between the main plot and subplots and the time points measuring the response. Letting $Y_t(\mathbf{r}, \mathbf{s})$ be the response at time t from location $(\mathbf{r}, \mathbf{s})$ (perhaps the notation $(\mathbf{s}(\mathbf{r}), \mathbf{s})$ brings out the nesting more clearly, but there is no confusion if we always consider $\mathbf{r}$ to be the coordinates of the subplot inside a main plot with centroid $\mathbf{s}$) and $\mathbf{x}_t(\mathbf{r}, \mathbf{s})$ be a $p \times 1$ vector of covariates from this location, a general multi-resolution model is given by

$$Y_t(\mathbf{r}, \mathbf{s}) = \mathbf{x}_t^{\mathrm{T}}(r, s)\boldsymbol{\beta} + f[(\mathbf{r}, \mathbf{s}), t] + \epsilon_t(\mathbf{r}, \mathbf{s}). \tag{10.68}$$

Here $\epsilon_t(\mathbf{r}, \mathbf{s}) \overset{iid}{\sim} N(0, \tau^2)$ over any finite collection of replicates within plots $(\mathbf{r}, \mathbf{s})$, denotes the measurement error, and $f[(\mathbf{r}, \mathbf{s}), t]$ is a function capturing the effect of time on the response at time t in location $(\mathbf{r}, \mathbf{s})$. Even a linear specification, $f[(\mathbf{r}, \mathbf{s}), t] = \alpha_0(\mathbf{r}, \mathbf{s}) + \alpha_1(\mathbf{r}, \mathbf{s})t$, affords richness and introduces micro-level variation in the coefficient processes. Banerjee and Johnson [2006] provide details on constructing multi-resolution cross-covariances.

10.7.3   Multivariate spatially varying coefficient models

For the case of a $p \times 1$ multivariate covariate vector $\mathbf{X}(\mathbf{s})$ at location $\mathbf{s}$ where, for convenience, $\mathbf{X}(\mathbf{s})$ includes a 1 as its first entry to accommodate an intercept, we generalize (10.61) to

$$Y(\mathbf{s}) = \mathbf{X}^{\mathrm{T}}(\mathbf{s})\tilde{\boldsymbol{\beta}}(\mathbf{s}) + \epsilon(\mathbf{s}), \tag{10.69}$$

where $\tilde{\boldsymbol{\beta}}(\mathbf{s})$ is assumed to follow a p-variate spatial process model. With observed locations $\mathbf{s}_1, \mathbf{s}_2, \ldots, \mathbf{s}_n$, let X be $n \times np$ block diagonal having as block for the ith row $\mathbf{X}^{\mathrm{T}}(\mathbf{s}_i)$. Then we can write $\mathbf{Y} = X^{\mathrm{T}}\tilde{\mathbf{B}} + \boldsymbol{\epsilon}$ where $\tilde{\mathbf{B}}$ is $np \times 1$, the concatenated vector of the $\tilde{\boldsymbol{\beta}}(s)$, and $\boldsymbol{\epsilon} \sim N(0, \tau^2 I)$.

As above, in practice, to assume that the component processes of $\tilde{\boldsymbol{\beta}}(\mathbf{s})$ are independent is likely inappropriate. The dramatic improvement in model performance when dependence is incorporated is shown in Example 10.4. To formulate a multivariate Gaussian process for $\tilde{\boldsymbol{\beta}}(\mathbf{s})$ we require the mean and the cross-covariance function. For the former, following Subsection 10.7.1, we take this to be $\boldsymbol{\mu}_\beta = (\beta_1, \ldots, \beta_p)^{\mathrm{T}}$. For the latter we require a valid p-variate choice. In the following paragraphs we work with a separable form (Section 10.3), yielding

$$\tilde{\mathbf{B}} \sim N(\mathbf{1}_{n \times 1} \otimes \boldsymbol{\mu}_\beta, \, H(\phi) \otimes T). \tag{10.70}$$

If $\tilde{\mathbf{B}} = \mathbf{B} + \mathbf{1}_{n \times 1} \otimes \boldsymbol{\mu}_\beta$, then we can write (10.69) as

$$Y(\mathbf{s}) = \mathbf{X}^{\mathrm{T}}(\mathbf{s})\boldsymbol{\mu}_\beta + \mathbf{X}^{\mathrm{T}}(\mathbf{s})\boldsymbol{\beta}(\mathbf{s}) + \epsilon(\mathbf{s}). \tag{10.71}$$

In (10.71) the total error in the regression model is partitioned into $p+1$ pieces, each with an obvious interpretation. Following Subsection 10.7.1, using (10.69) and (10.70) we can integrate over $\boldsymbol{\beta}$ to obtain

$$L(\boldsymbol{\mu}_\beta, \tau^2, T, \phi; \mathbf{y}) = |X(H(\phi) \otimes T)X^{\mathrm{T}} + \tau^2 I|^{-\frac{1}{2}} \times \exp\{-\tfrac{1}{2}Q\}, \tag{10.72}$$

where $Q = (\mathbf{y} - X(\mathbf{1} \otimes \boldsymbol{\mu}_\beta))^{\mathrm{T}}(X(H(\phi) \otimes T)X^{\mathrm{T}} + \tau^2 I)^{-1}(\mathbf{y} - X(\mathbf{1} \otimes \boldsymbol{\mu}_\beta))$. This apparently daunting form still involves only $n \times n$ matrices.

The Bayesian model is completed with a prior $p(\boldsymbol{\mu}_\beta, \tau^2, T, \phi)$, which we assume to take the product form $p(\boldsymbol{\mu}_\beta)p(\tau^2)p(T)p(\phi)$. Below, these components will be normal, inverse gamma, inverse Wishart, and gamma, respectively.

With regard to prediction, $p(\tilde{\mathbf{B}} \mid \mathbf{y})$ can be sampled one for one with the posterior samples from $f(\boldsymbol{\mu}_\beta, \tau^2, T, \phi \mid \mathbf{y})$ using $f(\tilde{\mathbf{B}} \mid \boldsymbol{\mu}_\beta, \tau^2, T, \phi, y)$, which is $N(A\mathbf{a}, A)$ where $A = (X^{\mathrm{T}}X/\tau^2 + H^{-1}(\phi) \otimes T^{-1})^{-1}$ and $\mathbf{a} = X^{\mathrm{T}}\mathbf{y}/\tau^2 + (H^{-1}(\phi) \otimes T^{-1})(\mathbf{1} \otimes \boldsymbol{\mu}_\beta)$. Here A is $np \times np$ but, for sampling $\tilde{\boldsymbol{\beta}}$, only a Cholesky decomposition of A is needed, and only for the retained posterior samples. Prediction at a new location, say, $\mathbf{s}_{new}$, requires samples from $f\left(\tilde{\boldsymbol{\beta}}\left(\mathbf{s}_{new}\right) \mid \tilde{\mathbf{B}}, \boldsymbol{\mu}_\beta, \tau^2, T, \phi\right)$. Defining $\mathbf{h}_{new}\left(\phi\right)$ to be the $n \times 1$ vector with ith row entry $\rho\left(\mathbf{s}_i - \mathbf{s}_{new}; \phi\right)$, this distribution is normal with mean

$$\boldsymbol{\mu}_\beta + (\mathbf{h}_{new}^{\mathrm{T}}\left(\phi\right) \otimes T)\left(H^{-1}\left(\phi\right) \otimes T^{-1}\right)\left(\tilde{\mathbf{B}} - \mathbf{1}_{nx1} \otimes \boldsymbol{\mu}_\beta\right)$$
$$= \boldsymbol{\mu}_\beta + \left(\mathbf{h}_{new}^{\mathrm{T}}\left(\phi\right) H^{-1}\left(\phi\right) \otimes I\right)\left(\tilde{\mathbf{B}} - \mathbf{1}_{nx1} \otimes \boldsymbol{\mu}_\beta\right) ,$$

and covariance matrix

$$T - (\mathbf{h}_{new}^{\mathrm{T}}\left(\phi\right) \otimes T)\left(H^{-1}\left(\phi\right) \otimes T^{-1}\right)(\mathbf{h}_{new}\left(\phi\right) \otimes T) = \left(I - \mathbf{h}_{new}^{\mathrm{T}}\left(\phi\right) H^{-1}\left(\phi\right) \mathbf{h}_{new}\left(\phi\right)\right) T .$$

Finally, the predictive distribution for $Y\left(\mathbf{s}_{new}\right)$, namely $f\left(Y\left(\mathbf{s}_{new}\right) \mid \mathbf{y}\right)$, is sampled by composition, as usual.

We conclude this subsection by noting an extension of (10.69) when we have repeated measurements at location s. That is, suppose we have

$$Y(\mathbf{s}, l) = \mathbf{X}^{\mathrm{T}}(\mathbf{s}, l)\boldsymbol{\beta}(\mathbf{s}) + \epsilon(\mathbf{s}, l) , \tag{10.73}$$

where $l = 1, \ldots, L_\mathbf{s}$ with $L_\mathbf{s}$ the number of measurements at $\mathbf{s}$ and the $\epsilon\left(\mathbf{s}, l\right)$ still white noise. As an illustration, in the real estate context, $\mathbf{s}$ might denote the location for an apartment block and l might index apartments in this block that have sold, with the lth apartment having characteristics $\mathbf{X}\left(\mathbf{s}, l\right)$. Suppose further that $\mathbf{Z}\left(\mathbf{s}\right)$ denotes an $r \times 1$ vector of site-level characteristics. For an apartment block, these characteristics might include amenities provided or distance to the central business district. Then (10.73) can be extended to a multilevel model in the sense of Goldstein [2003] or Raudenbush and Bryk [2002]. In particular we can write

$$\boldsymbol{\beta}(s) = \begin{pmatrix} \mathbf{Z}^{\mathrm{T}}\left(\mathbf{s}\right)\boldsymbol{\gamma}_1 \\ \vdots \\ \mathbf{Z}^{\mathrm{T}}\left(\mathbf{s}\right)\boldsymbol{\gamma}_p \end{pmatrix} + \mathbf{w}\left(\mathbf{s}\right) . \tag{10.74}$$

In (10.74), $\boldsymbol{\gamma}_j, j = 1, \ldots, p$, is an $r \times 1$ vector associated with $\tilde{\beta}_j\left(\mathbf{s}\right)$, and $\mathbf{w}\left(\mathbf{s}\right)$ is a mean-zero multivariate Gaussian spatial process, for example, as above. In (10.74), if the $\mathbf{w}\left(\mathbf{s}\right)$ were independent we would have a usual multilevel model specification. In the case where $\mathbf{Z}\left(\mathbf{s}\right)$ is a scalar capturing just an intercept, we return to the initial model of this subsection.

10.7.4   Spatially varying coregionalization models

A possible extension of the LMC would replace A by $A(\mathbf{s})$ and thus define

$$\mathbf{Y}(\mathbf{s}) = A(\mathbf{s})\mathbf{w}(\mathbf{s}) . \tag{10.75}$$

We refer to the model in (10.75) as a *spatially varying LMC*. Following the notation in Section 10.6.1, let $T(\mathbf{s}) = A(\mathbf{s})A(\mathbf{s})^{\mathrm{T}}$. Again $A(\mathbf{s})$ can be taken to be lower triangular for convenience. Now $C(\mathbf{s}, \mathbf{s}')$ is such that

$$C(\mathbf{s}, \mathbf{s}') = \sum \rho_j(\mathbf{s} - \mathbf{s}')\mathbf{a}_j(\mathbf{s})\mathbf{a}_j(\mathbf{s}') , \tag{10.76}$$

with $\mathbf{a}_j(\mathbf{s})$ the jth column of $A(\mathbf{s})$. Letting $T_j(\mathbf{s}) = \mathbf{a}_j(\mathbf{s})\mathbf{a}_j^{\mathrm{T}}(\mathbf{s})$, again, $\sum T_j(\mathbf{s}) = T(\mathbf{s})$. We see from (10.76) that $\mathbf{Y}(\mathbf{s})$ is no longer stationary. Extending the intrinsic specification for $\mathbf{Y}(\mathbf{s})$, $C(\mathbf{s}, \mathbf{s}') = \rho(\mathbf{s} - \mathbf{s}')T(\mathbf{s})$, which is a multivariate version of the case of a spatial process with a spatially varying variance.

This motivates a natural definition of $A(\mathbf{s})$ through its one-to-one correspondence with $T(\mathbf{s})$ (again from Section 10.6.1) since $T(\mathbf{s})$ is the covariance matrix for $\mathbf{Y}(\mathbf{s})$. In the univariate case choices for $\sigma^2(\mathbf{s})$ include $\sigma^2(\mathbf{s}, \theta)$, i.e., a parametric function of location; $\sigma^2(x(\mathbf{s})) = g(x(\mathbf{s}))\sigma^2$ where $x(\mathbf{s})$ is some covariate used to explain $\mathbf{Y}(\mathbf{s})$ and $g(\cdot) > 0$ (then $g(x(\mathbf{s}))$ is typically $x(\mathbf{s})$ or $x^2(\mathbf{s})$); or $\sigma^2(\mathbf{s})$ is itself a spatial process (e.g., $\log \sigma^2(\mathbf{s})$ might be a Gaussian process). In practice, $T(\mathbf{s}) = g(x(\mathbf{s}))T$ will likely be easiest to work with.

Note that all of the discussion in Section 10.6.2 regarding the relationship between conditional and unconditional specifications is applicable here. Particularly, if $p = 2$ and $T(\mathbf{s}) = g(x(\mathbf{s}))T$ then (T_{11}, T_{12}, T_{22}) is equivalent to $(\sigma_1, \sigma_2, \alpha)$, and we have $a_{11}(\mathbf{s}) = \sqrt{g(x(\mathbf{s})}\sigma_1$, $a_{22}(\mathbf{s}) = \sqrt{g(x(\mathbf{s})}\sigma_2$, and $a_{21} = \sqrt{g(x(\mathbf{s})}\alpha\sigma_1$. See Gelfand et al. [2004b] for further discussion.

10.7.5   Model-fitting issues

This subsection starts by discussing the computational issues in fitting the joint multivariate model presented in Subsection 10.6.1. It will be shown that it is a challenging task to fit this joint model. On the other hand, making use of the equivalence of the joint and conditional models, as discussed in Section 10.6.2, we demonstrate that it is much simpler to fit the latter.

10.7.5.1   Fitting the joint model

Different from previous approaches that have employed the coregionalization model, our intent is to follow the Bayesian paradigm. For this purpose, the model specification is complete only after assigning prior distributions to all unknown quantities in the model. The posterior distribution of the set of parameters is obtained after combining the information about them in the likelihood (see equation (10.51)) with their prior distributions.

Observing equation (10.51), we see that the parameter vector defined as $\boldsymbol{\theta}$ consists of $\{\boldsymbol{\beta}_j\}, D, \{\rho_j\}, \mathbf{T}, j = 1, \cdots, p$. Adopting a prior that assumes independence across j we take $p(\boldsymbol{\theta}) = \prod_j p(\boldsymbol{\beta}_j) p(\rho_j) p(\tau_j^2) p(T)$. Hence $p(\boldsymbol{\theta} \mid \mathbf{y})$ is given by

$$p(\boldsymbol{\theta} \mid \mathbf{y}) \propto p(\mathbf{y} \mid \{\boldsymbol{\beta}_j\}, D, \{\rho_j\}, T) \, p(\boldsymbol{\theta}) \, .$$

For the elements of $\boldsymbol{\beta}_j$, a normal mean-zero prior distribution with large variance can be assigned, resulting in a full conditional distribution that will also be normal. Inverse gamma distributions can be assigned to the elements of D, the variances of the p white noise processes. If there is no information about such variances, the means of these inverse gammas could be based on the least squares estimates of the independent models with large variances. Assigning inverse gamma distributions to τ_j^2 will result in inverse gamma full conditionals. The parameters of concern are the elements of ρ_j and T. Regardless of what prior distributions we assign, the full conditional distributions will not have a standard form. For example, if we assume that ρ_j is the exponential correlation function, $\rho_j(h) = \exp(-\phi_j h)$, a gamma prior distribution can be assigned to the ϕ_j's. In order to obtain samples of the ϕ_j's we can use the Metropolis-Hastings algorithm with, for instance, log-normal proposals centered at the current $\log \phi_j$.

We now consider how to sample T, the covariance matrix among the responses at each location $\mathbf{s}$. Due to the one-to-one relationship between T and the lower triangular A, one can either assign a prior to the elements of A, or set a prior on the matrix T. The latter seems to be more natural since T is interpreted as the covariance matrix of the elements of $\mathbf{Y}(\mathbf{s})$. As T must be positive definite, we use an inverse Wishart prior distribution with ν degrees of freedom and mean D^*, i.e., the scale matrix is $(\nu - p - 1)(D^*)^{-1}$. If there is no information about the prior mean structure of T, rough estimates of the elements of the diagonal of D^* can be obtained using ordinary least squares estimates based on the independent spatial models for each $Y_j(\mathbf{s})$, $j = 1, \cdots, p$. A small value of $\nu(> p+1)$ would be assigned to provide high uncertainty in the resulting prior distribution.

To sample from the full conditional for T, Metropolis-Hastings updates are a place to start. In our experience, random walk Wishart proposals do not work well, and importance sampled Wishart proposals have also proven problematic. Instead, we recommend updating the elements of T individually. In fact, it is easier to work in the unconstrained space of the components of A, so we would reparametrize the full conditional from T to A. Random walk normal proposals for the $\mathbf{a}$'s with suitably tuned variances will mix well, at least for $p = 2$ or 3. For larger p, repeated decomposition of T to A may prove too costly.

10.7.5.2 *Fitting the conditional model*

Section 10.6.2 showed the equivalence of conditional and unconditional specifications in terms of $\mathbf{v}(\mathbf{s})$. Here we write the multivariate model for $\mathbf{Y}(\mathbf{s})$ in its conditional parametrization and see that the inference procedure is simpler than for the multivariate parametrization. Following the discussion in Section 10.6.2, for a general p, the conditional parametrization is

$$
\begin{aligned}
Y_1(\mathbf{s}) &= \mathbf{X}_1^{\mathrm{T}}(\mathbf{s})\boldsymbol{\beta}_1 + \sigma_1 w_1(\mathbf{s}) \\
Y_2(\mathbf{s}) \mid Y_1(\mathbf{s}) &= \mathbf{X}_2^{\mathrm{T}}(\mathbf{s})\boldsymbol{\beta}_2 + \alpha^{2|1}Y_1(\mathbf{s}) + \sigma_2 w_2(\mathbf{s}) \\
&\vdots \\
Y_p(\mathbf{s}) \mid Y_1(\mathbf{s}), \cdots, Y_p(\mathbf{s}) &= \mathbf{X}_p^{\mathrm{T}}(\mathbf{s})\boldsymbol{\beta}_p + \alpha^{p|1}Y_1(\mathbf{s}) \\
&\quad + \cdots + \alpha^{p|p-1}Y_{p-1}(\mathbf{s}) + \sigma_p w_p(\mathbf{s}) .
\end{aligned}
\tag{10.77}
$$

In (10.77), the set of parameters to be estimated is $\boldsymbol{\theta}_c = \{\boldsymbol{\beta}, \boldsymbol{\alpha}, \boldsymbol{\sigma}^2, \boldsymbol{\phi}\}$, where $\boldsymbol{\alpha}^{\mathrm{T}} = (\alpha^{2|1}, \alpha^{3|1}, \alpha^{3|2}, \cdots, \alpha^{p|p-1})$, $\boldsymbol{\beta}^{\mathrm{T}} = (\boldsymbol{\beta}_1, \cdots, \boldsymbol{\beta}_p)$, $\boldsymbol{\sigma}^2 = (\sigma_1^2, \cdots, \sigma_p^2)$, and $\boldsymbol{\phi}$ is as defined in Subsection 10.6.2. The likelihood is given by

$$
f_c(\mathbf{Y} \mid \boldsymbol{\theta}_c) = f(\mathbf{Y}_1 \mid \boldsymbol{\theta}_{c_1}) f(\mathbf{Y}_2 \mid \mathbf{Y}_1, \boldsymbol{\theta}_{c_2}) \cdots f(\mathbf{Y}_p \mid \mathbf{Y}_1, \cdots, \mathbf{Y}_{p-1}, \boldsymbol{\theta}_{c_p}) .
$$

If $\pi(\boldsymbol{\theta}_c)$ is taken to be $\prod_{j=1}^{p} \pi(\boldsymbol{\theta}_{c_j})$ then this equation implies that the conditioning yields a factorization into p models each of which can be fitted separately. Prior specification of the parameters was discussed in Subsection 10.6.2.2. With those forms, standard univariate spatial models that can be fit using the `GeoBUGS` package arise.

Example 10.3 *(Commercial real estate example).* The selling price of commercial real estate, for example an apartment property, is theoretically the expected income capitalized at some (risk-adjusted) discount rate. (See Kinnard, 1971, Lusht, 1997, for general discussions of the basics of commercial property valuation theory and practice.) Here we consider a data set consisting of 78 apartment buildings, with 20 additional transactions held out for prediction of the selling price based on four different models. The locations of these buildings are shown in Figure 10.6. The aim here is to fit a joint model for selling price and net income and obtain a spatial surface associated with the risk, which, for any transaction, is given by net income/price. For this purpose we fit a model using the following covariates: average square feet of a unit within the building (sqft), the age of the building (age), the number of units within the building (unit), the selling price of the transaction (P), and the net income (I). Figure 10.7 shows the histograms of these variables on the log scale. Using the conditional parametrization, the model is

$$
\begin{aligned}
I(\mathbf{s}) &= sqft(\mathbf{s})\beta_{I1} + age(\mathbf{s})\beta_{I2} + unit(\mathbf{s})\beta_{I3} + \sigma_1 w_1(\mathbf{s}) \\
P(\mathbf{s}) \mid I(\mathbf{s}) &= sqft(\mathbf{s})\beta_{P1} + age(\mathbf{s})\beta_{P2} + unit(\mathbf{s})\beta_{P3} \\
&\quad + I(\mathbf{s})\alpha^{(2|1)} + \sigma_2 w_2(\mathbf{s}) + \epsilon(\mathbf{s}) .
\end{aligned}
\tag{10.78}
$$

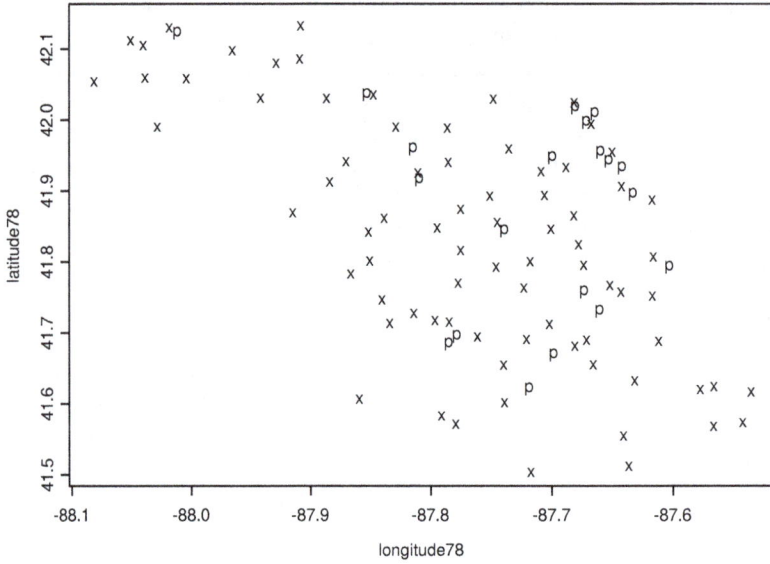

Figure 10.6 *Locations of the 78 sites* (x) *used to fit the (price, income) model, and the 20 sites used for prediction* (p).

Notice that $I(\mathbf{s})$ is considered to be purely spatial since, adjusted for building characteristics, we do not anticipate a microscale variability component. The need for white noise in the price component results from the fact that two identical properties at essentially the same location need not sell for the same price due to the motivation of the seller, the buyer, the brokerage process, etc. (If a white noise component for $I(\mathbf{s})$ were desired, we would fit the joint model as described near the beginning of Subsection 10.7.5.) The model in (10.78) is in accordance with the conditional parametrization in Subsection 10.6.2.2. The prior distributions were assigned as follows. For all the coefficients of the covariates, including $\alpha^{(2|1)}$, we assigned a normal 0 mean distribution with large variance. For σ_1^2 and σ_2^2 we used inverse gammas with infinite variance. We use exponential correlation functions and the decay parameters ϕ_j, $j = 1, 2$ have a gamma prior distribution arising from a mean range of one half the maximum interlocation distance, with infinite variance. Finally, τ_2^2, the variance of $\epsilon(\cdot)$, has an inverse gamma prior centered at the ordinary least squares variance estimate obtained from an independent model for log selling price given log net income.

Table 10.4 presents the posterior summaries of the parameters of the model. For the income model the age coefficient is significantly negative, and the coefficient for number of units is significantly positive. Notice further that the correlation between net income and price is very close to 1. Nevertheless, for the conditional price model age is still significant. Also we see that price shows a bigger range than net income. Figure 10.8 shows the spatial surfaces associated with the three processes: net income, price, and risk. It is straightforward to show that the logarithm of the spatial surface for risk is obtained through $(1 - \alpha^{(2|1)})\sigma_1 w_1(\mathbf{s}) - \sigma_2 w_2(\mathbf{s})$. Therefore, based on the posterior samples of $\alpha^{(2|1)}$, $w_1(\mathbf{s})$, σ_1, σ_2, and $w_2(\mathbf{s})$ we are able to obtain samples for the spatial surface for risk. From Figure 10.8(c), we note that the spatial risk surface tends to have smaller values than the other surfaces. Since $\log R(\mathbf{s}) = \log I(\mathbf{s}) - \log P(\mathbf{s})$ with $R(\mathbf{s})$ denoting the risk at location $\mathbf{s}$, the strong association between $I(\mathbf{s})$ and $P(\mathbf{s})$ appears to result in some cancellation of spatial effect for log risk. Actually, we can obtain the posterior distribution of the variance of the spatial process for $\log R(\mathbf{s})$. It is $(1 - \alpha^{(2|1)})^2\sigma_1^2 + \sigma_2^2$. The posterior mean of this variance is 0.036 and the 95% credible interval is given by $(0.0087, 0.1076)$ with a median equal to

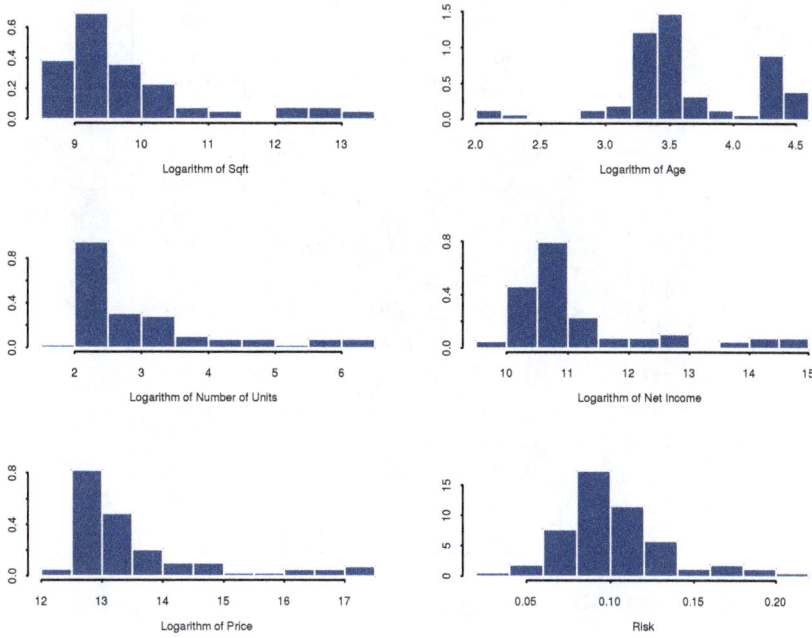

Figure 10.7 *Histograms of the logarithm of the variables.*

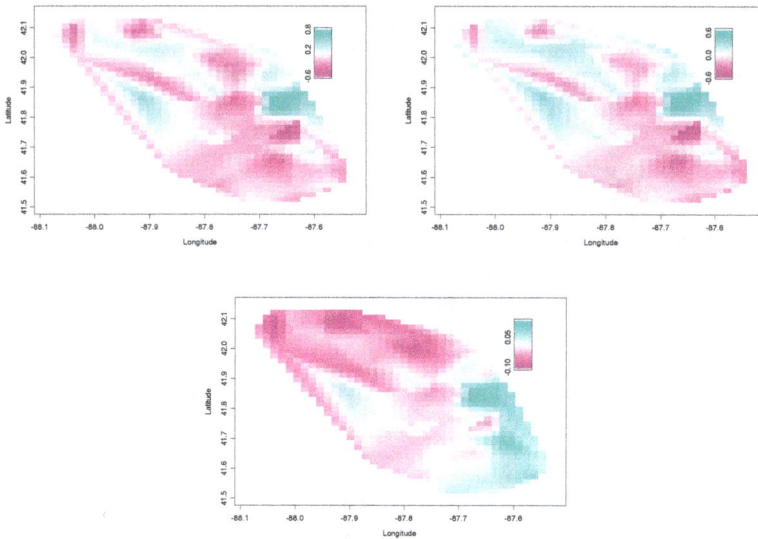

Figure 10.8 *Image plots of the spatial processes of (a) net income, (b) price, and (c) risk.*

0.028. The posterior variance of the noise term is given by τ_2^2, which is in Table 10.4. If we compare the medians of the posteriors of the variance of the spatial process of the risk and the variance of the white noise, we see that the spatial process presents a smaller variance; the variability of the risk process is being more explained by the residual component.

Parameter	Mean	2.50%	Median	97.50%
β_{I1}	0.156	-0.071	0.156	0.385
β_{I2}	-0.088	-0.169	-0.088	-0.008
β_{I3}	0.806	0.589	0.804	1.014
β_{P1}	0.225	0.010	0.229	0.439
β_{P2}	-0.092	-0.154	-0.091	-0.026
β_{P3}	-0.150	-0.389	-0.150	0.093
$\alpha^{(2|1)}$	0.858	0.648	0.856	1.064
σ_1^2	0.508	0.190	0.431	1.363
σ_2^2	0.017	0.006	0.014	0.045
τ_2^2	0.051	0.036	0.051	0.071
ϕ_I	3.762	1.269	3.510	7.497
ϕ_P	1.207	0.161	1.072	3.201
range$_I$	0.969	0.429	0.834	2.291
range$_P$	1.2383	0.554	1.064	2.937
corr(I,P)	0.971	0.912	0.979	0.995
T_{II}	0.508	0.190	0.431	1.363
T_{IP}	0.435	0.158	0.369	1.136
T_{PP}	0.396	0.137	0.340	1.000

Table 10.4 *Posterior summaries, joint model of price and income.*

Model	$\sum_{j=1}^{20} e_j^2$	$\sum_{j=1}^{20} Var(P(\mathbf{s}_j) \mid \mathbf{y})$
Independent, nonspatial	2.279	3.277
Independent, spatial	1.808	2.963
Conditional, nonspatial	0.932	1.772
Conditional, spatial	0.772	1.731

Table 10.5 *Squared error and the sum of the variances of the predictions for the 20 sites left out in the fitting of the model.*

In order to examine the comparative performance of the model proposed above we ran four different models for the selling price using each one to predict the locations marked with p in Figure 10.6. For all these models we used the same covariates as described before. Model 1 comprises an independent model for price, i.e., without a spatial component or net income. Model 2 has a spatial component and is not conditioned on net income. In Model 3 the selling price is conditioned on the net income but without a spatial component, and Model 4 has net income as a covariate and also a spatial component. Table 10.5 shows both $\sum_{j=1}^{20} e_j^2$, where $e_j = P(\mathbf{s}_j) - E(P(\mathbf{s}_j) \mid \mathbf{y}, \text{model})$ and $P(\mathbf{s}_j)$ is the observed log selling price for the jth transaction, and $\sum_{j=1}^{20} Var(P(\mathbf{s}_j) \mid \mathbf{y}, \text{model})$. Recall from equation (5.67) that the former is a measurement of predictive goodness of fit, while the latter is a measure of predictive variability. It is clear from the table that the model conditioned on net income and with a spatial component is best, both in terms of fit and predictive variability. ∎

10.8   Other constructive approaches ⋆

Here we consider two additional constructive strategies for building valid cross-covariance functions. The first is referred to as a moving average approach in Ver Hoef and Barry [1998]. It is a multivariate version of the kernel convolution development of Subsection 3.2.2

that convolves process variables to produce a new process. The second approach convolves valid covariance functions to produce a valid cross-covariance function.

For the first approach, expressions (3.11) and (3.13) suggest several ways to achieve multivariate extension. Again with $\mathbf{Y}(\mathbf{s}) = (Y_1(\mathbf{s}), \ldots, Y_p(\mathbf{s}))^{\mathrm{T}}$, define

$$Y_\ell(\mathbf{s}) = \int_{\Re^2} k_\ell(\mathbf{u}) Z(\mathbf{s} + \mathbf{u}) d\mathbf{u}, \quad \ell = 1, \ldots, p. \tag{10.79}$$

In this expression, $Z(\cdot)$ is a mean 0 stationary process with correlation function $\rho(\cdot)$, and k_ℓ is a kernel associated with the ℓth component of $\mathbf{Y}(\mathbf{s})$. In practice, $k_\ell(\mathbf{u})$ would be parametric, i.e., $k_\ell(\mathbf{u}; \boldsymbol{\theta}_\ell)$. The resulting cross-covariance matrix for $\mathbf{Y}(\mathbf{s})$ has entries

$$C_{\ell,\ell'}(\mathbf{s}, \mathbf{s}') = \sigma^2 \int_{\Re^2} \int_{\Re^2} k_\ell(\mathbf{s} - \mathbf{s}' + \mathbf{u}) k_{\ell'}(\mathbf{u}') \rho(\mathbf{u} - \mathbf{u}') d\mathbf{u} d\mathbf{u}'. \tag{10.80}$$

This cross-covariance matrix is necessarily valid. It is stationary and, as may be easily verified, is symmetric, i.e. $\mathrm{Cov}(Y_\ell(\mathbf{s}), Y_{\ell'}(\mathbf{s}')) = C_{\ell\ell'}(\mathbf{s} - \mathbf{s}') = C_{\ell'\ell}(\mathbf{s} - \mathbf{s}') = \mathrm{Cov}(Y_{\ell'}(\mathbf{s}), Y_\ell(\mathbf{s}'))$. Since the integration in (10.80) will not be possible to do explicitly except in certain special cases, finite sum approximation of (10.79) is an alternative.

An alternative extension to (10.65) introduces *lags* $\mathbf{h}_\ell$, defining

$$Y_\ell(\mathbf{s}) = \int_{\Re^2} k(\mathbf{u}) Z(\mathbf{s} + \mathbf{h}_\ell + \mathbf{u}) d\mathbf{u}, \quad \ell = 1, \ldots, p.$$

Now

$$C_{\ell,\ell'}(\mathbf{s}, \mathbf{s}') = \sigma^2 \int_{\Re^2} \int_{\Re^2} k(\mathbf{s} - \mathbf{s}' + \mathbf{u}) k(\mathbf{u}') \rho(\mathbf{h}_\ell - \mathbf{h}_{\ell'} + \mathbf{u} - \mathbf{u}') d\mathbf{u} d\mathbf{u}'$$

Again the resulting cross-covariance matrix is valid; again the process is stationary. However now it is easy to verify that the cross-covariance matrix is not symmetric. Whether a lagged relationship between the variables is appropriate in a purely spatial specification would depend upon the application. However, in practice the $\mathbf{h}_\ell$ would be unknown and would be considered as model parameters. A fully Bayesian treatment of such a model has not yet been discussed in the literature.

For the second approach, suppose $C_\ell(\mathbf{s})$, $\ell = 1, \ldots, p$ are each squared integrable stationary covariance functions valid in two-dimensional space. We now show $C_{\ell\ell}(\mathbf{s}) = \int_{R^2} C_\ell(\mathbf{s} - \mathbf{u}) C_\ell(\mathbf{u}) d\mathbf{u}$, the convolution of C_ℓ with itself, is again a valid covariance function. Writing $\widehat{C}_\ell(\mathbf{w}) = \int e^{-i\mathbf{w}^{\mathrm{T}} \mathbf{h}} C_\ell(\mathbf{h}) d\mathbf{h}$, by inversion, $C_\ell(\mathbf{s}) = \int e^{i\mathbf{w}^{\mathrm{T}} \mathbf{s}} \frac{\widehat{C}_\ell(\mathbf{w})}{(2\pi)^2} d\mathbf{w}$. But also, from (3.2), $\widehat{C}_{\ell\ell}(\mathbf{w}) \equiv \int e^{-i\mathbf{w}^{\mathrm{T}} \mathbf{s}} C_{\ell\ell}(\mathbf{s}) d\mathbf{s} = \int e^{-i\mathbf{s}^{\mathrm{T}} \mathbf{s}} \int C_\ell(\mathbf{s} - \mathbf{u}) C_\ell(\mathbf{u}) d\mathbf{u} d\mathbf{s} = \int \int e^{i\mathbf{w}^{\mathrm{T}} (\mathbf{s} - \mathbf{u})} C_\ell(\mathbf{s} - \mathbf{u}) e^{i\mathbf{w}^{\mathrm{T}} \mathbf{u}} C_\ell(\mathbf{u}) d\mathbf{u} d\mathbf{s} = (\widehat{C}_\ell(\mathbf{w}))^2$. Self-convolution of C_ℓ produces the square of the Fourier transform. However, since $C_\ell(\cdot)$ is valid, Bochner's Theorem (Subsection 3.1.2) tells us that $\widehat{C}_\ell(\mathbf{w})/(2\pi)^2 C(0)$ is a spectral density symmetric about 0. But then due to the squared integrability assumption, up to proportionality, so is $(\widehat{C}_\ell(w))^2$, and thus $C_{\ell\ell}(\cdot)$ is valid.

The same argument ensures that

$$C_{\ell\ell'}(\mathbf{s}) = \int_{R^2} C_\ell(\mathbf{s} - \mathbf{u}) C_{\ell'}(\mathbf{u}) d\mathbf{u} \tag{10.81}$$

is also a valid stationary covariance function; cross-convolution provides a valid covariance function. (Now $\widehat{C}_{\ell\ell'}(w) = \widehat{C}_\ell(w) \widehat{C}_{\ell'}(w)$.) Moreover, it can be shown that $C(\mathbf{s} - \mathbf{s}')$ defined by $(C(\mathbf{s} - \mathbf{s}'))_{\ell\ell'} = C_{\ell\ell'}(\mathbf{s} - \mathbf{s}')$ is a valid $p \times p$ cross-covariance function [see Majumdar and Gelfand, 2007]. It is also the case that if each C_ℓ is isotropic, then so is $C(\mathbf{s} - \mathbf{s}')$. To see this, suppose $\|\mathbf{h}_1\| = \|\mathbf{h}_2\|$. We need only show that $C_{\ell\ell'}(\mathbf{h}_1) = C_{\ell\ell'}(\mathbf{h}_2)$. But $\mathbf{h}_1 = P\mathbf{h}_2$ where

P is orthogonal. Hence, $C_{\ell\ell'}(\mathbf{h}_1) = \int C_\ell(\mathbf{h}_1 - \mathbf{u})C_{\ell'}(\mathbf{u})d\mathbf{u} = \int C_\ell(P(\mathbf{h}_1 - \mathbf{u}))C_{\ell'}(P\mathbf{u})d\mathbf{u} = \int C_\ell(\mathbf{h}_2 - \tilde{\mathbf{u}})C_{\ell'}(\tilde{\mathbf{u}})d\tilde{\mathbf{u}} = C_{\ell\ell'}(\mathbf{h}_2)$.

We note that the range associated with $C_{\ell\ell}$ is not the same as that for C_ℓ but that if the C_ℓ's have distinct ranges then so will the components, $Y_\ell(\mathbf{s})$. Computational issues associated with using $C(\mathbf{s} - \mathbf{s}')$ in model-fitting are also discussed in Majumdar and Gelfand [2007]. We note that (10.81) can in most cases be conveniently computed by transformation to polar coordinates and then using Monte Carlo integration. We leave this calculation to an exercise.

Other constructive approaches for cross-covariance functions include notable developments in Apanasovich and Genton [2010] [also see Apanasovich et al., 2012] , which builds extremely flexible classes of cross-covariance functions, drawing upon the constructions of Gneiting [2002a], as mentioned briefly in Section 12.3. The basic idea is to build valid forms using covariance functions involving latent dimensions. General expressions are complex, involving many parameters, suggesting potential identifiability problems in model fitting. Here, we offer a simple illustration. Recall that, under coregionalization, in the two-dimensional case, with exponential covariance functions, we create $C_{11}(\mathbf{h}) = a_{11}^2 \exp(-\alpha_1\|\mathbf{h}\|)$, $C_{22}(\mathbf{h}) = a_{21}^2 \exp(-\alpha_1\|\mathbf{h}\|) + a_{22}^2 \exp(-\alpha_2\|\mathbf{h}\|)$, and $C_{12}(\mathbf{h}) = a_{11}a_{21}\exp(-\alpha_1\|\mathbf{h}\|)$. Apanasovich et al. [2012] show that, if we retain $C_{11}(\mathbf{h})$ and $C_{22}(\mathbf{h})$ as they are but generalize $C_{12}(\mathbf{h})$ to $C_{12}(\mathbf{h}) = \frac{a_{11}a_{21}}{\delta_{12}+1} \exp(-\frac{\alpha_1\|\mathbf{h}\|}{(\delta_{12}+1)^{\beta/2}})$, we still obtain a valid cross-covariance function with added flexibility of the parameters $\delta_{12} \geq 0$ and $\beta \geq 0$.

This is easily generalized to higher dimensions. If $\mathbf{Y}(\mathbf{s}) = (Y_1(\mathbf{s}), \ldots, Y_q(\mathbf{s}))^\mathrm{T}$ is a $q \times 1$ spatial process and we seek a $q \times q$ valid cross-covariance matrix $C(\mathbf{s}, \mathbf{s}') = \mathrm{Cov}(\mathbf{Y}(\mathbf{s}), \mathbf{Y}(\mathbf{s}')) = [C_{ij}(\mathbf{s}, \mathbf{s}')]$ for $i, j = 1, \ldots, q$ and $\mathbf{s}, \mathbf{s}' \in \mathbb{R}^d$. Apanasovich and Genton [2010] propose building this cross-covariance matrix as $C_{ij}(\mathbf{s}, \mathbf{s}') = K((\mathbf{s}, \boldsymbol{\zeta}_i), (\mathbf{s}', \boldsymbol{\zeta}_j))$, where $\boldsymbol{\zeta}_i$ and $\boldsymbol{\zeta}_j$ are *latent* points in $\mathbb{R}^k$ for $1 \leq k \leq q$ and $K((\mathbf{s}, \boldsymbol{\zeta}_i), (\mathbf{s}', \boldsymbol{\zeta}_j))$ is a valid covariance function on an augmented latent space of dimension $\mathbb{R}^{d+k}$. This ensures that $C(\mathbf{s}, \mathbf{s}') = [C_{ij}(\mathbf{s}, \mathbf{s}')]$ is a valid cross-covariance function. An example [Apanasovich and Genton, 2010, Genton and Kleiber, 2015] is

$$C_{ij}(\mathbf{s}, \mathbf{s}') = \frac{\sigma_i \sigma_j}{\|\boldsymbol{\Delta}_{ij}\| + 1} \exp\left(-\frac{\phi\|\mathbf{s} - \mathbf{s}'\|}{(\|\boldsymbol{\Delta}_{ij}\| + 1)^{\beta/2}}\right), \tag{10.82}$$

where $\boldsymbol{\Delta}_{ij} = \boldsymbol{\zeta}_i - \boldsymbol{\zeta}_j$ is the difference between the two points in the latent dimension, σ_i and σ_j are marginal standard deviations and $\phi > 0$ controls rate of decay in spatial correlation as the distance between spatial locations increases. When $i = j$, then $\boldsymbol{\Delta}_{ij} = \mathbf{0}$, i.e., the two latent points coincide, and (10.82) reduces to the exponential covariance function for $Y_i(\mathbf{s})$.

One could also borrow ideas from multi-group Gaussian processes [Li et al., 2025] that are used in machine learning applications to model heterogeneous data containing multiple known discrete subgroups of samples. Li et al. [2025] construct Gaussian processes over $\mathbb{R}^d \times \mathcal{C}$, where $\mathcal{C}$ is a finite set of indices or labels for groups. While machine learning applications have employed multi-group Gaussian processes for capturing associations among groups, the connections to multivariate spatial processes are clear. In particular, Li et al. [2025] establish connections between multi-group covariance functions and multivariate cross-covariance functions. Li et al. [2025] devise flexible classes of multi-group covariance functions that can also serve as elements of a valid cross-covariance matrix. If the elements of $\mathcal{C}$ are labels for elements in a $q \times 1$ process $\mathbf{Y}(\mathbf{s})$ and if $Z(\mathbf{s}; c)$ is a univariate latent multi-group process over $\mathbb{R}^d \times \mathcal{C}$ with valid multi-group covariance function $K((\mathbf{s}; c_i), (\mathbf{s}; c_j))$, then we arrive at a valid cross-covariance matrix $C(\mathbf{s}, \mathbf{s}')$ with elements $C_{ij}(\mathbf{s}, \mathbf{s}') = \mathrm{Cov}(Z(\mathbf{s}; c_i), Z(\mathbf{s}'; c_j)) = K((\mathbf{s}; c_i), (\mathbf{s}; c_j))$.

The latent dimension and multi-group approaches induce a valid cross-covariance function from a valid univariate covariance functions. The former constructs the spatial process over an augmented Euclidean space with latent points $\boldsymbol{\zeta}$, while the latter uses a latent multi-group process $Z(\mathbf{s}; c)$. The properties of the univariate covariance functions define characteristics of the multivariate process. For example, if the covariance function $K((\mathbf{s}; \boldsymbol{\zeta}_i), (\mathbf{s} + \mathbf{h}; \boldsymbol{\zeta}_j))$ over the latent dimension $\mathbb{R}^{d+k}$ is stationary or isotropic, then so is the cross-covariance function $C_{ij}(\mathbf{s}, \mathbf{s} + h) := K(\mathbf{h}; \boldsymbol{\zeta}_i - \boldsymbol{\zeta}_j)$. Porcu and Zastavnyi [2011] build on similar ideas to construct compactly supported cross-covariance functions that can be used in multivariate covariance-tapering models for large datasets.

Finally, we point out some neat constructions of asymmetric cross-covariance functions by Li and Zhang [2011] using latent dimensions. While the latent dimension approach directly yields symmetric stationary cross-covariance functions $C_{ij}(\mathbf{h}) := C_{ij}(\mathbf{s}, \mathbf{s} + \mathbf{h}) = C_{ji}(\mathbf{h})$ (hence symmetric cross-covariance matrices $C(\mathbf{h}) = [C_{ij}(\mathbf{h})] = C(\mathbf{h})^{\mathrm{T}}$), it can be verified that $\tilde{C}_{ij}(\mathbf{h}) := C_{ij}(\mathbf{h} + \boldsymbol{\Delta}_{ij})$ is a valid asymmetric cross-covariance function for any $\boldsymbol{\Delta}_{ij} := \mathbf{r}_i - \mathbf{r}_j \neq \mathbf{0}$, for fixed set of q vectors $\mathbf{r}_i \in \mathbb{R}^d$ for $i = 1, \ldots, q$. In fact, if $\mathbf{Y}(\mathbf{s})$ has $q \times q$ cross-covariance matrix $C(\mathbf{h}) = [C_{ij}(\mathbf{h})]$, then the process $\tilde{\mathbf{Y}}(\mathbf{s}) = (Y_1(\mathbf{s} - \mathbf{r}_1), \ldots, Y_q(\mathbf{s} - \mathbf{r}_q))^{\mathrm{T}}$ has cross-covariance matrix $\tilde{C}(\mathbf{h}) = [\tilde{C}_{ij}(\mathbf{h})]$. This corresponds to spatial delay effects that introduce asymmetry in cross-covariances and generalizes two-dimensional spatial delay processes that could be constructed by regression of one element, say $Y_1(\mathbf{s})$ on $Y_2(\mathbf{s} - \mathbf{r})$ with one fixed latent point $\mathbf{r}$.

Example 10.4 *(Baton Rouge housing prices)*. We analyze a sample from a database of real estate transactions in Baton Rouge, LA, during the eight-year period 1985–1992. Here, we focus on the static case. In particular, we focus on modeling the log selling price of single-family homes. In real estate modeling it is customary to work with log selling price in order to achieve better approximate normality. A range of house characteristics are available. We use four of the most common choices: age of house, square feet of living area, square feet of other area (e.g., garages, carports, storage), and number of bathrooms. For the static spatial case, a sample of 237 transactions was drawn from 1992. Figure 10.9 shows the parish of Baton Rouge and the locations contained in an encompassing rectangle within the parish.

We fit a variety of models, where in all cases the correlation function is from the Matérn class. We used priors that are fairly noninformative and comparable across models as sensible. First, we started with a spatially varying intercept and one spatially varying slope coefficient (the remaining coefficients do not vary), requiring a bivariate process model. There are four such models, and using D_K, the balanced predictive loss criterion (5.67), the model with a spatially varying living area coefficient emerges as best. Next, we introduced two spatially varying slope coefficient processes along with a spatially varying intercept, requiring a trivariate process model. There are six models here; the one with spatially varying age and living area is best. Finally, we allowed five spatially varying processes: an intercept and all four coefficients, using a five-dimensional process model. We also fit a model with five independent processes. From Table 10.6 the five-dimensional dependent process model is far superior and the independent process model is a dismal last, supporting our earlier intuition.

The prior specification used for the five-dimensional dependent process model is as follows. We take vague $N\left(\mathbf{0}, 10^5 I\right)$ for $\boldsymbol{\mu}_\beta$, a five-dimensional inverse Wishart, $IW\left(5, Diag\left(0.001\right)\right)$, for T, and an inverse gamma $IG\left(2, 1\right)$ for τ^2 (mean 1, infinite variance). For the Matérn correlation function parameters ϕ and ν we assume gamma priors $G\left(2, 0.1\right)$ (mean 20 and variance 200). For all the models three parallel chains were run to assess convergence. Satisfactory mixing was obtained within 3000 iterations for all the models; 2000 further samples were generated and retained for posterior inference.

The resulting posterior inference summary is provided in Table 10.7. We note a significant negative overall age coefficient with significant positive overall coefficients for the other three covariates, as expected. The contribution to spatial variability from the components of $\boldsymbol{\beta}$ is captured through the diagonal elements of the T matrix scaled by the corresponding covariates following the discussion at the end of Subsection 10.7.1. We see that the spatial

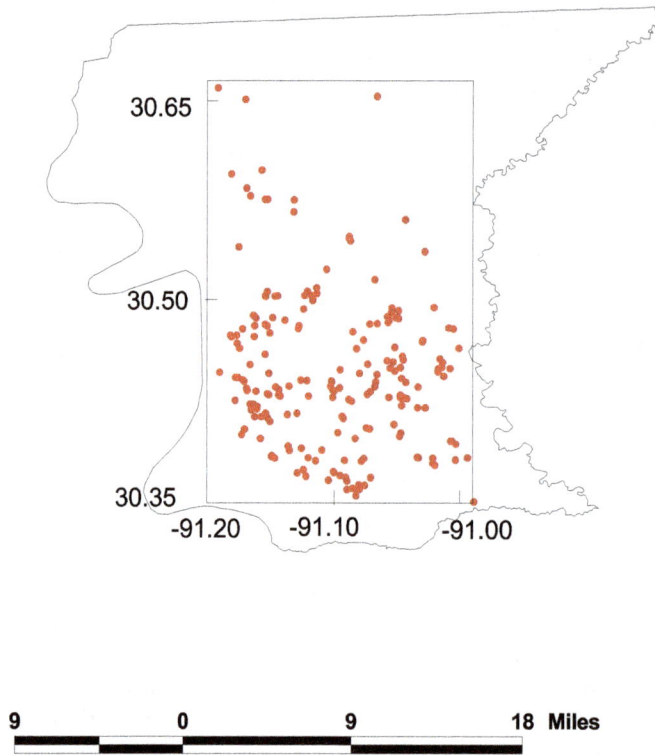

Figure 10.9 *Locations sampled within the parish of Baton Rouge for the static spatial models.*

Model	Fit	Variance penalty	D_K
Five-dimensional	42.21	36.01	78.22
Three-dimensional (best)	61.38	47.83	109.21
Two-dimensional (best)	69.87	46.24	116.11
Independent process	94.36	59.34	153.70

Table 10.6 *Values of posterior predictive model choice criterion (over all models).*

intercept process contributes most to the error variability with, perhaps surprisingly, the "bathrooms" process second. Clearly spatial variability overwhelms the pure error variability τ^2, showing the importance of the spatial model.

The dependence between the processes is evident in the posterior correlation between the components. We find the anticipated negative association between the intercept process and the slope processes (apart from that with the "other area" process). Under the Matérn correlation function, by inverting $\rho(\cdot;\phi) = 0.05$ for a given value of the decay parameter γ and the smoothing parameter ν, we obtain the range, i.e., the distance beyond which spatial association becomes negligible. Posterior samples of (γ, ν) produce posterior samples for the range. The resulting posterior median is roughly 4 km over a somewhat sprawling parish that is roughly 22 km × 33 km. The smoothness parameter suggests processes with mean square differentiable realizations ($\nu > 1$). Contour plots of the posterior mean spatial surfaces for each of the processes (not shown) are quite different. ■

Parameter	2.5%	50%	97.5%
β_0 (intercept)	9.908	9.917	9.928
β_1 (age)	−0.008	−0.005	−0.002
β_2 (living area)	0.283	0.341	0.401
β_3 (other area)	0.133	0.313	0.497
β_4 (bathrooms)	0.183	0.292	0.401
T_{11}	0.167	0.322	0.514
$\bar{x}_1^2 T_{22}$	0.029	0.046	0.063
$\bar{x}_2^2 T_{33}$	0.013	0.028	0.047
$\bar{x}_3^2 T_{44}$	0.034	0.045	0.066
$\bar{x}_4^2 T_{55}$	0.151	0.183	0.232
$T_{12}/\sqrt{T_{11}T_{22}}$	−0.219	−0.203	−0.184
$T_{13}/\sqrt{T_{11}T_{33}}$	−0.205	−0.186	−0.167
$T_{14}/\sqrt{T_{11}T_{44}}$	0.213	0.234	0.257
$T_{15}/\sqrt{T_{11}T_{55}}$	−0.647	−0.583	−0.534
$T_{23}/\sqrt{T_{22}T_{33}}$	−0.008	0.011	0.030
$T_{24}/\sqrt{T_{22}T_{44}}$	0.061	0.077	0.098
$T_{25}/\sqrt{T_{22}T_{55}}$	−0.013	0.018	0.054
$T_{34}/\sqrt{T_{33}T_{44}}$	−0.885	−0.839	−0.789
$T_{35}/\sqrt{T_{33}T_{55}}$	−0.614	−0.560	−0.507
$T_{45}/\sqrt{T_{44}T_{55}}$	0.173	0.232	0.301
ϕ (decay)	0.51	1.14	2.32
ν (smoothness)	0.91	1.47	2.87
range (in km)	2.05	4.17	9.32
τ^2	0.033	0.049	0.077

Table 10.7 *Inference summary for the five-dimensional multivariate spatially varying coefficients model.*

10.8.1   Generalized linear model setting

We briefly consider a generalized linear model version of (10.69), replacing the Gaussian first stage with

$$f\left(y\left(\mathbf{s}_i\right) \mid \theta\left(\mathbf{s}_i\right)\right) = h\left(y\left(\mathbf{s}_i\right)\right) \exp\left(\theta\left(\mathbf{s}_i\right) y\left(\mathbf{s}_i\right) - b\left(\theta\left(\mathbf{s}_i\right)\right)\right) , \qquad (10.83)$$

where, using a canonical link, $\theta\left(\mathbf{s}_i\right) = \mathbf{X}^{\mathrm{T}}\left(\mathbf{s}_i\right)\tilde{\boldsymbol{\beta}}\left(\mathbf{s}_i\right)$. In (10.83) we could include a dispersion parameter with little additional complication.

The resulting first-stage likelihood becomes

$$L\left(\tilde{\boldsymbol{\beta}}; \mathbf{y}\right) = \exp\left\{\sum y\left(\mathbf{s}_i\right)\mathbf{X}^{\mathrm{T}}\left(\mathbf{s}_i\right)\tilde{\boldsymbol{\beta}}\left(\mathbf{s}_i\right) - b\left(\mathbf{X}^{\mathrm{T}}\left(\mathbf{s}_i\right)\tilde{\boldsymbol{\beta}}\left(\mathbf{s}_i\right)\right)\right\}. \qquad (10.84)$$

Taking the prior on $\tilde{\boldsymbol{\beta}}$ in (10.70), the Bayesian model is completely specified with a prior on on ϕ, T and $\boldsymbol{\mu}_\beta$.

This model can be fit using a conceptually straightforward Gibbs sampling algorithm, which updates the components of $\boldsymbol{\mu}_\beta$ and $\tilde{\boldsymbol{\beta}}$ using adaptive rejection sampling. With an inverse Wishart prior on T, the resulting full conditional of T is again inverse Wishart. Updating ϕ is usually very awkward because it enters in the Kronecker form in (10.70). Slice sampling is not available here since we cannot marginalize the spatial effects; Metropolis updates are difficult to design but offer perhaps the best possibility. Also problematic is the repeated componentwise updating of $\tilde{\boldsymbol{\beta}}$. This hierarchically centered parametrization [Gelfand et al., 1995, 1996] is preferable to working with $\boldsymbol{\mu}_\beta$ and $\boldsymbol{\beta}$, but in our experience the algorithm still exhibits serious autocorrelation problems.

10.9   Spatial factor models

With larger numbers of dependent outcomes, modeling the cross-covariance becomes challenging. Even for stationary cross-covariance functions, where we assume that the associations among the variables do not change over space and the spatial association depends only on the difference of two positions, matters become computationally challenging. Spatial factor models address this issue and have been explored quite extensively by Wang and Wall [2003], Lopes et al. [2008], Ren and Banerjee [2013], Taylor-Rodríguez et al. [2017], Taylor-Rodriguez et al. [2019], Shirota et al. [2019], Zhang and Banerjee [2022] and Davies et al. [2022]. Lopes et al. [2008], in particular, provide an extensive discussion from the perspective of dynamic hierarchical models.

Spatial factor models arise as modifications of LMC but with a smaller number of latent processes than the number of outcomes in order to achieve dimension reduction. Let $\mathbf{y}(\mathbf{s}) = (y_1(\mathbf{s}), \ldots, y_q(\mathbf{s}))^{\mathrm{T}} \in \mathbb{R}^q$ denote the $q \times 1$ vector of dependent outcomes in location $\mathbf{s} \in \mathcal{D} \subset \mathbb{R}^d$, $\mathbf{x}(\mathbf{s}) = (x_1(\mathbf{s}), \ldots, x_p(\mathbf{s}))^{\mathrm{T}} \in \mathbb{R}^p$ be the corresponding explanatory variables, and β be a $p \times q$ regression coefficient matrix in the multivariate spatial model

$$\mathbf{y}(\mathbf{s}) = \beta^{\mathrm{T}}\mathbf{x}(\mathbf{s}) + \Lambda^{\mathrm{T}}\mathbf{f}(\mathbf{s}) + \boldsymbol{\epsilon}(\mathbf{s}) , \ \mathbf{s} \in \mathcal{D} , \tag{10.85}$$

where $\mathbf{f}(\mathbf{s}) = (f_1(\mathbf{s}), f_2(\mathbf{s}), \ldots, f_K(\mathbf{s}))^{\mathrm{T}}$ is a $K \times 1$ vector of independent spatial processes, $K < q$ to achieve dimension reduction, and Λ is a $q \times K$ matrix of factor loadings. For a finite collection of n locations $\mathcal{S} = \{\mathbf{s}_1, \ldots, \mathbf{s}_n\}$, we assume $\boldsymbol{\epsilon}(\mathbf{s}_i) \stackrel{ind}{\sim} N(\mathbf{0}, \Sigma)$.

A convenient way to describe hierarchical spatial factor models is with conjugate matrix-normal and inverse-Wishart distributions (recall Section 10.4) as

$$\beta \mid \Sigma \sim \mathrm{MN}(\mu_\beta, V_\beta, \Sigma) ; \ \Lambda \mid \Sigma \sim \mathrm{MN}(\mu_\Lambda, V_\Lambda, \Sigma) ; \ \Sigma \sim \mathrm{IW}(\Psi, \nu) \quad , \tag{10.86}$$

where μ_Λ is $q \times K$ and V_Λ is $K \times K$ and positive definite. The model in (10.85) applied to the locations in $\mathcal{S}$ yields the spatial factor model

$$Y_{n\times q} = X_{n\times p}\beta_{p\times q} + F_{n\times K}\Lambda_{K\times q} + E_{n\times q} , \quad E \sim \mathrm{MN}(O_{n\times q}, I_{n\times n}, \Sigma_{q\times q}) \tag{10.87}$$

where $Y = \mathbf{y}(\mathcal{S}) = [\mathbf{y}(\mathbf{s}_1) : \cdots : \mathbf{y}(\mathbf{s}_n)]^{\mathrm{T}}$ is the $n \times q$ response matrix, $X = \mathbf{x}(\mathcal{S}) = [\mathbf{x}(\mathbf{s}_1) : \cdots : \mathbf{x}(\mathbf{s}_n)]^{\mathrm{T}}$ is the corresponding design matrix, F is $n \times K$ with j-th column being the $n \times 1$ vector comprising $f_j(\mathbf{s}_i)$'s and E is $n \times q$ with rows $\boldsymbol{\epsilon}(\mathbf{s}_i)^{\mathrm{T}}$ for $i = 1, 2, \ldots, n$.

The parameters Λ and F are not jointly identified in factor models and some constraints are required to ensure identifiability [Lopes and West, 2004, Ren and Banerjee, 2013]. In LMC's, [Schmidt and Gelfand, 2003, Gelfand et al., 2004b, Finley et al., 2008], a lower-triangular square matrix Λ with positive diagonal elements identifies the covariances among the outcomes within a location because $C(\mathbf{0}) = \Lambda\Lambda^{\mathrm{T}}$ and Λ identifies with the Cholesky square-root since $C(\mathbf{0})$ is positive definite. However, with Λ no longer square in factor models this identification is lost. Finding proper constraints on Λ to be identifiable is equivalent to finding transformations that retain the statistical properties of the model.

These constraints are not without problems. For example, a lower-trapezoidal specification for Λ imposes possibly unjustifiable conditional independence on the spatial processes. Ren and Banerjee [2013] propose ordering the spatial range parameters to ensure identifiability, but this creates difficulties in computation and interpretation. Taylor-Rodriguez et al. [2019] and Shirota et al. [2019] investigate modeling the loading rows of the loading matrix using Dirichlet process mixtures, while Zhang and Banerjee [2022] suggest avoiding constraints and restricting inference to $\omega = F\Lambda$. This parametrization yields conditional conjugate distributions and, therefore, efficient posterior sampling.

We do not offer further details here (the reader is encouraged to refer to the foregoing publications) except to say that transitioning from LMCs to factor models, while conceptually simple, entails difficulties in computation arising from identifiability issues that are not entirely straightforward to obviate. Nevertheless, their conceptual simplicity and elegance of interpretation connected to traditional factor analysis or principal components analysis (PCA) have seen spatial factor models being applied to diverse applications such as studying biodiversity through the distribution and abundance of species; mapping remotely sensed Light Detection and Ranging (LiDAR) data in studying the health of forests; and mapping of cancer mortality data to ascertain whether a common spatial factor explains the associations among the diseases.

We conclude this section with two brief remarks. First, recent applications of spatial factor models, including in most of the foregoing literature, have been implemented in the context of large or massive spatial datasets with the number of spatial locations in the order or $\sim 10^4+$ or higher. This means that the process $\mathbf{f}(\mathbf{s})$ in (10.85) itself needs to be devised in a manner that can scale up to the massive datasets. We will turn to such processes in Chapter 13. Second, as in traditional factor analysis or PCA, factor models are effective when the variation among the associated outcomes can be reasonably captured by a relatively fewer number of factors. This begs the question about the size of q (the number of outcomes) for which factor models are effective. In fact, it is likely that the effectiveness of factor models as a tool for joint inference may be weakened, perhaps considerably so, in high-dimensional spaces when q is very large. In such settings the number of factors, K, may also need to be large to adequately capture the variability and associations among the outcomes which could severely compromise with the computational gains expected from factor models. Cases with q very large, say $\sim 10^2$ or higher, are referred to as *highly-multivariate* for which sparsity-inducing graphical models may be a preferable option. This is the topic of the next section.

10.10   Graphical models for multivariate point-referenced data

10.10.1   *Gaussian graphical models for point-referenced spatial data analysis*

While LMCs and spatial factor models are convenient choices for building multivariate spatial process models when the number of spatially dependent variables, their effectiveness, both in terms of inference and computational expense, is severely compromised when jointly modeling a very large number, q, of point-referenced variables (say exceeding 100). Keeping the number of factors too small relative to the number of variables could lead to poorer inference resulting from excessive smoothing of the spatial random field, while larger numbers of factors exacerbate the computational burden due to the estimation of the loading matrix in not very low-dimensional spaces. Furthermore, multivariate spatial analysis using LMC prohibits interpretation of spatial structures of each variable. For example, LMC endows each process with the same smoothness (the smoothness of the roughest latent process). This is implausible in most applications because different spatial variables typically exhibit very different degrees of smoothness. Here we turn to an alternative option where a graphical dependence model among the variables is used to model the spatial processes jointly.

Graphical models are widely used in contemporary multivariate data analysis in order to model complex dependencies among variables [see Spiegelhalter et al., 1993, Cox and Wermuth, 1993, 1996, Lauritzen, 1996, and references therein]. Such models are based upon joint probability distributions over directed acyclic graphs (DAGs or Bayesian networks) and undirected graphs (left and right panels in Figure 10.10, respectively). In fact, graphical models are especially attractive when modeling a very large number of variables. Later, in Section 13.6.2 of Chapter 13, we will see how a DAG is effectively used to construct nearest neighbor Gaussian Processes (NNGP) Datta et al. [2016a], Finley et al. [2019] that can scale inference to massive data sets by exploiting sparsity introduced by the underlying graph. In this section, we consider specifying multivariate spatial processes where the relationships

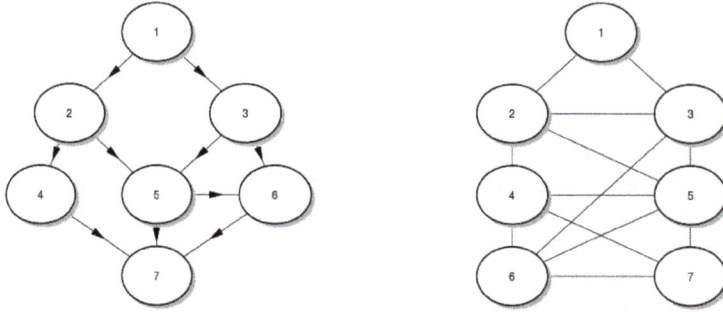

Figure 10.10 *Left: A directed graph (Bayesian network) with arrows pointing from parents to children. Right: An undirected conditional independence graph where the absence of edges between two nodes represent conditional independence given all other nodes.*

between individual processes are posited by an undirected graph. Such processes can scale inference to the highly multivariate setting with the number of variables in the order of $q \sim O(10^2)$ or more.

As in the preceding sections, we seek to build multivariate spatial processes where dependencies among the variables are specified by an undirected graph. Again, let $\mathbf{z}(\mathbf{s}) = (z_1(\mathbf{s}), \ldots, z_q(\mathbf{s}))^{\mathrm{T}}$ be a $q \times 1$ dimensional spatial process, where $z_i(\mathbf{s})$ is the univariate spatial process corresponding to variable i. Let $\mathbf{z}_i = (z_i(\mathbf{s}_1), z_i(\mathbf{s}_2), \ldots, z_i(\mathbf{s}_n))^{\mathrm{T}}$ be $n \times 1$ corresponding to the realizations of the i-th variable, $z_i(\mathbf{s})$, over n locations for each $i = 1, 2, \ldots, q$. We seek to build a multivariate spatial process $\mathbf{z}(\mathbf{s})$ such that the dependencies among $z_i(\mathbf{s})$ are posited by an undirected graph $\mathcal{G} = \{\mathcal{V}, \mathcal{E}\}$, where $\mathcal{V}$ is the set of q vertices corresponding to the q univariate processes and $\mathcal{E}$ is a set of edges. If an edge is absent between two nodes, then $\mathbf{z}_i$ and $\mathbf{z}_j$ are conditionally independent given all the remaining variables (denoted $\mathbf{z}_i \perp \mathbf{z}_j \mid \mathbf{z}_{-(i,j)}$) or, equivalently, the full conditional distribution $p(\mathbf{z}_i \mid \mathbf{z}_{-i})$ does not depend on $\mathbf{z}_j$.

Restricting ourselves, for now, to a fixed finite set of spatial locations $\mathcal{S} = \{\mathbf{s}_1, \ldots, \mathbf{s}_n\}$, it is tempting for us (given our developments in Chapter 4 on MRFs) to model the vectors $\mathbf{z}_i$ using a multivariate MRF. Here, two differences from Chapter 4 are evident. First, the graph now specifies dependencies among variables and not areal units. Second, each node represents a finite realization of a spatial process. Nevertheless, as in Chapter 4, we specify the following sequence of full conditional distributions,

$$\mathbf{z}_i \mid \mathbf{z}_{(-i)} \sim N\left(\sum_{j=1}^{q} A_{ij}\mathbf{z}_j, \Gamma_i\right), \quad i = 1, 2, \ldots, q, \tag{10.88}$$

where $\mathbf{z}_{-(i)} = \{\mathbf{z}_1, \mathbf{z}_2, \ldots, \mathbf{z}_q\} \setminus \{\mathbf{z}_i\}$, i.e., the collection of $\mathbf{z}_j$'s for $j = 1, 2, \ldots, q$ but excluding $\mathbf{z}_i$, A_{ij}'s are fixed $n \times n$ matrices, $A_{ii} = O$ (the matrix of zeroes), and Γ_i's are fixed positive definite matrices. Since each variable can exhibit its own spatial dependence structure, the Γ_i varies by variable.

A multivariate version of Brook's Lemma [recall Section 4.2 for the univariate case; also see Mardia, 1988, for applications to multivariate areal data, which we discuss in Chapter 10] provides a straightforward method for deriving the joint density from (10.88) using the identity

$$p(\mathbf{z}_1, \mathbf{z}_2, \ldots, \mathbf{z}_q) = \prod_{i=1}^{q} \frac{p(\mathbf{z}_i \mid \tilde{\mathbf{z}}_1, \ldots, \tilde{\mathbf{z}}_{i-1}, \mathbf{z}_{i+1}, \ldots, \mathbf{z}_q)}{p(\tilde{\mathbf{z}}_i \mid \tilde{\mathbf{z}}_1, \ldots, \tilde{\mathbf{z}}_{i-1}, \mathbf{z}_{i+1}, \ldots, \mathbf{z}_q))} \times p(\tilde{\mathbf{z}}_1, \tilde{\mathbf{z}}_2, \ldots, \tilde{\mathbf{z}}_q) , \qquad (10.89)$$

where $\tilde{\mathbf{z}} = (\tilde{\mathbf{z}}_1^{\mathrm{T}}, \tilde{\mathbf{z}}_2^{\mathrm{T}}, \ldots, \tilde{\mathbf{z}}_q^{\mathrm{T}})^{\mathrm{T}}$ is any fixed point in the support of $p(\mathbf{z})$ and we assume that the joint density $p(\cdot) > 0$ over its entire support. The proof is a straightforward verification proceeding from the last element of the right hand side (i.e., the joint density on the right hand side). We begin with the observation

$$\frac{p(\mathbf{z}_q \mid \tilde{\mathbf{z}}_1, \ldots, \tilde{\mathbf{z}}_{q-1})}{p(\tilde{\mathbf{z}}_q \mid \tilde{\mathbf{z}}_1, \ldots, \tilde{\mathbf{z}}_{q-1})} \times p(\tilde{\mathbf{z}}_1, \tilde{\mathbf{z}}_2, \ldots, \tilde{\mathbf{z}}_q) = p(\tilde{\mathbf{z}}_1, \tilde{\mathbf{z}}_2, \ldots, \tilde{\mathbf{z}}_{q-1}, \mathbf{z}_q) .$$

Proceeding as above will continue to replace $\tilde{\mathbf{z}}_i$ with $\mathbf{z}_i$ in the joint density on the right hand side of (10.89) for each $i = q - 1, q - 2, \ldots, 1$ and we eventually arrive at $p(\mathbf{z}_1, \mathbf{z}_2, \ldots, \mathbf{z}_q)$.

Applying (10.89) to the full conditional distributions in (10.88) with $\tilde{\mathbf{z}}_i = 0$ for each $i = 1, \ldots, q$ and assuming that $\Gamma_i^{-1} A_{ij} = \left(\Gamma_j^{-1} A_{ji}\right)^{\mathrm{T}} = A_{ji}^{\mathrm{T}} \Gamma_j^{-1}$ allows us to write

$$p(\mathbf{z}) \propto \exp\left\{ -\frac{1}{2}\left(\sum_{i=1}^{q} \mathbf{z}_i^{\mathrm{T}} \Gamma_i^{-1} \mathbf{z}_i - \sum_{i=1}^{q}\sum_{j \neq i}^{q} \mathbf{z}_i^{\mathrm{T}} \Gamma_i^{-1} A_{ij} \mathbf{z}_j \right) \right\} \propto \exp\left(-\frac{1}{2}\mathbf{z}^{\mathrm{T}} Q \mathbf{z} \right) , \qquad (10.90)$$

where $\mathbf{z} = (\mathbf{z}_1^{\mathrm{T}}, \ldots, \mathbf{z}_q^{\mathrm{T}})^{\mathrm{T}}$ is $nq \times 1$ and $Q = M^{-1}(I - A)$ is $nq \times nq$, $M = \oplus \Gamma_i$ is block-diagonal with (i, i)-th block Γ_i for $i = 1, \ldots, q$ and $A = (A_{ij})$ is $nq \times nq$ with A_{ij} as the (i, j)th block. For (10.90) to be a valid density, Q needs to be symmetric and positive definite and $\mathbf{z} \sim N(\mathbf{0}, Q^{-1})$ in (10.90). The condition $\Gamma_i^{-1} A_{ij} = \left(\Gamma_j^{-1} A_{ji}\right)^{\mathrm{T}}$ used to derive (10.90) ensures that Q is symmetric, but how can we ensure that Q is positive definite while also respecting the conditional independence relationships among the variables from a given undirected graph?

If two distinct nodes i and j in the graph do not have an edge, then the block $Q_{ij} = -\Gamma_i^{-1} A_{ij} = O$, i.e., the (i, j)-th block of Q must be an $n \times n$ matrix of zeros, so that $\mathbf{z}_i$ and $\mathbf{z}_j$ are conditionally independent given all other variables. Given the inter-variable graph $\mathcal{G} = \{\mathcal{V}, \mathcal{E}\}$, let $\Lambda = (\lambda_{ij}) = D - \rho W$ be the $q \times q$ graph Laplacian, where $W = (w_{ij})$ is the adjacency matrix with nonzero w_{ij} only if there is an edge between i and j, $D = (d_{ii})$ is a $q \times q$ diagonal matrix with the sum of each row of W along the diagonal, i.e., $d_{ii} = \sum_{j=1}^{q} w_{ij}$, and ρ is a scalar parameter that ensures positive-definiteness of Λ as long as $\rho \in (1/\zeta_{min}, \zeta_{max})$, where ζ_{min} and ζ_{max} are the minimum and maximum eigenvalues of $D^{-1/2} W D^{-1/2}$, respectively.

Let $C_i(\mathbf{s}, \mathbf{s}')$ be the spatial covariance function for $z_i(\mathbf{s})$ so that $\mathbf{z}_i \sim N(\mathbf{0}, C_{ii})$, where $C_{ii} = (C_i(\mathbf{s}_k, \mathbf{s}_l))$ is the $n \times n$ spatial covariance matrix for the i-th process. Let R_i be any nonsingular $n \times n$ factor (e.g., an upper-triangular Cholesky) of the positive definite precision matrix of $\mathbf{z}_i$, i.e., $(\mathrm{var}(\mathbf{z}_i))^{-1} = C_{ii}^{-1} = R_i^{\mathrm{T}} R_i$ for each $i = 1, 2, \ldots, q$. We set $\Gamma_i^{-1} = \lambda_{ii} R_i^{\mathrm{T}} R_i$ and $A_{ij} = -(\lambda_{ij}/\lambda_{ii}) R_i^{-1} R_j$ in (10.88). Then $Q_{ij} = \lambda_{ij} R_i^{\mathrm{T}} R_j$ and $Q = \tilde{R}^{\mathrm{T}}(\Lambda \otimes I)\tilde{R}$, where $\tilde{R} = \oplus_{i=1}^{q} R_i$. Since each R_i is nonsingular and Λ is positive definite, it follows that Q is positive definite and, hence, $\det(Q) = \prod_{i=1}^{q}(\det(R_i))^2 (\det(\Lambda))^n$.

The above construction yields proper densities in (10.90) that (i) model spatial dependence for each variable using its specified covariance function, and (ii) conform to conditional independence among the variables specified by the undirected graph. Sparser graphs imply sparser Q, which can accrue significant computational gains when jointly modeling a large number (q) of variables. The special case where each $R_i = R$ for $i = 1, 2, \ldots, q$ yields the separable model $Q = \Lambda \otimes (R^{\mathrm{T}} R)$.

However, it is challenging to characterize the marginal distributions of such a multivariate process parsimoniously in terms of a few parameters. Even if Γ_i is chosen from an interpretable parametric family of covariances like the Matérn family, the construction does not ensure that z_i will follow a Matérn distribution. Similarly, the cross-covariances will not correspond to standard families of valid, cross-covariance functions Genton and Kleiber [2015]. Furthermore, the construction of Q described above is essentially finite-dimensional and the notion of conditional independence is restricted to the finite set of locations $s_1, \ldots, s_n$. It is unclear if this construction is able to produce a multivariate spatial process over the entire domain $\mathcal{D}$ that conforms to process-level conditional independence.

10.10.2 *Graphical Gaussian Processes*

Dey et al. [2022] resolve the above issues by building a class of Graphical Gaussian Processes (GGPs). Rather than building a Gaussian graphical model whose nodes are finite-dimensional random vectors, GGP builds a graphical model whose nodes are the component Gaussian processes of a multivariate spatial GP and edges represent process-level conditional dependence between the incident nodes given all other nodes. A q-variate process $z(\mathbf{s})$ is a GGP with respect to an inter-variable graph $\mathcal{G} = \{\mathcal{V}, \mathcal{E}\}$ if $z_i(\cdot)$ and $z_j(\cdot)$ are conditionally independent given the remaining processes whenever $(i, j) \notin \mathcal{E}$. We first clarify what it means for two spatial process to be conditionally independent. Let the q nodes of $\mathcal{G}$ represent q different spatial processes, $\{z_i(\mathbf{s}) : s \in \mathcal{D}\}$ for $i = 1, 2, \ldots, q$. Two distinct processes $z_i(\cdot)$ and $z_j(\cdot)$ are conditionally independent given the remaining $q - 2$ processes, which we denote by $z_i(\cdot) \perp z_j(\cdot) \mid z_{-(ij)}(\cdot)$, if the covariance between $z_i(\mathbf{s})$ and $z_j(\mathbf{s}')$ is zero for any pair of locations $\mathbf{s}, \mathbf{s}' \in \mathcal{D}$ conditional on full realizations of all of the remaining processes. More formally, $z_i(\cdot)$ and $z_j(\cdot)$ are conditionally independent, given the remaining processes $\{z_k(\cdot) \mid k \in \mathcal{V} \setminus \{i, j\}\}$ if $\text{Cov}(\epsilon_{iB}(\mathbf{s}), \epsilon_{jB}(\mathbf{s}')) = 0$ for all $\mathbf{s}, \mathbf{s}' \in \mathcal{D}$ and $B = \mathcal{V} \setminus \{i, j\}$, where $\epsilon_{kB}(\mathbf{s}) = z_k(\mathbf{s}) - \text{E}[z_k(\mathbf{s}) \mid \sigma(\{z_j(\mathbf{s}') : j \in B, \ \mathbf{s}' \in \mathcal{D}\})]$ and $\sigma(\cdot)$ is the σ-algebra generated by its argument. This ensures that the conditional covariance between $z_i(\mathbf{s})$ and $z_j(\mathbf{s}')$ given any arbitrary finite realization $\{z_k(\mathbf{s}) : k \in \mathcal{V} \setminus \{i, j\}, s \in \mathcal{S}\}$ is zero for *any* arbitrary finite set of locations $\mathcal{S} \subset \mathcal{D}$ whenever there is no edge between i and j in the graph.

With the above definition, we ask the following question: Given an inter-variable graph and a cross-covariance matrix $C(\mathbf{s}, \mathbf{s}') = (C_{ij}(\mathbf{s}, \mathbf{s}'))$, can we construct a multivariate Gaussian process that conforms to $\mathcal{G}$ in terms of conditional dependencies and also offers some manner of a best approximation to a multivariate GP specified by $C(\mathbf{s}, \mathbf{s}')$? In this regard, we recall a seminal paper by Dempster [1972a] on covariance selection. Let $\mathcal{G} = \{\mathcal{V}, \mathcal{E}\}$ be any undirected graph with q nodes and let $H = (h_{ij})$ be any positive definite covariance matrix whose elements h_{ij} give the covariance between random variables at nodes i and j. Then, the best approximation of H (in terms of the Kullback-Leibler distance) among the class of covariance matrices that satisfy the conditional independence relationships specified by $\mathcal{G}$ is a unique positive definite matrix $\tilde{H} = (\tilde{h}_{ij})$ such that:

1. $\tilde{h}_{ii} = h_{ii}$;
2. $\tilde{h}_{ij} = h_{ij}$ for all $(i, j) \in \mathcal{E}$; and
3. $(\tilde{H}^{-1})_{ij} = 0$ for $(i, j) \notin \mathcal{E}$.

The first two conditions state that the optimal graphical model preserves all marginal variances and cross-covariances for edges included, while the third condition ensures the conditional dependencies specified by $\mathcal{G}$. Covariance selection does not explicitly require Markovian assumptions, nor does it require modeling full conditionals such as in (10.88).

Covariance selection, as described above, applies to finite-dimensional inference over a specified set of n spatial locations. If $C = (C(\mathbf{s}_i, \mathbf{s}_j))$ is an $nq \times nq$ spatial covariance matrix constructed from the $q \times q$ cross-covariance function $C(\mathbf{s}_i, \mathbf{s}_j) = (C_{k,l}(\mathbf{s}_i, \mathbf{s}_j))$ with $C_{k,l}(\mathbf{s}_i, \mathbf{s}_j)$ being the value of the cross-covariance between values of the k-th process at $\mathbf{s}_i$ and l-th process at $\mathbf{s}_j$, then an iterative proportional scaling (IPS) algorithm [Speed et al., 1986, Xu et al., 2011] can compute the unique $\tilde{C}$ preserving marginal variances and covariances and conforming to $\mathcal{G}$. However, this does not immediately extend to the process-level formulation of conditional dependence in a multivariate Gaussian process.

Dey et al. [2022] extend covariance selection to spatial processes to allow predictive inference at arbitrary spatial locations. We begin the construction with a given inter-variable graph $\mathcal{G}$, a $q \times q$ cross-covariance function $C(\mathbf{s}, \mathbf{s}')$, and a finite set of spatial locations $\mathcal{S}$. We define an $nq \times 1$ latent random variable

$$\tilde{\mathbf{z}} \sim N(\mathbf{0}, \tilde{C}) \,, \tag{10.91}$$

where $\tilde{\mathbf{z}} = (\tilde{\mathbf{z}}_1^\mathrm{T}, \dots, \tilde{\mathbf{z}}_q^\mathrm{T})^\mathrm{T}$, $\tilde{C} = (\tilde{C}_{ij})$ is the $nq \times nq$ positive definite matrix obtained from C using covariance selection and $\tilde{C}_{ij} = \mathrm{Cov}(\tilde{\mathbf{z}}_i, \tilde{\mathbf{z}}_j)$ gives the (i,j)-th block of $\tilde{C}$ for $i, j = 1, \dots, q$. Covariance selection ensures $\tilde{\mathbf{z}}_i \sim N(\mathbf{0}, C_{ii})$, $\mathrm{Cov}(\tilde{\mathbf{z}}_i, \tilde{\mathbf{z}}_j) = C_{ij}$ for all $(i.j) \in \mathcal{E}$, and $\mathrm{Cov}(\tilde{\mathbf{z}}_i, \tilde{\mathbf{z}}_j \mid \mathbf{z}_{-ij}) = (\tilde{C}^{-1})_{ij} = O$ for all $(i,j) \notin \mathcal{E}$. Therefore each element of $\tilde{\mathbf{z}}$ retains the marginal covariances as specified by C, marginal cross-covariances for variable pairs included in the graph, and conditional dependencies specified by the graph. To extend to a multivariate graphical process, we write $z_i(\mathbf{s})$ with covariance function $C_i(\mathbf{s}, \mathbf{s}')$ as

$$z_i(\mathbf{s}) = \mathbf{b}_i(\mathbf{s})^\mathrm{T} \tilde{\mathbf{z}}_i + r_i(\mathbf{s}), \quad i = 1, \dots, q \tag{10.92}$$

where $\mathbf{b}_i(\mathbf{s})$ is any $n \times 1$ vector of basis functions such that the process $\mathbf{b}_i(\mathbf{s})^\mathrm{T} \tilde{\mathbf{z}}_i = z_i(\mathbf{s})$ for all $\mathbf{s} \in \mathcal{S}$ and $r_i(\mathbf{s})$ is a *residual* Gaussian process independent of $\tilde{\mathbf{z}}_i$. There are plenty of choices for $\mathbf{b}_i(\mathbf{s})$ to ensure spatial interpolation (recall the discussion in Section 2.4), but a particularly simple choice is the kriging coefficients, $\mathbf{b}_i(\mathbf{s})^\mathrm{T} = \mathbf{c}_i(\mathbf{s})^\mathrm{T} C_{ii}(\mathcal{S})^{-1}$, where $\mathbf{c}_i(\mathbf{s})^\mathrm{T}$ is the $1 \times n$ vector with j-th element $C_i(\mathbf{s}, \mathbf{s}_j)$ for each location $\mathbf{s}_j \in \mathcal{S}$, $j = 1, \dots, n$, and $C_{ii}(\mathcal{S}) = (C_i(\mathbf{s}_j, \mathbf{s}_k))$ is $n \times n$ for $j, k = 1, \dots, n$. This choice for $\mathbf{b}_i(\mathbf{s})$ corresponds to a *predictive process* [Banerjee et al., 2008, which we discuss in Section 13.4.1 of Chapter 13]. Each $\mathbf{r}_i(\mathbf{s})$ is an independent (across $i = 1, \dots, q$) zero-centered Gaussian process with covariance function $R_{ii}(\mathbf{s}, \mathbf{s}') = C_i(\mathbf{s}, \mathbf{s}') - \mathbf{c}_i(\mathbf{s})^\mathrm{T} C_{ii}(\mathcal{S})^{-1} \mathbf{c}_i(\mathbf{s}')$ [this corresponds to a "modified" predictive process Finley et al., 2011a, which is also discussed in Section 13.4.3].

Each $z_i(\mathbf{s}) \sim GP(0, C_i(\cdot, \cdot))$ thereby preserving the marginals. Cross-covariances for $(i, j) \in \mathcal{E}$ are also preserved exactly on $\mathcal{S}$ and approximately elsewhere when choosing $\mathcal{S}$ to be sufficiently representative of the domain. Finally, the conditional independence of the elements of $\tilde{\mathbf{z}}$ with respect to $\mathcal{G}$ induced by covariance selection together with the component-wise independent residual processes $r_i(\cdot)$ ensure that the resulting multivariate process $\mathbf{z}(\mathbf{s}) = (z_1(\mathbf{s}), \dots, z_q(\mathbf{s}))^\mathrm{T}$ conforms to process-level conditional dependencies according to $\mathcal{G}$. This completes the construction of our GGP. Dey et al. [2022] refer to this construction *stitching*. The multiple processes are envisioned to be multiple layers of a fabric that are connected to each other only at the locations $\mathcal{S}$ via the edges of $\mathcal{G}$, serving as threads.

Given any non-graphical stationary multivariate GP and any inter-variable graph, the GGP constructed above is the unique and optimal multivariate Gaussian Process [in terms of integrated spectral Kullback-Leibler divergence from the original process; see Theorem 1 in Dey et al., 2022] while conforming to process-level conditional dependencies specified by the undirected graph. The GGP also maintains the univariate marginal distributions from the original process.

To summarize the above development, a $q \times 1$ multivariate process $\mathbf{z}(\mathbf{s}) = (z_1(\mathbf{s}), \dots, z_q(\mathbf{s}))^\mathrm{T}$ is a Graphical Gaussian Process (GGP) on $\mathcal{D}$ given a valid $q \times q$ cross-covariance function $C(\mathbf{s}, \mathbf{s}') = (C_{ij}(\mathbf{s}, \mathbf{s}'))$ on $\mathcal{D} \subseteq \mathbb{R}^d$ and a graph $\mathcal{G} = (\mathcal{V}, E)$, such that it satisfies the following properties.

1. Retains univariate marginal distributions: $z_j(\cdot) \sim GP(0, C_{jj}(\cdot, \cdot))$ for all $1 \le j \le q$.

2. Retains edge-specific cross-covariances: $\mathrm{Cov}(z_i(\mathbf{s}), z_j(\mathbf{s}')) = C_{ij}(\mathbf{s}, \mathbf{s}')$ for all $(i, j) \in E$ and $\mathbf{s}, \mathbf{s}' \in \mathcal{D}$.

3. Encodes process-level conditional dependencies exactly from the graph $\mathcal{G}$: $z_i(\cdot) \perp z_j(\cdot) \mid \mathbf{z}_{-ij}(\cdot)$ for all $(i, j) \notin E$.

A GGP is specified by a mean function $\mu(\cdot)$, an undirected graph $\mathcal{G}$ and a valid (parent) cross-covariance function $C(\cdot, \cdot)$ from which we derive $\tilde{C}$ using covariance selection. We denote a GGP as $z(\cdot) \sim GGP(\mu(\cdot), C(\cdot, \cdot), \mathcal{G})$ or simply as $z(\cdot) \sim GGP(C(\cdot, \cdot), \mathcal{G})$ if it is zero-centered with $\mu(\cdot) = 0$.

10.10.3 Hierarchical GGP models

We offer a brief overview of hierarchical modeling using the GGP we constructed above [see Dey et al., 2022, 2023, for further details and examples]. Let $\mathcal{S}$ be the set of n observed locations and let $\mathbf{y}(\mathbf{s}) = (y_1(\mathbf{s}), \ldots, y_q(\mathbf{s}))$ denote the $q \times 1$ vector of measurements at location $\mathbf{s} \in \mathcal{S}$ and let $\mathcal{L} \subset \mathcal{D} \setminus \mathcal{S}$ be a finite set of locations disjoint from $\mathcal{S}$ where we seek inference for the process. The hierarchical model we seek to fit is

$$y_i(\mathbf{s}) \sim \mathbf{x}_i(\mathbf{s})^{\mathrm{T}} \boldsymbol{\beta}_i + z_i(\mathbf{s}) + \epsilon_i(\mathbf{s}) \text{ for } i = 1, \ldots, q,$$

$$\epsilon_i(\mathbf{s}) \stackrel{iid}{\sim} N(0, \tau_i^2) \text{ for } i = 1, \ldots, q, s \in \mathcal{D}$$

$$z(\cdot) = (z_1(\cdot), \ldots, z_q(\cdot))^{\mathrm{T}} \sim GGP(0, C(\cdot, \cdot), \mathcal{G}), \quad (10.93)$$

where each $\mathbf{x}_i(\mathbf{s})$ is $p_i \times 1$ with values of predictors specific to the variable i at location $\mathbf{s}$, $\epsilon_i(\mathbf{s})$ captures measurement error at that location, τ_i^2 is the measurement error variance at $\mathbf{s}$, and $\mathbf{z}(\mathbf{s})$ is a GGP constructed from a parent cross-covariance function $C(\cdot, \cdot)$ and graph $\mathcal{G}$.

The GGP can be regarded as a two-step construction: first we specify $\tilde{\mathbf{z}}$ over $\mathcal{S}$ using (10.91) and then constructing $\mathbf{z}(\mathbf{s})$ using (10.92) while noting that $\mathbf{z}(\mathbf{s}) = \tilde{\mathbf{z}}(\mathbf{s})$ for all $\mathbf{s} \in \mathcal{S}$. It is possible that not all q variables have been observed over the same set of locations. Let $\mathcal{S}_i \subset \mathcal{S}$ denote the set of locations providing measurements on variable i. With a valid process specification, we embed the process within the hierarchical model

$$p(\boldsymbol{\beta}, \tau, \theta) \times N(\tilde{\mathbf{z}} \mid \mathbf{0}, \ddot{C}_\theta(\mathcal{S}, \mathcal{S})) \times \prod_{i=1}^{q} N(\mathbf{z}_i(\mathcal{L}_i) \mid B(\mathcal{L}_i) \ddot{\mathbf{z}}_i, R_{ii}(\mathcal{L}_i))$$

$$\times \prod_{i=1}^{q} N(\mathbf{y}(\mathcal{S}_i) \mid X(\mathcal{S}_i) \boldsymbol{\beta}_i + \mathbf{z}(\mathcal{S}_i), \tau_i^2 I), \quad (10.94)$$

where $\mathbf{y}(\mathcal{S}_i)$, $X(\mathcal{S}_i)$ and $\mathbf{z}(\mathcal{S}_i)$ are $\mathbf{y}(\mathbf{s})$, $X(\mathbf{s})$ and $\mathbf{z}(\mathbf{s})$ stacked over locations $\mathbf{s} \in \mathcal{S}_i$, $\mathbf{z}_i(\mathcal{L}_i)$ is the vector with elements $z_i(\mathbf{s})$, $B_i(\mathcal{L}_i)$ is block diagonal with $\mathbf{b}_i(\mathbf{s})^{\mathrm{T}}$ as diagonal blocks and $R_{ii}(\mathcal{L}_i)$ is the spatial covariance matrix of $r_i(\mathbf{s})$ over all $\mathbf{s} \in \mathcal{L}_i$; and θ is the set of all parameters in the cross-covariance function and $p(\beta, \tau, \theta)$ is a prior distribution on model parameters. The matrix $\tilde{C}_\theta$ is obtained from covariance selection as in (10.91).

The computational bottleneck in (10.94) arises from manipulating the $O(nq) \times O(nq)$ matrix $\tilde{C}$, which requires applying the IPS algorithm in each iteration of an MCMC algorithm as $\tilde{C}_\theta$ changes with every update of θ. In general, the IPS will require $O(q^2)$ floating point operations or flops [Xu et al., 2011]. For highly multivariate settings, where q is large, fitting (10.94) with a general undirected graph can be computationally prohibitive.

Fitting Bayesian graphical models is cumbersome for non-decomposable graphs [Roverato, 2002, Atay-Kayis and Massam, 2005], but matters are significantly improved when considering decomposable (or chordal) graphs. For a triplet (A, B, O) of disjoint subsets $\mathcal{V}$, O is said to *separate* A from B if every path from a vertex in A to a vertex in B passes through a vertex in O. If $\mathcal{V} = A \cup B \cup O$, and O is a complete subset of $\mathcal{V}$, then (A, B, O) is said to decompose $\mathcal{G}_\mathcal{V}$. The graph is said to be decomposable if it is either complete (meaning there is an edge between every pair of vertices) or if there exists a proper decomposition (A, B, O) into decomposable subgraphs $\mathcal{G}_{A \cup O}$ and $\mathcal{G}_{B \cup O}$. More generally, if a graph is non-decomposable, it can be embedded in a larger decomposable graph. Such graphs are popular in Bayesian graphical models [see, e.g., Dobra et al., 2003, Wang and West, 2009] because they correspond to several commonly encountered dependence structures, such as from low-rank models, factor models and autoregressive models [see Section 4 in Dey et al., 2022].

Decomposable graphs can be represented in terms of a set of cliques K and a set of separators S such that the likelihood for the GGP on $\mathcal{L}$ can be decomposed as

$$N(\mathbf{z}(\mathcal{L}) \mid \mathbf{0}, \tilde{C}) = \frac{\prod_{A \in K} N(\mathbf{z}_A(\mathcal{L}) \mid \mathbf{0}, C)}{\prod_{A \in S} N(\mathbf{z}_A(\mathcal{L}) \mid \mathbf{0}, C)} \tag{10.95}$$

where for any $A \subset \{1, \ldots, q\}$, $\mathbf{z}_A(\mathcal{L})$ denotes the subset of $\mathbf{z}(\mathcal{L})$ corresponding to the indices in A. This circumvents the use of the IPS algorithm as the right hand side can be written in terms of the multivariate densities based on the parent covariance C. The maximum dimension of any matrix involved is $O(c^3)$ where c denotes the largest clique-size and there would be at most $O(q)$ such matrices. Also the resulting likelihood only depends on the cross-covariances C_{ij} for either $i = j$ or $(i, j) \in \mathcal{E}$. Dey et al. [2022] and Dey et al. [2023] develop a chromatic Gibbs sampler [Gonzalez et al., 2011] for efficiently updating (10.94). Dey et al. [2022] devise reversible jump MCMC algorithms for GGPs that also infer about the underlying graph $\mathcal{G}$ using reversible jump MCMC for decomposable graphs [Green and Thomas, 2013].

10.10.4 Functional Graphical Gaussian models

A different class of graphical models for modeling multivariate spatial processes introduces conditional independence through the precision matrices of coefficient vectors in its basis expansion. Let $z_j(\mathbf{s}) = \sum_{l=1}^{\infty} \theta_{jl} \phi_{jl}(\mathbf{s})$, where $\{\phi_{jl}(\mathbf{s})\}$ is a countably infinite collection of orthogonal basis functions and θ_{jl} are the coefficients. The coefficients θ_{jl} are uniquely determined by $z_j(\mathbf{s})$ because $\theta_{jl} = \langle z_j, \phi_{jl} \rangle$, where $\langle \cdot, \cdot \rangle$ is the inner product (e.g., $\langle \phi_{jl}, \phi_{jl'} \rangle = \int \phi_{jl}(\mathbf{s}) \phi_{jl'}(\mathbf{s}) ds$) with respect to which $\langle \phi_{jl}, \phi_{jl'} \rangle = 0$. These are referred to as *Functional Graphical Gaussian Models* (FGGMs) as they have been originally developed for analyzing functional data. Inference using such processes have been explored by Zhu et al. [2016] and Zapata et al. [2021] in the context of functional data analysis, while Krock et al. [2023] develop such processes for highly multivariate spatial data. Zhu et al. [2016] show that modeling the coefficients θ_{jl} as Gaussian and assuming a sparsity structure between $\theta^{(j)} = \{\theta_{jl} : l = 1, \cdots, \infty\}$ imposes the corresponding process-level conditional independence among the component processes of $z(\cdot)$.

For modeling purposes, we can truncate the expansion of $z_j(\cdot)$ to m_j terms and fit a Graphical Gaussian model on the $M = \sum_{j=1}^{q} m_j$ dimensional vector of coefficients. In this case, we can write $z_j(\mathbf{s}) = \boldsymbol{\phi}_j(\mathbf{s})^{\mathrm{T}} \boldsymbol{\theta}_j$, where $\boldsymbol{\phi}_j(\mathbf{s})$ and $\boldsymbol{\theta}_j$ are $m \times 1$ vectors with elements $\phi_{jl}(\mathbf{s})$ and θ_{jl}, respectively. Modeling the conditional dependencies among $\boldsymbol{\theta}_j$'s according to a graph will induce the same dependencies among the elements of $z(\mathbf{s})$. The approach involves inverting an $M \times M$ matrix, thereby requiring $O(M^3)$ floating point operations (or $O(q^3 m^3)$ when $m_j = m$ for all j). The computational expense is cubic in the number of basis functions m, which can be computationally burdensome for large values for m and restricts the richness of the function class.

Zapata et al. [2021] and Krock et al. [2023] construct *partially separable* FGGMs of the form $z_j(\mathbf{s}) = \sum_{l=1}^{\infty} \theta_{jl} \phi_l(\mathbf{s})$ using a common set of orthonormal basis function, i.e., $\phi_{jl} = \phi_l$ for all $j = 1, \cdots, q$ and modeling the coefficients corresponding to different basis functions independently. Collecting the univariate $z_j(\mathbf{s})$ into the $q \times 1$ process $z(\mathbf{s})$, we write

$$\mathbf{z}(\mathbf{s}) = \sum_{l=1}^{\infty} \boldsymbol{\theta}_l \phi_l(\mathbf{s}), \quad \text{where} \quad \boldsymbol{\theta}_l = (\theta_{1l}, \ldots, \theta_{ql})^{\mathrm{T}} \overset{ind}{\sim} N(\mathbf{0}, \Sigma_l), \tag{10.96}$$

where each $\boldsymbol{\theta}_l = (\theta_{1l}, \ldots, \theta_{ql})^{\mathsf{T}}$ is $q \times 1$ consisting of basis coefficients for $\phi_l(\mathbf{s})$ from the q processes and Σ_l is the covariance matrix of $\boldsymbol{\theta}_l$. The basis function expansion is truncated to m terms in practical implementation and we only need m inversions of $q \times q$ matrices $(Cov(\boldsymbol{\theta}_j), j = 1, \ldots, m)$. This considerably reduces computational complexity to $O(mq^3)$, thereby allowing m to be larger. Using penalized likelihood maximization with a graphical Lasso penalty on the estimated covariance of the coefficient vectors introduces sparsity and can enable inference on an unknown graph in a computationally feasible manner. Recently, Li and Solea [2018], Lee et al. [2022] proposed a class of non-parametric graphical models for functional data that performs better than FGGMs when the relation between variables is non-linear or heteroskedastic in nature.

Dey et al. [2025] investigate the relationships among GGPs and FGGMs. The key difference between these two approaches is that the former builds the process from a graph and a cross-covariance function, while the latter uses basis functions and a graphical model on the coefficients. Let $C(\mathbf{s}, \mathbf{s}') = \text{Cov}(\mathbf{z}(\mathbf{s}), \mathbf{z}(\mathbf{s}'))$ be the cross-covariance function of a partially separable process $\mathbf{z}(\mathbf{s})$ in (10.96). Given a fixed graph $\mathcal{G}$, if we apply covariance selection to obtain $\tilde{\Sigma}_l$ from each Σ_l in (10.96), then we obtain $\tilde{\mathbf{z}}(\mathbf{s}) \sim\sim GGP(C, \mathcal{G})$ as defined in Section 10.10.2. In fact, any partially separable process $\tilde{\mathbf{z}}(\mathbf{s})$ admitting the representation in (10.96) is a $GGP(C, \mathcal{G})$ *only if* $\tilde{\Sigma}_l$ is obtained by applying covariance selection to Σ_l with respect to $\mathcal{G}$ for each l. In other words, starting with a partially separable FGGM, the optimal GGP can be obtained by applying Dempster's covariance selection on each covariance matrix Σ_l of the basis coefficients. And this graphical GP is a partially separable functional GGM. Dey et al. [2025] combine these connections between GGPs and FGGMs to investigate inferential performance and computational efficiency using empirical experiments and an application to functional modeling of neuroimaging data using the connectivity graph among regions of the brain.

10.11 Computer tutorials

Computer programs illustrating Bayesian analysis of multivariate spatial data using different libraries in R and its interfaces with Bayesian modeling environments are supplied in the folder titled Chapter 10 from https://github.com/sudiptobanerjee/BGC_2023.

In particular, we analyze a soil nutrient data which was collected at the La Selva Biological Station, Costa Rica [analyzed in greater detail by Guhaniyogi et al., 2013]. The data consists of measurements on soil nutrient concentrations of calcium (Ca), potassium (K) and magnesium (Mg) from $n = 80$ soil cores over a sparse grid centered on a more intensively sampled transect. These nutrient concentrations show a high positive correlation as seen in (10.97):

$$\begin{pmatrix} 1 & & \\ 0.7 & 1 & \\ 0.7 & 0.8 & 1 \end{pmatrix} \tag{10.97}$$

suggesting that we might build a richer model by explicitly accounting for spatial association among the $q = 3$ outcome variables. Our objective is to predict these nutrients at a fine resolution over the study plot. Ultimately, posterior predictive samples will serve as input to a vegetation competition model.

Figure 10.11 depicts the nutrient concentration surfaces. These patterns can be more formally examined using empirical semivariograms. We fit an exponential variogram model to each of the soil nutrients. The resulting variogram estimates are offered in Figure 10.12. Here the upper and lower horizontal lines are the *sill* and *nugget*, respectively, and the vertical line is the effective range (i.e., that distance at which the correlation drops to 0.05). Despite the patterns of spatial dependence seen in Figure 10.11, the variograms do not show much spatial variation. Changing the number of bins (bins) and maximum distance

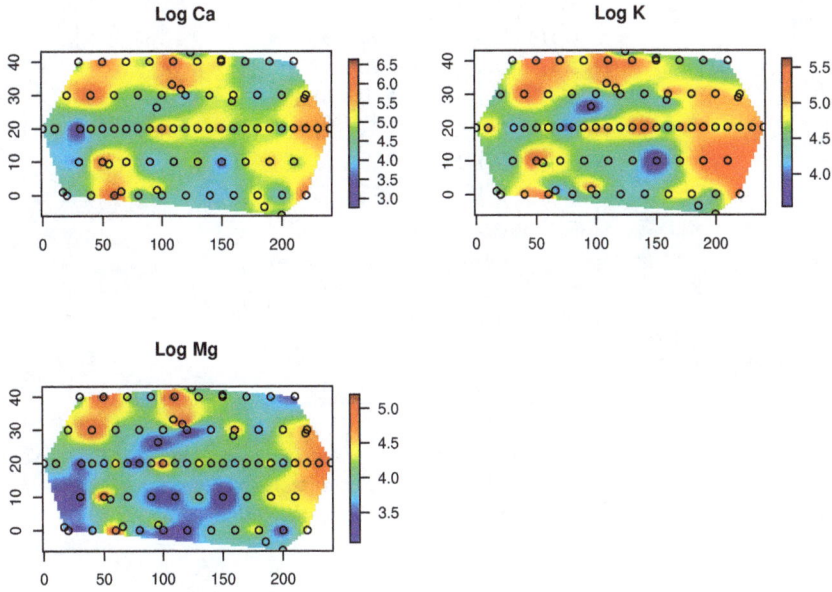

Figure 10.11 *Soil nutrient concentrations and sample array.*

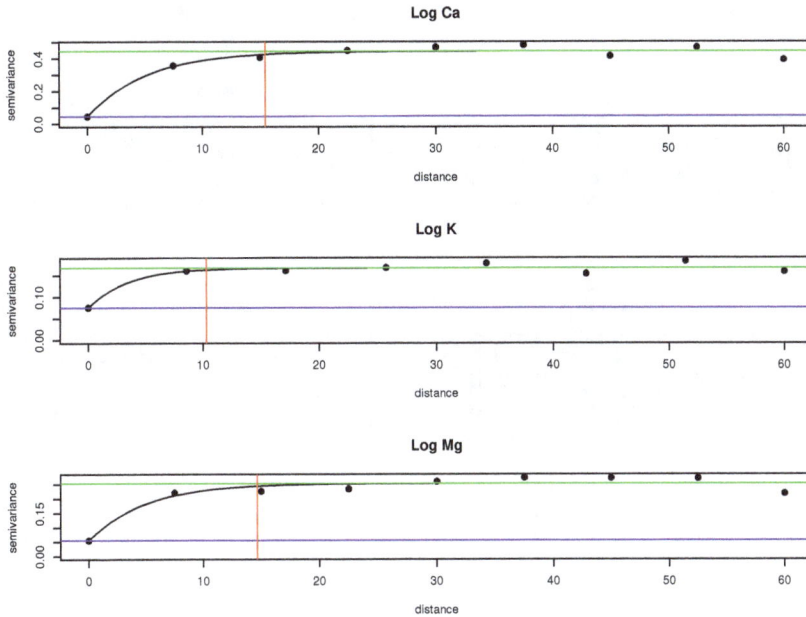

Figure 10.12 *Isotropic semivariograms for log nutrient concentrations.*

considered (`max`) produces effective spatial ranges of less than 20m for each of the nutrients. The signal is weak, likely due to the paucity of samples.

We fit the Bayesian LMC in (10.50), where each of the 3 nutrients is modeled with their own intercept, $y_j(\mathbf{s}) = \beta_j + v_j(\mathbf{s}) + \epsilon_j(\mathbf{s})$ for $j = 1, 2, 3$, a multivariate spatial process $v(\mathbf{s}) = Aw(\mathbf{s})$ (not spatially-varying) and $\epsilon_j(\mathbf{s}) \sim N(0, \tau_j^2)$. Each element of $w(\mathbf{s})$

	2.5%	50%	97.5%
K_{11}	0.299	0.448	0.614
K_{21}	0.136	0.216	0.301
K_{31}	0.209	0.320	0.430
K_{22}	0.092	0.134	0.195
K_{32}	0.112	0.169	0.231
K_{33}	0.179	0.255	0.355
τ_1^2	0.022	0.048	0.099
τ_2^2	0.023	0.042	0.063
τ_3^2	0.010	0.024	0.055
ϕ_1	0.100	0.199	0.294
ϕ_2	0.045	0.156	0.279
ϕ_3	0.036	0.133	0.279

Table 10.8 *Posterior quantiles of the elements of the non-spatial component of the covariance among the nutrients ($AA^{\mathrm{T}} = K$), the variance of the measurement errors for each of the nutrients (τ_j^2) and the rate of decay of spatial correlation (ϕ_j) for each of the nutrients.*

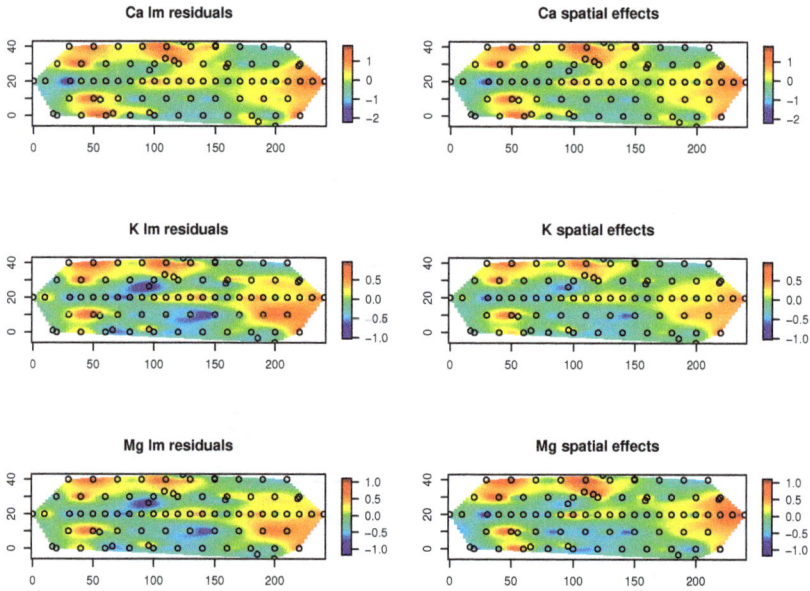

Figure 10.13 *Interpolated surface of the non-spatial model residuals and the mean of the random spatial effects posterior distribution.*

is an independent zero-centered spatial process with an exponential correlation function $\rho_j(\phi_j, d) = \exp(-\phi_j d)$. The posterior summaries of the entries of $AA^{\mathrm{T}} = K$, the τ_j^2's and ϕ_j's are presented in Table 10.8

We also recover the residual spatial effects. Figure 10.13 shows the nutrient concentration random spatial effects and compares them with the residual image plots from a non-spatial regression. With a sparse sample array, an estimated mean effective range of ~ 20, and no predictor variables, we cannot expect predictions to differ much from a constant mean concentration over the domain. We can produce interpolated image plots for the posterior predictive means and standard deviations for our three nutrients. These are displayed in

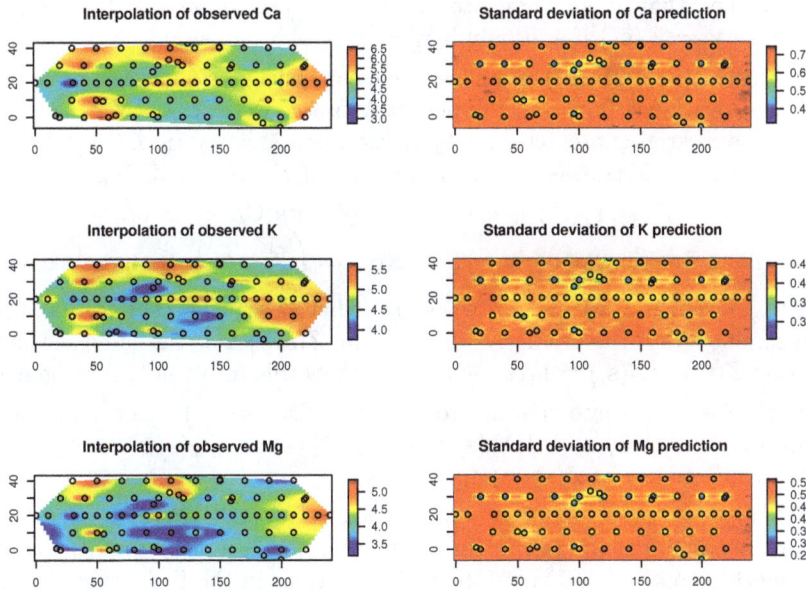

Figure 10.14 *Interpolated surface of observed log nutrient concentrations and standard deviation of each pixel's posterior predictive distribution.*

Figure 10.14. With such a small spatial range, increased precision does not extend far from the sample locations.

10.12   Exercises

1. Compute the coherence (generalized correlation) in (10.12):

   (a) for the cross-covariance in (10.45), and

   (b) for the cross-covariance in (10.49).

2. Show that the product of separable cross-covariance functions is a valid cross-covariance function.

3. Suppose $Z_j(\mathbf{s}) \sim \texttt{Poi}(\lambda_j(\mathbf{s})), j = 1, 2$ where $\lambda_j(\mathbf{s}) = N_j(\mathbf{s})e^{\beta_j + w_j(\mathbf{s})}$ and $\mathbf{w}(\mathbf{s}) = \begin{pmatrix} w_1(\mathbf{s}) \\ w_2(\mathbf{s}) \end{pmatrix} = A \begin{pmatrix} v_1(\mathbf{s}) \\ v_2(\mathbf{s}) \end{pmatrix}$ with $v_1(\mathbf{s})$ and $v_2(\mathbf{s})$ independent mean 0 Gaussian processes. Obtain $\text{Cov}(Z_1(\mathbf{s}), Z_2(\mathbf{s}'))$.

4. Let $Y(\mathbf{s}) = (Y_1(\mathbf{s}), Y_2(\mathbf{s}))^{\mathsf{T}}$ be a bivariate process with a stationary cross-covariance matrix function

$$C(\mathbf{s} - \mathbf{s}') = \begin{pmatrix} c_{11}(\mathbf{s} - \mathbf{s}') & c_{12}(\mathbf{s} - \mathbf{s}') \\ c_{12}(\mathbf{s}' - \mathbf{s}) & c_{22}(\mathbf{s} - \mathbf{s}') \end{pmatrix},$$

   and a set of covariates $\mathbf{x}(\mathbf{s})$. Let $\mathbf{y} = (\mathbf{y}_1^{\mathsf{T}}, \mathbf{y}_2^{\mathsf{T}})^{\mathsf{T}}$ be the $2n \times 1$ data vector, with $\mathbf{y}_1 = (y_1(\mathbf{s}_1), \ldots, y_1(\mathbf{s}_n))^{\mathsf{T}}$ and $\mathbf{y}_2 = (y_2(\mathbf{s}_1), \ldots, y_2(\mathbf{s}_n))^{\mathsf{T}}$.

   (a) Show that the cokriging predictor has the form

$$E[Y_1(\mathbf{s}_0) \mid \mathbf{y}] = \mathbf{x}^{\mathsf{T}}(\mathbf{s}_0)\boldsymbol{\beta} + \boldsymbol{\gamma}^{\mathsf{T}}\Sigma^{-1}(\mathbf{y} - X\boldsymbol{\beta}),$$

   i.e., as in (2.22), but with appropriate definitions of $\boldsymbol{\gamma}$ and Σ.

(b) Show further that if $\mathbf{s}_k$ is a site where $y_l(\mathbf{s}_k)$ is observed, then for $l = 1, 2$, $E[Y_l(\mathbf{s}_k) \mid \mathbf{y}] = y_l(\mathbf{s}_k)$ if and only if $\tau_l^2 = 0$.

5. Suppose $\mathbf{Y}(\mathbf{s}) = (Y_1(\mathbf{s}), Y_2(\mathbf{s}))$ is a constant mean, bivariate Gaussian process with separable cross covariance function. Suppose we observe data $\mathbf{Y} = \{\mathbf{Y}(\mathbf{s}_i), i = 1, 2, ..., n\}$. Show that the co-kriging predictor for $Y_1(\mathbf{s}_0)$ at a new location $\mathbf{s}_0$, $E(Y_1(\mathbf{s}_0) \mid \mathbf{Y})$ depends only upon the set of $Y_1(\mathbf{s}_i)$. Is the result still true if $E(Y_j(\mathbf{s}) = \mathbf{X}^{\mathrm{T}}(\mathbf{s})\boldsymbol{\beta}_j, j = 1, 2$?

6. Suppose $\mathbf{Y}(\mathbf{s}) = (Y_1(\mathbf{s}), Y_2(\mathbf{s}))$ is a mean 0, bivariate Gaussian process with separable cross-covariance function, $\rho(\mathbf{s} - \mathbf{s}'; \phi)T$ where $T = \begin{pmatrix} 1 & \gamma \\ \gamma & 1 \end{pmatrix}$. Suppose we observe data $\mathbf{Y} = \{\mathbf{Y}(\mathbf{s}_i), i = 1, 2, ..., n\}$ and we wish to predict $Z(\mathbf{s}_0) = Y_1(\mathbf{s}_0) + Y_2(\mathbf{s}_0)$ at the new location $\mathbf{s}_0$. Show that the predictive distribution $Z(\mathbf{s}_0) \mid \mathbf{Y}$ is the same as the predictive distribution $Z(\mathbf{s}_0) \mid \{Z(\mathbf{s}_i) = Y_1(\mathbf{s}_i) + y_2(\mathbf{s}_i)\}$. Does this result hold for a general T?

7. Suppose $\mathbf{Y}(\mathbf{s})$ is a bivariate spatial process as in Exercise 4 In fact, suppose $\mathbf{Y}(\mathbf{s})$ is a Gaussian process. Let $Z_1(\mathbf{s}) = I(Y_1(\mathbf{s}) > 0)$, and $Z_2(\mathbf{s}) = I(Y_2(\mathbf{s}) > 0)$. Approximate the cross-covariance matrix of $\mathbf{Z}(\mathbf{s}) = (Z_1(\mathbf{s}), Z_2(\mathbf{s}))^{\mathrm{T}}$.

8. The data in https://github.com/sudiptobanerjee/BGC_2023/data/ColoradoLMC.dat record maximum temperature (in tenths of a degree Celsius) and precipitation (in cm) during the month of January 1997 at 50 locations in the U.S. state of Colorado.

(a) Let X denote temperature and Y denote precipitation. Following the model of Example 10.3, fit an LMC model to these data using the conditional approach, fitting X and then $Y \mid X$.

(b) Repeat this analysis, but this time fitting Y and then $X \mid Y$. Show that your new results agree with those from part (a) up to simulation variability.

9. If C_l and $C_{l'}$ are isotropic, obtain $C_{ll'}(\mathbf{s})$ in (10.81) by transformation to polar coordinates.

10. More on the coregionalization asymmetry. Following up on the discussion surrounding (10.44), suppose we have the usual bivariate coregionalization model $\mathbf{W}(\mathbf{s}) = AV(\mathbf{s})$ with A lower triangular and say, $V_1(\mathbf{s})$ and $V_2(\mathbf{s})$ are independent mean 0 Gaussian processes. Clarify that $W_1(\mathbf{s})$ and $W_2(\mathbf{s}')$ are conditionally independent given $W_1(\mathbf{s}')$ but $W_1(\mathbf{s}')$ and $W_2(\mathbf{s})$ are not conditionally independent given $W_2(\mathbf{s}')$. (Of course, the asymmetry disappears if A is not lower triangular.)

Chapter 11

Models for multivariate areal data

In this chapter we explore the extension of univariate CAR methodology (Sections 4.3 and 6.8.3) to the multivariate setting. Such models can be employed to introduce multiple, dependent spatial random effects associated with areal units (as standard CAR models do for a single set of random effects). In this regard, Kim et al. [2001] presented a "twofold CAR" model to model counts for two different types of disease over each areal unit. Similarly, Knorr-Held and Best [2002] developed a "shared component" model for the above purpose, but their methodology is also specific to the bivariate situation. Knorr-Held and Rue [2002] illustrate sophisticated MCMC blocking approaches in a model placing three conditionally independent CAR priors on three sets of spatial random effects in a shared component model setting.

We mention some noteworthy contributions of MacNab who provides recent developmental multivariate CAR work. In particular, Macnab [2018] offers a lengthy review. Macnab [2022b] focuses on implementations using coregionalization (as we do here). Macnab [2022a] reviews multivariate MRF modeling in the context of disease mapping.

Multivariate CAR (MCAR) models can also provide coefficients in a multiple regression setting that are dependent and spatially varying at the areal unit level. For example, Gamerman et al. [2003] investigate a Gaussian Markov random field (GMRF) model (a multivariate generalization of the pairwise difference IAR model) and compare various MCMC blocking schemes for sampling from the posterior that results under a Gaussian multiple linear regression likelihood. They also investigate a "pinned down" version of this model that resolves the impropriety problem by centering the ϕ_i vectors around some mean location. These authors also place the spatial structure on the spatial regression coefficients themselves, instead of on extra intercept terms (that is, in (6.40) we would drop the ϕ_i, and replace β_1 by β_{1i}, which would now be assumed to have a CAR structure). Assunção et al. [2002] refer to these models as *space-varying coefficient* models and illustrate in the case of estimating fertility schedules. Assunção [2003] offers a nice review of the work in this area up to that time. Also working with areal units, Sain and Cressie [2007] offer multivariate GMRF models, proposing a generalization that permits asymmetry in the spatial conditional cross-correlation matrix. They use this approach to jointly model the counts of white and minority persons residing in the census block groups of St. James Parish, LA, a region containing several hazardous waste sites.

We point out that multivariate CAR models are not the only option available for analyzing multivariate areal data. Zhang et al. [2009] develop an arguably much simpler alternative approach building upon the techniques of smoothed ANOVA [SANOVA; see Hodges et al., 2007]. Instead of simply shrinking effects without any structure, these authors propose SANOVA to smooth spatial random effects by taking advantage of the spatial structure. The underlying idea is to extend SANOVA to cases in which one factor is a spatial lattice, which is smoothed using a CAR model, and a second factor is, for example, the type of disease. Datasets routinely lack enough information to identify the additional structure of MCAR. SANOVA offers a simpler and more intelligible structure than the MCAR while

DOI: 10.1201/9781003401728-11

performing as well. Nevertheless, the MCAR and more general CAR-based approaches provide a diverse and rich class of models suitable for capturing complex spatial associations. We focus on these approaches in the remainder of this section.

11.1 The multivariate CAR (MCAR) distribution

For a vector of univariate variables $\boldsymbol{\phi} = (\phi_1, \phi_2, \ldots, \phi_n)$, zero-centered CAR specifications were detailed in Section 4.3. For the MCAR model we instead let $\boldsymbol{\phi}^{\mathrm{T}} = (\boldsymbol{\phi}_1, \boldsymbol{\phi}_2, \ldots, \boldsymbol{\phi}_n)$ where each $\boldsymbol{\phi}_i = (\phi_{i1}, \phi_{i2}, \ldots, \phi_{ip})^{\mathrm{T}}$ is $p \times 1$. Most multivariate CAR models are members of the family developed by Mardia [1988]. Analogous to the univariate case, the joint distribution is derived from the full conditional distributions. Under the MRF assumption, we can specify these conditional distributions as

$$p(\boldsymbol{\phi}_i \mid \boldsymbol{\phi}_{j \neq i}, \Gamma_i) \;=\; N\left(\sum_{i \sim j} B_{ij}\boldsymbol{\phi}_j \,,\, \Gamma_i\right), \quad i, j = 1, \ldots, n, \tag{11.1}$$

where Γ_i and B_{ij} are $p \times p$ matrices. Mardia [1988] proved, using a multivariate analogue of Brook's lemma, that the full conditional distributions in (11.1) yield a joint distribution of the form

$$p(\boldsymbol{\phi} \mid \{\Gamma_i\}) \propto \exp\left\{-\frac{1}{2}\boldsymbol{\phi}^{\mathrm{T}}\Gamma^{-1}(I - \tilde{B})\boldsymbol{\phi}\right\},$$

where Γ is block-diagonal with blocks Γ_i, and $\tilde{B}$ is $np \times np$ with (i, j)-th block B_{ij}.

As in the univariate case, symmetry of $\Gamma^{-1}(I - \tilde{B})$ is required. A convenient special case sets $B_{ij} = b_{ij}I_{p \times p}$, yielding the symmetry condition $b_{ij}\Gamma_j = b_{ji}\Gamma_i$, analogous to (4.14). If as in Subsection 4.3.1 we take $b_{ij} = w_{ij}/w_{i+}$ and $\Sigma_i = w_{i+}^{-1}\Sigma$, then the symmetry condition is satisfied.

Kronecker product notation simplifies the form of $\Gamma^{-1}(I - \tilde{B})$. That is, setting $\tilde{B} = B \otimes I$ with B as in (4.13) and $\Gamma = D^{-1} \otimes \Sigma$ so

$$\Gamma^{-1}(I - \tilde{B}) = (D \otimes \Sigma^{-1})(I - B \otimes I) = (D - W) \otimes \Sigma^{-1}.$$

Again, the singularity of $D - W$ implies that $\Gamma^{-1}(I - B)$ is singular. We denote this distribution by $MCAR(1, \Sigma)$.

To consider remedies to the impropriety, Mardia [1988] proposed rewriting (11.1) as

$$p(\boldsymbol{\phi}_i \mid \boldsymbol{\phi}_{j \neq i}, \Gamma_i) \;=\; N\left(R_i \sum_{i \sim j} B_{ij}\boldsymbol{\phi}_j \,,\, \Gamma_i^{-1}\right), \quad i, j = 1, \ldots, n,$$

where R_i is $p \times p$. Now $\Gamma^{-1}(I - \tilde{B})$ is revised to $\Gamma^{-1}(I - \tilde{B}_R)$ where $\tilde{B}_R$ has (i, j)th block $R_i B_{ij}$. In general, then, the symmetry condition becomes

$$(\Gamma_i^{-1}R_iB_{ij})^{\mathrm{T}} = \Gamma_j^{-1}R_jB_{ji} \;\text{ or }\; \Gamma_j B_{ij}^{\mathrm{T}}R_i^{\mathrm{T}} = R_jB_{ji}\Gamma_i.$$

See Mardia [1988] expression (2.4) in this regard. If, in addition, $\Gamma^{-1}(I - \tilde{B}_R)$ is positive definite, then the conditional distributions uniquely determine the joint distribution

$$\boldsymbol{\phi} \sim N\left(\mathbf{0}, \, [\Gamma(I - \tilde{B}_R)]^{-1}\right). \tag{11.2}$$

In particular, if $B_{ij} = b_{ij}I_{p \times p}$ and $b_{ij} = w_{ij}/w_{i+}$, the symmetry condition simplifies to

$$w_{j+}\Gamma_j R_i^{\mathrm{T}} = w_{i+}R_j\Gamma_i.$$

Finally, if in addition we take $\Gamma_i = w_{i+}^{-1}\Lambda$, we obtain $\Lambda R_i^{\mathsf{T}} = R_j\Lambda$, which reveals that we must have $R_i = R_j = R$, and thus

$$\Lambda R^{\mathsf{T}} = R\Lambda . \tag{11.3}$$

For any arbitrary positive definite Λ, a generic solution to (11.3) is $R = \rho\Lambda^t$. Hence, regardless of t, (11.3) introduces a total of $\binom{p+1}{2} + 1$ parameters. Thus, without loss of generality, we can set $t = 0$, hence $R = \rho I$. Calculations as above yield

$$\Sigma^{-1} = \Gamma^{-1}(I - \tilde{B}_R) = (D - \rho W) \otimes \Lambda^{-1} , \tag{11.4}$$

where Σ is the $np \times np$ variance-covariance matrix of ϕ. Therefore, Σ has a *separable* structure and is nonsingular under the same restriction to ρ as in the univariate case, and we have

$$\phi \sim N\left(0, \ [(D - \rho W) \otimes \Lambda]^{-1}\right) . \tag{11.5}$$

Following Gelfand and Vounatsou [2003] and Carlin and Banerjee [2003], we denote this model by $MCAR(\rho, \Sigma)$.

The separable MCAR model in (11.5) is obtained under the assumption that $R_i = \rho I_{p\times p}, i = 1, \ldots, n$. While the positive definiteness condition of the covariance matrix follows immediately from its Kronecker product form, which also has computational benefits, the assumption of a common ρ for $j = 1, \ldots, p$ may well be too strong. To elucidate further, suppose $p = 2$ (e.g., two diseases in each county), and define $\phi_1' = (\phi_{11}, \ldots, \phi_{n1})$ and $\phi_2' = (\phi_{12}, \ldots, \phi_{n2})$. Then the MCAR formulation (11.5) can be written as

$$\begin{pmatrix} \phi_1 \\ \phi_2 \end{pmatrix} \sim N\left(\begin{pmatrix} 0 \\ 0 \end{pmatrix}, \begin{pmatrix} (D - \rho W)\Lambda_{11} & (D - \rho W)\Lambda_{12} \\ (D - \rho W)\Lambda_{12} & (D - \rho W)\Lambda_{22} \end{pmatrix}^{-1}\right) , \tag{11.6}$$

where $\Lambda_{ij}, i = 1, 2, j = 1, 2$ are the elements of Λ. More generally, we may need three different ρ_i parameters in (11.6) to explain the correlation between the two types of cancer and across the counties that neighbor each other [Kim et al., 2001]. The covariance matrix Σ would then be revised to

$$\Sigma = \begin{pmatrix} (D - \rho_1 W)\Lambda_{11} & (D - \rho_3 W)\Lambda_{12} \\ (D - \rho_3 W)\Lambda_{12} & (D - \rho_2 W)\Lambda_{22} \end{pmatrix}^{-1} , \tag{11.7}$$

where ρ_1 and ρ_2 are the smoothing parameters for the two cancer types, and ρ_3 is the "bridging" or "linking" parameter associating ϕ_{i1} with $\phi_{j2}, i \neq j$. Unfortunately, with this general covariance matrix, it is difficult to derive the conditions for positive definiteness as they depend upon the unknown Λ matrix.

Carlin and Banerjee [2003] and Gelfand and Vounatsou [2003] generalize the separable MCAR model by allowing two different ρ parameters (say, ρ_1 and ρ_2), and denote this model as $MCAR(\rho_1, \rho_2, \Lambda)$. They write the precision matrix Σ^{-1} as

$$\begin{pmatrix} U_1^{\mathsf{T}}U_1\Lambda_{11} & U_1^{\mathsf{T}}U_2\Lambda_{12} \\ U_2^{\mathsf{T}}U_1\Lambda_{12} & U_2^{\mathsf{T}}U_2\Lambda_{22} \end{pmatrix} = \begin{pmatrix} U_1^{\mathsf{T}} & 0 \\ 0 & U_2^{\mathsf{T}} \end{pmatrix} (\Lambda \otimes I_{n\times n}) \begin{pmatrix} U_1 & 0 \\ 0 & U_2 \end{pmatrix} , \tag{11.8}$$

where $U_k' U_k = D - \rho_k W, k = 1, 2$. Carlin and Banerjee [2003] take U_k to be the Cholesky decomposition of $D - \rho_k W$ so that U_k is an upper-triangular matrix, while Gelfand and Vounatsou [2003] employ a spectral decomposition, i.e., $U_k = \text{Diag}(1 - \rho_k\lambda_i)^{\frac{1}{2}}P'D^{\frac{1}{2}}P$, where the λ_i are the eigenvalues of $D^{-\frac{1}{2}}WD^{-\frac{1}{2}}$ and P is an orthogonal matrix with the corresponding eigenvectors as its columns. Either way, this generalization of the MCAR model permits different smoothing parameters ρ_k for each k (e.g., different strengths of spatial correlation for each type of cancer). As before, Λ controls the nonspatial correlation among cancers at any given location.

The conditions for the covariance matrix to be positive definite are easy to find as long as the Cholesky or spectral decompositions exist and Λ is positive definite. For the $p = 2$ case, these reduce to $|\rho_1| < 1$ and $|\rho_2| < 1$. The spectral approach may be better in terms of Bayesian computing since it does not requires the calculation of a Cholesky decomposition at each MCMC iteration, a substantial burden, particularly for a data set with many spatial regions. Neither of these MCAR structures allow a smoothing parameter ρ on the off-diagonal of the precision matrix as in (11.7); we cannot model the off-diagonal, since it is determined by the diagonal. Finally, since the decomposition of $D - \rho_k W$ is not unique, we can have different MCAR models with the covariance structure (11.8).

Note that the covariance structure (11.8) easily generalizes to an arbitrary number, say p, of diseases. We write

$$\boldsymbol{\phi} \ \sim \ N_{np}\Big(\mathbf{0}, \ [\mathrm{Diag}(U_1^{\mathrm{T}},\dots,U_p^{\mathrm{T}})(\Lambda \otimes I_{n\times n})\mathrm{Diag}(U_1,\dots,U_p)]^{-1}\Big) , \tag{11.9}$$

where $U_j^{\mathrm{T}}U_j = D - \rho_j W$, $j = 1,\dots,p$. We denote the distribution in (11.9) by $MCAR(\rho_1,\dots,\rho_p, \Lambda)$. Note that the off-diagonal block matrices (the U_i's) in the precision matrix in (11.9) are completely determined by the diagonal blocks. Thus, the spatial precision matrices for each disease induce the cross-covariance structure in (11.9).

Kim et al. [2001] proposed a multivariate CAR model in the bivariate ($p = 2$) case, which they dub the "twofold conditionally autoregressive" model, and which we notate as $2fCAR(\rho_0, \rho_1, \rho_2, \rho_3, \tau_1, \tau_2)$. They specify the moments of the full conditional distributions as

$$\mathrm{E}(\phi_{ik} \mid \phi_{il},\phi_{jk},\phi_{jl}) = \frac{1}{2m_i+1}\left(\rho_k\sum_{j\sim i}\phi_{jk} + \rho_3\sqrt{\frac{\tau_l}{\tau_k}}\sum_{j\sim i}\phi_{jl} + \rho_0\sqrt{\frac{\tau_l}{\tau_k}}\phi_{il}\right)$$

and

$$\mathrm{Var}(\phi_{ik} \mid \phi_{il},\phi_{jk},\phi_{jl}) = \frac{\tau_k^{-1}}{2m_i+1} , \quad i,j = 1,\dots n, \ \ l,k = 1,2, \ \ l \neq k,$$

where $j \sim i$ again means that region j is a neighbor of region i. Adding the Gaussian MRF structure, they derive the joint distribution arising from these full conditional distributions:

$$\begin{pmatrix}\phi_1\\\phi_2\end{pmatrix} \sim N\left(\begin{pmatrix}\mathbf{0}\\\mathbf{0}\end{pmatrix}, \begin{pmatrix}(2D+I-\rho_1 W)\tau_1 & -(\rho_0 I+\rho_3 W)\sqrt{\tau_1\tau_2} \\ -(\rho_0 I+\rho_3 W)\sqrt{\tau_1\tau_2} & (2D+I-\rho_2 W)\tau_2\end{pmatrix}^{-1}\right) , \tag{11.10}$$

where again $\boldsymbol{\phi}_1' = (\phi_{11},\dots,\phi_{n1})$, $\boldsymbol{\phi}_2' = (\phi_{12},\dots,\phi_{n2})$, $D = Diag(m_i)$, and W is the adjacency matrix. This model has the same number of parameters in the covariance structure (six) as the general formulation (11.7) in the bivariate case, so they are related to each other. In (11.10), ρ_1 and ρ_2 are the smoothing parameters, while ρ_0 and ρ_3 are the bridging parameters associating ϕ_{i1} with ϕ_{i2} and ϕ_{j2}, $j \neq i$, respectively. Unfortunately, this MCAR model is only designed for the bivariate case ($p = 2$) and seems difficult to generalize to higher dimensions. Also, under this approach it is hard to find conditions that guarantee a positive definite covariance matrix in (11.10). The conditions $|\rho_l| < 1$, $l = 0,1,2,3$ given by Kim et al. [2001] are sufficient but not necessary, and may be overly restrictive for some data sets since they restrict the correlation of ϕ_{i1} with ϕ_{i2} and ϕ_{j2}, $j \neq i$. Finally, this generalization comes at a significant price in terms of computing, since it requires many matrix multiplications, determinant evaluations, and inverses at each MCMC iteration, so can be very time-consuming even when working on a relatively small spatial domain.

11.2   Modeling with a proper, non-separable MCAR distribution

The MCAR specifications of the previous section are employed in models for spatial random effects arising in a hierarchical model. For instance, suppose we have a linear model with continuous data $\mathbf{Y}_{ik}, i = 1, \ldots, n, k = 1, \ldots, m_i$, where $\mathbf{Y}_{ik}$ is a $p \times 1$ vector denoting the kth response at the ith areal unit. The mean of the $\mathbf{Y}_{ik}$ is $\boldsymbol{\mu}_{ik}$ where $\mu_{ikj} = (\mathbf{X}_{ik})_j \boldsymbol{\beta}^{(j)} + \phi_{ij}, j = 1, \ldots, p$. Here $\mathbf{X}_{ik}$ is a $p \times s$ matrix with covariates associated with $\mathbf{Y}_{ik}$ having jth row $(\mathbf{X}_{ik})_j$, $\boldsymbol{\beta}^{(j)}$ is an $s \times 1$ coefficient vector associated with the jth component of the $\mathbf{Y}_{ik}$'s, and ϕ_{ij} is the jth component of the $p \times 1$ vector $\boldsymbol{\phi}_i$. Given $\left\{ \boldsymbol{\beta}^{(j)} \right\}, \{ \boldsymbol{\phi}_i \}$ and V, the $\mathbf{Y}_{ik}$ are conditionally independent $N(\boldsymbol{\mu}_{ik}, V)$ variables. Adding a prior for $\left\{ \boldsymbol{\beta}^{(j)} \right\}$ and V and one of the MCAR models from Subsection 11.1 for the $\boldsymbol{\phi}_i$ completes the second stage of the specification. Finally, a hyperprior on the MCAR parameters completes the model.

Alternatively, we might change the first stage to a multinomial. Here k disappears and $\mathbf{Y}_i$ is assumed to follow a multinomial distribution with sample size n_i and with $(p+1) \times 1$ probability vector $\boldsymbol{\pi}_i$. Working on the logit scale, using cell $p + 1$ as the baseline, we could set $\log \left(\frac{\pi_{ij}}{\pi_{i,p+1}} \right) = \mathbf{X}_i^{\mathrm{T}} \boldsymbol{\beta}^{(j)} + \phi_{ij}, j = 1, \ldots, p$, with $\mathbf{X}$'s, $\boldsymbol{\beta}$'s and $\boldsymbol{\phi}$'s interpreted as in the previous paragraph. Many other multivariate first stages could also be used, such as other multivariate exponential family models.

Regardless, model-fitting is most easily implemented using a Gibbs sampler with Metropolis updates where needed. The full conditionals for the $\boldsymbol{\beta}$'s will typically be normal (under a normal first-stage model) or else require Metropolis, slice, or adaptive rejection sampling (Gilks and Wild, 1992). For the $MCAR(1, \Sigma)$ and $MCAR(\rho, \Sigma)$ models, the full conditionals for the $\boldsymbol{\phi}_i$'s will be likelihood-adjusted versions of the conditional distributions that define the MCAR and are updated as a block. For the $MCAR(\boldsymbol{\rho}, \Sigma)$ model, we can work with either the $\boldsymbol{\phi}$ or the $\boldsymbol{\psi}$ parametrization. With a non-Gaussian first stage, it will be awkward to pull the transformed effects out of the likelihood in order to do the updating. However, with a Gaussian first stage, it may well be more efficient to work on the transformed scale. Under the Gaussian first stage, the full conditional for V will be seen to follow an inverse Wishart, as will Σ. The ρ's do not follow standard distributions; in fact, discretization expedites computation, avoiding Metropolis steps.

We have chosen an illustrative prior for ρ in the ensuing example following three criteria. First, we insist that $\rho < 1$ to ensure propriety but allow $\rho = 0.99$. Second, we do not allow $\rho < 0$ since this would violate the similarity of spatial neighbors that we seek. Third, since even moderate spatial dependence requires values of ρ near 1 (recall the discussion in Subsection 4.3.1) we place prior mass that favors the upper range of ρ. In particular, we put equal mass on the following 31 values: $0, 0.05, 0.1, \ldots, 0.8, 0.82, 0.84, \ldots, 0.90, 0.91, 0.92, \ldots, 0.99$.

Finally, model choice arises here only in selecting among MCAR specifications. That is, we do not alter the mean vector in these investigations; our interest here lies solely in comparing the spatial explanations. For illustration, multivariate versions of the Gelfand and Ghosh [1998] criterion (5.67) for multivariate Gaussian data are employed.

Example 11.1 (*Analysis of the child growth data*). Child growth is usually monitored using anthropometric indicators such as height adjusted for age (HAZ), weight adjusted for height (WHZ), and weight adjusted for age (WAZ). Independent analysis of each of these indicators is normally carried out to identify factors influencing growth that may range from genetic and environmental factors (e.g., altitude, seasonality) to differences in nutrition and social deprivation. Substantial variation in growth is common within as well as between populations. Recently, geographical variation in child growth has been thoroughly investigated in the country of Papua New Guinea in Mueller et al. [2001]. Independent spatial analyses for each of the anthropometric growth indicators identified complex geographical patterns of child growth finding areas where children are taller but skinnier than average, others where they are heavier but shorter, and areas where they are both short and light. These geographical patterns could be linked to differences in diet and subsistence agriculture, leading to the analysis presented here; see Gelfand and Vounatsou [2003] for further discussion.

The data for our illustration comes from the 1982–1983 Papua New Guinea National Nutrition Survey (NNS) [Heywood et al., 1988]. The survey includes anthropometric measures (age, height, weight) of approximately 28,000 children under 5 years of age, as well as dietary, socioeconomic, and demographic data about those children and their families. Dietary data include the type of food that respondents had eaten the previous day. Subsequently, the data were coded to 14 important staples and sources of protein. Each child was assigned to a village and each village was assigned to one of 4566 environmental zones (resource mapping units, or RMUs) into which Papua New Guinea has been divided for agriculture planning purposes. A detailed description of the data is given in Mueller et al. [2001].

The nutritional scores, height adjusted for age (HAZ), and weight adjusted for age (WAZ) that describe the nutritional status of a child were obtained using the method of Cole and Green [1992], which yields age-adjusted standard normal deviate Z-scores. The data set was collected at 537 RMUs. To overcome sparseness and to facilitate computation, we collapsed to 250 spatial units. In the absence of digitized boundaries, Delaunay tessellations were used to create the neighboring structure in the spatial units.

Because of the complex, multidimensional nature of human growth, a bivariate model that considers differences in height and weight jointly might be more appropriate for analyzing child growth data in general and to identify geographical patterns of growth in particular. We propose the use of Bayesian hierarchical spatial models and *multivariate CAR* (MCAR) specifications to analyze the bivariate pairs of indicators, HAZ and WAZ, of child growth. Our modeling reveals bivariate spatial random effects at the RMU level, justifying the MCAR specification.

Recalling the discussion of Subsection 11.1, it may be helpful to provide explicit expressions, with obvious notation, for the modeling and the resulting association structure. We have, for the jth child in the ith RMU,

$$\mathbf{Y}_{ij} = \left(\begin{array}{c} (HAZ)_{ij} \\ (WAZ)_{ij} \end{array} \right) = \mathbf{X}_{ij}^{\mathrm{T}} \left(\begin{array}{c} \boldsymbol{\beta}^{(H)} \\ \boldsymbol{\beta}^{(W)} \end{array} \right) + \left(\begin{array}{c} \phi_i^{(H)} \\ \phi_i^{(W)} \end{array} \right) + \left(\begin{array}{c} \epsilon_{ij}^{(H)} \\ \epsilon_{ij}^{(W)} \end{array} \right) .$$

In this setting, under say the $MCAR(\rho, \Sigma)$ model,

$$\mathrm{Cov}((HAZ)_{ij}, (HAZ)_{i^{\mathrm{T}} j^{\mathrm{T}}} \mid \boldsymbol{\beta}^{(H)}, \boldsymbol{\beta}^{(W)}, \rho, \Sigma, V)$$
$$= \mathrm{Cov}(\phi_i^{(H)}, \phi_{i^{\mathrm{T}}}^{(H)}) + V_{11} I_{i=i^{\mathrm{T}}, j=j^{\mathrm{T}}} ,$$
$$\mathrm{Cov}((WAZ)_{ij}, (WAZ)_{i^{\mathrm{T}} j^{\mathrm{T}}} \mid \boldsymbol{\beta}^{(H)}, \boldsymbol{\beta}^{(W)}, \rho, \Sigma, V)$$
$$= \mathrm{Cov}(\phi_i^{(W)}, \phi_{i^{\mathrm{T}}}^{(W)}) + V_{22} I_{i=i^{\mathrm{T}}, j=j^{\mathrm{T}}} ,$$
$$\text{and} \quad \mathrm{Cov}((HAZ)_{ij}, (WAZ)_{i^{\mathrm{T}} j^{\mathrm{T}}} \mid \boldsymbol{\beta}^{(H)}, \boldsymbol{\beta}^{(W)}, \rho, \Sigma, V)$$
$$= \mathrm{Cov}(\phi_i^{(H)}, \phi_{i^{\mathrm{T}}}^{(W)}) + V_{12} I_{i=i^{\mathrm{T}}, j=j^{\mathrm{T}}} ,$$

where $\mathrm{Cov}(\phi_i^{(H)}, \phi_{i^{\mathrm{T}}}^{(H)}) = (D_W - \rho W)_{ii^{\mathrm{T}}} \Sigma_{11}$, $\mathrm{Cov}(\phi_i^{(W)}, \phi_{i^{\mathrm{T}}}^{(W)}) = (D_W - \rho W)_{ii^{\mathrm{T}}} \Sigma_{22}$, and $\mathrm{Cov}(\phi_i^{(H)}, \phi_{i^{\mathrm{T}}}^{(W)}) = (D_W - \rho W)_{ii^{\mathrm{T}}} \Sigma_{12}$. The interpretation of the components of Σ and V (particularly Σ_{12} and V_{12}) is now clarified.

We adopted noninformative uniform prior specifications on $\boldsymbol{\beta}^{(H)}$ and $\boldsymbol{\beta}^{(W)}$. For Σ and V we use inverse Wishart priors, i.e., $\Sigma^{-1} \sim W(\Omega_1, c_1)$, $V^{-1} \sim W(\Omega_2, c_2)$ where Ω_1, Ω_2 are $p \times p$ matrices and c_1, c_2 are shape parameters. Since we have no prior knowledge regarding the nature or extent of dependence, we choose Ω_1 and Ω_2 diagonal; the data will inform about the dependence *a posteriori*. Since the $\mathbf{Y}_{ij}$'s are centered and scaled on each dimension, setting $\Omega_1 = \Omega_2 = I$ seems appropriate. Finally, we set $c_1 = c_2 = 4$ to provide low precision for these priors. For ρ_1 and ρ_2 we adopted the prior discussed in the previous

Model	G	P	D_∞
$MCAR(1,\Sigma)$	34300.69	33013.10	67313.79
$MCAR(\rho,\Sigma)$	34251.25	33202.86	67454.11
$MCAR(\boldsymbol{\rho},\Sigma)$	34014.46	33271.97	67286.43

Table 11.1 *Model comparison for child growth data.*

Covariate	Height (HAZ)			Weight (WAZ)		
	2.5%	50%	97.5%	2.5%	50%	97.5%
Global mean	−0.35	−0.16	−0.01	−0.48	−0.25	−0.15
Coconut	0.13	0.20	0.29	0.04	0.14	0.24
Sago	−0.16	−0.07	−0.00	−0.07	0.03	0.12
Sweet potato	−0.11	−0.03	0.05	−0.08	0.01	0.12
Taro	−0.09	0.01	0.10	−0.19	−0.09	0.00
Yams	−0.16	−0.04	0.07	−0.19	−0.05	0.08
Rice	0.30	0.40	0.51	0.26	0.38	0.49
Tinned fish	0.00	0.12	0.24	0.04	0.17	0.29
Fresh fish	0.13	0.23	0.32	0.08	0.18	0.28
Vegetables	−0.08	0.08	0.25	0.02	0.19	0.35
V_{11}, V_{22}	0.85	0.87	0.88	0.85	0.87	0.88
V_{12}	0.60	0.61	0.63			
Σ_{11}, Σ_{22}	0.30	0.37	0.47	0.30	0.39	0.52
Σ_{12}	0.19	0.25	0.35			
ρ_1, ρ_2	0.95	0.97	0.97	0.10	0.80	0.97

Table 11.2 *Posterior summaries of the dietary covariate coefficients, covariance components, and autoregression parameters for the child growth data using the most complex MCAR model.*

section. Simulation from the full conditional distributions of the $\boldsymbol{\beta}$'s and the $\boldsymbol{\psi}_i, i = 1, \ldots, n$ is straightforward as they are standard normal distributions. Similarly, the full conditionals for V^{-1} and Σ^{-1} are Wishart distributions. We implemented the Gibbs sampler with 10 parallel chains.

Table 11.1 offers a comparison of three MCAR models using (5.67), the Gelfand and Ghosh [1998] criterion. The most complex model is preferred, offering sufficient improvement in goodness of fit to offset the increased complexity penalty. Summaries of the posterior quantities under this model are shown in Table 11.2. These were obtained from a posterior sample of size 1,000, obtained after running a 10-chain Gibbs sampler for 30,000 iterations with a burn-in of 5,000 iterations and a thinning interval of 30 iterations. Among the dietary factors, high consumption of sago and taro is correlated with lighter and shorter children, while high consumption of rice, fresh fish, and coconut is associated with both heavier and taller children. Children from villages with high consumption of vegetables or tinned fish are heavier.

The posterior for the correlation associated with Σ, $\Sigma_{12}/\sqrt{\Sigma_{11}\Sigma_{22}}$, has mean of 0.67 with 95% credible interval $(0.57, 0.75)$, while the posterior for the correlation associated with V, $V_{12}/\sqrt{V_{11}V_{22}}$, has mean of 0.71 with 95% credible interval $(0.70, 0.72)$. In addition, ρ_1 and ρ_2 differ. ∎

11.3 Conditionally specified Generalized MCAR (GMCAR) distributions

Jin et al. [2005] expand upon this idea by building the joint distribution for a multivariate Markov random field (MRF) through specifications of simpler conditional and marginal models. The approach can be regarded as the analogue of the conditioning approach of Royle and Berliner [1999], see Section 10.5.5, for areal models.

This approach is best elucidated in the bivariate setting. We now assume the joint distribution of ϕ_1 and ϕ_2 is

$$\begin{pmatrix} \phi_1 \\ \phi_2 \end{pmatrix} \sim N \left(\begin{pmatrix} \mathbf{0} \\ \mathbf{0} \end{pmatrix}, \begin{pmatrix} \Sigma_{11} & \Sigma_{12} \\ \Sigma_{12}^{\mathrm{T}} & \Sigma_{22} \end{pmatrix} \right) ,$$

where the Σ_{kl}, $k, l = 1, 2$ are $n \times n$ covariance matrices. From standard multivariate normal theory, we have $\mathrm{E}(\phi_1 \mid \phi_2) = \Sigma_{12}\Sigma_{22}^{-1}\phi_2$ and $\mathrm{Var}(\phi_1 \mid \phi_2) = \Sigma_{11 \cdot 2} = \Sigma_{11} - \Sigma_{12}\Sigma_{22}^{-1}\Sigma_{12}^{\mathrm{T}}$. Now writing $A = \Sigma_{12}\Sigma_{22}^{-1}$, we can rewrite the joint distribution of ϕ_1 and ϕ_2 as

$$\begin{pmatrix} \phi_1 \\ \phi_2 \end{pmatrix} \sim N \left(\begin{pmatrix} \mathbf{0} \\ \mathbf{0} \end{pmatrix}, \begin{pmatrix} \Sigma_{11 \cdot 2} + A\Sigma_{22}A^{\mathrm{T}} & A\Sigma_{22} \\ (A\Sigma_{22})^{\mathrm{T}} & \Sigma_{22} \end{pmatrix} \right) . \tag{11.11}$$

The conditions that ensure the propriety of (11.11) are that Σ_{22} and $\Sigma_{11 \cdot 2}$ (Schur's complement) are positive definite [see, e.g., Harville, 2000, Banerjee and Roy, 2014]. Since $\phi_1 \mid \phi_2 \sim N(A\phi_2, \Sigma_{11 \cdot 2})$ and $\phi_2 \sim N(0, \Sigma_{22})$, we can construct $p(\phi) = p(\phi_1 \mid \phi_2)p(\phi_2)$ where $\phi^{\mathrm{T}} = (\phi_1^{\mathrm{T}}, \phi_2^{\mathrm{T}})$. For the joint distribution of ϕ, then, we need to specify the matrices $\Sigma_{11 \cdot 2}$, Σ_{22}, and A.

Jin et al. [2005] propose specifying the conditional distribution for $\phi_1 \mid \phi_2$ as $\phi_1 \mid \phi_2 \sim N\left(A\phi_2, [(D - \rho_1 W)\tau_1]^{-1}\right)$, and the marginal distribution of ϕ_2 as $\phi_2 \sim N\left(\mathbf{0}, [(D - \rho_2 W)\tau_2]^{-1}\right)$, where ρ_1 and ρ_2 are the smoothing parameters associated with the conditional distribution of $\phi_1 \mid \phi_2$ and the marginal distribution of ϕ_2 respectively, and τ_1 and τ_2 scale the precision of $\phi_1 \mid \phi_2$ and ϕ_2, respectively. The induced joint distribution will always be proper as long as these two CAR distributions are valid, so the positive definiteness of the covariance matrix in (11.11) is easily verified. If $D = Diag(m_i)$ and W be the adjacency matrix, then the positive definiteness conditions require only that $|\rho_1| < 1$ and $|\rho_2| < 1$. Further restricting these parameters between 0 and 1 $0 < \rho_1 < 1$ avoid negative spatial autocorrelation.

Regarding the A matrix, since $\mathrm{E}(\phi_1 \mid \phi_2) = A\phi_2$, we assume its elements are of the form

$$a_{ij} = \begin{cases} \eta_0 & \text{if } j = i \\ \eta_1 & \text{if } j \in N_i \text{ (i.e., if region } j \text{ is a neighbor of region } i) \\ 0 & \text{otherwise} \end{cases} .$$

Thus $A = \eta_0 I + \eta_1 W$ and $\mathrm{E}(\phi_1 \mid \phi_2) = (\eta_0 I + \eta_1 W)\phi_2$. Here η_0 and η_1 are the bridging parameters associating ϕ_{i1} with ϕ_{i2} and $\phi_{j2}, j \neq i$. One could easily augment A with another bridging parameter η_2 associated with the *second-order* neighbors (neighbors of neighbors) in each region, but we do not pursue this generalization here.) Under these assumptions, the covariance matrix in the joint distribution (11.11) can be written as $\Sigma = \begin{pmatrix} \Sigma_{11} & \Sigma_{12} \\ \Sigma_{12}^{\mathrm{T}} & \Sigma_{22} \end{pmatrix}$, where

$$\begin{aligned} \Sigma_{11} &= [\tau_1(D - \rho_1 W)]^{-1} + (\eta_0 I + \eta_1 W)[\tau_2(D - \rho_2 W)]^{-1}(\eta_0 I + \eta_1 W) \\ \Sigma_{12} &= (\eta_0 I + \eta_1 W)[\tau_2(D - \rho_2 W)]^{-1} \\ \Sigma_{22} &= [\tau_2(D - \rho_2 W)]^{-1} . \end{aligned} \tag{11.12}$$

Jin et al. [2005] denote this new model by $GMCAR(\rho_1, \rho_2, \eta_1, \eta_2, \tau_1, \tau_2)$. This bivariate GMCAR model has the same number of parameters as the twofold CAR model in (11.10) and has one more parameter than the $MCAR(\rho_1, \rho_2, \Lambda)$ model in (11.8).

Setting $\rho_1 = \rho_2 = \rho$ and $\eta_1 = 0$ in (11.12), and using a standard result from matrix theory [see, e.g., Harville, 2000, Banerjee and Roy, 2014], produces the separable precision matrix $\Sigma^{-1} = \Lambda \otimes (D - \rho W)$, where $\tau_1 = \Lambda_{11}$, $\tau_2 = \Lambda_{22} - \frac{\Lambda_{12}^2}{\Lambda_{11}}$, and $\eta_0 = -\frac{\Lambda_{12}}{\Lambda_{11}}$. Further assuming $\rho = 1$ produces an improper MIAR (Multivariate Intrinsic Autoregressive) model. If we assume $\rho_1 \neq \rho_2$ and $\eta_0 = \eta_1 = 0$, then we ignore dependence between the multivariate components, and the model turns out to be equivalent fitting two separate univariate CAR models. Finally, if we instead assume $\rho_1 = \rho_2 = 0$, $\eta_0 \neq 0$, and $\eta_1 = 0$, the model becomes an i.i.d. bivariate normal model.

The MCAR model in (11.6) has $\mathrm{E}(\phi_1 \mid \phi_2) = -\frac{\Lambda_{12}}{\Lambda_{11}}\phi_2$, which reveals that the conditional mean is merely a scale multiple of ϕ_2. Since $\mathrm{Var}(\phi_1 \mid \phi_2) = [\Lambda_{11}(D - \rho_1 W)]^{-1}$, which is free of ϕ_2, the distribution of the random variable at a particular site in one field is independent of neighbor variables in another field *given* the value of the related variable at the same area. The extended MCAR model (11.8) has

$$\mathrm{E}(\phi_1 \mid \phi_2) = -\frac{\Lambda_{12}}{\Lambda_{11}}(D - \rho_1 W)^{-\frac{1}{2}}(D - \rho_2 W)^{\frac{1}{2}}\phi_2$$

and $\mathrm{Var}(\phi_1 \mid \phi_2)$ identical to that of model (11.6). Therefore, the distribution of the random variable at a particular site in one field is no longer conditionally independent of neighboring variables in another field. However, this dependence is determined implicitly by ρ_1 and ρ_2 and is difficult to interpret.

By contrast, the GMCAR model has $\mathrm{E}(\phi_1 \mid \phi_2) = (\eta_0 I + \eta_1 W)\phi_2$ and $\mathrm{Var}(\phi_1 \mid \phi_2) = [\tau_1(D - \rho_1 W)]^{-1}$. Thus, while the conditional variance remains free of ϕ_2, the GMCAR allows spatial information (via the W matrix) to enter the conditional mean in an intuitive way, with a free parameter (η_1) to model the weights. That is, the GMCAR models the conditional mean of ϕ_1 for a given region as a sensible weighted average of the values of ϕ_2 for that region *and* a neighborhood of that region.

The GMCAR also allows us to incorporate different weighted adjacency matrices in the $MCAR(\rho, \Lambda)$ distribution. Suppose, we wish to extend the precision matrix in model (11.6) to

$$\Sigma^{-1} = \begin{pmatrix} (D_1 - \rho W^{(1)})\Lambda_{11} & (D_3 - \rho W^{(3)})\Lambda_{12} \\ (D_3 - \rho W^{(3)})\Lambda_{12} & (D_2 - \rho W^{(2)})\Lambda_{22} \end{pmatrix} , \tag{11.13}$$

where $D_k = Diag\left(\sum_{j=1}^n W_{1j}^{(k)}, \ldots, \sum_{j=1}^n W_{nj}^{(k)}\right)$ and $W^{(k)}$ is the weighted adjacency matrix with ij-element $W_{ij}^{(k)}$, $k = 1, 2, 3$, and $i, j = 1, \ldots, n$. The conditions for the precision matrix in (11.13) to be positive definite are less obvious. But in our GMCAR case, we obtain

$$\phi_1 \mid \phi_2 \quad \sim \quad N\left((\eta_0 I + \eta_1 W^{(3)})\phi_2, \, [\tau_1(D_1 - \rho_1 W^{(1)})]^{-1}\right) ,$$

$$\text{and} \quad \phi_2 \quad \sim \quad N\left(0, \, [\tau_2(D_2 - \rho_2 W^{(2)})]^{-1}\right) .$$

Sufficient conditions for positive definiteness can be easily seen to be $|\alpha_1| < 1$ and $|\alpha_2| < 1$ using the fact that diagonally dominant matrices are always positive definite.

Since we specify the joint distribution for a multivariate MRF directly through specification of simpler conditional and marginal distributions, an inherent problem with these methods is that their conditional specification imposes a potentially arbitrary order on the variables being modeled, as they lead to different marginal distributions depending upon the conditioning sequence (i.e., whether to model $p(\phi_1 \mid \phi_2)$ and then $p(\phi_2)$, or $p(\phi_2 \mid \phi_1)$ and then $p(\phi_1)$). This problem is somewhat mitigated in certain (e.g., medical and environmental) contexts where a *natural* order is reasonable, but in many disease mapping contexts, this is not the case. Although Jin et al. [2005] suggest using model comparison techniques to decide upon the proper modeling order since all possible permutations of the variables would need to be considered this seems feasible only with relatively few variables. In any case, the principle of choosing among conditioning sequences using model comparison metrics is perhaps not uncontroversial.

We note that, while the previous theory helps to illuminate structure in specifying multivariate CAR models, from a practical point of view, the reader may simply choose to implement an analogue of coregionalization to create a multivariate dependence model for areal data. In particular, suppose we assume a common proximity specification for each component of the random effects vector, $\boldsymbol{\phi}$. Then, we could write $\boldsymbol{\phi} = A\boldsymbol{\psi}$ where $\boldsymbol{\psi}_j$, the jth component of $\boldsymbol{\psi}$, is a univariate intrinsic CAR with precision parameter τ_j^2, and each of the component CAR models is independent. As above, we can take A to be lower triangular, with prior specifications discussed earlier. The resulting multivariate CAR model is, of course, improper which is fine as a prior for random effects. Moreover, it is easy to work with since we will only fit the model in the space of the independent CAR models, as we do with the coregionalization model fitting for point-referenced data. We could make the prior proper by making each of the components of $\boldsymbol{\psi}$ proper CAR's, should we wish. Jin et al. [2007] develop such coregionalized MCAR distributions, which we discuss in greater detail in Section 11.6. Similar modeling, both in the bivariate and trivariate cases, has been presented in Sang and Gelfand [2009] in the context of modeling temperature extremes (see Section 17.1).

11.4   Modeling using the GMCAR distribution

The $GMCAR(\rho_1, \rho_2, \eta_1, \eta_2, \tau_1, \tau_2)$ models are straightforwardly implemented in a Bayesian framework using MCMC methods. Matters are especially simple with Gaussian likelihoods in the first stage. As a specific example, consider the model

$$Y_{ij} \overset{ind}{\sim} N(Z_{ij}, \sigma^2), \ i = 1, \ldots, n, \ j = 1, 2. \tag{11.14}$$

Assume that $Z_{ij} = \beta_j + \phi_{ij}$, where the ϕ_{ij}'s follow the GMCAR distribution in (11.12). This means that the Z_{ij}'s follow the GMCAR distribution in (11.12) but with $\mathrm{E}[Z_{ij}] = \beta_j$, rather than zero. Then, we easily derive conditional distribution for $\mathbf{Z}_1 \mid \mathbf{Z}_2$

$$\mathbf{Z}_1 \mid \mathbf{Z}_2 \sim N\left(\beta_1 \mathbf{1} + (\eta_0 I + \eta_1 W)(\mathbf{Z}_2 - \beta_2 \mathbf{1}), \ [\tau_1(D - \rho_1 W)]^{-1}\right),$$

and the marginal distribution $\mathbf{Z}_2 \sim N\left(\beta_2 \mathbf{1}, \ [\tau_2(D - \rho_2 W)]^{-1}\right)$, where $\mathbf{Z}_1 = (Z_{11}, \ldots, Z_{n1})^{\mathrm{T}}$ and $\mathbf{Z}_2 = (Z_{12}, \ldots, Z_{n2})^{\mathrm{T}}$. Therefore, the joint distribution of $\mathbf{Z}^{\mathrm{T}} = (\mathbf{Z}_1^{\mathrm{T}}, \mathbf{Z}_2^{\mathrm{T}})$ $p(\mathbf{Z} \mid \boldsymbol{\beta}, \boldsymbol{\tau}, \boldsymbol{\alpha}, \boldsymbol{\eta})$ is proportional to

$$\begin{aligned} \tau_1^{\frac{n}{2}} |D - \rho_1 W|^{\frac{1}{2}} &\exp\{-\frac{\tau_1}{2}[\mathbf{Z}_1 - \beta_1 \mathbf{1} - (\eta_0 I + \eta_1 W)(\mathbf{Z}_2 - \beta_2 \mathbf{1})]' \\ &\times (D - \rho_1 W)[\mathbf{Z}_1 - \beta_1 \mathbf{1} - (\eta_0 I + \eta_1 W)(\mathbf{Z}_2 - \beta_2 \mathbf{1})]\} \\ &\times \tau_2^{\frac{n}{2}} |D - \rho_2 W|^{\frac{1}{2}} \exp\left[-\frac{\tau_2}{2}(\mathbf{Z}_2 - \beta_2 \mathbf{1})'(D - \rho_2 W)(\mathbf{Z}_2 - \beta_2 \mathbf{1})\right], \end{aligned} \tag{11.15}$$

where $\boldsymbol{\beta} = (\beta_1, \beta_2)$, $\boldsymbol{\tau} = (\tau_1, \tau_2)$, $\boldsymbol{\eta} = (\eta_0, \eta_1)$, and $\boldsymbol{\alpha} = (\rho_1, \rho_2)$.

The joint posterior distribution $p\left(\boldsymbol{\beta}, \sigma^2, \mathbf{Z}, \boldsymbol{\tau}, \boldsymbol{\alpha}, \boldsymbol{\eta} \mid \mathbf{Y}_1, \mathbf{Y}_2\right)$ is proportional to

$$L\left(\mathbf{Y}_1, \mathbf{Y}_2 \mid \mathbf{Z}, \sigma^2\right) p\left(\mathbf{Z} \mid \boldsymbol{\beta}, \boldsymbol{\tau}, \boldsymbol{\alpha}, \boldsymbol{\eta}\right) p(\boldsymbol{\beta}) p(\boldsymbol{\tau}) p(\boldsymbol{\alpha}) p(\boldsymbol{\eta}) p(\sigma^2), \tag{11.16}$$

where $\mathbf{Y}_1 = (Y_{11}, \ldots, Y_{n1})^{\mathrm{T}}$, $\mathbf{Y}_2 = (Y_{12}, \ldots, Y_{n2})^{\mathrm{T}}$, $L\left(\mathbf{Y}_1, \mathbf{Y}_2 \mid \mathbf{Z}, \sigma^2\right)$ is the likelihood

$$\sigma^{-2n} \exp\left\{-\frac{1}{2\sigma^2}[(\mathbf{Y}_1 - \mathbf{Z}_1)'(\mathbf{Y}_1 - \mathbf{Z}_1) + (\mathbf{Y}_2 - \mathbf{Z}_2)'(\mathbf{Y}_2 - \mathbf{Z}_2)]\right\},$$

$p(\mathbf{Z} \mid \boldsymbol{\beta}, \boldsymbol{\tau}, \boldsymbol{\alpha}, \boldsymbol{\eta})$ is given by (11.15), and the remaining terms in (11.16) are the prior distributions on $(\boldsymbol{\beta}, \boldsymbol{\tau}, \boldsymbol{\alpha}, \boldsymbol{\eta}, \sigma^2)$.

For the remaining terms, flat priors are chosen for β_1 and β_2, while σ^2 is assigned a vague inverse gamma prior, i.e. an $IG(1, 0.1)$ where we parametrize the $IG(a, b)$ so that $E(\sigma^2) = b/(a-1)$. Next, τ_1 and τ_2 are assigned vague gamma priors, specifically a $G(1, 0.1)$, which has mean of 10 and variance of 100. Finally, ρ_1, ρ_2 are given $Unif(0, 1)$ priors while η_0 and η_1 are given $N(0, \sigma_1^2)$ and $N(0, \sigma_2^2)$ priors, respectively. Jin et al. [2005] present several simulation studies using these specifications and report robust inference for varying hyperparameter values.

The Gibbs sampler is natural for updating the parameters in this setting because it can take advantage of the conditional specification of the GMCAR model. Each of the full conditional distributions required by the Gibbs sampler must be proportional to (11.16). Furthermore, no matrix inversion is required and only calculations on rather special (e.g. diagonal) n-dimensional matrices are required, regardless of the dimension p ($p = 2$ in our case). To calculate the determinant in (11.15), we have the fact that

$$
\begin{aligned}
|D - \rho_k W| &= |D^{\frac{1}{2}}(1 - \rho_k D^{-\frac{1}{2}} W D^{-\frac{1}{2}}) D^{\frac{1}{2}}| = |D| \prod_{i=1}^{n}(1 - \rho_k \lambda_i) \\
&\propto \prod_{i=1}^{n}(1 - \rho_k \lambda_i), \ k = 1, 2,
\end{aligned}
$$

where $\lambda_i, i = 1, \ldots n$ are the eigenvalues of the matrix $D^{-\frac{1}{2}} W D^{-\frac{1}{2}}$. The λ_i may be calculated prior to any MCMC iteration. Hence posterior computation for the GMCAR model is simpler and faster than that for existing MCAR models, especially for large areal data sets.

All of the parameters in (11.16) except $\boldsymbol{\eta}$ and $\boldsymbol{\alpha}$ have closed-form full conditionals, and so may be directly updated. For these two remaining parameters, Metropolis-Hastings steps with bivariate Gaussian proposals are convenient (though for $\boldsymbol{\alpha}$, a preliminary logit transformation, having Jacobian $\prod_{k=1}^{2} \rho_k(1 - \rho_k)$, is required). In practice, the ρ_k must be bounded away from 1 (say, by insisting $0 < \rho_k < 0.999$, $k = 1, 2$) to maintain identifiability and hence computational stability.

11.5   Illustration: Fitting conditional GMCAR to Minnesota cancer data

We now use GMCAR distributions as specifications for second-stage random effects in a hierarchical areal data model with a non-Gaussian first stage. Following Jin et al. [2005], we consider modeling the number of deaths due to cancers of the lung and esophagus in the years from 1991 to 1998 at the county level in Minnesota. We write the model as

$$
Y_{ij} \overset{ind}{\sim} Poisson(E_{ij} e^{Z_{ij}}), \ i = 1, \ldots, 87, \ j = 1, 2, \tag{11.17}
$$

where Y_{ij} is the observed number of deaths due to cancer j in county i, and E_{ij} is the corresponding expected number of deaths (assumed known). To calculate E_{ij}, we account for each county's age distribution by calculating the expected *age-adjusted* number of deaths due to cancer j in county i as

$$
E_{ij} = \sum_{k=1}^{m} \omega_j^k N_i^k, i = 1, \ldots, 87, \ j = 1, 2,
$$

where $\omega_j^k = (\sum_{i=1}^{87} D_{ij}^k)/(\sum_{i=1}^{87} N_i^k)$ is the age-specific death rate due to cancer j for age group k over all Minnesota counties, D_{ij}^k is the number of deaths in age group k of county i due to cancer j, and N_i^k is the total population at risk in county i, age group k.

The county-level maps of the raw standardized mortality ratios (i.e., $\text{SMR}_{ij} = Y_{ij}/E_{ij}$) shown in Figure 11.1 exhibit evidence of correlation both across space and between cancers,

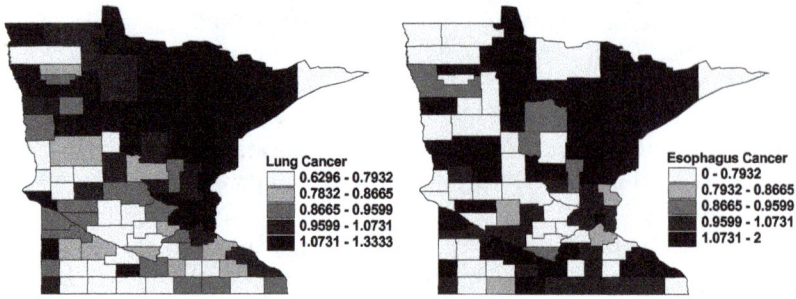

Figure 11.1 *Maps of raw standard mortality ratios (SMR) of lung and esophagus cancer in Minnesota.*

motivating the use of our proposed GMCAR models. Regarding the selection of the proper order in which to model the two cancers, Figure 11.2 gives a helpful data-based exploratory plot. We first obtain crude data-based estimates of the spatial random effects as $\hat{\phi}_{i1} = \log(\text{SMR}_{i1})$ and $\hat{\phi}_{i2} = \log(\text{SMR}_{i2})$. Next, recall the linearity of the conditional GMCAR mean for a given ordering (say, lung given esophagus), i.e.,

$$\text{E}(\phi_1 \mid \phi_2) = A\phi_2 = A(\eta_0, \eta_1)\phi_2 = (\eta_0 I + \eta_1 W)\phi_2 \; .$$

This motivates obtaining least-squares estimates $\hat{\eta}_0$ and $\hat{\eta}_1$ by minimizing $(\hat{\phi}_1 - A(\eta_0, \eta_1)\hat{\phi}_2)'(\hat{\phi}_1 - A(\eta_0, \eta_1)\hat{\phi}_2)$ as a function of η_0 and η_1. Finally, we plot $A(\hat{\eta}_0, \hat{\eta}_1)\hat{\phi}_2$ versus $\hat{\phi}_1$, and investigate how well the linearity assumption is supported by the data. Repeating this entire process for the reverse order (here, esophagus given lung) produces a second plot, which may be compared in quality to the first. In our case, Figure 11.2(a) (lung given esophagus) indicates more support for linearity, both in its appearance and in its higher sample correlation and regression t statistic.

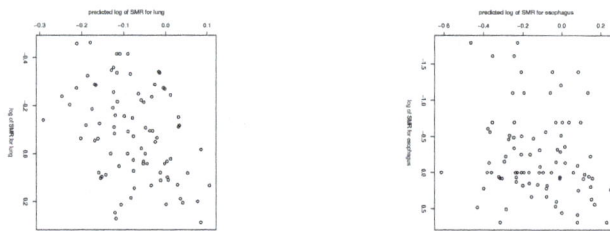

Figure 11.2 *Exploratory plot to help select modeling order: (a) [lung | esophagus], sample correlation 0.394, regression $t = 3.956$; (b) [esophagus | lung], sample correlation 0.193, regression $t = 1.813$.*

Using the likelihood in (11.17), we model the random effects Z_{ij} using the $GMCAR(\rho_1, \rho_2, \eta_1, \eta_2, \tau_1, \tau_2)$ with mean β. In what follows we compare the GMCAR with other existing MCAR models using DIC. In Table 11.3, Models 1–3 are members of our proposed GMCAR class. Specifically, in Model 1, we have the full model with all six parameters, and the conditioning order of the cancers is [lung | esophagus]. Model 2 assumes $\eta_1 = 0$ and uses the same conditioning order as Model 1. In Model 3, we switch the conditioning order to [esophagus | lung] and return to a full model. To compare the GMCAR to existing MCAR models, we take the $MCAR(\rho_1, \rho_2, \Lambda)$ using the Cholesky method for the U_k as Model 4, the same model but using the spectral decomposition for the U_k as Model 5, and the $2fCAR(\rho_0, \rho_1, \rho_2, \rho_3, \tau_1, \tau_2)$ as Model 6. We choose the prior distributions for each parameter as discussed in Section 11.4, and use Metropolis-Hastings and Gibbs sampling to update all parameters. We use 5,000 iterations as the pre-convergence burn-in period, and then a further 20,000 iterations as our production run for posterior summarization.

	Model	$\overline{D}$	p_D	DIC
1	GMCAR (full)	483.4	58.2	541.6
2	GMCAR (reduced; $\eta_1 = 0$)	483.0	63.8	546.8
3	GMCAR (full, reverse order)	480.6	63.3	543.9
4	MCAR (Cholesky decomposition)	483.6	61.3	544.9
5	MCAR (spectral decomposition)	483.8	60.6	544.4
6	2fCAR	482.6	65.1	547.7

Table 11.3 *Model comparison using DIC statistics, Minnesota cancer data analysis.*

Fit measures $\overline{D}$, effective numbers of parameters p_D, and DIC scores for each model are seen in Table 11.3. Model 1 has the smallest p_D and DIC values, so our $GMCAR(\rho_1, \rho_2, \eta_0, \eta_1, \tau_1, \tau_2)$ full model with the conditioning order [lung | esophagus] emerges as best for this data set. The reduced GMCAR Model 2 does less well, suggesting the need to account for bivariate spatial structure in these data. The two MCAR methods perform similarly to each other and the reduced GMCAR model, while the 2fCAR model does less well, largely because it does not seem to allow sufficient smoothing of the random effects (larger p_D score). Note that effective degrees of freedom may actually be smaller for apparently more complex models that allow more complicated forms of shrinkage, such as Model 1 in this case. We note that our "focus" parameter is the same for each model (both fixed and random effects are in focus), and the Poisson likelihood is also not changing across models. Also, our priors are all noninformative or quite vague (e.g., uniform priors for all ρ parameters). All of this suggests the DIC comparison in Table 11.3 is fair across models. Moreover, the resulting DIC scores were robust to the moderate changes in the prior distributions.

Regarding estimation of the fixed effects, under Model 1 we obtained point and 95% equal-tail interval estimates of 0.602 and (0.0267, 0.979) for ρ_1, and 0.699 and (0.0802, 0.973) for ρ_2. Recall these are spatial association parameters, but while their values are between 0 and 1 they are not "correlations" in the usual sense; the moderate point estimates and wide confidence intervals suggest a relatively modest degree of spatial association in the random effects. It is also important to remember that in this setup, ρ_2 measures spatial association in the esophagus random effects ϕ_2, while ρ_1 measures spatial association in the lung random effects ϕ_1 *given* the esophagus random effects ϕ_2. Thus the interpretation of the ρ_k would be different for Model 3 (due to the different conditioning order), and much different for Models 4 or 5. Note that for the MCAR model, $E(\phi_1 \mid \phi_2)$ and $E(\phi_2 \mid \phi_1)$ both depend on both ρ_1 and ρ_2. But for the GMCAR, $E(\phi_1 \mid \phi_2)$ is free of both ρ_1 and ρ_2, while of course $E(\phi_2) = 0$. Thus for this model, ρ_1 and ρ_2 unambiguously control only their corresponding variance matrices, and can be set without altering the mean structure.

Turning to τ_1 and τ_2, under Model 1 we obtained 32.65, (16.98, 66.71) and 13.73, (4.73, 38.05) as our point and interval estimates, respectively. Since these parameters measure spatial precision for each disease, they suggest slightly more variability in the esophagus random effects, although again comparison is difficult here since τ_2 is a *marginal* precision for ϕ_2 while τ_1 is a *conditional* precision for ϕ_1 given ϕ_2. Along these lines, Figure 11.3 shows estimated posteriors of the conditional variances $\sigma_1^2 = 1/\tau_1$ for several candidate multivariate spatial models. Panel (a) shows the situation for two separate CAR models, a model that ignores any possibility of a connection between the cancers. The remaining panels consider the $MCAR(\rho_1, \rho_2, \Lambda)$ model, the reduced $GMCAR(\rho_1, \rho_2, \eta_0, \tau_1, \tau_2)$ model, and the full $GMCAR(\rho_1, \rho_2, \eta_0, \eta_1, \tau_1, \tau_2)$ model. The reduction of uncertainty in ϕ_1 given

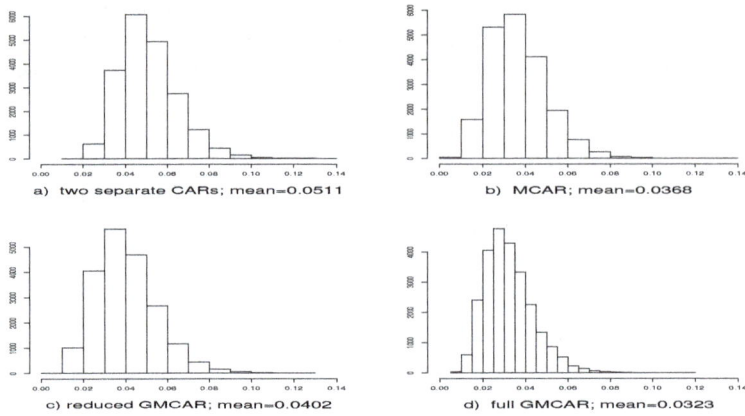

Figure 11.3 *Posterior samples of conditional variances $\sigma_1^2 = 1/\tau_1$ for various models: (a) two separate CAR models; (b) MCAR model; (c) reduced GMCAR model; (d) full GMCAR model.*

ϕ_2 in these more complex models is a measure of the information content between the cancers and is readily apparent from the histograms and their empirical means.

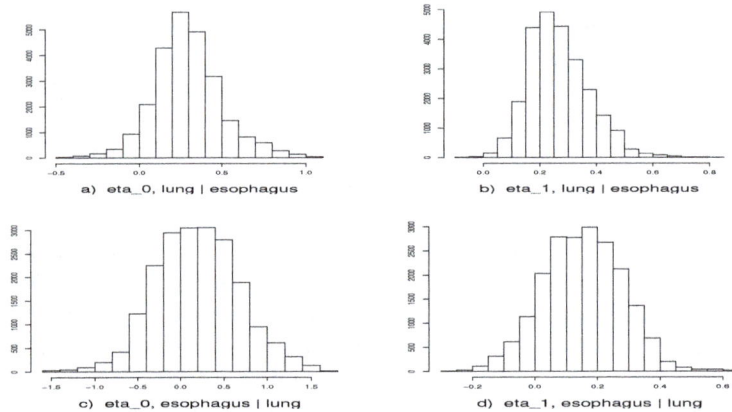

Figure 11.4 *Posterior samples of η_0 and η_1 using the full GMCAR model with two conditioning orders: (a) estimated posterior for η_0, [lung | esophagus]; (b) estimated posterior for η_1, [lung | esophagus]; (c) estimated posterior for η_0, [esophagus | lung]; (d) estimated posterior for η_1, [esophagus | lung].*

The slight preference of DIC for Model 1 is consistent with the estimated posteriors of the linking parameters η_0 and η_1 shown in Figure 11.4. The inclusion of 0 within the 95% credible interval for η_1 under the reverse ordering, but not under the natural ordering, is yet further evidence against the former. Note also that the linking parameters η_0 and η_1 have mostly positive support, meaning that the two cancers have positive spatial correlation. This is also evident from the maps of the posterior means of the SMRs for the two cancers under the full model shown in Figure 11.5. Clearly incidence of the two cancers is strongly correlated, with higher fitted ratios extending from the Twin Cities metro area (eastern side, about one-third of the way up) to the mining- and tourism-oriented north and northeast, regions where conventional wisdom suggests cigarette smoking may be more common.

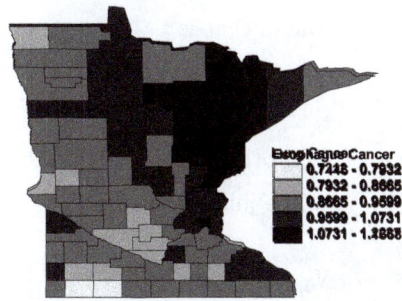

Figure 11.5 *Maps of posterior means of standardized mortality ratios (SMR) of lung and esophagus cancer in Minnesota from the full GMCAR model with conditioning order [lung | esophagus].*

11.6   Coregionalized MCAR distributions

As mentioned in Section 11.3, generalizations of GMCAR to settings with a large number of diseases are encumbered by the dependence of the joint distribution on the sequence of ordering in the hierarchy. To obviate this issue, Jin et al. [2007] develop an order-free framework for multivariate areal modeling that allows versatile spatial structures, yet is computationally feasible for many outcomes. This approach is based upon an adaptation of the *linear model of coregionalization* (LMC) to areal data.

The essential idea is to develop richer spatial association models using linear transformations of much simpler spatial distributions. The objective is to allow explicit smoothing of cross-covariances while at the same time not being hampered by conditional ordering. The most natural model here would parametrize the cross-covariances themselves as $D - \gamma_{ij}W$, instead of using the U_k's as in (11.6). Unfortunately, except in the separable model with only one smoothing parameter ρ, constructing such dispersion structures is not trivial and leads to identifiability issues on the γ's [see, e.g., Gelfand and Vounatsou, 2003]. Kim et al. [2001] resolve these identifiability issues in the bivariate setting using diagonal dominance but recognize the difficulty in extending this to the multivariate setting. We address this problem using a *linear model of coregionalization* (LMC).

It is worth pointing out that our use of the LMC here is somewhat different from what is usually encountered in geostatistics. In geostatistics we typically transform independent latent effects, which suffices in meeting the primary goal of introducing a different spatial range for each variable. This is akin to introducing different smoothing parameters for each variable and indeed, as we show below in Section 11.6.2, independent latent effects produce the $MCAR(\alpha_1, \ldots, \alpha_p; \Lambda)$ in (11.9).

However, to explicitly smooth the cross-covariances with identifiable parameters, we will relax the independence of latent effects. Still, in our ensuing parametrization, we are able to derive conditions that yield valid joint distributions. To be precise, let $\boldsymbol{\phi} = (\boldsymbol{\phi}_1^{\mathrm{T}}, \ldots, \boldsymbol{\phi}_p^{\mathrm{T}})^{\mathrm{T}}$ be an $np \times 1$ vector, where each $\boldsymbol{\phi}_j = (\phi_{1j}, \ldots, \phi_{nj})^{\mathrm{T}}$ is $n \times 1$ representing the spatial effects corresponding to disease j. We can write $\boldsymbol{\phi} = (A \otimes I_{n \times n})\mathbf{u}$, where $\mathbf{u} = (\mathbf{u}_1^{\mathrm{T}}, \ldots, \mathbf{u}_p^{\mathrm{T}})^{\mathrm{T}}$ is $np \times 1$ with each $\mathbf{u}_j$ being an $n \times 1$ areal process. Indeed, a proper distribution for $\mathbf{u}$ ensures a proper distribution for $\boldsymbol{\phi}$ subject only to the non-singularity of A. The flexibility of this approach is apparent: we obtain different multivariate lattice models with rich spatial covariance structures by making different assumptions about the p spatial processes $\mathbf{u}_j$.

11.6.1   *Case 1: Independent and identical latent CAR variables*

First, we will assume that the random spatial processes $\mathbf{u}_j$, $j = 1, \ldots, p$, are independent and identical. Since each spatial process $\mathbf{u}_j$ is a univariate process over areal units, we might adopt a CAR structure for each of them, that is

$$\mathbf{u}_j \; \sim \; N_n\left(\mathbf{0}, \; (D - \alpha W)^{-1}\right), \; j = 1, \ldots, p. \tag{11.18}$$

Since the $\mathbf{u}_j$'s are independent of each other, the joint distribution of $\mathbf{u} = (\mathbf{u}_1', \ldots, \mathbf{u}_p')'$ is $\mathbf{u} \sim N_{np}\left(\mathbf{0}, \; I_{p \times p} \otimes (D - \alpha W)^{-1}\right)$. The joint distribution of $\boldsymbol{\phi} = (A \otimes I_{n \times n})\mathbf{u}$ is

$$\boldsymbol{\phi} \; \sim \; N_{np}\left(\mathbf{0}, \; \Sigma \otimes (D - \alpha W)^{-1}\right), \tag{11.19}$$

defining $\Sigma = AA^{\mathrm{T}}$. We denote the distribution in (11.19) by $MCAR(\alpha, \Sigma)$. Note that the joint distribution of (11.19) is identifiable up to $\Sigma = AA'$, and is independent of the choice of A. Thus, without loss of generality, we can specify the matrix A as the upper-triangular Cholesky decomposition of Σ.

Since $\boldsymbol{\phi} = (A \otimes I_{n \times n})\mathbf{u}$, a valid joint distribution of $\boldsymbol{\phi}$ requires valid joint distributions of the $\mathbf{u}_j$, which happens if and only if $\frac{1}{\xi_{min}} < \alpha < \frac{1}{\xi_{max}}$, where ξ_{min} and ξ_{max} are the minimum and maximum eigenvalues of $D^{-\frac{1}{2}}WD^{-\frac{1}{2}}$. Note if $\alpha = 1$ in CAR structure (11.18), which is an ICAR, the joint distribution of $\boldsymbol{\phi}$ in (11.19) becomes the multivariate intrinsic CAR (Gelfand and Vounatsou, 2003).

Currently, BUGS offers an implementation of the $MCAR(\alpha = 1, \Sigma)$ distribution (using its mv.car distribution), but not the $MCAR(\alpha, \Sigma)$. However, through the LMC approach, we still can fit the $MCAR(\alpha, \Sigma)$ in BUGS by writing $\boldsymbol{\phi} = (A \otimes I_{n \times n})\mathbf{u}$ and assigning proper CAR priors (via the car.proper distribution) for each $\mathbf{u}_j$, $j = 1, \ldots, p$ with a common smoothing parameter α. Regarding the prior on A, note that since $AA' = \Sigma$ and A is the Cholesky decomposition of Σ, there is a one-to-one relationship between the elements of Σ and A. In Section 11.7, we argue that assigning a prior to Σ is computationally preferable.

11.6.2 Case 2: Independent but not identical latent CAR variables

Here, we continue to assume that the $\mathbf{u}_j$ are independent but relax them being identically distributed. Adopting the CAR structure, we assume

$$\mathbf{u}_j \; \sim \; N_n\left(\mathbf{0}, \; (D - \alpha_j W)^{-1}\right), \; j = 1, \ldots, p, \tag{11.20}$$

where α_j is the smoothing parameter for the jth spatial process. Since the $\mathbf{u}_j$'s are independent of each other and $\boldsymbol{\phi} = (A \otimes I_{n \times n})\mathbf{u}$, the joint distribution of $\boldsymbol{\phi}$ is

$$\boldsymbol{\phi} \; \sim \; N_{np}\left(\mathbf{0}, \; (A \otimes I_{n \times n})\Gamma^{-1}(A \otimes I_{n \times n})^{\mathrm{T}}\right), \tag{11.21}$$

where $\Sigma = AA^{\mathrm{T}}$ and Γ is an $np \times np$ block diagonal matrix with $n \times n$ diagonal entries $\Gamma_j = D - \alpha_j W, j = 1, \ldots, p$. We denote the distribution in (11.21) by $MCAR(\alpha_1, \ldots, \alpha_p, \Sigma)$.

It follows from (11.21) that different joint distributions of $\boldsymbol{\phi}$ having different covariance matrices emerge under different linear transformation matrices A. To ensure A is identifiable, we could again specify it to be the upper-triangular Cholesky decomposition of Σ, although this might not be the best choice computationally. Through the LMC approach in this case, the distribution in (11.21) is similar to the $MCAR(\alpha_1, \ldots, \alpha_p, \Lambda)$ structure (11.9), developed in Carlin and Banerjee (2003) and Gelfand and Vounatsou (2003). All of these have the same number of parameters, and there is no unique joint distribution for $\boldsymbol{\phi}$ with the $MCAR(\alpha_1, \ldots, \alpha_p, \Lambda)$ structure since there is not a unique R_j matrix such that $R_j R_j^{\mathrm{T}} = R_j PP^{\mathrm{T}} R_j^{\mathrm{T}} = D - \alpha_j W$ (P being an arbitrary orthogonal matrix).

Again, a valid joint distribution in (11.21) requires p valid distributions for $\mathbf{u}_j$, i.e. $\frac{1}{\xi_{min}} < \alpha_j < \frac{1}{\xi_{max}}$, $j = 1, \ldots, p$. Through the LMC approach, we can also fit the data with the $MCAR(\alpha_1, \ldots, \alpha_p, \Sigma)$ prior distribution (11.21) on ϕ in WinBUGS as in the previous subsection by writing $\phi = (A \otimes I_{n \times n})\mathbf{u}$ and assigning proper CAR priors (via the car.proper distribution) with a distinct smoothing parameter α_j for each $\mathbf{u}_j$, $j = 1, \ldots, p$. As mentioned in the preceding section, we assign a prior to $AA^{\mathrm{T}} = \Sigma$ (e.g., an inverse Wishart), and determine A from the one-to-one relationship between the elements of Σ and A; Section 11.7 below provides details.

11.6.3   Case 3: Dependent and not identical latent CAR variables

Finally, in this case we will assume that the random spatial processes $\mathbf{u}_j = (u_{1j}, \ldots, u_{nj})^{\mathrm{T}}$, $j = 1, \ldots, p$ are neither independent nor identically distributed. We now assume that u_{ij} and $u_{i, l \neq j}$ are independent given $u_{k \neq i, j}$ and $u_{k \neq i, l \neq j}$, where $l, j = 1, \ldots, p$ and $i, k = 1, \ldots, n$ implying that latent effects for different diseases in the same region are conditionally independent given those for diseases in the neighboring regions. Based upon the Markov property and similar to the conditional distribution in the univariate case, we specify the ij^{th} conditional distribution as Gaussian with mean

$$\mathrm{E}(u_{ij} \mid u_{k \neq i, j}, \, u_{i, l \neq j}, \, u_{k \neq i, l \neq j}) = b_{jj} \left(\sum_{k \sim i} u_{kj}/m_i \right) + \sum_{l \neq j} \left[b_{jl} \left(\sum_{k \sim i} u_{kl}/m_i \right) \right],$$

and conditional variance $\mathrm{Var}(u_{ij} \mid u_{k \neq i, j}, \, u_{i, l \neq j}, \, u_{k \neq i, l \neq j}) \propto 1/m_i$, where b_{jj} denotes the spatial autocorrelation for the random spatial process $\mathbf{u}_j$ while b_{jl} ($l \neq j$, $l, j = 1, \ldots, p$) denotes the cross spatial correlation between the random spatial process $\mathbf{u}_j$ and $\mathbf{u}_l$. Putting these conditional distributions together reveals the joint distribution of $\mathbf{u} = (\mathbf{u}_1^{\mathrm{T}}, \ldots, \mathbf{u}_p^{\mathrm{T}})^{\mathrm{T}}$ to be

$$\mathbf{u} \; \sim \; N_{np} \left(\mathbf{0}, \; (I_{p \times p} \otimes D - B \otimes W)^{-1} \right), \tag{11.22}$$

where I is a $p \times p$ identity matrix and B is a $p \times p$ symmetric matrix with the elements $b_{jl}, j, l = 1, \ldots, p$. As long as the dispersion matrix in (11.22) is positive-definite, which boils down to $(I_{p \times p} \otimes D - B \otimes W)$ being positive definite, (11.22) is itself a valid model. To assess non-singularity, note $I_{p \times p} \otimes D - B \otimes W = (I_{p \times p} \otimes D)^{\frac{1}{2}} \left(I_{pn \times pn} - B \otimes D^{-\frac{1}{2}} W D^{-\frac{1}{2}} \right) (I_{p \times p} \otimes D)^{\frac{1}{2}}$. Denoting the eigenvalues for $D^{-\frac{1}{2}} W D^{-\frac{1}{2}}$ as ξ_i, $i = 1, \ldots, n$, and the eigenvalues for B as ζ_j, $j = 1, \ldots, p$, one finds [see, e.g., Harville, 2000, Theorem 21.11.1] the eigenvalues for $B \otimes (D^{-\frac{1}{2}} W D^{-\frac{1}{2}})$ as $\xi_i \times \zeta_j$, $i = 1, \ldots, n$, $j = 1, \ldots, p$. Hence, the conditions for $I_{p \times p} \otimes D - B \otimes W$ being positive definite become $\xi_i \zeta_j < 1$, i.e., $\frac{1}{\xi_{min}} < \zeta_j < \frac{1}{\xi_{max}}$, $i = 1, \ldots, n$, $j = 1, \ldots, p$, where ξ_{min} and ξ_{max} are the minimum and maximum eigenvalues of $D^{-\frac{1}{2}} W D^{-\frac{1}{2}}$. Thus, $\frac{1}{\xi_{min}} < \zeta_j < 1$, $j = 1, \ldots, p$, ensures the positive definiteness of the matrix $I_{p \times p} \otimes D - B \otimes W$ and, hence, the validity of the distribution of $\mathbf{u}$ given in (11.22). In fact, $\xi_{max} = 1$ and $\xi_{min} < 0$, which makes this formulation easier to work with in practice (e.g. in choosing priors; see Section 11.7) than the alternative parametrization $\frac{1}{\zeta_{min}} < \xi_j < \frac{1}{\zeta_{max}}$.

The model in (11.22) introduces smoothing parameters in the cross-covariance structure through the matrix B, but unlike the $MCAR$ models in Sections 11.6.1 and 11.6.2 does not have the Σ matrix to capture non-spatial variances. To remedy this, we model $\phi = (A \otimes I_{n \times n})\mathbf{u}$ so that the joint distribution for the random effects ϕ is

$$\phi \; \sim \; N_{np} \left(\mathbf{0}, \; (A \otimes I_{n \times n}) (I_{p \times p} \otimes D - B \otimes W)^{-1} (A \otimes I_{n \times n})^{\mathrm{T}} \right). \tag{11.23}$$

Since $\phi = (A \otimes I_{n \times n})\mathbf{u}$, it is immediate that the validity of (11.22) ensures a valid joint distribution for (11.23). We denote distribution (11.23) by $MCAR(B, \Sigma)$, where $\Sigma = AA^{\mathrm{T}}$. Again, A identifies with the upper-triangular Cholesky square-root of Σ. Note that with $\Sigma = I$ we recover (11.22), which we henceforth denote as $MCAR(B, I)$.

To see the generality of (11.23), we find the joint distribution of $\boldsymbol{\phi}$ reduces to the $MCAR(\alpha_1, \ldots, \alpha_p, \Sigma)$ distribution (11.21) if $b_{jl} = 0$ and $b_{jj} = \alpha_j$, or the $MCAR(\alpha, \Sigma)$ distribution (11.19) if $b_{jl} = 0$ and $b_{jj} = \alpha$, in both cases for $j, l = 1, \ldots, p$. Also note that the distribution in (11.23) is invariant to orthogonal transformations (up to a reparametrization of B) in the following sense: let $T = AP$ with P being a $p \times p$ orthogonal matrix such that $TT^\mathrm{T} = APP^\mathrm{T}A^\mathrm{T} = \Sigma$. Then the covariance matrix in (11.23) can be expressed as $(A \otimes I_{n \times n})(I_{p \times p} \otimes D - B \otimes W)^{-1}(A \otimes I_{n \times n})^\mathrm{T} = (T \otimes I_{n \times n})(I_{p \times p} \otimes D - C \otimes W)^{-1}(T \otimes I_{n \times n})^\mathrm{T}$, where $C = P^\mathrm{T}BP$. Without loss of generality, then, we can choose the matrix A as the upper-triangular Cholesky decomposition of Σ.

To understand the features of the $MCAR(B, \Sigma)$ distribution (11.23), we illustrate in the bivariate case ($p = 2$). Define

$$(AA^\mathrm{T})^{-1} = \Sigma^{-1} = \begin{pmatrix} \Lambda_{11} & \Lambda_{12} \\ \Lambda_{12} & \Lambda_{22} \end{pmatrix}$$

and $B = A^\mathrm{T} \begin{pmatrix} \gamma_1\Lambda_{11} & \gamma_{12}\Lambda_{12} \\ \gamma_{12}\Lambda_{12} & \gamma_2\Lambda_{22} \end{pmatrix} A$, where $A = \begin{pmatrix} a_{11} & a_{12} \\ 0 & a_{22} \end{pmatrix}$. For convenience, we will denote the entries of B as b_{ij}. Note that the γ's are not identifiable from the matrix Λ and our reparametrization in terms of B must be used to conduct posterior inference on B and Λ (see Section 11.7), from which the cross-covariances may be recovered. The above expression does allow the $MCAR(B, \Sigma)$ distribution (11.23) to be rewritten as

$$\boldsymbol{\phi} \sim N_{2n}\left(\mathbf{0}, \; \begin{pmatrix} (D - \gamma_1 W)\Lambda_{11} & (D - \gamma_{12}W)\Lambda_{12} \\ (D - \gamma_{12}W)\Lambda_{12} & (D - \gamma_2 W)\Lambda_{22} \end{pmatrix}^{-1}\right), \tag{11.24}$$

which is precisely the general dispersion structure we set out to achieve. Jin et al. [2007] provide explicit expressions for the conditional means and variances, which offer further insight into how the parameters in (11.24) affect smoothing.

11.7   Modeling with coregionalized MCARs

The $MCAR(B, \Sigma)$ model is straightforwardly implemented in a Bayesian framework using MCMC methods. As in Section 11.6.3, we write $\boldsymbol{\phi} = (A \otimes I_{n \times n})\mathbf{u}$, where $\mathbf{u} = (\mathbf{u}_1^\mathrm{T}, \mathbf{u}_2^\mathrm{T})$ and $\mathbf{u}_j = (u_{1j}, \ldots, u_{nj})^\mathrm{T}$. The joint posterior distribution is $p\left(\boldsymbol{\beta}, \sigma^2, \mathbf{u}, A, B \mid \mathbf{Y}_1, \mathbf{Y}_2\right)$, which is proportional to

$$L\left(\mathbf{Y}_1, \mathbf{Y}_2 \mid \mathbf{u}, \sigma^2, A\right) p(\mathbf{u} \mid B) \, p(B)p(\boldsymbol{\beta})p(A)p(\sigma^2) , \tag{11.25}$$

where $\mathbf{Y}_1 = (Y_{11}, \ldots, Y_{n1})^\mathrm{T}$ and $\mathbf{Y}_2 = (Y_{12}, \ldots, Y_{n2})^\mathrm{T}$, $L\left(\mathbf{Y}_1, \mathbf{Y}_2 \mid \mathbf{u}, \sigma^2, A\right)$ is the data likelihood and $p(\mathbf{u} \mid B) = N_{np}\left(\mathbf{0}, \; (I_{p \times p} \otimes D - B \otimes W)^{-1}\right)$. As mentioned in Section 11.6.3, propriety of this distribution requires the eigenvalues ζ_j of B to satisfy $\frac{1}{\xi_{min}} < \zeta_j < 1$ ($j = 1, \ldots, p$). When p is large, it is hard to determine the intervals over the elements of B that result in $\frac{1}{\xi_{min}} < \zeta_j < 1$, and thus designing priors for B that guarantee this condition is awkward. In principle, one might impose the constraint numerically by assigning a flat prior or a normal prior with a large variance for the elements of B, and then simply check whether the eigenvalues of the corresponding B matrix are in that range during a random-walk Metropolis-Hastings (MH) update. If the resulting eigenvalues are out of range, the values are thrown out since they correspond to prior probability 0; otherwise we perform the standard MH comparison step. In our experience, however, this does not work well, especially when p is large.

Instead, here we outline a different strategy to update the matrix B. Our approach is to represent B using the spectral decomposition, which we write as $B = P\Delta P^\mathrm{T}$, where P is the corresponding orthogonal matrix of eigenvectors and Δ is a diagonal matrix of ordered eigenvalues, $\zeta_1, \ldots, \zeta_p$. We parameterize the $p \times p$ orthogonal matrix P in terms of the $p(p-1)/2$ Givens angles θ_{ij} for $i = 1, \ldots, p-1$ and $j = i+1, \ldots, p$ (Daniels and Kass, 1999). The matrix P is written as the product of $p(p-1)/2$ matrices, each one associated with a Givens angle. Specifically, $P = G_{12}G_{13} \ldots G_{1p} \ldots G_{(p-1)p}$ where i and j are distinct and G_{ij} is the $p \times p$ identity matrix with the ith and jth diagonal elements replaced by $\cos(\theta_{ij})$, and the (i,j)-th and (j,i)-th elements replaced by $\pm \sin(\theta_{ij})$, respectively. Since the Givens angles θ_{ij} are unique with a domain $(-\pi/2, \pi/2)$ and the eigenvalues ζ_j of B are in the range $(\frac{1}{\xi_{min}}, 1)$, we then put a Uniform$(-\pi/2, \pi/2)$ prior on the θ_{ij} and a Uniform$(\frac{1}{\xi_{min}}, 1)$ prior on the ζ_j. To update θ_{ij}'s or ζ_j's using random-walk Metropolis-Hastings steps with Gaussian proposals, we need to transform them to have support equal to the whole real line. A straightforward solution here is to use $g(\theta_{ij}) = \log(\frac{\pi/2 + \theta_{ij}}{\pi/2 - \theta_{ij}})$, a transformation having Jacobian $\prod_{i=1}^{p-1} \prod_{j=i+1}^{p} (\pi/2 + \theta_{ij})(\pi/2 - \theta_{ij})$. In practice, the ζ_j must be bounded away from 1 (say, by insisting $\frac{1}{\xi_{min}} < \zeta_j < 0.999$, $j = 1, \ldots, p$) to maintain identifiability and hence computational stability. In fact, with our approach it is also easy to calculate the determinant of the precision matrix, that is, $|I_{p \times p} \otimes D - B \otimes W| \propto \prod_{i=1}^{n} \prod_{j=1}^{p} (1 - \xi_i \zeta_j)$, where ξ_i are the eigenvalues of $D^{-\frac{1}{2}}WD^{-\frac{1}{2}}$, which can be calculated prior to any MCMC iteration. For the special case of the $MCAR(\alpha_1, \ldots, \alpha_p, \Sigma)$ models, one could assign each $\alpha_i \sim U(0,1)$, which would be sufficient to ensure a valid model (e.g. Carlin and Banerjee, 2002). However, we also investigated with more informative priors on the α_i's such as the $Beta(2,18)$ that centers the smoothing parameters closer to 1 and leads to greater smoothing.

With respect to the prior distribution $p(A)$ on the right-hand side of (11.25), we can put independent priors on the individual elements of A, such as inverse gamma for the square of the diagonal elements of A and normal for the off-diagonal elements. In practice, we cannot assign non-informative priors here, since then MCMC convergence is poor. In our experience it is easier to assign a vague (i.e., weakly informative) prior on Σ than to put such priors on the elements of A in terms of letting the data drive the inference and obtaining good convergence. Since Σ is a positive definite covariance matrix, the inverse Wishart prior distribution renders itself as a natural choice, that is, $\Sigma^{-1} \sim Wishart\left(\nu, (\nu R)^{-1}\right)$ (see e.g. Carlin and Louis, 2000, p.328). Hence, we instead place a prior directly on Σ, and then use the one-to-one relationship between the elements of Σ and the Cholesky factor A. Then, the prior distribution $p(A)$ becomes

$$p(A) \propto |AA^{\mathrm{T}}|^{-\frac{\nu+4}{2}} exp\left\{-\frac{1}{2}tr[\nu D(AA^{\mathrm{T}})^{-1}]\right\} \left|\frac{\partial \Sigma}{\partial a_{ij}}\right|,$$

where $\left|\frac{\partial \Sigma}{\partial a_{ij}}\right|$ is the Jacobian $2^p \prod_{i=1}^{p} a_{ii}^{p-i+1}$. For example, when $p = 2$, the Jacobian is $4a_{22}^2 a_{11}$. Rather than updating Σ as a block using a Wishart proposal, updating the elements a_{ij} of A offers better control. These are updated via a random-walk Metropolis, using log-normal proposals for the diagonal elements and normal proposals for the off-diagonal elements. With regard to choosing ν and R in the $Wishart\left(\nu, (\nu R)^{-1}\right)$, since $E(\Sigma^{-1}) = R^{-1}$, if there is no information about the prior mean structure of Σ, a diagonal matrix R can be chosen, with the scale of the diagonal elements being judged using ordinary least squares estimates based on independent models for each response variable. While this leads to a data-dependent prior, typically the Wishart prior lets the data drive the results, leading to robust posterior inference. In this study we adopt $\nu = 2$ (i.e., the smallest value for which this Wishart prior is proper) and $R = Diag(0.1, 0.1)$. Finally, for the remaining terms on the right hand side of (11.25), flat priors are chosen for β_1 and β_2, while σ^2 is assigned a vague inverse gamma prior, i.e. a $IG(1, 0.01)$ parameterized so that $E(\sigma^2) = b/(a-1)$. In this study, $\boldsymbol{\beta}$ and σ^2 have closed-form full conditionals, and so can be directly updated using Gibbs sampling.

11.8   Illustrating coregionalized MCAR models with 3 cancers from Minnesota

Jin et al. [2007] estimate different regionalized MCAR model methods with a data set consisting of the numbers of deaths due to cancers of the lung, larynx, and esophagus in the years from 1990 to 2000 at the county level in Minnesota. The larynx and esophagus are sites of the upper aerodigestive tract, so they are closely related anatomically. Epidemiological evidence shows a strong and consistent relationship between exposure to alcohol and tobacco and the risk of cancer at these two sites [Baron et al., 1993]. Meanwhile, lung cancer is the leading cause of cancer death for both men and women. An estimated 160,440 Americans will die in 2004 from lung cancer, accounting for 28% of all cancer deaths. It has long been established that tobacco, and particularly cigarette smoking, is the major cause of lung cancer. More than 87% of lung cancers are smoking-related (http://www.lungcancer.org).

Following Jin et al. [2007], we estimate the model

$$Y_{ij} \stackrel{ind}{\sim} Po(E_{ij}e^{\mu_{ij}}), \ i = 1, \ldots, n, \ j = 1, 2, 3 \tag{11.26}$$

where Y_{ij} is the observed number of cases of cancer type j (one of three types) in region i, $\log \mu_{ij} = \beta_j + \phi_{ij}$ with the β_j's being cancer-specific intercepts and the ϕ_{ij}'s being spatial random effects that are distributed according to some version of the coregionalized MCAR's we discussed earlier. To calculate the expected counts E_{ij}, we have to take each county's age distribution (over the 18 age groups) into account. To do so, we calculate the expected *age-adjusted* number of deaths due to cancer j in county i as $E_{ij} = \sum_{k=1}^{m} \omega_j^k N_i^k$, $i = 1, \ldots, 87$, $j = 1, 2, 3$, $k = 1, \ldots, 18$, where $\omega_j^k = (\sum_{i=1}^{87} D_{ij}^k)/(\sum_{i=1}^{87} N_i^k)$ is the age-specific death rate due to cancer j for age group k over all Minnesota counties, D_{ij}^k is the number of deaths in age group k of county i due to cancer j, and N_i^k is the total population at risk in county i, age group k, which we assume to be the same for each type of cancer.

The county-level maps of the raw age-adjusted standardized mortality ratios (i.e., $\text{SMR}_{ij} = Y_{ij}/E_{ij}$) shown in Figure 11.6 exhibit evidence of correlation across space and among the cancers, motivating use of our proposed multivariate lattice model. Using the likelihood in (11.26), we model the random effects ϕ_{ij} using our proposed $MCAR(B, \Sigma)$ model (11.23). In what follows we compare it with other MCAR models, including the $MCAR(\alpha, \Sigma)$ and $MCAR(1, \Sigma)$ from Section 11.6.1, a "three separate CARs" model ignoring correlation between cancers, and a trivariate i.i.d. model ignoring correlations of any kind. We also compare one of the $MCAR(\alpha_1, \alpha_2, \alpha_3, \Sigma)$ models given in (11.21) of Section 11.6.2 by choosing the matrix A as the upper-triangular Cholesky decomposition of Σ. Note that we do not consider the order-specific GMCAR model (Section 11.3), since with no natural causal order for these three cancers, it is hard to choose among the six possible conditioning orders.

For priors, we follow the guidelines outlined earlier and use the same specifications as in Jin et al. [2007]. Since $p = 3$ in this example, we choose the inverse Wishart distribution with $\nu = 3$ and $R = Diag(0.1, 0.1, 0.1)$ for Σ. For a model comparison using DIC, we retain the same "focus" parameters and likelihood across the models. We used 20,000 pre-convergence burn-in iterations followed by a further 20,000 production iterations for posterior summary. To see the relative performance of these models, we use DIC. As in the previous section, the deviance is the same for the models we wish to compare since they differ only in their random effect distributions $p(\phi \mid B, \Sigma)$.

In what follows, Models 1–6 are multivariate lattice models with different assumptions about the smoothing parameters. Model 1 is the full model $MCAR(B, \Sigma)$ (with a 3×3 matrix B whose elements are the six smoothing parameters) while Model 2 is the $MCAR(B, I)$ model. Model 2 is the $MCAR(\alpha_1, \alpha_2, \alpha_3; \Sigma)$ model (11.21) with a different smoothing parameter for each cancer. Model 3 assumes a common smoothing parameter α and Model 4

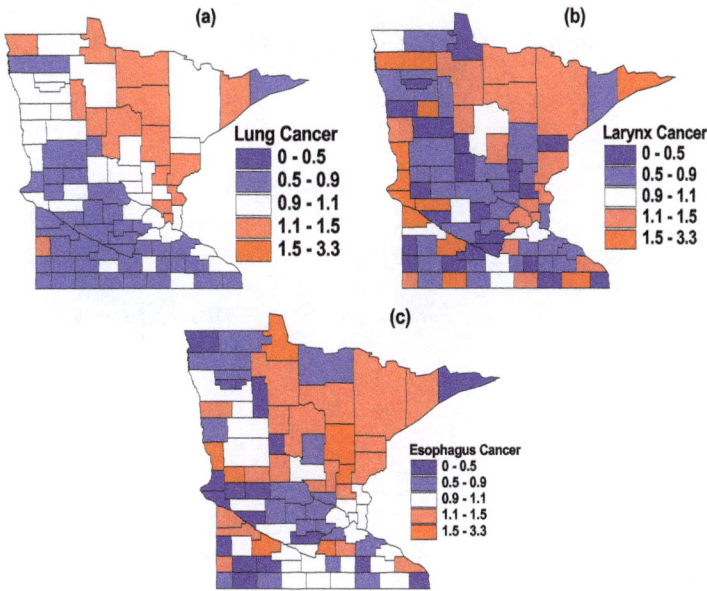

Figure 11.6 *Maps of raw standardized mortality ratios (SMR) of lung, larynx and esophagus cancer in the years from 1990 to 2000 in Minnesota.*

	Model	$\overline{D}$	p_D	DIC
1	$MCAR(B, \Sigma)$	138.8	82.5	221.3
2	$MCAR(B, I)$	147.6	81.4	229.0
3	$MCAR(\alpha_1, \alpha_2, \alpha_3, \Sigma)$	139.6	86.4	226.0
4	$MCAR(\alpha, \Sigma)$	143.4	81.9	225.3
5	separate CAR	147.6	82.8	230.4
6	trivariate i.i.d.	146.8	91.3	238.1
7	$MCAR(B, \Sigma)$ + trivariate I.I.D	129.6	137.6	267.2
8	$MCAR(B, I)$ + trivariate I.I.D	139.5	155.2	294.7
9	$MCAR(\alpha_1, \alpha_2, \alpha_3, \Sigma)$ + trivariate I.I.D	137.4	155.0	292.4
10	$MCAR(\alpha, \Sigma)$ + trivariate I.I.D	138.2	151.0	289.2
11	separate CAR + trivariate i.i.d.	139.2	162.8	302.0

Table 11.4 *Model comparison using DIC statistics, Minnesota cancer data analysis.*

fits the three separate univariate CAR model, while Model 6 is the trivariate i.i.d. model. Fit measures $\overline{D}$, effective numbers of parameters p_D, and DIC scores for each model are seen in Table 11.4. We find that the $MCAR(B, \Sigma)$ model has the smallest $\overline{D}$ and DIC values for this data set. The $MCAR(B, I)$ model again disappoints, excelling over the non-spatial model and the separate CAR models only (very marginally over the latter). The $MCAR(\alpha, \Sigma)$ and $MCAR(\alpha_1, \alpha_2, \alpha_3, \Sigma)$ models perform slightly worse than the $MCAR(B, \Sigma)$ model, suggesting the need for different spatial autocorrelation and cross spatial correlation parameters for this data set. Note that the effective numbers of parameters p_D in Model 3 is a little larger than in Model 1, even though the latter has three extra parameters. Finally, the MCAR models do better than the separate CAR model or the i.i.d. trivariate model, suggesting that it is worth taking account of the correlations both across counties and among cancers. Model 6 exhibits a large p_D score, suggesting it does not seem to allow sufficient smoothing of the random effects. This is what we might have expected since the spatial correlations are missed by this model.

	Lung median $(2.5\%, 97.5\%)$	Larynx median $(2.5\%, 97.5\%)$	Esophagus mean $(2.5\%, 97.5\%)$
$\beta_1, \beta_2, \beta_3$	$-0.093\ (-0.179, -0.006)$	$-0.128\ (-0.316, 0.027)$	$-0.080\ (-0.194, 0.025)$
$\Sigma_{11}, \Sigma_{22}, \Sigma_{33}$	$0.048\ (0.030, 0.073)$	$0.173\ (0.054, 0.395)$	$0.107\ (0.044, 0.212)$
ρ_{12}, ρ_{13}		$0.277\ (-0.112, 0.643)$	$0.378\ (-0.022, 0.716)$
ρ_{23}			$0.337\ (-0.311, 0.776)$
b_{11}, b_{22}, b_{33}	$0.442\ (-0.302, 0.921)$	$0.036\ (-0.830, 0.857)$	$0.312\ (-0.526, 0.901)$
b_{12}, b_{13}		$0.323\ (-0.156, 0.842)$	$0.389\ (-0.028, 0.837)$
b_{23}			$0.006\ (-0.519, 0.513)$

Table 11.5 *Posterior summaries of parameters in $MCAR(B, \Sigma)$ model for Minnesota cancer data.*

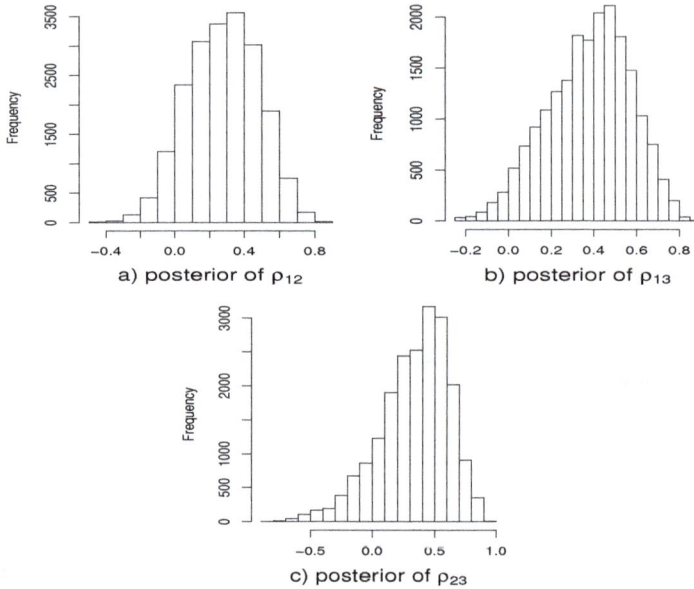

Figure 11.7 *Posterior samples of ρ_{12}, ρ_{13} and ρ_{23} in the Minnesota cancer data analysis using the $MCAR(B, \Sigma)$ model: (a) estimated posterior for correlation ρ_{12} between lung and larynx; (b) estimated posterior for correlation ρ_{13} between lung and esophagus; (c) estimated posterior for correlation ρ_{23} between larynx and esophagus.*

Models 7–11 are the convolution prior models corresponding to Models 1–5 formed by adding i.i.d. effects (following $N(0, \tau^2)$) to the ϕ_{ij}'s. Here the distinctions between the models are somewhat more pronounced due to the added variability in the models caused by the i.i.d. effects. The relative performances of the models remain the same with the $MCAR(B, \Sigma)$+ i.i.d. model emerging as best. Interestingly, none of the convolution models perform better than their purely spatial counterparts as the improvements in $\bar{D}$ in the former are insignificant compared to the increase in the effective dimensions brought about. This is indicative of the dominance of the spatial effects over the i.i.d. effects whence the convolution models seem to be rendering overparametrized models.

We summarize our results from the $MCAR(B, \Sigma)$, which is Model 1 in Table 11.4. Table 11.5 provides posterior means and associated standard deviations for the parameters β, Σ and b_{ij} in this model, where b_{ij} is the element of the symmetric matrix B. Instead of reporting Σ_{12}, Σ_{13} and Σ_{23}, we provide the mean and associated standard deviations for the correlation parameters ρ_{12}, ρ_{13} and ρ_{23}, which are calculated as $\rho_{ij} = \Sigma_{ij}/\sqrt{\Sigma_{ii}\Sigma_{jj}}$. We also plot histograms of the posterior samples ρ_{ij} in Figure 11.7, and histograms of the posterior samples b_{ij} in Figure 11.8.

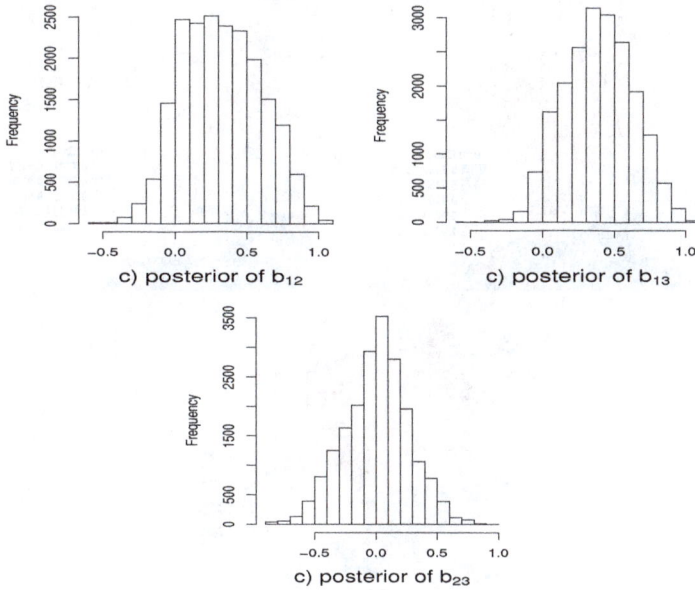

Figure 11.8 *Posterior samples of b_{12}, b_{13} and b_{23} in the Minnesota cancer data analysis using the $MCAR(B, \Sigma)$ model: (a) estimated posterior for b_{12}; (b) estimated posterior for b_{13}; (c) estimated posterior for b_{23}.*

Table 11.5 and Figure 11.7 reveal correlations between cancers, in particular a strong correlation between lung and esophagus (ρ_{13}). This might explain why the DIC scores for Models 1–4 in Table 11.5 are smaller than those under the separate CAR model. The b_{ij} in Table 11.5 are spatial autocorrelation and cross spatial correlation parameters for the latent spatial processes $\mathbf{u}_j$, $j = 1, 2, 3$. Figure 11.8 shows most of the b_{12} and b_{13} posterior samples are positive; the means of these two parameters are 0.323 and 0.389, respectively. Consistent with the DIC results in Table 11.4, these suggest it is worth fitting our proposed $MCAR(B, \Sigma)$ model to these data.

Turning to geographical summaries, Figure 11.9 maps the posterior means of the fitted standard mortality ratios (SMR) of lung, larynx and esophagus cancer from our $MCAR(B, \Sigma)$ model. From Figure 11.9, the correlation among the cancers is apparent, with higher fitted ratios extending from the Twin Cities metro area to the north and northeast (an area where previous studies have suggested smoking may be more common). In Figure 11.6, the range of the raw SMRs is seen to be from 0 to 3.3, while in Figure 11.9, the range of the fitted SMRs is from 0.7 to 1.3, due to spatial shrinkage in the random effects.

11.9 Modeling for multivariate disease mapping

We conclude this chapter with a brief review of what is called M modeling for multivariate disease mapping. We summarize work contained in Martinez-Beneito [2013] and Botella-Rocamora et al. [2015] as well as the recent book, Martínez-Beneito and Botella-Rocamora [2019]. This framework has, arguably, now become the standard in the field due to its convenient interpretation and rapid computation. Furthermore, it essentially includes all of the modeling we have presented in the previous eight sections.

Again, multivariate disease mapping addresses the joint mapping of multiple diseases from data aggregated to regions. The primary challenge is to map multiple diseases, accounting for any correlations among them. Martinez-Beneito [2013] opened up this modeling

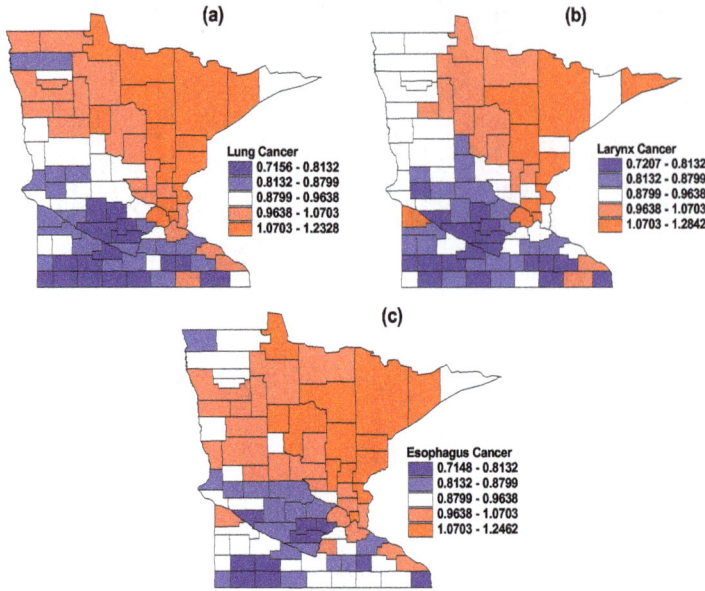

Figure 11.9 *Maps of posterior means of the fitted standard mortality ratios (SMR) of lung, larynx and esophagus cancer in the years from 1990 to 2000 in Minnesota from $MCAR(B, \Sigma)$ model.*

approach by providing a unifying framework for multivariate disease mapping. It incorporates a variety of existing statistical models for mapping multiple diseases but it and other existing approaches are computationally burdensome and preclude the multivariate analysis of moderate to large numbers of diseases. Botella-Rocamora et al. [2015] propose an alternative reformulation that offers substantial computational benefits, enabling the joint mapping of tens of diseases. Furthermore, it subsumes almost all existing classes of multivariate disease mapping models while enabling substantial insight into the properties of the statistical disease mapping models.

Continuing our previous notation, with say J diseases and n areal units, we need to specify the Poisson model for the set of observed counts $\{O_{ij}\}$ which requires the joint distribution of the $J \times n$ matrix of random effects, ϕ. Martinez-Beneito [2013] proposes that we write $\theta^{\mathrm{T}} = \phi P^{\mathrm{T}} \tilde{\Gamma}^{\mathrm{T}}$ where θ is a $J \times n$ matrix and the jth row of θ follows a multivariate normal distribution with covariance Σ_j. Further P is a $J \times J$ random orthogonal matrix and $\tilde{\Gamma}$ is the $J \times J$ upper triangular square root of the within areal unit covariance matrix, Γ.

While this notation seems a bit awkward, we can immediately see the linear model of coregionalization as a special case. That is, suppose we think about the random effects at unit i as $\boldsymbol{\eta}_i$ and write them as $\boldsymbol{\eta}_i = A\omega_i$ with A a $J \times J$ lower triangular matrix and ω_i as a $J \times 1$ vector such that the jth row is a realization from a ρ_j CAR model at unit i (Section 4.3). Collect the entire specification into matrix form as $\eta_{J \times n} = A_{J \times J} \omega_{J \times n}$. Then, we see that $\phi = \boldsymbol{\eta}^{\mathrm{T}}$, $\theta = \omega^{\mathrm{T}}$, $\Gamma = AA^{\mathrm{T}}$, and $P = I$. In particular, in the separable case, i.e., a common ρ across j, using vec notation, $\text{vec}(\phi)_{Jn \times 1} \sim N_{Jn}(\mathbf{0}, \Gamma \otimes \Sigma_\omega)$. (See for example, expression (11.19).)

Botella-Rocamora et al. [2015] realized that the term $P^{\mathrm{T}} \tilde{\Gamma}^{\mathrm{T}}$ in the expression above could, equivalently, be replaced with a $J \times J$ nonsingular matrix M and refer to this as M modeling for multivariate disease mapping. This dramatically simplifies computation and also enables interpretation of the specification as a factor model, i.e., to interpret the θ's in terms (factor loadings) of the ϕ's. Botella-Rocamora et al. [2015] offer some theoretical elaboration and exemplification of M modeling.

11.10   Computer tutorials

Computer programs illustrating Bayesian analysis of multivariate spatial data using different libraries in R and its interfaces with Bayesian modeling environments are supplied in the folder titled Chapter 11 from https://github.com/sudiptobanerjee/BGC_2023.

11.11   Exercises

1. The usual and generalized (but still proper) MCAR models may be constructed using linear transformations of some nonspatially correlated variables. Consider a vector blocked by components, say $\phi = (\phi_1^T, \phi_2^T)^T$, where each ϕ_i is $n \times 1$, n being the number of areal units. Suppose we look upon these vectors as arising from linear transformations

$$\phi_1 = A_1 \mathbf{v}_1 \text{ and } \phi_2 = A_2 \mathbf{v}_2 \,,$$

   where A_1 and A_2 are any $n \times n$ matrices, $\mathbf{v}_1 = (v_{11}, \ldots, v_{1n})^T$ and $\mathbf{v}_2 = (v_{21}, \ldots, v_{2n})^T$ with covariance structure

$$\text{Cov}(v_{1i}, v_{1j}) = \lambda_{11} I_{[i=j]}, \text{ Cov}(v_{1i}, v_{2j}) = \lambda_{12} I_{[i=j]},$$
$$\text{and Cov}(v_{2i}, v_{2j}) = \lambda_{22} I_{[i=j]} \,,$$

   where $I_{[i=j]} = 1$ if $i = j$ and 0 otherwise. Thus, although $\mathbf{v}_1$ and $\mathbf{v}_2$ are associated, their nature of association is nonspatial in that covariances remain same for every areal unit, and there is no association between variables in different units.

   (a) Show that the dispersion matrix $\Sigma_{(\mathbf{v}_1, \mathbf{v}_2)}$ equals $\Lambda \otimes I$, where $\Lambda = (\lambda_{ij})_{i,j=1,2}$.

   (b) Show that setting $A_1 = A_2 = A$ yields a separable covariance structure for ϕ. What choice of A would render a separable MCAR model, analogous to (11.4)?

   (c) Show that appropriate (different) choices of A_1 and A_2 yield the generalized MCAR model with a covariance matrix given by (11.7).

2. Derive the covariance matrix in (11.12) for the bivariate $GMCAR(\rho_1, \rho_2, \eta_1, \eta_2, \tau_1, \tau_2)$ model.

3. Show that the covariance matrix in (11.22) can be expressed as:

$$(I_{p\times p} \otimes D)^{\frac{1}{2}} \left(I_{pn \times pn} - B \otimes D^{-\frac{1}{2}} W D^{-\frac{1}{2}} \right) (I_{p\times p} \otimes D)^{\frac{1}{2}} \,.$$

4. Prove that $|I_{p\times p} \otimes D - B \otimes W| \propto \prod_{i=1}^{n} \prod_{j=1}^{p} (1 - \xi_i \zeta_j)$, where ξ_i's are the eigenvalues of $D^{-\frac{1}{2}} W D^{-\frac{1}{2}}$ and ζ_i's are the eigenvalues of B.

Chapter 12

Spatiotemporal modeling

In both theoretical and applied work, spatiotemporal modeling has received dramatically increased attention in the past few years. The reason is evident: the proliferation of data sets that are both spatially and temporally indexed, and the attendant need to understand them. For example, in studies of air pollution, we are interested not only in the spatial nature of a pollutant surface, but also in how this surface changes over time. Customarily, ongoing temporal measurements (e.g., hourly, daily, three-day average, etc.) are collected at monitoring sites yielding long time series of data. Similarly, with climate data we may be interested in spatial patterns of temperature or precipitation at a given time, but also in dynamic patterns in weather. With real estate markets, we might be interested in how the single-family home sales market changes on a quarterly or annual basis. Here an additional wrinkle arises in that we do not observe the *same* locations for each time period; the data are cross-sectional, rather than longitudinal.

Applications with areal unit data are also commonplace. For instance, we may look at annual lung cancer rates by county for a given state over a number of years to judge the effectiveness of a cancer control program. Or we might consider daily asthma hospitalization rates by zip code, over a period of several months.

So here we focus on geostatistical and discrete spatial data. A well-cited reference in this area is the book of Cressie and Wikle [2011]. We note that Section 9.7 devoted attention to spatiotemporal modeling for point patterns.

From a methodological point of view, the introduction of time into such spatial modeling brings a substantial increase in the scope of our work. We must incorporate decisions regarding spatial correlation, temporal correlation, and how space and time interact in our data. Such modeling will also carry an obvious associated increase in notational and computational complexity.

As in previous chapters, we make a distinction between the cases where the geographical aspect of the data is at point level versus where it is at areal unit level. Again the former case is typically handled via Gaussian process models, while the latter often uses CAR specifications. A parallel distinction could be drawn for the temporal scale: is time viewed as continuous (say, over $\Re^+$ or some subinterval thereof) or discrete (hourly, daily, etc.)? In the former case there is a conceptual measurement at each moment t. But in the latter case, we must determine whether each measurement should be interpreted as a block average over some time interval (analogous to block averaging in space), or whether it should be viewed merely as a measurement, e.g., a count attached to an associated time interval (and thus analogous to an areal unit measurement). Relatedly, when time is discretized, are we observing a time series of spatial data, e.g., the same points or areal units in each time period (as would be the case in our climate and pollution examples)? Or are we observing cross-sectional data, where the locations change with time period (as in our real estate setting)? In the case of time series, we could regard the data as a multivariate measurement vector at each location or areal unit. We could then employ multivariate spatial data models

DOI: 10.1201/9781003401728-12

as in the previous chapter. With short series, this might be reasonable; with longer series, we would likely want to introduce aspects of usual time series modeling.

The nature and location of missing data is another issue that we have faced before, yet becomes doubly complicated in the spatiotemporal setting. The major goal of our earlier kriging methods is to impute missing values at locations for which no data have been observed. Now we may encounter time points for which we lack spatial information, locations for which information is lacking for certain (possibly future) time points, or combinations thereof. Some of these combinations will be extrapolations (e.g., predicting future values at locations for which no data have been observed) that are statistically riskier than others (e.g., filling in missing values at locations for which we have data at some times but not others). Here the Bayesian hierarchical approach is particularly useful, since it not only helps organize our thinking about the model, but also fully accounts for all sources of uncertainty, and properly delivers wider confidence intervals for predictions that are "farther" from the observed data (in either space or time).

Our Atlanta data set (Figure 7.2) illustrates the sort of misalignment problem we face in many spatiotemporal settings. Here the number of ozone monitoring stations is small (just 8 or 10), but the amount of data collected from these stations over time (92 summer days for each of three years) is substantial. In this case, under suitable modeling assumptions, we may not only learn about the temporal nature of the data, but also enhance our understanding of the spatial process.

In the next few sections we consider the case of point-level spatial data, so that point-point and point-block realignment can be contemplated as in Section 7.1. We initially focus on relatively simple *separable* forms for the space-time correlation, but also consider more complex forms that do not impose the strong restrictions on space-time interaction that separability implies. We subsequently move on to spatiotemporal modeling for data where the spatial component can only be thought of as areal (block) level.

12.1 General modeling formulation

12.1.1 *Preliminary analysis*

Before embarking on a general spatiotemporal modeling formulation, consider the case of point-referenced data where time is discretized to customary integer-spaced intervals. We may look at a spatiotemporally indexed datum $Y(\mathbf{s}, t)$ in two ways. Writing $Y(\mathbf{s}, t) = Y_\mathbf{s}(t)$, it is evident that we have a spatially varying time series model. Writing $Y(\mathbf{s}, t) = Y_t(\mathbf{s})$, we instead have a temporally varying spatial model.

In fact, with locations $\mathbf{s}_i$, $i = 1, \ldots, n$ and time points $t = 1, \ldots, T$, we can collect the data into Y, an $n \times T$ matrix. Column averages of Y produce a space-averaged time series, while row averages yield a time-averaged spatial realization. In fact, suppose we center each column of Y by the vector of row averages and call the resulting matrix $\widetilde{Y}_{rows}$. Then clearly $\widetilde{Y}_{rows}\mathbf{1}_T = \mathbf{0}$, but also $\frac{1}{T}\widetilde{Y}_{rows}\widetilde{Y}_{rows}^{\mathrm{T}}$ is an $n \times n$ matrix that is the sample spatial covariance matrix. Similarly, suppose we center each row of Y by the vector of column averages and call the resulting matrix $\widetilde{Y}_{cols}$. Now $\mathbf{1}_n^{\mathrm{T}}\widetilde{Y}_{cols} = \mathbf{0}$ and $\frac{1}{n}\widetilde{Y}_{cols}^{\mathrm{T}}\widetilde{Y}_{cols}$ is the $T \times T$ sample autocorrelation matrix.

One could also center Y by the grand mean of the $Y(\mathbf{s}, t)$. Indeed, to examine residual spatiotemporal structure, adjusted for the mean, one could fit a suitable OLS regression to the $Y(\mathbf{s}, t)$ and examine $\widehat{E}$, the matrix of residuals $\widehat{e}(\mathbf{s}, t)$. As above, $\frac{1}{T}\widehat{E}\widehat{E}^{\mathrm{T}}$ is the residual spatial covariance matrix while $\frac{1}{n}\widehat{E}^{\mathrm{T}}\widehat{E}$ is the residual autocorrelation matrix.

We can create the singular value decomposition [see, e.g., Banerjee and Roy, 2014] for any of the foregoing matrices. Using E which would be most natural in practice to consider spatio-temporal structure, we can write

$$E = UDV^{\mathrm{T}} = \sum_{l=1}^{\min(n,T)} d_l \mathbf{u}_l \mathbf{v}_l^{\mathrm{T}} \,, \tag{12.1}$$

where U is an $n \times n$ orthogonal matrix with columns $\mathbf{u}_l$, V is a $T \times T$ orthogonal matrix with columns $\mathbf{v}_l$, and D is an $n \times T$ matrix of the form $\binom{\Delta}{0}$ where Δ is $T \times T$ diagonal with diagonal entries d_l, $l = 1, \ldots, T$. Without loss of generality, we can assume the d_l's are arranged in decreasing order of their absolute values. Then, $\mathbf{u}_l \mathbf{v}_l^{\mathrm{T}}$ is referred to as the lth *empirical orthogonal function* (EOF) since $\mathbf{u}_l \mathbf{v}_l^{\mathrm{T}} \perp \mathbf{u}_m \mathbf{v}_m^{\mathrm{T}}, l \neq m$ and $(\mathbf{u}_l \mathbf{v}_l^{\mathrm{T}})^{\mathrm{T}} \mathbf{u}_l \mathbf{v}_l^{\mathrm{T}} = 1$.

Thinking of $\mathbf{u}_l = (u_l(\mathbf{s}_1), \ldots, u_l(\mathbf{s}_n))^{\mathrm{T}}$ and $\mathbf{v}_l = (v_l(1), \ldots, v_l(T))^{\mathrm{T}}$, the expression in (12.1) represents the observed data as a sum of products of spatial and temporal variables, i.e., $E(\mathbf{s}_i, t) = \sum d_l u_l(\mathbf{s}_i) v_l(t)$. Suppose we approximate E by its first EOF, that is, $E \approx d_1 \mathbf{u}_1 \mathbf{v}_1^{\mathrm{T}}$. Then we are saying that $E(\mathbf{s}_i, t) \approx d_1 u_1(\mathbf{s}_i) v_1(t)$, i.e., the spatiotemporal process can be approximated by a product of a spatial process and a temporal process. If the u_1 and v_1 processes are mean 0 (as they would be for modeling residuals) and independent, this implies a *separable* covariance function for $E(\mathbf{s}, t)$ (see (12.18)).[1] Indeed, if the first term in the sum in (12.1) explains much of the residual matrix E, this is often taken as evidence for specifying a separable model. In any event, it does yield a reduction in dimension, introducing $n + T$ variables to represent E, rather than nT. Adding the second EOF yields the approximation $E(\mathbf{s}_i, t) \approx d_1 u_1(\mathbf{s}_i) v_1(t) + d_2 u_2(\mathbf{s}_i) v_2(t)$, a representation involving only $2(n + T)$ variables, and so on.

Note that, if say $T < n$,

$$EE^{\mathrm{T}} = UDD^{\mathrm{T}}U^{\mathrm{T}} = U \begin{pmatrix} \Delta^2 & 0 \\ 0 & 0 \end{pmatrix} U^{\mathrm{T}} = \sum_{l=1}^{T} d_l^2 \mathbf{u}_l \mathbf{u}_l^{\mathrm{T}} \,,$$

clarifying the interpretation of the d_l's. (Of course, $EE^{\mathrm{T}} = V^{\mathrm{T}}D^{\mathrm{T}}DV = V^{\mathrm{T}}\Delta^2 V$ as well.) Altogether, when applicable, EOFs provide an exploratory tool for learning about spatial structure and suggesting models, in the spirit of the tools described in Section 2.3. For full inference, however, we require a full spatiotemporal model specification, the subject to which we now turn.

12.1.2   Model formulation

Modeling for spatiotemporal data can be given a fairly general formulation which naturally extends that of Chapter 3. Consider point-referenced locations and continuous time. Let $Y(\mathbf{s}, t)$ denote the measurement at location $\mathbf{s}$ at time t. Extending (6.1), for continuous data assumed to be roughly normally distributed, we can write the general form

$$Y(\mathbf{s}, t) = \mu(\mathbf{s}, t) + e(\mathbf{s}, t) \,, \tag{12.2}$$

where $\mu(\mathbf{s}, t)$ denotes the mean structure and $e(\mathbf{s}, t)$ denotes the residual. If $\mathbf{x}(\mathbf{s}, t)$ is a vector of covariates associated with $Y(\mathbf{s}, t)$ then we can set $\mu(\mathbf{s}, t) = \mathbf{x}(\mathbf{s}, t)^{\mathrm{T}} \boldsymbol{\beta}(\mathbf{s}, t)$. Note that this form allows spatiotemporally varying coefficients (in the spirit of Section 10.7), which is likely more general than we would want than the data can handle. If t is discretized, $\boldsymbol{\beta}(\mathbf{s}, t) = \boldsymbol{\beta}_t$ might be adopted and might be appropriate if there were enough time points to suggest a temporal change in the coefficient vector. Similarly, setting $\boldsymbol{\beta}(\mathbf{s}, t) = \boldsymbol{\beta}(\mathbf{s})$ yields spatially varying coefficients, again following Section 10.7. Finally, $e(\mathbf{s}, t)$ would typically be rewritten as $w(\mathbf{s}, t) + \epsilon(\mathbf{s}, t)$, where $\epsilon(\mathbf{s}, t)$ is a Gaussian white noise process and $w(\mathbf{s}, t)$ is a mean-zero spatiotemporal process.

---

[1] It is routine to see that this will not be the case if the u_1 process and the v_1 process are not independent.

We can therefore view (12.2) as a hierarchical model with a conditionally independent first stage given $\{\mu(\mathbf{s}, t)\}$ and $\{w(\mathbf{s}, t)\}$. But then, in the spirit of Section 6.6, we can consider generalized linear model versions. We can replace the Gaussian first stage with another first-stage model (say, an exponential family model) and write $Y(\mathbf{s}, t) \sim f(y(\mathbf{s}, t) \mid \mu(\mathbf{s}, t), w(\mathbf{s}, t))$, where

$$f(y(\mathbf{s}, t) \mid \mu(\mathbf{s}, t), w(\mathbf{s}, t)) = h(y(\mathbf{s}, t)) \exp\{\gamma[\eta(\mathbf{s}, t)y(\mathbf{s}, t) - \chi(\eta(\mathbf{s}, t))]\} , \quad (12.3)$$

where γ is a positive dispersion parameter. In (12.3), $g(\eta(\mathbf{s}, t)) = \mu(\mathbf{s}, t) + w(\mathbf{s}, t)$ for some link function g.

For areal unit data with discrete time, let Y_{it} denote the measurement for unit i at time period t. (In some cases we might obtain replications at i or t, e.g., the jth cancer case in county i, or the the jth property sold in census tract i.) Analogous to (12.2) we can write

$$Y_{it} = \mu_{it} + e_{it} . \quad (12.4)$$

Now $\mu_{it} = \mathbf{x}_{it}^{\mathrm{T}}\boldsymbol{\beta}_t$ (or perhaps just $\boldsymbol{\beta}$), and $e_{it} = w_{it} + \epsilon_{it}$ where the ϵ_{it} are unstructured heterogeneity terms and the w_{it} are spatiotemporal random effects, typically associated with a spatiotemporal CAR specification. Choices for this latter part of the model will be presented in Section 12.7.

Since areal unit data are often non-Gaussian (e.g., sparse counts), again we would view (12.4) as a hierarchical model and replace the first stage Gaussian specification with, say, a Poisson model. We could then write $Y_{it} \sim f(y_{it}|\mu_{it}, w_{it})$, where

$$f(y_{it} \mid \mu_{it}, w_{it}) = h(y_{it}) \exp\{\gamma[\eta_{it}y_{it} - \chi(\eta_{it})]\} , \quad (12.5)$$

with γ again a dispersion parameter, and $g(\eta_{it}) = \mu_{it} + w_{it}$ for some suitable link function g. With replications, we obtain Y_{ijt} hence $\mathbf{x}_{ijt}, \mu_{ijt}$, and η_{ijt}. Now we can write $g(\eta_{ijt}) = \mu_{ijt} + w_{ijt} + \epsilon_{ijt}$, enabling separation of spatial and heterogeneity effects.

Returning to the point-referenced data model (12.2), spatiotemporal richness is captured by extending $e(\mathbf{s}, t)$ beyond $\epsilon(\mathbf{s}, t)$, a white noise process, as noted above. As a result, we need forms for $w(\mathbf{s}, t)$. Below, α's denote temporal effects and w's denote spatial effects. Following Gelfand et al. [2004a] with t discretized, consider the following forms for $w(\mathbf{s}, t)$:

$$w(\mathbf{s}, t) = \alpha(t) + w(\mathbf{s}) , \quad (12.6)$$
$$w(\mathbf{s}, t) = \alpha_{\mathbf{s}}(t) , \quad (12.7)$$
$$\text{and } w(\mathbf{s}, t) = w_t(\mathbf{s}) . \quad (12.8)$$

The given forms avoid specification of space-time interactions. With regard to (12.6), (12.7), and (12.8), the $\epsilon(\mathbf{s}, t)$ are i.i.d. $N(0, \sigma_{\epsilon}^2)$ and independent of the other processes. This pure error is viewed as a residual adjustment to the spatiotemporal explanation. (One could allow $\mathrm{Var}(\epsilon(\mathbf{s}, t)) = \sigma_{\epsilon,t}^2$, i.e., an error variance that changes with time. Modification to the details below is straightforward.)

Expression (12.6) provides an additive form in temporal and spatial effects. In fact, we can also introduce a multiplicative form, $\alpha(t)w(\mathbf{s})$ which would, of course, become additive on the log scale. Expression (12.7) provides temporal evolution at each site; temporal effects are nested within sites. Expression (12.8) provides spatial evolution over time; spatial effects are nested within time. Spatiotemporal modeling beyond (12.6), (12.7), and (12.8) (particularly if t is continuous) necessitates the choice of a specification to connect the space and time scales; this is the topic of Section 12.2.

Next, we consider the components in (12.6), (12.7), and (12.8) in more detail. In (12.6), if t were continuous we could model $\alpha(t)$ as a one-dimensional stationary Gaussian process. In particular, for the set of times, $\{t_1, t_2, ..., t_m\}$, $\boldsymbol{\alpha} = (\alpha(t_1), ..., \alpha(t_m))^{\mathrm{T}} \sim N(\mathbf{0}, \sigma_{\alpha}^2 \Sigma(\phi))$ where

$(\Sigma(\phi))_{rs} = Corr(\alpha(t_r), \alpha(t_s)) = \rho(| t_r - t_s |; \phi)$ for ρ a valid one-dimensional correlation function. A typical choice for ρ would be the exponential, $\exp(-\phi | t_r - t_s |)$ though other forms, analogous to the spatial forms in Table 2.1 are possible.

With t confined to an indexing set, $t = 1, 2, \ldots T$, we can simply view $\alpha(1), \ldots, \alpha(T)$ as the coefficients associated with a set of time dummy variables. With this assumption for the $\alpha(t)$'s, suppose in (12.6), $w(\mathbf{s})$ is set to zero, $\boldsymbol{\beta}(t)$ is assumed constant over time and $\mathbf{X}(\mathbf{s}, t)$ is assumed constant over t. Then, upon differencing, we find models described in the real estate literature, e.g., the seminal model for repeat property sales given in Bailey et al. [1963]. Also within these assumptions but restoring $\boldsymbol{\beta}$ to $\boldsymbol{\beta}(t)$, we obtain the extension of Knight et al. [1995]. In other work [Paci et al., 2013] the multiplicative form is used, with differencing, to implement real-time ozone forecasting.

Alternatively, we might set $\alpha(t+1) = \rho\alpha(t) + \eta(t)$ where $\eta(t)$ are i.i.d. $N(0, \sigma_\alpha^2)$. If $\rho < 1$ we have the familiar stationary $AR(1)$ time series, a special case of the continuous time model of the previous paragraph. If $\rho = 1$ the $\alpha(t)$ follow a random walk. With a finite set of times, time-dependent coefficients are handled analogously to the survival analysis setting [see, e.g., Cox and Oakes, 1984, Ch. 8].

The autoregressive and random walk specifications are naturally extended to provide a model for the $\alpha_\mathbf{s}(t)$ in (12.7). That is, we assume $\alpha_\mathbf{s}(t + 1) = \rho\alpha_\mathbf{s}(t) + \eta_\mathbf{s}(t)$ where again the $\eta_\mathbf{s}(t)$ are all i.i.d. Thus, there is no spatial modeling; rather, we imagine independent conceptual time series at each location. With spatial time series we can fit this model. With cross-sectional data, there is no information in the data about ρ so the likelihood can only identify the stationary variance $\sigma_\alpha^2/(1 - \rho^2)$ but not σ_α^2 or ρ. The case $\rho < 1$ with $\boldsymbol{\beta}(t)$ constant over time provides the models proposed in Hill et al. [1997] and in Hill et al. [1999]. If $\rho = 1$ with $\boldsymbol{\beta}(t)$ and $\mathbf{X}(\mathbf{s}, t)$ constant over time, upon differencing we obtain the widely used model dating to Case and Shiller [1989]. In application, it will be difficult to learn about the $\alpha_\mathbf{s}$ processes with typically one or at most two observations for each $\mathbf{s}$. The $w(\mathbf{s})$ are modeled as a Gaussian process following Section 3.1.

For $w_t(\mathbf{s})$ in (12.8), assuming t restricted to an index set, we can view the $w_t(\mathbf{s})$ as a collection of independent spatial processes. That is, rather than defining a dummy variable at each t, we conceptualize a separate spatial dummy *process* at each t. The components of $\mathbf{w}_t$ correspond to the sites at which measurements were observed in the time interval denoted by t. Thus, we capture the dynamics of location in a very general fashion. In particular, comparison of the respective process parameters reveals the nature of spatial evolution over time.

With a single time dummy variable at each t, assessment of temporal effects would be provided through inference associated with these variables. For example, a plot of the point estimates against time would clarify size and trend for the effects. With distinct spatial processes, how can we see such temporal patterns? A convenient reduction of each spatial process to a univariate random variable is the block average (see expression (7.1)).

To shed the independence assumption for the $w_t(\mathbf{s})$, we could instead assume that $w_t(\mathbf{s}) = \sum_{j=1}^{t} v_j(\mathbf{s})$ where the $v_j(\mathbf{s})$ are i.i.d. processes, again of one of the foregoing forms. Now, for $t < t^*$, $\mathbf{w}_t$ and $\mathbf{w}_{t^*}$ are not independent but $\mathbf{w}_t$ and $\mathbf{w}_{t^*} - \mathbf{w}_t$ are. This leads us to dynamic spatiotemporal models that are the focus of Section 12.4.

12.1.3 Associated distributional results

We begin by developing the likelihood under model (12.2) using (12.6), (12.7), or (12.8). Assuming t belongs to the set $\{1, 2, \ldots, T\}$, it is convenient to first obtain the joint distribution for $\mathbf{Y}^\mathrm{T} = (\mathbf{Y}_1^\mathrm{T}, \ldots, \mathbf{Y}_T^\mathrm{T})$ where $\mathbf{Y}_t^\mathrm{T} = (Y(\mathbf{s}_1, t), \ldots, Y(\mathbf{s}_n, t))$. That is, each $\mathbf{Y}_t$ is $n \times 1$ and $\mathbf{Y}$ is $Tn \times 1$. This joint distribution will be multivariate normal. Thus, the joint distribution for the observed $Y(\mathbf{s}, t)$ requires only pulling off the appropriate entries from the mean vector and appropriate rows and columns from the covariance matrix. This simplifies the computational bookkeeping, though care is still required.

In the constant $\boldsymbol{\beta}$ case, associate with $\mathbf{Y}_t$ the matrix X_t whose ith row is $\mathbf{X}(\mathbf{s}_i, t)^{\mathrm{T}}$. Let $\boldsymbol{\mu}_t = X_t\boldsymbol{\beta}$ and $\boldsymbol{\mu}^{\mathrm{T}} = (\boldsymbol{\mu}_1^{\mathrm{T}}, ..., \boldsymbol{\mu}_T^{\mathrm{T}})$. In the time-dependent parameter case we merely set $\boldsymbol{\mu}_t = X_t\boldsymbol{\beta}(t)$.

Under (12.6), let $\boldsymbol{\alpha}^{\mathrm{T}} = (\alpha(1), \ldots, \alpha(T))$, $\mathbf{w}^{\mathrm{T}} = (\mathbf{w}(\mathbf{s}_1), \ldots, \mathbf{w}(\mathbf{s}_n))$ and $\boldsymbol{\epsilon}^{\mathrm{T}} = (\epsilon(\mathbf{s}_1, 1),$ $\epsilon(\mathbf{s}_1, 2), ..., \epsilon(\mathbf{s}_n, T)$. Then,

$$\mathbf{Y} = \boldsymbol{\mu} + \boldsymbol{\alpha} \otimes \mathbf{1}_{n \times 1} + \mathbf{1}_{T \times 1} \otimes \mathbf{w} + \boldsymbol{\epsilon} \tag{12.9}$$

where $\otimes$ denotes the Kronecker product. Hence, given $\boldsymbol{\beta}$ along with the temporal and spatial effects,

$$\mathbf{Y} \mid \boldsymbol{\beta}, \boldsymbol{\alpha}, \mathbf{w}, \sigma_\epsilon^2 \sim N(\boldsymbol{\mu} + \boldsymbol{\alpha} \otimes \mathbf{1}_{n \times 1} + \mathbf{1}_{T \times 1} \otimes \mathbf{w}, \, \sigma_\epsilon^2 I_{Tn \times Tn}). \tag{12.10}$$

Let $\mathbf{w} \sim N(\mathbf{0}, \sigma_w^2 H(\delta))$. Suppose the $\alpha(t)$ follow an $AR(1)$ model, so that $\boldsymbol{\alpha} \sim N(\mathbf{0}, \sigma_\alpha^2 A(\rho))$ where $(A(\rho))_{ij} = \rho^{|i-j|}/(1 - \rho^2)$. Hence, if $\boldsymbol{\alpha}, \mathbf{w}$ and $\boldsymbol{\epsilon}$ are independent, marginalizing over $\boldsymbol{\alpha}$ and $\mathbf{w}$, i.e., integrating (12.10) with regard to the prior distribution of $\boldsymbol{\alpha}$ and $\mathbf{w}$, we obtain

$$
\begin{aligned}
\mathbf{Y} \mid \boldsymbol{\beta}, \sigma_\epsilon^2, \sigma_\alpha^2, \rho, \sigma_w^2, \delta & \\
\sim N\left(\boldsymbol{\mu}, \, \sigma_\alpha^2 A(\rho) \otimes \mathbf{1}_{n \times 1}\mathbf{1}_{n \times 1}^{\mathrm{T}} + \sigma_w^2 \mathbf{1}_{T \times 1}\mathbf{1}_{T \times 1}^{\mathrm{T}} \otimes H(\delta) + \sigma_\epsilon^2 I_{Tn \times Tn}\right) & .
\end{aligned} \tag{12.11}
$$

If the $\alpha(t)$ are coefficients associated with dummy variables (now $\boldsymbol{\beta}$ does not contain an intercept) we only marginalize over $\mathbf{w}$ to obtain

$$
\begin{aligned}
\mathbf{Y} \mid \boldsymbol{\beta}, \boldsymbol{\alpha}, \sigma_\epsilon^2, \sigma_w^2, \delta & \\
\sim N\left(\boldsymbol{\mu} + \boldsymbol{\alpha} \otimes \mathbf{1}_{n \times 1}, \, \sigma_w^2 \mathbf{1}_{T \times 1}\mathbf{1}_{T \times 1}^{\mathrm{T}} \otimes H(\delta) + \sigma_\epsilon^2 I_{Tn \times Tn}\right) & .
\end{aligned} \tag{12.12}
$$

The likelihood resulting from (12.10) arises as a product of independent normal densities by virtue of conditional independence. This can facilitate model fitting but at the expense of a very high-dimensional posterior distribution. Marginalizing to (12.11) or (12.12) results in a much lower-dimensional posterior. Note, however, that while the distributions in (12.11) and (12.12) can be determined, evaluating the likelihood (joint density) requires evaluation of a high-dimensional quadratic form and determinant calculation.

Turning to (12.7), if $\boldsymbol{\alpha}^{\mathrm{T}}(t) = (\alpha_{s_1}(t), ..., \alpha_{s_n}(t))$ and we also define $\boldsymbol{\alpha}^{\mathrm{T}} = (\boldsymbol{\alpha}^{\mathrm{T}}(1), \ldots, \boldsymbol{\alpha}^{\mathrm{T}}(T))$ with $\boldsymbol{\epsilon}$ as above, then

$$\mathbf{Y} = \boldsymbol{\mu} + \boldsymbol{\alpha} + \boldsymbol{\epsilon}.$$

Now

$$\mathbf{Y} \mid \boldsymbol{\beta}, \boldsymbol{\alpha}, \sigma_\epsilon^2 \sim N\left(\boldsymbol{\mu} + \boldsymbol{\alpha}, \, \sigma_\epsilon^2 I_{Tn \times Tn}\right).$$

If the $\alpha_{s_i}(t)$ follow an $AR(1)$ model independently across i, then marginalizing over $\boldsymbol{\alpha}$,

$$\mathbf{Y} \mid \boldsymbol{\beta}, \sigma_\epsilon^2, \sigma_\alpha^2, \rho \sim N(\boldsymbol{\mu}, \, A(\rho) \otimes I_{Tn \times Tn} + \sigma_\epsilon^2 I_{Tn \times Tn}). \tag{12.13}$$

For (12.8), let $\mathbf{w}_t^{\mathrm{T}} = (w_t(\mathbf{s}_1), \ldots, w_t(\mathbf{s}_n))$ and $\mathbf{w}^{\mathrm{T}} = (\mathbf{w}_1^{\mathrm{T}}, \ldots, \mathbf{w}_T^{\mathrm{T}})$. Then with $\boldsymbol{\epsilon}$ as above,

$$\mathbf{Y} = \boldsymbol{\mu} + \mathbf{w} + \boldsymbol{\epsilon} \tag{12.14}$$

and

$$\mathbf{Y} \mid \boldsymbol{\beta}, \mathbf{w}, \sigma_\epsilon^2 \sim N\left(\boldsymbol{\mu} + \mathbf{w}, \, \sigma_\epsilon^2 I_{Tn \times Tn}\right). \tag{12.15}$$

If $\mathbf{w}_t \sim N(\mathbf{0}, \sigma_w^{2(t)} H(\delta^{(t)}))$ independently for $t = 1, \ldots, T$, then, marginalizing over $\mathbf{w}$,

$$\mathbf{Y} \mid \boldsymbol{\beta}, \sigma_\epsilon^2, \boldsymbol{\sigma}_w^2, \delta \sim N(\boldsymbol{\mu}, \, D(\boldsymbol{\sigma}_w^2, \delta) + \sigma_\epsilon^2 I_{Tn \times Tn}), \tag{12.16}$$

where $\boldsymbol{\sigma}_w^{2\prime} = (\sigma_w^{2(1)}, ..., \sigma_w^{2(T)})$, $\boldsymbol{\delta}^{\mathrm{T}} = (\delta^{(1)}, ..., \delta^{(T)})$, and $D(\boldsymbol{\sigma}_w^2, \boldsymbol{\delta})$ is block diagonal with the tth block being $\sigma_w^{2(t)}(H(\delta^{(t)}))$. Because D is block diagonal, likelihood evaluation associated with (12.16) is less of an issue than for (12.11) and (12.12).

We note that with either (12.7) or (12.8), $e(\mathbf{s}, t)$ is comprised of two sources of error that the data cannot directly separate. However, by incorporating a stochastic assumption on the $\alpha_{\mathbf{s}}(t)$ or on the $w_t(\mathbf{s})$, we can learn about the processes that guide the error components, as (12.13) and (12.16) reveal.

12.1.4   Prediction and forecasting

We now turn to forecasting under (12.2) with models (12.6), (12.7), or (12.8). Such forecasting involves prediction at location $\mathbf{s}_0$ and time t_0, i.e., of $Y(\mathbf{s}_0, t_0)$. Here $\mathbf{s}_0$ may correspond to an already observed location, perhaps to a new location. However, typically $t_0 > T$ is of interest. Such prediction requires specification of an associated vector of characteristics $\mathbf{X}(\mathbf{s}_0, t_0)$. Also, prediction for $t_0 > T$ is available in the fixed coefficients case. For the time-varying coefficients case, we would need to specify a temporal model for $\boldsymbol{\beta}(t)$.

In general, within the Bayesian framework, prediction at $(\mathbf{s}_0, t_0)$ follows from the posterior predictive distribution of $f(Y(\mathbf{s}_0, t_0) \mid \mathbf{Y})$ where $\mathbf{Y}$ denotes the observed data vector. Assuming $\mathbf{s}_0$ and t_0 are new, and for illustration, taking the form in (12.6),

$$
\begin{aligned}
f(Y(\mathbf{s}_0, t_0) \mid \mathbf{Y}) = \int & f(Y(\mathbf{s}_0, t_0) \mid \boldsymbol{\beta}, \sigma_\epsilon^2, \alpha(t_0), w(\mathbf{s}_0)) \\
& \times dF(\boldsymbol{\beta}, \boldsymbol{\alpha}, \mathbf{w}, \sigma_\epsilon^2, \sigma_\alpha^2, \rho, \sigma_w^2, \delta, \alpha(t_0), w(\mathbf{s}_0) \mid \mathbf{Y}) .
\end{aligned}
\tag{12.17}
$$

Using (12.17), given a random draw $(\boldsymbol{\beta}^*, \sigma_\epsilon^{2*}, \alpha(t_0)^*, w(\mathbf{s}_0)^*)$ from the posterior $f(\boldsymbol{\beta}, \sigma_\epsilon^2, \alpha(t_0), w(\mathbf{s}_0) \mid \mathbf{Y})$, if we draw $Y^*(\mathbf{s}_0, t_0)$ from $N(X^{\mathrm{T}}(\mathbf{s}_0, t_0)\boldsymbol{\beta}^* + \alpha(t_0)^* + w(\mathbf{s}_0)^*, \sigma_e^{2*})$, marginally, $Y^*(\mathbf{s}_0, t_0) \sim f(Y(\mathbf{s}_0, t_0) \mid \mathbf{Y})$.

Using sampling-based model fitting and working with (12.10), we obtain samples $(\boldsymbol{\beta}^*, \sigma_\epsilon^{2*}, \sigma_\alpha^{2*}, \rho^*, \sigma_w^{2*}, \delta^*, \boldsymbol{\alpha}^*, \mathbf{w}^*)$ from the posterior distribution, $p(\boldsymbol{\beta}, \sigma_\epsilon^2, \sigma_\alpha^2, \rho, \sigma_w^2, \delta, \boldsymbol{\alpha}, \mathbf{w} \mid \mathbf{Y})$. But $f(\boldsymbol{\beta}, \sigma_\epsilon^2, \sigma_\alpha^2, \rho, \sigma_w^2, \delta, \boldsymbol{\alpha}, \mathbf{w}, \alpha(t_0), w(\mathbf{s}_0) \mid \mathbf{Y}) = f(\alpha(t_0) \mid \boldsymbol{\alpha}, \sigma_\alpha^2, \rho) \cdot f(w(\mathbf{s}_0) \mid \mathbf{w}, \sigma_w^2, \delta) \cdot f(\boldsymbol{\beta}, \sigma_\epsilon^2, \sigma_\alpha^2, \rho, \sigma_w^2, \delta, \boldsymbol{\alpha}, \mathbf{w} \mid \mathbf{Y})$. If, e.g., $t_0 = T + 1$, and $\alpha(t)$ is modeled as a time series, $f(\alpha(T+1) \mid \boldsymbol{\alpha}, \sigma_\alpha^2, \rho)$ is $N(\rho\alpha(T), \sigma_\alpha^2)$. If the $\alpha(t)$ are coefficients associated with dummy variables, setting $\alpha(T+1) = \alpha(T)$ is, arguably, the best one can do. The joint distribution of $\mathbf{w}$ and $w(\mathbf{s}_0)$ is a multivariate normal from which $f(w(\mathbf{s}_0) \mid \mathbf{w}, \sigma_w^2, \delta)$ is a univariate normal. So if $\alpha(t_0)^* \sim f(\alpha(t_0) \mid \boldsymbol{\alpha}^*, \sigma_\alpha^{2*}, \rho^*)$ and $w(\mathbf{s}_0)^* \sim f(w(\mathbf{s}_0) \mid \mathbf{w}^*, \sigma_w^{2*}, \delta^*)$, along with $\boldsymbol{\beta}^*$ and σ_ϵ^{2*} we obtain a draw from $f(\boldsymbol{\beta}, \sigma_\epsilon^2, \alpha(t_0), w(\mathbf{s}_0) \mid \mathbf{Y})$. (If $t_0 \epsilon \{1, 2, ..., T\}$, $\alpha(t_0)$ is a component of $\boldsymbol{\alpha}$, then $\alpha(t_0)^*$ is a component of $\boldsymbol{\alpha}^*$. If $\mathbf{s}_0$ is one of the $\mathbf{s}_1, \mathbf{s}_2, ..., \mathbf{s}_n$, $w^*(\mathbf{s}_0)$ is a component of $\mathbf{w}^*$.) Alternatively, one can work with (12.13). Now, having marginalized over $\boldsymbol{\alpha}$ and $\mathbf{w}$, $Y(\mathbf{s}, t)$ and $\mathbf{Y}$ are no longer independent. They have a multivariate normal distribution from which $f(Y(\mathbf{s}, t) \mid \mathbf{Y}, \boldsymbol{\beta}, \sigma_\epsilon^2, \sigma_\alpha^2, \rho, \sigma_w^2, \delta)$ must be obtained. Note that for multiple predictions, $w(\mathbf{s}_0)$ is replaced by a vector, say $\mathbf{w}_0$. Now $f(\mathbf{w}_0 \mid \mathbf{w}, \sigma_w^2, \delta)$ is a multivariate normal distribution. No additional complications arise.

Example 12.1 *(Baton Rouge home sales).* We present a portion of the data analysis developed in Gelfand et al. [2004a] for sales of single-family homes drawn from two regions in the city of Baton Rouge, LA. The two areas are known as Sherwood Forest and Highland Road. These regions are approximately the same size and have similar levels of transaction activity; they differ chiefly in the range of neighborhood characteristics and house amenities found within. Sherwood Forest is a large, fairly homogeneous neighborhood located east, southeast of downtown Baton Rouge. Highland Road, on the other hand, is a major thoroughfare connecting downtown with the residential area to the southeast. Rather than being one homogeneous neighborhood, the Highland Road area consists, instead, of heterogeneous subdivisions. Employing two regions makes a local isotropy assumption more comfortable and allows investigation of possibly differing time effects and location dynamics.

Year	Highland		Sherwood	
	Repeat	Single	Repeat	Single
1985	25	40	32	29
1986	20	35	32	39
1987	27	32	27	37
1988	16	26	20	34
1989	21	25	24	35
1990	42	29	27	37
1991	29	30	25	31
1992	33	38	39	27
1993	24	40	31	40
1994	26	35	20	34
1995	26	35	21	32
Total	289	365	298	375

Table 12.1 *Sample size by region, type of sale, and year.*

Variable	Highland		Sherwood	
	Repeat	Single	Repeat	Single
Age	11.10	12.49	14.21	14.75
	(8.15)	(11.37)	(8.32)	(10.16)
Bathrooms	2.18	2.16	2.05	2.02
	(0.46)	(0.56)	(0.36)	(0.40)
Living area	2265.4	2075.8	1996.0	1941.5
	(642.9)	(718.9)	(566.8)	(616.2)
Other area	815.1	706.0	726.0	670.6
	(337.7)	(363.6)	(258.1)	(289.2)

Table 12.2 *Mean (standard deviation) for house characteristics by region and type of sale.*

For these regions, a subsample of all homes sold only once during the period 1985 through 1995 (single-sale transactions) and a second subsample of homes sold more than once (repeat-sale transactions) were drawn. These two samples can be studied separately to assess whether the population of single-sale houses differs from that of repeat-sale houses. The sample sizes are provided by year in Table 12.1. The location of each property is defined by its latitude and longitude coordinates, rescaled to UTM projection. In addition, a variety of house characteristics, to control for physical differences among the properties, are recorded at the time of sale. We use age, living area, other area (e.g., patios, garages, and carports) and number of bathrooms as covariates in our analysis. Summary statistics for these attributes appear in Table 12.2. We see that the homes in the Highland Road area are somewhat newer and slightly larger than those in the Sherwood area. The greater heterogeneity of the Highland Road homes is borne out by the almost uniformly higher standard deviations for each covariate. In fact, we have more than 20 house characteristics in our data set, but elaborating the mean with additional features provides little improvement in R^2 and introduces multicollinearity problems. So, we confine ourselves to the four explanatory variables above and turn to spatial modeling to explain a portion of the remaining

Variable	Repeat	Single
Highland region:		
intercept (β_0)	11.63 (11.59, 11.66)	11.45 (11.40, 11.50)
age (β_1)	−0.04 (−0.07, −0.02)	−0.08 (−0.11, −0.06)
bathrooms (β_2)	0.02 (−0.01, 0.04)	0.02 (−0.01, 0.05)
living area (β_3)	0.28 (0.25, 0.31)	0.33 (0.29, 0.37)
other area (β_4)	0.08 (0.06, 0.11)	0.07 (0.04, 0.09)
Sherwood region:		
intercept (β_0)	11.33 (11.30, 11.36)	11.30 (11.27, 11.34)
age (β_1)	−0.06 (−0.07, −0.04)	−0.05 (−0.07, −0.03)
bathrooms (β_2)	0.05 (0.03, 0.07)	0.00 (−0.02, 0.02)
living area (β_3)	0.19 (0.17, 0.21)	0.22 (0.19, 0.24)
other area (β_4)	0.02 (0.01, 0.04)	0.06 (0.04, 0.08)

Table 12.3 *Parameter estimates (median and 95% interval estimates) for house characteristics.*

variability. Empirical semivariograms (2.9) offer evidence of spatial association, after adjusting for house characteristics.

We describe the results of fitting the model with mean $\mu(\mathbf{s}) = \mathbf{x}(\mathbf{s})^{\mathrm{T}}\boldsymbol{\beta}$ and the error structure in (12.8). This is also the preferred model using the predictive model choice approach of Gelfand and Ghosh (5.67); we omit details. Fixed coefficients were justified by the shortness of the observation period. Again, an exponential isotropic correlation function was adopted.

To complete the Bayesian specification, we adopt rather noninformative priors in order to resemble a likelihood/least squares analysis. In particular, we assume a flat prior on the regression parameter $\boldsymbol{\beta}$ and inverse gamma (a, b) priors for σ_ϵ^2, $\sigma_w^{2(t)}$ and $\delta^{(t)}$, $t = 1, \ldots T$. The shape parameter for these inverse gamma priors was fixed at 2, implying an infinite prior variance. We choose the inverse gamma scale parameter for all $\delta^{(t)}$'s to be equal, i.e., $b_{\delta(1)} = b_{\delta(2)} = \ldots = b_{\delta(T)} = b_\delta$, say, and likewise for $\sigma_w^{2(t)}$. Furthermore, we set $b_{\sigma_\epsilon} = b_{\sigma_w^2}$ reflecting uncertain prior contribution from the nugget to the sill. Finally, the exact values of b_{σ_ϵ}, $b_{\sigma_w^2}$ and b_δ vary between region and type of sale reflecting different prior beliefs about these characteristics.

Inference for the house characteristic coefficients is provided in Table 12.3 (point and 95% interval estimates). Age, living areas, and other areas are significant in all cases; number of bathrooms is significant only in Sherwood repeat sales. Significance of living area is much stronger in Highland than in Sherwood. The Highland sample is composed of homes from several heterogeneous neighborhoods. As such, living area not only measures differences in house size, but may also serve as a partial proxy for construction quality and for neighborhood location within the sample. The greater homogeneity of homes in Sherwood implies less variability in living areas (as seen in Table 12.2) and reduces the importance of these variables in explaining house price.

Turning to the error structure, the parameters of interest for each region are the $\sigma_w^{2(t)}$, the $\delta^{(t)}$, and σ_ϵ^2. The sill at time t is $\mathrm{Var}(Y(s,t)) = \sigma_w^{2(t)} + \sigma_\epsilon^2$. Figure 12.1 plots the posterior medians of these sills. We see considerable difference in variability over the groups and over time, providing support for distinct spatial models at each t. Variability is highest for Highland single sales, lowest for Sherwood repeats. The additional insight is the effect of time. Variability is generally increasing over time.

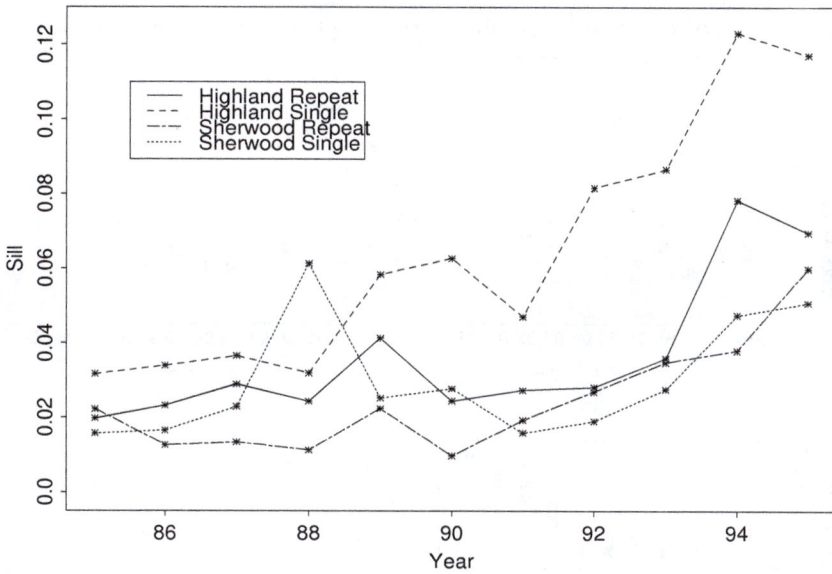

Figure 12.1 *Posterior median sill by year.*

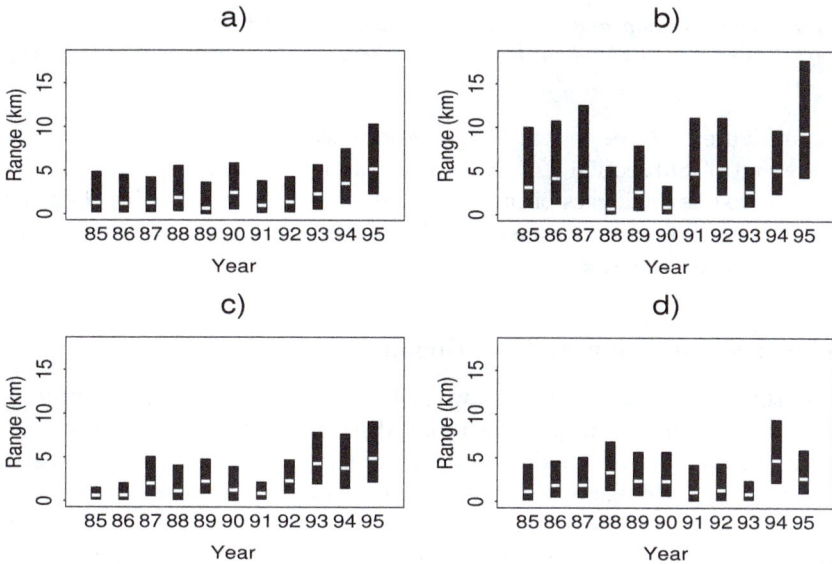

Figure 12.2 *Posterior median and 95% interval estimates for the range by year for (a) Highland repeat sales, (b) Highland single sales, (c) Sherwood repeat sales, and (d) Sherwood single sales.*

We can obtain posterior median and interval estimates for $\sigma_w^{2(t)}/(\sigma_e^2 + \sigma_w^{2(t)})$, the proportion of spatial variance to total. The strength of the spatial story is considerable; 40 to 80% of the variability is spatial.

In Figure 12.2 we provide point and interval estimates for the range. The ranges for the repeat sales are quite similar for the two regions, showing some tendency to increase in the later years of observation. By contrast, the range for the Highland single sales is much different from that for Sherwood. It is typically greater and much more variable. The latter again is a reflection of the high variability in the single-sale home prices in Highland. The resulting posteriors are more dispersed.

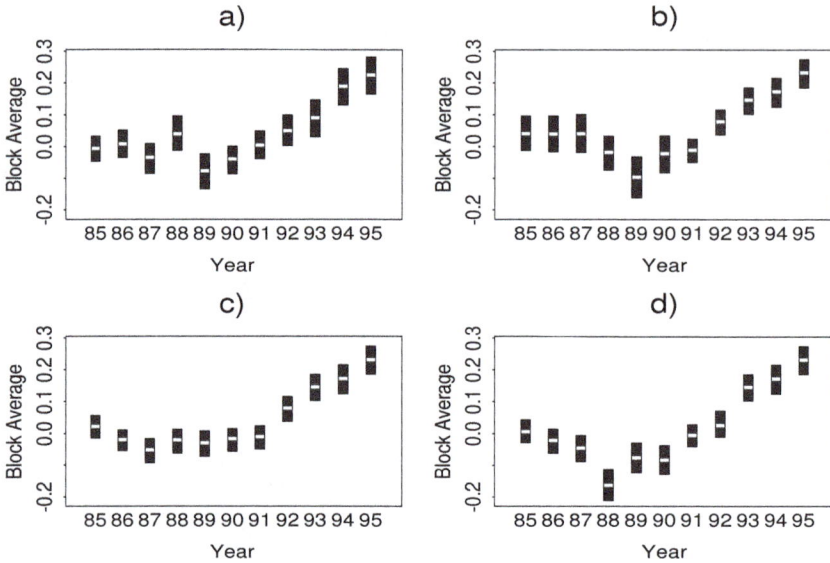

Figure 12.3 *Posterior median and 95% interval estimates for the block averages by year for (a) Highland repeat sales, (b) Highland single sales, (c) Sherwood repeat sales, and (d) Sherwood single sales.*

Finally, in Figure 12.3, we present the posterior distribution of the block averages, mentioned at the end of Subsection 12.1.2, for each of the four analyses. Again, these block averages are viewed as analogues of more familiar time dummy variables. Time effects are evident. In all cases, we witness somewhat of a decline in magnitude in the 1980s and an increasing trend in the 1990s. ∎

12.2   Point-level modeling with continuous time

Suppose now that $\mathbf{s} \in \Re^2$ and $t \in \Re^+$ and we seek to define a spatiotemporal process $Y(\mathbf{s}, t)$. As in Subsection 3.1 we have to provide a joint distribution for an uncountable number of random variables. Again, we do this through arbitrary finite dimensional distributions. Confining ourselves to the Gaussian case, we only need to specify a valid spatiotemporal covariance function. Here, "valid" means that for any set of locations and any set of time points, the covariance matrix for the resulting set of random variables is positive definite. An important point here is that it is not sensible to combine $\mathbf{s}$ and t and propose a valid correlation function on $\Re^3$. This is because distance in space has nothing to do with "distance" on the time scale.

As a result, a stationary spatiotemporal covariance specification is assumed to take the form $cov(Y(\mathbf{s}, t), Y(\mathbf{s}', t')) = c(\mathbf{s} - \mathbf{s}', t - t')$. An isotropic form sets $cov(Y(\mathbf{s}, t), Y(\mathbf{s}', t')) = c(\|\mathbf{s} - \mathbf{s}'\|, |t - t'|)$. A frequently used choice is the *separable* form

$$cov(Y(\mathbf{s}, t), Y(\mathbf{s}', t')) = \sigma^2 \rho^{(1)}(\mathbf{s} - \mathbf{s}'; \boldsymbol{\phi}) \, \rho^{(2)}(t - t'; \boldsymbol{\psi}) , \qquad (12.18)$$

where $\rho^{(1)}$ is a valid two-dimensional correlation function and $\rho^{(2)}$ is a valid one-dimensional correlation function. Expression (12.18) shows that dependence attenuates in a multiplicative manner across space and time. Forms such as (12.18) have a long history in spatiotemporal modeling; see, e.g., Mardia and Goodall [1993] and references therein for an early perspective.

Why is (12.18) valid? For locations $\mathbf{s}_1,\ldots,\mathbf{s}_I$ and times $t_1,\ldots,t_J$, collecting the variables a vector $\mathbf{Y}_s^{\mathrm{T}} = (\mathbf{Y}^{\mathrm{T}}(\mathbf{s}_1),\ldots,\mathbf{Y}^{\mathrm{T}}(\mathbf{s}_I))$ where $\mathbf{Y}(\mathbf{s}_i) = (Y(\mathbf{s}_i,t_1),\ldots,Y(\mathbf{s}_i,t_J))^{\mathrm{T}}$, the covariance matrix of $\mathbf{Y}_s$ is

$$\Sigma_{\mathbf{Y}_s}(\sigma^2,\boldsymbol{\phi},\boldsymbol{\psi}) = \sigma^2 H_s(\boldsymbol{\phi}) \otimes H_t(\boldsymbol{\psi}) , \qquad (12.19)$$

where "$\otimes$" again denotes the Kronecker product. In (12.19), $H_s(\boldsymbol{\phi})$ is $I \times I$ with $(H_s(\boldsymbol{\phi}))_{ii'} = \rho^{(1)}(\mathbf{s}_i - \mathbf{s}_i';\boldsymbol{\theta})$, and $H_t(\boldsymbol{\psi})$ is $J \times J$ with $(H_t(\boldsymbol{\psi}))_{jj'} = \rho^{(2)}(t_j - t_{j'};\boldsymbol{\psi})$. Expression (12.19) clarifies that $\Sigma_{\mathbf{Y}_s}$ is positive definite, following the argument below (10.11). So, $\mathbf{Y}_s$ will be IJ-dimensional multivariate normal with, in obvious notation, mean vector $\boldsymbol{\mu}_s(\boldsymbol{\beta})$ and covariance matrix (12.19).

Given a prior for $\boldsymbol{\beta},\sigma^2,\boldsymbol{\phi}$, and $\boldsymbol{\psi}$, the Bayesian model is completely specified. Simulation-based model fitting can be carried out similarly to the static spatial case by noting the following. The log-likelihood arising from $\mathbf{Y}_s$ is

$$-\tfrac{1}{2}\log\left|\sigma^2 H_s(\boldsymbol{\phi}) \otimes H_t(\boldsymbol{\psi})\right|$$
$$-\tfrac{1}{2\sigma^2}(\mathbf{Y}_s - \boldsymbol{\mu}_s(\boldsymbol{\beta}))^{\mathrm{T}}(H_s(\boldsymbol{\phi}) \otimes H_t(\boldsymbol{\psi}))^{-1}(\mathbf{Y}_s - \boldsymbol{\mu}_s(\boldsymbol{\beta})) .$$

But in fact $\left|\sigma^2 H_s(\boldsymbol{\phi}) \otimes H_t(\boldsymbol{\psi})\right| = (\sigma^2)^{IJ}|H_s(\boldsymbol{\phi})|^J|H_t(\boldsymbol{\psi})|^I$ and $(H_s(\boldsymbol{\phi}) \otimes H_t(\boldsymbol{\psi}))^{-1} = H_s^{-1}(\boldsymbol{\phi}) \otimes H_t^{-1}(\boldsymbol{\psi})$ by properties of Kronecker products. In other words, even though (12.19) is $IJ \times IJ$, we need only the determinant and inverse for an $I \times I$ and a $J \times J$ matrix, expediting likelihood evaluation and hence Gibbs sampling.

With regard to prediction, first consider new locations $\mathbf{s}_1',\ldots,\mathbf{s}_K'$ with interest in inference for $Y(\mathbf{s}_k',t_j)$. As with the observed data, we collect the $Y(\mathbf{s}_k',t_j)$ into vectors $\mathbf{Y}(\mathbf{s}_k')$, and the $\mathbf{Y}(\mathbf{s}_k')$ into a single $KJ \times 1$ vector $\mathbf{Y}_{s'}$. Even though we may not necessarily be interested in every component of $\mathbf{Y}_{s'}$, the simplifying forms that follow suggest that, with regard to programming, it may be easiest to simulate draws from the entire predictive distribution $f(\mathbf{Y}_{s'} \mid \mathbf{Y}_s)$ and then retain only the desired components.

Since $f(\mathbf{Y}_{s'}|\mathbf{Y}_s)$ has a form analogous to (7.3), given posterior samples $(\boldsymbol{\beta}_g^*,\sigma_g^{2*},\boldsymbol{\phi}_g^*,\boldsymbol{\psi}_g^*)$, we draw $\mathbf{Y}_{s',g}^*$ from $f(\mathbf{Y}_{s'} \mid \mathbf{Y}_s,\boldsymbol{\beta}_g^*,\sigma_g^{2*},\boldsymbol{\phi}_g^*,\boldsymbol{\psi}_g^*)$, $g = 1,\ldots,G$. Analogous to (7.4),

$$f\left(\begin{pmatrix}\mathbf{Y}_s\\\mathbf{Y}_{s'}\end{pmatrix}\middle|\boldsymbol{\beta},\sigma^2,\boldsymbol{\phi},\boldsymbol{\psi}\right) = N\left(\begin{pmatrix}\boldsymbol{\mu}_s(\boldsymbol{\beta})\\\boldsymbol{\mu}_{s^{\mathrm{T}}}(\boldsymbol{\beta})\end{pmatrix}, \Sigma_{\mathbf{Y}_s,\mathbf{Y}_{s'}}\right) \qquad (12.20)$$

where

$$\Sigma_{\mathbf{Y}_s,\mathbf{Y}_{s'}} = \sigma^2\begin{pmatrix} H_s(\boldsymbol{\phi}) \otimes H_t(\boldsymbol{\psi}) & H_{s,s'}(\boldsymbol{\phi}) \otimes H_t(\boldsymbol{\psi}) \\ H_{s,s'}^{\mathrm{T}}(\boldsymbol{\phi}) \otimes H_t(\boldsymbol{\psi}) & H_{s'}(\boldsymbol{\phi}) \otimes H_t(\boldsymbol{\psi}) \end{pmatrix} ,$$

with obvious definitions for $H_{s'}(\boldsymbol{\phi})$ and $H_{s,s'}(\boldsymbol{\phi})$. But then the conditional distribution $\mathbf{Y}_{s'} \mid \mathbf{Y}_s,\boldsymbol{\beta},\sigma^2,\boldsymbol{\phi},\boldsymbol{\psi}$ is also normal, with mean

$$\boldsymbol{\mu}_{s'}(\boldsymbol{\beta}) + (H_{s,s'}^{\mathrm{T}}(\boldsymbol{\phi}) \otimes H_t(\boldsymbol{\phi}))(H_s(\boldsymbol{\phi}) \otimes H_t(\boldsymbol{\psi}))^{-1}(Y_s - \boldsymbol{\mu}_s(\boldsymbol{\beta}))$$
$$= \boldsymbol{\mu}_{s'}(\boldsymbol{\beta}) + (H_{s,s'}^{\mathrm{T}}(\boldsymbol{\phi})H_s^{-1}(\boldsymbol{\phi}) \otimes I_{J\times J})(\mathbf{Y}_s - \boldsymbol{\mu}_s(\boldsymbol{\beta})) , \qquad (12.21)$$

and covariance matrix

$$H_{s'}(\boldsymbol{\phi}) \otimes H_t(\boldsymbol{\psi})$$
$$-(H_{s,s'}^{\mathrm{T}} \otimes H_t(\boldsymbol{\psi}))(H_s(\boldsymbol{\phi}) \otimes H_t(\boldsymbol{\psi}))^{-1}(H_{s,s'}(\boldsymbol{\phi}) \otimes H_t(\boldsymbol{\psi}))$$
$$= (H_{s'}(\boldsymbol{\phi}) - H_{s,s'}^{\mathrm{T}}(\boldsymbol{\phi})H_s^{-1}(\boldsymbol{\phi})H_{s,s}(\boldsymbol{\phi})) \otimes H_t(\boldsymbol{\psi}) , \qquad (12.22)$$

using standard properties of Kronecker products.

In (12.21), time disappears apart from $\boldsymbol{\mu}_{s'}(\boldsymbol{\beta})$, while in (12.22), time "factors out" of the conditioning. Sampling from this normal distribution usually employs the inverse square root of the conditional covariance matrix, but conveniently, this is

$$(H_{s'}(\boldsymbol{\phi}) - H_{s,s'}^{\mathrm{T}}(\boldsymbol{\phi})H_s^{-1}(\boldsymbol{\phi})H_{s,s'}(\boldsymbol{\phi}))^{-\frac{1}{2}} \otimes H_t^{-\frac{1}{2}}(\boldsymbol{\psi}) ,$$

so the only work required beyond that in (7.5) is obtaining $H_t^{-\frac{1}{2}}(\psi)$, since $H_t^{-1}(\psi)$ will already have been obtained in evaluating the likelihood, following the discussion above.

For prediction not for points but for areal units (blocks) $B_1,\ldots,B_K$, we would set $\mathbf{Y}^{\mathrm{T}}(B_k) = (Y(B_k,t_1),\ldots,Y(B_k,t_J))$ and then further set $\mathbf{Y}_B^{\mathrm{T}} = (\mathbf{Y}^{\mathrm{T}}(B_1),\ldots,\mathbf{Y}^{\mathrm{T}}(B_K))$. Analogous to (7.6) we seek to sample $f(\mathbf{Y}_B \mid \mathbf{Y}_s)$, so we require $f(\mathbf{Y}_B \mid \mathbf{Y}_s,\boldsymbol{\beta},\sigma^2,\boldsymbol{\phi},\boldsymbol{\psi})$. Analogous to (12.20), this can be derived from the joint distribution $f((\mathbf{Y}_s,\mathbf{Y}_B)^{\mathrm{T}} \mid \boldsymbol{\beta},\sigma^2,\boldsymbol{\phi},\boldsymbol{\psi})$, which is

$$N\left(\begin{pmatrix}\boldsymbol{\mu}_s(\boldsymbol{\beta})\\\boldsymbol{\mu}_B(\boldsymbol{\beta})\end{pmatrix}, \sigma^2\begin{pmatrix}H_s(\boldsymbol{\phi})\otimes H_t(\boldsymbol{\psi}) & H_{s,B}(\boldsymbol{\phi})\otimes H_t(\boldsymbol{\psi})\\H_{s,B}^{\mathrm{T}}(\boldsymbol{\phi})\otimes H_t(\boldsymbol{\psi}) & H_B(\boldsymbol{\phi})\otimes H_t(\boldsymbol{\psi})\end{pmatrix}\right),$$

with $\boldsymbol{\mu}_B(\boldsymbol{\beta})$, $H_B(\boldsymbol{\phi})$, and $H_{s,B}(\boldsymbol{\phi})$ defined as in Section 7.1.2. Thus the distribution $f(\mathbf{Y}_B|\mathbf{Y}_s,\boldsymbol{\beta},\sigma^2,\boldsymbol{\phi},\boldsymbol{\psi})$ is again normal with mean and covariance matrix as given in (12.21) and (12.22), but with $\boldsymbol{\mu}_B(\boldsymbol{\beta})$ replacing $\boldsymbol{\mu}_{s'}(\boldsymbol{\beta})$, $H_B(\boldsymbol{\phi})$ replacing $H_{s'}(\boldsymbol{\phi})$, and $H_{s,B}(\boldsymbol{\phi})$ replacing $H_{s,s'}(\boldsymbol{\phi})$. Using the same Monte Carlo integrations as proposed in Section 7.1.2 leads to sampling the resultant $\widehat{f}(\mathbf{Y}_B \mid \mathbf{Y}_s,\boldsymbol{\beta},\sigma^2,\boldsymbol{\phi},\boldsymbol{\psi})$, and the same technical justification applies.

If we started with block data, $Y(B_i,t_j)$, then following (7.11) and (12.19),

$$f(\mathbf{Y}_B \mid \boldsymbol{\beta},\sigma^2,\boldsymbol{\phi},\boldsymbol{\psi}) = N(\mu_B(\boldsymbol{\beta}), \sigma^2(H_B(\boldsymbol{\phi})\otimes H_t(\boldsymbol{\psi}))). \tag{12.23}$$

Given (12.23), the path for prediction at new points or at new blocks is clear, following the above and the end of Section 7.1.2; we omit the details.

Note that the association structure in (12.18) allows *forecasting* of the spatial process at time t_{J+1}. This can be done at observed or unobserved points or blocks following the foregoing development. To retain the above simplifying forms, we would first simulate the variables at t_{J+1} associated with observed points or blocks (with no change of support). We would then revise $H_t(\boldsymbol{\phi})$ to be $(J+1)\times(J+1)$ before proceeding as above.

Example 12.2 To illustrate the methods above, we use a spatiotemporal version of the Atlanta ozone data set. As mentioned in Section 7.1, we actually have ozone measurements at the 10 fixed monitoring stations shown in Figure 1.3 over the 92 summer days in 1995. Figure 12.4 shows the daily 1-hour maximum ozone reading for the sites during July of this same year. There are several sharp peaks, but little evidence of a weekly (7-day) period in the data. The mean structure appears reasonably constant in space, with the ordering of the site measurements changing dramatically for different days. Moreover, with only 10 "design points" in the metro area, any spatial trend surface we fit would be quite speculative over much of the study region (e.g., the northwest and southwest metro; see Figure 1.3). The temporal evolution of the series is not inconsistent with a constant mean autoregressive error model; indeed, the lag 1 sample autocorrelation varies between .27 and .73 over the 10 sites, strongly suggesting the need for a model accounting for both spatial and temporal correlations.

We thus fit our spatiotemporal model with mean $\boldsymbol{\mu}(\mathbf{s},t;\boldsymbol{\beta}) = \mu$, but with spatial and temporal correlation functions $\rho^{(1)}(\mathbf{s}_i - \mathbf{s}_{i'};\phi) = e^{-\phi\|\mathbf{s}_i-\mathbf{s}_{i'}\|}$ and $\rho^{(2)}(t_j - t_{j'};\psi) = \psi^{|j-j'|}$. Hence our model has four parameters: we use a flat prior for μ, an $IG(3,0.5)$ prior for σ^2, a $G(0.003,100)$ prior for ϕ, and a $U(0,1)$ prior for ψ (thus eliminating the implausible possibility of *negative* autocorrelation in our data, but favoring no positive value over any other). To facilitate our Gibbs-Metropolis approach, we transform to $\theta = \log\phi$ and $\lambda = \log(\psi/(1-\psi))$, and subsequently use Gaussian proposals on these transformed parameters.

Posterior medians and 95% equal-tail credible intervals for the four parameters are as follows: for μ, 0.068 and (0.057, 0.080); for σ^2, 0.11 and (0.08, 0.17); for ϕ, 0.06 and (0.03, 0.08); and for ψ, 0.42 and (0.31, 0.52). The rather large value of ψ confirms the strong temporal autocorrelation suspected in the daily ozone readings.

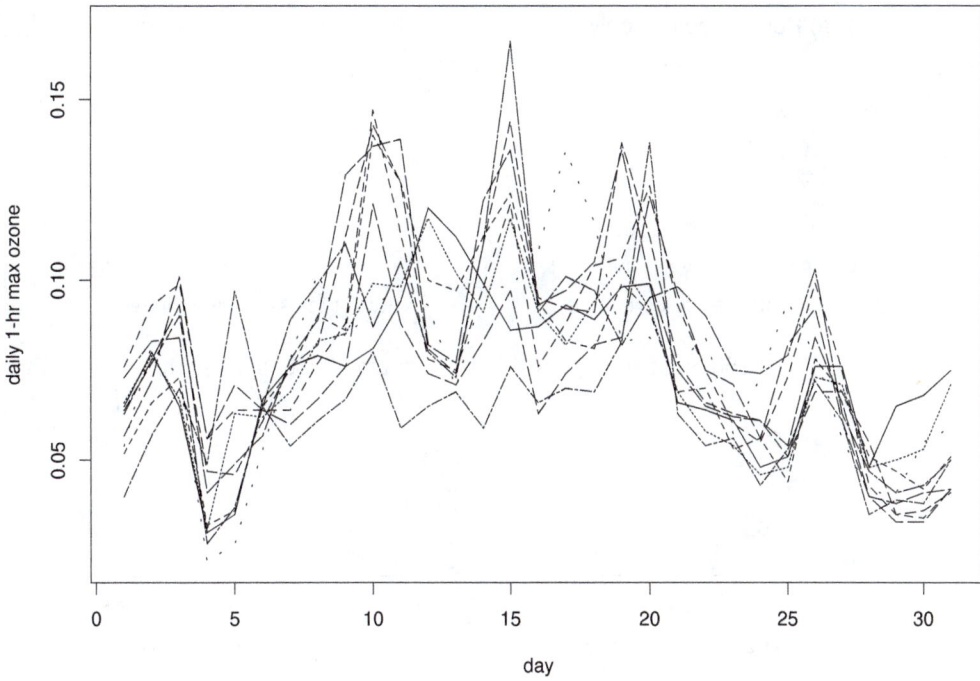

Figure 12.4 *Observed 1-hour maximum ozone measurement by day, July 1995, 10 Atlanta monitoring sites.*

| | Spatial only | | Spatiotemporal | |
	Point	95% Interval	Point	95% Interval
Point A	.125	(.040, .334)	.139	(.111, .169)
Point B	.116	(.031, .393)	.131	(.098, .169)
Zip 30317 (east-central)	.130	(.055, .270)	.138	(.121, .155)
Zip 30344 (south-central)	.123	(.055, .270)	.135	(.112, .161)
Zip 30350 (north)	.112	(.040, .283)	.109	(.084, .140)

Table 12.4 *Posterior medians and 95% equal-tail credible intervals for ozone levels at two points, and for average ozone levels over three blocks (zip codes), purely spatial model versus spatiotemporal model, Atlanta ozone data for July 15, 1995.*

Comparison of the posteriors for σ^2 and ϕ with those obtained for the static spatial model in Example 7.1 is not sensible, since these parameters have different meanings in the two models. Instead, we make this comparison in the context of point-point and point-block prediction. Table 12.4 provides posterior predictive summaries for the ozone concentrations for July 15, 1995, at points A and B (see Figure 1.3), as well as for the block averages over 3 selected Atlanta city zips: 30317, an east-central city zip very near two monitoring sites; 30344, the south-central zip containing points A and B; and 30350, the northernmost city zip. Results are shown for both the spatiotemporal model of this subsection and for the static spatial model previously fit in Example 7.1. Note that all the posterior medians are a bit higher under the spatiotemporal model, except for that for the northern zip, which remains low. Also note the significant increase in precision afforded by this model, which makes use of the data from all 31 days in July 1995, instead of only that from July 15. Figure 12.5 shows the estimated posteriors giving rise to the first and last rows in Table 12.4 (i.e., corresponding to the July 15, 1995, ozone levels at point A and the block average over

a) Point A, spatial only

Median: 0.125, 95% CI: (0.040, 0.334)

b) Point A, spatiotemporal

Median: 0.139, 95% CI: (0.111, 0.169)

c) Zip 30350, spatial only

Median: 0.112, 95% CI: (0.040, 0.283)

d) Zip 30350, spatiotemporal

Median: 0.109, 95% CI: (0.084, 0.140)

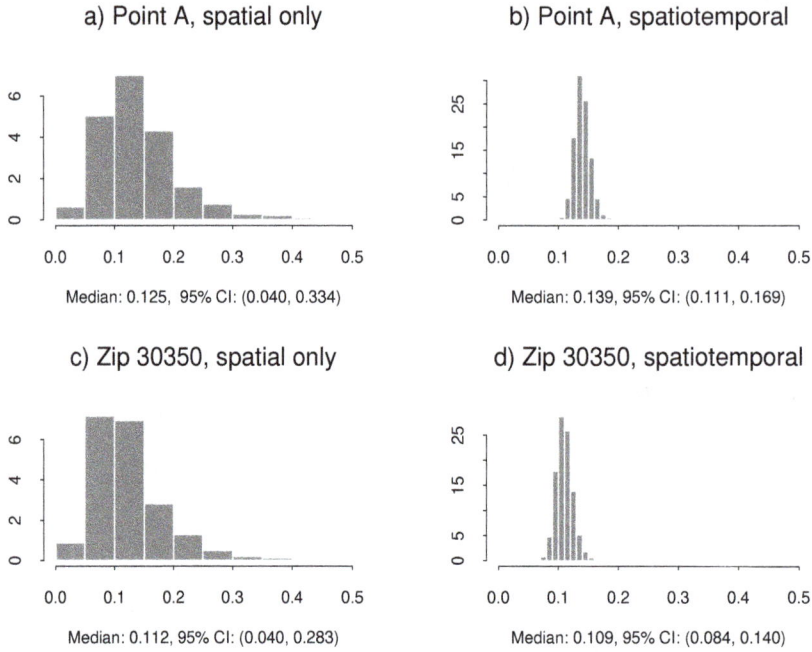

Figure 12.5 *Posterior predictive distributions for ozone concentration at point A and the block average over zip 30350, purely spatial model versus spatiotemporal model, Atlanta ozone data for July 15, 1995.*

the northernmost city zip, 30350). The ability of the Bayesian approach to reflect differing amounts of predictive uncertainty for the two models is clearly evident.

Finally, Figure 12.6 plots the posterior medians and upper and lower 0.025 quantiles produced by the spatiotemporal model by day for the ozone concentration at point A, as well as those for the block average in zip 30350. Note that the overall temporal pattern is quite similar to that for the data shown in Figure 12.4. Since point A is rather nearer to several data observation points, the confidence bands associated with it are often a bit narrower than those for the northern zip, but this pattern is not perfectly consistent over time. Also note that the relative positions of the bands for July 15 are consistent with the data pattern for this day seen in Figure 1.3, when downtown ozone exposures were higher than those in the northern metro. Finally, the day-to-day variability in the predicted series is substantially larger than the predictive variability associated with any given day. ∎

12.3 Nonseparable spatiotemporal models ⋆

The separable form for the spatiotemporal covariance function in (12.18) is convenient for computation and offers attractive interpretation. However, its form limits the nature of space-time interaction. Additive forms, arising from $w(\mathbf{s}, t) = w(\mathbf{s}) + \alpha(t)$ with $w(\mathbf{s})$ and $\alpha(t)$ independent may be even more unsatisfying.

A simple way to extend (12.18) is through *mixing*. For instance, suppose $w(\mathbf{s}, t) = w_1(\mathbf{s}, t) + w_2(\mathbf{s}, t)$ with w_1 and w_2 independent processes, each with a separable spatiotemporal covariance function, say $c_\ell(\mathbf{s} - \mathbf{s}', t - t) = \sigma_\ell^2 \rho_\ell^{(1)}(\mathbf{s} - \mathbf{s}') \rho_\ell^{(2)}(t - t')$, $\ell = 1, 2$. Then the covariance function for $w(\mathbf{s}, t)$ is evidently the sum and is not separable. Building covariance functions in this way is easy to interpret but yields an explosion of parameters with finite

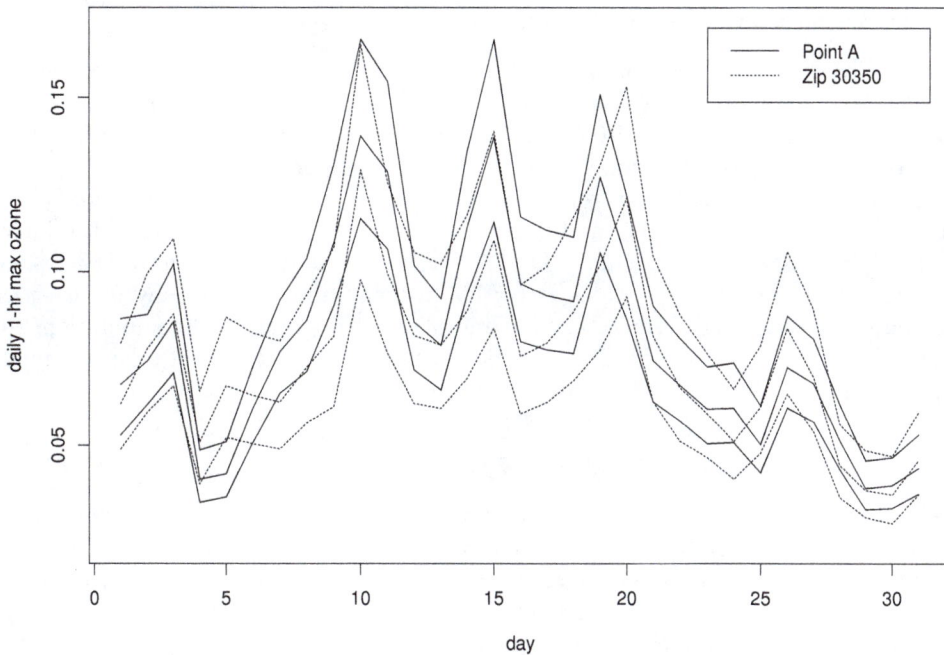

Figure 12.6 *Posterior medians and upper and lower .025 quantiles for the predicted 1-hour maximum ozone concentration by day, July 1995; solid lines, point A; dotted lines, block average over zip 30350 (northernmost Atlanta city zip).*

mixing. Continuous parametric mixing, e.g.,

$$c(\mathbf{s} - \mathbf{s}', t - t) = \sigma^2 \int \rho^{(1)}(\mathbf{s} - \mathbf{s}', \boldsymbol{\phi}) \rho^{(2)}(t - t', \boldsymbol{\psi}) \, G_{\boldsymbol{\gamma}}(d\boldsymbol{\phi}, d\boldsymbol{\psi}) \,, \qquad (12.24)$$

yields a function that depends only on σ^2 and $\boldsymbol{\gamma}$. Extensions of these ideas are developed in De Iaco et al. [2002]. However, these forms have not received much attention in the literature to date.

Cressie and Huang [1999] introduce a flexible class of nonseparable stationary covariance functions that allow for space-time interaction. However, they work in the spectral domain and require that $c(\mathbf{s} - \mathbf{s}', t - t')$ can be computed explicitly, i.e., the Fourier inversion can be obtained in closed-form. Unfortunately this occurs only in very special cases. Gneiting [2002a] adopts a similar approach but obtains very general classes of valid space-time models that do not rely on closed form Fourier inversions. One simple example is the class $c(\mathbf{s} - \mathbf{s}', t - t') = \sigma^2(|t - t'| + 1)^{-1} \exp(-\|\mathbf{s} - \mathbf{s}'\|(|t - t'| + 1)^{-\beta/2})$. Here, β is a space-time interaction parameter; $\beta = 0$ provides a separable specification.

Stein [2005] also works in the spectral domain, providing a class of spectral densities whose resulting spatiotemporal covariance function is nonseparable with flexible analytic behavior. These spectral densities extend the Matérn form as in (3.5).

In particular, the spectral density is

$$\widehat{c}(\mathbf{w}, v) \propto [c_1(\alpha_1^2 + \|\mathbf{w}\|^2)^{\alpha_1} + c_2(\alpha_2 + v^2)^{\alpha_2}]^{-v} \,.$$

Unfortunately, the associated covariance function cannot be computed explicitly; fast Fourier transforms offer the best computational prospects. Also, unlike Gneiting's class, separability does not arise as a special or limiting case. For further discussion of the above see Chapter 23 of Gelfand et al. [2010]. We also mention related work using "blurring" discussed in Brown et al. [2002].

12.4   Dynamic spatiotemporal models ⋆

In this section we follow the approach taken in Gelfand et al. [2005b], viewing the data
as arising from a time series of spatial processes. In particular, we work in the setting of
dynamic models [West and Harrison, 1997], describing the temporal evolution in a latent
space. We achieve a class of dynamic models for spatiotemporal data.

Here, there is a growing literature. Non-Bayesian approaches include Huang and Cressie
[1996], Wikle and Cressie [1999], and Mardia et al. [1998]. Bayesian approaches include
Tonellato [1998], Sansó and Guenni [1999], Stroud et al. [2001], and Huerta et al. [2004].
The paper by Stroud et al. [2001] is attractive in being applicable to any data set that
is continuous in space and discrete in time and allows straightforward computation using
Kalman filtering. Banerjee et al. [in press] demonstrate dynamic spatial-temporal models
for emulation and calibration of mechanistic systems.

12.4.1   Brief review of dynamic linear models

Dynamic linear models, often referred to as state-space models in the time-series literature,
offer a versatile framework for fitting several time-varying models [West and Harrison, 1997].
We briefly outline the general dynamic linear modeling framework. Thus, let $\mathbf{Y}_t$ be a $m \times 1$
vector of observables at time t. $\mathbf{Y}_t$ is related to a $p \times 1$ vector, $\boldsymbol{\theta}_t$, called the state vector,
through a *measurement equation*. In general, the elements of $\boldsymbol{\theta}_t$ are not observable, but are
generated by a first-order Markovian process, resulting in a *transition equation*. Therefore,
we can describe the above framework as

$$\mathbf{Y}_t = F_t \boldsymbol{\theta}_t + \boldsymbol{\epsilon}_t, \; \boldsymbol{\epsilon}_t \sim N\left(\mathbf{0}, \Sigma_t^\epsilon\right).$$
$$\boldsymbol{\theta}_t = G_t \boldsymbol{\theta}_{t-1} + \boldsymbol{\eta}_t, \; \boldsymbol{\eta}_t \sim N\left(\mathbf{0}, \Sigma_t^\eta\right),$$

where F_t and G_t are $m \times p$ and $p \times p$ matrices, respectively. The first equation is the measurement equation, where $\boldsymbol{\epsilon}_t$ is a $m \times 1$ vector of serially uncorrelated Gaussian variables with
mean $\mathbf{0}$ and an $m \times m$ covariance matrix, Σ_t^ϵ. The second equation is the transition equation
with $\boldsymbol{\eta}_t$ being a $p \times 1$ vector of serially uncorrelated zero-centered Gaussian disturbances
and Σ_t^η the corresponding $p \times p$ covariance matrix. Note that under (12.25), the association
structure can be computed explicitly across time, e.g., $\text{Cov}\left(\boldsymbol{\theta}_t, \boldsymbol{\theta}_{t-1}\right) = G_t \text{Var}\left(\boldsymbol{\theta}_{t-1}\right)$ and
$\text{Cov}\left(\mathbf{Y}_t, \mathbf{Y}_{t-1}\right) = F_t G_t \text{Var}\left(\boldsymbol{\theta}_{t-1}\right) F_t^\mathrm{T}$.

F_t (in the measurement equation) and G_t (in the transition equation) are referred to as
system matrices that may change over time. F_t and G_t may involve unknown parameters
but, given the parameters, temporal evolution is in a predetermined manner. The matrix
F_t is usually specified by the design of the problem at hand, while G_t is specified through
modeling assumptions; for example, $G_t = I_p$, the $p \times p$ identity matrix would provide a
random walk for $\boldsymbol{\theta}_t$. Regardless, the system is linear, and for any time point t, $\mathbf{Y}_t$ can be
expressed as a linear combination of the present $\boldsymbol{\epsilon}_t$ and the present and past $\boldsymbol{\eta}_t$'s.

12.4.2   Formulation for spatiotemporal models

In this section we adapt the above dynamic modeling framework to univariate spatiotemporal models with spatially varying coefficients. For this we consider a collection of sites
$S = \{\mathbf{s}_1, ..., \mathbf{s}_{N_s}\}$, and time-points $T = \{t_1, ..., t_{N_t}\}$, yielding observations $Y(\mathbf{s}, t)$, and covariate vectors $\mathbf{x}(\mathbf{s}, t)$, for every $(\mathbf{s}, t) \in S \times T$.

The response, $Y(\mathbf{s}, t)$, is first modeled through a measurement equation, which incorporates the measurement error, $\epsilon(\mathbf{s}, t)$, as serially and spatially uncorrelated zero-centered
Gaussian disturbances. The transition equation now involves the regression parameters
(slopes) of the covariates. The slope vector, say $\tilde{\boldsymbol{\beta}}(\mathbf{s}, t)$, is decomposed into a purely temporal component, $\boldsymbol{\beta}_t$, and a spatiotemporal component, $\boldsymbol{\beta}(\mathbf{s}, t)$. Both these are generated
through transition equations, capturing their Markovian dependence in time. While the

transition equation of the purely temporal component is as in usual state-space modeling, the spatiotemporal component is generated by a multivariate Gaussian spatial process. Thus, we may write the spatiotemporal modeling framework as

$$Y(\mathbf{s},t) = \mu(\mathbf{s},t) + \epsilon(\mathbf{s},t)\,;\; \epsilon(\mathbf{s},t) \overset{ind}{\sim} N\left(0,\sigma^2\right), \qquad (12.25)$$

$$\mu(\mathbf{s},t) = \mathbf{x}^{\mathrm{T}}(\mathbf{s},t)\,\tilde{\boldsymbol{\beta}}(\mathbf{s},t),$$

$$\tilde{\boldsymbol{\beta}}(\mathbf{s},t) = \boldsymbol{\beta}_t + \boldsymbol{\beta}(\mathbf{s},t), \qquad (12.26)$$

$$\boldsymbol{\beta}_t = \boldsymbol{\beta}_{t-1} + \boldsymbol{\eta}_t,\; \boldsymbol{\eta}_t \overset{ind}{\sim} N_p\left(\mathbf{0},\Sigma_{\boldsymbol{\eta}}\right),$$

$$\text{and } \boldsymbol{\beta}(\mathbf{s},t) = \boldsymbol{\beta}(\mathbf{s},t-1) + \mathbf{w}(\mathbf{s},t)\,.$$

In (12.26), we introduce a linear model of regionalization (Section 10.6) for $\mathbf{w}(\mathbf{s},t)$, i.e., $\mathbf{w}(\mathbf{s},t) = A\mathbf{v}(\mathbf{s},t)$, with $\mathbf{v}(\mathbf{s},t) = (v_1(\mathbf{s},t),...,v_p(\mathbf{s},t))^{\mathrm{T}}$, yielding $\Sigma_w = AA^{\mathrm{T}}$. The $v_l(\mathbf{s},t)$ are serially independent replications of a Gaussian process with unit variance and correlation function $\rho_l(\,\cdot\,;\phi_l)$, henceforth denoted by $GP(0,\rho_l(\,\cdot\,;\phi_l))$, for $l = 1,\ldots,p$ and independent across l. In the current context, we assume that A does not depend upon $(\mathbf{s},t)$. Nevertheless, this still allows flexible modeling for the spatial covariance structure, as we discuss below.

Moreover, allowing a spatially varying coefficient $\boldsymbol{\beta}(\mathbf{s},t)$ to be associated with $\mathbf{x}(\mathbf{s},t)$ provides an arbitrarily rich explanatory relationship for the x's with regard to the Y's (see Section 10.7 in this regard). By comparison, in Stroud et al. [2001] at a given t, a locally weighted mixture of linear regressions is proposed and only the purely temporal component of $\tilde{\boldsymbol{\beta}}(\mathbf{s},t)$ is used. Such a specification requires both the number of basis functions and the number of mixture components.

Returning to our specification, note that if $v_l(\cdot,t) \overset{ind}{\sim} GP(0,\rho(\cdot;\phi))$, we have the intrinsic or separable model for $w(\mathbf{s},t)$. Allowing different correlation functions and decay parameters for the $v_l(\mathbf{s},t)$, i.e., $v_l(\cdot,t) \overset{ind}{\sim} GP(0,\rho_l(\cdot;\phi_l))$ yields the linear model of coregionalization (Section 10.6).

Following Section 12.4.1, we can compute the general association structure for the Y's under (12.25) and (12.26). For instance, we have the result that

$$\text{Cov}\left(Y(\mathbf{s},t),Y(\mathbf{s}',t-1)\right) = \mathbf{x}^{\mathrm{T}}(\mathbf{s},t)\,\Sigma_{\tilde{\boldsymbol{\beta}}(\mathbf{s},t),\tilde{\boldsymbol{\beta}}(\mathbf{s}',t-1)}\,\mathbf{x}(\mathbf{s},t-1)\,,$$

where $\Sigma_{\tilde{\boldsymbol{\beta}}(\mathbf{s},t),\tilde{\boldsymbol{\beta}}(\mathbf{s}',t-1)} = (t-1)\left(\Sigma_{\boldsymbol{\eta}} + \sum_{l=1}^{p}\rho_l(\mathbf{s}-\mathbf{s}';\phi_l)\,\mathbf{a}_l\mathbf{a}_l^{\mathrm{T}}\right)$. Furthermore,

$$\text{Var}\left(Y(\mathbf{s},t)\right) = \mathbf{x}^{\mathrm{T}}(\mathbf{s},t)\,t\left[\Sigma_{\boldsymbol{\eta}} + AA^{\mathrm{T}}\right]\mathbf{x}(\mathbf{s},t)$$

with the result that $\text{Corr}\left(Y(\mathbf{s},t),Y(\mathbf{s}',t-1)\right) = O(1)$ as $t \to \infty$.

A Bayesian hierarchical model for (12.25) and (12.26) may be completed by prior specifications such as

$$\boldsymbol{\beta}_0 \sim N(\mathbf{m}_0,C_0) \text{ and } \boldsymbol{\beta}(\cdot,0) \equiv 0. \qquad (12.27)$$

$$\Sigma_{\boldsymbol{\eta}} \sim IW\left(a_{\boldsymbol{\eta}},B_{\boldsymbol{\eta}}\right),\; \Sigma_{\mathbf{w}} \sim IW\left(a_{\mathbf{w}},B_{\mathbf{w}}\right) \text{ and } \sigma_{\epsilon}^2 \sim IG\left(a_{\epsilon},b_{\epsilon}\right),$$

$$\mathbf{m}_0 \sim N(\mathbf{0},\Sigma_0)\,;\; \Sigma_0 = 10^5 \times I_p\,,$$

where B_{η} and $B_{\mathbf{w}}$ are $p \times p$ precision (hyperparameter) matrices for the inverted Wishart distribution.

Consider now data, in the form $(Y(\mathbf{s}_i,t_j))$ with $i = 1,2,...,N_s$ and $j = 1,2,...,N_t$. Let us collect, for each time point, the observations on all the sites. That is, we form, $\mathbf{Y}_t = (Y(\mathbf{s}_1,t),...,Y(\mathbf{s}_{N_s},t))^{\mathrm{T}}$ and the $N_s \times N_s p$ block diagonal matrix $F_t = (\mathbf{x}^{\mathrm{T}}(\mathbf{s}_1,t),\mathbf{x}^{\mathrm{T}}(\mathbf{s}_2,t),...,\mathbf{x}^{\mathrm{T}}(\mathbf{s}_N,t))$ for $t = t_1,...,t_{N_t}$. ^{T}. Analogously we form the $N_s p \times 1$

vector $\boldsymbol{\theta}_t = \mathbf{1}_{N_s} \otimes \boldsymbol{\beta}_t + \boldsymbol{\beta}_t^*$, where $\boldsymbol{\beta}_t^* = (\boldsymbol{\beta}\,(\mathbf{s}_1, t), \ldots, \boldsymbol{\beta}\,(\mathbf{s}_{N_s}, t))^{\mathrm{T}}$, $\boldsymbol{\beta}_t = \boldsymbol{\beta}_{t-1} + \boldsymbol{\eta}_t$, $\boldsymbol{\eta}_t \stackrel{ind}{\sim}$ $N_p\,(\mathbf{0}, \Sigma_{\boldsymbol{\eta}})$; and, with $\mathbf{w}_t = (\mathbf{w}^{\mathrm{T}}\,(\mathbf{s}_1, t), \ldots, \mathbf{w}^{\mathrm{T}}\,(\mathbf{s}_{N_s}, t))^{\mathrm{T}}$,

$$\boldsymbol{\beta}_t^* = \boldsymbol{\beta}_{t-1}^* + \mathbf{w}_t, \ \mathbf{w}_t \stackrel{ind}{\sim} N\left(\mathbf{0}, \sum_{l=1}^{p} \left(R_l\,(\phi_l) \otimes \Sigma_{\mathbf{w}, l}\right)\right),$$

where $[R_l(\phi_l)]_{ij} = \rho_l(\mathbf{s}_i - \mathbf{s}_j; \phi_l)$ is the correlation matrix for $v_l(\cdot, t)$. We then write the data equation for a dynamic spatial model as

$$\mathbf{Y}_t = F_t \boldsymbol{\theta}_t + \boldsymbol{\epsilon}_t; \ t = 1, \ldots, N_t; \ \boldsymbol{\epsilon}_t \sim N\left(\mathbf{0}, \sigma_\epsilon^2 I_{N_s}\right).$$

With the prior specifications in (12.27), we can design a Gibbs sampler with Gaussian full conditionals for the temporal coefficients $\{\boldsymbol{\beta}_t\}$, the spatiotemporal coefficients $\{\boldsymbol{\beta}_t^*\}$, inverted Wishart for $\Sigma_{\boldsymbol{\eta}}$, and Metropolis steps for ϕ and the elements of $\Sigma_{\mathbf{w}, l}$. Updating of $\Sigma_{\mathbf{w}} = \sum_{l=1}^{p} \Sigma_{\mathbf{w}, l}$ is most efficiently done by reparametrizing the model in terms of the matrix square root of $\Sigma_{\mathbf{w}}$, say A, and updating the elements of the lower triangular matrix A. To be precise, consider the full conditional distribution,

$$f\,(\Sigma_{\mathbf{w}} \mid \boldsymbol{\gamma}, \phi_1, \phi_2) \propto f\,(\Sigma_{\mathbf{w}} \mid a_\gamma, B_\gamma)\, \frac{1}{\left|\sum_{l=1}^{p} R_l(\phi_l) \otimes \Sigma_{\mathbf{w}, l}\right|}$$
$$\times \exp\left(-\tfrac{1}{2}\boldsymbol{\beta}^{*T}\left(J^{-1} \otimes \left(\sum_{l=1}^{p} R_l\,(\phi_l) \otimes \Sigma_{\mathbf{w}, l}\right)^{-1}\right)\boldsymbol{\beta}^*\right).$$

The one-to-one relationship between elements of $\Sigma_{\mathbf{w}}$ and the Cholesky square root A is well known [Harville, 2000, Banerjee and Roy, 2014, see, e.g.,]. So, we reparametrize the above full conditional as

$$f\,(A \mid \boldsymbol{\gamma}, \phi_1, \phi_2) \propto f\,(h\,(A) \mid a_\gamma, B_\gamma)\left|\frac{\partial h}{\partial a_{ij}}\right| \frac{1}{\left|\sum_{l=1}^{p} R_l(\phi_l) \otimes (\mathbf{a}_l \mathbf{a}_l^{\mathrm{T}})\right|}$$
$$\times \exp\left(-\tfrac{1}{2}\boldsymbol{\beta}^{*T}\left(J^{-1} \otimes \left(\sum_{l=1}^{p} R_l\,(\phi_l) \otimes (\mathbf{a}_l \mathbf{a}_l^{\mathrm{T}})\right)^{-1}\right)\boldsymbol{\beta}^*\right).$$

Here, h is the function taking the elements of A, say a_{ij}, to those of the symmetric positive definite matrix $\Sigma_{\mathbf{w}}$. In the 2×2 case we have

$$h\,(a_{11}, a_{21}, a_{22}) = \left(a_{11}^2, a_{11}a_{21}, a_{21}^2 + a_{22}^2\right),$$

and the Jacobian is $4a_{11}^2 a_{22}$. Now, the elements of A are updated with univariate random-walk Metropolis proposals: lognormal or gamma for a_{11} and a_{22}, and normal for a_{21}. Additional computational burden is created, since now the likelihood needs to be computed for each of the three updates, but the chains are much better tuned (by controlling the scale of the univariate proposals) to move around the parameter space, thereby leading to better convergence behavior.

Example 12.3 (*Modeling temperature given precipitation*). Our spatial domain, shown in Figure 12.7 along with elevation contours (in 100-m units), provides a sample of 50 locations (indicated by "+") in the state of Colorado. Each site provides information on monthly maximum temperature, and monthly mean precipitation. We denote the temperature summary in location $\mathbf{s}$ at time t, by $Y\,(\mathbf{s}, t)$, and the precipitation by $x\,(\mathbf{s}, t)$. Forming a covariate vector $\mathbf{x}^{\mathrm{T}}\,(\mathbf{s}, t) = (1, x\,(\mathbf{s}, t))$, we analyze the data using a coregionalized dynamic model, as outlined in Subsection 12.4.2. As a result, we have an intercept process $\tilde{\beta}_0\,(\mathbf{s}, t)$ and a slope process $\tilde{\beta}_1\,(\mathbf{s}, t)$, and the two processes are dependent.

Figure 12.8 displays the time-varying intercepts and slopes (coefficient of precipitation). As expected, the intercept is higher in the summer months and lower in the winter months, highest in July, lowest in December. In fact, the gradual increase from January to July, and the subsequent decrease toward December is evident from the plot. Precipitation seems to

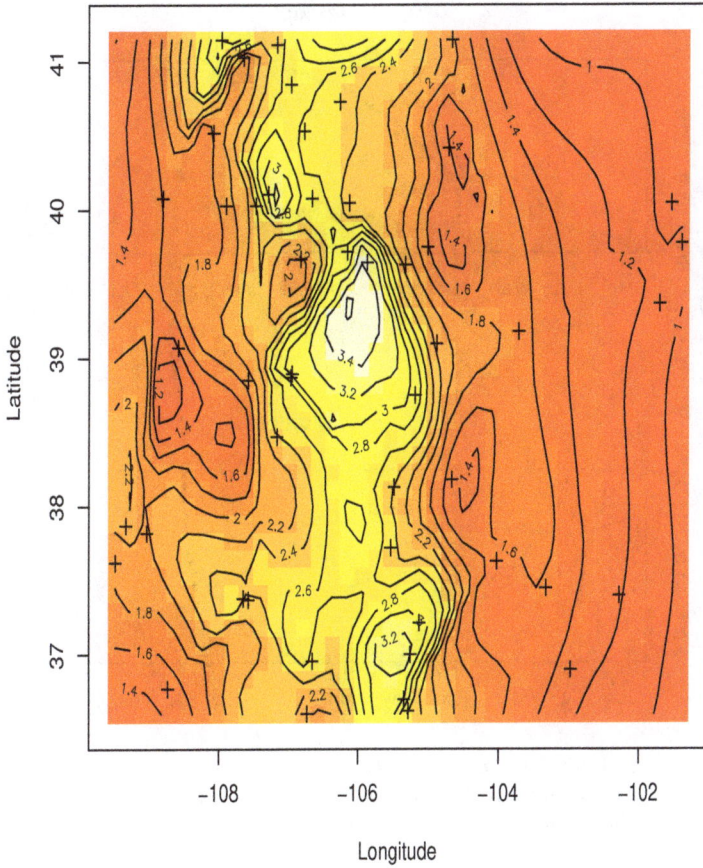

Figure 12.7 *Map of the region in Colorado that forms the spatial domain. The data for the illustrations come from 50 locations, marked by "+" signs in this region.*

have a negative impact on temperature, although this seems to be significant only in the months of January, March, May, June, November, and December, i.e., seasonal pattern is retrieved although no such structure is imposed.

Table 12.5 displays the credible intervals for elements of the $\Sigma_{\boldsymbol{\eta}}$ matrix. Rows 1 and 2 show the medians and credible intervals for the respective *variances*; while Row 3 shows the *correlation*. The corresponding results for the elements of $\Sigma_{\mathbf{w}}$ are given in Table 12.6. A significant negative correlation is seen between the intercept and the slope processes, justifying our use of dependent processes. Next, in Table 12.7, we provide the measurement error variances for temperature along with the estimates of the spatial correlation parameters for the intercept and slope process. Also presented are the ranges implied by ϕ_1 and ϕ_2 for the marginal intercept process, $w_1(\mathbf{s})$, and the marginal slope process, $w_2(\mathbf{s})$. The first range is computed by solving for the distance d, $\rho_1(\phi_1, d) = 0.05$, while the second range is obtained by solving $\left(a_{21}^2 \exp\left(-\phi_1 d\right) + a_{22}^2 \exp\left(-\phi_2 d\right)\right) / \left(a_{21}^2 + a_{22}^2\right) = 0.05$. The ranges are presented in units of 100 km with the maximum observed distance between our sites being approximately 742 km.

Finally, Figure 12.9 displays the time-sliced image-contour plots for the slope process; similar figures can be drawn for the intercept process. For both processes, the spatial variation is better captured in the central and western edges of the domain. In Figure 12.9, all the months display broadly similar spatial patterns, with denser contour variations toward the west than the east. However, the spatial pattern does seem to be more pronounced in

intercept

precipitation

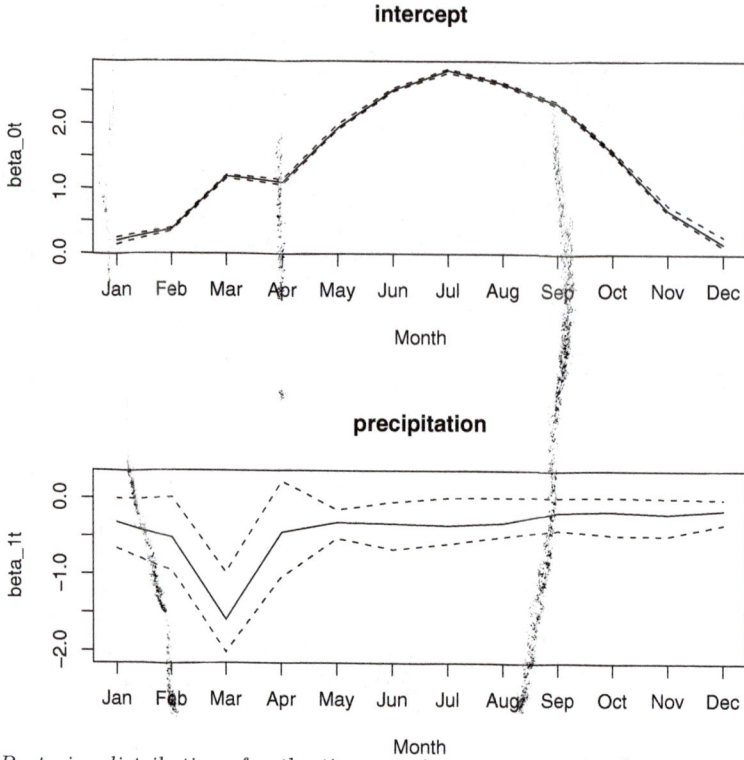

Figure 12.8 *Posterior distributions for the time-varying parameters in the temperature given precipitation example. The top graph corresponds to the intercept, while the lower one is the coefficient of precipitation. Solid lines represent the medians while the dashed lines correspond to the upper and lower credible intervals.*

Σ_η	Median (2.5%, 97.5%)
$\Sigma_\eta\,[1,1]$	0.296 (0.130, 0.621)
$\Sigma_\eta\,[2,2]$	0.786 (0.198, 1.952)
$\Sigma_\eta\,[1,2]\,/\sqrt{\Sigma_\eta\,[1,1]\,\Sigma_\eta\,[2,2]}$	-0.562 (-0.807, -0.137)

Table 12.5 *Estimates of the variances and correlation from Σ_η, dynamic spatiotemporal modeling example.*

$\Sigma_{\mathbf{w}}$	Median (2.5%, 97.5%)
$\Sigma_{\mathbf{w}}\,[1,1]$	0.017 (0.016, 0.019)
$\Sigma_{\mathbf{w}}\,[2,2]$	0.026 (0.0065, 0.108)
$\Sigma_{\mathbf{w}}\,[1,2]\,/\sqrt{\Sigma_{\mathbf{w}}\,[1,1]\,\Sigma_{\mathbf{w}}\,[2,2]}$	-0.704 (-0.843, -0.545)

Table 12.6 *Estimates of the variances and correlation from $\Sigma_{\mathbf{w}}$, dynamic spatiotemporal modeling example.*

the months with more extreme weather, namely in the winter months of November through January and the summer months of June through August. ■

12.4.3 Spatiotemporal data

A natural extension of the modeling of the previous sections is the case where we have data correlated at spatial locations across time. If, as in Section 12.2, we assume that time is

Parameters	Median (2.5%, 97.5%)
σ_ϵ^2	0.134 (0.106, 0.185)
ϕ_1	1.09 (0.58, 2.04)
ϕ_2	0.58 (0.37, 1.97)
Range for intercept process	2.75 (1.47, 5.17)
Range for slope process	4.68 (1.60, 6.21)

Table 12.7 *Nugget effects and spatial correlation parameters, dynamic spatiotemporal modeling example.*

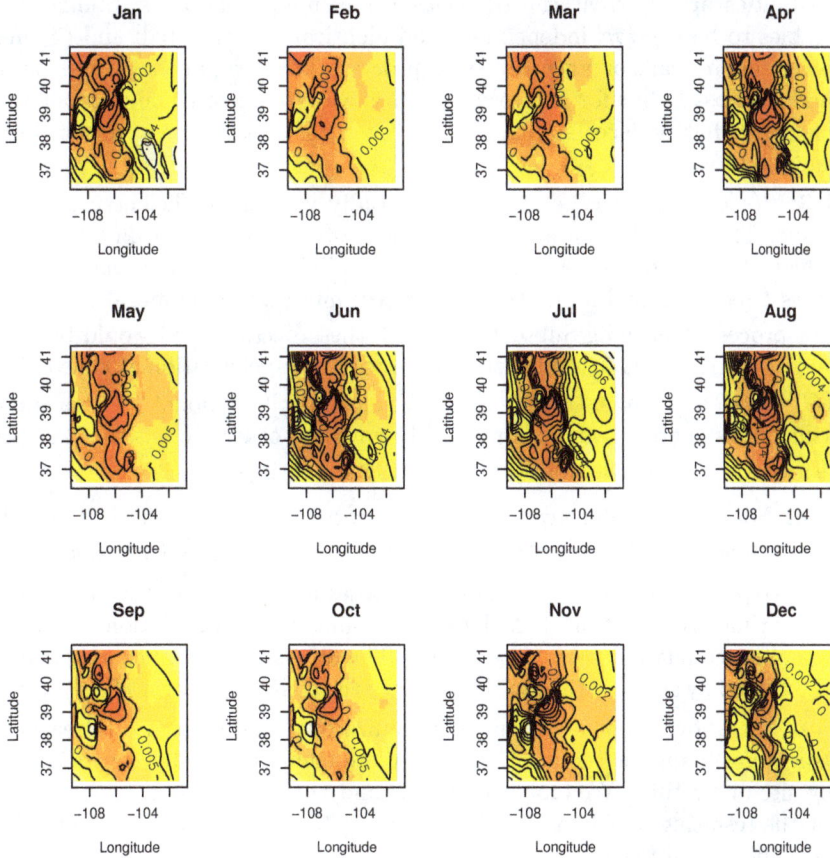

Figure 12.9 *Time-sliced image-contour plots displaying the posterior mean surface of the spatial residuals corresponding to the slope process in the temperature given precipitation model.*

discretized to a finite set of equally spaced points on a scale, we can conceptualize a time series of spatial processes that are observed only at the spatial locations $\mathbf{s}_1, \ldots, \mathbf{s}_n$.

Adopting a general notation that parallels (10.61), let

$$Y(\mathbf{s}, t) = \mathbf{X}^{\mathrm{T}}(\mathbf{s}, t)\, \tilde{\boldsymbol{\beta}}(\mathbf{s}, t) + \epsilon(\mathbf{s}, t) \ , \quad t = 1, 2, \ldots, M \ . \qquad (12.28)$$

That is, we introduce spatiotemporally varying intercepts and spatiotemporally varying slopes. Alternatively, if we write $\tilde{\boldsymbol{\beta}}(\mathbf{s}, t) = \boldsymbol{\beta}(\mathbf{s}, t) + \boldsymbol{\mu}_\beta$, we are partitioning the total error into $p + 1$ spatiotemporal intercept pieces including $\epsilon(\mathbf{s}, t)$, each with an obvious interpretation. So we continue to assume that $\epsilon(\mathbf{s}, t) \overset{iid}{\sim} N(0, \tau^2)$, but need to specify a model for $\tilde{\boldsymbol{\beta}}(\mathbf{s}, t)$. Regardless, (12.28) defines a nonstationary process having moments $E(Y(\mathbf{s},t)) = \mathbf{X}^{\mathrm{T}}(\mathbf{s},t)\tilde{\boldsymbol{\beta}}(s,t)$, $\mathrm{Var}(Y(\mathbf{s},t)) = \mathbf{X}^{\mathrm{T}}(\mathbf{s},t)\Sigma_{\tilde{\boldsymbol{\beta}}_{(,t)}}\mathbf{X}(\mathbf{s},t) + \tau^2$, and $\mathrm{Cov}(Y(\mathbf{s},t), Y(\mathbf{s}^{\mathrm{T}},t^{\mathrm{T}})) = \mathbf{X}^{\mathrm{T}}(\mathbf{s},t)\Sigma_{\tilde{\beta}(\mathbf{s},t),\tilde{\beta}(\mathbf{s}^{\mathrm{T}},t^{\mathrm{T}})}\mathbf{X}(\mathbf{s}',t^{\mathrm{T}})$.

Section 12.4 handled (12.28) using a dynamic model. Here we consider four alternative specifications for $\boldsymbol{\beta}(\mathbf{s},t)$. Paralleling the customary assumption from longitudinal data modeling (where the time series are usually short), we could set

- **Model 1:** $\boldsymbol{\beta}(\mathbf{s},t) = \boldsymbol{\beta}(\mathbf{s})$, where $\boldsymbol{\beta}(\mathbf{s})$ is modeled as in the previous sections. This model can be viewed as a locally linear growth curve model.

- **Model 2:** $\boldsymbol{\beta}(\mathbf{s},t) = \boldsymbol{\beta}(\mathbf{s}) + \boldsymbol{\alpha}(t)$, where $\boldsymbol{\beta}(\mathbf{s})$ is again as in Model 1. In modeling $\boldsymbol{\alpha}(t)$, two possibilities are (i) treat the $\alpha_k(t)$ as time dummy variables, taking this set of pM variables to be *a priori* independent and identically distributed; and (ii) model the $\boldsymbol{\alpha}(t)$ as a random walk or autoregressive process. The components could be assumed independent across k, but for greater generality, we take them to be dependent, using a separable form that replaces $\mathbf{s}$ with t and takes ρ to be a valid correlation function in just one dimension.

- **Model 3:** $\boldsymbol{\beta}(\mathbf{s},t) = \boldsymbol{\beta}^{(t)}(\mathbf{s})$, i.e., we have spatially varying coefficient processes nested within time. This model is an analogue of the nested effects areal unit specification in Waller et al. [1997]; see also Gelfand et al. [2004a]. The processes are assumed independent across t (essentially dummy time processes) and permit temporal evolution of the coefficient process. Following Subsection 10.7.3, the process $\boldsymbol{\beta}^{(t)}(\mathbf{s})$ would be mean-zero, second-order stationary Gaussian with cross-covariance specification at time t, $C^{(t)}(\mathbf{s},\mathbf{s}')$ where $\left(C^{(t)}(\mathbf{s},\mathbf{s}')\right)_{lm} = \rho\left(\mathbf{s}-\mathbf{s}';\phi^{(t)}\right)\tau^{(t)}_{lm}$. We have specified Model 3 with a common $\boldsymbol{\mu}_{\beta}$ across time. This enables some comparability with the other models we have proposed. However, we can increase flexibility by replacing $\boldsymbol{\mu}_{\beta}$ with $\boldsymbol{\mu}_{\beta}^{(t)}$.

- **Model 4:** For $\rho^{(1)}$ a valid two-dimensional correlation function, $\rho^{(2)}$ a valid one-dimensional choice, and T positive definite symmetric, $\boldsymbol{\beta}(\mathbf{s},t)$ such that $\Sigma_{[\boldsymbol{\beta}(\mathbf{s},t),\boldsymbol{\beta}(\mathbf{s}',t')]} = \rho^{(1)}(\mathbf{s}-\mathbf{s}';\phi)\rho^{(2)}(t-t';\gamma)T$. This model proposes a separable covariance specification in space and time, as in Section 12.2. Here $\rho^{(1)}$ obtains spatial association as in earlier subsections that is attenuated across time by $\rho^{(2)}$. The resulting covariance matrix for the full vector $\boldsymbol{\beta}$, blocked by site and time within site has the convenient form $H_2(\gamma)\otimes H_1(\phi)\otimes T$.

In each of the above models we can marginalize over $\boldsymbol{\beta}(\mathbf{s},t)$ as we did earlier in this section. Depending upon the model it may be more computationally convenient to block the data by site or by time. We omit the details and notice only that, with n sites and T time points, the resulting likelihood will involve the determinant and inverse of an $nT \times nT$ matrix (typically a large matrix).

Note that all of the foregoing modeling can be applied to the case of cross-sectional data where the set of observed locations varies with t. This is the case, for instance, with our real estate data. We only observe a selling price at the time of a transaction. With n_t locations in year t, the likelihood for all but Model 3 will involve a $\sum n_t \times \sum n_t$ matrix.

Example 12.4 *(Baton Rouge housing prices (contd.))*. We now turn to the dynamic version of Baton Rouge dataset presented in Example 10.4. From the Baton Rouge database we drew a sample of 120 transactions at distinct spatial locations for the years 1989, 1990, 1991, and 1992. We compare Models 1–4. In particular, we have two versions of Model 2; 2a has the $\boldsymbol{\alpha}(t)$ as four i.i.d. time dummies, while 2b uses the multivariate temporal process model for $\boldsymbol{\alpha}(t)$. We also have two versions of Model 3; 3a has a common $\boldsymbol{\mu}_{\beta}$ across t, while 3b uses $\boldsymbol{\mu}_{\beta}^{(t)}$. In all cases the five-dimensional spatially varying coefficient model for $\boldsymbol{\beta}$'s was employed. Table 12.8 shows the results. Model 3, where space is nested within time, turns out to be the best with Model 4 following closely behind. We omit the posterior inference

Model	Independent process			Dependent process		
	G	P	D_∞	G	P	D_∞
1	88.58	56.15	144.73	54.54	29.11	83.65
2a	77.79	50.65	128.44	47.92	26.95	74.87
2b	74.68	50.38	125.06	43.38	29.10	72.48
3a	59.46	48.55	108.01	43.74	20.63	64.37
3b	57.09	48.41	105.50	42.35	21.04	63.39
4	53.55	52.98	106.53	37.84	26.47	64.31

Table 12.8 *Model choice criteria for various spatiotemporal process models.*

summary for Model 3b, noting only that the overall coefficients $\left(\boldsymbol{\mu}_\beta^{(t)}\right)$ do not change much over time. However, there is some indication that spatial range is changing over time. ∎

12.5 Fitting dynamic spatiotemporal models

Let us explore a relatively simple, but rather flexible, univariate version of the dynamic models discussed in the preceding section. Suppose, $y_t(\mathbf{s})$ denotes the observation at location $\mathbf{s}$ and time t. We model $y_t(\mathbf{s})$ through a *measurement equation* that provides a regression specification with a space-time varying intercept and serially and spatially uncorrelated zero-centered Gaussian disturbances as measurement error $\epsilon_t(\mathbf{s})$. Next a *transition equation* introduces a $p \times 1$ coefficient vector, say $\boldsymbol{\beta}_t$, which is a purely temporal component (i.e., time-varying regression parameters), and a spatio-temporal component $u_t(\mathbf{s})$. Both these are generated through transition equations, capturing their Markovian dependence in time. While the transition equation of the purely temporal component is akin to usual state-space modeling, the spatio-temporal component is generated using Gaussian spatial processes. The overall model, for $t = 1, 2, \ldots, N_t$, is written as

$$
\begin{aligned}
y_t(\mathbf{s}) &= \mathbf{x}_t(\mathbf{s})^\top \boldsymbol{\beta}_t + u_t(\mathbf{s}) + \epsilon_t(\mathbf{s}), \quad \epsilon_t(\mathbf{s}) \overset{ind.}{\sim} N(0, \tau_t^2) ; \\
\boldsymbol{\beta}_t &= \boldsymbol{\beta}_{t-1} + \boldsymbol{\eta}_t, \quad \boldsymbol{\eta}_t \overset{i.i.d.}{\sim} N(0, \Sigma_\eta) ; \\
u_t(\mathbf{s}) &= u_{t-1}(\mathbf{s}) + w_t(\mathbf{s}), \quad w_t(\mathbf{s}) \overset{ind.}{\sim} GP\left(0, C_t(\cdot, \boldsymbol{\theta}_t)\right) ,
\end{aligned}
\tag{12.29}
$$

where the abbreviations *ind.* and *i.i.d* are *independent* and *independent and identically distributed*, respectively. Here $\mathbf{x}_t(\mathbf{s})$ is a $p \times 1$ vector of predictors and $\boldsymbol{\beta}_t$ is a $p \times 1$ vector of coefficients. In addition to an intercept, $\mathbf{x}_t(\mathbf{s})$ can include location specific variables useful for explaining the variability in $y_t(\mathbf{s})$. The $GP(\mathbf{0}, C_t(\cdot, \boldsymbol{\theta}_t))$ denotes a spatial Gaussian process with covariance function $C_t(\cdot; \boldsymbol{\theta}_t)$. We customarily specify $C_t(\mathbf{s}_1, \mathbf{s}_2; \boldsymbol{\theta}_t) = \sigma_t^2 \rho(\mathbf{s}_1, \mathbf{s}_2; \phi_t)$, where $\boldsymbol{\theta}_t = \{\sigma_t^2, \phi_t\}$ and $\rho(\cdot; \phi)$ is a *correlation function* with ϕ controlling the correlation decay and σ_t^2 represents the spatial variance component. We further assume $\boldsymbol{\beta}_0 \sim N(\mathbf{m}_0, \Sigma_0)$ and $u_0(\mathbf{s}) \equiv 0$, which completes the prior specifications leading to a well-identified Bayesian hierarchical model with reasonable dependence structures. In practice, estimation of model parameters are robust to these hyper-prior specifications. Also note that (12.29) reduces to a simple spatial regression model for $t = 1$.

We consider settings where the inferential interest lies in spatial prediction or interpolation over a region for a set of discrete time points. We also assume that the same locations are monitored for each time point resulting in a space-time matrix whose rows index the locations and columns index the time points, i.e., the (i, j)-th element is $y_j(\mathbf{s}_i)$. Our algorithm will accommodate the situation where some cells of the space-time data matrix may have missing observations, as is common in monitoring environmental variables.

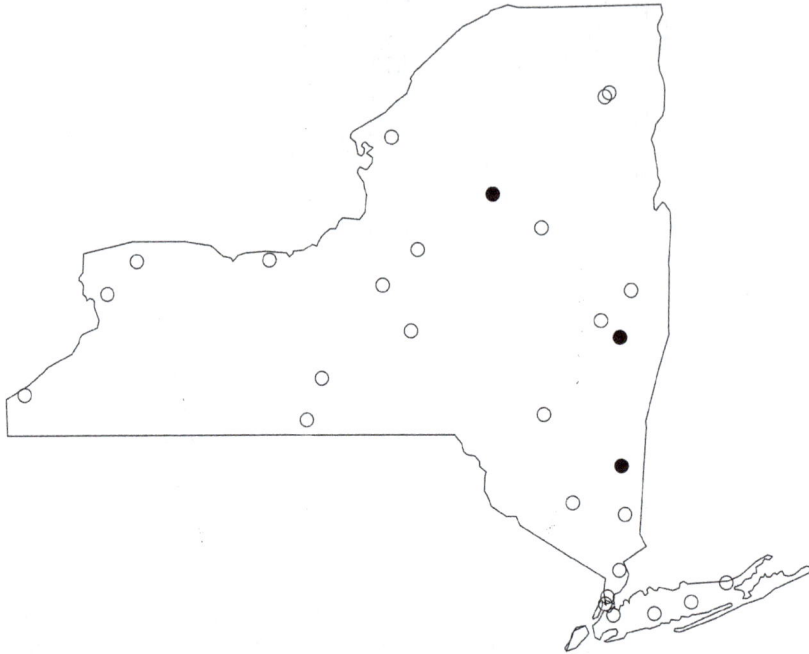

Figure 12.10 *Open and filled circle symbols indicate the location of 28 ozone monitoring stations across New York State. Filled circle symbols identify those stations that have half of the daily ozone measurements withheld to assess model predictive performance.*

The dynamic model (12.29) is also easily adapted to a computationally efficient low-rank version using the *predictive process* (see Section 13.4). Here we illustrate the full rank dynamic model using an ozone monitoring dataset that was previously analyzed by Sahu and Bakar [2012]. This is a relatively small dataset and does not require dimension reduction.

The dataset comprises 28 Environmental Protection Agency monitoring stations that recorded ozone from July 1 to August 31, 2006. The outcome is daily 8-hour maximum average ozone concentrations (parts per billion; O3.8HRMAX), and predictors include maximum temperature (Celsius; cMAXTMP), wind speed (knots; WDSP), and relative humidity (RM). Of the $1,736$ possible observations, i.e., $n=28$ locations times $N_t=62$ daily O3.8HRMAX measurements, 114 are missing. In this illustrative analysis we use the predictors cMAXTMP, WDSP, and RM as well as the spatially and temporally structured residuals to predict missing O3.8HRMAX values. To gain a better sense of the dynamic model's predictive performance, we withheld half of the observations from the records of three stations for subsequent validation. Figure 12.10 shows the monitoring station locations and identifies those stations where data were withheld.

To fit the model we define the station coordinates as well as starting, tuning, and prior distributions for the model parameters. Exploratory data analysis using time step specific variograms can be helpful for defining starting values and prior support for parameters in $\boldsymbol{\theta}_t$ and τ_t^2. To avoid cluttering the code, we specify the same prior for the ϕ_t's, σ_t^2's, and τ_t^2's. One can choose among several popular spatial correlation functions including the exponential, spherical, Gaussian and Matérn. We illustrate with the exponential function.

Time series plots of parameters' posterior summary statistics are often useful for exploring the temporal evolution of the parameters. In the case of the regression coefficients, these plots describe the time-varying trend in the outcome and impact of covariates. For example,

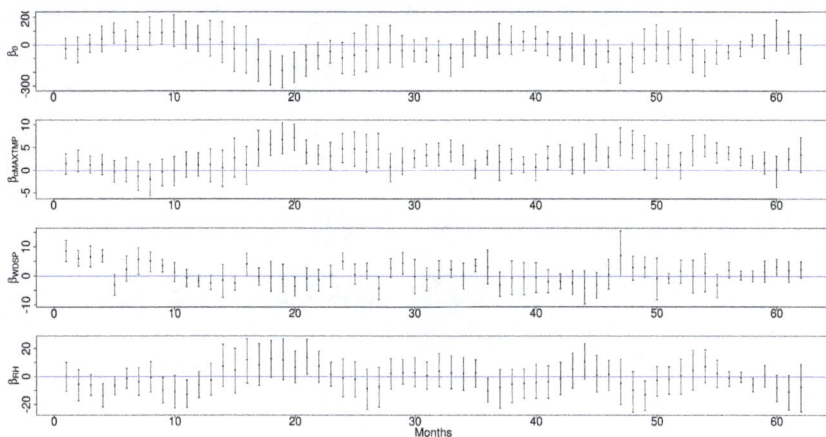

Figure 12.11 *Posterior distribution medians and 95% credible intervals for model intercept and predictors.*

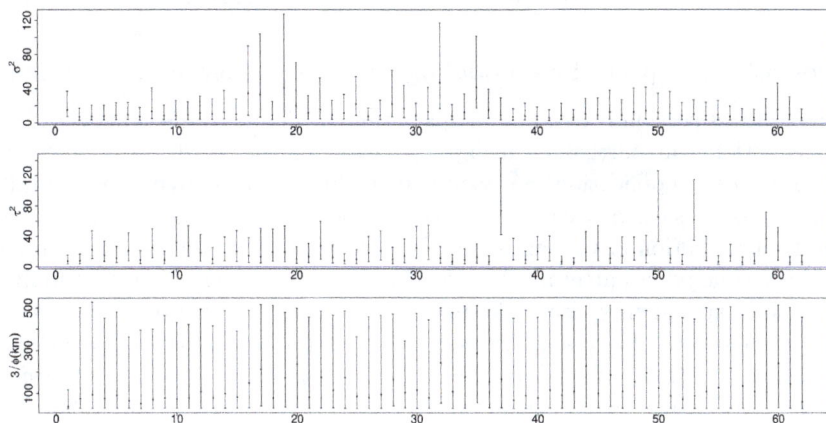

Figure 12.12 *Posterior distribution medians and 95% credible intervals for $\boldsymbol{\theta}$ and τ^2.*

the sinusoidal pattern in the model intercept, β_0, seen in Figure 12.11, correlates strongly with both cMAXTMP, RM, and to a lesser degree with WDSP. With only a maximum of 28 observations within each time step, there is not much information to inform estimates of $\boldsymbol{\theta}$. As seen in Figure 12.12, this paucity of information is reflected in the imprecise CI's for the ϕ's and small deviations from the priors on σ^2 and τ^2. There are, however, noticeable trends in the variance components over time.

Figure 12.13 shows the observed and predicted values for the three stations used for validation. Here, open circle symbols indicate those observations used for parameter estimation and filled circles identify hold-out observations. The posterior predicted median and 95% CIs are overlaid using solid and dashed lines, respectively. Three of the 36 hold-out measurements fell outside of their 95% predicted CI, a $\sim$92% coverage rate. As noted in Sahu and Bakar [2012], there is a noticeable reduction in ozone levels in the last two weeks in August.

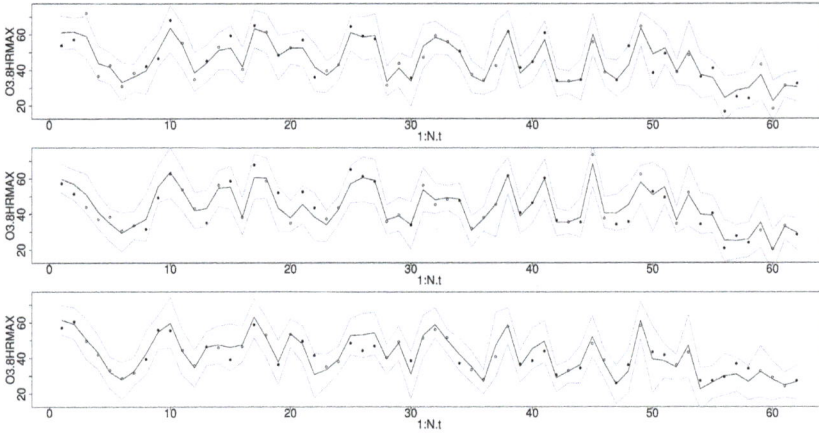

Figure 12.13 *Posterior predicted distribution medians and 95% credible intervals, solid and dashed lines respectively, for three stations. Open circle symbols indicate those observations used for model parameter estimation and filled circle symbols indicate those observations withheld for validation.*

12.6   Geostatistical space-time modeling driven by differential equations

The objective of this section is to consider space-time modeling in the context of stochastic differential equations, in particular, using stochastic diffusion processes. Such processes arise frequently in the application. For example, we find various environmental diffusions: for emerging diseases such as avian or H1N1 flu, for the progression of invasive species, and for the transformation of the landscape. In Section 8.8.3, we find a diffusion model to describe a space-time point pattern, urban development with regard to single-family homes. Here, we consider spread in space and time in the geostatistical setting, with associated uncertainty, with potential explanatory covariates. Our starting point is a deterministic integro-difference equation or partial differential equation. Many of the ideas in this section arise from the work of Wikle and colleagues. See, e.g., Wikle and Hooten (2006, 2010) and references therein. In general, differential equations that have analytical solutions are too simple to capture what we seek in practice. So, to accommodate more flexible forms, we adopt a strategy which carries out discretization in time. In fact, it may be argued that we should begin with a temporally discretized version, incorporating the features we seek, rather than attempting to frame a particular SDE.

We continue to work within our hierarchical paradigm,

$$[data \mid process, parameters][process \mid parameters][parameters]$$

We continue to work within the Bayesian framework, employing structured dependence in space and time. So, model fitting using MCMC will be challenging and we will typically have to resort to dimension reduction techniques (see Chapter 13).

More precisely, our discretization envisions continuous space with discrete time, i.e., $w_t(\mathbf{s})$. Without loss of generality we can take time to be $t \in \{1, 2, ..., T\}$. The customary terminology here refers to $w_t(\mathbf{s})$ as a *dynamical* process. In fact, we simplify to a first order Markov process, i.e., for the finite set of locations $\mathbf{s}_1, \mathbf{s}_2, ..., \mathbf{s}_n$, let $\mathbf{w}_t = (w_t(\mathbf{s}_1), w_t(\mathbf{s}_2), ..., w_t(\mathbf{s}_n))^\mathsf{T}$. Then $[\mathbf{w}_t \mid \mathbf{w}_0, \mathbf{w}_1, ..., \mathbf{w}_{t-1}] = [\mathbf{w}_t \mid \mathbf{w}_{t-1}]$. For example, a linear update would be

$$\mathbf{w}_t = H\mathbf{w}_{t-1} + \boldsymbol{\eta}_t \tag{12.30}$$

where $\eta_t(\mathbf{s})$ incorporates spatial structure. In the literature, this is referred to as a vector AR(1) model and H is called the propagator matrix. The modeling challenge is the specification of H.

We can look at several cases. For instance, consider $H = I$. Note that this form will not provide stationary behavior but, more importantly, there is no interaction across space and time, and so this choice would not be realistic for most dynamic processes of interest. Next, consider $H = \text{Diag}(h)$ where $\text{Diag}(h)$ has diagonal elements $0 < h_i < 1$. Now, we achieve stationarity but still we have no space-time interaction. A more general form is an integro-difference equation (IDE),

$$w_t(\mathbf{s}) = \int h(\mathbf{s}, \mathbf{r}; \phi) w_{t-1}(\mathbf{r}) d\mathbf{r} + \eta_t(\mathbf{s}). \tag{12.31}$$

We see that (12.31) does enable the dynamics we seek. In particular, in (12.31), h is a "redistribution kernel," providing redistribution in space which determines the rate of *diffusion* and the *advection*. If we require $w > 0$, then we could work with $\log w_t(\mathbf{s}) = \log(\int h(\mathbf{s}, \mathbf{r}; \phi) w_{t-1}(\mathbf{r}) d\mathbf{r}) + \eta_t(\mathbf{s})$. Again, we resort to discretization in order to supply the H matrix. In this regard, we might begin with forms for h in (12.31); would we want a stationary choice, $h(\mathbf{s}, \mathbf{r}; \phi)$ or would a time-dependent choice, $h_t(\mathbf{s}, \mathbf{r}; \phi)$ be more appropriate? Might ϕ depend upon $\mathbf{r}$?

Recall the linear partial differential equation (PDE), $\frac{dw(\mathbf{s},t)}{dt} = h(\mathbf{s})w(\mathbf{s}, t)$. Applying finite differencing yields $w(\mathbf{s}, t+\Delta t) - w(\mathbf{s}, t) = h(\mathbf{s})w(\mathbf{s}, t)\Delta t$, i.e., $w(\mathbf{s}, t+1) \approx \tilde{h}(\mathbf{s})w(\mathbf{s}, t)$. We see that the linear PDE suffers the same problems as we noted above. There is no space-time interaction; there is no redistribution over space. So, we need more general PDE's which, as noted above, can motivate an IDE, can illuminate the choice of H.

In this regard, it may be useful to note the "forward" vs. "backward" perspective associated with an IDE. In one sense, we can think of (12.31) as moving forward in time, taking us from a current "state," $w_t(\mathbf{s}), \mathbf{s} \in D$ to a new state $w_{t+1}(\mathbf{s}), \mathbf{s} \in D$. In another sense, we can look backwards, thinking of (12.31) as clarifying how the "state" $w_t(\mathbf{r}), \mathbf{r} \in D$ contributed to give us the current state, $w_{t+1}(\mathbf{s}), \mathbf{s} \in D$. Depending upon the process we are modeling, specification of h may emerge more naturally under one perspective rather than the other. Also, IDE's can be specified directly without using PDE's. That is, $h(\mathbf{s}, \mathbf{r})$ can be developed using process-based assumptions, e.g., as a sum of a survival/growth term plus a birth/replenishment term as in Ghosh et al. [2012].

Now, consider a diffusion in one dimension. Fick's Law of diffusion [Fick, 1995, Crank, 1979] asserts that the diffusive flux from *high* concentration to *low* is $-\delta \frac{\partial w(x,t)}{\partial x}$ with δ being the diffusion coefficient. More flexible dynamics arise with a location varying diffusion coefficient $\delta(x)$ supplying a location varying diffusion. The associated diffusion equation is $\partial w/\partial t = -\partial \texttt{flux}/\partial x$, i.e., $\frac{\partial w(x,t)}{\partial t} = \frac{\partial}{\partial x}(\delta(x)\frac{\partial w(x,t)}{\partial x})$ Applying the chain rule, the one dimensional diffusion equation is

$$\frac{\partial w(x,t)}{\partial t} = \delta^{\mathrm{T}}(x)\frac{\partial w(x,t)}{\partial x} + \delta(x)\frac{\partial^2 w(x,t)}{\partial x^2}. \tag{12.32}$$

Moving to two-dimensional space, writing $\mathbf{s} = (x, y)$, the diffusive flux is $-\delta(x,y)\nabla w(x, y, t)$ where $\nabla w(x, y, t)$ is the concentration gradient at time t. Now, the resulting diffusion PDE is

$$\frac{\partial w(x,y,t)}{\partial t} = \frac{\partial}{\partial x}(\delta(x,y)\frac{\partial w(x,y,t)}{\partial x}) + \frac{\partial}{\partial y}(\delta(x,y)\frac{\partial w(x,y,t)}{\partial y}). \tag{12.33}$$

We can complete the chain rule calculation to explicitly obtain the spatial diffusion equation, noting that it will involve the second-order partial derivatives, $\frac{\partial^2 w}{\partial x^2}$ and $\frac{\partial^2 w}{\partial y^2}$. Now, suppose we introduce Δt, Δx, Δy and replace ∂'s with finite differences (first forward and second order centered). The resulting expressions are elaborate and messy but are developed in careful detail in Hooten and Wikle (2007). The critical point is that, after the smoke clears, we obtain the propagator matrix H to insert into $\mathbf{w}_{t+\Delta t} = H\mathbf{w}_t$. Again, we would add independent spatial noise at each time point, $\boldsymbol{\eta}_t$. Evidently, we are back to our earlier redistribution form in (12.30).

Hooten and Wikle [2008] illustrate with data from the U.S. Breeding Bird Survey, focusing on the Eurasian collared dove. For the years $1986 - 2003$, the data consist of recorded bird counts by sight (for three minutes) over a collection of routes each roughly of length 40kms with 50 stops per route. The counts are attached to grid boxes i in year t and are given a Poisson specification reflecting the number of visits to the site in a given year and a model for the associated expected counts (intensities, see Chapter 8) per visit. The H matrix is a function of the vector of local diffusion coefficients. Since the number of sites is large, dimension reduction is applied to the vector of w's (see Chapter 13). We do not offer further detail here, encouraging the reader to consult the Hooten and Wikle (2007) paper. Instead we present a different geostatistical example below.

The foregoing dynamics redistribute the existing population spatially over time. However, in many situations it would be the case that there is growth or decline in the population. For instance, with housing stock, while some new homes are built, others are torn down. With species populations, change in population size is *density dependent*; competition may encourage or discourage population growth. Hence, we might attempt to add a growth rate to the model. An illustrative choice, which we employ below, is the logistic differential equation (see, e.g., Kot, 2001),

$$\frac{\partial w(\mathbf{s}, t)}{\partial t} = r w(\mathbf{s}, t) \left(1 - \frac{w(\mathbf{s}, t)}{K} \right). \tag{12.34}$$

Here, r is the growth rate and K is the carrying capacity. In practice, we would imagine a spatially varying growth rate, $r(\mathbf{s})$ and perhaps even a spatially varying capacity, $K(\mathbf{s})$.

Turning to more general structures, suppose $w(\mathbf{s}, t)$ is a mean (second stage) specification for a space-time geostatistical model or GLM (or perhaps an intensity for a space-time point pattern, as in Section 10.3.3). A general deterministic diffusion PDE for $w(\mathbf{s}, t)$ looks like

$$\frac{\partial w(\mathbf{s}, t)}{\partial t} = a(w(\mathbf{s}, t), v(\mathbf{s}, t); \theta) \tag{12.35}$$

where $v(\mathbf{s}, t)$ includes other potential variables; in its simplest form, $v(\mathbf{s}, t) = t$. Furthermore, we might extend θ to $\theta(\mathbf{s})$ or perhaps to $\theta(\mathbf{s}, t)$.

To think about adding uncertainty, ignore location $\mathbf{s}$ for the moment and consider a usual nonlinear differential equation, $d\mu(t) = a(\mu(t), t, \theta)dt$ with $\mu(0) = \mu_0$. A simple way to add stochasticity is to make θ random. However, this imposes the likely unreasonable assumption that the functional form of the equation is *true*. Instead, we might assume $d\mu(t) = a(\mu(t), t, \theta)dt + b(\mu(t), t, \theta)dZ(t)$ where $Z(t)$ is variance 1 Brownian motion over R^1 (Section 3.2) with a and b the "drift" and "volatility" respectively. Now we obtain a *stochastic* differential equation (SDE) in which we would still assume θ to be random. A bit more generality is achieved with the form, $d\mu(t) = a(\mu(t), t, \theta(t))dt$ where $\theta(t)$ is driven by an SDE, $d\theta(t) = g(\theta(t), t, \beta)dt + h(\theta(t), \sigma)dZ(t)$ where, again, $Z(t)$ is variance 1 Brownian motion.

Now, we add space. We write $d\mu(\mathbf{s}, t) = a(\mu(\mathbf{s}, t), t, \theta(\mathbf{s}))dt$ with $\mu(\mathbf{s}, 0)) = \mu_0(\mathbf{s})$, a partial differential equation (PDE). We add randomness through $\theta(\mathbf{s})$, a process realization, resulting in a stochastic process of differential equations. Again, we don't believe the form of the PDE is true. So, we write

$$d\mu(\mathbf{s}, t) = a(\mu(\mathbf{s}, t), t, \theta(\mathbf{s}))dt + b(\mu(\mathbf{s}, t), t, \theta(\mathbf{s}))dZ(\mathbf{s}, t), \tag{12.36}$$

where we need to extend our modeling of Brownian motion to $Z(\mathbf{s}, t)$. For a fixed finite set of spatial locations we customarily assume independent Brownian motion at each location, allowing the $\theta(\mathbf{s})$ process to provide the spatial dependence. Again, we can work with the more general form, $d\mu(\mathbf{s}, t) = a(\mu(\mathbf{s}, t), t, \theta(\mathbf{s}, t))dt$ where, simplifying the volatility, $d\theta(\mathbf{s}, t) = g(\theta(\mathbf{s}, t))dt + bdZ(\mathbf{s}, t)$. When $d\theta(\mathbf{s}, t) = \alpha(\theta(\mathbf{s}, t) - \theta(\mathbf{s}))dt + bdZ(\mathbf{s}, t)$ with $\theta(\mathbf{s})$ a process realization as above, we have characterized $\theta(\mathbf{s}, t)$ through an infinite dimensional SDE and it can be shown that the associated covariance function of the space time process is separable.

The usual space-time "geostatistics" setting with observations at locations and times in the foregoing context takes the form $Y(\mathbf{s}, t) = w(\mathbf{s}, t) + \epsilon(\mathbf{s}, t)$ where now, $w(\mathbf{s}, t)$ is modeled through a stochastic PDE. In particular, the logistic equation now extended to space and time becomes

$$\frac{\partial w(\mathbf{s}, t)}{\partial t} = r(\mathbf{s}, t)w(\mathbf{s}, t)\left(1 - \frac{w(\mathbf{s}, t)}{K(\mathbf{s})}\right). \tag{12.37}$$

Again, with time discretized to intervals Δt, indexed as $t_j, j = 0, 1, 2, ...J$ and locations $\mathbf{s}_i$, our data takes the form $\{Y(\mathbf{s}_i, t_j)\}$ with resulting dynamic model $Y(\mathbf{s}_i, t_j) = w(\mathbf{s}_i, t_j) + \varepsilon(\mathbf{s}_i, t_j)$. Using Euler's approximation yields the difference equation:

$$\Delta w(\mathbf{s}, t_j) = r(\mathbf{s}, t_j)w(\mathbf{s}, t_{j-1})\left[1 - \frac{w(\mathbf{s}, t_{j-1})}{K(\mathbf{s})}\right]\Delta t,$$

$$w(\mathbf{s}, t_j) \approx w(\mathbf{s}, 0) + \sum_{l=1}^{j}\Delta w(\mathbf{s}, t). \tag{12.38}$$

For $r(\mathbf{s}, t)$, we adopt the model given above for $\theta(\mathbf{s}, t)$, i.e., $dr(\mathbf{s}, t) = \alpha(r(\mathbf{s}, t) - r(\mathbf{s}))dt + bdZ(\mathbf{s}, t)$. For the initial positive $w(\mathbf{s}, 0)$ and $K(\mathbf{s})$ we can use log-Gaussian spatial processes with regression forms for the means:

$$\log w(\mathbf{s}, 0) = \mu_w(X_w(\mathbf{s}), \beta_w) + \eta_w(\mathbf{s})$$

with $\eta_w(\mathbf{s}) \sim \text{GP}\left(0, \sigma_w^2\rho_w(\mathbf{s} - \mathbf{s}'; \phi_w)\right)$ and

$$\log K(\mathbf{s}) = \mu_K(X_K(\mathbf{s}), \beta_K) + \eta_K(\mathbf{s})$$

with $\eta_K(\mathbf{s}) \sim \text{GP}\left(0, \sigma_K^2\rho_K(\mathbf{s} - \mathbf{s}'; \phi_K)\right)$. In fact, to simplify here, we assume $K(s)$ known and set to 1, yielding an interpretation of w on a *percent* scale with K as 100% or full capacity. We add similar modeling for $\mu_r(\mathbf{s})$. Altogether, we have specified a hierarchical model. After specifying priors, we fit with MCMC cumulatively in w over time. We consider the usual prediction questions—(i) interpolating the past at new locations and (ii) forecasting the future at current and new locations

For our specific example, we use a 10×10 study region introducing 44 locations over 30 time periods with 4 sites retained as hold out for validation (see Figure 12.14). To illustrate with a different choice of covariance function, we adopted the Matérn with smoothness parameter $\nu = 3/2$ for both $w_0(\mathbf{s})$ and $r(\mathbf{s})$. Hence, the resulting space time covariance function for $r(\mathbf{s}, t)$ becomes

$$\sigma_r^2 \exp\left(-\alpha|t_{j_1} - t_{j_2}|\right)\left(1 + \phi_r||\mathbf{s}_{i_1} - \mathbf{s}_{i_2}||\right)\exp\left(-\phi_r|\mathbf{s}_{i_1} - \mathbf{s}_{i_2}|\right).$$

We use the simulated $\mathbf{r}$ and $\mathbf{w}_0$ with the transition equation recursively to obtain $\boldsymbol{\Delta}\mathbf{w}_j$ and $\mathbf{w}_j$ for each of the 30 periods. The observed data are sampled as mutually independent given $\mathbf{w}_j$ with the random noise $\boldsymbol{\varepsilon}_j$. The data at four selected locations marked as 1, 2, 3, and 4 in Figure 12.14 are shown as small circles. We leave out the data at four randomly chosen locations (shown in a diamond shape and marked as A, B, C and D in Figure 12.14 for spatial prediction and out-of-sample validation for our model.

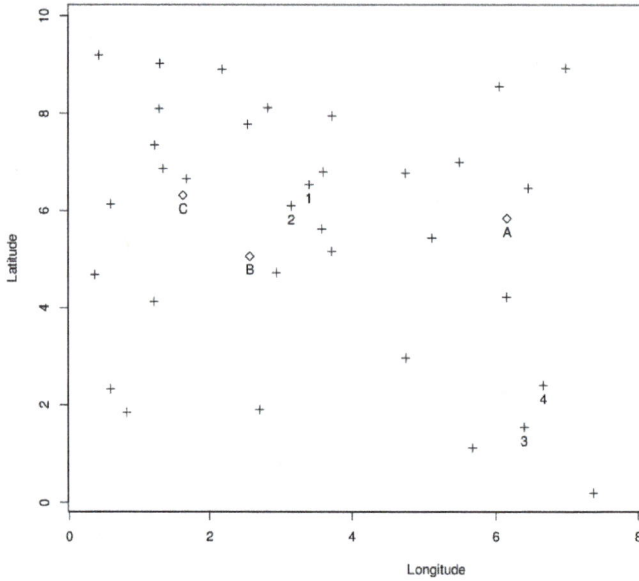

Figure 12.14 *Simulated data at four selected locations marked as 1, 2, 3, and 4 are shown as small circles. Hold-out data at four randomly chosen locations for out-of-sample validation is shown in diamond shape and marked as A, B, C and D.*

We fit the foregoing model to the data at the remaining 40 locations (hence a 40×30 spatio-temporal data set). We use very vague priors for the constant means: $\pi(\mu_w) \sim N\left(0, 10^8\right)$ and $\pi(\mu_r) \sim N\left(0, 10^8\right)$. We use conjugate gamma priors for the precision parameters of r and w_0: $\pi\left(1/\sigma_r^2\right) \sim Gamma(1, 1)$ and $\pi\left(1/\sigma_w^2\right) \sim Gamma(1, 1)$. The positive parameter for the temporal correlation of r also has a vague log-normal prior: $\pi(\alpha) \sim log\text{-}N\left(0, 10^8\right)$. Because the spatial range parameters ϕ_r and ϕ_w are only weakly identified, we only use informative and discrete prior for them. Indeed, we have chosen 20 values (from 0.1 to 2.0) and assume uniform priors over them for both ϕ_r and ϕ_w.

We use the random-walk Metropolis-Hastings algorithm to simulate posterior samples of $\mathbf{r}$ and $\mathbf{w_0}$. We draw the entire vector of $\mathbf{w_0}$ for all forty locations as a single block in every iteration. Because $\mathbf{r}$ is very high-dimensional (a 40×30 matrix concatenated into a vector), we cannot draw the entire matrix of $\mathbf{r}$ as one block and achieve a satisfactory acceptance rate. So, we partition $\mathbf{r}$ into 40 row blocks (location-wise) in every odd-numbered iteration and 30 column blocks (period-wise) in every even numbered iteration. Each block is drawn in one Metropolis step. The posterior samples start to converge after about 30,000 iterations. Given the sampled $\mathbf{r}$ and $\mathbf{w_0}$, the mean parameters μ_r, μ_w and the precision parameters $1/\sigma_r^2$ and $1/\sigma_A^2$ all have conjugate priors, and therefore their posterior samples are drawn directly. ϕ_r and ϕ_w have discrete priors and therefore are also directly sampled. We use the random-walk Metropolis-Hastings algorithm to draw α.

We obtain 200,000 samples from the algorithm and discard the first 100,000 as burn-in. For the posterior inference, we use 4,000 subsamples from the remaining 100,000 samples, with a thinning equal to 25. The posterior means and 95% equal-tail Bayesian posterior predictive intervals for the model parameters are presented in Table 12.9. Evidently we are recovering the true parameter values very well.

Figure 12.15 displays the posterior mean of the growth curves and 95% Bayesian predictive intervals for the four locations which were used in the fitting (1, 2, 3 and 4), compared with the actual latent growth curve $w(\mathbf{s}, t)$ and observed data. Up to the uncertainty in the

Model Parameters	True Value	Posterior Mean	95% Equal-tail Interval
μ_w	-4.2	-4.14	(-4.88, -3.33)
σ_w	1.0	0.91	(0.62, 1.46)
ϕ_w	0.7	0.77	(0.50, 1.20)
σ_ε	0.05	0.049	(0.047, 0.052)
μ_r	0.24	0.24	(0.22, 0.26)
σ_r	0.08	0.088	(0.077, 0.097)
ϕ_r	0.7	0.78	(0.60, 1.10)
α	0.6	0.64	(0.51, 0.98)

Table 12.9 *Parameters and their posterior inference for the simulated example.*

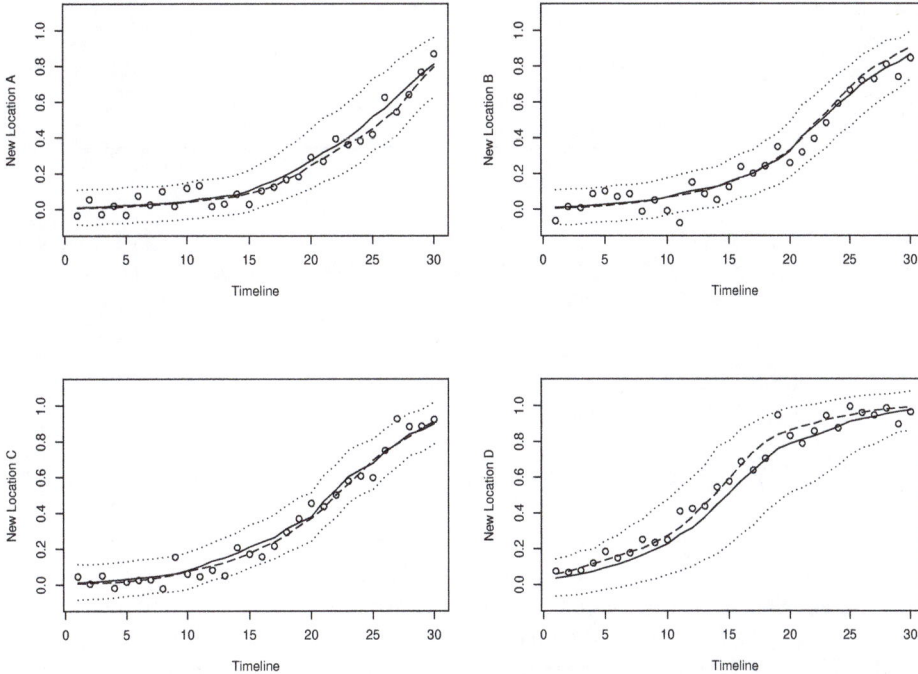

Figure 12.15 *Hold-out space-time geostatistical data at 4 locations, actual (dashed line) and predicted mean growth curves (solid line) and 95% predictive intervals (dotted line) by our model (12.36) for the simulated data example.*

model we approximate the actual curves very well. The fitted mean growth curves almost perfectly overlap with the actual simulated growth curves. The empirical coverage of the Bayesian predictive bounds is 93.4%.

Interpolation yields the predictive growth curve for the four hold-out locations (A, B, C and D). In Figure 12.15 we display the means of the predicted curves and 95% Bayesian predictive intervals, together with the hold-out data. We can see the spatial prediction captures the patterns of the hold-out data very well. The predicted mean growth curves overlap with the actual simulated growth curves very well except for location D (because location D is rather far from all the observed locations). The empirical coverage of the Bayesian predictive intervals is 95.8%.

Finally, for comparison and in the absence of covariates, we also fit the following customary process realization model with space-time random effects to the simulated data set

$$\mathbf{y}_j = \mu \mathbf{1} + \boldsymbol{\xi}_j + \boldsymbol{\epsilon}_j \; ; \; \boldsymbol{\epsilon}_j \sim N\left(0, \sigma_\epsilon^2 I_n\right), \quad \text{for } j = 0, 1, \ldots, J \qquad (12.39)$$

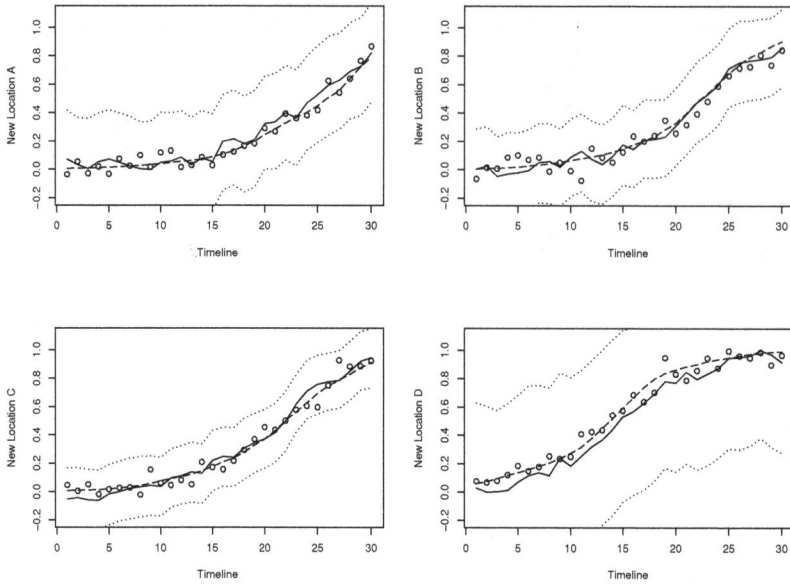

Figure 12.16 *Hold-out space-time geostatistical data at 4 locations, actual (dashed line) and pre-dicted mean growth curves (solid line) and 95% predictive intervals (dotted line) by the benchmark model (12.39) for the simulated data example.*

where the random effects $\boldsymbol{\xi} = [\boldsymbol{\xi}_0, \ldots, \boldsymbol{\xi}_J]$ come from a Gaussian process with a separable spatio-temporal correlation of the form:

$$C_\xi(t - t', s - s') = \sigma_\xi^2 \exp\left(-\alpha_\xi |t - t'|\right) \left(\phi_\xi |s - s'|\right)^\nu \kappa_\nu \left(\phi_\xi |s - s'|\right), \ \nu = \frac{3}{2} \ . \quad (12.40)$$

Comparison of model performance between the SPDE model and this model is done using spatial prediction at the 4 new locations. In Figure 12.16 we display the means of the predicted curves and 95% Bayesian predictive intervals, together with the hold-out data. For the four hold-out sites, the average mean square error for the SPDE model is 1.75×10^{-3} versus 3.34×10^{-3} for the standard model. The average length of the 95% predictive intervals for the SPDE model is 0.29 versus 0.72 for the standard model. It is evident that, when we have an SPDE driving the data, discretized as above, we can learn about it and will do better in terms of prediction than using a standard model.

12.7   Areal unit space-time modeling

We now return to spatiotemporal modeling for areal unit data, following the discussion of equations (12.4) and (12.5) in Section 12.1. Recall that we have briefly discussed general space-time SAR modeling in Section 4.4.2. Here, we focus on the spatio-temporal disease mapping setting.

12.7.1   Aligned data

In the aligned data case, matters are relatively straightforward. Consider for example the spatiotemporal extension of the standard disease mapping setting described in Section 6.8.1. Here we would have $Y_{i\ell t}$ and $E_{i\ell t}$, the observed and expected disease counts in county i and demographic subgroup ℓ (race, gender, etc.) during time period t (without loss of generality we let t correspond to years in what follows). Again the issue of whether the $E_{i\ell t}$ are internally or externally standardized arises; in the more common former case we would use $n_{i\ell t}$, the number of persons at risk in county i during year t, to compute $E_{i\ell t} = n_{i\ell t}(\sum_{i\ell t} Y_{i\ell t}/\sum_{i\ell t} n_{i\ell t})$. That is, $E_{i\ell t}$ is the number of cases we would expect if the grand disease rate (all regions, subgroups, and years) were in operation throughout. The extension of the basic Section 6.8.1 Poisson regression model is then

$$Y_{i\ell t} \mid \mu_{i\ell t} \overset{ind}{\sim} Po\left(E_{i\ell t}\, e^{\mu_{i\ell t}}\right) ,$$

where $\mu_{i\ell t}$ is the log-relative risk of disease for region i, subgroup ℓ, and year t.

It now remains to specify the main effect and interaction components of $\mu_{i\ell t}$, and corresponding prior distributions. First the main effect for the demographic subgroups can be taken to have ordinary linear regression structure, i.e., $\varepsilon_\ell = \mathbf{x}'_\ell \boldsymbol{\beta}$, with a flat prior for $\boldsymbol{\beta}$. Next, the main effects for time (say, δ_t) can be assigned flat priors (if we wish to view them as fixed effects, i.e., temporal dummy variables), or an $AR(1)$ specification (if we wish them to reflect temporal autocorrelation). In some cases an even simpler structure (say, $\delta_t = \gamma t$) may be appropriate.

Finally, the main effects for space are similar to those assumed in the nontemporal case. Specifically, we might let

$$\psi_i = \mathbf{z}'_i \boldsymbol{\omega} + \theta_i + \phi_i ,$$

where $\boldsymbol{\omega}$ has a flat prior, the θ_i capture *heterogeneity* among the regions via the i.i.d. specification,

$$\theta_i \overset{iid}{\sim} N(0\,,\, 1/\tau) ,$$

and the ϕ_i capture regional *clustering* via the CAR prior,

$$\phi_i \mid \phi_{j\neq i} \sim N(\bar{\phi}_i\,,\, 1/(\lambda m_i)) .$$

As usual, m_i is the number of neighbors of region i, and $\bar{\phi}_i = m_i^{-1}\Sigma_{j\in\partial_i} \phi_j$.

Turning to spatiotemporal interactions, suppose for the moment that demographic effects are not affected by region and year. Consider then the *nested* model,

$$\theta_{it} \overset{iid}{\sim} N(0\,,\, 1/\tau_t) \ \text{ and } \ \phi_{it} \sim CAR(\lambda_t) , \tag{12.41}$$

where $\tau_t \overset{iid}{\sim} G(a,b)$ and $\lambda_t \overset{iid}{\sim} G(c,d)$. Provided these hyperpriors are not too informative, this allows "shrinkage" of the year-specific effects toward their grand mean, and in a way that allows the data to determine the amount of shrinkage.

Thus our most general model for $\mu_{i\ell t}$ is

$$\mu_{i\ell t} = \mathbf{x}'_\ell \boldsymbol{\beta} + \delta_t + \mathbf{z}'_i \boldsymbol{\omega} + \theta_{it} + \phi_{it} ,$$

with corresponding joint posterior distribution proportional to

$$L(\boldsymbol{\beta}, \delta, \boldsymbol{\omega}, \boldsymbol{\theta}, \boldsymbol{\phi}; \mathbf{y}) p(\delta) p(\boldsymbol{\theta} \mid \tau) p(\boldsymbol{\phi} \mid \lambda) p(\tau) p(\lambda) .$$

Computation via univariate Metropolis and Gibbs updating steps is relatively straightforward (and readily available in this aligned data setting. However, convergence can be rather slow due to the weak identifiability of the joint parameter space. As a possible remedy, consider the simple space-only case again for a moment. We may transform from $(\boldsymbol{\theta}, \boldsymbol{\phi})$ to $(\boldsymbol{\theta}, \boldsymbol{\eta})$ where $\eta_i = \theta_i + \phi_i$. Then $p(\boldsymbol{\theta}, \boldsymbol{\eta} \mid \mathbf{y}) \propto L(\boldsymbol{\eta}; \mathbf{y}) p(\boldsymbol{\theta}) p(\boldsymbol{\eta}-\boldsymbol{\theta})$, so that

$$p(\eta_i \mid \eta_{j\neq i}, \boldsymbol{\theta}, \mathbf{y}) \propto L(\eta_i; y_i)\, p(\eta_i - \theta_i \mid \{\eta_j - \theta_j\}_{j\neq i})$$

and

$$p(\theta_i \mid \theta_{j\neq i}, \boldsymbol{\eta}, \mathbf{y}) \propto p(\theta_i)\, p(\eta_i - \theta_i \mid \{\eta_j - \theta_j\}_{j\neq i}) .$$

This simple transformation improves matters since each η_i full conditional is now well identified by the data point Y_i, while the weakly identified (indeed, "Bayesianly unidentified") θ_i now emerges in closed form as a normal distribution (since the nonconjugate Poisson likelihood no longer appears).

Demographic subgroup	Contribution to ε_{jk}	Fitted relative risk
White males	0	1
White females	α	0.34
Nonwhite males	β	1.02
Nonwhite females	$\alpha + \beta + \xi$	0.28

Table 12.10 *Fitted relative risks, four sociodemographic subgroups in the Ohio lung cancer data.*

Example 12.5 The study of the trend of risk for a given disease in space and time may provide important clues in exploring the underlying causes of the disease and helping to develop environmental health policy. Waller et al. [1997] consider the following data set on lung cancer mortality in Ohio. Here Y_{ijkt} is the number of lung cancer deaths in county i during year t for gender j and race k in the state of Ohio. The data are recorded for $J = 2$ genders (male and female, indexed by s_j) and $K = 2$ races (white and nonwhite, indexed by r_k) for each of the $I = 88$ Ohio counties over $T = 21$ years (1968–1988).

We adopt the model,

$$\mu_{ijkt} = s_j\alpha + r_k\beta + s_jr_k\xi + \theta_{it} + \phi_{it} , \qquad (12.42)$$

where $s_j = 1$ if $j = 2$ (female) and 0 otherwise, and $r_k = 1$ if $k = 2$ (nonwhite) and 0 otherwise. That is, there is one subgroup (white males) for which there is no contribution to the mean structure (12.51). For our prior specification, we select

$$\theta_{it} \stackrel{ind}{\sim} N\left(0, \tfrac{1}{\tau_t}\right) \quad \text{and} \quad \phi_{it} \sim CAR(\lambda_t) ;$$
$$\alpha, \beta, \xi \sim \text{flat} ;$$
$$\tau_t \stackrel{iid}{\sim} G(1, 100) \quad \text{and} \quad \lambda_t \stackrel{iid}{\sim} G(1, 7) ,$$

where the relative sizes of the hyperparameters in these two gamma distributions were selected following guidance given in Bernardinelli et al. [1995a] ; see also Best et al. [1999] and Eberly and Carlin [2000].

Histograms of the sampled values showed θ_{it} distributions centered near 0 in most cases, but ϕ_{it} distributions typically removed from 0, suggesting that the heterogeneity effects are not really needed in this model. Plots of $E(\tau_t \mid \mathbf{y})$ and $E(\lambda_t \mid \mathbf{y})$ versus t suggest increasing clustering and slightly increasing heterogeneity over time. The former might be the result of flight from the cities to suburban "collar counties" over time, while the latter is likely due to the elevated mean levels over time (for the Poisson, the variance increases with the mean).

Fitted relative risks obtained by Waller et al. [1997] for the four main demographic subgroups are shown in Table 12.10. The counterintuitively positive fitted value for nonwhite females may be an artifact of the failure of this analysis to age-standardize the rates prior to modeling (or at least to incorporate age group as another demographic component in the model). To remedy this, consider the following revised and enhanced model, described by Xia and Carlin [1998], where we assume that

$$Y^*_{ijkt} \sim Poisson(E_{ijkt}\exp(\mu_{ijkt})) , \qquad (12.43)$$

where again Y^*_{ijkt} denotes the observed age-adjusted deaths in county i for sex j, race k, and year t, and E_{ijkt} are the expected death counts. We also incorporate an ecological level smoking behavior covariate into our log-relative risk model, namely,

$$\mu_{ijkt} = \mu + s_j\alpha + r_k\beta + s_jr_k\xi + p_i\rho + \gamma t + \phi_{it} , \qquad (12.44)$$

Demographic subgroup	Contribution to ε_{jk}	Fitted log- relative risk	Fitted relative risk
White males	0	0	1
White females	α	-1.06	0.35
Nonwhite males	β	0.18	1.20
Nonwhite females	$\alpha + \beta + \xi$	-1.07	0.34

Table 12.11 *Fitted relative risks, four sociodemographic subgroups in the Ohio lung cancer data.*

where p_i is the true smoking proportion in county i, γ represents the fixed time effect, and the ϕ_{it} capture the random spatial effects over time, wherein clustering effects are nested within time. That is, writing $\boldsymbol{\phi}_t = (\phi_{1t}, \ldots, \phi_{It})'$, we let $\boldsymbol{\phi}_t \sim CAR(\lambda_t)$ where $\lambda_t \overset{iid}{\sim} G(c,d)$. We assume that the sociodemographic covariates (sex and race) do not interact with time or space. Following the approach of Bernardinelli et al. [1995b], we introduce both sampling error and spatial correlation into the smoking covariate. Let

$$q_i \mid p_i \sim N(p_i, \sigma_q^2), \; i = 1, \ldots, I, \text{ and} \tag{12.45}$$

$$\mathbf{p} \sim CAR(\lambda_p) \iff p_i \mid p_{j \neq i} \sim N(\mu_{p_i}, \sigma_{p_i}^2), \; i = 1, \ldots, I, \tag{12.46}$$

where q_i is the current smoking proportion observed in a sample survey of county i (an imperfect measurement of p_i), $\mu_{p_i} = \sum_{j \neq i} w_{ij} p_j / \sum_{j \neq i} w_{ij}$, and $\sigma_{p_i}^2 = (\lambda_p \sum_{j \neq i} w_{ij})^{-1}$. Note that the amount of smoothing in the two CAR priors above may differ, since the smoothing is controlled by different parameters λ_ϕ and λ_p. Like λ_ϕ, λ_p is also assigned a gamma hyperprior, namely, a $Gamma(e, f)$.

Using a Gibbs-Metropolis algorithm for fitting, we obtained the 95% posterior credible sets $[-1.14, -0.98]$, $[0.07, 0.28]$, and $[-0.37, -0.01]$ for α, β, and ξ, respectively. Note that all 3 fixed effects are significantly different from 0, in contrast to our Table 12.10 results, which failed to uncover a main effect for race. The corresponding point estimates are translated into the fitted relative risks for the four sociodemographic subgroups in Table 12.11. Nonwhite males experience the highest risk, followed by white males, with females of both races having much lower risks. ■

12.7.2 *Misalignment across years*

In this subsection we develop a spatiotemporal model to accommodate the situation of Figure 12.17, wherein the response variable and the covariate are spatially aligned within any given timepoint, but not across timepoints (due to periodic changes in the regional grid). Assuming that the observed disease count Y_{it} for zip i in year t is conditionally independent of the other zip-level disease counts given the covariate values, we have the model,

$$Y_{it} \mid \mu_{it} \overset{ind}{\sim} Po(E_{it} \exp(\mu_{it})), \; i = 1, \ldots, I_t, \; t = 1, \ldots, T,$$

where the expected count for zip i in year t, E_{it}, is proportional to the population count. In our case, we set $E_{it} = R n_{it}$, where n_{it} is the population count in zip i at year t and $R = (\sum_{it} Y_{it}) / (\sum_{it} n_{it})$, the grand asthma hospitalization rate (i.e., the expected counts assume homogeneity of disease rates across all zips and years). The log-relative risk is modeled as

$$\mu_{it} = x_{it} \beta_t + \delta_t + \theta_{it} + \phi_{it}, \tag{12.47}$$

where x_{it} is the zip-level exposure covariate (traffic density) depicted for 1983 in Figure 12.17, β_t is the corresponding main effect, δ_t is an overall intercept for year t, and

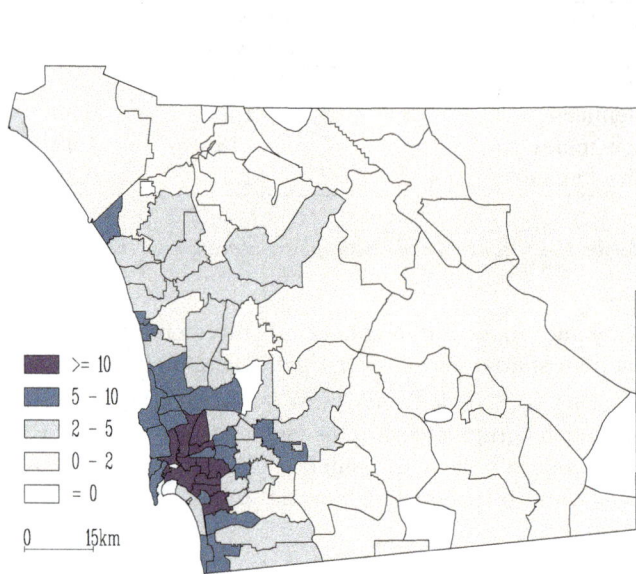

Figure 12.17 *Traffic density (average vehicles per km of major roadway) in thousands by zip code for 1983, San Diego County.*

θ_{it} and ϕ_{it} are zip- and year-specific heterogeneity and clustering random effects, analogous to those described in Section 12.7.1. The changes in the zip grid over time cloud the interpretation of these random effects (e.g., a particular region may be indexed by different i in different years), but this does not affect the interpretation of the main effects β_t and δ_t; it is simply the analogue of unbalanced data in a longitudinal setting. In the spatiotemporal case, the distributions on these effects become

$$\boldsymbol{\theta}_t \stackrel{ind}{\sim} N\left(0, \frac{1}{\tau_t} I\right) \ \text{ and } \ \boldsymbol{\phi}_t \stackrel{ind}{\sim} CAR(\lambda_t) \,, \tag{12.48}$$

where $\boldsymbol{\theta}_t = (\theta_1, \ldots, \theta_{I_t})'$, $\boldsymbol{\phi}_t = (\phi_1, \ldots, \phi_{I_t})'$, and we encourage similarity among these effects across years by assuming $\tau_t \stackrel{iid}{\sim} G(a,b)$ and $\lambda_t \stackrel{iid}{\sim} G(c,d)$, where G again denotes the gamma distribution. Placing flat (uniform) priors on the main effects β_t and δ_t completes the model specification. Note that the constraints $\sum_i \phi_{it} = 0$, $t = 1, \ldots, T$ must be added to identify the year effects δ_t, due to the location invariance of the CAR prior.

Example 12.6 Asthma is the most common chronic disease diagnosis for children in the U.S. (National Center for Environmental Health, 1996). A large number of studies have shown a correlation between known products and byproducts of auto exhaust (such as ozone, nitrogen dioxide, and particulate matter) and pediatric asthma ER visits or hospitalizations. Several studies [e.g., Tolbert et al., 2000b, Zidek et al., 1998, Best et al., 2000] have used hierarchical Bayesian methods in such investigations. An approach taken by some authors is to use proximity to major roadways (or some more refined measure of closeness to automobile traffic) as an omnibus measure of exposure to various asthma-inducing pollutants. We also adopt this approach and use the phrase "exposure" in what follows, even though in fact our traffic measures are really surrogates for true exposure.

Our data set arises from San Diego County, CA, the region pictured in Figure 12.17. The city of San Diego is located near the southwestern corner of the map; the map's western boundary is the Pacific Ocean, while Mexico forms its southern boundary. The subregions pictured are the zip codes as defined in 1983; as mentioned earlier this grid changes over time. Specifically, during the course of our 8-year (1983–1990) study period, the zip code boundaries changed four times: in 1984, 1987, 1988, and 1990.

The components of our data set are as follows. First, for a given year, we have the number of discharges from hospitalizations due to asthma for children aged 14 and younger by zip code (California Office of Statewide Health Planning and Development, 1997). The primary diagnosis was asthma based on the International Classification of Diseases, code 493 (U.S. Department of Health and Human Services, 1989). Assuming that patient records accurately report the correct zip code of residence, these data can be thought of as error-free.

Second, we have zip-level population estimates (numbers of residents aged 14 and younger) for each of these years [see Zhu et al., 2000]. These estimates were obtained in ARC/INFO using the following process. First, a land-use covariate was used to assist in a linear interpolation between the 1980 and 1990 U.S. Census figures, to obtain estimates at the census block group level. Digitized hard-copy U.S. Postal Service maps or suitably modified street network files provided by the San Diego Association of Governments (SANDAG) were then used to reallocate these counts to the zip code grid for the year in question. To do this, the GIS first created a subregional grid by intersecting the block group and zip code grids. The block group population totals were allocated to the subregions per a combination of subregional area and population density (the latter again based on the land-use covariate). Finally, these imputed subregional counts were reaggregated to the zip grid. While there are several possible sources of uncertainty in these calculations, we ignore them in our initial round of modeling, assuming these population counts to be fixed and known.

Finally, for each of the major roads in San Diego County, we have mean yearly traffic counts on each road segment in our map. Here "major" roads are defined by SANDAG to include interstate highways or equivalent, major highways, access or minor highways, and arterial or collector routes. The sum of these numbers within a given zip divided by the total length of its major roads provides an aggregate measure of traffic exposure for the zip. These zip-level *traffic densities* are plotted for 1983 in Figure 12.17; this is the exposure measure we use in the following text.

We set $a = 1$, $b = 10$ (i.e., the τ_t have prior mean and standard deviation both equal to 10) and $c = 0.1$, $d = 10$ (i.e., the λ_t have prior mean 1, standard deviation $\sqrt{10}$). These are fairly vague priors designed to let the data dominate the allocation of excess spatial variability to heterogeneity and clustering. (As mentioned near equation (6.39), simply setting these two priors equal to each other would not achieve this, since the prior for the θ_{it} is specified *marginally*, while that for the ϕ_{it} is specified *conditionally* given the neighboring ϕ_{jt}.)

Plots of the posterior medians and 95% equal-tail Bayesian confidence intervals for β_t (not shown) makes clear that, with the exception of that for 1986, all of the β_t's are significantly greater than 0. Hence, the traffic exposure covariate in Figure 12.17 is positively associated with increased pediatric asthma hospitalization in seven of the eight years of our study. To interpret these posterior summaries, recall that their values are on the *log*-relative risk scale. Thus a zip having a 1983 traffic density of 10 thousand cars per km of roadway would have median relative risk $e^{10(.065)} = 1.92$ times higher than a zip with essentially no traffic exposure, with a corresponding 95% confidence interval of $(e^{10(.000)}, e^{10(.120)}) = (1.00, 3.32)$. There also appears to be a slight weakening of the traffic-asthma association over time.

Figure 12.18 provides ARC/INFO maps of the crude and fitted asthma rates (per thousand) in each of the zips for 1983. The crude rates are of course given by $r_{it} = Y_{it}/n_{it}$, while the fitted rates are given by $R \exp(\hat{\mu}_{it})$, where R is again the grand asthma rate across all zips and years and $\hat{\mu}_{it}$ is obtained by plugging in the estimated posterior means for the various components in equation (12.47). The figure clearly shows the characteristic Bayesian shrinkage of the crude rates toward the grand rate. In particular, no zip is now assigned a rate of exactly zero, and the rather high rates in the thinly populated eastern part of the

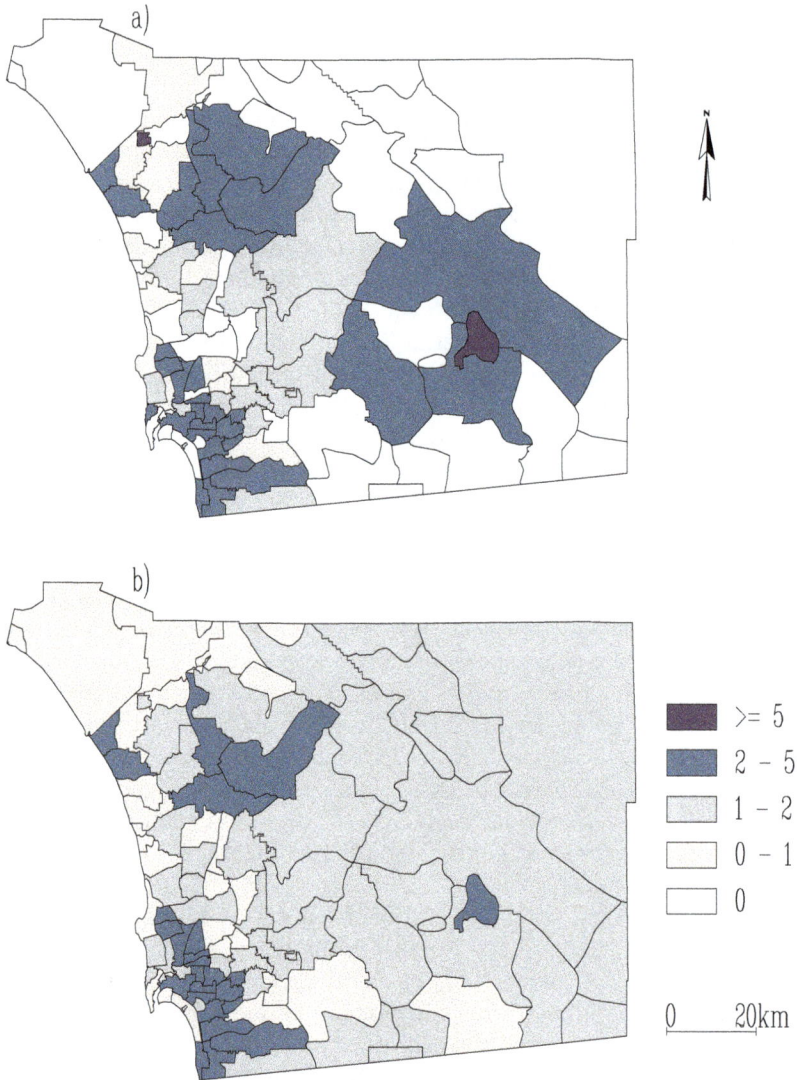

Figure 12.18 *Pediatric asthma hospitalization rate (per thousand children) by zip code for 1983, San Diego County: (a) crude rate, (b) temporally misaligned model fitted rate.*

map have been substantially reduced. However, the high observed rates in urban San Diego continue to be high, as the method properly recognizes the much higher sample sizes in these zips. There also appears to be some tendency for clusters of similar crude rates to be preserved, the probable outcome of the CAR portion of our model. ■

12.7.3   *Nested misalignment both within and across years*

In this subsection we extend our spatiotemporal model to accommodate the situation of Figure 12.19, wherein the covariate is available on a grid that is a refinement of the grid for which the response variable is available (i.e., nested misalignment within years, as well as misalignment across years). Letting the subscript j index the subregions (which we also refer to as *atoms*) of zip i, our model now becomes

$$Y_{ijt} \mid \mu_{ijt} \sim Po(E_{ijt} \exp(\mu_{ijt})), \; i = 1, \ldots, I_t, \; j = 1, \ldots, J_{it}, \; t = 1, \ldots, T,$$

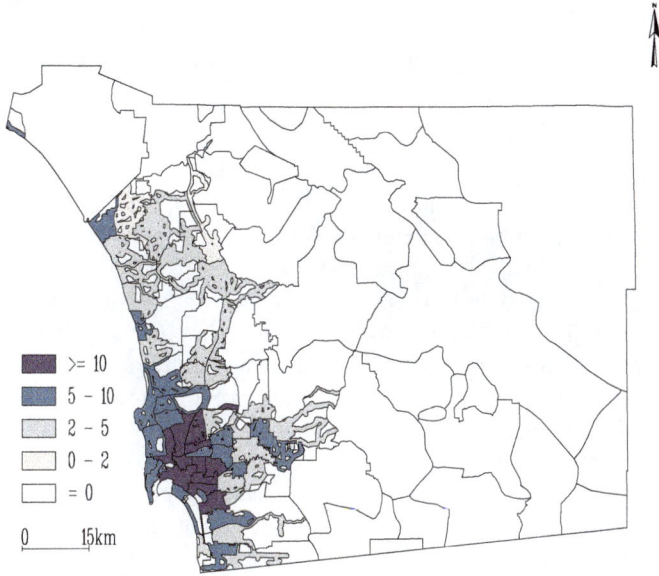

Figure 12.19 *Adjusted traffic density (average vehicles per km of major roadway) in thousands by zip code subregion for 1983, San Diego County.*

where the expected counts E_{ijt} are now Rn_{ijt}, with the grand rate R as before. The population of atom ijt is not known, and so we determine it by areal interpolation as $n_{ijt} = n_{it}(\text{area of atom } ijt)/(\text{area of zip } it)$. The log-relative risk in atom ijt is then modeled as

$$\mu_{ijt} = x_{ijt}\beta_t + \delta_t + \theta_{it} + \phi_{it}, \tag{12.49}$$

where x_{ijt} is now the atom-level exposure covariate (depicted for 1983 in Figure 12.19), but β_t, δ_t, θ_{it} and ϕ_{it} are as before. Thus our prior specification is exactly that of the previous subsection; priors for the $\boldsymbol{\theta}_t$ and $\boldsymbol{\phi}_t$ as given in equation (12.48), exchangeable gamma hyperpriors for the τ_t and λ_t with $a = 1$, $b = 10$, $c = 0.1$, and $d = 10$, and flat priors for the main effects β_t and δ_t.

Since only the zip-level hospitalization totals Y_{it} (and not the atom-level totals Y_{ijt}) are observed, we use the additivity of conditionally independent Poisson distributions to obtain

$$Y_{it} \mid \beta_{it}, \delta_t, \theta_{it}, \phi_{it} \sim Po\left(\sum_{j=1}^{J_{it}} E_{ijt}\exp(\mu_{ijt})\right),\ i = 1,\ldots,I_t,\ t = 1,\ldots,T. \tag{12.50}$$

Using expression (12.50), we can obtain the full Bayesian model specification for the observed data as

$$\left[\prod_{t=1}^{T}\prod_{i=1}^{I_t} p\left(y_{it} \mid \beta_t, \delta_t, \theta_{it}, \phi_{it}\right)\right]\left[\prod_{t=1}^{T} p(\boldsymbol{\theta}_t \mid \tau_t)p(\boldsymbol{\phi}_t \mid \lambda_t)p(\tau_t)p(\lambda_t)\right] \tag{12.51}$$

As in the previous section, only the τ_t and λ_t parameters may be updated via ordinary Gibbs steps, with Metropolis steps required for the rest.

Note that model specification (12.51) makes use of the atom-level covariate values x_{ijt}, but only the zip-level hospitalization counts Y_{it}. Of course, we might well be interested in *imputing* the values of the missing subregional counts Y_{ijt}, whose full conditional distribution is multinomial, namely,

$$(Y_{i1t}, \dots, Y_{iJ_it}) \mid Y_{it}, \beta_t, \delta_t, \theta_{it}, \phi_{it} \sim Mult(Y_{it}, \{q_{ijt}\}) \,, \qquad (12.52)$$

$$\text{where} \quad q_{ijt} = \frac{E_{ijt}e^{\mu_{ijt}}}{\sum_{j=1}^{J_{it}} E_{ijt}e^{\mu_{ijt}}} \,.$$

Since this is a purely predictive calculation, Y_{ijt} values need not be drawn as part of the MCMC sampling order, but instead at the very end, conditional on the post-convergence samples.

Zhu et al. [2003] use Figure 12.19 to refine the definition of exposure used in Example 12.6 by subdividing each zip into subregions based on whether or not they are closer than 500 m to a major road. This process involves creating "buffers" around each road and subsequently overlaying them in a GIS, and has been previously used in several studies of vehicle emissions. This definition leads to some urban zips becoming "entirely exposed," as they contain no point further than 500 m from a major road; these are roughly the zips with the darkest shading in Figure 12.17 (i.e., those having traffic densities greater than 10,000 cars per year per km of major roadway). Analogously, many zips in the thinly populated eastern part of the county contained at most one major road, suggestive of little or no traffic exposure. As a result, we (somewhat arbitrarily) defined those zips in the two lightest shadings (i.e., those having traffic densities less than 2,000 cars per year per km of roadway) as being "entirely unexposed." This typically left slightly less than half the zips (47 for the year shown, 1983) in the middle range, having some exposed and some unexposed subregions, as determined by the intersection of the road proximity buffers. These subregions are apparent as the lightly shaded regions in Figure 12.19; the "entirely exposed" regions continue to be those with the darkest shading, while the "entirely unexposed" regions have no shading.

The fitted rates obtained by Zhu et al. [2000] provide a similar overall impression as those in Figure 12.18, except that the newer map is able to show subtle differences within several "partially exposed" regions. These authors also illustrate the interpolation of missing subregional counts Y_{ijt} using equation (12.52). Analogous to the block-block FMPC imputation in Subsection 7.2, the sampling-based hierarchical Bayesian method produces more realistic estimates of the subregional hospitalization counts, with associated confidence limits emerging as an automatic byproduct.

12.7.4 Nonnested misalignment and regression

In this subsection we consider spatiotemporal *regression* in the misaligned data setting motivated by our Atlanta ozone data set. Recall that the first component of this data set provides ozone measurements X_{itr} at between 8 and 10 fixed monitoring sites i for day t of year r, where $t = 1, \dots, 92$ (the summer days from June 1 through August 31) and $r = 1, 2, 3$, corresponding to years 1993, 1994, and 1995 For example, Figure 1.3 shows the 8-hour daily maximum ozone measurements (in parts per million) at the 10 monitoring sites for a particular day (July 15, 1995), along with the boundaries of the 162 zip codes in the Atlanta metropolitan area.

A *second* component of this data set (about which we so far have said far less) provides relevant health outcomes, but only at the zip code level. Specifically, for each zip l, day t, and year r, we have the number of pediatric emergency room (ER) visits for asthma, Y_{ltr}, as well as the total number of pediatric ER visits, n_{ltr}. These data come from a historical records-based investigation of pediatric asthma emergency room visits to seven major emergency care centers in the Atlanta metropolitan statistical area during the same three summers. Our main substantive goal is an investigation of the relationship between ozone and pediatric ER visits for asthma in Atlanta, controlling for a range of sociodemographic covariates. Potential covariates (available only as zip-level summaries in our data set) include average age, percent male, percent black, and percent using Medicaid for payment (a crude surrogate for socioeconomic status). Clearly an investigation of the relationship between ozone exposure and pediatric ER visit count cannot be undertaken until the mismatch in the support of the (point-level) predictor and (zip-level) response variables is resolved.

A naive approach would be to average the ozone measurements belonging to a specific zip code, then relate this average ozone measurement to the pediatric asthma ER visit count in this zip. In fact, there are few monitoring sites relative to the number of zip codes; Figure 1.3 shows most of the zip codes contain no sites at all, so that most of the zip-level ER visit count data would be discarded. An alternative would be to aggregate the ER visits over the entire area and model them as a function of the average of the ozone measurements (that is, eliminate the spatial aspect of the data and fit a temporal-only model). Using this idea in a Poisson regression, we obtained a coefficient for ozone of 2.48 with asymptotic standard error 0.71 (i.e., significant positive effect of high ozone on ER visit rates). While this result is generally consistent with our findings, precise comparison is impossible for a number of reasons. First, this approach requires use of data from the entire Atlanta metro area (due to the widely dispersed locations of the monitoring stations), not data from the city only as our approach allows. Second, it does not permit use of available covariates (such as race and SES) that were spatially but not temporally resolved in our data set. Third, standardizing using expected counts E_i (as in equation (12.53) below) must be done only over days (not regions), so the effect of including them is now merely to adjust the model's intercept.

We now describe the disease component of our model, and subsequently assemble the full Bayesian hierarchical modeling specification for our spatially misaligned regression. Similar to the model of Subsection 12.7.3, we assume the zip-level asthma ER visit counts, Y_{ltr} for zip l during day t of summer r, follow a Poisson distribution,

$$Y_{ltr} \sim Poisson\left(E_{ltr}\exp(\lambda_{ltr})\right),$$ (12.53)

where the E_{ltr} are expected asthma visit counts, determined via internal standardization as $E_{ltr} = n_{ltr}(\sum_{ltr} Y_{ltr} / \sum_{ltr} n_{ltr})$, where n_{ltr} is the total number of pediatric ER visits in zip code l on day t of year r. Thus E_{ltr} is the number of pediatric ER asthma visits we would expect from the given zip and day if the proportion of such visits relative to the total pediatric ER visit rate was homogeneous across all zips, days, and years. Hence λ_{ltr} in (12.53) can be interpreted as a log-relative risk of asthma among those children visiting the ER in group ltr. Our study design is thus a *proportional admissions model* [Breslow et al., 1980, pp. 153–155].

We do not take n_{ltr} equal to the total number of children *residing* in zip l on day t of year r, since this standardization would implicitly presume a constant usage of the ER for pediatric asthma management across all zips, which seems unlikely (children from more affluent zips are more likely to have the help of family doctors or specialists in managing their asthma, and so would not need to rely on the ER; see Congdon and Best [2000] for a solution to the related problem of adjusting for patient referral practices). Note however that this in turn means that our disease (pediatric asthma visits) is not particularly "rare" relative to the total (all pediatric visits). As such, our use of the Poisson distribution in (12.53) should not be thought of as an approximation to a binomial distribution for a rare event, but merely as a convenient and sensible model for a discrete variable.

For the log-relative risks in group ltr, we begin with the model,

$$\lambda_{ltr} = \beta_0 + \beta_1 X_{l,t-1,r} + \sum_{c=1}^{C} \alpha_c Z_{cl} + \sum_{d=1}^{D} \delta_d W_{dt} + \theta_l.$$ (12.54)

Here, β_0 is an intercept term, and β_1 denotes the effect of ozone exposure $X_{l,t-1,r}$ in zip l during day $t-1$ of year r. Note that we model pediatric asthma ER visit counts as a function of the ozone level on the *previous* day, in keeping with the most common practice in the epidemiological literature [Tolbert et al., 2000b]. This facilitates next-day predictions for pediatric ER visits given the current day's ozone level, with our Bayesian approach permitting full posterior inference (e.g., 95% prediction limits). However, it also means we have only $(J-1) \times 3 = 273$ days worth of usable data in our sample. Also, $\mathbf{Z}_l = (Z_{1l},\ldots,Z_{Cl})^{\mathrm{T}}$ is a vector of C zip-level (but not time-varying) sociodemographic covariates with corresponding coefficient vector $\boldsymbol{\alpha} = (\alpha_1,\ldots,\alpha_C)^{\mathrm{T}}$, and $\mathbf{W}_t = (W_{1t},\ldots,W_{Dt})^{\mathrm{T}}$ is a vector of D day-level (but not spatially varying) temporal covariates with corresponding coefficient vector $\boldsymbol{\delta} = (\delta_1,\ldots,\delta_D)^{\mathrm{T}}$. Finally, θ_l is a zip-specific random effect designed to capture extra-Poisson variability in the observed ER visitation rates. These random effects may simply be assumed to be exchangeable draws from a $N(0,1/\tau)$ distribution (thus modeling overall *heterogeneity*), or may instead be assumed to vary spatially using a conditionally autoregressive (CAR) specification.

Of course, model (12.53)–(12.54) is not fittable as stated, since the zip-level previous-day ozone values $X_{l,t-1,r}$ are not observed. Fortunately, we may use the methods of Section 7.1 to perform the necessary point-block realignment. To connect our equation (12.54) notation with that used in Section 7.1, let us write $\mathbf{X}_{B,r} \equiv \{X_{l,t-1,r},\ l=1,\ldots,L, t=2,\ldots,J\}$ for the unobserved block-level data from year r, and $\mathbf{X}_{s,r} \equiv \{X_{itr},\ i=1,\ldots,I, t=1,\ldots,J\}$ for the observed site-level data from year r. Then, from equations (12.21) and (12.22) and assuming no missing ozone station data for the moment, we can find the conditional predictive distribution $f(\mathbf{X}_{B,r} \mid \mathbf{X}_{s,r},\boldsymbol{\gamma}_r,\sigma_r^2,\boldsymbol{\phi}_r,\boldsymbol{\rho}_r)$ for year r. However, for these data some components of the $\mathbf{X}_{s,r}$ will be missing, and thus replaced with imputed values $\mathbf{X}_{s,r}^{(m)}$, $m = 1,\ldots,M$, for some modest number of imputations M (say, $M=3$). (In a slight abuse of notation here, we assume that any *observed* component of $\mathbf{X}_{s,r}^{(m)}$ is simply set equal to that observed value for all m.)

Thus, the full Bayesian hierarchical model specification is given by

$$
\begin{aligned}
& \left[\prod_r \prod_t \prod_l f(Y_{ltr} \mid \boldsymbol{\beta},\boldsymbol{\alpha},\boldsymbol{\delta},\boldsymbol{\theta}, X_{l,t-1,r}) \right] p(\boldsymbol{\beta},\boldsymbol{\alpha},\boldsymbol{\delta},\boldsymbol{\theta}) \\
& \times \left[\prod_r f(\mathbf{X}_{B,r} \mid \mathbf{X}_{s,r}^{(m)},\boldsymbol{\gamma}_r,\sigma_r^2,\boldsymbol{\phi}_r,\boldsymbol{\rho}_r) \right. \\
& \left. \qquad \times f(\mathbf{X}_{s,r}^{(m)} \mid \boldsymbol{\gamma}_r,\sigma_r^2,\boldsymbol{\phi}_r,\boldsymbol{\rho}_r) p(\boldsymbol{\gamma}_r,\sigma_r^2,\boldsymbol{\phi}_r,\boldsymbol{\rho}_r) \right] ,
\end{aligned}
\tag{12.55}
$$

where $\boldsymbol{\beta} = (\beta_0,\beta_1)^{\mathrm{T}}$, and $\boldsymbol{\gamma}_r,\sigma_r^2,\boldsymbol{\phi}_r$ and $\boldsymbol{\rho}_r$ are year-specific versions of the parameters in (7.6). Note that there is a posterior distribution for each of the M imputations. Model (12.55) assumes the asthma-ozone relationship does not depend on year; the misalignment parameters are year-specific only to permit year-by-year realignment.

Zhu et al. [2003] offer a reanalysis of the Atlanta ozone and asthma data by fitting a version of model (12.54), namely,

$$
\lambda_{ltr} = \beta_0 + \beta_1 X_{l,t-1,r}^{*(m,v)} + \alpha_1 Z_{1l} + \alpha_2 Z_{2l} + \delta_1 W_{1t} + \delta_2 W_{2t} + \delta_3 W_{3t} + \delta_4 W_{4t} ,
\tag{12.56}
$$

where $X_{l,t-1,r}^{*(m,v)}$ denotes the (m,v)th imputed value for the zip-level estimate of the 8-hour daily maximum ozone measurement on the previous day $(t-1)$. Our zip-specific covariates are Z_{1l} and Z_{2l}, the percent high socioeconomic status and percent black race of those pediatric asthma ER visitors from zip l, respectively. Of the day-specific covariates, W_{1t} indexes day of summer ($W_{1t} = t \bmod 91$) and $W_{2t} = W_{1t}^2$, while W_{3t} and W_{4t} are indicator variables for days in 1994 and 1995, respectively (so that 1993 is taken as the reference year). We include both linear and quadratic terms for day of summer in order to capture the rough U-shape in pediatric ER asthma visits, with June and August higher than July.

Parameter	Effect	Posterior median	95% Posterior credible set	Fitted relative risk
β_0	intercept	−0.4815	(−0.5761, −0.3813)	—
β_1	ozone	0.7860	(−0.7921, 2.3867)	1.016†
α_1	high SES	−0.5754	(−0.9839, −0.1644)	0.562
α_2	black	0.5682	(0.3093, 0.8243)	1.765
δ_1	day	−0.0131	(−0.0190, −0.0078)	—
δ_2	day^2	0.00017	(0.0001, 0.0002)	—
δ_3	year 1994	0.1352	(0.0081, 0.2478)	1.145
δ_4	year 1995	0.4969	(0.3932, 0.5962)	1.644

Table 12.12 *Fitted relative risks for the parameters of interest in the Atlanta pediatric asthma ER visit data, full model. (†This is the posterior median relative risk predicted to arise from a .02 ppm increase in ozone.)*

The analysis of Zhu et al. [2003] is only approximate, in that they run *separate* MCMC algorithms on the portions of the model corresponding to the two lines of model (12.55). In the spirit of the multiple imputation approach to the missing (point-level) ozone observations, they also retain $V = 3$ post-convergence draws from each of our $M = 3$ imputed data sets, resulting in $MV = 9$ zip-level approximately imputed ozone vectors $\mathbf{X}_{B,r}^{*(m,v)}$.

The results of this approach are shown in Table 12.12. The posterior median of β_1 (.7860) is positive, as expected. An increase of .02 ppm in 8-hour maximum ozone concentration (a relatively modest increase, as seen from Figure 1.3) thus corresponds to a fitted relative risk of exp(.7860 × .02) ≈ 1.016, or a 1.6% increase in relative risk of a pediatric asthma ER visit. However, the 95% credible set for β_1 does include 0, meaning that this positive association between ozone level and ER visits is not "Bayesianly significant" at the 0.05 level. Using a more naive approach but data from all 162 zips in the Atlanta metro area, Carlin et al. [1999] estimate the above relative risk as 1.026, marginally significant at the .05 level (that is, the lower limit of the 95% credible set for β_1 was precisely 0).

Regarding the demographic variables, the effects of both percent high SES and percent black emerge as significantly different from 0. The relative risk for a zip made entirely of high SES residents would be slightly more than half that of a comparable all-low SES zip, while a zip with a 100% black population would have a relative risk nearly 1.8 times that of a 100% nonblack zip. As for the temporal variables, day of summer is significantly negative and its square is significantly positive, confirming the U-shape of asthma relative risks over a given summer. Both year 1994 and year 1995 show higher relative risk compared with year 1993, with estimated increases in relative risk of about 15% and 64%, respectively.

12.8   Areal-level continuous time modeling

Quick et al. [2013] address the less common setting where space is discrete and time is continuous. This can be envisioned in situations where a collection of N_s *spatially associated* functions of time over N_s regions are posited. Put another way, functions arising from neighboring regions are believed to resemble each other. The functional data analysis literature [see Ramsay and Silverman, 2005, and references therein] deals almost exclusively with kernel smoothers and roughness-penalty type (spline) models; recent discrete-space, continuous time examples using spline-based methods include the works by MacNab and Gustafson [2007] and Ugarte et al. [2010]. Baladandayuthapani et al. [2008] consider spatially correlated functional data modeling for point-referenced data by treating space as continuous. Delicado et al. [2010] provides a review of spatially associated functional modeling of time.

Quick et al. [2013] propose a class of Bayesian space-time models based upon a dynamic MRF evolves continuously over time. This accommodates spatial processes that are posited to be spatially indexed over a geographical map with a well-defined system of neighbors. Rather than modeling time using simple parametric forms, as is often done in longitudinal contexts, these authors employ a stochastic process, enhancing the model's adaptability to the data.

The benefits of using a continuous-time model over a discrete-time model are pronounced when investigators (e.g. public health officials) seek to understand the local effects of temporal impact at a resolution finer than that at which the data were sampled. For instance, despite collecting data monthly, there may be interest in interpolating over a particular week or even at a given day of that month. Dynamic space-time models that treat time discretely can offer statistically legitimate inferences only at the level of the data. In addition, the modeling also allows us to subsequently carry out inference on temporal gradients; that is, the rate of change of the underlying process over time (see Chapter 14 for inference on spatial gradients). Quick et al. [2013] show how such inference can be carried out in fully model-based fashion using exact posterior predictive distributions for the gradients at any arbitrary time point.

12.8.1 Areally referenced temporal processes

Here we provide a brief overview of the approach proposed by Quick et al. [2013]. Consider a map of a geographical region comprising N_s regions that are delineated by well-defined boundaries, and let $Y_i(t)$ be the outcome arising from region i at time t. For every region i, we believe that $Y_i(t)$ exists, at least conceptually, at every time point. However, the observations are collected not continuously but at discrete time points, say $\mathcal{T} = \{t_1, t_2, \ldots, t_{N_t}\}$. For simplicity, let us assume that the data comes from the same set of time points in $\mathcal{T}$ for each region. This is not strictly necessary for the ensuing development, but will facilitate the notation.

A spatial random effect model that treats space as continuous and time as discrete assumes that

$$Y_i(t) = \mu_i(t) + Z_i(t) + \epsilon_i(t), \quad \epsilon_i(t) \overset{ind}{\sim} N(0, \tau_i^2) \text{ for } i = 1, 2, \ldots, N_s, \tag{12.57}$$

where $\mu_i(t)$ captures large scale variation or trends, for example using a regression model, and $Z_i(t)$ is an underlying areally-referenced stochastic process over time that captures smaller-scale variations in the time scale while also accommodating spatial associations. Each region also has its own variance component, τ_i^2, which captures residual variation not captured by the other components.

The process $Z_i(t)$ specifies the probability distribution of correlated space-time random effects while treating space as discrete and time as continuous. We seek a specification that will allow temporal processes from neighboring regions to be more alike than from non-neighbors. As regards spatial associations, we will respect the discreteness inherent in the aggregated outcome. Rather than model an underlying response surface continuously over the region of interest, we want to treat the $Z_i(t)$'s as functions of time that are smoothed across neighbors.

The neighborhood structure arises from a discrete topology comprising a list of neighbors for each region. This is described using an $N_s \times N_s$ adjacency matrix $W = \{w_{ij}\}$, where $w_{ij} = 0$ if regions i and j are not neighbors and $w_{ij} = c \neq 0$ when regions i and j are neighbors, denoted by $i \sim j$. By convention, the diagonal elements of W are all zero. To account for spatial association in the $Z_i(t)$'s, a temporally evolving MRF for the areal units at any arbitrary time point t specifies the full conditional distribution for $Z_i(t)$ as depending only upon the neighbors of region i,

$$p(Z_i(t) \mid \{Z_{j \neq i}(t)\}) \sim N\left(\sum_{j \sim i} \alpha \frac{w_{ij}}{w_{i+}} Z_j(t), \frac{\sigma^2}{w_{i+}}\right),$$

where $w_{i+} = \sum_{j \sim i} w_{ij}$, $\sigma^2 > 0$, and α is a propriety parameter described below. This means that the $N_s \times 1$ vector $\mathbf{Z}(t) = (Z_1(t), Z_2(t), \ldots, Z_{N_s}(t))^{\mathrm{T}}$ follows a multivariate normal distribution with zero mean and a precision matrix $\frac{1}{\sigma^2}(D - \alpha W)$, where D is a diagonal matrix with w_{i+} as its i-th diagonal elements. The precision matrix is invertible as long as $\alpha \in (1/\lambda_{(1)}, 1/\lambda_{(n)})$, where $\lambda_{(1)}$ (which can be shown to be negative) and $\lambda_{(n)}$ (which can be shown to be 1) are the smallest (i.e., most negative) and largest eigenvalues of $D^{-1/2}WD^{-1/2}$, respectively, and this yields a proper distribution for $\mathbf{Z}(t)$ at each timepoint t.

The MRF in (12.58) does not allow temporal dependence; the $\mathbf{Z}(t)$'s are independently and identically distributed as $N\left(\mathbf{0}, \sigma^2(D - \alpha W)^{-1}\right)$. We could allow time-varying parameters σ_t^2 and α_t so that $\mathbf{Z}(t) \overset{ind}{\sim} N\left(\mathbf{0}, \sigma_t^2(D - \alpha_t W)^{-1}\right)$ for every t. If time were treated discretely, then we could envision dynamic autoregressive priors for these time-varying parameters, or some transformations thereof. However, there are two reasons why we do not pursue this further. First, we do not consider time as discrete because that would preclude inference on temporal gradients, which, as we have mentioned, is a major objective here. Second, time-varying hyperparameters, especially the α_t's, in MRF models are usually weakly identified by the data; they permit very little prior-to-posterior learning and often lead to over-parametrized models that impair predictive performance over time.

Quick et al. [2013] prefer to jointly build spatial-temporal associations into the model using a multivariate process specification for $\mathbf{Z}(t)$. A highly flexible and computationally tractable option is to assume that $\mathbf{Z}(t)$ is a zero-centered multivariate Gaussian process, $GP(\mathbf{0}, K_Z(\cdot, \cdot))$, where the matrix-valued covariance function (e.g., "*cross-covariance* matrix function", Cressie, 1993) $K_Z(t, u) = \mathrm{Cov}\{\mathbf{Z}(t), \mathbf{Z}(u)\}$ is defined to be the $N_s \times N_s$ matrix with (i, j)-th entry $\mathrm{Cov}\{Z_i(t), Z_j(u)\}$ for any $(t, u) \in \Re^+ \times \Re^+$. Thus, for any two positive real numbers t and u, $K_Z(t, u)$ is an $N_s \times N_s$ matrix with (i, j)-th element given by the covariance between $Z_i(t)$ and $Z_j(u)$. These multivariate processes are *stationary* when the covariances are functions of the separation between the time-points, in which case we write $K_Z(t, u) = K_Z(\Delta)$, and *fully symmetric* when $K_Z(t, u) = K_Z(|\Delta|)$, where $\Delta = t - u$.

To ensure valid joint distributions for process realizations, we use a constructive approach similar to that used in *linear models of coregionalization* (LMC) and, more generally, belonging to the class of multivariate latent process models. We assume that $\mathbf{Z}(t)$ arises as a (possibly temporally-varying) linear transformation $\mathbf{Z}(t) = A(t)\mathbf{v}(t)$ of a simpler process $\mathbf{v}(t) = (v_1(t), v_2(t), \ldots, v_{N_s}(t))^{\mathrm{T}}$, where the $v_i(t)$'s are univariate temporal processes, independent of each other, and with unit variances. This differs from the conventional LMC approach based on *spatial* processes, which treats space as continuous. The matrix-valued covariance function for $\mathbf{v}(t)$, say $K_{\mathbf{v}}(t, u)$, thus has a simple diagonal form and $K_Z(t, u) = A(t)K_{\mathbf{v}}(t, u)A(u)^{\mathrm{T}}$. The dispersion matrix for $\mathbf{Z}$ is $\Sigma_Z = \mathcal{A}\Sigma_{\mathbf{v}}\mathcal{A}^{\mathrm{T}}$, where $\mathcal{A}$ is a block-diagonal matrix with $A(t_j)$'s as blocks, and $\Sigma_{\mathbf{v}}$ is the dispersion matrix constructed from $K_{\mathbf{v}}(t, u)$. Constructing simple valid matrix-valued covariance functions for $\mathbf{v}(t)$ automatically ensures valid probability models for $\mathbf{Z}(t)$. Also note that for $t = u$, $K_{\mathbf{v}}(t, t)$ is the identity matrix so that $K_Z(t, t) = A(t)A(t)^{\mathrm{T}}$ and $A(t)$ is a square-root (e.g. obtained from the triangular Cholesky factorization) of the matrix-valued covariance function at time t.

The above framework subsumes several simpler and more intuitive specifications. One particular specification that we pursue here assumes that each $v_i(t)$ follows a stationary Gaussian Process $GP(0, \rho(\cdot, \cdot; \phi))$, where $\rho(\cdot, \cdot; \phi)$ is a positive definite correlation function parametrized by ϕ (e.g. Stein, 1999), so that $\mathrm{Cov}(v_i(t), v_i(u)) = \rho(t, u; \phi)$ for every

$i = 1, 2, \ldots, N_s$ for all non-negative real numbers t and u. Since the $v_i(t)$ are independent across i, $\mathrm{Cov}\{v_i(t), v_j(u)\} = 0$ for $i \neq j$.

The matrix-valued covariance function for $\mathbf{Z}(t)$ becomes $K_Z(t, u) = \rho(t, u; \boldsymbol{\phi})A(t)A(u)^{\mathrm{T}}$. If we further assume that $A(t) = A$ is constant over time, then the process $\mathbf{Z}(t)$ is stationary if and only if $\mathbf{v}(t)$ is stationary. Further, we obtain a *separable* specification, so that $K_Z(t, u) = \rho(t, u; \boldsymbol{\phi})AA^{\mathrm{T}}$. Letting A be some square-root (e.g. Cholesky) of the $N_s \times N_s$ dispersion matrix $\sigma^2(D - \alpha W)^{-1}$ and $R(\boldsymbol{\phi})$ be the $N_t \times N_t$ temporal correlation matrix having (i, j)-th element $\rho(t_i, t_j; \boldsymbol{\phi})$ yields

$$K_Z(t, u) = \sigma^2 \rho(t, u; \boldsymbol{\phi})(D - \alpha W)^{-1} \text{ and } \Sigma_Z = R(\boldsymbol{\phi}) \otimes \sigma^2(D - \alpha W)^{-1} . \qquad (12.58)$$

It is straightforward to show that the marginal distribution from this constructive approach for each $\mathbf{Z}(t_i)$ is $N\left(\mathbf{0}, \sigma^2(D - \alpha W)^{-1}\right)$, the same marginal distribution as the temporally independent MRF specification in (12.58). Therefore, our constructive approach ensures a valid space-time process, where associations in space are modeled discretely using a MRF, and those in time through a continuous Gaussian process.

This separable specification is easily interpretable because it factorizes the dispersion into a spatial association component (areal) and a temporal component. Another significant practical advantage is its computational feasibility. Estimating more general space-time models usually entails matrix factorizations with $O(N_s^3 N_t^3)$ computational complexity. The separable specification allows us to reduce this complexity substantially by avoiding factorizations of $N_s N_t \times N_s N_t$ matrices. One could design algorithms to work with matrices whose dimension is the smaller of N_s and N_t, thereby accruing massive computational gains. More general models using this approach are introduced and discussed in the online supplement [Quick et al., 2013], but since they do not offer anything new in terms of temporal gradients, we do not pursue them further.

12.8.2 Hierarchical modeling

Following Quick et al. [2013], we build a hierarchical modeling framework using the likelihood from our spatial random effects model in (12.57) and the distributions emerging from the temporal Gaussian process discussed in Section 12.8.1. The mean $\mu_i(t)$ in (12.57) is often indexed by a parameter vector $\boldsymbol{\beta}$, for example a linear regression with regressors indexed by space and time so that $\mu_i(t; \boldsymbol{\beta}) = \mathbf{x}_i(t)^{\mathrm{T}}\boldsymbol{\beta}$.

The posterior distribution is

$$
\begin{aligned}
p(\boldsymbol{\theta}, \mathbf{Z} \mid \mathbf{Y}) \quad \propto \quad & p(\boldsymbol{\phi}) \times IG(\sigma^2 \mid a_\sigma, b_\sigma) \times \left(\prod_{i=1}^{M} IG(\tau_i^2 \mid a_\tau, b_\tau) \right) \\
& \times N(\boldsymbol{\beta} \mid \mu_\beta, \Sigma_\beta) \times Beta(\alpha \mid a_\alpha, b_\alpha) \\
& \times N(\mathbf{Z} \mid \mathbf{0}, R(\phi) \otimes \sigma^2(D - \alpha W)^{-1}) \\
& \times \prod_{j=1}^{N_t} \prod_{i=1}^{N_s} N(Y_i(t_j) \mid \mathbf{x}_i(t_j)^{\mathrm{T}}\boldsymbol{\beta} + Z_i(t_j), \tau_i^2) , \qquad (12.59)
\end{aligned}
$$

where $\boldsymbol{\theta} = \{\phi, \alpha, \sigma^2, \beta, \tau_1^2, \tau_2^2, \ldots, \tau_{N_s}^2\}$ and $\mathbf{Y}$ is the vector of observed outcomes defined analogous to $\mathbf{Z}$. The parametrizations for the standard densities are as in Gelman et al. [2013]. We assume all the other hyperparameters in (12.59) are known.

Recall the separable matrix-valued covariance function in (12.58). The correlation function $\rho(\cdot; \boldsymbol{\phi})$ determines process smoothness and we choose it to be a fully symmetric Matérn correlation function Markov chain Monte Carlo (MCMC) can be used to evaluate the joint posterior in (12.59), using Metropolis steps for updating ϕ and Gibbs steps for all other parameters; details are available in the supplemental article [Quick et al., 2013]. Sampling-based Bayesian inference seamlessly delivers inference on the residual spatial effects. Specifically, if t_0 is an arbitrary unobserved timepoint, then, for any region i, we sample from the posterior predictive distribution

$$p(Z_i(t_0) \mid \mathbf{Y}) = \int p(Z_i(t_0) \mid \mathbf{Z}, \boldsymbol{\theta}) \, p(\boldsymbol{\theta}, \mathbf{Z} \mid \mathbf{Y}) d\boldsymbol{\theta} \, d\mathbf{Z} \, .$$

This is achieved using *composition sampling*: for each sampled value of $\{\boldsymbol{\theta}, \mathbf{Z}\}$, we draw $Z_i(t_0)$, one for one, from $p(Z_i(t_0) \mid \mathbf{Z}, \boldsymbol{\theta})$, which is Gaussian. Also, our sampler easily adapts to situations where $Y_i(t)$ is missing (or not monitored) for some of the time points in region i. We simply treat such variables as missing values and update them, from their associated full conditional distributions, which of course are $N(\mathbf{x}_i(t)^{\mathsf{T}}\boldsymbol{\beta} + Z_i(t), \tau_i^2)$. We assume that all predictors in $\mathbf{x}_i(t)$ will be available in the space-time data matrix, so this temporal interpolation step for missing outcomes is straightforward and inexpensive.

Model checking is facilitated by simulating *independent* replicates for each observed outcome: for each region i and observed timepoint t_j, we sample from $p(Y_{rep,i}(t_j) \mid \mathbf{Y})$, which is equal to

$$\int N(Y_{rep,i}(t_j) \mid \mathbf{x}_i(t_j)^{\mathsf{T}}\boldsymbol{\beta} + Z_i(t_j), \tau_i^2) \, p(\boldsymbol{\beta}, Z_i(t_j), \tau_i^2 \mid \mathbf{Y}) d\boldsymbol{\beta} dZ_i(t_j) d\tau_i^2 \, ,$$

where $p(\boldsymbol{\beta}, Z_i(t_j), \tau_i^2 \mid \mathbf{Y})$ is the marginal posterior distribution of the unknowns in the likelihood. Sampling from the posterior predictive distribution is straightforward, again, using composition sampling.

Example 12.7 Quick et al. [2013] analyze a dataset consisting of monthly asthma hospitalization rates in the 58 counties of California over an 18-year period. As such, $N_t = 12 \times 18 = 216$, and we can simply set $t_j = j = 1, 2, \ldots, N_t$. The covariates in this model include population density, ozone level, the percent of the county under 18, and percent black. Population-based covariates are calculated for each county using the 2000 U.S. Census, so they do not vary temporally. In order to accommodate seasonality in the data, monthly fixed effects are included, using January as a baseline. Thus, after accounting for the monthly fixed effects and the four covariates of interest, $\mathbf{x}_i(t)$ is a 16×1 vector.

We compare the model in (12.29) with three alternative models using the DIC criterion. These models are all still of the form

$$Y_i(t) = \mathbf{x}_i(t)'\boldsymbol{\beta} + Z_i(t) + \epsilon_i(t), \quad \epsilon_i(t) \stackrel{ind}{\sim} N(0, \tau_i^2) \text{ for } i = 1, 2, \ldots, N_s \, , \qquad (12.60)$$

but with different $Z_i(t)$. Our first model is a simple linear regression model which ignores both the spatial and the temporal autocorrelation, i.e., $Z_i(t) = 0 \, \forall \, i, t$. The second model allows for a random intercept and random temporal slope, but ignores the spatial nature of the data, i.e., here $Z_i(t) = \alpha_{0i} + \alpha_{1i}t$, where $\alpha_{ki} \stackrel{iid}{\sim} N(0, \sigma_k^2)$, for $k = 0, 1$. In this model, to preserve model identifiability, we must remove the global intercept from our design matrix, $\mathbf{x}_i(t)$. Our third model builds upon the second, but introduces spatial autocorrelation by letting $\boldsymbol{\alpha}_k = (\alpha_{k1}, \ldots, \alpha_{kN_s})' \sim CAR(\sigma_k^2), k = 0, 1$. The results of the model comparison can be seen in Table 12.13, which indicates that our Gaussian process model has the lowest DIC value, and is thus the preferred model and the only one we consider henceforth. The surprisingly large p_D for the areally referenced Gaussian process model arises due to the very large size of the dataset (58 counties $\times$ 216 timepoints).

The estimates for our model parameters can be seen in Table 12.14. The coefficients for the monthly covariates indicate decreased hospitalization rates in the summer months, a trend which is consistent with previous findings. The coefficients for population density, percent under 18, and percent black are all significantly positive, also as expected. There

	p_D	DIC*
Simple Linear Regression	79	9,894
Random Intercept and Slope	165	4,347
CAR Model	117	7,302
Areally Referenced Gaussian Process	5,256	0

Table 12.13 *Comparisons between our areally referenced Gaussian process model and the three alternatives. p_D is a measure of model complexity, as it represents the effective number of parameters. Smaller values of DIC indicate a better trade-off between in sample model fit and model complexity. * DIC is standardized relative to the areally referenced Gaussian process model.*

Parameter	Median (95% CI)	Parameter	Median (95% CI)
β_0 (Intercept)	9.17 (8.93, 9.42)	β_{11} (August)	-3.58 (-4.02, -3.13)
β_1 (Pop Den)	0.60 (0.49, 0.70)	β_{12} (September)	-1.96 (-2.37, -1.54)
β_3 (% Under 18)	1.24 (1.15, 1.34)	β_{13} (October)	-1.36 (-1.73, -1.00)
β_4 (% Black)	1.12 (1.01, 1.24)	β_{14} (November)	-0.71 (-1.02, -0.42)
β_5 (February)	-0.25 (-0.46, -0.04)	β_{15} (December)	0.63 (0.41, 0.86)
β_6 (March)	-0.21 (-0.48, 0.07)	ϕ	0.90 (0.84, 0.97)
β_7 (April)	-1.47 (-1.81, -1.12)	α	0.77 (0.71, 0.80)
β_8 (May)	-1.17 (-1.53, -0.8)	σ^2	21.52 (20.18, 23.06)
β_9 (June)	-2.79 (-3.21, -2.4)	$\bar{\tau}^2$	3.32 (0.18, 213.16)
β_{10} (July)	-3.78 (-4.21, -3.37)		

Table 12.14 *Parameter estimates for asthma hospitalization data, where estimates for $\bar{\tau}^2$ represent the median (95% CI) of the $\tau_i^2, i = 1, \ldots, N_s = 58$.*

is a large range of values for the county-specific residual variance parameters, τ_i^2. Perhaps not surprisingly, the magnitude of these terms seems to be negatively correlated with the population of the given counties, demonstrating the effect a (relatively) small denominator can have when computing and modeling rates. The strong spatial story seen in the maps is reflected by the size of σ^2 compared to the majority of the τ_i^2. There is also relatively strong temporal correlation, with $\phi = 0.9$ corresponding to $\rho(t_i, t_j; \phi) \geq 0.4$ for $|t_j - t_i|$ less than 2 months.

Maps of the yearly (averaged across month) spatiotemporal random effects can be seen in Figure 12.20. Since here we are dealing with the *residual* curve after accounting for a number of mostly non-time-varying covariates, it comes as no surprise that the spatiotemporal random effects capture most of the variability in the model, including the striking decrease in yearly hospitalization rates over the study period. It also appears that our model is providing a better fit to the data in the years surrounding 2000, perhaps indicating that we could improve our fit by allowing our demographic covariates to vary temporally. Our model also appears to be performing well in the central counties, where asthma hospitalization rates remained relatively stable for much of the study period. ∎

12.9   Computer tutorials

Computer programs illustrating Bayesian analysis of spatial-temporal data using different libraries in R are supplied in the folder titled Chapter 12 from https://github.com/sudiptobanerjee/BGC_2023.

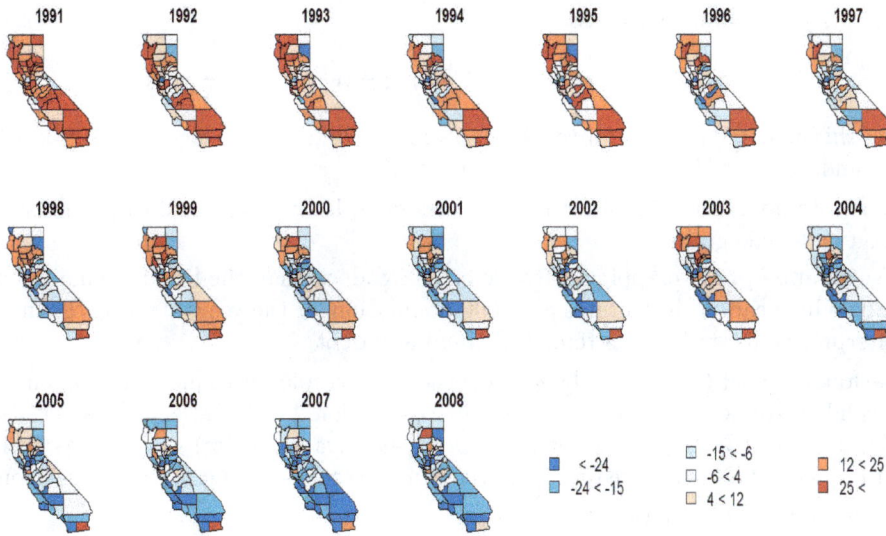

Figure 12.20 *Spatial random effects for asthma hospitalization data, by year.*

12.10 Exercises

1. Suppose $\text{Var}(\epsilon(\mathbf{s},t))$ in (12.6), (12.7), and (12.8) is revised to $\sigma_\epsilon^{2(t)}$.

   (a) Revise expressions (12.11), (12.13), and (12.16), respectively.

   (b) How would these changes affect simulation-based model fitting?

2. The data https://github.com/sudiptobanerjee/BGC_2023/data/ColoradoS-T.dat contain the maximum monthly temperatures (in tenths of a degree Celcius) for 50 locations over 12 months in 1997. The elevation at each of the 50 sites is also given.

   (a) Treating month as the discrete time unit, temperature as the dependent variable, and elevation as a covariate, fit the additive space-time model (12.6) to this data. Provide posterior estimates of the important model parameters, and draw image-contour plots for each month.

   (b) Compare a few sensible models (changing the prior for the ϕ_t, including/excluding the covariate, etc.).

   (c) Repeat part (a) assuming the error structures (12.7) and (12.8).

3. Suppose $Y(\mathbf{s}_i, t_j)$, $i = 1, \ldots, n$, $j = 1, \ldots, m$ arise from a mean-zero stationary spatiotemporal process. Let $a_{ii'} = \sum_{j=1}^{m} Y(\mathbf{s}_i, t_j) Y(\mathbf{s}_{i'}, t_j)/m$, let $b_{jj'} = \sum_{i=1}^{n} Y(\mathbf{s}_i, t_j) Y(\mathbf{s}_i, t_{j'})/n$, and let $c_{ii',jj'} = Y(\mathbf{s}_i, t_j) Y(\mathbf{s}_{i'}, t_{j'})$.

   (a) Obtain $E(a_{ii'})$, $E(b_{jj'})$, and $E(c_{ii',jj'})$.

   (b) Argue that if we plot $c_{ii',jj'}$ versus $a_{ii'} \cdot b_{jj'}$, under a separable covariance structure, we can expect the plotted points to roughly lie along a straight line. (As a result, we might call this a *separability plot*.) What is the slope of this theoretical line?

   (c) Create a separability plot for the data in Exercise 2 Was the separability assumption there justified?

4. Consider again the data and model of Example 12.5, the former located at https://github.com/sudiptobanerjee/BGC_2023/data/ColoradoS-T.dat. Fit the Poisson spatiotemporal disease mapping model (12.43), but where we discard the smoking covariate, and also reverse the gender scores ($s_j = 1$ if male, 0 if female) so that the log-relative

risk (12.44) is reparametrized as

$$\mu_{ijkt} = \mu + s_j\alpha + r_k\beta + s_jr_k(\xi - \alpha - \beta) + \gamma t + \phi_{it} \,.$$

Under this model, β now unequivocally captures the difference in log-relative risk between white and nonwhite females.

(a) Compute point and 95% interval estimates of β. Is there any real difference between the two female groups?

(b) Use an appropriate mapping software or GIS code to map the fitted median nonwhite female lung cancer death rates per 1000 population for the years 1968, 1978, and 1988. Interpret your results. Is a temporal trend apparent?

5. In the following, let C_1 be a valid two-dimensional isotropic covariance function and let C_2 be a valid one-dimensional isotropic covariance function. Let $C_A(\mathbf{s}, t) = C_1(\mathbf{s}) + C_2(t)$ and $C_M(\mathbf{s}, t) = C_1(\mathbf{s})C_2(t)$. C_A is referred to as an *additive* (or *linear*) space-time covariance function, while C_M is referred to as a *multiplicative* space-time covariance function.

(a) Why are C_A and C_M valid?

(b) Comment on the behavior of C_A and C_M as $||\mathbf{s} - \mathbf{s}', t - t'|| \to 0$ (local limit), and as $||\mathbf{s} - \mathbf{s}', t - t'|| \to \infty$ (global limit).

6. Suppose we observe a constant mean space-time process, $Y(\mathbf{s}, t)$ at equally spaced time points over a regular lattice. How might we obtain simple sample estimates for the covariance functions, $C(\mathbf{s}, t)$, $C(\mathbf{s})$, and $C(t)$? How might we use these to do some exploratory data analysis with regard to separability of the covariance function?

7. Suppose a simple dynamic model $Y_t(\mathbf{s}) = \gamma Y_{t-1}(\mathbf{s}) + \eta_t(\mathbf{s})$ where $|\gamma| < 1$ and the $\eta_t(\mathbf{s})$ are independent and identically distributed stationary mean 0 Gaussian processes for $\mathbf{s} \in D$. Show that the $Y_t(\mathbf{s})$ has a separable covariance function in space and time. Show that this is not the case if γ depends upon $\mathbf{s}$. More generally, consider the dynamical process with integro-difference equation $Y_t(\mathbf{s}) = \int h(\mathbf{s} - \mathbf{s}')Y_{t-1}(\mathbf{s}')d\mathbf{s}' + \eta_t(\mathbf{s})$ with $\eta(\mathbf{s})$ as above. Show that the process is not separable unless $h(\mathbf{0}) = \gamma \neq 0$ and $h(\mathbf{u}) = 0$ for almost all $\mathbf{u} \neq 0$.

Chapter 13

Modeling large spatial and spatiotemporal datasets

13.1 Introduction

Implementing Gibbs sampling or other MCMC algorithms requires repeated evaluation of various full conditional density functions. In the case of hierarchical models built from random effects using Gaussian processes, this requires repeated evaluation of the likelihood and/or joint or conditional densities arising under the Gaussian process. In particular, such computation requires evaluation of quadratic forms involving the inverse of covariance matrix and also the determinant of that matrix. Strictly speaking, we do not have to obtain the inverse in order to compute the quadratic form. Letting $\mathbf{z}^{\mathrm{T}} A^{-1} \mathbf{z}$ denote a general object of this sort, if we obtain $A^{\frac{1}{2}}$ (e.g. a triangular Cholesky factorization) and solve $\mathbf{z} = A^{\frac{1}{2}} \mathbf{v}$ for $\mathbf{v}$, then $\mathbf{v}^{\mathrm{T}} \mathbf{v} = \mathbf{z}^{\mathrm{T}} A^{-1} \mathbf{z}$. Still, with large n, computation associated with resulting $n \times n$ matrices can be unstable, and repeated computation (as for simulation-based model fitting) can be very slow, perhaps infeasible. After all, spatial covariance matrices, in general, are dense and the Cholesky factorization requires about $O(n^3/3)$ flops. We refer to this situation informally as *"the big n problem."*

Extension to multivariate models with, say, p measurements at a location leads to $np \times np$ matrices (see Section 10.6). Extension to spatiotemporal models (say, spatial time series at T time points) leads to $nT \times nT$ matrices (see Section 12.2). Of course, there may be modeling strategies that will simplify to, say, T $n \times n$ matrices or an $n \times n$ and a $T \times T$ matrix, but the problem will still persist if n is large. The objective of this section is thus to review approaches for handling spatial process models in this case.

Even a cursory review reveals significant literature on statistical methods for massive spatial datasets, which is too vast to be summarized here [see, e.g., Banerjee, 2017, Heaton et al., 2019, and references therein]. Within the hierarchical setup $[data \mid process] \times [process \mid parameters] \times [parameters]$, inference proceeds from spatial processes that scale massive data sets. Examples range from reduced-rank processes or subsets of regression approaches [Quiñonero-Candela and Rasmussen, 2005, Cressie and Johannesson, 2008, Banerjee et al., 2008, Wikle, 2010, and references therein], multi-resolution approaches [Nychka et al., 2015, Katzfuss, 2017], and graph-based models Vecchia [1988], Datta et al. [2016a], Katzfuss and Guinness [2021], Dey et al. [2022], Sauer et al. [2023a]. Full inference typically requires efficient Markov chain Monte Carlo [Finley et al., 2019], variational approximations [Ren et al., 2011, Wu et al., 2022, Cao et al., 2023] and a significant body of literature on Gaussian Markov random field approximations [Rue et al., 2009, Lindgren et al., 2011, and references therein] in conjunction with integrated nested Laplace approximations (see Section 5.5.4.1) for computing the marginal distributions of the process at given locations.

Broadly speaking, the approaches for tackling the big n problem can be classified as those that seek approximations of the exact likelihood and those that develop models which can handle fitting with large values of n. There is, also, an expanding literature on the use of machine learning methods based on dividing and conquering massive data sets [Guhaniyogi and Banerjee, 2018, 2019, Guhaniyogi et al., 2019, 2022, 2023], some of which we briefly discuss in Section 13.7, and on using random data compression [Guhaniyogi and Scheffler, 2021, Guhaniyogi et al., 2025], which we do not discuss in this text. We first describe the approximate likelihood approaches and then turn to the arguably, richer option of devising models for large spatial datasets.

13.2 Approximate likelihood approaches

13.2.1 Spectral methods

A rich, and theoretically attractive, option is to work in the spectral domain (as advocated by Stein [1999a] and Fuentes [2002a]. The idea is to transform to the space of frequencies, develop a periodogram (an estimate of the spectral density), and utilize the Whittle likelihood [Whittle, 1954, Guyon, 1995] in the spectral domain as an approximation to the data likelihood in the original space. The Whittle likelihood requires no matrix inversion so, as a result, computation is very rapid. In principle, inversion back to the original space is straightforward.

The practical concerns here are the following. First, there is discretization to implement a fast Fourier transform. Then, there is a certain arbitrariness to the development of a periodogram. Empirical experience is employed to suggest how many low frequencies should be discarded. Also, there is concern regarding the performance of the Whittle likelihood as an approximation to the exact likelihood. Some empirical investigation we have attempted suggests that this approximation is reasonably well centered, but does a less than satisfactory job in the tails (thus leading to poor estimation of model variances). Lastly, with non-Gaussian first stages, we will be doing all of this with random spatial effects that are never observed, making the implementation impossible. In summary, use of the spectral domain with regard to handling large n is limited in its application, and requires considerable familiarity with spectral analysis (discussed briefly in Subsection 3.1.2).

13.2.2 Lattice and conditional independence methods

Though Gaussian Markov random fields have received a great deal of recent attention for modeling areal unit data, they were originally introduced for points on a regular lattice. In fact, using inverse distance to create a proximity matrix, we can immediately supply a joint spatial distribution for variables at an arbitrary set of locations. As in Section 4.2, this joint distribution will be defined through its full conditional distribution. The joint density is recaptured using Brook's Lemma (4.7). The inverse of the covariance matrix is directly available, and the joint distribution can be made proper through the inclusion of an autocorrelation parameter. Other than the need to sample a large number of full conditional distributions, there is no big n problem. Indeed, many practitioners immediately adopt Gaussian Markov random field models as the spatial specification due to the computational convenience.

The disadvantages arising with the use of Gaussian Markov random fields should by now be familiar. First, and perhaps most importantly, we do not model association directly, which precludes the specification of models exhibiting desired correlation behavior. The joint distribution of the variables at two locations depends not only on their joint distribution given the rest of the variables, but also on the joint distribution of the rest of the variables. In fact, the relationship between entries in the inverse covariance matrix and the actual covariance matrix is very complex and highly nonlinear. Besag and Kooperberg [1995] (1995) showed, using a fairly small n that entries in the covariance matrix resulting from a Gaussian Markov random field specification need not behave as desired. They need not be positive nor decay with distance. With large n, the implicit transformation from inverse covariance matrix to covariance matrix is even more ill behaved [Wall, 2004].

In addition, with a Gaussian Markov random field there is no notion of a stochastic process, i.e., a collection of variables at all locations in the region of interest with joint distributions determined through finite dimensional distributions. In particular, we cannot write down the distribution of the variable at a selected location in the region. Rather, the best we can do is determine a conditional distribution for this variable given the variables at some prespecified number of and set of locations. Also, introduction of nonspatial error is confusing. The conditional variance in the Gaussian Markov random field cannot be aligned in magnitude with the marginal variance associated with a white noise process. Also, as we clarified in 4.3, Markov random field models preclude valid interpolation diminishing their utility as approximations .

Some authors have proposed approximating a Gaussian process with a Gaussian Markov random field. More precisely, a given set of spatial locations $s_1, ..., s_n$ along with a choice of correlation function yields an $n \times n$ covariance matrix Σ_1. How might we specify a Gaussian Markov random field with full rank inverse matrix Σ_2^{-1} such that $\Sigma_2 \approx \Sigma_1$? That is, unlike the previous paragraph where we start with a Gaussian Markov random field, here we start with the Gaussian spatial process.

A natural metric in this setting is Kullback-Liebler distance [see Besag and Kooperberg, 1995] 1995). If $f_1 \sim N(0, \Sigma_1)$ and $f_2 \sim N(0, \Sigma_2)$, the Kullback-Leibler distance of f_2 from f_1 is

$$KL(f_1, f_2) = \int f_1 \log(f_1/f_2) = -\frac{1}{2} \log \left| \Sigma_2^{-1} \Sigma_1 \right| + \frac{1}{2} tr(\Sigma_2^{-1} \Sigma_1 - I). \tag{13.1}$$

Hence, we only need Σ_1 and Σ_2^{-1} to compute (13.1). Using an algorithm originally proposed by Dempster [1972a], Besag and Kooperberg [1995] provide approximation based upon making (13.1) small. Rue and Tjelmeland [2002] note that this approach does not well approximate the correlation function of the Gaussian process. In particular, it will not do well when spatial association decays slowly. Rue and Tjelmeland [2002] propose a "matched correlation" criterion that accommodates both local and global behavior.

13.2.3   INLA

Laplace approximation provided an early approach to handling challenging Bayesian computation [Tierney et al., 1989, Kass et al., 1990]. It was particularly successful in the context of so-called conditionally independent hierarchical models, models with conditionally independent first stage specifications and with exchangeable parameters at the second stage. Its application diminished with the arrival of MCMC model fitting but, recently, it has enjoyed a rejuvenation through the popular Integrated Nested Laplace Approximation (INLA) package [Rue et al., 2009]. INLA handles spatial analysis through Markov random field approximation on regular grids. The software runs very rapidly and the Laplace approximation under the hood is very well done. However, overall inference is limited and application to more challenging multi-level models may be difficult.

The previous section offered some insight into how we might develop an approximation to a Gaussian process using a Gaussian Markov random field (GMRF). INLA adopts a very attractive choice through the use of the stochastic partial differential equation approach (SPDE), Lindgren et al. [2011]. This approach offers an explicit link between a Gaussian process and a GMRF. The SPDE approach extends work of Besag [1981], who proposed to approximate a Gaussian process when $\nu \to 0$ in the Matérn correlation function. This approximation imagines a regular two-dimensional lattice where the number of sites tends to infinity with local conditional distributions having $E(Y_{ij} \mid Y_{-ij}) = \frac{1}{a}(Y_{i-1,j} + Y_{i+1,j} +$

$Y_{i,j-1} + Y_{i,j+1})$ and $\text{Var}(Y_{ij} \mid Y_{-ij}) = \frac{1}{a}$ for $|a| > 4$. In the precision matrix, for site (i,j), we have the value a at (i,j) and the values -1 at each of the four immediate $N, E, S,$ and W neighbors of (i,j).

In Lindgren et al. [2011] it is noted that a Gaussian process $X(\mathbf{s})$ with the Matérn covariance is a solution to the linear fractional stochastic partial differential equation (SPDE) $(\kappa^2 - \Delta)^{\alpha/2} X(\mathbf{s}) = W(\mathbf{s})$ where Δ is the Laplacian, $W(\mathbf{s})$ is a white noise process, $\alpha = \nu + d/2$ with $\kappa > 0$, $\nu > 0$ (the usual Matérn smoothness parameter) and, in the spatial setting, $d = 2$. When $\nu = 1$ we achieve an extension of Besag, adding 8 more neighbors, expanding from 4 to 12 neighbors, providing local entries in the precision matrix as follows: $4 + a^2$ at $i, j)$, $-2a$ at each of the immediate $N, E, S,$ and W neighbors, 2 at the NW, NE, SE and SW neighbors and 1 at the two-away $N, E, S,$ and W neighbors. Extension to $\nu = 2$ adds 12 more neighbors yielding a total of 24.

Intuition suggests that if we have larger ν in the Matérn correlation function, i.e., a more smooth process realization, we need more non-zero neighbors sites in the GMRF representation. If the spatial locations are on an irregular grid, it is necessary to use a second result in Lindgren et al. [2011] which overlays a regular grid and employs the finite element method for an interpolation of the locations of observations to the nearest grid point. A basis function representation with random Gaussian weights provides the interpolation.

13.2.4 Approximate likelihood

Evidently, the big n challenge arises because we have a high-dimensional joint distribution in the likelihood. So, a natural approximation would be to attempt some sort of pseudo-likelihood approximation, replacing the joint density $f(y(\mathbf{s}_1), y(\mathbf{s}_2), ..., y(\mathbf{s}_n))$ with a product approximation, using a conditional density for $Y(\mathbf{s}_i)$ given a subset of the remaining y's. This idea dates at least to Vecchia [1988] who proposed this as an early spatial analysis computational trick. However, the approach suffers many problems. First, it is not formally defined. Second, it will typically be sequence dependent because there is no natural ordering of the spatial locations. Most troubling is the arbitrariness in the number of and choice of "neighbors." Moreover, perhaps counter-intuitively, we can not merely select locations close to $\mathbf{s}_i$.

Specifically, Stein et al. [2004] pointed out that we need locations at larger distances from each of the $\mathbf{s}_i$ (as we would have with the full data likelihood) in order to learn about the spatial decay in dependence for the process. So, altogether, we do not see such approximations as a useful approach.

For hierarchical models with a non-Gaussian first stage, the foregoing forms the basis for coarse-fine coupling as in Higdon et al. [2003]. The idea here is, with a non-Gaussian first stage, if spatial random effects (say, $\theta(\mathbf{s}_1), ..., \theta(\mathbf{s}_n)$) are introduced at the second stage, then, as in Subsection 6.6, the set of $\theta(\mathbf{s}_i)$ will have to be updated at each iteration of a Gibbs sampling algorithm.

Suppose n is large and that a "fine" chain does such updating. This chain will proceed very slowly. Suppose, concurrently, we run a "coarse" chain using a much smaller subset n' of the $\mathbf{s}_i$'s. The coarse chain will update very rapidly. Since the process for $\theta(\cdot)$ is the same in both chains it will be the case that the coarse one will explore the posterior more rapidly. However, we need realizations from the fine chain to fit the model using all of the data.

The coupling idea is to let both the fine and coarse chains run, and after a specified number of updates of the fine chain (and many more updates of the coarse chain, of course) we attempt a "swap;" i.e., we propose to swap the current value of the fine chain with that of the coarse chain. The swap attempt ensures that the equilibrium distributions for both chains are not compromised [see Higdon et al., 2003]. For instance, given the values of the θ's for the fine iteration, we might just use the subset of θ's at the locations for the coarse chain. Given the values of the θ's for the coarse chain, we might do an appropriate kriging to obtain the θ's for the fine chain.

With regard to specifying the coarse chain, one could employ a subsample of the sampled locations. Subsampling can be formalized into a model-fitting approach following the ideas of Pardo-Igúzquiza and Dowd [1997]. Specifically, for observations $Y(\mathbf{s}_i)$ arising from a Gaussian process with parameters $\boldsymbol{\theta}$, they propose replacing the joint density of $\mathbf{Y} = (Y(\mathbf{s}_1), \dots, Y(\mathbf{s}_n))^\mathrm{T}$, $f(\mathbf{y} \mid \boldsymbol{\theta})$, by

$$\prod_{i=1}^{n} f(y(\mathbf{s}_i) \mid y(\mathbf{s}_j), \; \mathbf{s}_j \in \partial\mathbf{s}_i) , \tag{13.2}$$

where $\partial\mathbf{s}_i$ defines some neighborhood of $\mathbf{s}_i$. For instance, it might be all $\mathbf{s}_j$ within some specified distance of $\mathbf{s}_i$, or perhaps the m $\mathbf{s}_j$'s closest to $\mathbf{s}_i$ for some integer m. Pardo-Igúzquiza and Dowd [1997] suggest the latter, propose $m = 10$ to 15, and check for stability of the inference about $\boldsymbol{\theta}$. Formal argument for approximating $f(\mathbf{y} \mid \boldsymbol{\theta})$ by (13.2) is essentially from Vecchia [1988] above. In light of the above, we view this approach as purely an algorithm for large datasets rather than a recommendable modeling approach.

13.2.5   *Covariance tapering*

A wide class of likelihood approximations rely upon sparse approximations. One such approach comes from tapered processes. Recall that we discussed covariance tapering in Section 3.4 in the context of using product form for constructing valid covariance functions. Returning to that theme, tapered processes offer an alternative means to dimension reduction, by producing sparse spatial covariance matrices (e.g., Furrer et al., 2006, Kaufman et al., 2008, Du et al., 2009). The underlying idea is to use a compactly supported covariance function (Wendland, 1995, Gneiting, 2002b) as a *tapering kernel* $C_\nu(\mathbf{s}_1, \mathbf{s}_2)$, which is a positive-definite function satisfying

$$C_\nu(\mathbf{s}_1, \mathbf{s}_2) = 0 \quad \text{if} \quad ||\mathbf{s}_1 - \mathbf{s}_2|| > \nu , \tag{13.3}$$

where ν is the distance beyond which the covariance becomes zero. Tapering introduces a sparse structure for the spatial covariance matrix from the Gaussian process model. Let T_ν be the $n \times n$ matrix with (i, j)-th element $C_\nu(\mathbf{s}_i, \mathbf{s}_j)$. Clearly, the matrix T will have zero entries for any pair of locations separated by more than ν units and, therefore, is sparse. There are choices aplenty for the tapering kernel, but the more widely used kernels use the Wendland family of tapered covariance functions (Wendland, 1995, Furrer et al., 2006). One particularly popular choice is given by

$$C_\nu(\mathbf{s}_1, \mathbf{s}_2) = \left(1 - \frac{h}{\nu}\right)_+^4 \left(1 + 4\frac{h}{\nu}\right) , \tag{13.4}$$

where $h = ||\mathbf{s}_1 - \mathbf{s}_2||$. Note that ν is typically not estimated, but fixed to achieve the desired degree of sparsity in the spatial covariance matrix (Kaufman et al., 2008). In a Bayesian context, we can estimate ν using some prior distribution, but such priors will need to be strongly informative for ν to be identified, which may not be straightforward.

Tapering a covariance function $C(\mathbf{s}_1, \mathbf{s}_2; \boldsymbol{\theta}_1)$ yields

$$C_{tap}(\mathbf{s}_1, \mathbf{s}_2; \boldsymbol{\theta}_1) = C_\nu(\mathbf{s}_1, \mathbf{s}_2) C(\mathbf{s}_1, \mathbf{s}_2, \boldsymbol{\theta}_1) .$$

Tapered covariances have been used effectively for analyzing large spatial datasets (e.g. Furrer et al. 2009) as its process realizations yield a sparse dispersion matrix $C(\boldsymbol{\theta}_1) \odot T_\nu$, where $\odot$ is the elementwise matrix product (or the Hadamard product). A standard property of the Hadamard product ensures that $C(\boldsymbol{\theta}_1) \odot T_\nu$ will be positive definite because $C(\boldsymbol{\theta}_1)$ and T_ν are (Banerjee and Roy, 2014, section 14.9). Since T_ν is sparse, so is $C(\boldsymbol{\theta}_1) \odot T_\nu$; therefore, sparse matrix algorithms can be employed to estimate tapered spatial process models.

13.3   Models for large spatial data: low rank models

A popular way of dealing with large spatial datasets is to devise models that bring about dimension reduction. The essential idea is to replace the spatial process $w(\mathbf{s})$ with $\tilde{w}(\mathbf{s})$, where the latter is a dimension-reducing process. How can we construct such dimension-reducing processes? A reduced rank *low rank* or *reduced rank* specification is typically based upon a representation in terms of the realizations of some latent process over a smaller set of coordinates called *knots*. To be precise,

$$\tilde{w}(\mathbf{s}) = \sum_{j=1}^{m} l(\mathbf{s}, \mathbf{s}_j^*) Z(\mathbf{s}_j^*), \tag{13.5}$$

where $Z(\mathbf{s})$ is a well defined process. The surface/process realization for $\tilde{w}(\mathbf{s})$ is completely determined by the function $l(\cdot, \cdot)$ and the set of variables, $\{Z(\mathbf{s}_j^*), j = 1, 2, ..., m\}$. The collection of $\mathbf{s}_j^*$'s are the knots. For a collection of locations, with associated vector denoted by $\tilde{\mathbf{w}} = (\tilde{w}(\mathbf{s}_1), \tilde{w}(\mathbf{s}_2), ..., \tilde{w}(\mathbf{s}_n))^\mathrm{T}$, we write

$$\tilde{\mathbf{w}} = \mathbf{L}\mathbf{z}^*, \tag{13.6}$$

where $\mathbf{L}$ is the $n \times m$ matrix with (i, j)-th element $l(\mathbf{s}_i, \mathbf{s}_j^*)$ and $\mathbf{z}^*$ is the $m \times 1$ vector with entries $Z(\mathbf{s}_j^*)$ with $m < n$.

Equation (13.6) immediately reveals dimension reduction: despite there being n $\tilde{w}(\mathbf{s}_i)$'s, we will only have to work with m $Z(\mathbf{s}_j^*)$'s. Since we anticipate $m << n$, the consequential dimension reduction is evident and, since we will write the model in terms of the Z's (with the $\tilde{w}$'s being deterministic from the Z's, given $l(\cdot, \cdot)$), the associated matrices we work with will be $m \times m$. Evidently, $\tilde{w}(\mathbf{s})$ as defined in (13.5) spans only an m-dimensional space; we create an uncountable number of variables through a finite number of variables. When $n > m$, the joint distribution of $\tilde{\mathbf{w}}$ is singular. However, we do create a valid stochastic process. In particular, the valid covariance function is

$$\mathrm{cov}(\tilde{w}(\mathbf{s}), \tilde{w}(\mathbf{s}')) = \mathbf{l}(\mathbf{s})^\mathrm{T} \Sigma_{\mathbf{z}^*} \mathbf{l}(\mathbf{s}') \tag{13.7}$$

where $\mathbf{l}(\mathbf{s})$ is the $m \times 1$ vector with entries $l(\mathbf{s}, \mathbf{s}_j^*)$. From (13.7), we see that, even if $l(\cdot, \cdot)$ is stationary, i.e., of the form $l(\cdot - \cdot)$, the induced covariance function is not. Also, if the Z's are Gaussian, then $\tilde{w}(\mathbf{s})$ is a Gaussian process.

How do we view the Z's? Are they a collection of w's from a process of interest, whence the $\tilde{w}$'s provide an approximation to enable computational tractability? Or are they merely a specification to provide a spatial process model? Also, are the $\mathbf{s}_j^*$ a subset of the observed $\mathbf{s}$'s or chosen otherwise? We offer a few more words about the first three questions. The last question takes us to the design problem.

The most prevalent specification for the Z's is i.i.d. normal with mean 0 and variance σ^2, i.e., $Z(\mathbf{s})$ is a white noise process, whence (13.7) simplifies to $\sigma^2 \mathbf{l}(\mathbf{s})^\mathrm{T} \mathbf{l}(\mathbf{s}')$. This form appears in Barry and Ver Hoef [1996] and in a series of papers by Higdon and collaborators [e.g., in Higdon, 1998, 2002c], the former calling it a "moving average" model, the latter, "kernel convolution." In particular, a natural choice for l is a kernel function, say $K(\mathbf{s} - \mathbf{s}')$ which puts more weight on $\mathbf{s}'$ near $\mathbf{s}$. The kernel would have parameters (which induces a parametric covariance function) and might be spatially varying [Higdon, 2002c, Paciorek and Schervish, 2006]. The reduced rank form can be viewed as a discretization of a process specification of the form $\tilde{w}(\mathbf{s}) = \int_{R^2} K(\mathbf{s} - \mathbf{s}') Z(\mathbf{s}') d\mathbf{s}'$. Gaussian kernels are frequently used though they lead to Gaussian covariance functions which, typically, yield process realizations too smooth to be satisfactory in practice [Stein, 1999a, Paciorek and Schervish, 2006]. Moreover, the scope of processes that can be obtained through kernel convolution is limited; for instance the widely used exponential covariance function does not arise from kernel convolution.

A different approach to specification for the Z's is to endow them with a stochastic process model having a selected covariance function. Again, from (13.7), this will impart a covariance function to the $\tilde{w}$'s. Reversing the perspective, if we have a particular covariance function that we wish for the $\tilde{w}(\mathbf{s})$, what covariance function shall we choose for the Z's? We argue in Section 13.4 that, in some sense, the *predictive process* provides an *optimal* choice.

13.3.1 Kernel-based dimension reduction

Recall the idea of kernel convolution (see Section 3.1.1) where we represent the process $Y(s)$ by

$$Y(\mathbf{s}) = \int k(\mathbf{s} - \mathbf{s}')z(\mathbf{s}')d\mathbf{s}' , \qquad (13.8)$$

where k is a kernel function (which might be parametric, and might be spatially varying) and $z(\mathbf{s})$ is a stationary spatial process (which might be white noise, that is, $\int_A z(\mathbf{s})d\mathbf{s} \sim N(0, \sigma^2 A)$ and $cov(\int_A z(\mathbf{s})d\mathbf{s}, \int_B z(\mathbf{s})d\mathbf{s}) = \sigma^2|A \cap B|)$. A finite version of (13.8) yields

$$Y(\mathbf{s}) = \sum_{j=1}^{J} k(\mathbf{s} - \mathbf{s}_j^*)z(\mathbf{s}_j^*) . \qquad (13.9)$$

Expression (13.9) shows that given k, every variable in the region is expressible as a linear combination of the set $\{z(\mathbf{s}), j = 1, ..., J\}$. Hence, no matter how large n is, working with the z's, we never have to handle more than a $J \times J$ matrix. The richness associated with the class in (13.8) suggests reasonably good richness associated with (13.9). Versions of (13.9) to accommodate multivariate processes and spatiotemporal processes can be readily envisioned.

Concerns regarding the use of (13.9) involve two issues. First, how does one determine the number of and choice of the $\mathbf{s}_j^*$'s? How sensitive will inference be to these choices? Also, the joint distribution of $\{Y(\mathbf{s}_i), i = 1, ..., n\}$ will be singular for $n > J$. While this does not mean that $Y(\mathbf{s}_i)$ and $Y(\mathbf{s}_i')$ are perfectly associated, it does mean that specifying $Y(\cdot)$ at J distinct locations determines the value of the process at all other locations. As a result, such modeling may be more attractive for spatial random effects than for the data itself.

A variant of this strategy is a conditioning idea. Suppose we partition the region of interest into M subregions so that we have the total of n points partitioned into n_m in subregion m with $\sum_{m=1}^{M} n_m = n$. Suppose we assume that $Y(\mathbf{s})$ and $Y(\mathbf{s}')$ are conditionally independent given $\mathbf{s}$ lies in subregion m and $\mathbf{s}'$ lies in subregion m'. However, suppose we assign random effects $\gamma(\mathbf{s}_1^*), ..., \gamma(\mathbf{s}_M^*)$ with $\gamma(\mathbf{s}_m^*)$ assigned to subregion m. Suppose the $\mathbf{s}_M^*$'s are "centers" of the subregions (using an appropriate definition) and that the $\gamma(\mathbf{s}_M^*)$ follows a spatial process that we can envision as a *hyper*spatial process. There are obviously many ways to build such multilevel spatial structures, achieving a variety of spatial association behaviors. We do not elaborate here but note that we will now have $n_m \times n_m$ matrices with an $M \times M$ matrix rather than a single $n \times n$.

13.3.2 The Karhunen-Loéve representation of Gaussian processes

Reduced rank approaches approximate the parent process $w(\mathbf{s})$ by a process $\tilde{w}(\mathbf{s})$ that lies in a fixed, finite-dimensional space. In seeking such approximations, one can consider the Karhunen-Loeve theorem (named after Kari Karhunen and Michel Loéve) that represents a stochastic process as an infinite linear combination of orthogonal functions, analogous to a Fourier series representation of a function on a bounded interval. In the case of a zero-centered spatial process $w_D = \{w(\mathbf{s}) : \mathbf{s} \in D\}$ with covariance function $C(\mathbf{s}_1, \mathbf{s}_2)$, where D

is a compact subset of $\Re^d$, the Karhunen-Loéve expansion can be written as

$$w(\mathbf{s}) = \sum_{i=1}^{\infty} \sqrt{\lambda_i} \phi_i(\mathbf{s}) Z_i \; , \tag{13.10}$$

where Z_i's are a sequence of independent and identically distributed $N(0,1)$ random variables, λ_i's are the (positive) eigen-values, often arranged in non-increasing order $\lambda_1 \geq \lambda_2 \geq \cdots$, of the symmetric positive definite function $C(\mathbf{s}_1, \mathbf{s}_2)$, and the $\phi_i(\mathbf{s})$'s are the corresponding eigenfunctions. The $\phi_i(\mathbf{s})$'s are continuous real-valued functions on D, which are mutually orthogonal in the L^2 space of functions over D. They form a set of "basis" functions to represent $w(\mathbf{s})$ with Z_i being the random coefficients with respect to this basis and λ_i's providing a scale adjustment to the random coefficients. Using standard inversion techniques, we have $Z_i = \frac{1}{\sqrt{\lambda_i}} \int_D w(\mathbf{s}) \phi_i(\mathbf{s}) d\mathbf{s}$, $i = 1, 2, \ldots$. The eigen-pairs $\{\lambda_i, \phi_i(\mathbf{s})\}$ satisfy the integral equation

$$\int_D C(\mathbf{s}, \mathbf{u}) \phi_i(\mathbf{u}) d\mathbf{u} = \lambda_i \phi_i(\mathbf{s}) \; . \tag{13.11}$$

Note that $\sqrt{\lambda_i}$'s are often referred to as the *singular values* of the process $w(\mathbf{s})$. The underlying theme in reduced rank approaches is that only the leading terms in the K-L expansion capture the main feature of the process, so the remaining terms can be dropped from the expansion in (13.10) to yield a reasonable reduced rank approximation of the process. Assuming that $\lambda_i \approx 0$ for $i = m+1, m+2, \ldots$, we retain only the first m terms in (13.10) to arrive at a rank-m approximation:

$$w(\mathbf{s}) \approx \tilde{w}(\mathbf{s}) = \sum_{i=1}^{m} \sqrt{\lambda_i} \phi_i(\mathbf{s}) Z_i \; . \tag{13.12}$$

The covariance function for the rank-m process $\tilde{w}(\mathbf{s})$ is given by $\tilde{C}(\mathbf{s}_1, \mathbf{s}_2) = \boldsymbol{\phi}(\mathbf{s}_1)^{\mathrm{T}} \Lambda \boldsymbol{\phi}(\mathbf{s}_2)$, where $\boldsymbol{\phi}(\mathbf{s}) = (\phi_1(\mathbf{s}), \phi_2(\mathbf{s}), \ldots, \phi_m(\mathbf{s}))^{\mathrm{T}}$ and Λ is an $m \times m$ diagonal matrix with λ_i as its i-th diagonal entry. Note that irrespective of how many locations we have, the rank of the matrix Λ will always remain fixed at m. Therefore, the process $\tilde{w}(\mathbf{s})$ is a *degenerate* Gaussian process whose (partial) realizations yield singular (rank-deficient) normal distributions over sets with more than m locations.

Curiously, the reduced-rank representation in (13.12) does not explicitly depend upon knots as does the representation (13.5). After all, knots, or any subset of locations, do not arise in constructing (13.12). Instead, (13.12) truncates a basis expansion based upon the magnitude of the eigenvalues of the parent covariance function. However, the predictive process (Banerjee et al. 2008), which is our topic of discussion in the next section, emerges as a special case of the reduced-rank representation in (13.12). To see how, consider implementing (13.12) in practice. This will entail computing the m eigen-pairs $\{\lambda_i, \phi_i(\mathbf{s})\}$ by solving (13.11). One way to solve this is to discretize (13.11) using an approximate linear system. The discretized system will use the "knots," say $\mathcal{S}^* = \{\mathbf{s}_1^*, \mathbf{s}_2^*, \ldots, \mathbf{s}_m^*\}$, as the arguments in the integrand in (13.11), so that

$$\frac{1}{m} \sum_{i=1}^{m} C(\mathbf{s}, \mathbf{s}_i^*) \phi_i(\mathbf{s}_i^*) = \lambda_i \phi_i(\mathbf{s}) \Rightarrow \mathbf{c}^{\mathrm{T}}(\mathbf{s}) \boldsymbol{\phi}_i^* = m \lambda_i \phi_i(\mathbf{s}) \; , \quad i = 1, 2, \ldots, m \; . \tag{13.13}$$

where $\mathbf{c}(\mathbf{s}) = (C(\mathbf{s}, \mathbf{s}_1^*), C(\mathbf{s}, \mathbf{s}_2^*), \ldots, C(\mathbf{s}, \mathbf{s}_m))^{\mathrm{T}}$ is the $m \times 1$ vector with $C(\mathbf{s}, \mathbf{s}_i)'$ as the i-th element and $\boldsymbol{\phi}_i^* = (\phi_i(\mathbf{s}_1^*), \phi_i(\mathbf{s}_2^*), \ldots, \phi_i(\mathbf{s}_m^*))^{\mathrm{T}}$. Furthermore, substituting $\mathbf{s}_i^*$ for $\mathbf{s}$ in (13.13) leads to the following $m \times m$ non-singular system:

$$\mathbf{c}^{\mathrm{T}}(\mathbf{s}_i^*) \boldsymbol{\phi}_i^* = m \lambda_i \phi_i(\mathbf{s}_i^*) \; , \quad i = 1, 2, \ldots, m \Rightarrow \mathbf{C}^* \boldsymbol{\phi}_i^* = m \lambda_i \boldsymbol{\phi}_i^* \; , \tag{13.14}$$

where $\mathbf{C}^*$ is the $m \times m$ matrix with $C(\mathbf{s}_i^*, \mathbf{s}_j^*)$ as its (i,j)-th element. This implies that $(m\lambda_i, \boldsymbol{\phi}_i^*)$'s are eigen-pairs for the full-rank $m \times m$ matrix $\mathbf{C}^*$.

Using (13.13) we write $\sqrt{\lambda_i}\phi_i(\mathbf{s}) = \dfrac{1}{m\sqrt{\lambda_i}}\mathbf{c}(\mathbf{s})'\boldsymbol{\phi}_i^*$ and using (13.14) we we can write $\boldsymbol{\phi}_i^* = m\lambda_i \mathbf{C}^{*-1}\boldsymbol{\phi}_i^*$. Substituting these in (13.12), we can write

$$
\begin{aligned}
\tilde{w}(\mathbf{s}) &= \sum_{i=1}^{m} \sqrt{\lambda_i}\phi_i(\mathbf{s})Z_i = \sum_{i=1}^{m} \frac{1}{m\sqrt{\lambda_i}}\mathbf{c}(\mathbf{s})'\boldsymbol{\phi}_i^* Z_i \\
&= \sum_{i=1}^{m} \frac{1}{m\sqrt{\lambda_i}}(m\lambda_i)\mathbf{c}^{\mathrm{T}}(\mathbf{s})\mathbf{C}^{*-1}\boldsymbol{\phi}_i^* Z_i \sum_{i=1}^{m} \sqrt{\lambda_i}\mathbf{c}^{\mathrm{T}}(\mathbf{s})\mathbf{C}^{*-1}\boldsymbol{\phi}_i^* Z_i \\
&= \mathbf{c}^{\mathrm{T}}(\mathbf{s})\mathbf{C}^{*-1}\sum_{i=1}^{m} \sqrt{\lambda_i}\boldsymbol{\phi}_i^* Z_i \approx \mathbf{c}^{\mathrm{T}}(\mathbf{s})\mathbf{C}^{*-1}\mathbf{w}^* ,
\end{aligned}
\tag{13.15}
$$

where $\mathbf{w}^* = (w(\mathbf{s}_1^*), w(\mathbf{s}_2^*), \ldots, w(\mathbf{s}_m^*))^{\mathrm{T}}$ and the last approximation follows from the low-rank Karhunen-Loéve representation $w(\mathbf{s}_i^*) \approx \sum_{i=1}^{m} \sqrt{\lambda_i}\phi(\mathbf{s}_i^*)Z_i$. The final expression in (13.15) is precisely the predictive process of Banerjee et al. [2008]; see Section 13.4. In fact, the predictive process offers a "closed-form" expression for (13.12) by circumventing the difficult problem of computing functional eigen-pairs.

As a last comment here, once we start down the path of basis function representations for processes, we can similarly consider basis representations for surfaces. By now, this is a standard literature where we can flexibly represent surfaces using say spline bases or wavelet bases. A general version would take the form $g(\mathbf{s}) = \sum_{l=1}^{L} a_l f_l(\mathbf{s})$ where we have chosen L basis functions (typically orthonormal), the f_l's, and we would estimate the coefficients, the a_l's, in order to fit a surface. We would face the challenge of specifying choice of and number of functions. Typically, these choices are made using knots, e.g., we use local cubic functions. More commonly, these functions are over one dimensional space rather than two dimensional space. In this setting, if the coefficients are random, the surface is random. If the coefficients are, say i.i.d. normal, we have an analogue of the kernel-based dimension reduction. Throughout this book we prefer to work with random surfaces captured through realizations of stochastic processes over $\Re^2$, rather than through linear combination of functions.

A useful point in this regard is the following. Basis representations of functions provide an explicit function to evaluate at any location of interest while process representations require interpolation to infer about arbitrary locations; a process realization is not a function. However, process realizations are truly nonparametric while basis representations, though sometimes described as "nonparametric" are, in fact, parametric, based upon a finite set of coefficients.

13.4   Predictive process models

13.4.1   The predictive process

Banerjee et al. [2008] propose a class of models based upon the idea of a spatial predictive process (motivated from kriging ideas). Consider a set of "knots" $\mathcal{S}^* = \{\mathbf{s}_1^*, \ldots, \mathbf{s}_m^*\}$, which may but need not be a subset of the entire collection of observed locations in $\mathcal{S} = \{\mathbf{s}_1, \mathbf{s}_2, \ldots, \mathbf{s}_n\}$. All we require is that m be much smaller than n. Assume that $w(\mathbf{s}) \sim GP(0, C(\cdot; \boldsymbol{\theta}))$ and let $\mathbf{w}^*$ be a realization of $w(\mathbf{s})$ over $\mathcal{S}^*$. That is, $\mathbf{w}^*$ is $m \times 1$ with entries $w(\mathbf{s}_i^*)$ and $\mathbf{w}^* \sim MVN(\mathbf{0}, C^*(\boldsymbol{\theta}))$, where $C^*(\boldsymbol{\theta})$ is the associated $m \times m$ covariance matrix with entries $C(\mathbf{s}_i^*, \mathbf{s}_j^*; \boldsymbol{\theta})$.

The spatial interpolant (that leads to "kriging") at a site $\mathbf{s}_0$ is given by

$$\tilde{w}(\mathbf{s}_0) = \mathbb{E}[w(\mathbf{s}_0) \mid \mathbf{w}^*] = \mathbf{c}^{\mathrm{T}}(\mathbf{s}_0; \boldsymbol{\theta}) C^{*-1}(\boldsymbol{\theta}) \mathbf{w}^* , \qquad (13.16)$$

where $\mathbf{c}(\mathbf{s}_0; \boldsymbol{\theta})$ is $m \times 1$ with entries $C(\mathbf{s}_0, \mathbf{s}_j^*; \boldsymbol{\theta})$. This single site interpolator, in fact, defines a spatial process $\tilde{w}(\mathbf{s}) \sim GP(0, \tilde{C}(\cdot))$ with covariance function,

$$\tilde{C}(\mathbf{s}, \mathbf{s}'; \boldsymbol{\theta}) = \mathbf{c}^{\mathrm{T}}(\mathbf{s}; \boldsymbol{\theta}) C^{*-1}(\boldsymbol{\theta}) \mathbf{c}(\mathbf{s}', \boldsymbol{\theta}), \qquad (13.17)$$

where $\mathbf{c}(\mathbf{s}; \boldsymbol{\theta}) = [C(\mathbf{s}, \mathbf{s}_j^*; \boldsymbol{\theta})]_{j=1}^m$. We refer to $\tilde{w}(\mathbf{s})$ as the *predictive process* derived from the *parent process* $w(\mathbf{s})$. The realizations of $\tilde{w}(\mathbf{s})$ are precisely the kriged predictions conditional upon a realization of $w(\mathbf{s})$ over $\mathcal{S}^*$. The process is completely specified given the covariance function of the parent process and $\mathcal{S}^*$. So, to be precise, we should write $\tilde{w}_{\mathcal{S}^*}(\mathbf{s})$, but we suppress this implicit dependence. From (13.17), this process is nonstationary regardless of whether $w(\mathbf{s})$ is. The connection with (13.5) is clear: we take $Z(\mathbf{s})$ to be the parent process and $\mathbf{l}(\mathbf{s}) = \mathbf{c}(\mathbf{s}; \boldsymbol{\theta})^T C^{*-1}(\boldsymbol{\theta})$.

Therefore, every spatial process induces a predictive process model (in fact, arbitrarily many of them). The latter models project process realizations of the former to a lower-dimensional subspace, thereby reducing the computational burden. For example, consider the customary spatial regression model

$$Y(\mathbf{s}) = \mathbf{x}^{\mathrm{T}}(\mathbf{s})\boldsymbol{\beta} + w(\mathbf{s}) + \epsilon(\mathbf{s}). \qquad (13.18)$$

Replacing $w(\mathbf{s})$ in (13.18) with $\tilde{w}(\mathbf{s})$, we obtain the predictive process model,

$$Y(\mathbf{s}) = \mathbf{x}^{\mathrm{T}}(\mathbf{s})\boldsymbol{\beta} + \tilde{w}(\mathbf{s}) + \epsilon(\mathbf{s}). \qquad (13.19)$$

Since $\tilde{w}(\mathbf{s}) = \mathbf{c}^{\mathrm{T}}(\mathbf{s}) C^{*-1}(\boldsymbol{\theta}) \mathbf{w}^*$, $\tilde{w}(\mathbf{s})$ is a spatially varying linear transformation of $\mathbf{w}^*$. The dimension reduction is seen immediately. In fitting the model in (13.19), the n random effects $\{w(\mathbf{s}_i), i = 1, 2, ..., n\}$ are replaced with only the m random effects in $\mathbf{w}^*$; we can work with an m dimensional joint distribution involving only $m \times m$ matrices. Evidently, the model in (13.19) is different from that in (13.18). Hence, though we introduce the same set of parameters in both models, they will not be identical in both models.

Knot-based linear combinations such as $\sum_{i=1}^m a_i(\mathbf{s}) w(\mathbf{s}_i^*)$ resemble other process approximation approaches. For instance, motivated by an integral representation of (certain) stationary processes as a kernel convolution of Brownian motion on $\Re^2$, Higdon [2002b] proposes a finite approximation to the parent process of the form $\sum_{i=1}^m a_i(\mathbf{s}; \boldsymbol{\theta}) u_i$ where u_i's are i.i.d. $N(0,1)$ and $a_i(\mathbf{s}; \boldsymbol{\theta}) = k(\mathbf{s}, \mathbf{s}_i^*; \boldsymbol{\theta})$ with $k(\cdot; \boldsymbol{\theta})$ being a Gaussian *kernel* function. Evidently, Gaussian kernels only capture Gaussian processes with Gaussian covariance functions [see Paciorek and Schervish, 2006]. Xia and Gelfand [2006] suggest extensions to capture more general classes of stationary Gaussian processes by aligning kernels with covariance functions. However, the class of stationary Gaussian process models admitting a kernel representation is limited (see Section 3.1.4). Sikorski et al. [2024] investigate stationary process approximations for large data sets using fast algorithms to "normalize" basis functions.

In the representation of Higdon [2002b], the $k(\mathbf{s}, \mathbf{s}_i^*; \boldsymbol{\theta})$ can be spatially varying, as in Higdon et al. [1999]. The u_i can be replaced with realizations from a stationary Gaussian process on $\Re^2$, say, $\mathbf{w}^*$. Then, the original realizations are projected onto an m-dimensional subspace generated by the columns of the $n \times m$ matrix K with entries $k(\mathbf{s}_i, \mathbf{s}_j^*; \boldsymbol{\theta})$, where $\tilde{\mathbf{w}} = K\mathbf{w}^*$. Alternatively, one could project as $\tilde{\mathbf{w}} = Z\mathbf{u}$, where $\mathbf{u} \sim N(\mathbf{0}, I)$, and Z is $n \times m$ with i-th row $\mathbf{c}^{\mathrm{T}}(\mathbf{s}_i; \boldsymbol{\theta}) C^{*-1/2}(\boldsymbol{\theta})$ to yield the same joint distribution as the predictive process model in (13.19). This approach has been used in "low-rank kriging" methods [Kammann and Wand, 2003]. More general low-rank spline models are also discussed in Ruppert et al. [2003, Ch 13] and Lin et al. [2000].

We regard the predictive process as a competing model specification with, computational advantages, but induced by an underlying full rank process. In fact, these models are, in some sense, *optimal* projections as we clarify in the next subsection. Also, $\tilde{w}(\mathbf{s})$ does not arise as a discretization of an integral representation of a process and we only require a valid covariance function to induce it.

Recall the discussion regarding fixed rank kriging [see also Section 3.2 and Cressie and Johannesson, 2008]. Again, letting $\mathbf{g}(\mathbf{s})$ be a $k \times 1$ vector of specified basis functions on $\Re^2$, the proposed covariance function is $C(\mathbf{s}, \mathbf{s}') = \mathbf{g}(\mathbf{s})^{\mathrm{T}} K \mathbf{g}(\mathbf{s}')$ with K an unknown positive definite $k \times k$ matrix that is estimated from the data using a method of moments approach. Such an approach may be challenging for the hierarchical models we envision here. We will be providing spatial modeling with random effects at the second stage of the specification. We have no "data" to provide an empirical covariance function. In contrast, we focus upon fitting hierarchical models (including, but not limited to, kriging models) to large spatial datasets. Depending upon where the spatial process $w(\mathbf{s})$ arises in the hierarchy it may be completely unobserved, unlike in classical kriging where the process is typically partially observed, whence empirical data-based estimators of the spatial dispersion matrix will be unavailable. Instead of regarding induced covariance structure of the predictive process as an approximation to an empirical gold standard, we consider them as models that are, in some sense, *optimal* projections as we clarify in the next subsection. Also, $\tilde{w}(\mathbf{s})$ does not arise as a discretization of an integral representation of a process and we only require a valid covariance function to induce it.

Lastly, we can draw a connection to recent work in spatial dynamic factor analysis [see, e.g., Lopes et al., 2008, and references therein], where K is viewed as an $n \times m$ matrix of factor loadings. Neither K nor $\mathbf{w}^*$ are known but replication over time in the form of a dynamic model is introduced to enable the data to separate them and to infer about them. In our case, the entries in K are "known" given the covariance function C.

13.4.2 *Properties of the predictive process*

The predictive process $\tilde{w}(\mathbf{s})$ is, in some sense, an *optimal* projection process as we clarify below. First, note that $\tilde{w}(\mathbf{s}_0)$ is an orthogonal projection of $w(\mathbf{s}_0)$ on to a particular linear subspace (e.g. Stein [1999a]. Let $\mathcal{H}_{m+1}$ be the Hilbert space generated by $w(\mathbf{s}_0)$ and the m random variables in $\mathbf{w}^*$ (with $\mathcal{H}_m$ denoting the space generated by the latter); hence, $\mathcal{H}_{m+1}$ comprises all linear combinations of these $m+1$ zero-centered, finite variance random variables along with their mean square limit points. If we seek the element in $\tilde{w}(\mathbf{s}_0) \in \mathcal{H}_m$ closest to $w(\mathbf{s}_0)$ in terms of the inner product norm induced by $\mathrm{E}[w(\mathbf{s})w(\mathbf{s}')]$, we obtain the linear system $\mathrm{E}[(w(\mathbf{s}_0) - \tilde{w}(\mathbf{s}_0))w(\mathbf{s}_j^*)] = 0$, $j = 1,\ldots,m$ with the unique solution $\tilde{w}(\mathbf{s}_0) = \mathbf{c}^{\mathrm{T}}(\mathbf{s}_0)C^{*-1}(\boldsymbol{\theta})\mathbf{w}^*$. Being a conditional expectation, it immediately follows that $\tilde{w}(\mathbf{s}_0)$ minimizes $\mathrm{E}[w(\mathbf{s}_0) - f(\mathbf{w}^*) \mid \mathbf{w}^*]$ over all real-valued functions $f(\mathbf{w}^*)$. In this sense, the predictive process is the best approximation for the parent process.

Also, $\tilde{w}(\mathbf{s}_0)$ *deterministically* interpolates $w(\mathbf{s})$ over $\mathcal{S}^*$. Indeed, if $\mathbf{s}_0 = \mathbf{s}_j^* \in \mathcal{S}^*$ we have

$$\tilde{w}(\mathbf{s}_j^*) = \mathbf{c}^{\mathrm{T}}(\mathbf{s}_j^*; \boldsymbol{\theta})C^{*-1}(\boldsymbol{\theta})\mathbf{w}^* = w(\mathbf{s}_j^*) \tag{13.20}$$

since $\mathbf{e}_j^{\mathrm{T}}C^*(\boldsymbol{\theta}) = \mathbf{c}^{\mathrm{T}}(\mathbf{s}_j^*, \boldsymbol{\theta})$, where $\mathbf{e}_j$ denotes the vector with 1 in the j-th position and 0 elsewhere. So, (13.20) shows that $\mathrm{E}[\tilde{w}(\mathbf{s}_j^*) \mid \mathbf{w}^*] = w(\mathbf{s}_j^*)$ and $\mathrm{Var}(\tilde{w}(\mathbf{s}_j^*) \mid \mathbf{w}^*) = 0$ (a property of "kriging"). At the other extreme, suppose $\mathcal{S}$ and $\mathcal{S}^*$ are disjoint. Then $\mathbf{w} \mid \mathbf{w}^* \sim MVN(c^{\mathrm{T}}C^{*-1}\mathbf{w}^*, C - c^{\mathrm{T}}C^{*-1}c)$ where c is the $m \times n$ matrix whose columns are the $\mathbf{c}(\mathbf{s}_i)$ and C is the $n \times n$ covariance matrix of $\mathbf{w}$. We can write $\mathbf{w} = \tilde{\mathbf{w}} + (\mathbf{w} - \tilde{\mathbf{w}})$ and the choice of $\mathbf{w}^*$ determines the (conditional) variability in the second term on the right side, i.e., how close $\Sigma_{\mathbf{w}}$ is to $\Sigma_{\tilde{\mathbf{w}}}$. It also reveals that there will be less variability in the predictive process than in the parent process as n variables are determined by $m < n$ random variables.

Kullback-Leibler based justification for $\tilde{w}(\mathbf{s}^*)$ is discussed in Csató [2002] and Seeger et al. [2003]. The former offers a general theory for Kullback-Leibler projections and proposes sequential algorithms for computations. As a simpler and more direct argument, let us assume that $\mathcal{S}^*$ and $\mathcal{S}$ are disjoint and let $\mathbf{w}_a = (\mathbf{w}^*, \mathbf{w})^{\mathrm{T}}$ be the $(m+n) \times 1$ vector of realizations over $\mathcal{S}^* \cup \mathcal{S}$. In (13.18), assuming all other model parameters are fixed, the posterior distribution for $p(\mathbf{w}_a \mid \mathbf{Y})$ is proportional to $p(\mathbf{w}_a)p(\mathbf{Y} \mid \mathbf{w})$ since $p(\mathbf{Y} \mid \mathbf{w}_a) = p(\mathbf{Y} \mid \mathbf{w})$. The corresponding posterior in (13.19) replaces $p(\mathbf{Y} \mid \mathbf{w})$ with a density $q(\mathbf{Y} \mid \mathbf{w}^*)$. Letting $\mathcal{Q}$ be the class of all probability densities satisfying $q(\mathbf{Y} \mid \mathbf{w}_a) = q(\mathbf{Y} \mid \mathbf{w}^*)$, suppose we seek the density $q \in \mathcal{Q}$ that minimizes the reverse Kullback-Leibler divergence $KL(q,p) = \int q \log(q/p)$. Banerjee et al. [2008] argue that $KL(q(\mathbf{w}_a \mid \mathbf{Y}), p(\mathbf{w}_a \mid \mathbf{Y}))$ is minimized when $q(\mathbf{Y} \mid \mathbf{w}^*) \propto \exp\left(\mathrm{E}_{\mathbf{w} \mid \mathbf{w}^*}[\log p(\mathbf{Y} \mid \mathbf{w}_a)]\right)$. Subsequent calculations from standard multivariate normal theory reveal this to be the Gaussian likelihood corresponding to the predictive process model.

Turning briefly to local smoothness of $\tilde{w}(\mathbf{s})$, often useful in modeling spatial gradients, note that $\tilde{w}(\mathbf{s})$ depends upon $\mathbf{s}$ only through $\mathbf{c}(\mathbf{s})$, hence $C(\mathbf{s}, \mathbf{s}')$, which is a *deterministic* function of $\mathbf{s}$. Thus smoothness of the predictive process amounts to investigating the analyticity of the parent covariance function. In fact, most stationary covariance functions (e.g. the Matérn family) admit infinite (multivariate) Taylor expansions in D outside a neighborhood of $\mathbf{0}$ leading to infinite differentiability in the almost sure sense everywhere in the domain, except possibly for $\mathbf{s} \in \mathcal{S}^*$. Mean square differentiability of the predictive process, defined in terms of Taylor expansions in the L^2 metric, follows from a similar argument. For further details on process smoothness see Chapter 14.

Turning to the smoothness properties of the process $\tilde{w}(\mathbf{s})$, let $C(\mathbf{s}; \boldsymbol{\theta})$ be a stationary covariance function for the parent process $w(\mathbf{s})$ such that $C(\mathbf{s}; \boldsymbol{\theta})$ is infinitely differentiable whenever $\|\mathbf{s}\| > 0$. This assumption is commonly satisfied by most covariance functions (e.g. the Matérn function; see Stein, 1999). Then, $\tilde{w}(\mathbf{s})$ inherits its smoothness properties from the elements of $\mathbf{c}^{\mathrm{T}}(\mathbf{s})$, i.e., from the C's. Hence it is infinitely differentiable in the almost sure sense everywhere in the domain, except possibly for $\mathbf{s} \in \mathcal{S}^*$. Mean square differentiability is also immediate as $C(\mathbf{s}; \boldsymbol{\theta})$ admits infinite Taylor expansions for every $\mathbf{s} \neq \mathbf{0}$ and since $\mathbf{w}^*$ is Gaussian with a finite variance. Consequently, for say, the Matérn correlation family, even with $\nu < 1$ (e.g. with the exponential correlation function) so the parent process $w(\mathbf{s})$ is *not* mean square differentiable, the predictive process still is.

13.4.3   *Biases in low-rank models and the bias-adjusted modified predictive process*

Irrespective of their precise specifications, low-rank models tend to underestimate uncertainty (since they are driven by a finite number of random variables), hence, overestimate the residual variance. In other words, this arises from systemic over-smoothing or model under-specification by the low-rank model when compared to the parent model. In fact, this becomes especially transparent from writing the parent likelihood and low-rank likelihood as mixed linear models. To elucidate, suppose, without much loss of generality, that $\mathcal{S} \cap \mathcal{S}^* = \mathcal{S}^*$ with the first m locations in $\mathcal{S}$ acting as the knots. Note that the Gaussian likelihood with the parent process in (13.18) can be written as $N(\mathbf{y} \mid X\boldsymbol{\beta} + Z(\boldsymbol{\theta})\mathbf{u}, \tau^2 I)$, where $Z(\boldsymbol{\theta})$ is the $n \times n$ lower-triangular Cholesky square-root of $C(\boldsymbol{\theta})$ and $\mathbf{u} = (u_1, u_2, \ldots, u_n)^{\mathrm{T}}$ is now an $n \times 1$ vector such that $u_i \overset{iid}{\sim} N(0,1)$. Writing $Z(\boldsymbol{\theta}) = [Z_1(\boldsymbol{\theta}) : Z_2(\boldsymbol{\theta})]$, a low-rank model would work with the likelihood $N(\mathbf{y} \mid X\boldsymbol{\beta} + Z_1(\boldsymbol{\theta})\mathbf{u}_1, \tau^2 I)$, where $\mathbf{u}_1$ is an $m \times 1$ vector whose components are independently and identically distributed $N(0,1)$ variables. Dimension reduction occurs because estimating the low-rank likelihood requires $m \times m$ (instead of $n \times n$) matrix decompositions. The parent and low-rank likelihoods can now be written as

Parent likelihood: $\qquad \mathbf{y} = X\boldsymbol{\beta} + Z_1(\boldsymbol{\theta})\mathbf{u}_1 + Z_2(\boldsymbol{\theta})\mathbf{u}_2 + \boldsymbol{\epsilon}_1$

Low rank likelihood: $\qquad \mathbf{y} = X\boldsymbol{\beta} + Z_1(\boldsymbol{\theta})\mathbf{u}_1 + \boldsymbol{\epsilon}_2 \,,$

where $\boldsymbol{\epsilon}_i \sim N(0, \tau_i^2 I)$ for $i = 1, 2$. For fixed $\boldsymbol{\beta}$ and $\boldsymbol{\theta}$, the basis functions forming the columns of $\mathbf{Z}_2(\boldsymbol{\theta})$ in the parent likelihood are absorbed into the residual error in the low rank likelihood, leading to an upward bias in the estimate of the nugget. Put another way, being smoother than the parent process, the low rank process tends to have lower variance which, in turn, inflates the residual variability often manifested as an overestimation of τ^2.

Let us be a bit more precise. Let $P_Z = Z(Z^{\mathrm{T}}Z)^{-1}Z^{\mathrm{T}}$ be the orthogonal projection matrix (or "hat" matrix) into the column space of Z. Customary linear model calculations reveal that the magnitude of the residual vector from the parent model is given by $\mathbf{y}^{\mathrm{T}}(I - P_Z)\mathbf{y}$, while that from the low-rank model is given by $\mathbf{y}^{\mathrm{T}}(I - P_{Z_1})\mathbf{y}$. Using the fact that $P_Z = P_{Z_1} + P_{[(I-P_{Z_1})Z_2]}$ (see exercises), we find the excess residual variability in the low-rank likelihood is summarized by

$$(\mathbf{y} - X\boldsymbol{\beta})^{\mathrm{T}} P_{[(I-P_{Z_1})Z_2]}(\mathbf{y} - X\boldsymbol{\beta}) \,.$$

Although this excess residual variability can be quantified as above, it is less clear how the low-rank spatial likelihood could be modified to compensate for this oversmoothing without adding significantly to the computational burden. Matters are complicated by the fact that expressions for the excess variability will involve the unknown process parameters $\boldsymbol{\theta}$, which must be estimated.

In fact, not all low-rank models lead to a straightforward quantification for this bias. For instance, low-rank models based upon kernel convolutions (Higdon, 2002) approximate $w(\mathbf{s})$ with $w_{KC}(\mathbf{s}) = \sum_{j=1}^{n^*} k(\mathbf{s} - \mathbf{s}_j^*, \boldsymbol{\theta}_1)u_j$, where $k(\cdot, \boldsymbol{\theta}_1)$ is some kernel function and $u_j \overset{iid}{\sim} N(0, 1)$, assumed to arise from a Brownian motion $U(\mathbf{v})$ on $\Re^2$. So $w(\mathbf{s}) - w_{KC}(\mathbf{s})$ is

$$\int k(\mathbf{s} - \mathbf{v}, \boldsymbol{\theta})dU(\mathbf{v}) - \sum_{j=1}^{n^*} k(\mathbf{s} - \mathbf{s}_j^*, \boldsymbol{\theta})u_j \approx \sum_{j=n^*+1}^{\infty} k(\mathbf{s} - \mathbf{s}_j^*, \boldsymbol{\theta})u_j \,, \tag{13.21}$$

which does not, in general, render a closed form and may be difficult to compute accurately. Furthermore, Gaussian kernels only capture Gaussian processes with Gaussian covariance functions [Paciorek and Schervish, 2006]; beyond this class, there is no theoretical assurance that $\mathrm{var}\{w(\mathbf{s})\} - \mathrm{var}\{w_{KC}(\mathbf{s})\} = C(\mathbf{s}, \mathbf{s}; \boldsymbol{\theta}) - \sum_{j=1}^{n^*} k^2(\mathbf{s} - \mathbf{s}_j^*, \boldsymbol{\theta})$ will be positive. In fact, the predictive process, being a conditional expectation, orthogonally decomposes the spatial variance so that $\mathrm{var}\{w(\mathbf{s}) - \tilde{w}(\mathbf{s})\} = \mathrm{var}\{w(\mathbf{s})\} - \mathrm{var}\{\tilde{w}(\mathbf{s})\}$; such orthogonal decompositions do not arise naturally in kernel convolution approximations.

The predictive process has a distinct advantage here. Being a conditional expectation, it is a projection onto a Hilbert space [e.g., Banerjee et al., 2008] and, unlike other existing low-rank methods, renders a closed form for the residual process arising from subtracting the predictive process from the parent process. In fact, the following inequality holds for any fixed $\mathcal{S}^*$ and for any spatial process $w(\mathbf{s})$:

$$\mathrm{var}\{w(\mathbf{s})\} = \mathrm{var}\{\mathrm{E}[w(\mathbf{s}) \mid \mathbf{w}^*]\} + \mathrm{E}\{\mathrm{var}[w(\mathbf{s}) \mid \mathbf{w}^*]\} \geq \mathrm{var}\{\mathrm{E}[w(\mathbf{s}) \mid \mathbf{w}^*]\} \,, \tag{13.22}$$

which implies that $\mathrm{var}\{w(\mathbf{s})\} \geq \mathrm{var}\{\tilde{w}(\mathbf{s})\}$. If the process is Gaussian then standard multivariate normal calculations yields the following closed form for this difference at any arbitrary location $\mathbf{s}$,

$$\delta^2(\mathbf{s}) = \mathrm{E}\{\mathrm{var}[w(\mathbf{s}) \mid \mathbf{w}^*]\} = C(\mathbf{s}, \mathbf{s}; \boldsymbol{\theta}_1) - \mathbf{c}(\mathbf{s}; \boldsymbol{\theta}_1)'\mathbf{C}^*(\boldsymbol{\theta}_1)^{-1}\mathbf{c}(\mathbf{s}, \boldsymbol{\theta}_1) \,. \tag{13.23}$$

	μ	σ^2	τ^2	RMSPE
True	1	1	1	
$m = 49$				
PP	1.37 (0.29,2.61)	1.37 (0.65,2.37)	1.18 (1.07,1.23)	1.21
MPP	1.36 (0.51,2.39)	1.04 (0.52,1.92)	0.94 (0.68.1,14)	1.20
$m = 144$				
PP	1.36 (0.52,2.32)	1.39 (0.76,2.44)	1.09 (0.96, 1.24)	1.17
MPP	1.33 (0.50,2.24)	1.14 (0.64,1.78)	0.93 (0.76,1.22)	1.17
$m = 900$				
PP	1.31 (0.23, 2.55)	1.12 (0.85,1.58)	0.99 (0.85,1.16)	1.17
MPP	1.31 (0.23,2.63)	1.04 (0.76,1.49)	0.98 (0.87,1.21)	1.17

Table 13.1 *Parameter estimates for the predictive process (PP) and modified predictive process (MPP) models in the univariate simulation.*

One simple remedy for the bias in the predictive process model [Finley et al., 2009b,a, Banerjee et al., 2010] is to use the process

$$\tilde{w}_\epsilon(\mathbf{s}) = \tilde{w}(\mathbf{s}) + \tilde{\epsilon}(\mathbf{s}) \ , \tag{13.24}$$

where $\tilde{\epsilon}(\mathbf{s}) \overset{iid}{\sim} N(0, \delta^2(\mathbf{s}))$ and $\tilde{\epsilon}(\mathbf{s})$ is independent of $\tilde{w}(\mathbf{s})$. We call this the *modified predictive process*. Now, the variance of $\tilde{w}_\epsilon(\mathbf{s})$ equals that of the parent process $w(\mathbf{s})$ and the remedy is computationally efficient—adding an independent space-varying nugget does not incur substantial computational expense. To summarize, we do not recommend the use of *just* a reduced/low rank model. To improve performance, it is necessary to approximate the residual process and, in this regard, the predictive process is especially attractive since the residual process is available explicitly

We present a brief simulation example revealing the benefit of the modified predictive process. We generate 2000 locations within a $[0, 100] \times [0, 100]$ square and then generate the dependent variable from model (13.18) with an intercept as the regressor, an exponential covariance function with range parameter $\phi = 0.06$ (i.e., such that the spatial correlation is ~ 0.05 at 50 distance units), scale $\sigma^2 = 1$ for the spatial process, and with nugget variance $\tau^2 = 1$. We then fit the predictive process and modified predictive process models using a holding out set of randomly selected sites, along with a separate set of regular lattices for the knots ($m = 49$, 144 and 900). Table 13.1 shows the posterior estimates and the square roots of MSPE based on the prediction for the hold-out data set. The overestimation of τ^2 by the unmodified predictive process is apparent and we also see how the modified predictive process is able to adjust for the τ^2. Not surprisingly, the RMSPE is essentially the same under either process model.

The remedy for the bias, as presented above, is effective in most practical settings and also leads to improved predictive performance by compensating for oversmoothing. However, it may be less effective in capturing small-scale spatial variation as it may not account for the spatial dependence in the residual process. Sang et al. [2011] and Sang and Huang [2012] propose to *taper* the bias adjustment process $\tilde{\epsilon}(\mathbf{s})$. This yields $\tilde{w}_{\tilde{\epsilon}_2}(\mathbf{s}) = \tilde{w}(\mathbf{s}) + \tilde{\epsilon}_2(\mathbf{s})$, where $\tilde{\epsilon}_2(\mathbf{s})$ is now a Gaussian process with covariance function

$$C_{\tilde{\epsilon}_2}(\mathbf{s}_1, \mathbf{s}_2, \boldsymbol{\theta}) = C_{\tilde{\epsilon}}(\mathbf{s}_1, \mathbf{s}_2, \boldsymbol{\theta}) C_\nu(\mathbf{s}_1, \mathbf{s}_2)$$

While existing remedies only adjust for the variability in the residual process, the tapered approach, on the other hand, accounts for the residual spatial association. Since the residual process mainly captures the small scale dependence and the tapering has little impact on such dependence other than introducing sparsity, the resulting error of the new approximation is expected to be small. Sang and Huang (2012) refer to this approach as *full-scale approximation* because of its capability of providing high-quality approximations at both the small and large spatial scales. In this sense, the proposed tapering approach will enrich the approximation to the underlying parent process.

Katzfuss [2017] has extended the tapering ideas in Sang et al. [2011] with regard to predictive process modeling. He specifies spatial error using two terms, combining a low rank component to capture both medium-to-long-range dependence with a tapered residual component to capture local dependence. Nonstationary Matérn covariance functions are adopted to provide increased flexibility.

13.4.4   Selection of knots

For a given set of observations, Finley et al. (2009) proposed a knot selection strategy designed to improve the induced predictive process as an approximation to the parent process. For a selected set of knots, $\tilde{w}(\mathbf{s}) = \mathrm{E}[w(\mathbf{s}) \mid \mathbf{w}^*]$ is considered as an approximation to the parent process. Given $\boldsymbol{\theta}_1$, the associated predictive variance of $w(\mathbf{s})$ conditional on $\mathbf{w}^*$, denoted $V_{\boldsymbol{\theta}_1}(\mathbf{s}, \mathcal{S}^*)$, is

$$\mathrm{var}[w(\mathbf{s}) \mid \mathbf{w}(\cdot), \mathcal{S}^*, \boldsymbol{\theta}_1] = \mathbf{C}(\mathbf{s}, \mathbf{s}; \boldsymbol{\theta}_1) - \mathbf{c}(\mathbf{s}, \boldsymbol{\theta}_1)^{\mathrm{T}} \mathbf{C}^{*-1}(\boldsymbol{\theta}_1) \mathbf{c}(\mathbf{s}, \boldsymbol{\theta}_1) , \qquad (13.25)$$

which measures how well we approximate $w(\mathbf{s})$ by the predictive process $\tilde{w}(\mathbf{s})$. This measure is in the spirit of work by Zidek and colleagues [Le and Zidek, 1992, Zidek et al., 2000], i.e., measuring knot *value* in terms of conditional variance. There, the best knot selection maximizes conditional variance given the previously selected knots. Here, we measure the effectiveness of a given collection of selected knots through small conditional variance.

In particular, the knot selection criterion is then defined as a function of $V_{\boldsymbol{\theta}_1}(\mathbf{s}, \mathcal{S}^*)$. One commonly used criterion is:

$$V_{\boldsymbol{\theta}_1}(\mathcal{S}^*) = \int_D V_{\boldsymbol{\theta}_1}(\mathbf{s}, \mathcal{S}^*) g(\mathbf{s}) d\mathbf{s} = \int_D \mathrm{var}[w(\mathbf{s}) \mid \mathbf{w}^*, \boldsymbol{\theta}_1] g(\mathbf{s}) d\mathbf{s} \qquad (13.26)$$

where $g(\mathbf{s})$, integrable over D, is the weight assigned to location $\mathbf{s}$ Zidek et al. [2000] and Diggle and Lophaven [2006]. Here, we only consider the simple case for which $g(\mathbf{s}) \equiv 1$. $V_{\boldsymbol{\theta}_1}(\mathcal{S}^*)$ can be regarded as a spatially averaged predictive variance. The integral in (13.26) is analytically intractable and discrete approximations such as numerical quadrature or Monte Carlo integration will be required. We use the discrete approximation which computes the spatially averaged prediction variance over all the observed locations,

$$V_{\boldsymbol{\theta}_1}(\mathcal{S}^*) \approx \frac{\sum_{i=1}^n \mathrm{var}[w(\mathbf{s}_i) \mid \mathbf{w}^*, \boldsymbol{\theta}_1]}{n} \qquad (13.27)$$

We ultimately reduce the problem of knot performance to the minimization of a design criterion, which is the function $V_{\boldsymbol{\theta}_1}(\mathcal{S}^*)$.

The following facts are easily verified for $V_{\boldsymbol{\theta}_1}(\mathcal{S}^*)$ defined in (13.26):

- $V_{\boldsymbol{\theta}_1}(\{\mathcal{S}^*, \mathbf{s}_0\}) - V_{\boldsymbol{\theta}_1}(\mathcal{S}^*) < 0$ for a new site $\mathbf{s}_0$;
- $V_{\boldsymbol{\theta}_1}(\{\mathcal{S}^*, \mathbf{s}_0\}) - V_{\boldsymbol{\theta}_1}(\mathcal{S}^*) \longrightarrow 0$ when $\|\mathbf{s}_0 - \mathbf{s}_i^*\| \longrightarrow 0$, where $\mathbf{s}_i^*$ is any member of $\mathcal{S}^*$, and
- $V_{\boldsymbol{\theta}_1}(\mathcal{S}) = 0$, where $\mathcal{S} = \{\mathbf{s}_1, \ldots, \mathbf{s}_n\}$ are the original observed locations.

The variance-covariance matrix under the parent process model is $\Sigma_{\mathbf{Y}} = \mathbf{C}(\boldsymbol{\theta}_1) + \tau^2 \mathbf{I}$, while that from the corresponding predictive process is given by $\tilde{\Sigma}_{\mathbf{Y}} = \mathcal{C}(\boldsymbol{\theta}_1)^{\mathrm{T}} \mathbf{C}^{*-1}(\boldsymbol{\theta}_1) \mathcal{C}(\boldsymbol{\theta}_1) + \tau^2 \mathbf{I}$. The Frobenius norm between $\Sigma_{\mathbf{Y}}$ and $\tilde{\Sigma}_{\mathbf{Y}}$ is $\|\Sigma_{\mathbf{Y}} - \tilde{\Sigma}_{\mathbf{Y}}\|_F \equiv \mathrm{tr}\left([\mathbf{C}(\boldsymbol{\theta}_1) - \mathcal{C}(\boldsymbol{\theta}_1)^{\mathrm{T}} \mathbf{C}^{*-1}(\boldsymbol{\theta}_1) \mathcal{C}(\boldsymbol{\theta}_1)]^2\right)$. Since $\mathbf{C}(\boldsymbol{\theta}_1) - \mathcal{C}(\boldsymbol{\theta}_1)^{\mathrm{T}} \mathbf{C}^{*-1}(\boldsymbol{\theta}_1) \mathcal{C}(\boldsymbol{\theta}_1)$ is positive definite, the norm $\|\Sigma_{\mathbf{Y}} - \tilde{\Sigma}_{\mathbf{Y}}\|_F \equiv \sum \lambda_i^2$, where λ_i is the i-th eigenvalue of $\Sigma_{\mathbf{Y}} - \tilde{\Sigma}_{\mathbf{Y}}$. Also, the averaged predictive variance is given by $\bar{V} = \frac{1}{n} \mathrm{tr}(\Sigma_{\mathbf{Y}} - \tilde{\Sigma}_{\mathbf{Y}}) = \frac{1}{n} \sum \lambda_i$.

Note that, even after discretization, we can not evaluate $V_{\boldsymbol{\theta}_1}(\mathcal{S}^*)$ since it depends upon the unknown $\boldsymbol{\theta}_1$. Available options to accommodate this include obtaining parameter estimates by using a subset of the original data or more fully Bayesian strategies which place a prior on $\boldsymbol{\theta}_1$ and then minimizes $\mathrm{E}_{\boldsymbol{\theta}_1}(V_{\boldsymbol{\theta}_1}(\mathcal{S}^*))$ [see Diggle and Lophaven, 2006]. In fact, we might naturally use the same prior as we would use to fit the model.

Regardless of which of these strategies we adopt, how shall we proceed to find a good $\mathcal{S}^*$? Suppose the values of the parameters and the knot size m are given. The following sequential search algorithm finds an approximately optimal design:

- Initialization: As in all cases where the domain is continuous, for implementation of an optimal design, we need to reduce the possible sampling locations to a finite set. Natural choices include a fine grid set, the observed set of locations or the union of these two sets.

- Specify an initial set of locations of size $m_0 \ll m$ as starting points for knot selection; possible choices include a coarse grid, or a subset of the observed locations, chosen randomly or deterministically.

- At step $t+1$,
  - For each sample point $\mathbf{s}_i$ in the allowable sample set, evaluate $V(\{\mathcal{S}^{*(t)}, \mathbf{s}_i\})$.
  - Remove the sample point with maximum decrease in $\bar{V}$ from the allowable sample set and add it to the knot set.

- Repeat the above procedure until we obtain m knots.

The sequential evaluation of $\bar{V}$ is achieved using a very efficient routine incorporating block matrix computation. Utilizing this, we have successfully implemented the sequential algorithm in a simulation study presented in Section 13.4.5.

We remark that the sequential algorithm does not necessarily achieve the global optimization solution. Alternative computational approaches are available to us in finding approximately optimal designs such as stochastic search and block selection [see Xia et al., 2006]. As to the choice of m, the obvious answer is "as large as possible." Evidently, this is governed by computational cost and sensitivity to choice. So, for the former, we will have to implement the analysis over different choices of m to consider run time. For the latter, we look for stability of predictive inference as m increases. We measure this by the value of minimized $\bar{V}$ under different choices of m. Unlike more formal sampling design contexts, our goal here is to achieve "good" knot selection to enable model fitting. We find coarse progression in m to be adequate. Finally, we can implement the analysis in two steps by combining this knot selection procedure with the modified predictive process in the obvious way: (1) choose a set of knots to minimize the averaged predictive variances; (2) then use the modified process in the model fitting. For further discussion of this design problem see Gelfand et al. [2012]. Here we present some examples.

13.4.5 A simulation example using the two step analysis

We generated 1,000 data points in a $[0, 100] \times [0, 100]$ square and then generated the dependent variable from model (13.18) with an intercept $\mu = 1$ as a regressor, an exponential covariance function with range parameter $\phi = 0.06$ (i.e., an effective range of $\sim$50 units), scale $\sigma = 1$ for the spatial process, and with nugget variance $\tau^2 = 1$. We illustrate a comparison among three design strategies, including regular grids, sequential search over all the observed locations and sequential search over a fine regular lattice. In Figure 13.1, we plot the averaged predictive variances under each strategy. The sequential search algorithm is clearly better than choosing a regular grid as knots. For instance, with 180 sites selected, sequential search over the observed locations yielded an averaged predictive variance approximately 0.15. For the regular grids, roughly 150 *additional* sites are needed to achieve the same level of performance.

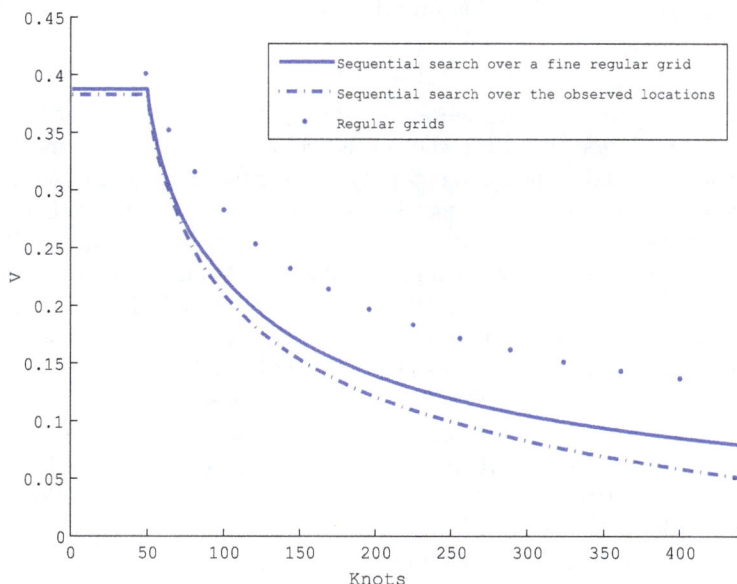

Figure 13.1 *Averaged prediction variance (V) versus number of knots (m). Solid dots denote results for regular grids; dash-dot line denotes results for the sequential search over the observed data locations (starting with 49 randomly chosen sites from the observed locations), and; solid line denotes results for the sequential search over a 60×60 regular grid (starting with a 7×7 regular grid).*

13.4.6 Non-Gaussian first stage models

There are two typical non-Gaussian first stage settings: (i) binary response at locations modeled using logit or probit regression and (ii) count data at locations modeled using Poisson regression. Diggle et al. [1998] unify the use of generalized linear models in spatial data contexts. See also Section 6.6 as well as Lin et al. [2000] and Lin and Zhang [1999] and Kammann and Wand [2003]. Essentially we replace (13.18) with the assumption that $E[Y(\mathbf{s})]$ is linear on a transformed scale, i.e., $\eta(\mathbf{s}) \equiv g(E(Y(\mathbf{s}))) = \mathbf{x}^T(\mathbf{s})\boldsymbol{\beta} + w(\mathbf{s})$ where $g(\cdot)$ is a suitable link function. With the Gaussian first stage, we can marginalize over the w's to achieve the covariance matrix $\tau^2 I + \mathbf{c}^T C^{*-1} \mathbf{c}$. Though this matrix is $n \times n$, using the Sherman-Woodbury result [Harville, 2000, Banerjee and Roy, 2014, also see Section 13.5, inversion only requires C^{*-1}]. With, say, a binary or Poisson first stage, such marginalization is precluded; we have to update the w's in running our Gibbs sampler. Using the predictive process, we only have to update the $m \times 1$ vector $\mathbf{w}^*$.

A bit more clarification may be useful. As described in the previous paragraph, the resulting model would take the form

$$\prod_i p(Y(\mathbf{s}_i) \mid \boldsymbol{\beta}, \mathbf{w}^*, \phi) p(\mathbf{w}^* \mid \sigma^2, \phi) p(\boldsymbol{\beta}, \phi, \sigma^2)$$

Though $\mathbf{w}^*$ is only $m \times 1$, making draws of this vector from its full conditional distribution will require Metropolis-Hastings updates.

13.4.7 Spatiotemporal versions

There are various spatiotemporal contexts in which predictive processes can be introduced to render computation feasible. We illustrate three of them here. First, we generalize (13.18) to

$$Y(\mathbf{s}, t) = \mathbf{x}(\mathbf{s}, t)\boldsymbol{\beta} + w(\mathbf{s}, t) + \epsilon(\mathbf{s}, t) \tag{13.28}$$

for $\mathbf{s} \in D$ and $t \in [0, T]$. In (13.28), the ϵ's are again, pure error terms, the $\mathbf{x}(\mathbf{s}, t)$ are local space-time covariate vectors, and $\boldsymbol{\beta}$ is a coefficient vector, here assumed constant over space and time but can be spatially and/or temporally varying (see Section 3). We have replaced the spatial random effects, $w(\mathbf{s})$, with space-time random effects, $w(\mathbf{s}, t)$ that come from a Gaussian process with covariance function $cov(w(\mathbf{s}, t), w(\mathbf{s}', t')) \equiv C(\mathbf{s}, \mathbf{s}'; t, t')$. There has been recent discussion regarding valid nonseparable space-time covariance functions; see Section 12.3. Now, we assume data $Y(\mathbf{s}_i, t_i), i = 1, 2, ...n$, where n can be very large because there are many distinct locations, times or both. In any event, the predictive process will be defined analogous to that above—$\tilde{w}(\mathbf{s}, t) = c(\mathbf{s}, t)^{\mathrm{T}} C^{*-1} \mathbf{w}^*$ where now $\mathbf{w}^*$ is an $m \times 1$ vector associated with m knots over $D \times [0, T]$ having covariance matrix C^* and $c(\mathbf{s}, t)$ is the vector of covariances of $w(\mathbf{s}, t)$ with the entries in $\mathbf{w}^*$. The spatiotemporal predictive process model, $\tilde{w}(\mathbf{s}, t)$ will enjoy the same properties as $\tilde{w}(s)$. Now, the issue of knot selection arises over $D \times [0, T]$. Knot selection over D follows the ideas above. Knot selection over $[0, T]$ is much easier due to the ordering in one dimension. For example, with annual data, we might use monthly knots, with monthly data we might use weekly knots.

Next, suppose we discretize time to say, $t = 1, 2, ...T$. Now, we would write the response as $Y_t(\mathbf{s})$ and the random effects as $w_t(\mathbf{s})$. Dynamic evolution of $w_t(\mathbf{s})$ is natural, leading to a spatial dynamic model as discussed in, e.g., Gelfand, Banerjee, and Gamerman (2005). In one scenario the data may arise as a time series of spatial processes, i.e., there is a conceptual time series at each location $\mathbf{s} \in D$. Alternatively, it may arise as cross-sectional data, i.e., there is a set of locations associated with each time point and these can differ from time point to time point. In the latter case, we can anticipate an explosion of locations as time goes on. Use of predictive process modeling, defined through a dynamic sequence of $\mathbf{w}_t^*$'s sharing the same knots enables us to handle this. A detailed explanation can be found in Finley et al. [2012].

It is also not uncommon to find space-time datasets with a very large number of distinct time points, possibly with different time points observed at different locations (e.g., real estate transactions). Predictive processes can be used to improve the applicability of a class of dynamic space-time models proposed by Gelfand et al. [2005b] by alleviating a computational bottleneck without sacrificing model flexibility and with minimal loss of information. Finley et al. [2012] focused on the common setting where space is considered continuous but time is taken to be discrete. Here, data is viewed as arising from a time series of spatial processes. Some examples of data that fit this description include: US Environmental Protection Agency's Air Quality System which reports pollutants' mean, minimum, and maximum at 8 and 24 hour intervals; climate model outputs of weather variables generated on hourly or daily intervals, and; remotely sensed landuse/landcover change recorded at annual or decadal time steps.

Finally, predictive processes offer an alternative to the dimension-reduction approach to space-time Kalman filtering presented by Wikle and Cressie [1999]. With time discretized, they envision evolution through a discretized integro-differential equation with a spatially structured noise. That is, $w_t(\mathbf{s}) = \int h_\mathbf{s}(\mathbf{u})w_{t-1}(\mathbf{u})d\mathbf{u} + \eta_t(\mathbf{s})$ with $h_\mathbf{s}$ a location interaction function and η a *spatially-colored* error process. $w_t(\mathbf{s})$ is decomposed as $\sum_{k=1}^{K} \phi_k(\mathbf{s})a_{kt}$ where the $\phi_k(\mathbf{s})$'s are deterministic orthonormal basis functions and the a's are mean 0 time series. Then, each $h_\mathbf{s}$ has a basis representation in the ϕ's, i.e., $h_\mathbf{s}(\mathbf{u}) = \sum_{l=1}^{\infty} b_l(\mathbf{s})\phi_l(\mathbf{u})$ where the b's are unknown. A dynamic model for the $k \times 1$ vector a_t driven by a linear transformation of the spatial noise process $\eta(\mathbf{s})$ results. Instead of the above decomposition for $w_t(\mathbf{s})$, we would introduce a predictive process model using $\mathbf{w}_t^*$. We replace the projection onto an arbitrary basis with a projection based upon a desired covariance specification.

13.4.8   Multivariate predictive process models

The predictive process immediately extends to multivariate Gaussian process settings. For a $q \times 1$ multivariate Gaussian parent process, $\mathbf{w}(\mathbf{s})$, the corresponding predictive process is

$$\tilde{\mathbf{w}}(\mathbf{s}) = \text{Cov}(\mathbf{w}(\mathbf{s}), \mathbf{w}^*)\text{Var}^{-1}(\mathbf{w}^*)\mathbf{w}^* = \mathcal{C}^{\mathrm{T}}(\mathbf{s}; \boldsymbol{\theta})\mathcal{C}^{*-1}(\boldsymbol{\theta})\mathbf{w}^* , \quad (13.29)$$

where $\mathcal{C}^{\mathrm{T}}(\mathbf{s}; \boldsymbol{\theta}) = [\Gamma_\mathbf{w}(\mathbf{s}, \mathbf{s}_1^*; \boldsymbol{\theta}), \ldots, \Gamma_\mathbf{w}(\mathbf{s}, \mathbf{s}_m^*; \boldsymbol{\theta})]$ is $q \times mq$, $\Gamma_\mathbf{w}(\mathbf{s}, \mathbf{s}')$ is the *cross-covariance* matrix (see Chapter 5), and $\mathcal{C}^*(\boldsymbol{\theta}) = [\Gamma_\mathbf{w}(\mathbf{s}_i^*, \mathbf{s}_j^*; \boldsymbol{\theta})]_{i,j=1}^m$ is the $mq \times mq$ covariance matrix of $\mathbf{w}^* = (\mathbf{w}(\mathbf{s}_1^*)^{\mathrm{T}}, \mathbf{w}(\mathbf{s}_2^*)^{\mathrm{T}}, \ldots, \mathbf{w}(\mathbf{s}_m)^{\mathrm{T}})^{\mathrm{T}}$. Equation (13.29) shows $\tilde{\mathbf{w}}(\mathbf{s})$ is a zero mean $q \times 1$ predictive process with cross-covariance matrix $\Gamma_{\tilde{\mathbf{w}}}(\mathbf{s}, \mathbf{s}') = \mathcal{C}^{\mathrm{T}}(\mathbf{s}; \boldsymbol{\theta})\mathcal{C}^{*-1}(\boldsymbol{\theta})\mathcal{C}(\mathbf{s}'; \boldsymbol{\theta})$. This is especially important for the applications we consider, where each location $\mathbf{s}$ yields observations on q dependent variables given by a $q \times 1$ vector $\mathbf{Y}(\mathbf{s}) = [Y_l(\mathbf{s})]_{l=1}^q$. For each $Y_l(\mathbf{s})$, we also observe a $p_l \times 1$ vector of regressors $\mathbf{x}_l(\mathbf{s})$. Thus, for each location we have q univariate spatial regression equations which can be combined into the following multivariate regression model:

$$\mathbf{Y}(\mathbf{s}) = X^{\mathrm{T}}(\mathbf{s})\boldsymbol{\beta} + \mathbf{w}(\mathbf{s}) + \boldsymbol{\epsilon}(\mathbf{s}) , \quad (13.30)$$

where $X^{\mathrm{T}}(\mathbf{s})$ is a $q \times p$ matrix $(p = \sum_{l=1}^q p_l)$ having a block diagonal structure with its l-th diagonal being the $1 \times p_l$ vector $\mathbf{x}_l^{\mathrm{T}}(\mathbf{s})$. Note that $\boldsymbol{\beta} = (\boldsymbol{\beta}_1, \ldots, \boldsymbol{\beta}_p)^{\mathrm{T}}$ is a $p \times 1$ vector of regression coefficients with $\boldsymbol{\beta}_l$ being the $p_l \times 1$ vector of regression coefficients corresponding to $\mathbf{x}_l^{\mathrm{T}}(\mathbf{s})$. Likelihood evaluation from (13.30) involves $nq \times nq$ matrices which can be reduced to $mq \times mq$ matrices by simply replacing $\mathbf{w}(\mathbf{s})$ in (13.30) by $\tilde{\mathbf{w}}(\mathbf{s})$.

Further computational gains in computing $\mathcal{C}^{*-1}(\boldsymbol{\theta})$ can be achieved by adopting the linear model of coregionalization, which specifies $\Gamma_\mathbf{w}(\mathbf{s}, \mathbf{s}') = A(\mathbf{s})\text{Diag}[\rho_l(\mathbf{s}, \mathbf{s}'; \boldsymbol{\theta})]_{l=1}^q A^{\mathrm{T}}(\mathbf{s}')$, where each $\rho_l(\mathbf{s}, \mathbf{s}'; \boldsymbol{\phi})$ is a univariate correlation function satisfying $\rho_l(\mathbf{s}, \mathbf{s}'; \boldsymbol{\phi}) \to 1$ as $\mathbf{s} \to \mathbf{s}'$. Note that $\Gamma_\mathbf{w}(\mathbf{s}, \mathbf{s}) = A(\mathbf{s})A^{\mathrm{T}}(\mathbf{s})$, hence $A(\mathbf{s}) = \Gamma_\mathbf{w}^{1/2}(\mathbf{s}, \mathbf{s})$ can be taken as any square-root of $\Gamma_\mathbf{w}(\mathbf{s}, \mathbf{s})$. Often we assume $A(\mathbf{s}) = A$ and assign an inverse-Wishart prior on AA^{T} with A a computationally efficient square-root (e.g., Cholesky or spectral). It now easily follows that $\mathcal{C}^*(\boldsymbol{\theta}) = (I_m \otimes A)\Sigma^*(\boldsymbol{\theta})(I_m \otimes A^{\mathrm{T}})$, where $\Sigma^*(\boldsymbol{\theta})$ is an $mq \times mq$ matrix partitioned into $q \times q$ blocks, whose (i,j)-th block is the diagonal matrix $\text{Diag}[\rho_l(\mathbf{s}_i^*, \mathbf{s}_j^*; \boldsymbol{\theta})]_{l=1}^q$. This yields a sparse structure and can be computed efficiently using specialized sparse matrix algorithms. Alternatively, we can write Σ^* as an orthogonally transformed matrix of $m \times m$ block diagonal matrix, $P^{\mathrm{T}}[\oplus_{l=1}^q [\rho_l(\mathbf{s}_i^*, \mathbf{s}_j^*; \theta_l)]_{i,j=1}^m]P$, where $\oplus$ is the block diagonal operator and P is a permutation (hence orthogonal) matrix. Since $P^{-1} = P^{\mathrm{T}}$, we need to invert q $m \times m$ symmetric correlation matrices rather than a single $qm \times qm$ matrix. Constructing the $nq \times mq$ matrix $\tilde{\Sigma}(\boldsymbol{\theta}) = [\text{Diag}[\rho_l(\mathbf{s}_i, \mathbf{s}_j^*; \boldsymbol{\theta})]_{l=1}^q]_{i,j=1}^{n,m}$, we further have

$$\text{Var}(\tilde{\mathbf{w}}) = \mathcal{C}^{\mathrm{T}}(\boldsymbol{\theta})\mathcal{C}^{*-1}(\boldsymbol{\theta})\mathcal{C}(\boldsymbol{\theta}) = (I_n \otimes A)\tilde{\Sigma}(\boldsymbol{\theta})\Sigma^{*-1}(\boldsymbol{\theta})\tilde{\Sigma}^{\mathrm{T}}(\boldsymbol{\theta})(I_m \otimes A^{\mathrm{T}}), \quad (13.31)$$

where the Kronecker structures and sparse matrices render easier computations. For multivariate spatial processes, Banerjee et al. [2010] show that

$$
\begin{aligned}
\text{Var}\{\mathbf{w}(\mathbf{s})\} - \text{Var}\{\tilde{\mathbf{w}}(\mathbf{s})\} &= C_\mathbf{w}(\mathbf{s}, \mathbf{s}) - \mathcal{C}^{\mathrm{T}}(\mathbf{s}; \boldsymbol{\theta})\Sigma^{*-1}(\boldsymbol{\theta})\mathcal{C}(\mathbf{s}, \boldsymbol{\theta}) \\
&= \text{var}\{\mathbf{w}(\mathbf{s}) \mid \mathbf{w}^*\} \succeq 0,
\end{aligned} \quad (13.32)
$$

where $\text{Var}\{\cdot\}$ denotes the variance-covariance matrix and $\succeq 0$ indicates non-negative definiteness. Equality holds only when $\mathbf{s} \in \mathcal{S}^*$, whereupon the predictive process coincides with the parent process.

Equation (13.32) is analogous to (13.22). The bias-adjustment for the multivariate predictive process is analogous to the univariate case. More precisely, we introduce $\tilde{\mathbf{w}}_\epsilon(\mathbf{s}) =$

$\tilde{\mathbf{w}}(\mathbf{s}) + \tilde{\boldsymbol{\epsilon}}(\mathbf{s})$, where $\tilde{\boldsymbol{\epsilon}}(\mathbf{s}) \sim N(\mathbf{0}, \Gamma_{\mathbf{w}}(\mathbf{s}, \mathbf{s}) - \mathcal{C}^{\mathrm{T}}(\mathbf{s}, \boldsymbol{\theta})\mathcal{C}^{*-1}(\boldsymbol{\theta})\mathcal{C}(\mathbf{s}, \boldsymbol{\theta}))$. Notice that $\tilde{\boldsymbol{\epsilon}}(\mathbf{s})$ acts as a nonstationary adjustment to the "residual process" (i.e. the difference between the parent process and the low-rank process) eliminating the systematic bias in variance components.

13.5   Modeling with the predictive process

We outline the implementation details for estimating a modified predictive process version of a multivariate spatial regression model

$$\mathbf{Y}(\mathbf{s}) = X^{\mathrm{T}}(\mathbf{s})\boldsymbol{\beta} + \mathbf{w}(\mathbf{s}) + \boldsymbol{\epsilon}(\mathbf{s}), \tag{13.33}$$

where $X^{\mathrm{T}}(\mathbf{s})$ is a $q \times p$ matrix ($p = \sum_{l=1}^{q} p_l$) having a block diagonal structure with its l-th diagonal being the $1 \times p_l$ vector $\mathbf{x}_l^{\mathrm{T}}(\mathbf{s})$. Note that $\boldsymbol{\beta} = (\boldsymbol{\beta}_1, \ldots, \boldsymbol{\beta}_p)^{\mathrm{T}}$ is a $p \times 1$ vector of regression coefficients with $\boldsymbol{\beta}_l$ being the $p_l \times 1$ vector of regression coefficients corresponding to $\mathbf{x}_l^{\mathrm{T}}(\mathbf{s})$. Likelihood evaluation from (13.33) that involves $nq \times nq$ matrices can be reduced to $mq \times mq$ matrices by simply replacing $\mathbf{w}(\mathbf{s})$ in (13.33) by $\tilde{\mathbf{w}}(\mathbf{s})$.

The modified predictive process model derived from (13.33) has the data likelihood

$$\mathbf{Y} = X\boldsymbol{\beta} + \mathcal{C}^{\mathrm{T}}(\boldsymbol{\theta})\mathcal{C}^{*-1}(\boldsymbol{\theta})\mathbf{w}^* + \tilde{\boldsymbol{\epsilon}} + \boldsymbol{\epsilon}; \ \tilde{\boldsymbol{\epsilon}} \sim N(\mathbf{0}, \Sigma_{\tilde{\epsilon}}), \ \boldsymbol{\epsilon} \sim N(\mathbf{0}, I_q \otimes \Psi), \tag{13.34}$$

where $\mathbf{Y} = [\mathbf{Y}(\mathbf{s}_i)]_{i=1}^n$ is the $nq \times 1$ response vector, $X = [X^{\mathrm{T}}(\mathbf{s}_i)]_{i=1}^n$ is the $nq \times p$ matrix of regressors, $\boldsymbol{\beta}$ is the $p \times 1$ vector of regression coefficients and $\mathcal{C}^{\mathrm{T}}(\boldsymbol{\theta}) = [\Gamma_{\mathbf{w}}(\mathbf{s}_i, \mathbf{s}_j^*; \boldsymbol{\theta})]_{i,j=1}^{n,m}$ is $nq \times mq$. In addition, $\Sigma_{\tilde{\epsilon}} = \mathrm{Diag}[\Gamma_{\mathbf{w}}(\mathbf{s}_i, \mathbf{s}_i) - \mathcal{C}^{\mathrm{T}}(\mathbf{s}_i, \boldsymbol{\theta})\mathcal{C}^{*-1}\mathcal{C}(\mathbf{s}_i, \boldsymbol{\theta})]_{i=1}^n$.

Given priors, model fitting employs a Gibbs sampler with Metropolis-Hastings steps using the marginalized likelihood

$$\mathbf{Y} \mid \boldsymbol{\beta}, \boldsymbol{\theta} \sim N(X\boldsymbol{\beta}, \mathcal{C}^{\mathrm{T}}(\boldsymbol{\theta})\mathcal{C}^{*-1}(\boldsymbol{\theta})\mathcal{C}(\boldsymbol{\theta}) + \Sigma_{\tilde{\epsilon}+\epsilon}(\boldsymbol{\theta})) \,,$$

where $\Sigma_{\tilde{\epsilon}+\epsilon}(\boldsymbol{\theta}) = \mathrm{Diag}[\Psi + \Gamma_{\mathbf{w}}(\mathbf{s}_i, \mathbf{s}_i) - \mathcal{C}^{\mathrm{T}}(\mathbf{s}_i, \boldsymbol{\theta})\mathcal{C}^{*-1}\mathcal{C}(\mathbf{s}_i, \boldsymbol{\theta})]_{i=1}^n$. Computing this marginalized likelihood now requires the inverse and determinant of $\mathcal{C}^{\mathrm{T}}(\boldsymbol{\theta})\mathcal{C}^{*-1}(\boldsymbol{\theta})\mathcal{C}(\boldsymbol{\theta}) + \Sigma_{\tilde{\epsilon}+\epsilon}(\boldsymbol{\theta})$. The inverse is computed using the Sherman-Woodbury-Morrison formula,

$$\Sigma_{\tilde{\epsilon}+\epsilon}^{-1}(\boldsymbol{\theta}) - \Sigma_{\tilde{\epsilon}+\epsilon}^{-1}(\boldsymbol{\theta})\mathcal{C}^{\mathrm{T}}(\boldsymbol{\theta})\left[\mathcal{C}^*(\boldsymbol{\theta}) + \mathcal{C}(\boldsymbol{\theta})\Sigma_{\tilde{\epsilon}+\epsilon}^{-1}(\boldsymbol{\theta})\mathcal{C}^{\mathrm{T}}(\boldsymbol{\theta})\right]^{-1}\mathcal{C}(\boldsymbol{\theta})\Sigma_{\tilde{\epsilon}+\epsilon}^{-1}(\boldsymbol{\theta}) \,, \tag{13.35}$$

requiring $mq \times mq$ inversions instead of $nq \times nq$ inversions, while the determinant is computed as

$$|\Sigma_{\tilde{\epsilon}+\epsilon}(\boldsymbol{\theta})||\mathcal{C}^*(\boldsymbol{\theta}) + \mathcal{C}(\boldsymbol{\theta})\Sigma_{\tilde{\epsilon}+\epsilon}(\boldsymbol{\theta})\mathcal{C}^{\mathrm{T}}(\boldsymbol{\theta})|/|\mathcal{C}^*(\boldsymbol{\theta})| \,. \tag{13.36}$$

In particular, with coregionalized models, $\mathcal{C}^{\mathrm{T}}(\boldsymbol{\theta})\mathcal{C}^{*-1}(\boldsymbol{\theta})\mathcal{C}(\boldsymbol{\theta})$ can be expressed as in (13.31), while $\Sigma_{\tilde{\epsilon}}(\boldsymbol{\theta})$ is given by

$$\mathrm{Diag}[AA^{\mathrm{T}} - (\mathbf{1}_m^{\mathrm{T}} \otimes A)[\oplus_{j=1}^m \Gamma(\mathbf{s}_i, \mathbf{s}_j^*; \boldsymbol{\theta})]\Sigma^{*-1}(\boldsymbol{\theta})[\oplus_{j=1}^m \Gamma^{\mathrm{T}}(\mathbf{s}_i, \mathbf{s}_j^*; \boldsymbol{\theta})](\mathbf{1}_m \otimes A^{\mathrm{T}})].$$

To complete the hierarchical specifications, customarily we assign a flat prior for $\boldsymbol{\beta}$, while Ψ could be assigned an inverse-Wishart prior. More commonly, independence of pure error for the different responses at each site is adopted, yielding a diagonal $\Psi = \mathrm{Diag}(\tau_i^2)_{i=1}^q$ with $\tau_i^2 \sim IG(a_i, b_i)$. Also, we model AA^{T} with an inverse Wishart prior. The priors for $\boldsymbol{\theta}$ will be assigned using customary specifications. Recall that, with the Matérn, the spatial decay parameters are generally weakly identifiable so reasonably informative priors are needed for satisfactory MCMC behavior. Priors for the decay parameters are set relative to the size of $\mathcal{D}$, e.g., prior means that imply the spatial ranges to be a chosen fraction of the maximum distance. The smoothness parameter ν is typically assigned a prior support of $(0, 2)$ as the data can rarely inform about smoothness of higher orders.

We obtain L samples, say $\{\Omega^{(l)}\}_{l=1}^{L}$, from

$$p(\Omega \mid Data) \propto p(\boldsymbol{\beta})p(A)p(\boldsymbol{\theta})p(\mathbf{Y} \mid \boldsymbol{\beta}, A, \boldsymbol{\theta}, \Psi) , \qquad (13.37)$$

where $\Omega = \{\boldsymbol{\beta}, A, \boldsymbol{\theta}, \Psi\}$. In each iteration, we update $\boldsymbol{\beta}$ by drawing from a $N(B\mathbf{b}, B)$ distribution, where $B = [\Sigma_{\boldsymbol{\beta}}^{-1} + (X^{\mathrm{T}}\mathcal{C}^{\mathrm{T}}(\boldsymbol{\theta})\mathcal{C}^{*-1}(\boldsymbol{\theta})\mathcal{C}(\boldsymbol{\theta}) + \Sigma_{\tilde{\epsilon}+\epsilon})^{-1}X]^{-1}$ and $\mathbf{b} = X^{\mathrm{T}}(\mathcal{C}^{\mathrm{T}}(\boldsymbol{\theta})\mathcal{C}^{*-1}(\boldsymbol{\theta})\mathcal{C}(\boldsymbol{\theta}) + \Sigma_{\tilde{\epsilon}+\epsilon})^{-1}\mathbf{Y}$. The remaining parameters are updated using Metropolis steps, possibly with block-updates (e.g. all the parameters in Ψ in one block and those in A in another). Typically, random walk Metropolis with (multivariate) normal proposals is adopted; since all parameters with positive support are converted to their logarithms, some Jacobian computation is needed. For instance, while we assign an inverted Wishart prior to AA^{T}, in the Metropolis update we update A, which requires transforming the prior by the Jacobian $2^{k} \prod_{i=1}^{k} a_{ii}^{k-i+1}$. Uniform priors on the spatial decay parameters will require a Hastings step due to the asymmetry in the priors.

Once the posterior samples from $P(\Omega \mid Data)$, $\{\Omega^{(l)}\}_{l=1}^{L}$, have been obtained, posterior samples from $P(\mathbf{w}^{*} \mid Data)$ are drawn by sampling $\mathbf{w}^{*(l)}$ for each $\Omega^{(l)}$ from $P(\mathbf{w}^{*} \mid \Omega^{(l)}, Data)$. This composition sampling is routine because $P(\mathbf{w}^{*} \mid \Omega, Data)$ is Gaussian (follows from (13.34) with mean $B\mathbf{b}$ and covariance matrix B, where

$$\mathbf{b} = \mathcal{C}^{*-1}(\boldsymbol{\theta})\mathcal{C}(\boldsymbol{\theta})\Sigma_{\tilde{\epsilon}+\epsilon}^{-1}(\mathbf{Y} - X\boldsymbol{\beta}) \text{ and}$$

$$B = (\mathcal{C}^{*-1}(\boldsymbol{\theta}) + \mathcal{C}^{*-1}(\boldsymbol{\theta})\mathcal{C}(\boldsymbol{\theta})\Sigma_{\tilde{\epsilon}+\epsilon}^{-1}\mathcal{C}^{\mathrm{T}}(\boldsymbol{\theta})\mathcal{C}^{*-1}(\boldsymbol{\theta}))^{-1} \text{ respectively.}$$

In some instances (e.g., prediction) it may be desirable to recover $\tilde{\boldsymbol{\epsilon}}$, in which case we again use composition sampling to draw $\tilde{\boldsymbol{\epsilon}}^{(l)}$ from the distribution $N(B\mathbf{b}, B)$, where $\mathbf{b} = (I_n \otimes \Psi^{-1})(\mathbf{Y} - X\boldsymbol{\beta} - \mathcal{C}^{\mathrm{T}}(\boldsymbol{\theta})\mathcal{C}^{*-1}(\boldsymbol{\theta})\mathbf{w}^{*})$ and $B = (\Sigma_{\tilde{\epsilon}}^{-1} + (I_n \otimes \Psi^{-1}))^{-1}$. Once $\mathbf{w}^{*}$ and $\tilde{\boldsymbol{\epsilon}}$ are recovered, prediction is carried out by drawing $\mathbf{Y}^{(l)}(\mathbf{s}_0)$, for each $l = 1, \ldots, L$ from a $q \times 1$ multivariate normal distribution with mean $X^{\mathrm{T}}(\mathbf{s}_0)\boldsymbol{\beta}^{(l)} + \mathcal{C}^{\mathrm{T}}(\boldsymbol{\theta}^{(l)})\mathcal{C}^{*-1}(\boldsymbol{\theta}^{(l)})\mathbf{w}^{*(l)} + \tilde{\boldsymbol{\epsilon}}^{(l)}$ and and variance $\Psi^{(l)}$.

Example 13.1 Forest biomass prediction and mapping. Spatial modeling of forest biomass and other variables related to measurements of current carbon stocks and flux have recently attracted much attention for quantifying the current and future ecological and economic viability of forest landscapes. Interest often lies in detecting how biomass changes across the landscape (as a continuous surface) by forest tree species. We consider point-referenced biomass (log-transformed) data observed at 437 forest inventory plots across the USDA Forest Service Bartlett Experimental Forest (BEF) in Bartlett, New Hampshire. Each location yields measurements of metric tons of above-ground biomass per hectare for American beech (BE), eastern hemlock (EH), red maple (RM), sugar maple (SM), and yellow birch (YB) and five covariates: TC1, TC2, and TC3 tasseled cap components derived from a spring date of mid-resolution Landsat 7 ETM+ satellite imagery from the National Land Cover Database and elevation (ELEV) and slope (SLOPE) derived from a digital elevation model data (see `http://seamless.usgs.gov` for metadata).

Figure 13.2 offers interpolated surfaces of the response variables. Covariates were measured on a 30×30 m pixel grid and are available for every location across the BEF. Interest lies in producing pixel-level prediction of biomass by species across large geographic areas. Because data layers such as these serve as input variables to subsequent forest carbon estimation models, it is crucial that each layer also provides a pixel-level measure of uncertainty in prediction.

Obtaining predictions from a predictive process model could substantially reduce the time necessary to estimate the posterior predictive distributions over a large array of pixels. A similar analysis was conducted by Finley et al. (2008); however, due to computational limitations they were only able to fit models using half of the available data and pixel-level prediction was still infeasible.

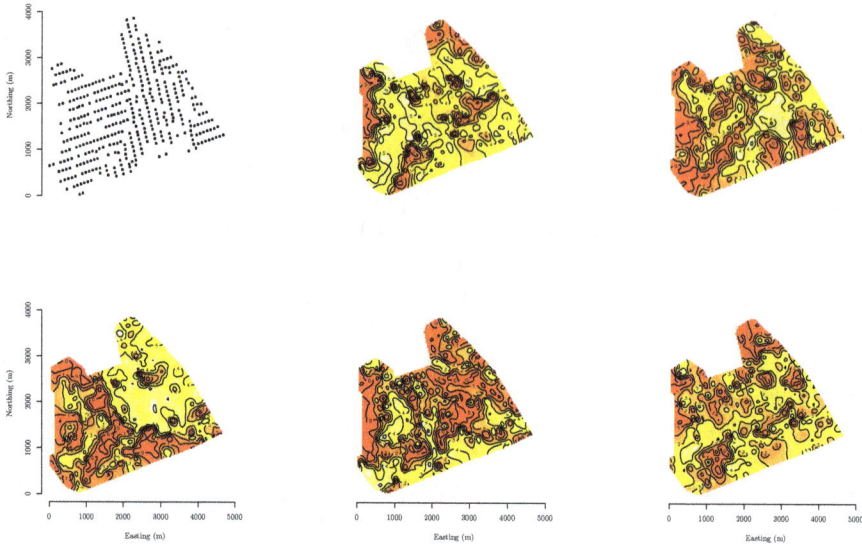

Figure 13.2 *Interpolation surfaces of log-transformed metric tons of biomass per hectare by species measured on forest inventory plots across the BEF. Response variables ordered BE, EH top row and RM, SM, YB bottom row. The set of 437 forest inventory plots are represented as points in the top left panel.*

Here we considered sub-models of (13.33) including the non-spatial and spatial non-separable models with the modified predictive process and three knot intensities of 51, 126, and 206. For all models Ψ and Γ_w are considered full $q \times q$ cross-covariance matrices where $q = 5$. Predictive process knots were located on a uniform grid within the BEF. We judge the performance of these models based on prediction of a hold-out set of 37 inventory plots, and visual similarity between the predicted and observed response surfaces.

We assigned a flat prior to each of the 30 β parameters (i.e., $p = \sum_{l=1}^{5} p_l = 30$ with each p_l including an intercept, TC1, TC2, TC3, ELEV, and SLOPE) . The cross-covariance matrices Ψ and Γ_w each receive an inverse-Wishart, $IW(df, S)$, with the degrees of freedom set to $q + 1 = 6$. Again, diagonal elements in the IW hyperprior scale matrix for Ψ and Γ_w were taken from univariate semi-variograms fit to the residuals of the non-spatial multivariate model. The decay parameter ϕ in the Matérn correlation function spatial follows a $U(0.002, 0.06)$ which corresponds to an effective spatial range between 50 and 1,500 m. Again, the smoothness parameter, ν, was fixed at 0.5, which reduces to the exponential correlation function. For each model, we ran three initially over-dispersed chains for 35,000 iterations. Unlike in the simulation analysis, substantial effort was required to select tuning values that achieved acceptable Metropolis acceptance rates. Ultimately, we resorted to univariate updates of elements in $\Psi^{1/2}$ and $\Gamma_w^{1/2}$ to gain the control necessary to maintain an acceptance of approximately 20%. Convergence diagnostics revealed 5,000 iterations to be sufficient for initial burn-in so the remaining 30,000 samples from each chain were used for posterior inference. The 206 knot model required approximately 2 hours to complete the MCMC sampling with the 106 and 51 knot models requiring substantial less time to collect the specified number of samples.

For the three knot intensities, there was negligible difference among the β parameter estimates. The estimated diagonal elements of Ψ and Γ_w for the three models were also nearly identical. Further, all of the 95% credible intervals for the off-diagonal elements in Ψ and Γ_w overlapped between the 126 and 206 knot models; however, the 206 knot model had several more significant off-diagonal elements (i.e., indicated by a credible interval that does not include zero). For the 51 knot model, off-diagonal elements of Γ_w were general closer to zero and the corresponding elements in Ψ were significantly different than zero, suggesting that the coarseness of this knot grid could not capture the residual spatial process.

Parameter	50% (2.5%, 97.5%)	Parameter	50% (2.5%, 97.5%)
$\Gamma_{w;1,1}$	1.97 (1.93, 2.02)	$\Psi_{1,1}$	1.95 (1.92, 1.98)
$\Gamma_{w;1,2}$	0.0044 (-0.0029, 0.019)	$\Psi_{1,2}$	-0.01 (-0.031, -0.0002)
$\Gamma_{w;1,3}$	-0.014 (-0.034, -0.004)	$\Psi_{1,3}$	-0.0069 (-0.018, 0.001)
$\Gamma_{w;1,4}$	0.011 (-0.0004, 0.027)	$\Psi_{1,4}$	0.01 (-0.0026, 0.019)
$\Gamma_{w;1,5}$	0.012 (0.0009, 0.018)	$\Psi_{1,5}$	-0.0048 (-0.022, 0.013)
$\Gamma_{w;2,2}$	1.96 (1.89, 2.00)	$\Psi_{2,2}$	1.92 (1.88, 1.97)
$\Gamma_{w;2,3}$	0.017 (0.0043, 0.032)	$\Psi_{2,3}$	0.0081 (-0.0001, 0.015)
$\Gamma_{w;2,4}$	0.0032 (-0.01, 0.013)	$\Psi_{2,4}$	-0.0048 (-0.012, 0.0019)
$\Gamma_{w;2,5}$	0.0031 (-0.0058, 0.041)	$\Psi_{2,5}$	0.011 (0.0042, 0.038)
$\Gamma_{w;3,3}$	1.98 (1.9, 2.01)	$\Psi_{3,3}$	1.97 (1.95, 1.98)
$\Gamma_{w;3,4}$	-0.0058 (-0.015, 0.012)	$\Psi_{3,4}$	-0.013 (-0.045, -0.0002)
$\Gamma_{w;3,5}$	0.016 (-0.0017, 0.029)	$\Psi_{3,5}$	0.0018 (-0.0089, 0.016)
$\Gamma_{w;4,4}$	2.03 (1.99, 2.065)	$\Psi_{4,4}$	1.94 (1.90, 1.98)
$\Gamma_{w;4,5}$	0.0064 (-0.0091, 0.016)	$\Psi_{4,5}$	0.0044 (-0.003, 0.012)
$\Gamma_{w;5,5}$	1.91 (1.84, 2.026)	$\Psi_{5,5}$	1.96 (1.93, 1.98)
ϕ_{w_1}	0.0056 (0.0033, 0.01)	Range_{w_1}	536.75 (296.06, 903.66)
ϕ_{w_2}	0.0048 (0.0037, 0.0144)	Range_{w_2}	624.72 (208.76, 806.32)
ϕ_{w_3}	0.0028 (0.0021, 0.0053)	Range_{w_3}	1085.68 (563.5, 1453.63)
ϕ_{w_4}	0.0051 (0.0035, 0.0085)	Range_{w_4}	586.02 (350.93, 846.24)
ϕ_{w_5}	0.0059 (0.0032, 0.0102)	Range_{w_5}	506.06 (293.25, 934.23)

Table 13.2 *BEF biomass parameter estimates for the 126 knot modified predictive process model. Subscripts 1-6 correspond to BE, EH, RM, SM, and YB species.*

Table 13.2 presents the parameter estimates of Γ_w, Ψ, and ϕ for the 126 knot model. For brevity we have omitted β estimates but note that 15 were significant at the 0.05 level. Those significant off-diagonal elements of Γ_w in Table 13.2 are apparent in the interpolated surface of $\tilde{w}$ depicted in Figure 13.3, where positive residual spatial correlation can be seen between BE and YB and between EH and RM.

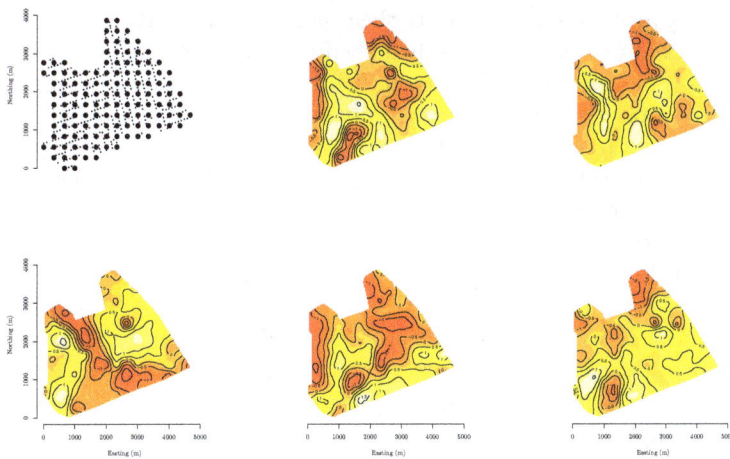

Figure 13.3 *Interpolated surfaces of the 126 knot model's median $\tilde{w}$ at each inventory plot. Top left panel shows forest inventory plots (small points) under the 126 knots (large points). The order of response variables in the subsequent panels correspond to Figure 13.2.*

Turning to prediction, it appears that the covariates and spatial proximity of observed inventory plots explain a significant portion of the variation in the response variables, perhaps leading to overfitting. We note that for our 37 hold-out plots the 95% prediction intervals are quite broad yielding a 100% empirical coverage for all three knot intensities. Finally, comparing the surface of pixel-level prediction for 1,000 randomly selected pixels (Figure 13.4) to the observed (Figure 13.2) we see that the model can capture landscape-level variation in biomass and spatial patterns in biomass by species. ∎

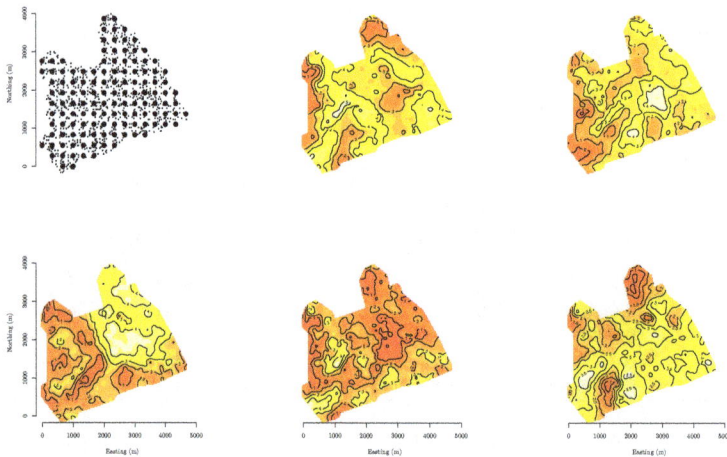

Figure 13.4 *Interpolated surfaces of the 126 knot model's median predicted response value over a random subset of 1,000 pixels in the BEF. Top left panel shows the subset of prediction pixels (small points) under the 126 knots (large points). The order of response variables in the subsequent panels correspond to Figure 13.2.*

13.6   Sparse Gaussian graphical spatial process models

Rather than working with approximations to the process, we can construct sparsity-inducing Gaussian processes that can be conveniently embedded within Bayesian hierarchical models and deliver full Bayesian inference for random fields at arbitrary resolutions. One such approach is covariance tapering that induces sparsity in the covariance matrix (Section 13.2.5). An alternate approach is to introduce sparsity in the precision matrices for graphical Gaussian models by exploiting the relationship between the Cholesky decomposition of a positive definite matrix and conditional independence. Readers will be reminded of SAR and CAR models based on adjacency matrices. Such models are well suited for finite collections of regions or sites, but are not easily extended to continuous spatial processes. The link between Gaussian Markov random fields (GMRFs) and continuous Gaussian random fields (GRF) has been characterized through stochastic partial differential equations in the seminal discussion paper by [Lindgren et al., 2011] and constitutes the underlying methodology for the popular Integrated Nested Laplace Approximation [INLA, Rue et al., 2009, also see Section 5.5.4.1 in Chapter 5] of Gaussian random fields using GMRFs.

An alternate approach, which can be regarded as a special case of a GMRF but offers a simpler and more flexible route to a spatial process is with the use of directed acyclic graphs (DAGs) and graphical Gaussian models. The result will be a Gaussian process whose finite-dimensional realizations will have sparse precision matrices. We call them Nearest Neighbor Gaussian Processes (NNGP). Section 13.6.3 outlines how the process can be embedded within hierarchical models and presents some brief simulation examples demonstrating aspects of inference from NNGP models. We will provide a brief overview here on the lines of Banerjee [2017] while referring the reader to relevant articles for further details.

13.6.1 Graphical Gaussian models for spatial data

Consider an expensive prior density $N(w \mid 0, K_\theta)$ with a dense covariance matrix K_θ. We wish to obtain a covariance matrix $\tilde{K}_\theta$ such that $\tilde{K}_\theta^{-1}$ is sparse and, importantly, its determinant is available cheaply. What would be an effective way of achieving this? One approach would be to consider *modeling* the Cholesky decomposition of the precision matrix so that it is sparse [Pourahmadi, 1999]. For example, forcing some elements in the dense half of the triangular Cholesky factor to be zero will introduce sparsity in the precision matrix. To precisely set out which elements should be made zero in the Cholesky factor, we borrow some fundamental notions of sparsity from graphical (Gaussian) models.

The underlying idea is prevalent in graphical models or Bayesian networks [see, e.g., Lauritzen, 1996, Bishop, 2006, Murphy, 2012]. The joint distribution for a random vector w can be regarded as a directed acyclic graph (DAG), where each node is a random variable w_i. We write the joint distribution as

$$p(w_1) \prod_{i=2}^{n} p(w_i \mid w_1, \ldots, w_{i-1}) = \prod_{i=1}^{n} p(w_i \mid w_{\mathrm{Pa}[i]}) \,,$$

where $\mathrm{Pa}[1]$ is the empty set and $\mathrm{Pa}[i] = \{1, 2, \ldots, i-1\}$ for $i = 2, 3, \ldots, n-1$ is the set of parent nodes with directed edges to i. This model is specific to the ordering (sometimes called "topological ordering") of the nodes. The DAG corresponding to this factorization is shown in Figure 13.5(a) for $n = 7$ nodes. This is a full graphical model, so called because $\mathrm{Pa}[i]$ comprises all nodes preceding i in the topological order, which defines the joint distribution. Shrinking $\mathrm{Pa}[i]$ from the set of all nodes preceding i to a smaller subset of parent nodes yields a different, but still valid, joint distribution. In spatial settings, each of the nodes in the DAG have associated spatial coordinates. Thus, the parents for any node i can be chosen to include a certain fixed number of "nearest neighbors," say based upon their distance from node i. For example, Figure 13.5(b) shows the DAG when some of the edges are deleted so as to retain at most 3 nearest neighbors in the conditional probabilities. The resulting joint density is

$$p(w_1) \times p(w_2 \mid w_1) \times p(w_3 \mid w_1, w_2) \times p(w_4 \mid w_1, w_2, w_3) \times p(w_5 \mid \cancel{w_1}, w_2, w_3, w_4)$$
$$\times p(w_6 \mid w_1, \cancel{w_2}, \cancel{w_3}, w_4, w_5) \times p(w_7 \mid w_1, w_2, \cancel{w_3}, \cancel{w_4}, \cancel{w_5}, w_6) \,.$$

The above model posits that any node i, given its parents, is conditionally independent of any other node that is neither its parent nor its child.

Multivariate Gaussian densities are especially useful in manifesting the explicit connection between conditional independence in DAGs and sparsity. Consider an $n \times 1$ random vector $\mathbf{w}$ distributed as $N(\mathbf{0}, K_\theta)$. Writing $N(\mathbf{w} \mid \mathbf{0}, K_\theta)$ as $p(w_1) \prod_{i=2}^{n} p(w_i \mid w_1, w_2, \ldots, w_{i-1})$ is equivalent to the following set of linear models,

$$w_1 = 0 + \eta_1 \quad \text{and} \quad w_i = a_{i1} w_1 + a_{i2} w_2 + \cdots + a_{i,i-1} w_{i-1} + \eta_i \text{ for } i = 2, \ldots, n \,,$$

or, more compactly, simply $\mathbf{w} = A\mathbf{w} + \boldsymbol{\eta}$, where A is $n \times n$ strictly lower-triangular with elements $a_{ij} = 0$ whenever $j \geq i$ and $\boldsymbol{\eta} \sim N(\mathbf{0}, D)$ and D is diagonal with diagonal entries $d_{11} = \mathrm{var}\{w_1\}$ and $d_{ii} = \mathrm{var}\{w_i \mid w_j : j < i\} > 0$ for $i = 2, \ldots, n$.

Clearly $I - A$ is nonsingular since it is a lower-triangular matrix with 1's along the diagonal and $K_\theta = (I - A)^{-1} D (I - A)^{-\mathrm{T}}$ is positive definite since D has positive diagonal entries. The possibly nonzero elements of A and D are completely determined by K_θ. Let

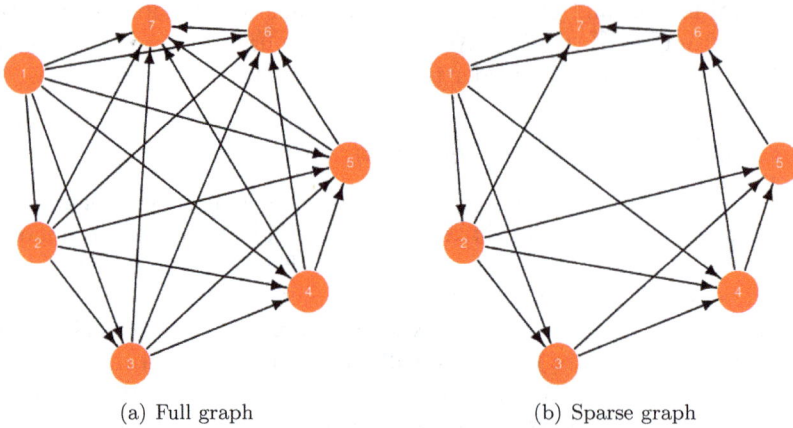

(a) Full graph   (b) Sparse graph

Figure 13.5 *Sparsity using directed acyclic graphs.*

a[i,j], d[i,j] and K[i,j] denote the (i,j)-th entries of A, D and K_θ, respectively. Note that d[1,1] = K[1,1] and the first row of A is 0. A pseudo-code to compute the remaining elements of A and D is:

```
for(i in 1:(n-1)) {
    a[i+1,1:i] = solve(K[1:i,1:i], K[1:i,i+1])
    d[i+1,i+1] = K[i+1,i+1] - dot(K[i+1,1:i],a[i+1,1:i])
}.
```
(13.38)

Here a[i+1,1:i] denotes the $1 \times$ i row vector comprising the possibly nonzero elements of the i+1-th row of A, K[1:i,1:i] is the i $\times$ i leading principal submatrix of K_θ, K[1:i, i] is the i $\times$ 1 row vector with the first i elements in the i-th column of K_θ, K[i, 1:i] is the $1 \times$ i row vector formed by the first i elements in the i-th row of K_θ, solve(B,b) computes the solution for the linear system Bx = b, and dot(u,v) provides the inner product between vectors u and v. The determinant of K_θ is obtained with almost no additional cost: it is simply $\prod_{i=1}^{n}$ d[i,i].

The above algorithm delivers the Cholesky decomposition of K_θ. If $K_\theta = LDL^T$ is the Cholesky decomposition, then $L = (I - A)^{-1}$. However, there is no apparent gain from the preceding computations since we need to solve increasingly larger linear systems as the loop runs into higher values of i. Nevertheless, we can exploit sparsity if we set some of the elements in the lower triangular part of A to be zero. For example, suppose we set at most m elements in each row of A to be nonzero. Let N[i] be the set of indices j < i such that a[i,j] $\neq$ 0. We can compute the nonzero elements of A and the diagonal elements of D much more efficiently as:

```
for(i in 1:(n-1) {
    Pa = N[i+1] # neighbors of i+1
    a[i+1,Pa] = solve(K[Pa,Pa], K[(i+1),Pa])
    d[i+1,i+1] = K[i+1,i+1] - dot(K[(i+1),Pa], a[i+1,Pa])
}.
```
(13.39)

In (13.39) we solve n-1 linear systems of size at most m $\times$ m. This can be performed in $\sim$ nm^3 flops, whereas (13.38) for the dense model requires $\sim$ n^3 flops. These computations can be performed in parallel as each iteration of the loop is independent of the others.

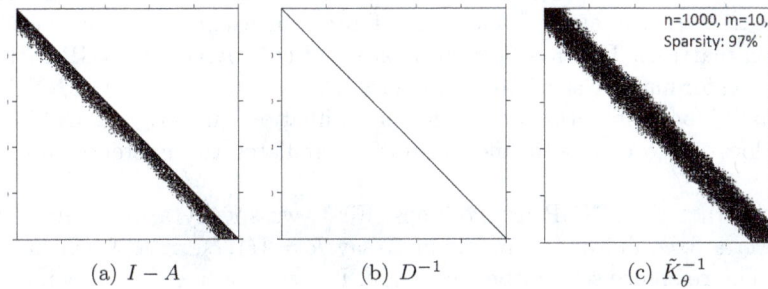

(a) $I - A$ (b) D^{-1} (c) $\tilde{K}_\theta^{-1}$

Figure 13.6 *Structure of the factors making up the sparse $\tilde{K}_\theta^{-1}$ matrix.*

We have arrived at a very useful strategy for introducing sparsity in a precision matrix. Let K_θ and K_θ^{-1} both be dense $n \times n$ positive definite matrices. Suppose we use (13.39) with $\mathtt{K} = K_\theta$ to construct a sparse strictly lower-triangular matrix A with no more than m non-zero entries in each row, where m is considerably smaller than n, and the diagonal matrix D. The resulting matrix $\tilde{K}_\theta = (I - A)^{-1}D(I - A)^{-\mathrm{T}}$ is a covariance matrix whose inverse $\tilde{K}_\theta^{-1} = (I - A^{\mathrm{T}})D^{-1}(I - A)$ is sparse. Figure 13.6 presents a visual representation of the sparsity. While $\tilde{K}_\theta$ need not be sparse, the density $N(w \mid 0, \tilde{K}_\theta)$ is cheap to compute since $\tilde{K}_\theta^{-1}$ is sparse and $\det(\tilde{K}_\theta) = \det(D) = \prod_{i=1}^n \mathtt{d[i,i]}$ is calculated from (13.39). Therefore, we can scale models such as (6.5) to massive data sets by assuming that $w \sim N(0, \tilde{K}_\theta)$ as the prior instead of $w \sim N(0, K_\theta)$.

13.6.2   Nearest-Neighbor Gaussian Processes (NNGP)

If we are interested in estimating the spatial or spatiotemporal process parameters from a finite collection of random variables, then we can use the approach in Section 13.6.1 with $w_i := w(\ell_i)$. In spatial settings, matters are especially convenient as we can delete the edges in the DAG based upon the distances among ℓ_i's. In fact, one can decide to retain at most m of the nearest neighbors for each location and delete all remaining edges. This implies that the (i, j)-th element of A in Section 13.6.1 will be nonzero only if ℓ_j is one of the m nearest neighbors of ℓ_i. In fact, this idea has been effectively used to construct approximate likelihoods for Gaussian process models by Vecchia [1988], Stein et al. [2004] and Stroud et al. [2017].

Localized Gaussian process regression based on few nearest neighbors has also been used to obtain fast kriging estimates. Emery [2009] provides fast updates for kriging equations after adding a new location to the input set. Iterative application of their algorithm yields a localized kriging estimate based on a small set of locations (including few nearest neighbors). The local estimate often provides an excellent approximation to the global kriging estimate which uses data observed at all the locations to predict at a new location. However, this assumes that the parameters associated with the mean and covariance of the GP are known or already estimated. Local Approximation GP, or LAGP [Gramacy and Apley, 2015, Gramacy and Haaland, 2016, Gramacy, 2016], extends this further to estimate the parameters at each new location, essentially providing a non-stationary local approximation to a Gaussian Process at every predictive location and can be used to interpolate or smooth the observed data.

If posterior predictive inference is sought at arbitrary spatiotemporal resolutions, i.e., for the entire process $\{w(\ell) : \ell \in \mathcal{L}\}$, then the ideas in Section 13.6.1 need to be extended to process-based models. In this regard, Datta et al. [2016a] proposed a Nearest Neighbor

Gaussian Process (NNGP) for modeling large spatial data. The NNGP is a well defined Gaussian Process over a domain $\mathcal{L}$ and yields finite dimensional Gaussian densities with sparse precision matrices. This has been also extended to a dynamic NNGP with dynamic neighbor selection for massive spatiotemporal data [Datta et al., 2016b]. The NNGP delivers massive scalability both in terms of parameter estimation and kriging. Unlike low rank processes, it does not oversmooth and accurately emulates the inference from full rank GPs.

We will construct the NNGP in two steps. First, we specify a multivariate Gaussian distribution over a fixed finite set r points in $\mathcal{L}$, say $\mathcal{R} = \{\ell_1^*, \ell_2^*, \ldots, \ell_r^*\}$, which we call the *reference set*. The reference set can be very large. It can be a fine grid of points over $\mathcal{L}$ or one can simply take $r = n$ and let $\mathcal{R}$ be the set of observed points in $\mathcal{L}$. We require that the inverse of the covariance matrix be sparse and computationally efficient. Therefore, we specify that $\mathbf{w}_{\mathcal{R}} \sim N(\mathbf{0}, \tilde{K}_\theta)$, where $\mathbf{w}_{\mathcal{R}}$ is $r \times 1$ with elements $w(\ell_i^*)$ and $\tilde{K}_\theta$ is a covariance matrix such that $\tilde{K}_\theta^{-1}$ is sparse. The matrix $\tilde{K}_\theta$ is constructed from a dense covariance matrix K_θ as described in Section 13.6.1. This provides a highly effective approximation [Vecchia, 1988, Stein et al., 2004] as below:

$$N(\mathbf{w}_{\mathcal{R}} \mid \mathbf{0}, K_\theta) = \prod_{i=1}^{r} p(w(\ell_i^*) \mid \mathbf{w}_{H(\ell_i^*)}) \approx \prod_{i=1}^{r} p(w(\ell_i^*) \mid \mathbf{w}_{N(\ell_i^*)}) = N(\mathbf{w}_{\mathcal{R}} \mid \mathbf{0}, \tilde{K}_\theta) , \quad (13.40)$$

where *history sets* $H(\ell_i^*)$ are defined so that $H(\ell_1^*)$ is empty and $H(\ell_i^*) = \{\ell_1^*, \ell_2^*, \ldots, \ell_{i-1}^*\}$ for $i = 2, 3, \ldots, r$ and we have much smaller *neighbor sets* $N(\ell_i^*) \subseteq H(\ell_i^*)$ for each ℓ_i^* in $\mathcal{R}$. We have legitimate probability models for any choice of $N(\ell_i^*)$'s as long as $N(\ell_i^*) \subseteq H(\ell_i^*)$. One easy specification is to define $N(\ell_i^*)$ as the set of m nearest neighbors of ℓ_i^* among the points in $\mathcal{R}$. Therefore,

$$N(\ell_i) = \begin{cases} \text{empty set for } i = 1 \\ H(\ell_i^*) = \{\ell_1^*, \ell_2^*, \ldots, \ell_{i-1}^*\} \text{ for } i = 2, 3, \ldots, m \\ m \text{ nearest neighbors of } \ell_i^* \text{ among } H(\ell_i^*) \text{ for } i = m+1, \ldots, n \end{cases} .$$

If $m(<< r)$ denotes the limiting size of the neighbor sets $N(\ell)$, then $\tilde{K}_\theta^{-1}$ has at most $O(rm^2)$ non-zero elements. Hence, the approximation in (13.40) produces a sparsity-inducing proper prior distribution for random effects over $\mathcal{R}$ that closely approximates a $N(\mathbf{0}, K_\theta)$ density.

To construct the NNGP we extend the above model to arbitrary locations. We define neighbor sets $N(\ell)$ for any $\ell \in \mathcal{L}$ as the set of m nearest neighbors of ℓ in $\mathcal{R}$. Thus, $N(\ell) \subseteq \mathcal{R}$ and the process can be derived from $p(\mathbf{w}_{\mathcal{R}}, w(\ell) \mid \theta) = N(\mathbf{w}_{\mathcal{R}} \mid \mathbf{0}, \tilde{K}_\theta) \times p\left(w(\ell) \mid \mathbf{w}_{N(\ell)}, \theta\right)$ or, equivalently, by writing

$$w(\ell) = \sum_{i=1}^{r} a_i(\ell) w(\ell_i^*) + \eta(\ell) \text{ for any } \ell \notin \mathcal{R} , \quad (13.41)$$

where $a_i(\ell) = 0$ whenever $\ell_i^* \notin N(\ell)$, $\eta(\ell) \overset{ind}{\sim} N(0, \delta^2(\ell))$ is a process independent of $w(\ell)$, $\text{Cov}\{\eta(\ell), \eta(\ell')\} = 0$ for any two distinct points in $\mathcal{L}$, and

$$\delta^2(\ell) = K_\theta(\ell, \ell) - K_\theta(\ell, N(\ell)) K_\theta^{-1}(N(\ell), N(\ell)) K_\theta(N(\ell), \ell) .$$

Taking conditional expectations in (13.41) yields $\mathbb{E}[w(\ell) \mid \mathbf{w}_{N(\ell)}] = \sum_{i:\ell_i \in N(\ell)} a_i(\ell) w(\ell_i^*)$, which implies that for each ℓ the nonzero $a_i(\ell)$'s are obtained by solving an $m \times m$ linear system. The above construction ensures that $w(\ell)$ is a legitimate Gaussian process whose realizations over any finite collection of arbitrary points in $\mathcal{L}$ will have a multivariate normal distribution with a sparse precision matrix. More formal developments and technical details in the spatial and spatiotemporal settings can be found in Datta et al. [2016a] and Datta et al. [2016b], respectively.

One point worth considering is the definition of "neighbors." The neighbor sets can be fixed before the model fitting exercise. There is some flexibility here. In the spatial setting, the correlation functions usually decay with increasing inter-site distance, so the set of nearest neighbors based on the inter-site distances represents locations exhibiting highest correlation with the given locations. For example, on the plane one could simply use the Euclidean metric to construct neighbor sets, although Stein et al. [2004] recommends including a few points that are farther apart. In-depth investigations into the effects of neighbors on inference is provided by Guinness [2018], who also offer a permutation-based algorithm for choosing neighbors.

In spatiotemporal settings, matters are more complicated. Spatiotemporal covariances between two points typically depend on the spatial as well as the temporal lag between the points. Non-separable isotropic spatiotemporal covariance functions can be written as $K_\theta((s_1, t_1), (s_2, t_2)) = K_\theta(h, u)$ where $h = \|s_1 - s_2\|$ and $u = |t_1 - t_2|$. This often precludes defining any universal distance function $d : (\mathcal{S} \times \mathcal{T})^2 \to \Re^+$ such that $K_\theta((s_1, t_1), (s_2, t_2))$ will be monotonic with respect to $d((s_1, t_1), (s_2, t_2))$ for all choices of θ. This makes it difficult to define universal nearest neighbors in spatiotemporal domains. To obviate this hurdle, Datta et al. [2016b] define "nearest neighbors" in a spatiotemporal domain using the spatiotemporal covariance function itself as a proxy for distance. This can work for arbitrary domains. For any three points ℓ_1, ℓ_2 and ℓ_3, we say that ℓ_1 is nearer to ℓ_2 than to ℓ_3 if $K_\theta(\ell_1, \ell_2) > K_\theta(\ell_1, \ell_3)$. Subsequently, this definition of "distance" is used to find m nearest neighbors for any location. Prediction at any arbitrary location $\ell \notin \mathcal{R}$ is performed by sampling from the posterior predictive distribution.

However, for every point ℓ_i, its neighbor set $N_\theta(\ell)$ will now depend on θ and can change from iteration to iteration in the estimation algorithm. If θ were known, one could have simply evaluated the pairwise correlations between any point ℓ_i^* in $\mathcal{R}$ and all points in its history set $H(\ell_i^*)$ to obtain $N_\theta(\ell_i^*)$—the set of m true nearest neighbors. In practice, however, θ is unknown and for every new value of θ in an iterative algorithm, we need to search for the neighbor sets within the history sets. Since the history sets are quite large, searching the entire space for nearest neighbors in each iteration will be computationally unfeasible. Datta et al. [2016b] offer some smart strategies for selecting spatiotemporal neighbors. They propose restricting the search for the neighbor sets to carefully constructed small subsets of the history sets. These small *eligible sets* $E(\ell_i^*)$ are constructed in such a manner that, despite being much smaller than the history sets, they are guaranteed to contain the true nearest neighbor sets. This strategy works when we choose m to be a perfect square and the original nonseparable covariance function $K_\theta(h, u)$, where $h = \|\mathbf{s} - \mathbf{s}'\|$ and $u = |t - t'|$ satisfies *natural monotonicity*, i.e. $K_\theta(h, u)$ is decreasing in h for fixed u and decreasing in u for fixed h. All Matérn-based space-time separable covariances and many non-separable classes of covariance functions possess this property [Stein, 2013, Omidi and Mohammadzadeh, 2015].

13.6.3 Hierarchical NNGP models

We briefly turn to model fitting and estimation. For the approximation in (13.40) to be effective, the size of the reference set, r, needs to be large enough to represent the spatial domain. However, this does not impede computations involving NNGP models because the storage and number of floating point operations are always linear in r. The reference set $\mathcal{R}$ can, in principle, be any finite set of locations in the study domain. A particularly convenient choice, in practice, is to simply take $\mathcal{R}$ to be the set of observed locations in the dataset. Datta et al. [2016a] demonstrate through extensive simulation experiments and a real application that this simple choice seems to be very effective.

Since the NNGP is a proper Gaussian process, we can use it as a prior for the spatial random effects in any hierarchical model. We write $w(\ell) \sim NNGP(0, \tilde{K}_\theta(\cdot, \cdot))$, where

$\tilde{K}_\theta(\ell, \ell')$ is the covariance function for the NNGP [see Datta et al., 2016a, for a closed form expression]. For example, with $r = n$ and $\mathcal{R}$ the set of observed locations, one can build a scalable Bayesian hierarchical model exactly as with a usual spatial process, but assigning an NNGP to the spatial random effects. Here is a simple hierarchical NNGP spatial model:

$$Y(\ell) \mid g(\cdot), \boldsymbol{\beta}, w(\ell) \stackrel{ind}{\sim} P_\tau \quad \text{(exponential family)}, \quad g(\mathbb{E}[Y(\ell)]) = \mathbf{x}(\ell)^{\mathrm{T}}\boldsymbol{\beta} + w(\ell) ; \quad (13.42)$$

$$w(\ell) \sim NNGP(0, \tilde{K}_\theta(\cdot, \cdot)) ; \quad \{\theta, \boldsymbol{\beta}, \tau\} \sim p(\theta, \boldsymbol{\beta}, \tau) , \quad (13.43)$$

where P_τ is an exponential family distribution with link function $g(\cdot)$. Posterior sampling from (13.42) is customarily performed using MCMC. Computational benefits emerge from the full conditional distribution $p(w(\ell_i) \mid \mathbf{w}_\mathcal{R}, \theta, \boldsymbol{\beta}, \tau) = p(w(\ell_i) \mid \mathbf{w}_{N(\ell_i)}, \theta, \boldsymbol{\beta}, \tau)$, where $\mathbf{w}_{N(\ell_i)}$ is an $m \times 1$ subset of $\mathbf{w}_\mathcal{R}$. Prediction at any arbitrary location $\ell \notin \mathcal{R}$ is performed by sampling from the posterior predictive distribution. For each draw of $\{\mathbf{w}_\mathcal{R}, \boldsymbol{\beta}, \theta, \tau\}$ from $p(\mathbf{w}_\mathcal{R}, \boldsymbol{\beta}, \tau, \theta \mid \mathbf{y})$, we draw a $w(\ell)$ from $N(\mathbf{a}(\ell)^{\mathrm{T}}\mathbf{w}_{N(\ell)}, \delta^2(\ell))$ and $y(\ell)$ from $p(y(\ell) \mid \boldsymbol{\beta}, w(\ell), \tau)$, where $\mathbf{y}$ is the vector of observed outcomes and $a(\ell)$ is a vector of the nonzero $a_j(\ell)$'s in (13.41).

Another, even simpler, example could model a continuous outcome itself as an NNGP. Let the desired full GP specification be $Y(\ell) \sim GP(\mathbf{x}^{\mathrm{T}}(\ell)\boldsymbol{\beta}, K_\theta(\cdot, \cdot))$. We derive the NNGP from this K_θ and obtain

$$Y(\ell) \sim NNGP(\mu(\ell), \tilde{K}_\theta(\cdot, \cdot)) ; \quad \mu(\ell) = \mathbf{x}^{\mathrm{T}}(\ell)\boldsymbol{\beta} ; \quad \{\theta, \boldsymbol{\beta}\} \sim p(\theta, \boldsymbol{\beta}) . \quad (13.44)$$

The above model is extremely fast. The likelihood is of the form $y \sim N(X\boldsymbol{\beta}, \tilde{K}_\theta)$, where $\tilde{K}_\theta^{-1} = (I - A^{\mathrm{T}})D^{-1}(I - A)$ is sparse and A and D are obtained from (13.39) efficiently in parallel. The parameter space of interest is $\{\theta, \boldsymbol{\beta}\}$, which is much smaller than for (13.42) where the latent spatial process also was unknown. While (13.44) does not separate the residuals into a spatial process and a measurement error process, one can still include measurement error variance, or the nugget, in (13.44). Here, one would absorb the nugget into θ. For example, we could write the likelihood as $N(\mathbf{y} \mid X\boldsymbol{\beta}, K_\theta)$, where $K_\theta = \sigma^2 R_\phi + \tau^2 I_n$, R_ϕ is a spatial correlation matrix and $\theta = \{\sigma^2, \phi, \tau^2\}$. These will also feature in the derived NNGP covariance matrix $\tilde{K}_\theta$. We can predict the outcome at an arbitrary point ℓ by sampling from the posterior predictive distribution as follows: for each draw of $\{\boldsymbol{\beta}, \theta\}$ from $p(\boldsymbol{\beta}, \theta \mid \mathbf{y})$, we draw a $y(\ell)$ from $N(y(\ell) \mid \mathbf{x}^{\mathrm{T}}(\ell)\boldsymbol{\beta}, \delta^2(\ell))$. There is, however, no latent smooth process $w(\ell)$ in (13.44) and inference on the latent spatial process is precluded.

Likelihood computations in NNGP models usually involve $O(nm^3)$ flops. One does not need to store $n \times n$ matrices, only $m \times m$ matrices which leads to storage $\sim nm^2$. Substantial computational savings accrue because m is usually very small. Datta et al. [2016a] demonstrate that fitting NNGP models to the simulated data with number of neighbors as less as $m = 10$ produce posterior estimates of the spatial surface indistinguishable from those fitted with a full Gaussian process model with a dense covariance matrix. In fact, simulation experiments in Datta et al. [2016a] and Datta et al. [2016b] also affirm that m can usually be taken to be very small compared to r; there seems to be no inferential advantage to taking m to exceed 15, even for datasets with over 10^5 spatial locations. For example, Figure 13.7 shows the 95% posterior credible intervals for a series of 10 simulation experiments where the true effective range was fixed at values from 0.1 to 1.0 in increments of 0.1. Each dataset comprised 2500 points. Even with $m = 10$ neighbors, the credible intervals for the effective spatial range from the NNGP model were very consistent with those from the full GP model. Datta et al. [2016a] present simulations using the Matérn and other covariance functions revealing very similar behavior.

Another important point to note is that $\tilde{K}_\theta$ is not invariant to the order in which we define $H(\ell_1) \subseteq H(\ell_2) \subseteq \cdots \subseteq H(\ell_r)$ (i.e., the topological order). Vecchia [1988] and Stein et al. [2004] both assert that the approximation in (13.40) is not sensitive to this ordering.

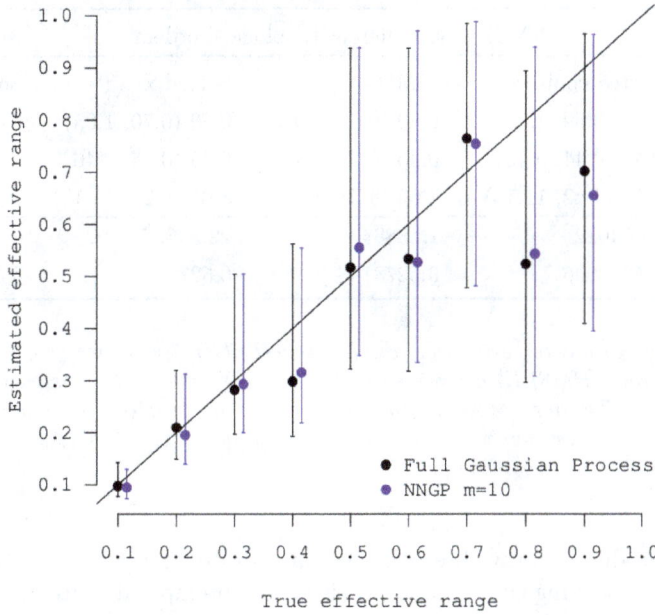

Figure 13.7 *95% credible intervals for the effective spatial range from an NNGP model with $m = 10$ and a full GP model fitted to 10 different simulated datasets with true effective range fixed at values between 0.1 and 1.0 in increments of 0.1.*

This is corroborated by simulation experiments by Datta et al. [2016a], although Guinness [2018] has indicated sensitivity to the ordering in terms of model deviance and recommends a maximum-minimum distance (MMD) algorithm. In fact, there is also interest in choosing the number of neighbors. First, it should be intuitively clear that DAGs constructed using shrunk neighbor sets will yield probability models farther away from the full model as the neighbor sets get smaller. Following Banerjee [2020], consider a random vector $\mathbf{w} = (\mathbf{w}_A^\mathsf{T}, \mathbf{w}_B^\mathsf{T})^\mathsf{T}$, where A and B are mutually exclusive sets containing indices for the elements of $\mathbf{w}$, and let $p(\mathbf{w}) = p(\mathbf{w}_A)p(\mathbf{w}_B \mid \mathbf{w}_A)$ denote the joint probability density for $\mathbf{w}$. Consider two sub-models $p_1(\mathbf{w}) = p(\mathbf{w}_A) \times p(\mathbf{w}_B \mid \mathbf{w}_{N_{1B}})$ and $p_2(\mathbf{w}) = p(\mathbf{w}_A) \times p(\mathbf{w}_B \mid \mathbf{w}_{N_{2B}})$, where $N_{2B} \subset N_{1B} \subset A$. The model p_2 is farther than p_1 from p in terms of KL divergence:

$$
\begin{aligned}
KL(p\|\|p_2) - KL(p\|\|p_1) &= \int \left\{ \log\left(\frac{p(\mathbf{w})}{p_2(\mathbf{w})}\right) - \log\left(\frac{p(\mathbf{w})}{p_1(\mathbf{w})}\right) \right\} p(\mathbf{w})d\mathbf{w} \\
&= \int \log\left(\frac{p_1(\mathbf{w})}{p_2(\mathbf{w})}\right) p(\mathbf{w})d\mathbf{w} = \int \log\left(\frac{p(\mathbf{w}_B \mid \mathbf{w}_{N_{1B}})}{p(\mathbf{w}_B \mid \mathbf{w}_{N_{2B}})}\right) p(\mathbf{w})d\mathbf{w} \\
&= \int \log\left(\frac{p(\mathbf{w}_B \mid \mathbf{w}_{N_{1B}})}{p(\mathbf{w}_B \mid \mathbf{w}_{N_{2B}})}\right) p(\mathbf{w}_B \mid \mathbf{w}_{N_{1B}})p(\mathbf{w}_{N_{1B}})d\mathbf{w}_B d\mathbf{w}_{N_{1B}} \\
&= \int \left\{ \int \log\left(\frac{p(\mathbf{w}_B \mid \mathbf{w}_{N_{1B}})}{p(\mathbf{w}_B \mid \mathbf{w}_{N_{2B}})}\right) p(\mathbf{w}_B \mid \mathbf{w}_{N_{1B}})d\mathbf{w}_B \right\} p(\mathbf{w}_{N_{1B}})d\mathbf{w}_{N_{1B}} \geq 0 \,,
\end{aligned}
\tag{13.45}
$$

where we have used the fact that $A \setminus N_{1B}$ is mutually exclusive of N_{1B} and, crucially, also of N_{2B} (since $N_{2B} \subset N_{1B}$) to legitimately integrate out $w_{A \setminus N_{1B}}$. The final conclusion follows from a customary application of Jensen's inequality to show that the inner integral in the last equation is non-negative. Equation (13.45) provides an alternate distribution-free proof of a result for Gaussian likelihoods [see Theorem 1 in Guinness, 2018, for the proof with Gaussian likelihoods]. These results also indicate that the ordering of the variables to construct the approximation can affect model performance and certain designs to determine

	True	Sorted coord(x+y)	MMD	Sorted x	Sorted y
		NNGP from different topological orders			
σ	1	0.79 (0.69, 1.04)	0.80 (0.69, 1.02)	0.80 (0.70, 1.05)	0.83 (0.69, 1.08)
τ	0.45	0.45 (0.44, 0.46)	0.45 (0.44, 0.47)	0.45 (0.44, 0.46)	0.45 (0.44, 0.47)
ϕ	5	8.11 (4.42, 11.10)	7.63 (4.58, 10.97)	8.01 (4.26, 11.18)	7.12 (4.06, 11.03)
KL-D	–	24.04022	13.88847	22.30667	21.59174
RMSPE	–	0.5278996	0.5278198	0.527912	0.527807

Table 13.3 *Posterior parameter estimates, the Kullback-Leibler divergence (KL-D) and root mean square predictive errors (RMSPE) are presented for four NNGP models constructed from different topological orderings. The four orderings from left to right are "sorted on the sum of vertical and horizontal coordinate," maximum-minimum distance [Guinness, 2018], sorted on horizontal coordinate and sorted on vertical coordinate.*

the ordering can produce improved results [as demonstrated in Guinness, 2018]. Datta et al. [2016b] argued against fixing the neighborhoods in spatiotemporal contexts (since neighbors in space and neighbors in time may not align) and demonstrate a computationally efficient method to learn about neighbors in spatiotemporal domains.

Datta et al. [2016a] conducted some preliminary investigations to investigate the effect of the topological order. In one simple experiment, we generated data from the "true" model $y(\ell_i) = w(\ell_i) + \epsilon(\ell_i)$ for $i = 1, \ldots, n$ with $n = 6400$ spatial locations arranged over an 80×80 grid. The covariance function was specified as $K_\theta(\ell_i, \ell_j) = \sigma^2 \exp(-\phi\|\ell_i - \ell_j\|)$, and $\epsilon(\ell_i) \overset{iid}{\sim} N(0, \tau^2)$ with the true values of σ^2, ϕ and τ^2 given in the second column of Table 13.3. Four different NNGP models corresponding to (13.44) with $\tilde{K}_\theta$ derived from $K_\theta = \sigma^2 R_\phi + \tau^2 I$ and R_ϕ having elements $\exp(-\phi\|\ell_i - \ell_j\|)$, were fitted to the simulated data. Each of these models were constructed with $m = 10$ nearest neighbors, but with different ordering of the points $\ell = (x, y)$. These were performed according to the sum of the coordinates $x + y$, a maximum-minimum distance (MMD) proposed by Guinness [2018], the x coordinate, and the y coordinate. Table 13.3 presents a comparison of these NNGP models. Irrespective of the ordering of the points, the inference with respect to parameter estimates and predictive performance is extremely robust and effectively indistinguishable from each other. However, the posterior mean of the Kullback-Leibler divergence of these models from the true generating model revealed that the metric proposed by Guinness [2018] is indeed less than the other three.

It is worth pointing out that there is a substantial array of methods developed on the ideas laid out by Vecchia [1988] for approximate likelihoods and Datta et al. [2016a] for sparsity-inducing spatial processes. Notable extensions include Katzfuss et al. [2020] and Katzfuss and Guinness [2021] where the models are being derived using DAGs over the expanded set of observations and process realizations; Peruzzi et al. [2022] who develop meshed NNGPs by partitioning the spatial domain and modeling the regions in the partition using a sparsity-inducing directed acyclic graph (DAG); and developments of Vecchia-approximated deep Gaussian processes by Sauer et al. [2023a]. In more computationally oriented developments, Finley et al. [2019] develop and compare different algorithms and models for computing hierarchical NNGP model, Zhang et al. [2019] offer some practical strategies for fitting NNGPs using conjugate gradient solvers and without using MCMC, while Wu et al. [2022] and Cao et al. [2023] explore variational inference for NNGPs and general Vecchia processes, respectively.

13.7 Spatial Meta-Kriging

A different approach toward BIG DATA problems relies upon divide and conquer methods. The idea here is divide and conquer by pooling posterior inference across a partition of data subsets. The concept of delegating Bayesian calculations to a group of independent datasets is intuitive and has been explored from diverse perspectives. Examples include Consensus Monte Carlo [Scott et al., 2016] and its kernel-based adaptations [Rendell et al., 2021], Bayesian meta-analysis in clinical applications [see, e.g., Chapter 4 in Parmigiani, 2002] and in spatially-temporally structured data [Bell et al., 2005, Kang et al., 2011]; and, more recently, approaches such as "meta-kriging" using geometric medians of posterior distributions [Srivastava et al., 2015, Minsker et al., 2017, Guhaniyogi and Banerjee, 2018, 2019, Guhaniyogi et al., 2019].

Recall the conjugate Bayesian linear regression model in Section 5.3,

$$p(\boldsymbol{\beta}, \sigma^2 \mid \mathbf{y}) \propto IG(\sigma^2 \mid a_\sigma, b_\sigma) \times N(\boldsymbol{\beta} \mid \boldsymbol{\mu}_\beta, \sigma^2 V_\beta) \times N(\mathbf{y} \mid X\boldsymbol{\beta}, \sigma^2 V_y) \,, \tag{13.46}$$

where $\mathbf{y}$ is $N \times 1$, X is $N \times p$, $\boldsymbol{\beta}$ is $p \times 1$, V_y is a fixed $N \times N$ covariance matrix, $\boldsymbol{\mu}_\beta$ is a fixed $p \times 1$ vector and V_β is a fixed $p \times p$ matrix. The joint posterior density $p(\boldsymbol{\beta}, \sigma^2 \mid \mathbf{y})$ is available in closed form as $p(\boldsymbol{\beta}, \sigma^2 \mid \mathbf{y}) = p(\sigma^2 \mid \mathbf{y}) \times p(\boldsymbol{\beta} \mid \sigma^2, \mathbf{y})$, where the marginal posterior density $p(\sigma^2 \mid \mathbf{y}) = IG(\sigma^2 \mid a^*, b^*)$ and the conditional posterior density $p(\boldsymbol{\beta} \mid \sigma^2, \mathbf{y}) = N(\boldsymbol{\beta} \mid M\mathbf{m}, \sigma^2 M)$ with $a^* = a_\sigma + N/2$, $b^* = b_\sigma + c/2$, $\mathbf{m} = V_\beta^{-1}\boldsymbol{\mu}_\beta + X^\mathsf{T} V_y^{-1}\mathbf{y}$, $M^{-1} = V_\beta^{-1} + X^\mathsf{T} V_y^{-1} X$ and $c = \boldsymbol{\mu}_\beta^\mathsf{T} V_\beta^{-1}\boldsymbol{\mu}_\beta + \mathbf{y}^\mathsf{T} V_y^{-1}\mathbf{y} - \mathbf{m}^\mathsf{T} M\mathbf{m}$. We sample from (13.46) by first sampling σ^2 from $IG(a^*, b^*)$ and then sampling $\boldsymbol{\beta}$ from $N(M\mathbf{m}, \sigma^2 M)$ for each sampled value of σ^2. This results in samples from $p(\boldsymbol{\beta}, \sigma^2 \mid \mathbf{y})$. Besides the fixed hyperparameters in the prior distributions, this exercise requires computing $\mathbf{m}$, M and c.

Now consider a situation where N is large enough so that memory requirements for computing (13.46) is impractical. Therefore, we split the $N \times 1$ vector $\mathbf{y} = (\mathbf{y}_1^\mathsf{T}, \ldots, \mathbf{y}_K^\mathsf{T})^\mathsf{T}$ into K vectors with $\mathbf{y}_k$ as the $n_k \times 1$ vector, where $\sum_{k=1}^K n_k = N$. The size of the k-th subset is n_k. These sizes need not be the same across k, but will be chosen in a manner so that each of the subsets can be fitted easily with the computational resources available. Also, let X_k be the $n_k \times p$ matrix of predictors corresponding to $\mathbf{y}_k$ and let V_{y_k} be the marginal correlation matrix for $\mathbf{y}_k$ for each $k = 1, \ldots, K$. If V_y in (13.46) has a block diagonal structure with V_{y_k} as its diagonal blocks, then $\mathbf{y}_k = X_k\boldsymbol{\beta} + \boldsymbol{\epsilon}_k$, where $\boldsymbol{\epsilon}_k \overset{ind}{\sim} N(\mathbf{0}, \sigma^2 V_{y_k})$. The Bayesian specification is completed by assigning priors to σ^2 and $\boldsymbol{\beta}$ as in (13.46). If we distribute the analysis to K different computing cores, where the k-th core fits $N(\mathbf{y}_k \mid X_k\boldsymbol{\beta}, \sigma^2 V_{y_k})$, then the quantities needed for sampling from the full $p(\boldsymbol{\beta}, \sigma^2 \mid \mathbf{y})$ can be computed entirely using quantities obtained from the individual subsets of the data. For each $k = 1, 2, \ldots, K$ we independently compute $\mathbf{m}_k = V_\beta^{-1}\boldsymbol{\mu}_\beta + X_k^\mathsf{T} V_{y_k}^{-1}\mathbf{y}_k$ and $M_k^{-1} = V_\beta^{-1} + X_k^\mathsf{T} V_{y_k}^{-1} X_k$ based upon the k-th subset of the data. We then combine them to obtain

$$\mathbf{m} = \sum_{k=1}^K (\mathbf{m}_k - (1 - 1/K) V_\beta^{-1}\boldsymbol{\mu}_\beta) \quad \text{and} \quad M^{-1} = \sum_{k=1}^K (M_k^{-1} - (1 - 1/K) V_\beta^{-1}) \,. \tag{13.47}$$

Subsequently, we compute $c = \boldsymbol{\mu}_\beta^\mathsf{T} V_\beta^{-1}\boldsymbol{\mu}_\beta + \sum_{k=1}^K \mathbf{y}_k^\mathsf{T} V_{y_k}^{-1}\mathbf{y}_k - \mathbf{m}^\mathsf{T} M\mathbf{m}$.

Hence, sampling from the posterior distribution of $\boldsymbol{\beta}$ and σ^2 in (13.46) given the entire dataset can be achieved using quantities computed independently from each of the K smaller subsets of the data. There is no need to interact between the subsets and one does not require to store or compute with large objects based upon the entire dataset. This computation can also be done sequentially. We first obtain the posterior distribution $p(\boldsymbol{\beta}, \sigma^2 \mid \mathbf{y}_1)$ based only upon the first data set. This posterior becomes the prior for the next step and we obtain $p(\boldsymbol{\beta}, \sigma^2 \mid \mathbf{y}_1, \mathbf{y}_2) \propto p(\boldsymbol{\beta}, \sigma^2 \mid \mathbf{y}_1) \times p(\mathbf{y}_2 \mid \boldsymbol{\beta}, \sigma^2)$ and so on until we arrive at $p(\boldsymbol{\beta}, \sigma^2 \mid \mathbf{y}_1, \mathbf{y}_2, \ldots, \mathbf{y}_K) \propto p(\boldsymbol{\beta}, \sigma^2 \mid \mathbf{y}_1, \mathbf{y}_2, \ldots, \mathbf{y}_{K-1}) \times p(\mathbf{y}_K \mid \boldsymbol{\beta}, \sigma^2)$.

Clearly such exact recovery of the full posterior crucially depends on the conditional independence across the different data blocks (e.g., $p(\mathbf{y}_k \mid \boldsymbol{\beta}, \sigma^2, \mathbf{y}_1, \ldots, \mathbf{y}_{k-1}) = p(\mathbf{y}_k \mid \boldsymbol{\beta}, \sigma^2)$ for each $k = 2, \ldots, K$). This works for uncorrelated outcomes, as in standard linear regression, or uncorrelated blocks of outcomes. Exact recovery is precluded for spatial and spatiotemporal process models and, more generally, for correlated data. Nevertheless, we can develop a general approximation framework for obtaining the full posterior from posterior densities calculated over smaller subsets. One general way to pool information across these individual posteriors is to use the unique *Geometric Median* (GM) of the subset posteriors, as developed by [Minsker, 2015]. Assume that the individual posterior densities $p_k \equiv p(\Omega \mid \mathbf{y}_k)$ reside on a Banach space $\mathcal{H}$ equipped with norm $\|\cdot\|$.

The GM is defined as $\pi^*(\cdot \mid \mathbf{y}) = \arg\min_{\pi \in \mathcal{H}} \sum_{k=1}^{K} \|p_k - \pi\|_\rho$, where $\mathbf{y} = (\mathbf{y}_1^{\mathrm{T}}, \mathbf{y}_2^{\mathrm{T}}, \ldots, \mathbf{y}_K^{\mathrm{T}})^{\mathrm{T}}$.
The norm quantifies the distance between any two posterior densities $\pi_1(\cdot)$ and $\pi_2(\cdot)$ as
$\|\pi_1 - \pi_2\|_\rho = \left\| \int \rho(\Omega, \cdot) d(\pi_1 - \pi_2)(\Omega) \right\|$, where $\rho(\cdot)$ is a positive-definite kernel function.
Assume $\rho(\mathbf{z}_1, \mathbf{z}_2) = \exp(-\|\mathbf{z}_1 - \mathbf{z}_2\|^2)$. The GM is unique and lies in the convex hull of the individual posteriors, so $\pi^*(\Omega \mid \mathbf{y})$ is a legitimate probability density. Specifically, $\pi^*(\Omega \mid \mathbf{y}) = \sum_{k=1}^{K} \alpha_{\rho,k}(\mathbf{y}) p_k, \sum_{k=1}^{K} \alpha_{\rho,k}(\mathbf{y}) = 1$, each $\alpha_{\rho,k}(\mathbf{y})$ being a function of $\rho, \mathbf{y}$, so that $\int_\Omega \pi^*(\Omega \mid \mathbf{y}) d\Omega = 1$. Computing the GM $\pi^* \equiv \pi^*(\Omega \mid \mathbf{y})$ is achieved by an iterative algorithm that estimates $\alpha_{\rho,k}(\mathbf{y})$ from the subset posteriors p_k for each $k = 1, 2, \ldots, K$. To further elucidate, we use a well known result that the GM π^* satisfies $\pi^* = \frac{\sum_{k=1}^{K} \|p_k - \pi^*\|_\rho^{-1} p_k}{\sum_{k=1}^{K} \|p_k - \pi^*\|_\rho^{-1}}$,
so that $\alpha_{\rho,k}(\mathbf{y}) = \frac{\|p_k - \pi^*\|_\rho^{-1}}{\sum_{j=1}^{K} \|p_k - \pi^*\|_\rho^{-1}}$. There is no apparent closed-form solution for $\alpha_{\rho,k}(\mathbf{y})$ satisfying this equation, so Weiszfeld's algorithm [Minsker et al., 2017] is used to estimate these functions.

This approach has been extended to spatial process settings by Guhaniyogi and Banerjee [2018, 2019]. The advantage here is that one can use existing Bayesian geostatistical software to sample from the posterior distributions of the different subsets. This can be performed either in parallel over multiple cores or across different machines altogether. One then needs to save only the post burn-in samples and execute Weiszfeld's algorithm to these samples. Weiszfeld's algorithm is extremely fast and easy to program.

13.8   Computer tutorials

Computer programs illustrating predictive process and NNGP models using the spBayes and spNNGP libraries in R as well as an NNGP implementation in Stan are supplied in the folder titled Chapter 13 from https://github.com/sudiptobanerjee/BGC_2023.

13.9   Exercises

1. Consider the Sherman-Woodbury-Morrison formula (assuming the inverses exist and the matrices have conformable dimensions):

$$(A + BDC)^{-1} = A^{-1} - A^{-1}BMCA^{-1}, \quad \text{where} \quad M = (D^{-1} + CA^{-1}B)^{-1}.$$

Derive this in the following two ways:

(a) Multiply the right hand side by $(A + BDC)$ and show that it equals to the identity matrix.

(b) Consider the following block of linear equations:

$$\begin{pmatrix} A & -B \\ C & D^{-1} \end{pmatrix} \begin{pmatrix} X_1 \\ X_2 \end{pmatrix} = \begin{pmatrix} I \\ O \end{pmatrix} \tag{13.48}$$

Eliminate X_2 from the first block of equations in (13.48) to show that $X_1 = (A + BDC)^{-1}$. Now eliminate X_1 from the second block of equations, solve for X_2 and then substitute this solution in the first system to solve for X_1. Show that this gives $X_1 = A^{-1} - A^{-1}BMCA^{-1}$, where $M = (D^{-1} + CA^{-1}B)^{-1}$. These two different expressions for X_1 establishes the Sherman-Woodbury-Morrison identity.

2. Consider the Bayesian hierarchical model:

$$N(\mathbf{x} \mid \mathbf{0}, D) \times N(\mathbf{y} \mid B\mathbf{x}, A) , \qquad (13.49)$$

where $\mathbf{y}$ is $n \times 1$, $\mathbf{x}$ is $p \times 1$, A and D are positive definite matrices of sizes $n \times n$ and $p \times p$ respectively, and B is $n \times p$. By computing the marginal covariance matrix $\mathrm{Var}(\mathbf{y})$ in two ways, show that:

$$(A + BDB')^{-1} = A^{-1} - A^{-1}B(D^{-1} + B'A^{-1}B)^{-1}B'A^{-1} .$$

3. Let $Z = [Z_1 : Z_2]$ be a matrix of full column rank and let $P_Z = Z(Z'Z)^{-1}Z'$. Prove that $P_Z = P_{Z_1} + P_W$, where $W = (I - P_{Z_1})Z_2$.

4. Use the spBayes library in R to fit a predictive process model with 10, 20 and 30 knots to the Colorado temperature and precipitation dataset available from `https://github.com/sudiptobanerjee/BGC_2023/data/ColoradoLMC.dat`. First fit a univariate predictive process model with temperature as the dependent variable and precipitation as the explanatory variable. Next fit a bivariate predictive process model with temperature and precipitation modeled jointly with their own intercepts and using elevation as a shared predictor.

5. Repeat the above analysis in Problem 4 by programming a predictive process model in a Bayesian modeling language such as rjags or nimble. Compare the speed of these programs with spBayes.

6. Write a variational Bayes algorithm to estimate a full geostatistical (parent) and a predictive process model for the BEF data in spBayes.

7. Use the spNNGP library in R to fit a Nearest Neighbor Gaussian Process (NNGP) model to the BCEF forest canopy data available in the spNNGP library. First obtain exact inference by fixing the parameters range (or decay), smoothness and the ratio between the nugget and spatial variance component to obtain conjugate closed-form posterior distributions using the spConjNNGP function. Next use the spNNGP function to perform spatial linear regression with priors on all unknown parameters in the process.

8. Use the spNNGP library in R to fit a Nearest Neighbor Gaussian Process (NNGP) model to the Colorado temperature dataset available from `https://github.com/sudiptobanerjee/BGC_2023/data/ColoradoLMC.dat` using precipitation and elevation as predictors. As in Problem 7, analyze the data using the conjugate model in spConjN-NGP and the full Bayesian spatial regression model in the spNNGP function.

9. Repeat the above analysis in Problems 7 and 8 by programming an NNGP model in rstan (see `https://mc-stan.org/learn-stan/case-studies/nngp.html` for the code). Compare the speed of your program the output of your inference with that from the Use the spNNGP library in R.

10. Consider the setup in Section 13.7 and the conjugate Bayesian linear regression model in (13.46). Assume that the data arrives in a sequential stream of K independent blocks $\mathbf{y} = (\mathbf{y}_1^{\mathrm{T}}, \ldots, \mathbf{y}_K^{\mathrm{T}})^{\mathrm{T}}$ such that $p(\mathbf{y} \mid \boldsymbol{\beta}, \sigma^2) = N(\mathbf{y} \mid X\boldsymbol{\beta}, \sigma^2 V_y) = \prod_{k=1}^{K} N(\mathbf{y}_k \mid X_k\boldsymbol{\beta}, \sigma^2 V_{y_k})$. Suppose that the computer memory available to us can load and analyze only one block of data at a time. Suppose the posterior distribution including and up to block k is

$$p(\boldsymbol{\beta}, \sigma^2 \mid \mathbf{y}_1, \ldots, \mathbf{y}_k) = NIG(\boldsymbol{\beta}, \sigma^2 \mid M_k\mathbf{m}_k, M_k, a_k^*, b_k^*) , \qquad (13.50)$$

where $NIG(\boldsymbol{\beta}, \sigma^2 \mid M_k\mathbf{m}_k, M_k, a_k^*, b_k^*)$ is the density in (5.22) (recall Section 5.3).

(a) Show that the posterior distribution the data including and up to block $k+1$ is

$$p(\boldsymbol{\beta}, \sigma^2 \mid \mathbf{y}_1, \ldots, \mathbf{y}_k, \mathbf{y}_{k+1}) = NIG(\boldsymbol{\beta}, \sigma^2 \mid M_{k+1}\mathbf{m}_{k+1}, M_{k+1}, a_{k+1}^*, b_{k+1}^*), \quad (13.51)$$

where the parameters in the the NIG density can all be updated using the parameters in (13.50) and the data associated with block $k+1$ ($\{\mathbf{y}_{k+1}, X_{k+1}, V_{y_{k+1}}\}$).

(b) Prove that this sequential updating will result in the same posterior distribution as $p(\boldsymbol{\beta}, \sigma^2 \mid \mathbf{y}) = NIG(\boldsymbol{\beta}, \sigma^2 \mid M\mathbf{m}, \mathbf{m}, a^*, b^*)$ described in Section 13.7, i.e., where $\mathbf{m}$ and M are as in (13.47) and

$$a^* = a_\sigma + \frac{N}{2}; \quad b^* = b_\sigma + \frac{\boldsymbol{\mu}_\beta^{\mathrm{T}} V_\beta^{-1} \boldsymbol{\mu}_\beta + \sum_{k=1}^{K} \mathbf{y}_k^{\mathrm{T}} V_{y_k}^{-1} \mathbf{y}_k - \mathbf{m}^{\mathrm{T}} M\mathbf{m}}{2}.$$

(c) Write down an efficient "divide and conquer" algorithm to implement the above updating scheme that arrives at the full data analysis while only loading objects from a single data block at each step.

Chapter 14

Spatial gradients and wombling

14.1 Introduction

Much of this text has focused upon spatial process models that presume, for a region of study $\mathcal{D}$, a collection of random variables $\{Y(\mathbf{s}) : \mathbf{s} \in \mathcal{D}\}$, which can be viewed as a randomly realized surface over the region. In practice, this surface is only observed at a finite set of locations and inferential interest typically resides in estimation of the process parameters as well as in spatial interpolation of the process at the unobserved locations.

Once such an interpolated surface has been obtained, investigation of rapid change on the surface may be of interest. In such contexts, interest often lies in the rate of change of a spatial surface at a given location in a given direction. Such slopes or gradients are of interest in so-called digital terrain models for exploring surface roughness. They would also be of interest in meteorology to recognize temperature or rainfall gradients or in environmental monitoring to understand pollution gradients. Such rates of change may also be sought for unobservable or latent spatial processes. For instance, in understanding real estate markets supplying house prices for single family homes, spatial modeling of residuals provides adjustment to reflect desirability of location, controlling for the characteristics for the home and property. Directional gradients at a given location illuminate potential investment decision-making.

Evidently, inferring about rapid change is often equivalent to investigating smoothness of process realizations. Scale is clearly of crucial importance in capturing roughness. For instance, in terrain modeling, at low resolution, roughness recognizes global features such as hills and valleys. At high resolution, we would be identifying more local features. Indeed, to characterize local rates of change without having to specify a scale we can conceptualize infinitesimal gradients. Under suitable conditions, we can formally define directional derivatives and the existence of directional derivative processes [Banerjee and Gelfand, 2003].

Two points are critical here. First, it is evident that the finitely sampled data cannot visually inform about the smoothness of such realizations. Rather, such smoothness is captured in the specification of the process and hence would be motivated by mechanistic considerations associated with the process yielding the data. Second, the choice of the process covariance function, i.e., the function which provides the covariance between $Y(\mathbf{s})$ and $Y(\mathbf{s}')$ for arbitrary pairs of locations $\mathbf{s}$ and $\mathbf{s}'$ in $\mathcal{D}$ characterizes the smoothness of process realizations. For example, Kent (1989) pursues the notion of almost sure (a.s.) continuity while Stein (1999a) follows the path of mean square continuity (and more generally, mean square differentiability). Banerjee and Gelfand (2002) clarifies and extends these ideas in various ways.

Such local assessments of spatial surfaces are not restricted to points, but are often desired for curves and boundaries. For instance, environmental scientists are interested in ascertaining whether natural boundaries (e.g. mountains, forest edges etc.) represent a zone of rapid change in weather, ecologists are interested in determining curves that delineate differing zones of species abundance, while public health officials want to identify change in health care delivery across municipal boundaries, counties or states. The above objectives

DOI: 10.1201/9781003401728-14

require the notion of gradients and, in particular, assigning gradients to curves (*curvilinear gradients*) in order to identify curves that track a path through the region where the surface is rapidly changing [Banerjee and Gelfand, 2006]. Such boundaries are commonly referred to as difference boundaries or *wombling boundaries*, named after a seminal paper by Womble [Womble, 1951], who discussed their importance in understanding scientific phenomena [also see Fagan et al., 2003]

As a concept, wombling is useful because it attempts to quantify spatial information in objects such as curves and paths which is not easy to model using regressors. It is similar to image analysis in that it also seeks to capture lurking "spatial effects" on curves. However, unlike images, where edges and lines represent discontinuities or breaks, wombling boundaries capture rapid surface change; cutting across a wombling boundary should tend to reveal a steep change in elevation or, equivalently, a sharp gradient. Evidently, gradients are central to wombling and the spatial surfaces under investigation must be sufficiently smooth. This precludes methods such as wavelets that have been employed in detecting image discontinuities, such as ridges and cliffs [e.g., Csillag and Kabos, 2002] but do not admit gradients.

Visual assessment of the surface over D often proceeds from contour and image plots of the surface fitted from the data using surface interpolators. Surface representation and contouring methods range from tensor-product interpolators for gridded data [e.g., Cohen et al., 2001] to more elaborate adaptive control-lattice or tessellation based interpolators for scattered data [Akima, 1996, Lee et al., 1997, Finley et al., 2024a]. Mitas and Mitasova [1999] provide a review of several such methods available in GIS software (e.g. GRASS: https://grass.osgeo.org/). These methods are often fast and simple to implement and produce contour maps that reveal topographic features. However, they do not account for association and uncertainty in the data. Contrary to being competitive with statistical methods, they play a complementary role creating descriptive plots from the raw data in the pre-modeling stage and providing visual displays of estimated response or residual surfaces in the post-modeling stage. It is worth pointing out that while contours often provide an idea about the local topography, they are not the same as wombling boundaries. Contour lines connect points with the same spatial elevation and may or may not track large gradients, so they may or may not correspond to wombling boundaries.

Existing wombling methods for point referenced data concentrate upon finding points that have large gradients and attempt to connect them in an algorithmic fashion, which then defines a boundary. Such algorithms have been employed widely in computational ecology, anthropology and geography. For example, Barbujani et al. [1990] and Barbujani et al. [1997] used wombling on red blood cell markers to identify genetic boundaries in Eurasian human populations by different processes restricting gene flow; Bocquet-Appel and Bacro [1994] investigated genetic, morphometric and physiologic boundaries; Fortin [1994] and Fortin [1997] delineated boundaries related to specific vegetation zones; Fortin and Drapeau [1995] applied wombling on real environmental data, and Jacquez and Greiling [2003] estimated boundaries for breast, lung, and colorectal cancer rates in males and females in Nassau, Suffolk, and Queens counties in New York. This last application is somewhat different from the others in that the data were areally referenced with counties. Unlike image pixels, these counties are not regularly spaced but still have a well-defined neighborhood structure (a topological graph) and the image analysis methods can be applied directly. The gradient is not explicitly modeled; boundary effects are looked upon as edge effects and modeled using Markov random field specifications. A Bayesian framework for areal boundary analysis has been provided by Lu and Carlin [2005].

Banerjee and Gelfand [2006] formulated a Bayesian framework for point-referenced curvilinear gradients or boundary analysis, a conceptually harder problem due to the lack of definitive candidate boundaries. Spatial process models help in estimating not only response surfaces, but residual surfaces after covariate and systematic trends have been accounted. Depending upon the scientific application, boundary analysis may be desirable on either. Algorithmic methods treat statistical estimates of the surface as "data" and apply interpolation-based wombling to obtain boundaries. Although such methods produce useful descriptive surface plots, they preclude formal statistical inference. Indeed, boundary assessment using such reconstructed surfaces will suffer from inaccurate estimation of uncertainty.

In the next section, we present a formal development of inference for directional finite difference processes and directional derivative processes. We then move from points to curves and assign a meaningful gradient to a curve. For a point, if the gradient in a particular direction is large (positive or negative) then the surface is rapidly increasing or decreasing in that direction. For a curve, if the gradients in the direction orthogonal to the curve tend to be large then the curve tracks a path through the region where the surface is rapidly changing. We obtain complete distribution theory results under the assumptions of a stationary Gaussian process model either for the data or for spatial random effects. We present inference under a Bayesian framework which, in this setting, offers several advantages.

14.2   Process smoothness revisited ⋆

We confine ourselves to smoothness properties of a univariate spatial process, say, $\{Y(\mathbf{s}), \mathbf{s} \in \Re^d\}$; for a discussion of multivariate processes, see Banerjee and Gelfand [2003]. In our investigation of smoothness properties we look at two types of continuity, continuity in the L_2 sense and continuity in the sense of process realizations. Unless otherwise noted, we assume the processes to have 0 mean and finite second-order moments.

Definition 14.1 A process $\{Y(\mathbf{s}), \mathbf{s} \in \Re^d\}$ is L_2 continuous at $\mathbf{s}_0$ if and only if $\lim_{\mathbf{s} \to \mathbf{s}_0} E[Y(\mathbf{s}) - Y(\mathbf{s}_0)]^2 = 0$. Continuity in the L_2 sense is also referred to as *mean square continuity*, and will be denoted by $Y(\mathbf{s}) \xrightarrow{L_2} Y(\mathbf{s}_0)$.

Definition 14.2 A process $\{Y(\mathbf{s}), \mathbf{s} \in \Re^d\}$ is *almost surely continuous* at $\mathbf{s}_0$ if $Y(\mathbf{s}) \longrightarrow Y(\mathbf{s}_0)$ *a.s.* as $\mathbf{s} \longrightarrow \mathbf{s}_0$. If the process is almost surely continuous for every $\mathbf{s}_0 \in \Re^d$ then the process is said to have continuous realizations.

In general, one form of continuity does not imply the other since one form of convergence does not imply the other. However, if $Y(\mathbf{s})$ is a bounded process then a.s. continuity implies L_2 continuity. Of course, each implies that $Y(\mathbf{s}) \xrightarrow{P} Y(\mathbf{s}_0)$.

Example 14.1 Almost sure continuity does not imply mean square continuity. To see this, let $t \in [0,1]$ with $\omega \sim U(0,1)$ and define

$$Y(t;\omega) = \begin{cases} \left(t - \frac{1}{2}\right)^{-1} I_{\left(\frac{1}{2},t\right)}(\omega) & \text{if } t \in \left(\frac{1}{2},1\right] \\ 0 & \text{if } t \in \left[0,\frac{1}{2}\right] \end{cases}.$$

Then $Y(t;\omega) \longrightarrow 0$ *a.s.* as $t \longrightarrow \frac{1}{2}$. But $E[Y^2(t;\omega)] \longrightarrow \infty$ as $t \longrightarrow \frac{1}{2}$ if $t \in \left(\frac{1}{2},1\right]$ and $E[Y^2(t;\omega)] = 0$ if $t \in \left[0,\frac{1}{2}\right]$. Thus the process does not converge in L_2 although it does so almost surely. ∎

Example 14.2 Mean square continuity does not imply almost sure continuity. To see this, construct a process over $t \in \Re^+$ defined through $\omega \sim U(0,1)$ as follows. Let $Y\left(\frac{1}{t};\omega\right) = 0$, if t is not a positive integer, $Y(1;\omega) = I_{\left(0,\frac{1}{2}\right)}(\omega)$, $Y\left(\frac{1}{2};\omega\right) = I_{\left(\frac{1}{2},1\right)}(\omega)$, $Y\left(\frac{1}{3};\omega\right) = I_{\left(0,\frac{1}{3}\right)}(\omega)$, $Y\left(\frac{1}{4};\omega\right) = I_{\left(\frac{1}{3},\frac{2}{3}\right)}(\omega)$, $Y\left(\frac{1}{5};\omega\right) = I_{\left(\frac{2}{3},1\right)}(\omega)$, and so on. That is, we construct the process as a sequence of moving indicators on successively finer arithmetic divisions of the unit interval. We see here that $E\left[Y^2\left(\frac{1}{t};\omega\right)\right] \longrightarrow 0$ as $t \longrightarrow 0$, so that $Y\left(\frac{1}{t};\omega\right) \xrightarrow{L_2} 0$. However the process is not continuous almost surely since $Y\left(\frac{1}{t};\omega\right)$ is equal to one infinitely often. ∎

The above definitions apply to any stochastic process (possibly nonstationary). Cramer and Leadbetter [2005] and Hoel et al. [1972] outline conditions on the covariance function for mean square continuity for processes on the real line. For a process on $\Re^d$, we denote the covariance function, as usual, by $C\left(\mathbf{s}, \mathbf{s}'\right) = \operatorname{Cov}\left(Y\left(\mathbf{s}\right), Y\left(\mathbf{s}'\right)\right)$, so that the definition of mean square continuity is equivalent to $\lim_{\mathbf{s}' \to \mathbf{s}}\left[C\left(\mathbf{s}', \mathbf{s}'\right) - 2C(\mathbf{s}', \mathbf{s}) + C(\mathbf{s}, \mathbf{s})\right] = 0$. It follows that continuity in $\mathbf{s}$ and $\mathbf{s}'$ serve as sufficient conditions for mean square continuity. For a (weakly) stationary process, mean square continuity is equivalent to the covariance function $C\left(\mathbf{s}\right)$ being continuous at $\mathbf{0}$. This follows easily since $E\left[Y\left(\mathbf{s}'\right) - Y\left(\mathbf{s}\right)\right]^2 = 2(C\left(\mathbf{0}\right) - C\left(\mathbf{s}' - \mathbf{s}\right))$ for a weakly stationary process and enables a simple practical check for mean square continuity.

Kent [1989] investigates continuous process realizations through a Taylor expansion of the covariance function. Let $\{Y\left(\mathbf{s}\right), \mathbf{s} \in \Re^d\}$ be a real-valued stationary spatial process on $\Re^d$. Kent proves that if $C\left(\mathbf{s}\right)$ is d-times continuously differentiable and $C_d\left(\mathbf{s}\right) = C\left(\mathbf{s}\right) - P_d\left(\mathbf{s}\right)$, where $P_d\left(\mathbf{s}\right)$ is the Taylor polynomial of degree d for $C\left(\mathbf{s}\right)$ about $\mathbf{0}$, satisfies the condition,

$$|C_d\left(\mathbf{s}\right)| = O\left(||\mathbf{s}||^{d+\beta}\right)$$

for some $\beta > 0$, then there exists a version of the spatial process $\{Y\left(\mathbf{s}\right), \mathbf{s} \in \Re^d\}$ with continuous realizations. If $C\left(\mathbf{s}\right)$ is d-times continuously differentiable then it is of course continuous at $\mathbf{0}$ and so, from the previous paragraph, the process is mean square continuous.

Let us suppose that $f : L_2 \longrightarrow \Re^1$ (L_2 is the usual Hilbert space of random variables induced by the L_2 metric) is a continuous function. Let $\{Y\left(\mathbf{s}\right), \mathbf{s} \in \Re^d\}$ be a process that is continuous almost surely. Then the process $Z\left(\mathbf{s}\right) = f\left(Y\left(\mathbf{s}\right)\right)$ is almost surely continuous, being the composition of two continuous functions. The validity of this statement is direct and does not require checking Kent's conditions. However, the process $Z(\mathbf{s})$ need not be stationary even if $Y\left(\mathbf{s}\right)$ is. Moreover, the existence of the covariance function $C\left(\mathbf{s}, \mathbf{s}'\right) = E\left[f\left(Y\left(\mathbf{s}\right)\right) f\left(Y\left(\mathbf{s}'\right)\right)\right]$, via the Cauchy-Schwartz inequality, requires $Ef^2\left(Y\left(\mathbf{s}\right)\right) < \infty$.

While almost sure continuity of the new process $Z\left(\mathbf{s}\right)$ follows routinely, the mean square continuity of $Z\left(\mathbf{s}\right)$ is not immediate. However, from the remark below Definition 14.2, if $f : \Re^1 \longrightarrow \Re^1$ is a continuous function that is bounded and $Y\left(\mathbf{s}\right)$ is a process that is continuous almost surely, then the process $Z\left(\mathbf{s}\right) = f\left(Y\left(\mathbf{s}\right)\right)$ (a process on $\Re^d$) is mean square continuous.

More generally suppose f is a continuous function that is Lipschitz of order 1, and $\{Y\left(\mathbf{s}\right), \mathbf{s} \in \Re^d\}$ is a process which is mean square continuous. Then the process $Z\left(\mathbf{s}\right) = f\left(Y\left(\mathbf{s}\right)\right)$ is mean square continuous. To see this, note that since f is Lipschitz of order 1 we have $|f\left(Y\left(\mathbf{s}+\mathbf{h}\right)\right) - f\left(Y\left(\mathbf{s}\right)\right)| \leq K\left|Y\left(\mathbf{s}+\mathbf{h}\right) - Y\left(\mathbf{s}\right)\right|$ for some constant K. It follows that $E[f\left(Y\left(\mathbf{s}+\mathbf{h}\right)\right) - f\left(Y\left(\mathbf{s}\right)\right)]^2 \leq K^2 E[Y\left(\mathbf{s}+\mathbf{h}\right) - Y\left(\mathbf{s}\right)]^2$, and the mean square continuity of $Z\left(\mathbf{s}\right)$ follows directly from the mean square continuity of $Y\left(\mathbf{s}\right)$.

We next formalize the notion of a mean square differentiable process. Our definition is motivated by the analogous definition of total differentiability of a function of $\Re^d$ in a nonstochastic setting. In particular, $Y\left(\mathbf{s}\right)$ is mean square differentiable at $\mathbf{s}_0$ if there exists a vector $\nabla_Y\left(\mathbf{s}_0\right)$, such that, for any scalar h and any unit vector $\mathbf{u}$,

$$Y\left(\mathbf{s}_0 + h\mathbf{u}\right) = Y\left(\mathbf{s}_0\right) + h\mathbf{u}^{\mathsf{T}}\nabla_Y\left(\mathbf{s}_0\right) + r\left(\mathbf{s}_0, h\mathbf{u}\right), \tag{14.1}$$

where $r\left(\mathbf{s}_0, h\mathbf{u}\right) \to 0$ in the L_2 sense as $h \to 0$. That is, we require for any unit vector $\mathbf{u}$,

$$\lim_{h \to 0} E\left(\frac{Y\left(\mathbf{s}_0 + h\mathbf{u}\right) - Y\left(\mathbf{s}_0\right) - h\mathbf{u}^{\mathsf{T}}\nabla_Y\left(\mathbf{s}_0\right)}{h}\right)^2 = 0. \tag{14.2}$$

The first-order linearity condition for the process is required to ensure that mean square differentiable processes are mean square continuous. A counterexample when this condition does not hold is the following.

Example 14.3 Let $Z \sim N(0,1)$ and consider the process $\{Y(\mathbf{s}) : \mathbf{s} = (s_1, s_2) \in \Re^2\}$ defined as follows:

$$Y(\mathbf{s}) = \begin{cases} \frac{s_1 s_2^2}{s_1^2 + s_2^4} Z & \text{if } \mathbf{s} \neq \mathbf{0}, \\ 0 & \text{if } \mathbf{s} = \mathbf{0}. \end{cases}$$

Then, the finite difference process $Y_{\mathbf{u},h}(\mathbf{0})$ is

$$Y_{\mathbf{u},h}(\mathbf{0}) = \frac{Y(\mathbf{0} + h\mathbf{u}) - Y(\mathbf{0})}{h} = \frac{Y(h\mathbf{u})}{h} = \frac{u_1 u_2^2}{u_1^2 + h^2 u_2^4}$$

for any $\mathbf{u} = (u_1, u_2)^{\mathsf{T}}$. Therefore, $D_{\mathbf{u}}Y(\mathbf{0}) = \lim_{h \to 0} Y_{h\mathbf{u}}(\mathbf{0}) = (u_2^2/u_1)Z$ for every $\mathbf{u}$ with $u_1 \neq 0$ and $D_{\mathbf{u}}Y(\mathbf{0}) = 0$ for any direction $\mathbf{u}$ with $u_1 = 0$, where the limits are taken in the L^2 sense. However, the above process is not mean square continuous at $\mathbf{0}$ as can be seen by considering the path $\{(s_2^2, s_2) : s_2 \in \Re\}$, along which $\mathrm{E}\{Y(\mathbf{s}) - Y(\mathbf{0})\}^2 = 1/4$. ∎

The above example shows that, even though the directional derivatives exist in all directions at $\mathbf{0}$, the process is not mean-square differentiable because there does not exist the required linear function of $\mathbf{u}$. On the other hand, if $Y(\mathbf{s})$ is a mean square differentiable process on $\Re^d$, then $Y(\mathbf{s})$ is mean square continuous as well. That is, any direction $\mathbf{u}$ can be taken to be of the form $h\mathbf{v}$ where $\mathbf{v}$ is the unit vector giving the direction of $\mathbf{u}$ and h is a scalar denoting the magnitude of $\mathbf{u}$. The point here is the assumed existence of the surface, $\nabla_Y(\mathbf{s})$ under a mean square differentiable process. And, it is important to note that this surface is random; $\nabla_Y(\mathbf{s})$ is not a function.

14.3 Directional finite difference and derivative processes

The focus of this subsection is to formally address the problem of the rate of change of a spatial surface at a given point in a given direction. As noted in the introduction to this chapter, such slopes or gradients are of interest in so-called digital terrain models for exploring surface roughness. They would also arise in meteorology to recognize temperature or rainfall gradients or in environmental monitoring to understand pollution gradients. With spatial computer models, where the process generating the $Y(\mathbf{s})$ is essentially a black box and realizations are costly to obtain, inference regarding local rates of change becomes important. The application we study here considers rates of change for unobservable or latent spatial processes. For instance, in understanding real estate markets, i.e., house prices for single-family homes, spatial modeling of residuals provides adjustment to reflect desirability of location, controlling for the characteristics of the home and property. Suppose we consider the rate of change of the residual surface in a given direction at, say, the central business district. Transportation costs to the central business district vary with direction. Increased costs are expected to reduce the price of housing. Since transportation cost information is not included in the mean, directional gradients to the residual surface can clarify this issue.

Spatial gradients are customarily defined as finite differences [see, e.g., Greenwood, 1984, Meyer et al., 2001]. Evidently the scale of resolution will affect the nature of the resulting gradient (as we illustrate in Example 14.2). To characterize local rates of change without having to specify a scale, infinitesimal gradients may be preferred. Ultimately, the nature of the data collection and the scientific questions of interest would determine preference for an infinitesimal or a finite gradient. For the former, gradients (derivatives) are quantities of basic importance in geometry and physics. Researchers in the physical sciences (e.g., geophysics, meteorology, oceanography) often formulate relationships in terms of gradients. For the latter, differences, viewed as discrete approximations to gradients, may initially seem less attractive. However, in applications involving spatial data, scale is usually a critical question (e.g., in environmental, ecological, or demographic settings). Infinitesimal local rates of change may be of less interest than finite differences at the scale of a map of interpoint distances.

Following the discussion of Section 14.2, with $\mathbf{u}$ a unit vector, let

$$Y_{\mathbf{u},h}(\mathbf{s}) = \frac{Y(\mathbf{s}+h\mathbf{u})-Y(\mathbf{s})}{h} \qquad (14.3)$$

be the finite difference at $\mathbf{s}$ in direction $\mathbf{u}$ at scale h. Clearly, for a fixed $\mathbf{u}$ and h, $Y_{\mathbf{u},h}(\mathbf{s})$ is a well-defined process on $\Re^d$, which we refer to as the finite difference process at scale h in direction $\mathbf{u}$.

Next, let $D_{\mathbf{u}}Y(\mathbf{s}) = \lim_{h\to 0} Y_{\mathbf{u},h}(\mathbf{s})$ if the limit exists. We see that if $Y(\mathbf{s})$ is a mean square differentiable process in $\Re^d$, i.e., (14.2) holds for every $\mathbf{s}_0$ in $\Re^d$, then for each $\mathbf{u}$,

$$\begin{aligned}
D_{\mathbf{u}}Y(\mathbf{s}) &= \lim_{h\to 0}\frac{Y(\mathbf{s}+h\mathbf{u})-Y(\mathbf{s})}{h}\\
&= \lim_{h\to 0}\frac{h\mathbf{u}^{\mathrm{T}}\boldsymbol{\nabla}_Y(\mathbf{s})+r(\mathbf{s},h\mathbf{u})}{h} = \mathbf{u}^{\mathrm{T}}\boldsymbol{\nabla}_Y(\mathbf{s})\ .
\end{aligned}$$

So $D_{\mathbf{u}}Y(\mathbf{s})$ is a well-defined process on $\mathbb{R}^d$, which we call the *directional derivative process* in the direction $\mathbf{u}$. If unit vectors $\mathbf{e}_1,\mathbf{e}_2,\ldots,\mathbf{e}_d$ form an orthonormal basis set for $\mathbb{R}^d$, then $\mathbf{u} = \sum_{i=1}^d w_i\mathbf{e}_i$ with $w_i = \mathbf{u}^{\mathrm{T}}\mathbf{e}_i$ and $\sum_{i=1}^d w_i^2 = 1$. This yields

$$D_{\mathbf{u}}Y(\mathbf{s}) = \mathbf{u}^{\mathrm{T}}\boldsymbol{\nabla}_Y(\mathbf{s}) = \sum_{i=1}^d w_i\mathbf{e}_i^{\mathrm{T}}\boldsymbol{\nabla}_Y(\mathbf{s}) = \sum_{i=1}^d w_i D_{\mathbf{e}_i}Y(\mathbf{s})\ . \qquad (14.4)$$

Hence, to study directional derivative processes in arbitrary directions we need only work with a basis set of directional derivative processes. Also from (14.4) it is clear that $D_{-\mathbf{u}}Y(\mathbf{s}) = -D_{\mathbf{u}}Y(\mathbf{s})$. Applying the Cauchy-Schwarz inequality to (14.4), for every unit vector $\mathbf{u}$, $D_{\mathbf{u}}^2Y(\mathbf{s}) \leq \sum_{i=1}^d D_{\mathbf{e}_i}^2Y(\mathbf{s})$. Hence, $\sum_{i=1}^d D_{\mathbf{e}_i}^2Y(\mathbf{s})$ is the maximum over all directions of $D_{\mathbf{u}}^2Y(\mathbf{s})$. At location $\mathbf{s}$, this maximum is achieved in the direction $\mathbf{u} = \boldsymbol{\nabla}_Y(\mathbf{s})/\|\boldsymbol{\nabla}_Y(\mathbf{s})\|$, and the maximizing value is $\|\boldsymbol{\nabla}_Y(\mathbf{s})\|$.

In the following text we work with the customary orthonormal basis defined by the coordinate axes so that $\mathbf{e}_i$ is a $d\times 1$ vector with all 0's except for a 1 in the ith row. In fact, with this basis, $\boldsymbol{\nabla}_Y(\mathbf{s}) = (D_{\mathbf{e}_1}Y(\mathbf{s}),\ldots,D_{\mathbf{e}_d}Y(\mathbf{s}))^{\mathrm{T}}$. The result in (14.4) is a limiting result as $h\to 0$. From (14.3), the presence of h shows that to study finite difference processes at scale h in arbitrary directions we have no reduction to a basis set.

A useful comment is that the directional gradients for $g(Z(\mathbf{s}))$, with g continuous and monotonic increasing, are immediately $D_{\mathbf{u}}g(Z(\mathbf{s})) = g'(Z(\mathbf{s}))D_{\mathbf{u}}Z(\mathbf{s})$, by use of the chain rule. This enables us to study gradient behavior of mean surfaces under link functions.

Formally, finite difference processes require less assumption for their existence. To compute differences we need not worry about a *degree* of smoothness for the realized spatial surface. However, issues of numerical stability can arise if h is too small. Also, with directional derivatives in, say, two-dimensional space, following the discussion below (14.4), we only need to work with north and east directional derivatives processes in order to study directional derivatives in arbitrary directions.

14.4   Distribution theory for finite differences and directional gradients

If $E(Y(\mathbf{s})) = 0$ for all $\mathbf{s}\in\Re^d$ then $E(Y_{\mathbf{u},h}(\mathbf{s})) = 0$ and $E(D_{\mathbf{u}}Y(\mathbf{s})) = 0$. Let $C_{\mathbf{u}}^{(h)}(\mathbf{s},\mathbf{s}')$ and $C_{\mathbf{u}}(\mathbf{s},\mathbf{s}')$ denote the covariance functions associated with the process $Y_{\mathbf{u},h}(\mathbf{s})$ and $D_{\mathbf{u}}Y(\mathbf{s})$, respectively. If $\boldsymbol{\Delta} = \mathbf{s}-\mathbf{s}'$ and $Y(\mathbf{s})$ is (weakly) stationary we immediately have

$$C_{\mathbf{u}}^{(h)}(\mathbf{s},\mathbf{s}') = \frac{(2C(\boldsymbol{\Delta})-C(\boldsymbol{\Delta}+h\mathbf{u})-C(\boldsymbol{\Delta}-h\mathbf{u}))}{h^2}\ , \qquad (14.5)$$

whence $\text{Var}(Y_{\mathbf{u},h}(\mathbf{s})) = 2(C(\mathbf{0}) - C(h\mathbf{u}))/h^2$. If $Y(\mathbf{s})$ is isotropic and we replace $C(\mathbf{s}, \mathbf{s}')$ by $\widetilde{C}(||\mathbf{s} - \mathbf{s}'||)$, we obtain

$$C_{\mathbf{u}}^{(h)}(\mathbf{s}, \mathbf{s}') = \frac{\left(2\widetilde{C}(||\boldsymbol{\Delta}||) - \widetilde{C}(||\boldsymbol{\Delta} + h\mathbf{u}||) - \widetilde{C}(||\boldsymbol{\Delta} - h\mathbf{u}||)\right)}{h^2}. \tag{14.6}$$

Expression (14.6) shows that even if $Y(\mathbf{s})$ is isotropic, $Y_{\mathbf{u},h}(\mathbf{s})$ is only stationary. Also $\text{Var}(Y_{\mathbf{u},h}(\mathbf{s})) = 2\left(\widetilde{C}(0) - \widetilde{C}(h)\right)/h^2 = \gamma(h)/h^2$ where $\gamma(h)$ is the familiar variogram of the $Y(\mathbf{s})$ process (Subsection 2.1.2).

Similarly, if $Y(\mathbf{s})$ is stationary we may show that if all second-order partial and mixed derivatives of C exist and are continuous, the limit of (14.5) as $h \to 0$ is

$$C_{\mathbf{u}}(\mathbf{s}, \mathbf{s}') = -\mathbf{u}^{\mathsf{T}}\Omega(\boldsymbol{\Delta})\mathbf{u}, \tag{14.7}$$

where $(\Omega(\boldsymbol{\Delta}))_{ij} = \partial^2 C(\boldsymbol{\Delta})/\partial\boldsymbol{\Delta}_i\partial\boldsymbol{\Delta}_j$. By construction, (14.7) is a valid covariance function on $\Re^d$ for any $\mathbf{u}$. Also, $\text{Var}(D_{\mathbf{u}}Y(\mathbf{s})) = -\mathbf{u}^{\mathsf{T}}\Omega(0)\mathbf{u}$. If $Y(\mathbf{s})$ is isotropic, using standard chain rule calculations, we obtain

$$C_{\mathbf{u}}(\mathbf{s}, \mathbf{s}') = -\left\{\left(1 - \frac{(\mathbf{u}^{\mathsf{T}}\boldsymbol{\Delta})^2}{||\boldsymbol{\Delta}||^2}\right)\frac{\widetilde{C}'(||\boldsymbol{\Delta}||)}{||\boldsymbol{\Delta}||} + \frac{(\mathbf{u}^{\mathsf{T}}\boldsymbol{\Delta})^2}{||\boldsymbol{\Delta}||^2}\widetilde{C}''(||\boldsymbol{\Delta}||)\right\}. \tag{14.8}$$

Again, if $Y(\mathbf{s})$ is isotropic, $D_{\mathbf{u}}Y(\mathbf{s})$ is only stationary. In addition, we have $\text{Var}(D_{\mathbf{u}}Y(\mathbf{s})) = -\widetilde{C}''(0)$ which also shows that, provided $\widetilde{C}$ is twice differentiable at 0, $\lim_{h \to 0}\gamma(h)/h^2 = -\widetilde{C}''(0)$, i.e., $\gamma(h) = O(h^2)$ for h small.

For $Y(\mathbf{s})$ stationary we can also calculate

$$\text{Cov}(Y(\mathbf{s}), Y_{\mathbf{u},h}(\mathbf{s}')) = (C(\boldsymbol{\Delta} - h\mathbf{u}) - C(\boldsymbol{\Delta}))/h,$$

from which $\text{Cov}(Y(\mathbf{s}), Y_{\mathbf{u},h}(\mathbf{s})) = (C(h\mathbf{u}) - C(\mathbf{0}))/h$. But then,

$$\begin{aligned}\text{Cov}(Y(\mathbf{s}), D_{\mathbf{u}}Y(\mathbf{s}')) &= \lim_{h \to 0}(C(\boldsymbol{\Delta} - h\mathbf{u}) - C(\boldsymbol{\Delta}))/h \\ &= -D_{\mathbf{u}}C(\boldsymbol{\Delta}) = D_{\mathbf{u}}C(-\boldsymbol{\Delta}),\end{aligned}$$

since $C(\boldsymbol{\Delta}) = C(-\boldsymbol{\Delta})$. In particular, we have that $\text{Cov}(Y(\mathbf{s}), D_{\mathbf{u}}Y(\mathbf{s})) = \lim_{h \to 0}(C(h\mathbf{u}) - C(\mathbf{0}))/h = D_{\mathbf{u}}C(\mathbf{0})$. The existence of the directional derivative process ensures the existence of $D_{\mathbf{u}}C(\mathbf{0})$. Moreover, since $C(h\mathbf{u}) = C(-h\mathbf{u})$, $C(h\mathbf{u})$ (viewed as a function of h) is even, so $D_{\mathbf{u}}C(\mathbf{0}) = 0$. Thus, $Y(\mathbf{s})$ and $D_{\mathbf{u}}Y(\mathbf{s})$ are uncorrelated. Intuitively, this is sensible. The level of the process at a particular location is uncorrelated with the directional derivative in any direction at that location. This is not true for directional differences. Also, in general, $\text{Cov}(Y(\mathbf{s}), D_{\mathbf{u}}Y(\mathbf{s}'))$ will not be 0.

Under isotropy,

$$\text{Cov}(Y(\mathbf{s}), Y_{\mathbf{u},h}(\mathbf{s}')) = \frac{\widetilde{C}(||\boldsymbol{\Delta} - h\mathbf{u}||) - \widetilde{C}(||\boldsymbol{\Delta}||)}{h}.$$

Now $\text{Cov}(Y(\mathbf{s}), Y_{\mathbf{u},h}(\mathbf{s})) = \left(\widetilde{C}(h) - \widetilde{C}(0)\right)/h = \gamma(h)/2h$, so this means $\text{Cov}(Y(\mathbf{s}), D_{\mathbf{u}}Y(\mathbf{s})) = \widetilde{C}^{\mathsf{T}}(0) = \lim_{h \to 0}\gamma(h)/2h = 0$ since, as above, if $\widetilde{C}''(0)$ exists, $\gamma(h) = O(h^2)$.

Suppose we consider the bivariate process $\mathbf{Z}_{\mathbf{u}}^{(h)}(\mathbf{s}) = (Y(\mathbf{s}), Y_{\mathbf{u},h}(\mathbf{s}))^{\mathsf{T}}$. It is clear that this process has mean 0 and, if $Y(\mathbf{s})$ is stationary, cross-covariance matrix $V_{\mathbf{u},h}(\boldsymbol{\Delta})$ given by

$$\begin{pmatrix} C(\boldsymbol{\Delta}) & \frac{C(\boldsymbol{\Delta} - h\mathbf{u}) - C(\boldsymbol{\Delta})}{h} \\ \frac{C(\boldsymbol{\Delta} + h\mathbf{u}) - C(\boldsymbol{\Delta})}{h} & \frac{2C(\boldsymbol{\Delta}) - C(\boldsymbol{\Delta} + h\mathbf{u}) - C(\boldsymbol{\Delta} - h\mathbf{u})}{h^2} \end{pmatrix}. \tag{14.9}$$

Since $\mathbf{Z}_\mathbf{u}^{(h)}(\mathbf{s})$ arises by linear transformation of $Y(\mathbf{s})$, (14.9) is a valid cross-covariance matrix in $\Re^d$. But since this is true for every h, letting $h \to 0$,

$$V_\mathbf{u}(\boldsymbol{\Delta}) = \begin{pmatrix} C(\boldsymbol{\Delta}) & -D_\mathbf{u}C(\boldsymbol{\Delta}) \\ D_\mathbf{u}C(\boldsymbol{\Delta}) & -\mathbf{u}^\mathrm{T}\Omega(\boldsymbol{\Delta})\mathbf{u} \end{pmatrix} \tag{14.10}$$

is a valid cross-covariance matrix in $\Re^d$. In fact, $V_\mathbf{u}$ is the cross-covariance matrix for the bivariate process $\mathbf{Z}_\mathbf{u}(s) = \begin{pmatrix} Y(\mathbf{s}) \\ D_\mathbf{u}Y(\mathbf{s}) \end{pmatrix}$.

Suppose we now assume that $Y(\mathbf{s})$ is a stationary Gaussian process. Then, it is clear, again by linearity, that $Y_{\mathbf{u},h}(\mathbf{s})$ and, in fact, $\mathbf{Z}_\mathbf{u}^h(\mathbf{s})$ are both stationary Gaussian processes. But then, by a standard limiting moment generating function argument, $\mathbf{Z}_\mathbf{u}(\mathbf{s})$ is a stationary bivariate Gaussian process and thus $D_\mathbf{u}Y(\mathbf{s})$ is a stationary univariate Gaussian process. As an aside, we note that for a given $\mathbf{s}$, $D_{\frac{\nabla_Y(\mathbf{s})}{||\nabla_Y(\mathbf{s})||}}Y(\mathbf{s})$ is not normally distributed, and in fact the set $\{D_{\frac{\nabla_Y(\mathbf{s})}{||\nabla_Y(\mathbf{s})||}}Y(\mathbf{s}) : \mathbf{s} \in \Re^d\}$ is not a spatial process.

Extension to a pair of directions with associated unit vectors $\mathbf{u}_1$ and $\mathbf{u}_2$ results in a trivariate Gaussian process $\mathbf{Z}(\mathbf{s}) = (Y(\mathbf{s}), D_{\mathbf{u}_1}Y(\mathbf{s}), D_{\mathbf{u}_2}Y(\mathbf{s}))^\mathrm{T}$ with associated cross-covariance matrix $V_\mathbf{Z}(\boldsymbol{\Delta})$ given by

$$\begin{pmatrix} C(\boldsymbol{\Delta}) & -(\boldsymbol{\nabla}C(\boldsymbol{\Delta}))^\mathrm{T} \\ \boldsymbol{\nabla}C(\boldsymbol{\Delta}) & -\Omega(\boldsymbol{\Delta}) \end{pmatrix}. \tag{14.11}$$

At $\boldsymbol{\Delta} = 0$, (14.11) becomes a diagonal matrix.

We conclude this subsection with a useful example. Recall the power exponential family of isotropic covariance functions of the previous subsection, $\widetilde{C}(||\boldsymbol{\Delta}||) = \alpha\exp\left(-\phi||\boldsymbol{\Delta}||^\nu\right)$, $0 < \nu \le 2$. It is apparent that $\widetilde{C}''(0)$ exists only for $\nu = 2$. The Gaussian covariance function is the only member of the class for which directional derivative processes can be defined. However, as we have noted in Subsection 3.1.4, the Gaussian covariance function produces process realizations that are too smooth to be attractive for practical modeling.

Turning to the Matérn class, $\widetilde{C}(||\boldsymbol{\Delta}||) = \alpha(\phi||\boldsymbol{\Delta}||)^\nu K_\nu(\phi||\boldsymbol{\Delta}||)$, ν is a smoothness parameter controlling the extent of mean square differentiability of process realizations [Stein, 1999a]. At $\nu = 3/2$, $\widetilde{C}(||\boldsymbol{\Delta}||)$ takes the closed form $\widetilde{C}(||\boldsymbol{\Delta}||) = \sigma^2(1 + \phi||\boldsymbol{\Delta}||)\exp\left(-\phi||\boldsymbol{\Delta}||\right)$ where σ^2 is the process variance. This function is exactly twice differentiable at 0. We have a (once but not twice) mean square differentiable process, which therefore does not suffer the excessive smoothness implicit with the Gaussian covariance function.

For this choice one can show that $\boldsymbol{\nabla}\widetilde{C}(||\boldsymbol{\Delta}||) = -\sigma^2\phi^2\exp\left(-\phi||\boldsymbol{\Delta}||\right)\boldsymbol{\Delta}$, that $\left(H_{\widetilde{C}}(||\boldsymbol{\Delta}||)\right)_{ii} = -\sigma^2\phi^2\exp\left(-\phi||\boldsymbol{\Delta}||\right)\left(1 - \phi\Delta_i^2/||\boldsymbol{\Delta}||\right)$, and also that $\left(H_{\widetilde{C}}(||\boldsymbol{\Delta}||)\right)_{ij} = \sigma^2\phi^2\exp\left(-\phi||\boldsymbol{\Delta}||\right)\Delta_i\Delta_j/||\boldsymbol{\Delta}||$. In particular, $V_\mathbf{u}(\mathbf{0}) = \sigma^2\mathrm{BlockDiag}(1, \phi^2I)$.

14.5 Directional derivative processes in modeling

We work in $d = 2$-dimensional space and can envision the following types of modeling settings in which directional derivative processes would be of interest. For $Y(\mathbf{s})$ purely spatial with constant mean, we would seek $D_\mathbf{u}Y(\mathbf{s})$. In the customary formulation $Y(\mathbf{s}) = \mu(\mathbf{s}) + w(\mathbf{s}) + \epsilon(\mathbf{s})$ we would instead want $D_\mathbf{u}w(\mathbf{s})$. In the case of a spatially varying coefficient model $Y(\mathbf{s}) = \beta_0(\mathbf{s}) + \beta_1(\mathbf{s})X(\mathbf{s}) + \epsilon(\mathbf{s})$ such as in Section 10.7, we would examine $D_\mathbf{u}\beta_0(\mathbf{s})$, $D_\mathbf{u}\beta_1(\mathbf{s})$, and $D_\mathbf{u}EY(\mathbf{s})$ with $EY(\mathbf{s}) = \beta_0(\mathbf{s}) + \beta_1(\mathbf{s})X(\mathbf{s})$.

Consider the constant mean purely spatial process for illustration, where we have $Y(\mathbf{s})$ a stationary process with mean μ and covariance function $C(\boldsymbol{\Delta}) = \sigma^2\rho(\boldsymbol{\Delta})$ where ρ is a valid two-dimensional correlation function. For illustration we work with the general Matérn class parametrized by ϕ and ν, constraining $\nu > 1$ to ensure the (mean square) existence of the directional derivative processes. Letting $\boldsymbol{\theta} = \left(\mu, \sigma^2, \phi, \nu\right)$, for locations $\mathbf{s}_1, \mathbf{s}_2, ..., \mathbf{s}_n$, the likelihood $L\left(\boldsymbol{\theta}; \mathbf{Y}\right)$ is proportional to

$$\left(\sigma^2\right)^{-n/2} \left|R\left(\phi, \nu\right)\right|^{-1/2} \exp\left\{-\frac{1}{2\sigma^2} \left(\mathbf{Y} - \mu\mathbf{1}\right)^{\mathrm{T}} R^{-1}\left(\phi, \nu\right)\left(\mathbf{Y} - \mu\mathbf{1}\right)\right\}. \tag{14.12}$$

In (14.12), $\mathbf{Y}^{\mathrm{T}} = \left(Y\left(\mathbf{s}_1\right), ..., Y\left(\mathbf{s}_n\right)\right)^{\mathrm{T}}$ and $\left(R\left(\phi, \nu\right)\right)_{ij} = \rho\left(\mathbf{s}_i - \mathbf{s}_j; \phi, \nu\right)$.

We have discussed prior specification for such a model in Chapter 3. Customary choices include a vague normal (perhaps flat) prior for μ; we would deal with the identifiability issue for σ^2 and ϕ through a centered inverse gamma on the former and a weakly informative prior for the latter. (Below, we work with an inverse gamma and a gamma prior respectively). With regard to the prior on ν, Banerjee et al. [2003a] follow the suggestion of Stein [1999a] and others, who observe that, in practice, it will be very difficult to distinguish $\nu = 2$ from $\nu > 2$ and so, adopt a $U\left(1, 2\right)$ prior. The algorithms discussed in Chapter 3, and available in the spBayes library, can be used for fitting the Bayesian model. We can also use slice sampling [Agarwal and Gelfand, 2005] as an easily programmable alternative.

We note that all gradient analysis is a post-model fitting activity, employing posterior samples of the model parameters to obtain samples from posterior predictive distributions. In particular, a contour or a grey-scale plot of the posterior mean surface is of primary interest in providing a smoothed display of spatial pattern and of areas where the process is elevated or depressed. To handle finite differences at scale h, in the sequel we work with the vector of eight compass directions, $N, NE, E, \ldots$. At $\mathbf{s}_i$, we denote this vector by $\mathbf{Y}_h\left(\mathbf{s}_i\right)$ and let $\mathbf{Y}_h = \left\{\mathbf{Y}_h\left(\mathbf{s}_i\right), i = 1, 2, ..., n\right\}$. With directional derivatives we only need $\mathbf{D}\left(\mathbf{s}_i\right)^{\mathrm{T}} = \left(D_{(1,0)}Y\left(\mathbf{s}_i\right), D_{(0,1)}Y\left(\mathbf{s}_i\right)\right)$ and let $\mathbf{D} = \left\{\mathbf{D}\left(\mathbf{s}_i\right), i = 1, 2, ..., n\right\}$. We seek samples from the predictive distribution $f\left(\mathbf{Y}_h \mid \mathbf{Y}\right)$ and $f\left(\mathbf{D} \mid \mathbf{Y}\right)$. In $\mathbf{Y}_h\left(\mathbf{s}_i\right)$, $Y\left(\mathbf{s}_i\right)$ is observed, hence fixed in the predictive distribution. So we can replace $Y_{u,h}\left(\mathbf{s}_i\right)$ with $Y\left(\mathbf{s}_i + h\mathbf{u}\right)$; posterior predictive samples of $Y\left(\mathbf{s}_i + h\mathbf{u}\right)$ are immediately converted to posterior predictive samples of $Y_{u,h}\left(\mathbf{s}_i\right)$ by linear transformation. Hence, the directional finite differences problem is merely a large Bayesian kriging problem requiring spatial prediction at the set of $8n$ locations $\left\{Y\left(\mathbf{s}_i + h\mathbf{u}_r\right), i = 1, 2, ..., n; r = 1, 2, ..., 8\right\}$. Denoting this set by $\widetilde{\mathbf{Y}}_h$, we require samples from $f\left(\widetilde{\mathbf{Y}}_h \mid \mathbf{Y}\right)$. From the relationship, $f\left(\widetilde{\mathbf{Y}}_h \mid \mathbf{Y}\right) = \int f\left(\widetilde{\mathbf{Y}}_h \mid \mathbf{Y}, \boldsymbol{\theta}\right) f\left(\boldsymbol{\theta} \mid \mathbf{Y}\right) d\boldsymbol{\theta}$ this can be done one for one with the $\boldsymbol{\theta}_l^*$'s by drawing $\widetilde{\mathbf{Y}}_{h,l}^*$ from the multivariate normal distribution $f\left(\widetilde{\mathbf{Y}}_h \mid \mathbf{Y}, \boldsymbol{\theta}_l^*\right)$, as detailed in Section 6.1. Similarly $f\left(\mathbf{D} \mid \mathbf{Y}\right) = \int f\left(\mathbf{D} \mid \mathbf{Y}, \boldsymbol{\theta}\right) f\left(\boldsymbol{\theta} \mid \mathbf{Y}\right) d\boldsymbol{\theta}$. The cross-covariance function in (14.11) allows us to immediately write down the joint multivariate normal distribution of $\mathbf{Y}$ and $\mathbf{D}$ given $\boldsymbol{\theta}$ and thus, at $\boldsymbol{\theta}_l^*$, the conditional multivariate normal distribution $f\left(\mathbf{D} \mid \mathbf{Y}, \boldsymbol{\theta}_l^*\right)$. At a specified new location $\mathbf{s}_0$, with finite directional differences we need to add spatial prediction at the nine new locations, $Y\left(\mathbf{s}_0\right)$ and $Y\left(\mathbf{s}_0 + h\mathbf{u}_r\right)$, $r = 1, 2, ..., 8$. With directional derivatives, we again can use (14.9) to obtain the joint distribution of $\mathbf{Y}, \mathbf{D}$, $Y\left(\mathbf{s}_0\right)$ and $\mathbf{D}\left(\mathbf{s}_0\right)$ given θ and thus the conditional distribution $f\left(\mathbf{D}, Y\left(\mathbf{s}_0\right), \mathbf{D}\left(\mathbf{s}_0\right) \mid \mathbf{Y}, \boldsymbol{\theta}\right)$.

Turning to the random spatial effects model we now assume that

$$Y\left(\mathbf{s}\right) = \mathbf{x}^{\mathrm{T}}\left(\mathbf{s}\right)\boldsymbol{\beta} + w\left(\mathbf{s}\right) + \epsilon\left(\mathbf{s}\right). \tag{14.13}$$

In (14.13), $\mathbf{x}\left(\mathbf{s}\right)$ is a vector of location characteristics, $w\left(\mathbf{s}\right)$ is a mean 0 stationary Gaussian spatial process with parameters σ^2, ϕ, and ν as above, and $\epsilon\left(\mathbf{s}\right)$ is a Gaussian white noise process with variance τ^2, intended to capture measurement error or microscale variability. Such a model is appropriate for the real estate example mentioned in Subsection 14.3, where $Y\left(\mathbf{s}\right)$ is the log selling price and $\mathbf{x}\left(\mathbf{s}\right)$ denotes associated house and property characteristics. Here $w\left(\mathbf{s}\right)$ measures the spatial adjustment to log selling price at location $\mathbf{s}$ reflecting relative desirability of the location. $\epsilon\left(\mathbf{s}\right)$ is needed to capture microscale variability. Here such variability arises because two identical houses arbitrarily close to each other need not sell for essentially the same price due to unobserved differences in buyers, sellers, and brokers across transactions.

For locations $\mathbf{s}_1, \mathbf{s}_2, ..., \mathbf{s}_n$, with $\boldsymbol{\theta} = (\boldsymbol{\beta}, \tau^2, \sigma^2, \phi, \nu)$ the model in (14.13) produces a marginal likelihood $L(\boldsymbol{\theta}; Y)$ (integrating over $\{w(\mathbf{s}_i)\}$) proportional to

$$|\Sigma(\boldsymbol{\gamma})|^{-1/2} \exp\left\{ -\frac{1}{2}(\mathbf{Y} - X\boldsymbol{\beta})^{\mathrm{T}} \Sigma(\boldsymbol{\gamma})^{-1}(\mathbf{Y} - X\boldsymbol{\beta}) \right\},$$

where $\boldsymbol{\gamma} = \{\tau^2, \sigma^2, \phi, \nu\}$ and $\Sigma(\boldsymbol{\gamma}) = \sigma^2 R(\phi, \nu) + \tau^2 I$. Priors for $\boldsymbol{\theta}$ can be prescribed as for (14.12). Again, the procedures discussed in Chapter 3, or even slice sampling, provides an efficient fitting mechanism and is available through the `spLM` function in the `spBayes` library.

Further inference with regard to (14.13) focuses on the spatial process itself. That is, we would be interested in the posterior spatial effect surface and in rates of change associated with this surface. The former is usually handled with samples of the set of $w(\mathbf{s}_i)$ given $\mathbf{Y}$ along with, perhaps, a grey-scaling or contouring routine. The latter would likely be examined at new locations. For instance in the real estate example, spatial gradients would be of interest at the central business district or at other externalities such as major roadway intersections, shopping malls, airports, or waste disposal sites but not likely at the locations of the individual houses.

As below (14.12) with $\mathbf{w}^{\mathrm{T}} = (w(\mathbf{s}_1), \ldots, w(\mathbf{s}_n))$, we sample $f(\mathbf{w} \mid Y)$ one for one with the $\boldsymbol{\theta}_l^*$'s using $f(\mathbf{w} \mid \mathbf{Y}) = \int f(\mathbf{w} \mid \mathbf{Y}, \boldsymbol{\theta}) f(\boldsymbol{\theta} \mid \mathbf{Y}) d\boldsymbol{\theta}$, as described in Section 6.1. But also, given $\boldsymbol{\theta}$, the joint distribution of $\mathbf{w}$ and $\mathbf{V}(\mathbf{s}_0)$ where $\mathbf{V}(\mathbf{s}_0)$ is either $\mathbf{w}_h(\mathbf{s}_0)$ or $\mathbf{D}(\mathbf{s}_0)$ is multivariate normal. For instance, with $D(\mathbf{s}_0)$, the joint normal distribution can be obtained using (14.11) and as a result so can the conditional normal distribution $f(\mathbf{V}(\mathbf{s}_0) \mid \mathbf{w}, \boldsymbol{\theta})$. Lastly, since

$$f(\mathbf{V}(\mathbf{s}_0) \mid \mathbf{Y}) = \int f(\mathbf{V}(\mathbf{s}_0) \mid \mathbf{w}, \boldsymbol{\theta}) f(\mathbf{w} \mid \boldsymbol{\theta}, \mathbf{Y}) f(\boldsymbol{\theta} \mid \mathbf{Y}) d\boldsymbol{\theta} d\mathbf{w},$$

we can also obtain samples from $f(\mathbf{V}(\mathbf{s}_0) \mid \mathbf{Y})$ one for one with the $\boldsymbol{\theta}_l^*$'s.

14.6   Illustration: Inference for differences and gradients

A simulation example adapted from Banerjee et al. [2003a] is provided to illustrate inference on finite differences and directional gradients. We generate data from a Gaussian random field with constant mean μ and a covariance structure specified through the Matèrn ($\nu = 3/2$) covariance function, $\sigma^2(1 + \phi d)\exp(-\phi d)$. This will yield a realized field that will be mean-square differentiable exactly once. The field is observed on a randomly sampled set of points within a 10x10 square. We set $\mu = 0$, $\sigma^2 = 1.0$ and $\phi = 1.05$. In the subsequent illustration our data consists of $n = 100$ observations at the randomly selected sites shown in the left panel of Figure 14.1. The maximum observed distance in our generated field is approximately 13.25 units. The value of $\phi = 1.05$ provides an effective isotropic range of about 4.5 units. We also perform a Bayesian kriging on the data to develop a predicted field. The right panel of Figure 14.1 shows a grey-scale plot with contour lines displaying the topography of the "kriged" field. We will see below that our predictions of the spatial gradients at selected points are consistent with the topography around those points, as depicted in the right panel of Figure 14.1. Adopting a flat prior for μ, an $\mathrm{IG}(2, 0.1)$ (mean $= 10$, infinite variance) prior for σ^2, a $G(2, 0.1)$ prior (mean=20, variance=200) for ϕ, and a uniform on $(1, 2)$ for ν, we obtain the posterior estimates for our parameters shown in Table 14.1.

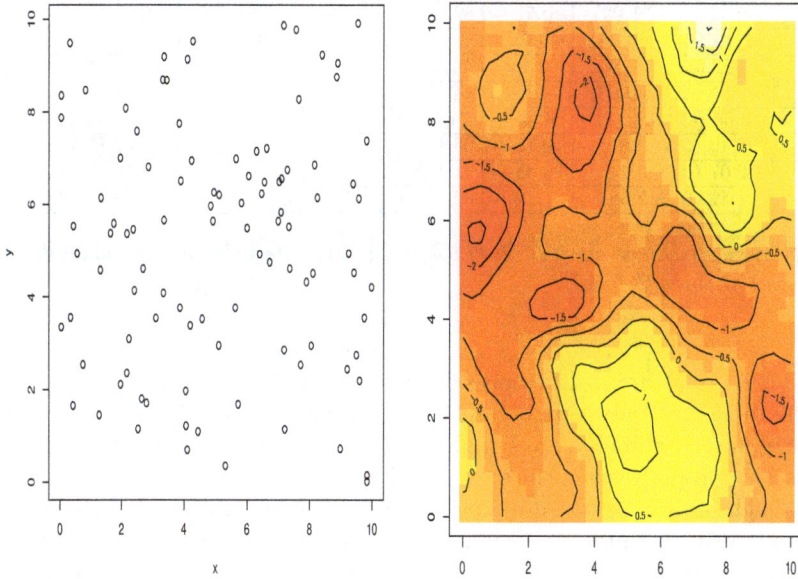

Figure 14.1 *Left panel: Location of the 100 sites where the random field has been observed. Right panel: A grey scale plot with contour lines showing the topography of the random field in the simulation example.*

We next predict the directional derivatives and directional finite differences for the unit vectors corresponding to angles of 0, 45, 90, 135, 180, 225, 270, and 315 degrees with the horizontal axis in a counter-clockwise direction at the point. For the finite differences we consider $h = 1.0, 0.1$ and 0.01. Recall that $D_{-\mathbf{u}}Y(\mathbf{s}) = -D_{\mathbf{u}}Y(\mathbf{s})$.

Parameter	50% (2.5%,97.5%)
μ	-0.39 (-0.91,0.10)
σ^2	0.74 (0.50,1.46)
ϕ	1.12 (0.85,1.41)
ν	1.50 (1.24,1.77)

Table 14.1 *Posterior estimates for model parameters.*

Table 14.2 presents the resulting posterior predictive inference for the point (3.5, 3.5) in Figure 14.1. We see that (3.5, 3.5) seems to be in a rather interesting portion of the surface, with many contour lines nearby. It is clear from the contour lines that there is a negative northern gradient (downhill) and a positive southern gradient (uphill) around this point. On the other hand, there does not seem to be any significant EW gradient around that point as seen from the contour lines through that point running EW. This is brought out very clearly in column 1 of Table 14.2. The angles of 0 and 180 degrees which correspond to the EW gradients are not at all significant. The NS gradients are indeed pronounced as seen by the 90 and 270 degree gradients. The directional derivatives along the diagonals also indicate presence of a gradient. There is a significant downhill gradient towards the NE and (therefore) a significant uphill gradient towards the SW. Hence the directional derivative process provides inference consistent with features captured descriptively and visually in Figure 14.1.

Angle	$D_{\mathbf{u}}Y(\mathbf{s})$ $(h=0)$	$h=1.0$	$h=0.1$	$h=0.01$
0	-0.06 (-1.12,1.09)	0.51 (-0.82,1.81)	-0.08 (-1.23,1.20)	-0.07 (-1.11,1.10)
45	-1.49 (-2.81,-0.34)	-0.01 (-1.29,1.32)	-1.55 (-2.93,-0.56)	-1.53 (-2.89,-0.49)
90	-2.07 (-3.44,-0.66)	-0.46 (-1.71,0.84)	-2.13 (-3.40,-0.70)	-2.11 (-3.41,-0.69)
135	-1.42 (-2.68,-0.23)	-0.43 (-1.69,0.82)	-1.44 (-2.64,-0.23)	-1.43 (-2.70,-0.23)
180	0.06 (-1.09,1.12)	-0.48 (-1.74,0.80)	0.08 (-1.19,1.23)	0.06 (-1.10,1.12)
225	1.49 (0.34,2.81)	0.16 (-1.05,1.41)	1.61 (0.52,3.03)	1.52 (0.48,2.90)
270	2.07 (0.66,3.44)	0.48 (-0.91,1.73)	2.12 (0.68,3.43)	2.10 (0.68,3.42)
315	1.42 (0.23,2.68)	1.12 (-0.09,2.41)	1.44 (0.24,2.68)	1.42 (0.23,2.70)

Table 14.2 *Posterior medians and (2.5%,97.5%) predictive intervals for directional derivatives and finite differences at point (3.5,3.5).*

For the directional finite differences in columns 2, 3 and 4 of Table 14.2, note, for instance, the difference between column 2 and columns 3 and 4. In the former, none of the directional finite differences are significant. The low resolution (large h) fails to capture local topographic properties. On the other hand the latter very much resemble column 1. As expected, at high resolution, the directional finite difference process results match those of the directional derivative process. Computational simplicity and stability (difficulties may arise with very small h in the denominator of (2)) encourage the use of the latter [see Banerjee et al., 2003a, for details].

14.7 Curvilinear gradients and wombling

We now extend the developments in the previous section to an inferential framework for gradients along curves. The conceptual challenge in moving from points to curves is the construction of a sensible measure to associate with a curve in order to assess whether it can be declared a wombling boundary. In this regard, we can consider open or closed curves. In Section 14.7.1 we formally develop the notion of an average gradient to associate with a curve and in Section 14.7.2 we are able to offer a formal definition of a wombling boundary.

We use differential geometric notions for parametric boundaries [as developed, e.g., in Rudin et al., 1976, Spivak, 1979, Frankel, 1997]. Since most spatial modeling is done on domains in $\Re^2$ we restrict our attention to this case, focusing on a real-valued process $Y(\mathbf{s})$ with the spatial domain as an open subset of $\Re^2$. Thus, we offer an independent development of gradients along planar curves without resorting to geometry on manifolds. For hyper-curves in general $\Re^d$, the theory is more complicated (especially if $d > 3$) and must involve development of calculus on abstract manifolds.

14.7.1 Gradients along curves

Let C be an open curve in $\Re^2$ and suppose it is desired to ascertain whether such a curve is a wombling boundary with regard to $Y(\mathbf{s})$. To do so we seek to associate an average gradient with C. In particular, for each point $\mathbf{s}$ lying on C, we let $D_{n(\mathbf{s})}Y(\mathbf{s})$ be the directional derivative in the direction of the unit normal $n(\mathbf{s})$.[1] We can define the *wombling measure* of the curve either as the total gradient along C,

$$\int_C D_{n(\mathbf{s})}Y(\mathbf{s})\,d\nu = \int_C \langle \nabla Y(\mathbf{s}), n(\mathbf{s})\rangle d\nu, \qquad (14.14)$$

---

[1]Again, the rationale for the choice of direction normal to the curve is that, for a curve tracking rapid change in the spatial surface, lines orthogonal to the curve should reveal sharp gradients.

or perhaps as the average gradient along C,

$$\frac{1}{\nu(C)} \int_C D_{n(\mathbf{s})} Y(\mathbf{s}) \, d\nu = \frac{1}{\nu(C)} \int_C \langle \nabla Y(\mathbf{s}), n(\mathbf{s}) \rangle d\nu, \tag{14.15}$$

where $\nu(\cdot)$ is an appropriate measure. For (14.14) and (14.15), ambiguity arises with respect to the choice of measure. For example, $\nu(C) = 0$ if we take ν as two-dimensional Lebesgue measure and, indeed, this is true for any ν which is mutually absolutely continuous with respect to Lebesgue measure. Upon reflection, an appropriate choice for ν turns out to be arc-length. This can be made clear by a parametric treatment of the curve C.

In particular, a curve C in $\Re^2$ is a set parametrized by a single parameter $t \in \Re^1$ where $C = \{\mathbf{s}(t) : t \in \mathcal{T}\}$, with $\mathcal{T} \subset \Re^1$. We call $\mathbf{s}(t) = (s_{(1)}(t), s_{(2)}(t)) \in \Re^2$ the position vector of the curve—$\mathbf{s}(t)$ traces out C as t spans its domain. Then, assuming a differentiable curve with non-vanishing derivative $\mathbf{s}'(t) \neq 0$ (such a curve is often called *regular*), we obtain the (component-wise) derivative $\mathbf{s}'(t)$ as the "velocity" vector, with unit velocity (or tangent) vector $\mathbf{s}'(t)/\|\mathbf{s}'(t)\|$. Letting $n(\mathbf{s}(t))$ denote the parametrized unit normal vector to C, again if C is sufficiently smooth, then $\langle \mathbf{s}'(t), n(\mathbf{s}(t)) \rangle = 0, a.e.\mathcal{T}$. In $\Re^2$ we see that

$$n(\mathbf{s}(t)) = \frac{\left(s_{(2)}'(t), -s_{(1)}'(t) \right)}{\|\mathbf{s}'(t)\|}. \tag{14.16}$$

Under the above parametrization (and the regularity assumption) the arc-length measure ν can be defined as

$$\nu(\mathcal{T}) = \int_{\mathcal{T}} \|\mathbf{s}'(t)\| \, dt. \tag{14.17}$$

In fact, $\|\mathbf{s}'(t)\|$ is analogous to the "speed" (the norm of the velocity) at "time" t, so the above integral is interpretable as the distance traversed or, equivalently, the arc-length $\nu(C)$ or $\nu(\mathcal{T})$. In particular, if $\mathcal{T}$ is an interval, say $[t_0, t_1]$, we can write

$$\nu(\mathcal{T}) = \nu_{t_0}(t_1) = \int_{t_0}^{t_1} \|\mathbf{s}'(t)\| \, dt.$$

Thus we have $d\nu_{t_0}(t) = \|\mathbf{s}'(t)\| \, dt$ and, taking ν as the arc-length measure for C, we have the wombling measures in (14.14) (total gradient) and (14.15) (average gradient) respectively as

$$\Gamma_{Y(\mathbf{s})}(\mathcal{T}) = \int_C \langle \nabla Y(\mathbf{s}), n(\mathbf{s}) \rangle d\nu = \int_{\mathcal{T}} \langle \nabla Y(\mathbf{s}(t)), n(\mathbf{s}(t)) \rangle \|\mathbf{s}'(t)\| \, dt$$

$$\text{and } \bar{\Gamma}_{Y(\mathbf{s})}(\mathcal{T}) = \frac{1}{\nu(\mathcal{T})} \Gamma_{Y(\mathbf{s})}(\mathcal{T}). \tag{14.18}$$

This result is important since we want to take ν as the arc-length measure, but it will be easier to use the parametric representation and work in t space. Also, it is a consequence of the implicit mapping theorem in mathematical analysis (see, e.g., Rudin, 1976) that any other parametrization $\mathbf{s}^*(t)$ of the curve C is related to $\mathbf{s}(t)$ through a differentiable mapping g such that $\mathbf{s}^*(t) = \mathbf{s}(g(t))$. This immediately implies (using (14.18)) that our proposed wombling measure is invariant to the parametrization of C and, as desired, a feature of the curve itself.

For some simple curves the wombling measure can be evaluated quite easily. For instance, when C is a segment of length 1 of the straight line through the point $\mathbf{s}_0$ in the direction $\mathbf{u} = (u_{(1)}, u_{(2)})$, then we have $C = \{\mathbf{s}_0 + t\mathbf{u} : t \in [0, 1]\}$. Under this parametrization, $\mathbf{s}'(t)^{\mathrm{T}} = (u_{(1)}, u_{(2)})$, $\|\mathbf{s}'(t)\| = 1$, and $\nu_{t_0}(t) = t$. Clearly, $n(\mathbf{s}(t)) = (u_{(2)}, -u_{(1)})$, (independent of t), which we write as $\mathbf{u}^{\perp}$—the normal direction to $\mathbf{u}$. Therefore $\Gamma_{Y(\mathbf{s})}(\mathcal{T})$ in (14.18) becomes

$$\int_0^1 \langle \nabla Y(\mathbf{s}(t)), n(\mathbf{s}(t)) \rangle dt = \int_0^1 D_{\mathbf{u}^{\perp}} Y(\mathbf{s}(t)) \, dt.$$

Another example is when C is the arc of a circle with radius r. For example suppose C is traced out by $\mathbf{s}(t) = (r \cos t, r \sin t)$ as $t \in [0, \pi/4]$. Then, since $\|\mathbf{s}'(t)\| = r$, the average gradient is more easily computed as

$$\frac{1}{\nu(C)} \int_0^{\pi/4} \langle \nabla Y(\mathbf{s}(t)), n(\mathbf{s}(t)) \rangle r dt \;=\; \frac{4}{r\pi} \int_0^{\pi/4} \langle \nabla Y(\mathbf{s}(t)), n(\mathbf{s}(t)) \rangle r dt$$

$$= \frac{4}{\pi} \int_0^{\pi/4} \langle \nabla Y(\mathbf{s}(t)), n(\mathbf{s}(t)) \rangle dt.$$

In either case, $n(\mathbf{s}(t))$ is given by (14.16).

Note that while the normal component, $D_{n(\mathbf{s})} Y(\mathbf{s})$, seems to be more appropriate for assessing whether a curve provides a wombling boundary, one may also consider the tangential direction, $\mathbf{u}(t) = \mathbf{s}'(t) / \|\mathbf{s}'(t)\|$ along a curve C. In this case, the average gradient will be given by

$$\frac{1}{\nu(C)} \int_C \langle \nabla Y(\mathbf{s}(t)), \mathbf{u}(t) \rangle \|\mathbf{s}'(t)\| \, dt = \frac{1}{\nu(C)} \int_C \langle \nabla Y(\mathbf{s}(t)), \mathbf{s}'(t) \rangle dt.$$

In fact, we have

$$\int_C \langle \nabla Y(\mathbf{s}(t)), \mathbf{s}'(t) \rangle dt \;=\; \int_{t_0}^{t_1} \langle \nabla Y(\mathbf{s}(t)), \mathbf{s}'(t) \rangle dt$$

$$= \int_{\mathbf{s}_0}^{\mathbf{s}_1} \langle \nabla Y(\mathbf{s}), d\mathbf{s} \rangle = Y(\mathbf{s}_1) - Y(\mathbf{s}_0),$$

where $\mathbf{s}_1 = \mathbf{s}(t_1)$ and $\mathbf{s}_0 = \mathbf{s}(t_0)$ are the endpoints of C. That is, unsatisfyingly, the average directional gradient in the tangential direction is independent of the path C, depending only upon the endpoints of the curve C. Furthermore, Banerjee and Gelfand [2006] show that for a closed path C, the average gradient in the tangential direction is zero. These considerations motivate us to define a "wombling boundary" (see below) with respect to the direction normal to the curve (i.e. perpendicular to the tangent).

14.7.2 Wombling boundary

With the above formulation in place, we now offer a formal definition of a curvilinear *wombling boundary*:

Definition: A curvilinear wombling boundary is a curve C that reveals a large wombling measure, $\Gamma_{Y(\mathbf{s})}(\mathcal{T})$ or $\bar{\Gamma}_{Y(\mathbf{s})}(\mathcal{T})$ (as given in (14.18)) in the direction normal to the curve.

Were the surface fixed, we would have to set a threshold to determine what "large", say in absolute value, means. Since the surface is a random realization, $\Gamma_{Y(\mathbf{s})}(\mathcal{T})$ and $\bar{\Gamma}_{Y(\mathbf{s})}(\mathcal{T})$ are random. Hence, we declare a curve to be a wombling boundary if say a 95% credible set for $\bar{\Gamma}_{Y(\mathbf{s})}(\mathcal{T})$ does not contain 0. It is worth pointing out that while one normal direction (as defined in (14.16)) is used in (14.18), $-n(\mathbf{s}(t))$ would also have been a valid choice. Since $D_{-n(\mathbf{s}(t))} Y(\mathbf{s}(t)) = -D_{n(\mathbf{s}(t))} Y(\mathbf{s}(t))$, we note that the wombling measure with respect to one is simply the negative of the other. Thus, in the above definition large positive as well as large negative values of the integral in (14.18) would signify a wombling boundary. Being a local concept, across a curve, an uphill gradient is equivalent to a downhill gradient.

We also point out that, being a continuous average (or sum) of the directional gradients, the wombling measure may "cancel" the overall gradient effect. For instance, imagine a curve C that exhibits a large positive gradient in the $n(\mathbf{s})$ direction for the first half of its length and a large negative gradient for the second half, thereby cancelling the total or average gradient effect. A potential remedy is to redefine the wombling measure using absolute gradients, $|D_{n(\mathbf{s})}Y(\mathbf{s})|$, in (14.14) and (14.15). The corresponding development does not entail any substantially new ideas, but would destroy the attractive distribution theory in Section 14.8 below and make the computation less tractable. In particular, it will make calibration of the resulting measure with regard to significance much more difficult; how do we select a threshold? Moreover, in practice a descriptive contour representation is usually available where sharp gradients will usually reflect themselves and one could instead compute the wombling measure for appropriate sub-curves of C. Though somewhat subjective, identifying such sub-curves is usually unambiguous and leads to robust scientific inference. More fundamentally, in certain applications a signed measure may actually be desirable: one might want to classify a curve as a wombling boundary if it reflects either an overall "large positive" or a "large negative" gradient effect across it. For these reasons, we confine ourselves to working with $D_{n(\mathbf{s})}Y(\mathbf{s})$ and turn to the distribution theory for the wombling measure in the next section.

14.8 Distribution theory for curvilinear gradients

Curvilinear wombling amounts to performing predictive inference for a line integral parametrized over $\mathcal{T}$. Let us suppose that $\mathcal{T}$ is an interval, $[0, T]$, which generates the curve $C = \{\mathbf{s}(t) : t \in [0, T]\}$. For any $t^* \in [0, T]$ let $\nu(t^*)$ denote the arc length of the associated curve C_{t^*}. The line integrals for total gradient and average gradient along C_{t^*} are given by $\Gamma_{Y(\mathbf{s})}(t^*)$ and $\bar{\Gamma}_{Y(\mathbf{s})}(t^*)$ respectively as:

$$\Gamma_{Y(\mathbf{s})}(t^*) = \int_0^{t^*} D_{n(\mathbf{s}(t))}Y(\mathbf{s}(t))\|\mathbf{s}'(t)\|dt \text{ and } \bar{\Gamma}_{Y(\mathbf{s})}(t^*) = \frac{1}{\nu(t^*)}\Gamma_{Y(\mathbf{s})}(t^*). \quad (14.19)$$

We seek to infer about $\Gamma_{Y(\mathbf{s})}(t^*)$ based upon data $Y = (Y(\mathbf{s}_1), \ldots, Y(\mathbf{s}_n))$. Since $D_{n(\mathbf{s}(t))}Y(\mathbf{s}(t)) = \langle \nabla Y(\mathbf{s}(t)), n(\mathbf{s}(t)) \rangle$ is a Gaussian process (from Section 2), $\Gamma_{Y(\mathbf{s})}(t^*)$ is a Gaussian process on $[0, T]$, equivalently on the curve C. Note that although $D_{n(\mathbf{s})}Y(\mathbf{s})$ is a process on $\Re^d$, our parametrization of the coordinates by $t \in \mathcal{T} \subseteq \Re^1$ induces a valid process on $\mathcal{T}$. In fact, $\Gamma_{Y(\mathbf{s})}(t^*)$ is a Gaussian process with mean and covariance functions

$$\mu_{\Gamma_{Y(\mathbf{s})}}(t^*) = \int_0^{t^*} D_{n(\mathbf{s}(t))}\mu((\mathbf{s}(t)))\|\mathbf{s}'(t)\|dt$$

$$K_{\Gamma_{Y(\mathbf{s})}}(t_1^*, t_2^*) = \int_0^{t_1^*} \int_0^{t_2^*} q_{n(\mathbf{s})}(t_1, t_2)\|\mathbf{s}'(t_1)\|\|\mathbf{s}'(t_2)\|dt_1 dt_2,$$

where $q_{n(\mathbf{s})}(t_1, t_2) = n^{\mathsf{T}}(\mathbf{s}(t_1))H_K(\Delta(t_1, t_2))n(\mathbf{s}(t_2))$ and $\Delta(t_1, t_2) = \mathbf{s}(t_2) - \mathbf{s}(t_1)$. In particular, $\text{Var}(\Gamma_{Y(\mathbf{s})}(t^*)) = K_{\Gamma_{Y(\mathbf{s})}}(t^*, t^*)$ is

$$\int_0^{t^*} \int_0^{t^*} n^{\mathsf{T}}(\mathbf{s}(t_1))H_K(\Delta(t_1, t_2))n(\mathbf{s}(t_2))\|\mathbf{s}'(t_1)\|\,\|\mathbf{s}'(t_2)\|\,dt_1 dt_2.$$

Evidently, $\Gamma_{Y(\mathbf{s})}(t^*)$ is mean square continuous. But, from the above, note that even if $Y(\mathbf{s})$ is a stationary process, $\Gamma_{Y(\mathbf{s})}(t^*)$ is not. For any $\mathbf{s}_j$ in the domain of Y,

$$\text{Cov}(\Gamma_{Y(\mathbf{s})}(t^*), Y(\mathbf{s}_j)) = \int_0^{t^*} Cov(D_{n(\mathbf{s}(t))}Y(\mathbf{s}(t)), Y(\mathbf{s}_j))\|\mathbf{s}'(t)\|dt$$

$$= \int_0^{t^*} D_{n(\mathbf{s}(t))}K(\Delta_j(t))\|\mathbf{s}'(t)\|dt, \quad (14.20)$$

where $\Delta_j(t) = \mathbf{s}(t) - \mathbf{s}_j$. Based upon data $\mathbf{Y} = (Y(\mathbf{s}_1), \ldots, Y(\mathbf{s}_n))$, we seek the predictive distribution $P(\Gamma_{Y(\mathbf{s})}(t^*) \mid \mathbf{Y})$, but note that $Y(\mathbf{s})$ and $\Gamma_{Y(\mathbf{s})}(t^*)$ are processes on different domains—the former is over a connected region in $\Re^2$, while the latter is on a parametrized curve, $\mathbf{s}(t)$, indexed by $\mathcal{T}$. Nevertheless, $\Gamma_{Y(\mathbf{s})}(t^*)$ is derived from $Y(\mathbf{s})$ and we have a valid *joint distribution* $(\mathbf{Y}, \Gamma_{Y(\mathbf{s})}(t^*))$ for any $t^* \in \mathcal{T}$, given by

$$N_{n+1} \left(\begin{pmatrix} \boldsymbol{\mu} \\ \mu_{\Gamma_{Y(s)}}(t^*) \end{pmatrix}, \begin{pmatrix} \Sigma_Y & \gamma_{\Gamma,\mathbf{Y}}(t^*) \\ \gamma_{\Gamma,\mathbf{Y}}^{\mathrm{T}}(t^*) & K_\Gamma(t^*,t^*) \end{pmatrix} \right). \tag{14.21}$$

Here, $\boldsymbol{\mu} = (\mu(\mathbf{s}_1), \ldots, \mu(\mathbf{s}_n))$ and

$$\gamma_{\Gamma,\mathbf{Y}}^{\mathrm{T}}(t^*) = \left(Cov\left(Y(\mathbf{s}_1), \Gamma_{Y(\mathbf{s})}(t^*) \right), \ldots, Cov\left(Y(\mathbf{s}_n), \Gamma_{Y(\mathbf{s})}(t^*) \right) \right),$$

each component being evaluated from (14.20).

Suppose $\mu(\mathbf{s}; \boldsymbol{\beta})$ and $K(\cdot; \boldsymbol{\eta})$ are indexed by regression parameters $\boldsymbol{\beta}$ and covariance parameters $\boldsymbol{\eta}$ respectively. For now, assume that $\mu(\mathbf{s}; \boldsymbol{\beta})$ is a smooth function in $\mathbf{s}$ (as would be needed to do prediction for $Y(\mathbf{s})$). Using MCMC, these model parameters, $\boldsymbol{\theta} = (\boldsymbol{\beta}, \boldsymbol{\eta})$, are available to us as samples, $\{\boldsymbol{\theta}_l\}$, from their posterior distribution $P(\boldsymbol{\theta} \mid \mathbf{Y})$. Therefore, $P(\Gamma_{Y(\mathbf{s})}(t^*) \mid \mathbf{Y}) = \int P(\Gamma_{Y(\mathbf{s})}(t^*) \mid \mathbf{Y}, \boldsymbol{\theta}) P(\boldsymbol{\theta} \mid \mathbf{Y}) d\boldsymbol{\theta}$ will be obtained by sampling, for each $\boldsymbol{\theta}_l$, $\Gamma_{Y(\mathbf{s})}^l(t^*)$ from $P(\Gamma_{Y(\mathbf{s})}(t^*) \mid \mathbf{Y}, \boldsymbol{\theta}_l)$, which, using (14.21), is normally distributed with mean and variance given by

$$\begin{aligned} &\mu_{\Gamma_{Y(s)}}(t^*; \boldsymbol{\beta}_l) - \gamma_{\Gamma,\mathbf{Y}}^{\mathrm{T}}(t^*; \boldsymbol{\eta}_l) \Sigma_Y^{-1}(\boldsymbol{\eta}_l)(\mathbf{Y} - \boldsymbol{\mu}) \text{ and} \\ &K_{\Gamma_{Y(s)}}(t^*, t^*; \boldsymbol{\eta}_l) - \gamma_{\Gamma,\mathbf{Y}}(t^*; \boldsymbol{\eta}_l)^{\mathrm{T}} \Sigma_Y^{-1}(\boldsymbol{\eta}_l) \gamma_{\Gamma,\mathbf{Y}}(t^*; \boldsymbol{\eta}_l) \text{ respectively}. \end{aligned} \tag{14.22}$$

In particular, when $C_{t^*} = \{\mathbf{s}_0 + t\mathbf{u} : t \in [0, t^*]\}$, is a line segment of length t^* joining $\mathbf{s}_0$ and $\mathbf{s}_1 = \mathbf{s}_0 + t^*\mathbf{u}$, we have seen below (14.18) that $\Gamma_{Y(s)}(t^*)$ equals $\int_0^{t^*} D_{\mathbf{u}^\perp} Y(\mathbf{s}(t)) dt$. Thus, defining $\Delta_{0j} = \mathbf{s}_0 - \mathbf{s}_j$, we have

$$\begin{aligned} \mu_{\Gamma_{Y(s)}}(t^*; \boldsymbol{\beta}) &= \int_0^{t^*} \langle \mathbf{u}^\perp, \nabla\mu(\mathbf{s}(t); \boldsymbol{\beta}) \rangle dt; \\ (\gamma_{\Gamma,\mathbf{Y}}(t^*; \boldsymbol{\eta}))_j &= \int_0^{t^*} D_{\mathbf{u}^\perp} K(\Delta_{0j} + t\mathbf{u}; \boldsymbol{\eta}) dt, \ j = 1, \ldots, n; \\ K_{\Gamma_{Y(s)}}(t^*, t^*; \boldsymbol{\eta}) &= \int_0^{t^*} \int_0^{t^*} -(\mathbf{u}^\perp)^{\mathrm{T}} H_{K(\boldsymbol{\eta})}(\Delta(t_1, t_2)) \mathbf{u}^\perp dt_1 dt_2. \end{aligned}$$

These integrals need to be computed for each $\boldsymbol{\theta}_l = (\boldsymbol{\beta}_l, \boldsymbol{\eta}_l)$. Though they may not be analytically tractable (depending upon our choice of $\mu(\cdot)$ and $K(\cdot)$), they are one or two-dimensional integrals that can be efficiently computed using quadrature. Furthermore, since the $\boldsymbol{\theta}_l$'s will already be available, the quadrature calculations (for each $\boldsymbol{\theta}_l$) can be performed ahead of the predictive inference, perhaps using a separate quadrature program, and the output stored in a file for use in the predictive program. The only needed inputs are $\mathbf{s}_0$, $\mathbf{u}$, and the value of t^*. For a specified line segment, we will know these. In fact, for a general curve C, its representation on a given map is as a polygonal curve. As a result, the total or average gradient for C can be obtained through the $\Gamma_{Y(\mathbf{s})}(t^*)$ associated with the line segments that comprise C.

Specifically, with GIS software we can easily extract (at high resolution) the coordinates along the boundary, thus approximating C by line segments connecting adjacent points. Thus, $C = \cup_{k=1}^M C_k$ where C_k's are virtually disjoint (only one common point at the "join") line segments, and $\nu(C) = \sum_{k=1}^M \nu(C_k)$. If we parametrize each line segment as above and compute the line integral along each C_k by the above steps, the total gradient is the sum of the piece-wise line integrals. To be precise, if $\Gamma_k(\mathbf{s}(t_k^*))$ is the line-integral process on the linear segment C_k, we will obtain predictive samples, $\Gamma_k^{(l)}\left(\mathbf{s}\left(t_k^*\right)\right)$ from each $P\left(\Gamma_k\left(\mathbf{s}\left(t_k^*\right)\right) \mid \mathbf{Y}\right)$, $k=1,\ldots,M$. Inference on the average gradient along C will stem from posterior samples of

$$\frac{1}{\sum_{k=1}^{M} \nu\left(C_k\right)} \sum_{k=1}^{M} \Gamma_k^{(l)}\left(\mathbf{s}\left(t_k^*\right)\right).$$

Thus, with regard to boundary analysis, the wombling measure reduces a curve to an average gradient and inference to the examination of the posterior of the average gradient.

When $Y(\mathbf{s})$ is a Gaussian process with constant mean, $\mu(\mathbf{s})=\mu$ and an isotropic correlation function $K\left(\|\Delta\|; \sigma^2, \phi\right)=\sigma^2 \exp\left(-\phi\|\Delta\|^2\right)$, calculations simplify. We have $\nabla \mu(\mathbf{s})=0$, $\mu_{\Gamma}\left(t^*\right)=0$, and

$$H_K(\Delta)=-2\sigma^2 \phi \exp\left(-\phi\|\Delta\|^2\right)\left(I-2\phi \Delta\Delta^{\mathrm{T}}\right).$$

Further calculations reveal that $\left(\gamma_{YZ}\left(t^*; \sigma^2, \phi\right)\right)_j$ can be computed as,

$$c\left(\sigma^2, \phi, \mathbf{u}^{\perp}, \Delta_{0j}\right)\left(\Phi\left(\sqrt{2\phi}\left(t^*+\left\langle\mathbf{u}, \Delta_{0j}\right\rangle\right)\right)-\Phi\left(\sqrt{2\phi}\left\langle\mathbf{u}, \Delta_{0j}\right\rangle\right)\right), \tag{14.23}$$

where $\Phi()$ is the standard Gaussian cumulative distribution function, and

$$c\left(\sigma^2, \phi, \mathbf{u}^{\perp}, \Delta_{0j}\right)=-2\sigma^2 \sqrt{\pi\phi}\left\langle\mathbf{u}^{\perp}, \Delta_{0j}\right\rangle \exp\left(-\phi|\left\langle\mathbf{u}^{\perp}, \Delta_{0j}\right\rangle|^2\right).$$

These computations can be performed using the Gaussian cdf function, and quadrature is needed only for $\operatorname{Var}\left(\Gamma_{Y(\mathbf{s})}\left(t^*\right)\right)$.

Returning to the model $Y(\mathbf{s})=\mathbf{x}^{\mathrm{T}}(\mathbf{s})\boldsymbol{\beta}+w(\mathbf{s})+\epsilon(\mathbf{s})$ with $\mathbf{x}(\mathbf{s})$ a general covariate vector, $w(\mathbf{s}) \sim GP\left(0, \sigma^2 \rho(\cdot, \phi)\right)$ and $\epsilon(\mathbf{s})$ a zero-centered white-noise process with variance τ^2, consider boundary analysis for the residual surface $w(\mathbf{s})$. In fact, boundary analysis on the spatial residual surface is feasible in generalized linear modeling contexts with exponential families, where $w(\mathbf{s})$ may be looked upon as a non-parametric latent structure in the mean of the parent process. Denoting by $\Gamma_{w(\mathbf{s})}(t)$ and $\bar{\Gamma}_{w(\mathbf{s})}(t)$ as the total and average gradient processes (as defined in (14.18)) for $w(\mathbf{s})$, we seek the posterior distributions $P\left(\Gamma_{w(\mathbf{s})}\left(t^*\right) \mid \mathbf{Y}\right)$ and $P\left(\bar{\Gamma}_{w(\mathbf{s})}\left(t^*\right) \mid \mathbf{Y}\right)$. Note that

$$P\left(\Gamma_{w(\mathbf{s})}\left(t^*\right) \mid \mathbf{Y}\right)=\int P\left(\Gamma_{w(\mathbf{s})}\left(t^*\right) \mid \mathbf{w}, \boldsymbol{\theta}\right) P(\mathbf{w} \mid \boldsymbol{\theta}, \mathbf{Y}) P(\boldsymbol{\theta} \mid \mathbf{Y}) d\boldsymbol{\theta} d\mathbf{w}, \tag{14.24}$$

where $\mathbf{w}=\left(w\left(s_1\right),\ldots,w\left(s_n\right)\right)$ denotes a realization of the residual process and $\boldsymbol{\theta}=\left(\boldsymbol{\beta}, \sigma^2, \phi, \tau^2\right)$. Sampling of this distribution again proceeds in a posterior-predictive fashion using posterior samples of $\boldsymbol{\theta}$, and is expedited in a Gaussian setting since $P(\mathbf{w} \mid \boldsymbol{\theta}, \mathbf{Y})$ and $P\left(\Gamma_{w(\mathbf{s})}\left(t^*\right) \mid \mathbf{w}, \boldsymbol{\theta}\right)$ are both Gaussian distributions.

Formal inference for a wombling boundary is done more naturally on the residual surface $w(\mathbf{s})$, i.e. for $\Gamma_{w(\mathbf{s})}\left(t^*\right)$ and $\bar{\Gamma}_{w(\mathbf{s})}\left(t^*\right)$, because $w(\mathbf{s})$ is the surface containing any non-systematic spatial information on the parent process $Y(\mathbf{s})$. Since $w(\mathbf{s})$ is a zero-mean process, $\mu_{\Gamma_{w(\mathbf{s})}}\left(t^*; \boldsymbol{\beta}\right)=0$, and thus one needs to check for the inclusion of this null value in the resulting 95% credible intervals for $\Gamma_{w(\mathbf{s})}\left(t^*\right)$ or, equivalently, for $\bar{\Gamma}_{w(\mathbf{s})}\left(t^*\right)$. Again, this clarifies the issue of the normal direction mentioned in Section 14.7.2; significance using $n(\mathbf{s}(t))$ is equivalent to significance using $-n(\mathbf{s}(t))$. One only needs to select and maintain a particular orthogonal direction relative to the curve. In accord with our remarks concerning absolute gradients in Section 14.7.2, we could compute (14.19) using $\left|D_{n(\mathbf{s}(t))} Y(\mathbf{s}(t))\right|$ using a Riemann sum, but would be computationally expensive and would not offer a Gaussian calibration of significance.

14.9   Illustration: Spatial boundaries for invasive plant species

Banerjee and Gelfand [2006] consider data collected from 603 locations in Connecticut with presence/absence and abundance indicators for several invasive plant species, plus environmental covariates. The covariates are available only at the sample locations, not on a grid. The response variable $Y(\mathbf{s})$ is a presence-absence binary indicator (0 for absence) for one species *Celastrus orbiculatus* at location $\mathbf{s}$. There are four categorical covariates: habitat class (representing the current state of the habitat) of four different types, land use, and land cover (LULC) types (Land use/cover history of the location; e.g. always forest, formerly pasture now forest, etc.) at five levels and a 1970 category number (LULC at one point in the past: 1970; e.g. forest, pasture, residential, etc.) with six levels. In addition, we have an ordinal covariate, canopy closure percentage (percent of the sky that is blocked by "canopy" of leaves of trees. A location under mature forest would have close to 100% canopy closure while a forest edge would have closer to 25%) with four levels in increasing order, a binary variable for heavily managed points (0 if "no;" "heavy management" implies active landscaping or lawn mowing) and a continuous variable measuring the distance from the forest edge in the logarithm scale. Figure 14.2 is a digital terrain image of the study domain, with the labeled curves indicating forest edges extracted using GIS software (https://www.esri.com/en-us/home). Ecologists are interested in evaluating spatial gradients along these ten natural curves and identifying them as wombling boundaries.

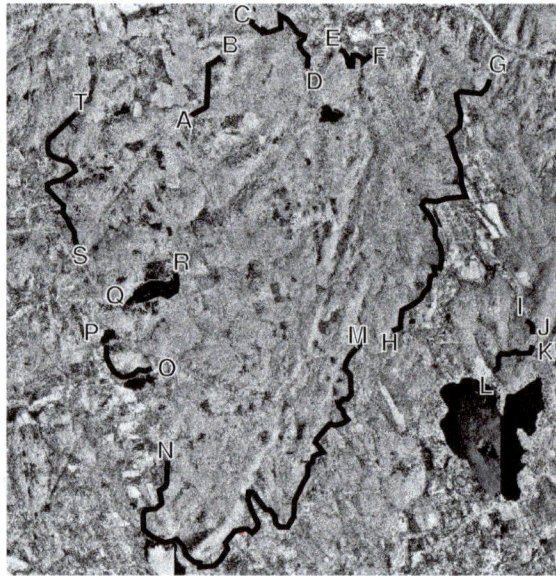

Figure 14.2 *A digital image of the study domain in Connecticut indicating the forest edges as marked curves. These are assessed for significant gradients. Note: Eastings range from 699148 to 708961; Northings range from 4604089 to 4615875 for the picture.*

We fit a logistic regression model with spatial random effects,

$$\log\left(\frac{P(Y(\mathbf{s})=1)}{P(Y(\mathbf{s})=0)}\right) = \mathbf{x}^{\mathrm{T}}(\mathbf{s})\boldsymbol{\beta} + w(\mathbf{s}),$$

where $\mathbf{x}(\mathbf{s})$ is the vector of covariates observed at location $\mathbf{s}$ and $w(\mathbf{s}) \sim GP(0, \sigma^2\rho(\cdot; \phi, \nu))$ is a Gaussian process with $\rho(\cdot; \phi, \nu)$ as a Matérn correlation function. While $Y(\mathbf{s})$ is a binary surface that does not admit gradients, conducting boundary analysis on $w(\mathbf{s})$ is perfectly legitimate. The residual spatial surface reflects unmeasured or unobservable environmental features in the mean surface. Again, attaching curvilinear wombling boundaries to the residual surface tracks the rapid change in the departure from the mean surface.

We adopt a flat prior for β, an inverse gamma $IG(2, 0.001)$ prior for σ^2 and the Matérn correlation function with a Gamma prior for the correlation decay parameter, ϕ, specified so that the prior spatial range has a mean of about half of the observed maximum inter-site distance (the maximum distance is 11887 meters based on a UTM projection), and a $U(1, 2)$ prior to the smoothness parameter ν. Three parallel MCMC chains were run for 15000 iterations each and $5000 \times 3 = 15,000$ samples were retained after adequate mixing for posterior analysis.

Parameters	50% (2.5%,97.5%)
Intercept	0.983 (-2.619, 4.482)
Habitat Class (Baseline: Type 1)	
Type 2	-0.660 (-1.044,-0.409)
Type 3	-0.553 (-1.254, 0.751)
Type 4	-0.400 (-0.804,-0.145)
Land use Land Cover Types (Baseline: Level 1)	
Type 2	0.591 (0.094, 1.305)
Type 3	1.434 (0.946, 2.269)
Type 4	1.425 (0.982, 1.974)
Type 5	1.692 (0.934, 2.384)
1970 Category Types (Baseline: Category 1)	
Category 2	-4.394 (-6.169,-3.090)
Category 3	-0.104 (-0.504, 0.226)
Category 4	1.217 (0.864, 1.588)
Category 5	-0.039 (-0.316, 0.154)
Category 6	0.613 (0.123, 1.006)
Canopy Closure	0.337 (0.174, 0.459)
Heavily Managed Points (Baseline: No)	
Yes	-1.545 (-2.027,-0.975)
Log Edge Distance	-1.501 (-1.891,-1.194)
σ^2	8.629 (7.005, 18.401)
ϕ	1.75E-3 (1.14E-3, 3.03E-3)
ν	1.496 (1.102, 1.839)
Range (in meters)	1109.3 (632.8, 1741.7)

Table 14.3 *Parameter estimates for the logistic spatial regression example.*

Table 14.3 presents the posterior estimates of the model parameters. We do not have a statistically significant intercept, but most of the categorical variables reveal significance: Types 2 and 4 for habitat class have significantly different effects from Type 1; all the four types of LULC show significant departure from the baseline Type 1; for the 1970 category number, category 2 shows a significant negative effect, while categories 4 and 6 show significant positive effects compared to category 1. Canopy closure is significantly positive, implying higher presence probabilities of *Celastrus orbiculatus* with higher canopy blockage, while points that are more heavily managed appear to have a significantly lower probability of species presence as does the distance from the nearest forest edge. Posterior summaries of the spatial process parameters are also presented and the effective spatial range is approximated to be around 1109.3 meters approximately. These produce the mean posterior surface of $w(\mathbf{s})$, shown in Figure 14.3 with the twenty endpoints of the forest edges from Figure 14.2 labeled to connect the figures.

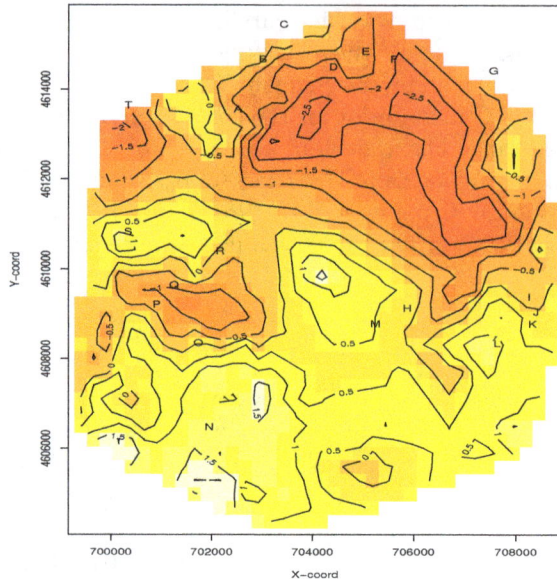

Figure 14.3 *The spatial residual surface from the presence-absence application for the 603 observed locations (not shown) in Connecticut. Also shown are the twenty endpoints for the ten curves in Figure 14.2 to connect the figures.*

Finally, in Table 14.4, we present the formal curvilinear gradient analysis for the ten forest edges in Figure 14.2. We find that six out of the ten edge curves (with the exception of CD, EF, KL, and MN) are formally tested to be wombling boundaries. Our methodology proves useful here since some of these edge curves meander along the terrain for substantially long distances. Indeed, while the residual surface in Figure 14.3 reveals a general pattern of spatial variation (higher residuals in the South), it is difficult to make visual assessments on the size (and significance) of average gradients for the longer curves. Furthermore, with non-gridded data as here, the surface interpolators [in this case the Akima, 1996, interpolator in R] often find it difficult to extrapolate beyond a convex hull of the site locations. Consequently, parts of the curve (e.g. endpoints C, G, and (almost) T) lie outside the fitted surface, making local visual assessment on them impossible.

Quickly and reliably identifying forest edges could be useful in determining boundaries between areas of substantial anthropogenic activity and minimally managed forest habitats. Such boundaries are important because locations at which forest blocks have not been invaded by exotic plant species may be subject to significant seed rain from these species. These boundaries thus might form important "front lines" for efforts to monitor or control invasive species.

Curve	Average Gradient	Curve	Average Gradient
AB	1.021 (0.912, 1.116)	KL	0.036 (-0.154, 0.202)
CD	0.131 (-0.031, 0.273)	MN	0.005 (-0.021, 0.028)
EF	0.037 (-0.157, 0.207)	OP	0.227 (0.087, 0.349)
GH	1.538 (1.343, 1.707)	QR	0.282 (0.118, 0.424)
IJ	0.586 (0.136, 0.978)	ST	0.070 (0.017, 0.117)

Table 14.4 *Curvilinear gradient assessment for the ten forest edges labelled in Figure 14.2 for the logistic regression example.*

14.10   Extensions and related ideas

Boundary analysis is related to the problem of spatial cluster analysis. However, the latter is focused on detecting clusters of homogeneous regions, identifying their shape and range, and often on identifying whether a particular region is or is not part of a particular "cluster" (say, of infected or exposed counties). By contrast, boundary analysis is more focused on detecting and locating rapid changes, which are typically thought of as "lines" or "curves" on the spatial surface. Substantive interest focuses on the boundary itself and what distinguishes the regions on either side, rather than on any particular region. As such, methods for spatial clustering (as summarized for instance by Lawson and Denison [2002] are not expected to be directly applicable here.

In the previous sections, we have primarily focused on inference on spatial gradients to detect points and curves that represent rapid change. We have confined ourselves to curves that track zones of rapid change. However, as we have alluded to above, zones of rapid change are areal notions; description by a curve may be an unsatisfying simplification. Describing zones as areal quantities, i.e., as sets of nonzero Lebesgue measure in $\Re^2$ is an alternative. To proceed, the crucial issue is to formalize shape-oriented definitions of a wombling boundary. There are opportunities to adapt ideas from statistical shape analysis and its applications in imaging to evaluate their effectiveness in boundary detection. Alternatively, we can recast boundary detection as a problem of identifying a set of boundaries on a map of areal units. This, in some sense, requires less mathematical sophistication as one no longer defines boundaries using the smoothness of the random field. Instead, we seek inference over a finite, but possibly large, set of political boundaries delineating areal units and identify neighboring regions with significantly different outcomes. This is the topic of Section 14.11.

Other possibilities include the use of formal differential geometry and calculus of variations to conduct inference on the local smoothness of surfaces using fundamental forms of the surface. A few extensions are worth mentioning. Halder et al. [2024a] extend Bayesian inference to higher-order gradients by constructing directional curvature processes to analyze differential behavior in responses. Perhaps a more significant challenge is addressed by Halder et al. [2024b] who extend wombling methods to spatial-temporal processes and consider settings where the boundary of interest is static over time (e.g., a physical feature such as a mountain or a river) or where it changes over time (e.g., in imaging applications where a boundary of interest on the brain may change with the progression of a disease). One might also examine the boundary analysis problem from an alternative modeling perspective, where curves or zones arise as random processes. Possibilities include line processes, random tessellations, and random areal units.

Notable developments in boundary detection are also available in Qu et al. [2021], who propose a multiscale representation of the directional derivative Karhunen–Loéve expansion to perform directionally based boundary detection. They seek alternative, and perhaps simpler to compute, metrics to identify wombling boundaries using the concept of curvilinear boundary fallacy (CBF) error. They adopt a multi-scale spatial model to define the CBF error, which can be regarded as a boundary detection counterpart to ecological fallacy is often studied in spatial change of support literature. Qu et al. [2021] also propose a directionally based multi-scale curvilinear boundary error criterion to quantify CBFs and compute a boundary aggregation error (BAGE) to perform boundary detection.

14.11   Areal wombling

Areal wombling refers to ascertaining boundaries on *areally* referenced data. Such methods are valuable in determining boundaries for data sets that, perhaps due to confidentiality concerns, are available only in ecological (aggregated) format, or are only collected this way (e.g., delivery of health care or cost information). In this case, since we lack smooth realizations of spatial surfaces, areal wombling cannot employ spatial gradients. Here the gradient is not explicitly modeled; instead, boundary effects are looked upon as edge effects and modeled using Markov random field specifications. Boundaries in areal wombling are just a collection of segments (or *arcs*, in geographic information systems (GIS) parlance) dually indexed by ij, corresponding to the two adjacent regions i and j the segment separates. In the fields of image analysis and pattern recognition, there has been much research in using statistical models for capturing "edge" and "line" effects [see, e.g., Besag, 1974, Geman and Geman, 1984, 1986, Geman and McClure, 1987, Helterbrand et al., 1994]; see also [Cressie, 1993, Sec. 7.4] and references therein. Such models are based upon probability distributions such as Gibbs distributions or Markov random fields that model pixel intensities as conditional dependencies using the neighborhood structure [see, e.g., Chellappa and Jain, 1992]. Modeling objectives include identification of edges based on distinctly different image intensities in adjacent pixels.

As seen earlier, in areal models, local spatial dependence between the observed image characteristics is captured by a *neighborhood structure*, where a pixel is independent of the rest of the pixels, given the values of its neighbors. Various neighborhood structures are possible, but all propose stronger statistical dependence between data values from areas that are spatially closer, thus inducing local smoothing. However, in the context of areal wombling this leads to a new problem: when real discontinuities (boundaries) exist between neighboring pixels, MRF models tend to smooth across them, thus blurring the very edges we hope to detect.

Although the boundary analysis problem for public health data resembles the edge-detection problem in image processing, significant differences exist. Unlike image pixels, geographical maps that form the domain of most public health data are not regularly spaced but still have a well-defined neighborhood structure. Furthermore, there are usually far fewer of these areas than the number of pixels that would arise in a typical image restoration problem, so we have far less data. Finally, the areal units (polygons) are often quite different in size, shape, and number of neighbors, leading for example to different degrees of smoothing in urban and rural regions, as well as near the external boundary of the study region.

In this section, after a brief review of existing algorithmic techniques, we propose a variety of fully model-based frameworks for areal wombling, using Bayesian hierarchical models with posterior summaries computed using MCMC methods. We explore the suitability of various existing hierarchical and spatial software packages (notably R and BUGS/JAGS) to the task and indicate the approaches' superiority over existing non-stochastic alternatives. We also illustrate our methods (as well as the solution of advanced modeling issues such as simultaneous inference) using county-level cancer surveillance and zip code-level hospice utilization data in the state of Minnesota.

Areal wombling (also known as *polygonal* wombling) has less development in the literature than point or raster wombling, but some notable papers exist. Oden et al. [1993] provide a wombling algorithm for multivariate categorical data defined on a lattice. The statistic chosen is the average proportion of category mismatches at each pair of neighboring sites, with significance relative to an independence or particular spatial null distribution judged by a randomization test. Csillag and Kabos [2002] developed a procedure for characterizing the strength of boundaries examined at the neighborhood level. In this method, a topological or a metric distance δ defines a the neighborhood of the candidate set of polygons (say, p_i). A weighted local statistic is attached to each p_i. The difference statistic calculated as the squared difference between any two sets of polygons' local statistic and its quantile measure are used as a relative measure of the distinctiveness of the boundary at the scale of neighborhood size δ. Jacquez and Greiling [2003] estimate boundaries of rapid change for colorectal, lung, and breast cancer incidence in Nassau, Suffolk, and Queens counties in New York.

Classical boundary analysis research often proceeds by selecting a *dissimilarity metric* (say, Euclidean distance) to measure the difference in response between the values at (say) adjacent polygon centroids. An absolute (dissimilarity metrics greater than C) or relative (dissimilarity metrics in the top $k\%$) threshold then determines which borders are considered actual barriers, or parts of the boundary. The relative (top $k\%$) thresholding method for determining boundary elements is easily criticized, since for a given threshold, a fixed number of boundary elements will always be found regardless of whether or not the responses separated by the boundary are statistically different. Jacquez and Maruca (1998) suggest the use of both local and global statistics to determine where statistically significant boundary elements are, and a randomization test (with or without spatial constraints) for whether the boundaries for the entire surface are statistically unusual.

These older areal wombling approaches, including the algorithm implemented in the **BoundarySeer** software (https://biomedware.com/), tend to be algorithmic, rather than model-based. That is, they do not involve a probability distribution for the data, and therefore permit statements about the "significance" of a detected boundary only relative to predetermined, often unrealistic null distributions. We prefer a hierarchical statistical modeling framework for areal wombling, to permit direct estimation of the probability that two geographic regions are separated by the wombled boundary. Such models can account for spatial and/or temporal association and permit the borrowing of strength across different levels of the model hierarchy.

14.11.1 Simple MRF-based areal wombling

Suppose we have regions $i = 1, \ldots, N$ along with the areal adjacency matrix A, and we have observed a response Y_i (e.g., a disease count or rate) for the ith region. Traditional areal wombling algorithms assign a *boundary likelihood value* (BLV) to each areal boundary using a Euclidean distance metric between neighboring observations. This distance is taken as the dissimilarity metric, calculated for each pair of adjacent regions. Thus, if i and j are neighbors, the BLV associated with the edge (i, j) is

$$D_{ij} = \|Y_i - Y_j\| \,,$$

where $\| \cdot \|$ is a distance metric. Locations with higher BLV's are more likely to be a part of a difference boundary, since the variable changes rapidly there.

The wombling literature and attendant software further distinguishes between *crisp* and *fuzzy* wombling. In the former, BLV's exceeding specified thresholds are assigned a *boundary membership value* (BMV) of 1, so the wombled boundary is $\{(i, j) : \|Y_i - Y_j\| > c, i \text{ adjacent to } j\}$ for some $c > 0$. The resulting edges are called *boundary elements* (BEs). In the latter (fuzzy) case, BMVs can range between 0 and 1 (say, $\|Y_i - Y_j\| / \max_{ij}\{\|Y_i - Y_j\|\}$) and indicate *partial* membership in the boundary.

For our hierarchical modeling framework, suppose we employ the usual Poisson log-linear form for observed and expected disease counts Y_i and E_i,

$$Y_i \sim Poisson(\mu_i) \text{ where } \log \mu_i = \log E_i + \mathbf{x}_i'\boldsymbol{\beta} + \phi_i \,. \tag{14.25}$$

This model allows a vector of region-specific covariates $\mathbf{x}_i$ (if available), and a random effect vector $\boldsymbol{\phi} = (\phi_1, \ldots, \phi_N)'$ that is given a conditionally autoregressive (CAR) specification. As seen earlier, the intrinsic CAR, or IAR, form has an improper joint distribution, but conditional distributions of the form

$$\phi_i \mid \boldsymbol{\phi}_{j \neq i} \sim N(\bar{\phi}_i \,, 1/(\tau m_i)) \,, \tag{14.26}$$

where N denotes the normal distribution, $\bar{\phi}_i$ is the average of the $\phi_{j \neq i}$ that are adjacent to ϕ_i, and m_i is the number of these adjacencies. Finally, τ is typically set equal to some fixed value, or assigned a distribution itself (usually a relatively vague gamma distribution).

As we have seen, MCMC samples $\mu_i^{(g)}$, $g = 1, \ldots, G$ from the marginal posterior distribution $p(\mu_i \mid \mathbf{y})$ can be obtained for each i, from which corresponding samples of the (theoretical) standardized morbidity ratio,

$$\eta_i = \frac{\mu_i}{E_i}, \ i = 1, \ldots, N,$$

are immediately obtained. We may then define the BLV for boundary (i, j) as

$$\Delta_{ij} = |\eta_i - \eta_j| \quad \text{for all } i \text{ adjacent to } j \ . \tag{14.27}$$

Crisp and fuzzy wombling boundaries are then based on the posterior distribution of the BLVs. In the crisp case, we might define ij to be part of the boundary if and only if $\mathrm{E}(\Delta_{ij} \mid \mathbf{y}) > c$ for some constant $c > 0$, or if and only if $P(\Delta_{ij} \geq c \mid \mathbf{y}) > c^*$ for some constant $0 < c^* < 1$.

Model (14.25)–(14.26) can be easily implemented in the **BUGS/JAGS** software package. The case of *multivariate* response variables requires multivariate CAR (MCAR) models (see Chapter 11). Posterior draws $\{\eta_i^{(g)}, \ g = 1, \ldots, G\}$ and their sample means $\widehat{E}(\eta_i \mid \mathbf{y}) = \frac{1}{G} \sum_{g=1}^{G} \eta_i^{(g)}$ are easily obtained for our problem. Bayesian areal wombling would naturally obtain posterior draws of the BLV's in (14.27) by simple transformation as $\Delta_{ij}^{(g)} = |\eta_i^{(g)} - \eta_j^{(g)}|$, and then base the boundaries on their empirical distribution. For instance, we might estimate the posterior means as

$$\widehat{E}(\Delta_{ij} \mid \mathbf{y}) = \frac{1}{G} \sum_{g=1}^{G} \Delta_{ij}^{(g)} = \frac{1}{G} \sum_{g=1}^{G} |\eta_i^{(g)} - \eta_j^{(g)}| \,, \tag{14.28}$$

and take as our wombling boundaries the borders corresponding to the top 20% or 50% of these values.

Turning to fuzzy wombling, suppose we select a cutoff c such that, were we *certain* a particular BLV exceeded c, we would also be certain the corresponding segment was part of the boundary. Since our statistical model (14.25)–(14.26) delivers the full posterior distribution of every Δ_{ij}, we can compute $P(\Delta_{ij} > c \mid \mathbf{y})$, and take this probability as our fuzzy BMV for segment ij. In fact, the availability of the posterior distribution provides another benefit: a way to directly assess the *uncertainty* in our fuzzy BMVs. Our Monte Carlo estimate of $P(\Delta_{ij} > c \mid \mathbf{y})$ is

$$\hat{p}_{ij} \equiv \widehat{P}(\Delta_{ij} > c \mid \mathbf{y}) = \frac{\#\Delta_{ij}^{(g)} > c}{G} \ . \tag{14.29}$$

This is nothing but a binomial proportion; were its components independent, basic binomial theory implies an approximate standard error for it would be

$$\widehat{se}(\hat{p}_{ij}) = \sqrt{\frac{\hat{p}_{ij}(1 - \hat{p}_{ij})}{G}} \ . \tag{14.30}$$

Of course, our Gibbs samples Δ_{ij} are *not* independent in general, since they arise from a Markov chain, but we can make them approximately so by subsampling, retaining only every M^{th} sample. Note that this subsampling does *not* remove the *spatial* dependence among the Δ_{ij}, so repeated use of formula (14.30) would not be appropriate if we wanted to make a *joint* probability statement involving more than one of the Δ_{ij} at the same time; Subsection 14.11.3 revisits this "simultaneous inference" issue.

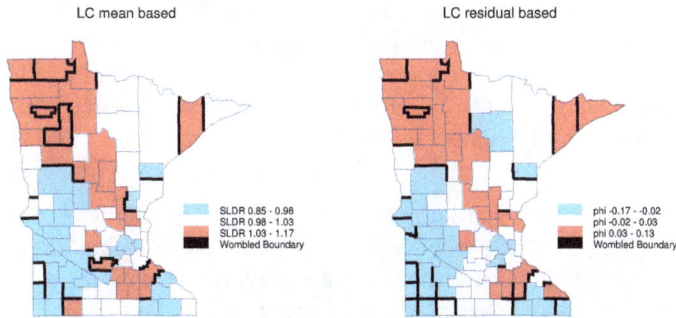

Figure 14.4 *Lu and Carlin crisp (top 20%) wombled maps based on SLDRs η_i (left) and residuals ϕ_i (right).*

Example 14.4 Minnesota breast cancer late detection data. As an illustration, we consider boundary analysis for a dataset recording the rate of late detection of several cancers collected by the Minnesota Cancer Surveillance System (MCSS), a population-based cancer registry maintained by the Minnesota Department of Health. The MCSS collects information on geographic location and stage at detection for colorectal, prostate, lung, and breast cancers. For each county, the late detection rate is defined as the number of regional or distant case detections divided by the total cases observed, for the years 1995 to 1997. Since higher late detection rates are indicative of possibly poorer cancer control in a county, a wombled boundary for this map might help identify barriers separating counties with different cancer control methods. Such a boundary might also motivate a more careful study of the counties that separate it, in order to identify previously unknown covariates (population characteristics, dominant employment type, etc.) that explain the difference.

In this example, we consider n_i, the total number of breast cancers occurring in county i, and Y_i, the number of these that were detected late (i.e., at the regional or distant stage). To correct for the differing number of detections (i.e., the total population) in each county, we womble not on the Y scale, but on the *standardized late detection ratio* (SLDR) scale, $\eta_i = \mu_i/E_i$, where the expected counts E_i are computed via internal standardization as $E_i = n_i\bar{r}$, where $\bar{r} = \sum_i Y_i / \sum_i n_i$, the statewide late detection rate.

As discussed above, areal wombling is naturally based here on the absolute SLDR differences $\Delta_{ij} = |\eta_i - \eta_j|$; we refer to this as *mean-based* wombling. But one might redefine the BLVs $\Delta_{ij} = |\phi_i - \phi_j|$, resulting in *residual-based* wombling. Comparison of the mean and residual-based wombling maps may provide epidemiologists with information regarding barriers that separate regions with different cancer prevention and control patterns. For instance, if segment ij is picked up as a BE in the mean-based map but not in the residual-based map, this suggests that the difference between the two regions in the response counts may be due to the differences in their fixed effects. If on the other hand segment ij is a BE in the residual-based map but not the mean-based map, this may suggest a boundary between region i and j caused by excess spatial heterogeneity, possibly indicating missing spatial covariates.

Our model assumes the mean structure $\log \mu_i = \log E_i + \beta_1(x_i - \bar{x}) + \phi_i$, where x_i is the average annual age-adjusted cancer mortality rate in county i for 1993–1997, and with ϕ following the usual CAR prior with fixed adjacency matrix. Regarding prior distributions for this model, we selected a $N(0,1)$ prior for β_1 and a $G(0.1, 0.1)$ prior for τ, choices designed to be vague enough to allow the data to dominate the determination of the posterior.

Figure 14.4 shows the resulting crisp wombling maps based on the top 20% of the BLVs. The county shading (blue for low values and red for high values) indicates the posterior mean SLDR η_i or residual ϕ_i, while the boundary segments are shown as dark lines. The residual-based map indicates many boundaries in the southern borders of the state and also in the more thinly-populated north region of the state, an area with large farms and native American reservation land. County 63 (Red Lake, a T-shaped county in the northwest) seems "isolated" from its two neighbors, a finding also noted by Lu and Carlin [2005]. The mean map seems to identify the regions with differing fixed effects for mortality. ∎

Note that wombling on the spatial residuals ϕ_i instead of the fitted SLDRs η_i changes not only the scale of the c cutoff (to difference in *log*-relative risk) but the interpretation of the results as well. Specifically, we can borrow an interpretation often mentioned in spatial epidemiology regarding spatially oriented covariates $\mathbf{x}_i$ still missing from the model (14.25). Since boundaries based on the ϕ_i separate regions that differ in their unmodeled spatial heterogeneity, a careful comparison of such regions identified by the wombled map should prove the most fruitful in any missing covariate search.

14.11.1.1 *Adding covariates*

Lu et al. [2007] develop randomly weighted hierarchical areal wombling models that allow the data to help determine the degree and nature of spatial smoothing to perform areal wombling. Specifically, they propose a model that allows the choice of the neighborhood structure to be determined by the value of the process in each region and by variables determining the similarity of two regions. This approach is natural for Bayesian areal wombling since detecting boundary membership can be viewed as the dual problem of regional estimation using adjacency information. Such a method permits the data (and perhaps other observed covariate information) to help determine the neighborhood structure and the degree and nature of spatial smoothing, while simultaneously offering a new stochastic definition of the boundary elements.

Recall that spatial random effects $\boldsymbol{\phi}$ following a CAR prior have conditional distributions

$$\phi_i \mid \boldsymbol{\phi}_{(-i)} \sim N\left(\frac{\sum_j w_{ij}\phi_j}{\sum_j w_{ij}}, \frac{1}{\tau \sum_j w_{ij}}\right), \tag{14.31}$$

where N denotes the normal distribution. Here the neighborhood weights w_{ij} are traditionally fixed at 1 or 0 based on whether regions i and j share a common geographic boundary; this is the approach taken above. Now, rather than fix the w_{ij}, we model them as

$$w_{ij} \mid p_{ij} \stackrel{indep}{\sim} \text{Bernoulli}(p_{ij}), \text{ where } \log\left(\frac{p_{ij}}{1-p_{ij}}\right) = \mathbf{z}'_{ij}\boldsymbol{\gamma}. \tag{14.32}$$

Here $\mathbf{z}_{ij}$ is a set of known features of the $(i,j)^{th}$ pair of regions, with corresponding parameter vector $\boldsymbol{\gamma}$. Regions i and j are thus considered to be neighbors with probability p_{ij} provided they share a common geographical boundary; $w_{ij} = 0$ for all non-adjacent regions. Now that the w_{ij} are random, one can also draw crisp wombled maps based on the $1 - w_{ij}$ posteriors, offering a third, clearly variance-based alternative to the mean- and residual-based methods illustrated in Figure 14.4.

A wide variety of covariates might be considered for $\mathbf{z}_{ij}$ in (14.32), containing information about the difference between regions i and j that might impact neighbor status. For instance, $\mathbf{z}_{ij}$ could be purely map-based, such as the distance between the centroids of regions i and j, or the difference between the percentage of common boundaries shared by the two regions among their own total geographical boundaries. Auxiliary topological information, such as the presence of a mountain range or river across which travel is difficult, might also be important here. Alternatively, $\mathbf{z}_{ij}$ may include sociodemographic information, such as the difference between the regions' percentage of urban area, or the absolute difference of some regional covariate (say, the percent of residents who are smokers, or even the region's expected age-adjusted disease count).

Vague priors are typically chosen for $\boldsymbol{\beta}$ in the mean structure, and a conventional gamma prior can be used for the precision parameter τ. However, it is not clear how to specify the prior for $\boldsymbol{\gamma}$. If the covariates $\mathbf{z}_{ij}$ measure dissimilarity between regions i and j, common sense might suggest an informative prior on $\boldsymbol{\gamma}$ that favors negative values, because the more dissimilar two regions are, the less likely that they should be thought of as neighbors. The induced prior on p_{ij} could be fairly flat even if $\boldsymbol{\gamma}$'s prior has a small variance, due to the logit transformation. If the proper CAR prior is used, i.e., $\boldsymbol{\phi} \sim N(\mathbf{0}, \tau^2(D - \rho W)^{-1})$ with positive definite precision matrix $D - \rho W$, W being the $N \times N$ matrix of weights w_{ij} and $D = \text{Diag}(\sum_{j \sim i}^{m_i} w_{ij})$, either a noninformative (e.g., Unif$(0,1)$) or an informative (e.g., Beta$(18,2)$) prior could be used for ρ, depending in part on one's tolerance for slow MCMC convergence. The covariate effect $\boldsymbol{\gamma}$ is identified by the data, even under a noninformative prior distribution for $\boldsymbol{\gamma}$; however, as with α, moderately informative priors are often used.

14.11.2   Joint site-edge areal wombling

The method of Lu and Carlin [2005] is very easy to understand and implement but suffers from the oversmoothing problems mentioned above. It also generally fails to produce the long series of connected boundary segments often desired by practitioners. In this subsection, we suggest a variety of hierarchical models for areal boundary analysis that hierarchically or jointly parameterize *both* the areas and the edge segments. The approach uses a compound Gaussian Markov random field model that adopts an *Ising* distribution as the prior on the edges. This leads to conceptually appealing solutions for our data that remain computationally feasible. While our approaches parallel similar developments in statistical image restoration using Markov random fields, important differences arise due to the irregular nature of our lattices, the sparseness and high variability of our data, the existence of important covariate information, and most importantly, our desire for full posterior inference on the boundary. We will illustrate the performance of these methods compared to the basic Lu and Carlin approach in the context of a zip code-level hospice utilization dataset. Our description basically follows that in Ma et al. [2010], but will bring in other references as appropriate.

14.11.2.1   Edge smoothing and random neighborhood structure

We begin by extending our notion of adjacency to the edge domain. Figure 14.5(a) illustrates this neighborhood structure on an idealized regular lattice. The dark square (Region 7) has 4 neighbors (Regions 3, 6, 8, and 11, shaded light gray). In this case we have $w_{i+} = m_i$, the number of neighbors for region i, so the conditional distribution has mean $\bar{\phi}_i$, the average of the neighboring ϕ_j's, and variance inversely proportional to m_i.

Ma et al. [2010] proposed direct modeling in the edge domain, where the basic data elements are assumed to arise on the edge segments themselves. A CAR model for the edge segments is adopted to favor connected boundaries. For example, in Figure 14.5(b), the thick black boundary corresponding to edge (7,11) has six "neighboring" edges, highlighted as thick gray lines. Thus edge segments are adjacent if and only if they connect to one another. Note that edges (6,10) and (8,12) are adjacent to edges (7,11) even though these segments have no areal units in common.

14.11.2.2   Two-level CAR model

As mentioned in the previous subsection, the edge elements in the adjacency matrix can be modeled as random, potentially offering a natural framework for areal wombling. Since we prefer connected boundaries, given that a particular edge segment is part of the boundary,

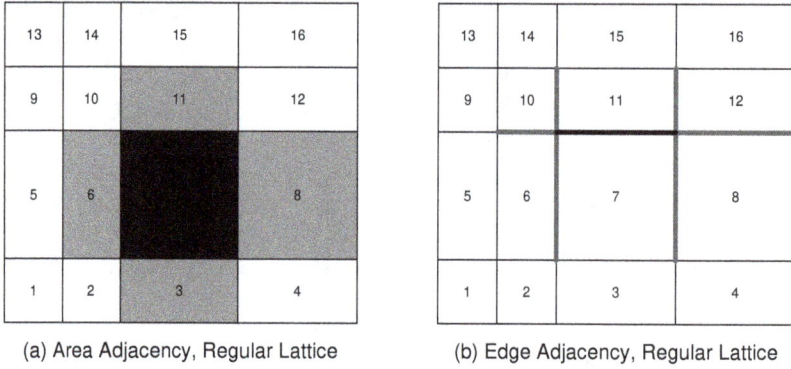

(a) Area Adjacency, Regular Lattice   (b) Edge Adjacency, Regular Lattice

Figure 14.5 *Illustration of area and edge domain neighborhood structures: (a) areal neighborhood structure, regular idealized map; (b) edge neighborhood structure, regular idealized map. In panel (b), note that edges (such as (6,10) and (7,11)) can be neighbors even though their index sets have no areal units in common.*

we would like our model to favor the inclusion of neighboring edge segments in the boundary as well. The standard, 0-1 adjacency-based CAR model appears naturally suited to this task: all we require is a *second* CAR model on the edge space (in addition to the original CAR on the areal unit space) with edge adjacency matrix W^* determined by the regional map as illustrated in Figure 14.5.

Similar to LC, the *two-level hierarchical CAR* (CAR2) model follows

$$Y_i \mid \boldsymbol{\beta}, \phi_i \overset{ind}{\sim} Poisson(\mu_i)$$
$$\text{where } \log(\mu_i) = \log(E_i) + \mathbf{x}_i' \boldsymbol{\beta} + \phi_i , \quad i = 1, \dots, n , \tag{14.33}$$
$$\text{and } p(\boldsymbol{\phi} \mid \tau_\phi, W) = C(\tau_\phi, W) \exp\left\{-\frac{\tau_\phi}{2} \boldsymbol{\phi}'(D_w - W)\boldsymbol{\phi}\right\} , \tag{14.34}$$

where $C(\tau_\phi, W)$ is an unknown normalizing constant. We then augment model (14.32) to

$$w_{ij} \mid p_{ij} \sim Bernoulli(p_{ij}) \text{ and } logit(p_{ij}) = \mathbf{z}_{ij}' \boldsymbol{\gamma} + \theta_{ij} , \tag{14.35}$$

where θ_{ij} is a spatial random effect associated with the edge separating areas i and j. Note that in our random W setting, if two regions i and j are neighbors (i.e., $w_{ij} = 1$) then they must also be adjacent, but the converse need not be true. Because of the symmetry of W, we need only be concerned with its upper triangle.

We reorder the w_{ij} into the singly-indexed vector $\boldsymbol{\xi} = (\xi_1, \dots \xi_K)^\mathrm{T}$, where K is the number of regional adjacencies in the map. We also carry out a corresponding reordering of the θ_{ij} into a vector $\boldsymbol{\psi} = (\psi_1, \dots \psi_K)^\mathrm{T}$. We then place the second level CAR as the prior on the edge random effects, i.e.,

$$\boldsymbol{\psi} \mid \tau_\psi \sim CAR(\tau_\psi, W^*) , \tag{14.36}$$

so that ψ_k has conditional distribution $N(\bar{\psi}_k , 1/(\tau_\psi w_{k+}^*))$, where $\tau_\psi > 0$ and W^* is the fixed $K \times K$ 0-1 adjacency matrix for $\boldsymbol{\psi}$, determined as in Figure 14.5(b).

Equations (14.34)–(14.36) comprise the CAR2 model. Vague conjugate gamma prior distributions for the precision hyperparameters τ_ϕ and τ_ψ, along with normal or flat priors for $\boldsymbol{\beta}$ and $\boldsymbol{\gamma}$, complete the hierarchical specification. The posterior distribution of the parameters can be estimated via MCMC techniques; the "inhomogeneous model" of Aykroyd [1998] is a Gaussian-response variant of the CAR2 for image data over a regular grid.

14.11.2.3 Site-edge (SE) models

A primary issue in implementing the CAR2 method is the determination of good "discrepancy" covariates $\mathbf{z}_{ij}$. Although $\boldsymbol{\gamma}$ is estimable even under a noninformative prior distribution, these second-level regression coefficients are often hard to estimate. Also, p_{ij} (and correspondingly w_{ij}) can be sensitive to the prior specification of $\boldsymbol{\gamma}$. Since the edge parameters enter the model only to specify the variances of the first-level random effects, they may be "too far away from the data" in the hierarchical model. This motivates a model with fewer levels or more direct modeling of edge effects.

As such, in this subsection we now consider "site-edge" (SE) models, where both the areal units (sites) *and* the edges between them contribute random effects to the mean structure. Let $\mathcal{G} = (\mathcal{S}, \mathcal{E})$, where $\mathcal{S} = \{1, \ldots, n\}$ is a set of sites/areas, and $\mathcal{E} = \{(i, j) : i \sim j\}$ is a set of edges, where $\sim$ indicates the symmetric "adjacency" relation. Suppose the data follow an exponential family likelihood, and let $\boldsymbol{\phi} = (\boldsymbol{\phi}^S, \boldsymbol{\phi}^E)$ be a vector of site- and edge-level effects, respectively. The general class of SE models is then given by $g(\mu_i) = f(\mathbf{x}_i, \boldsymbol{\beta}, \boldsymbol{\phi})$ and $\boldsymbol{\phi} = (\boldsymbol{\phi}^S, \boldsymbol{\phi}^E) \sim MRF$, where μ_i is the mean of the likelihood and $g(\cdot)$ stands for the canonical link function (e.g., the log link for the Poisson).

Here we take f to be linear, but this is not required for posterior propriety. To facilitate parameter identification and data information flow while encouraging sensible interaction between the areal and edge random effects, we now propose the hierarchical model,

$$
\begin{aligned}
Y_i \mid \boldsymbol{\beta}, \phi_i^S &\sim Poisson(\mu_i), \text{ with} \\
\log \mu_i &= \log E_i + \mathbf{x}_i' \boldsymbol{\beta} + \phi_i^S, \quad i = 1, \ldots, n\,,
\end{aligned}
$$

$$
p(\boldsymbol{\phi}^S \mid \boldsymbol{\phi}^E, \tau_\phi) = C(\tau_\phi, \boldsymbol{\phi}^E) \exp\left\{ -\frac{\tau_\phi}{2} \sum_{i \sim j} (1 - \phi_{ij}^E)(\phi_i^S - \phi_j^S)^2 \right\} \quad (14.37)
$$

$$
\text{and} \quad p(\boldsymbol{\phi}^E) \propto \exp\left\{ -\nu \sum_{ij \sim kl} \phi_{ij}^E \phi_{kl}^E \right\}, \quad (14.38)
$$

where $\phi_i^S \in \Re$ as before, but now $\phi_{ij}^E \in \{0, 1\}$ for all edges $(i, j) \in \mathcal{E}$. The conditional distribution in (14.37) is IAR, with $(1 - \phi_{ij}^E)$ playing the roles of the w_{ij} in (14.34). That is, $\phi_{ij}^E = 1$ if edge (i, j) is a boundary element, and 0 otherwise. Thus smoothing of neighboring ϕ_i^S and ϕ_j^S is only encouraged if there is no boundary between them. The prior for $\boldsymbol{\phi}^E$ in (14.38) is an *Ising* model with tuning parameter ν, often used in image restoration [e.g., Geman and Geman, 1984, p.725]. This prior yields a binary MRF that allows binary variables (the ϕ_{ij}^E's) to directly borrow strength across their neighbors, avoiding the need for a link function to introduce continuous spatial effects as in (14.35). The ν here is interpreted as measuring "binding strength" between the edges; smaller values of ν lead to more connected boundary elements, hence more separated areal units. We recommend comparing results under different fixed ν values. We refer to model (14.37)–(14.38) as a *SE-Ising* model. Strictly speaking, this model is not an MRF. Based upon our discussion in Chapter 4, an MRF would require not only $\phi_{ij}^E \phi_{kl}^E$, but also $(1 - \phi_{ij}^E)(1 - \phi_{kl}^E)$. The latter is absent in $p(\boldsymbol{\phi}^E)$. Nevertheless, the SE-Ising model is a legitimate probability model that supplies legitimate posteriors.

The improper CAR prior for $\boldsymbol{\phi}^S$ makes the joint prior $p(\boldsymbol{\phi}^S, \boldsymbol{\phi}^E) \equiv p(\boldsymbol{\phi}^S \mid \boldsymbol{\phi}^E) p(\boldsymbol{\phi}^E)$ improper regardless of the choice of $p(\boldsymbol{\phi}^E)$, but the joint *posterior* of these parameters will still be proper. To see this, note that $p(\boldsymbol{\phi}^S \mid \boldsymbol{\phi}^E, \mathbf{y})$ is proper under the usual improper CAR prior, and the discrete support of $p(\boldsymbol{\phi}^E)$ in (14.38) means it too is proper by construction. Since $p(\boldsymbol{\phi}^S, \boldsymbol{\phi}^E \mid \mathbf{y}) \propto p(\boldsymbol{\phi}^S \mid \boldsymbol{\phi}^E, \mathbf{y}) p(\boldsymbol{\phi}^E)$, the joint posterior is proper as well.

While the SE-Ising model is quite sensible for boundary analysis, it does not explicitly encourage long strings of connected boundary segments of the sort that would be needed to separate a hospice service area from an unserved area. As such, we further propose a *penalized* SE-Ising distribution

$$p(\phi^E) \propto \exp\left\{-\nu \sum_{ij\sim kl} \phi_{ij}^E \phi_{kl}^E + \kappa M\right\}, \tag{14.39}$$

where M is the number of strings of connected "on" edges ($\phi_{ij}^E = 1$) and $\kappa < 0$ is a second tuning parameter. Adding this additional penalty on edge arrangements that do not favor series of connected boundary segments helps to impose the kind of structure we want on our fitted boundaries.

Example 14.5 Minnesota Medicare hospice utilization data. Ma et al. [2010] consider the problem of identifying unserved areas of two particular cancer hospice systems in northeastern Minnesota in the state of Minnesota. Their data consist of ZIP code area-level Medicare beneficiary death counts from 2000 to 2002, as well as the number of these deaths among patients served by each hospice, both based on Medicare billing records. The use of ZIP code areas as our zonal system has inherent problems [Grubesic, 2008], including that it evolves over time at the whim of the US Postal Service. Ma et al. [2010] focus on the two hospice systems headquartered in the city of Duluth that serve rural northeast and north-central Minnesota, St. Luke's and St. Mary's/Duluth Clinic (SMDC). Figures 14.6(a) and (c) give raw data maps for St. Luke's, while those for SMDC appear in Figures 14.6(b) and (d). The first row of the figure maps the numbers of hospice deaths during the three-year period by ZIP code for the two hospice systems, while the second row maps the internally standardized mortality ratios, i.e., actual hospice death count divided by expected deaths (taken as proportional to the total Medicare death count) in each ZIP code area. Using either definition of "service," St. Luke's service area appears much smaller and more tightly clustered than SMDC's.

Determining the "service area" for each hospice system based only on the ZIP code area-specific hospice and total death counts is not as easy as simply drawing boundaries that separate ZIP code areas with zero counts from those with nonzero counts, since a patient's actual and billing addresses may not coincide. While calling every hospice in the country is infeasible, one might wonder if this would at least provide a "gold standard" to help validate a statistical model in a few cases. To check this, the two hospices were contacted and lists of ZIP code areas that each said it served were obtained. These results are shown in Figures 14.6(e) and (f) for St. Luke's and SMDC, respectively. The former self-reported service area appears congruent with the observed data for St. Luke's, a smaller hospice focusing on ZIP codes in or just northeast of Duluth (indicated with a "D" in panels (e) and (f)). But this is not the case for SMDC, a fast-developing hospice system with a home base in Duluth and two satellite bases in Grand Rapids ("GR") and Virginia ("V"), both north of Duluth. Comparing the SMDC self-report to the actual hospice death counts in Figure 14.6(b), it does not appear that its service region extended quite as far south or west as claimed during the three years covered by our data.

Ma et al. [2010] apply the LC, CAR2, SE-Ising, and penalized SE-Ising models to these data. With the latter three models, we use edge correction, while for all four methods, we use a *thresholding* approach designed to detect a boundary only when differences in the means of adjacent ZIP codes lie on opposite sides of some predetermined minimum service level. Their analysis considers a single covariate x_i, the intercentroidal (geodetic) distance from the patient's ZIP code area to the nearest relevant hospice home base ZIP code area (again see Figures 14.6(e) and (f) for locations). Since hospice services are provided in the

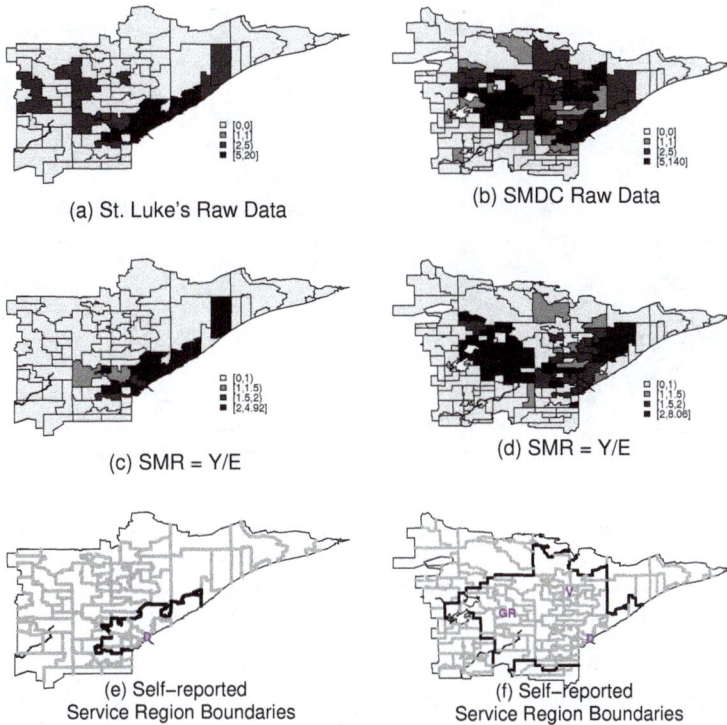

Figure 14.6 *St. Luke's and SMDC hospice system usage data, northeastern Minnesota ZIP codes: (a) St. Luke's hospice death counts; (b) SMDC hospice death counts; (c) St. Luke's internally standardized mortality ratios; (d) SMDC internally standardized mortality ratios; (e) St. Luke's self-reported service area boundaries; (f) SMDC self-reported service area boundaries. In (e) and (f), hospice home bases are marked (D = Duluth, GR = Grand Rapids, V = Virginia). Note that the self-reported service area for St. Luke's is entirely contained by the self-reported service area for SMDC.*

patient's home, increasing this distance should decrease the probability of that ZIP code area being served.

Ma et al. [2010] report that the SE-Ising models perform best with respect to DIC. The corresponding effects are significant: for example, using the penalized SE-Ising model, the posterior medians and 95% equal-tail credible sets for their regression slope of the distance between the centroids of the patient's ZIP code area to that of the nearest relevant hospice home base ZIP code area are –2.93 (–4.26,–1.58) for St. Luke's and –3.06 (–5.03, –1.50) for SMDC. The negative signs indicate that the farther away a ZIP code is from the nearest hospice home base, the less likely it is to be served by that hospice. Figure 14.7(a)–(b) show μ-based boundary maps for St. Luke's, while Figure 14.7(c)–(d) give them for SMDC. All four panels in these figures are based on absolute posterior medians of $\Delta_{\mu,ij} = \mu_i - \mu_j$. Panels (a) and (c) give results from the LC model, which appears to do a credible job for both hospices. However, even in the easier St. Luke's case, the LC map does include some "clutter" (identified boundary segments apparently internal to the service area) near the bottom of the map.

Panels (b) and (d) in Figure 14.7 give the hierarchically smoothed boundaries from the penalized SE-Ising model. For St. Luke's, the penalized SE-Ising boundaries in Figure 14.7(b) are quite satisfactory, showing essentially no internal clutter and offering a good match with the self-reported boundaries in Figure 14.6(e). However, for SMDC the

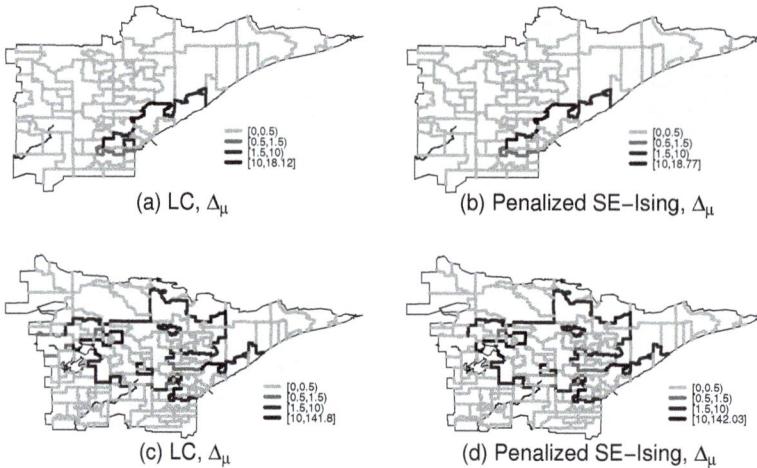

Figure 14.7 *Maps of St. Luke's and SMDC's service area boundaries: (a) St. Luke's service area boundaries given by the LC (usual CAR) model; (b) St. Luke's service area boundaries given by the penalized SE-Ising model; (c) SMDC's service area boundaries given by the LC (usual CAR) model; (d) SMDC's service area boundaries given by the penalized SE-Ising model.*

boundaries are less well connected, perhaps owing to the more complex nature of the data, which features a much larger service region and comprises three offices shown in Figure 14.6(f).

The wombled boundaries in Figures 14.7(c) and (d) are quite similar, as the μ_i are fairly well-estimated by any reasonable spatial model. However, none of our SMDC wombled maps provide a very good match to the self-reported boundaries in the south, since the data do not support the claim of service coverage there. This disagreement between our results and the self-report could be the result of reporting lags or migration in and out of service areas but is more likely due to the focus of some hospices (especially larger ones, like SMDC) on urban patients, at the expense of harder-to-reach rural ones. ∎

Ma et al. [2010] also consider boundaries based not on the $\Delta_{\mu,ij}$ but on the $\Delta_{\phi^S,ij} \equiv \phi_i^S - \phi_j^S$ or on the ϕ_{ij}^E themselves, to provide information about boundaries separating areas having significantly different *residuals*. This analysis reveals a diagonal edge separating urban Duluth from the rest of the service area, and a potential need for an indicator of whether a ZIP code area is a largely uninhabited, protected area to be added as an areal covariate $\mathbf{x}_i$ in (14.37)–(14.38).

14.11.3 Bayesian model-based areal wombling using False Discovery Rates (FDR)

Fully model-based areal wombling typically proceeds along one of two lines. The first builds stochastic models for the adjacency matrices [Lu et al., 2007, Ma et al., 2010, Lee and Mitchell, 2012, Liang, 2019, Corpas-Burgos and Martinez-Beneito, 2020] that, in principle, can accommodate information from explanatory variables in ascertaining the presence of edges. These approaches, including some discussed in the previous subsections, do not reckon with the multiplicity issues afflicting inference from marginal posterior estimates.

The second approach forgoes introducing variables in the adjacency but formulates the problem as one of multiple testing of different boundary hypotheses. A boundary hypothesis posits whether a pair of neighbors have equal spatial random effects or not. We want to test, for each pair of adjacent geographical regions in a map, a null model that posits equal spatial effects for the two regions against an alternative model that allows unconstrained, but spatially correlated, regional effects. As such, we will have as many hypotheses as there are geographical boundary segments on our map. Each hypothesis corresponds to a two-component mixture distribution that assigns a point mass to the null hypothesis and distributes the remaining mass to the alternative. Other model-based approaches aiming for full probabilistic uncertainty quantification have also received attention and include, but are not limited to, developments in Li et al. [2015], Hanson et al. [2015], Corpas-Burgos and Martinez-Beneito [2020], Gianella et al. [2023], Gao et al. [2023], and Pavani and Quintana [2024]. The idea is to detect edges using multiple comparisons by estimating the spatial random effects and, subsequently, testing how many such pairs are significantly different.

When multiple hypotheses are tested simultaneously, classical inference is usually concerned with controlling the overall Type I error rate. Benjamini and Hochberg [1995] introduced the False Discovery Rate (FDR) as an error criterion in multiple testing and described procedures to control it. The FDR is the expected proportion of falsely rejected null hypotheses among all rejected null hypotheses. Bayesian versions of FDR have been proposed and discussed by several authors including Storey [2002], Storey [2003], Genovese and Wasserman [2004], and Newton et al. [2004]. Müller et al. [2004] used a decision-theoretic perspective and set up decision problems that lead to the use of FDR-based rules and generalizations. Specifically for spatial data analysis, we note the work by Perone Pacifico et al. [2004] for testing an uncountable set of hypothesis tests on Gaussian random fields and FDR smoothing developed by Tansey et al. [2018] that exploits spatial structure within a multiple-testing problem. A Bayesian approach is appealing as the posterior distribution of spatial effects produce an exact Bayesian FDR [Müller et al., 2004] without recourse to asymptotic assumptions that are inappropriate in areal settings.

Formulating areal wombling as a multiple testing problem requires adjustments to modeling spatial effects that allow us to consider probabilities $P(\phi_i = \phi_j \mid i \sim j)$, where $\{\phi_i\}$ are spatial effects and $i \sim j$ indicates i and j are neighbors. Clearly, continuous priors for the ϕ_i's do not work since $P(\phi_i = \phi_j \mid i \sim j) = 0$. This prompted Li et al. [2012] to depart from the more traditional conditionally autoregressive (CAR) and simultaneous autoregressive (SAR) models and build classes of parametric univariate discrete spatial moving average models (SMA). Bayesian nonparametric models for areal data have been developed using "Areally-Referenced Dirichlet Process" (ARDP) and "Areally-Referenced Stick Breaking Process" (ARSB) that introduce spatial dependence using CAR models as baseline distributions of Dirichlet or more general stick-breaking processes [Li et al., 2015, Hanson et al., 2015]. Gao et al. [2023] extend them to analyze multiple correlated diseases and achieve probabilistic estimation for difference boundaries by embedding a multivariate areal model within a hierarchical Dirichlet process model.

The basic framework for using Bayesian nonparametric models to detect wombling boundaries is as follows. We consider a multivariate setting with q outcomes recorded in each areal unit. Let Y_{id} be the random variable denoting the measurement of outcome d in areal unit i. We construct a generalized linear mixed model using a distribution from the exponential family with a canonical link,

$$g(\mathrm{E}(Y_{id})) = \mathbf{x}_{id}^{\mathsf{T}}\boldsymbol{\beta}_d + \phi_{id}, \quad \text{for} \quad i = 1, \ldots, n \,; \ d = 1, \ldots, q \,, \tag{14.40}$$

where $\mathbf{x}_{id}$ is a vector of explanatory variables and $\boldsymbol{\beta}_d$ is the corresponding vector of slopes specific to outcome d, and ϕ_{id} is the spatial random effect for outcome d in areal unit i. For detecting difference boundaries, we seek $P(\phi_{id} \neq \phi_{jd} \mid \mathbf{y}, i \sim j)$, where $\mathbf{y}$ is the collection of observed y_{id}'s and $i \sim j$ means that regions i and j are neighbors.

The Dirichlet Process (DP) offers a natural way to cluster regions by endowing possibly positive mass on $P(\phi_{id} = \phi_{jd} \mid \mathbf{y}, i \sim j)$. The DP is especially appealing here as it still captures spatial associations in ϕ_{id} through the baseline covariance (or precision) matrix. More generally, stick-breaking process priors can achieve the same result [Li et al., 2015, Hanson et al., 2015]. Let $\boldsymbol{\phi} = (\boldsymbol{\phi}_1^{\mathrm{T}}, \ldots, \boldsymbol{\phi}_q^{\mathrm{T}})^{\mathrm{T}}$ be the $N \times 1$, where $N = nq$ and $\boldsymbol{\phi}_d = (\phi_{1d}, \ldots, \phi_{nd})^{\mathrm{T}}$ for each $d = 1, \ldots, q$ and let $\{1, \ldots, n, \ldots, (q-1)n + 1, \ldots, N\}$ be an enumeration of the pairwise (i, d) indices corresponding to the ordering of $\boldsymbol{\phi}$. We model $\boldsymbol{\phi} \sim G_N$, where G_N is an unknown distribution further specified by $G_N = \sum_{u_1, \ldots, u_N} \pi_{u_1, \ldots, u_N} \delta_{(\theta_{u_1}, \ldots, \theta_{u_N})}$, where $u_1, \ldots, u_N \in \{1, \ldots, K\}$, $\theta_k \mid \tau \overset{iid}{\sim} N\left(0, \tau^2\right)$ for $k = 1, \ldots, K$, $\delta_{(\theta_{u_1}, \ldots, \theta_{u_N})}$ is the Dirac measure located at $(\theta_{u_1}, \ldots, \theta_{u_N})$ and

$$\pi_{u_1, \ldots, u_N} = P\left(\sum_{k=1}^{u_1 - 1} p_k < F^{(1)}(\gamma_1) < \sum_{k=1}^{u_1} p_k, \ldots, \sum_{k=1}^{u_N - 1} p_k < F^{(N)}(\gamma_N) < \sum_{k=1}^{u_N} p_k\right) \quad (14.41)$$

with $\boldsymbol{\gamma} = (\gamma_1, \ldots, \gamma_N)^{\mathrm{T}} \sim N(\mathbf{0}, \Sigma_\gamma)$. The choice of the independent and identical distributions for the atoms base distribution accommodates the possibility of ties across both regions and outcomes. For example, if we were to assign different means to different outcomes, we could still accommodate ties within each outcome, wherein different regions' random effects assume the same value. However, we would be unable to establish ties across diseases as the underlying distribution from which the atoms are sampled would differ.

Regression coefficients and variance parameters can follow customary specifications such as $\boldsymbol{\beta}_d \overset{ind}{\sim} N(\mathbf{0}, \Sigma_{\beta_d})$ and $\tau^2 \sim IG(a, b)$. Stick-breaking weights are customarily defined as $p_1 = V_1$ and $p_j = V_j \prod_{k<j}(1 - V_k)$ for $j = 2, \ldots, K$, where $V_k \overset{iid}{\sim} \mathrm{Beta}(1, \alpha)$, while $F^{(i)}(\cdot)$'s are cumulative distribution functions of the marginal distribution of γ_i for $i = 1, \ldots, N$. The weights in the distribution G_N are constructed by utilizing the marginal cumulative distribution function (CDF) of the elements of $\boldsymbol{\gamma}$. There is ample scope for richness and flexibility here [see, e.g., Petrone et al., 2009, Rodríguez et al., 2010, Hjort et al., 2010, and several references therein on stock-breaking processes and their adaptations for modeling]. Marginally, each $F^{(\cdot)}(\gamma.)$ is $\mathcal{U}(0, 1)$ but they are dependent through the joint distribution of $\boldsymbol{\gamma}$, and the distribution of each spatial random effect $\phi.$ is a regular univariate DP, with the spatial dependence between these DPs introduced using a copula representation for the weights. Spatial dependence is introduced in Σ_γ, which can be specified using any appropriate multivariate spatial covariance matrix. Gao et al. [2023] build Σ_γ using MCAR (Chapter 11) and constructions of multivariate DAGAR (Section 4.5) distributions using coregionalization, while Aiello and Banerjee [2023] incorporate graphical models for multivariate outcomes alongside spatial associations in Σ_γ.

The above modeling framework equips us with the posterior probability of $\phi_{id} = \phi_{jd'}$ and its complement $\phi_{id'} \neq \phi_{jd'}$ for region-outcome pairs (i, d) and (j, d'). We compute these probabilities for each pair of adjacent regions, i.e., $i \sim j$, and each outcome d and designate the edge (i, j) as a difference boundary if the posterior probability of $\phi_{id} \neq \phi_{jd'}$ given $\mathbf{y}$ surpasses a predefined threshold t. Joint modeling in the multivariate outcome settings enable us to report the *cross-difference* boundary based upon $v_{(i,d)(j,d')} = P(\phi_{id} \neq \phi_{jd'}, \phi_{id'} \neq \phi_{jd} \mid \mathbf{y})$. We derive a threshold t that controls the false discovery rate (FDR) below a level $\zeta = 0.05$. To this end, we define

$$\mathrm{FDR}_{d,d'}(t) = \frac{\sum_{i \sim j} I(\phi_{id} = \phi_{jd'}) I(v_{(i,d)(j,d')} > t)}{\sum_{i \sim j} I(v_{(i,d)(j,d')} > t)} \quad (14.42)$$

every disease pair d and d'. The quantity in (14.42) is evaluated as

$$\widehat{\mathrm{FDR}}_{d,d'}(t) = \mathrm{E}[\mathrm{FDR}_{d,d'} \mid \mathbf{y}] = \frac{\sum_{i \sim j}(1 - v_{(i,d)(j,d')}) I(v_{(i,d)(j,d')} > t)}{\sum_{i \sim j} I(v_{(i,d)(j,d')} > t)}. \quad (14.43)$$

Following Müller et al. [2004], we define

$$t^* = \sup\left\{t : \widehat{\mathrm{FDR}}_{d,d'}(t) \leq \zeta\right\}, \quad (14.44)$$

which is based upon the optimal decision that minimizes the estimated false negative rate (FNR), $\widehat{\text{FNR}}_{d,d'}(t) = \frac{\sum_{i\sim j} v_{(i,d)(j,d')}(1-I(v_{(i,d)(j,d')}>t))}{m-\sum_{i\sim j} I(v_{(i,d)(j,d')}>t)}$, where m is the total number of geographic boundaries, subject to $\widehat{\text{FDR}} \leq \zeta$ [also see Sun et al., 2015, who proffer a similar approach]. This decision is predicated on a bivariate loss function $L_{2R} = (\widehat{\text{FDR}}, \widehat{\text{FNR}})$. The single disease setting is obtained by substituting $v_{i,j}^{(d)} = P(\phi_{id} \neq \phi_{jd'}, \phi_{id'} \neq \phi_{jd} \mid \mathbf{y})$ in the above expressions.

Li et al. [2012, 2015], Gao et al. [2023], and Aiello and Banerjee [2023] pursue Bayesian nonparametric models to investigate the effectiveness of FDR-based decision rules in detecting wombling boundaries for areal data, with the latter two articles focusing on multivariate areal data. Wu and Banerjee [2024] also use FDR to detect boundaries but define difference boundaries differently from the Bayesian nonparametric settings. These authors pursue multiple testing within a Bayesian linear regression framework to exploit analytically accessible posterior distributions and accelerate computation significantly over Bayesian nonparametric approaches. This approach uses continuous priors on spatial random effects (to exploit closed-form distributions) and defines difference boundaries in terms of the difference between neighboring spatial effects exceeding a threshold. This threshold is determined using an entropy loss function. This approach scales effectively for analyzing large datasets.

14.12   Wombling with point process data

In disease mapping and public health, a spatial layer for which boundary analysis would be of considerable interest is the pattern of disease incidence. In particular, we would seek to identify the transition from areas with low incidence to areas with elevated incidence. For cases aggregated to counts for areal units, e.g., census blocks or zip codes (in order to protect confidentiality), this would require obtaining standardized incidence rates for the units. Wombling for such data is discussed in Jacquez and Greiling [2003] as well as in the previous section.

With observations as points, what would wombling for a spatial point pattern mean? Following the development in Chapter 8, we assume that the point pattern is driven by an intensity surface, $\lambda(\mathbf{s})$. So, wombling for the observed point pattern would be achieved by wombling a realization of $\lambda(\mathbf{s})$, analogous to wombling a realization of a Gaussian process. Hence, with the previous sections in this chapter along with the machinery of Chapter 8, we have the tools to go forward. That is, for an LGCP, with an intensity of the form, $\lambda(\mathbf{s}) = \exp(\mathbf{X}(\mathbf{s})^{\mathsf{T}}\boldsymbol{\beta} + Z(\mathbf{s}))$ (see 8.4.2) with $Z(\mathbf{s})$ a Gaussian process, upon model fitting, we can study gradient behavior of $Z(\mathbf{s})$, hence, with smooth $\mathbf{X}(\mathbf{s})^{\mathsf{T}}$, for $\log\lambda(\mathbf{s})$. Since, from Section 14.3 the gradient for $g(Z(\mathbf{s}))$, with g continuous and monotonic increasing, results in $D_{\mathbf{u}}g(Z(\mathbf{s})) = g'(Z(\mathbf{s}))D_{\mathbf{u}}Z(\mathbf{s})$, we can directly explore the gradient surface of $\exp(Z(\mathbf{s}))$ and therefore $\lambda(\mathbf{s})$, as well. Details, including an extension to *marked point processes* is discussed in Liang et al. [2009a].

14.13   Directional velocities

Here, we briefly review an extension of the use of spatial gradients to the development of velocities. We draw on the work in Schliep et al. [2015]. The context is with regard to the velocity of climate change. This velocity is defined as an instantaneous rate of change needed to maintain a constant climate. It is developed as the ratio of the temporal gradient of climate change over the spatial gradient of climate change. A fully stochastic hierarchical model is employed that allows for the inherent relationship between climate, time, and space. Space-time processes are employed to capture the spatial correlation in both the climate variable and the rate of change in climate over time. Directional derivative processes yield spatial and temporal gradients and, thus, the resulting velocities of a climate variable.

The gradients and velocities can be obtained at an arbitrary location and direction for any given time. In fact, maximum gradients and their directions can be obtained, hence minimum velocities. The work in Schliep et al. [2015] applied the modeling to average annual temperature across the eastern United States for the years 1963 to 2012.

Working with temperature, the notion of *velocity of climate change* was introduced in the work of Loarie et al. [2009]. The basic idea is to derive an index of velocity of temperature change in km/yr over a large spatial region. It is developed from spatial gradients of temperature change in °C/km and temporal rates of temperature change in °C/yr. Dividing the latter by the former produces a *velocity*. The ratio is interpreted as the "instantaneous local velocity along the earth's surface needed to maintain constant temperatures" [Loarie et al., 2009].

14.13.1 Modeling details

Schliep et al. [2015] model annual average temperature using a linear mixed model with spatially correlated random effects. The model is inherently hierarchical as it combines two sources of data, annual average temperature and elevation. Let $T(x,y,t)$ be the annual average temperature for location $\mathbf{s} = (x,y)$ at time t where x is the easting coordinate and y is the northing coordinate. Further, let $E(x,y)$ be elevation at location (x,y). Then, model $T(x,y,t)$ and $\mathrm{E}(x,y)$ as

$$T(x,y,t) = \beta_0 + \beta_1 t + \beta_2 y + \beta_3 Z(x,y) + \beta_0(x,y) + \beta_1(x,y)t + \epsilon(x,y,t) \qquad (14.45a)$$

$$E(x,y) = \mu + Z(x,y) + \eta(x,y) \qquad (14.45b)$$

where $\epsilon(x,y,t) \sim N(0,\sigma_T^2)$ and $\eta(x,y) \sim N(0,\sigma_E^2)$.

Both $\beta_0(x,y)$ and $\beta_1(x,y)$ in (14.45a) are spatial random effects that account for the remaining spatial variation in annual average temperature and rate of change in annual temperature over time, respectively, that is not accounted for by latitude and elevation. In other words, we have spatially varying intercepts and slopes in our regression on temperature (see Section 10.6). Schliep et al. [2015] do not introduce a longitudinal term with a coefficient as the temperature gradient in longitude is not significant over the study region. The spatial random effect, $Z(x,y)$, in (14.45b) accounts for the spatial variation in elevation and enters as a covariate not connected to t in the temperature model. To enable desired differentiation, the three spatial random effects, $\beta_0(x,y)$, $\beta_1(x,y)$, and $Z(x,y)$ must be sufficiently smooth differentiable surfaces. As such, these processes allow for gradients and velocities to be calculable at arbitrary locations. The specification of $E(x,y)$ does not assume the observed elevation surface to be smooth, but rather, specifies the "centering" surface, $Z(x,y)$ to be smooth.

There are two approaches for modeling the three spatial processes, $\beta_0(x,y)$, $\beta_1(x,y)$, and $Z(x,y)$ that enable stochastic gradient calculation. One is to model them as customary Gaussian processes, possibly dependent as in the previous sections of this chapter. See Schliep and Gelfand [2019] for an illustration of this approach for velocity calculation in the context of point patterns of crime, using the nearest neighbor Gaussian process [Datta et al., 2016a]. A second approach is to model them using dimension reduction, i.e., to express the surfaces as parametric linear transformations of a finite set of random variables at fixed locations. With suitably differentiable functions, such a representation enables explicit gradient calculation. Schliep et al. [2015] adopt this latter approach due to computational necessity as they consider temperature at more than 21,000 gridded locations. They use the predictive process for dimension reduction for each of the three spatial processes (Section 13.4). Anticipating customary association between slope and intercept, they employ coregionalization to connect the slope and intercept processes in the mean temperature model. The latent elevation process is treated as independent of these two processes.

14.13.2   Gradient and velocity calculation

Turning to the needed gradient calculations, for the annual temperature model, the temporal gradient of temperature change is the *expected* change in temperature per year whereas the spatial gradient is the *expected* change in temperature per kilometer. A spatial gradient can be calculated in an arbitrary direction and time. Attractively, the gradient in any direction can be calculated from the gradient in the easting direction ($\partial E(T(x,y,t))/\partial x$) and in the northing direction ($\partial E(T(x,y,t))/\partial y$). Let

$$\nabla E(T(x,y,t)) = \begin{pmatrix} \partial E(T(x,y,t))/\partial x \\ \partial E(T(x,y,t))/\partial y \end{pmatrix}. \tag{14.46}$$

Then, the gradient in the direction $\mathbf{u}$, where $\mathbf{u} = (u_x, u_y)'$ is a unit vector, is $\nabla E(T(x,y,t))^\mathsf{T}\mathbf{u}$. Evidently, the gradient in direction $-\mathbf{u}$ is the negative of the gradient in direction $\mathbf{u}$ but the magnitudes will be the same. Furthermore, following the calculation in Section 14.3, the direction of the maximum gradient is $\nabla E(T(x,y,t))/\|\nabla ET((x,y,t))\|$ and the magnitude of the maximum gradient is $\|\nabla E(T(x,y,t))\|$. The explicit gradient calculations using predictive processes are supplied in Schliep et al. [2015].

Finally, velocities can be obtained from the spatial and temporal gradients. A climate velocity for annual temperature is the ratio of the temporal gradient to the spatial gradient and is measured in km/yr. The velocity in the direction $\mathbf{u}$ is

$$\frac{\frac{\partial E(T(x,y,t))}{\partial t}}{\nabla E(T(x,y,t))^\mathsf{T}\mathbf{u}} = \frac{\frac{\partial E(T(x,y,t))}{\partial t}}{\frac{\partial E(T(x,y,t))}{\partial x}u_1 + \frac{\partial E(T(x,y,t))}{\partial y}u_2}. \tag{14.47}$$

In an absolute sense, the minimum velocity is the velocity in the direction of the maximum gradient and is $\frac{\partial E(T(x,y,t))/\partial t}{\|\nabla E(T(x,y,t))\|}$. Ecologically, the direction and magnitude of the minimum velocity are meaningful as they capture what will be required for survival at that location and time. The direction of maximum velocity is not meaningful mathematically nor ecologically as there will be a direction(s) where the spatial gradient of temperature is essentially 0, yielding essentially infinite velocity in that direction.

By way of summary, large velocities indicate large changes in temperature over time relative to changes in temperature across space. Small velocities, on the other hand, indicate large changes in temperature across space relative to changes in time. Locations with a steep elevation gradient may see very small climate velocities since short distances lead to large changes in elevation, and thus, may result in large changes in temperature.

14.14   Computer tutorials

Computer programs illustrating Bayesian wombling for point-referenced and areal data in R and its interfaces with Bayesian modeling environments are supplied in the folder titled Chapter 14 from https://github.com/sudiptobanerjee/BGC_2023.

14.15   Exercises

1. Let $X(t) \sim GP(0, K(\cdot,\cdot)$ be a Gaussian process on the real line with mean 0 and covariance function $K(t,t')$. Define the finite-difference process,

$$X_h(t) = \frac{X(t+h) - X(t)}{h},$$

where h is a fixed positive real number.

(a) Prove that $X_h(t)$ is also a Gaussian process. What is its covariance function?

(b) Establish conditions on $K(\cdot, \cdot)$ for $\lim_{h \to 0} \mathrm{Cov}(X_h(t), X_h(t'))$ to exist. Prove that if the above limit exists, then the function is a valid covariance function.

(c) Are the above conditions sufficient for $dX(t)/dt = \lim_{h \to 0} X_h(t)$ to be a valid Gaussian process?

2. Let $\{Y(\mathbf{s}) : \mathbf{s} \in \mathbb{R}^d\}$ be a random field with an isotropic covariance function $K(\mathbf{s}, \mathbf{s}') = \sigma^2 \exp(-\phi \|\boldsymbol{\Delta}\|)$, where $\boldsymbol{\Delta} = \mathbf{s} - \mathbf{s}'$. Let $\{\mathbf{s}_1^*, \ldots, \mathbf{s}_m^*\}$ be a set of m fixed spatial locations. Let $\tilde{Y}(\mathbf{s})$ be the *predictive process* (Section 13.4.1),

$$\tilde{Y}(\mathbf{s}) = \mathrm{E}[Y(\mathbf{s}) \mid \mathbf{y}^*] = \mathbf{c}(\mathbf{s})^{\mathsf{T}} C^{*-1} \mathbf{y}^* \,,$$

where $\mathbf{y}^*$ is $m \times 1$ with elements $Y(\mathbf{s}_j^*)$, $\mathbf{c}(\mathbf{s})$ is $m \times 1$ with entries $C(\mathbf{s}, \mathbf{s}_j^*)$ and C^* is $m \times m$ with entries $K(\mathbf{s}_i^*, \mathbf{s}_j^*)$. Describe the smoothness properties of $\tilde{Y}(\mathbf{s})$ in the mean-square sense. Is it mean-square continuous? Mean-square differentiable? Is it "too" smooth?

3. Let $\mathcal{R} = \{\mathbf{s}_1^*, \ldots, \mathbf{s}_r^*\}$ be a set of r spatial locations forming a *reference set* toward the construction of the NNGP described in Section 13.6.2. Define *history sets* $H(\mathbf{s}_i^*)$ so that $H(\mathbf{s}_1^*)$ is empty and $H(\mathbf{s}_i^*) = \{\mathbf{s}_1^*, \mathbf{s}_2^*, \ldots, \mathbf{s}_{i-1}^*\}$ for $i = 2, 3, \ldots, r$. Let $m << r$ be a fixed positive integer and define *neighbor sets*

$$N(\mathbf{s}) = \begin{cases} \text{empty set if } \mathbf{s} = \mathbf{s}_1 \\ H(\mathbf{s}_i^*) = \{\mathbf{s}_1^*, \mathbf{s}_2^*, \ldots, \mathbf{s}_{i-1}^*\} \text{ if } \mathbf{s} = \mathbf{s}_i^*, \text{ for } i = 2, 3, \ldots, m \\ m \text{ nearest neighbors of } \mathbf{s}_i^* \text{ among } H(\mathbf{s}_i^*) \text{ if } \mathbf{s} = \mathbf{s}_i^*, \text{ for } i = m+1, \ldots, n \\ m \text{ nearest neighbors of } \mathbf{s} \text{ among } \mathcal{R} \text{ if } \mathbf{s} \notin \mathcal{R} \,. \end{cases}$$

Let $Y(\mathbf{s}) \sim GP(0, K(\mathbf{s}, \mathbf{s}'))$ be a spatial process with covariance function $K(\mathbf{s}, \mathbf{s}')$ and let $\mathbf{y}_{\mathcal{R}} = (y(\mathbf{s}_1^*), \ldots, y(\mathbf{s}_r^*))^{\mathsf{T}} \sim N(\mathbf{0}, \tilde{K})$, where $\tilde{K}$ is the covariance matrix derived from the Vecchia approximation $\prod_{i=1}^{r} p(Y(\mathbf{s}_i^*) \mid \mathbf{Y}_{N(\mathbf{s}_i^*)})$ (recall (13.40)). For any two finite sets of locations A and B with n_A and n_B locations, respectively, let $K(A, B)$ denote the $n_A \times n_B$ matrix with (i, j)th element $K(\mathbf{a}_i, \mathbf{b}_j)$, where $\mathbf{a}_i \in A$ and $\mathbf{b}_j \in B$. Construct the NNGP

$$\tilde{Y}(\mathbf{s}) = \sum_{i=1}^{r} a_i(\mathbf{s}) y(\mathbf{s}_j^*) + \eta(\mathbf{s}) \,, \quad \eta(\mathbf{s}) \overset{ind}{\sim} N(0, \delta^2(\mathbf{s})) \tag{14.48}$$

where $\delta^2(\mathbf{s}) = K(\mathbf{s}, \mathbf{s}) - K(\mathbf{s}, N(\mathbf{s})) K(N(\mathbf{s}), N(\mathbf{s}))^{-1} K(N(\mathbf{s}), \mathbf{s})$,

$$a_i(\mathbf{s}) = \begin{cases} 0 & \text{if } \mathbf{s} \notin N(\mathbf{s}_i^*) \\ \left[K(N(\mathbf{s}), N(\mathbf{s}))^{-1} K(N(\mathbf{s}), \mathbf{s}) \right]_i & \text{if } \mathbf{s} \in N(\mathbf{s}_i^*) \,, \end{cases}$$

$\left[K(N(\mathbf{s}), N(\mathbf{s}))^{-1} K(N(\mathbf{s}), \mathbf{s}) \right]_i$ denotes the i-th element of the solution vector of $K(N(\mathbf{s}), N(\mathbf{s})) \mathbf{x} = K(N(\mathbf{s}), \mathbf{s})$. Investigate the smoothness of $\tilde{Y}(\mathbf{s})$ defined in (14.48) in terms of mean-square continuity or differentiability. What impact, if any, do the smoothness properties of $K(\mathbf{s}, \mathbf{s}')$ have on the smoothness of $\tilde{Y}(\mathbf{s})$ constructed in (14.48)?

4. Let $\{Y(\mathbf{s}) : \mathbf{s} \in \mathbb{R}^d\}$ be a random field with a covariance function $K(\mathbf{s}, \mathbf{s}')$. Define,

$$Y_{\mathbf{u}, \mathbf{v}, h}^{(2)}(\mathbf{s}_0) = (Y(\mathbf{s}_0 + h(\mathbf{u} + \mathbf{v})) - Y(\mathbf{s}_0 + h\mathbf{u}) - Y(\mathbf{s}_0 + h\mathbf{v}) + Y(\mathbf{s}_0))/h^2 \,,$$

where $h > 0$ is a positive real number, $\mathbf{u}$ and $\mathbf{v}$ are unit vectors. Using a second-order Taylor expansion of $Y(\mathbf{s}_0 + h\mathbf{u})$ (i.e., extend (14.1) to a second order-term), investigate the existence of the process $D_{\mathbf{u}, \mathbf{v}}^{(2)} Y(\mathbf{s}_0) = \lim_{h \to 0} Y_{\mathbf{u}, \mathbf{v}, h}^{(2)}(\mathbf{s}_0)$. If the process exists, show that it can be expressed as $D_{\mathbf{u}, \mathbf{v}}^{(2)} Y(\mathbf{s}_0) = \mathbf{u}^{\mathsf{T}} [\nabla_Y^2(\mathbf{s}_0)] \mathbf{v}$, where $[\nabla_Y^2(\mathbf{s}_0)]$ is a $d \times d$ symmetric matrix. Describe the elements of $[\nabla_Y^2(\mathbf{s}_0)]$ in terms of partial derivatives of $Y(\mathbf{s}_0)$.

5. Let $\{Y(\mathbf{s}) : \mathbf{s} \in \mathbb{R}^d\}$ be a stationary random field with a covariance function $K(\mathbf{s}, \mathbf{s}') = K(\boldsymbol{\Delta})$, where $\boldsymbol{\Delta} = \mathbf{s} - \mathbf{s}'$. Define

$$\mathbf{Z}(\mathbf{s}) = \begin{pmatrix} Y(\mathbf{s}) \\ \nabla Y(\mathbf{s}) \\ \mathrm{vech}(\nabla^2 Y(\mathbf{s})) \end{pmatrix},$$

where $\mathrm{vech}(\nabla^2 Y(\mathbf{s}))$ is the half-vectorization operator on the symmetric matrix $[\nabla^2_Y(\mathbf{s}_0)]$ found in (4), which stacks the columns of the lower-triangular part of $[\nabla^2_Y(\mathbf{s}_0)]$.

(a) What is the dimension of the vector $\mathbf{Z}(\mathbf{s})$?

(b) Assuming all required derivatives of $K(\boldsymbol{\Delta})$ to exist, write down the cross-covariance function of $\mathbf{Z}(\mathbf{s})$. Is $\mathbf{Z}(\mathbf{s})$ stationary?

6. Consider the dataset https://github.com/sudiptobanerjee/BGC_2023/data/rate_lung_cancer_with_smoking.csv which includes the age-adjusted lung cancer rates reported from the $n = 58$ counties in California. Fit the following model with Y_i as cancer rate x_i as the smoking rate for county i,

$$Y_i = \beta_0 + \beta_1 x_i + \phi_i + \epsilon_i, \quad \epsilon_i \overset{iid}{\sim} N(0, \sigma^2), \quad i = 1, \ldots, n,$$

where the spatial random effects $\boldsymbol{\phi} = (\phi_1, \ldots, \phi_n)^\mathrm{T}$ follow a CAR model, i.e., $\boldsymbol{\phi} \sim N(\mathbf{0}, \sigma^2(D - \rho W)^{-1})$, where $W = [w_{ij}]$ is the $n \times n$ binary adjacency matrix and D is $n \times n$ diagonal with a number of neighbors of each region along the diagonal; $\tau^2 \sim IG(a_\tau, b_\tau)$ and $\sigma^2 \sim IG(a_\sigma, b_\sigma)$ with $a_\tau = a_\sigma = 2$ and $b_\sigma = b_\tau = 1.0$ and fix $\rho = 0.98$. Define the Boundary Likelihood Value (BLV) $\Delta_{ij} = |\phi_i - \phi_j|$ for all neighboring pairs (i, j). Carry out crisp and fuzzy wombling using the posterior distribution of this BLV.

7. Repeat the analysis of the dataset https://github.com/sudiptobanerjee/BGC_2023/data/rate_lung_cancer_with_smoking.csv using a univariate Bayesian nonparametric model derived from (14.40) with a Gaussian distribution $Y_i \sim N(\mu_i, \sigma^2)$, $\mu_i = \mathbf{x}_i^\mathrm{T} \boldsymbol{\beta} + \phi_i$, $\boldsymbol{\beta} \sim N(\mathbf{0}, (\sigma^2/n_0)I)$, where n_0 is a fixed real number to control the precision of the prior, $\sigma^2 \sim IG(a_\sigma, b_\sigma)$ and the nonparametric prior on $\{\phi_i\}$ described in the construction of (14.41) but setting $d = 1$ (for only one outcome). Use CAR or DAGAR for the spatial dependence in Σ_γ. Use (14.43) and (14.44) to obtain difference boundaries across the counties in California.

8. Extract data for four cancers: (i) lung, (ii) esophageal, (iii) larynx, and (iv) colorectal for the 58 counties of California from the data set in https://github.com/sudiptobanerjee/BGC_2023/data/age_adjusted.csv. Extract the smoking rates for the counties in California from https://github.com/sudiptobanerjee/BGC_2023/data/smoking.csv. Write a program to implement the multivariate areal stick-breaking model described in (14.41) specifying a Gaussian likelihood for (14.40) to analyze the data. Use MCAR or MDAGAR for spatial dependence in Σ_γ, Using (14.43) and (14.44) obtain the cross-difference boundaries for the four cancers across the counties in California.

Chapter 15

Spatial survival models

The use of survival models involving a random effect or "frailty" term is now common. Usually, the random effects are assumed to represent different clusters, and clusters are assumed to be independent. In this chapter, we consider random effects corresponding to clusters that are spatially arranged, such as clinical sites or geographical regions. That is, we might suspect that random effects corresponding to strata in closer proximity to each other might also be similar in magnitude.

Survival models have a long history in the biostatistical and medical literature [see, e.g., Cox and Oakes, 1984]. Very often, time-to-event data will be grouped into *strata* (or *clusters*), such as clinical sites, geographic regions, and so on. In this setting, a hierarchical modeling approach using stratum-specific parameters called *frailties* is often appropriate. Introduced by Vaupel et al. [1979], this is a mixed model with random effects (the frailties) that correspond to a stratum's overall health status.

To illustrate, let t_{ij} be the time to death or censoring for subject j in stratum i, $j = 1, \ldots, n_i$, $i = 1, \ldots, I$. Let $\mathbf{x}_{ij}$ be a vector of individual-specific covariates. The usual assumption of proportional hazards $h(t_{ij}; \mathbf{x}_{ij})$ enables models of the form

$$h(t_{ij}; \mathbf{x}_{ij}) = h_0(t_{ij}) \exp(\boldsymbol{\beta}^\mathrm{T} \mathbf{x}_{ij}) , \tag{15.1}$$

where h_0 is the *baseline hazard*, which is affected only multiplicatively by the exponential term involving the covariates. In the frailty setting, model (15.1) is extended to

$$\begin{aligned} h(t_{ij}; x_{ij}) &= h_0(t_{ij}) \, \omega_i \exp(\boldsymbol{\beta}^\mathrm{T} \mathbf{x}_{ij}) \\ &= h_0(t_{ij}) \exp(\boldsymbol{\beta}^\mathrm{T} \mathbf{x}_{ij} + W_i) , \end{aligned} \tag{15.2}$$

where $W_i \equiv \log \omega_i$ is the stratum-specific frailty term, designed to capture differences among the strata. Typically a simple i.i.d. specification for the W_i is assumed, e.g.,

$$W_i \overset{iid}{\sim} N(0, \sigma^2) . \tag{15.3}$$

With the advent of MCMC computational methods, the Bayesian approach to fitting hierarchical frailty models such as these has become increasingly popular [see, e.g., Carlin and Louis, 2008, Sec. 7.6]. Perhaps the simplest approach is to assume a *parametric* form for the baseline hazard h_0. While a variety of choices (gamma, lognormal, etc.) have been explored in the literature, in Section 15.1 we adopt the Weibull, which seems to represent a good trade-off between simplicity and flexibility. This then produces

$$h(t_{ij}; x_{ij}) = \rho t_{ij}^{\rho-1} \exp(\boldsymbol{\beta}^\mathrm{T} \mathbf{x}_{ij} + W_i) . \tag{15.4}$$

Now, placing prior distributions on $\rho, \boldsymbol{\beta}$, and σ^2 completes the Bayesian model specification. Such models are by now a standard part of the literature, and easily fit (at least in the univariate case) using a Bayesian modeling language (BUGS/JAGS/NIMBLE). Carlin and Hodges [1999] consider further extending the model (15.4) to allow stratum-specific baseline

DOI: 10.1201/9781003401728-15

hazards, i.e., by replacing ρ by ρ_i. MCMC fitting is again routine given distribution for these new random effects, say, $\rho_i \overset{iid}{\sim} Gamma(\alpha, 1/\alpha)$, so that the ρ_i have mean 1 (corresponding to a constant hazard over time) but variance $1/\alpha$.

A richer but somewhat more complex alternative is to model the baseline hazard *non-parametrically*. In this case, letting γ_{ij} be a death indicator (0 if alive, 1 if dead) for patient ij, we may write the likelihood for our model $L(\beta, \mathbf{W}; \mathbf{t}, \mathbf{x}, \boldsymbol{\gamma})$ generically as

$$\prod_{i=1}^{I} \prod_{j=1}^{n_i} \{h(t_{ij}; \mathbf{x}_{ij})\}^{\gamma_{ij}} \exp\{-H_{0i}(t_{ij}) \exp(\boldsymbol{\beta}^{\mathrm{T}} \mathbf{x}_{ij} + W_i)\} \ ,$$

where $H_{0i}(t) = \int_0^t h_{0i}(u) \, du$, the integrated baseline hazard. A frailty distribution parametrized by λ, $p(\mathbf{W}|\lambda)$, coupled with prior distributions for λ, β, and the hazard function h complete the hierarchical Bayesian model specification.

In this chapter we consider both parametric and semiparametric hierarchical survival models for data sets that are spatially arranged. Such models might be appropriate anytime we suspect that frailties W_i corresponding to strata in closer proximity to each other might also be similar in magnitude. This could arise if, say, the strata corresponded to hospitals in a given region, to counties in a given state, and so on. The basic assumption here is that "expected" survival times (or hazard rates) will be more similar in proximate regions, due to underlying factors (access to care, the willingness of the population to seek care, etc.) that vary spatially. We remind the reader that this does not imply that the observed survival times from subjects in proximate regions must be similar since they include an extra level of randomness arising from their variability around their (spatially correlated) underlying model quantities.

15.1   Parametric models

15.1.1   *Univariate spatial frailty modeling*

While it is possible to identify centroids of geographic regions and employ spatial process modeling for these locations, the effects in our examples are more naturally associated with areal units. As such, we work exclusively with CAR models for these effects, i.e., we assume that

$$\mathbf{W} \mid \lambda \sim CAR(\lambda) \, . \tag{15.5}$$

Also, we note that the resulting model for, say, (15.2) is an extended example of a generalized linear model for areal spatial data (Section 6.9). That is, (15.2) implies that

$$f(t_{ij}|\boldsymbol{\beta}, x_{ij}, W_i) = h_0(t_{ij}) e^{\boldsymbol{\beta}^{\mathrm{T}} \mathbf{x}_{ij} + W_i} e^{-H_0(t_{ij}) \exp(\boldsymbol{\beta}^{\mathrm{T}} \mathbf{x}_{ij} + W_i)} \, . \tag{15.6}$$

In other words, $U_{ij} = H_0(t_{ij}) \sim \text{Exponential}(\exp[-(\boldsymbol{\beta}^{\mathrm{T}}\mathbf{x}_{ij} + W_i)])$ so $-\log EH_0(t_{ij}) = \boldsymbol{\beta}^{\mathrm{T}}\mathbf{x}_{ij} + W_i$. The analogy with (6.42) and $g(\eta_i)$ is clear. The critical difference is that in Section 6.9 the link g is assumed known; here the link to the linear scale requires h_0, which is unknown (and will be modeled parametrically or nonparametrically).

Finally, we remark that it would certainly be possible to include both spatial and nonspatial frailties, which as already seen (Subsection 6.8.3) is now common practice in areal data modeling. Here, this would mean supplementing our spatial frailties W_i with a collection of nonspatial frailties, say, $V_i \overset{iid}{\sim} N(0, 1/\tau)$. The main problem with this approach is again that the frailties now become identified only by the prior, and so the proper choice of priors for τ and λ (or $\boldsymbol{\theta}$) becomes problematic. Another problem is the poor model fitting performance resulting from the addition of so many additional, weakly identified parameters.

15.1.1.1   Bayesian implementation

As already mentioned, the models outlined above are straightforwardly implemented in a Bayesian framework using MCMC methods. In the parametric case, say (15.4), the joint posterior distribution of interest is

$$p\left(\boldsymbol{\beta}, \mathbf{W}, \rho, \lambda \mid \mathbf{t}, \mathbf{x}, \boldsymbol{\gamma}\right) \propto L\left(\boldsymbol{\beta}, \mathbf{W}, \rho\,;\, \mathbf{t}, \mathbf{x}, \boldsymbol{\gamma}\right) p\left(\mathbf{W} \mid \lambda\right) p(\boldsymbol{\beta}) p\left(\rho\right) p(\lambda)\,, \qquad (15.7)$$

where the first term on the right-hand side is the Weibull likelihood, the second is the CAR distribution of the random frailties, and the remaining terms are prior distributions. In (15.7), $\mathbf{t} = \{t_{ij}\}$ denotes the collection of times to death, $\mathbf{x} = \{\mathbf{x}_{ij}\}$ the collection of covariate vectors, and $\boldsymbol{\gamma} = \{\gamma_{ij}\}$ the collection of death indicators for all subjects in all strata.

For our investigations, we retain the parametric form of the baseline hazard given in (15.4). Thus $L\left(\boldsymbol{\beta}, \mathbf{W}, \rho\,;\, \mathbf{t}, \mathbf{x}, \boldsymbol{\gamma}\right)$ is proportional to

$$\prod_{i=1}^{I} \prod_{j=1}^{n_i} \left\{ \rho t_{ij}^{\rho-1} \exp\left(\boldsymbol{\beta}^{\mathrm{T}} \mathbf{x}_{ij} + W_i\right) \right\}^{\gamma_{ij}} \exp\left\{ -t_{ij}^{\rho} \exp\left(\boldsymbol{\beta}^{\mathrm{T}} \mathbf{x}_{ij} + W_i\right) \right\}. \qquad (15.8)$$

The model specification in the Bayesian setup is completed by assigning prior distributions for $\boldsymbol{\beta}, \rho$, and λ. Typically, a flat (improper uniform) prior is chosen for $\boldsymbol{\beta}$, while vague but proper priors are chosen for ρ and λ, such as a $G(\alpha, 1/\alpha)$ prior for ρ and a $G(a,b)$ prior for λ. Hence the only extension beyond the disease mapping illustrations of Section 6.8 is the need to update ρ.

Example 15.1 *(Application to Minnesota infant mortality data).* We apply the methodology above to the analysis of infant mortality in Minnesota, originally considered by Banerjee et al. [2003b]. The data were obtained from the linked birth-death records data registry kept by the Minnesota Department of Health. The data comprise 267,646 live births occurring during the years 1992–1996 followed through the first year of life, together with relevant covariate information such as birth weight, sex, race, mother's age, and the mother's total number of previous births. Because of the careful linkage connecting infant death certificates with birth certificates (even when the death occurs in a separate state), we assume that each baby in the data set that is not linked with death must have been alive at the end of one year. Of the live births, only 1,547 babies died before the end of their first year. The number of days they lived is treated as the response t_{ij} in our models, while the remaining survivors were treated as "censored," or in other words, alive at the end of the study period. In addition to this information, the mother's Minnesota county of residence prior to the birth is provided. We implement the areal frailty model (15.5), the nonspatial frailty model (15.3), and a simple nonhierarchical ("no-frailty") model that sets $W_i = 0$ for all i.

For all of our models, we adopt a flat prior for $\boldsymbol{\beta}$, and a $G(\alpha, 1/\alpha)$ prior for ρ, setting $\alpha = 0.01$. Metropolis random walk steps with Gaussian proposals were used for sampling from the full conditionals for $\boldsymbol{\beta}$, while Hastings independence steps with gamma proposals were used for updating ρ. As for λ, in our case we are fortunate to have a data set that is large relative to the number of random effects to be estimated. As such, we simply select a vague (mean 1, variance 1000) gamma specification for λ, and rely on the data to overwhelm the priors.

Table 15.1 compares our three models in terms of DIC and effective model size p_D. For the no-frailty model, we see a p_D of 8.72, very close to the actual number of parameters, 9 (8 components of $\boldsymbol{\beta}$ plus the Weibull parameter ρ). The random effects models have substantially larger p_D values, though much smaller than their actual parameter counts (which would include the 87 random frailties W_i); apparently, there is substantial shrinkage of the frailties toward their grand mean. The DIC values suggest that each of these models

Model	p_D	DIC
No-frailty	8.72	511
Nonspatial frailty	39.35	392
CAR frailty	34.52	371

Table 15.1 *DIC and effective number of parameters p_D for competing parametric survival models.*

Covariate	2.5%	50%	97.5%
Intercept	-2.135	-2.024	-1.976
Sex (boys $= 0$)			
girls	-0.271	-0.189	-0.105
Race (white $= 0$)			
black	-0.209	-0.104	-0.003
Native American	0.457	0.776	1.004
unknown	0.303	0.871	1.381
Mother's age	-0.005	-0.003	-0.001
Birth weight in kg	-1.820	-1.731	-1.640
Total births	0.064	0.121	0.184
ρ	0.411	0.431	0.480
σ	0.083	0.175	0.298

Table 15.2 *Posterior summaries for the nonspatial frailty model.*

is substantially better than the no-frailty model, despite their increased size. Though the spatial frailty model has the best DIC value, plots of the full estimated posterior deviance distributions (not shown) suggest substantial overlap. On the whole, we seem to have modest support for the spatial frailty model over the ordinary frailty model.

Tables 15.2 and 15.3 provide 2.5, 50, and 97.5 posterior percentiles for the main effects in our two frailty models. In both cases, all of the predictors are significant at the .05 level. Since the reference group for the sex variable is boys, we see that girls have a lower hazard of death during the first year of life. The reference group for the race variables is white; the Native American beta coefficient is rather striking. In the CAR model, this covariate increases the posterior median hazard rate by a factor of $e^{0.782} = 2.19$. The effect of "unknown" race is also significant, but more difficult to interpret: in this group, the race of the infant was not recorded on the birth certificate. Separate terms for Hispanics, Asians, and Pacific Islanders were also originally included in the model but were eliminated after emerging as not significantly different from zero. Note that the estimate of ρ is quite similar across models, and suggests a decreasing baseline hazard over time. This is consistent with the fact that a high proportion (495, or 32%) of the infant deaths in our data set occurred in the first *day* of life: the force of mortality (hazard rate) is very high initially, but drops quickly and continues to decrease throughout the first year.

A benefit of fitting the spatial CAR structure is seen in the reduction of the length of the 95% credible intervals for the covariates in the spatial models compared to the i.i.d. model. As we might expect, there are modest efficiency gains when the model that better specifies the covariance structure of its random effects is used. That is since the spatial dependence priors for the frailties are in better agreement with the likelihood than is the independence prior, the prior-to-posterior learning afforded by Bayes' Rule leads to smaller posterior variances in the former cases. Most notably, the 95% credible set for the effect of "unknown" race is (0.303, 1.381) under the nonspatial frailty model (Table 15.2), but (0.351, 1.165) under the CAR frailty model (Table 15.3), a reduction in length of roughly 25%.

Covariate	2.5%	50%	97.5%
Intercept	−2.585	−2.461	−2.405
Sex (boys = 0)			
girls	−0.224	−0.183	−0.096
Race (white = 0)			
black	−0.219	−0.105	−0.007
Native American	0.455	0.782	0.975
unknown	0.351	0.831	1.165
Mother's age	−0.005	−0.004	−0.003
Birth weight in kg	−1.953	−1.932	−1.898
Total births	0.088	0.119	0.151
ρ	0.470	0.484	0.497
λ	12.62	46.07	100.4

Table 15.3 *Posterior summaries for the CAR frailty model.*

I.I.D. model (with covariates)

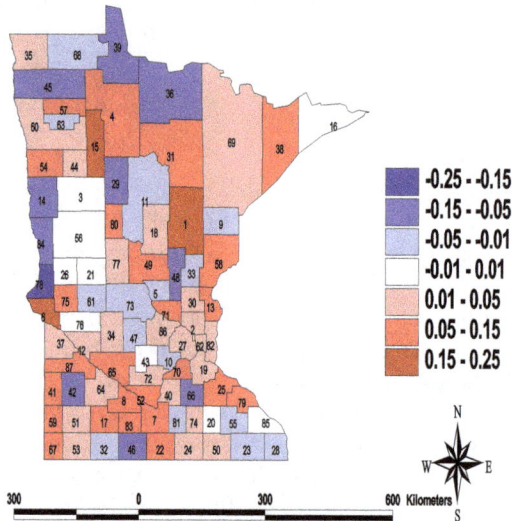

Figure 15.1 *Posterior median frailties, i.i.d. model with covariates, Minnesota county-level infant mortality data.*

Figures 15.1 and 15.2 map the posterior medians of the W_i under the nonspatial (i.i.d. frailties) and CAR models, respectively, where the models include all of the covariates listed in Tables 15.2 and 15.3. As expected, no clear spatial pattern is evident in the i.i.d. map, but from the CAR map we are able to identify two clusters of counties having somewhat higher hazards (in the southwest following the Minnesota River, and in the northeast "arrowhead" region), and two clusters with somewhat lower hazards (in the northwest, and the southeastern corner). Thus, despite the significance of the covariates now in these models, Figure 15.2 suggests the presence of some still-missing, spatially varying covariate(s) relevant to infant mortality. Such covariates might include location of birth (home or hospital), overall quality of available health or hospital care, mother's economic status, and mother's number of prior abortions or miscarriages.

CAR model (with covariates)

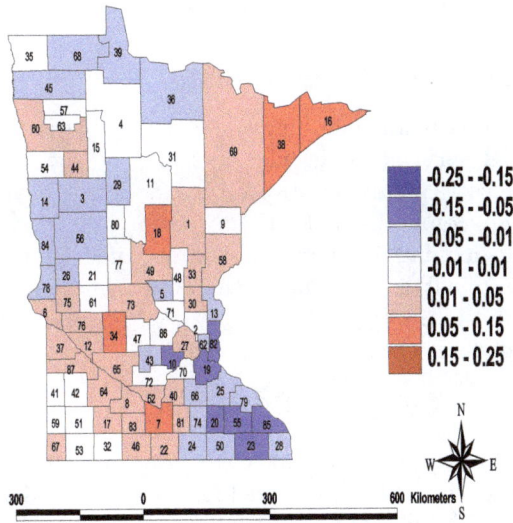

Figure 15.2 *Posterior median frailties, CAR model with covariates, Minnesota county-level infant mortality data.*

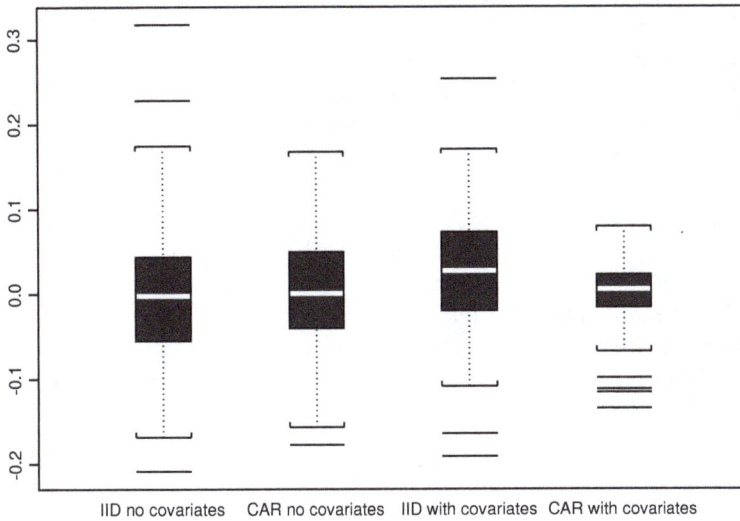

Figure 15.3 *Boxplots of posterior median frailties, i.i.d. and CAR models with and without covariates.*

In addition to the improved appearance and epidemiological interpretation of Figure 15.2, another reason to prefer the CAR model is provided in Figure 15.3, which shows boxplots of the posterior median frailties for the two cases corresponding to Figures 15.1 and 15.2, plus two preliminary models in which *no* covariates **x** are included. The tightness of the full CAR boxplot suggests this model is best at reducing the need for the frailty terms. This is as it should be since these terms are essentially spatial residuals, and capture lingering lack of fit in our spatial model (although they may also account for some excess *non*spatial variability since our current models do not include nonspatial frailty terms).

Note that all of the full CAR residuals are in the range (–0.15, 0.10), or (0.86, 1.11) on the hazard scale, suggesting that missing spatially varying covariates have only a modest (10 to 15%) impact on the hazard; from a practical standpoint, this model fits quite well. ∎

15.1.2 Spatial frailty versus logistic regression models

In many contexts (say, a clinical trial enrolling and following patients at spatially proximate clinical centers), a spatial survival model like ours may be the only appropriate model. However, since the Minnesota infant mortality data does not have any babies censored because of loss to followup, competing risks, or any reason other than the end of the study, there is no ambiguity in defining a *binary* survival outcome for use in a random effects logistic regression model. That is, we replace the event time data t_{ij} with an indicator of whether the subject did ($Y_{ij} = 0$) or did not ($Y_{ij} = 1$) survive the first year. Letting $p_{ij} = Pr(Y_{ij} = 1)$, our model is then

$$logit(p_{ij}) = \widetilde{\boldsymbol{\beta}}^{\mathrm{T}} \mathbf{x}_{ij} + \widetilde{W}_i \,, \tag{15.9}$$

with the usual flat prior for $\widetilde{\boldsymbol{\beta}}$ and an i.i.d. or CAR prior for the $\widetilde{W}_i$. As a result, (15.9) is exactly an example of a generalized linear model for areal spatial data.

Other authors [Doksum and Gasko, 1990, Ingram and Kleinman, 1989] have shown that in this case of no censoring before followup (and even in cases of equal censoring across groups), it is possible to get results for the $\widetilde{\boldsymbol{\beta}}$ parameters in the logistic regression model very similar to those obtained in the proportional hazards model (15.1), except of course for the differing interpretations (log odds versus log relative risk, respectively). Moreover when the probability of death is very small, as it is in the case of infant mortality, the log odds and log relative risk become even more similar. Since it uses more information (i.e., time to death rather than just a survival indicator), intuitively, the proportional hazards model should make gains over the logistic model in terms of power to detect significant covariate effects. Yet, consistent with the simulation studies performed by Ingram and Kleinman [1989], our experience with the infant mortality data indicate that only a marginal increase in efficiency (decrease in variance) is exhibited by the posterior distributions of the parameters.

On the other hand, we did find some differences in terms of the estimated random effects in the logistic model compared to the proportional hazards model. Figure 15.4 shows a scatterplot of the estimated posterior medians of W_i versus $\widetilde{W}_i$ for each county obtained from the models where there were no covariates, and the random effects were assumed to i.i.d. The sample correlation of these estimated random effects is 0.81, clearly indicating that they are quite similar. Yet there are still some particular counties that result in rather different values under the two models. One way to explain this difference is that the hazard functions are not exactly proportional across the 87 counties of Minnesota. A close examination of the counties that had differing $\widetilde{W}_i$ versus W_i shows that they had different average times at death compared to other counties with similar overall death rates. Consider for example County 70, an outlier circled in Figure 15.4, and its comparison to circled Counties 73, 55, and 2, which have similar death rates (and hence roughly the same horizontal position in Figure 15.4). We find County 70 has the smallest mean age at death, implying that it has more early deaths, explaining its smaller frailty estimate. Conversely, County 14 has a higher average time at death but overall death rates similar to Counties 82, 48, and 5 (again note the horizontal alignment in Figure 15.4), and as a result has higher estimated frailty. A lack of proportionality in the baseline hazard rates across counties thus appears to manifest as a departure from linearity in Figure 15.4.

We conclude this subsection by noting that previous work by Carlin and Hodges [1999] suggests a generalization of our basic model (15.4) to

$$h(t_{ij}; x_{ij}) = \rho_i t_{ij}^{\rho_i - 1} \exp(\boldsymbol{\beta}^{\mathrm{T}} \mathbf{x}_{ij} + W_i) \,.$$

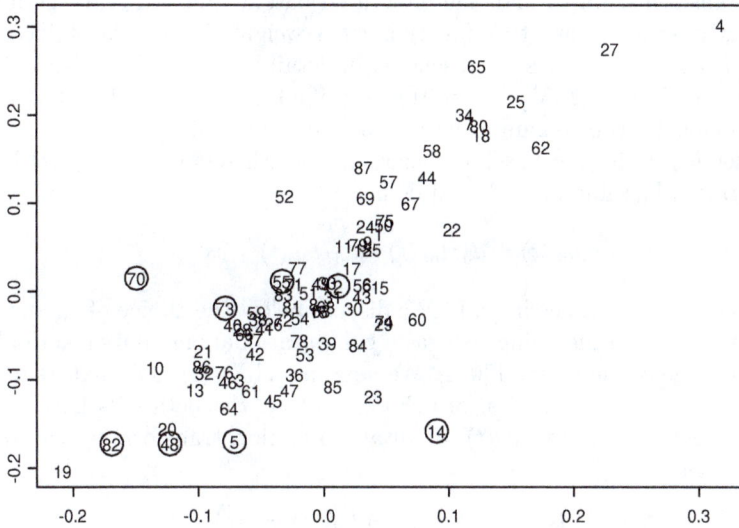

Figure 15.4 *Posterior medians of the frailties W_i (horizontal axis) versus posterior medians of the logistic random effects $\widetilde{W}_i$ (vertical axis). Plotting character is county number; significance of circled counties is described in the text.*

That is, we allow two sets of random effects: the existing frailty parameters W_i, and a new set of shape parameters ρ_i. This then allows both the overall level and the shape of the hazard function over time to vary from county to county. Either i.i.d. or CAR priors could be assigned to these two sets of random effects, which could themselves be correlated within county. In the latter case, this might be fit using the MCAR model of Section 11.1; see [Jin et al., 2005, 2007] as well as Section 15.4.

15.2   Semiparametric models

While parametric models are easily interpretable and often afford a good fit to survival data, many practitioners continue to prefer the additional richness of the nonparametric baseline hazard offered by the celebrated Cox model. In this section we turn to nonparametric models for the baseline hazard. Such models are often referred to as *semiparametric*, since we continue to assume proportional hazards of the form (15.1) in which the covariate effects are still modeled parametrically, While Li and Ryan [2002] address this problem from a classical perspective, in this section we follow the hierarchical Bayesian approach of Banerjee and Carlin [2002].

Within the Bayesian framework, several authors have proposed treating the Cox partial likelihood as a full likelihood, to obtain a posterior distribution for the treatment effect. However, this approach does not allow fully hierarchical modeling of stratum-specific baseline hazards (with stratum-specific frailties) because the baseline hazard is implicit in the partial likelihood computation. In the remainder of this section, we describe two possible methodological approaches to modeling the baseline hazard in Cox regression, which thus lead to two semiparametric spatial frailty techniques. We subsequently revisit the Minnesota infant mortality data.

15.2.1   Beta mixture approach

Our first approach uses an idea of Gelfand and Mallick [1995] that flexibly models the integrated baseline hazard as a mixture of monotone functions. In particular, these authors use a simple transformation to map the integrated baseline hazard onto the interval $[0, 1]$, and subsequently approximate this function by a weighted mixture of incomplete beta functions. Implementation issues are discussed in detail by Gelfand and Mallick [1995] and also by Carlin and Hodges [1999] for stratum-specific baseline hazards. The likelihood and Bayesian hierarchical setup remain exactly as above.

Thus, we let $h_{0i}(t)$ be the baseline hazard in the ith region and $H_{0i}(t)$ be the corresponding integrated baseline hazard, and define

$$J_{0i}(t) = a_0 H_{0i}(t) / [a_0 H_{0i}(t) + b_0] \ ,$$

which conveniently takes values in $[0, 1]$. We discuss below the choice of a_0 and b_0 but note that this is not as much a modeling issue as a computational one, important only to ensure appropriate coverage of the interval $[0, 1]$. We next model $J_{0i}(t)$ as a mixture of $Beta(r_l, s_l)$ cdfs, for $l = 1, \ldots, m$. The r_l and s_l are chosen so that the beta cdfs have evenly spaced means and are centered around $\widetilde{J_0}(t)$, a suitable function transforming the time scale to $[0, 1]$. We thus have

$$J_{0i}(t) = \sum_{l=1}^{m} v_{il} \, IB\left(\widetilde{J_0}(t); r_l, s_l\right) \ ,$$

where $\sum_{l=1}^{m} v_{il} = 1$ for all i, and $IB(\cdot; a, b)$ denotes the incomplete beta function (i.e., the cdf of a $Beta(a, b)$ distribution). Since any distribution function on $[0, 1]$ can be approximated arbitrarily well by a finite mixture of beta cdfs, the same is true for J_{0i}, an increasing function that maps $[0, 1]$ onto itself. Thus, working backward, we find the following expression for the cumulative hazard in terms of the above parameters:

$$H_{0i}(t) = \frac{b_0 \sum_{l=1}^{m} v_{il} \, IB\left(\widetilde{J_0}(t); r_l, s_l\right)}{a_0 \left\{1 - \sum_{l=1}^{m} v_{il} \, IB\left(\widetilde{J_0}(t); r_l, s_l\right)\right\}} \ .$$

Taking derivatives, we have for the hazard function,

$$h_{0i}(t) = \frac{b_0 \frac{\partial}{\partial t} \widetilde{J_0}(t) \sum_{l=1}^{m} v_{il} Beta\left(\widetilde{J_0}(t); r_l, s_l\right)}{a_0 \left\{1 - \sum_{l=1}^{m} v_{il} \, IB\left(\widetilde{J_0}(t); r_l, s_l\right)\right\}^2} \ .$$

Typically m, the number of mixands of the beta cdfs, is fixed, as are the $\{(r_l, s_l)\}_{l=1}^{m}$, so chosen that the resulting beta densities cover the interval $[0, 1]$. For example, we might fix $m = 5$, $\{r_l\} = (1, 2, 3, 4, 5)$ and $\{s_l\} = (5, 4, 3, 2, 1)$, producing five evenly-spaced beta cdfs.

Regarding the choice of a_0 and b_0, we note that it is intuitive to specify $\widetilde{J_0}(t)$ to represent a plausible central function around which the J_{0i}'s are distributed. Thus, if we consider the cumulative hazard function of an exponential distribution to specify $\widetilde{J_0}(t)$, then we get $\widetilde{J_0}(t) = a_0 t / (a_0 t + b_0)$. In our Minnesota infant mortality data set, since the survival times ranged between 1 day and 365 days, we found $a_0 = 5$ and $b_0 = 100$ lead to values for $\widetilde{J_0}(t)$ that largely cover the interval $[0, 1]$, and so fixed them as such. The likelihood is thus a function of the regression coefficients $\boldsymbol{\beta}$, the stratum-specific weight vectors $\mathbf{v}_i = (v_{i1}, \ldots, v_{im})^{\mathrm{T}}$, and the spatial effects W_i. It is natural to model the $\mathbf{v}_i$'s as draws from a Dirichlet$(\phi_1, \ldots, \phi_m)$ distribution, where for simplicity we often take $\phi_1 = \cdots = \phi_m = \phi$.

15.2.2   Counting process approach

The second nonparametric baseline hazard modeling approach we investigate is that of Clayton [1991] and Clayton [1994]. Here we give only the essential ideas, referring the reader to Andersen and Gill [1982], Clayton [1991], or Fleming and Harrington [1991] for a more complete treatment. The underlying idea is that the number of failures up to time t is assumed to arise from a *counting process* $N(t)$. The corresponding *intensity process* is defined as

$$I(t)\,dt = \mathrm{E}\left(dN\left(t\right)|F_{t-}\right)\,,$$

where $dN(t)$ is the increment of N over the time interval $[t, t+dt)$, and F_{t-} represents the available data up to time t. For each individual, $dN(t)$ therefore takes the value 1 if the subject fails in that interval, and 0 otherwise. Thus $dN(t)$ may be thought of as the "death indicator process," analogous to γ in the model of the previous subsection. For the jth subject in the ith region, under the proportional hazards assumption, the intensity process (analogous to our hazard function $h\left(t_{ij}; \mathbf{x}_{ij}\right)$) is modeled as

$$I_{ij}(t) = Y_{ij}(t)\lambda_0\left(t\right)\exp\left(\boldsymbol{\beta}^{\mathrm{T}}\mathbf{x}_{ij} + W_i\right)\,,$$

where $\lambda_0\left(t\right)$ is the baseline hazard function and $Y_{ij}\left(t\right)$ is an indicator process taking the value 1 or 0 according to whether or not subject i is observed at time t. Under the above formulation and keeping the same notation as above for $\mathbf{W}$ and $\mathbf{x}$, a Bayesian hierarchical model may be formulated as:

$$
\begin{aligned}
dN_{ij}\left(t\right) &\sim Poisson\left(I_{ij}\left(t\right)dt\right)\,, \\
I_{ij}\left(t\right)dt &= Y_{ij}(t)\exp\left(\boldsymbol{\beta}^{\mathrm{T}}\mathbf{x}_{ij} + W_i\right)d\Lambda_0\left(t\right)\,, \\
d\Lambda_0\left(t\right) &\sim Gamma\left(c\,d\Lambda_0^*\left(t\right), c\right)\,.
\end{aligned}
$$

As before, priors $p\left(\mathbf{W}|\lambda\right)$, $p\left(\lambda\right)$, and $p(\boldsymbol{\beta})$ are required to completely specify the Bayesian hierarchical model. Here, $d\Lambda_0\left(t\right) = \lambda_0\left(t\right)dt$ may be looked upon as the increment or jump in the integrated baseline hazard function occurring during the time interval $[t, t+dt)$. Since the conjugate prior for the Poisson mean is the gamma distribution, $\Lambda_0\left(t\right)$ is conveniently modeled as a process whose increments $d\Lambda_0\left(t\right)$ are distributed according to gamma distributions. The parameter c in the above setup represents the degree of confidence in our prior guess for $d\Lambda_0\left(t\right)$, given by $d\Lambda_0^*\left(t\right)$. Typically, the prior guess $d\Lambda_0^*\left(t\right)$ is modeled as $r\,dt$, where r is a guess at the failure rate per unit time. The LeukFr example in the BUGS examples manual offers an illustration of how to code the above formulation.

Example 15.2 *(Application to Minnesota infant mortality data, continued)*. We now apply the methodology above to the reanalysis of our Minnesota infant mortality data set. For both the CAR and nonspatial models we implemented the Cox model with the two semiparametric approaches outlined above. We found very similar results, and so in our subsequent analysis, we present only the results with the beta mixture approach (Subsection 15.2.1). For all of our models, we adopt vague Gaussian priors for $\boldsymbol{\beta}$. Since the full conditionals for each component of $\boldsymbol{\beta}$ are log-concave, adaptive rejection sampling was used for sampling from the $\boldsymbol{\beta}$ full conditionals. As in Section 15.1, we again simply select a vague $G(0.001, 1000)$ (mean 1, variance 1000) specification for CAR smoothness parameter λ, though we maintain more informative priors on the other variance components.

Table 15.4 compares our three models in terms of DIC and effective model size p_D. For the no-frailty model, we see a p_D of 6.82, reasonably close to the actual number of parameters, 8 (the components of $\boldsymbol{\beta}$). The other two models have substantially larger p_D values, though much smaller than their actual parameter counts (which would include the 87 random frailties W_i); apparently, there is substantial shrinkage of the frailties toward their grand mean. The DIC values suggest that both of these models are substantially better than the no-frailty model, despite their increased size. As in Table 15.1, the spatial frailty model has the best DIC value.

Model	p_D	DIC
No-frailty	6.82	507
Nonspatial frailty	27.46	391
CAR frailty	32.52	367

Table 15.4 *DIC and effective number of parameters p_D for competing nonparametric survival models.*

Covariate	2.5%	50%	97.5%
Intercept	−2.524	−1.673	−0.832
Sex (boys = 0)			
girls	−0.274	−0.189	−0.104
Race (white = 0)			
black	−0.365	−0.186	−0.012
Native American	0.427	0.737	1.034
unknown	0.295	0.841	1.381
Mother's age	−0.054	−0.035	−0.014
Birth weight in kg	−1.324	−1.301	−1.280
Total births	0.064	0.121	0.184

Table 15.5 *Posterior summaries for the nonspatial semiparametric frailty model.*

Covariate	2.5%	50%	97.5%
Intercept	−1.961	−1.532	−0.845
Sex (boys = 0)			
girls	−0.351	−0.290	−0.217
Race (white = 0)			
black	−0.359	−0.217	−0.014
Native American	0.324	0.599	0.919
unknown	0.365	0.863	1.316
Mother's age	−0.042	−0.026	−0.013
Birth weight in kg	−1.325	−1.301	−1.283
Total births	0.088	0.135	0.193

Table 15.6 *Posterior summaries for the CAR semiparametric frailty model.*

Tables 15.5 and 15.6 provide 2.5, 50, and 97.5 posterior percentiles for the main effects in our two frailty models, respectively. In both tables, all of the predictors are significant at the .05 level. Overall, the results are broadly similar to those from our earlier parametric analysis in Tables 15.2 and 15.3. For instance, the effect of being in the Native American group is again noteworthy. Under the CAR model, this covariate increases the posterior median hazard rate by a factor of $e^{0.599} = 1.82$. The benefit of fitting the spatial CAR structure is also seen again in the reduction of the length of the 95% credible intervals for the spatial model compared to the i.i.d. model. Most notably, the 95% credible set for the effect of "mother's age" is $(-0.054, -0.014)$ under the nonspatial frailty model (Table 15.5), but $(-0.042, -0.013)$ under the CAR frailty model (Table 15.6), a reduction in length of roughly 28%. Thus overall, adding spatial structure to the frailty terms appears to be reasonable and beneficial. Maps analogous to Figures 15.1 and 15.2 (not shown) reveal a very similar story. ∎

15.3 Spatiotemporal models

In this section we follow Carlin and Banerjee [2003] to develop a semiparametric (Cox) hierarchical Bayesian frailty model for capturing spatiotemporal heterogeneity in survival data. We then use these models to describe the pattern of breast cancer in the 99 counties of Iowa while accounting for important covariates, spatially correlated differences in the hazards among the counties, and possible space-time interactions.

We begin by extending the framework of the preceding section to incorporate temporal dependence. Here we have t_{ijk} as the response (time to death) for the jth subject residing in the ith county who was diagnosed in the kth year, while the individual-specific vector of covariates is now denoted by $\mathbf{x}_{ijk}$, for $i = 1, 2, ..., I$, $k = 1, ..., K$, and $j = 1, 2, ..., n_{ik}$. We note that "time" is now being used in two ways. The measurement or response is a survival time, but these responses are themselves observed at different areal units *and* different times (years). Furthermore, the spatial random effects W_i in the preceding section are now modified to W_{ik}, to represent spatiotemporal frailties corresponding to the ith county for the kth diagnosis year. Our spatial frailty specification in (15.1) now becomes

$$h\left(t_{ijk}; \mathbf{x}_{ijk}\right) = h_{0i}\left(t_{ijk}\right) \exp\left(\boldsymbol{\beta}^{\mathrm{T}} \mathbf{x}_{ijk} + W_{ik}\right) . \tag{15.10}$$

Our CAR prior would now have conditional representation $W_{ik} \mid \ W_{(i' \neq i)k} \ \sim \ N(\overline{W}_{ik}, 1/(\lambda_k m_i))$.

Note that we can account for temporal correlation in the frailties by assuming that the λ_k are themselves identically distributed from a common hyperprior (Subsection 12.7.1). A gamma prior (usually vague but proper) is often selected here since this is particularly convenient for MCMC implementation. A flat prior for $\boldsymbol{\beta}$ is typically chosen since this still admits a proper posterior distribution. Adaptive rejection [Gilks and Wild, 1992] or Metropolis-Hastings sampling are usually required to update the $\mathbf{W}_k$ and $\boldsymbol{\beta}$ parameters in a Gibbs sampler.

We remark that it would certainly be possible to include both spatial and nonspatial frailties, as mentioned in Subsection 15.1.1. This would mean supplementing our spatial frailties W_{ik} with a collection of nonspatial frailties, say $V_{ik} \overset{iid}{\sim} N(0, 1/\tau_k)$. We summarize our full hierarchical model as follows:

$$
\begin{aligned}
L\left(\beta, \mathbf{W}; \mathbf{t}, \mathbf{x}, \gamma\right) \ &\propto \ \prod_{k=1}^{K} \prod_{i=1}^{I} \prod_{j=1}^{n_{ik}} \left\{h_{0i}\left(t_{ijk}; \mathbf{x}_{ijk}\right)\right\}^{\gamma_{ijk}} \\
&\quad \times \exp\left\{-H_{0i}\left(t_{ijk}\right) \exp\left(\boldsymbol{\beta}^{\mathrm{T}} \mathbf{x}_{ijk} + W_{ik} + V_{ik}\right)\right\}, \\
\text{where } p(\mathbf{W}_k|\lambda_k) \ &\sim \ CAR(\lambda_k) \quad p(\mathbf{V}_k|\tau_k) \sim N_I\left(\mathbf{0}, \tau_k \mathbf{I}\right) \\
\text{and } \lambda_k \ &\sim \ G\left(a, b\right) \ , \ \ \tau_k \sim G\left(c, d\right) \ \text{ for } \ k = 1, 2, ..., K \ .
\end{aligned}
$$

In the sequel we adopt the beta mixture approach of Subsection 15.2.1 to model the baseline hazard functions $H_{0i}(t_{ijk})$ nonparametrically.

Example 15.3 (*Analysis of Iowa SEER breast cancer data*). The National Cancer Institute's SEER program (`seer.cancer.gov`) is the most authoritative source of cancer data in the U.S., offering county-level summaries on a yearly basis for several states in various parts of the country. In particular, the database provides a cohort of 15,375 women in Iowa who were diagnosed with breast cancer starting in 1973, and have been undergoing treatment and have been progressively monitored since. Only those who have been identified as having died from metastasis of cancerous nodes in the breast are considered to have failed, while the rest (including those who might have died from metastasis of other types of cancer, or from other causes of death) are considered censored. By the end of 1998, 11,912 of the patients had died of breast cancer while the remaining were censored, either because they

Covariate	2.5%	50%	97.5%
Age at diagnosis	0.0135	0.0148	0.0163
Number of primaries	−0.43	−0.40	−0.36
Race (white = 0)			
black	−0.14	0.21	0.53
other	−2.25	−0.30	0.97
Stage (local = 0)			
regional	0.30	0.34	0.38
distant	1.45	1.51	1.58

Table 15.7 *Posterior summaries for the spatiotemporal frailty model.*

survived until the end of the study period, dropped out of the study, or died of causes other than breast cancer. For each individual, the data set records the time in months (1 to 312) that the patient survived, and her county of residence at diagnosis. Several individual-level covariates are also available, including race (black, white, or other), age at diagnosis, number of primaries (i.e., the number of other types of cancer diagnosed for this patient), and the stage of the disease (local, regional, or distant).

15.3.0.1   *Results for the full model*

We begin by summarizing our results for the spatiotemporal frailty model described above, i.e., the full model having both spatial frailties W_{ik} and nonspatial frailties V_{ik}. We chose vague $G(0.01, 0.01)$ hyperpriors for the λ_k and τ_k (having mean 1 but variance 100) in order to allow maximum flexibility in the partitioning of the frailties into spatial and nonspatial components. Best et al. [1999] suggest that a higher variance prior for the τ_k (say, a $G(0.001, 0.001)$) may lead to better prior "balance" between the spatial and nonspatial random effects, but there is controversy on this point and so we do not pursue it here. While overly diffuse priors may result in weak identifiability of these parameters, their posteriors remain proper, and the impact of these priors on the posterior for the well-identified subset of parameters (including β and the log-relative hazards themselves) should be minimal [Daniels and Kass, 1999, Eberly and Carlin, 2000].

Table 15.7 provides 2.5, 50, and 97.5 posterior percentiles for the main effects (components of β) in our model. All of the predictors *except* those having to do with race are significant at the .05 level. Since the reference group for the stage variable is local, we see that women with regional and distant (metastasized) diagnoses have higher and much higher hazard of death, respectively; the posterior median hazard rate increases by a factor of $e^{1.51} = 4.53$ for the latter group. Higher age at diagnosis also increases the hazard, but a larger number of primaries (the number of other types of cancer a patient is suffering from) actually leads to a *lower* hazard, presumably due to the competing risk of dying from one of these other cancers.

Figure 15.5 maps the posterior medians of the frailties $W_{ik} + V_{ik}$ for the representative year 1986. We see clusters of counties with lower median frailties in the north-central and south-central parts of the state, and also clusters of counties with higher median frailties in the central, northeastern, and southeastern parts of the state.

Maps for other representative years showed very similar patterns, as well as an overall decreasing pattern in the frailties over time [see Carlin and Banerjee, 2003, for details]. Figure 15.6 clarifies this pattern by showing boxplots of the posterior medians of the W_{ik} over time (recall our full model does not have year-specific intercepts; the average of the W_{ik} for year k plays this role). We see an essentially horizontal trend during roughly the first half of our observation period, followed by a decreasing trend that seems to be accelerating.

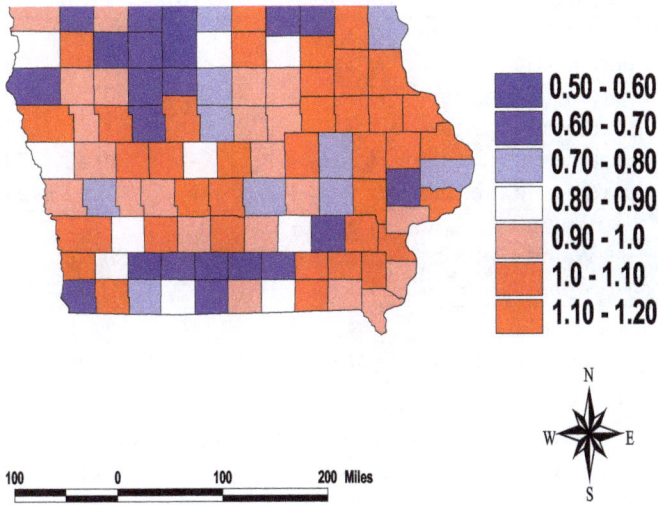

Figure 15.5 *Fitted spatiotemporal frailties, Iowa counties, 1986.*

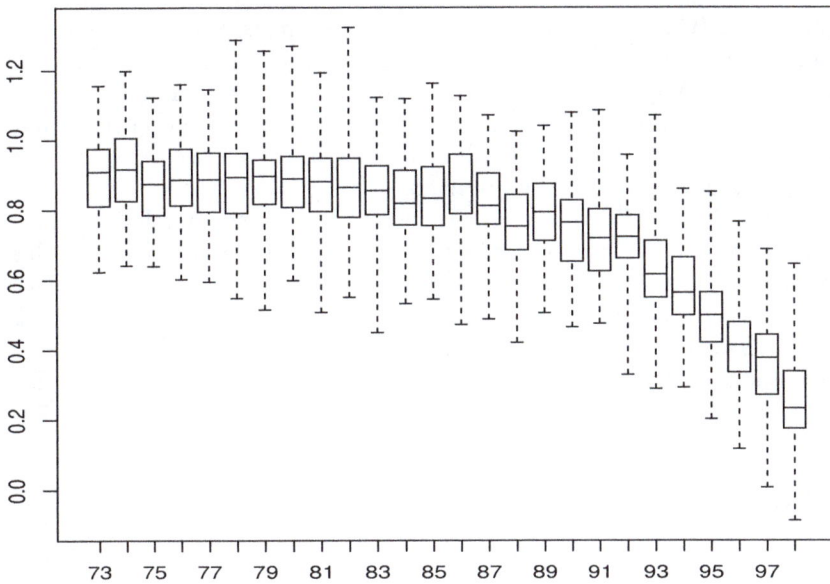

Figure 15.6 *Boxplots of posterior medians for the spatial frailties W_{ik} over the Iowa counties for each year, $k=1973, \ldots, 1998$.*

Overall the total decrease in median log hazard is about 0.7 units, or about a 50% reduction in hazard over the observation period. A cancer epidemiologist would likely be unsurprised by this decline since it coincides with the recent rise in the use of mammography by American women.

15.3.0.2 Bayesian model choice

For model choice, we again turn to the DIC criterion. The first six lines of Table 15.8 provide p_D and DIC values for our full model and several simplications thereof. Note the full model (sixth line) is estimated to have only just over 150 effective parameters, a substantial reduction (recall there are $2 \times 99 \times 26 = 5148$ random frailty parameters alone). Removing

Baseline hazard	Log-relative hazard	p_D	DIC
Semiparametric mixture	$\boldsymbol{\beta}^{\mathrm{T}}\mathbf{x}_{ijk}$	6.17	780
Semiparametric mixture	$\boldsymbol{\beta}^{\mathrm{T}}\mathbf{x}_{ijk} + \alpha_k$	33.16	743
Semiparametric mixture	$\boldsymbol{\beta}^{\mathrm{T}}\mathbf{x}_{ijk} + \alpha_k + W_i$	80.02	187
Semiparametric mixture	$\boldsymbol{\beta}^{\mathrm{T}}\mathbf{x}_{ijk} + W_{ik}$	81.13	208
Semiparametric mixture	$\boldsymbol{\beta}^{\mathrm{T}}\mathbf{x}_{ijk} + V_{ik}$	149.45	732
Semiparametric mixture	$\boldsymbol{\beta}^{\mathrm{T}}\mathbf{x}_{ijk} + W_{ik} + V_{ik}$	151.62	280
Weibull	$\boldsymbol{\beta}^{\mathrm{T}}\mathbf{x}_{ijk} + \alpha_k + W_i$	79.22	221
Weibull	$\boldsymbol{\beta}^{\mathrm{T}}\mathbf{x}_{ijk} + W_{ik}$	80.75	239
Weibull	$\boldsymbol{\beta}^{\mathrm{T}}\mathbf{x}_{ijk} + W_{ik} + V_{ik}$	141.67	315

Table 15.8 *DIC and effective number of parameters p_D for the competing models.*

the spatial frailties W_{ik} from the log-relative hazard has little impact on p_D, but substantial negative impact on the DIC score. By contrast, removing the nonspatial frailties V_{ik} reduces (i.e., improves) both p_D and DIC, consistent with our findings in the previous subsection. Further simplifying the model to having a single set of spatial frailties W_i that do not vary with time (but now also reinserting year-specific intercepts α_k) has little effect on p_D but does improve DIC a bit more (though this improvement appears only slightly larger than the order of Monte Carlo error in our calculations). Even more drastic simplifications (eliminating the W_i, and perhaps even the α_k) lead to further drops in p_D, but at the cost of unacceptably large increases in DIC. Thus our county-level breast cancer survival data seem to have strong spatial structure that is still unaccounted for by the covariates in Table 15.7, but structure that is fairly similar for all diagnosis years.

The last three lines of Table 15.8 reconsider the best three log-relative hazard models above, but where we now replace the semiparametric mixture baseline hazard with a Weibull hazard having region-specific baseline hazards $h_{0i}(t_{ijk}; \rho_i) = \rho_i t_{ijk}^{\rho_i - 1}$ (note the spatial frailties play the role of the second parameter customarily associated with the Weibull model). These fully parametric models offer small advantages in terms of parsimony (smaller p_D), but these gains are apparently more than outweighed by a corresponding degradation in fit (much larger DIC score). ∎

15.4 Multivariate models ⋆

In this section we extend to multivariate spatial frailty modeling, using the MCAR model introduced in Subsection 11.1. In particular, we use a semiparametric model and consider MCAR structure on both residual (spatial frailty) and regression (space-varying coefficient) terms. We also extend to the spatiotemporal case by including temporally correlated cohort effects (say, one for each year of initial disease diagnosis) that can be summarized and plotted over time. Example 15.4 illustrates the utility of our approach in an analysis of survival times of patients suffering from one or more types of cancer. We obtain posterior estimates of key fixed effects, smoothed maps of both frailties and spatially varying coefficients and compare models using the DIC criterion.

15.4.0.1   Static spatial survival data with multiple causes of death

Consider the following multivariate survival setting. Let t_{ijk} denote the time to death or censoring for the kth patient having the jth type of primary cancer living in the ith county, $i = 1, \ldots, n$, $j = 1, \ldots, p$, $k = 1, \ldots, s_{ij}$, and let γ_{ijk} be the corresponding death indicator. Let us write $\mathbf{x}_{ijk}$ as the vector of covariates for the above individual, and let $\mathbf{z}_{ijk}$ denote the vector of cancer indicators for this individual. That is, $\mathbf{z}_{ijk} = (z_{ijk1}, z_{ijk2}, \ldots, z_{ijkp})^{\mathrm{T}}$

where $z_{ijkl} = 1$ if patient ijk suffers from cancer type l, and 0 otherwise (note that $z_{ijkj} = 1$ by definition). Then we can write the likelihood of our proportional hazards model $L(\boldsymbol{\beta}, \boldsymbol{\theta}, \boldsymbol{\Phi}; \mathbf{t}, \mathbf{x}, \boldsymbol{\gamma})$ as

$$\prod_{i=1}^{n} \prod_{j=1}^{p} \prod_{k=1}^{s_{ij}} \{h(t_{ijk}; \mathbf{x}_{ijk}, \mathbf{z}_{ijk})\}^{\gamma_{ijk}} \times \exp\left\{-H_{0i}(t_{ijk}) \exp\left(\mathbf{x}_{ijk}^{\mathrm{T}}\boldsymbol{\beta} + \mathbf{z}_{ijk}^{\mathrm{T}}\boldsymbol{\theta} + \phi_{ij}\right)\right\}, \tag{15.11}$$

where

$$h(t_{ijk}; \mathbf{x}_{ijk}, \mathbf{z}_{ijk}) = h_{0i}(t_{ijk}) \exp\left(\mathbf{x}_{ijk}^{\mathrm{T}}\boldsymbol{\beta} + \mathbf{z}_{ijk}^{\mathrm{T}}\boldsymbol{\theta} + \phi_{ij}\right). \tag{15.12}$$

Here, $H_{0i}(t_{ijk}) = \int_0^{t_{ijk}} h_{0i}(u)\, du$, $\boldsymbol{\phi}_i = (\phi_{i1}, \phi_{i2}, ..., \phi_{in})^{\mathrm{T}}$, $\boldsymbol{\beta}$ and $\boldsymbol{\theta}$ are given flat priors, and

$$\boldsymbol{\Phi} \equiv (\boldsymbol{\phi}_1^{\mathrm{T}}, \ldots, \boldsymbol{\phi}_n^{\mathrm{T}})^{\mathrm{T}} \sim MCAR(\rho, \boldsymbol{\Sigma}),$$

using the notation of Section 11.1. The region-specific baseline hazard functions $h_{0i}(t_{ijk})$ are modeled using the beta mixture approach (Subsection 15.2.1) in such a way that the intercept in $\boldsymbol{\beta}$ remains estimable. We note that we could extend to a county *and* cancer-specific baseline hazard h_{0ij}; however, preliminary exploratory analyses of our data suggest such generality is not needed here.

Several alternatives to model formulation (15.12) immediately present themselves. For example, we could convert to a space-varying coefficients model [Assunção, 2003], replacing the log-relative hazard $\mathbf{x}_{ijk}^{\mathrm{T}}\boldsymbol{\beta} + \mathbf{z}_{ijk}^{\mathrm{T}}\boldsymbol{\theta} + \phi_{ij}$ in (15.12) with

$$\mathbf{x}_{ijk}^{\mathrm{T}}\boldsymbol{\beta} + \mathbf{z}_{ijk}^{\mathrm{T}}\boldsymbol{\theta}_i, \tag{15.13}$$

where $\boldsymbol{\beta}$ again has a flat prior, but $\boldsymbol{\Theta} \equiv (\boldsymbol{\theta}_1^{\mathrm{T}}, \ldots, \boldsymbol{\theta}_n^{\mathrm{T}})^{\mathrm{T}} \sim MCAR(\rho, \boldsymbol{\Sigma})$. In Example 15.4 we apply this method to our cancer data set; we defer mention of still other log-relative hazard modeling possibilities until after this illustration.

15.4.0.2 MCAR specification, simplification, and computing

To efficiently implement the $MCAR(\rho, \boldsymbol{\Sigma})$ as a prior distribution for our spatial process, suppose that we are using the usual 0-1 adjacency weights in W. Then recall from equation (11.4) that we may express the MCAR precision matrix $\mathbf{B} \equiv \boldsymbol{\Sigma}_{\boldsymbol{\phi}}^{-1}$ in terms of the $n \times n$ adjacency matrix W as

$$\mathbf{B} = (Diag(m_i) - \rho W) \otimes \Lambda,$$

where we have added a propriety parameter ρ. Note that this is a Kronecker product of an $n \times n$ and a $p \times p$ matrix, thereby rendering $\mathbf{B}$ as $np \times np$ as required. In fact, $\mathbf{B}$ may be looked upon as the Kronecker product of two partial precision matrices: one for the spatial components, $(Diag(m_i) - \rho W)$ (depending upon their adjacency structure and number of neighbors), and another for the variation across diseases, given by Λ. We thus alter our notation slightly to $MCAR(\rho, \Lambda)$.

Also as a consequence of this form, a sufficient condition for positive definiteness of the dispersion matrix for this MCAR model becomes $|\rho| < 1$ (as in the univariate case). Negative smoothness parameters are not desirable, so we typically take $0 < \rho < 1$. We can now complete the Bayesian hierarchical formulation by placing appropriate priors on ρ (say, a $Unif(0,1)$ or $Beta(18,2)$) and Λ (say, a $Wishart(\rho, \Lambda_0)$).

The Gibbs sampler is the MCMC method of choice here, particularly because, as in the univariate case, it takes advantage of the MCAR's conditional specification. Adaptive rejection sampling may be used to sample the regression coefficients $\boldsymbol{\beta}$ and $\boldsymbol{\theta}$, while Metropolis steps with (possibly multivariate) Gaussian proposals may be employed for the spatial effects $\boldsymbol{\Phi}$. The full conditional for ρ is nicely suited for slice sampling (see Subsection 5.5.1.5), given its bounded support. Finally, the full conditional for Λ^{-1} emerges in closed form as an inverse Wishart distribution.

We conclude this subsection by recalling that our model can be generalized to admit different propriety parameters ρ_j for different diseases (recall Section 11.1 and, in particular, the discussion surrounding (11.8)) and (11.9). We notate this model as $MCAR(\boldsymbol{\rho}, \Lambda)$, where $\boldsymbol{\rho} = (\rho_1, \ldots, \rho_p)^T$.

15.4.0.3 Spatiotemporal survival data

Here we extend our model to allow for cohort effects. Let r index the year in which patient ijk entered the study (i.e., the year in which the patient's primary cancer was diagnosed). Extending the model (15.12) we obtain the log-relative hazard,

$$\mathbf{x}_{ijkr}^T \boldsymbol{\beta} + \mathbf{z}_{ijkr}^T \boldsymbol{\theta} + \phi_{ijr} , \qquad (15.14)$$

with the obvious corresponding modifications to the likelihood (15.11). Here, $\phi_{ir} = (\phi_{i1r}, \phi_{i2r}, ..., \phi_{ipr})^T$ and $\boldsymbol{\Phi}_r = (\phi_{1r}^T, \ldots, \phi_{nr}^T)^T \overset{ind}{\sim} MCAR(\rho_r, \Lambda_r)$. This permits addition of an exchangeable prior structure,

$$\rho_r \overset{iid}{\sim} Beta(a, b) \text{ and } \Lambda_r \overset{iid}{\sim} Wishart(\rho, \Lambda_0) ,$$

where we may choose fixed values for a, b, ρ, and Λ_0, or place hyperpriors on them and estimate them from the data. Note also the obvious extension to disease-specific ρ_{jr}, as mentioned at the end of the previous subsection.

Example 15.4 *(Application to Iowa SEER multiple cancer survival data).* We illustrate the approach with an analysis of SEER data on 17,146 patients from the 99 counties of the state of Iowa who have been diagnosed with cancer between 1992 and 1998, and who have a well-identified primary cancer. Our covariate vector $\mathbf{x}_{ijk}$ consists of a constant (intercept), a gender indicator, the age of the patient, indicators for race with "white" as the baseline, indicators for the stage of the primary cancer with "local" as the baseline, and indicators for the year of primary cancer diagnosis (cohort) with the first year (1992) as the baseline. The vector $\mathbf{z}_{ijk}$ comprises the indicators of which cancers the patient has; the corresponding parameters will thus capture the effect of these cancers on the hazards regardless of whether they emerge as primary or secondary.

With regard to modeling details, we used five separate (cancer-specific) propriety parameters ρ_j having an exchangeable $Beta(18, 2)$ prior, and a vague $Wishart(\rho, \Lambda_0)$ for Λ, where $\rho = 5$ and $\Lambda_0 = 0.01 I_{5 \times 5}$. (Results for $\boldsymbol{\beta}, \boldsymbol{\theta}$, and $\boldsymbol{\Phi}$ under a $U(0, 1)$ prior for the ρ_j were broadly similar.) Table 15.9 gives posterior summaries for the main effects $\boldsymbol{\beta}$ and $\boldsymbol{\theta}$; note that $\boldsymbol{\theta}$ is estimable despite the presence of the intercept since many individuals have more than one cancer. No race or cohort effects emerged as significantly different from zero, so they have been deleted; all remaining effects are shown here. All of these effects are significant and in the directions one would expect. In particular, the five cancer effects are consistent with results of previous modeling of this and similar data sets, with pancreatic cancer emerging as the most deadly (posterior median log relative hazard 1.701) and colorectal and small intestinal cancer relatively less so (.252 and .287, respectively).

Table 15.10 gives posterior variance and correlation summaries for the frailties ϕ_{ij} among the five cancers for two representative counties, Dallas (urban; Des Moines area) and Clay (rural northwest). Note that the correlations are as high as 0.528 (pancreas and stomach in Dallas County), suggesting the need for the multivariate structure inherent in our MCAR frailty model. Note also that summarizing the posterior distribution of Λ^{-1} would be inappropriate here, since, despite the Kronecker structure here, Λ^{-1} cannot be directly interpreted as a primary cancer covariance matrix across counties.

Variable	2.5%	50%	97.5%
Intercept	0.102	0.265	0.421
Sex (female = 0)	0.097	0.136	0.182
Age	0.028	0.029	0.030
Stage of primary cancer (local = 0)			
regional	0.322	0.373	0.421
distant	1.527	1.580	1.654
Type of primary cancer			
colorectal	0.112	0.252	0.453
gallbladder	1.074	1.201	1.330
pancreas	1.603	1.701	1.807
small intestine	0.128	0.287	0.445
stomach	1.005	1.072	1.141

Table 15.9 *Posterior quantiles for the fixed effects in the MCAR frailty model.*

Dallas County	Colo-rectal	Gall-bladder	Pancreas	Small intestine	Stomach
Colorectal	0.852	0.262	0.294	0.413	0.464
Gallbladder		1.151	0.314	0.187	0.175
Pancreas			0.846	0.454	0.528
Small intestine				1.47	0.413
Stomach					0.908

Clay County	Colo-rectal	Gall-bladder	Pancreas	Small intestine	Stomach
Colorectal	0.903	0.215	0.273	0.342	0.352
Gallbladder		1.196	0.274	0.128	0.150
Pancreas			0.852	0.322	0.402
Small intestine				1.515	0.371
Stomach					1.068

Table 15.10 *Posterior variances and correlation summaries, Dallas and Clay Counties, MCAR spatial frailty model. Diagonal elements are estimated variances, while off-diagonal elements are estimated correlations.*

Turning to geographic summaries, Figure 15.7 shows ArcView maps of the posterior means of the MCAR spatial frailties ϕ_{ij}. Recall that in this model, the ϕ_{ij} play the role of spatial residuals, capturing any spatial variation not already accounted for by the spatial main effects β and θ. The lack of spatial pattern in these maps suggests there is little additional spatial "story" in the data beyond what is already being told by the fixed effects. However, the map scales reveal that one cancer (gallbladder) is markedly different from the others, both in terms of total range of the mean frailties (rather broad) and their center (negative; the other four are centered near 0).

Next, we change from the MCAR spatial frailty model to the MCAR spatially varying coefficients model (15.13). This model required a longer burn-in period but otherwise, our prior and MCMC control parameters remain unchanged. Figure 15.8 shows ArcView maps of the resulting posterior means of the spatially varying coefficients θ_{ij}. Unlike the ϕ_{ij} in the previous model, these parameters are not "residuals," but the effects of the presence of primary cancer indicated on the death rate in each county. Clearly, these maps show a strong spatial pattern, with (for example) southwestern Iowa counties having relatively high fitted values for pancreatic and stomach cancer, while southeastern counties fare relatively

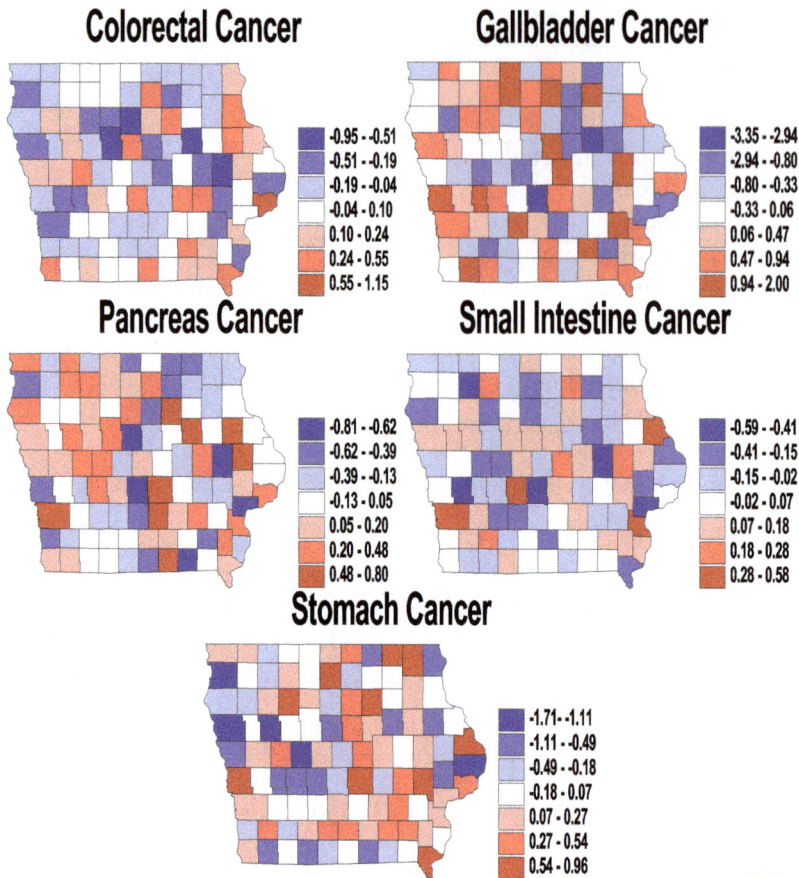

Figure 15.7 *Posterior mean spatial frailties, Iowa cancer data, static spatial MCAR model.*

poorly with respect to colorectal and small intestinal cancer. The overall levels for each cancer are consistent with those given for the corresponding fixed effects θ in Table 15.9 for the spatial frailty model.

Table 15.11 gives the effective model sizes p_D and DIC scores for a variety of spatial survival models. The first two listed (fixed effects only and standard CAR frailty) have few effective parameters, but also poor (large) DIC scores. The MCAR spatial frailty models (which place the MCAR on $\mathbf{\Phi}$) fare better, especially when we add the disease-specific ρ_j (the model summarized in Table 15.9, Table 15.10, and Figure 15.7). However, adding heterogeneity effects ϵ_{ij} to this model adds essentially no extra effective parameters, and is actually harmful to the overall DIC score (since we are adding complexity for little or no benefit in terms of fit). Finally, the two spatially varying coefficients models enjoy the best (smallest) DIC scores, but only by a small margin over the best spatial frailty model.

Finally, we fit the spatiotemporal extension (15.14) of our MCAR frailty model to the data where the cohort effect (year of study entry r) is taken into account. Year-by-year box-plots of the posterior median frailties (Figure 15.9) reveal the expected steadily decreasing trend for all five cancers, though it is not clear how much of this decrease is simply an artifact of the censoring of survival times for patients in more recent cohorts. The spatiotemporal extension of the spatially varying coefficients model (15.13) (i.e., $\mathbf{x}_{ijkr}^{T}\boldsymbol{\beta} + \mathbf{z}_{ijkr}^{T}\boldsymbol{\theta}_{ir}$) might well produce results that are temporally more interesting in this case. Incorporating change points, cancer start date measurement errors, and other model enhancements (say, interval censoring) might also be practically important model enhancements here. ∎

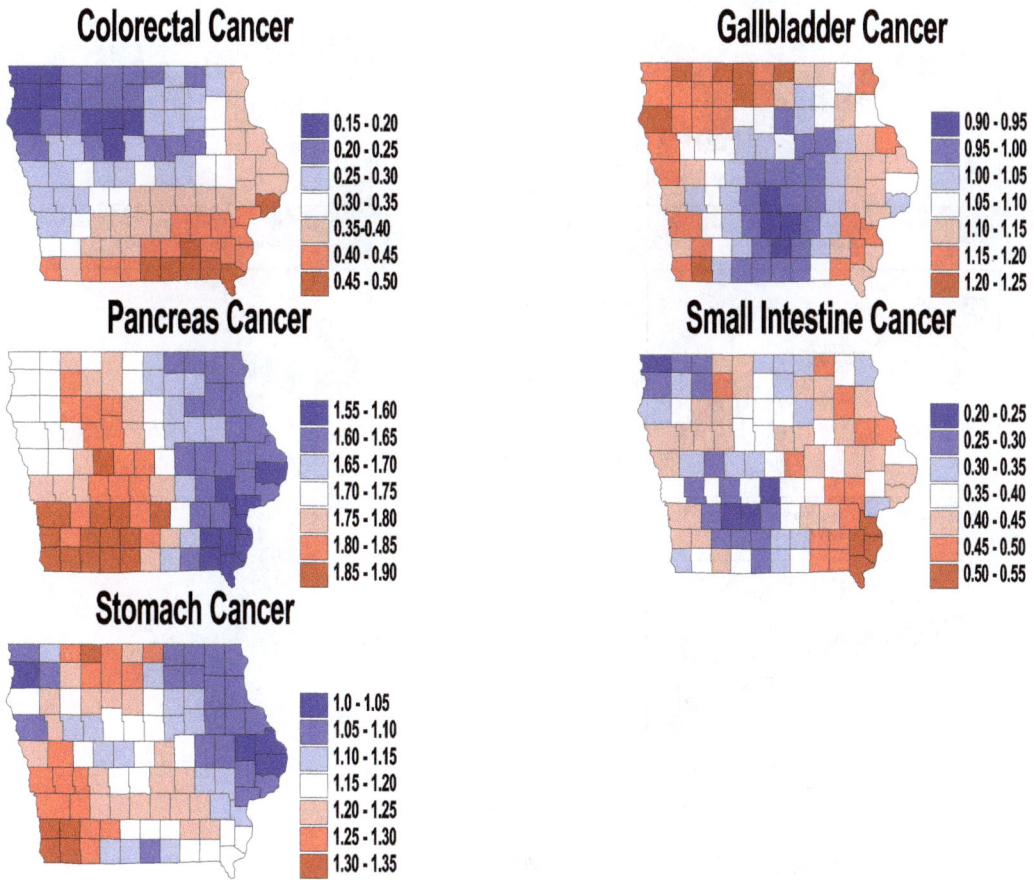

Figure 15.8 *Posterior mean spatially varying coefficients, Iowa cancer data, static spatial MCAR model.*

Log-relative hazard model	p_D	DIC
$\mathbf{x}_{ijk}^{\mathrm{T}}\boldsymbol{\beta} + \mathbf{z}_{ijk}^{\mathrm{T}}\boldsymbol{\theta}$	10.97	642
$\mathbf{x}_{ijk}^{\mathrm{T}}\boldsymbol{\beta} + \mathbf{z}_{ijk}^{\mathrm{T}}\boldsymbol{\theta} + \phi_i,\ \boldsymbol{\phi} \sim CAR(\rho, \lambda)$	103.95	358
$\mathbf{x}_{ijk}^{\mathrm{T}}\boldsymbol{\beta} + \mathbf{z}_{ijk}^{\mathrm{T}}\boldsymbol{\theta} + \phi_{ij},\ \boldsymbol{\Phi} \sim MCAR(\rho = 1, \Lambda)$	172.75	247
$\mathbf{x}_{ijk}^{\mathrm{T}}\boldsymbol{\beta} + \mathbf{z}_{ijk}^{\mathrm{T}}\boldsymbol{\theta} + \phi_{ij},\ \boldsymbol{\Phi} \sim MCAR(\rho, \Lambda)$	172.40	246
$\mathbf{x}_{ijk}^{\mathrm{T}}\boldsymbol{\beta} + \mathbf{z}_{ijk}^{\mathrm{T}}\boldsymbol{\theta} + \phi_{ij},\ \boldsymbol{\Phi} \sim MCAR(\rho_1, \ldots, \rho_5, \Lambda)$	175.71	237
$\mathbf{x}_{ijk}^{\mathrm{T}}\boldsymbol{\beta} + \mathbf{z}_{ijk}^{\mathrm{T}}\boldsymbol{\theta} + \phi_{ij} + \epsilon_{ij},\ \boldsymbol{\Phi} \sim MCAR(\rho_1, \ldots, \rho_5, \Lambda),$ $\epsilon_{ij} \stackrel{iid}{\sim} N(0, \tau^2)$	177.25	255
$\mathbf{x}_{ijk}^{\mathrm{T}}\boldsymbol{\beta} + \mathbf{z}_{ijk}^{\mathrm{T}}\boldsymbol{\theta}_i,\ \boldsymbol{\Theta} \sim MCAR(\rho, \Lambda)$	169.42	235
$\mathbf{x}_{ijk}^{\mathrm{T}}\boldsymbol{\beta} + \mathbf{z}_{ijk}^{\mathrm{T}}\boldsymbol{\theta}_i,\ \boldsymbol{\Theta} \sim MCAR(\rho_1, \ldots, \rho_5, \Lambda)$	171.46	229

Table 15.11 *DIC comparison, spatial survival models for the Iowa cancer data.*

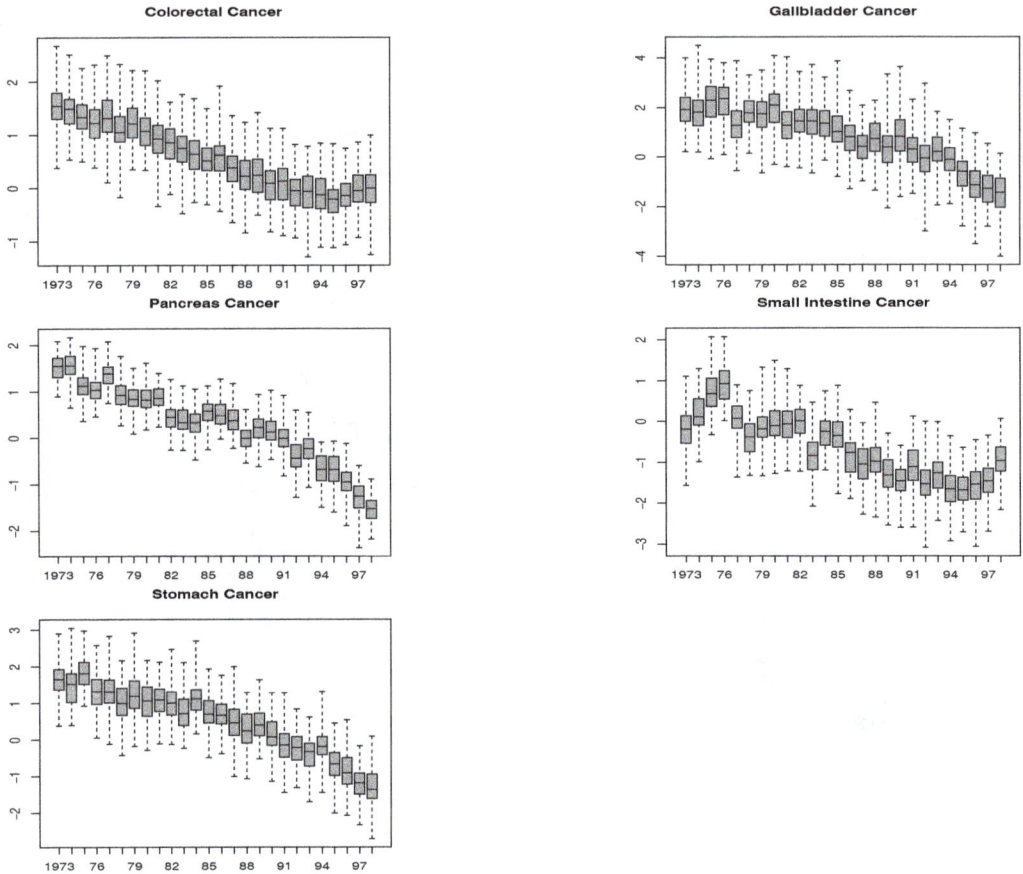

Figure 15.9 *Boxplots of posterior medians for the spatial frailties ϕ_{ijr} over the 99 Iowa counties for each year, $r = 1973, \ldots, 1998$.*

15.5 Spatial cure rate models $\star$

In Section 15.1 we investigated spatially correlated frailties in traditional parametric survival models, choosing a random effects distribution to reflect the spatial structure in the problem. Sections 15.2 and 15.3 extended this approach to spatial and spatiotemporal settings within a semiparametric model.

In this section our ultimate goal is the proper analysis of a geographically referenced smoking cessation study, in which we observe subjects periodically through time to check for relapse following an initial quit attempt. Each patient is observed once each year for five consecutive years, whereupon the current average number of cigarettes smoked at each visit is recorded, along with the zip code of residence and several other potential explanatory variables. This data set requires us to extend the work of Carlin and Hodges [1999] in a number of ways. The primary extension involves the incorporation of a *cure fraction* in our models. In investigating the effectiveness of quitting programs, data typically reveal many former smokers having successfully given up smoking, and as such may be thought of as "cured" of the deleterious habit. Incorporating such cure fractions in survival models leads to *cure rate models*, which are often applied in survival settings where the endpoint is a particular disease (say, breast cancer) which the subject may never reexperience. These models have a long history in the biostatistical literature, dating to Berkson and Gage [1952]. This model has been extensively studied in the statistical literature by a number of authors, including Farewell [1982] and Farewell [1986], Goldman [1984] and Ewell and Ibrahim [1997]. Cure rates have been studied in more general settings by Chen et al. [1999] and Cooner et al. [2007] following earlier work by Yakovlev et al. [1996].

In addition, while this design can be analyzed as an ordinary right-censored survival model (with relapse to smoking as the endpoint), the data are perhaps more accurately viewed as *interval-censored*, since we actually observe only approximately annual intervals within which a failed quitter resumed smoking. We will consider both right- and interval-censored models, where in the former case we simply approximate the time of relapse by the midpoint of the corresponding time interval. Finally, we capture spatial variation through zip code-specific spatial random effects in the cure fraction or the hazard function, which in either case may act as spatial frailties. We find that incorporating the covariates and frailties into the hazard function is most natural (both intuitively and methodologically), especially after adopting a Weibull form for the baseline hazard.

15.5.1 Models for right- and interval-censored data

15.5.1.1 Right-censored data

Our cure rate models are based on those of Chen et al. [1999] and derived assuming that some latent biological process is generating the observed data. Suppose there are I regions and n_i patients in the ith region. We denote by T_{ij} the random variable for time to event (relapse, in our case) of the jth person in the ith region, where $j = 1, 2, ..., n_i$ and $i = 1, 2, ..., I$. (While acknowledging the presence of the regions in our notation, we postpone explicit spatial modeling to the next section.) Suppose that the (i, j)th individual has N_{ij} potential latent (unobserved) risk factors, the presence of any of which (i.e., $N_{ij} \geq 1$) will ultimately lead to the event. For example, in cancer settings these factors may correspond to metastasis-competent tumor cells within the individual. Typically, there will be a number of subjects who do not undergo the event during the observation period and are therefore considered censored. Thus, letting U_{ijk}, $k = 1, 2, ..., N_{ij}$ be the time to an event arising from the kth latent factor for the (i, j)th individual, the observed time to event for an uncensored individual is generated by $T_{ij} = \min \{U_{ijk}, k = 1, 2, ..., N_{ij}\}$. If the (i, j)th individual is right-censored at time t_{ij}, none of the latent factors have led to an event by that time, and clearly $T_{ij} > t_{ij}$ (and in fact $T_{ij} = \infty$ if $N_{ij} = 0$).

Given N_{ij}, the U_{ijk}'s are independent with survival function $S(t|\Psi_{ij})$ and corresponding density function $f(t|\Psi_{ij})$. The parameter Ψ_{ij} is a collection of all the parameters (including possible regression parameters) that may be involved in a parametric specification for the survival function S. In this section we will work with a two-parameter Weibull distribution specification for the density function $f(t|\Psi_{ij})$, where we allow the Weibull scale parameter ρ to vary across the regions, and η, which may serve as a link to covariates in a regression setup, to vary across individuals. Therefore $f(t|\rho_i, \eta_{ij}) = \rho_i t^{\rho_i - 1} \exp(\eta_{ij} - t^{\rho_i} \exp(\eta_{ij}))$.

In terms of the hazard function h, $f(t|\rho_i, \eta_{ij}) = h(t|\rho_i, \eta_{ij}) S(t|\rho_i, \eta_{ij})$, with $h(t; \rho_i, \eta_{ij}) = \rho_i t^{\rho_i - 1} \exp(\eta_{ij})$ and $S(t|\rho_i, \eta_{ij}) = \exp(-t^{\rho_i} \exp(\eta_{ij}))$. Note we implicitly assume proportional hazards, with baseline hazard function $h_0(t|\rho_i) = \rho_i t^{\rho_i - 1}$. Thus an individual ij who is censored at time t_{ij} before undergoing the event contributes $(S(t_{ij}|\rho_i, \eta_{ij}))^{N_{ij}}$ to the likelihood, while an individual who experiences the event at time t_{ij} contributes $N_{ij} (S(t_{ij}|\rho_i, \eta_{ij}))^{N_{ij}-1} f(t_{ij}|\rho_i, \eta_{ij})$. The latter expression follows from the fact that the event is experienced when any one of the latent factors occurs. Letting ν_{ij} be the observed event indicator for individual ij, this person contributes

$$L(t_{ij}|N_{ij}, \rho_i, \eta_{ij}, \nu_{ij})$$
$$= (S(t_{ij}|\rho_i, \eta_{ij}))^{N_{ij}(1-\nu_{ij})} \left(N_{ij} S(t_{ij}|\rho_i, \eta_{ij})^{N_{ij}-1} f(t_{ij}|\rho_i, \eta_{ij})\right)^{\nu_{ij}},$$

and the joint likelihood for all the patients can now be expressed as

$$
\begin{aligned}
& L\left(\{t_{ij}\} \mid \{N_{ij}\},\{\rho_i\},\{\eta_{ij}\},\{\nu_{ij}\}\right) \\
&= \prod_{i=1}^{I} \prod_{j=1}^{n_i} L\left(t_{ij} \mid N_{ij}, \rho_i, \eta_{ij}, \nu_{ij}\right) \\
&= \prod_{i=1}^{I} \prod_{j=1}^{n_i}\left(S\left(t_{ij} \mid \rho_i, \eta_{ij}\right)\right)^{N_{ij}(1-\nu_{ij})} \\
& \quad \times\left(N_{ij} S\left(t_{ij} \rho_i, \eta_{ij}\right)^{N_{ij}-1} f\left(t_{ij} \mid \rho_i, \eta_{ij}\right)\right)^{\nu_{ij}} \\
&= \prod_{i=1}^{I} \prod_{j=1}^{n_i}\left(S\left(t_{ij} \mid \rho_i, \eta_{ij}\right)\right)^{N_{ij}-\nu_{ij}}\left(N_{ij} f\left(t_{ij} \mid \rho_i, \eta_{ij}\right)\right)^{\nu_{ij}}.
\end{aligned}
$$

This expression can be rewritten in terms of the hazard function as

$$
\prod_{i=1}^{I} \prod_{j=1}^{n_i}\left(S\left(t_{ij} \mid \rho_i, \eta_{ij}\right)\right)^{N_{ij}}\left(N_{ij} h\left(t_{ij} \mid \rho_i, \eta_{ij}\right)\right)^{\nu_{ij}}. \tag{15.15}
$$

A Bayesian hierarchical formulation is completed by introducing prior distributions on the parameters. We will specify independent prior distributions $p\left(N_{ij} \mid \theta_{ij}\right)$, $p\left(\rho_i \mid \psi_\rho\right)$, and $p\left(\eta_{ij} \mid \psi_\eta\right)$ for $\{N_{ij}\}$, $\{\rho_i\}$, and $\{\eta_{ij}\}$, respectively. Here, ψ_ρ, ψ_η, and $\{\theta_{ij}\}$ are appropriate hyperparameters. Assigning independent hyperpriors $p\left(\theta_{ij} \mid \psi_\theta\right)$ for $\{\theta_{ij}\}$ and assuming the hyperparameters $\psi=\left(\psi_\rho, \psi_\eta, \psi_\theta\right)$ to be fixed, the posterior distribution for the parameters, $p\left(\{\theta_{ij}\},\{\eta_{ij}\},\{N_{ij}\},\{\rho_i\} \mid \{t_{ij}\},\{\nu_{ij}\}\right)$, is easily found (up to a proportionality constant) using (15.15) as

$$
\begin{aligned}
\prod_{i=1}^{I} & \left\{p\left(\rho_i \mid \psi_\rho\right) \prod_{j=1}^{n_i}\left[S\left(t_{ij} \mid \rho_i, \eta_{ij}\right)\right]^{N_{ij}}\left[N_{ij} h\left(t_{ij} \mid \rho_i, \eta_{ij}\right)\right]^{\nu_{ij}}\right. \\
& \left. \times p\left(N_{ij} \mid \theta_{ij}\right) p\left(\eta_{ij} \mid \psi_\eta\right) p\left(\theta_{ij} \mid \psi_\theta\right)\right\}.
\end{aligned}
$$

Chen et al. [1999] assume that the N_{ij}'s are are distributed as independent Poisson random variables with mean θ_{ij}, i.e., $p\left(N_{ij} \mid \theta_{ij}\right)$ is *Poisson* $\left(\theta_{ij}\right)$. In this setting it is easily seen that the survival distribution for the (i, j)th patient, $P\left(T_{ij} \geq t_{ij} \mid \rho_i, \eta_{ij}\right)$, is given by $\exp \left\{-\theta_{ij}\left(1-S\left(t_{ij} \mid \rho_i, \eta_{ij}\right)\right)\right\}$. Since $S\left(t_{ij} \mid \rho_i, \eta_{ij}\right)$ is a proper survival function (corresponding to the latent factor times U_{ijk}, as $t_{ij} \rightarrow \infty$, $P\left(T_{ij} \geq t_{ij} \mid \rho_i, \eta_{ij}\right) \rightarrow \exp \left(-\theta_{ij}\right)>0$. Thus we have a subdistribution for T_{ij} with a *cure fraction* given by $\exp \left(-\theta_{ij}\right)$. Here a hyperprior on the θ_{ij}'s would have support on the positive real line.

While there could certainly be multiple latent factors that increase the risk of smoking relapse (age started smoking, occupation, amount of time spent driving, tendency toward addictive behavior, etc.), this is rather speculative and certainly not as justifiable as in the cancer setting for which the multiple factor approach was developed (where $N_{ij}>1$ is biologically motivated). As such, we instead form our model using a single, omnibus, "propensity for relapse" latent factor. In this case, we think of N_{ij} as a *binary* variable and specify $p\left(N_{ij} \mid \theta_{ij}\right)$ as Bernoulli $\left(1-\theta_{ij}\right)$. In this setting it is easier to look at the survival distribution after marginalizing out the N_{ij}. In particular, note that

$$
P\left(T_{ij} \geq t_{ij} \mid \rho_i, \eta_{ij}, N_{ij}\right)=\left\{\begin{array}{cl}
S\left(t_{ij} \mid \rho_i, \eta_{ij}\right), & N_{ij}=1 \\
1, & N_{ij}=0
\end{array}\right. .
$$

That is, if the latent factor is absent, the subject is cured (does not experience the event). Marginalizing over the Bernoulli distribution for N_{ij}, we obtain for the (i, j)th patient the survival function $S^*\left(t_{ij} \mid \theta_{ij}, \rho_i, \eta_{ij}\right) \equiv P\left(T_{ij} \geq t_{ij} \mid \rho_i, \eta_{ij}\right)=\theta_{ij}+\left(1-\theta_{ij}\right) S\left(t_{ij} \mid \rho_i, \eta_{ij}\right)$, which is the classic cure-rate model attributed to Berkson and Gage [1952] with cure fraction θ_{ij}. Now we can write the likelihood function for the data marginalized over $\{N_{ij}\}$, $L\left(\{t_{ij}\} \mid \{\rho_i\},\{\theta_{ij}\},\{\eta_{ij}\},\{\nu_{ij}\}\right)$, as

$$
\begin{aligned}
\prod_{i=1}^{I} & \prod_{j=1}^{n_i}\left[S^*\left(t_{ij} \mid \theta_{ij}, \rho_i, \eta_{ij}\right)\right]^{1-\nu_{ij}}\left(-\frac{d}{dt_{ij}} S^*\left(t_{ij} \mid \theta_{ij}, \rho_i, \eta_{ij}\right)\right)^{\nu_{ij}} \\
&= \prod_{i=1}^{I} \prod_{j=1}^{n_i}\left[S^*\left(t_{ij} \mid \theta_{ij}, \rho_i, \eta_{ij}\right)\right]^{1-\nu_{ij}}\left[\left(1-\theta_{ij}\right) f\left(t_{ij} \mid \rho_i, \eta_{ij}\right)\right]^{\nu_{ij}},
\end{aligned}
$$

which in terms of the hazard function becomes

$$\prod_{i=1}^{I} \prod_{j=1}^{n_i} \left[S^* \left(t_{ij}|\theta_{ij}, \rho_i, \eta_{ij}\right)\right]^{1-\nu_{ij}} \left[(1-\theta_{ij}) S\left(t_{ij}|\rho_i, \eta_{ij}\right) h\left(t_{ij}|\rho_i, \eta_{ij}\right)\right]^{\nu_{ij}}, \qquad (15.16)$$

where the hyperprior for θ_{ij} has support on $(0,1)$. Now the posterior distribution of the parameters is proportional to

$$L\left(\{t_{ij}\} \mid \{\rho_i\}, \{\theta_{ij}\}, \{\eta_{ij}\}, \{\nu_{ij}\}\right) \prod_{i=1}^{I} \left\{ p\left(\rho_i|\psi_\rho\right) \prod_{j=1}^{n_i} p\left(\eta_{ij}|\psi_\eta\right) p\left(\theta_{ij}|\psi_\theta\right) \right\}. \qquad (15.17)$$

Turning to the issue of incorporating covariates, in the general setting with N_{ij} assumed to be distributed Poisson, Chen et al. [1999] propose their introduction in the cure fraction through a suitable link function g, so that $\theta_{ij} = g\left(\mathbf{x}_{ij}^{\mathrm{T}}\widetilde{\boldsymbol{\beta}}\right)$, where g maps the entire real line to the positive axis. This is sensible when we believe that the risk factors affect the probability of an individual being cured. Proper posteriors arise for the regression coefficients $\widetilde{\boldsymbol{\beta}}$ even under improper priors. Unfortunately, this is no longer true when N_{ij} is Bernoulli (i.e., in the Berkson and Gage model). Vague but proper priors may still be used, but this makes the parameters difficult to interpret, and can often lead to poor MCMC convergence.

Since a binary N_{ij} seems most natural in our setting, we instead introduce covariates into $S\left(t_{ij}|\rho_i, \eta_{ij}\right)$ through the Weibull link η_{ij}, i.e., we let $\eta_{ij} = \mathbf{x}_{ij}^{\mathrm{T}}\boldsymbol{\beta}$. This seems intuitively more reasonable anyway since now the covariates influence the underlying factor that brings about the smoking relapse (and thus the rapidity of this event). Also, proper posteriors arise here for $\boldsymbol{\beta}$ under improper posteriors even though N_{ij} is binary. As such, henceforth we will only consider the situation where the covariates enter the model in this way (through the Weibull link function). This means we are unable to separately estimate the effect of the covariates on both the *rate* of relapse and the *ultimate level* of relapse, but "fair" estimation here (i.e., allocating the proper proportions of the covariates' effects to each component) is not clear anyway since flat priors could be selected for $\boldsymbol{\beta}$, but not for $\widetilde{\boldsymbol{\beta}}$. Finally, all of our subsequent models also assume a constant cure fraction for the entire population (i.e., we set $\theta_{ij} = \theta$ for all i, j).

Note that the posterior distribution in (15.17) is easily modified to incorporate covariates. For example, with $\eta_{ij} = \mathbf{x}_{ij}^{\mathrm{T}}\boldsymbol{\beta}$, we replace $\prod_{ij} p\left(\eta_{ij}|\psi_\eta\right)$ in (15.17) with $p\left(\boldsymbol{\beta}|\psi_\beta\right)$, with ψ_β as a fixed hyperparameter. Typically a flat or vague Gaussian prior may be taken for $p\left(\boldsymbol{\beta}|\psi_\beta\right)$.

15.5.1.2 *Interval-censored data*

The formulation above assumes that our observed data are right-censored. This means that we are able to observe the actual relapse time t_{ij} when it occurs prior to the final office visit. In reality, our study (like many others of its kind) is only able to determine patient status at the office visits themselves, meaning we observe only a time *interval* (t_{ijL}, t_{ijU}) within which the event (in our case, smoking relapse) is known to have occurred. For patients who did not resume smoking prior to the end of the study, we have $t_{ijU} = \infty$, returning us to the case of right-censoring at time point t_{ijL}. Thus we now set $\nu_{ij} = 1$ if subject ij is interval-censored (i.e., experienced the event), and $\nu_{ij} = 0$ if the subject is right-censored.

Following Finkelstein [1986], the general interval-censored cure rate likelihood, $L\left(\{(t_{ijL}, t_{ijU})\} \mid \{N_{ij}\}, \{\rho_i\}, \{\eta_{ij}\}, \{\nu_{ij}\}\right)$, is given by

$$\prod_{i=1}^{I} \prod_{j=1}^{n_i} [S(t_{ijL}|\rho_i, \eta_{ij})]^{N_{ij}-\nu_{ij}} \{N_{ij}[S(t_{ijL}|\rho_i, \eta_{ij}) - S(t_{ijU}|\rho_i, \eta_{ij})]\}^{\nu_{ij}}$$

$$= \prod_{i=1}^{I} \prod_{j=1}^{n_i} [S(t_{ijL}|\rho_i, \eta_{ij})]^{N_{ij}} \left\{ N_{ij}\left(1 - \frac{S(t_{ijU}|\rho_i, \eta_{ij})}{S(t_{ijL}|\rho_i, \eta_{ij})}\right) \right\}^{\nu_{ij}}.$$

As in the previous section, in the Bernoulli setup after marginalizing out the $\{N_{ij}\}$ the foregoing becomes $L\left(\{(t_{ijL}, t_{ijU})\} \mid \{\rho_i\}, \{\theta_{ij}\}, \{\eta_{ij}\}, \{\nu_{ij}\}\right)$, and can be written as

$$\prod_{i=1}^{I} \prod_{j=1}^{n_i} S^*(t_{ijL}|\theta_{ij}, \rho_i, \eta_{ij}) \left\{ 1 - \frac{S^*(t_{ijU}|\theta_{ij}, \rho_i, \eta_{ij})}{S^*(t_{ijL}|\theta_{ij}, \rho_i, \eta_{ij})} \right\}^{\nu_{ij}}. \tag{15.18}$$

We omit details (similar to those in the previous section) arising from the Weibull parametrization and subsequent incorporation of covariates through the link function η_{ij}.

15.5.2 Spatial frailties in cure rate models

The development of the hierarchical framework in the preceding section acknowledged the data as coming from I different geographical regions (clusters). Such clustered data are common in survival analysis and often modeled using cluster-specific frailties ϕ_i. As with the covariates, we will introduce the frailties ϕ_i through the Weibull link as intercept terms in the log-relative risk; that is, we set $\eta_{ij} = \mathbf{x}_{ij}^{\mathrm{T}}\boldsymbol{\beta} + \phi_i$.

Here we allow the ϕ_i to be spatially correlated across the regions; similarly, we would like to permit the Weibull baseline hazard parameters, ρ_i, to be spatially correlated. A natural approach in both cases is to use a univariate CAR prior. While one may certainly employ separate, independent CAR priors on $\boldsymbol{\phi}$ and $\boldsymbol{\zeta} \equiv \{\log \rho_i\}$, another option is to allow these two spatial priors to themselves be correlated. In other words, we may want a bivariate spatial model for the $\delta_i = (\phi_i, \zeta_i)^{\mathrm{T}} = (\phi_i, \log \rho_i)^{\mathrm{T}}$. As mentioned in Sections 11.1 and 15.4, we may use the MCAR distribution for this purpose. In our setting, the MCAR distribution on the concatenated vector $\boldsymbol{\delta} = (\boldsymbol{\phi}^{\mathrm{T}}, \boldsymbol{\zeta}^{\mathrm{T}})^{\mathrm{T}}$ is Gaussian with mean $\mathbf{0}$ and precision matrix $\Lambda^{-1} \otimes (Diag(m_i) - \rho W)$, where Λ is a 2×2 symmetric and positive definite matrix, $\rho \in (0, 1)$, and m_i and W remain as above. In the current context, we may also wish to allow different smoothness parameters (say, ρ_1 and ρ_2) for $\boldsymbol{\phi}$ and $\boldsymbol{\zeta}$, respectively, as in Section 15.4. Henceforth, in this section we will denote the proper MCAR with a common smoothness parameter by $MCAR(\rho, \Lambda)$, and the multiple smoothness parameter generalized MCAR by $MCAR(\rho_1, \rho_2, \Lambda)$. Combined with independent (univariate) CAR models for $\boldsymbol{\phi}$ and $\boldsymbol{\zeta}$, these offer a broad range of potential spatial models.

15.5.3 Model comparison

Suppose we let Ω denote the set of all model parameters, so that the deviance statistic (recall (5.65) in Section 5.6.2) becomes

$$D(\Omega) = -2\log f(\mathbf{y}|\Omega) + 2\log h(\mathbf{y}). \tag{15.19}$$

When DIC is used to compare nested models in standard exponential family settings, the unnormalized likelihood $L(\Omega; \mathbf{y})$ is often used in place of the normalized form $f(\mathbf{y} \mid \Omega)$ in (15.19), since in this case the normalizing function $m(\Omega) = \int L(\Omega; \mathbf{y}) d\mathbf{y}$ will be free of Ω and constant across models, hence contribute equally to the DIC scores of each (and thus have no impact on model selection). However, in settings where we require comparisons across different likelihood distributional forms, it appears one must be careful to use the properly scaled joint density $f(\mathbf{y}|\Omega)$ for each model.

We argue that use of the usual proportional hazards likelihood (which of course is not a joint density function) *is* in fact appropriate for DIC computation here, provided we make a fairly standard assumption regarding the relationship between the survival and censoring mechanisms generating the data. Specifically, suppose the distribution of the censoring times is independent of that of the survival times *and* does not depend upon the survival model parameters (i.e., independent, noninformative censoring). Let $g(t_{ij})$ denote the density of the censoring time for the ijth individual, with the corresponding survival (1-cdf) function $R(t_{ij})$. Then the right-censored likelihood (15.16) can be extended to the joint likelihood specification,

$$\prod_{i=1}^{I} \prod_{j=1}^{n_i} [S^* (t_{ij}|\theta_{ij}, \rho_i, \eta_{ij})]^{1-\nu_{ij}}$$
$$\times [(1 - \theta_{ij}) S (t_{ij}|\rho_i, \eta_{ij}) h (t_{ij}|\rho_i, \eta_{ij})]^{\nu_{ij}} [R (t_{ij})]^{\nu_{ij}} [g (t_{ij})]^{1-\nu_{ij}} ,$$

as for example in [Le, 1997, pp. 69–70]. While not a joint probability density, this likelihood is still an everywhere nonnegative and integrable function of the survival model parameters Ω, and thus suitable for use with the Kullback-Leibler divergences that underlie DIC [Spiegelhalter et al., 2002, p. 586]. But by assumption, $R(t)$ and $g(t)$ do not depend upon Ω. Thus, like an $m(\Omega)$ that is free of Ω, they may be safely ignored in both the p_D and DIC calculations. Note this same argument implies that we can use the unnormalized likelihood (15.16) when comparing not only nonnested parametric survival models (say, Weibull versus gamma), but even parametric and semiparametric models (say, Weibull versus Cox) provided our definition of "likelihood" is comparable across models.

Note also that here our "focus" [in the nomenclature of Spiegelhalter et al., 2002] is solely on Ω. An alternative would be instead to use a missing data formulation, where we include the likelihood contribution of $\{s_{ij}\}$, the collection of latent survival times for the right-censored individuals. Values for both Ω and the $\{s_{ij}\}$ could then be imputed along the lines given by [Cox and Oakes, 1984, pp. 165–166] for the EM algorithm or [Spiegelhalter et al., 1995, the "mice" example] for the Gibbs sampler. This would alter our focus from Ω to $(\Omega, \{s_{ij}\})$, and p_D would reflect the correspondingly larger effective parameter count.

Turning to the interval censored case, here matters are only a bit more complicated. Converting the interval-censored likelihood (15.18) to a joint likelihood specification yields

$$\prod_{i=1}^{I} \prod_{j=1}^{n_i} S^* (t_{ijL}|\theta_{ij}, \rho_i, \eta_{ij}) \left(1 - \frac{S^* (t_{ijU}|\theta_{ij}, \rho_i, \eta_{ij})}{S^* (t_{ijL}|\theta_{ij}, \rho_i, \eta_{ij})} \right)^{\nu_{ij}}$$

$$\times [R (t_{ijL})]^{\nu_{ij}} \left(1 - \frac{R (t_{ijU})}{R (t_{ijL})} \right)^{\nu_{ij}} [g (t_{ijL})]^{1-\nu_{ij}} .$$

Now $[R (t_{ijL})]^{\nu_{ij}} (1 - R (t_{ijU}) / R (t_{ijL}))^{\nu_{ij}} [g (t_{ijL})]^{1-\nu_{ij}}$ is the function absorbed into $m(\Omega)$, and is again free of Ω. Thus again, use of the usual form of the interval-censored likelihood presents no problems when comparing models within the interval-censored framework (including nonnested parametric models, or even parametric and semiparametric models).

Note that it does *not* make sense to compare a particular right-censored model with a particular interval-censored model. The form of the available data is different; model comparison is only appropriate to a given data set.

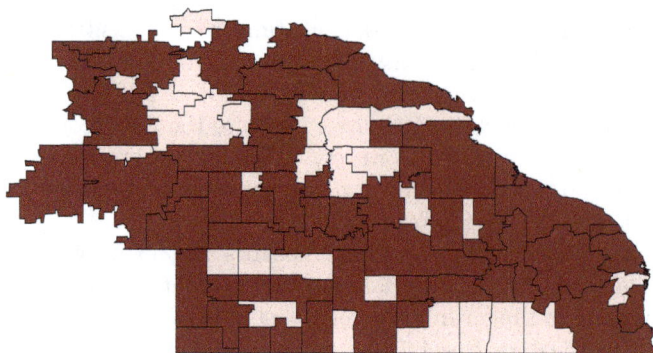

Figure 15.10 *Map showing missingness pattern for the smoking cessation data: lightly shaded regions are those having no responses.*

Model	Log-relative risk	pD	DIC
1	$\mathbf{x}_{ij}^{\mathrm{T}}\boldsymbol{\beta} + \phi_i;\ \phi_i \overset{iid}{\sim} N(0,\tau_\phi),\ \rho_i = \rho\ \forall\ i$	10.3	438
2	$\mathbf{x}_{ij}^{\mathrm{T}}\boldsymbol{\beta} + \phi_i;\ \{\phi_i\} \sim CAR(\lambda_\phi),\ \rho_i = \rho\ \forall\ i$	9.4	435
3	$\mathbf{x}_{ij}^{\mathrm{T}}\boldsymbol{\beta} + \phi_i;\ \phi_i \overset{iid}{\sim} N(0,\tau_\phi),\ \zeta_i \overset{iid}{\sim} N(0,\tau_\zeta)$	13.1	440
4	$\mathbf{x}_{ij}^{\mathrm{T}}\boldsymbol{\beta} + \phi_i;\ \{\phi_i\} \sim CAR(\lambda_\phi),\ \{\zeta_i\} \sim CAR(\lambda_\zeta)$	10.4	439
5	$\mathbf{x}_{ij}^{\mathrm{T}}\boldsymbol{\beta} + \phi_i;\ (\{\phi_i\},\{\zeta_i\}) \sim MCAR(\rho,\Lambda)$	7.9	434
6	$\mathbf{x}_{ij}^{\mathrm{T}}\boldsymbol{\beta} + \phi_i;\ (\{\phi_i\},\{\zeta_i\}) \sim MCAR(\rho_\phi,\rho_\zeta,\Lambda)$	8.2	434

Table 15.12 *DIC and p_D values for various competing interval-censored models.*

Example 15.5 *(Smoking cessation data).* We illustrate our methods using the aforementioned study of smoking cessation, a subject of particular interest in studies of lung health and primary cancer control. Described more fully by Murray et al. [1998], the data consist of 223 subjects who reside in 53 zip codes in the southeastern corner of Minnesota. The subjects, all of whom were smokers at study entry, were randomized into either a smoking intervention (SI) group, or a usual care (UC) group that received no special antismoking intervention. Each subject's smoking habits were monitored at roughly annual visits for five consecutive years. The subjects we analyze are actually the subset who are known to have quit smoking at least once during these five years, and our event of interest is whether they relapse (resume smoking) or not. Covariate information available for each subject includes sex, years as a smoker, and the average number of cigarettes smoked per day just prior to the quit attempt.

To simplify matters somewhat, we actually fit our spatial cure rate models over the 81 contiguous zip codes shown in Figure 15.10, of which only the 54 dark-shaded regions are those contributing patients to our data set. This enables our models to produce spatial predictions even for the 27 unshaded regions in which no study patients actually resided.

Table 15.12 provides the DIC scores for a variety of random effects cure rate models in the interval-censored case. Models 1 and 2 have only random frailty terms ϕ_i with i.i.d. and CAR priors, respectively. Models 3 and 4 add random Weibull shape parameters $\zeta_i = \log \rho_i$, again with i.i.d. and CAR priors, respectively, independent of the priors for the ϕ_i. Finally, Models 5 and 6 consider the full MCAR structure for the (ϕ_i, ζ_i) pairs, assuming common and distinct spatial smoothing parameters, respectively. The DIC scores do not suggest that the more complex models are significantly better; apparently, the data encourage a high degree of shrinkage in the random effects (note the low p_D scores). In what follows we

Parameter	Median	(2.5%, 97.5%)
Intercept	−2.720	(−4.803, −0.648)
Sex (male = 0)	0.291	(−0.173, 0.754)
Duration as smoker	−0.025	(−0.059, 0.009)
SI/UC (usual care = 0)	−0.355	(−0.856, 0.146)
Cigarettes smoked per day	0.010	(−0.010, 0.030)
θ (cure fraction)	0.694	(0.602, 0.782)
ρ_ϕ	0.912	(0.869, 0.988)
ρ_ζ	0.927	(0.906, 0.982)
Λ_{11} (spatial variance component, ϕ_i)	0.005	(0.001, 0.029)
Λ_{22} (spatial variance component, ζ_i)	0.007	(0.002, 0.043)
$\Lambda_{12}/\sqrt{\Lambda_{11}\Lambda_{22}}$	0.323	(−0.746, 0.905)

Table 15.13 *Posterior quantiles, full model, interval-censored case.*

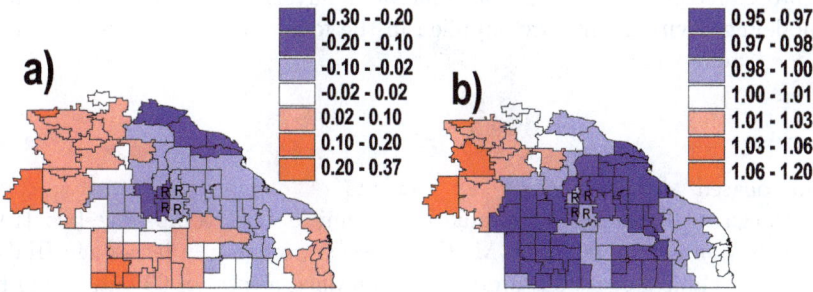

Figure 15.11 *Maps of posterior means for the ϕ_i (a) and the ρ_i (b) in the full spatial MCAR model, assuming the data to be interval-censored.*

present results for the "full" model (Model 6) in order to preserve complete generality, but emphasize that any of the models in Table 15.12 could be used with equal confidence.

Table 15.13 presents estimated posterior quantiles (medians, and upper and lower .025 points) for the fixed effects β, cure fraction θ, and hyperparameters in the interval-censored case. The smoking intervention does appear to produce a decrease in the log relative risk of relapse, as expected. Patient sex is also marginally significant, with women more likely to relapse than men, a result often attributed to the (real or perceived) risk of weight gain following smoking cessation. The number of cigarettes smoked per day does not seem important, but duration as a smoker is significant, and in a possibly counterintuitive direction: shorter-term smokers relapse sooner. This may be due to the fact that people are better able to quit smoking as they age (and are thus confronted more clearly with their own mortality).

The estimated cure fraction in Table 15.13 is roughly .70, indicating that roughly 70% of smokers in this study who attempted to quit have in fact been "cured." The spatial smoothness parameters ρ_ϕ and ρ_ζ are both close to 1, again suggesting we would lose little by simply setting them both equal to 1 (as in the standard CAR model). Finally, the last lines of both tables indicate only a moderate correlation between the two random effects, again consistent with the rather weak case for including them in the model at all.

We compared our results to those obtained from the R function `survreg` using a Weibull link, and also to Bayesian Weibull regression models using `BUGS/JAGS`. While neither of these alternatives featured a cure rate (and only the `BUGS/JAGS` analysis included spatial random effects), both produced fixed effect estimates quite consistent with those in Table 15.13.

Turning to graphical summaries, Figure 15.11 maps the posterior medians of the frailty (ϕ_i) and shape (ρ_i) parameters in the full spatial MCAR (Model 6) case. The maps reveal some interesting spatial patterns, though the magnitudes of the differences appear relatively small across zip codes. The south-central region seems to be of some concern, with its high values for both ϕ_i (high overall relapse rate) and ρ_i (increasing baseline hazard over time). By contrast, the four zip codes comprising the city of Rochester, MN (home of the Mayo Clinic, and marked with an "R" in each map) suggest slightly better than average cessation behavior. Note that a nonspatial model cannot impute anything other than the "null values" ($\phi_i = 0$ and $\rho_i = 1$) for any zip code contributing no data (all of the unshaded regions in Figure 15.10). Our spatial model however is able to impute nonnull values here, in accordance with the observed values in neighboring regions. ∎

15.6 Computer tutorials

Computer programs illustrating spatial survival analysis in R and its interfaces with Bayesian modeling environments are supplied in the folder titled Chapter 15 from https://github.com/sudiptobanerjee/BGC_2023.

15.7 Exercises

1. The data located at https://github.com/sudipt%obanerjee/BGC_2023/data/MAC.dat also shown in Table 15.14, summarize a clinical trial comparing two treatments for *Mycobacterium avium* complex (MAC), a disease common in late-stage HIV-infected persons. Eleven clinical centers ("units") have enrolled a total of 69 patients in the trial, 18 of which have died; see Cohn et al. [1999] and Carlin and Hodges [1999] for full details regarding this trial.

   As in Section 15.1, let t_{ij} be the time to death or censoring and x_{ij} be the treatment indicator for subject j in stratum i ($j = 1, \ldots, n_i$, $i = 1, \ldots, k$). With proportional hazards and a Weibull baseline hazard, stratum i's hazard is then

$$
\begin{aligned}
h(t_{ij}; x_{ij}) &= h_0(t_{ij})\omega_i \exp(\beta_0 + \beta_1 x_{ij}) \\
&= \rho_i t_{ij}^{\rho_i - 1} \exp(\beta_0 + \beta_1 x_{ij} + W_i),
\end{aligned}
$$

   where $\rho_i > 0$, $\boldsymbol{\beta} = (\beta_0, \beta_1)' \in \Re^2$, and $W_i = \log \omega_i$ is a clinic-specific frailty term.

   (a) Assume i.i.d. specifications for these random effects, i.e.,

$$
W_i \overset{iid}{\sim} N(0, 1/\tau) \quad \text{and} \quad \rho_i \overset{iid}{\sim} G(\alpha, \alpha).
$$

   Then,

$$
\mu_{ij} = \exp(\beta_0 + \beta_1 x_{ij} + W_i),
$$

   so that $t_{ij} \sim Weibull(\rho_i, \mu_{ij})$. Use a Bayesian modeling environment to obtain posterior summaries for the main and random effects in this model. Use vague priors on β_0 and β_1, a moderately informative $G(1,1)$ prior on τ, and set $\alpha = 10$. (You might also recode the drug covariate from (1,2) to (−1,1), in order to ease collinearity between the slope β_1 and the intercept β_0.)

   (b) From Table 15.15, we can obtain the latitude and longitude of each of the 11 sites, hence the distance d_{ij} between each pair. These distances are included in https://github.com/sudiptobanerjee/BGC_2023/data/MAC.dat note they have been scaled so that the largest (New York-San Francisco) equals 1. (Note that since sites F and H are virtually coincident (both in Detroit, MI), we have recoded them as a single clinic

Unit	Drug	Time	Unit	Drug	Time	Unit	Drug	Time
A	1	74+	E	1	214	H	1	74+
A	2	248	E	2	228+	H	1	88+
A	1	272+	E	2	262	H	1	148+
A	2	344				H	2	162
			F	1	6			
B	2	4+	F	2	16+	I	2	8
B	1	156+	F	1	76	I	2	16+
			F	2	80	I	2	40
C	2	100+	F	2	202	I	1	120+
			F	1	258+	I	1	168+
D	2	20+	F	1	268+	I	2	174+
D	2	64	F	2	368+	I	1	268+
D	2	88	F	1	380+	I	2	276
D	2	148+	F	1	424+	I	1	286+
D	1	162+	F	2	428+	I	1	366
D	1	184+	F	2	436+	I	2	396+
D	1	188+				I	2	466+
D	1	198+	G	2	32+	I	1	468+
D	1	382+	G	1	64+			
D	1	436+	G	1	102	J	1	18+
			G	2	162+	J	1	36+
E	1	50+	G	2	182+	J	2	160+
E	2	64+	G	1	364+	J	2	254
E	2	82						
E	1	186+	H	2	22+	K	1	28+
E	1	214+	H	1	22+	K	1	70+
						K	2	106+

Table 15.14 *Survival times (in half-days) from the MAC treatment trial, from Carlin and Hodges (1999). Here, "+" indicates a censored observation.*

(#6) and now think of this as a 10-site model.) Refit the model assuming the frailties to have spatial correlation following the isotropic exponential kriging model,

$$\mathbf{W} \sim N_k(\mathbf{0}, H), \text{ where } H_{ij} = \sigma^2 \exp(-\phi d_{ij}),$$

where as usual $\sigma^2 = 1/\tau$, and where we place a $G(3, 0.1)$ (mean 30) prior on ϕ.

2. The file https://github.com/sudiptobanerjee/BGC_2023/data/smoking.dat contains the southeastern Minnesota smoking cessation data discussed in Section 15.5. At each of up to five office visits, the smoking status of persons who had recently quit smoking was assessed. We define relapse to smoking as the endpoint, and denote the failure or censoring time of person j in county i by t_{ij}. The data set also contains the adjacency matrix for the counties in question.

(a) Assuming that smoking relapses occurred on the day of the office visit when they were detected, build a hierarchical spatial frailty model to analyze these data. Implement your model in a Bayesian modeling environment of your choice and summarize your results. Compare your analysis for a few competing prior or likelihood specifications and comment on the robustness of your inference with regard to the different specifications.

Unit	Number	City
A	1	Harlem (New York City), NY
B	2	New Orleans, LA
C	3	Washington, DC
D	4	San Francisco, CA
E	5	Portland, OR
F	6a	Detroit, MI (Henry Ford Hospital)
G	7	Atlanta, GA
H	6b	Detroit, MI (Wayne State University)
I	8	Richmond, VA
J	9	Camden, NJ
K	10	Albuquerque, NM

Table 15.15 *Locations of the clinical sites in the MAC treatment data set.*

(b) When we observe a subject who has resumed smoking, all we really know is that his failure (relapse) point occurred somewhere between his last office visit and this one. As such, improve your model from part (a) by building an interval-censored version.

3. Consider the extension of the Section 15.4 model in the single endpoint, multiple cause case to the *multiple* endpoint, multiple cause case—say, for analyzing times until diagnosis of each cancer (if any), rather than merely a single time until death. Write down a model, likelihood, and prior specification (including an appropriately specified MCAR distribution) to handle this case.

Chapter 16

Spatial data fusion (and preferential sampling)

16.1 Introduction

In this chapter we explore an important and rapidly growing area in spatial statistics which considers fusing data sources. Indeed, with an increased collection of spatial (and spatio-temporal) datasets, we often find multiple sources that are jointly capable of informing about features of a process of interest. It seems evident that, through suitable fusion of the data sources, we can learn at least as much about these features of interest as we can from any individual source. Our intent here is to consider the range of potential spatial data sources available for fusion and, accordingly, to propose suitable generative stochastic modeling to implement a fusion of these sources. Again, we consider only model-based approaches, only coherent models, i.e., only models that could have produced the data we have observed. Such modeling enables full inference both with regard to estimation and prediction, with implicit incorporation of uncertainty. Here, through hierarchical modeling, we introduce a suitable latent stochastic process which connects the data sources.

Further, in Section 16.5, we develop the ideas behind preferential sampling, i.e., the presence of sampling bias in a set of geostatistical locations and its impact on spatial interpolation. Preferential sampling also motivates a hierarchical model specification which enables fusion of geostatistical data with spatial point pattern data in Section 16.6.

The idea of fusing data sources is well established in the literature and, even with a focus exclusively on spatial data, by now that literature is substantial. The term data fusion has been referred to under different names in the literature, e.g., data assimilation, data aggregation, data melding, and data integration (references are supplied below). In this regard, we are not thinking of fusion in terms of increasing sample size, combining directly comparable datasets/databases. Rather, in the case of say two sources, we are imagining data collected under two different sampling regimes. We are also not thinking in terms of a regression setting where one data source to explain the other. Rather, there can be covariate information associated with each data source but the response information is what we seek to fuse and the nature of the response information need not be the same for both sources. Further, we are also not thinking about the fusion of multivariate spatial data vectors collected from a single source. In general, such data is best modeled through joint modeling of the components of the vectors collected at the spatial locations (Chapter 10). Consistent with our philosophy in this volume, we will consider our fusion objective within a Bayesian framework, eschewing algorithmic and pseudo-statistical approaches.

The range of fusion applications spans all of spatial data. However, a very common theme in the literature addresses environmental and ecological applications, and so our examples here will reflect this. Often, the literature introduces simple aggregation, e.g., ad hoc descriptive summaries, perhaps using geographic information system (GIS) tools. A well-discussed example, usually referred to as data assimilation, is in weather prediction. An illustration considers a rich literature on data aggregation from climate models [Berrocal, 2019]. Here, atmospheric and physical models are built on global or regional scales. Each model produces a scenario for climate over some common future time window. The aggregation fuses these scenarios to obtain a version of an "average" scenario and also computes the variability in these scenarios as a measure of uncertainty.

Other common fusions in the literature include data sources providing temperatures as well as environmental contaminant exposure. These sources incorporate monitoring station networks, satellite data collection, and computer model outputs (see Berrocal [2019]). Ecological settings often seek to fuse data sources on species distribution and species abundance. Here, say in the context of plants, sources can include both designed data collection as well as "citizen science" or museum data base/non-systematic data collection (see Gelfand and Shirota [2019] and references therein). In the context of animal movement, for say terrestrial animals, we can envision fusing well-established distance sampling data collection [Buckland et al., 2007] with say, camera trap data. For marine mammals, we can envision again, fusing distance sampling (by ship or by air) with acoustic monitoring data as well as opportunistic sighting data.

All of this work originates, in a sense, from the bigger world of data assimilation with computer model output [Poole and Raftery, 2000]. Again, here the overall goal is to synthesize multiple data sources that are informing about a common spatial or spatio-temporal process. The anticipated benefit will be improved kriging and, in the temporal case, improved short-term forecasting. It will almost always be the case that the data layers are misaligned in space and in time, e.g., areal units for one, point-referenced locations for another; hourly availability for one, daily collection for another. This recalls the misalignment discussion of Chapter 7. The misalignment can be more challenging, e.g., in space, not a common region, and in time, data collected in differing weeks, months, and years. Taking this further, for one source, we may have geostatistical data, explicitly point-referenced or perhaps, collected as averages or proportions for areal units. For a second source, we may have a point pattern of occurrences or perhaps, areal counts where the points have been aggregated to specified areal units.

16.1.1 Algorithmic and pseudo-statistical approaches in weather prediction

We briefly review algorithmic and pseudo-statistical approaches to spatial data assimilation in the context of numerical weather prediction. Kalnay [2003] provides a development of this material. Early work created local polynomial interpolations using quadratic trend surfaces in locations in order to interpolate observed values to grid values. Eventually, what emerged in the meteorology community was the recognition that a first guess (or background field or prior information) was needed (Bergthórsson and Döös [1955]), supplying the *initial conditions*. The climatological intuition here is worth articulating. Over "data-rich" areas, the observational data dominates while in "data-poor" regions the forecast facilitates the transport of information from the data-rich areas.

We review several numerical approaches using temperature as an illustrative variable of interest. At time t, we let $T_{obs}(t)$ be an observed measurement, $T_b(t)$ a background level, $T_a(t)$ an assimilated value, and $T_{true}(t)$ the true value. An early scheme is known as the successive corrections method (SCM) which obtains $T_{i,a}(t)$ iteratively through $T_{i,a}^{(r+1)}(t) = T_{i,a}^{(r)}(t) + \frac{\sum_k w_{ik}(T_{k,obs}(t) - T_{k,a}^{(r)}(t))}{\sum_k w_{ik} + \epsilon^2}$. Here, i indexes the grid cells for the interpolation while k indexes the observed data locations. $T_{k,a}^{(r)}(t)$ is the value of the assimilator at the r-th iteration at the observation point k (obtained from interpolating the surrounding grid points). The weights can be defined in various ways but usually as a decreasing function of the distance between the grid point and the observation point. They can vary with iteration, perhaps becoming increasingly local. (See, e.g., Cressman, 1959, and Bratseth, 1986.)

A second empirical approach is called *nudging* or Newtonian relaxation. Suppose, suppressing location, we think about a differential equation driving temperature, e.g.,

$\frac{dT(t)}{dt} = a(T(t), t, \theta(t))$. If we write $a(\cdot)$ as an additive form say $a(T(t), t) + \theta(t)$ and let $\theta(t) = (T_{obs}(t) - T(t))/\tau$ then τ controls the relaxation. Small τ implies that the $\theta(t)$ term dominates while large τ implies that the nudging effect will be negligible.

We also might consider a least squares approach. Again, suppressing location, suppose we assume that $T_{obs}^{(1)}(t) = T_{true}(t) + \epsilon_1(t)$ and $T_{obs}^{(2)}(t) = T_{true}(t) + \epsilon_2(t)$ where we envision two sources of observational data on the true temperature at t. The ϵ_l have mean 0 and variance $\sigma_l^2, l = 1, 2,$. Then, with the variances known, it is a familiar exercise to obtain the best unbiased estimator of $T_{true}(t)$ based on these two pieces of information. That is, $T_a(t) = a_1 T_{obs}^{(1)}(t) + a_2 T_{obs}^{(2)}(t)$ where $a_1 = \sigma_2^2/(\sigma_1^2 + \sigma_2^2)$ and $a_2 = \sigma_1^2/(\sigma_1^2 + \sigma_2^2)$. We obtain the same solution as the MLE if we use independent normal likelihoods for the $T_{obs}^{(l)}(t)$'s.

A last idea here is simple sequential assimilation and its connection to the Kalman filter. In the univariate case suppose we write $T_a(t) = T_b(t) + \gamma(T_{obs}(t) - T_b(t))$. Here, $T_{obs}(t) - T_b(t)$ is referred to as the observational innovation or observational increment relative to the background. The optimal weight $\gamma = \sigma_{obs}^2/(\sigma_{obs}^2 + \sigma_b^2)$, analogous to the previous paragraph. Hence, we only need an estimate of the ratio of the observational variance to the background variance in order to obtain $T_a(t)$. To make this scheme dynamic, suppose the background is updated through the assimilation, i.e., $T_b(t+1) = h(T_a(t))$ where $h(\cdot)$ denotes some choice of forecast model. Then we will also need a revised background variance; this is usually taken to be a scalar (> 1) multiple of the variance of $T_a(t)$.

16.2 Types of fusions

A fusion which rarely occurs in practice would envision $Y(B_{11}), ..., Y(B_{1k_1})$ and $Y(B_{21}), ..., Y(B_{2k_2})$, different partitions of essentially the same region (nested, if the same, nonnested if not). This fusion is essentially an analogue to the more common modifiable areal unit problem (MAUP) discussed in Sections 7.2–7.4 where the goal was an interpolation of the first set of Y's to the second set of Y's. Elementary areal interpolation was proposed along with formal modeling using say, a Poisson regression for counts or perhaps a probit or logit for proportions. The regression assumes shared covariates associated with the observations from both sources.

No further discussion is warranted here except to raise the general question of what does the random variable $Y(B)$ mean? Here, there has been confusion in the literature. Is it the value of the variable at some location in B. Is the process constant over B at this value? Is it an average of some values over locations in B? Is it the (conceptual) average over all locations in B? Suitable modeling for $Y(B)$ should depend on its interpretation. We are clear regarding the interpretation in what follows.

Our general elaboration of the spatial fusion framework envisions the process operating at the point level enabling the spatial data to be viewed in the setting of points and marks (as in Chapter 8), jointly modeled as $[points][marks|points]$, points in $\mathcal{D}$, marks in $\mathcal{Y}$. In this regard, the process can model the points themselves, the marks themselves (ignoring any randomness in the points), or the points and marks jointly. This leads to four spatial data types: (i) a spatial point pattern, $\mathcal{S} = (\mathbf{s}_1, \mathbf{s}_2, ..., \mathbf{s}_n)$; (ii) a collection of disjoint sets with a count of the number of points observed in each set, $\{N(B_k), k = 1, 2, .., K\}$ where $N(B)$ is a number of points in B or, as a variant, $N(s_k^*), k = 1, 2, ..., K$ where the s_k^* are a set of points in $\mathcal{D}$ and $N(s_k^*)$ is a count associated with location s_k^*; (iii) a geostatistical dataset with observations at a finite set of point referenced locations, $\{Y(\mathbf{s}_i), i = 1, 2, ..., n\}$, (iv) a set of areal data observations, $\{Y(B_1), Y(B_2), ..., Y(B_k)\}$ with the B's usually disjoint and the Y's viewed as arising as block averages. Note that we do not consider fusion with marked point pattern data. We are unaware of any such work in the literature.

In this chapter, we present modeling detail for two data sources; the approach to extension to more sources is hopefully evident but clearly more computationally demanding.

The two sources can each be any one of the four data types. So, potentially, we might have 16 different pairs. For our discussion, we split the scenarios into three groups—mark/mark, point/point, and point/mark.

Regardless of how the data are observed, we imagine the process of interest operates at the point level. We seek to learn about a process at the point level; conceptually, an event can be observed at any point $\mathbf{s}$ in our study domain, i.e., a point can arise or a mark can be observed at any location in $\mathcal{D}$. Point level specification need not be viewed as a restriction. Data collected at areal scale, e.g., an average or proportion can be viewed as a block average over geostatistical data collected at points (see Chapter 7). Abundance data at areal scale can be viewed as counting/aggregating point-level occurrences. Moreover, point-level modeling (the highest spatial resolution) is the most informative way to learn about the process of interest. A point-level stochastic process over $\mathcal{D}$ that links the two data sources provides a natural bridge.

We present these fusions in three subsequent sections. Section 16.3 ignores modeling the point pattern(s) associated with the data, only modeling the marks. That is, both sources may be viewed as "geostatistical." This scenario provides the most familiar version of spatial data fusion in the literature. Section 16.4 presumes no marks and focuses solely on sources arising from point patterns. Fusing two point pattern-based sources also finds some literature; we focus on fusion of spatially varying thinning of a latent unobserved point pattern. Section 16.6 assumes both the points and the marks are random. Again, it enriches preferential sampling, which we elaborate in Section 16.5. There seems to be little work and some controversy in this context.

We remind the reader about the common challenge which arise in any of these formal fusions, i.e., spatial and temporal misalignment of the data sources. We don't discuss this issue here since it is usually application specific and, when considered, it is usually dealt with in an ad hoc fashion.

16.3   Geostatistical-geostatistical fusion

Here, again there are three possibilities. All assume that there is a latent "true" process realization operating, $Y_{true}(\mathbf{s}), \mathbf{s} \in \mathcal{D}$, driving the observations of the process. The cases are:

(i) $\mathcal{Y}_1 = (Y(\mathbf{s}_{11}), Y(\mathbf{s}_{12}), ..., Y(\mathbf{s}_{1n_1}), \mathcal{Y}_2 = (Y(\mathbf{s}_{21}), Y(\mathbf{s}_{22}), ..., Y(\mathbf{s}_{2n_2})$. Typically, we assume one with just measurement error, the other needing calibration, usually spatially varying;

(ii) $\mathcal{Y}_1 = (Y(\mathbf{s}_{11}), Y(\mathbf{s}_{12}), ..., Y(\mathbf{s}_{1n_1}), Y(B_1), ..., Y(B_{k_2})$. Here, the $Y(B)$'s are assumed to be block averages of the truth plus error. Again, one source is assumed to be observed with measurement error. Naturally, it is the point-referenced source, with the block averaged data needing calibration. Here, we can consider upscaling, the data melding approach, as well as downscaling [Berrocal, 2019];

(iii) $Y(B_{11}), ..., Y(B_{1k_1})$ and $Y(B_{21}), ..., Y(B_{2k_2})$. Both sets of Y's are assumed to be block averages. This setting is customarily referred to as the change of support problem or the modifiable areal unit problem. Interest would presumably be in interpolation to unobserved B's. It need not be viewed at point level and is discussed in detail in Chapter 7 with an explicit assimilation example in Section 7.4. We do not consider it further here.

The notions of fusion and calibration arise here from the well-discussed issue of measurement error (or errors in variables) in the Statistics literature [see, e.g., Fuller, 1987, Carroll et al., 2006]. The idea is that neither data source provides the "true" response. Each is observed with error. Typically, one source has just simple a measurement error, with the observed measurements varying around the true ones with some, essentially known, variance. The other requires calibration, with error, relative to the truth. There is already a substantial published literature on stochastic spatial or spatio-temporal data fusion modeling for environmental and ecological data adopting this approach. For example, Fuentes and

Raftery [2005] consider fusion of air pollution data sources (see below). In addition, Liu et al. [2011] and Zidek et al. [2012] consider a fusion for ozone sources. Foley and Fuentes [2008] present wind data modeling fusion. Finley et al. [2011a] offer a fusion for agricultural crop yields. Liu et al. [2016] investigate fusion to predict spatial tracks for marine mammals. In other work [Villejo et al., 2023] introduces the INLA-SPDE approach for spatial data fusion and Banerjee et al. [in press] develop Bayesian learning frameworks for spatial-temporal mechanistic systems.

16.3.1 Fusion modeling using stochastic integration

The fusion approach proposed by Fuentes and Raftery [2005] works with block averaging (Chapter 7) and builds upon earlier work referred to as Bayesian melding in Poole and Raftery [2000]. It conceptualizes a true exposure surface, say $Y_{true}(\mathbf{s})$, and views the two sources as monitoring station data along with computer model output data each varying in a suitable way around the true surface. In particular, the true average exposure in a grid cell A differs from the exposure at any particular location $\mathbf{s}$. The so called change of support problem (see Chapter 7) in the context here addresses converting the point level $Y_{true}(\mathbf{s})$ to the grid level $Y_{true}(A)$ through the stochastic integral,

$$Y_{true}(A) = \frac{1}{|A|} \int_A Y_{true}(\mathbf{s}) \, d\mathbf{s}, \tag{16.1}$$

where $|A|$ denotes the area of grid cell A and $Y_{true}(A)$ is the block average (see Chapter 7).

Let $Y(\mathbf{s})$ denote the observed exposure at a station $\mathbf{s}$. The measurement error model assumption is

$$Y(\mathbf{s}) = Y_{true}(\mathbf{s}) + \epsilon(\mathbf{s}) \tag{16.2}$$

where $\epsilon(\mathbf{s}) \sim N(0, \sigma_\epsilon)$ represents the measurement error at location $\mathbf{s}$. The true exposure process is assumed to be

$$Y_{true}(\mathbf{s}) = \mu(\mathbf{s}) + \eta(\mathbf{s}) \tag{16.3}$$

where $\mu(\mathbf{s})$ provides the spatial mean surface, typically characterized by a trend, by elevation, etc. The error term $\eta(\mathbf{s})$ is a spatial process assumed to be a zero mean Gaussian process (GP) with a specified covariance function. Inserting (16.3) into (16.2), we obtain the familiar geostatistical model (See Chapter 6) where one proposed interpretation for $\epsilon(\mathbf{s})$ was measurement error.

The second source in Fuentes and Raftery [2005] is output from a computer model at areal unit scale. Such output is assumed to be biased/uncalibrated. Hence, the conceptual point level output from this model, denoted by $Q(\mathbf{s})$, usually is related to the true surface as

$$Q(\mathbf{s}) = a(\mathbf{s}) + b(\mathbf{s}) Y_{true}(\mathbf{s}) + \delta(\mathbf{s}) \tag{16.4}$$

where $a(\mathbf{s})$ denotes the additive bias and $b(\mathbf{s})$ denotes the multiplicative bias.[1] The error term, $\delta(\mathbf{s})$, is assumed to be a white noise process given by $N(0, \sigma_\delta^2)$. Then, since the computer model output is provided on a grid, $A_1, \ldots, A_J$, the point level process is converted to grid level by stochastic integration of (16.4), i.e.,

$$Q(A_j) = \int_{A_j} a(\mathbf{s}) \, d\mathbf{s} + \int_{A_j} b(\mathbf{s}) Y_{true}(\mathbf{s}) \, d\mathbf{s} + \int_{A_j} \delta(\mathbf{s}) \, d\mathbf{s}.$$

We note that if $a(\mathbf{s})$ and $b(\mathbf{s})$ are smooth Gaussian process realizations, their integrals exist. However, since $\delta(\mathbf{s})$ is everywhere discontinuous, formally, the third integral doesn't exist. We imagine it as denoting a Gaussian random variable distributed as $N(0, |A_j|\sigma_\delta^2)$.

---

[1] Recall the spatially varying coefficient modeling from Section 10.7

It has been observed that unstable model fitting (identifiability problems) may accrue to the case where we have spatially varying $b(\mathbf{s})$ so $b(\mathbf{s}) = b$ may need to be adopted. Spatial prediction at a new location $\mathbf{s}_0$ is done through the posterior predictive distribution $f(Y_{true}(\mathbf{s}_0) \mid \mathbf{Y}, \mathbf{Q})$ where $\mathbf{Y}$ denotes all the station data and $\mathbf{Q}$ denote all the grid-level computer output $Q(A_1), \ldots, Q(A_J)$.

This fusion strategy starts to become potentially computationally infeasible in the setting below, where we have more than $40,000$ grid cells for the computer model output over the Eastern U.S. We have a very large number of grid cells providing computer model output (CMAQ data, see below) with a relatively sparse number of monitoring sites (station data). An enormous amount of stochastic integration is required. A dynamic implementation over many time periods becomes even more challenging. This motivates the suggestion that we try scaling down rather than integrating up. This so-called downscaling approach is developed in the following subsection.

While the Fuentes and Raftery [2005] approach models at the point level, the strategy in McMillan et al. [2010] models at the grid cell level. In this fashion, computation is simplified, and fusion with space-time data is manageable. In particular, suppose that we have say n monitoring stations with say J grid cells. Let $Q(A_j)$ denote the CMAQ output value for cell A_j while $Y_{A_j}(\mathbf{s}_k)$ denotes the station data for site $\mathbf{s}_k$ within cell A_j. Of course, for most of the j's, k will be 0 since $n << J$. Let $Y_{true}(A_j)$ denote the true value for cell A_j.

Then, paralleling (16.2) and (16.4),

$$Y_{A_j}(\mathbf{s}_k) = Y_{true}(A_j) + \epsilon_{A_j}(\mathbf{s}_k) \tag{16.5}$$

and

$$Q(A_j) = Y_{true}(A_j) + b(A_j) + \gamma(A_j). \tag{16.6}$$

In (16.6), the CMAQ output is modeled as varying around the true value with a bias term, denoted by $b(A_j)$, specified using a B-spline model. Also, the ϵ's are assumed to be independently and identically distributed and so are the γ's, each with a respective variance component. So, the station data and the CMAQ data are conditionally independent given the true surface. Finally, the true surface is modeled analogously to (16.3) but at a grid scale. Operating at areal scale, the η's are given a CAR specification (see Chapter 4). For space-time data, McMillan et al. [2010] offer a dynamic version of this approach, assuming a dynamic CAR specification for the η's.

Wikle and Berliner [2005] are concerned with two sources of wind data—daily wind satellite data and computer model output data supplied by a weather center. The satellite data are at 0.5^o resolution, not on a regular grid while the the computer model output is on a 2.5^o regular grid. The objective is to predict a windstream surface at 1.0^o resolution. They work within the familiar hierarchical framework,

[data | process][process | parameters][parameters] .

Here, the process is the true underlying wind surface and the two data sources are assumed observed with measurement error.

Their data structure implies three scales: subgrid (0.5^o resolution), supergrid (2.5^o resolution) and prediction grid (1.0^o resolution). They use block averaging on each of the three scales, with the aforementioned measurement error, to infer about the blocks on the prediction grid from the "observed" blocks on the super and subgrids. Formally, the model introduces a latent Gaussian process to do the block averaging and local scaling factors to move from one grid to another. With temporal wind data, Wikle et al. [2001] implement a space-time data fusion. Here, they account for temporal dependence by using dynamic coefficients in the process model in order to avoid the computation of stochastic integrals.

16.3.2   The downscaler

The downscaler model was proposed as an alternative to Bayesian melding in Berrocal et al. [2010a]. First, we show a static spatial version which can be used at any temporal scale, e.g., for daily data or for annual averages. Then, we present several ways in which the static model can be extended to handle spatio-temporal data.

Though it is more broadly applicable, for ease of interpretation, we present it with regard to an ozone application. So, we consider two sources of ground-level ozone data (daily 8-hour maxima in ppb). The first one comes from the National Air Monitoring Stations/State and Local Air Monitoring Stations

(NAMS/SLAMS; *http://www.epa.gov/monitor/programs/namsslam.html*)

network for the Eastern US. Figure 16.1 shows a map of the NAMS/SLAMS monitoring sites used in our analysis. Panel (a) displays all the monitoring sites (N=803) used to fit and validate the spatio-temporal version of our downscaling model. Panel (b) shows the monitoring sites (N=69) enclosed in the smaller region highlighted by the square in panel (a) that have been used to enable comparison of the static version of our downscaling model with the Bayesian melding model of Fuentes and Raftery [2005] and with ordinary kriging.

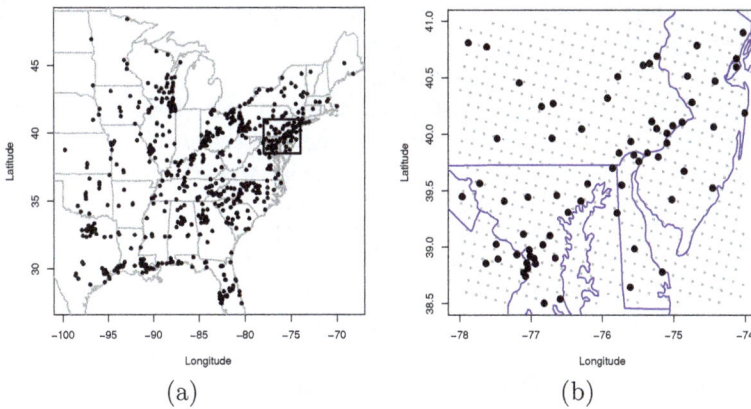

(a)   (b)

Figure 16.1 *(a): Ozone monitoring sites in the Eastern US; (b): Subset region used to compare ordinary kriging, Bayesian melding, and the downscaler. The black points represent monitoring sites, and the black dots represent the centroids of the CMAQ grid cells.*

The second source of data in our study is the Models-3/Community Multiscale Air Quality (CMAQ; *http://www.epa.gov/asmdnerl/CMAQ*) model, mentioned above. CMAQ is a numerical model that estimates the daily 8-hour maximum concentration of ozone by integrating information coming from three different components: a meteorology component which accounts for the state and the evolution in time of the atmosphere, an emissions component which deals with emissions injected in the atmosphere by both chemical plants and natural processes, and a component that accounts for the chemical and physical interactions occurring in the atmosphere. We use daily CMAQ predictions gridded to 12km spatial resolution. There are 40,044 grid cells covering the portion of the eastern US displayed in Figure 16.1(a) resulting in $40,044 \times 168 = 6,727,392$ daily modeled output measurements. 651 grid cells cover the smaller region shown in Figure 16.1(b). There, gray dots display the centroids of each 12km CMAQ grid cell. Again, we have 69 monitoring sites in the smaller region. We immediately see that we have many more grid cells than monitoring sites; the enormous amount of block averaging that would be needed to implement the fusion model of Fuentes and Raftery [2005] makes it computationally overwhelming.

With regard to normality assumptions, it is better to model ozone on the square root scale. Therefore, we denote with $Y(\mathbf{s})$ the square root of the observed ozone concentration at $\mathbf{s}$. We use $x(B)$ to denote the square root of the numerical model output over grid cell B. Each point $\mathbf{s}$ is associated with the CMAQ grid cell B in which it lies. So, all the points $\mathbf{s}$ falling in the same 12-km square region are assigned the same CMAQ output value. This is the usual interpretation of the CMAQ model output, i.e., it is viewed as a tiled surface at grid cell resolution, providing a tile height for each grid cell.

We relate the observed data to the CMAQ output in the following way. For each $\mathbf{s}$ in B, we assume that

$$Y(\mathbf{s}) = \tilde{\beta}_0(\mathbf{s}) + \tilde{\beta}_1(\mathbf{s})\, x(B) + \epsilon(\mathbf{s}), \qquad \epsilon(\mathbf{s}) \overset{ind}{\sim} \mathrm{N}(0, \tau^2) \qquad (16.7)$$

where

$$\begin{aligned} \tilde{\beta}_0(\mathbf{s}) &= \beta_0 + \beta_0(\mathbf{s}) \\ \tilde{\beta}_1(\mathbf{s}) &= \beta_1 + \beta_1(\mathbf{s}), \end{aligned} \qquad (16.8)$$

and $\epsilon(\mathbf{s})$ is a white noise process with nugget variance τ^2. β_0 and β_1 represent the overall/global intercept and slope bias of the CMAQ model, while $\beta_0(\mathbf{s})$ and $\beta_1(\mathbf{s})$ are local adjustments to the additive and multiplicative bias, respectively. The spatially-varying coefficients $\beta_0(\mathbf{s})$ and $\beta_1(\mathbf{s})$ are in turn modeled as bivariate mean-zero Gaussian spatial processes using the method of coregionalization (see Chapter 5).

Therefore, we suppose that there exist two mean-zero unit-variance independent Gaussian processes $w_0(\mathbf{s})$ and $w_1(\mathbf{s})$ such that, for convenience, $\mathrm{Cov}(w_j(\mathbf{s}), w_j(\mathbf{s}')) = \exp(-\phi_j|\mathbf{s} - \mathbf{s}'|)$, i.e., ϕ_j is the spatial decay parameter for the Gaussian process $w_j(\mathbf{s})$, $j = 0, 1$. So,

$$\begin{pmatrix} \beta_0(\mathbf{s}) \\ \beta_1(\mathbf{s}) \end{pmatrix} = A \begin{pmatrix} w_0(\mathbf{s}) \\ w_1(\mathbf{s}) \end{pmatrix} \qquad (16.9)$$

where the unknown A matrix in (16.9) is assumed to be a lower-triangular. We note that there is no identifiability problem introduced when we have, say multiple $\mathbf{s}$'s within a given B since the associated $Y(\mathbf{s})$'s will vary over the $\mathbf{s}$'s in B. Hence, the data informs locally about the $\beta_0(\mathbf{s})$ and $\beta_1(\mathbf{s})$ surfaces. Then, the spatial dependence introduced by (16.9) enables interpolation of these surfaces over the region of interest.

We complete the specification of the Bayesian hierarchical model with the following prior specifications. We use a bivariate normal distribution for the overall bias terms β_0 and β_1, lognormal distributions for the two diagonal entries, a_{11} and a_{22}, of the coregionalization matrix A, a normal distribution for the off-diagonal entry, a_{21}, of A, and an inverse gamma distribution for τ^2. It is not possible to estimate consistently all of the covariance parameters (Zhang, 2004, Tang et al., 2021); under weak prior specifications we find weak identifiability of these parameters in the MCMC chains. Hence, we use discrete uniform priors on m values for the decay parameters ϕ_j, $j = 0, 1$.

Our model specification is rather simple yet it provides calibration at the local level and endows the spatial process $Y(\mathbf{s})$ with a flexible non-stationary covariance structure. The model is much easier to fit than Bayesian melding since we eliminate the need to handle stochastic integrals. Moreover, to fit the model we only need to work with the responses associated with the $Y(\mathbf{s}_i)$, i.e., with the set of monitoring sites, a much smaller number compared to the number of grid cells. Arguably, our model is also preferable to that of McMillan et al. [2010] since we downscale to the point level rather than upscale to the grid cell level. Spatial interpolation to a new location is based upon the predictive distribution, i.e., at location $\mathbf{s}_0$, we sample $f(Y(\mathbf{s}_0)|\{Y(\mathbf{s}_i)\}, \{x(B_j)\})$.

16.3.3   Spatio-temporal versions

The foregoing downscaling model can be extended to accommodate data collected over time in several ways. With time denoted by t, $t = 1, \ldots, T$, let $Y(\mathbf{s}, t)$ be the square root of observed daily 8-hour maximum ozone concentration at $\mathbf{s}$ at time t and let $x(B, t)$ be the square root of the CMAQ predicted average ozone concentration for grid cell B at time t. Again, we associate with each point $\mathbf{s}$ the CMAQ grid cell B in which it lies.

We start by assuming that the overall bias terms, β_0 and β_1, change in time, while the local adjustments to β_0 and β_1, the spatially varying coefficients $\beta_0(\mathbf{s})$ and $\beta_1(\mathbf{s})$, remain constant in time. This means that

$$Y(\mathbf{s}, t) = \beta_{0t} + \beta_{1t} x(B, t) + \beta_0(\mathbf{s}) + \beta_1(\mathbf{s}) x(B, t) + \epsilon(\mathbf{s}, t) \tag{16.10}$$

where $\epsilon(\mathbf{s}, t) \stackrel{ind}{\sim} N(0, \tau^2)$. There are two customary ways in which the β_{0t} and β_{1t} terms could be specified. The first is to assume that β_{0t} and β_{1t} are nested within time, or, in other words, they are independent across time:

$$\begin{pmatrix} \beta_{0t} \\ \beta_{1t} \end{pmatrix} \stackrel{ind}{\sim} \text{MVN}_2(\mu, V) \tag{16.11}$$

The second is to assume that the two overall bias terms β_{0t} and β_{1t} evolve dynamically in time. That is,

$$\begin{pmatrix} \beta_{0t} \\ \beta_{1t} \end{pmatrix} = \begin{pmatrix} \rho_0 & 0 \\ 0 & \rho_1 \end{pmatrix} \begin{pmatrix} \beta_{0\,t-1} \\ \beta_{1\,t-1} \end{pmatrix} + \begin{pmatrix} \eta_{0t} \\ \eta_{1t} \end{pmatrix} \tag{16.12}$$

where $\begin{pmatrix} \eta_{0t} \\ \eta_{1t} \end{pmatrix} \stackrel{ind}{\sim} \text{MVN}_2(\mathbf{0}, V_\eta)$ and $\begin{pmatrix} \beta_{00} \\ \beta_{10} \end{pmatrix} \sim \text{MVN}_2(\mu_0, V_0)$.

A more general way to introduce time in our downscaling model is by assuming that both the overall bias terms β_0 and β_1, and the local adjustments, $\beta_0(\mathbf{s})$ and $\beta_1(\mathbf{s})$, vary with time, that is,

$$Y(\mathbf{s}, t) = \beta_{0t} + \beta_{1t} x(B, t) + \beta_0(\mathbf{s}, t) + \beta_1(\mathbf{s}, t) x(B, t) + \epsilon(\mathbf{s}, t) \tag{16.13}$$

As with β_{0t} and β_{1t} in model (16.10), there are two ways in which we can specify the $\beta_0(\mathbf{s}, t)$ and $\beta_1(\mathbf{s}, t)$ terms. In the first case, we can have

$$\begin{pmatrix} \beta_0(\mathbf{s}, t) \\ \beta_1(\mathbf{s}, t) \end{pmatrix} = A \begin{pmatrix} w_{0t}(\mathbf{s}) \\ w_{1t}(\mathbf{s}) \end{pmatrix} \tag{16.14}$$

with A lower-triangular. In (16.14) $w_{0t}(\mathbf{s})$ and $w_{1t}(\mathbf{s})$ are serially independent replicates of two independent unit-variance Gaussian processes with mean zero and exponential covariance function having spatial decay parameters ϕ_0 and ϕ_1, respectively.

In the second case, that is, if $\beta_0(\mathbf{s}, t)$ and $\beta_1(\mathbf{s}, t)$ evolve dynamically in time, we follow Section 12.4. Therefore,

$$\begin{aligned} \begin{pmatrix} \beta_0(\mathbf{s}, t) \\ \beta_1(\mathbf{s}, t) \end{pmatrix} &= \begin{pmatrix} \rho_0 & 0 \\ 0 & \rho_1 \end{pmatrix} \begin{pmatrix} \beta_0(\mathbf{s}, t-1) \\ \beta_1(\mathbf{s}, t-1) \end{pmatrix} + \begin{pmatrix} v_{0t}(\mathbf{s}) \\ v_{1t}(\mathbf{s}) \end{pmatrix} \\ \begin{pmatrix} v_{0t}(\mathbf{s}) \\ v_{1t}(\mathbf{s}) \end{pmatrix} &= A \begin{pmatrix} w_{0t}(\mathbf{s}) \\ w_{1t}(\mathbf{s}) \end{pmatrix} \end{aligned} \tag{16.15}$$

where $\begin{pmatrix} \beta_0(\mathbf{s}, 0) \\ \beta_1(\mathbf{s}, 0) \end{pmatrix} \equiv \begin{pmatrix} 0 \\ 0 \end{pmatrix}$. Again, A is lower-triangular and does not depend on t, while $w_{0t}(\mathbf{s})$ while $w_{1t}(\mathbf{s})$ are as above.

16.3.4   An illustration

Ozone poses a threat primarily during summer months, i.e., the ozone season, May 1—October 15, 2001, when it is present at higher concentrations. So, to demonstrate the predictive performance of the static downscaler on ozone data, we consider three illustrative dates, randomly chosen: one in June, one in July, and one in August. For each day, we fit the downscaler using ozone observations coming from 69 monitoring sites and estimates of average ozone concentration over the associated grid cell provided by CMAQ. For comparison, we fit Bayesian melding using the 69 observations and CMAQ model output for the 651 grid cells covering the subset region displayed in Figure 1(b). For ordinary kriging, we estimate parameters of the exponential covariance function using only the 69 observations. The mean and standard deviation of the observed daily 8 hour maximum ozone concentration for the three days were, respectively, 69.94 and 14.28 ppb, for June 21 2001, 72.23 and 9.61 ppb, for July 10 2001, and 41.78 and 13.96 ppb for August 11 2001.

Day	Method	MSE	MAE	Empirical coverage of 90% PI	Average length of 90% PI	Average predictive variance
06/21/2001	Ordinary kriging	37.38	4.87	89.9%	20.52	39.81
	Bayesian melding	35.35	4.81	78.26%	14.22	19.36
	Downscaler	9.79	2.45	98.55%	19.34	35.47
07/10/2001	Ordinary kriging	39.15	4.66	95.65%	22.75	48.49
	Bayesian melding	32.26	4.26	88.41%	16.70	26.36
	Downscaler	16.33	2.90	97.10%	21.29	42.75
08/11/2001	Ordinary kriging	30.70	4.38	90.90%	20.53	41.44
	Bayesian melding	14.94	2.99	78.79%	11.74	13.39
	Downscaler	6.06	1.89	96.97 %	17.49	29.48

Table 16.1 *Mean Square Error (MSE), Mean Absolute Error (MAE), empirical coverage, average length of 90% predictive intervals (PI) and predictive variance for ordinary kriging, Bayesian melding and the downscaler method for three days in the 2001 high-ozone season.*

After estimating the model parameters, we proceeded to predict ozone concentration at all of the 69 monitoring sites using the three different methods. We evaluate the quality of the predictions by computing the MSE and MAE of the predictions, and by looking at the empirical coverage and average length of the 90% predictive intervals. Table 16.1 reports these summary statistics for all three selected days. We see that the downscaler produces predictions that are far better calibrated than the other methods. Moreover, the empirical coverage of the downscaler is always above the nominal value, thus indicating that it is more conservative than Bayesian melding, whose empirical coverage is always below 90%. Alternatively, the predictive intervals constructed using the downscaler are wider than those obtained using Bayesian melding, which are too narrow.

Ordinary kriging performs as well as the downscaler model in terms of coverage and average length of the 90% predictive intervals, but its predictions are much more biased than the ones produced using our downscaling model. This might be expected since ordinary kriging does not exploit the additional information contained in the high-resolution CMAQ output, relying only on the 69 ozone observations.

We conclude by noting that the downscaler was extended to accommodate bivariate data at locations in Berrocal et al. [2010b]. Also, there can be measurement error concerns when downscaling. Berrocal et al. [2012] propose new models that are intended to address two such concerns with the computer model output. One is potential spatial displacement in the computer model values assigned to a grid cell. Possibly, this output is appropriate for a displacement of the grid cell. The second recognizes that, with regard to improving

	RMSE	MAE	CRPS
Model 1	1.72	1.40	1.02
Model 2	1.45	1.17	0.85

Table 16.2 *Performance metrics (RMSE, MAE, and CRPS) for the two models summarized by their average across 100 simulations.*

the predictive performance of the fusion at a location, there may be useful information in the outputs for grid cells that are neighbors of the one in which the location lies.

16.3.5 A useful simulation

In the measurement error fusion setting, it is challenging to demonstrate the benefit of calibration and fusion with real data. Specifically, in the context of more than one source informing about the true value of a variable, neither source provides observations that are the true values. Each source has measurement error and, in the absence of external validation data, the true value of the variable is not observed.

In a regression setting, whether a predictor or the response (or both) is observed with measurement error, learning about the true regression relationship is the objective. Therefore, in the absence of the true values, using the observed values implies increased uncertainty in this inference objective. Ignoring measurement error leads to potentially inaccurate assessment of precision and accuracy. In order to demonstrate the improvement in inference by incorporating measurement error into a model specification in the form of calibration and fusion, we present a simple simulation example where the truth is known.

We generated 100 points inside the unit square, as realizations from $Y_{true}(\mathbf{s})$. We captured spatial dependence with a mean $\mu = 4$ Gaussian process (GP) arising from an exponential covariance function with variance $\sigma_{true}^2 = 4$ and decay parameter $\phi_{true} = 3/0.5$, so the spatial range is 0.5. Then, we sampled 80 locations at random. We randomly assigned half of these locations to source 1 and the other half to source 2. For source 1, we obtain the responses using $Y_1(\mathbf{s}) = Y_{true}(\mathbf{s}) + \epsilon_1(\mathbf{s})$ where the $\epsilon_1(\mathbf{s})$ are i.i.d. $N(0, \sigma_1^2 = 4)$. This is the measurement error sample. For source 2, we obtain the responses using $Y_2(\mathbf{s}) = \lambda_0 + \lambda_1 Y_{true}(\mathbf{s}) + \epsilon_2(\mathbf{s})$ where the $\epsilon_2(\mathbf{s})$ are i.i.d. $N(0, \sigma_2^2 = 1)$ for fixed $\lambda_0 = 2$ and $\lambda_1 = 2$. This is the calibration sample. The remaining 20 locations are viewed as "out of sample" for both sources, employed for validation. We predict the $Y_{true}(\mathbf{s})$ for these points using kriging. We compare models using by now familiar criteria: root mean square error (RMSE), mean absolute error (MAE), and continuous rank probability score (CRPS).

We fit Model 1 using just the source 1 data in a standard spatial linear regression model (Section 6.1), $Y_1(\mathbf{s}) = \mu + w(\mathbf{s}) + e(\mathbf{s})$ where $w(\mathbf{s})$ is a mean-zero spatial GP (with known decay parameter) and $e(\mathbf{s})$ is pure error. So, this model has three parameters, $(\mu, \sigma_w^2, \sigma_e^2)$. It is useful to note that this, by now familiar geostatistical model, is in fact a measurement error model since we can write it as $Y_1(\mathbf{s}) = Y_{true}(\mathbf{s}) + \epsilon(\mathbf{s})$ with $Y_{true}(\mathbf{s}) = \mu + w(\mathbf{s})$. In fact, in the introduction to Chapter 6, we commented that measurement error was one of the possible interpretations of $\epsilon(\mathbf{s})$. Further, the measurement error is identifiable in this modeling because of the dependence structure associated with $w(\mathbf{s})$. Model 2 uses the source 1 and source 2 data. This model specifies $Y_1(\mathbf{s})$ and $Y_2(\mathbf{s})$ as above. It has six parameters, $(\mu, \sigma_{true}^2, \sigma_1^2, \lambda_0, \lambda_1, \sigma_2^2)$. Under each model, we do posterior prediction for the 20 hold-out true values. We repeat the entire simulation 100 times. Table 16.2 summarizes across simulations, using the predictive performance metrics above. In this simple illustration, Model 2 improves Model 1 by roughly 15% to 20% based on all metrics, demonstrating the benefits of the fusion in terms of predictive performance.

16.4   Point pattern-point pattern fusion

Assuming two point pattern sources, we have to be explicit with regard to how we imagine each of the point patterns is observed in order to develop an appropriate a fusion. One scenario would propose that each source provides a full realization of a point pattern and that the focus of the modeling is to learn about the nature of the intensity that is driving each of the realizations. Such a fusion would be sensible if the intensities for the two-point patterns may be assumed to be common, up to varying covariate levels according to the different locations for the two point patterns. This might occur say if the point patterns were collected at different times for the same region or perhaps if the point patterns were the locations of T-cells (part of the immune system) in tissue samples from the same patient. With a log Gaussian Cox process (Chapter 8), both point patterns arise from intensity with a shared Gaussian process. Regardless, both point patterns are viewed as conditionally independent given the intensity. As a result, again following Chapter 8, the hierarchical modeling is straightforward with a product form for the two point pattern likelihoods at the first stage; approaches for fitting are presented in Section 9.4.2. This setting is analogous to increasing sample size for the fusion. It may also be viewed as a version of a superposition (Section 9.3.6.4). So, we do not pursue this fusion further here.

An alternative, arguably, more challenging fusion arises if we assume that there is a single true point pattern which yields our observed point patterns. That is, we assume that each observed point pattern data source is a partial realization, arising in some form, of the full point pattern. The realizations are *partial* in the sense that, due to, e.g., sampling effort or detection, only a portion of the true point pattern was seen. Here, we assume a model for the true intensity, $\lambda_{true}(\mathbf{s})$ which has generated a point pattern realization, $\mathcal{S}$ over $\mathcal{D}$. In the most direct case, what we observe are $\mathcal{S}_1 = \{\mathbf{s}_1, ..., \mathbf{s}_{n_1}\}$ and $\mathcal{S}_2 = \{\mathbf{s}_1, ..., \mathbf{s}_{n_2}\}$, each a *thinned* version (Section 9.3.6.2) of $\mathcal{S}$, each with its own thinning mechanism.

More generally, we could have $\lambda_{1,true}(\mathbf{s})$ and $\lambda_{2,true}(\mathbf{s})$ where the log intensities share a common GP but otherwise may have different covariates. In either case, we have a dependence on the intensities but conditionally independent realizations given the intensities. Examples here could be two sets of distance sampling data over the same region say, for different days, yielding different covariates. This might arise from flying different sets of trajectories over the region on different days.

Extending this, for data source 2 we can have $\mathcal{S}_2$ in the form $N(B_k), k = 1, 2, ..., K$, where $N(B_k)$ is the number of points in set B_k and the $B_k \subset \mathcal{D}$ with data source 1 as above. The sets $\{B_k\}$ are disjoint but their union will not be $\mathcal{D}$, again reflecting thinning. Another variant would replace the sets B_k with points. For example, we could observe a count of the number of individuals observed at say, a monitoring location, $\mathbf{s}_k^*$, obtaining data, $\{N(\mathbf{s}_k^*), k = 1, 2, ..., K\}$. Finally, we can have both data sources observed gathered into counts for sets (or points), with differing collections of disjoint subsets (points) of $\mathcal{D}$. Regardless, we have a shared $\lambda_{true}(\mathbf{s})$ driving both sources. Again, we have a thinning mechanism associated with each source.

For the direct case of two thinned point patterns, with interest in the actual point pattern which is operating over $\mathcal{D}$, we assume that data source 1 arises from a thinning mechanism $p_1(\mathbf{s})$; similarly, $p_2(\mathbf{s})$ is the thinning mechanism for source 2. These thinning mechanisms are applied independently to a latent realization $\mathcal{S}$ from $\lambda_{true}(\mathbf{s})$ to provide the observed point patterns for each source, respectively. This is the sampling model and the resulting observed point patterns are conditionally independent given the thinning mechanisms and the latent point pattern. Under such assumption, the likelihood for the sampling model arises as a product form.

The fitting model specification also employs both thinning mechanisms but with an intensity model. The thinning functions are specified, each appropriately capturing sampling effort, location availability, and detection (or misclassification). The intensity, $\lambda(\mathbf{s})$ will arise from a log Gaussian Cox process (LGCP) model. The resulting *operating* intensities in the fitting model become $p_1(\mathbf{s})\lambda(\mathbf{s})$ and $p_2(\mathbf{s})\lambda(\mathbf{s})$. See, e.g., Chakraborty et al. [2011] and Section 9.8.2 for an illustration of this version of thinning. Note that the thinned observations could be sampled directly using the thinned/degraded intensity with a gridding of the operating intensity and a version of simulation using the method of Lewis and Shedler [1979] (Section 9.3.2).

As is always the case in practice, the fitting model will not be the sampling model. The fitting model treats the observed thinned point patterns as conditionally independent given the thinning mechanisms and the intensity. Then, the primary inference objective will be to back out this intensity from both thinnings. This intensity enables inference regarding the spatial variation of the point pattern process as well as expected counts in subsets of the region. If there are parameters in the thinning mechanisms, then identifiability of these parameters along with those of $\lambda(\mathbf{s})$ becomes a concern. Some prior information regarding them will be needed in order to provide an *anchor/scale* for the true intensity. Alternatively, knowledge regarding total expected number over the region can anchor the individual thinning mechanisms. Also, there can be interest in features of the realization $\mathcal{S}$ which has been thinned, e.g., the number of points in $\mathcal{D}$ or in a subset of $\mathcal{D}$.

For the case of thinned points, we obtain the familiar LGCP likelihood (Section 8.4). For the case of counts, they are assumed to arise from Poisson distributions, e.g., $N(B_k) \sim Po(\lambda(B_k))$ where $\lambda(B_k) = \int_{B_k} p(\mathbf{s})\lambda(\mathbf{s})ds$. The fact that the sets are disjoint immediately enables the likelihood. Regardless, discretization for the stochastic integrals following Section 8.4 will be required. Again, the sources are conditionally independent given the thinning mechanisms and the true intensities. The case with counts for both sources, $N(B_{1k}), k = 1, 2, ..., K_1$ and $N(B_{2k}), k = 1, 2, ..., K_2$, all $B \subset \mathcal{D}$ and the sets for each source disjoint can be handled similarly. Again, the disjointedness of the sets enables an immediate likelihood for each source after which the model fitting and inference play out as above.

We note that, in principle, we can remove the disjointedness requirement by creating latent counts associated with a disjoint partitioning of the union of all of the $N_1(B_k)$. Then, customary Gibbs looping (Section 5.3) would be implemented to alternate between updating the model parameters given the latent counts and then updating the counts given the parameters. We are unaware of an application in the literature of this version.

Viewing the data collected from the fixed monitoring locations as conditionally independent counts given the fitting model, again, enables an immediate likelihood with model fitting and inference similar to the above. In fact, we illustrate this below, following Schliep et al. [2024] .

The fusion is expected to benefit both inference regarding $\lambda(\mathbf{s})$, the nondegraded intensity, and $\mathcal{S}$ the actual full realization from that intensity. We can offer some informal support for the benefit. Suppose we have two point patterns, $\mathcal{S}_1 = \{\mathbf{s}_{1,1}, ..., \mathbf{s}_{1,n_1}\}$ and $\mathcal{S}_2 = \{\mathbf{s}_{2,1}, ..., \mathbf{s}_{2,n_2}\}$. Suppose that $\mathcal{S}_1$ is a realization from an NHPP (or LGCP) with intensity $p_1\lambda(\mathbf{s})$ and $\mathcal{S}_2$ is a realization from an NHPP (or LGCP) with intensity $p_2\lambda(\mathbf{s})$ both over the region $\mathcal{D}$, i.e., we adopt constant thinning. Suppose we seek to estimate $\lambda(\mathcal{D})$ which is $E(N(\mathcal{S}))$ if $\mathcal{S}$ is a realization from an NHPP (or LGCP) with intensity $\lambda(\mathbf{s})$. Let $N(\mathcal{S}_1)$ be the number of points in $\mathcal{S}_1$ and $N(\mathcal{S}_2)$ be the number of points in $\mathcal{S}_2$. Then, since $E(N(\mathcal{S}_1)) = p_1\lambda(\mathcal{D})$ and $E(N(\mathcal{S}_2)) = p_2\lambda(\mathcal{D})$, $N(\mathcal{S}_1)/p_1$ and $N(\mathcal{S}_2)/p_2$ are both unbiased estimators of $\lambda(\mathcal{D})$. However, if $p_1 < p_2$, the variance of the latter is smaller than the variance of the former; the less we thin, the better we can estimate $\lambda(\mathcal{D})$, in accord with intuition. In our fusion setting, the joint thinning of $\mathcal{S}$ yields a thinning probability $1 - (1 - p_1)(1 - p_2) > \max(p_1, p_2)$. We thin less under the fusion than under either source; we expect to better estimate $EN(\mathcal{S})$ and better predict $\mathcal{S}$.

We can extend this argument to the case of spatially varying thinning, $p_1(\mathbf{s})$ and $p_2(\mathbf{s})$. Suppose we partition $\mathcal{D}$ into arbitrarily fine disjoint sets B_k such that if $\mathbf{s} \in B_k$, $p_1(\mathbf{s}) = p_{1k}$

and $p_2(\mathbf{s}) = p_{2k}$ where $p_{1k} < p_{2k} \; \forall k$. We can consider the two unbiased estimates for $\lambda(\mathcal{D})$, $\sum_k N_{1k}/p_{1k}$ and $\sum_k N_{2k}//p_{2k}$, where N_{ik} denotes the random number of points from $\mathcal{S}_i$ in the set B_k. Since N_{ik} and $N_{ik'}$ are independent for $k \neq k'$, $\mathrm{Var}\left(\sum_k N_{1k}/p_{1k}\right) = \sum_k \lambda(B_k)/p_{1k} > \sum_k \lambda(B_k)/p_{2k} = \mathrm{Var}\left(\sum_k N_{2k}/p_{2k}\right)$. Again, this implies that we can better estimate $\lambda(\mathcal{D})$ using the data fusion given the joint thinning of $\mathcal{S}$.

16.4.1 A challenging example for whale abundance estimation

Abundance estimation for marine mammal populations is generally very difficult since, unlike terrestrial animals, marine mammals typically spend only a portion of their time at or near the surface where they are visible. Since current abundance estimation practice is built upon some version of (aerial) distance sampling (Mayo et al. [2018] and Ganley et al. [2019]), potential sightings are limited to the window when an individual is essentially at the surface. It is further limited by whether the collection is going on at the time and is in sufficient proximity to detect a surfacing. Adding an additional data source, in the illustration here, acoustic data, improves our understanding of the spatial distribution as well as the abundance for these mammals.

The example in Schliep et al. [2024] considers the time of observation as fixed, over a specified region. What we conceive of is a latent, unobserved set of locations for all of the individuals in the region at the fixed time. That is, the truth is viewed as a conceptual *snapshot* of a point pattern. In fact, we have such a conceptual snapshot at any time within a window of interest. This differs from the standard *spatio-temporal* point process setting where a point pattern is realized over a window of time (Section 9.7) and abundance arises by integrating over space as well as time.

Associated with the unobservable point pattern is an unobservable "true" latent intensity surface which *explains/drives* this set of locations. With temporal data collection, the resulting objectives are (i) at a given time, to estimate abundance locally (or globally) by integrating the intensity over subregions (or the entire study region) and (ii) to learn about spatial variation in intensity, including comparison and evolution over time.

Here we are in a version of the second case of the previous section. One data source is aerial distance sampling data, collected irregularly in a given year during the first half of the year. Flights are scheduled roughly every 10 days to two weeks, weather permitting. On any particular day, a partially observed realization of the full point pattern is obtained. The collection is made through a designed sequence of airplane trajectories over a few-hour period. In what follows this window is assumed short enough to view the collection as a portion of the foregoing snapshot of the entire set of animal presences. For each trajectory, what is observed is a degraded/thinned point pattern due to time on the surface, sampling effort, and detection probability.

The second data source is referred to as passive acoustic monitoring (PAM) data. Its form provides the number of whale calls received hourly at each hydrophone over a network of hydrophone monitors. Evidently, hourly collection is at substantially higher temporal resolution than the aerial data. Here, for any hour the set of number of calls over the hydrophone network is also viewed as arising from a snapshot of animal presences. The additional challenge is that, for this data, no animals are ever observed. The acoustic monitors are set on the ocean floor and detect calls throughout the water column; animals need not be on the surface in order to be detected through their calls. The strategy here is to employ an acoustic detection function as well as to adjust call rates to individuals. Again, we have a degraded intensity such that the expected number of calls received by a monitor is viewed as the integration of this degraded intensity relative to the geo-coded location of the monitor. In other words, from this degraded source we can back out the expected number of individuals detected over the study region for each PAM, using assumptions about calling rates. This also enables us to learn about the latent intensity, assuming the expected number of calls is the integrated degraded intensity over the study area. Altogether, both sources degrade the true intensity surface but each informs about this true intensity.

As a last point, lacking whale depth data, it is assumed that the true latent intensity operates only on the (x, y) plane; we imagine a conceptual three-dimensional intensity integrated over depth. For the distance sampling source, the true intensity operating at the surface is the true intensity multiplied by the proportion of time at the surface. See Schliep et al. [2024] for the full development of this example.

16.5   Preferential sampling

Before elaborating a proposed data fusion for a point-referenced dataset and a point pattern dataset, we digress to review the ideas of preferential sampling which motivate a natural model specification for such fusion. Preferential sampling was introduced into the literature in the seminal paper of Diggle et al. [2010]. Subsequently, there has been considerable followup research. Two useful methodological papers in this regard are Pati et al. [2011] and Cecconi et al. [2016]. Preferential sampling has been found to be useful to consider in areas such as environmental exposure, adaptive geostatistical design, prevalence mapping, veterinary parasitology, missing data, species distribution modeling, and selling prices of properties.

A standard illustration arises in geostatistical modeling where we seek to infer about an environmental exposure surface. If environmental monitors tend to be placed in locations where environmental levels tend to be high, then interpolation based upon observations from these locations will necessarily tend to produce only high predictions. A natural remedy lies in suitable spatial design of the locations, e.g., a random or space-filling design for locations over the region of interest [Nychka and Saltzman, 1998] is expected to preclude such bias. However, in practice, the sampling may be designed in order to learn about areas of high exposure.

In the context of preferential sampling, the set of sampling locations may not have been developed randomly but it is studied/modeled as if it was a realization of a spatial point process. That is, it may be designed/specified in some fashion but not necessarily with the intention of being roughly uniformly distributed over $\mathcal{D}$. Then, the question becomes a stochastic one: is the realization of the responses independent of the realization of the locations? If no, then we have what is called preferential sampling. The key point is that the dependence here is **stochastic** dependence. We clearly have **functional** dependence; the responses are associated with the locations.

Below we will offer two illustrative settings. One takes up hedonic modeling for selling price of properties. Specifically, hedonic modeling is a regression model specification at the level of the individual sale. It customarily introduces features associated with the property sold at the given location and time. Common property features include size, age, number of bedrooms, number of baths, and other amenities. Other covariates include features of the location, e.g., neighborhood safety, quality of schools, and access to public transportation, access to green spaces. Contemporary hedonic modeling also includes spatial and perhaps spatio-temporal random effects [Gelfand et al., 1998]. In such modeling, locations and times of transactions are random. That is, the number of transactions is random and, given this number, the locations and times of the transactions are random. So, the collection of property sales is a realization of a random point pattern over space (the sales market) and time (the period). The geostatistical response at a location is the selling price, modeled conditionally on location and time. Randomness is introduced only in the regression given predictors at the location and time. Random effects are also introduced, associated with locations and times that, again, are viewed as fixed. The randomness in the transaction locations has been ignored in the hedonic modeling literature apart from the work of Paci et al. [2020].

The second setting considers species distribution modeling. Here, since one of the most important ecological questions to ask is where species are and why, there is an enormous literature with presence/absence or abundance recorded at locations. Spatial sampling bias can arise because ecologists will tend to sample locations where they expect to find individuals (referred to as sampling effort). So, with presence/absence data, we may find an oversampling of presences. Further, in this setting and to complete this chapter, we show how preferential sampling can be extended to a data fusion where we have both presence/absence data and presence-only data. The presence-only data adds random locations to those already in the presence/absence collection. And, the presence-only set of data locations will be *extremely* biased since only presences are observed at these locations. Here, we review the work of Gelfand and Shirota [2019]

Preferential sampling opens up three inference issues with regard to potential bias in the collection of sampling locations; they can affect the predictive performance of a geostatistical model. These issues are: (i) can we identify the occurrence of a preferential sampling effect? (ii) can we revise modeling in the presence of preferential sampling? (iii) when can such adjustment improve predictive performance over a customary geostatistical model?

Let us formalize, probabilistically, the preferential sampling setting. We adopt the general notation: $\boldsymbol{\eta}$, $\mathcal{S}$, and $\mathcal{Y}$ to denote a latent spatial process, the design (location) data, and the measurement (geostatistical) data, respectively. A factorization of the joint distribution of $\boldsymbol{\eta}$, $\mathcal{S}$, and $\mathcal{Y}$ is

$$[\boldsymbol{\eta}, \mathcal{S}, \mathcal{Y}] = [\boldsymbol{\eta}][\mathcal{S} \mid \boldsymbol{\eta}][\mathcal{Y} \mid \mathcal{S}, \boldsymbol{\eta}]. \tag{16.16}$$

Integration with respect to $\boldsymbol{\eta}$ gives the log-likelihood function based on the data $\mathcal{S}$ and $\mathcal{Y}$ as

$$\log L(\mathcal{S}, \mathcal{Y}) = \log \int [\mathcal{S} \mid \boldsymbol{\eta}][\mathcal{Y} \mid \mathcal{S}, \boldsymbol{\eta}][\boldsymbol{\eta}] d\boldsymbol{\eta} \tag{16.17}$$

The above factorization is most natural from a modeling perspective because (i) the latent process drives both data mechanisms and (ii) $(\mathcal{S}, \mathcal{Y})$ can be viewed as a marked point process and, in the geostatistical setting, we would model locations and then response given location (see Section 8.6.3). Following Diggle et al. [2010], under non-preferential sampling, $[\mathcal{S}|\boldsymbol{\eta}] = [\mathcal{S}]$ and the log likelihood becomes

$$\log L(\mathcal{S}, \mathcal{Y}) = \log \int [\mathcal{Y} \mid \mathcal{S}, \boldsymbol{\eta}][\boldsymbol{\eta}] d\boldsymbol{\eta} + \log[\mathcal{S}] \tag{16.18}$$

The stochastic variation in $\mathcal{S}$ can be ignored for inference about $\boldsymbol{\eta}$ or $\mathcal{Y}$.

Conventional geostatistical modeling (Chapter 6), treats the design as a fixed set of locations $\mathbf{s}_i : i = 1, ..., n$. Typically, it assumes that the measurements $Y(\mathbf{s}_i)$ are conditionally independent given the corresponding $\eta(\mathbf{s}_i)$, hence

$$[\mathcal{Y} \mid \mathcal{S}, \boldsymbol{\eta}] = \prod_{i=1}^{n}[Y(\mathbf{s}_i) \mid \eta(\mathbf{s}_i)]. \tag{16.19}$$

Diggle et al. [2010] refer to the specification, $[\mathcal{S}|\boldsymbol{\eta}]$ as *strong* preferential sampling. Under strongly preferential sampling, no factorization of $[\boldsymbol{\eta}, \mathcal{S}, \mathcal{Y}]$ separates $\mathcal{S}$ and $\boldsymbol{\eta}$ Valid inference requires the stochastic nature of $\mathcal{S}$ to be taken into account, i.e., $\mathcal{S}$ is not ignorable. This leads us to the preferential sampling model in Diggle et al. [2010]. Suppose the geostatistical model for $Y(\mathbf{s})$ is of the form:

$$Y(\mathbf{s}) = \mu(\mathbf{s}) + \eta(\mathbf{s}) + \epsilon(\mathbf{s}) \tag{16.20}$$

Diggle et al. [2010] assume that $\eta(\mathbf{s})$ is a stationary Gaussian process and, conditional on $\boldsymbol{\eta}$, $\mathcal{S}$ is a nonhomogeneous Poisson process with intensity $\lambda(\mathbf{s}) = \exp\{\alpha + \beta\eta(\mathbf{s})\}$. Unconditionally,

S is a log Gaussian Cox process (Section 8.4). They refer to this as a "shared process" $(\eta(\mathbf{s}))$ model. We have non-preferential sampling when $\beta = 0$, and strong preferential sampling otherwise.

Perhaps a more useful inferential form is presented in Pati et al. [2011] who extend the above by adding a second Gaussian process and reparametrizing. Now assume S is a log-Gaussian Cox process with intensity

$$\lambda(\mathbf{s}) = \exp\{\alpha + \eta(\mathbf{s})\} \tag{16.21}$$

where $\eta(\mathbf{s})$ is a Gaussian process with mean zero, variance σ_1^2 and correlation function $\rho(\mathbf{h}; \phi_1)$. Further, assume that measurements $Y(\mathbf{s}_i) : i = 1, ..., n$ at locations $\mathbf{s}_i$ follow the "geostatistical" model (Chapter 6)

$$Y(\mathbf{s}) = \mu(\mathbf{s}) + \delta \log \lambda(\mathbf{s}) + w(\mathbf{s}) + \epsilon(\mathbf{s}) \tag{16.22}$$

where $w(\mathbf{s})$ is a Gaussian process, independent of $\eta(\mathbf{s})$, with mean zero, variance σ_2^2 and correlation function $\rho(\mathbf{h}; \phi_2)$. The process $w(\mathbf{s})$ allows for a component of the spatial variation in the measurement process that is not linked to the sampling process. The $\epsilon(\mathbf{s})$ are pure error $N(0, \tau^2)$ variables. As a result, we have a *geostatistical* model with the intensity as a *regressor*; δ controls the degree of "preferentiality" in the sampling of the $Y(\mathbf{s})$. We are back to the general geostatistical model when $\beta = 0$. Note that, equivalently, we can replace $\delta \log \lambda(\mathbf{s})$ with $\delta\eta(\mathbf{s})$

Now, we present a collection of stochastic specifications that illuminate preferential sampling for a region $\mathcal{D}$. We consider two cases for the intensity associated with the point pattern of sampling locations, $\mathcal{S}$:

(i) $\log \lambda(\mathbf{s}) = \mathbf{W}^{\mathrm{T}}(\mathbf{s})\boldsymbol{\beta}$, i.e., a nonhomogeneous Poisson process (NHPP) and

(ii) $\log \lambda(\mathbf{s}) = \mathbf{W}^{\mathrm{T}}(\mathbf{s})\boldsymbol{\beta} + \eta(\mathbf{s})$, a log Gaussian Cox process (LGCP).

Here, $\mathbf{W}(\mathbf{s})$ is a vector of predictors with associated regression coefficients $\boldsymbol{\beta}$, and $\eta(\mathbf{s})$ is a mean 0 GP (below, for convenience, with an exponential covariance function). See Chapter 8 for a discussion of NHPPs and LGCPs. In the sequel we only work with (ii).

Turning to the geostatistical data, $Y(\mathbf{s})$, we start with a simple spatial regression,

(a) $Y(\mathbf{s}) = \mathbf{X}^{\mathrm{T}}(\mathbf{s})\boldsymbol{\alpha} + \epsilon(\mathbf{s})$,

where the predictors in $\mathbf{X}(\mathbf{s})$ and those in $\mathbf{W}(\mathbf{s})$ need not be identical and $\epsilon(\mathbf{s})$ is a pure error with homogeneous variance. Extension to a customary geostatistical model for $Y(\mathbf{s})$ becomes

(b) $Y(\mathbf{s}) = \mathbf{X}^{\mathrm{T}}(\mathbf{s})\boldsymbol{\alpha} + \omega(\mathbf{s}) + \epsilon(\mathbf{s})$,

adding $\omega(\mathbf{s})$ as a mean 0 GP (also with an exponential covariance function for convenience), independent of $\eta(\mathbf{s})$ above. So, using (ii) to model S with (b) to model $Y(\mathbf{s})$, S and $\mathcal{Y}$ are probabilistically independent; there is no preferential sampling.

As noted above, Pati et al. [2011] attempt to interpret $\eta(\mathbf{s})$ as a regressor to add to the geostatistical model for $\mathcal{Y}$. That is, they extend (b) to model

(c): $Y(\mathbf{s}) = \mathbf{X}^{\mathrm{T}}(\mathbf{s})\boldsymbol{\alpha} + \delta\eta(\mathbf{s}) + \omega(\mathbf{s}) + \epsilon(\mathbf{s})$.

Here, the coefficient δ plays a preferential sampling role. For example, suppose the design S oversamples locations in $\mathcal{D}$ where we observe presences, where $Y(\mathbf{s})$ tends to be high. Then, $\eta(\mathbf{s})$ will tend to be high around those locations. Therefore, $\eta(\mathbf{s})$ can be a significant

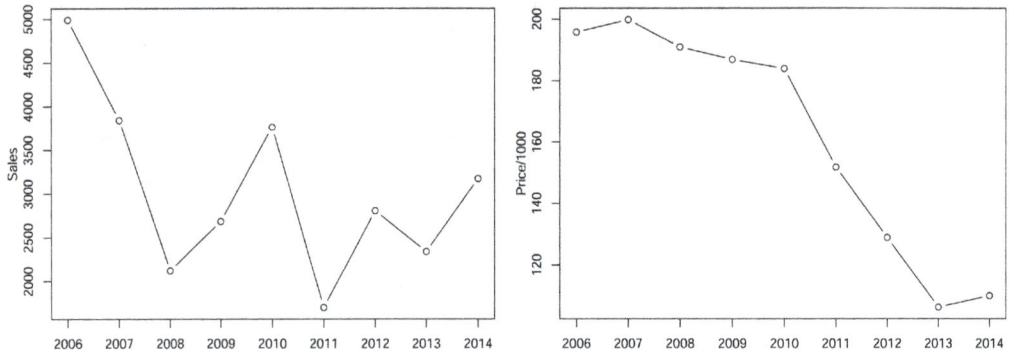

Figure 16.2 *Number of transactions (left panel) and average price (right panel) per year.*

predictor for $Y(\mathbf{s})$ with $\delta > 0$. (A similar argument applies when $\delta < 0$.) With (ii) and (c), $\eta(\mathbf{s})$ is the shared process.

A further shared process model for $\mathcal{Y}$, that can be explored in this regard, extends (a) to

(d): $Y(\mathbf{s}) = \mathbf{X}^{\mathrm{T}}(\mathbf{s})\boldsymbol{\alpha} + \delta\eta(\mathbf{s}) + \epsilon(\mathbf{s}).$

Here, interest is in comparing (d) and (ii) with (a) and (ii); is $\delta \neq 0$, i.e., we have a shared process model? Diggle et al. [2010] focus on comparing (b) and (i) with (b) and (iii). Pati et al. [2011] focus on comparing (ii) and (b) with (ii) and (c).

16.5.1   Two preferential sampling examples

We return to the two examples mentioned above. First, we consider selling prices of properties. Paci et al. [2020] consider residential sales in Zaragoza, Spain, during the 9-year period $2006 - 2014$ for Zaragoza, a city situated in the northeast part of Spain with a population of $664,953$ inhabitants as of 2014. For every sale, the dataset contains information on location (UTM coordinates), selling price, living area, year of building construction, and the exact day of the transaction.

The area of the municipality is large, with the town hall dividing the city into 89 urban polygons. Some of them, considered rural neighborhoods/areas, are distant from the city center and, in some cases, are disconnected from the rest of the city. For this study, 52 of those polygons are employed, all of them around the city center, following the dense distribution of population in old cities in Spain. Apartments are the dominant type of dwelling everywhere in the city. Other properties are discarded; family houses, townhouses, etc. are located only in a few polygons. Large areas with no housing within the polygons have been removed (e.g. parks, the University of Zaragoza campus, hospital areas, etc.). What is retained is, essentially, a full census of apartment sale transactions in and around the city center.

Figure 16.2 shows the average price per year and the number of transactions per year. For the period analyzed in this work, it is important to keep in mind the deep economic financial crisis that, starting in 2008, sank the economy in Spain and many other countries, leading to devastating consequences, especially in the real estate market. A clear drop in the prices is shown starting from 2010. Paci et al. [2017] analyzed the spatio-temporal point pattern of sales over the 9-year period. Paci et al. [2020] bring in the selling price with the objective, as described above, to assess preferential sampling with the regard to the selling

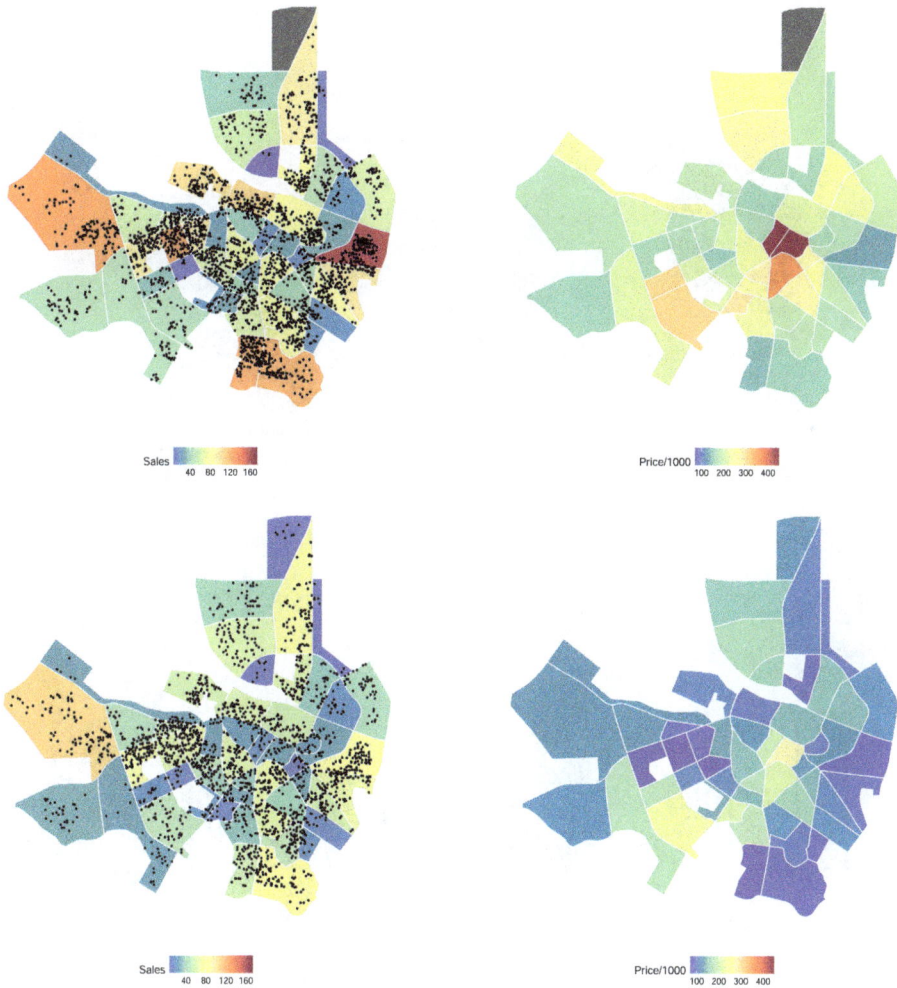

Figure 16.3 *Number of sales (left panels) and average price (right panels) by urban polygons in 2007 (top panels) and 2012 (bottom panels). Grey color denotes no sales.*

price. They focus on relating the spatial point pattern of sales to the spatial surface of selling price over the city. They analyze selling prices for two different years, 2007 and 2012, pre- and post-crisis, respectively. Figure 16.3 shows the number of sales and the average price by urban polygons in 2007 (top panels) and 2012 (bottom panels); there are $2,969$ sales in 2007 and $2,186$ in 2012.[2]

Features of the sold properties are available to explain the selling price, including the size of the property, its age, and a binary regressor which records the presence of *accessory elements* (AE). Accessory elements are considered as complements to the apartment such as a garage space or a storage area. Specifically, $AE = 1$ if *no* accessories, $AE = 0$ if there are. Figure 16.4 displays the scatterplot of log size and log price for 2007 (left panel) and 2012 (right panel), showing the expected strong relationship. Also considered is the age of the property, defined as the difference between the year of transaction and the year of building construction. Then, a logarithmic transformation is applied and the scaled variable

---

[2]While we see spatial pattern in the selling prices, it is important to note that they have not yet been adjusted to reflect differences in the regressors associated with each sale.

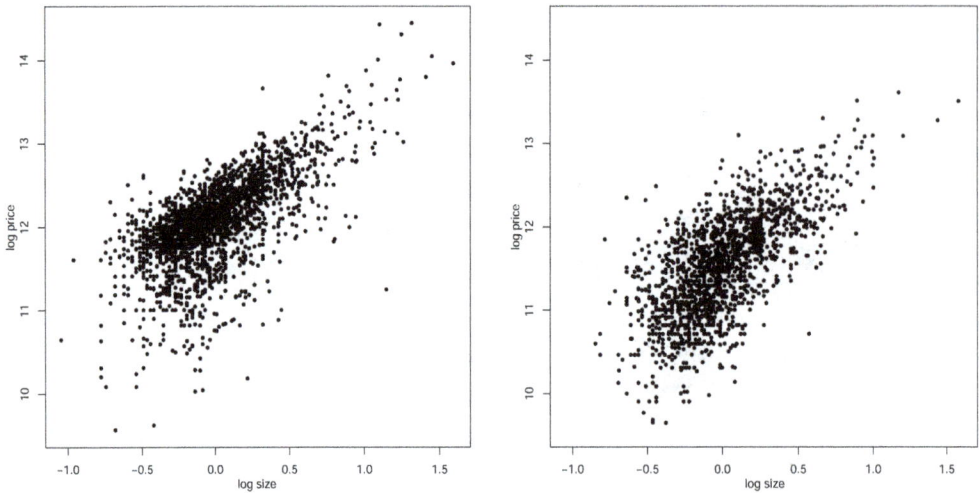

Figure 16.4 *Scatterplot of log size and log price for* 2007 *(left panel) and* 2012 *(right panel).*

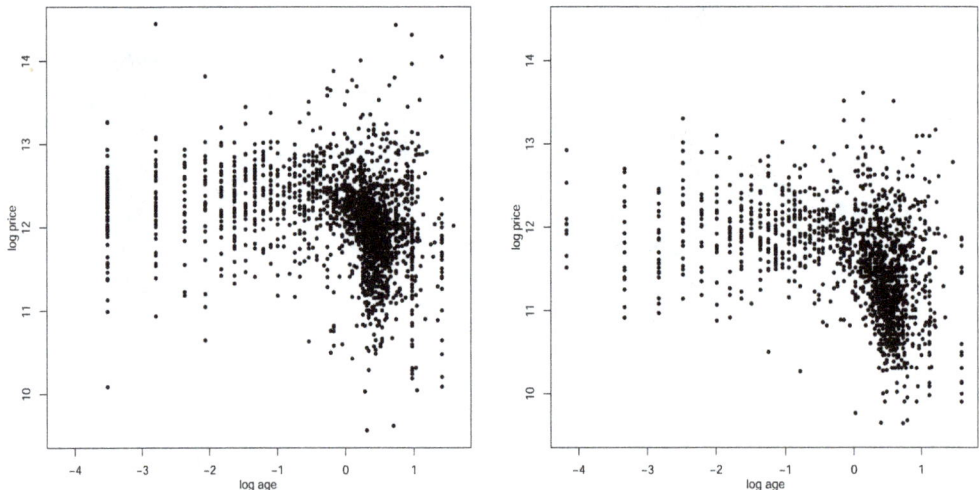

Figure 16.5 *Scatterplot of log age and log price for* 2007 *(left panel) and* 2012 *(right panel).*

is retained. The weak relationship between log price and log age is displayed in Figure 16.5. Usually, economic variables are employed in hedonic modeling, such as interest rates, unemployment rates, and consumer price indexes. However, such variables do not vary over the city and so they cannot be used for studying the spatial variation of selling price. Finally, demographic covariates are introduced at spatial (polygon) level, i.e., total population, percentage of youth population (≤ 14 years old), elderly population (≥ 65 years old), and the percentage of foreign population.

Let $\mathcal{Y}$ denote the set of selling prices and let $\mathcal{S}$ denote the spatial point pattern of sales. In Paci et al. [2017], $\mathcal{S}$ is modeled using an LGCP driven by an intensity $\lambda(\mathbf{s})$, i.e., $[\mathcal{S}|\lambda(\mathbf{s})], \mathbf{s} \in \mathcal{D}$, where $\mathcal{D}$ is the spatial domain. Note that, if we simply overlay on this model a hedonic model for $\mathcal{Y}$, e.g. $[\mathcal{Y}|\mathcal{S}, parameters]$, then we can fit both models separately. $\mathcal{S}$ is viewed as fixed in the hedonic model; hence, $\mathcal{S}$ does not inform about $\mathcal{Y}$ and we do not introduce the notion of preferential sampling. Paci et al. [2020] consider the foregoing models for $\mathcal{S}$ and those for the selling prices, $Y(\mathbf{s})$.

They note the following:

- Along with a suitable set of regressors, the flexibility of model (b) may enable it to predict well ignoring the information in $\mathcal{S}$ if $\mathcal{S}$ does not show strong bias with regard to locations of sales over $\mathcal{D}$. In other words, a shared process model for selling price need not be better than a geostatistical model for predicting selling price. The geostatistical model allows the selling price to have its own random effects process (GP) while the shared process model forces $\mathcal{Y}$ to inherit the same GP as for $\mathcal{S}$.

- Combining model (2) with models (c) or (d) creates a dependence between $\mathcal{Y}$ and $\mathcal{S}$ in the sense that $\eta(\mathbf{s})$ is shared in the model for each. When there is stochastic dependence between $\mathcal{Y}$ and $\mathcal{S}$ then not only does $\mathcal{S}$ inform about the mean surface for $y(\mathbf{s})$ but also $\mathcal{Y}$ informs about the intensity, $\lambda(\mathbf{s})$.

- Though the $\eta(\mathbf{s})$ and $\phi(\mathbf{s})$ processes are assumed independent, model (2) with the model (c) specifies a joint model for sales activity and a selling price that provides spatial dependence through a linear model of coregionalization (See Chapter 10). That is if we write the shared component in the log intensity for the sales locations as $\eta(\mathbf{s}) = \sigma_\eta U_1(\mathbf{s})$ and we write the hedonic model in the form $y(\mathbf{s}) = \mathbf{W}(\mathbf{s})^\mathrm{T}\boldsymbol{\alpha} + \delta\sigma_\eta U_1(\mathbf{s}) + \sigma_\phi U_2(\mathbf{s}) + \epsilon(\mathbf{s})$, where $\phi(\mathbf{s}) = \sigma_\phi U_2(\mathbf{s})$, then we have a coregionalization model with coregionalization matrix $A = \begin{pmatrix} \sigma_\eta & 0 \\ \delta\sigma_\eta & \sigma_\phi \end{pmatrix}$. We again see that $\delta = 0$ endows the locations model and the selling price model with spatial random effects but that the models are independent.

- The need to approximate model (2) with representative points along with the fairly large number of properties in the product term in model (2) results in a high-dimensional multivariate normal vector of random effects as part of the model fitting. The additional GP for the selling prices included in models (b) and (c) brings an additional high-dimensional multivariate normal vector of random effects. To expedite computations Paci et al. [2020] employed a nearest neighbor Gaussian process (NNGP) for model fitting (see Chapter 13).

Posterior summaries for the models (a), (b), (c), and (d), each fitted with (2), for each of the two years, are presented in Paci et al. [2020]. They are not shown here but reveal that the price is higher in less populated neighborhoods, with a higher percentage of elderly people and a lower proportion of foreigners. However, the coefficients associated with the demographic covariates are not significant when spatial random effects are included in the hedonic model, with the only exception being the percent of foreigners. The size of the apartment and the age of the property are on the log scale; in this way, the corresponding coefficients estimate the elasticity of the price with respect to these covariates. The coefficient of the size of the property is positive, indicating that the bigger the property, the higher the price. Conversely, the negative coefficient of the age of the property reflects an age penalization on the price of the dwelling. The AE variable presents a negative coefficient in 2012, revealing, as anticipated, that presenting accessories yields an increase in the expected price.

With regard to the preferential sampling coefficient δ under model (d), it is significantly negative in 2007 and positive in 2012. These effects have to be interpreted with care. In 2007, before the crisis, there was high demand with relatively higher sales activity away from the center where properties were developed (some with subsidies) with relatively lower prices. So, $\delta < 0$ for this year reflects the effect of the log intensity to pull down prices where there was high activity. By contrast, in 2012, just after the crisis, overall demand decreased but there was relatively higher sales activity in the center, where prices were relatively higher.

	2007		2012	
	PMSE	CRPS	PMSE	CRPS
model (a)	0.1487	0.2066	0.1247	0.1947
model (d)	0.1346	0.1943	0.1121	0.1832
model (c)	0.1328	0.1940	0.1098	0.1808
model (b)	0.1327	0.1939	0.1124	0.1834

Table 16.3 *Model comparison based on predictive mean square error (PMSE) and continuous ranked probability score (CRPS).*

So, $\delta > 0$ for this year reflects the effect of the log intensity to push up prices where there was high intensity.

Under model (c), the coefficient δ is not significant for both 2007 and 2012, revealing that the sales activity does not affect the selling prices in the presence of flexible GP random effects in the hedonic model. Again, when analyzing the full dataset, there is no expectation of strong sampling bias, hence there is no reason to expect a significant preferential sampling effect in model (c).

Paci et al. [2020] focused on model comparison only for the prediction of selling prices since the motivation for this work is to assess the effect of preferential sampling on the prediction of such prices. That is, they hold out a proportion of the selling prices and obtain their associated posterior predictive distributions. Predictive mean square error (PMSE) and continuous ranked probability score (CRPS) are used to compare performance across the four choices of hedonic models. Results of the model comparison are presented in Table 16.3. Comparing the simple spatial regression model with the shared component model we can appreciate the benefit of the correction introduced by the preferential sampling in the hedonic model. The comparison of model (d) with model (c) tells us that the geostatistical model slightly outperforms the preferential sampling model. However, the closeness in performance between these two models suggests that the effects of omitted variables are captured roughly equally as well by $\eta(\mathbf{s})$ as $\phi(\mathbf{s})$. Finally, adding a shared component to the geostatistical hedonic model, model (b) does not provide any improvement in terms of predictive performance. Again, we have anticipated this in our discussion above.

In summary, there is a preferential sampling effect in the absence of the spatial GP in the hedonic model and there is substantial improvement in prediction when a shared process is introduced. However, with the GP included in the hedonic model, such benefit disappears. Moreover, with a full inventory of transactions, the prediction of selling prices using model (b) is not beneficial.

Now, we consider presence/absence data for species distributions. The intent of this example is to investigate preferential sampling for a geostatistical model with binary— presence/absence response. Much of it is drawn from Gelfand and Shirota [2019]. First, it is necessary to clarify exactly what a presence/absence event means. If one asks different ecologists one may get different answers. Does it mean going to a geo-coded location and recording whether or not the species is there? Or, is it a binary summary for an areal unit; was the species observed on the unit or not? In order to bridge with preferential sampling ideas, we need to conceptualize presence/absence at the point-level. In this way, we can formalize a "probability of presence" surface which provides the presence/absence probability for the species at each location in the region, driven by environmental features at the location. If we add to this surface a conceptual Bernoulli trial at every location, we obtain a realization of a presence/absence surface for the species, a binary map with a 1 or 0 at each point in the region (See Section 10.4). This surface is only partially observed through the data collection. In terms of "seeing" this surface, at best we can display it with a high-resolution grid of points.

Coherent modeling for these two surfaces which enables a *generative* probabilistic model for presence/absence data is a primary objective. Presence/absence modeling in the context of areal units does not permit modeling of a probability of presence surface; in this case presence/absence probability will depend upon the size, shape, orientation, etc. of the areal units. Focusing on plants, for a given species, we can ask what the *true* realization of a presence/absence surface over a fixed region at a *fixed* time would look like. At any location in the region, this surface must take on the value of either 1 if the species is present there or 0 if it is not. If this surface is specified to be 1 for some areal units and 0 for others, then, as above, the realization of the surface will be dependent upon the selection of the units, their size, their shape, their orientation. This would seem to conflict with how presence/absence arises in nature. Such incoherence can be avoided if presence/absence is viewed at *point* level.

More explicitly, from a point-level perspective, we can "see" the realization of the presence/absence surface over the entire region of interest, and thus, over any subset of the region without imposing any areal scales. However, presence/absence data is frequently associated with areal units, e.g., described as presence/absence over a grid cell, e.g., a plot or a quadrat. Depending upon the size of the region relative to the size of the areal units, the unit may be considered as a point in the region. In any event, formally, presence/absence is *never* observed at a point. Even at fine resolution, a point is only specified with regard to a number of significant decimal places so, implicitly, it is an area due to rounding. The idea of a point-level process specification is accepted as conceptual.

When presence/absence data is recorded at areal units, presence is customarily declared if the species is found anywhere in the unit. But then, it seems necessary to model the probability of presence considering the size of the unit. Moreover, this definition would ignore the abundance of the unit. Should the presence associated with one individual in a unit be the same as the presence associated with ten individuals in a unit of the same size? Shouldn't there be implications for the probability of presence in the unit? Again, coherence finds us wanting to think of presence/absence in a *unitless* fashion.

The set of presences of a species over a region may be viewed as a point pattern. That is, there are a random, finite, number of individuals randomly located in the region. Again, scale enters here. At a suitable spatial scale, a presence will be recorded as a point; however, in a unitless sense, a presence is not at a point but over some, perhaps very small, area. In this regard, thinking about presence through point patterns suggests a perspective that is associated with presence-only data, as we develop below. From the point pattern perspective, an intensity would typically be prescribed as the generator of the point pattern realization, following Chapter 8. Intensity surfaces can be normalized to density surfaces. Such density surfaces reflect the relative chance of observing a species at a given location compared with other locations in the region. They have nothing to do with providing a probability of presence surface.

If we scale a realization of a presence/absence surface as a binary map to an areal unit then it makes sense to think about the average of the realization over that unit, i.e., the proportion of 1's over the unit. This proportion is the *empirical* chance of finding the species present at a randomly selected location within the unit. In fact, the proportion of 1's over the entire region can be interpreted as the prevalence of the species over the region. Similarly, with a modeled probability of presence surface, if we scale this surface over the unit, we obtain an average probability over the unit. This average conveys the modeled probability of finding a presence at a randomly selected location within the unit. The issue is that we need not think in terms of areal units in order to model presence/absence. If we want to investigate units then we can scale accordingly.

Let's offer a more explicit discussion regarding what an observed presence means and the associated implications. Repeating, the issue is whether presence/absence is viewed as an event at the point level or at the areal level. Is it a Bernoulli trial at say location $\mathbf{s}$ or is it the probability that the number of individuals of a species in a set, say A, is ≥ 1?

If we model presence/absence at point level, then $Y(\mathbf{s}) = 1$ is a Bernoulli trial at location $\mathbf{s}$. However, what does $Y(A)$ mean? A coherent probabilistic definition specifies it as a block average (see Chapter 7), i.e., a realization of $Y(A)$ is $Y(A) = \int_A 1(Y(\mathbf{s}) = 1)ds/|A|$ (where $|A|$ is the area of A). It is the proportion of the $Y(\mathbf{s})$ in A that equal 1; it is not a Bernoulli trial and $P(Y(A) = 1) = 0$ since the probability that almost every Bernoulli trial in A results in a 1 equals 0. We can calculate $\mathrm{E}(Y(A)) = \int_A p(\mathbf{s})ds/|A|$ with $p(\mathbf{s})$ as below. That is, $\mathrm{E}(Y(A))$ becomes the average probability of presence over A. It is the probability that, at a randomly selected location in A, the species is present.

Now, suppose we consider the locations of all individuals in a study region as a random point pattern. Then, if $N(A)$ is the number of individuals in set A, $P(presence\ in\ A) = P(N(A) \geq 1)$. Here, assuming a nonhomogeneous Poisson process (NHPP) or, more generally a log Gaussian Cox process (LGCP) with intensity $\lambda(\mathbf{s})$ (see Chapter 8), $N(A) \sim Po(\lambda(A))$ where $\lambda(A) = \int_A \lambda(s)ds$. Then, taking the areal unit definition of a presence in A, we seek $P(Y(A) = 1) = P(N(A) \geq 1) = 1 - e^{-\lambda(A)}$. Since presence-only data samples the point pattern (although likely not fully but, rather, up to sampling effort over the region (Chakraborty et al. [2011] as well as Renner et al. [2015]), it is compatible with this definition of presence/absence. However, the probability of a presence in A is only defined with regard to the size of A and will vary with A, a concern raised in Hastie and Fithian [2013]. As a result, it is unclear how to specify a meaningful probability of presence surface. Furthermore, the definition of probability of presence as "one or more" observations of the species in A yields local distortion to any such surface; $N(A) = 1$ or $N(A) = 11$ are treated the same with regard to probability of presence in A. Moreover, even if we ignore the size of A and return to a grid of cells over $\mathcal{D}$, then it is clear that $p \equiv P(Y(A_i) = 1)$ has nothing to do with p_i defined in the previous paragraph.

The two foregoing definitions associated with $P(presence\ in\ A)$ are incompatible and the fundamental difference between them has often been missed in the literature. The conceptualization for the first choice is that we go to fixed "point" locations and see what is there; we are not sampling a point pattern. We model a surface over a domain $\mathcal{D}$ which captures the probability of presence at every location in $\mathcal{D}$. The conceptualization for the second is that we identify an area of interest $\mathcal{D}$ and, conceptually, we census it completely for all of the occurrences of the point pattern. We model an intensity which, using the definition above, provides a probability of presence for A.

The need for care is clear in terms of formalizing the notion of the presence of a species as well as the challenge of fusing presence/absence and presence-only data. In the literature to date, ignoring the incompatibility associated with the scaling issue is the way that presence-only data has been used to provide presence/absence probabilities and also the way presence-only data has customarily been fused with presence/absence data [see, e.g., Pacifici et al., 2017]. Below we provide a probabilistically coherent remediation for this incompatibility.

Some presence/absence modeling

Viewing the visited sites as points and therefore modeling at point scale, $Y(\mathbf{s})$ would be taken as

$$Y(\mathbf{s}) \sim Bernoulli(p(\mathbf{s})), \tag{16.23}$$

analogously relating the probability that the species occurs in site $\mathbf{s}$, $p(\mathbf{s})$, to the set of environmental variables as a logit $(\log\left(\frac{p(\mathbf{s})}{1-p(\mathbf{s})}\right))$ or probit $(\Phi^{-1}(p(\mathbf{s}))) = \mathbf{w}^{\mathrm{T}}(\mathbf{s})\boldsymbol{\beta}$. Such modeling requires that we have covariate levels $\mathbf{w}(\mathbf{s})$ for each site. This model is referred to as a spatial regression in the sense that the regressors are spatially referenced.

We bring in spatial dependence between points based on their relative locations using Gaussian processes, creating geostatistical models. We would model $Y(\mathbf{s})$ given $p(\mathbf{s})$ and augment the explanation of $p(\mathbf{s})$ through the form

$$\log \frac{p(\mathbf{s})}{1 - p(\mathbf{s})} = \mathbf{w}^{\mathrm{T}}(\mathbf{s})\boldsymbol{\beta} + \omega(\mathbf{s}). \tag{16.24}$$

Here, $\omega(\mathbf{s})$ is the spatial random effect associated with point $\mathbf{s}$, arising as a realization of a Gaussian process. A suitable covariance function would be selected. With binary response, this model is a spatial generalized linear model (GLM) as Section 6.2. Reviewing, the model has two levels: the first or data-level specification is a Bernoulli trial and the second or process level presents the probability of presence surface.

Assuming data of the form, $(\mathbf{w}(\mathbf{s}_i), Y(\mathbf{s}_i))$ for sites $i = 1, 2, ..., n$ and adopting the hierarchical (multi-level) regression with say, a probit link, $P(Y(\mathbf{s}) = 1) \equiv p(\mathbf{s}) = \Phi(\mathbf{w}^{\mathrm{T}}(\mathbf{s})\boldsymbol{\beta} + \omega(\mathbf{s}))$. That is, $P(Y(\mathbf{s}) = 1) = P(Z(\mathbf{s}) > 0)$ where $Z(\mathbf{s}) = \mathbf{w}^{\mathrm{T}}(\mathbf{s})\boldsymbol{\beta} + \omega(\mathbf{s}) + \epsilon(\mathbf{s})$, a geostatistical model. Here, $\epsilon(\mathbf{s})$ is a pure error, i.e., $\epsilon \sim N(0, 1)$ and $\omega(\mathbf{s})$ is a mean 0 Gaussian process with a suitable correlation function. Under this model, the $Y(\mathbf{s})$ are drawn as conditionally independent Bernoulli trials given $p(\mathbf{s})$ (and the associated $Z(\mathbf{s})$ are conditionally independent normals). As a result, even if the probability of presence surface $p(\mathbf{s})$ is smooth, realizations of the presence/absence surface are everywhere discontinuous. In the context of preferential sampling, we can consider the models (a)-(d) for $Z(\mathbf{s})$ supplied above. Hence, we can study preferential sampling in the context of binary response data.

Some presence-only data modeling

As noted above, there has been considerable recent growth in the analysis of presence-only data, e.g., citizen science and museum databases, and other non-systematic surveys. It is noteworthy that presence-only data is not *inferior* to presence/absence data. In principle, presence-only data offer a complete census while presence/absence data, confined to a specified set of sampling sites, contains less information. However, in practice, a complete census of individuals is rarely achieved. Sampling effort usually exceeds available resources.

In any event, to explain this partially observed point pattern, a formal point pattern modeling approach, say a logGaussian Cox process (LGCP) would be adopted, specifying an associated intensity in terms of the available environments with spatial structure through spatial random effects. The observed point pattern is biased through (i) anthropogenic processes, e.g., transformation of the landscape
(ii) non-uniform (typically, very irregular) sampling effort
(iii) the possibility of missed detection or mis-classification. As a result, it is necessary to adjust the *potential* species intensity to a *realized* intensity, a *degradation or thinning* of the former to explain the realized/observed point pattern.

An illustrative specific model is presented in Chakraborty et al. [2011]. Let $\lambda(\mathbf{s})$ be the "potential" intensity surface where modeling for $\lambda(\mathbf{s})$ is a LGCP,

$$\log \lambda(\mathbf{s}) \quad = \quad \mathbf{w}^{\mathrm{T}}(\mathbf{s})\beta + \omega(\mathbf{s}) \tag{16.25}$$

with $\omega(\mathbf{s})$ a zero-mean GP to capture residual spatial association in the $\lambda(\mathbf{s})$ surface across grid cells. For degradation, a binary availability surface, $U(\mathbf{s})$; $U(\mathbf{s}) = 1$ or 0 according to whether or not $\mathbf{s}$ is untransformed/available. So, $\lambda(\mathbf{s})U(\mathbf{s})$ can only be $\lambda(\mathbf{s})$ or 0, $\mathbf{s}$ available or not. Then, a binary sampling effort surface, $T(\mathbf{s})$, is introduced such that $T(\mathbf{s})U(\mathbf{s}) = 1$ indicates that $\mathbf{s}$ is both available and sampled. As a result, $\lambda(\mathbf{s})U(\mathbf{s})T(\mathbf{s})$ becomes the degraded/operating intensity. Where no locations were sampled, the operating intensity is 0.

16.5.2 Application: invasive plant data from New England

First, consider the presence/absence data. From Gelfand and Shirota [2019] consider a sub-region of six New England states (Connecticut, Rhode Island, Massachusetts, Vermont, New Hampshire, Maine) in the U.S. The presence/absence dataset comes from the Invasive Plant Atlas of New England (IPANE). IPANE is a citizen science organization; volunteers using rigorous sampling protocols. There are 4314 unique sampling sites across New England. Each site is provided with a location (latitude, longitude) and has been classified with regard to each focal species as a presence or an absence. We examine seven common invasive plant species in the IPANE database: multiflora rose (MR), oriental bittersweet (OB), Japanese barberry (JB), glossy buckthorn (GB), autumn olive (AO), burning bush (BB), and garlic mustard (GM).

The presence-only dataset comes from the Global Biodiversity Information Facility (GBIF) which is a data aggregator for biological collections worldwide. The number of observations will vary from species to species. We consider the same set of species as above, within GBIF. Longitude and latitude are transformed to eastings and northings, and rescaled from km units to 10 km units. All of upper Maine is removed; both the IPANE and the GBIF data are too sparse there. Table 16.4 summarizes the species and sample sizes for each dataset. Figure 16.6 shows, for the IPANE data the distribution of presence (blue) and absence (green) points for each species across the study region. Figure 16.7, for the GBIF data, shows the distribution of presence-only points (red) for each species across the study region.

Common name	symbol	IPANE presences	IPANE absences	GBIF presences
multiflora rose	MR	1230	3084	249
oriental bittersweet	OB	1106	3208	305
Japanese barberry	JB	1012	3302	399
glossy buckthorn	GB	755	3559	223
autumn olive	AO	386	3928	193
burning bush	BB	336	3978	257
garlic mustard	GM	279	4035	440

Table 16.4 *Study species and sample sizes.*

Gelfand and Shirota [2019] employ seven covariates:

(1) mean diurnal range (mDR, mean of monthly (max temp-min temp)),

(2) max temperature of warmest month (maxTWM),

(3) min temperature of the coldest month (minTCM),

(4) mean temperature of the driest quarter (meanTDQ),

(5) precipitation of wettest month (PWM),

(6) precipitation seasonality (PS, the standard deviation of the monthly precipitation estimates expressed as a percentage of the mean of those estimates, that is, the annual mean),

(7) precipitation of warmest quarter (PWQ).

Figure 16.8 shows the standardized spatial surfaces for these covariates.

Applying the preferential sampling modeling above, Table 16.5 displays the estimation results for δ for models (c) and (d). For MR, JB, GB, and AO, the results for model (d) suggest significant preferential sampling effects; the means for δ are significantly different from 0. When $\delta > 0$, this implies that, in the selection of the presence/absence locations for

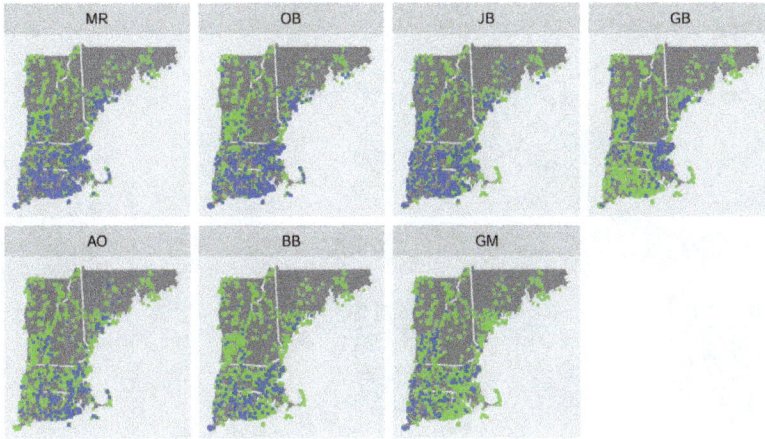

Figure 16.6 *IPANE: the distribution of presence (blue) and absence (green) points for each species across the study region.*

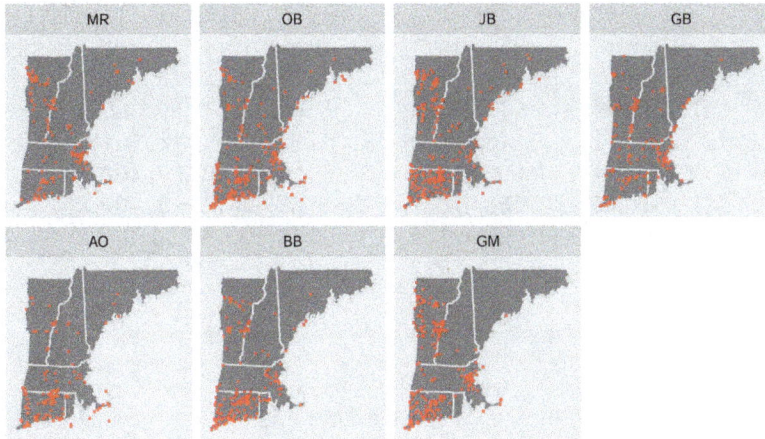

Figure 16.7 *GBIF: the distribution of presence-only points (red) for each species across the study region.*

the species, presences were oversampled. When $\delta < 0$, this means that in the selection of the presence/absence locations for the species, presences were undersampled. This insight is useful and can help in predicting the probability of presence at unobserved locations. Furthermore, failing to include the $\eta(\mathbf{s})$ in the modeling might lead to misinterpretation of the effects of the regressors.

The $\eta(\mathbf{s})$ also provide improved prediction of presence/absence (see below). However, with the inclusion of the $\omega(\mathbf{s})$ surface (model (c)), the δ coefficients become insignificant. The flexibility of the $\omega(\mathbf{s})$ surface seems to remove the benefit of using the $\eta(\mathbf{s})$ surface as a predictor.

Table 16.6 displays the estimation results for models (a) and (d) for the vector of regression coefficients, $\boldsymbol{\alpha}$, for MR, JB, GB, and AO (each has a significant δ with model (d)). Introducing the $\eta(\mathbf{s})$ surface affects the estimation results for $\boldsymbol{\alpha}$. For example, the estimated

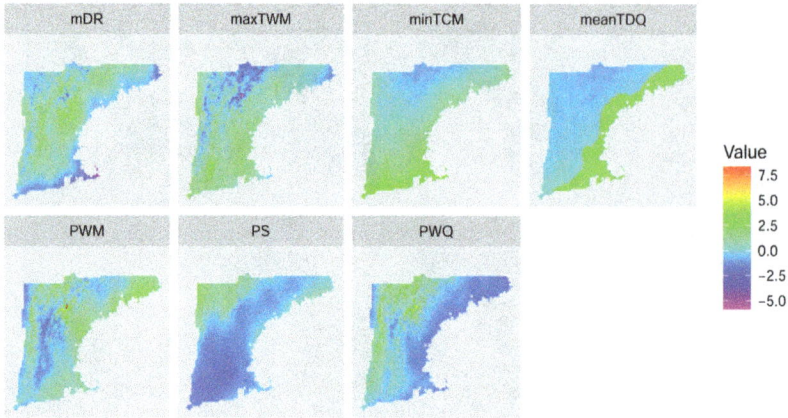

Figure 16.8 *The standardized covariate surfaces for seven selected covariates across the study region.*

	Model (c)		Model (d)	
	Mean	95% Int	Mean	95% Int
MR	0.028	[-0.029, 0.092]	0.037	**[0.023, 0.071]**
OB	-0.014	[-0.072, 0.044]	-0.027	[-0.063, 0.006]
JB	0.024	[-0.043, 0.093]	0.085	**[0.048, 0.122]**
GB	0.075	[-0.052, 0.194]	0.225	**[0.179, 0.274]**
AO	-0.049	[-0.133, 0.039]	-0.076	**[-0.120, -0.030]**
BB	0.064	[-0.025, 0.164]	0.013	[-0.033, 0.062]
GM	0.041	[-0.100, 0.213]	-0.036	[-0.085, 0.015]

Table 16.5 *Estimation results for δ with models (c) and (d).*

α of meanTDQ for GB is significantly negative for model (a) but becomes insignificant for model (d).

With binary response, to demonstrate improved prediction we consider misclassification error using the Tjur R^2 coefficient of determination [T, 2009]. This measure prefers a model with high probability of presence when presence is observed and low probability of presence when absence is observed. For species j, this quantity is given by $TR_j = (\hat{\pi}_j(1) - \hat{\pi}_j(0))$ where $\hat{\pi}_j(1)$ and $\hat{\pi}_j(0)$ are the average probabilities of presence for the observed ones and zeros associated with the j-th species across the locations. The larger the TR_j, the better the discrimination.

20% (879) of the presence/absence locations were held out, chosen at random, for the seven species. Table 16.7 displays the results for the TR measure for models (a)—(d). For all species, the models including $\omega(\mathbf{s})$, (b), and (c), outperform those without, (a) and (d). Model (c) with (ii) tends to be better than model (b) (which ignores (ii)) particularly for AO, BB, and GM. Model (d) with (ii) is at least as good as model (a) (which ignores (ii)) but is really only consequentially better for species GB which has the largest δ coefficient under model (d).

	MR		JB		GB		AO	
Model(a)	Mean	95% Int	Mean	95% Int	Mean	95% Int	Mean	95% Int
const	-1.774	[-1.951, -1.602]	-1.507	[-1.676, -1.344]	-1.906	[-2.126, -1.693]	-2.360	[-2.634, -2.102]
mDR	0.193	[0.027, 0.362]	0.383	[0.213, 0.556]	0.110	[-0.082, 0.305]	0.381	[0.161, 0.603]
maxTWM	-0.020	[-0.217, 0.175]	-0.236	[-0.435, -0.035]	0.463	[0.223, 0.704]	-0.554	[-0.813, -0.300]
meanTDQ	0.715	[0.460, 0.970]	0.762	[0.498, 1.028]	-0.324	[-0.619, -0.025]	0.859	[0.532, 1.196]
minTCM	-0.070	[-0.149, 0.010]	-0.065	[-0.146, 0.015]	0.415	[0.309, 0.522]	0.029	[-0.074, 0.136]
PWM	0.178	[0.092, 0.262]	0.063	[-0.024, 0.151]	-0.139	[-0.240, -0.038]	-0.182	[-0.300, -0.066]
PS	-0.142	[-0.246, -0.040]	-0.083	[-0.182, 0.017]	-0.030	[-0.141, 0.081]	-0.162	[-0.320, -0.010]
PWQ	-0.062	[-0.166, 0.042]	0.204	[0.103, 0.307]	-0.371	[-0.502, -0.239]	-0.156	[-0.308, -0.001]
Model(d)	Mean	95% Int	Mean	95% Int	Mean	95% Int	Mean	95% Int
const	-1.931	[-2.156, -1.705]	-1.845	[-2.072, -1.620]	-2.985	[-3.331, -2.654]	-2.066	[-2.391, -1.751]
mDR	0.215	[0.043, 0.386]	0.423	[0.246, 0.597]	**0.352**	**[0.124, 0.578]**	0.362	[0.141, 0.585]
maxTWM	-0.049	[-0.247, 0.145]	-0.308	[-0.514, -0.102]	**0.149**	**[-0.135, 0.433]**	-0.507	[-0.760, -0.252]
meanTDQ	0.792	[0.524, 1.054]	0.900	[0.622, 1.174]	**0.206**	**[-0.145, 0.550]**	0.756	[0.418, 1.089]
minTCM	-0.063	[-0.146, 0.021]	-0.051	[-0.137, 0.034]	0.480	[0.350, 0.612]	0.020	[-0.087, 0.129]
PWM	0.161	[0.073, 0.250]	0.022	[-0.070, 0.114]	-0.225	[-0.371, -0.096]	-0.148	[-0.269, -0.026]
PS	-0.132	[-0.238, -0.026]	-0.061	[-0.165, 0.040]	0.046	[-0.083, 0.183]	-0.174	[-0.333, -0.017]
PWQ	-0.028	[-0.141, 0.083]	0.267	[0.153, 0.381]	-0.218	[-0.399, -0.032]	-0.195	[-0.357, -0.038]
δ	0.037	[0.023, 0.071]	0.085	[0.048, 0.122]	0.225	[0.179, 0.274]	-0.076	[-0.120, -0.030]

Table 16.6 *Estimation results for MR, JB, GB, and AO for models (a) and (d). The bold font suggests the change of significance.*

	Model (a)		Model (b)		Model (c)		Model (d)	
	Mean	95% Int	Mean	95% Int	Mean	95% Int	Mean	95% Int
MR	0.104	[0.094, 0.114]	0.168	[0.145, 0.191]	**0.176**	**[0.157, 0.202]**	0.105	[0.096, 0.114]
OB	0.099	[0.088, 0.109]	**0.183**	**[0.163, 0.200]**	**0.183**	**[0.158, 0.203]**	0.101	[0.092, 0.111]
JB	0.072	[0.063, 0.081]	**0.201**	**[0.180, 0.230]**	0.198	[0.169, 0.227]	0.075	[0.066, 0.083]
GB	0.126	[0.112, 0.139]	0.405	[0.372, 0.434]	**0.412**	**[0.382, 0.451]**	0.162	[0.147, 0.175]
AO	0.034	[0.027, 0.043]	0.095	[0.068, 0.129]	**0.115**	**[0.073, 0.156]**	0.039	[0.033, 0.051]
BB	0.057	[0.048, 0.068]	0.112	[0.078, 0.152]	**0.131**	**[0.097, 0.168]**	0.057	[0.049, 0.066]
GM	0.026	[0.020, 0.032]	0.135	[0.111, 0.166]	**0.164**	**[0.118, 0.206]**	0.027	[0.022, 0.033]

Table 16.7 *Estimation results for the TR measure for preferential sampling.*

16.6 Geostatistical and point pattern data fusion

We consider this fusion through the following presence/absence and presence-only data example. The fusion objective here considers the two types of plant data collection discussed in detail above—survey sampled presence/absence data with presence-only citizen science/museum databases/non-systematic data collection. We take as the modeling goal the development of a probability of presence surface over the study region. As detailed above, this fusion has been widely discussed and abused in the literature; a probabilistically coherent joint modeling specification is presented which enables remediation of the abuse.

Having clarified the different stochastic mechanisms which are associated with point-referenced presence/absence data and presence-only data we can offer a formal prescription for a data fusion to enhance learning about the probability of presence surface. The foregoing preferential sampling development motivates the path. Further, consider the general objective of improved learning about the mean geostatistical response surface (perhaps transformed through a suitable link function). We assert that having additional data in the form of a marked point pattern with the mark being the geostatistical response, an analogous data fusion will respond to the general objective.

Again, the data fusion here differs from geostatistical x geostatistical setting. We don't have multiple data sources informing about a common response. We have two different data types and a different type of fusion is needed. Again, $\mathcal{S}_{PA}$ may exhibit sampling bias; ecologists will tend to sample where they expect to find (or not find) individuals. As above, *preferential sampling* treats $\mathcal{S}_{PA}$ as a point pattern, models it using a LGCP, and introduces the intensity as a *regressor* in the P/A model to improve the $p(\mathbf{s})$ surface. The extra information available for data fusion from the presence-only data is $\mathcal{S}_{PO}$, the set of observed presence-only locations. Importantly, the locations in $\mathcal{S}_{PO}$ are severely biased; they are locations where we see only 1's. We are severely oversampling presences with $\mathcal{S}_{PO}$. Suppose we model them also using an LGCP and also introduce the associated intensity as a *regressor* in the P/A model. Extracting a preferential sampling effect should improve inference regarding the $p(\mathbf{s})$ surface.

Formally, what information does $\mathcal{S}_{PO}$ bring with regard to learning about the probability of presence surface? Suppose $\mathcal{S}_{PO}$ is a complete census. Associate with $\mathcal{S}_{PO} = \{\mathbf{s}_1^*, ..., \mathbf{s}_m^*\}$ an intensity $\lambda_{PO}(\mathbf{s}) = \mathbf{w}^{\mathrm{T}}(\mathbf{s})\boldsymbol{\beta}_{PO} + \eta_{PO}(\mathbf{s})$, like above, using the same predictors as with the presence/absence modeling. We expect $\lambda_{PO}(\mathbf{s})$ to be elevated near these observations. Hence, if inserted into a presence-absence model, we expect to increase the probability of presence around the $\mathbf{s}_j^*$'s.

So, in order to capture the influence of $\mathcal{S}_{PO}$ on the $p(\mathbf{s})$ surface associated with $\mathcal{Y}_{PA}$ (the presence/absence data), we could add $\delta_{PO}\eta_{PO}(\mathbf{s})$ to the mean for $Z(\mathbf{s})$, i.e., we could have a $\delta_{PA}\eta_{PA}(\mathbf{s})$ term and a $\delta_{PO}\eta_{PO}(\mathbf{s})$ term in order to improve prediction of presence/absence. Each dataset enables a possible preferential sampling source. We might insist that $\delta_{PO} > 0$. Then, from the presence-only data, the probability of presence will be increased around the $\mathbf{s}_j^*$'s and decreased away from them.

We now have four potential models for $\lambda_{PO}(\mathbf{s})$, parallel to those for $\lambda_{PA}(\mathbf{s})$, to combine with the model for $Y(\mathbf{s})$. We focus on an LGCP for $\mathcal{S}_{PO}$ analogous to an LGCP for $\mathcal{S}_{PA}$. Then, we can add a $\delta_{PO}\eta_{PO}(\mathbf{s})$ term to the mean of $Z(\mathbf{s})$ under (b), (c), or (d). The full model takes the form

$$[\mathcal{Y}|\mathcal{S}_{PA}, \boldsymbol{\alpha}, \boldsymbol{\eta}_{PA,Y}, \delta_{PA}, \boldsymbol{\eta}_{PO,Y}, \delta_{PO}]$$
$$[\mathcal{S}_{PA}|\boldsymbol{\beta}_{PA}, \boldsymbol{\eta}_{PA,D}][\mathcal{S}_{PO}|\boldsymbol{\beta}_{PO}, \boldsymbol{\eta}_{PO,D}]$$

With a partial realization of the presence-only point pattern, we **need** to degrade $\lambda_{PO}(\mathbf{s})$ in the model fitting.

Explicitly, $P(Y(\mathbf{s}) = 1) = P(Z(\mathbf{s}) > 0)$, with $Z(\mathbf{s})$ a latent GP. Each dataset enables a possible preferential sampling source. We have the following models for data fusion:

(a): $Z(\mathbf{s}) = \mathbf{x}^{\mathrm{T}}(\mathbf{s})\boldsymbol{\alpha} + \delta_{PO}\eta_{PO}(\mathbf{s}) + \epsilon(\mathbf{s})$.

(b): $Z(\mathbf{s}) = \mathbf{x}^{\mathrm{T}}(\mathbf{s})\boldsymbol{\alpha} + \delta_{PO}\eta_{PO}(\mathbf{s}) + \omega(\mathbf{s}) + \epsilon(\mathbf{s})$.

In addition, we have two models which also include preferential sampling associated with the presence/absence data, $\delta_{PA}\eta_{PA}(\mathbf{s})$:

(c): $Z(\mathbf{s}) = \mathbf{x}^{\mathrm{T}}(\mathbf{s})\boldsymbol{\alpha} + \delta_{PA}\eta_{PA}(\mathbf{s}) + \delta_{PO}\eta_{PO}(\mathbf{s}) + \epsilon(\mathbf{s})$

(d): $Z(\mathbf{s}) = \mathbf{x}^{\mathrm{T}}(\mathbf{s})\boldsymbol{\alpha} + \delta_{PA}\eta_{PA}(\mathbf{s}) + \delta_{PO}\eta_{PO}(\mathbf{s}) + \omega(\mathbf{s}) + \epsilon(\mathbf{s})$

We briefly summarize the results of the fusion. We note that Gelfand and Shirota [2019] adopt the sampling effort surface $T(\mathbf{s})$ for each grid cell so that $T(\mathbf{s}) = 1$ for all cells where at least one presence-only point is observed, $T(\mathbf{s}) = 0$ otherwise. Again, δ_{PO} is expected to be positive, i.e., the probability of presence will be increased around the $\mathbf{s}_j^*$'s and decreased away from them, so a uniform prior on $[0, 100]$ was adopted. *None* of the δ_{PA} are significantly different from 0. *All* of the δ_{PO} are far from 0; evidently, presence-only sites reveal preferential sampling! They clearly improve the presence/absence surface. Omitting details, performance is essentially indistinguishable across the models (all have the PO intensity as a regressor). However, model (b) emerges as the best.

We conclude by noting that with the rapidly increasing extent of data collection, we anticipate that data fusion will play an increasingly important role in the future of statistical data analysis. Evidently, this will be comparably important within the analysis of spatial and spatio-temporal data. We suggest that the model-based fusions presented in this chapter can be foundational in this evolution.

Chapter 17

Special topics in spatial process modeling

Earlier chapters have developed the basic theory and the general hierarchical Bayesian modeling approach for handling spatial and spatiotemporal point-referenced data. In this chapter, we consider some special topics that are finding increasing interest in the context of such models.

17.1 Space-time modeling for extremes

Extreme value analysis is frequently applied to environmental science data, see e.g., Tawn et al. [2018], particularly, extremes in exposure to environmental contaminants are of interest with regard to public health outcomes. Extremes in weather are of interest with regard to the performance of plants and animals. In particular, for plants, it is suggested that extreme weather events, such as drought, heavy rainfall, and very high or low temperatures, might be more significant factors in explaining plant performance with regard to survival, growth, reproductivity, etc., than trends in the mean climate.

There is no unique perspective on what modeling extremes means. For instance, focusing on temperatures at, say, a daily maximum scale, we can consider mixture models for mean temperatures (see, e.g., Schliep et al. [2021] and discussion therein), we can consider quantile regression and autoregression to move to the upper tails of the temperatures (see, e.g., Castillo-Mateo et al. [2023a]), we can consider record breaking of the daily temperatures (Castillo-Mateo et al. [2025]), we can explore heat waves or, more or less equivalently, extreme heat events in terms of duration and magnitude [Schliep et al., 2021]. Here, we confine ourselves to the most written-about modeling approach utilizing Generalized Extreme Value (GEV) distributions and the associated peaks over thresholds (POT) modeling. We present this in the context of spatial extremes so the data we work with is customarily a collection of time series (at a suitable temporal scale) over a set of spatial locations.

We present an illustration of the analysis of spatio-temporal weather extremes with data derived from precipitation surfaces in the Cape Floristic Region (CFR), the smallest but, arguably, the richest of the world's six floral kingdoms, encompassing a region of roughly $90,000 km^2$ in southwestern South Africa. The daily precipitation surfaces we employ arise via interpolation to grid cells at 10km resolution based on records reported by up to 3000 stations in South Africa over the period from 1950-1999. We have 50 derived surfaces of annual maxima surfaces from daily rainfalls for 1332 grid cells. Eventually, we hold-out 1999 for model validation. Figure 17.1 shows the CFR and the data for the year 1999.

17.1.1 Possibilities for modeling maxima

Let us consider possibilities for modeling maxima. To make things concrete, suppose at site $\mathbf{s}$, daily data j in year t, e.g., precipitation or temperature, ozone or $PM_{2.5}$, That is, we envision the collection of surfaces $W_{t,j}(\mathbf{s})$ which are observed at $\{s_i, i = 1, 2, ..., n\}$. We seek inference regarding $Y_t(\mathbf{s}) = \max_j W_{t,j}(s)$. To be more explicit, we are considering maxima

DOI: 10.1201/9781003401728-17

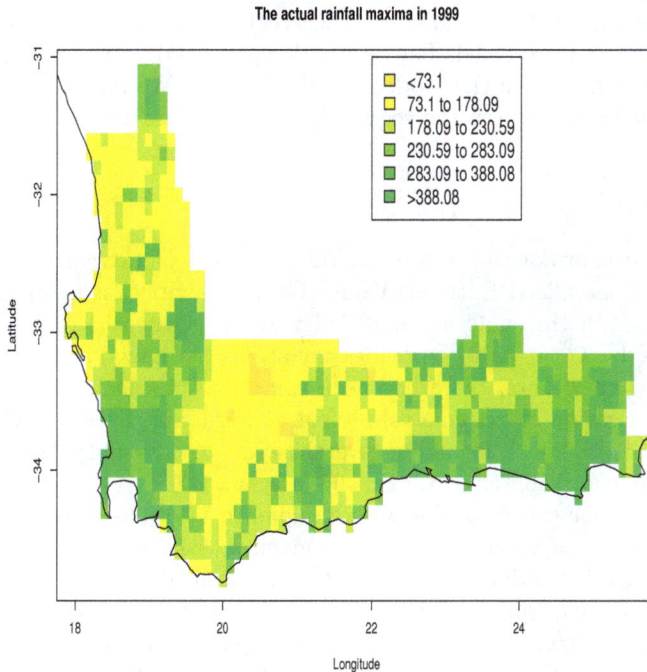

Figure 17.1 *The Cape Floristic Region (CFR) and the data for the year 1999.*

over time at a particular location, not maxima over space at a given time. And, with interest in a surface of extremes, we are evidently in the setting of multivariate extreme value theory. It seems clear that spatial dependence in the maxima will be weaker than spatial dependence in the daily data. Inference focuses on parametric issues, e.g., trends in time, and dependence in space, as well as on prediction at new locations and at future (one-step ahead) times. Also of interest is risk assessment with regard to return time. That is, if $P(Y_t(\mathbf{s}) > y) = p$, the expected return time until an exceedance of y is $1/p$ time units.

With regard to stochastic specification, in principle, we could model the W's, which will induce inference for the Y's. In this case, we have an enormous literature, using hierarchical space-time modeling that has been much discussed in this book. Summarily, let θ denote all of the parameters in a particular model. Suppose we fit the model and obtain the posterior, $[\theta|\{W_{t,j}(\mathbf{s}_i)\}]$. At a new $\mathbf{s}_0$, we can obtain the predictive distribution for $W_{t,j}(\mathbf{s}_0)$ in the form of posterior samples. These posterior samples can be converted into posterior samples of $Y_t(\mathbf{s}_0)$, i.e., "derived" quantities. We can do similarly for projection to time $T+1$. However, there are several reasons why this path is not of interest here. First, it is not modeling or explaining the maxima, and second, it is extremely computationally demanding, generating hundreds and hundreds of daily data samples and discarding them all, retaining just the maxima as a summary. Second, say for modeling daily max temperatures, while there is substantial literature on building mean models for such data, they are well known to perform poorly with regard to inference for the extreme tails. Third, there will be greater uncertainty in predictions made using derived quantities than in predictions doing direct modeling of the maxima.

There are still two paths to directly model the maxima: we can model the maxima or model the "process" that drives the maxima. More precisely, under the first path, the usual assumptions are that the $Y_t(\mathbf{s})$ are spatially dependent but temporally independent, i.e., they are viewed as replicates. The argument here is that there will be strong temporal dependence at the scale of days, that is, in the $W_{t,j}(\mathbf{s})$'s but temporal dependence will be negligible between yearly values, that is, for the $Y_t(\mathbf{s})$'s. Usually, this is not an explanatory model, no covariates are introduced. The alternative or second path is to provide space-time modeling for the parameters in the models for the $\{Y_t(\mathbf{s})\}$. We refer to this as the process path and it is what we develop in the sequel. First, we review a little bit of basic extreme value theory.

17.1.2   Review of Extreme Value Theory

By now there is an enormous literature on the modeling of extremes through the standard approach of Generalized Extreme Value (GEV) distribution families. Alternatively, exceedances for a given threshold are modeled with the Generalized Pareto Distribution (GPD). The book by Coles [2001] provides an accessible introduction. Relevant to the subject matter presented here, there has been some work focusing on spatial (or spatio-temporal) characterization of extreme values [e.g., Kharin and Zwiers, 2005, Cooley et al., 2007, Sang and Gelfand, 2009], including several papers discussing spatial interpolation for extreme values [see, e.g., Ribatet, 2013, Buishand et al., 2008].

Formal extreme value theory begins with a sequence $Y_1, Y_2, \ldots$ of independent and identically distributed random variables and, for a given n, asks about parametric models for $M_n = \max(Y_1, \ldots, Y_n)$. If the distribution of the Y_i is specified, the exact distribution of M_n is known. In the absence of such specification, extreme value theory considers the existence of $\lim_{n\to\infty} Pr((M_n - b_n)/a_n \le y) \equiv F(y)$ for two sequences of real numbers $a_n > 0, b_n$. If $F(y)$ is a non degenerate distribution function, it belongs to either the Gumbel, the Fréchet or the Weibull class of distributions, which can all be usefully expressed under the umbrella of the GEV. That is,

$$G(y; \mu, \sigma, \xi) = \exp\left\{-\left[1 + \xi\left(\frac{y - \mu}{\sigma}\right)\right]^{-1/\xi}\right\} \tag{17.1}$$

for $\{y : 1 + \xi(y - \mu)/\sigma > 0\}$. Here, $\mu \in \Re$ is the location parameter, $\sigma > 0$ is the scale parameter, and $\xi \in \Re$ is the shape parameter. The "residual" $V = (Y - \mu)/\sigma$ follows a GEV$(0, 1, \xi)$. Let $Z = (1 + \xi V)^{\frac{1}{\xi}} \Leftrightarrow V = \frac{Z^{\xi} - 1}{\xi}$. Then Z follows a standard Fréchet distribution, with the distribution function $\exp(-z^{-1})$. (For us, this distribution is more familiar as an inverse gamma, $IG(1, 1)$.) The GEV distribution is heavy-tailed and its probability density function decreases at a slow rate when the shape parameter, ξ, is positive. On the other hand, the GEV distribution has a bounded upper tail for a negative shape parameter. Note that n is not specified; the GEV is viewed as an approximate distribution to model the maximum of a sufficiently long sequence of random variables. In our setting, we will assign daily precipitation into annual blocks and, initially, assume the maxima are conditionally independent (but not identically distributed) across years given their respective, parametrically modeled, μ, σ, and ξ.

Closely related (and, in fact, derivable from the GEVs) are the generalized Pareto distributions (GPDs) which are a companion to the GEV distributions. From expression (17.1), with $\mu = 0$ and $\sigma = 1$, we consider the cdf,

$$P(Y < y) = 1 - ((1 - \xi y)^{-\frac{1}{\xi}} \qquad \xi \ne 0 \qquad P(Y < y) = 1 - \exp(-y) \qquad \xi = 0. \tag{17.2}$$

The support for Y is $y \ge 0$ for $\xi \ge 0$ and $(0 \le y \le -\frac{1}{\xi})$ for $\xi < 0$. Again, ξ is a shape parameter. The GPD specifies the exceedance distribution, $P(Y \le y + u \mid y > u) = \frac{P(Y \le y + u) - P(Y < u)}{P(Y > u)}$ for a specified threshold u.

Next, we turn to the work of Sang and Gelfand [2009, 2010] which develops hierarchical models for rainfall that reflect dependence on space and time. In particular, they use the GEV which is characterized by a location, a scale, and a shape parameter. Conceptually, each of these could vary in space and time and they could be mutually dependent. As a result, one can envision a range of such models, fitting them, and comparing them. Initially, we discuss the grid cell data, using Markov random field models. Then, we move to a dataset at the station level and work with Gaussian processes.

We introduce the GEV distribution as a first-stage model for annual precipitation maxima, specifying μ, σ, and ξ at the second stage to reflect the underlying spatio-temporal structure. In particular, working with grid cells, let $Y_{i,t}$ denote the annual maximum of daily rainfall for cell i in year t. We assume the $Y_{i,t}$ are conditionally independent, each following a GEV distribution with parameters $\mu_{i,t}$, $\sigma_{i,t}$, and $\xi_{i,t}$, respectively. Attention focuses on the specification of the models for $\mu_{i,t}$, $\sigma_{i,t}$, and $\xi_{i,t}$.

The conditional independence assumption is interpreted as an interest in smoothing the surfaces around which the interpolated data is centered rather than smoothing the data surface itself. As a formal assumption, it is defendable in time since, at a site, the annual maxima likely occur with sufficient time between them to be assumed independent. In space, we would expect small-scale dependence on the data at a given time. However, with observations assigned to grid cells at 10 km resolution, we can not hope to learn about fine scale dependence. Below, when we work at the point level, we will restore this spatial dependence.

Exploratory analysis for the foregoing CFR precipitation data, calculating MLE's through a suitable package, suggests $\xi_{i,t} = \xi$ for all i and t and $\sigma_{i,t} = \sigma_i$ for all t so modeling focuses on the $\mu_{i,t}$. But, in addition, we want the $\mu_{i,t}$, and σ_i to be dependent at the same site. This requires the introduction of an association model for a collection of spatially co-varying parameters over a collection of grid cells. We adopt the coregionalization method, as developed in Chapter 9, making a random linear transformation of conditionally autoregressive (CAR) models in order to greatly reduce the computational burden in model fitting.

With regard to modeling the $\mu_{i,t}$, we have many options. With spatial covariates $\mathbf{X}_i$, it is natural to specify a regression model with random effects, $[\mu_{i,t} \mid \boldsymbol{\beta}, W_{i,t}, \tau^2] = N(\mathbf{X}_i^{\mathsf{T}}\boldsymbol{\beta} + W_{i,t}, \tau^2)$. Here, for example, the $\mathbf{X}_i$ could include altitude or specify a trend surface, with coefficient vector $\boldsymbol{\beta}$ while $W_{i,t}$ is a spatio-temporal random effect.

Illustrative possibilities for modeling $W_{i,t}$ include: (i) an additive form, **Model A**: $W_{i,t} = \psi_i + \delta_t$, $\delta_t = \phi\delta_{t-1} + \omega_t$, where $\omega_t \sim N(0, W_0^2)$ $i.i.d$; (ii) a linear form in time with spatial random effects, **Model B**: $W_{i,t} = \psi_i + \rho(t - t_0)$; (iii) a linear form in time with local slope, **Model C**: $W_{i,t} = \psi_i + (\rho + \rho_i)(t - t_0)$; (iv) a multiplicative form in space and time, **Model D**: $W_{i,t} = \psi_i\delta_t$, $\delta_t = \phi\delta_{t-1} + \omega_t$, where $\omega_t \sim N(0, W_0^2)$ $i.i.d.$

The additive form in Model A might appear to over-simplify spatio-temporal structure. However, the data may not be rich enough to find space-time interaction in the $\mu_{i,t}$. Model B and Model C provide evaluations of temporal trends in terms of global and local assessments respectively. The coefficient $\rho + \rho_i$ in Model C represents the spatial trend in location parameters, where ρ could be interpreted as the global change level in the CFR per year. Finally, Model D provides a multiplicative representation of $W_{i,t}$, similar in spirit to the work of G. Huerta and B. Sansó [2007]. Models A and D yield special cases of a dynamic linear model (West and Harrison, 1997). Again, we want to model the dependence between location and scale parameters in the GEV model. In models A, B, and D, we do this by specifying $\log\sigma_i$ and ψ_i to be dependent. We work with $\sigma_i = \sigma_0\exp(\lambda_i)$ and a coregionalized bivariate CAR model. In model C, we specify $\log\sigma_i, \psi_i$, and ρ_i to be dependent and use a coregionalized trivariate CAR specification.

In the foregoing models, temporal evolution in the extreme rainfalls is taken into account in the model for $W_{i,t}$. More precisely, each of the models enables prediction for any grid cell for any year. In fact, with the CFR precipitation data, Sang and Gelfand [2009] held out the annual maximum rainfalls in 1999 (Figure 17.1) for validation purposes, in order to compare models in terms of the predictive performance; see the paper for model fitting details. Posterior medians are adopted as the point estimates of the predicted annual maxima because of the skewness of the predictive distributions. Predictive performance was assessed first by computing the averaged absolute predictive errors (AAPE) for each model. A second comparison among models is to study the proportion of the true annual maximum rainfalls in the year 1999 which lie in the estimated 95% credible intervals for each model. A third model selection criterion which is easily calculated from the posterior samples is the deviance information criterion (DIC). Using these criteria, Model C emerged as the best (we omit details here).

17.1.3   A continuous spatial process model

Again, we focus on spatial extremes for precipitation events. However, now the motivating data are annual maxima of daily precipitations derived from daily station records at 281 sites over South Africa in 2006. Often, extreme climate events are driven by multi-scale spatial forcings, say, large regional forcing and small scale local forcing. Therefore, attractive modeling in such cases would have the potential to characterize the multi-scale dependence between locations for extreme values of the spatial process. Additionally, with point-referenced station data, spatial interpolation is of interest to learn about the predictive distribution of the unobserved extreme value at unmonitored locations.

So, we extend the hierarchical modeling approach developed in the previous subsection to accommodate a collection of point-referenced extreme values in order to achieve multi-scale dependence along with spatial smoothing for realizations of the surface of extremes. In particular, we assume annual maxima follow GEV distributions, with parameters μ, σ , and ϕ specified in the latent stage to reflect the underlying spatial structure. We relax the conditional independence assumption previously imposed on the first stage hierarchical specifications. Again, in space, despite the fact that large scale spatial dependence may be accounted for in the latent parameter specifications, there may still remain unexplained small scale spatial dependence in the extreme data. So, we propose a (mean square) continuous spatial process model for the actual extreme values to account for spatial dependence which is unexplained by the latent spatial specifications for the GEV parameters. In other words, we imagine a scale of space-time dependence which is captured at a second stage of a hierarchical model specification with additional proposed first stage spatial smoothing at a much shorter range. This first stage process model is created through a copula approach where the copula idea is applied to a Gaussian spatial process using suitable transformation.

More formally, the first stage of the hierarchical model can be written as:

$$Y(\mathbf{s}) \;=\; \mu(\mathbf{s}) + \frac{\sigma(\mathbf{s})}{\xi(\mathbf{s})}(Z(\mathbf{s})^{\xi(\mathbf{s})} - 1) \tag{17.3}$$

where $Z(\mathbf{s})$ follows a standard Fréchet distribution. We may view $Z(\mathbf{s})$ as the "standardized residual" in the first stage GEV model. The conditional independence assumption is equivalent to the assumption that the $Z(\mathbf{s})$ are $i.i.d.$ So, again, even if the surface for each model parameter is smooth, realizations of the predictive surface will be everywhere discontinuous under the conditional independence assumption. We present a Gaussian processes approach using copulas.

Using copulas, we seek to modify the assumption that the $z(\mathbf{s})$ are i.i.d. Fréchet. We wish to introduce spatial dependence to the $z(\mathbf{s})$ while retaining Fréchet marginal distributions. A frequent strategy for introducing dependence subject to specified marginal distributions is through copula models. We need to apply the copula approach to a stochastic process. The Gaussian process, which is determined through its finite dimensional distributions, offers the most convenient mechanism for doing this. With a suitable choice of the correlation function, mean square continuous surface realizations result for the Gaussian process, hence for the transformed surfaces under monotone transformation. In other words, through transformation of a Gaussian process, we can obtain a continuous spatial process of extreme values with standard Fréchet marginal distributions.

SPACE-TIME MODELING FOR EXTREMES

Copulas have received much attention and application in the past two decades (see, Nelsen, 2006, for a review). Consider a random two-dimensional vector distributed according to a standard bivariate Gaussian distribution with correlation ρ. The Gaussian copula function is defined as follows: $C_\rho(u,v) = \Phi_\rho(\Phi^{-1}(u), \Phi^{-1}(v))$ where the $u, v \in [0,1]$, Φ denotes the standard normal cumulative distribution function and Φ_ρ denotes the cumulative distribution function of the standard bivariate Gaussian distribution with correlation ρ. The bivariate random vector (X, Y) having GEV distributions as marginals, denoted as G_x and G_y respectively, can be given a bivariate extreme value distribution using the Gaussian copula as follows. Let $(X, Y) = (G_x^{-1}(\Phi(X')), G_y^{-1}(\Phi(Y')))$, where G_x^{-1} and G_y^{-1} are the inverse marginal distribution functions for X and Y and $(X', Y') \sim \Phi_\rho$. Then the distribution function of (X,Y) is given by $H(X, Y; \rho) = C_\rho(\Phi(X'), \Phi(Y'))$ and the marginal distributions of X and Y remain to be G_x and G_y.

Now, we can directly propose a spatial extreme value process which is transformed from a standard spatial Gaussian process with mean 0, variance 1, and correlation function $\rho(\mathbf{s}, \mathbf{s}'; \theta)$. The standard Fréchet spatial process is the transformed Gaussian process defined as $z(\mathbf{s}) = G^{-1}(\Phi(z'(\mathbf{s})))$ where now G is the distribution function of a standard Fréchet distribution. It is clear that z(s) is a valid stochastic process since it is induced by a strictly monotone transformation of a Gaussian process. Indeed, this standard Fréchet process is completely determined by $\rho(\mathbf{s}, \mathbf{s}; \theta)$. More precisely, suppose we observe extreme values at a set of sites $\{s_i, i = 1, 2, ..., n\}$. The realizations $\mathbf{z}^T = (z^T(\mathbf{s}_1), ..., z^T(\mathbf{s}_n))$ follow a multivariate normal distribution which is determined by ρ and induces the joint distribution for $\mathbf{z} = (z(\mathbf{s}_1), ..., z(\mathbf{s}_n))$.

Properties of the transformed Gaussian process include: (i) joint, marginal and conditional distributions for $\mathbf{z}$ are all immediate. Dependence in the Fréchet process is inherited through $\rho(\cdot)$, (ii) a Matérn with smoothness parameter greater than 0 assures mean square continuous realizations, (iii) efficient model fitting approaches for Gaussian processes can be utilized after inverse transformation, (iv) since the $z^T(\mathbf{s})$ process is strongly stationary, then so is the $z(\mathbf{s})$ process, (v) despite the inherited dependence through ρ, G has no moments so non-moment based dependence metrics for $z(\mathbf{s})$ are needed, e.g., $\int ([z(\mathbf{s}), z(\mathbf{s}')] - [z(\mathbf{s})][z(\mathbf{s}')])^2 ds ds'$, and (vi) evidently, the transformed Gaussian approach is not limited to extreme value analysis; it can create other classes of means square continuous processes.

17.1.4 *Hierarchical modeling for spatial extreme values*

Returning to (2), we now assume that the $z(\mathbf{s})$ follow a standard Fréchet process. That is, we have a hierarchical model in which the first stage conditional independence assumption is removed. Now, we turn to the specification of the second stage, creating *latent* spatial models, following Coles and Tawn [1996]. Specifications for $\mu(\mathbf{s})$, $\sigma(\mathbf{s})$, and $\xi(\mathbf{s})$ have to be made with care. A simplification to facilitate model fitting is to assume there is spatial dependence for the $\mu(\mathbf{s})$ but that $\sigma(\mathbf{s})$ and $\xi(\mathbf{s})$ are constant across the study region. In fact, the data is not likely to be able to inform about processes for $\sigma(\mathbf{s})$ and for $\xi(\mathbf{s})$.

More precisely, suppose we propose the specification $\mu(\mathbf{s}) = X(\mathbf{s})^T \beta + W(\mathbf{s})$. $X(\mathbf{s})$ is the site-specific vector of potential explanatory variables. The $W(\mathbf{s})$ are spatial random effects, capturing the effect of unmeasured or unobserved covariates with large operational scale spatial patterns. A natural specification for $W(\mathbf{s})$ is a zero-centered Gaussian process determined by a valid covariance function $C(\mathbf{s}_i, \mathbf{s}_j)$. Here C is apart from ρ introduced at the first stage. In fact, from (2), plugging in the model for $\mu(\mathbf{s})$ we obtain

$$Y(\mathbf{s}) = \mathbf{x}^T(\mathbf{s})\boldsymbol{\beta} + W(\mathbf{s}) + \frac{\sigma}{\xi}(z(\mathbf{s})^\xi - 1) \tag{17.4}$$

585

and $z(\mathbf{s}) = G^{-1}\Phi(z^{'}(\mathbf{s}))$. We clearly see the two sources of spatial dependence, the $z(\mathbf{s})$ and the $W(\mathbf{s})$ with two associated scales of dependence. The foregoing interpretation as well as the need for identifiability imply that we assign the shorter-range spatial dependence to the $z(\mathbf{s})$ process. Of course, we can compare this specification with the case where we have $z(\mathbf{s})$ i.i.d. Fréchet.

Again, we often have space-time data over long periods of time, e.g., many years, and we seek to study say, annual spatial maxima. Now, we are given a set of extremes $\{Y(\mathbf{s}_i,t), i = 1, ..., n; t = 1, ..., T\}$. Now, the first stage of the hierarchical model specifies a space time standard Fréchet process for the $z(\mathbf{s},t)$, built from a space-time Gaussian process. If time is viewed as continuous, we need only specify a valid space-time covariance function. If time is discretized then we need only provide a dynamic Gaussian process model (see Chapter 10). In fact, we might assume $z_t(\mathbf{s})$ and $z_{t'}(\mathbf{s})$ are two independent Gaussian processes when $t \neq t'$. That is, it may be plausible to assume temporal independence since the annual block size may be long enough to yield approximately independent annual maximum observation surfaces. Model specifications for $\mu(\mathbf{s},t)$, $\sigma(\mathbf{s},t)$, and $\xi(\mathbf{s},t)$ should account for dependence structures both within and across location and time. In Section 13.2.2 above, we offered suggested forms with regard to these latent parameter specifications.

Finally, Sang and Gelfand (2011) work with the foregoing annual maxima of daily rainfalls from the station data in South Africa. They consider 200 monitoring sites and fit each of the years 1956, 1976, 1996, and 2006 separately; see the paper for model fitting details. They offer a comparison between the conditionally independent first stage specification and the smoothed first stage using the Gaussian copula for the annual maximum rainfall station data for the year 2006. The smoothed first stage specification was superior, achieving a 20% reduction in AAPE and more accurate empirical coverage.

17.1.5 *Max stable spatial processes and generalized Pareto spatial processes*

In the extreme value literature there is a general class of multivariate extreme value processes which introduce desired dependence between pairs of maxima, the so-called max-stable processes, dating to the seminal work of De Haan [1984]. From Coles [2001], let $W_1(\mathbf{s}), W_2(\mathbf{s}), ...,$ be i.i.d. replicates of a process on R^2. Let the process $Y^{(n)}(\mathbf{s}) = \max_{j=1,...,n} W_j(\mathbf{s}) \sim nW_1(\mathbf{s}) \; \forall n$. There is a limiting characterization of this class of processes due to De Haan and specific versions have been presented by Smith and by Schlather [see discussion and references in Coles, 2001, Schlather, 2001, 2002, Schlather and Tawn, 2003, De Haan and Ferreira, 2006, Ballani and Schlather, 2011, for further insights and technical details].

For instance, for suitable normalizing sequences $a_n(\mathbf{s})$ and $b_n(\mathbf{s})$, under certain conditions [see, e.g., Ribatet, 2013], there exists a spatial process such that $Z(\mathbf{s}) = \lim_{n\to\infty}(Y^{(n)}(\mathbf{s}) - b_n(\mathbf{s}))/a_n(\mathbf{s})$ will have a GEV distribution. Unfortunately, model fitting using these max-stable processes is challenging since joint distributions are intractable for more than two locations—in practice, we would have hundreds of locations requiring explicit forms for high-dimensional joint distributions in order to work with likelihoods. Currently, two approximation approaches have emerged: the use of pairwise likelihoods and an approximate Bayesian computation approach [see, e.g., Zhang et al., 2022, for discussion regarding large spatial datasets].

However, a very recent paper [Huser et al., 2024] argues that the future for spatial modeling of block maxima does not lie with max stable processes. Max stable processes suffer limitations including the fact that: (i) they implicitly assume that the max at location $\mathbf{s}$ occurs at the same time as the max at location $\mathbf{s}^{'}$. This may be approximately true, yielding a reasonable approximation for the joint distribution of the block maxima over space but it is not going to be true with real data, (ii) the dependence structure is independent of the size of the blocks. 10, 100, or 1000 year maxima all are captured with the same dependence; dependence doesn't change with the "severity" of the event, and (iii) spatial dependence does not tend to 0 as locations become further apart. So-called "bulk" modeling strategies, i.e., mixture models with a component for the bulk of the data along with a component for the upper tail [and perhaps the lower tail; see, e.g., Naveau et al., 2016, and discussion therein] can more realistically capture the behavior of extremes.

Alternatively, peaks-over threshold modeling has been around for many years [again, see, e.g., Coles, 2001]. These approaches focus on modeling extremes of the original spatial events that effectively took place. They are potentially more meaningful from a practical perspective and allow customization of the definition of a 'spatial extreme event' to the specific problem of interest. In particular, peaks-over-threshold approaches do not only provide valid probability approximations when all variables are simultaneously large (which rarely occurs in practice), but they can be adapted to events where only a subset of variables are extreme. Discussion of the existence of Generalized Pareto processes with a view toward application and simulation can be found in Ferrieira and De Haan [2014]. Pareto processes are mathematically simpler than max stable processes and have the potential advantage of incorporating all relevant extreme events, by generalizing the notion of a univariate exceedance. de Fondeville and Davison [2018] provides details as well as an application to rainfall on a grid of 3600 locations.

17.2 Spatial CDFs and extents

In this section, we review the essentials of spatial cumulative distribution functions (SCDFs), including a hierarchical modeling approach for inference. We then extend the basic definition to allow covariate weighting of the SCDF estimate, as well as versions arising under a bivariate random process. Finally, we extend the idea of an SCDF to a spatial *extent* to consider a random spatial realization relative to a specified spatial surface.

17.2.1 Basic definition of a spatial CDF and motivating data sets

Suppose that $X(\mathbf{s})$ is the log-ozone concentration at location $\mathbf{s}$ over a particular time period. Thinking of $X(\mathbf{s})$, $\mathbf{s} \in D$ as a spatial process providing continuous realizations, we might wish to find the proportion of area in D that has ozone concentration below some level w (say, a level above which exposure is considered to be unhealthful). This proportion is the random variable,

$$F(w) = Pr\left[\mathbf{s} \in D : X(\mathbf{s}) \le w\right] = \frac{1}{|D|} \int_D Z_w(\mathbf{s})d\mathbf{s} , \qquad (17.5)$$

where $|D|$ is the area of D, and $Z_w(\mathbf{s}) = 1$ if $X(\mathbf{s}) \le w$, and 0 otherwise. It is worth commenting that the integral in (17.5) exists in the sense that for any realization, $\int_D X(\mathbf{s})d\mathbf{s}$ is a Reimann integral (though it can not be evaluated since $X(\mathbf{s})$ is not a function). Similarly, $\frac{1}{|D|} \int_D Z_w(\mathbf{s})d\mathbf{s}$ exists since the indicator function applied to $X(\mathbf{s})$ will be continuous almost everywhere. Since $X(\mathbf{s})$, $\mathbf{s} \in D$ is random, (17.5) is a random function of $w \in \Re$ (a random Reimann integral) that increases from 0 to 1 and is right-continuous. Thus, while $F(w)$ is not the usual cumulative distribution function (CDF) of X at $\mathbf{s}$ (which would be given by $Pr[X(\mathbf{s}) \le x]$, and is not random), it does have all the properties of a CDF, and so is referred to as the *spatial cumulative distribution function*, or SCDF. For a constant mean stationary process, all $X(\mathbf{s})$ have the same marginal distribution, whence $\mathrm{E}[F(w)] = Pr\left(X(\mathbf{s}) \le w\right)$. It is also easy to show that $\mathrm{Var}[F(w)] = \frac{1}{|D|^2} \int_D \int_D Pr\left(X(\mathbf{s}) \le w, X(\mathbf{s}') \le w\right)d\mathbf{s}d\mathbf{s}' - [Pr\left(X(\mathbf{s}) \le w\right)]^2$. Overton (1989) introduced the idea of an SCDF, and used it to analyze data from the National Surface Water Surveys. Lahiri et al. [1999] developed a subsampling method that provides (among other things) large-sample prediction bands for the SCDF, which they show to be useful in assessing the foliage condition of red maple trees in the state of Maine.

The *empirical* SCDF based upon data $\mathbf{X}_s = (X(\mathbf{s}_1), \ldots, X(\mathbf{s}_n))'$ at w is the proportion of the $X(\mathbf{s}_i)$ that take values less than or equal to w. Large-sample investigation of the behavior of the empirical SCDF requires care to define the appropriate asymptotics; see Lahiri et al. [1999] and Zhu et al. [2002] for details. When n is not large, as is the case in our applications, the empirical SCDF may become less attractive. Stronger inference can be achieved if one is willing to make stronger distributional assumptions regarding the process $X(\mathbf{s})$. For instance, if $X(\mathbf{s})$ is assumed to be a Gaussian process, the joint distribution of $\mathbf{X}_s$ is multivariate normal. Given a suitable prior specification, a Bayesian framework provides the predictive distribution of $F(w)$ given $X(\mathbf{s})$.

Though (17.5) can be studied analytically, since it is a stochastic integral, it can never be observed and it can never be calculated exactly. (See Section 7.1.2 in this regard.) However, approximation of (17.5) via Monte Carlo integration is available (or creating a grid of points over D)[1], i.e., replacing $F(w)$ by

$$\widehat{F}(w) = \frac{1}{L} \sum_{\ell=1}^{L} Z_w(\tilde{\mathbf{s}}_\ell) , \qquad (17.6)$$

where the $\tilde{\mathbf{s}}_\ell$ are chosen randomly in D, and $Z_w(\tilde{\mathbf{s}}_\ell) = 1$ if $X(\tilde{\mathbf{s}}_\ell) \leq w$, and 0 otherwise. Suppose we seek a realization of $\widehat{F}(w)$ from the predictive distribution of $\widehat{F}(w)$ given $\mathbf{X}_s$. In Section 7.1 we showed how to sample from the predictive distribution $p(\mathbf{X}_{\tilde{s}} \mid \mathbf{X}_s)$ for $\mathbf{X}_{\tilde{s}}$ arising from new locations $\tilde{\mathbf{s}} = (\tilde{\mathbf{s}}_1, \ldots, \tilde{\mathbf{s}}_L)'$. In fact, samples $\{\mathbf{X}_{\tilde{s}}^{(g)}, g = 1, \ldots, G\}$ from the posterior predictive distribution,

$$p(\mathbf{X}_{\tilde{s}} \mid \mathbf{X}_s) = \int p(\mathbf{X}_{\tilde{s}} \mid \mathbf{X}_s, \boldsymbol{\beta}, \boldsymbol{\theta}) p(\boldsymbol{\beta}, \boldsymbol{\theta} \mid \mathbf{X}_s) d\boldsymbol{\beta} d\boldsymbol{\theta} ,$$

may be obtained one for one from posterior samples by composition.

The predictive distribution of (17.5) can be sampled at a given w by obtaining $X(\tilde{\mathbf{s}}_\ell)$ using the above algorithm, hence $Z_w(\tilde{\mathbf{s}}_\ell)$, and then calculating $\widehat{F}(w)$ using (17.6). Since interest is in the entire function $F(w)$, we would seek realizations of the approximate function $\widehat{F}(w)$. These are most easily obtained, up to an interpolation, using a grid of w values $\{w_1 < \cdots < w_k < \cdots < w_K\}$, whence each $(\boldsymbol{\beta}^{(g)}, \boldsymbol{\theta}^{(g)})$ gives a realization at grid point w_k,

$$\widehat{F}^{(g)}(w_k) = \frac{1}{L} \sum_{\ell=1}^{L} Z_{w_k}^{(g)}(\tilde{\mathbf{s}}_\ell) , \qquad (17.7)$$

where now $Z_{w_k}^{(g)}(\tilde{\mathbf{s}}_\ell) = 1$ if $X^{(g)}(\tilde{\mathbf{s}}_\ell) \leq w_k$, and 0 otherwise. Handcock [1999] describes a similar Monte Carlo Bayesian approach to estimating SCDFs in his discussion of Lahiri et al. [1999].

Expression (17.7) suggests placing all of our $\widehat{F}^{(g)}(w_k)$ values in a $K \times G$ matrix for easy summarization. For example, a histogram of all the $\widehat{F}^{(g)}(w_k)$ in a particular row (i.e., for a given grid point w_k) provides an estimate of the predictive distribution of $\widehat{F}(w_k)$. On the other hand, each column (i.e., for a given Gibbs draw g) provides (again up to, say, linear interpolation) an approximate draw from the predictive distribution of the SCDF. Hence, averaging these columns provides, with interpolation, essentially the posterior predictive mean for F and can be taken as an *estimated* SCDF. But also, each draw from the predictive distribution of the SCDF can be inverted to obtain any quantile of interest (e.g., the median exposure $d_{.50}^{(g)}$). A histogram of these inverted values in turn provides an estimate

---

[1]There is some delicacy regarding whether, according to the choice of function $h(X(\mathbf{s}))$ used in the integration, the discrete approximation converges to the stochastic integral. However, for the indicator functions here, there is no problem.

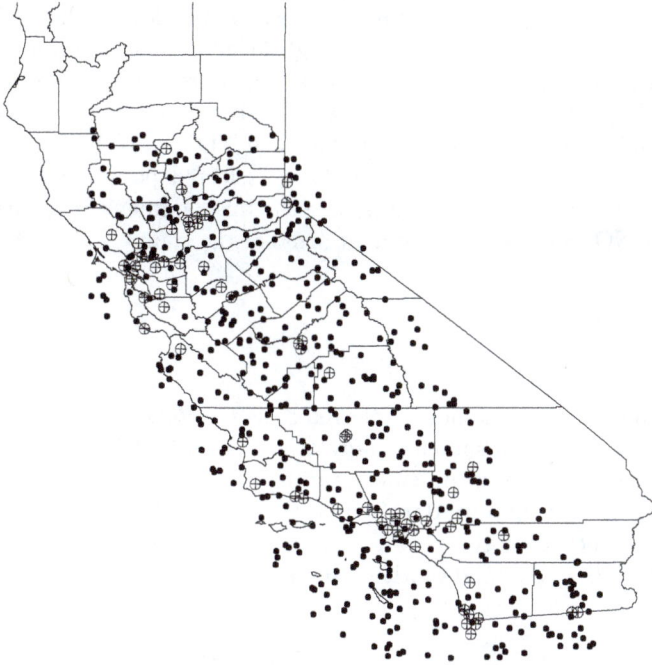

Figure 17.2 *Locations of 67 NO and NO$_2$ monitoring sites, California air quality data; 500 randomly selected target locations are also shown as dots.*

of the posterior distribution of this quantile (in this case, $\widehat{p}(d_{.50}|\mathbf{X}_s)$). While this algorithm provides general inference for SCDFs, for most data sets it will be computationally very demanding since a large L will be required to make (17.7) sufficiently accurate.

This methodology is motivated by two environmental data sets; we describe both here but only present the inference for the second. The first is the Atlanta eight-hour maximum ozone data, which exemplifies the case of an air pollution variable measured at points, with a demographic covariate measured at a block level. Recall that its first component is a collection of ambient ozone levels in the Atlanta, GA, metropolitan area, as reported by Tolbert et al. (2000). Ozone measurements X_{itr} are available at between 8 and 10 fixed monitoring sites i for day t of year r, where $t = 1, \ldots, 92$ (the summer days from June 1 through August 31) and $r = 1, 2, 3$, corresponding to years 1993, 1994, and 1995. The reader may wish to flip back to Figure 1.3, which shows the 8-hour daily maximum ozone measurements in parts per million at the 10 monitoring sites for one of the days (July 15, 1995). This figure also shows the boundaries of the 162 zip codes in the Atlanta metropolitan area, with the 36 zips falling within the city of Atlanta encircled by the darker boundary on the map. An environmental justice assessment of exposure to potentially harmful levels of ozone would be clarified by examination of the predictive distribution of a *weighted* SCDF that uses the racial makeups of these city zips as the weights. This requires generalizing our SCDF simulation in (17.7) to accommodate covariate weighting in the presence of misalignment between the response variable (at point-referenced level) and the covariate (at areal-unit level).

SCDFs adjusted with point-level covariates present similar challenges. Consider the spatial data setting of Figure 17.2, recently presented and analyzed by Gelfand, Schmidt, and Sirmans (2002). These are the locations of several air pollutant monitoring sites in central and southern California, all of which measure ozone, carbon monoxide, nitric oxide (NO), and nitrogen dioxide (NO_2). For a given day, suppose we wish to compute an SCDF for the log of the daily median NO exposure adjusted for the log of the daily median NO_2 level (since the health effects of exposure to high levels of one pollutant may be exacerbated by further exposure to high levels of the other). Here the data are all point level, so that Bayesian kriging methods of the sort described above may be used. However, we must still tackle the problem of *bivariate* kriging (for both NO and NO_2) in a computationally demanding setting (say, to the $L = 500$ randomly selected points shown as dots in Figure 17.2). In some settings, we must also resolve the misalignment in the data itself, which arises when NO or NO_2 values are missing at some of the source sites.

17.2.2   Derived-process spatial CDFs

17.2.2.1   Point- versus block-level spatial CDFs

The spatial CDF in (17.5) is customarily referred to as *the* SCDF associated with the spatial process $X(\mathbf{s})$. In fact, we can formulate many other useful SCDFs under this process. We proceed to elaborate on choices of possible interest.

Suppose for instance that our data arrive at the areal unit level, i.e., we observe $X(B_j)$, $j = 1,\ldots,J$ such that the B_j are disjoint with union D, the entire study region. Let $Z_w(B_j) = 1$ if $X(B_j) \leq w$, and 0 otherwise. Then

$$\widetilde{F}(w) = \frac{1}{|D|} \sum_{j=1}^{J} |B_j|\, Z_w(B_j) \tag{17.8}$$

again has the properties of a CDF and thus can also be interpreted as a spatial CDF. In fact, this CDF is a step function recording the proportion of the area of D that (at block-level resolution) lies below w. Suppose in fact that the $X(B_j)$ can be viewed as block averages of the process $X(\mathbf{s})$, i.e., $X(B_j) = \frac{1}{|B_j|} \int_{B_j} X(\mathbf{s})ds$. Then (17.5) and (17.8) can be compared: write (17.5) as $\frac{1}{|D|} \sum_j |B_j| \left[\frac{1}{|B_j|} \int_{B_j} I(X(\mathbf{s}) \leq w)ds \right]$ and (17.8) as $\frac{1}{|D|} \sum_j |B_j|\, I\left[\left(\frac{1}{|B_j|} \int_{B_j} X(\mathbf{s})ds \right) \leq w \right]$. Interpreting $\mathbf{s}$ to have a uniform distribution on B_j, the former is $\frac{1}{|D|} \sum_j |B_j|\, E_{B_j} [I(X(\mathbf{s}) \leq w)]$ while the latter is $\frac{1}{|D|} \sum_j |B_j|\, I[E_{B_j}(X(\mathbf{s})) \leq w]$. In fact, if $X(\mathbf{s})$ is stationary, while $E[F(w)] = P(X(\mathbf{s}) \leq w)$, $E[\widetilde{F}(w)] = \frac{1}{|D|} \sum_j |B_j|\, P(X(B_j) \leq w)$. For a Gaussian process, under weak conditions $X(B_j)$ is normally distributed with mean $E[X(\mathbf{s})]$ and variance $\frac{1}{|B_j|^2} \int_{B_j} \int_{B_j} c(\mathbf{s}-\mathbf{s}';\boldsymbol{\theta})dsds'$, so $E[\widetilde{F}(w)]$ can be obtained explicitly. Note also that since $\frac{1}{|B_j|} \int_{B_j} I(X(\mathbf{s}) \leq w)ds$ is the customary spatial CDF for region B_j, then by the alternate expression for (17.5) above, $F(w)$ is an areally weighted average of *local* SCDFs.

Thus (17.5) and (17.8) differ, but (17.8) should neither be viewed as "incorrect" nor as an approximation to (17.5). Rather, it is an alternative SCDF derived under the $X(\mathbf{s})$ process. Moreover, if only the $X(B_j)$ has been observed, it is arguably the most sensible empirical choice. Indeed, the Multiscale Advanced Raster Map (MARMAP) analysis system project (`www.stat.psu.edu/~gpp/marmap_system_partnership.htm`) is designed to work with "empirical cell intensity surfaces" (i.e., the tiled surface of the $X(B_j)$'s over D) and calculates the "upper level surfaces" (variants of (17.8)) for description and inference regarding multicategorical maps and cellular surfaces.

Next, we seek to introduce covariate weights to the spatial CDF, as motivated in Subsection 17.2.1. For a nonnegative function $r(\mathbf{s})$ that is integrable over D, define the SCDF associated with $X(\mathbf{s})$ weighted by r as

$$F_r(w) = \frac{\int_D r(\mathbf{s})Z_w(\mathbf{s})d\mathbf{s}}{\int_D r(\mathbf{s})d\mathbf{s}} . \tag{17.9}$$

Evidently (17.9) satisfies the properties of a CDF and generalizes (17.5) (i.e., (17.5) is restored by taking $r(\mathbf{s}) \equiv 1$). But as (17.5) suggests expectations with respect to a uniform density for $\mathbf{s}$ over D, (17.9) suggests expectation with respect to the density $r(\mathbf{s})/\int_D r(\mathbf{s})d\mathbf{s}$. Under a stationary process, $\mathrm{E}[F(w)] = P(X(\mathbf{s}) \le w)$ and $Var[F(w)]$ is

$$\frac{1}{(\int_D r(\mathbf{s})d\mathbf{s})^2} \int_D \int_D r(\mathbf{s})r(\mathbf{s}')P(X(\mathbf{s}) \le w, X(\mathbf{s}') \le w)d\mathbf{s}d\mathbf{s}'$$
$$-[P(X(\mathbf{s}) \le w)]^2 .$$

There is an empirical SCDF associated with (17.9) that extends the empirical SCDF in Subsection 17.2.1 using weights $r(\mathbf{s}_i)/\sum_i r(\mathbf{s}_i)$ rather than $1/n$. This random variable is mentioned in p87 of Lahiri et al. [1999]. Following Subsection 17.2.1, we adopt a Bayesian approach and seek a predictive distribution for $F_r(w)$ given $\mathbf{X}_s$. This is facilitated by Monte Carlo integration of (17.9), i.e.,

$$\widehat{F}_r(w) = \frac{\sum_{\ell=1}^{L} r(\mathbf{s}_\ell)Z_w(\mathbf{s}_\ell)}{\sum_{\ell=1}^{L} r(\mathbf{s}_\ell)} . \tag{17.10}$$

17.2.2.2 Covariate weighted SCDFs for misaligned data

In the environmental justice application described in Subsection 17.2.1, the covariate is only available (indeed, only meaningful) at an areal level, i.e., we observe only the population density associated with B_j. How can we construct a covariate weighted SCDF in this case? Suppose we make the assignment $r(\mathbf{s}) = r_j$ for all $\mathbf{s} \in B_j$, i.e., that the density surface is constant over the areal unit (so that $r_j|B_j|$ is the observed population density for B_j). Inserting this into (17.9) we obtain

$$F_r^*(w) = \frac{\sum_{j=1}^{J} r_j|B_j| \left[\frac{1}{|B_j|} \int_{B_j} Z_w(\mathbf{s})d\mathbf{s} \right]}{\sum_{j=1}^{J} r_j|B_j|} . \tag{17.11}$$

As a special case of (17.9), (17.11) again satisfies the properties of a CDF and again has mean $P(X(\mathbf{s}) \le w)$. Moreover, as below (17.8), the bracketed expression in (17.11) is the spatial CDF associated with $X(\mathbf{s})$ restricted to B_j. Monte Carlo integration applied to (17.11) can use the same set of $\mathbf{s}_\ell$'s chosen randomly over D as in Subsection 17.2.1 or as in (17.10). In fact (17.10) becomes

$$\widehat{F}_r(w) = \frac{\sum_{j=1}^{J} r_j L_j \left[\frac{1}{L_j} \sum_{\mathbf{s}_\ell \in B_j} Z_w(\mathbf{s}_\ell) \right]}{\sum_{j=1}^{J} r_j L_j} , \tag{17.12}$$

where L_j is the number of $\mathbf{s}_\ell$ falling in B_j. Equation (17.11) suggests the alternative expression,

$$\widehat{F}_r^*(w) = \frac{\sum_{j=1}^{J} r_j|B_j| \left[\frac{1}{L_j} \sum_{\mathbf{s}_\ell \in B_j} Z_w(\mathbf{s}_\ell) \right]}{\sum_{j=1}^{J} r_j|B_j|} . \tag{17.13}$$

Expression (17.13) may be preferable to (17.12) since it uses the exact $|B_j|$ rather than the random L_j.

17.2.3   Randomly weighted SCDFs

If we work solely with the r_j's, we can view (17.10)–(17.13) as *conditional* on the r_j's. However, if we work with $r(\mathbf{s})$'s, then we will need a probability model for $r(\mathbf{s})$ in order to interpolate to $r(\mathbf{s}_\ell)$ in (17.10). Since $r(\mathbf{s})$ and $X(\mathbf{s})$ are expected to be associated, we may conceptualize them as arising from a spatial process, and develop, say, a bivariate Gaussian spatial process model for both $X(\mathbf{s})$ and $h(r(\mathbf{s}))$, where h maps the weights onto $\Re^1$.

Let $\mathbf{Y}(\mathbf{s}) = (X(\mathbf{s}), h(r(\mathbf{s}))^{\mathrm{T}}$ and $\mathbf{Y} = (\mathbf{Y}(\mathbf{s}_1), \dots, \mathbf{Y}(\mathbf{s}_n))^{\mathrm{T}}$. Analogous to the univariate situation in Subsection 17.2.1, we need to draw samples $\mathbf{Y}_{\tilde{s}}^{(g)}$ from $p(\mathbf{Y}_{\tilde{s}} \mid \mathbf{Y}, \boldsymbol{\beta}^{(g)}, \boldsymbol{\theta}^{(g)}, T^{(g)})$. Again this is routinely done via composition from posterior samples. Since the $\mathbf{Y}_{\tilde{s}}^{(g)}$ samples have marginal distribution $p(\mathbf{Y}_{\tilde{s}} \mid \mathbf{Y})$, we may use them to obtain predictive realizations of the SCDF, using either the unweighted form (17.7) or the weighted form (17.10).

The bivariate structure also allows for the definition of a *bivariate SCDF*,

$$F_{U,V}(w_u, w_v) = \frac{1}{|D|} \int_D I(U(\mathbf{s}) \leq w_u, V(\mathbf{s}) \leq w_v) d\mathbf{s} , \qquad (17.14)$$

which gives $Pr\,[\mathbf{s} \in D : U(\mathbf{s}) \leq w_u, V(\mathbf{s}) \leq w_v]$, the proportion of the region having values below the given thresholds for, say, two pollutants. Finally, a sensible *conditional* SCDF might be

$$\begin{aligned} F_{U|V}(w_u|w_v) &= \frac{\int_D I(U(\mathbf{s}) \leq w_u, V(\mathbf{s}) \leq w_v) d\mathbf{s}}{\int_D I(V(\mathbf{s}) \leq w_v) d\mathbf{s}} \\ &= \frac{\int_D I(U(\mathbf{s}) \leq w_u) I(V(\mathbf{s}) \leq w_v) d\mathbf{s}}{\int_D I(V(\mathbf{s}) \leq w_v) d\mathbf{s}} . \end{aligned} \qquad (17.15)$$

This expression gives $Pr\,[\mathbf{s} \in D : U(\mathbf{s}) \leq w_u \mid V(\mathbf{s}) \leq w_v]$, the proportion of the region having second pollutant values below the threshold w_v that *also has* first pollutant values below the threshold w_u. Note that (17.15) is again a weighted SCDF, with $r(\mathbf{s}) = I(V(\mathbf{s}) \leq w_v)$. Note further that we could easily alter (17.15) by changing the directions of either or both of its inequalities, if conditional statements involving high (instead of low) levels of either pollutant were of interest.

Example 17.1 *(California air quality data)*. We illustrate in the case of a bivariate Gaussian process using data collected by the California Air Resources Board, available at `www.arb.ca.gov/aqd/aqdcd/aqdcddld.htm`. The particular subset we consider are the mean NO and NO_2 values for July 6, 1999, as observed at the 67 monitoring sites shown as solid dots in Figure 17.2. Recall that in our notation, U corresponds to log(mean NO) while V corresponds to log(mean NO_2). A WSCDF based on these two variables is of interest since persons already at high NO risk may be especially vulnerable to elevated NO_2 levels. Figure 17.3 shows interpolated perspective, image, and contour plots of the raw data. The association of the pollutant levels is apparent; in fact, the sample correlation coefficient over the 67 pairs is 0.74.

We fit a separable, Gaussian bivariate model using the simple exponential spatial covariance structure $\rho(d_{ii'}, \boldsymbol{\theta}) = \exp(-\lambda d_{ii'})$, so that $\boldsymbol{\theta} \equiv \lambda$; no σ^2 parameter is required (nor identifiable) here due to the multiplicative presence of the T matrix. For prior distributions, we first assumed $T^{-1} \sim W((\nu R)^{-1}, \nu)$ where $\nu = 2$ and $R = 4I$. This is a reasonably vague specification, both in its small degrees of freedom ν and in the relative size of R (roughly the prior mean of T), since the entire range of the data (for both log-NO and log-NO_2) is only about 3 units. Next, we assume $\lambda \sim G(a, b)$, with the parameters chosen so that the effective spatial range is half the maximum diagonal distance M in Figure 17.2 (i.e., $3/\mathrm{E}(\lambda) = .5M$), and the standard deviation is one half of this mean. Finally, we assume constant means $\mu_U(\mathbf{s}; \boldsymbol{\beta}) = \beta_U$ and $\mu_V(\mathbf{s}; \boldsymbol{\beta}) = \beta_V$, and let β_U and β_V have vague normal priors (mean 0, variance 1000).

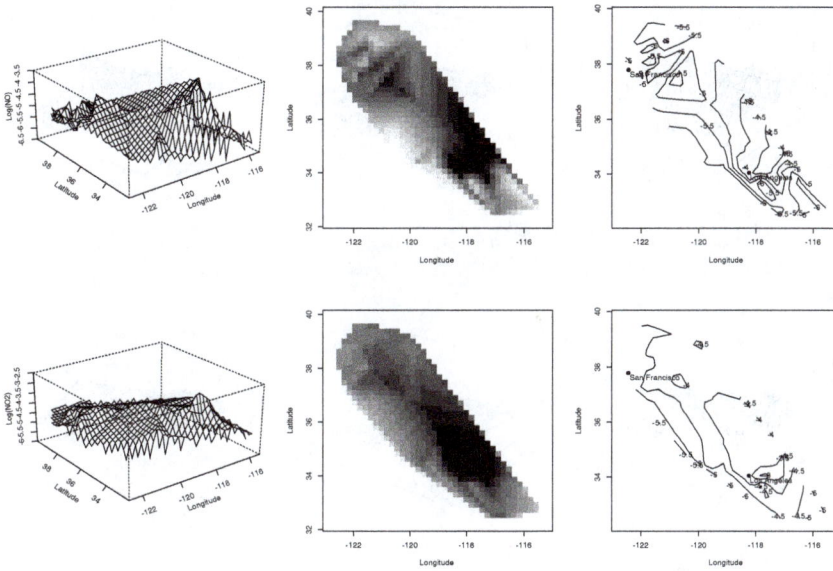

Figure 17.3 *Interpolated perspective, image, and contour plots of the raw log-NO (first row) and log-NO₂ (second row), California air quality data, July 6, 1999.*

Our initial Gibbs algorithm sampled over β, T^{-1}, and λ. The first two of these may be sampled from closed-form full conditionals (normal and inverse Wishart, respectively) while λ is sampled using Hastings independence chains with $G(\frac{2}{3}, \frac{3}{2})$ proposals. We used 3 parallel chains to check convergence, followed by a "production run" of 2000 samples from a single chain for posterior summarization. Histograms (not shown) of the posterior samples for the bivariate kriging model are generally well behaved and consistent with the results in Figure 17.3.

Figure 17.4 shows perspective plots of raw and kriged log-NO and log-NO₂ surfaces, where the plots in the first column are the (interpolated) raw data (as in the first column of Figure 17.3), those in the second column are based on a single Gibbs sample, and those in the third column represent the average over 2000 post-convergence Gibbs samples. The plots in this final column are generally consistent with those in the first, except that they exhibit the spatial smoothness we expect of our posterior means.

Figure 17.5 shows several SCDFs arising from samples from our bivariate kriging algorithm. First, the solid line shows the ordinary SCDF (17.7) for log-NO. Next, we computed the weighted SCDF for two choices of weight function in (17.10). In particular, we weight log-NO exposure U by $h^{-1}(V)$ using $h^{-1}(V) = \exp(V)$ and $h^{-1}(V) = \exp(V)/(\exp(V)+1)$ (the exponential and inverse logit functions). Since V is log-NO₂ exposure, this amounts to weighting by NO₂ itself, and by NO₂/(NO₂+1). The results from these two h^{-1} functions turn out to be visually indistinguishable and are shown as the dotted line in Figure 17.5. This line is shifted to the right from the unweighted version, indicating higher harmful exposure when the second (positively correlated) pollutant is accounted for.

Also shown as dashed lines in Figure 17.5 are several WSCDFs that result from using a particular indicator of whether log-NO₂ remains below a certain threshold. These WSCDFs are thus also conditional SCDFs, as in equation (17.15). Existing EPA guidelines and expertise could be used to inform the choice of clinically meaningful thresholds; here we simply demonstrate the procedure's behavior for a few illustrative thresholds. For example, when the threshold is set to −3.0 (a rather high value for this pollutant on the log scale), nearly all of the weights equal 1, and the WSCDF differs little from the unweighted SCDF. However,

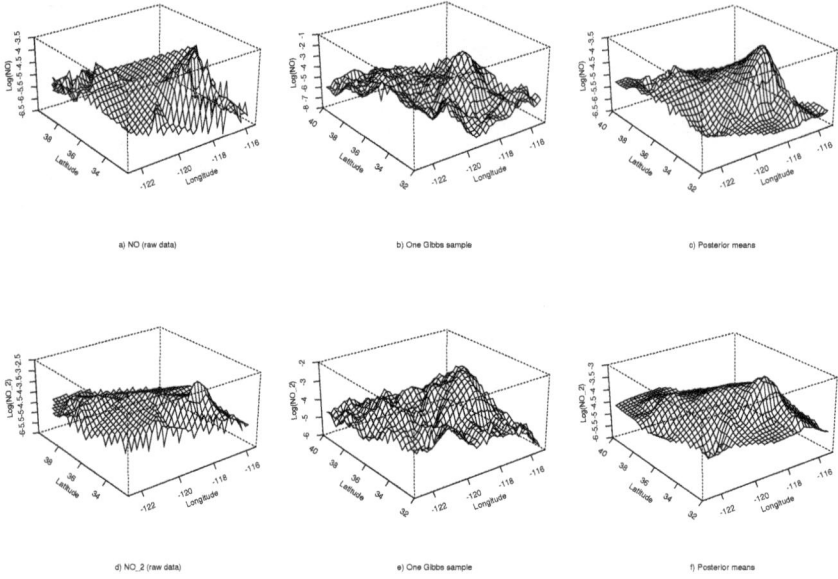

Figure 17.4 *Perspective plots of kriged log-NO and log-NO$_2$ surfaces, California air quality data. First column, raw data; second column, based on a single Gibbs sample; third column, average over 2000 post-convergence Gibbs samples.*

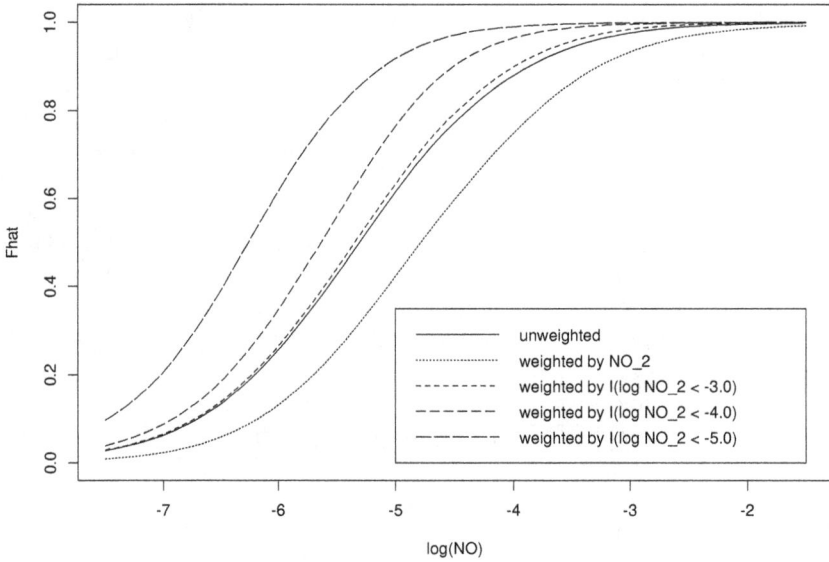

Figure 17.5 *Weighted SCDFs for the California air quality data: solid line, ordinary SCDF for log-NO; dotted line, weighted SCDF for log-NO using NO$_2$ as the weight; dashed lines, weighted SCDF for log-NO using various indicator functions of log-NO$_2$ as the weights.*

as this threshold moves lower (to −4.0 and −5.0), the WSCDF moves further to the left of its unweighted counterpart. This movement is understandable since our indicator functions are *decreasing* functions of log-NO$_2$; movement to the right could be obtained simply by reversing the indicator inequality. ∎

17.2.4   Spatial extents

Here, we consider an extension of the idea of a spatial cdf to a spatial extent. In particular, we are motivated by the extent of an extreme heat event (EHE) and follow ideas presented in Cebrian et al. [2022]. Interest in in assessing such extent is clear. However, essentially all of the work in the literature analyzes the extent using descriptive approaches and using observed or gridded data with no formal inference from probabilistic modeling. Some more formal definitions related to the concept of area under extreme conditions have been introduced in the statistical literature. For example, French and Sain [2013] present a method for constructing confidence regions for Gaussian processes that contain the true exceedance regions with some predefined probability. Extending this methodology, Hazra and Huser [2021] obtain confidence regions that contain joint threshold exceedances of surface sea temperatures, using a semiparametric Bayesian spatial mixed-effects linear model. Bolin and Lindgren [2015] consider excursion sets, which are sets of points in an area where a spatial function is above a given threshold. Sommerfeld et al. [2018] develop confidence regions for these spatial excursion sets with an application to climate. Cebrian et al. [2022] defined the notion of the extent of an extreme heat event as a stochastic object and used it to calculate daily, seasonal, and decadal averages.

We define the object of interest here, a spatial extent, in the context of temperature exceedance. Let $Y_t(\mathbf{s})$ denote the daily max temp for day t at location $\mathbf{s}$. Suppose we consider a subregion $B \subseteq \mathcal{D}$ of the study domain. Then, for a given w, the extent of the EHE in subregion B on day t is:

$$Ext_t(w; B) = \frac{1}{|B|} \int_B \mathbf{1}(Y_t(\mathbf{s}) - q(\mathbf{s}) \geq w) d\mathbf{s}. \qquad (17.16)$$

Here, $q(\mathbf{s})$ is a specified threshold surface over B. As a result, $Ext_t(w; B)$ is the proportion of B which is experiencing extreme heat at least w degrees above or below (according to the sign of w) associated local thresholds on day t. It is a block average (Section 7.1) in this case of indicator functions. Evidently, we can choose B as we wish; we might consider $B = \mathcal{D}$. EHE is applicable to the extent when $w = 0$, probably of greatest interest, but we can also look at the extent of more ($w > 0$) or less ($w < 0$) extreme heat events.

As an aside, Cebrian et al. [2022] obtain the first and second moments under an illustrative Gaussian spatial first order autoregression model:

$$Y_t(\mathbf{s}) = \mu_t(\mathbf{s}) + \eta(\mathbf{s}) + \rho(Y_{t-1}(\mathbf{s}) - (\mu_{t-1}(\mathbf{s}) + \eta(\mathbf{s}))) + \epsilon_t(\mathbf{s}) \qquad (17.17)$$

where $\mu_t(\mathbf{s})$ is a spatio-temporal drift term. Here, $\eta(\mathbf{s})$ is a mean 0 Gaussian process with covariance $cov(\eta(\mathbf{s}), \eta(\mathbf{s}')) = \sigma^2 h(\mathbf{s} - \mathbf{s}')$ providing local spatial adjustment to the drift terms as well as spatial dependence across locations. The $\epsilon_t(\mathbf{s})$ are pure errors, independent and identically distributed as $N(0, \tau^2)$.

We note a caveat here. Formally, $Ext_t(w : B)$ applied to $Y_t(\mathbf{s})$ does not exist. That is, with the inclusion of pure error in 17.17, a realization of $Y_t(\mathbf{s})$ over D will be everywhere discontinuous; the proposed Reimann integral does not exist. However, the quantity which is the proportion of B that is experiencing extreme heat at least w degrees above or below (according to the sign of w) associated local thresholds on day t" is well defined and we use the integral notation symbolically.

The posterior predictive distribution for $Ext_t(w; B)$ is needed for inference. We use a discretization to obtain an approximate realization of it by computing:

$$\widetilde{Ext}_t(w; B) = \frac{1}{m} \sum_{j=1}^{m} \mathbf{1}(Y_t(\mathbf{s}_j) - q(\mathbf{s}_j) \geq w). \qquad (17.18)$$

Here, for a selected set of m locations in B with an associated set of thresholds, $\{q(\mathbf{s}_j), j = 1, 2, ..., m\}$, $\{Y_t(\mathbf{s}_j), j = 1, 2, ..., m\}$ is a posterior predictive realization of daily max temperatures for day t at the locations, $\mathbf{s}_j$. If we have a collection of these realizations, then we can obtain posterior samples of $\widetilde{Ext}_t(w; B)$ for any choices of w. In this way, with arbitrarily many posterior predictive realizations, we can learn arbitrarily well about the posterior predictive distribution for $Ext_t(w; B)$. We can obtain such posterior predictive realizations using composition sampling.

As a last remark, it is possible to create an empirical extent (analogous to the empirical SCDF above), employing the form in (17.18) but using only the available observed sites that are within B. With few monitoring stations, only a few will be in B, yielding only a point estimate that can assume only a few discrete values and with no uncertainty. In implementing the Monte Carlo integrations we can obtain an arbitrarily large m, yielding much smoother extents as well as replication, enabling us to see the distribution.

17.3 Spatial quantile regression and autoregression

Quantile regression (QR) has a rich history by now, dating to Koenker [1978], with much seminal work by Koenker and colleagues, [see, e.g., Koenker and Machado, 1999, Koenker, 2005, Koenker and Xiao, 2006]. Many facets are considered in the literature including choice of optimization function (equivalently error distribution), dependence through autoregression, and quantile crossing.

Quantile regression is useful in studying other features of the conditional distribution of an outcome variable given covariates. This contrasts with mean regression, in which the conditional mean of the outcome variable is the object of interest relative to the covariates. Quantile regression supplements mean regression by supplying information about the relationships between the covariates and the tails of the distribution. Quantile regression can also illustrate the differing effects of the covariates on the outcome according to various quantiles of the distribution.

There are two modeling approaches for QR in the literature. The first follows the original ideas by Koenker [1978] and offers a separate regression model for each of the quantiles of interest. This approach is usually called *multiple* QR, and inference typically proceeds by minimizing a check loss function or assuming an asymptotic Laplace (AL) error term. Examples of multiple QR with AL errors appear in Yu and Moyeed [2001] while Kozumi and Kobayashi [2011] present a Gibbs sampler for a Bayesian QR model. The second approach, which is usually called *joint* QR, specifies an appropriate joint model for all quantiles, see, e.g., [Tokdar and Kadane, 2012, Yang and Tokdar, 2017, Das and Ghosal, 2017b]. Broad implementation for joint QR has proven challenging.

17.3.1 Modeling spatial dependence

Spatial dependence is captured through spatially varying quantiles which are analogous to introducing spatially varying coefficients in spatial linear regression, and dependent quantile levels which are analogous to introducing dependence through the errors in the linear regression.

We can also classify the models in terms of whether they incorporate temporal, spatial, or spatio-temporal dependence. Koenker and Xiao [2006] established the basis for joint quantile autoregression (QAR) models in time series. A detailed overview of the different strands of time series QR modeling can be found in Peters [2018]. Recently, spatial quantiles have been an active area of research. Hallin et al. [2009] introduce spatial multiple QR that is nonparametric, focusing on asymptotic behavior using assumptions associated with time series asymptotics. Reich et al. [2011] develop a spatial joint QR model that incorporates spatial dependence through spatially varying regression coefficients, which are expressed as a weighted sum of Bernstein basis polynomials where the weights are constrained spatial GPs. Lum and Gelfand [2012] consider spatial multiple QR with AL errors and then extend it to capture spatial dependence by introducing the AL process. Yang and He [2015] consider a nonparametric approach based on Bayesian spatial QR using empirical likelihood as a working likelihood and spatial priors. Chen and Tokdar [2021] specify a spatial joint QR based on the so-called constraint-free reparametrization by generalizing the model of Yang and Tokdar [2017] and characterizing spatial dependence via a Gaussian or t copula process on the underlying quantile levels of the observation units. Spatio-temporal quantile models are the most challenging and little work has been done in that regard. For example, Reich [2012] follows Reich et al. [2011], but allows for residual correlation via a spatio-temporal copula model. Neelon et al. [2015] propose a multiple QR model for areal data. They model the random effects via intrinsic conditionally autoregressive priors, and they adopt the Bayesian approach based on the AL errors. Das and Ghosal [2017b] develop a joint QR model with a single explanatory variable following the representation of quantile functions given by Tokdar and Kadane [2012] and Das and Ghosal [2017a]. The explanatory variable is a linear trend over time and spatial dependence is captured by a B-spline basis expansion prior.

There has been recent work in the spatial non-quantile crossing setting, e.g., Chen and Tokdar [2021] present a spatial quantile regression implementation using covariates but with global regression coefficients. Castillo-Mateo et al. [2024] also take up the non-quantile crossing spatial quantile autoregression challenge in the auto-regressive case. We present some brief details. Let $Y_t(\mathbf{s})$ denote the observation for time $t = 1, \ldots, T$ at location $\mathbf{s} \in D$, where $D \subset \mathbb{R}^2$ is the study region. They have a time series at each of the locations, $\{\mathbf{s}_1, \ldots, \mathbf{s}_n\}$, say, the locations of monitoring stations. The joint spatial QAR model is given by

$$Y_t(\mathbf{s}) = \theta_0(U_t(\mathbf{s}); \mathbf{s}) + \theta_1(U_t(\mathbf{s}); \mathbf{s})Y_{t-1}(\mathbf{s}), \tag{17.19}$$

where the θ functions are quantile and spatially varying. Chen and Tokdar [2021] propose to model the spatial dependence of the realizations in their QR model using a spatial copula process. Generalizing this, in Castillo-Mateo et al. [2024] , the vectors $(U_t(\mathbf{s}_1), \ldots, U_t(\mathbf{s}_n))^{\mathrm{T}}$ follow an independent copula distribution for every t. We see that they introduce spatially varying coefficients rather than global coefficients. As a consequence, they have a dependence in the time series realizations as well as spatially varying quantile functions.

17.3.2   A quick review of basic quantile regression and the asymmetric Laplace distribution

In its most elementary form, quantile regression imagines a model for observation Y_i, such that $Y_i = \mathbf{X}_i^{\mathrm{T}}\boldsymbol{\beta} + \epsilon_i$ with the ϵ_i independent error variables. However, unlike usual mean regression where the ϵ_i have mean 0, here, if $P(\epsilon_i < 0) = \tau$, with $\tau \in (0, 1)$, then $P(Y_i < \mathbf{X}^{\mathrm{T}}\boldsymbol{\beta}) = \tau$ and $\mathbf{X}^{\mathrm{T}}\boldsymbol{\beta}$ is referred to as the τ quantile of the distribution of Y_i, denoted by say $q_{Y_i}(\tau|\mathbf{X}^{\mathrm{T}}\boldsymbol{\beta})$. As τ varies over $(0, 1)$, $q_{Y_i}(\tau|\mathbf{X}^{\mathrm{T}}\boldsymbol{\beta})$ is referred to as the quantile function associated with Y_i. In fact, it is the inverse of the cdf of Y_i in that $F_{Y_i}(q_{Y_i}(\tau|\mathbf{X}^{\mathrm{T}}\boldsymbol{\beta})) = \tau$. Given a sample, $\{(Y_i, \mathbf{X}_i), i = 1, 2, ..., n\}$, to estimate $\boldsymbol{\beta}$ for a given τ, customarily a minimization using a check loss function is adopted, i.e., we seek $argmin_{\boldsymbol{\beta}} \sum_{i=1}^n \rho_\tau(Y_i - \mathbf{X}^{\mathrm{T}}\boldsymbol{\beta})$ where $\rho_\tau(u) = u(\tau - 1(u < 0))$. A plot of $\rho_\tau(u)$ for choices of τ motivates the name check loss!

It then becomes natural to specify a distribution for the ϵ_i which captures quantiles in the form of the check loss function as an error distribution for conditional quantile regression (in the same way that squared error loss captures the error distribution for the mean). This distribution is the asymmetric Laplace distribution.

We introduce the AL distribution as an error distribution for multiple QR models using the following parametrization. We denote by $\epsilon \sim AL(\mu, \sigma, \tau)$ a random variable with probability density function (pdf),

$$f(\epsilon \mid \mu, \sigma, \tau) = \sigma\tau(1 - \tau) \begin{cases} \exp\{-(1 - \tau)\sigma|\epsilon - \mu|\}, & \text{if } \epsilon < \mu, \\ \exp\{-\tau\sigma|\epsilon - \mu|\}, & \text{if } \epsilon \geq \mu. \end{cases}$$

The cumulative distribution function is

$$F(\epsilon \mid \mu, \sigma, \tau) = \begin{cases} \tau\exp\{-(1 - \tau)\sigma|\epsilon - \mu|\}, & \text{if } \epsilon < \mu, \\ 1 - (1 - \tau)\exp\{-\tau\sigma|\epsilon - \mu|\}, & \text{if } \epsilon \geq \mu. \end{cases}$$

Here, μ is a location parameter, $\sigma > 0$ is a scale parameter, and $\tau \in (0, 1)$ is an asymmetry parameter. In particular, it is easily checked that μ is the τ quantile of the distribution and we will typically set $\mu = 0$ so that $P(\epsilon \leq 0) = \tau$.

To make the connection with the check loss function, the pdf above can be rewritten as $f(\epsilon \mid \mu, \sigma, \tau) = \sigma\tau(1 - \tau)\exp\{-\sigma\rho_\tau(\epsilon - \mu)\}$ where, as above, $\rho_\tau(u) = u(\tau - \mathbf{1}(u < 0))$ is the *check loss* function [Koenker, 1978]. Again, for a sample $\{Y_i : i = 1, \ldots, n\}$, finding $argmin_\mu \sum \delta_\tau(Y_i - \mu)$ returns the τ empirical quantile. Just as minimizing the sum of squares loss is associated with normal errors, minimizing check loss is associated with AL errors.

A convenient strategy for generating $\epsilon \sim AL(0, \sigma, \tau)$ variables is to use the following representation proven by comparing moment generating functions, see, e.g., We can express ϵ in terms of

$$\epsilon = \sqrt{\frac{2U}{\sigma^2\tau(1 - \tau)}}Z + \frac{1 - 2\tau}{\sigma\tau(1 - \tau)}U$$

where $Z \sim N(0, 1)$ and $U \sim Exp(1)$. So,

$$\epsilon \mid \sigma, U \sim N\left(\frac{1 - 2\tau}{\sigma\tau(1 - \tau)}U, \frac{2U}{\sigma^2\tau(1 - \tau)}\right) \tag{17.20}$$

is normally distributed enabling us to use all of the familiar Gaussian theory.

The asymmetric Laplace distribution as the error term achieves stochastic ordering for quantiles across τ. If we let $\epsilon_\tau \sim AL(\tau, 0, \sigma)$ and $\epsilon_{\tau'} \sim AL(\tau', 0, \tau)$ for $\tau < \tau'$, then it is straightforward to show that ϵ_τ is stochastically larger than $\epsilon_{\tau'}$. Hence, for fixed Y with two quantile models, $Y = \mu_\tau + \epsilon_\tau$ and $Y = \mu_{\tau'} + \epsilon_{\tau'}$, $\epsilon_\tau \preceq \epsilon_{\tau'}$. This implies that μ_τ is stochastically larger than $\mu_{\tau'}$.

17.3.3   The asymmetric Laplace process (ALP)

In order to incorporate spatial structure in the model, we introduce the asymmetric Laplace process. The mixture representation of the asymmetric Laplace distribution suggests a straightforward way to create a spatial process model for quantiles. By replacing the univariate standard normal Z with a mean zero, variance one Gaussian process (GP), $Z(\mathbf{s})$, we create a process model in which each location has an asymmetric Laplace marginal distribution, provided that $U(\mathbf{s})$ is independent of $Z(\mathbf{s})$ and is marginally exponentially distributed. Thus, our spatial quantile process model can be written as

$$\epsilon_\tau(\mathbf{s}) = \sqrt{\frac{2U(\mathbf{s})}{\sigma^2\tau(1 - \tau)}}Z(\mathbf{s}) + \frac{1 - 2\tau}{\sigma\tau(1 - \tau)}U(\mathbf{s})$$

$$Z(\mathbf{s}) \sim GP(\mathbf{0}, \rho_Z(\mathbf{s}, \mathbf{s}'; \boldsymbol{\theta})),$$

where $\rho_Z(\mathbf{s}, \mathbf{s}'; \boldsymbol{\theta})$ is a valid correlation function and we address a joint specification of $U(\mathbf{s})$ below.[2] Because $U(\mathbf{s})$ is marginally exponentially distributed with rate 1, because $Z(\mathbf{s})$ is marginally a standard normal, and because $U(\mathbf{s})$ is conditionally independent of $Z(\mathbf{s})$, at any given location, $\mathbf{s}_0$, $\epsilon_\tau(\mathbf{s}_0)|U(\mathbf{s}_0) \sim N\left(\frac{1-2\tau}{\tau(1-\tau)}U(\mathbf{s}_0), \frac{2U(\mathbf{s}_0)}{\sigma\tau(1-\tau)}\right)$, i.e., an $AL(\tau, 0, \sigma)$. So, the marginals of this process are distributed asymmetric Laplace. Lum and Gelfand [2012] consider three options for modeling the $U(\mathbf{s})$: (i) common $U(\mathbf{s})$, (ii) i.i.d $U(\mathbf{s}_i)$, and (iii) $U(\mathbf{s}_i)$ spatially structured. However, regardless of this specification, the GP model for $Z(\mathbf{s})$ ensures spatial dependence for the $\epsilon_\tau(\mathbf{s})$.

17.3.4 The spatial quantile regression model

A spatial quantile regression is generally given by $Y(\mathbf{s}) - \mu_\tau(\mathbf{s}) = \epsilon_\tau(\mathbf{s})$, where $\epsilon_\tau(\mathbf{s})$ must satisfy the constraint that $\Pr(\epsilon_\tau(\mathbf{s}) \leq 0) = \tau$. This equation reveals a different opportunity for including a spatial component in this model. Why not let $\mu_\tau(\mathbf{s}) = \mathbf{X}(\mathbf{s})^\mathrm{T}\boldsymbol{\beta}_\tau + w_\tau(\mathbf{s})$ with $w_\tau(\mathbf{s})$ a spatial Gaussian process? The problem is that then $\mathbf{X}^\mathrm{T}(\mathbf{s})\boldsymbol{\beta}_\tau + w_\tau(\mathbf{s})$ would correspond to the τth conditional quantile of $Y(\mathbf{s})$ rather than $\mathbf{X}^\mathrm{T}(\mathbf{s})\boldsymbol{\beta}_\tau$. In the case of spatial mean regression, we may interpret the $w_{mean}(\mathbf{s})$ as a local spatial adjustment to the mean. That is, $\mathrm{E}[Y(\mathbf{s})|w_{mean}(\mathbf{s})] = \mathbf{X}^\mathrm{T}(\mathbf{s})\boldsymbol{\beta}_{mean} + w_{mean}(\mathbf{s})$ will be a better estimator in the mean squared error sense than the marginal expectation, $\mathrm{E}[Y(\mathbf{s})] = \mathbf{X}^\mathrm{T}(\mathbf{s})\boldsymbol{\beta}_{mean}$. This argument does not extend to quantile processes or the asymmetric Laplace distribution as specified above.

Analogous to the spatial mean model, because of the way in which the spatial component is embedded within $\epsilon_\tau(\mathbf{s})$ in the ALP, we retain the interpretation of $\boldsymbol{\beta}_\tau$ as a global quantile regression coefficient. Suppose we enrich $Z(\mathbf{s})$ such that $Z(\mathbf{s}) = \sqrt{1-\alpha}w(\mathbf{s}) + \sqrt{\alpha}\delta(\mathbf{s})$ where $w(\mathbf{s})$ is a mean 0 variance 1 GP, $\delta(\mathbf{s})$ is a pure error process, and $\alpha \in [0,1]$. Then, $Z(\mathbf{s}) \sim N(0,1)$ and we can pull apart a spatial adjustment to the marginal quantile. That is, now a spatially adjusted conditional quantile using the ALP is $q_\tau(Y(\mathbf{s})|w(\mathbf{s})) = \mathbf{X}^\mathrm{T}(\mathbf{s})\boldsymbol{\beta}_\tau + w(\mathbf{s})\sqrt{\frac{\pi}{2\sigma^2\tau(1-\tau)}}$, which is anticipated to be better in terms of check loss than the marginal quantile $q_\tau(Y(\mathbf{s}))$. Notice that, for each τ, the $q_\tau(Y(\mathbf{s})|w(\mathbf{s}))$ surface will be smooth if $\mathbf{X}^\mathrm{T}(\mathbf{s})$ and $w(\mathbf{s})$ are. In summary, to specify a quantile model that can produce spatial adjustments for more accurate local quantile regressions, the spatial structure must be embedded within the error process to produce a quantile regression at the marginal level.

Hence, we arrive at a preferred quantile regression model:

$$Y(\mathbf{s}) = \mu_\tau(\mathbf{s}) + \epsilon_\tau(\mathbf{s}) \tag{17.21}$$

$$\epsilon_\tau(\mathbf{s}) = \sqrt{\frac{2U(\mathbf{s})}{\sigma^2\tau(1-\tau)}}Z(\mathbf{s}) + \frac{1-2\tau}{\sigma\tau(1-\tau)}U(\mathbf{s}) \tag{17.22}$$

$$Z(\mathbf{s}) = \sqrt{1-\alpha}w(\mathbf{s}) + \sqrt{\alpha}\delta(\mathbf{s}), \tag{17.23}$$

with all of the model components defined above. Evidently, the parameter α dictates what proportion of the variance of $Z(\mathbf{s})$ is due to the spatial component.

17.3.5 A spatial quantile autoregression example

With time series data across spatial locations, a spatial quantile autoregression analysis may be of interest. An example is presented in Castillo-Mateo et al. [2023b] which considers daily maximum temperature ($^\circ$C) data at $n = 18$ sites around the Comunidad Autónoma de Aragón provided by the Agencia Estatal de Meteorología (AEMET) in Spain. The data

---

[2]In the above, since we specify a model for each τ, we introduce a $Z_\tau(\mathbf{s})$ and $U_\tau(\mathbf{s})$ for each τ. In the sequel, we suppress the subscript τ's.

are available on a daily scale from 1956 to 2015, but the focus of the analyses is in the warm months of June, July, and August (denoted as JJA); in this regard, they fit the models with data in an extended period from May 1 to September 30 to avoid boundary issues.

Turning to the modeling, let $\tau \in (0,1)$ denote a quantile order, where each quantile is modeled separately. The general form for a spatio-temporal τ-QAR with two time scales, years (t) and days within year(ℓ) is given by

$$
\begin{aligned}
Y_{t\ell}(\mathbf{s}) &= Q_{Y_{t\ell}(\mathbf{s})}(\tau \mid Y_{t,\ell-1}(\mathbf{s})) + \epsilon_{t\ell}^{\tau}(\mathbf{s}) \\
&= q_{t\ell}^{\tau}(\mathbf{s}) + \rho^{\tau}(\mathbf{s})\left(Y_{t,\ell-1}(\mathbf{s}) - q_{t,\ell-1}^{\tau}(\mathbf{s})\right) + \epsilon_{t\ell}^{\tau}(\mathbf{s})
\end{aligned}
\tag{17.24}
$$

where $Q_{Y_{t\ell}(\mathbf{s})}(\tau \mid Y_{t,\ell-1}(\mathbf{s}))$ is the τ conditional quantile of $Y_{t\ell}(\mathbf{s})$ given $Y_{t,\ell-1}(\mathbf{s})$ and the error term is $\epsilon_{t\ell}^{\tau}(\mathbf{s}) \sim$ ind. $AL(0, \sigma^{\tau}(\mathbf{s}), \tau)$. Here, $q_{t\ell}^{\tau}(\mathbf{s})$ contains fixed and random effects as below. In addition, $\rho^{\tau}(\mathbf{s})$ is a spatially varying autoregression coefficient and $\sigma^{\tau}(\mathbf{s})$ is a spatially varying pure error scale parameter at location $\mathbf{s}$.

Castillo-Mateo et al. [2023a] adopt an analogue of the spatio-temporal mean autoregression model in Castillo-Mateo et al. [2022]. Here, $Y_{t\ell}(\mathbf{s})$ denotes the daily maximum temperature for day ℓ, $\ell = 2, \ldots, L$ of year t, $t = 1, \ldots, T$ at location $\mathbf{s}$, $\mathbf{s} \in \mathcal{D}$, the study region. They specify $\rho^{\tau}(\mathbf{s})$ to capture spatial autoregession dependence through the GP $Z_{\rho}^{\tau}(\mathbf{s}) = \log\{(1 + \rho^{\tau}(\mathbf{s}))/(1 - \rho^{\tau}(\mathbf{s}))\}$ with mean Z_{ρ}^{τ} and exponential covariance function having variance parameter $\sigma_{\rho}^{2,\tau}$ and decay parameter ϕ_{ρ}^{τ}. In the same manner, they specify $\sigma^{\tau}(\mathbf{s})$ to capture spatial scale dependence through the GP $Z_{\sigma}^{\tau}(\mathbf{s}) = \log\{\sigma^{\tau}(\mathbf{s})\}$ with mean Z_{σ}^{τ} and exponential covariance function having variance parameter $\sigma_{\sigma}^{2,\tau}$ and decay parameter ϕ_{σ}^{τ}.

As for $q_{t\ell}^{\tau}(\mathbf{s})$, they adopt

$$
q_{t\ell}^{\tau}(\mathbf{s}) = \beta_0^{\tau} + \alpha^{\tau} t + \beta_1^{\tau} \sin(2\pi\ell/365) + \beta_2^{\tau} \cos(2\pi\ell/365) + \beta_3^{\tau} elev(\mathbf{s}) + \gamma_t^{\tau}(\mathbf{s}),
$$

where $\gamma_t^{\tau}(\mathbf{s}) = \beta_0^{\tau}(\mathbf{s}) + \alpha^{\tau}(\mathbf{s})t + \psi_t^{\tau} + \eta_t^{\tau}(\mathbf{s})$. The *fixed effects* are given by β_0^{τ}, a global intercept, $\alpha^{\tau} t$, a global long-term linear trend, sin and cos terms that provide the annual seasonal component, and $elev(\mathbf{s})$, the elevation at $\mathbf{s}$. The *random effects* given by $\gamma_t^{\tau}(\mathbf{s})$ capture space-time dependence through GPs. In particular, $\beta_0^{\tau}(\mathbf{s})$ is a GP with zero mean and exponential covariance function having variance parameter $\sigma_{\beta_0}^{2,\tau}$ and decay parameter $\phi_{\beta_0}^{\tau}$, and it provides a local spatial adjustment to the intercept. The $\alpha^{\tau}(\mathbf{s})$ are a GP, with zero mean and exponential covariance function having variance parameter $\sigma_{\alpha}^{2,\tau}$ and decay parameter ϕ_{α}^{τ}, to provide a local slope adjustment to the linear trend. Together, $\gamma_t(\mathbf{s})$ supplies a *locally linear* trend, an exceptionally rich spatial specification. With the inclusion of seasonality, it is difficult to imagine that the data could inform about a higher order local choice. Continuing, $\psi_t^{\tau} \sim$ i.i.d. $N(0, \sigma_{\psi}^{2,\tau})$ provides annual intercepts to allow for yearly shifts (i.e., for hotter or colder years).

17.4   Spatial functional regression

Functional data analysis is by now a well-established field with an enormous literature, dating to the seminal work of Ramsay and Silverman [Ramsay and Silverman, 2005]. Here, our interest is in spatial functional regression. The functional regression literature is by now also substantial. See, e.g., the excellent review paper of Morris [2015] still thorough though now nearly a decade old. The objective can be either (i) *functional predictor regression*, to employ a function on the right side of a regression model as a predictor for say a scalar or multivariate response vector, or (ii) *functional response regression*, to place the function on the left side of this regression as a response and attempt to explain an entire function using scalar predictors. Under (i), the predictor will be the function. It will be observed at a finite set of arguments. However, conceptually the entire function is unknown but driving the regression across a continuous domain for the argument. Under (ii), the observed response is a multivariate vector at a set of arguments for the function. However, we seek to learn the behavior of the entire function, again over a continuous domain, as a response to predictors. It is worth noting that any dataset having a scalar variable as well as a functional variable can be placed in either setting (i) or (ii) according to which variable is viewed as a predictor and which is viewed as a response. Function on function regression, i.e., a functional predictor with a functional response, has received little attention, perhaps because of the paucity of suitable applications. Typical examples of a functional variable in the literature consider independent observations of individuals, with the function being, say, growth curves, proteomic spectra, progesterone levels, or diffusion fission imaging scans. With spectrometry data, the associated scalar variable might be an indicator of cancer or not. With imaging scans, the associated scalar variable might be a measure of disease progression.

In the nonspatial setting, we imagine independent replicates for fitting the regression, $i = 1, 2, ...n$, providing the scalar values $\{Y_i\}$ and the function, $X_i(u)$ observed at a finite set of values for the function, u_j, $j = 1, 2, ...p$. In this simplest version, the functional predictor setting envisions a regression of the form $Y_i = \beta_0 + \int_{\mathcal{U}} \beta(u) X_i(u) du + \epsilon_i$. In the functional response setting, the regression is of the form $X_i(u) = \beta_0 + \beta(u) X_i + \epsilon_i$. In either case, the coefficient function, $\beta(u)$, is of interest. We elaborate all of this below.

The spatial functional regression literature imagines functions observed at spatially referenced locations and seeks to model spatial dependence between the functions rather than assuming independent replicates as above. In the predictor setting, such modeling would seek to capture spatial dependence between the predictor functions over the spatial domain. This literature is sparse. If the function is a response, then we are seeking spatial dependence between the response functions. There is some literature here; see, e.g., Bouka et al. [2018], Rimalova et al. [2022], Kang et al. [2023]. We also note the contributions on geostatistics, point processes, and areal data with functional observations, as well as examples incorporating spatial dependence, presented in Delicado et al. [2010]. At present, geostatistics for functional data is the most developed.

We offer examples of flexible spatial functional response models below. Evidently, this setting extends multivariate spatial response to a continuous response curve. In fact, in practice, gridding the functional response to a high dimensional response vector takes us back to the multivariate setting, albeit in a very high dimension. It provides an example which is handled through dimension reduction.

Environmental functional data examples in the literature include temperature vs. depth profiles say for data collected in bodies of water or for plant reflectance spectra associated with a band of wavelengths (the profile of light reflected by leaves across different wavelengths). Both of these illustrations are naturally spatial, i.e., the temperature depth profiles will be collected at recorded locations in a body of water, and the plant reflectances will be collected at sampled locations in a study area. Further, the reflectance setting provides an opportunity to consider both (i) and (ii) above. For (i), we can use the reflectance spectra to explain traits observed for the leaves. Examples of some continuous traits include leaf water content (LWC), leaf mass per area (LMA), percent Nitrogen (pN), and leaf succulence (LS). For (ii), we can use environmental features to explain the reflectance spectra. We illustrate this case below.

Leaf reflectance is viewed as a function over wavelength $w \in [450, 950]$ nanometers, observed as a 500 dimensional reflectance vector, by the nanometer. In fact, to further elaborate the inference we might attempt to jointly model the continuous trait levels with the functional reflectance over wavelength and seek to learn about the dependence between trait levels and reflectances. This is undertaken in White et al. [2023] using a South Africa dataset to jointly explain plant traits and reflectance, illustrative environmental covariates include (i) elevation, (ii) annual precipitation, (iii) rainfall concentration, and (iv) minimum average temperature in January, the peak of the austral summer.

We first remind the reader of functional predictor regression in the nonspatial case. This case assumes the predictor functions are independent across say, individuals, $i = 1, 2, \ldots, n$. Functional data differs from time series data in the sense that the function need not be a function of time but, in any event, that its support is a continuous domain. The above examples reveal this.

Consider the model

$$Y_i = \mathbf{Z}_i^{\mathrm{T}} \boldsymbol{\beta} + \int_{\mathcal{X}} \beta(x) w_i(x) dx + \epsilon_i. \tag{17.25}$$

Here, $\mathbf{Z}_i$ is a vector of scalar covariates, $w_i(x)$ is a functional predictor associated with, e.g., individual i, $\mathcal{X}$ is a suitable interval of support for x, and the ϵ_i are pure errors. Note that we have replaced a usual sum with an integral in (17.25). As a result, we have a coefficient function, $\beta(x)$. Evidently, such a model will fit the data with no errors since we have more parameters than observations. So, some form of *regularization* is required in order to make the model useful. A customary choice is through basis functions, e.g., splines, wavelets, etc., here taking the form $\beta(x) = \sum_{l=1}^{L} \alpha_l h_l(x) \equiv \boldsymbol{\alpha}^{\mathrm{T}} \mathbf{h}(x)$ where the $h_l(x)$ are a set of basis functions over $x \in \mathcal{X}$. These basis function representations for functions are often referred to as *nonparametric* but, in fact, in practice, they are parametric with the $\{\alpha_l\}$ providing the L parameters which define the function. The function becomes random if the set $\{\alpha_l\}$ is assumed random.

Continuing, inserting the basis function representation for $\beta(x)$ into (17.25), we obtain $\int_{\mathcal{X}} \sum_{l=1}^{L} \alpha_l h_l(x) w_i(x) dx = \sum_{l=1}^{L} \alpha_l \int_{\mathcal{X}} h_l(x) w_i(x) dx$. Implementing a numerical integration for $\tilde{w}_{il} \equiv \int_{\mathcal{X}} h_l(x) w_i(x) dx$ since the h's are known and the w_i's are observed over $\mathcal{X}$, (17.25) becomes $Y_i = \mathbf{Z}_i^{\mathrm{T}} \boldsymbol{\beta} + \sum_{l=1}^{L} \alpha_l \tilde{w}_{il} + \epsilon_l$. This is now a finite dimensional linear regression model which can be fitted within a standard likelihood framework or a Bayesian framework. In particular, the estimate of the $\boldsymbol{\alpha}$ vector enables estimation of the coefficient function at any x, i.e., the coefficient of $w_i(x)$ as a predictor in the functional regression. Further, we can display the coefficient function vs. x.

An alternative specification for $\beta(x)$ is as a realization of a stochastic process over D, e.g., a Gaussian process. Convenient choices include kernel convolution or a predictive process (see Chapter 13), providing reduction to a finite dimensional predictor. For instance, the kernel convolution version would specify $\beta(x) = \sum_{l=1}^{L} K(x - x_l^*; \boldsymbol{\theta}) \eta(x_l^*)$ where K is a kernel function with parameters $\boldsymbol{\theta}$, the x_l^* are a grid over $\mathcal{X}$ and $\eta(x)$ is a realization of a mean 0 Gaussian process on R^1. Again, plugging in $\beta(x)$ into $\int_{\mathcal{X}} \beta(x) w_i(x) dx$ and exchanging sum and integration, we obtain $\sum_{l=1}^{L} \eta(x_l^*) \int_{\mathcal{X}} K(x - x_l^*; \boldsymbol{\theta}) w_i(x) dx$. Again, with a numerical integration of the integral, we obtain $\sum_{l=1}^{L} \eta(x_l^*) \tilde{w}_{il}$ as the functional predictor. Given the partial realization, $\{\eta(x_l^*)\}$, we can obtain the coefficient $\beta(x)$ for any $x \in \mathcal{X}$.

Evidently, with either specification above, we can also offer a generalized functional linear model. That is, we introduce a suitable link function, e.g., logit, probit, or log such that $\mathrm{E}(Y_i) = g(\mathbf{Z}_i^{\mathrm{T}} \boldsymbol{\beta} + \int_{\mathcal{X}} \beta(x) w_i(x) dx)$ and we drop the ϵ_i's since they are redundant with the stochastic mechanism providing the Y_i's, e.g., binomial or Poisson.

How can we adapt this functional predictor setting to spatial modeling? Now, we have a response $Y(\mathbf{s})$ for $\mathbf{s} \in D$. We would propose the geostatistical model

$$Y(\mathbf{s}) = \mathbf{X}(\mathbf{s})^{\mathrm{T}} \boldsymbol{\beta} + \int_{\mathcal{X}} \beta(x) w(x, \mathbf{s}) dx + \eta(\mathbf{s}) + \epsilon(\mathbf{s}). \tag{17.26}$$

This is our familiar geostatistical model with the only novelty being the integral term as a spatial predictor. Using say the basis representation for $\beta(x)$ above, the integral is replaced with $\sum_l \alpha_l \tilde{w}_l(\mathbf{s})$ where $\tilde{w}_l(\mathbf{s}) = \int_{\mathcal{X}} h_l(x) w(x, \mathbf{s}) dx$. Again, we can fit in a classical way or use our preferred Bayesian approach as developed in Chapter 6. Again, the estimate of the $\boldsymbol{\alpha}$ vector enables the estimation of the coefficient function at any x, i.e., the coefficient of $w(x, \mathbf{s})$) as a predictor in the functional regression. Analogous versions using kernel convolution or predictive processes can be readily written down; details are omitted.

Next, we turn to spatial functional response modeling. Now, we imagine a response function $Y_i(x)$ which, for each i, is observed at a finite set of x's, not necessarily the same set for each i. To obtain a functional form to explain the functional response, we would write

$$Y_i(x) = \mu_i(x) + \epsilon_i(x) \tag{17.27}$$

If we have $\mu_i(x)$ as a mean given scalar regressors, we could write $\mu_i(x) = \sum_{p=1}^{P} \beta_p(x) Z_{ip}$. We would view $\beta_p(x)$ as the *partial* effect of predictor Z_p on Y at level x. Using a basis function representation for each $\beta_p(x)$, we obtain $\mu_i(x) = \sum_{p=1}^{P} Z_{ip} \sum_{l=1}^{L} \alpha_{pl} h_l(x) = \sum_{l=1}^{L} \alpha_{pl} \sum_{p=1}^{P} h_l(x) Z_{ip}$. We have a linear regression with each vector $\boldsymbol{\alpha}_p$ an $L \times 1$ vector assumed i.i.d. normal. Typically, the $\epsilon_i(x)$ are assumed to be Gaussian process realizations over $\mathcal{X}$ with a suitable covariance function on R^1. Thus, if $\mu_i(x)$ is continuous and the realization, $\epsilon_i(x)$ is continuous, the modeled function will be continuous.

With interest in estimating the response functions in the absence of covariates, we could propose $\mu_i(x) = \mu_i + \eta_i(x)$ with μ_i being an individual level intercept and the $\epsilon_i(x)$ supplying curve to curve residual error deviations. Then, it is attractive to view the $\eta_i(x)$ as curve level random effect functions, again using a basis function representation, i.e., $\eta_i(x) = \sum_{l=1}^{L} \alpha_{il} h_l(x)$. Here, each $\boldsymbol{\alpha}_i$ vector is $L \times 1$ and is an independent realization from an L dimensional mean 0 multivariate normal with either a specified covariance matrix, Σ or given as a simple parametric form. We might prefer to use say a kernel representation for the $\mu_i(x)$, i.e., $\mu_i(x) = \sum_{l=1}^{L} K(x - x_l^*; \boldsymbol{\theta}) \gamma_{il}$ where K is a suitable kernel function with parameter $\boldsymbol{\theta}$ and the γ_{il} are i.i.d. normal variables.

Turning to the spatial version for the response function, we would now imagine a "geostatistical" model of the form

$$Y(x, \mathbf{s}) = \mu(x, \mathbf{s}) + \eta(x, \mathbf{s}) + \epsilon(x, \mathbf{s}). \tag{17.28}$$

With spatially referenced covariates, e.g., a $p \times 1$ vector $\mathbf{Z}(\mathbf{s})$, we could write $\mu(x, \mathbf{s}) = \mathbf{Z}^{\mathrm{T}}(\mathbf{s}) \boldsymbol{\beta}(x)$. We could model the $\boldsymbol{\beta}(x)$ vectors as above. For the pure error term, we could model the $\epsilon(x, \mathbf{s})$ as $N(0, \sigma^2(x))$. Here, $\sigma^2(x)$ could be a constant or perhaps a function of the form $g(x)\sigma^2$ for a suitable choice of g.

Interesting in this regard is a proposed model for the process $\eta(x, \mathbf{s})$ over $x \in \mathcal{X}$ and $\mathbf{s} \in D$. We want smooth curves at each location and we want $\eta(x, \mathbf{s})$ close to $\eta(x, \mathbf{s}')$ when $||\mathbf{s} - \mathbf{s}'||$ is small. An easy to use process representation is through a generalization of "kernel convolution," i.e., $\eta(x, \mathbf{s}) = \sum_{l=1}^{L} K(x - x_l^*; \boldsymbol{\theta}) \omega_l(\mathbf{s})$. As above, K is a suitable kernel function, e.g., Laplace or Gaussian, and the $\omega_l(\mathbf{s})$ are independent Gaussian processes each over D. All sorts of analytical work regarding dependence structure, mean square difference between curves at a level x, and integrated mean square difference over $\mathcal{X}$, can be developed. Details are omitted.

Finally, we comment that we have bypassed the large Bayesian nonparametric functional data analysis literature here. A typical version provides a hierarchical model that allows simultaneous estimation of multiple curves nonparametrically using dependent Dirichlet Process mixtures of Gaussians to characterize the joint distribution of predictors and outcomes. Function estimates are then induced through the conditional distribution of the outcome given the predictors. See, e.g., Rodriguez et al. [2009] and references therein. However, the spatial modeling proposed above is expected to provide rich enough specifications for most practical applications. In fact, we illustrate such an application using leaf reflectance as a response function in the next subsection.

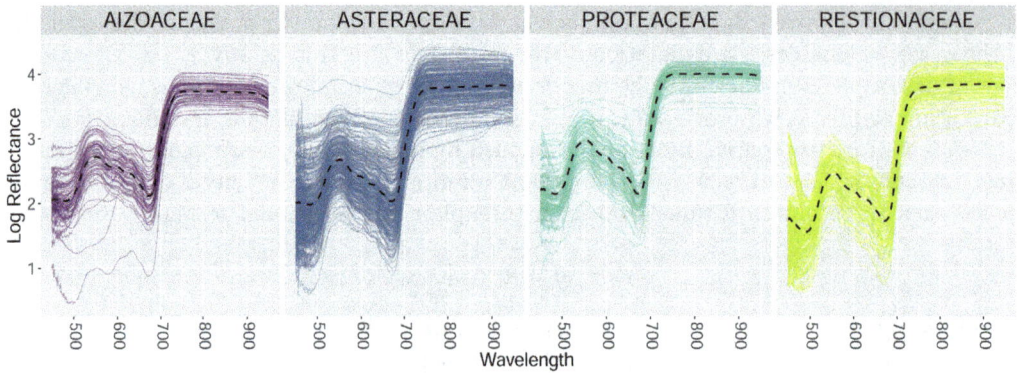

Figure 17.6 *Some leaf reflectance curves.*

17.4.1 A functional response example

The reflectance of the surface of a material is its effectiveness in reflecting radiant energy. It is the fraction of incident electromagnetic power that is reflected at the boundary. Reflectance is a component of the response of the electronic structure of the material to the electromagnetic field of light, and, in general, is a function of the frequency, or wavelength, of the light. The dependence of reflectance on the wavelength is called a reflectance spectrum or spectral reflectance curve. We consider leaf spectra as a "response" to be explained spatially, by genus within a family. They are functions over a wavelength band, as above, and can be viewed as an "uber" trait which is expected to respond to the environment.

We follow the work in White et al. [2022] whose contribution is the development of extremely rich spatial models for explaining leaf spectrum reflectance data using a South African dataset. Reflectance spectra are examined for genera within families. The set of wavelength responses for an individual leaf is viewed as a function of wavelength so reflectance is explained using functional data modeling. Local spatial covariates/environmental features are employed as regressors.

Formal spatial modeling enables the prediction of leaf spectra for genera at unobserved locations with known environmental features. Spatial dependence as well as wavelength dependence are expressed, both through random effects. Further, space-wavelength interaction (in the spirit of space-time interaction) is introduced along with heterogeneity of variance in response across wavelength.

The leaf reflectance data were collected from the Cape Floristic Region (CFR) in South Africa. Four families that are prevalent in this area are Aizoaceae, Asteraceae, Proteaceae, and Restionaceae. These families have broad overlaps in their spectral reflectances (Figure 17.6). However, a linear discriminant analysis to predict these families based on their reflectance spectra yields a clear separation of the groups indicating that their spectral reflectances can be used to effectively predict taxonomic differences. Also, in Figure 17.6, some reflectance curves are plotted at a site where all three families are observed to show the variability in reflectance.

Through EDA, the authors are led to four modeling considerations employing functional data analysis: (i) the need to account for family and genus differences, (ii) the need for a heterogeneous model for the reflectance spectrum because the within-curve variability changes as a function of wavelength (iii) between-curve variability attributable to geographic differences that can be captured through spatial modeling and/or environmental variables, and (iv) reflectances at lower wavelengths (< 500 nm) appear to be more volatile, suggesting a heteroscedastic error model. The families are modeled separately at the genus level, treating species within the genus as replicates.

For a given family, let i denote genera within the particular family, and let j denote replicates within the genus. Again, these replicates are associated with different species. Let $\mathbf{s}$ denote spatial location and t denote wavelength. There is a severe imbalance in the data. The genera observed varies across locations and the number of replicates/species observed within a genus varies considerably across the locations. Altogether, the most general model for *log* reflectance takes the form:

$$Y_{ij}(\mathbf{s},t) = \mu_i(\mathbf{s},t) + \gamma_i(t) + \alpha_i(\mathbf{s}) + \eta(\mathbf{s},t) + \epsilon_{ij}(\mathbf{s},t) \qquad (17.29)$$

Specifically, with regard to the site level covariates, $\mathbf{X}(\mathbf{s})$, the authors write $\mu_i(\mathbf{s},t) = \alpha_i + \mathbf{X}^{\mathrm{T}}(\mathbf{s})\boldsymbol{\beta}(t)$. That is, they introduce a genus level random effect but family level coefficients, allowing these coefficients to vary with wavelength. So, a first model choice clarification is whether constant coefficients are adequate or whether wavelength varying coefficients are needed. Two further model choice comparisons are whether the γ's and whether the α's should be genus specific. Additionally, to allow the functional reflectance spectrum model to vary more adaptively over space, the $\eta(\mathbf{s},t)$ term is needed. Finally, the heterogeneity in variance is captured through the $\epsilon_{ij}(\mathbf{s},t)$ terms, setting $\mathrm{var}(\epsilon_{ij}(\mathbf{s},t)) = \sigma^2(t)$.

The specification for each $\alpha_i(\mathbf{s})$ is a mean 0 genus level Gaussian process with exponential covariance function. The GPs are conditionally independent across genera given a shared decay and scale parameter. The $\gamma_i(t)$ are specified using hierarchical process convolutions of normal random variables, adopting a set of wavelength knots $t_1^\gamma, ..., t_{J_\gamma}^\gamma$, spaced every 25 nm from 437.5-962.5 nm (22 in total). For model components $\alpha_i(\mathbf{s})$ and $\gamma_i(t)$, when the subscript i is dropped, all genera share, respectively, a common spatial or wavelength random effect.

The authors let $\gamma_i(t) = \sum_{j=1}^{J_\gamma} K_{t_j^\gamma}(t - t_j^\gamma; \theta_{t_j^\gamma}^{(\gamma)})\gamma_i^*(t_j^\gamma)$ where $\gamma_i^*(t_j)$ are independent, normally distributed, and centered on a common $\gamma^*(t_j)$. They use Gaussian kernels for $K_{t_j}(\cdot; \theta_{t_j}^{(\gamma)})$ with bandwidths $\theta_{t_j}^{(\gamma)}$ (standard deviation of the Gaussian pdf) varying over wavelength. They assume that the log-bandwidths follow a multivariate normal distribution with global log-bandwidth and $\mathrm{Cov}\left(\theta_{t_j}^{(\gamma)}, \theta_{t_{j'}}^{(\gamma)}\right) = v_\gamma^2 \exp\left(-|t_j - t_{j'}|/\phi_\gamma\right)$, yielding a nonstationary process because of the heterogeneous bandwidth. For model components $\alpha_i(\mathbf{s})$ and $\gamma_i(t)$, when the subscript i is dropped, all genera share, respectively, a common spatial or wavelength random effect.

Turning to $\eta(\mathbf{s},t)$, given the relatively simple shape of the log-reflectance spectra and the evident wavelength heterogeneity, as well as possible nonstationarity and nonseparability, They use wavelength kernel convolutions of spatially-varying variables. That is, They consider low-rank but heterogeneous and nonstationary (in the wavelength domain) specifications. Specifically, upon selecting a set of wavelength knots $t_1^\eta, ..., t_{J_\eta}^\eta$, spaced every 25 nm from 437.5-962.5 nm (22, in total), they define

$$\eta(\mathbf{s},t) = \mathbf{K}(t)^{\mathrm{T}}\mathbf{z}(\mathbf{s}) = \sum_{j=1}^{J_\eta} K_{t_j^\eta}(t - t_j^\eta; \boldsymbol{\theta}^{(\eta)})\omega_{t_j^\eta}(\mathbf{s}), \qquad (17.30)$$

where $\omega_{t_j^\eta}(\mathbf{s})$ are spatially-varying random variables associated with Gaussian wavelength kernels $K_{t_j^\eta}(\cdot; \boldsymbol{\theta}_{t_j^\eta})$. Dimension reduction for the $\boldsymbol{\omega}$ vector employs the linear model of coregionalization (see Chapter 10).

Model selection using mean square error, mean absolute error, and continuous rank probability scores was implemented. The model with (1) a global wavelength random effect, (2) a spatially-varying genus-specific intercept, (3) functional regression coefficients, and (4) a space-wavelength random effect specified through the wavelength kernel convolution of a multivariate spatial process constructed by 10 latent spatial GPs with different decay parameters was adopted for the ensuing inference.

17.5   Bayesian modeling of spatially dependent finite populations

Finite population survey sampling concerns statistical modeling and inference on finite populations from sampling designs [see, e.g., classic texts such as Kish, 1965, Cochran, 1977]. Bayesian inference for finite population survey sampling has been discussed from diverse perspectives, for example, in Ericson [1969], Rao and Ghangurde [1972], Arora et al. [1997], Ghosh and Meeden [1997], Little and Rubin [2002] Gelman [2007], Little [2004] and Gelman et al. [2013], while an excellent appraisal of classical and Bayesian approaches is offered by Rao [2011].

Let $\mathcal{N} = \{1, 2, \ldots, N\}$ denote a finite set of integers labeling the units in a finite population. Given a sample $\{i_1, i_2, \ldots, i_n\}$ of size n, which is a subset of $\mathcal{N}$, units in the population and sample are described by $\{(Y_t, Z_t) : t \in \mathcal{N}\}$, where Y_t denotes the value of population unit labeled t and Z_t is a binary variable indicating if unit t is in the sample or not. The binary vector $\mathbf{Z} = (Z_1, Z_2, \ldots, Z_N)^{\mathrm{T}}$ completely encodes the sampled units in the population and, analogously, $\mathbf{Y} = (Y_1, Y_2, \ldots, Y_N)^{\mathrm{T}}$ denotes the values of the all units in the finite population. The sampling design determines the joint distribution over $\mathbf{Y}$ and $\mathbf{Z}$ by specifying $p(\mathbf{Z} \mid \mathbf{Y})$. Bayesian inference proceeds from $p(\mathbf{Y} \mid \mathcal{D}, \mathbf{Z}) \propto p(\mathbf{Y})p(\mathbf{Z} \mid \mathbf{Y})p(\mathcal{D} \mid \mathbf{Y}, \mathbf{Z})$, where $\mathcal{D}$ is the sampled data. For simplicity (which follows from "exchangeable" distributions), we denote $\mathcal{D}$ by $\mathbf{y} = (y_1, y_2, \ldots, y_n)^{\mathrm{T}}$ as the values of the n sampled units and $\mathbf{Y}_u = (Y_{n+1}, \ldots, Y_N)^{\mathrm{T}}$ as the unknown values of units not sampled. We will primarily focus upon settings where the units in the population are spatially dependent, but first consider the following simple example of a Bayesian model for finite populations.

17.5.1   Simple random sampling without replacement

In simple random sampling without replacement (SRSWOR) from a population of N units, $p(\mathbf{Z} \mid \mathbf{Y}) = 1/\binom{N}{n}$ whenever $\sum_{i=1}^{N} Z_i = n$ and 0 otherwise. Since this distribution does not depend upon $\mathbf{Y}$, we can write $p(\mathbf{Y}, \mathbf{Z}) = p(\mathbf{Y})p(\mathbf{Z})$ and we seek the posterior distribution $p(\mathbf{Y} \mid \mathcal{D}, \mathbf{Z}) \propto p(\mathbf{Y})p(\mathcal{D} \mid \mathbf{Y}, \mathbf{Z})$. Inference on the population is based solely on the prior distribution for $\mathbf{Y}$, which induces the distribution $p(\mathcal{D} \mid \mathbf{Y}, \mathbf{Z})$. If the probability model assumes that the units are exchangeable, then we can reorder the labels as necessary and denote $\mathcal{D}$ by $\mathbf{y} = (y_1, y_2, \ldots, y_n)^{\mathrm{T}}$ as the values of the n sampled units and $\mathbf{Y}_u = (Y_{n+1}, \ldots, Y_N)^{\mathrm{T}}$ as the unknown values of units not sampled.

Example 17.2 (Inference for finite populations) Consider the simple hierarchical model for population units:

$$Y_i \overset{iid}{\sim} N(\mu, \sigma^2), \; i = 1, 2, \ldots, N; \quad \mu \sim N(\theta, \sigma^2/n_0); \quad \sigma^2 \sim IG(a_0, b_0) , \tag{17.31}$$

where θ, n_0, a_0 and b_0 are known scalars. The joint posterior distribution $p(\mu, \sigma^2 \mid y)$ factorizes as in (5.12) Conditional on σ^2, it is easy to derive the joint posterior distribution $p(\mathbf{Y}_u, \mu \mid \sigma^2, \mathbf{y}) = N(\mu \mid Mm, \sigma^2 M) \times N(\mathbf{Y}_u \mid \mathbf{1}_{N-n}\mu, \sigma^2 I_{N-n})$, where $m = n_0\theta + n\bar{y}$ and $M = 1/(n + n_0)$. The marginal posterior distribution $p(\mathbf{Y}_u \mid \sigma^2, \mathbf{y})$ is easily obtained (without requiring explicit integration) by substituting $\mu = Mm + \omega$, where $\omega \sim N(0, \sigma^2 M)$, as $N(\mathbf{Y}_u \mid \mathbf{1}_{N-n}(n_0\theta + n\bar{y})/(n + n_0), \sigma^2(\mathbf{1}_{N-n}\mathbf{1}_{N-n}^{\mathrm{T}}/(n + n_0) + I_{N-n}/n))$.

The posterior distribution for any linear function of the population units, say $p(\mathbf{a}^{\mathrm{T}}\mathbf{Y} \mid \sigma^2, \mathbf{y})$, where $\mathbf{a} = (a_1, \ldots, a_N)^{\mathrm{T}}$, is again Normal with mean

$$\sum_{i=1}^{n} a_i y_i + \sum_{i=n+1}^{N} a_i \mathbb{E}[Y_{u,i} \mid \sigma^2, y] = \sum_{i=1}^{n} \left(a_i + \frac{\sum_{i=n+1}^{N} a_i}{n_0 + n} \right) y_i + \frac{n_0 \sum_{i=n+1}^{N} a_i}{n_0 + n}\theta . \tag{17.32}$$

The posterior covariance matrix of $\mathbf{Y}_u$ is the sum of $\mathbb{E}[\mathbb{V}(\mathbf{Y}_u \mid \sigma^2, \mu, \mathbf{y})]$ and $\mathbb{V}(\mathbb{E}[\mathbf{Y}_u \mid \sigma^2, \mu, \mathbf{y}])$, which yields $\sigma^2(\mathbf{1}_{N-n}\mathbf{1}_{N-n}^{\mathrm{T}}/(n+n_0) + I_{N-n})$. Therefore,

$$\mathbb{V}\left(\mathbf{a}^{\mathrm{T}}\mathbf{Y}_u \mid \sigma^2, \mathbf{y}\right) = \sigma^2 \left(\frac{\left(\sum_{i=n+1}^{N} a_i\right)^2}{n+n_0} + \sum_{i=n+1}^{N} a_i^2 \right). \tag{17.33}$$

For the population total, $a_i = 1$ for each $i = 1, 2, \ldots, N$ so the mean and variance of $p(\sum_{i=1}^{N} Y_i \mid \sigma^2, \mathbf{y})$ is given by $\sum_{i=1}^{n}(n_0 + N)y_i/(n_0 + n)$ and the variance is $\sigma^2\left((N-n)(n_0+N)/(n_0+n)\right)$. Letting $f = n/N$ and $f_0 = n_0/N$, the posterior distribution of the population mean, $p(\bar{Y} \mid \sigma^2, \mathbf{y})$, can be expressed as

$$\left[\frac{\bar{Y} - \left\{ f\bar{y} + (1-f)\frac{n_0\theta + n\bar{y}}{n_0+n} \right\}}{\sigma\sqrt{\frac{(1-f)(1+f_0)}{n+n_0}}} \,\middle|\, \sigma^2, \mathbf{y} \right] \sim N(0,1). \tag{17.34}$$

If the prior distribution on μ becomes vague, i.e., $n_0 \to 0$, then the posterior mean and variance in (17.34) are $\bar{y}$ and $(1-f)\sigma^2/n$, respectively. Further, as $n_0 \to 0$ and $a_0 \to -1/2$ (corresponds to a uniform prior over the real line for $(\mu, \log \sigma^2)$), the resulting posterior distribution $p(\sigma^2 \mid \mathbf{y}) \to IG((n-1)/2, (n-1)s^2/2)$ and the marginal posterior distribution for the population mean is given by $\left[\frac{\bar{Y} - \bar{y}}{s\sqrt{(1-f)/n}} \,\middle|\, \mathbf{y} \right] \sim T_{n-1}$, where T_{n-1} denotes the Student's t distribution with $n-1$ degrees of freedom. This, numerically, reproduces the inference from the classical theory of design-based estimators [Cochran, 1977] while retaining the appealing interpretation of direct probability statements about the population mean and without recourse to the specialized mathematical developments of Central Limit Theorems for SRSWOR sampling from finite populations. ∎

The posterior distribution of the population total, $N\bar{Y}$, has mean $\sum_{i=1}^{n} y_i/\pi_i$, where $\pi_i = n/N$ is the inclusion probability of the i-th unit in the sample. This is precisely the design-based Horvitz-Thompson estimator [Narain, 1951, Horvitz and Thompson, 1952] of the population total. Bayesian hierarchical models that yield posterior means equaling the Horvitz-Thompson estimator have been discussed in Ghosh and Sinha [1990], Ghosh and Meeden [1997], Little [2004], Ghosh [2012], and Banerjee [2024]. One simple example is the following model,

$$Y_i = \pi_i\beta + \pi_i\epsilon_i; \quad \epsilon_i \overset{ind}{\sim} N(0, \sigma_i^2), \ i = 1, 2, \ldots, N; \quad p(\beta) \propto 1, \tag{17.35}$$

where the σ_i^2's are assumed to be fixed constants. The posterior mean of β is

$$\mathbb{E}[\beta \mid \mathbf{y}] = (\boldsymbol{\pi}^{\mathrm{T}}\mathrm{diag}(1/(\sigma_i\pi_i)^2)\boldsymbol{\pi})^{-1}\boldsymbol{\pi}^{\mathrm{T}}\mathrm{diag}(1/(\sigma_i\pi_i)^2)\mathbf{y} = \sum_{i=1}^{n} w_i(y_i/\pi_i),$$

where $\boldsymbol{\pi} = (\pi_1, \ldots, \pi_N)^{\mathrm{T}}$ and $w_i = (1/\sigma_i^2)/\left(\sum_{i=1}^{n} 1/\sigma_i^2\right)$. This yields the posterior mean of the population total as

$$\mathbb{E}\left[\sum_{i=1}^{N} Y_i \,\middle|\, \mathbf{y}\right] = \sum_{i=1}^{n} y_i + \left(\sum_{i=n+1}^{N} \pi_i\right)\mathbb{E}[\beta \mid \mathbf{y}] = \sum_{i=1}^{n} y_i + \left(\sum_{i=n+1}^{N} \pi_i\right)\left(\sum_{i=1}^{n} w_i(y_i/\pi_i)\right)$$

$$= \sum_{i=1}^{n} \left(1 + \left(n - \sum_{i=1}^{n} \pi_i\right) w_i/\pi_i\right) y_i = \sum_{i=1}^{n} \tilde{w}_i y_i, \tag{17.36}$$

where $\tilde{w}_i = 1 + (n - \sum_{i=1}^n \pi_i)\, w_i/\pi_i$. We obtain the Horvitz-Thompson estimator in the last expression if $\tilde{w}_i = 1/\pi_i$. Letting $\sigma_i^2 = 1/(1-\pi_i)$ yields $w_i = (1-\pi_i)/(n - \sum_{i=1}^n \pi_i)$ so that $\tilde{w}_i = 1 + (1-\pi_i)/\pi_i = 1/\pi_i$. Therefore, setting $\sigma_i^2 = 1/(1-\pi_i)$ in (17.35) yields a Bayesian model where the posterior expectation of the population total exactly agrees with the Horvitz-Thompson estimator [Ghosh and Sinha, 1990]. See Ghosh [2012] for generalizations of this result to exponential families and other families of priors.

17.5.2　Incorporating information in population units

A key advantage of the model-based approach to survey sampling is its ability to incorporate information, either known or posited, about the population units. The ability to endow probability laws over the finite population units makes the Bayesian framework substantially richer than design-based or classical model-based frameworks. Subsequent inference about the finite population quantities follows from the probability distribution of unobserved quantities while accounting for the dependence structure in the population [Rubin, 1976, Gelman et al., 2013]. As earlier, let $\mathbf{Y} = (y_1, \ldots, y_N)^\mathrm{T}$ be the units in the population and let $\mathbf{Z} = (Z_1, \ldots, Z_N)^\mathrm{T}$ be the sampling inclusion indicators. A probability model for the entire finite population is given by $p(\mathbf{Y}, \mathbf{Z} \mid \theta, \phi) = p(\mathbf{Y} \mid \theta) \times p(\mathbf{Z} \mid \mathbf{Y}, \phi)$, where θ and ϕ are unknown parameters specifying the distribution of $\mathbf{Y}$ and the conditional distribution of $\mathbf{Z}$ given $\mathbf{Y}$, respectively. We sample from

$$p(\mathbf{Y}_u \mid \mathbf{y}, \mathbf{Z}) \propto \int p(\mathbf{Y}_u \mid \mathbf{y}, \mathbf{Z}, \theta, \phi) \times p(\theta, \phi \mid \mathbf{y}, \mathbf{Z})\, d\theta\, d\phi \,, \qquad (17.37)$$

where $\mathbf{Y} = \{\mathbf{Y}_u, \mathbf{y}\}$ with $\mathbf{y}$ denoting values realized in the observed samples and $\mathbf{Y}_u$ unknown values in population units outside of the sample. In order to sample from (17.37), we note that $p(\mathbf{Y}_u, \theta, \phi \mid \mathbf{y}, \mathbf{Z}) = p(\mathbf{Y}_u \mid \mathbf{y}, \mathbf{Z}, \theta, \phi) \times p(\theta, \phi \mid \mathbf{y}, \mathbf{Z})$. Therefore, we first sample values of θ and ϕ from their joint posterior distribution $p(\theta, \phi \mid \mathbf{y}, \mathbf{Z})$ and, subsequently, we draw one value of $\mathbf{Y}_u$ from $p(\mathbf{Y}_u \mid \mathbf{y}, \mathbf{Z}, \theta, \phi)$ using each sampled value of $\{\theta, \phi\}$.

17.5.3　Ignorable designs

The conditional distribution $p(\mathbf{Z} \mid \mathbf{Y}, \phi)$ *models* the sampling design and is accounted for in the finite population inference. Following Rubin [1976], if $p(\mathbf{Z} \mid \mathbf{y}, \phi) = p(\mathbf{Z})$ does not depend on the sampled values and on unknown design parameters ϕ, then the design is referred to as *ignorable*. Purely model-based approaches can reproduce the Horvitz-Thompson estimator without requiring any information on the sampling design. Inference for finite population units from an ignorable design are, therefore, essentially a model-based prediction or Bayesian imputation problem. Linear regression models with dependent population units are formulated as

$$\begin{bmatrix} \mathbf{y} \\ \mathbf{Y}_u \end{bmatrix} = \begin{bmatrix} X_s \\ X_u \end{bmatrix} \beta + \begin{bmatrix} \epsilon_s \\ \epsilon_u \end{bmatrix} ; \quad \begin{bmatrix} \epsilon_s \\ \epsilon_u \end{bmatrix} \sim N\left(\begin{bmatrix} \mathbf{0} \\ \mathbf{0} \end{bmatrix}, \begin{bmatrix} V_s(\theta) & V_{su}(\theta) \\ V_{us}(\theta) & V_u(\theta) \end{bmatrix} \right) ;$$
$$\beta = A\mu + \eta ; \quad \eta \sim N(\mathbf{0}, V_\beta) ; \quad \theta \sim p(\theta) , \qquad (17.38)$$

where X_s and X_u are fixed design matrices corresponding to sampled and non-sampled units, respectively, $\begin{bmatrix} V_s(\theta) & V_{su}(\theta) \\ V_{us}(\theta) & V_u(\theta) \end{bmatrix}$ is a positive definite covariance matrix partitioned accordingly, and A and μ are assumed fixed. Inference proceeds from the following steps: (i) Draw samples from the posterior distribution $p(\beta, \theta \mid \mathbf{y})$; (ii) Draw samples from $p(\mathbf{Y}_u \mid \mathbf{y})$ so that $\mathbb{E}_{\beta,\theta \mid y}\left[N\left(\mathbf{Y}_u \mid X_u\beta + V_{us}V_s^{-1}(\mathbf{y} - X_s\beta), V_u - V_{us}V_s^{-1}V_{su}\right)\right]$. The posterior samples $\mathbf{Y}_u$ and the observations $\mathbf{y}$ are then substituted into desired functions of the population units to obtain the posterior samples of the desired functions. The posterior estimator obtained from (17.38) is consistent in the sense that $\mathbb{E}[\mathbf{Y}_u \mid \mathbf{y}, \boldsymbol{\beta}, \theta] = X_u\boldsymbol{\beta} + V_{us}V_s^{-1}(\mathbf{y} - X_s\boldsymbol{\beta})$ equals $\mathbf{y}$ if the entire finite population is sampled whereupon $X_u = X_s$ and $V_{us} = V_s$ and the expression becomes free of model parameters $\{\boldsymbol{\beta}, \theta\}$. Furthermore, the conditional variance $\mathbb{V}(\mathbf{Y}_u \mid \mathbf{y}, \boldsymbol{\beta}, \theta)$ equals 0 when all units of the finite population are sampled. In spatial contexts, this reveals an interpolation property for finite populations using (17.38).

For ignorable designs, the framework in (17.38) is sufficient to accommodate spatial dependence using the covariance matrix in (17.38). For example, there is substantial literature on small area estimation for regionally aggregated data [see, e.g., Rao and Molina, 2015, Clayton and Kaldor, 1987, Datta and Ghosh, 1991, Ghosh et al., 1998, Arora and Lahiri, 1997, Ghosh and Rao, 1994], where interest lies in modeling dependencies across regions such as counties or districts within states, states within a country, or census-tracts or postal codes within a state or city. If the population units correspond to areal regions, then we can introduce dependencies using Markov random fields, where the covariance matrix in (17.38) may be specified using a proper CAR or SAR model.

Modeling considerations are different when we consider quantities that, at least conceptually, exist in a continuum over the entire domain, even if they are recorded at a finite number of point-referenced locations. Now, dependence among the population units is induced from a spatial process, which assigns a probability law to an uncountable subset within a d-dimensional Euclidean domain. Here, the covariance matrix in (17.38) is modeled using a spatial covariance function. The literature on finite population sampling in spatial process settings is sparser than small area estimation. Here, Ver Hoef [2002] discusses geostatistical models and classical design-based sampling and develops methods for executing model-based block "kriging." Cicchitelli and Montanari [2012] present a spline regression model-assisted, design-based estimator of the mean for use on a random sample from both finite and infinite spatial populations, while Bruno et al. [2013] use linear spatial interpolation to create a design-based predictor of values at unobserved locations. Chan-Golston et al. [2020] use (17.38) to introduce spatial processes in multi-stage sampling designs.

17.5.4 *Non-ignorable spatial designs*

Spatial survey sampling accommodates non-ignorable sampling designs using (17.37). Let $\mathcal{L} \subseteq \mathbb{R}^2$ and let $\mathcal{L}_{FP} = \{\mathbf{s}_1, \ldots, \mathbf{s}_N\}$ be a finite collection of N spatial locations representing units in a finite population referenced by locations that are sampled, denoted by $\mathcal{L}_s$, and those that are not, denoted by $\mathcal{L}_u$. Hence, $\mathcal{L}_s \cup \mathcal{L}_u = \mathcal{L}_{FP}$. Spatial applications may also require a "structural zero" subset, denoted by $\mathcal{L}_0 \subset \mathcal{L} \setminus \mathcal{L}_{FP}$, which consists of locations where sampling is precluded due to geographic considerations (e.g., lakes and rivers are excluded when modeling household economic surveys; heavily urbanized metropolitan areas are excluded from studies concerning biomass interpolation from a finite population of trees).

We model the set $\{(Y(\mathbf{s}), Z(\mathbf{s})) : \mathbf{s} \in \mathcal{L}\}$ using a spatial process, which endows the finite population $\{(Y(\mathbf{s}), Z(\mathbf{s})) : \mathbf{s} \in \mathcal{L}_{FP}\}$ with a probability law. We consider three underlying stochastic processes generating the data: (i) a possibly vector-valued latent spatial process, $\{\mathbf{w}(\mathbf{s}) : \mathbf{s} \in \mathcal{L}\}$, which models the variable(s) of interest at $\mathbf{s}$; (ii) a sampling indicator process, $\{Z(\mathbf{s}) : \mathbf{s} \in \mathcal{L}\}$, which indicates whether a location is sampled or not; and possibly (iii) a structural zero indicator process (partially observed), $\{u(\mathbf{s}) : \mathbf{s} \in \mathcal{L}\}$, which indicates whether $\mathbf{s} \in \mathcal{L}_{FP}$ or not. A joint hierarchical model, conceptually, is constructed as

$$[Y(\cdot) \mid \mathbf{w}(\cdot), u(\cdot)] \times [Z(\cdot) \mid Y(\cdot), u(\cdot)] \times [\mathbf{w}(\cdot) \mid u(\cdot)] \times [u(\cdot)]. \qquad (17.39)$$

We turn to specifying probability models for each of these components in the spirit of Finley et al. [2011b], who predicted continuous forest variables at new locations accounting for uncertainty in whether the new locations were forested or not.

17.5.4.1 Modeling $[Y(\cdot) \mid \mathbf{w}(\cdot), u(\cdot)]$

Let $u(\mathbf{s}) = 1$ if $\mathbf{s} \in \mathcal{L}_{FP}$ and $u(\mathbf{s}) = 0$ if $\mathbf{s} \notin \mathcal{L}_{FP}$. The first component in (17.39) is modeled as a customary spatial (or spatial-temporal) regression model for point-referenced data if $u(\mathbf{s}) = 1$ and is distributed as the Dirac measure, δ_0, with a point mass at 0 if $u(\mathbf{s}) = 0$. Thus,

$$Y(\mathbf{s}) = u(\mathbf{s})\{\mathbf{x}(\mathbf{s})^{\mathsf{T}}\boldsymbol{\beta} + \tilde{\mathbf{x}}(\mathbf{s})^{\mathsf{T}}\mathbf{w}(\mathbf{s}) + \epsilon(\mathbf{s})\} + (1 - u(\mathbf{s}))\delta_0 \; ; \quad \epsilon(\mathbf{s}) \stackrel{iid}{\sim} N(0, \sigma^2)$$

where $\mathbf{x}(\mathbf{s})$ is a vector of explanatory (perhaps including design) variables at $\mathbf{s}$ with regression coefficients $\boldsymbol{\beta}$, $\tilde{\mathbf{x}}(\mathbf{s})$ is a subset of $\mathbf{x}(\mathbf{s})$ whose regression coefficients, $\mathbf{w}(\mathbf{s})$, vary over space, that is, the impact of the predictors on the population units vary over space and $\mathbf{w}(\mathbf{s})$ is a multivariate Gaussian process with a valid cross-covariance matrix function or a vector of independent univariate processes. The measurement error process $\epsilon(\mathbf{s})$ captures micro-scale variation existing at fine scales (smaller than the minimum distances among locations in the finite population) that is attributed to measurement errors in surveys.

17.5.4.2 Modeling $[Z(\cdot) \mid Y(\cdot), \omega(\cdot), u(\cdot)]$

While the sampling indicator process $Z(\cdot)$ "disappears" from the likelihood or the posterior for ignorable designs, we can model non-ignorable designs using such a process. For example, we consider the model

$$Z(\mathbf{s}) \sim u(\mathbf{s})\text{Ber}(\pi(\mathbf{s})) + (1 - u(\mathbf{s}))\delta_0 \; ; \quad \text{logit}(\pi(\mathbf{s})) = \beta_{0Z} + \beta_{1Z}Y(\mathbf{s}) + \mathbf{x}_Z^{\mathsf{T}}\boldsymbol{\beta}_{2Z} \; ,$$

where, for an arbitrary location $\mathbf{s}$, the indicator process is distributed as a Bernoulli distribution with probability $\pi(\mathbf{s})$ if $u(\mathbf{s}) = 1$, that is, if $\mathbf{s} \in \mathcal{L}_{FP}$, and is set equal to zero if $u(\mathbf{s}) = 0$, that is, if $\mathbf{s} \notin \mathcal{L}_{FP}$. We introduce external predictors, $\mathbf{x}_Z$, along with the finite population units $Y(\mathbf{s})$ to inform the sampling indicator process. The slope for $Y(\mathbf{s})$ is β_{1Z}, while $\boldsymbol{\beta}_{2Z}$ is the vector of slopes corresponding to $\mathbf{x}_Z$, In fact, one can conceive of introducing another spatial process

$$\text{logit}(\pi(\mathbf{s})) = \beta_{0Z} + \beta_{1Z}Y(\mathbf{s}) + \mathbf{x}_Z^{\mathsf{T}}\boldsymbol{\beta}_Z + \omega(\mathbf{s}) \; ; \quad \omega(\mathbf{s}) \sim GP(0, C_{\theta_\omega}(\cdot)) \; ,$$

where $\omega(\mathbf{s})$ is a zero centered Gaussian process with a specified covariance kernel $C_{\theta_\omega}(\cdot)$.

17.5.4.3 Modeling $[\mathbf{w}(\cdot) \mid u(\cdot)] = [\mathbf{w}(\cdot)]$ and $[u(\cdot)]$

The spatial process $w(\mathbf{s})$ is assumed to be independent of the population indicator $u(\mathbf{s})$ and is modeled as a multivariate Gaussian process (see Chapter 10). The quantity $u(\mathbf{s})$ is a population indicator process modeled by

$$u(\mathbf{s}) \sim \text{Ber}(\pi(\mathbf{s})) \; ; \quad \text{logit}(\pi(\mathbf{s})) = \mathbf{x}_u(\mathbf{s})^{\mathsf{T}}\boldsymbol{\beta}_u + v(\mathbf{s}) \; ; \quad v(\mathbf{s}) \sim GP(0, C_{\theta_v}(\cdot)) \; ,$$

where $v(\mathbf{s})$ is a spatial process with covariance function $C_{\theta_v}(\cdot)$. Here, the model for $\pi(\mathbf{s}) = P(u(\mathbf{s}) = 1)$ is a logistic regression accommodating explanatory variables that inform about a location being a part of the finite population or not.

17.5.5 Examples

We present an example from Chan-Golston et al. [2020] who built a hierarchical model from a two-stage cluster sampling design to study ground-water nitrate levels in California's Central Valley. Our finite population consists of 6,117 unique wells situated in the Tulare Lake Basin (TLB) among 63 zip codes, where health officials seek estimates of mean nitrate levels. The spatial locations of these wells are available. The health implications of high

Model	Finite Population Mean in mg/L (95% CI)	WAIC
(i) Two-Stage	26.6 (19.2, 35.1)	4701.6
(ii) Spatial	32.6 (27.2, 38.7)	4858.4
(iii) Two-Stage + Spatial	31.2 (24.2, 37.3)	2697.0
(iv) Regional Spatial	26.2 (18.7, 34.4)	4536.6

Table 17.1 *Results of California nitrate data analysis.*

nitrate levels in wells can be serious. At high levels, infants and pregnant women are more susceptible to nitrate poisoning, which makes it more difficult for oxygen to be distributed to the body and can be fatal to infants less than six months old.

Twenty-one zip codes were randomly chosen in the first stage and 50%–90% of the wells in each zip code were randomly sampled in the second stage. The nitrate levels in the sample wells were measured in milligrams per liter (mg/L). The following models were used to analyze the sampled data: (i) a two-stage hierarchical cluster model; (ii) a spatial random field model over the entire region that ignores the structure of the cluster sampling design (so a single set of spatial process parameters for the entire region); (iii) a two-stage model with an additive global spatial process; and (iv) a regional spatial process model where each cluster has its own mean parameter and spatial process. All of these models can, in fact, be subsumed into (17.38) with appropriately constructed design and covariance matrices [see Chan-Golston et al., 2020, for details].

Table 17.1 presents the posterior mean and 95% credible intervals for the finite population means obtained from each model. There are some discrepancies in these estimates among the different models. Hence, it will be appropriate to The WAIC scores for each of the models are also presented. Some key features are worth pointing out. Notably, models that either ignore the two-stage structure of the sampled data (models ii and iv) or that ignore the spatial dependence (model i) appear to perform considerably worse (in terms of WAIC) than the model that accommodates both the sampling structure and the spatial dependence (model iii). Our recommendation is to report the analysis from model (iii). Also, while the credible intervals around the finite population mean from different models largely overlap, the point estimates from (i) and (iv) are lower than (ii) or (iii). This is, perhaps, attributed to the former two models not having a single spatial random field over the entire domain. Finally, the spatial clustering introduced in the latter two models results in shorter credible intervals.

We turn to a second example from a study to estimate the percentage of annual reported income spent on fruits and vegetables (PIFV) in two low-income communities in Los Angeles, California [see Chan-Golston et al., 2022, and references therein for a detailed description of the study]. This study follows the framework described in Section 17.5.4. There are $N = 635$ locations in $\mathcal{L}_{FP}$ and within each location, we have a number of households comprising the population units. The total number of population units is 2015. We randomly sampled $n = 555$ locations and then sampled households within each of these locations to obtain a sample of 1294 population units. These units reported amounts spent on fruits and vegetables on weekly, bi-weekly, or monthly scale. These values were multiplied by 52, 26, and 12, respectively, to reflect the annual amount spent on fruits and vegetables in a household. Reported yearly incomes in the population units were also collected.

We consider a simpler version by setting $u(\cdot) = 1$. We assume that the PIFV of the j-th household within location $\mathbf{s}_i$ is modeled as $Y_j(\mathbf{s}_i) = \mathbf{x}_j(\mathbf{s}_i)^{\mathrm{T}}\boldsymbol{\beta} + w(\mathbf{s}_i) + \epsilon(\mathbf{s}_i)$, $\mathbf{x}_j(\mathbf{s}_i)$ is a vector of fixed explanatory variables, $w(\mathbf{s}_i)$ is a spatial random effect and $\epsilon(\mathbf{s}_i)$ is white noise representing measurement errors and micro-scale variation. This is the model for $[Y(\cdot) \mid w(\cdot)]$ in Section 17.5.4 with $u(\mathbf{s}) = 1$. Based upon this, we consider four models: (i) a simple linear regression model that sets $w(\mathbf{s}) = 0$ for all locations (yields an ignorable

	Model 1	Model 2	Model 3	Model 4
PIFV %	17% (16%, 19%)	26% (22%, 31%)	35% (28%, 43%)	26% (22%, 32%)
D	3527.8	3692.2	3701.2	3482
GRS	-1841.2	-1903.6	-1863	-1767.8

Table 17.2 *Results of regression models predicting the percentage of income spent on fruits and vegetables.*

response); (ii) a model with nonignorable response that models $[Z(\cdot) \mid Y(\cdot), \omega(\cdot)]$ in addition to $[Y(\cdot) \mid w(\cdot)]$, but with no spatial effect (hence, $w(\cdot) = 0$ and $\omega(\cdot) = 0$); (iii) spatial random effects in $[Y(\cdot) \mid w(\cdot)]$, but no spatial effects in $[Z(\cdot) \mid Y(\cdot)]$; and (iv) fully spatial model with nonignorable responses with spatial effects $w(\cdot)$ and $\omega(\cdot)$ as described in Section 17.5.4.

Table 17.2 presents the posterior estimates of the finite population PIFV (first row) from the four models described above. The performance of these models was evaluated using D and GRS defined in (5.67) and (5.69), respectively. Based on these metrics, model (iii) is marginally preferred to model (iv), which suggests a lack of spatial dependence in the sampling inclusion model. Both the spatial models, (iii) and (iv), considerably outperform the non-spatial models including (ii) which still accommodates the nonignorable response. Model (i), which assumes ignorability, performs the worst. The credible interval from the ignorable model seems very tight, while those from the other models are wider because of the propagation of uncertainty from the additional parameters being estimated. The PIFV from the model (iii) is recommended for reporting given the model's superior performance.

These examples illustrate the benefits of Bayesian inference for finite populations in stochastic process settings, an area that has received relatively limited attention. A recent, more comprehensive, case study is available in Finley et al. [2024b] who developed a two-stage hierarchical Bayesian model to estimate forest biomass density and total given sparsely sampled LiDAR and georeferenced forest inventory plot measurements.

17.6   Spatial directional data analysis

Circular data, i.e., observations recorded as directions, arise in many different contexts. Examples include natural directions, such as wind directions (meteorology), animal movement directions (biology), and rock fracture orientations (geology). Another type of directional data arises by wrapping periodic time data with period L (say, daily or weekly) onto a circle with circumference L, e.g., times of crimes or hospital arrivals, and then rescaling the circumference to 2π, that of a unit circle. Directional data may be linked to other variables measured in space, time, or space and time, as in marine applications with significant waves height and direction, or in modeling wind fields (intensity and direction) resulting in circular data regression modeling (see Wang and Gelfand [2014].

Analysis of circular data can be challenging due to the restriction of the domain to the circle. With such data, customary statistical summaries can lose their meaning and should be replaced with their circular counterpart; for example, the mean is replaced by the circular mean [Jammalamadaka and Sengupta, 2001]. Most of the general inference tools have been adapted for application to circular variables, for example, see the books by Mardia and Jupp [2000], Fisher [1993], Jammalamadaka and Sengupta [2001] or the review paper by Lee [2010]. As with data on the real line, directional data can also arise in the space and space-time setting. Process modeling for such data is the goal of this brief review. As expected, we present a Bayesian/hierarchical modeling perspective. Inference focuses on population features and, with spatial or spatio-temporal data, on prediction in space and time (kriging). Model fitting is through MCMC which becomes straightforward through the introduction of latent variables. Much of what follows is drawn from the work of Jona Lasinio et al. [2012], Wang and Gelfand [2013], and Wang and Gelfand [2014]. As well, fully developed applications are supplied in these papers.

A circular distribution is a probability distribution whose entire mass is on the circumference of a unit circle. We work with the absolutely continuous case (w.r.t. Lebesgue measure on the circle) assuming cdf $F(\theta)$ and density $f(\theta)$ with the following properties: (i) for the cdf, $F(\theta + 2\pi) - F(\theta) = 1$, $\theta \in R^1$, (ii) for the density, $\int_0^{2\pi} f(\theta)d\theta = 1$ and $f(\theta + 2k\pi) = f(\theta)$ for any integer k (f is periodic).

Expectations under circular densities are hard to compute, e.g., for the Von Mises, wrapped distributions, and projected normals presented below. Working with the associated complex variable on the unit circle in the complex plane, $Z = e^{i\theta}$, is convenient. $E(Z^r) = E(e^{ir\theta}) \equiv \rho_r e^{i\mu_r}$, i.e, the value of the characteristic function at r. Thus, we have $\rho_r \cos(\mu_r) = E\cos(r\theta)$ and $\rho_r \sin(\mu_r) = E\sin(r\theta)$. When $r = 1$, we have $\rho\cos\mu = E(\cos\theta)$ and $\rho\sin\mu = E(\sin\theta)$. Solving for the **mean direction** $\mu = \text{atan2} \frac{E\sin(\theta)}{E\cos(\theta)}$.[3] We also have the **resultant/concentration** $\rho(\equiv c) = \sqrt{(E\cos(\theta))^2 + (E\sin(\theta))^2} \leq 1$.

These moment properties suggest exploratory data ideas. In particular, consider data $\{\theta_i, i = 1, 2, ...n\}$, equivalently $\{(\cos\theta_i, \sin\theta_i), i = 1, 2, ...n\}$. Let $\bar{C} = \frac{1}{n}\sum_i \cos\theta_i$, $\bar{S} = \frac{1}{n}\sum_i \sin\theta_i$. Then, we have a method of moments estimators using $\bar{C} = \hat{\rho}\cos\hat{\mu}$, $\bar{S} = \hat{\rho}\sin\hat{\mu}$. That is, $\hat{\mu} = \text{atan2} \frac{\bar{S}}{\bar{C}}$ and $\hat{\rho} = \sqrt{\bar{C}^2 + \bar{S}^2}$.

We note several challenges with circular data. The support restriction is not just $[0, 2\pi)$ but circularity, i.e., sensitivity to the starting point. An angle or, equivalently, a real number on $[0, 2\pi)$ arises given a fixed orientation, e.g., the positive x-axis. However, inference should not depend upon "choice of origin," i.e., it should be rotation invariant. Further, the circle is very different topologically from the line; the beginning coincides with the end. That is, "direction has no magnitude." There is no ordering or ranking. Also, is the direction 2-dimensional, e.g., an angle associated with the resultant of a N-S direction and an E-W direction? The customary sample mean and variance don't mean anything,—for the sample $1^o, 0^o, 359^o$, the sample mean is 120°, clearly meaningless; 0° is more sensible. Similarly, the sample variance is also silly. That's why we consider the mean direction and concentration, as above.

Continuing, what should the association between say, θ and θ' mean? What good properties should we require for a correlation between dependent directions when directions have no magnitude? In particular, in a spatial setting, how can we connect the covariance function of a linear Gausssian Process to that of an induced spatial process model for directional variables? The following are some desirable properties of a circular correlation coefficient: (i) $\rho_c(\theta_1, \theta_2)$ should not depend upon the "zero" direction, (ii) $\rho_c(\theta_1, \theta_2) = \rho_c(\theta_2, \theta_1)$, (iii) $|\rho_c(\theta_1, \theta_2)| \leq 1$, (iv) $\rho_c(\theta_1, \theta_2) = 0$ if θ_1, θ_2 are independent. Jammalamadaka and Sarma [1988] provide the following measure which satisfies the above properties:

$$\rho_c(\theta_1, \theta_2) = \frac{E(\sin(\theta_1 - \mu_1)\sin(\theta_2 - \mu_2))}{\sqrt{\text{Var}(\sin(\theta_1 - \mu_1))\text{Var}(\sin(\theta_2 - \mu_2))}} \tag{17.40}$$

Let's turn to the circular distribution modeling which we find in the literature. The *standard/most common* so-called intrinsic approach adopts the von Mises distribution $M(\mu, \kappa)$ with density

$$f(\theta; \mu, \kappa) = \frac{1}{2\pi I_0(\kappa)}e^{\kappa\cos(\theta - \mu)},$$

where μ is the mean direction, κ is the concentration, and I_0 is the modified Bessel function of the first kind of order 0. It is the circular analogue of normal distribution for linear data.

---

[3]atan2 returns the arc tangent of the two numbers x and y. It is similar to calculating the arc tangent of y/x, except that the signs of both arguments are used to determine the quadrant of the result. The result is an angle expressed in radians.

The density is, evidently, symmetric and unimodal; mixture models have been proposed to add flexibility. The von Mises distribution is not extensible for use with multivariate angular data. So, for spatial or space-time data, we find conditionally independent von Mises specifications with process modeling at the second stage for μ, κ. As data models, this approach may be unattractive.

A second family of models are the wrapped distributions. The idea is to wrap a linear variable, i.e., define $\theta = Y \bmod 2\pi$. If $g(y)$ is a density on R^1, the associated wrapped density becomes

$$f(\theta) = \sum_{k=-\infty}^{\infty} g(\theta + 2\pi k).$$

Obviously, we can rescale from $[0, L)$ to $[0, 2\pi)$.

Taking g as the unit normal density, we have the wrapped normal density, $WN(\mu, \sigma^2)$, in the form, for $0 \le \theta < 2\pi$:

$$f(\theta) = \sum_{k=-\infty}^{\infty} g(\theta + 2k\pi) = \frac{1}{\sigma\sqrt{2\pi}} \sum_{k=-\infty}^{\infty} \exp\left(-\frac{(\theta + 2k\pi - \mu)^2}{2\sigma^2}\right).$$

Then, $E(Z) = e^{-\sigma^2/2} e^{i\mu}$, i.e., μ is the linear mean with $\mu = \tilde{\mu} + 2\pi K_\mu$ with $\tilde{\mu} \in [0, 2\pi)$ the mean direction and $c = e^{-\sigma^2/2}$ is concentration. Here, θ is observed; $\theta + 2K\pi$ is the associated linear variable. The joint density for θ and K is $f(\theta, k) = g(\theta + 2k\pi) =$

$$\frac{1}{\sigma\sqrt{2\pi}} \exp\left(-\frac{(\theta + 2k\pi - \mu)^2}{2\sigma^2}\right), \quad 0 \le \theta < 2\pi, K \in \{0, \pm 1, \pm 2, ...\}$$

This form removes the doubly infinite sum and suggests introducing K as a latent variable.

In this regard, note that as $\sigma \to \infty$ the WN tends to a circular uniform. The mean direction does not exist for the circular uniform ($E\sin\theta = E\cos\theta = 0$) and $c = 0$. To address this, we can truncate the doubly infinite sum according to σ, e.g., if $\sigma < 2\pi/3$ then $P(K \in \{-1, 0, 1\}) > .997$; if $2\pi/3 < \sigma < 4\pi/3$ then $P(K \in \{-2, -1, 0, 1, 2\}) > .997$. So, K can be large i.f.f. σ large, suggesting an identifiability issue. That is, if σ large, it will be hard to distinguish from a circular uniform, and "close to" uniform implies inference problems for μ and c. Perhaps more importantly, we can use this truncation trick to sample latent K's from their full conditional distributions in MCMC model fitting, as described below.

A multivariate version (say p-dimensional) is easy to specify. With a multivariate density g on R^p,

$$f(\boldsymbol{\theta}) = \sum_{k_1=-\infty}^{\infty} \cdots \sum_{k_p=-\infty}^{\infty} g(\boldsymbol{\theta} + 2\pi\mathbf{k}).$$

A convenient choice, which takes us to Gaussian processes, is a multivariate normal. That is, a Gaussian process on $\mathbb{R}^2$ induces a wrapped Gaussian process on $\mathbb{R}^2$. In particular, the GP is specified through its finite dimensional distributions which in turn induce the finite dimensional distributions for the wrapped process, returning us to the multivariate wrapped distributional models. In particular, if $Y(\mathbf{s})$ is a GP with mean $\mu(\mathbf{s})$ and covariance function, say $\sigma^2 \rho(||\mathbf{s} - \mathbf{s}'||; \psi)$ where ψ is a decay parameter, then, for locations $\mathbf{s}_1, \mathbf{s}_2, ..., \mathbf{s}_n$, $\boldsymbol{\theta} = (\theta(\mathbf{s}_1), \theta(\mathbf{s}_2), ..., \theta(\mathbf{s}_n)) \sim WN(\boldsymbol{\mu}, \sigma^2 \mathbf{R}(\psi))$ where $\boldsymbol{\mu} = (\mu(\mathbf{s}_1), ..., \mu(\mathbf{s}_n))$ and $R(\psi)_{ij} = \rho(\mathbf{s}_i - \mathbf{s}_j; \psi)$.

Fitting a wrapped GP model: Model fitting for a wrapped GP within a Bayesian framework can be readily done using MCMC. First, suppose a linear GP model of the form $Y(\mathbf{s}_i) = \mu + w(\mathbf{s}_i), i = 1, 2, ..., n$ where $w(\mathbf{s}_i)$ is a mean 0 GP with covariance function

$\sigma^2 \rho(\mathbf{s} - \mathbf{s}'; \phi)$. Consider an exponential covariance and a prior on $\boldsymbol{\gamma} = (\mu, \sigma^2, \phi)$ of the form $[\mu][\sigma^2][\phi]$ which is normal, inverse Gamma, and uniform, respectively.

For the wrapped GP, the approach follows by introducing $\mathbf{K}$, a vector of K_i's. Again, the induced wrapped GP provides $\boldsymbol{\theta} \sim WN(\mu\mathbf{1}, \sigma^2 \mathbf{R}(\phi))$ where $R(\phi)_{jk} = \rho(s_j - s_k; \phi)$ so that the joint distribution of $\boldsymbol{\theta}, \mathbf{K}$ takes the form $N(\boldsymbol{\theta} + 2\pi\mathbf{K}|\mu\mathbf{1}, \sigma^2 \mathbf{R}(\phi))$. In fitting, the full conditionals for μ and σ^2 are familiar. The full conditional for ϕ is unpleasant because it is buried in the covariance matrix associated with the $n-$ variate normal distribution for $\boldsymbol{\theta} + 2\pi\mathbf{K}$. Generally, it is best to update σ^2 and ϕ as a pair on the log scale as in the linear case. In fact, with such joint sampling, very large values of σ^2 are rejected in the M-H step, so, in fact, we do not need to impose any truncation on the prior for σ^2. The full conditionals for the K_i arise from the conditional distribution of $Y_i = \theta_i + 2\pi k_i | \{Y_j = \theta_j + 2\pi K_j, j \neq i\}; \boldsymbol{\gamma}$. The form requires the conditional mean and variance μ_i and σ_i^2 which are functions of $\{\theta_j, j \neq i\}$, $\{K_j, j \neq i\}$, and $\boldsymbol{\gamma}$. The adaptive truncation approximation above can be employed.

A third modeling approach is an embedding of the unit circle within R^2. Let $\mathbf{U} = (U_1, U_2) \sim g(u_1, u_2)$, a density on R^2 Then, define $(V_1, V_2) \equiv (\frac{U_1}{||\mathbf{U}||}, \frac{U_2}{||\mathbf{U}||})$ where $||\mathbf{U}||$ is the length of $\mathbf{U}$, a point on the circle associated with angle $\theta = \mathrm{atan}2\frac{V_2}{V_1} = \mathrm{atan}2\frac{U_2}{U_1}$. In fact, $U_1 = R\cos\theta$ and $U_2 = R\sin\theta$, with R latent. That is, $R = ||\mathbf{U}||$, $V_1 = \cos\theta$, $V_2 = \sin\theta$ Again, the angular mean direction is $\mathrm{atan}2\frac{E\sin\theta}{E\cos\theta} = \frac{\mathrm{E}(V_2)}{\mathrm{E}(V_1)} \neq \frac{\mathrm{E}(U_2)}{\mathrm{E}(U_1)}$. The concentration is $||\mathrm{E}(\mathbf{V})|| \leq 1$.

This leads us to the projected normal distribution. Suppose the random vector $\mathbf{U} \sim N_2(\boldsymbol{\mu}, \boldsymbol{\Sigma})$, then $\theta \sim PN_2(\boldsymbol{\mu}, \boldsymbol{\Sigma})$ $\boldsymbol{\mu} = (\mu_1, \mu_2)^{\mathrm{T}}$ and covariance matrix $\boldsymbol{\Sigma} = \begin{pmatrix} \sigma_1^2 & \rho\sigma_1\sigma_2 \\ \rho\sigma_1\sigma_2 & \sigma_2^2 \end{pmatrix}$. It can be shown that θ has a density $f(\theta|\boldsymbol{\mu}, \boldsymbol{\Sigma})$, (see p.52 of Mardia [1972]) given by

$$
\begin{aligned}
f(\theta|\boldsymbol{\mu}, \boldsymbol{\Sigma}) &= \frac{\phi_2(\mu_1, \mu_2; \mathbf{0}, \boldsymbol{\Sigma})}{C(\theta)} \\
&+ \frac{aD(\theta)\Phi_1\{D(\theta)\}\phi_1\left[a\{C(\theta)\}^{-\frac{1}{2}}(\mu_1 \sin\theta - \mu_2 \cos\theta)\right]}{C(\theta)},
\end{aligned} \qquad (17.41)
$$

where ϕ_1 and Φ_1 are standard univariate normal pdf and cdf, ϕ_2 is the standard bivariate normal pdf, and

$$
a = \{\sigma_1\sigma_2\sqrt{1 - \rho^2}\}^{-1},
$$
$$
C(\theta) = a^2(\sigma_2^2 \cos^2\theta - \rho\sigma_1\sigma_2 \sin 2\theta + \sigma_1^2 \sin^2\theta),
$$
$$
D(\theta) = a^2\{C(\theta)\}^{-\frac{1}{2}}\{\mu_1\sigma_2(\sigma_2 \cos\theta - \rho\sigma_1 \sin\theta) + \mu_2\sigma_1(\sigma_1 \sin\theta - \rho\sigma_2 \cos\theta)\}.
$$

Evidently, this density will be unpleasant to work with. Rather, we introduce a latent variable $R = ||\mathbf{U}||$ and work with the joint distribution of R and Θ, easily obtained through changing variables to polar coordinates from the joint distribution of U_1 and U_2.

While other joint densities on $\Re^2$ can be projected onto the unit circle, the projected normal distribution is most commonly used. However, the literature using the projected normal model for angular data is small and initially, the special case, $PN_2(\boldsymbol{\mu}, \mathbf{I})$ was the focus [Presnell et al., 1998].

The projected normal distribution with the identity covariance matrix is also known as the *displaced normal* [Kendall, 1974]. Its corresponding density function is symmetric and unimodal so that $PN(\boldsymbol{\mu}, \mathbf{I})$ is comparable to other symmetric unimodal distributions, such as the von Mises. Wang and Gelfand [2013] study the projected normal family with a possibly non-identity covariance matrix $\boldsymbol{\Sigma}$ and refer to this richer class $PN(\boldsymbol{\mu}, \boldsymbol{\Sigma})$ as the

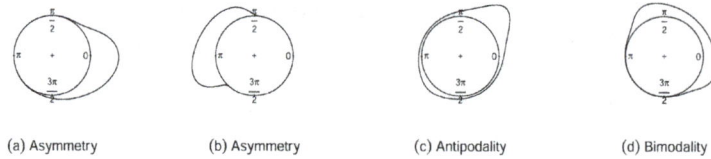

Figure 17.7 *Shapes for the general projected normal distribution.*

general projected normal distribution. They point out that this general version provides flexibility, such as asymmetry and possible bimodality. Figure 17.7 demonstrates some possible shapes of the general projected normal, other than unimodal and symmetric ones. The original specification of the general projected normal distribution is not fully identified; $\mathbf{U} = \mathbf{V}/\|\mathbf{V}\|$ is invariant to scale transformation. Wang and Gelfand [2013] set $\sigma_1 = \tau$ and $\sigma_2 = 1$ to ensure identifiability. Thus, the general projected normal is a four-parameter $(\mu_1, \mu_2, \tau, \rho)$ distribution. In this regard, comparable flexibility with the von Mises model would need to mix two distributions, resulting in a five parameter model with identifiability problems. With regard to mixtures, we note that the general projected normal distributions are dense in the class of all distributions on the circle. No such result is available for the von Mises family so mixing with projected normals may be more promising in practice. However, the mixture modeling will be difficult to work with for more than two components.

The above naturally suggests that one can create a stochastic spatial process of random variables taking values on a circle, by projecting a bivariate spatial process on $\mathcal{R}^2$. Changing to more customary notation, let $\mathbf{Y}(\mathbf{s}) = (Y_1(\mathbf{s}), Y_2(\mathbf{s}))^\mathsf{T}$ denote a 2×1 vector of random variables at location $\mathbf{s}$, $\mathcal{D}$ be the domain of interest, and $\{\mathbf{Y}(\mathbf{s}) : \mathbf{s} \in \mathcal{D}\}$ be a realization of a bivariate stochastic process. Letting $(\cos\Theta(\mathbf{s}), \sin\Theta(\mathbf{s})) = (Y_1(\mathbf{s}), Y_2(\mathbf{s}))/\|\mathbf{Y}(\mathbf{s})\|$, one obtains a circular process $\Theta(\mathbf{s})$. This projected process inherits some of the properties of the inline bivariate process, such as stationarity and isotropy. For example, straightforwardly, if the bivariate process $\mathbf{Y}(\mathbf{s})$ is strictly stationary, the induced circular projected process $\Theta(\mathbf{s})$ is strictly stationary. Letting $\rho_c(\Theta, \Theta')$ be the circular correlation, defined as in the expression (17.40), between two circular random variables Θ and Θ', we define a circular process $\Theta(\mathbf{s})$ to be *isotropic* if $\rho_c(\Theta(\mathbf{s}), \Theta(\mathbf{s}'))$ is a function of $\|\mathbf{s} - \mathbf{s}'\|$. If we assume $\mathbf{Y}(\mathbf{s})$ to be a bivariate Gaussian process, then, if the bivariate Gaussian process $\mathbf{Y}(\mathbf{s})$ is isotropic, the induced circular process $\Theta(\mathbf{s})$ is isotropic.

Finally, suppose $\mathbf{Y}(\mathbf{s})$ is a bivariate Gaussian process with mean $\boldsymbol{\mu}(\mathbf{s})$ and cross-covariance function $C(\mathbf{s}, \mathbf{s}') = \text{cov}(\mathbf{Y}(\mathbf{s}), \mathbf{Y}(\mathbf{s}'))$. We define the induced circular process upon projection to be the *projected Gaussian process*. In principle, we could consider non-Gaussian bivariate processes. However, the multivariate distributional convenience of the Gaussian process enables straightforward distribution theory for the projected Gaussian process. Moreover, with a single sample of angles, say $\Theta(\mathbf{s}_i)$, $i = 1, 2, ..., n$, it may be difficult for the data to criticize the normality assumption. Suppose we consider a separable choice of cross-covariance function, $C(\mathbf{s}, \mathbf{s}') = \varrho(\mathbf{s}, \mathbf{s}') \cdot T$, where ϱ is a valid correlation function and T is a 2×2 positive definite matrix, the homogeneous (over R^2) correlation between $Y_1(\mathbf{s})$ and $Y_2(\mathbf{s})$. Under the identifiability constraint, T is set to be $\begin{pmatrix} \tau^2 & \rho\tau \\ \rho\tau & 1 \end{pmatrix}$.

The marginal distribution for the random variable at each location $\Theta(\mathbf{s})$ is a univariate projected normal with shapes as suggested in Figure 17.7. To provide a feel for the scope of possible joint distributions, we select three illustrative sets of parameters, representing the marginals of unimodal and symmetric, unimodal and asymmetric, and bimodal, respectively. The joint density plots are shown in Figure 17.8; the rows represent different types of

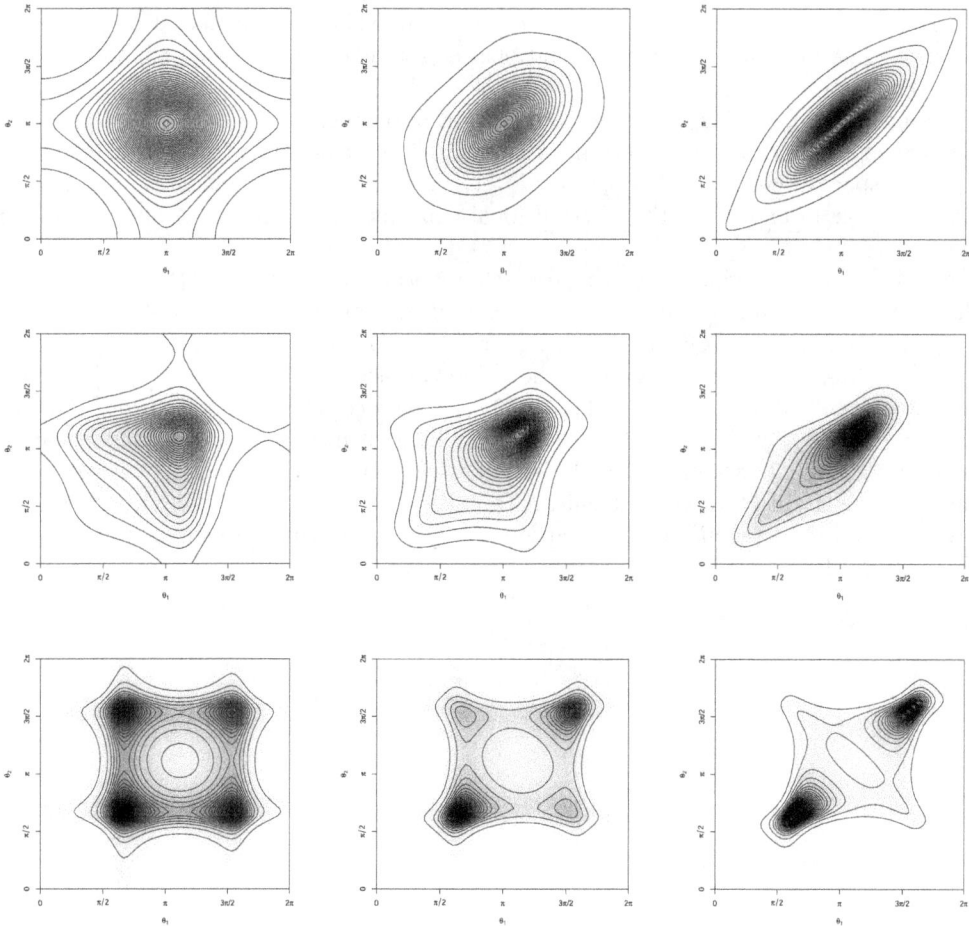

Figure 17.8 *Bivariate joint distribution of $\Theta(\mathbf{s})$ and $\Theta(\mathbf{s}')$ of three different marginals (rows) and three different levels of spatial dependence (columns).*

univariate marginals and the columns illustrate the increasing levels of spatial dependence. The richness, departing from a symmetric case as with independent von Mises or independent $PN(\boldsymbol{\mu}, \mathbf{I})$, is evident. It is noteworthy that, by column, $C(\|\mathbf{s} - \mathbf{s}'\|)$ is the same, by row, $\boldsymbol{\mu}$ and T are the same. So, the nature of the joint distribution is more affected by changes in the latter than in the former.

Model Fitting: Suppose we have a projected Gaussian spatial process model, $\Theta(\mathbf{s})$, which is induced from a linear bivariate process $\mathbf{Y}(\mathbf{s})$ with mean $\boldsymbol{\mu}(\mathbf{s})$ and the separable cross-covariance $C(\mathbf{s}, \mathbf{s}') = \varrho(\mathbf{s} - \mathbf{s}'; \phi) \cdot T$. Local covariates $X(\mathbf{s})$'s can be readily incorporated into the circular spatial process through the mean $\boldsymbol{\mu}(\mathbf{s})$. However, for simplicity, we only provide details of model fitting of the projected normal process with constant mean $\boldsymbol{\mu}(\mathbf{s}) \equiv \boldsymbol{\mu} = (\mu_1, \mu_2)'$ and an exponential correlation function in the cross-covariance denoted by $\varrho(\mathbf{s} - \mathbf{s}'; \phi) = e^{-\phi\|\mathbf{s} - \mathbf{s}'\|}$. As previously defined, $T = \begin{pmatrix} \tau^2 & \rho\tau \\ \rho\tau & 1 \end{pmatrix}$. Therefore, we have five parameters to estimate: $\boldsymbol{\Psi} = (\mu_1, \mu_2, \tau, \rho, \phi)$.

For data $\boldsymbol{\theta} = (\theta(\mathbf{s}_1), \dots, \theta(\mathbf{s}_n))^{\mathrm{T}}$, we seek to obtain the likelihood, which is the joint distribution $f(\theta(\mathbf{s}_1), \dots, \theta(\mathbf{s}_n))$. As noted above, in implementing MCMC to fit the model, it is convenient to introduce, at each $\mathbf{s}_i$, a latent variable $R(\mathbf{s}_i)$ associated with $\Theta(\mathbf{s}_i)$ and work in the space of $\Theta(\mathbf{s}_i)$'s and $R(\mathbf{s}_i)$'s for $i = 1, \ldots, n$. This data augmentation trick plays an essential role in exploiting the distributional convenience of the latent inline Gaussian process. For data $\boldsymbol{\theta} = (\theta(\mathbf{s}_1), \ldots, \theta(\mathbf{s}_n))'$, we seek to obtain the likelihood, which is the joint distribution $f(\theta(\mathbf{s}_1), \ldots, \theta(\mathbf{s}_n))$. As noted above, in implementing MCMC to fit the model, it is convenient to introduce, at each $\mathbf{s}_i$, a latent variable $R(\mathbf{s}_i)$ associated with $\Theta(\mathbf{s}_i)$ and work in the space of $\Theta(\mathbf{s}_i)$'s and $R(\mathbf{s}_i)$'s for $i = 1, \ldots, n$. This data augmentation trick plays an essential role in exploiting the distributional convenience of the latent inline Gaussian process.

According to the definition of the projected normal process, with the latent $R(\mathbf{s}_i)$, we have the corresponding "unobserved" latent linear variables $Y_1(\mathbf{s}_i) = R(\mathbf{s}_i)\cos\Theta(\mathbf{s}_i)$ and $Y_2(\mathbf{s}_i) = R(\mathbf{s}_i)\sin\Theta(\mathbf{s}_i)$. Denoting $\mathbf{Y}_1 = (Y_1(\mathbf{s}_1), \ldots, Y_1(\mathbf{s}_n))^{\mathrm{T}}$ and $\mathbf{Y}_2 = (Y_2(\mathbf{s}_1), \ldots, Y_2(\mathbf{s}_n))^{\mathrm{T}}$, the realizations of the inline Gaussian process $\mathbf{Y} = (\mathbf{Y}_1^{\mathrm{T}}, \mathbf{Y}_2^{\mathrm{T}})^{\mathrm{T}}$ follow a multivariate normal with mean $(\mu_1 \mathbf{1}_{1\times n}, \mu_2 \mathbf{1}_{1\times n})^{\mathrm{T}}$ and covariance matrix $\Sigma = T \otimes \mathbf{H}(\phi)$, where $\{\mathbf{H}(\phi)\}_{j,k} = \varrho(\mathbf{s}_j - \mathbf{s}_k; \phi)$, $j, k = 1, \ldots, n$. The joint distribution of Θ and $\mathbf{R}$ can be obtained by changing variables from the joint distribution of $\mathbf{Y}_1$ and $\mathbf{Y}_2$. However, block updating of the latent $\mathbf{R}$'s seems difficult. Alternatively, one can utilize the properties of inline GP to obtain the conditional distribution $\mathbf{Y}(\mathbf{s}_i)|\mathbf{Y}_{-\mathbf{s}_i}, \mathbf{\Psi}$, thus inducing the conditional for $R(\mathbf{s}_i)|R_{-\mathbf{s}_i}, \mathbf{\Psi}$.

Bibliography

P. Abrahamsen. Bayesian kriging for seismic depth conversion of a multi-layer reservoir. In A. Soares, editor, *Geostatistics Tróia '92: Volume 1*, pages 385–398. Springer Netherlands, Dordrecht, 1993. ISBN 978-94-011-1739-5. doi: 10.1007/978-94-011-1739-5_31. URL https://doi.org/10.1007/978-94-011-1739-5_31.

M. Abramowitz and I. A. Stegun, editors. *Handbook of mathematical functions.* Dover Books on Mathematics. Dover Publications, Mineola, NY, June 1965.

R. H. Affandi, E. B. Fox, and B. Taskar. Approximate inference in continuous determinantal processes. In *Advances in Neural Information Processing Systems 26 (NIPS)*, 2013.

R. H. Affandi, E. B. Fox, R. P. Adams, and B. Taskar. Learning the parameters of determinantal point process kernels. In *Thirty-First International Conference on Machine Learning (ICML)*, 2014.

D. K. Agarwal and A. E. Gelfand. Slice sampling for simulation based fitting of spatial data models. *Statistics and Computing*, 15(1):61–69, Jan 2005. ISSN 1573-1375. doi: 10.1007/s11222-005-4790-z. URL https://doi.org/10.1007/s11222-005-4790-z.

D. K. Agarwal, A. E. Gelfand, and J. A. Silander. Investigating tropical deforestation using two-stage spatially misaligned regression models. *Journal of Agricultural, Biological, and Environmental Statistics*, 7(3):420–439, Sep 2002. ISSN 1537-2693. doi: 10.1198/108571102348. URL https://doi.org/10.1198/108571102348.

A. Agresti. *Categorical data analysis.* John Wiley & Sons, Hokoken, NJ, second edition, 2002. ISBN 0-471-36093-7.

L. Aiello and S. Banerjee. Detecting spatial health disparities using disease maps, 2023. URL https://arxiv.org/abs/2309.02086.

H. Akaike. Information theory and an extension of the maximum likelihood principle. In E. Parzen, K. Tanabe, and G. Kitagawa, editors, *Selected Papers of Hirotugu Akaike*, pages 199–213. Springer New York, New York, NY, 1998. ISBN 978-1-4612-1694-0. doi: 10.1007/978-1-4612-1694-0_15. URL https://doi.org/10.1007/978-1-4612-1694-0_15.

H. Akima. A method of bivariate interpolation and smooth surface fitting for irregularly distributed data points. *ACM Transactions on Mathematical Software*, 4(2):148–159, jun 1978. ISSN 0098-3500. doi: 10.1145/355780.355786. URL https://doi.org/10.1145/355780.355786.

H. Akima. Algorithm 761: Scattered-data surface fitting that has the accuracy of a cubic polynomial. *ACM Transactions on Mathematical Software*, 22(3):362–371, Sept. 1996. ISSN 0098-3500. doi: 10.1145/232826.232856. URL https://doi.org/10.1145/232826.232856.

J. Albert. An introduction to R. In *Bayesian Computation with R*, pages 1–17. Springer New York, New York, NY, 2009.

M. A. Amaral Turkman, C. D. Paulino, and P. Müller. *Computational Bayesian statistics.* Institute of Mathematical Statistics Textbooks. Cambridge University Press, Cambridge, England, Feb. 2019.

E. Anderes. On the consistent separation of scale and variance for Gaussian random fields. *Annals of Statistics*, 38(2):870–893, 2010.

P. K. Andersen and R. D. Gill. Cox's Regression Model for Counting Processes: A Large Sample Study. *The Annals of Statistics*, 10(4):1100–1120, 1982. doi: 10.1214/aos/1176345976. URL https://doi.org/10.1214/aos/1176345976.

P. K. Andersen, O. Borgan, R. D. Gill, and N. Keiding. *Statistical models based on counting processes*. Springer series in statistics. Springer, New York, NY, 1 edition, Dec. 1996.

C. Andrieu and J. Thoms. A tutorial on adaptive MCMC. *Statistics and Computing*, 18(4):343–373, 2008.

T. V. Apanasovich and M. G. Genton. Cross-covariance functions for multivariate random fields based on latent dimensions. *Biometrika*, 97(1):15–30, 2010.

T. V. Apanasovich, M. G. Genton, and Y. Sun. A valid matérn class of cross-covariance functions for multivariate random fields with any number of components. *Journal of the American Statistical Association*, 107(497):180–193, 2012.

M. Armstrong and P. Diamond. Testing variograms for positive-definiteness. *Journal of the International Association for Mathematical Geology*, 16(4):407–421, May 1984. ISSN 1573-8868. doi: 10.1007/BF01029889. URL https://doi.org/10.1007/BF01029889.

M. Armstrong and R. Jabin. Variogram models must be positive-definite. *Journal of the International Association for Mathematical Geology*, 13(5):455–459, Oct 1981. ISSN 1573-8868. doi: 10.1007/BF01079648. URL https://doi.org/10.1007/BF01079648.

B. C. Arnold and D. J. Strauss. Bivariate distributions with conditionals in prescribed exponential families. *Journal of the Royal Statistical Society. Series B (Methodological)*, 53(2):365–375, 1991. ISSN 00359246. URL http://www.jstor.org/stable/2345747.

B. C. Arnold, E. Castillo, and J. M. Sarabia. *Conditional specification of statistical models*. Springer Series in Statistics. Springer, New York, NY, 1999 edition, Oct. 1999.

V. Arora and P. Lahiri. On the superiority of the Bayesian method over the blup in small area estimation problems. *Statistica Sinica*, 7(4):1053–1063, 1997. ISSN 10170405, 19968507. URL http://www.jstor.org/stable/24306172.

V. Arora, P. Lahiri, and K. Mukherjee. Empirical bayes estimation of finite population means from complex surveys. *Journal of the American Statistical Association*, 92(440):1555–1562, 1997.

R. M. Assunção. Space varying coefficient models for small area data. *Environmetrics*, 14(5):453–473, 2003. doi: https://doi.org/10.1002/env.599. URL https://onlinelibrary.wiley.com/doi/abs/10.1002/env.599.

R. M. Assunção, J. E. Potter, and S. M. Cavenaghi. A Bayesian space varying parameter model applied to estimating fertility schedules. *Statistics in Medicine*, 21(14):2057–2075, 2002. doi: https://doi.org/10.1002/sim.1153. URL https://onlinelibrary.wiley.com/doi/abs/10.1002/sim.1153.

A. Atay-Kayis and H. Massam. A Monte Carlo method for computing the marginal likelihood in nondecomposable Gaussian graphical models. *Biometrika*, 92(2):317–335, 2005.

H. Attias. Variational Bayesian framework for graphical models. In *In proceeding: Advances in Neural Information Processing Systems*. MIT Press, Cambridge, MA, 2000.

R. Aykroyd. Bayesian estimation for homogeneous and inhomogeneous Gaussian random fields. *IEEE Transactions on Pattern Analysis and Machine Intelligence*, 20(5):533–539, 1998. doi: 10.1109/34.682182.

A. Baddeley and R. Turner. Practical maximum pseudolikelihood for spatial point patterns. *Austrian and New Zealand Journal of Statistics*, 42:283–315, 2000.

A. Baddeley and R. Turner. spatstat: an R package for analyzing spatial point patterns. *Journal of Statistical Software*, 12(6):1–42, 2005.

A. Baddeley, R. Turner, J. Møller, and M. Hazelton. Residual analysis for spatial point processes (with discussion). *Journal of the Royal Statistical Society: Series B (Statistical Methodology)*, 67(5):617–666, 2005.

A. Baddeley, J. Møller, and A. G. Pakes. Properties of residuals for spatial point processes. *Annals of the Institute of Statistical Mathematics*, 60(3):627–649, 2008.

A. Baddeley, E. Rubak, and R. Turner. *Spatial Point Patterns: Methodology and Applications with R*. Chapman & Hall/CRC Interdisciplinary Statistics. Chapman & Hall/CRC Press, Oakville, MO, Nov. 2015.

A. J. Baddeley, J. Møller, and R. Waagepetersen. Non-and semi-parametric estimation of interaction in inhomogeneous point patterns. *Statistica Neerlandica*, 54(3):329–350, 2000.

M. J. Bailey, R. F. Muth, and H. O. Nourse. A regression method for real estate price index construction. *Journal of the American Statistical Association*, 58(304):933–942, 1963. ISSN 01621459, 1537274X. URL http://www.jstor.org/stable/2283324.

T. C. Bailey and A. C. Gatrell. *Interactive spatial data analysis*. Prentice Hall, Old Tappan, NJ, 1995.

V. Baladandayuthapani, B. Mallick, M. Hong, J. Lupton, N. Turner, and R. Carroll. Bayesian hierarchical spatially correlated functional data analysis with application to colon carcinogenesis. *Biometrics*, 64:64–73, 04 2008. doi: 10.1111/j.1541-0420.2007.00846.x.

F. Ballani and M. Schlather. A construction principle for multivariate extreme value distributions. *Biometrika*, 98(3):633–645, 09 2011. ISSN 0006-3444. doi: 10.1093/biomet/asr034. URL https://doi.org/10.1093/biomet/asr034.

S. Banerjee. Revisiting spherical trigonometry with orthogonal projectors. *The College Mathematics Journal*, 35(5):375–381, 2004. doi: 10.1080/07468342.2004.11922099. URL https://doi.org/10.1080/07468342.2004.11922099.

S. Banerjee. On geodetic distance computations in spatial modeling. *Biometrics*, 61(2):617–625, 2005. doi: https://doi.org/10.1111/j.1541-0420.2005.00320.x. URL https://onlinelibrary.wiley.com/doi/abs/10.1111/j.1541-0420.2005.00320.x.

S. Banerjee. High-Dimensional Bayesian Geostatistics. *Bayesian Analysis*, 12(2):583–614, 2017. doi: 10.1214/17-BA1056R. URL https://doi.org/10.1214/17-BA1056R.

S. Banerjee. Modeling massive spatial datasets using a conjugate Bayesian linear modeling framework. *Spatial Statistics*, 37:100417, 2020. ISSN 2211-6753. doi: https://doi.org/10.1016/j.spasta.2020.100417. URL https://www.sciencedirect.com/science/article/pii/S2211675320300117. Frontiers in Spatial and Spatio-temporal Research.

S. Banerjee. Finite population survey sampling: An unapologetic Bayesian perspective. *Sankhya A*, 86(1):95–124, Nov 2024. ISSN 0976-8378. doi: 10.1007/s13171-024-00348-8. URL https://doi.org/10.1007/s13171-024-00348-8.

S. Banerjee and B. P. Carlin. Spatial semi-parametric proportional hazards models for analyzing infant mortality rates in minnesota counties. In C. Gatsonis, R. E. Kass, A. Carriquiry, A. Gelman, D. Higdon, D. K. Pauler, and I. Verdinelli, editors, *Case Studies in Bayesian Statistics Volume 6*, pages 137–151. Springer, New York, 2002. doi: https://doi.org/10.1007/978-1-4612-2078-7.

S. Banerjee and A. Gelfand. On smoothness properties of spatial processes. *Journal of Multivariate Analysis*, 84(1):85–100, 2003. ISSN 0047-259X. doi: https://doi.org/10.1016/S0047-259X(02)00016-7. URL https://www.sciencedirect.com/science/article/pii/S0047259X02000167.

S. Banerjee and A. E. Gelfand. Prediction, interpolation and regression for spatially misaligned data. *Sankhyā: The Indian Journal of Statistics, Series A (1961-2002)*, 64(2): 227–245, 2002. ISSN 0581572X. URL http://www.jstor.org/stable/25051392.

S. Banerjee and A. E. Gelfand. Bayesian Wombling: Curvilinear Gradient Assessment Under Spatial Process Models. *Journal of the American Statistical Association*, 101(476):1487–1501, 2006.

S. Banerjee and G. A. Johnson. Coregionalized single- and multiresolution spatially varying growth curve modeling with application to weed growth. *Biometrics*, 62(3):864–876, 2006. doi: https://doi.org/10.1111/j.1541-0420.2006.00535.x. URL https://onlinelibrary.wiley.com/doi/abs/10.1111/j.1541-0420.2006.00535.x.

S. Banerjee and A. Roy. *Linear Algebra and Matrix Analysis for Statistics*. Chapman & Hall/CRC Texts in Statistical Science. Chapman & Hall/CRC, Philadelphia, PA, June 2014.

S. Banerjee, X. Chen, I. Frankenburg and D. Zhou. Dynamic Bayesian learning for spatiotemporal mechanistic models. *Journal of Machine Learning Research* (in press). URL https://arxiv.org/abs/2208.06528.

S. Banerjee, A. E. Gelfand, and W. Polasek. Geostatistical modelling for spatial interaction data with application to postal service performance. *Journal of Statistical Planning and Inference*, 90(1):87–105, 2000. ISSN 0378-3758. doi: https://doi.org/10.1016/S0378-3758(00)00111-7. URL https://www.sciencedirect.com/science/article/pii/S0378375800001117.

S. Banerjee, A. E. Gelfand, and C. Sirmans. Directional rates of change under spatial process models. *Journal of the American Statistical Association*, 98(464):946–954, 2003a.

S. Banerjee, M. M. Wall, and B. P. Carlin. Frailty modeling for spatially correlated survival data, with application to infant mortality in minnesota. *Biostatistics*, 4(1):123–142, 2003b.

S. Banerjee, A. E. Gelfand, J. R. Knight, and C. F. Sirmans. Spatial modeling of house prices using normalized distance-weighted sums of stationary processes. *Journal of Business & Economic Statistics*, 22(2):206–213, 2004. ISSN 07350015. URL http://www.jstor.org/stable/1392179.

S. Banerjee, A. E. Gelfand, A. O. Finley, and H. Sang. Gaussian predictive process models for large spatial data sets. *Journal of the Royal Statistical Society: Series B (Statistical Methodology)*, 70(4):825–848, 2008. doi: https://doi.org/10.1111/j.1467-9868.2008.00663.x. URL https://rss.onlinelibrary.wiley.com/doi/abs/10.1111/j.1467-9868.2008.00663.x.

S. Banerjee, A. O. Finley, P. Waldmann, and T. Ericcson. Hierarchical spatial process models for multiple traits in large genetic trials. *Journal of the American Statistical Association*, 105:506–521, 2010.

J. J. Barber, A. E. Gelfand, and J. A. Silander. Modelling map positional error to infer true feature location. *Canadian Journal of Statistics*, 34(4):659–676, 2006.

G. Barbujani, G. Jacquez, and L. Ligi. Diversity of some gene frequencies in european and asian populations. v. steep multilocus clines. *American Journal of Human Genetics*, 47: 867–75, 12 1990.

G. Barbujani, A. Magagni, E. Minch, and L. L. Cavalli-Sforza. An apportionment of human dna diversity. *Proceedings of the National Academy of Sciences*, 94(9):4516–4519, 1997. doi: 10.1073/pnas.94.9.4516. URL https://www.pnas.org/doi/abs/10.1073/pnas.94.9.4516.

G. Barnard. Studies in the history of probability and statistics: IX. Thomas Bayes's essay towards solving a problem in the doctrine of chances*: Reproduced with the permission of the Council of the Royal Society from The Philosophical Transactions (1763), 53, 370-418. *Biometrika*, 45(3-4):293–295, 12 1958. ISSN 0006-3444. doi: 10.1093/biomet/45.3-4.293. URL https://doi.org/10.1093/biomet/45.3-4.293.

A. E. Baron, S. Franceschi, S. Barra, R. Talamini, and C. La Vecchia. A comparison of the joint effects of alcohol and smoking on the risk of cancer across sites in the upper aerodigestive tract. *Cancer epidemiology, biomarkers & prevention: a publication of the American Association for Cancer Research, cosponsored by the American Society of Preventive Oncology*, 2(6):519–523, 1993.

R. Barry and J. Ver Hoef. Blackbox kriging: Spatial prediction without specifying variogram models. *Journal of Agricultural, Biological and Environmental Statistics*, 1:297–322, 1996.

M. J. Bayarri, J. O. Berger, and F. Liu. Modularization in Bayesian analysis, with emphasis on analysis of computer models. *Bayesian Analysis*, 4(1):119–150, 2009.

T. Bayes. LII. An essay towards solving a problem in the doctrine of chances. by the late Rev. Mr. Bayes, F.R.S. communicated by Mr. Price, in a letter to John Canton, A. M. F. R. S. *Philosophical Transactions of the Royal Society of London*, 53:370–418, 1763. doi: 10.1098/rstl.1763.0053. URL https://royalsocietypublishing.org/doi/abs/10.1098/rstl.1763.0053.

A. Baíllo and A. Grané. Local linear regression for functional predictor and scalar response. *Journal of Multivariate Analysis*, 100(1):102–111, 2009. ISSN 0047-259X. doi: https://doi.org/10.1016/j.jmva.2008.03.008. URL https://www.sciencedirect.com/science/article/pii/S0047259X08000973.

M. A. Beaumont. Approximate Bayesian computation. *Annual Review of Statistics and Its Application*, 6:379–403, 2019.

M. A. Beaumont, W. Zhang, and D. J. Balding. Approximate Bayesian computation in population genetics. *Genetics*, 162:2025–2035, 2002.

R. Becker and A. Wilks. Maps in s. *AT&T Bell Laboratories Statistics Research Report*, 03 1998.

R. A. Becker, A. R. Wilks, and M. Hill. Constructing a geographical database, 1997. URL https://api.semanticscholar.org/CorpusID:17258119.

M. L. Bell, F. Dominici, and J. M. Samet. A meta-analysis of time-series studies of ozone and mortality with comparison to the national morbidity, mortality, and air pollution study. *Epidemiology*, 16(4):436–45, 2005. ISSN 1044-3983. URL https://journals.lww.com/epidem/fulltext/2005/07000/a_meta_analysis_of_time_series_studies_of_ozone.4.aspx.

V. Benes, K. Bodlák, J. Møller, and R. P. Waagepetersen. Application of log Gaussian Cox processes in disease mapping. In *The ISI International Conference on Environmental Statistics and Health*, pages 95–105. University of Santiago de Compostela, 2003.

Y. Benjamini and Y. Hochberg. Controlling the false discovery rate: A practical and powerful approach to multiple testing. *Journal of the Royal Statistical Society. Series B (Methodological)*, 57(1):289–300, 1995. ISSN 00359246. URL http://www.jstor.org/stable/2346101.

J. O. Berger. *Statistical decision theory and Bayesian analysis*. Springer Series in Statistics. Springer, New York, NY, 2 edition, Mar. 1993.

J. O. Berger and L. R. Pericchi. The Intrinsic Bayes Factor for Linear Models[1]. In *Bayesian Statistics 5: Proceedings of the Fifth Valencia International Meeting*. Oxford University Press, 05 1996. ISBN 9780198523567. doi: 10.1093/oso/9780198523567.003.0002. URL https://doi.org/10.1093/oso/9780198523567.003.0002.

J. O. Berger, V. D. Oliveira, and B. Sansó. Objective Bayesian analysis of spatially correlated data. *Journal of the American Statistical Association*, 96(456):1361–1374, 2001. doi: 10.1198/016214501753382282. URL https://doi.org/10.1198/016214501753382282.

P. Bergthórsson and B. R. Döös. Numerical weather map analysis. *Tellus*, 7(3):329–340, 1955. doi: https://doi.org/10.1111/j.2153-3490.1955.tb01170.x. URL https://onlinelibrary.wiley.com/doi/abs/10.1111/j.2153-3490.1955.tb01170.x.

J. Berkson and R. P. Gage. Survival curve for cancer patients following treatment. *Journal of the American Statistical Association*, 47(259):501–515, 1952. ISSN 01621459, 1537274X. URL http://www.jstor.org/stable/2281318.

M. Berliner. Hierarchical Bayesian modeling in the environmental sciences. *AStA Advances in Statistical Analysis*, 2(84):141–153, 2000.

M. Berman and T. R. Turner. Approximating point process likelihoods with GLIM. *Applied Statistics*, 14(1):31–38, 1992.

L. Bernardinelli and C. Montomoli. Empirical Bayes versus fully Bayesian analysis of geographical variation in disease risk. *Statistics in Medicine*, 11(8):983–1007, 1992. doi: https://doi.org/10.1002/sim.4780110802. URL https://onlinelibrary.wiley.com/doi/abs/10.1002/sim.4780110802.

L. Bernardinelli, D. Clayton, and C. Montomoli. Bayesian estimates of disease maps: How important are priors? *Statistics in Medicine*, 14(21-22):2411–2431, 1995a. doi: https://doi.org/10.1002/sim.4780142111. URL https://onlinelibrary.wiley.com/doi/abs/10.1002/sim.4780142111.

L. Bernardinelli, D. Clayton, C. Pascutto, C. Montomoli, M. Ghislandi, and M. Songini. Bayesian analysis of space—time variation in disease risk. *Statistics in Medicine*, 14(21-22):2433–2443, 1995b. doi: https://doi.org/10.1002/sim.4780142112. URL https://onlinelibrary.wiley.com/doi/abs/10.1002/sim.4780142112.

J. M. Bernardo and A. F. M. Smith. *Bayesian Theory*. Wiley Series in Probability and Statistics. John Wiley & Sons, Chichester, England, Mar. 2000.

V. Berrocal. Data assimilation. In A. E. Gelfand, M. Fuentes, J. Hoeting, and R. L. Smith, editors, *The Handbook of Environmental and Ecological Statistics*. (Chapman Hall), Boca Raton, 2019.

V. J. Berrocal, A. E. Gelfand, and D. M. Holland. A spatio-temporal downscaler for output from numerical models. *Journal of Agricultural, Biological and Environmental Statistics*, 15(2):176–197, 2010a. doi: 10.1007/s13253-009-0004-z.

V. J. Berrocal, A. E. Gelfand, and D. M. Holland. A bivariate space–time downscaler under space and time misalignment. *The Annals of Applied Statistics*, 4(4):1942–1975, 2010b. doi: 10.1214/10-AOAS351. URL https://doi.org/10.1214/10-AOAS351.

V. J. Berrocal, A. E. Gelfand, and D. M. Holland. Space-time data fusion under error in computer model output: An application to modeling air quality. *Biometrics*, 68(3):837–848, 2012. doi: 10.1111/j.1541-0420.2011.01725.x.

K. K. Berthelsen and J. Møller. A primer on perfect simulation for spatial point processes. *Bulletin of the Brazilian Mathematical Society*, 33:351–367, 2002.

K. K. Berthelsen and J. Møller. Likelihood and non-parametric Bayesian MCMC inference for spatial point processes based on perfect simulation and path sampling. *Scandinavian Journal of Statistics*, 30(3):549–564, 2003.

K. K. Berthelsen and J. Møller. Non-parameteric Bayesian inference for inhomogeneous Markov point proceses. *Australian & New Zealand Journal of Statistics*, 50(3):257–272, 2008.

J. Besag. Spatial Interaction and the Statistical Analysis of Lattice Systems. *Journal of the Royal Statistical Society: Series B (Methodological)*, 36(2):192–225, 12 1974. ISSN 0035-9246. doi: 10.1111/j.2517-6161.1974.tb00999.x. URL https://doi.org/10.1111/j.2517-6161.1974.tb00999.x.

J. Besag. On a system of two-dimensional recurrence equations. *Journal of the Royal Statistical Society: Series B (Methodological)*, 43(3):302–309, 1981. ISSN 00359246. URL http://www.jstor.org/stable/2984940.

J. Besag. On the statistical analysis of dirty pictures. *Journal of the Royal Statistical Society: Series B (Statistical Methodology)*, 48:259–302, 1986.

J. Besag and P. J. Diggle. Simple Monte Carlo tests for spatial pattern. *Applied Statistics*, pages 327–333, 1977.

J. Besag and C. Kooperberg. On conditional and intrinsic autoregression. *Biometrika*, 82 (4):733–746, 1995. ISSN 00063444, 14643510. URL http://www.jstor.org/stable/2337341.

J. Besag, J. York, and A. Mollié. Bayesian image restoration, with two applications in spatial statistics. *Annals of the Institute of Statistical Mathematics*, 43(1):1–20, Mar 1991. ISSN 1572-9052. doi: 10.1007/BF00116466. URL https://doi.org/10.1007/BF00116466.

J. Besag, P. Green, D. Higdon, and K. Mengersen. Bayesian Computation and Stochastic Systems. *Statistical Science*, 10(1):3–41, 1995. doi: 10.1214/ss/1177010123. URL https://doi.org/10.1214/ss/1177010123.

J. E. Besag. Statistical analysis of non-lattice data. *The Statistician*, 24:179–195, 1975.

N. G. Best, R. A. Arnold, A. Thomas, L. A. Waller, and E. M. Conlon. Bayesian Models for Spatially Correlated Disease and Exposure Data. In *Bayesian Statistics 6: Proceedings of the Sixth Valencia International Meeting June 6-10, 1998*. Oxford University Press, 08 1999. ISBN 9780198504856. doi: 10.1093/oso/9780198504856.003.0006. URL https://doi.org/10.1093/oso/9780198504856.003.0006.

N. G. Best, K. Ickstadt, and R. L. Wolpert. Spatial poisson regression for health and exposure data measured at disparate resolutions. *Journal of the American Statistical Association*, 95(452):1076–1088, 2000. doi: 10.1080/01621459.2000.10474304. URL https://www.tandfonline.com/doi/abs/10.1080/01621459.2000.10474304.

P. Billingsley. *Probability and Measure*. Wiley Series in Probability and Statistics. Wiley, 2012. ISBN 9781118341919. URL https://books.google.com/books?id=a3gavZbxyJcC.

C. Bishop. *Pattern Recognition and Machine Learning*. Springer-Verlag, New York, NY, 2006.

R. S. Bivand, E. Pebesma, and V. Gomez-Rubio. *Applied Spatial Data Analysis with R*. Springer, New York, NY, second edition, 2013.

D. M. Blei, A. Kucukelbir, and J. D. McAuliffe. Variational inference: A review for statisticians. *Journal of the American Statistical Association*, 112(518):859–877, 2017. doi: 10.1080/01621459.2017.1285773. URL https://doi.org/10.1080/01621459.2017.1285773.

M. Blum, M. Nunes, D. Prangle, and S. Sisson. A comparative review of dimension reduction methods in approximate Bayesian computation. *Statistical Science*, 28:189–208, 2013.

M. G. B. Blum and V. C. Tran. HIV with contact tracing: a case study in approximate Bayesian computation. *Biostatistics*, 11:644–660, 2010.

J. P. Bocquet-Appel and J. N. Bacro. Generalized Wombling. *Systematic Biology*, 43(3): 442–448, 09 1994. ISSN 1063-5157. doi: 10.1093/sysbio/43.3.442. URL https://doi.org/10.1093/sysbio/43.3.442.

D. Bolin and F. Lindgren. Spatial models generated by nested stochastic partial differential equations, with an application to global ozone mapping. *The Annals of Applied Statistics*, 5(1):523–550, 2011. doi: 10.1214/10-AOAS383. URL https://doi.org/10.1214/10-AOAS383.

D. Bolin and F. Lindgren. Excursion and contour uncertainty regions for latent Gaussian models. *Journal of the Royal Statistical Society, B*, 77(1):85–106, 2015.

P. Botella-Rocamora, M. A. Martinez-Beneito, and S. Banerjee. A unifying modeling framework for highly multivariate disease mapping. *Statistics in Medicine*, 34:1548–1559, 2015.

S. Bouka, S. Dabo-Niang, and G. M. Nkiet. On estimation in a spatial functional linear regression model with derivatives. *Comptes Rendus Mathematique*, 356(5):558–562, 2018. ISSN 1631-073X. doi: https://doi.org/10.1016/j.crma.2018.02.013. URL https://www.sciencedirect.com/science/article/pii/S1631073X18301122.

G. E. P. Box and G. C. Tiao. *Bayesian inference in statistical analysis*. Wiley Classics Library. John Wiley & Sons, Nashville, TN, Mar. 1992.

A. M. Bratseth. Statistical interpolation by means of successive corrections. *Tellus A*, 38A(5):439–447, 1986. doi: https://doi.org/10.1111/j.1600-0870.1986.tb00476.x. URL https://onlinelibrary.wiley.com/doi/abs/10.1111/j.1600-0870.1986.tb00476.x.

N. E. Breslow and D. G. Clayton. Approximate inference in generalized linear mixed models. *Journal of the American Statistical Association*, 88(421):9–25, 1993. ISSN 01621459, 1537274X. URL http://www.jstor.org/stable/2290687.

N. E. Breslow, N. E. Day, and E. Heseltine. Statistical methods in cancer research volume i: The analysis of case-control studies, 1980.

D. Brook. On the distinction between the conditional probability and the joint probability approaches in the specification of nearest-neighbour systems. *Biometrika*, 51(3-4):481–483, 12 1964. ISSN 0006-3444. doi: 10.1093/biomet/51.3-4.481. URL https://doi.org/10.1093/biomet/51.3-4.481.

S. Brooks, A. Gelman, G. Jones, and X. Meng. *Handbook of Markov Chain Monte Carlo*. ISSN. CRC Press, 2011. ISBN 9781420079425. URL https://books.google.com/books?id=qfRsAIKZ4rIC.

S. P. Brooks and A. Gelman. General methods for monitoring convergence of iterative simulations. *Journal of Computational and Graphical Statistics*, 7(4):434–455, 1998. doi: 10.1080/10618600.1998.10474787. URL https://www.tandfonline.com/doi/abs/10.1080/10618600.1998.10474787.

P. E. Brown, G. O. Roberts, K. F. Karesen, and S. Tonellato. Blur-generated non-separable space–time models. *Journal of the Royal Statistical Society Series B: Statistical Methodology*, 62(4):847–860, 01 2002. ISSN 1369-7412. doi: 10.1111/1467-9868.00269. URL https://doi.org/10.1111/1467-9868.00269.

P. J. Brown, N. D. Le, and J. V. Zidek. Multivariate spatial interpolation and exposure to air pollutants. *The Canadian Journal of Statistics / La Revue Canadienne de Statistique*, 22(4):489–509, 1994. ISSN 03195724. URL http://www.jstor.org/stable/3315406.

F. Bruno, D. Cocchi, and A. Vagheggini. Finite population properties of individual predictors based on spatial pattern. *Environmental and Ecological Statistics*, 20(3):467–494, 2013.

S. Brush. History of the lenz-ising model. *Reviews of Modern Physics*, 39:883–893, Oct 1967. doi: 10.1103/RevModPhys.39.883. URL https://link.aps.org/doi/10.1103/RevModPhys.39.883.

S. T. Buckland, D. R. Anderson, K. P. Burnham, and L. Thomas. *Advanced Distance Sampling*. Oxford University Press, 2007. ISBN 978-0-19-922587-3.

L. Bugayevskiy and J. Snyder. *Map projections*. Taylor & Francis, London, England, June 1995.

T. A. Buishand, L. de Haan, and C. Zhou. On spatial extremes: With application to a rainfall problem. *The Annals of Applied Statistics*, 2(2):624 – 642, 2008. doi: 10.1214/ 08-AOAS159. URL https://doi.org/10.1214/08-AOAS159.

C. A. Calder. A dynamic process convolution approach to modeling ambient particulate matter concentrations. *Environmetrics*, 19(1):39–48, 2008.

J. Cao, M. Kang, F. Jimenez, H. Sang, F. T. Schaefer, and M. Katzfuss. Variational sparse inverse cholesky approximation for latent Gaussian processes via double kullback-leibler minimization. In A. Krause, E. Brunskill, K. Cho, B. Engelhardt, S. Sabato, and J. Scarlett, editors, *Proceedings of the 40th International Conference on Machine Learning*, volume 202 of *Proceedings of Machine Learning Research*, pages 3559–3576. PMLR, 23–29 Jul 2023. URL https://proceedings.mlr.press/v202/cao23b.html.

B. P. Carlin and S. Banerjee. Hierarchical multivariate car models for spatio-temporally correlated survival data. *Bayesian Statistics*, 7(7):45–63, 2003.

B. P. Carlin and J. S. Hodges. Hierarchical proportional hazards regression models for highly stratified data. *Biometrics*, 55(4):1162–1170, 1999. ISSN 0006341X, 15410420. URL http://www.jstor.org/stable/2533735.

B. P. Carlin and T. A. Louis. *Bayesian Methods for Data Analysis*. Chapman and Hall/CRC, Boca Raton, FL, third edition, 2008.

B. P. Carlin and M.-E. Pérez. Robust Bayesian analysis in medical and epidemiological settings. In D. R. Insua and F. Ruggeri, editors, *Robust Bayesian Analysis*, pages 351–372. Springer New York, New York, NY, 2000. ISBN 978-1-4612-1306-2. doi: 10.1007/ 978-1-4612-1306-2_19. URL https://doi.org/10.1007/978-1-4612-1306-2_19.

B. P. Carlin, H. Xia, O. Devine, P. Tolbert, and J. Mulholland. Spatio-temporal hierarchical models for analyzing atlanta pediatric asthma er visit rates. In C. Gatsonis, R. E. Kass, B. Carlin, A. Carriquiry, A. Gelman, I. Verdinelli, and M. West, editors, *Case Studies in Bayesian Statistics*, pages 303–320, New York, NY, 1999. Springer New York. ISBN 978-1-4612-1502-8.

B. Carpenter, A. Gelman, M. D. Hoffman, D. Lee, B. Goodrich, M. Betancourt, M. Brubaker, J. Guo, P. Li, and A. Riddell. Stan: A probabilistic programming language. *Journal of Statistical Software*, 76(1), 2017.

R. J. Carroll, D. Ruppert, L. A. Stefanski, and C. Crainiceanu. *Measurement Error in Nonlinear Models*. Chapman and Hall/CRC, New York, NY, 2 edition, 2006. doi: 10. 1201/9781420010138.

K. E. Case and R. J. Shiller. The efficiency of the market for single-family homes. *The American Economic Review*, 79(1):125–137, 1989. ISSN 00028282. URL http://www. jstor.org/stable/1804778.

G. Casella and E. I. George. Explaining the gibbs sampler. *The American Statistician*, 46(3): 167–174, 1992. doi: 10.1080/00031305.1992.10475878. URL https://www.tandfonline. com/doi/abs/10.1080/00031305.1992.10475878.

W. Caspary and R. Scheuring. Positional accuracy in spatial databases. *Computers, Environment and Urban Systems*, 17(2):103–110, 1993.

J. Castillo-Mateo, M. Lafuente Blasco, A. Gelfand, J. Asin, A. Cebrian, and J. Abaurrea. Spatial modeling of day-within-year temperature time series: an examination of daily maximum temperatures in aragon, spain. *Journal of Agricultural, Biological and Environmental Statistics*, 27:487–505, 2022.

J. Castillo-Mateo, A. Gelfand, J. Asin, A. Cebrian, and J. Abaurrea. Spatial quantile autoregression for season within year daily temperature data. *Annals of Applied Statistics*, 17:2305–2325, 2023a.

J. Castillo-Mateo, A. E. Gelfand, C. A. Hudak, C. A. Mayo, and R. S. Schick. Space-time multi-level modeling for zooplankton abundance employing double data fusion and calibration. *Environmental and Ecological Statistics*, 30:769–795, 2023b.

J. Castillo-Mateo, A. E. Gelfand, J. Asín, A. C. Cebrián, and J. Abaurrea. Bayesian joint quantile autoregression. *TEST*, 33:335–357, 2024.

J. Castillo-Mateo, Z. Tabuenca, A. Gelfand, J. Asin, and A. Cebrian. Spatio-temporal modeling for record-breaking temperature events in spain. *Journal of the Amer Stat Assoc*, page (forthcoming), 2025.

A. C. Cebrian, J. Asin, A. E. Gelfand, E. M. Schliep, J. Castillo-Mateo, M. A. Beamonte, and J. Abaurrea. Spatio-temporal analysis of the extent of an extreme heat event. *Stochastic Environmental Research and Risk Assessment*, 36:2737–2751, 2022.

L. Cecconi, L. Grisotto, D. Catelan, C. Lagazio, V. Berrocal, and A. Biggeri. Preferential sampling and Bayesian geostatistics: Statistical modeling and examples. *Statistical Methods in Medical Research*, 25(4):1224–1243, 2016.

G. Celeux, M. Hurn, and C. P. Robert. Computational and inferential difficulties with mixture posterior distributions. *Journal of the American Statistical Association*, 95(451): 957–970, 2000. doi: 10.1080/01621459.2000.10474285. URL https://www.tandfonline.com/doi/abs/10.1080/01621459.2000.10474285.

A. Chakraborty and A. E. Gelfand. Analyzing spatial point patterns subject to measurement error. *Bayesian Analysis*, 5(1):97–122, 2010. doi: 10.1214/10-BA504. URL https://doi.org/10.1214/10-BA504.

A. Chakraborty, A. E. Gelfand, A. M. Wilson, A. M. Latimer, and J. A. Silander Jr. Modeling large scale species abundance with latent spatial processes. *The Annals of Applied Statistics*, pages 1403–1429, 2010.

A. Chakraborty, A. E. Gelfand, A. M. Wilson, A. M. Latimer, and J. A. Silander. Point pattern modelling for degraded presence-only data over large regions. *Journal of the Royal Statistical Society: Series C (Applied Statistics)*, 60(5):757–776, 2011.

J. M. Chambers. *Graphical methods for data analysis*. Duxbury Resource Center, 1983.

A. M. Chan-Golston, S. Banerjee, and M. S. Handcock. Bayesian inference for finite populations under spatial process settings. *Environmetrics*, 31(3):e2606, 2020. doi: https://doi.org/10.1002/env.2606. URL https://onlinelibrary.wiley.com/doi/abs/10.1002/env.2606. e2606 env.2606.

A. M. Chan-Golston, S. Banerjee, T. R. Belin, S. E. Roth, and M. L. Prelip. Bayesian finite-population inference with spatially correlated measurements. *Japanese Journal of Statistics and Data Science*, 5:407–430, 2022. doi: https://doi.org/10.1007/s42081-022-00178-8. URL https://link.springer.com/article/10.1007/s42081-022-00178-8#citeas.

R. Chellappa and A. K. Jain, editors. *Markov random fields*. Academic Press, San Diego, CA, Dec. 1992.

M.-H. Chen, J. G. Ibrahim, and D. Sinha. A new Bayesian model for survival data with a surviving fraction. *Journal of the American Statistical Association*, 94(447):909–919, 1999. ISSN 01621459, 1537274X. URL http://www.jstor.org/stable/2670006.

M.-H. Chen, Q.-M. Shao, and J. G. Ibrahim. *Monte Carlo methods in Bayesian computation*. Springer Series in Statistics. Springer, New York, NY, 1 edition, Jan. 2000.

W. Chen, Y. Li, B. J. Reich, and Y. Sun. Deepkriging: Spatially dependent deep neural networks for spatial prediction, 2022. URL https://arxiv.org/abs/2007.11972.

X. Chen and S. T. Tokdar. Joint quantile regression for spatial data. *Journal of the Royal Statistical Society: Series B (Statistical Methodology)*, 83(4):826–852, 2021. doi: 10.1111/rssb.12467.

S. Cherry. An evaluation of a non-parametric method of estimating semi-variograms of isotropic spatial processes. *Journal of Applied Statistics*, 23(4):435–449, 1996. doi: 10. 1080/02664769624170. URL https://doi.org/10.1080/02664769624170.

S. Chib and E. Greenberg. Analysis of multivariate probit models. *Biometrika*, 85(2):347–361, 06 1998. ISSN 0006-3444. doi: 10.1093/biomet/85.2.347. URL https://doi.org/10.1093/biomet/85.2.347.

J. Chilès and P. Delfiner. *Geostatistics: Modeling Spatial Uncertainty.* John Wiley: New York., 1999.

J.-P. Chilès and P. Delfiner. *Geostatistics: Modeling Spatial Uncertainty.* Wiley Series in Probability and Statistics. Wiley-Blackwell, Hoboken, NJ, 2nd edition, Mar. 2012.

G. Christakos. On the problem of permissible covariance and variogram models. *Water Resources Research*, 20(2):251–265, 1984. doi: https://doi.org/10.1029/WR020i002p00251. URL https://agupubs.onlinelibrary.wiley.com/doi/abs/10.1029/WR020i002p00251.

R. Christensen, W. Johnson, A. Branscum, and T. Hanson. *Bayesian Ideas and Data Analysis: An Introduction for Scientists and Statisticians.* Chapman & Hall/CRC Texts in Statistical Science. Taylor & Francis, 2011. ISBN 9781439803554. URL https://books.google.com/books?id=qPERhCbePNcC.

Y. Chun and D. A. Griffith. *Spatial Statistics and Geostatistics: Theory and Applications for Geographic Information Science and Technology.* SAGE, London, UK, 2013.

G. Cicchitelli and G. E. Montanari. Model-assisted estimation of a spatial population mean. *International Statistical Review*, 80(1):111–126, 2012.

I. Clark, K. L. Basinger, and W. V. Harper. Muck: A novel approach to co-kriging. In B. Buxton, editor, *Proceedings of the Conference on Geostatistical, Sensitivity, and Uncertainty Methods for Ground-Water Flow and Radionuclide Transport Modeling.* attelle Press, Columbus, OH., 1989.

D. Clayton. Some approaches to the analysis of recurrent event data. *Statistical Methods in Medical Research*, 3(3):244–262, 1994. doi: 10.1177/096228029400300304. URL https://doi.org/10.1177/096228029400300304. PMID: 7820294.

D. G. Clayton. A Monte Carlo method for Bayesian inference in frailty models. *Biometrics*, 47(2):467–485, 1991. ISSN 0006341X, 15410420. URL http://www.jstor.org/stable/2532139.

D. G. Clayton and J. M. Kaldor. Empirical bayes estimates of age-standardized relative risks for use in disease mapping. *Biometrics*, 43 3:671–81, 1987. URL https://api.semanticscholar.org/CorpusID:21068137.

A. D. Cliff and J. K. Ord. *Spatial Autocorrelation.* Pion, London, England, Sept. 1973.

P. Clifford. Markov random fields in statistics. In G. Grimmett and D. Welsh, editors, *Disorder in physical systems.* Clarendon Press, Oxford, England, 1990.

M. Clyde and E. S. Iversen. Bayesian model averaging in the m-open framework. *Bayesian theory and applications*, 14(4):483–498, 2013. doi: http://dx.doi.org/10.1093/acprof:oso/9780199695607.003.0024.

W. G. Cochran. *Sampling Techniques.* John Wiley & Sons, Hoboken, NJ, 3rd edition, 1977.

E. Cohen, R. F. Riesenfeld, and G. Elber. *Geometric modeling with splines.* A K Peters, Natick, MA, July 2001.

D. L. Cohn, E. J. Fisher, G. T. Peng, J. S. Hodges, J. Chesnut, C. C. Child, B. Franchino, C. L. Gibert, W. El-Sadr, R. Hafner, J. Korvick, M. Ropka, L. Heifets, J. Clotfelter, D. Munroe, J. Horsburgh, C. Robert, and T. B. C. P. for Clinical Research on AIDS. A Prospective Randomized Trial of Four Three-Drug Regimens in the Treatment of Disseminated Mycobacterium avium Complex Disease in AIDS Patients: Excess Mortality Associated with High-Dose Clarithromycin. *Clinical Infectious Diseases*, 29(1):125–133, 07 1999. ISSN 1058-4838. doi: 10.1086/520141. URL https://doi.org/10.1086/520141.

T. J. Cole and P. J. Green. Smoothing reference centile curves: The lms method and penalized likelihood. *Statistics in Medicine*, 11(10):1305–1319, 1992. doi: https://doi.org/10.1002/sim.4780111005. URL https://onlinelibrary.wiley.com/doi/abs/10.1002/sim.4780111005.

S. Coles. *An introduction to statistical modeling of extreme values.* Springer Series in Statistics. Springer, London, England, 2001 edition, Aug. 2001.

S. G. Coles and J. A. Tawn. a Bayesian analysis of extreme rainfall data. *Journal of the Royal Statistical Society: Series C (Applied Statistics)*, 45(4):463–478, 1996. doi: https://doi.org/10.2307/2986068. URL https://rss.onlinelibrary.wiley.com/doi/abs/10.2307/2986068.

P. Congdon and N. Best. Small area variation in hospital admission rates: Bayesian adjustment for primary care and hospital factors. *Journal of the Royal Statistical Society. Series C (Applied Statistics)*, 49(2):207–226, 2000. ISSN 00359254, 14679876. URL http://www.jstor.org/stable/2680850.

D. Cooley, D. Nychka, and P. Naveau. Bayesian spatial modeling of extreme precipitation return levels. *Journal of the American Statistical Association*, 102(479):824–840, 2007. doi: 10.1198/016214506000000780. URL https://doi.org/10.1198/016214506000000780.

F. Cooner, S. Banerjee, B.P. Carlin and D. Sinha. Flexible cure rate modeling under latent activation schemes. *Journal of the American Statistical Association*, 102(478), 560–572, 2007. URL https://doi.org/10.1198/01621450700000011.

F. Corpas-Burgos and M. A. Martinez-Beneito. On the use of adaptive spatial weight matrices from disease mapping multivariate analyses. *Stochastic Environmental Research and Risk Assessment*, 34:531–544, 2020.

L. C. A. Corsten. Interpolation and optimal linear prediction. *Statistica Neerlandica*, 43:69–84, 1989. URL https://api.semanticscholar.org/CorpusID:120447658.

S. Coube-Sisqueille, S. Banerjee, and B. Liquet. Nonstationary spatial process models with spatially varying covariance kernels. *Journal of Computational and Graphical Statistics* (in press). URL https://doi.org/10.1080/10618600.2025.2516020. Nonstationary spatial process models with spatially varying covariance kernels, 2024. URL https://arxiv.org/abs/2203.11873.

M. K. Cowles and B. P. Carlin. Markov chain Monte Carlo convergence diagnostics: A comparative review. *Journal of the American Statistical Association*, 91(434):883–904, 1996. doi: 10.1080/01621459.1996.10476956. URL https://www.tandfonline.com/doi/abs/10.1080/01621459.1996.10476956.

D. Cox and N. Wermuth. *Multivariate Dependencies.* Chapman & Hall/CRC, Boca Raton, FL, 1996.

D. R. Cox and D. Oakes. *Analysis of survival data.* Chapman & Hall/CRC Monographs on Statistics and Applied Probability. Chapman & Hall/CRC, Philadelphia, PA, June 1984.

D. R. Cox and N. Wermuth. Linear Dependencies Represented by Chain Graphs. *Statistical Science*, 8(3):204 – 218, 1993. doi: 10.1214/ss/1177010887. URL https://doi.org/10.1214/ss/1177010887.

H. Cramér. On the theory of stationary random processes. *Annals of Mathematics*, 41:215, 1940. URL https://api.semanticscholar.org/CorpusID:123755454.

H. Cramer and R. Leadbetter. *Stationary and related stochastic processes*. Dover Books on Mathematics. Dover Publications, Mineola, NY, Mar. 2005.

J. Crank. *The mathematics of diffusion*. Oxford science publications. Oxford University Press, London, England, 2 edition, Mar. 1979.

N. Cressie. Change of support and the modifiable areal unit problem. *Geographical Systems*, 3:159–180, Jan 1996.

N. Cressie and N. H. Chan. Spatial modeling of regional variables. *Journal of the American Statistical Association*, 84(406):393–401, 1989. ISSN 01621459, 1537274X. URL http://www.jstor.org/stable/2289922.

N. Cressie and J. L. Davidson. Image analysis with partially ordered Markov models. *Computational Statistics and Data Analysis*, 29(1):1–26, 1998. ISSN 0167-9473. doi: http://dx.doi.org/10.1016/S0167-9473(98)00052-8. URL http://www.sciencedirect.com/science/article/pii/S0167947398000528.

N. Cressie and D. M. Hawkins. Robust estimation of the variogram: I. *Journal of the International Association for Mathematical Geology*, 12(2):115–125, Apr 1980. ISSN 1573-8868. doi: 10.1007/BF01035243. URL https://doi.org/10.1007/BF01035243.

N. Cressie and H. Huang. Classes of nonseparable, spatio-temporal stationary covariance functions. *Journal of the American Statistical Association*, 94:1330–1340, 1999.

N. Cressie and G. Johannesson. Fixed rank kriging for very large spatial data sets. *Journal of the Royal Statistical Society: Series B (Statistical Methodology)*, 70(1):209–226, 2008.

N. Cressie and J. Kornak. Spatial statistics in the presence of location error with an application to remote sensing of the environment. *Statistical Science*, pages 436–456, 2003.

N. Cressie and T. Read. Do sudden infant deaths come in clusters? *Statistics and Decisions Supplement Issue*, 2:333–349, 1985.

N. Cressie and C. K. Wikle. The variance-based cross-variogram: You can add apples and oranges. *Mathematical Geology*, 30(7):789–799, Oct 1998. ISSN 1573-8868. doi: 10.1023/A:1021770324434. URL https://doi.org/10.1023/A:1021770324434.

N. Cressie and C. K. Wikle. *Statistics for Spatio-Temporal Data*. Wiley-Blackwell, Chichester, England, Mar. 2011.

N. Cressie and A. Zammit-Mangion. Multivariate spatial covariance models: a conditional approach. *Biometrika*, 103(4):915–935, 12 2016. ISSN 0006-3444. doi: 10.1093/biomet/asw045. URL https://doi.org/10.1093/biomet/asw045.

N. A. Cressie. *Statistics for Spatial Data*. Probability & Statistics. John Wiley & Sons, Nashville, TN, 2nd edition, oct 1993.

G. P. Cressman. An operational objective analysis system. *Monthly Weather Review*, 87(10):367–374, 1959. doi: 10.1175/1520-0493(1959)087⟨0367:AOOAS⟩2.0.CO;2. URL https://journals.ametsoc.org/view/journals/mwre/87/10/1520-0493_1959_087_0367_aooas_2_0_co_2.xml.

L. Csató. Gaussian processes:iterative sparse approximations. March 2002. URL https://publications.aston.ac.uk/id/eprint/1327/.

F. Csillag and S. Kabos. Wavelets, boundaries, and the spatial analysis of landscape pattern. *Écoscience*, 9(2):177–190, 2002. doi: 10.1080/11956860.2002.11682704. URL https://doi.org/10.1080/11956860.2002.11682704.

K. Csillery, O. Francois, and M. G. B. Blum. abc: an r package for approximate Bayesian computation (abc). *Methods in Ecology and Evolution*, 2012. doi: 10.1111/j.2041-210X. 2011.00179.x.

D. J. Daley and D. Vere-Jones. *An introduction to the theory of point processes*. Probability and Its Applications. Springer, New York, NY, 2 edition, Nov. 2003.

D. Damian, P. D. Sampson, and P. Guttorp. Bayesian estimation of semi-parametric non-stationary spatial covariance structures. *Environmetrics*, 12(2):161–178, 2001. doi: https://doi.org/10.1002/1099-095X(200103)12:2⟨161::AID-ENV452⟩3.0.CO;2-G. URL https://onlinelibrary.wiley.com/doi/abs/10.1002/1099-095X%28200103%2912%3A2%3C161%3A%3AAID-ENV452%3E3.0.CO%3B2-G.

M. J. Daniels and R. E. Kass. Nonconjugate Bayesian estimation of covariance matrices and its use in hierarchical models. *Journal of the American Statistical Association*, 94(448): 1254–1263, 1999. ISSN 01621459, 1537274X. URL http://www.jstor.org/stable/2669939.

P. Das and S. Ghosal. Bayesian quantile regression using random B-spline series prior. *Computational Statistics & Data Analysis*, 109:121–143, 2017a. ISSN 0167-9473. doi: 10.1016/j.csda.2016.11.014.

P. Das and S. Ghosal. Analyzing ozone concentration by Bayesian spatio-temporal quantile regression. *Environmetrics*, 28(4):e2443, 2017b. doi: 10.1002/env.2443.

A. Datta, S. Banerjee, A. O. Finley, and A. E. Gelfand. Hierarchical nearest-neighbor Gaussian process models for large geostatistical datasets. *Journal of the American Statistical Association*, 111:800–812, 2016a. URL http://dx.doi.org/10.1080/01621459.2015.1044091.

A. Datta, S. Banerjee, A. O. Finley, N. A. S. Hamm, and M. Schaap. Non-separable dynamic nearest-neighbor Gaussian process models for large spatio-temporal data with an application to particulate matter analysis. *Annals of Applied Statistics*, 10:1286–1316, 2016b. URL http://dx.doi.org/10.1214/16-AOAS931.

A. Datta, S. Banerjee, J. S. Hodges, and L. Gao. Spatial Disease Mapping Using Directed Acyclic Graph Auto-Regressive (DAGAR) Models. *Bayesian Analysis*, 14(4):1221–1244, 2019. doi: 10.1214/19-BA1177. URL https://doi.org/10.1214/19-BA1177.

G. S. Datta and M. Ghosh. Bayesian prediction in linear models: Applications to small area estimation. *The Annals of Statistics*, pages 1748–1770, 1991.

D. M. Blei, A. Kucukelbir, and J. D. McAuliffe. Variational inference: A review for statisticians. *Journal of the American Statistical Association*, 112(518):859–877, 2017. doi: 10.1080/01621459.2017.1285773. URL https://doi.org/10.1080/01621459.2017.1285773.

T. M. Davies, S. Banerjee, A. P. Martin, and R. E. Turnbull. A nearest-neighbour Gaussian process spatial factor model for censored, multi-depth geochemical data. *Journal of the Royal Statistical Society Series C: Applied Statistics*, 71(4):1014–1043, 05 2022. ISSN 0035-9254. doi: 10.1111/rssc.12565. URL https://doi.org/10.1111/rssc.12565.

R. de Fondeville and A. Davison. High-dimensional peaks-over-threshold inference. *Biometrika*, 105:575–592, 2018.

L. De Haan. A spectral representation for max-stable processes. *The Annals of Probability*, 12(4):1194–1204, 1984. doi: 10.1214/aop/1176993148. URL https://doi.org/10.1214/aop/1176993148.

L. De Haan and A. Ferreira. *Extreme value theory: an introduction*. Springer, New York, NY., 2006.

S. De Iaco, D. Myers, and D. Posa. Nonseparable space-time covariance models: Some parametric families. *Mathematical Geology*, 34(1):23–42, Jan 2002. ISSN 1573-8868. doi: 10.1023/A:1014075310344. URL https://doi.org/10.1023/A:1014075310344.

M. D. DeGroot. *Optimal Statistical Decisions*. Wiley Classics Library. John Wiley & Sons, Nashville, TN, Apr. 2004.

P. Delicado, R. Giraldo, C. Comas, and J. Mateu. Statistics for spatial functional data: some recent contributions. *Environmetrics*, 21(3-4):224–239, 2010. doi: https://doi.org/10.1002/env.1003. URL https://onlinelibrary.wiley.com/doi/abs/10.1002/env.1003.

A. P. Dempster. Covariance selection. *Biometrics*, 28(1):157–175, 1972. ISSN 0006341X, 15410420. URL http://www.jstor.org/stable/2528966.

A. P. Dempster, N. M. Laird, and D. B. Rubin. Maximum likelihood from incomplete data via the em algorithm. *Journal of the Royal Statistical Society: Series B (Methodological)*, 39(1):1–22, 12 2018. ISSN 0035-9246. doi: 10.1111/j.2517-6161.1977.tb01600.x. URL https://doi.org/10.1111/j.2517-6161.1977.tb01600.x.

O. Devine, J. Qualters, J. Morrissey, and P. Wall. Estimation of the impact of the former feed materials production center (fmpc) on lung cancer mortality in the surrounding community, 1998. Technical report: Centers for Disease Control and Prevention. Atlanta, GA.

D. Dey, A. Datta, and S. Banerjee. Graphical Gaussian process models for highly multivariate spatial data. *Biometrika*, 109(4):993–1014, 12 2022. ISSN 1464-3510. doi: 10.1093/biomet/asab061. URL https://doi.org/10.1093/biomet/asab061.

D. Dey, A. Datta, and S. Banerjee. Modeling multivariate spatial dependencies using graphical models. *The New England Journal of Statistics in Data Science*, 1(2):283–295, 2023. ISSN 2693-7166. doi: 10.51387/23-NEJSDS47.

D. Dey, S. Banerjee, M. A. Lindquist, and A. Datta. Graph-constrained analysis for multivariate functional data. *Journal of Multivariate Analysis*, 207:105428, 2025. ISSN 0047-259X. doi: 10.1016/j.jmva.2025.105428. URL https://www.sciencedirect.com/science/article/pii/S0047259X25000235.

D. K. Dey, A. E. Gelfand, T. B. Swartz, and P. K. Vlachos. A simulation-intensive approach for checking hierarchical models. *Test*, 7(2):325–346, 1998.

P. J. Diggle. Point process modelling in environmental epidemiology. In V. Barnett and K. F. Turkman, editors, *Statistics for the environment: Volume 1*. John Wiley & Sons, 1993.

P. J. Diggle. *Statistical analysis of spatial and spatio-temporal point patterns*. CRC Press, 3 edition, 2013.

P. J. Diggle and P. J. Ribeiro. *Model-based Geostatistics*. Springer, New York, NY, 2007.

P. J. Diggle, V. Gómez-Rubio, P. E. Brown, A. G. Chetwynd, and S. Gooding. Second-order analysis of inhomogeneous spatial point processes using case–control data. *Biometrics*, 63(2):550–557, 2007.

P. J. Diggle and E. Gabriel. Spatio-temporal point processes. In A. Gelfand, P. Diggle, M. Fuentes, and P. Guttorp, editors, *Handbook of Spatial Statistics*, pages 449–461. CRC Press, 2010.

P. J. Diggle and S. Lophaven. Bayesian geostatistical design. *Scandinavian Journal of Statistics*, 33(1):53–64, 2006. doi: https://doi.org/10.1111/j.1467-9469.2005.00469.x. URL https://onlinelibrary.wiley.com/doi/abs/10.1111/j.1467-9469.2005.00469.x.

P. J. Diggle and P. J. Ribiero. Bayesian inference in Gaussian model-based geostatistics. *Geographical and Environmental Modelling*, 6(2):129–146, 2002. doi: 10.1080/1361593022000029467. URL https://doi.org/10.1080/1361593022000029467.

P. J. Diggle, J. A. Tawn, and R. A. Moyeed. Model-based geostatistics. *Journal of the Royal Statistical Society: Series C (Applied Statistics)*, 47(3):299–350, 1998. doi: https://doi.org/10.1111/1467-9876.00113. URL https://rss.onlinelibrary.wiley.com/doi/abs/10.1111/1467-9876.00113.

P. J. Diggle, R. Menezes, and T.-l. Su. Geostatistical inference under preferential sampling. *Journal of the Royal Statistical Society: Series C (Applied Statistics)*, 59(2): 191–232, 2010. doi: https://doi.org/10.1111/j.1467-9876.2009.00701.x. URL https://rss.onlinelibrary.wiley.com/doi/abs/10.1111/j.1467-9876.2009.00701.x.

P. J. Diggle, P. Heagerty, K. Liang, and S. Zeger. *Analysis of Longitudinal Data*. Oxford Statistical Science Series. OUP Oxford, 2013. ISBN 9780191664335. URL https://books.google.com/books?id=zAiK-gWUqDUC.

A. Dobra et al. Markov bases for decomposable graphical models. *Bernoulli*, 9(6):1093–1108, 2003.

K. A. Doksum and M. Gasko. On a correspondence between models in binary regression analysis and in survival analysis. *International Statistical Review / Revue Internationale de Statistique*, 58(3):243–252, 1990. ISSN 03067734, 17515823. URL http://www.jstor.org/stable/1403807.

C. C. Drovandi and A. N. Pettitt. Estimation of parameters for macroparasite population evolution using approximate Bayesian computation. *Biometrics*, 67:225–233, 2011.

J. Du, H. Zhang, and V. S. Mandrekar. Fixed-domain asymptotic properties of tapered maximum likelihood estimators. *The Annals of Statistics*, 37(6A):3330–3361, 2009. doi: 10.1214/08-AOS676. URL https://doi.org/10.1214/08-AOS676.

J. Duan and A. Gelfand. Finite mixture model of nonstationary spatial data. Technical Report, 2003. Technical report, Institute for Statistics and Decision Sciences, Duke University.

J. A. Duan, A. E. Gelfand, and C. F. Sirmans. Modeling space-time data using stochastic differential equations. *Bayesian Analysis*, 4(4):733–758, 2009. doi: 10.1214/09-BA427.

S. Duane, A. Kennedy, B. J. Pendleton, and D. Roweth. Hybrid Monte Carlo. *Physics Letters B*, 195(2):216–222, 1987. ISSN 0370-2693. doi: https://doi.org/10.1016/0370-2693(87)91197-X. URL https://www.sciencedirect.com/science/article/pii/037026938791197X.

R. Dutta, M. Schoengens, L. Pacchiardi, A. Ummadisingu, N. Widmer, P. Künzli, J.-P. Onnela, and A. Mira. Abcpy: A high-performance computing perspective to approximate Bayesian computation. *Journal of Statistical Software*, 100(7):1–38, 2021. doi: 10.18637/jss.v100.i07. URL https://www.jstatsoft.org/index.php/jss/article/view/v100i07.

L. E. Eberly and B. P. Carlin. Identifiability and convergence issues for Markov chain Monte Carlo fitting of spatial models. *Statistics in medicine*, 19(17-18):2279–94, 2000. URL https://api.semanticscholar.org/CorpusID:10881631.

M. Ecker and J. Heltshe. Geostatistical estimates of scallop abundance. In N. Lange, L. Ryan, L. Billard, D. Brillinger, L. Conquest, and J. Greenhouse, editors, *Case studies in biometry*, pages 107–124. John Wiley and Sons, New York, NY, 1994.

M. D. Ecker and A. E. Gelfand. Bayesian variogram modeling for an isotropic spatial process. *Journal of Agricultural, Biological, and Environmental Statistics*, 2(4):347–369, 1997. ISSN 10857117. URL http://www.jstor.org/stable/1400508.

M. D. Ecker and A. E. Gelfand. Bayesian modeling and inference for geometrically anisotropic spatial data. *Mathematical Geology*, 31:67–83, 1999.

M. D. Ecker and A. E. Gelfand. Spatial modeling and prediction under stationary non-geometric range anisotropy. *Environmental and Ecological Statistics*, 10(2):165–178, Jun 2003. ISSN 1573-3009. doi: 10.1023/A:1023600123559. URL https://doi.org/10.1023/A:1023600123559.

B. Efron. *Institute of mathematical statistics monographs: Large-scale inference: Empirical Bayes methods for estimation, testing, and prediction series number 1*. Institute of mathematical statistics monographs. Cambridge University Press, Cambridge, England, Nov. 2012.

X. Emery. The kriging update equations and their application to the selection of neighboring data. *Computational Geosciences*, 13(3):269–280, 2009. ISSN 1573-1499. doi: 10.1007/s10596-008-9116-8. URL http://dx.doi.org/10.1007/s10596-008-9116-8.

E. S. Epstein. A scoring system for probability forecasts of ranked categories. *Journal of Applied Meteorology*, 8(6):985–987, 1969.

R. J. Erhardt and R. L. Smith. Approximate Bayesian computing for spatial extremes. *Computational Statistics and Data Analysis*, 56:1468–1481, 2012.

W. A. Ericson. Subjective Bayesian models in sampling finite populations. *Journal of the Royal Statistical Society, Series B*, 31(2):195–233, 1969.

M. Ewell and J. G. Ibrahim. The large sample distribution of the weighted log rank statistic under general local alternatives. *Lifetime Data Analysis*, 3(1):5–12, Jan 1997. ISSN 1572-9249. doi: 10.1023/A:1009690200504. URL https://doi.org/10.1023/A:1009690200504.

W. F. Fagan, M.-j. Fortin, and C. Soykan. Integrating Edge Detection and Dynamic Modeling in Quantitative Analyses of Ecological Boundaries. *BioScience*, 53(8):730–738, 08 2003. ISSN 0006-3568. doi: 10.1641/0006-3568(2003)053[0730:IEDADM]2.0.CO;2. URL https://doi.org/10.1641/0006-3568(2003)053[0730:IEDADM]2.0.CO;2.

L. Fahrmeir, T. Kneib, S. Lang, and B. Marx. *Regression*. Springer, Berlin, Germany, 2013 edition, May 2013.

V. T. Farewell. The use of mixture models for the analysis of survival data with long-term survivors. *Biometrics*, 38(4):1041–1046, 1982. ISSN 0006341X, 15410420. URL http://www.jstor.org/stable/2529885.

V. T. Farewell. Mixture models in survival analysis: Are they worth the risk? *Canadian Journal of Statistics*, 14(3):257–262, 1986. doi: https://doi.org/10.2307/3314804. URL https://onlinelibrary.wiley.com/doi/abs/10.2307/3314804.

M. T. Farr, D. S. Green, K. E. Holekamp, and E. F. Zipkin. Integrating distance sampling and presence-only data to estimate species abundance. *Ecology*, 102(1):e03204, 2021.

P. Fearnhead and D. Prangle. Constructing summary statistics for approximate Bayesian computation: semi-automatic approximate Bayesian computation. *Journal of the Royal Statistical Society: Series B (Statistical Methodology)*, 74:419–474, 2012.

A. Ferrieira and L. De Haan. The generalized pareto process; with a view towards application and simulation. *Bernoulli*, 20:1717–1737, 2014.

A. Fick. On liquid diffusion. *Journal of Membrane Science*, 100(1):33–38, 1995. ISSN 0376-7388. doi: https://doi.org/10.1016/0376-7388(94)00230-V. URL https://www.sciencedirect.com/science/article/pii/037673889400230V. The early history of membrane science selected papers celebrating vol. 100.

D. M. Finkelstein. A proportional hazards model for interval-censored failure time data. *Biometrics*, 42(4):845–854, 1986. ISSN 0006341X, 15410420. URL http://www.jstor.org/stable/2530698.

A. Finley, S. Banerjee, and Øyvind Hjelle. *MBA: Multilevel B-Spline Approximation*, 2024a. URL https://CRAN.R-project.org/package=MBA. R package version 0.1-2.

A. O. Finley, S. Banerjee, A. R. Ek, and R. E. McRoberts. Bayesian multivariate process modeling for prediction of forest attributes. *Journal of Agricultural, Biological, and Environmental Statistics*, 1:60–83, 2008.

A. O. Finley, S. Banerjee, and R. E. McRoberts. Hierarchical spatial models for predicting tree species assemblages across large domains. *The Annals of Applied Statistics*, 3(3): 1052–1079, 09 2009a. doi: 10.1214/09-AOAS250. URL http://dx.doi.org/10.1214/09-AOAS250.

A. O. Finley, H. Sang, S. Banerjee, and A. E. Gelfand. Improving the performance of predictive process modeling for large datasets. *Computational statistics and data analysis*, 53(8):2873–2884, 2009b.

A. O. Finley, S. Banerjee, and B. Basso. Improving crop model inference through Bayesian melding with spatially varying parameters. *Journal of Agricultural, Biological, and Environmental Statistics*, 16:453–474, 2011a. doi: https://doi.org/10.1007/s13253-011-0070-x.

A. O. Finley, S. Banerjee, and D. W. MacFarlane. A hierarchical model for quantifying forest variables over large heterogeneous landscapes with uncertain forest areas. *Journal of the American Statistical Association*, 106(493):31–48, 2011b. doi: 10.1198/jasa.2011.ap09653. URL https://doi.org/10.1198/jasa.2011.ap09653. PMID: 26139950.

A. O. Finley, S. Banerjee, and A. E. Gelfand. Bayesian dynamic modeling for large space-time datasets using Gaussian predictive processes. *Journal of Geographical Systems*, 14: 29–47, 2012.

A. O. Finley, A. Datta, B. D. Cook, D. C. Morton, H.-E. Andersen, and S. Banerjee. Efficient algorithms for Bayesian nearest neighbor Gaussian processes. *Journal of Computational and Graphical Statistics*, 28(2):401–414, 2019.

A. O. Finley, H.-E. Andersen, C. Babcock, B. D. Cook, D. C. Morton, and S. Banerjee. Models to support forest inventory and small area estimation using sparsely sampled lidar: A case study involving g-liht lidar in tanana, alaska. *Journal of Agricultural, Biological and Environmental Statistics*, 29(4):695–722, Dec 2024b. ISSN 1537-2693. doi: 10.1007/s13253-024-00611-3. URL https://doi.org/10.1007/s13253-024-00611-3.

N. I. Fisher. *Statistical Analysis of Circular Data*. Cambridge, 1993.

T. Fleming and D. Harrington. *Counting Processes and Survival Analysis*. Wiley Series in Probability and Statistics. Wiley, 1991. ISBN 9780471522188. URL https://books.google.com/books?id=nIWSQgAACAAJ.

R. Flowerdew and M. Green. Statistical methods for inference between incompatible zonal systems. In M. F. Goodchild and S. Gopal, editors, *The accuracy of spatial databases*. North West Regional Research Laboratory, 1989.

R. Flowerdew and M. Green. Developments in areal interpolation methods and gis. *The Annals of Regional Science*, 26(1):67–78, Mar 1992. ISSN 1432-0592. doi: 10.1007/BF01581481. URL https://doi.org/10.1007/BF01581481.

K. M. Foley and M. Fuentes. A statistical framework to combine multivariate spatial data and physical models for Hurricane surface wind prediction. *Journal of Agricultural, Biological, and Environmental Statistics*, 13(1):37–59, 2008. doi: 10.1198/108571108X276473.

M.-J. Fortin. Edge detection algorithms for two-dimensional ecological data. *Ecology*, 75 (4):956–965, 1994. ISSN 00129658, 19399170. URL http://www.jstor.org/stable/1939419.

M.-J. Fortin. Effects of data types on vegetation boundary delineation. *Canadian Journal of Forest Research*, 27(11):1851–1858, 1997. doi: 10.1139/x97-156. URL https://doi.org/10.1139/x97-156.

M.-J. Fortin and P. Drapeau. Delineation of ecological boundaries: Comparison of approaches and significance tests. *Oikos*, 72(3):323–332, 1995. ISSN 00301299, 16000706. URL http://www.jstor.org/stable/3546117.

A. Fotheringham and P. Rogerson. The sage handbook of spatial analysis, 2009. URL https://doi.org/10.4135/9780857020130.

A. Fotheringham and P. Rogerson. *Spatial Analysis And GIS*. CRC Press, Apr 2013. ISBN 9781482272468. doi: 10.1201/9781482272468. URL https://doi.org/10.1201/9781482272468.

C. W. Fox and S. J. Roberts. A tutorial on variational Bayesian inference. *Artificial Intelligence Review*, 38(2):85–95, Aug 2012. ISSN 1573-7462. doi: 10.1007/s10462-011-9236-8. URL https://doi.org/10.1007/s10462-011-9236-8.

T. Frankel. *The Geometry of Physics (revised edition)*. Cambridge Univ. Press, Cambridge, 1997.

J. French and S. Sain. Spatio-temporal exceedance locations and confidence regions. *Annals of Applied Statistics*, 7, 2013.

J. Friedman, T. Hastie, and R. Tibshirani. Regularization paths for generalized linear models via coordinate descent. *Journal of Statistical Software*, 33:1–22, 2010.

M. Fuentes. A high frequency kriging approach for non-stationary environmental processes. *Environmetrics*, 12(5):469–483, 2001. doi: https://doi.org/10.1002/env.473. URL https://onlinelibrary.wiley.com/doi/abs/10.1002/env.473.

M. Fuentes. Interpolation of nonstationary air pollution processes: a spatial spectral approach. *Statistical Modelling*, 2(4):281–298, 2002a. doi: 10.1191/1471082x02st034oa. URL https://doi.org/10.1191/1471082x02st034oa.

M. Fuentes. Spectral methods for nonstationary spatial processes. *Biometrika*, 89(1):197–210, 2002b. ISSN 00063444. URL http://www.jstor.org/stable/4140567.

M. Fuentes and A. E. Raftery. Model evaluation and spatial interpolation by Bayesian combination of observations with outputs from numerical models. *Biometrics*, 61(1): 36–45, 2005. doi: 10.1111/j.0006-341X.2005.030821.x.

M. Fuentes and R. Smith. Modeling nonstationary processes as a convolution of local stationary processes. Technical Report, 2001. Technical report, Department of Statistics, North Carolina State University.

M. Fuentes and R. Smith. A new class of models for nonstationary processes. Technical Report, 2003. Technical report, Department of Statistics, North Carolina State University.

W. A. Fuller. *Measurement Error Models*. John Wiley & Sons, New York, NY, 1987.

R. Furrer, M. G. Genton, and D. Nychka. Covariance tapering for interpolation of large spatial datasets. *Journal of Computational and Graphical Statistics*, 15(3):502–523, 2006. ISSN 10618600. URL http://www.jstor.org/stable/27594195.

D. Gamerman and H. F. Lopes. *Markov chain Monte Carlo*. Chapman & Hall/CRC Texts in Statistical Science. Chapman & Hall/CRC, Philadelphia, PA, 2 edition, May 2006.

D. Gamerman, A. R. Moreira, and H. Rue. Space-varying regression models: specifications and simulation. *Computational Statistics & Data Analysis*, 42(3):513–533, 2003. ISSN 0167-9473. doi: https://doi.org/10.1016/S0167-9473(02)00211-6. URL https://www.sciencedirect.com/science/article/pii/S0167947302002116. Computational Ecometrics.

L. C. Ganley, S. Brault, and C. A. Mayo. What we see is not what there is: estimating North Atlantic right whale *Eubalaena glacialis* local abundance. *Endangered Species Research*, 38:101–113, 2019. doi: 10.3354/esr00938.

L. Gao, S. Banerjee, and B. Ritz. Spatial difference boundary detection for multiple outcomes using Bayesian disease mapping. *Biostatistics*, 24(4):922–944, 2023.

G. Gaspari and S. E. Cohn. Construction of correlation functions in two and three dimensions. *Quarterly Journal of the Royal Meteorological Society*, 125(554):723–757, 1999. doi: https://doi.org/10.1002/qj.49712555417. URL https://rmets.onlinelibrary.wiley.com/doi/abs/10.1002/qj.49712555417.

M. Gaudard, M. Karson, E. Linder, and D. Sinha. Bayesian spatial prediction. *Environmental and Ecological Statistics*, 6(2):147–171, Jun 1999. ISSN 1573-3009. doi: 10.1023/A:1009614003692. URL https://doi.org/10.1023/A:1009614003692.

A. E. Gelfand. H.-J. Kim. C. F. Sirmans, and S. Baneriee. Spatially modeling with spatially varying coefficient process. *Journal of the American Statistical Association*, 98 (462):387–396. 2003. doi: 10.1198/016214503000170. URL https://doi.org/10.1198/01621950300070.

A. E. Gelfand and S. K. Ghosh. Model choice: A minimum posterior predictive loss approach. *Biometrika*, 85(1):1–11, 03 1998. ISSN 0006-3444. doi: 10.1093/biomet/85.1.1. URL https://doi.org/10.1093/biomet/85.1.1.

A. E. Gelfand and B. K. Mallick. Bayesian analysis of proportional hazards models built from monotone functions. *Biometrics*, 51(3):843–852, 1995. ISSN 0006341X, 15410420. URL http://www.jstor.org/stable/2532986.

A. E. Gelfand and E. M. Schliep. *Bayesian Inference and Computing for Spatial Point Patterns*. NSF-CBMS Regional Conference Series in Probability and Statistics, Volume 10. Institute of Mathematical Statistics and American Statistical Association, 2018.

A. E. Gelfand and S. Shirota. Preferential sampling for presence/absence data and for fusion of presence/absence data with presence-only data. *Ecological Monographs*, 89(3): e01372, 2019.

A. E. Gelfand and A. F. M. Smith. Sampling-based approaches to calculating marginal densities. *Journal of the American Statistical Association*, 85(410):398–409, 1990. ISSN 01621459, 1537274X. URL http://www.jstor.org/stable/2289776.

A. E. Gelfand and P. Vounatsou. Proper multivariate conditional autoregressive models for spatial data analysis. *Biostatistics*, 4(1):11–25, 2003. URL https://api.semanticscholar.org/CorpusID:8618500.

A. E. Gelfand, S. K. Sahu, and B. P. Carlin. Efficient parametrisations for normal linear mixed models. *Biometrika*, 82(3):479–488, 1995. doi: 10.1093/biomet/82.3.479.

A. E. Gelfand, S. Sahu, and B. Carlin. Efficient parametrization for generalized linear mixed models. In J. Bernardo and et al., editors, *Bayesian Statistics 5*, pages 165–180. Clarendon Press, Oxford, 1996.

A. E. Gelfand, S. K. Ghosh, J. Knight, and C. F. Sirmans. Spatio-temporal modeling of residential sales markets. *Journal of Business and Economic Statistics*, 16:312–321, 1998.

A. E. Gelfand, M. D. Ecker, J. R. Knight, and C. F. Sirmans. The dynamics of location in home price. *The Journal of Real Estate Finance and Economics*, 29(2):149–166, Sep 2004a. ISSN 1573-045X. doi: 10.1023/B:REAL.0000035308.15346.0a. URL https://doi.org/10.1023/B:REAL.0000035308.15346.0a.

A. E. Gelfand, A. M. Schmidt, S. Banerjee, and C. F. Sirmans. Nonstationary multivariate process modeling through spatially varying coregionalization. *TEST*, 13(2):263–312, 2004b.

A. E. Gelfand, S. Banerjee, and D. Gamerman. Spatial process modelling for univariate and multivariate dynamic spatial data. *Environmetrics*, 16:465–479, 2005b.

A. E. Gelfand, A. Kottas, and S. MacEachern. Bayesian nonparametric spatial modeling with dirichlet process mixing. *Journal of the American Statistical Association*, 100(471): 1021–1035, 2005b. doi: 10.1198/016214504000002078. URL https://doi.org/10.1198/016214504000002078.

A. E. Gelfand, A. M. Schmidt, S. Wu, J. A. Silander, A. Latimer, and A. G. Rebelo. Modelling species diversity through species level hierarchical modelling. *Journal of the Royal Statistical Society: Series C (Applied Statistics)*, 54(1):1–20, 2005c.

A. E. Gelfand, P. J. Diggle, P. Guttorp, and M. Fuentes. Handbook of Spatial Statistics. Chapman & Hall/CRC Handbooks of Modern Statistical Methods. Taylor & Francis, 2010. ISBN 9781420072877. URL https://www.routledge.com/Handbook-of-Spatial-Statistics/Gelfand-Diggle-Guttorp-Fuentes/p/book/9781420072877.

A. E. Gelfand, S. Banerjee, and A. O. Finley. *Spatial Design for Knot Selection in Knot-Based Dimension Reduction Models*, chapter 7, pages 142–169. John Wiley & Sons, Ltd, 2012. ISBN 9781118441862. doi: https://doi.org/10.1002/9781118441862.ch7. URL https://onlinelibrary.wiley.com/doi/abs/10.1002/9781118441862.ch7.

I. M. Gelfand and S. Fomin. *Calculus of Variations*. Prentice-Hall, Inc, Englewood CliKs, New Jersey, 1963.

A. Gelman. Struggles with Survey Weighting and Regression Modeling. *Statistical Science*, 22(2):153–164, 2007.

A. Gelman and X.-L. Meng. Simulating normalizing constants: From importance sampling to bridge sampling to path sampling. *Statistical Science*, 13(2):163–185, 1998.

A. Gelman and D. B. Rubin. Inference from Iterative Simulation Using Multiple Sequences. *Statistical Science*, 7(4):457–472, 1992. doi: 10.1214/ss/1177011136. URL https://doi.org/10.1214/ss/1177011136.

A. Gelman, X.-L. Meng, and H. Stern. Posterior predictive assessment of model fitness via realized discrepancies. *Statistica Sinica*, 6(4):733–760, 1996a.

A. Gelman, G. Roberts, and W. Gilks. Efficient Metropolis Jumping Rules. In *Bayesian Statistics 5: Proceedings of the Fifth Valencia International Meeting*. Oxford University Press, 05 1996b. ISBN 9780198523567. doi: 10.1093/oso/9780198523567.003.0038. URL https://doi.org/10.1093/oso/9780198523567.003.0038.

A. Gelman, J. B. Carlin, H. S. Stern, D. B. Dunson, A. Vehtari, and D. B. Rubin. *Bayesian Data Analysis*. Chapman & Hall/CRC Texts in Statistical Science. Chapman & Hall/CRC, New York, NY, 3 edition, Nov. 2013.

A. Gelman, J. Hwang, and A. Vehtari. Understanding predictive information criteria for Bayesian models. *Statistics and Computing*, 24(6):997–1016, Nov 2014. ISSN 1573-1375. doi: 10.1007/s11222-013-9416-2. URL https://doi.org/10.1007/s11222-013-9416-2.

D. Geman and S. Geman. Bayesian image analysis. In *Disordered systems and biological organization*, pages 301–319. Springer, 1986.

S. Geman and D. Geman. Stochastic relaxation, gibbs distributions, and the Bayesian restoration of images. *IEEE Transactions on Pattern Analysis and Machine Intelligence*, PAMI-6(6):721–741, 1984. doi: 10.1109/TPAMI.1984.4767596.

S. Geman and D. E. McClure. Statistical methods for tomographic image reconstruction. *Bulletin of the International Statistical Institute*, 52:5–21, 1987.

C. Genovese and L. Wasserman. A stochastic process approach to false discovery control. *The Annals of Statistics*, 32(3):1035–1061, 2004. doi: 10.1214/009053604000000283. URL https://doi.org/10.1214/009053604000000283.

M. G. Genton and W. Kleiber. Cross-covariance functions for multivariate geostatistics. *Statistical Science*, pages 147–163, 2015.

H.-O. Georgii. Canonical and grand canonical Gibbs states for continuum systems. *Communications in Mathematical Physics*, 48(1):31–51, 1976.

C. J. Geyer. Practical Markov Chain Monte Carlo. *Statistical Science*, 7(4):473–483, 1992. doi: 10.1214/ss/1177011137. URL https://doi.org/10.1214/ss/1177011137.

C. J. Geyer and J. Møller. Simulation procedures and likelihood inference for spatial point processes. *Scandinavian Journal of Statistics*, 21(4):359–373, 1994.

C. J. Geyer and E. A. Thompson. Annealing Markov chain Monte Carlo with applications to ancestral inference. *Journal of the American Statistical Association*, 90(431):909–920, 1995.

S. Ghosal and A. van der Vaart. *Fundamentals of Nonparametric Bayesian Inference*. Cambridge Series in Statistical and Probabilistic Mathematics. Cambridge University Press, 2017.

M. Ghosh. Finite population sampling: A model-design synthesis. *Statistics in Transition new series*, 13(2):235–242, 2012.

M. Ghosh and G. Meeden. *Bayesian Methods for Finite Population Sampling*. Chapman & Hall, London, 1997.

M. Ghosh and J. N. K. Rao. Small Area Estimation: An Appraisal. *Statistical Science*, 9(1):55–93, 1994.

M. Ghosh and B. K. Sinha. On the consistency between model-and design-based estimators in survey sampling. *Communications in Statistics - Theory and Methods*, 19(2):689–702, 1990. doi: 10.1080/03610929008830226. URL https://doi.org/10.1080/03610929008830226.

M. Ghosh, K. Natarajan, T. W. F. Stroud, and B. P. Carlin. Generalized linear models for small-area estimation. *Journal of the American Statistical Association*, 93(441):273–282, 1998.

S. Ghosh, A. E. Gelfand, and J. S. Clark. Inference for size demography from point pattern data using integral projection models. *Journal of Agricultural, Biological, and Environmental Statistics*, 17(4):641–677, Dec 2012. ISSN 1537-2693. doi: 10.1007/s13253-012-0123-9. URL https://doi.org/10.1007/s13253-012-0123-9.

M. Gianella, M. Beraha, and A. Guglielmi. Bayesian nonparametric boundary detection for income areal data. *arXiv preprint arXiv:2312.13992*, 2023.

I. I. Gikhman and A. V. Skorokhod. *Stochastic Differential Equations*, pages 113–219. Springer Berlin Heidelberg, Berlin, Heidelberg, 2007. ISBN 978-3-540-49941-1. doi: 10.1007/978-3-540-49941-1_2. URL https://doi.org/10.1007/978-3-540-49941-1_2.

W. R. Gilks and P. Wild. Adaptive rejection sampling for gibbs sampling. *Journal of the Royal Statistical Society. Series C (Applied Statistics)*, 41(2):337–348, 1992. ISSN 00359254, 14679876. URL http://www.jstor.org/stable/2347565.

W. R. Gilks, A. Thomas, and D. J. Spiegelhalter. A language and program for complex Bayesian modelling. *Journal of the Royal Statistical Society. Series D (The Statistician)*, 43(1):169–177, 1994. ISSN 00390526, 14679884. URL http://www.jstor.org/stable/2348941.

W. R. Gilks, S. Richardson, and D. J. Spiegelhalter. *Markov Chain Monte Carlo in Practice*. Chapman & Hall/CRC Interdisciplinary Statistics. Chapman & Hall/CRC, Philadelphia, PA, Dec. 1995.

M. Girolami and B. Calderhead. Riemann manifold Langevin and Hamiltonian Monte Carlo methods. *Journal of the Royal Statistical Society: Series B (Statistical Methodology)*, 73(2):123–214, 2011.

T. Gneiting. Nonseparable, stationary covariance functions for space–time data. *Journal of the American Statistical Association*, 97:590–600, 2002a.

T. Gneiting. Compactly supported correlation functions. *Journal of Multivariate Analysis*, 83(2):493–508, 2002b. ISSN 0047-259X. doi: https://doi.org/10.1006/jmva.2001.2056. URL https://www.sciencedirect.com/science/article/pii/S0047259X01920561.

T. Gneiting and P. Guttorp. Continuous-parameter spatio-temporal processes. In A. Gelfand, P. Diggle, M. Fuentes, and P. Guttorp, editors, *Handbook of Spatial Statistics*, pages 427–436. CRC Press, 2010.

T. Gneiting and A. E. Raftery. Strictly proper scoring rules, prediction, and estimation. *Journal of the American Statistical Association*, 102(477):359–378, 2007.

A. I. Goldman. Survivorship analysis when cure is a possibility: A Monte Carlo study. *Statistics in Medicine*, 3(2):153–163, 1984. doi: https://doi.org/10.1002/sim.4780030208. URL https://onlinelibrary.wiley.com/doi/abs/10.1002/sim.4780030208.

A. I. Goldman, B. P. Carlin, L. R. Crane, C. Launer, J. A. Korvick, L. Deyton, and D. I. Abrams. Response of cd4 lymphocytes and clinical consequences of treatment using ddi or ddc in patients with advanced hiv infection. *JAIDS Journal of Acquired Immune Deficiency Syndromes*, 11(2):161–169, 1996. ISSN 1525-4135. URL https://journals.lww.com/jaids/fulltext/1996/02010/response_of_cd4_lymphocytes_and_clinical.7.aspx.

H. Goldstein. *Multilevel statistical models*. Kendall's library of statistics. Hodder Arnold, London, England, 3rd edition, Feb. 2003.

J. Goldstein, M. Haran, I. Simeonov, J. Fricks, and F. Chiaromonte. An attraction–repulsion point process model for respiratory syncytial virus infections. *Biometrics*, 71(2):376–385, 2015.

G. H. Golub and C. F. Van Loan. *Matrix Computations*. Johns Hopkins Studies in the Mathematical Sciences. Johns Hopkins University Press, Baltimore, MD, 4 edition, Feb. 2013.

F. B. Gonçalves and D. Gamerman. Exact Bayesian inference in spatiotemporal Cox processes driven by multivariate Gaussian processes. *Journal of the Royal Statistical Society: Series B (Statistical Methodology)*, 80(1):157–175, 2018.

J. Gonzalez, Y. Low, A. Gretton, and C. Guestrin. Parallel Gibbs sampling: From colored fields to thin junction trees. In *Proceedings of the Fourteenth International Conference on Artificial Intelligence and Statistics*, pages 324–332, 2011.

I. J. Good. Studies in the History of Probability and Statistics. XXXVII A. M. Turing's statistical work in World War II. *Biometrika*, 66(2):393–396, 08 1979. ISSN 0006-3444. doi: 10.1093/biomet/66.2.393. URL https://doi.org/10.1093/biomet/66.2.393.

C. A. Gotway and L. J. Young. Combining incompatible spatial data. *Journal of the American Statistical Association*, 97(458):632–648, 2002. doi: 10.1198/016214502760047140. URL https://doi.org/10.1198/016214502760047140.

R. J. B. Goudie, R. M. Turner, D. De Angelis, and A. Thomas. Multibugs: A parallel implementation of the bugs modeling framework for faster Bayesian inference. *Journal of Statistical Software*, 95(7):1–20, 2020. doi: 10.18637/jss.v095.i07. URL https://www.jstatsoft.org/index.php/jss/article/view/v095i07.

R. Gramacy. lagp: Large-scale spatial modeling via local approximate Gaussian processes in r. *Journal of Statistical Software*, 72(1):1–46, 2016. ISSN 1548-7660. doi: 10.18637/jss.v072.i01. URL https://www.jstatsoft.org/index.php/jss/article/view/v072i01.

R. B. Gramacy and D. W. Apley. Local Gaussian process approximation for large computer experiments. *Journal of Computational and Graphical Statistics*, 24(2):561–578,

2015.   doi: 10.1080/10618600.2014.914442.   URL http://dx.doi.org/10.1080/
10618600.2014.914442.

R. B. Gramacy and B. Haaland. Speeding up neighborhood search in local Gaussian process
prediction. *Technometrics*, 58(3):294–303, 2016. doi: 10.1080/00401706.2015.1027067.
URL http://dx.doi.org/10.1080/00401706.2015.1027067.

P. J. Green and S. Richardson. Hidden Markov models and disease mapping. *Journal of the American Statistical Association*, 97(460):1055–1070, 2002. doi: 10.1198/
016214502388618870. URL https://doi.org/10.1198/016214502388618870.

P. J. Green and A. Thomas. Sampling decomposable graphs using a Markov chain on
junction trees. *Biometrika*, 100(1):91–110, 2013.

J. Greenwood. A unified theory of surface roughness. *Proceedings of the Royal Society of
London. A. Mathematical and Physical Sciences*, 393(1804):133–157, 1984.

D. A. Griffith. *Advanced spatial statistics*. Advanced Studies in Theoretical and Applied
Econometrics. Kluwer Academic, Dordrecht, Netherlands, 1988 edition, Apr. 1988.

T. H. Grubesic. Zip codes and spatial analysis: Problems and prospects. *Socio-Economic
Planning Sciences*, 42(2):129–149, 2008.   ISSN 0038-0121.   doi: https://doi.org/10.
1016/j.seps.2006.09.001.   URL https://www.sciencedirect.com/science/article/
pii/S0038012106000516.

M. Grzebyk and H. Wackernagel. Multivariate analysis and spatial/temporal scales: Real
and complex models. In *Proceedings of the XVIIth International Biometrics Conference*,
pages 19–33, 1994.

Y. Guan. A composite likelihood approach in fitting spatial point process models. *Journal
of the American Statistical Association*, 101:1502–1512, 2006.

H. W. Guggenheimer. *Differential Geometry*. Dover Books on Mathematics. Dover Publications, Mineola, NY, June 1977.

R. Guhaniyogi and S. Banerjee. Meta-kriging: Scalable Bayesian modeling and inference
for massive spatial datasets. *Technometrics*, 60(4):430–444, 2018. doi: 10.1080/00401706.
2018.1437474.

R. Guhaniyogi and S. Banerjee.   Multivariate spatial meta kriging.   *Statistics &
Probability Letters*, 144:3–8, May 2019.   ISSN 0167-7152.   doi: https://doi.org/
10.1016/j.spl.2018.04.017. URL https://www.sciencedirect.com/science/article/
pii/S0167715218301718.

R. Guhaniyogi and A. Scheffler. Sketching in Bayesian high dimensional regression with big
data using Gaussian scale mixture priors, 2021. URL https://arxiv.org/abs/2105.
04795.

R. Guhaniyogi, A. O. Finley, S. Banerjee, and R. K. Kobe. Modeling complex spatial
dependencies: Low-rank spatially varying cross-covariances with application to soil nutrient data. *Journal of Agricultural, Biological, and Environmental Statistics*, 18(3):
274–298, Sep 2013. ISSN 1537-2693. doi: 10.1007/s13253-013-0140-3. URL https:
//doi.org/10.1007/s13253-013-0140-3.

R. Guhaniyogi, C. Li, T. D. Savitsky, and S. Srivastava. A divide-and-conquer Bayesian
approach to large-scale kriging. *arXiv preprint*, page arXiv:1712.09767, 2019. doi: 10.
48550/arXiv.1712.09767.

R. Guhaniyogi, C. Li, T. D. Savitsky, and S. Srivastava. Distributed Bayesian varying coefficient modeling using a Gaussian process prior. *Journal of Machine Learning Research*,
23(84):1–59, 2022. URL http://jmlr.org/papers/v23/20-543.html.

R. Guhaniyogi, C. Li, T. Savitsky, and S. Srivastava. Distributed Bayesian inference in
massive spatial data. *Statistical science*, 38(2):262–284, 2023.

R. Guhaniyogi, L. Baracaldo, and S. Banerjee. Bayesian data sketching for varying coefficient regression models. *Journal of Machine Learning Research*, 26(98):1-29, 2025. URL http://imlr.org/papers/v26/23-0505.html.

J. Guinness. Permutation and grouping methods for sharpening Gaussian process approximations. *Technometrics*, 60(4):415–429, 2018. doi: 10.1080/00401706.2018.1437476. URL https://doi.org/10.1080/00401706.2018.1437476.

X. Guo and B. P. Carlin. Separate and joint modeling of longitudinal and event time data using standard computer packages. *The American Statistician*, 58(1):16–24, 2004. doi: 10.1198/0003130042854. URL https://doi.org/10.1198/0003130042854.

I. Guttman. *Linear Models: An Introduction*. WILEY SERIES in PROBABILITY and STATISTICS: PROBABILITY and STATISTICS SECTION Series. Wiley, 1982. ISBN 9780471099154. URL https://books.google.com/books?id=MyHvAAAAMAAJ.

P. Guttorp and T. L. Thorarinsdottir. Bayesian inference for non-Markovian point processes. In *Advances and Challenges in Space-time Modelling of Natural Events*, pages 79–102. Springer, 2012.

X. Guyon. *Random fields on a network*. Probability and Its Applications. Springer, New York, NY, 1995 edition, June 1995.

O. Häggström, M.-C. N. Van Lieshout, and J. Møller. Characterization results and Markov chain Monte Carlo algorithms including exact simulation for some spatial point processes. *Bernoulli*, 5(4):641–658, 1999.

R. Haining. *Spatial Data Analysis in the Social and Environmental Sciences*. Cambridge University Press, 1990.

R. Haining. *Spatial Data Anaalysis: Theory and Practice*. Cambridge University Press, Cambridge, UK, 2003.

A. Halder, S. Banerjee, and D. K. Dey. Bayesian modeling with spatial curvature processes. *Journal of the American Statistical Association*, 119(546):1155–1167, 2024a. doi: 10.1080/01621459.2023.2177166. URL https://doi.org/10.1080/01621459.2023.2177166.

A. Halder, D. Li, and S. Banerjee. Bayesian spatiotemporal wombling, 2024b. URL https://arxiv.org/abs/2407.17804.

P. Hall, N. I. Fisher, and B. Hoffmann. On the Nonparametric Estimation of Covariance Functions. *The Annals of Statistics*, 22(4):2115–2134, 1994. doi: 10.1214/aos/1176325774. URL https://doi.org/10.1214/aos/1176325774.

M. Hallin, Z. Lu, and K. Yu. Local linear spatial quantile regression. *Bernoulli*, 15(3):659–686, 2009. doi: 10.3150/08-BEJ168.

M. S. Handcock. Comment on "prediction of spatial cumulative distribution functions using subsampling". *Journal of the American Statistical Association*, 94(445):100–102, 1999. doi: 10.1080/01621459.1999.10473824. URL https://doi.org/10.1080/01621459.1999.10473824.

M. S. Handcock and M. L. Stein. A Bayesian analysis of kriging. *Technometrics*, 35(4):403–410, 1993. doi: 10.1080/00401706.1993.10485354. URL https://www.tandfonline.com/doi/abs/10.1080/00401706.1993.10485354.

M. S. Handcock and J. R. Wallis. An approach to statistical spatial-temporal modeling of meteorological fields. *Journal of the American Statistical Association*, 89(426):368–378, 1994. doi: 10.1080/01621459.1994.10476754. URL https://doi.org/10.1080/01621459.1994.10476754.

T. Hanson, S. Banerjee, P. Li, and A. McBean. Spatial boundary detection for areal counts. In *Nonparametric Bayesian Inference in Biostatistics*, pages 377–399. Springer, 2015.

D. A. Harville. *Matrix Algebra from a Statistician's Perspective*. Springer, New York, NY, 1 edition, Nov. 2000.

T. Hastie and W. Fithian. Inference from presence-only data; the ongoing controversy. *Ecography*, 36(8):864–867, 2013.

W. K. Hastings. Monte Carlo sampling methods using Markov chains and their applications. *Biometrika*, 57(1):97–109, 04 1970. ISSN 0006-3444. doi: 10.1093/biomet/57.1.97. URL https://doi.org/10.1093/biomet/57.1.97.

A. Hazra and R. Huser. Estimating high-resolution red sea surface temperature hotspots, using a low-rank semiparametric spatial model. *Annals of Applied Statistics*, 15(2):572–596, 2021.

M. Heaton, A. Datta, A. Finley, R. Furrer, J. Guinness, R. Guhaniyogi, F. Gerber, R. Gramacy, D. Hammerling, M. Katzfuss, F. Lindgren, D. Nychka, F. Sun, and A. Zammit-Mangion. Methods for analyzing large spatial data: A review and comparison. *Journal of Agricultural, Biological and Environmental Statistics*, 24(3):398–425, 2019. doi: 10.1007/s13253-018-00348-w. URL https://doi.org/10.1007/s13253-018-00348-w.

M. J. Heaton and A. E. Gelfand. Spatial regression using kernel averaged predictors. *Journal of Agricultural, Biological, and Environmental Statistics*, 16(2):233–252, Jun 2011. ISSN 1537-2693. doi: 10.1007/s13253-010-0050-6. URL https://doi.org/10.1007/s13253-010-0050-6.

J. Heikkinen and H. Hogmander. Fully Bayesian approach to image restoration with an application in biogeography. *Journal of the Royal Statistical Society. Series C (Applied Statistics)*, 43(4):569–582, 1994. ISSN 00359254, 14679876. URL http://www.jstor.org/stable/2986258.

J. D. Helterbrand, N. Cressie, and J. L. Davidson. A statistical approach to identifying closed object boundaries in images. *Advances in Applied Probability*, 26(4):831–854, 1994. ISSN 00018678. URL http://www.jstor.org/stable/1427893.

R. Henderson, P. Diggle, and A. Dobson. Joint modelling of longitudinal measurements and event time data. *Biostatistics*, 1(4):465–480, 12 2000. ISSN 1465-4644. doi: 10.1093/biostatistics/1.4.465. URL https://doi.org/10.1093/biostatistics/1.4.465.

P. Heywood, N. Singleton, and J. Ross. Nutritional status of young children–the 1982/83 national nutrition survey. *Papua New Guinea Medical Journal*, 31(2):91–101, June 1988.

D. Higdon. A process-convolution approach to modeling temperatures in the north atlantic ocean. *Environmental and Ecological Statistics*, 5:173–190, 1998.

D. Higdon. Space and space-time modeling using process convolutions. In *Quantitative Methods for Current Environmental Issues*, pages 37–56. Springer, 2002a.

D. Higdon. Space and space time modeling using process convolutions. In C. Anderson, V. Barnett, P. Chatwin, and A. El-Shaarawi, editors, *Quantitative Methods for Current Environmental Issues*, pages 37–56. Springer, 2002b.

D. Higdon. Space and space time modeling using process convolutions. In C. Anderson, V. Barnett, P. Chatwin, and A. El-Shaarawi, editors, *Quantitative Methods for Current Environmental Issues*, pages 37–56. Springer, 2002c.

D. Higdon, J. Swall, and J. Kern. Non-stationary spatial modeling. In J. Bernardo, J. Berger, A. Dawid, and A. Smith, editors, *Bayesian Statistics 6*, pages 761–768. Oxford: Oxford University Press, 1999.

D. Higdon, H. Lee, and C. Holloman. Markov chain Monte Carlo-based approaches for inference in computationally intensive inverse problems. In *Bayesian Statistics 7: Proceedings of the Seventh Valencia International Meeting*. Oxford University Press, 07 2003. ISBN 9780198526155. doi: 10.1093/oso/9780198526155.003.0010. URL https://doi.org/10.1093/oso/9780198526155.003.0010.

R. Hill, C. Sirmans, and J. R. Knight. A random walk down main street? *Regional Science and Urban Economics*, 29(1):89–103, 1999. ISSN 0166-0462. doi: https://doi.org/10.1016/S0166-0462(98)00014-3. URL https://www.sciencedirect.com/science/article/pii/S0166046298000143.

R. C. Hill, J. R. Knight, and C. F. Sirmans. Estimating capital asset price indexes. *The Review of Economics and Statistics*, 79(2):226–233, 1997. ISSN 00346535, 15309142. URL http://www.jstor.org/stable/2951455.

N. L. Hjort, H. Omre, M. Frisén, F. Godtliebsen, J. Helgeland, J. Møller, E. B. V. Jensen, M. Rudemo, and H. Stryhn. Topics in spatial statistics (with discussion, comments and rejoinder). *Scandinavian Journal of Statistics*, 21(4):289–357, 1994. ISSN 03036898, 14679469. URL http://www.jstor.org/stable/4616322.

N. L. Hjort, C. Holmes, P. Muller, and S. G. Walker, editors. *Bayesian nonparametrics*. Cambridge series in statistical and probabilistic mathematics. Cambridge University Press, Cambridge, England, Apr. 2010.

D. C. Hoaglin and etc. *Exploring data tables, trends and shapes*. Series: Wiley Series in Probability & Mathematical Statistics. John Wiley & Sons, Nashville, TN, Sept. 1985.

D. C. Hoaglin, F. Mosteller, and J. W. Tukey, editors. *Understanding robust and exploratory data analysis*. Wiley Classics Library. John Wiley & Sons, Nashville, TN, May 2000.

J. P. Hobert, G. L. Jones, B. Presnell, and J. S. Rosenthal. On the applicability of regenerative simulation in Markov chain Monte Carlo. *Biometrika*, 89(4):731–743, 12 2002. ISSN 0006-3444. doi: 10.1093/biomet/89.4.731. URL https://doi.org/10.1093/biomet/89.4.731.

J. S. Hodges. *Richly Parameterized Linear Models: Additive, Time Series, and Spatial Models Using Random Effects*. Chapman & Hall/CRC Texts in Statistical Science. Chapman & Hall/CRC, Boca Raton, FL, 2013.

J. S. Hodges, Y. Cui, D. J. Sargent, and B. P. Carlin. Smoothing balanced single-error-term analysis of variance. *Technometrics*, 49(1):12–25, 2007.

P. G. Hoel, S. Port, and C. Stone. *Introduction to probability theory*. Houghton Mifflin, Boston, MA, July 1972.

J. A. Hoeting, M. Leecaster, and D. Bowden. An improved model for spatially correlated binary responses. *Journal of Agricultural, Biological, and Environmental Statistics*, 5(1):102–114, 2000. ISSN 10857117. URL http://www.jstor.org/stable/1400634.

P. D. Hoff. Introduction and examples. In *Springer Texts in Statistics*, Springer texts in statistics, pages 1–12. Springer New York, New York, NY, 2009.

M. D. Hoffman and A. Gelman. The no u-turn sampler: Adaptively setting path lengths in Hamiltonian Monte Carlo. *Journal of Machine Learning Research*, 15:1593–1623, 2014.

H. Hogmander and J. Møller. Estimating distribution maps from atlas data using methods of statistical image analysis. *Biometrics*, 51(2):393–404, 1995. ISSN 0006341X, 15410420. URL http://www.jstor.org/stable/2532928.

M. E. Hohn. *Geostatistics and Petroleum Geology*. Springer US, 1988. ISBN 9781461571063. doi: 10.1007/978-1-4615-7106-3. URL http://dx.doi.org/10.1007/978-1-4615-7106-3.

M. B. Hooten and C. K. Wikle. A hierarchical Bayesian non-linear spatio-temporal model for the spread of invasive species with application to the eurasian collared-dove. *Environmental and Ecological Statistics*, 15(1):59–70, Mar 2008. ISSN 1573-3009. doi: 10.1007/s10651-007-0040-1. URL https://doi.org/10.1007/s10651-007-0040-1.

R. A. Horn and C. R. Johnson. *Matrix Analysis*. Cambridge University Press, Cambridge, England, 2 edition, Oct. 2012.

D. G. Horvitz and D. J. Thompson. A Generalization of Sampling Without Replacement From a Finite Universe. *Journal of the American Statistical Association*, 47(260):663–685, 1952.

B. Hrafnkelsson and N. Cressie. Hierarchical modeling of count data with application to nuclear fall-out. *Environmental and Ecological Statistics*, 10(2):179–200, Jun 2003. ISSN 1573-3009. doi: 10.1023/A:1023674107629. URL https://doi.org/10.1023/A:1023674107629.

F. Huang and Y. Ogata. Improvements of the maximum pseudo-likelihood estimators in various spatial statistical models. *Journal of Computational and Graphical Statistics*, 8:510–530, 1999.

H.-C. Huang and N. Cressie. Spatio-temporal prediction of snow water equivalent using the kalman filter. *Computational Statistics & Data Analysis*, 22(2):159–175, 1996. ISSN 0167-9473. doi: https://doi.org/10.1016/0167-9473(95)00047-X. URL https://www.sciencedirect.com/science/article/pii/016794739500047X.

G. Huerta and B. Sansó. Time-varying models for extreme values. *Environmental and Ecological Statistics*, 14(3):285–299, Sep 2007. ISSN 1573-3009. doi: 10.1007/s10651-007-0014-3. URL https://doi.org/10.1007/s10651-007-0014-3.

G. Huerta, B. Sansó, and J. R. Stroud. A spatiotemporal model for mexico city ozone levels. *Journal of the Royal Statistical Society: Series C (Applied Statistics)*, 53(2):231–248, 2004. doi: https://doi.org/10.1046/j.1467-9876.2003.05100.x. URL https://rss.onlinelibrary.wiley.com/doi/abs/10.1046/j.1467-9876.2003.05100.x.

R. Huser, M. L. Stein, and P. Zhong. Vecchia likelihood approximation for accurate and fast inference with intractable spatial max-stable models. *Journal of Computational and Graphical Statistics*, 33(3):978–990, 2024. doi: 10.1080/10618600.2023.2285332. URL https://doi.org/10.1080/10618600.2023.2285332.

J. Illian, A. Penttinen, H. Stoyan, and D. Stoyan. *Statistical Analysis and Modelling of Spatial Point Patterns*. John Wiley & Sons, Ltd, Hoboken, NJ, 2007.

J. Illian, A. Penttinen, H. Stoyan, and D. Stoyan. *Statistical analysis and modelling of spatial point patterns*, volume 70. John Wiley & Sons, 2008.

J. B. Illian, S. H. Sørbye, and H. Rue. A toolbox for fitting complex spatial point process models using integrated nested Laplace approximation (INLA). *The Annals of Applied Statistics*, pages 1499–1530, 2012.

J. B. Illian, S. Martino, S. H. Sørbye, J. B. Gallego-Fernández, M. Zunzunegui, M. P. Esquivias, and J. M. Travis. Fitting complex ecological point process models with integrated nested Laplace approximation. *Methods in Ecology and Evolution*, 4(4):305–315, 2013.

D. D. Ingram and J. C. Kleinman. Empirical comparisons of proportional hazards and logistic regression models. *Statistics in Medicine*, 8(5):525–538, 1989. doi: https://doi.org/10.1002/sim.4780080502. URL https://onlinelibrary.wiley.com/doi/abs/10.1002/sim.4780080502.

E. H. Isaaks and M. R. Srivastava. *Applied Geostatistics*. Oxford University Press, New York, NY, Oct. 1990.

G. M. Jacquez and D. A. Greiling. Geographic boundaries in breast, lung and colorectal cancers in relation to exposure to air toxics in long island, new york. *International Journal of Health Geographics*, 2(1):4, Feb 2003. ISSN 1476-072X. doi: 10.1186/1476-072X-2-4. URL https://doi.org/10.1186/1476-072X-2-4.

S. R. Jammalamadaka and Y. R. Sarma. "A Correlation Coefficient for Angular Variables". In K. Matusita, editor, *Statistical Theory and Data Analysis II*, pages 349–364. North Holland, 1988.

S. R. Jammalamadaka and A. Sengupta. *Topics in Circular Statistics*. World Scientific, 2001.

H. Jeffreys. *The Theory of Probability*. Oxford Classic Texts in the Physical Sciences. OUP Oxford, 1998. ISBN 9780191589676. URL https://books.google.com/books?id=vh9Act9rtzQC.

J. L. Jensen and J. Møller. Pseudolikelihood for exponential family models of spatial point processes. *The Annals of Applied Probability*, 1(3):445–461, 1991.

X. Jin, B. P. Carlin, and S. Banerjee. Generalized Hierarchical Multivariate CAR Models for Areal Data. *Biometrics*, 61(4):950–961, 12 2005. ISSN 0006-341X. doi: 10.1111/j.1541-0420.2005.00359.x. URL https://doi.org/10.1111/j.1541-0420.2005.00359.x.

X. Jin, S. Banerjee, and B. P. Carlin. Order-Free Co-Regionalized Areal Data Models with Application to Multiple-Disease Mapping. *Journal of the Royal Statistical Society Series B: Statistical Methodology*, 69(5):817–838, 10 2007. ISSN 1369-7412. doi: 10.1111/j.1467-9868.2007.00612.x. URL https://doi.org/10.1111/j.1467-9868.2007.00612.x.

G. Jona Lasinio, A. E. Gelfand, and M. Jona Lasinio. Spatial analysis ofwave direction data using wrapped Gaussian processes. *Annals of Applied Statistics*, 6(4):1478–1498, 2012.

M. I. Jordan, Z. Ghahramani, T. S. Jaakkola, and L. K. Saul. An introduction to variational methods for graphical models. *Machine Learning*, 37(2):183–233, Nov 1999. ISSN 1573-0565. doi: 10.1023/A:1007665907178. URL https://doi.org/10.1023/A:1007665907178.

A. Journel and C. Huijbregts. *Mining Geostatistics*. Blackburn Press, 2003. ISBN 9781930665910. URL https://books.google.com/books?id=Id1GAAAAYAAJ.

A. G. Journel and R. Froidevaux. Anisotropic hole-effect modeling. *Journal of the International Association for Mathematical Geology*, 14(3):217–239, Jun 1982. ISSN 1573-8868. doi: 10.1007/BF01032885. URL https://doi.org/10.1007/BF01032885.

M. S. Kaiser and N. Cressie. The construction of multivariate distributions from Markov random fields. *Journal of Multivariate Analysis*, 73(2):199–220, 2000. ISSN 0047-259X. doi: https://doi.org/10.1006/jmva.1999.1878. URL https://www.sciencedirect.com/science/article/pii/S0047259X9991878X.

E. Kalnay. *Atmospheric Modeling, Data Assimilation and Predictability*. Cambridge University Press, 2002.

S. P. Kaluzny, S. C. Vega, T. P. Cardoso, and A. A. Shelly. *S+SpatialStats*. Springer, New York, NY, 1 edition, July 1999.

E. E. Kammann and M. P. Wand. Geoadditive models. *Applied Statistics*, 52:1–18, 2003.

E. L. Kang and N. Cressie. Bayesian inference for the spatial random effects model. *Journal of the American Statistical Association*, 106(495):972–983, 2011. doi: 10.1198/jasa.2011.tm09680. URL https://doi.org/10.1198/jasa.2011.tm09680.

H. B. Kang, Y. J. Jung, and J. Park. Fast Bayesian functional regression for non-Gaussian spatial data. *Bayesian Analysis*, 19(2):1–32, 2023.

J. Kang, T. D. Johnson, T. E. Nichols, and T. D. Wager. Meta analysis of functional neuroimaging data via Bayesian spatial point processes. *Journal of the American Statistical Association*, 106(493):124–134, 2011. doi: 10.1198/jasa.2011.ap09735.

R. E. Kass and A. E. Raftery. Bayes factors. *Journal of the American Statistical Association*, 90(430):773–795, 1995. doi: 10.1080/01621459.1995.10476572. URL https://www.tandfonline.com/doi/abs/10.1080/01621459.1995.10476572.

R. E. Kass, L. Tierney, and J. B. Kadane. The validity of posterior expansions based on laplaces method. In S. Geisser, J. Hodges, S. Press, and A. Zellner, editors, *Essays in Honor of George Barnard*. North–Holland, 1990.

R. E. Kass, B. P. Carlin, A. Gelman, and R. M. Neal. Markov chain Monte Carlo in practice: A roundtable discussion. *The American Statistician*, 52(2):93–100, 1998. ISSN 00031305, 15372731. URL http://www.jstor.org/stable/2685466.

M. Katzfuss. A multi-resolution approximation for massive spatial datasets. *Journal of the American Statistical Association*, 112:201–214, 2017. doi: 10.1080/01621459.2015. 1123632. URL http://dx.doi.org/10.1080/01621459.2015.1123632.

M. Katzfuss and J. Guinness. A general framework for Vecchia approximations of Gaussian processes. *Statistical Science*, 36(1):124–141, 2021. doi: 10.1214/19-STS755. URL https://doi.org/10.1214/19-STS755.

M. Katzfuss and F. Schäfer. Scalable Bayesian transport maps for high-dimensional non-Gaussian spatial fields. *Journal of the American Statistical Association*, 119(546):1409–1423, 2024. doi: 10.1080/01621459.2023.2197158. URL https://doi.org/10.1080/01621459.2023.2197158.

M. Katzfuss, J. Guinness, W. Gong, and D. Zilber. Vecchia approximations of Gaussian-process predictions. *Journal of Agricultural, Biological and Environmental Statistics*, 25 (3):383–414, Sep 2020. ISSN 1537-2693. doi: 10.1007/s13253-020-00401-7. URL https://doi.org/10.1007/s13253-020-00401-7.

C. G. Kaufman, M. J. Schervish, and D. W. Nychka. Covariance tapering for likelihood-based estimation in large spatial data sets. *Journal of the American Statistical Association*, 103(484):1545–1555, 2008. ISSN 01621459. URL http://www.jstor.org/stable/27640203.

F. P. Kelly and B. D. Ripley. A note on Strauss' model for clustering. *Biometrika*, 63: 357–360, 1976.

J. E. Kelsall and P. J. Diggle. Non-parametric estimation of spatial variation in relative risk. *Statistics in Medicine*, 14(21-22):2335–2342, 1995.

D. G. Kendall. Pole-seeking brownian motion and bird navigation. *Journal of the Royal Statistical Society. Series B*, 36(3):365–417, 1974.

W. S. Kendall and J. Møller. Perfect simulation using dominating processes on ordered spaces, with application to locally stable point processes. *Advances in Applied Probability*, 32:844–865, 2000.

M. C. Kennedy and A. O'Hagan. Bayesian Calibration of Computer Models. *Journal of the Royal Statistical Society Series B: Statistical Methodology*, 63(3):425–464, 01 2002. ISSN 1369-7412. doi: 10.1111/1467-9868.00294. URL https://doi.org/10.1111/1467-9868.00294.

J. T. Kent. Continuity Properties for Random Fields. *The Annals of Probability*, 17(4): 1432–1440, 1989. doi: 10.1214/aop/1176991163. URL https://doi.org/10.1214/aop/1176991163.

J. T. Kent and K. V. Mardia. *Spatial Analysis*. Springer, Berlin, Heidelberg, third edition, 2003.

V. V. Kharin and F. W. Zwiers. Estimating extremes in transient climate change simulations. *Journal of Climate*, 18(8):1156–1173, 2005. doi: 10.1175/JCLI3320.1. URL https://journals.ametsoc.org/view/journals/clim/18/8/jcli3320.1.xml.

G. Killough, M. Case, K. Meyer, R. Moore, S. Rope, D. Schmidt, B. Schleien, Sinclair, P. W.K., Voillequé, and J. Till. Task 6: Radiation doses and risk to residents from fmpc operations from 1951–1988, 1996. Draft report: Radiological Assessments Corporation, Neeses, SC.

H. Kim, D. Sun, and R. K. Tsutakawa. A bivariate bayes method for improving the estimates of mortality rates with a twofold conditional autoregressive model. *Journal of the American Statistical Association*, 96(456):1506–1521, 2001. doi: 10.1198/016214501753382408. URL https://doi.org/10.1198/016214501753382408.

R. King, J. B. Illian, S. E. King, G. F. Nightingale, and D. K. Hendrichsen. A Bayesian approach to fitting Gibbs processes with temporal random effects. *Journal of Agricultural, Biological, and Environmental Statistics*, 17(4):601–622, 2012a.

R. King, J. B. Illian, S. E. King, G. F. Nightingale, and D. K. Hendrichsen. A Bayesian approach to fitting Gibbs processes with temporal random effects. *Journal of Agricultural Biological and Environmental Statistics*, 17:601–622, 2012b.

W. N. Kinnard. *Income property valuation*. Study in Business, Industry & Technology. Free Press, New York, NY, Nov. 1971.

L. Kish. *Survey Sampling*. John Wiley & Sons, Inc., Hoboken, New Jersey, 1965.

J. R. Knight, J. Dombrow, and C. F. Sirmans. A varying parameters approach to constructing house price indexes. *Real Estate Economics*, 23(2):187–205, 1995. doi: https://doi.org/10.1111/1540-6229.00663. URL https://onlinelibrary.wiley.com/doi/abs/10.1111/1540-6229.00663.

L. Knorr-Held and N. G. Best. A Shared Component Model for Detecting Joint and Selective Clustering of Two Diseases. *Journal of the Royal Statistical Society Series A: Statistics in Society*, 164(1):73–85, 01 2002. ISSN 0964-1998. doi: 10.1111/1467-985X.00187. URL https://doi.org/10.1111/1467-985X.00187.

L. Knorr-Held and H. Rue. On block updating in Markov random field models for disease mapping. *Scandinavian Journal of Statistics*, 29(4):597–614, 2002. doi: https://doi.org/10.1111/1467-9469.00308. URL https://onlinelibrary.wiley.com/doi/abs/10.1111/1467-9469.00308.

I. Kobyzev, S. J. Prince, and M. A. Brubaker. Normalizing flows: An introduction and review of current methods. *IEEE Transactions on Pattern Analysis and Machine Intelligence*, 43(11):3964–3979, 2021. doi: 10.1109/TPAMI.2020.2992934.

R. Koenker. Regression quantiles. *Econometrica*, 46(1):33–50, 1978.

R. Koenker. *Quantile Regression*. Cambridge University Press, New York, NY, 2005. doi: 10.1017/CBO9780511754098.

R. Koenker and J. A. F. Machado. Goodness of fit and related inference processes for quantile regression. *Journal of the American Statistical Association*, 94(448):1296–1310, 1999. doi: 10.1080/01621459.1999.10473882.

R. Koenker and Z. Xiao. Quantile autoregression. *Journal of the American Statistical Association*, 101(475):980–990, 2006. doi: 10.1198/016214506000000672.

A. Kottas and B. Sansó. Bayesian mixture modeling for spatial Poisson process intensities, with applications to extreme value analysis. *Journal of Statistical Planning and Inference*, 137(10):3151–3163, 2007.

H. Kozumi and G. Kobayashi. Gibbs sampling methods for Bayesian quantile regression. *Journal of Statistical Computation and Simulation*, 81(11):1565–1578, 2011. doi: 10.1080/00949655.2010.496117.

M. Kulldorff. A spatial scan statistic. *Communications in Statistics - Theory and Methods*, 26(6):1481–1496, 1997. doi: 10.1080/03610929708831995. URL https://doi.org/10.1080/03610929708831995.

D. Krige. A statistical approach to some basic mine valuation problems on the witwatersrand. *Journal of the Southern African Institute of Mining and Metallurgy*, 52(6):119–139, 1951. doi: 10.10520/AJA0038223X_4792. URL https://journals.co.za/doi/abs/10.10520/AJA0038223X_4792.

M. L. Krock, W. Kleiber, D. Hammerling, and S. Becker. Modeling massive highly multivariate nonstationary spatial data with the basis graphical lasso. *Journal of Computational and Graphical Statistics*, 0(0):1–16, 2023. doi: 10.1080/10618600.2023.2174126. URL https://doi.org/10.1080/10618600.2023.2174126.

S. N. Lahiri, M. S. Kaiser, N. Cressie, and N.-J. Hsu. Prediction of spatial cumulative distribution functions using subsampling. *Journal of the American Statistical Association*, 94(445):86–97, 1999. doi: 10.1080/01621459.1999.10473821. URL https://www.tandfonline.com/doi/abs/10.1080/01621459.1999.10473821.

N. M. Laird and J. H. Ware. Random-effects models for longitudinal data. *Biometrics*, 38(4):963–974, 1982. ISSN 0006341X, 15410420. URL http://www.jstor.org/stable/2529876.

I. H. Langford, A. H. Leyland, J. Rasbash, and H. Goldstein. Multilevel Modelling of the Geographical Distributions of Diseases. *Journal of the Royal Statistical Society Series C: Applied Statistics*, 48(2):253–268, 01 2002. ISSN 0035-9254. doi: 10.1111/1467-9876.00153. URL https://doi.org/10.1111/1467-9876.00153.

S. L. Lauritzen. *Graphical Models*. Clarendon Press, Oxford, United Kingdom, 1996.

F. Lavancier, J. Møller, and E. Rubak. Determinantal point process models and statistical inference. *Journal of the Royal Statistical Society: Series B (Statistical Methodology)*, 77(4):853–877, 2015.

A. B. Lawson and D. G. Denison, editors. *Spatial Cluster Modelling*. Chapman & Hall/CRC, Philadelphia, PA, 2002.

C. T. Le. *Applied Survival Analysis*. Wiley Series in Probability and Statistics. John Wiley & Sons, Nashville, TN, Oct. 1997.

N. D. Le and J. V. Zidck. Interpolation with uncertain spatial covariances: A Bayesian alternative to kriging. *Journal of Multivariate Analysis*, 43(2):351–374, 1992. ISSN 0047-259X. doi: https://doi.org/10.1016/0047-259X(92)90040-M. URL https://www.sciencedirect.com/science/article/pii/0047259X9290040M.

N. D. Le, W. Sun, and J. V. Zidek. Bayesian multivariate spatial interpolation with data missing by design. *Journal of the Royal Statistical Society. Series B (Methodological)*, 59(2):501–510, 1997. ISSN 00359246. URL http://www.jstor.org/stable/2346059.

T. Le and B. Clarke. A Bayes interpretation of stacking for m-complete and m-open settings. *Bayesian Analysis*, 12(3):807–829, 2017.

A. J. Lee. Circular data. *Wiley Interdisciplinary Reviews: Computational Statistics*, 2(4):477–486, 2010.

D. Lee and R. Mitchell. Boundary detection in disease mapping studies. *Biostatistics*, 13(3):415–426, 2012.

K.-Y. Lee, L. Li, B. Li, and H. Zhao. Nonparametric functional graphical modeling through functional additive regression operator. *Journal of the American Statistical Association*, pages 1–15, 2022.

P. M. Lee. *Bayesian Statistics: An Introduction*. John Wiley & Sons, Nashville, TN, 4 edition, Aug. 2012.

S. Lee, G. Wolberg, and S. Shin. Scattered data interpolation with multilevel b-splines. *IEEE Transactions on Visualization and Computer Graphics*, 3(3):228–244, 1997. doi: 10.1109/2945.620490.

M. Leecaster. Geostatistical modeling of subsurface characteristics in the radioactive waste management complex region, operable unit 7-13/14. Technical report, Idaho National Engineering and Environmental Laboratory (INEEL), 2002.

T. J. Leininger and A. E. Gelfand. Bayesian inference and model assessment for spatial point patterns using posterior predictive samples. *Bayesian Analysis*, 12(1):1–30, 2017.

S. Lele. Inner product matrices, kriging, and nonparametric estimation of variogram. *Mathematical Geology*, 27(5):673–692, Jul 1995. ISSN 1573-8868. doi: 10.1007/BF02093907. URL https://doi.org/10.1007/BF02093907.

B. G. Leroux, X. Lei, and N. Breslow. Estimation of disease rates in small areas: A new mixed model for spatial dependence. In M. E. Halloran and D. Berry, editors, *Statistical Models in Epidemiology, the Environment, and Clinical Trials*, pages 179–191, New York, NY, 2000. Springer New York. ISBN 978-1-4612-1284-3.

P. Lewis and G. Shedler. Simulation of a nonhomogeneous Poisson process by thinning. *Naval Logistics Quarterly*, 26:403–413, 1979.

B. Li and E. Solea. A nonparametric graphical model for functional data with application to brain networks based on fmri. *Journal of the American Statistical Association*, 113 (524):1637–1655, 2018.

B. Li and H. Zhang. An approach to modeling asymmetric multivariate spatial covariance structures. *Journal of Multivariate Analysis*, 102(10):1445–1453, 2011. ISSN 0047-259X. doi: https://doi.org/10.1016/j.jmva.2011.05.010. URL https://www.sciencedirect.com/science/article/pii/S0047259X11000819.

C. Li. Bayesian fixed-domain asymptotics for covariance parameters in a Gaussian process model. *The Annals of Statistics*, 50(6):3334–3363, 2022. doi: 10.1214/22-AOS2230. URL https://doi.org/10.1214/22-AOS2230.

C. Li, S. Sun, and Y. Zhu. Fixed-domain posterior contraction rates for spatial Gaussian process model with nugget. *Journal of the American Statistical Association*, 119 (546):1336–1347, 2024. doi: 10.1080/01621459.2023.2191380. URL https://doi.org/10.1080/01621459.2023.2191380.

D. Li, W. Tang, and S. Banerjee. Inference for Gaussian processes with Matern covariogram on compact Riemannian manifolds. *Journal of Machine Learning Research*, 24(101):1–26, 2023. URL http://jmlr.org/papers/v24/22-0503.html.

D. Li, A. Jones, S. Banerjee, and B. E. Engelhardt. Bayesian multi-group Gaussian process models for heterogeneous group-structured data. *Journal of Machine Learning Research*, 26(30):1–34, 2025. URL http://jmlr.org/papers/v26/23-0291.html.

P. Li, S. Banerjee, B. P. Carlin, and A. M. McBean. Bayesian areal wombling using false discovery rates. *Statistics and its Interface*, 5(2):149–158, 2012.

P. Li, S. Banerjee, T. A. Hanson, and A. M. McBean. Bayesian models for detecting difference boundaries in areal data. *Statistica Sinica*, 25(1):385, 2015.

Y. Li and L. Ryan. Modeling spatial survival data using semiparametric frailty models. *Biometrics*, 58(2):287–297, 2002. ISSN 0006341X, 15410420. URL http://www.jstor.org/stable/3068467.

S. Liang, S. Banerjee, and B. P. Carlin. Bayesian wombling for spatial point processes. *Biometrics*, 65(4):1243–1253, 11, 2009a. ISSN 0006-341X. doi: 10.1111/j.1541-0420.2009.01203.x. URL https://doi.org/10.1111/j.1541-0420.2009.01203.x.

S. Liang, B. P. Carlin, and A. E. Gelfand. Analysis of minnesota colon and rectum cancer point patterns with spatial and nonspatial covariate information. *The Annals of Applied Statistics*, 3(3):943–962, 2009b.

Y. Liang. Graph-based multivariate conditional autoregressive models. *Statistical Theory and Related Fields*, 3(2):158–169, 2019.

K. Lim, P. Treitz, M. Wulder, B. St-Onge, and M. Flood. LiDAR remote sensing of forest structure. *Progress in Physical Geography*, 27(1):88–106, 2003.

X. Lin and D. Zhang. Inference in generalized additive mixed models by using smoothing splines. *Journal of the Royal Statistical Society. Series B (Statistical Methodology)*, 61(2):381–400, 1999. ISSN 13697412, 14679868. URL http://www.jstor.org/stable/2680648.

X. Lin, G. Wahba, D. Xiang, F. Gao, R. Klein, and B. Klein. Smoothing spline anova models for large data sets with bernoulli observations and the randomized gacv. *Annals of Statistics*, 28:1570–1600, 2000.

S. W. Linderman and R. P. Adams. Discovering latent network structure in point process data. *Proceedings of Machine Learning Research*, 32(2):1413–1421, 2014.

F. Lindgren, H. Rue, and J. Lindstrom. An explicit link between Gaussian fields and Gaussian Markov random fields: the stochastic partial differential equation approach. *Journal of the Royal Statistical Society: Series B (Statistical Methodology)*, 73(4):423–498, 2011. ISSN 1467-9868. doi: 10.1111/j.1467-9868.2011.00777.x. URL http://dx.doi.org/10.1111/j.1467-9868.2011.00777.x.

D. Lindley and A. Smith. Bayes estimates for the linear model. *Journal of the Royal Statistical society, Series B*, 34:1–41, 1972.

R. Little and D. Rubin. *Statistical Analysis with Missing Data*. John Wiley & Sons, Inc., Hoboken, New Jersey, 2002.

R. J. Little. To Model or Not To Model? Competing Modes of Inference for Finite Population Sampling. *Journal of the American Statistical Association*, 99(466):546–556, 2004.

J. S. Liu. *Monte Carlo strategies in scientific computing*. Springer Series in Statistics. Springer, New York, NY, 1 edition, Oct. 2002.

Y. Liu, J. V. Zidek, A. W. Trites, and B. C. Battaile. Bayesian data fusion approaches to predicting spatial tracks: Application to marine mammals. *The Annals of Applied Statistics*, 10(3):1517–1546, 2016. doi: 10.1214/16-AOAS945.

Z. Liu, N. D. Le, and J. V. Zidek. An empirical assessment of Bayesian melding for mapping ozone pollution. *Environmetrics*, 22(3):340–353, 2011. doi: 10.1002/env.1054.

S. R. Loarie, P. B. Duffy, H. Hamilton, G. P. Asner, C. B. Field, and D. D. Ackerly. The velocity of climate change. *Nature*, 462:1052–1055, 2009.

W.-L. Loh. Fixed-domain asymptotics for a subclass of Matérn-type Gaussian random fields. *The Annals of Statistics*, 33(5):2344–2394, 2005. doi: 10.1214/009053605000000516. URL https://doi.org/10.1214/009053605000000516.

H. F. Lopes and M. West. Bayesian model assessment in factor analysis. *Statistica Sinica*, 14:41–67, 2004.

H. F. Lopes, E. Salazar, and D. Gamerman. Spatial dynamic factor analysis. *Bayesian Analysis*, 3(4):759–792, 2008.

H. Lu and B. P. Carlin. Bayesian areal wombling for geographical boundary analysis. *Geographical Analysis*, 37(3):265–285, 2005.

H. Lu, C. S. Reilly, S. Banerjee, and B. P. Carlin. Bayesian areal wombling via adjacency modeling. *Environmental and ecological statistics*, 14(4):433–452, 2007.

K. Lum and A. E. Gelfand. Spatial quantile multiple regression using the asymmetric Laplace process. *Bayesian Analysis*, 7(2):235–258, 2012. doi: 10.1214/12-BA708.

J. Lund and M. Rudemo. Models for point processes observed with noise. *Biometrika*, 87(2):235–249, 06 2000. ISSN 0006-3444. doi: 10.1093/biomet/87.2.235. URL https://doi.org/10.1093/biomet/87.2.235.

J. Lund, A. Penttinen, and M. Rudemo. Bayesian analysis of spatial point patterns from noisy observations. *Report, Department of Mathematics and Physics*, The Royal Veterinary and Agricultural University, Copenhagen, 1999.

D. Lunn, D. Spiegelhalter, A. Thomas, and N. Best. The bugs project: Evolution, critique and future directions. *Statistics in Medicine*, 28(25):3049–3067, 2009. doi: https://doi.org/10.1002/sim.3680. URL https://onlinelibrary.wiley.com/doi/abs/10.1002/sim.3680.

D. J. Lunn, A. Thomas, N. Best, and D. Spiegelhalter. Winbugs - a Bayesian modelling framework: Concepts, structure, and extensibility. *Statistics and Computing*, 10(4):325–337, Oct 2000. ISSN 1573-1375. doi: 10.1023/A:1008929526011. URL https://doi.org/10.1023/A:1008929526011.

N. Lynch and F. Vaandrager. Forward and backward simulation. *Information and Computation*, 121(2):214–238, 1995.

K. Lusht. *Real Estate Valuation: Principles and Applications*. Irwin series in finance, insurance and real estate. Irwin, 1997. ISBN 9780256190595. URL https://books.google.com/books?id=vVawPAAACAAJ.

H. Ma, B. P. Carlin, and S. Banerjee. Hierarchical and joint site-edge methods for medicare hospice service region boundary analysis. *Biometrics*, 66(2):355–364, 2010.

S. N. Maceachern and L. M. Berliner. Subsampling the gibbs sampler. *The American Statistician*, 48(3):188–190, 1994. doi: 10.1080/00031305.1994.10476054. URL https://www.tandfonline.com/doi/abs/10.1080/00031305.1994.10476054.

Y. Macnab. Some recent work on multivariate Gaussian Markov random fields. *TEST*, 27, 09 2018. doi: 10.1007/s11749-018-0605-3.

Y. Macnab. Bayesian disease mapping: Past, present, and future. *Spatial Statistics*, 50: 100593, 01 2022a. doi: 10.1016/j.spasta.2022.100593.

Y. Macnab. On coregionalized multivariate Gaussian Markov random fields: construction, parameterization, and Bayesian estimation and inference. *TEST*, 09 2022b. doi: 10.1007/s11749-022-00832-z.

Y. C. MacNab and C. B. Dean. Parametric bootstrap and penalized quasi-likelihood inference in conditional autoregressive models. *Statistics in Medicine*, 19(17-18):2421–2435, 2000. doi: https://doi.org/10.1002/1097-0258(20000915/30)19:17/18⟨2421::AID-SIM579⟩3.0.CO;2-C. URL https://onlinelibrary.wiley.com/doi/abs/10.1002/1097-0258%2820000915/30%2919%3A17/18%3C2421%3A%3AAID-SIM579%3E3.0.CO%3B2-C.

Y. C. MacNab and P. Gustafson. Regression b-spline smoothing in Bayesian disease mapping: with an application to patient safety surveillance. *Statistics in Medicine*, 26(24): 4455–4474, 2007. doi: https://doi.org/10.1002/sim.2868. URL https://onlinelibrary.wiley.com/doi/abs/10.1002/sim.2868.

A. Majumdar and A. E. Gelfand. Multivariate spatial modeling for geostatistical data using convolved covariance functions. *Mathematical Geology*, 39:225–245, 2007. URL https://api.semanticscholar.org/CorpusID:121010917.

K. Mardia. Multi-dimensional multivariate Gaussian Markov random fields with application to image processing. *Journal of Multivariate Analysis*, 24(2):265–284, 1988.

K. Mardia and C. Goodall. Spatio-temporal analyses of multivariate environmental monitoring data. In G. Patil and C. Rao, editors, *Multivariate Environmental Statistics*, pages 347–386. Elsevier, 1993.

K. Mardia, J. Kent, and J. Bibby. *Multivariate Analysis*. Probability and Mathematical Statistics : a series of monographs and textbooks. Academic Press, 1979. ISBN 9780124712508. URL https://books.google.com/books?id=bxjvAAAAMAAJ.

K. Mardia, C. Goodall, E. Redfern, and F. Alonso. The kriged kalman filter. *TEST: An Official Journal of the Spanish Society of Statistics and Operations Research*, 7:217–282, 02 1998. doi: 10.1007/BF02565111.

K. V. Mardia. *Statistics of Directional Data*. Academic Press, New York, 1972.

K. V. Mardia and P. E. Jupp. *Directional Statistics*. Wiley, 2000.

J.-M. Marin, P. Pudlo, C. P. Robert, and R. J. Ryder. Approximate Baeysian computational methods. *Statistics and Computing*, 22:1167–1180, 2012.

E. Marinari and G. Parisi. Simulated tempering: a new Monte Carlo scheme. *Europhysics Letters*, 19(6):451, 1992.

J. S. Maritz and T. Lwin. *Empirical Bayes methods*. Routledge Library Editions: Econometrics. Routledge Member of the Taylor and Francis Group, New York, NY, June 2019.

P. Marjoram, J. Molitor, V. Plagnol, and S. Tavare. Markov chain Monte Carlo without likelihoods. *Proceedings of the National Academy of Sciences*, 100:15324–15328, 2003.

M. A. Martinez-Beneito. A general modeling framework for multivariate disease mapping. *Biometrika*, 100:539–553, 2013.

S. Martino, D. S. Pace, S. Moro, E. Casoli, D. Ventura, A. Frachea, M. Silvestri, A. Arcangeli, G. Giacomini, G. Ardizzone, et al. Integration of presence-only data from several sources: a case study on dolphins' spatial distribution. *Ecography*, 44(10):1533–1543, 2021.

M. Martínez-Beneito and P. Botella-Rocamora. *Disease mapping: from foundations to multidimensional modeling*. CRC Press, Boca Raton, FL, 2019.

Y. Marzouk, T. Moselhy, M. Parno, and A. Spantini. Sampling via measure transport: An introduction. In R. Ghanem, D. Higdon, and H. Owhadi, editors, *Handbook of Uncertainty Quantification*, pages 1–41. Springer International Publishing, Cham, 2016. ISBN 978-3-319-11259-6. doi: 10.1007/978-3-319-11259-6_23-1. URL https://doi.org/10.1007/978-3-319-11259-6_23-1.

S. Mase, J. Møller, D. Stoyan, R. P. Waagepetersen, and G. Döge. Packing densities and simulated tempering for hard core Gibbs point processes. *Annals of the Institute of Statistical Mathematics*, 53(4):661–680, 2001.

B. Matern. *Spatial Variation*. Lecture notes in statistics. Springer, New York, NY, 2 edition, Dec 1986. ISBN 9780387963655. doi: 10.1007/978-1-4615-7892-5. URL https://doi.org/10.1007/978-1-4615-7892-5.

G. Matheron. Principles of geostatistics. *Economic Geology*, 58(8):1246–1266, 12 1963. ISSN 0361-0128. doi: 10.2113/gsecongeo.58.8.1246. URL https://doi.org/10.2113/gsecongeo.58.8.1246.

G. Matheron. The intrinsic random functions and their applications. *Advances in Applied Probability*, 5(3):439–468, 1973. doi: 10.2307/1425829.

G. Matheron. Recherche de simplification dans un probleme de cokrigeage. Technical Report, 1979. Centre de Géostatistique, Fountainebleau, N-698.

G. Matheron. Pour une analyse krigeante des données régionalisées. *Centre de Géostatistique, Report N-132, Fontainebleau*, page 22, 1982.

T. Mattfeldt, S. Eckel, F. Fleischer, and V. Schmidt. Statistical modelling of the geometry of planar sections of prostatic capillaries on the basis of stationary Strauss hard-core processes. *Journal of Microscopy*, 228:272–81, 2007.

C. A. Mayo, L. Ganley, C. A. Hudak, S. Brault, M. K. Marx, E. Burke, and M. W. Brown. Distribution, demography, and behavior of North Atlantic right whales (*Eubalaena Glacialis*) in Cape Cod Bay, Massachusetts, 1998-2013. *Marine Mammal Science*, 34(4):979–996, 2018. ISSN 0824-0469. doi: 10.1111/mms.12511.

A. B. McBratney and R. Webster. Choosing functions for semi-variograms of soil properties and fitting them to sampling estimates. *Journal of Soil Science*, 37(4):617–639, 1986.

doi: https://doi.org/10.1111/j.1365-2389.1986.tb00392.x. URL https://bsssjournals.onlinelibrary.wiley.com/doi/abs/10.1111/j.1365-2389.1986.tb00392.x.

N. J. McMillan, D. M. Holland, M. Morara, and J. Feng. Combining numerical model output and particulate data using Bayesian space–time modeling. *Environmetrics*, 21(1):48–65, 2010. doi: https://doi.org/10.1002/env.984. URL https://onlinelibrary.wiley.com/doi/abs/10.1002/env.984.

K. L. Mengersen, C. P. Robert, and C. Guihenneuc-Jouyaux. MCMC Convergence Diagnostics: A Review. In *Bayesian Statistics 6: Proceedings of the Sixth Valencia International Meeting June 6-10, 1998*. Oxford University Press, 08 1999. ISBN 9780198504856. doi: 10.1093/oso/9780198504856.003.0018. URL https://doi.org/10.1093/oso/9780198504856.003.0018.

N. Metropolis, A. W. Rosenbluth, M. N. Rosenbluth, A. H. Teller, and E. Teller. Equation of state calculations by fast computing machines. *Journal of Chemical Physics*, 21(6): 1087–1092, June 1953.

T. H. Meyer, M. Eriksson, and R. C. Maggio. Gradient estimation from irregularly spaced data sets. *Mathematical Geology*, 33(6):693–717, Aug 2001. ISSN 1573-8868. doi: 10.1023/A:1011026732182. URL https://doi.org/10.1023/A:1011026732182.

S. Minsker. Geometric median and robust estimation in banach spaces. *Bernoulli*, 21: 2308–2335, 2015.

S. Minsker, S. Srivastava, L. Lin, and D. B. Dunson. Robust and scalable Bayes via a median of subset posterior measures. *Journal of Machine Learning Research*, 18(124): 1–40, 2017.

A. Mira and D. J. Sargent. A new strategy for speeding Markov chain Monte Carlo algorithms. *Statistical Methods and Applications*, 12(1):49–60, Feb. 2003.

L. Mitas and H. Mitasova. Spatial interpolation. In P. A. Longley, M. F. Goodchild, D. J. Maguire, and D. W. Rhind, editors, *Geographical information systems: principles, techniques, management and applications*, pages 481–492. Wiley Hoboken, NJ, USA, 2 edition, 1999.

G. O. Mohler, M. B. Short, P. J. Brantingham, F. P. Schoenberg, and G. E. Tita. Self-exciting point process modeling of crime. *Journal of the American Statistical Association*, 100:100–108, 2011.

J. Møller. Perfect simulation of conditionally specified models. *Journal of the Royal Statistical Society: Series B (Statistical Methodology)*, 61(1):251–264, 1999.

J. Møller. A review of perfect simulation in stochastic geometry. *IMS Lecture Notes-Monograph Series*, pages 333–355, 2001.

J. Møller. Aspects of spatial point process modelling and Bayesian inference. Presentation, 2012. Presentation at Bayes Lectures 2012, Edinburgh, UK.

J. Moller and R. P. Waagepetersen. *Statistical inference and simulation for spatial point processes*. Chapman & Hall/CRC Monographs on Statistics & Applied Probability. Chapman and Hall, London, England, Sept. 2003.

J. Møller and R. P. Waagepetersen. Modern statistics for spatial point processes. *Scandinavian Journal of Statistics*, 34(4):643–684, 2007.

J. Møller, A. N. Pettitt, R. Reeves, and K. K. Berthelsen. An efficient Markov chain Monte Carlo method for distributions with intractable normalising constants. *Biometrika*, 93 (2):451–458, 2006.

P. Moraga. *Spatial statistics for data science*. Chapman & Hall/CRC, Philadelphia, PA, Dec 2023. ISBN 9781032633510.

R. D. Morey, J.-W. Romeijn, and J. N. Rouder. The philosophy of bayes factors and the quantification of statistical evidence. *Journal of Mathematical Psychology*, 72: 6–18, 2016. ISSN 0022-2496. doi: https://doi.org/10.1016/j.jmp.2015.11.001. URL https://www.sciencedirect.com/science/article/pii/S0022249615000723. Bayes Factors for Testing Hypotheses in Psychological Research: Practical Relevance and New Developments.

J. S. Morris. Functional regression. *Annual Review of Statistics and Its Application*, 2(1): 321–359, 2015. doi: 10.1146/annurev-statistics-010814-020413.

I. Mueller, P. Vounatsou, B. J. Allen, and T. Smith. Spatial patterns of child growth in papua new guinea and their relation to environment, diet, socio-economic status and subsistence activities. *Annals of Human Biology*, 28(3):263–280, 2001. doi: 10.1080/030144601300119089. URL https://doi.org/10.1080/030144601300119089. PMID: 11393334.

A. S. Mugglin and B. P. Carlin. Hierarchical modeling in geographic information systems: Population interpolation over incompatible zones. *Journal of Agricultural, Biological, and Environmental Statistics*, 3(2):111–130, 1998. ISSN 10857117. URL http://www.jstor.org/stable/1400646.

A. S. Mugglin, B. P. Carlin, L. Zhu, and E. Conlon. Bayesian areal interpolation, estimation, and smoothing: An inferential approach for geographic information systems. *Environment and Planning A: Economy and Space*, 31(8):1337–1352, 1999. doi: 10.1068/a311337. URL https://doi.org/10.1068/a311337.

A. S. Mugglin, B. P. Carlin, and A. E. Gelfand. Fully model-based approaches for spatially misaligned data. *Journal of the American Statistical Association*, 95(451):877–887, 2000. doi: 10.1080/01621459.2000.10474279. URL https://www.tandfonline.com/doi/abs/10.1080/01621459.2000.10474279.

P. Müller, G. Parmigiani, C. Robert, and J. Rousseau. Optimal sample size for multiple testing: the case of gene expression microarrays. *Journal of the American Statistical Association*, 99(468):990–1001, 2004.

K. Murphy. *Machine Learning: A probabilistic perspective*. The MIT Press, Cambridge, MA, 2012.

I. Murray and R. P. Adams. Slice sampling covariance hyperparameters of latent Gaussian models. In *Neural Information Processing Systems*, 2010. URL https://api.semanticscholar.org/CorpusID:9152657.

I. Murray, Z. Ghahramani, and D. J. C. MacKay. MCMC for doubly-intractable distributions. In *Proceedings of the 22nd Annual Conference on Uncertainty in Artificial Intelligence (UAI)*. AUAI Press, 2006.

I. Murray, R. Adams, and D. MacKay. Elliptical slice sampling. In Y. W. Teh and M. Titterington, editors, *Proceedings of the Thirteenth International Conference on Artificial Intelligence and Statistics*, volume 9 of *Proceedings of Machine Learning Research*, pages 541–548, Chia Laguna Resort, Sardinia, Italy, 13–15 May 2010a. PMLR. URL https://proceedings.mlr.press/v9/murray10a.html.

I. Murray, R. P. Adams, and D. J. MacKay. Elliptical slice sampling. *Journal of Machine Learning Research: Workshop and Conference Proceedings (AISTATS)*, 9:541–548, 2010b.

R. P. Murray, N. R. Anthonisen, J. E. Connett, R. A. Wise, P. G. Lindgren, P. G. Greene, M. A. Nides, and for the. Effects of multiple attempts to quit smoking and relapses to smoking on pulmonary function. *Journal of Clinical Epidemiology*, 51(12):1317–1326, 1998. ISSN 0895-4356. doi: https://doi.org/10.1016/S0895-4356(98)00120-6. URL https://www.sciencedirect.com/science/article/pii/S0895435698001206.

D. E. Myers. Matrix formulation of co-kriging. *Journal of the International Association for Mathematical Geology*, 14(3):249–257, Jun 1982. ISSN 1573-8868. doi: 10.1007/BF01032887. URL https://doi.org/10.1007/BF01032887.

D. E. Myers. Pseudo-cross variograms, positive-definiteness, and cokriging. *Mathematical Geology*, 23(6):805–816, Jul 1991. ISSN 1573-8868. doi: 10.1007/BF02068776. URL https://doi.org/10.1007/BF02068776.

D. E. Myers. Kriging, cokriging, radial basis functions and the role of positive definiteness. *Computers & Mathematics With Applications*, 24:139–148, 1992. URL https://api.semanticscholar.org/CorpusID:119559150.

P. Mykland, L. Tierney, and B. Yu. Regeneration in Markov chain samplers. *Journal of the American Statistical Association*, 90(429):233–241, 1995. ISSN 01621459, 1537274X. URL http://www.jstor.org/stable/2291148.

R. Narain. On sampling without replacement with varying probabilities. *Journal of the Indian Society of Agricultural Statistics*, 3:169–175, 1951.

P. Naveau, R. Huser, P. Ribereau, and A. Hannart. Modeling jointly low, moderate, and heavy rainfall intensities without a threshold selection. *Water Resources Research*, 52(4): 2753–2769, 2016. doi: https://doi.org/10.1002/2015WR018552. URL https://agupubs.onlinelibrary.wiley.com/doi/abs/10.1002/2015WR018552.

R. Neal. Mcmc using Hamiltonian dynamics. In S. Brooks, A. Gelman, G. L. Jones, and X.-L. Meng, editors, *Handbook of Markov Chain Monte Carlo*, pages 113–162. Boca Raton, FL: CRC Press, 2011.

R. M. Neal. Probabilistic inference using Markov chain Monte Carlo methods. Technical report, Department of Computer Science, University of Toronto Toronto, Ontario, Canada, 1993a.

R. M. Neal. Bayesian learning via stochastic dynamics. In *Advances in neural information processing systems*, pages 475–482, 1993b.

R. M. Neal. An improved acceptance procedure for the hybrid Monte Carlo algorithm. *Journal of Computational Physics*, 111(1):194–203, 1994. ISSN 0021-9991. doi: https://doi.org/10.1006/jcph.1994.1054. URL https://www.sciencedirect.com/science/article/pii/S0021999184710540.

R. M. Neal. *Bayesian learning for neural networks*, volume 118. Springer, 1996.

R. M. Neal. Slice sampling. *The Annals of Statistics*, 31(3):705–767, 2003. doi: 10.1214/aos/1056562461. URL https://doi.org/10.1214/aos/1056562461.

B. Neelon, F. Li, L. F. Burgette, and S. E. Neelon. A spatiotemporal quantile regression model for emergency department expenditures. *Statistics in Medicine*, 34(17):2559–2575, 2015. doi: 10.1002/sim.6480.

R. B. Nelsen. *An Introduction to Copulas*. Springer Series in Statistics. Springer, New York, NY, 2 edition, Jan. 2006.

M. A. Newton, A. Noueiry, D. Sarkar, and P. Ahlquist. Detecting differential gene expression with a semiparametric hierarchical mixture method. *Biostatistics*, 5(2):155–176, 04 2004. ISSN 1465-4644. doi: 10.1093/biostatistics/5.2.155. URL https://doi.org/10.1093/biostatistics/5.2.155.

D. Nychka and N. Saltzman. Design of air-quality monitoring networks. *Case Studies in Environmental Statistics*, pages 51–76, 1998.

D. Nychka, S. Bandyopadhyay, D. Hammerling, F. Lindgren, and S. Sain. A multiresolution Gaussian process model for the analysis of large spatial datasets. *Journal of Computational and Graphical Statistics*, 24(2):579–599, 2015. doi: 10.1080/10618600.2014.914946. URL http://dx.doi.org/10.1080/10618600.2014.914946.

J. E. Oakley and A. O'Hagan. Probabilistic sensitivity analysis of complex models: A Bayesian approach. *Journal of the Royal Statistical Society: Series B (Statistical Methodology)*, 66(3):751–769, 2004. doi: https://doi.org/10.1111/j.1467-9868.2004.05304. x. URL https://rss.onlinelibrary.wiley.com/doi/abs/10.1111/j.1467-9868.2004.05304.x.

N. L. Oden, R. R. Sokal, M.-J. Fortin, and H. Goebl. Categorical wombling: Detecting regions of significant change in spatially located categorical variables. *Geographical Analysis*, 25(4):315–336, 1993. doi: https://doi.org/10.1111/j.1538-4632.1993.tb00301. x. URL https://onlinelibrary.wiley.com/doi/abs/10.1111/j.1538-4632.1993.tb00301.x.

A. O'Hagan. Fractional bayes factors for model comparison. *Journal of the Royal Statistical Society: Series B (Methodological)*, 57(1):99–118, 1995. doi: https://doi.org/10.1111/j.2517-6161.1995.tb02017.x. URL https://rss.onlinelibrary.wiley.com/doi/abs/10.1111/j.2517-6161.1995.tb02017.x.

A. O'Hagan, M. Kendall, and A. Stuart. *Advanced theory of statistics: Bayesian inference v. 2B*. Kendall's advanced statistics library. Hodder Arnold, London, England, 6 edition, June 1994.

V. D. Oliveira. Bayesian prediction of clipped Gaussian random fields. *Computational Statistics & Data Analysis*, 34(3):299–314, 2000. ISSN 0167-9473. doi: https://doi.org/10.1016/S0167-9473(99)00103-6. URL https://www.sciencedirect.com/science/article/pii/S0167947399001036.

V. D. Oliveira. Models for geostatistical binary data: Properties and connections. *The American Statistician*, 74(1):72–79, 2020. doi: 10.1080/00031305.2018.1444674. URL https://doi.org/10.1080/00031305.2018.1444674.

M. Omidi and M. Mohammadzadeh. A new method to build spatio-temporal covariance functions: Analysis of ozone data. *Statistical Papers*, pages 1–15, 2015. ISSN 0932-5026. doi: 10.1007/s00362-015-0674-2. URL http://dx.doi.org/10.1007/s00362-015-0674-2.

H. Omre. Bayesian kriging—merging observations and qualified guesses in kriging. *Mathematical Geology*, 19(1):25–39, Jan 1987. ISSN 1573-8868. doi: 10.1007/BF01275432. URL https://doi.org/10.1007/BF01275432.

H. Omre. A Bayesian approach to surface estimation. In C. F. Chung, A. G. Fabbri, and R. Sinding-Larsen, editors, *Quantitative Analysis of Mineral and Energy Resources*, pages 359–373. Springer Netherlands, Dordrecht, 1988. ISBN 978-94-009-4029-1. doi: 10.1007/978-94-009-4029-1_21. URL https://doi.org/10.1007/978-94-009-4029-1_21.

H. Omre and K. B. Halvorsen. The Bayesian bridge between simple and universal kriging. *Mathematical Geology*, 21(7):767–786, Oct 1989. ISSN 1573-8868. doi: 10.1007/BF00893321. URL https://doi.org/10.1007/BF00893321.

H. Omre, K. B. Halvorsen, and V. Berteig. A Bayesian approach to kriging. In M. Armstrong, editor, *Geostatistics*, pages 109–126, Dordrecht, 1989. Springer Netherlands. ISBN 978-94-015-6844-9.

R. Pace, R. Barry, O. W. Gilley, and C. Sirmans. A method for spatial–temporal forecasting with an application to real estate prices. *International Journal of Forecasting*, 16(2):229–246, 2000. ISSN 0169-2070. doi: https://doi.org/10.1016/S0169-2070(99)00047-3. URL https://www.sciencedirect.com/science/article/pii/S0169207099000473.

R. K. Pace and R. Barry. Sparse spatial autoregressions. *Statistics & Probability Letters*, 33(3):291–297, 1997a. ISSN 0167-7152. doi: https://doi.org/10.1016/S0167-7152(96)00140-X. URL https://www.sciencedirect.com/science/article/pii/S016771529600140X.

R. K. Pace and R. Barry. Fast spatial estimation. *Applied Economics Letters*, 4(5):337–341, 1997b. doi: 10.1080/758532605. URL https://doi.org/10.1080/758532605.

L. Paci, A. E. Gelfand, and D. M. Holland. Spatio-temporal modeling for real-time ozone forecasting. *Spatial Statistics*, 4:79–93, 2013. ISSN 2211-6753. doi: https://doi.org/10.1016/j.spasta.2013.04.003. URL https://www.sciencedirect.com/science/article/pii/S2211675313000195.

L. Paci, A. E. Gelfand, M. A. Beamonte, P. Gargallo, and M. Salvador. Analysis of residential property sales using space-time point patterns. *Spatial Statistics*, 21:149–165, 2017.

L. Paci, A. E. Gelfand, M. A. Beamonte, P. Gargallo, and M. Salvador. Spatial hedonic modeling adjusted for preferential sampling. *Journal of the Royal Statistical Society, Series A*, 183:169–192, 2020.

K. Pacifici, B. J. Reich, D. A. Miller, B. Gardner, G. Stauffer, S. Singh, A. McKerrow, and J. A. Collazo. Integrating multiple data sources in species distribution modeling: A framework for data fusion. *Ecology*, 98(3):840–850, 2017.

C. J. Paciorek and M. J. Schervish. Spatial modelling using a new class of nonstationary covariance functions. *Environmetrics*, 17:483–506, 2006.

S. Pan, L. Zhang, J. R. Bradley, and S. Banerjee. Bayesian inference for spatial-temporal non-Gaussian data using predictive stacking, 2025. URL https://arxiv.org/abs/2406.04655.

E. Pardo-Igúzquiza and P. A. Dowd. Amle3d: A computer program for the inference of spatial covariance parameters by approximate maximum likelihood estimation. *Computers & Geosciences*, 23(7):793–805, 1997. ISSN 0098-3004. doi: https://doi.org/10.1016/S0098-3004(97)00040-X. URL https://www.sciencedirect.com/science/article/pii/S009830049700040X.

G. Parmigiani. *Modeling in Medical Decision Making: A Bayesian Approach*. John Wiley & Sons, Hoboken, NJ, 2002.

D. Pati, B. J. Reich, and D. B. Dunson. Bayesian geostatistical modelling with informative sampling locations. *Biometrika*, 98:35–48, 2011.

J. Pavani and F. A. Quintana. A Bayesian multivariate model with temporal dependence on random partition of areal data. *arXiv preprint arXiv:2401.08303*, 2024.

F. Pearson II. *Map Projections: Theory and Applications*. CRC Press, Boca Raton, FL, second edition, Mar. 1990.

E. Pebesma and R. Bivand. *Spatial Data Science: With Applications in R*. Chapman and Hall/CRC, Boca Raton, 2023. doi: 10.1201/9780429459016. URL https://doi.org/10.1201/9780429459016.

M. Perone Pacifico, C. Genovese, I. Verdinelli, and L. Wasserman. False discovery control for random fields. *Journal of the American Statistical Association*, 99(468):1002–1014, 2004.

M. Peruzzi, S. Banerjee, and A. Finley. Highly Scalable Bayesian Geostatistical Modeling via Meshed Gaussian Processes on Partitioned Domains. *Journal of the American Statistical Association*, 117(538):969–982, 2022.

G. W. Peters. General quantile time series regressions for applications in population demographics. *Risks*, 6(3):97, 2018. doi: 10.3390/risks6030097.

S. Petrone, M. Guindani, and A. E. Gelfand. Hybrid dirichlet mixture models for functional data. *Journal of the Royal Statistical Society: Series B (Statistical Methodology)*, 71(4):755–782, 2009.

U. Picchini. Inference for SDE models via approximate Bayesian computation. *Journal of Computational and Graphical Statistics*, 23:1080–1100, 2013.

D. Poole and A. E. Raftery. Inference for deterministic simulation models: The Bayesian melding approach. *Journal of the American Statistical Association*, 95(452):1244–1255, 2000. doi: 10.1080/01621459.2000.10474324.

E. Porcu and V. Zastavnyi. Characterization theorems for some classes of covariance functions associated to vector valued random fields. *Journal of Multivariate Analysis*, 102 (9):1293–1301, 2011. ISSN 0047-259X. doi: https://doi.org/10.1016/j.jmva.2011.04.013. URL https://www.sciencedirect.com/science/article/pii/S0047259X11000698.

R. B. Potts. Some generalized order-disorder transformations. *Mathematical Proceedings of the Cambridge Philosophical Society*, 48(1):106–109, 1952. doi: 10.1017/S0305004100027419.

M. Pourahmadi. Joint mean-covariance models with applications to longitudinal data: unconstrained parameterisation. *Biometrika*, 86(3):677–690, 09 1999. ISSN 0006-3444. doi: 10.1093/biomet/86.3.677. URL https://doi.org/10.1093/biomet/86.3.677.

B. Presnell, S. P. Morrison, and R. C. Littell. Projected multivariate linear models for directional data. *Journal of the American Statistical Association*, 93(443):1068–1077, 1998.

J. Pritchard, M. Seielstad, A. Perez-Lezaun and M. Feldman. Population growth of human y chromosomes: a study of y chromosome microsatellites. *Molecular Biology and Evolution*, 16:1791–98, 1999.

J. Pritchard, M. Seielstad, A. Perez-Lezaun, and M. Feldman. Population growth of human Y chromosomes: a study of Y chromosome microsatellites. *Molecular Biology and Evolution*, 16:1791–1798, 1999.

J. G. Propp and D. B. Wilson. Exact sampling with coupled Markov chains and applications to statistical mechanics. *Random Structures and Algorithms*, 9(1-2):223–252, 1996.

J. G. Propp and D. B. Wilson. How to get a perfectly random sample from a generic Markov chain and generate a random spanning tree of a directed graph. *Journal of Algorithms*, 27(2):170–217, 1998.

K. Qu, J. R. Bradley, and X. Niu. Boundary detection using a Bayesian hierarchical model for multiscale spatial data. *Technometrics*, 63(1):64–76, 2021.

J. Quiñonero-Candela and C. E. Rasmussen. A unifying view of sparse approximate Gaussian process regression. *Journal of Machine Learning Research*, 6(65):1939–1959, 2005. URL http://jmlr.org/papers/v6/quinonero-candela05a.html.

H. Quick, S. Banerjee, and B. P. Carlin. Modeling temporal gradients in regionally aggregated California asthma hospitalization data. *The Annals of Applied Statistics*, 7(1):154–176, 2013. doi: 10.1214/12-AOAS600. URL https://doi.org/10.1214/12-AOAS600.

R Core Team. *R: A Language and Environment for Statistical Computing*. R Foundation for Statistical Computing, Vienna, Austria, 2021. URL https://www.R-project.org/.

S. T. Radev, U. K. Mertens, A. Voss, L. Ardizzone, and U. Köthe. BayesFlow: Learning complex stochastic models with invertible neural networks. *IEEE transactions on neural networks and learning systems*, 33(4):1452–1466, 2020.

S. T. Radev, M. Schmitt, V. Pratz, U. Picchini, U. Koethe, and P.-C. Buerkner. JANA: Jointly amortized neural approximation of complex Bayesian models. In *The 39th Conference on Uncertainty in Artificial Intelligence*, 2023a. URL https://openreview.net/forum?id=dS3wVICQrU0.

S. T. Radev, M. Schmitt, L. Schumacher, L. Elsemüller, V. Pratz, Y. Schälte, U. Köthe, and P.-C. Bürkner. Bayesflow: Amortized Bayesian workflows with neural networks, 2023b. URL https://arxiv.org/abs/2306.16015.

J. O. Ramsay and B. W. Silverman. *Functional Data Analysis*. Springer series in statistics. Springer, New York, NY, 2 edition, June 2005.

J. N. K. Rao. Impact of Frequentist and Bayesian Methods on Survey Sampling Practice: A Selective Appraisal. *Statistical Science*, 26(2):240–256, 2011. doi: 10.1214/10-STS346. URL https://doi.org/10.1214/10-STS346.

J. N. K. Rao and P. D. Ghangurde. Bayesian optimization in sampling finite populations. *Journal of the American Statistical Association*, 67(338):439–443, 1972. doi: 10.1080/01621459.1972.10482406. URL https://www.tandfonline.com/doi/abs/10.1080/01621459.1972.10482406.

J. N. K. Rao and I. Molina. *Small Area Estimation*. John Wiley & Sons, Hoboken, NJ, 2nd edition, 2015.

C. E. Rasmussen and C. K. I. Williams. *Gaussian processes for machine learning*. Adaptive Computation and Machine Learning series. MIT Press, Cambridge, MA, Nov. 2005.

J. G. Rasmussen. Bayesian inference for hawkes processes. *Methodolgy and Computing in Applied Probability*, 15:623–642, 2013.

S. W. Raudenbush and A. S. Bryk. *Hierarchical Linear Models: Applications and Data Analysis Methods*. SAGE, 2nd edition, 2002. ISBN 076191904X, 9780761919049.

S. U. Rehman and A. Shapiro. An integral transform approach to cross-variograms modeling. *Computational Statistics & Data Analysis*, 22(3):213–233, 1996. ISSN 0167-9473. doi: https://doi.org/10.1016/0167-9473(95)00052-6. URL https://www.sciencedirect.com/science/article/pii/0167947395000526.

B. J. Reich. Spatiotemporal quantile regression for detecting distributional changes in environmental processes. *Journal of the Royal Statistical Society: Series C (Applied Statistics)*, 61(4):535–553, 2012. doi: 10.1111/j.1467-9876.2011.01025.x.

B. J. Reich, J. Eidsvik, M. Guindani, A. J. Nail, and A. M. Schmidt. A class of covariate-dependent spatiotemporal covariance functions for the analysis of daily ozone concentration. *Annals of Applied Statistics*, 5:2425–2447, 2011.

A. Reinhart. A review of self-exciting spatio-temporal point processes and their applications. *Statistical Science*, 33(3):299–318, 2018.

Q. Ren and S. Banerjee. Hierarchical factor models for large spatially misaligned data: A low-rank predictive process approach. *Biometrics*, 69(1):19–30, 2013. doi: 10.1111/j.1541-0420.2012.01832.x. URL https://onlinelibrary.wiley.com/doi/abs/10.1111/j.1541-0420.2012.01832.x.

Q. Ren, S. Banerjee, A. O. Finley, and J. S. Hodges. Variational Bayesian methods for spatial data analysis. *Computational Statistics & Data Analysis*, 55(12):3197–3217, 2011. ISSN 0167-9473. doi: https://doi.org/10.1016/j.csda.2011.05.021. URL https://www.sciencedirect.com/science/article/pii/S0167947311002003.

L. J. Rendell, A. L. Adam M. Johansen, and N. Whiteley. Global consensus Monte Carlo. *Journal of Computational and Graphical Statistics*, 30(2):249–259, 2021. doi: 10.1080/10618600.2020.1811105.

I. W. Renner, J. Elith, A. Baddeley, W. Fithian, T. Hastie, S. J. Phillips, G. Popovic, and D. I. Warton. Point process models for presence-only analysis. *Methods in Ecology and Evolution*, 6(4):366–379, 2015.

M. Ribatet. Spatial extremes: Max-stable processes at work. *Journal de la société française de statistique*, 154(2):156–177, 2013. URL http://www.numdam.org/item/JSFS_2013_154_2_156_0/.

A. Riebler, S. Sorbye, D. Simpson, and H. Rue. An intuitive Bayesian spatial model for disease mapping that accounts for scaling. *Statistical Methods in Medical Research*, 25(4):1145–1165, 2016.

V. Rimalova, E. Fiserova, A. Menafoglio, and A. Pini. Inference for spatial regression models with functional response using a permutational approach. *Journal of Multivariate Analysis*, 189, 2022.

B. D. Ripley. Modelling spatial patterns. *Journal of the Royal Statistical Society. Series B (Methodological)*, 39(2):172–212, 1977.

B. D. Ripley. *Spatial statistics*. John Wiley & Sons, 1981.

M. D. Risser and C. A. Calder. Regression-based covariance functions for nonstationary spatial modeling. *Environmetrics*, 26:284–297, 2015.

C. Robert. *The Bayesian choice*. Springer Texts in Statistics. Springer, New York, NY, 2 edition, June 2007.

C. Robert and G. Casella. *Monte Carlo Statistical Methods*. Boca Raton, FL: CRC Press, second edition, 2004.

G. O. Roberts and J. S. Rosenthal. Optimal scaling of discrete approximations to Langevin diffusions. *Journal of the Royal Statistical Society: Series B (Statistical Methodology)*, 60 (1):255–268, 1998a.

G. O. Roberts and J. S. Rosenthal. Markov-chain Monte Carlo: Some practical implications of theoretical results. *Canadian Journal of Statistics*, 26(1):5–20, 1998b.

G. O. Roberts and J. S. Rosenthal. Coupling and ergodicity of adaptive Markov chain Monte Carlo algorithms. *Journal of Applied Probability*, 44(2):458–475, 2007.

G. O. Roberts and O. Stramer. Langevin diffusions and Metropolis-Hastings algorithms. *Methodology and Computing in Applied Probability*, 4(4):337–357, 2002.

G. O. Roberts and R. L. Tweedie. Exponential convergence of langevin distributions and their discrete approximations. *Bernoulli*, 2(4):341–363, 1996.

G. O. Roberts, J. S. Rosenthal, et al. Optimal scaling for various Metropolis-Hastings algorithms. *Statistical Science*, 16(4):351–367, 2001.

W. S. Robinson. Ecological correlations and the behavior of individuals. *American Sociological Review*, 15(3):351–357, 1950. ISSN 00031224, 19398271. URL http://www.jstor.org/stable/2087176.

A. Rodriguez, D. B. Dunson, and A. E. Gelfand. Bayesian nonparametric functional data analysis through density estimation. *Biometrika*, 96(1):149–162, 2009. ISSN 00063444, 14643510. URL http://www.jstor.org/stable/27798808.

A. Rodríguez, D. B. Dunson, and A. E. Gelfand. Latent stick-breaking processes. *Journal of the American Statistical Association*, 105(490):647–659, 2010.

J. F. Rogers and G. G. Killough. Historical dose reconstruction project: Estimating the population at risk. *Health Physics*, 72(2), 1997. ISSN 0017-9078. URL https://journals.lww.com/health-physics/fulltext/1997/02000/historical_dose_reconstruction_project__estimating.2.aspx.

A. Roverato. Hyper inverse Wishart distribution for non-decomposable graphs and its application to Bayesian inference for Gaussian graphical models. *Scandinavian Journal of Statistics*, 29(3):391–411, 2002.

J. A. Royle and L. M. Berliner. A hierarchical approach to multivariate spatial modeling and prediction. *Journal of Agricultural Biological and Environmental Statistics*, 4:29–56, 1999. URL https://api.semanticscholar.org/CorpusID:121346808.

D. B. Rubin. Inference and Missing Data. *Biometrika*, 63(3):581–592, 1976.

W. Rudin et al. *Principles of mathematical analysis*, volume 3. McGraw-hill New York, 1976.

H. Rue and L. Held. *Gaussian Markov Random Fields : Theory and Applications*. Monographs on statistics and applied probability. Chapman & Hall/CRC, Boca Raton, FL, 2005. ISBN 1-584-88432-0. URL http://opac.inria.fr/record=b1119989.

H. Rue and H. Tjelmeland. Fitting Gaussian Markov random fields to Gaussian fields. *Scandinavian Journal of Statistics*, 29(1):31–49, 2002. doi: https://doi.org/10.1111/1467-9469.00058. URL https://onlinelibrary.wiley.com/doi/abs/10.1111/1467-9469.00058.

H. Rue, S. Martino, and N. Chopin. Approximate Bayesian inference for latent Gaussian models by using integrated nested laplace approximations. *Journal of the Royal Statistical Society: Series B (Statistical Methodology)*, 71(2):319–392, 2009. ISSN 1467-9868. doi: 10.1111/j.1467-9868.2008.00700.x. URL http://dx.doi.org/10.1111/j.1467-9868.2008.00700.x.

D. Ruppert, M. Wand, and R. Carroll. *Semiparametric Regression*. Cambridge University Press, Cambridge, United Kingdom, 2003.

J. Sacks, W. J. Welch, T. J. Mitchell, and H. P. Wynn. Design and Analysis of Computer Experiments. *Statistical Science*, 4(4):409–423, 1989. doi: 10.1214/ss/1177012413. URL https://doi.org/10.1214/ss/1177012413.

A. Saha, S. Basu, and A. Datta. Randomforestsgls: an R package for random forests for dependent data. *Journal of Open Source Software*, 7(71), 2022.

A. Saha, S. Basu, and A. Datta. Random forests for spatially dependent data. *Journal of the American Statistical Association*, 118(541):665–683, 2023.

S. Sahu and K. Bakar. A comparison of Bayesian models for daily ozone concentration levels. *Statistical Methodology*, 9(1):144–157, 2012. ISSN 1572-3127. doi: https://doi.org/10.1016/j.stamet.2011.04.009. URL https://www.sciencedirect.com/science/article/pii/S1572312711000475. Special Issue on Astrostatistics + Special Issue on Spatial Statistics.

S. K. Sahu. *Bayesian modeling of spatio-temporal data with R*. Chapman and Hall/CRC, Boca Raton, Dec 2021. ISBN 9780429318443. doi: 10.1201/9780429318443. URL https://doi.org/10.1201/9780429318443.

S. R. Sain and N. Cressie. A spatial model for multivariate lattice data. *Journal of Econometrics*, 140(1):226–259, 2007. ISSN 0304-4076. doi: https://doi.org/10.1016/j.jeconom.2006.09.010. URL https://www.sciencedirect.com/science/article/pii/S0304407606002302. Analysis of spatially dependent data.

M. Sainsbury-Dale, A. Zammit-Mangion, and R. Huser. Likelihood-free parameter estimation with neural bayes estimators. *The American Statistician*, 78(1):1–14, 2024. doi: 10.1080/00031305.2023.2249522. URL https://doi.org/10.1080/00031305.2023.2249522.

P. D. Sampson and P. Guttorp. Nonparametric estimation of nonstationary spatial covariance structure. *Journal of the American Statistical Association*, 87(417):108–119, 1992. ISSN 01621459, 1537274X. URL http://www.jstor.org/stable/2290458.

H. Sang and A. E. Gelfand. Hierarchical modeling for extreme values observed over space and time. *Environmental and Ecological Statistics*, 16(3):407–426, Sep 2009. ISSN 1573-3009. doi: 10.1007/s10651-007-0078-0. URL https://doi.org/10.1007/s10651-007-0078-0.

H. Sang and A. E. Gelfand. Continuous spatial process models for spatial extreme values. *Journal of Agricultural, Biological, and Environmental Statistics*, 15(1):49–65, Mar 2010. ISSN 1537-2693. doi: 10.1007/s13253-009-0010-1. URL https://doi.org/10.1007/s13253-009-0010-1.

H. Sang and J. Z. Huang. A full scale approximation of covariance functions for large spatial data sets. *Journal of the Royal Statistical Society, Series B*, 74:111–132, 2012.

H. Sang, M. Jun, and J. Huang. Covariance approximation for large multivariate spatial datasets with an application to multiple climate model errors. *Annals of Applied Statistics*, 4:2519–2548, 2011.

B. Sansó and L. Guenni. Venezuelan rainfall data analysed by using a Bayesian space-time model. *Journal of the Royal Statistical Society, Series C (Applied Statistics)*, 48(3):345–362, 1999. ISSN 00359254, 14679876. URL http://www.jstor.org/stable/2680829.

T. J. Santner, B. J. Williams, and W. I. Notz. *The design and analysis of computer experiments the design and analysis of computer experiments*. Springer series in statistics. Springer, New York, NY, 2nd edition, Jan. 2019.

A. Sauer, A. Cooper, and R. B. Gramacy. Vecchia-approximated deep Gaussian processes for computer experiments. *Journal of Computational and Graphical Statistics*, 32(3):824–837, 2023a. doi: 10.1080/10618600.2022.2129662. URL https://doi.org/10.1080/10618600.2022.2129662.

A. Sauer, R. B. Gramacy, and D. Higdon. Active learning for deep Gaussian process surrogates. *Technometrics*, 65(1):4–18, 2023b. doi: 10.1080/00401706.2021.2008505. URL https://doi.org/10.1080/00401706.2021.2008505.

O. Schabenberger and C. A. Gotway. *Statistical Methods for Spatial Data Analysis*. Chapman and Hall/CRC, first edition, 2004.

M. Schlather. Examples for the coefficient of tail dependence and the domain of attraction of a bivariate extreme value distribution. *Statistics & Probability Letters*, 53(3):325–329, 2001. ISSN 0167-7152. doi: https://doi.org/10.1016/S0167-7152(01)00090-6. URL https://www.sciencedirect.com/science/article/pii/S0167715201000906.

M. Schlather. Models for stationary max-stable random fields. Extremes, 5:33–44, 2002.

M. Schlather and J. A. Tawn. A dependence measure for multivariate and spatial extreme values: Properties and inference. *Biometrika*, 90(1):139–156, 2003. ISSN 00063444. URL http://www.jstor.org/stable/30042025.

E. Schliep, A. Gelfand, J. Asin, A. Cebrian, M. Beamonte, and J. Abaurrea. Long-term spatial modeling for characteristics of extreme heat events. *Journal of the Royal Statistical Society, Series A*, 184:1070–1092, 2021.

E. M. Schliep and A. E. Gelfand. Velocities for space-time point patterns. *Spatial Statistics*, 29:204–229, 2019.

E. M. Schliep, A. E. Gelfand, and J. S. Clark. Stochastic modeling for the velocity of climate change. *Journal of Agricultural, Biological and Environmental Statistics*, 20:323–342, 2015.

E. M. Schliep, A. E. A E Gelfand, C. W. Clark, C. M. Mayo, B. McKenna, S. E. Parks, T. M. Yack, and R. S. Schick. Assessing marine mammal abundance: A novel data fusion. *Annals of Applied Statistics*, 18(4):3071–3090, 2024.

A. M. Schmidt and A. E. Gelfand. A Bayesian coregionalization approach for multivariate pollutant data. *Journal of Geophysical Research*, 108:D24, 2003.

A. M. Schmidt and P. Guttorp. Flexible spatial covariance functions. *Spatial Statistics*, 37:100416, 2020. ISSN 2211-6753. doi: https://doi.org/10.1016/j.spasta.2020.100416. URL https://www.sciencedirect.com/science/article/pii/S2211675320300105. Frontiers in Spatial and Spatio-temporal Research.

A. M. Schmidt and A. O'Hagan. Bayesian inference for non-stationary spatial covariance structure via spatial deformations. *Journal of the Royal Statistical Society: Series B*

(Statistical Methodology), 65(3):743–758, 2003. doi: https://doi.org/10.1111/1467-9868. 00413. URL https://rss.onlinelibrary.wiley.com/doi/abs/10.1111/1467-9868. 00413.

A. M. Schmidt, P. Guttorp, and A. O'Hagan. Considering covariates in the covariance structure of spatial processes. *Environmetrics*, 22:487–500, 2011.

I. J. Schoenberg. Positive definite functions on spheres. *Duke Mathematical Journal*, 9(1): 96–108, 1942. doi: 10.1215/S0012-7094-42-00908-6. URL https://doi.org/10.1215/S0012-7094-42-00908-6.

G. Schwarz. Estimating the Dimension of a Model. *The Annals of Statistics*, 6(2):461–464, 1978. doi: 10.1214/aos/1176344136. URL https://doi.org/10.1214/aos/1176344136.

S. L. Scott, F. V. B. Alexander W. Blocker, H. A. Chipman, E. I. George, and R. E. McCulloch. Bayes and big data: the consensus Monte Carlo algorithm. *International Journal of Management Science and Engineering Management*, 11(2):78–88, 2016. doi: 10.1080/17509653.2016.1142191.

M. W. Seeger, C. K. I. Williams, and N. D. Lawrence. Fast forward selection to speed up sparse Gaussian process regression. In C. M. Bishop and B. J. Frey, editors, *Proceedings of the Ninth International Workshop on Artificial Intelligence and Statistics*, volume R4 of *Proceedings of Machine Learning Research*, pages 254–261. PMLR, 03–06 Jan 2003. URL https://proceedings.mlr.press/r4/seeger03a.html. Reissued by PMLR on 01 April 2021.

A. Shapiro and J. Botha. Variogram fitting with a general class of conditionally non-negative definite functions. *Computational Statistics & Data Analysis*, 11(1):87–96, 1991. ISSN 0167-9473. doi: https://doi.org/10.1016/0167-9473(91)90055-7. URL https://www.sciencedirect.com/science/article/pii/0167947391900557.

S. Shirota and S. Banerjee. Scalable inference for space-time Gaussian cox processes. *Journal of Time Series Analysis*, 40:269–287, 2019.

S. Shirota and A. E. Gelfand. Approximate Bayesian computation and model assessment for repulsive spatial point processes. *Journal of Computational and Graphical Statistics*, 26(3):646–657, 2017.

S. Shirota, A. E. Gelfand, and S. Banerjee. Spatial joint species distribution modeling using Dirichlet processes. *Statistica Sinica*, 29(3):1127–1154, 2019. doi: https://doi.org/10.5705/ss.202017.0482.

P. Sidén and F. Lindsten. Deep Gaussian Markov random fields. In H. Daumé III and A. Singh, editors, *Proceedings of the 37th International Conference on Machine Learning*, volume 119 of *Proceedings of Machine Learning Research*, pages 8916–8926. PMLR, 13–18 Jul 2020. URL https://proceedings.mlr.press/v119/siden20a.html.

A. Sikorski, D. McKenzie, and D. Nychka. Normalizing basis functions: Approximate stationary models for large spatial data, 2024. URL https://arxiv.org/abs/2405.13821.

B. W. Silverman. *Density estimation for statistics and data analysis*. Chapman & Hall/CRC, 2018.

D. Simpson, J. B. Illian, F. Lindgren, S. H. Sørbye, and H. Rue. Going off grid: Computationally efficient inference for log-Gaussian Cox processes. *Biometrika*, 103(1):49–70, 2016.

S. A. Sisson and Y. Fan. Likelihood-free Markov chain Monte Carlo. In *Handbook of Markov Chain Monte Carlo*, pages 313–338. Chapman and Hall/CRC, 2011.

R. J. Smith. Estimating nonstationary spatial correlations, 1996. URL https://api.semanticscholar.org/CorpusID:118255697. Technical Report, Cambridge University and University of North Carolina, Chapel Hill.

J. P. Snyder. Map projections: A working manual, 1987. URL https://doi.org/10.3133/pp1395.

A. R. Solow. Mapping by simple indicator kriging. *Mathematical Geology*, 18(3):335–352, Apr 1986. ISSN 1573-8868. doi: 10.1007/BF00898037. URL https://doi.org/10.1007/BF00898037.

M. Sommerfeld, S. Sain, and A. Schwartzman. Confidence regions for spatial excursion sets from repeated random field observations, with an application to climate. *Journal of the American Statistical Association*, 113:1327–1340, 2018.

S. Soubeyrand, F. Carpentier, F. Guiton, and E. K. Klein. Approximate Bayesian computation with functional statistics. *Statistical Applications in Genetics and Molecular Biology*, 12(1):17–37, 2013.

T. P. Speed, H. T. Kiiveri, et al. Gaussian Markov distributions over finite graphs. *The Annals of Statistics*, 14(1):138–150, 1986.

D. J. Spiegelhalter, A. P. Dawid, S. L. Lauritzen, and R. G. Cowell. Bayesian Analysis in Expert Systems. *Statistical Science*, 8(3):219–247, 1993. doi: 10.1214/ss/1177010888. URL https://doi.org/10.1214/ss/1177010888.

D. J. Spiegelhalter, A. Thomas, N. G. Best, and W. R. Gilks. Bugs - Bayesian inference using gibbs sampling version 0.50, 1995. URL https://api.semanticscholar.org/CorpusID:115455672. Technical report, Medical Research Council Biostatistics Unit, Institute of Public Health, Cambridge University.

D. J. Spiegelhalter, N. G. Best, B. P. Carlin, and A. van der Linde. Bayesian measures of model complexity and fit. *Journal of the Royal Statistical Society B*, 64:583–639, 2002.

M. Spivak. *A Comprehensive Introduction To Differential Geometry*. Number v. 1–5 in A Comprehensive Introduction to Differential Geometry. Publish or Perish Inc., 1979.

S. Srivastava, V. Cevher, Q. Dinh, and D. Dunson. WASP: Scalable Bayes via barycenters of subset posteriors. In G. Lebanon and S. V. N. Vishwanathan, editors, *Proceedings of the Eighteenth International Conference on Artificial Intelligence and Statistics*, volume 38 of *Proceedings of Machine Learning Research*, pages 912–920, San Diego, California, USA, 09–12, May 2015. PMLR.

A. Stein and L. C. A. Corsten. Universal kriging and cokriging as a regression procedure. *Biometrics*, 47:575–587, 1991. URL https://api.semanticscholar.org/CorpusID:120428386.

A. Stein, A. C. van Eijnsbergen, and L. G. Barendregt. Cokriging nonstationary data. *Mathematical Geology*, 23(5):703–719, Jun 1991. ISSN 1573-8868. doi: 10.1007/BF02082532. URL https://doi.org/10.1007/BF02082532.

M. L. Stein. *Interpolation of spatial data: Some Theory for Kriging*. Springer Science & Business Media, 1999a.

M. L. Stein. Predicting random fields with increasing dense observations. *The Annals of Applied Probability*, 9(1):242–273, 1999b. ISSN 10505164. URL http://www.jstor.org/stable/2667300.

M. L. Stein. Space–time covariance functions. *Journal of the American Statistical Association*, 100:310–321, 2005.

M. L. Stein. On a class of space–time intrinsic random functions. *Bernoulli*, 19(2):387–408, 05 2013. doi: 10.3150/11-BEJ405. URL http://dx.doi.org/10.3150/11-BEJ405.

M. L. Stein, Z. Chi, and L. J. Welty. Approximating likelihoods for large spatial data sets. *Journal of the Royal Statistical society, Series B*, 66:275–296, 2004.

H. Stern and N. Cressie. Inference for extremes in disease mapping. In A. Lawson, A. Biggeri, D. Boehning, E. Lesaffre, J. Viel, and R. Bertollini, editors, *Disease Mapping and Risk Assessment for Public Health*. Wiley, 1999.

J. D. Storey. A direct approach to false discovery rates. *Journal of the Royal Statistical Society: Series B (Statistical Methodology)*, 64(3):479–498, 2002. doi: https://doi.org/ 10.1111/1467-9868.00346. URL https://rss.onlinelibrary.wiley.com/doi/abs/10. 1111/1467-9868.00346.

J. D. Storey. The positive false discovery rate: a Bayesian interpretation and the q-value. *The Annals of Statistics*, 31(6):2013–2035, 2003. doi: 10.1214/aos/1074290335. URL https://doi.org/10.1214/aos/1074290335.

D. Stoyan and P. Grabarnik. Second-order characteristics for stochastic structures connected with Gibbs point processes. *Mathematische Nachrichten*, 151(1):95–100, 1991.

D. Stoyan and A. Penttinen. Recent applications of point process methods in forestry statistics. *Statistical Science*, 15:61–78, 2000.

G. Strang and K. Borre. *Linear Algebra, Geodesy and GPS*. Wellesley-Cambridge Press, Wellesley, MA, Dec 1997. ISBN 9780961408862.

D. J. Strauss. A model for clustering. *Biometrika*, 62:467–475, 1975.

J. R. Stroud, P. Müller, and B. Sansó. Dynamic models for spatiotemporal data. *Journal of the Royal Statistical society, Series B*, 63:673–689, 2001.

J. R. Stroud, M. L. Stein, and S. Lysen. Bayesian and maximum likelihood estimation for Gaussian processes on an incomplete lattice. *Journal of Computational and Graphical Statistics*, 26(1):108–120, 2017. doi: 10.1080/10618600.2016.1152970. URL https://doi. org/10.1080/10618600.2016.1152970.

W. Sun, B. J. Reich, T. Tony Cai, M. Guindani, and A. Schwartzman. False discovery control in large-scale spatial multiple testing. *Journal of the Royal Statistical Society Series B: Statistical Methodology*, 77(1):59–83, 2015.

T. Tjur. Coefficients of determination in logistic regression models-a new proposal: the coefficient of discrimination. *The American Statistician*, 63:366–372, 2009.

M. A. Taddy. Autoregressive mixture models for dynamic spatial Poisson processes: Application to tracking intensity of violent crime. *Journal of the American Statistical Association*, 105(492):1403–1417, 2010.

M. A. Taddy and A. Kottas. Mixture modeling for marked Poisson processes. *Bayesian Analysis*, 7(2):335–362, 2012.

W. Tang, L. Zhang, and S. Banerjee. On identifiability and consistency of the nugget in Gaussian spatial process models. *Journal of the Royal Statistical Society: Series B (Statistical Methodology)*, 83(5):1044–1070, 2021. doi: https://doi.org/10.1111/rssb.12472. URL https://rss.onlinelibrary.wiley.com/doi/abs/10.1111/rssb.12472.

M. Tanner. *Tools for statistical inference*. Springer Series in Statistics. Springer, New York, NY, 3 edition, Aug. 1997.

W. Tansey, O. Koyejo, R. A. Poldrack, and J. G. Scott. False discovery rate smoothing. *Journal of the American Statistical Association*, 113(523):1156–1171, 2018.

J. Tawn, R. Shooter, R. Towe, and R. Lamb. Modelling spatial extreme events with environmental applications. *Spatial Statistics*, 28:39–58, 2018.

B. M. Taylor and P. J. Diggle. INLA or MCMC? a tutorial and comparative evaluation for spatial prediction in log-Gaussian Cox processes. *Journal of Statistical Computation and Simulation*, 84(10):2266–2284, 2014.

D. Taylor-Rodríguez, K. Kaufeld, E. M. Schliep, J. S. Clark, and A. E. Gelfand. Joint Species Distribution Modeling: Dimension Reduction Using Dirichlet Processes. *Bayesian Analysis*, 12(4):939–967, 2017. doi: 10.1214/16-BA1031. URL https://doi.org/10.1214/16-BA1031.

D. Taylor-Rodriguez, A. O. Finley, A. Datta, C. Babcock, H. E. Andersen, B. D. Cook, D. C. Morton, and S. Banerjee. Spatial factor models for high-dimensional and large spatial data: An application in forest variable mapping. *Statistica Sinica*, 29(3):1155–1180, 2019.

K. Thapa and J. Bossler. Accuracy of spatial data used information systems. *Photogrammetric Engineering & Remote Sensing*, 58(6):835–841, 1992.

R. B. Thompson. Global positioning system: The mathematics of gps receivers. *Mathematics Magazine*, 71(4):260–269, 1998. doi: 10.1080/0025570X.1998.11996650. URL https://doi.org/10.1080/0025570X.1998.11996650.

R. Tibshirani. Regression shrinkage and selection via the lasso. *Journal of the Royal Statistical Society: Series B (Statistical Methodology)*, 58:267–288, 1996.

L. Tierney, R. E. Kass, and J. B. Kadane. Fully exponential laplace approximations to expectations and variances of nonpositive functions. *Journal of the American Statistical Association*, 84(407):710–716, 1989. doi: 10.1080/01621459.1989.10478824. URL https://doi.org/10.1080/01621459.1989.10478824.

W. R. Tobler. Smooth pycnophylactic interpolation for geographical regions. *Journal of the American Statistical Association*, 74(367):519–530, 1979. doi: 10.1080/01621459.1979.10481647. URL https://www.tandfonline.com/doi/abs/10.1080/01621459.1979.10481647. PMID: 12310706.

I. Todhunter. *Spherical Trigonometry: For the Use of Colleges and Schools*. Macmillan and Company, 1863. URL https://books.google.com/books?id=8M02AAAAMAAJ.

S. T. Tokdar and J. B. Kadane. Simultaneous linear quantile regression: A semiparametric Bayesian approach. *Bayesian Analysis*, 7(1):51–72, 2012. doi: 10.1214/12-BA702.

P. E. Tolbert, J. A. Mulholland, D. L. Macintosh, F. Xu, D. Daniels, O. J. Devine, B. P. Carlin, M. Klein, A. J. Butler, D. F. Nordenberg, H. Frumkin, P. B. Ryan, and M. C. White. Air Quality and Pediatric Emergency Room Visits for Asthma and Atlanta, Georgia. *American Journal of Epidemiology*, 151(8):798–810, 04 2000a. ISSN 0002-9262. doi: 10.1093/oxfordjournals.aje.a010280. URL https://doi.org/10.1093/oxfordjournals.aje.a010280.

P. E. Tolbert, J. A. Mulholland, D. L. Macintosh, F. Xu, D. Daniels, O. J. Devine, B. P. Carlin, M. Klein, A. J. Butler, D. F. Nordenberg, H. Frumkin, P. B. Ryan, and M. C. White. Air Quality and Pediatric Emergency Room Visits for Asthma and Atlanta, Georgia. *American Journal of Epidemiology*, 151(8):798–810, 04 2000b. ISSN 0002-9262. doi: 10.1093/oxfordjournals.aje.a010280. URL https://doi.org/10.1093/oxfordjournals.aje.a010280.

S. F. Tonellato. A Bayesian approach to the analysis of spatial time series. In A. Rizzi, M. Vichi, and H.-H. Bock, editors, *Advances in Data Science and Classification*, pages 579–584, Berlin, Heidelberg, 1998. Springer Berlin Heidelberg. ISBN 978-3-642-72253-0.

M. D. Ugarte, T. Goicoa, and A. F. Militino. Spatio-temporal modeling of mortality risks using penalized splines. *Environmetrics*, 21(3-4):270–289, 2010. doi: https://doi.org/10.1002/env.1011. URL https://onlinelibrary.wiley.com/doi/abs/10.1002/env.1011.

G. Van Brummelen. *Heavenly mathematics*. Princeton University Press, Princeton, NJ, Apr. 2017.

M. van Lieshout. *Theory of Spatial Statistics: A Concise Introduction.* Chapman and Hall/CRC Press, Boca Raton, FL, 2019.

S. R. S. Varadhan. *Probability Theory.* Courant Lecture Notes. American Mathematical Society, Providence, RI, Sept. 2001.

J. Vargas-Guzmán, A. Warrick, and D. Myers. Coregionalization by linear combination of nonorthogonal components. *Mathematical Geology*, 34:405–419, 05 2002. doi: 10.1023/A:1015078911063.

J. W. Vaupel, K. G. Manton, and E. Stallard. The impact of heterogeneity in individual frailty on the dynamics of mortality. *Demography*, 16(3):439–454, 1979. ISSN 00703370, 15337790. URL http://www.jstor.org/stable/2061224.

A. V. Vecchia. Estimation and model identification for continuous spatial processes. *Journal of the Royal Statistical Society: Series B (Methodological)*, 50(2):297–312, 12 1988. ISSN 0035-9246. doi: 10.1111/j.2517-6161.1988.tb01729.x. URL https://doi.org/10.1111/j.2517-6161.1988.tb01729.x.

J. Ver Hoef. Sampling and geostatistics for spatial data. *Écoscience*, 9(2):152–161, 2002.

J. Ver Hoef and N. Cressie. Multivariable spatial prediction. *Mathematical Geology*, 25:219–240, 02 1993. doi: 10.1007/BF00893273.

J. M. Ver Hoef and R. P. Barry. Constructing and fitting models for cokriging and multivariable spatial prediction. *Journal of Statistical Planning and Inference*, 69(2):275–294, 1998. ISSN 0378-3758. doi: https://doi.org/10.1016/S0378-3758(97)00162-6. URL https://www.sciencedirect.com/science/article/pii/S0378375897001626.

J. M. Ver Hoef, E. M. Hanks, and M. B. Hooten. On the relationship between conditional (car) and simultaneous (sar) autoregressive models. *Spatial Statistics*, 25:68–85, 2018. ISSN 2211-6753. doi: https://doi.org/10.1016/j.spasta.2018.04.006. URL https://www.sciencedirect.com/science/article/pii/S2211675317302725.

B. K. Victor De Oliveira and D. A. Short. Bayesian prediction of transformed Gaussian random fields. *Journal of the American Statistical Association*, 92(440):1422–1433, 1997. doi: 10.1080/01621459.1997.10473663. URL https://doi.org/10.1080/01621459.1997.10473663.

S. J. Villejo, J. B. Illian, and B. Swallow. Data fusion in a two-stage spatio-temporal model using the INLA-SPDE approach. *Spatial Statistics*, 54:100744, 2023. doi: 10.1016/j.spasta.2023.100744.

R. Waagepetersen and Y. Guan. Two-step estimation for inhomogeneous spatial point processes. *Journal of the Royal Statistical Society: Series B (Statistical Methodology)*, 71(3):685–702, 2009. doi: https://doi.org/10.1111/j.1467-9868.2008.00702.x. URL https://rss.onlinelibrary.wiley.com/doi/abs/10.1111/j.1467-9868.2008.00702.x.

H. Wackernagel. *Multivariate geostatistics: an introduction with applications.* Springer, Berlin, Heidelberg, third edition, 2003.

M. J. Wainwright and M. I. Jordan. Graphical models, exponential families, and variational inference. *Foundations and Trends in Machine Learning*, 1(1–2):1–305, 2008. ISSN 1935-8237. doi: 10.1561/2200000001. URL http://dx.doi.org/10.1561/2200000001.

J. Wakefield. Sensitivity analyses for ecological regression. *Biometrics*, 59(1):9–17, 2003. doi: https://doi.org/10.1111/1541-0420.00002. URL https://onlinelibrary.wiley.com/doi/abs/10.1111/1541-0420.00002.

J. Wakefield. A critique of statistical aspects of ecological studies in spatial epidemiology. *Environmental and Ecological Statistics*, 11(1):31–54, Mar 2004. ISSN 1573-3009. doi: 10.1023/B:EEST.0000011363.12720.38. URL https://doi.org/10.1023/B:EEST.0000011363.12720.38.

J. Wakefield and R. Salway. A statistical framework for ecological and aggregate studies. *Journal of the Royal Statistical Society. Series A (Statistics in Society)*, 164(1):119–137, 2001. ISSN 09641998, 1467985X. URL http://www.jstor.org/stable/2680539.

J. C. Wakefield and S. E. Morris. The Bayesian modeling of disease risk in relation to a point source. *Journal of the American Statistical Association*, 96(453):77–91, 2001. doi: 10.1198/016214501750332992. URL https://doi.org/10.1198/016214501750332992.

M. M. Wall. A close look at the spatial structure implied by the car and sar models. *Journal of Statistical Planning and Inference*, 121(2):311–324, 2004. ISSN 0378-3758. doi: https://doi.org/10.1016/S0378-3758(03)00111-3. URL https://www.sciencedirect.com/science/article/pii/S0378375803001113.

L. Waller, B. Turnbull, L. Clark, and P. Nasca. Spatial pattern analyses to detect rare disease clusters. In N. Lange, L. Ryan, L. Billard, D. Brillinger, L. Conquest, and J. Greenhouse, editors, *Case Studies in Biometry*. John Wiley & Sons, Inc., New York, NY., 1994.

L. A. Waller and C. A. Gotway. *Applied spatial statistics for public health data*, volume 368. John Wiley & Sons, 2004.

L. A. Waller, B. P. Carlin, H. Xia, and A. E. Gelfand. Hierarchical spatio-temporal mapping of disease rates. *Journal of the American Statistical Association*, 92(438):607–617, 1997. doi: 10.1080/01621459.1997.10474012. URL https://doi.org/10.1080/01621459.1997.10474012.

F. Wang and A. E. Gelfand. Directional data analysis under the general projected normal distribution. *Statistical Methodology*, 10(1):113–127, 2013.

F. Wang and A. E. Gelfand. Modeling space and space-time directional data using projected Gaussian processes. *Journal of the American Statistical Association*, 109(508):1565–1580, 2014. doi: 10.1080/01621459.2014.934454. URL https://doi.org/10.1080/01621459.2014.934454.

F. Wang and M. M. Wall. Generalized common spatial factor model. *Biostatistics*, 4(4):569–582, Oct 2003. doi: 10.1093/biostatistics/4.4.569. URL http://dx.doi.org/10.1093/biostatistics/4.4.569.

H. Wang and M. West. Bayesian analysis of matrix normal graphical models. *Biometrika*, 96(4):821–834, 2009.

G. Ward, T. Hastie, S. Barry, J. Elith, and J. R. Leathwick. Presence-only data and the em algorithm. *Biometrics*, 65(2):554–563, 05 2009. ISSN 0006-341X. doi: 10.1111/j.1541-0420.2008.01116.x. URL https://doi.org/10.1111/j.1541-0420.2008.01116.x.

D. I. Warton and L. C. Shepherd. Poisson point process models solve the pseudo-absence problem for presence-only data in ecology. *Annals of Applied Statistics*, 4:1383–1402, 2010.

S. Watanabe and M. Opper. Asymptotic equivalence of bayes cross validation and widely applicable information criterion in singular learning theory. *Journal of machine learning research*, 11(12), 2010.

W. W. Wei. *Time series analysis: Univariate and Multivariate Methods*. Pearson, Upper Saddle River, NJ, 2 edition, July 2005. ISBN 0-321-32216-9.

H. Wendland. Piecewise polynomial, positive definite and compactly supported radial functions of minimal degree. *Advances in Computational Mathematics*, 4(1):389–396, Dec 1995. ISSN 1572-9044. doi: 10.1007/BF02123482. URL https://doi.org/10.1007/BF02123482.

M. West and J. Harrison. *Bayesian Forecasting and Dynamic Models*. Springer, 2 edition, 1997.

P. A. White, M. F. Christensen, A. E. Gelfand, H. A. Frye, and J. A. Silander. Spatial functional data modeling of plant reflectances. *Annals of Applied Statistics*, 16(3):1919–1936, 2022.

P. A. White, M. F. Christensen, A. E. Gelfand, H. A. Frye, and J.A. Silander. Joint multivariate and functional modeling for plant traits and reflectances. *Ecology and Environmental Statistics*, 30(3):501–528, 2023.

P. Whittle. On stationary processes in the plane. *Biometrika*, 41:434–449, 1954.

C. K. Wikle. Low-rank representations for spatial processes. *Handbook of Spatial Statistics*, pages 107–118, 2010. Gelfand, A. E., Diggle, P., Fuentes, M. and Guttorp, P., editors, Chapman and Hall/CRC, pp. 107-118.

C. K. Wikle and L. M. Berliner. Combining information across spatial scales. *Technometrics*, 47(1):80–91, 2005. doi: 10.1198/004017004000000572. URL https://doi.org/10.1198/004017004000000572.

C. K. Wikle and N. Cressie. A dimension reduced approach to space-time Kalman filtering. *Biometrika*, 86:815–829, 1999.

C. K. Wikle and A. Zammit-Mangion. Statistical deep learning for spatial and spatiotemporal data. *Annual Review of Statistics and Its Application*, 10 (Volume 10, 2023):247–270, 2023. ISSN 2326-831X. doi: https://doi.org/10.1146/annurev-statistics-033021-112628. URL https://www.annualreviews.org/content/journals/10.1146/annurev-statistics-033021-112628.

C. K. Wikle, R. F. Milliff, D. Nychka, and L. M. Berliner. Spatiotemporal hierarchical Bayesian modeling tropical ocean surface winds. *Journal of the American Statistical Association*, 96(454):382–397, 2001. doi: 10.1198/016214501753168109. URL https://doi.org/10.1198/016214501753168109.

C. K. Wikle, A. Zammit-Mangion, and N. Cressie. *Spatio-Temporal Statistics with R*. Chapman & Hall/CRC Press, Boca Raton, FL, 2019.

D. H. Wolpert. Stacked generalization. *Neural networks*, 5(2):241–259, 1992.

R. L. Wolpert and K. Ickstadt. Poisson/gamma random field models for spatial statistics. *Biometrika*, 85(2):251–267, 1998.

W. H. Womble. Differential Systematics. *Science*, 114(2961):315–322, 1951.

A. D. Woodbury. Bayesian updating revisited. *Mathematical Geology*, 21(3):285–308, Apr 1989. ISSN 1573-8868. doi: 10.1007/BF00893691. URL https://doi.org/10.1007/BF00893691.

K. L. Wu and S. Banerjee. Assessing spatial disparities: A Bayesian linear regression approach, 2024. URL https://arxiv.org/abs/2407.19171.

L. Wu, G. Pleiss, and J. P. Cunningham. Variational nearest neighbor Gaussian process. In K. Chaudhuri, S. Jegelka, L. Song, C. Szepesvari, G. Niu, and S. Sabato, editors, *Proceedings of the 39th International Conference on Machine Learning*, volume 162 of *Proceedings of Machine Learning Research*, pages 24114–24130. PMLR, 17–23 Jul 2022. URL https://proceedings.mlr.press/v162/wu22h.html.

N. X. Xanh and H. Zessin. Integral and differential characterizations of the Gibbs process. *Mathematische Nachrichten*, 88(1):105–115, 1979.

G. Xia and A. Gelfand. Stationary process approximation for the analysis of large spatial datasets. Technical report, Technical Report, ISDS, Duke University, 2006.

G. Xia, M. L. Miranda, and A. E. Gelfand. Approximately optimal spatial design approaches for environmental health data. *Environmetrics*, 17(4):363–385, 2006. doi: https://doi.org/10.1002/env.775. URL https://onlinelibrary.wiley.com/doi/abs/10.1002/env.775.

H. Xia and B. P. Carlin. Spatio-temporal models with errors in covariates: mapping ohio lung cancer mortality. *Statistics in Medicine*, 17(18):2025–2043, 1998. doi: https://doi.org/10.1002/(SICI)1097-0258(19980930)17:18⟨2025::AID-SIM865⟩ 3.0.CO;2-M. URL https://onlinelibrary.wiley.com/doi/abs/10.1002/%28SICI% 291097-0258%2819980930%2917%3A18%3C2025%3A%3AAID-SIM865%3E3.0.CO%3B2-M.

S. Xiao, A. Kottas, and B. Sansó. Modeling for seasonal marked point processes: An analysis of evolving hurricane occurrences. *The Annals of Applied Statistics*, 9(1):353–382, 2015.

P.-F. Xu, J. Guo, and X. He. An improved iterative proportional scaling procedure for Gaussian graphical models. *Journal of Computational and Graphical Statistics*, 20(2): 417–431, 2011.

A. Yaglom. *An Introduction to the Theory of Stationary Random Functions*. Dover books on intermediate and advanced mathematics. Dover Publications, Incorporated, 2004. ISBN 9780486495712. URL https://books.google.com/books?id=bnQKXWjGoSEC.

A. M. Yaglom. *Correlation theory of stationary and related random functions*. Springer series in statistics. Springer, New York, NY, 1987 edition, Nov. 1987.

A. Y. Yakovlev, A. D. Tsodikov, and B. Asselain. *Stochastic Models of Tumor Latency and Their Biostatistical Applications*. WORLD SCIENTIFIC, 1996. doi: 10.1142/2420. URL https://www.worldscientific.com/doi/abs/10.1142/2420.

Y. Yang and X. He. Quantile regression for spatially correlated data: An empirical likelihood approach. *Statistica Sinica*, 25(1):261–274, 2015.

Y. Yang and S. T. Tokdar. Joint estimation of quantile planes over arbitrary predictor spaces. *Journal of the American Statistical Association*, 112(519):1107–1120, 2017. doi: 10.1080/01621459.2016.1192545.

Y. Yao, A. Vehtari, D. Simpson, and A. Gelman. Using stacking to average Bayesian predictive distributions (with discussion). *Bayesian Analysis*, 13(3):917–1007, 2018.

Y. Yao, G. Pirs, A. Vehtari, and A. Gelman. Bayesian hierarchical stacking: Some models are (somewhere) useful. *Bayesian Analysis*, 1(1):1–29, 2021.

Y. Yao, A. Vehtari, and A. Gelman. Stacking for non-mixing Bayesian computations: The curse and blessing of multimodal posteriors. *Journal of Machine Learning Research*, 23 (79):1–45, 2022. URL http://jmlr.org/papers/v23/20-1426.html.

K. Yu and R. A. Moyeed. Bayesian quantile regression. *Statistics & Probability Letters*, 54 (4):437–447, 2001. doi: 10.1016/S0167-7152(01)00124-9.

A. Zammit-Mangion, T. L. J. Ng, Q. Vu, and M. Filippone. Deep compositional spatial models. *Journal of the American Statistical Association*, 117(540):1787–1808, 2022. doi: 10.1080/01621459.2021.1887741. URL https://doi.org/10.1080/01621459.2021. 1887741.

A. Zammit-Mangion, M. Sainsbury-Dale, and R. Huser. Neural methods for amortized inference, 2024. URL https://arxiv.org/abs/2404.12484.

J. Zapata, S. Y. Oh, and A. Petersen. Partial separability and functional graphical models for multivariate Gaussian processes. *Biometrika*, 10 2021. ISSN 0006-3444. doi: 10.1093/ biomet/asab046. URL https://doi.org/10.1093/biomet/asab046. asab046.

W. Zhan and A. Datta. Neural networks for geospatial data. *Journal of the American Statistical Association*, 1–21, 2024.

H. Zhang. Inconsistent estimation and asymptotically equal interpolations in model-based geostatistics. *Journal of the American Statistical Association*, 99(465):250–261, 2004. doi: 10.1198/016214504000000241. URL https://doi.org/10.1198/016214504000000241.

H. Zhang and D. L. Zimmerman. Towards reconciling two asymptotic frameworks in spatial statistics. *Biometrika*, 92(4):921–936, 12 2005. ISSN 0006-3444. doi: 10.1093/biomet/92. 4.921. URL https://doi.org/10.1093/biomet/92.4.921.

L. Zhang and S. Banerjee. Spatial factor modeling: A Bayesian matrix-normal approach for misaligned data. *Biometrics*, 78(2):560–573, 2022. doi: https://doi.org/10.1111/biom. 13452. URL https://onlinelibrary.wiley.com/doi/abs/10.1111/biom.13452.

L. Zhang, A. Datta, and S. Banerjee. Practical Bayesian modeling and inference for massive spatial datasets on modest computing environments. *Statistical Analysis and Data Mining: The ASA Data Science Journal*, 12(3):197–209, 2019. doi: 10.1002/sam.11413. URL https://doi.org/10.1002/sam.11413.

L. Zhang, B. A. Shaby, and J. L. Wadsworth. Hierarchical transformed scale mixtures for flexible modeling of spatial extremes on datasets with many locations. *Journal of the American Statistical Association*, 117(539):1357–1369, 2022. doi: 10.1080/01621459.2020. 1858838. URL https://doi.org/10.1080/01621459.2020.1858838.

L. Zhang, W. Tang, and S. Banerjee. Bayesian geostatistics using predictive stacking, 2025. URL https://arxiv.org/abs/2304.12414.

Y. Zhang, J. S. Hodges, and S. Banerjee. Smoothed ANOVA with spatial effects as a competitor to MCAR in multivariate spatial smoothing. *The Annals of Applied Statistics*, 3(4):1805 – 1830, 2009. doi: 10.1214/09-AOAS267. URL https://doi.org/10.1214/09-AOAS267.

Z. Zhou, D. S. Matteson, D. B. Woodard, S. G. Henderson, and A. C. Micheas. A spatio-temporal point process model for ambulance demand. *Journal of the American Statistical Association*, 110(509):6–15, 2015.

H. Zhu, N. Strawn, and D. B. Dunson. Bayesian graphical models for multivariate functional data. *Journal of Machine Learning Research*, 17(204):1–27, 2016.

J. Zhu, S. N. Lahiri, and N. Cressie. Asymptotic inference for spatial cdfs over time. *Statistica Sinica*, 12(3):843–861, 2002. ISSN 10170405, 19968507. URL http://www.jstor.org/stable/24306998.

L. Zhu, B. P. Carlin, P. English, and R. Scalf. Hierarchical modeling of spatio-temporally misaligned data: relating traffic density to pediatric asthma hospitalizations. *Environmetrics*, 11(1):43–61, 2000. doi: https://doi.org/10.1002/(SICI)1099-095X(200001/02)11: 1⟨43::AID-ENV380⟩3.0.CO;2-V. URL https://onlinelibrary.wiley.com/doi/abs/10.1002/%28SICI%291099-095X%28200001/02%2911%3A1%3C43%3A%3AAID-ENV380%3E3.0.CO%3B2-V.

L. Zhu, B. P. Carlin, and A. E. Gelfand. Hierarchical regression with misaligned spatial data: relating ambient ozone and pediatric asthma er visits in atlanta. *Environmetrics*, 14 (5):537–557, 2003. doi: https://doi.org/10.1002/env.614. URL https://onlinelibrary.wiley.com/doi/abs/10.1002/env.614.

J. V. Zidek, R. White, W. Sun, R. T. Burnett, and N. D. Le. Imputing unmeasured explanatory variables in environmental epidemiology with application to health impact analysis of air pollution. *Environmental and Ecological Statistics*, 5(2):99–105, Jun 1998. ISSN 1573-3009. doi: 10.1023/A:1009610720709. URL https://doi.org/10.1023/A:1009610720709.

J. V. Zidek, W. Sun, and N. D. Le. Designing and Integrating Composite Networks for Monitoring Multivariate Gaussian Pollution Fields. *Journal of the Royal Statistical Society Series C: Applied Statistics*, 49(1):63–79, 01 2000. ISSN 0035-9254. doi: 10.1111/1467-9876.00179. URL https://doi.org/10.1111/1467-9876.00179.

J. V. Zidek, N. D. Le, and Z. Liu. Combining data and simulated data for space–time fields: application to ozone. *Environmental and Ecological Statistics*, 19:37–56, 2012.

D. L. Zimmerman. Another look at anisotropy in geostatistics. *Mathematical Geology*, 25(4):453–470, May 1993. ISSN 1573-8868. doi: 10.1007/BF00894779. URL https://doi.org/10.1007/BF00894779.

Index

For Product Safety Concerns and Information please contact our EU
representative GPSR@taylorandfrancis.com
Taylor & Francis Verlag GmbH, Kaufingerstraße 24, 80331 München, Germany

www.ingramcontent.com/pod-product-compliance
Lightning Source LLC
Chambersburg PA
CBHW061928190326
41458CB00009B/2694

9 781032 508559